ESSENTIAL

UNIVERSITY PHYSICS

VOLUME 2 CHAPTERS 20–39

THIRD EDITION

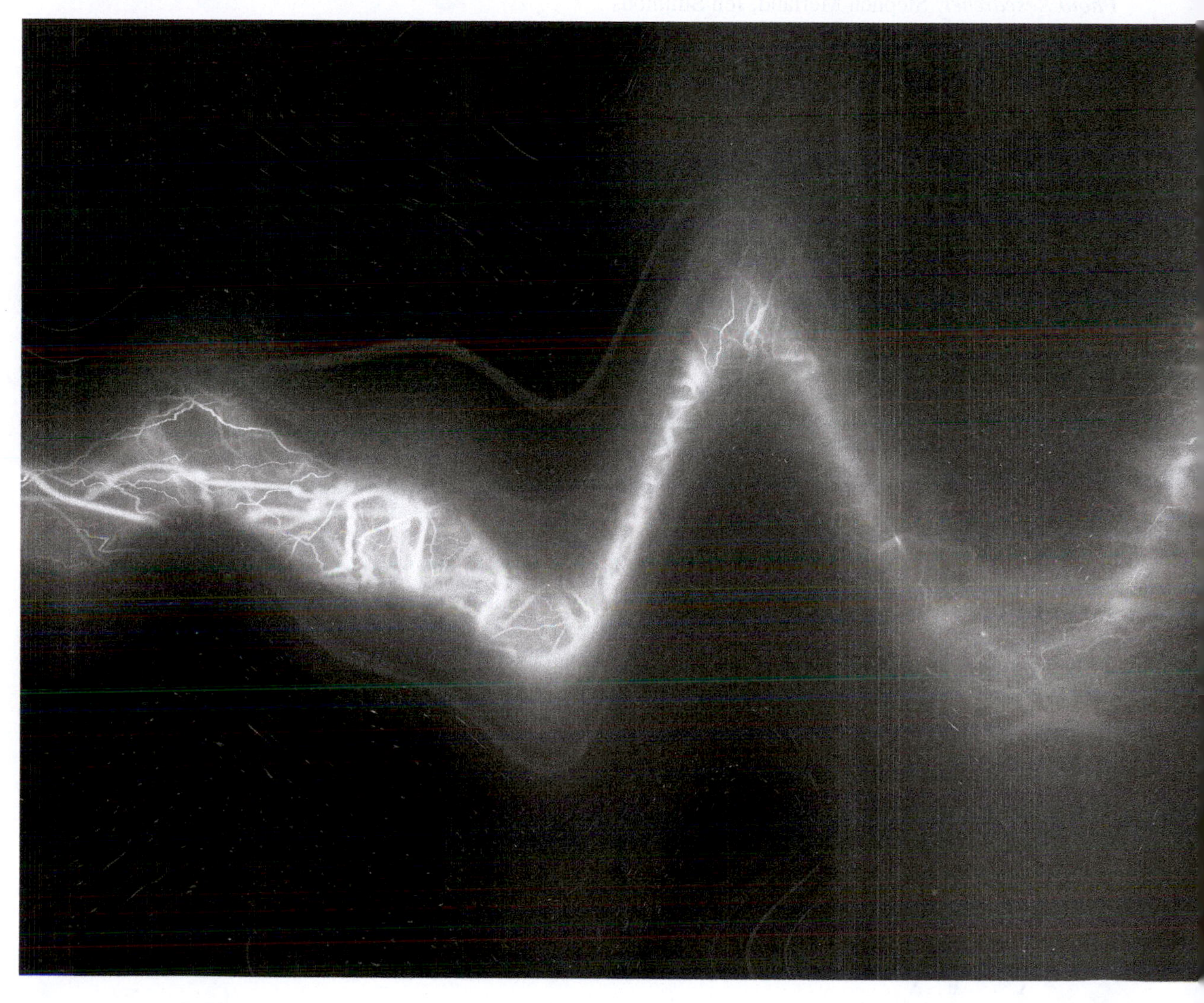

Richard Wolfson

Middlebury College

Executive Editor: Nancy Whilton
Project Manager: Katie Conley
Development Editors: John Murdzek, Matt Walker
Editorial Assistant: Sarah Kaubisch
Development Manager: Cathy Murphy
Project Management Team Lead: Kristen Flathman
Compositor: Lumina Datamatics, Inc.
Design Manager: Marilyn Perry
Illustrators: Rolin Graphics
Rights & Permissions Management: Timothy Nicholls
Photo Researcher: Stephen Merland, Jen Simmons
Manufacturing Buyer: Maura Zaldivar-Garcia
Marketing Manager: Will Moore
Cover Photo Credit: Argus/Shutterstock

CIP data is on file with the Library of Congress.

www.pearsonhighered.com

ISBN 10: 0-321-97642-8; ISBN 13: 978-0-321-97642-0

PhET Simulations

Available in the Pearson eText and in the Study Area of MasteringPhysics MP®

Video Tutor Demonstrations

Video tutor demonstrations can be accessed by scanning the QR codes in the textbook using a smartphone. They are also available in the Study Area and Instructor's Resource Area on MasteringPhysics and in the eText.

Brief Contents

About the Author

Richard Wolfson

Richard Wolfson is the Benjamin F. Wissler Professor of Physics at Middlebury College, where he has taught since 1976. He did undergraduate work at MIT and Swarthmore College, and he holds an M.S. degree from the University of Michigan and Ph.D. from Dartmouth. His ongoing research on the Sun's corona and climate change has taken him to sabbaticals at the National Center for Atmospheric Research in Boulder, Colorado; St. Andrews University in Scotland; and Stanford University.

Rich is a committed and passionate teacher. This is reflected in his many publications for students and the general public, including the video series *Einstein's Relativity and the Quantum Revolution: Modern Physics for Nonscientists* (The Teaching Company, 1999), *Physics in Your Life* (The Teaching Company, 2004), *Physics and Our Universe: How It All Works* (The Teaching Company, 2011), and *Understanding Modern Electronics* (The Teaching Company, 2014); books *Nuclear Choices: A Citizen's Guide to Nuclear Technology* (MIT Press, 1993), *Simply Einstein: Relativity Demystified* (W. W. Norton, 2003), and *Energy, Environment, and Climate* (W. W. Norton, 2012); and articles for *Scientific American* and the *World Book Encyclopedia.*

Outside of his research and teaching, Rich enjoys hiking, canoeing, gardening, cooking, and watercolor painting.

Preface to the Instructor

Introductory physics texts have grown ever larger, more massive, more encyclopedic, more colorful, and more expensive. *Essential University Physics* bucks that trend—without compromising coverage, pedagogy, or quality. The text benefits from the author's three decades of teaching introductory physics, seeing firsthand the difficulties and misconceptions that students face as well as the "Got It!" moments when big ideas become clear. It also builds on the author's honing multiple editions of a previous calculus-based textbook and on feedback from hundreds of instructors and students.

Goals of This Book

Physics is the fundamental science, at once fascinating, challenging, and subtle—and yet simple in a way that reflects the few basic principles that govern the physical universe. My goal is to bring this sense of physics alive for students in a range of academic disciplines who need a solid calculus-based physics course—whether they're engineers, physics majors, premeds, biologists, chemists, geologists, mathematicians, computer scientists, or other majors. My own courses are populated by just such a variety of students, and among my greatest joys as a teacher is having students who took a course only because it was required say afterward that they really enjoyed their exposure to the ideas of physics. More specifically, my goals include:

- Helping students build the analytical and quantitative skills and confidence needed to apply physics in problem solving for science and engineering.
- Addressing key misconceptions and helping students build a stronger conceptual understanding.
- Helping students see the relevance and excitement of the physics they're studying with contemporary applications in science, technology, and everyday life.
- Helping students develop an appreciation of the physical universe at its most fundamental level.
- Engaging students with an informal, conversational writing style that balances precision with approachability.

New to the Third Edition

The overall theme for this third-edition revision is to present a more unified view of physics, emphasizing "big ideas" and the connections among different topics covered throughout the book. We've also updated material and features based on feedback from instructors, students, and reviewers. A modest growth, averaging about one page per chapter, allows for expanded coverage of topics where additional elaboration seemed warranted. Several chapters have had major rewrites of key physics topics. We've also made a number of additions and modifications aimed at improving students' understanding, increasing relevancy, and offering expanded problem-solving opportunities.

- Chapter opening pages have been redesigned to include explicit connections, both textual and graphic, with preceding and subsequent chapters.
- The presentation of **energy and work** in Chapters 6 and 7 has been extensively rewritten with a clearer invocation of **systems concepts**. Internal energy is introduced much earlier in the book, and potential energy is carefully presented as a property not of objects but of systems. Two new sections in Chapter 7 emphasize the universality of energy conservation, including the role of internal energy in systems subject to dissipative forces. Forward references tie this material to the chapters on thermodynamics, electromagnetism, and relativity. The updated treatment of energy also allows the text to make a closer connection between the conservation laws for energy and momentum.

- The presentation of **magnetic flux and Faraday's law** in Chapter 27 has been recast so as to distinguish motional emf from emfs induced by changing magnetic fields—including Einstein's observation about induction, which is presented as a forward-looking connection to Chapter 33.
- There is more emphasis on calculus in earlier chapters, allowing instructors who wish to do so to use calculus approaches to topics that are usually introduced algebraically. We've also added more calculus-based problems. However, we continue to emphasize the standard approach in the main text for those who teach the course with a calculus corequisite or otherwise want to go slowly with more challenging math.
- A host of **new applications** connects the physics concepts that students are learning with contemporary technological and biomedical innovations, as well as recent scientific discoveries. A sample of new applications includes Inertial Guidance Systems, Vehicle Stability Control, Climate Modeling, Electrophoresis, MEMS (Microelectromechanical Systems), The Taser, Uninterruptible Power Supplies, Geomagnetic Storms, PET Scans, Noise-Cancelling Headphones, Femtosecond Chemistry, Windows on the Universe, and many more.
- Additional **worked examples** have been added in areas where students show the need for more practice in problem solving. Many of these are not just artificial textbook problems but are based on contemporary science and technology, such as the Mars *Curiosity* rover landing, the Fukushima accident, and the Chelyabinsk meteor. Following user requests, we've added an example of a collision in the center-of-mass reference frame.
- New GOT IT? boxes, now in nearly every section of every chapter, provide quick checks on students' conceptual understanding. Many of the GOT IT? questions have been formatted as Clicker questions, available on the Instructor's Resource DVD and in the Instructor's Resource Area in Mastering.
- End-of chapter problem sets have been extensively revised:
  - Each EOC problem set has at least 10 percent new or substantially revised problems.
  - More "For Thought and Discussion Questions" have been added.
  - Nearly every chapter has more intermediate-level problems.
  - More calculus-based problems have been added.
  - Every chapter now has at least one data problem, designed to help students develop strong quantitative reasoning skills. These problems present a data table and require students to determine appropriate functions of the data to plot in order to achieve a linear relationship and from that to find values of physical quantities involved in the experiment from which the data were taken.
  - New tags have been added to label appropriate problems. These include CH (challenge), ENV (environmental), and DATA, and they join the previous BIO and COMP (computer) problem tags.
- QR codes in margins allow students to use smartphones or other devices for immediate access to video tutor demonstrations that illustrate selected concepts while challenging students to interact with the video by predicting outcomes of simple experiments.
- References to PhET simulations appear in the margins where appropriate.
- As with earlier revisions, we've incorporated new research results, new applications of physics principles, and findings from physics education research.

Pedagogical Innovations

This book is *concise*, but it's also *progressive* in its embrace of proven techniques from physics education research and *strategic* in its approach to learning physics. Chapter 1 introduces the IDEA framework for problem solving, and every one of the book's subsequent **worked examples** employs this framework. IDEA—an acronym for Identify, Develop, Evaluate, Assess—is not a "cookbook" method for students to apply mindlessly, but rather a tool for organizing students' thinking and discouraging equation hunting. It begins with an interpretation of the problem and an identification of the key physics concepts involved; develops a plan for reaching the solution; carries out the mathematical evaluation; and assesses the solution to see that it makes sense, to compare the example with others, and to mine additional insights into physics. In nearly all of the text's worked examples, the Develop phase includes making a drawing, and most of these use a hand-drawn style to encourage students to make their own drawings—a step that research suggests they often skip. IDEA provides a common approach to all physics problem solving, an approach that emphasizes the conceptual unity of physics and helps break the typical student view of physics as a hodgepodge of equations and unrelated ideas. In addition to IDEA-based worked examples, other pedagogical features include:

- **Problem-Solving Strategy boxes** that follow the IDEA framework to provide detailed guidance for specific classes of physics problems, such as Newton's second law, conservation of energy, thermal-energy balance, Gauss's law, or multiloop circuits.
- **Tactics boxes** that reinforce specific essential skills such as differentiation, setting up integrals, vector products, drawing free-body diagrams, simplifying series and parallel circuits, or ray tracing.
- **QR codes** in the textbook allow students to link to video tutor demonstrations as they read, using their smartphones. These "Pause and predict" videos of key physics concepts ask students to submit a prediction before they see the outcome. The videos are also available in the Study Area of Mastering and in the Pearson eText.
- **GOT IT? boxes** that provide quick checks for students to test their conceptual understanding. Many of these use a multiple-choice or quantitative ranking format to probe student misconceptions and facilitate their use with classroom-response systems. Many new GOT IT? boxes have been added in the third edition, and now nearly every section of every chapter has at least one GOT IT? box.
- **Tips** that provide helpful problem-solving hints or warn against common pitfalls and misconceptions.
- **Chapter openers** that include a graphical indication of where the chapter lies in sequence as well as three columns of points that help make connections with other material throughout the book. These include a backward-looking "What You Know," "What You're Learning" for the present chapter, and a forward-looking "How You'll Use It." Each chapter also includes an opening photo, captioned with a question whose answer should be evident after the student has completed the chapter.
- **Applications**, self-contained presentations typically shorter than half a page, provide interesting and contemporary instances of physics in the real world, such as bicycle stability; flywheel energy storage; laser vision correction; ultracapacitors; noise-cancelling headphones; wind energy; magnetic resonance imaging; smartphone gyroscopes; combined-cycle power generation; circuit models of the cell membrane; CD, DVD, and Blu-ray technologies; radiocarbon dating; and many, many more.
- **For Thought and Discussion** questions at the end of each chapter designed for peer learning or for self-study to enhance students' conceptual understanding of physics.
- **Annotated figures** that adopt the research-based approach of including simple "instructor's voice" commentary to help students read and interpret pictorial and graphical information.
- **End-of-chapter** problems that begin with simpler exercises keyed to individual chapter sections and ramp up to more challenging and often multistep problems that synthesize chapter material. Context-rich problems focusing on real-world situations are interspersed throughout each problem set.
- **Chapter summaries** that combine text, art, and equations to provide a synthesized overview of each chapter. Each summary is hierarchical, beginning with the chapter's "big ideas," then focusing on key concepts and equations, and ending with a list of "applications"—specific instances or applications of the physics presented in the chapter.

Organization

This contemporary book is *concise*, *strategic*, and *progressive*, but it's *traditional* in its organization. Following the introductory Chapter 1, the book is divided into six parts. Part One (Chapters 2–12) develops the basic concepts of mechanics, including Newton's laws and conservation principles as applied to single particles and multiparticle systems. Part Two (Chapters 13–15) extends mechanics to oscillations, waves, and fluids. Part Three (Chapters 16–19) covers thermodynamics. Part Four (Chapters 20–29) deals with electricity and magnetism. Part Five (Chapters 30–32) treats optics, first in the geometrical optics approximation and then including wave phenomena. Part Six (Chapters 33–39) introduces relativity and quantum physics. Each part begins with a brief description of its coverage, and ends with a conceptual summary and a challenge problem that synthesizes ideas from several chapters.

Essential University Physics is available in two paperback volumes, so students can purchase only what they need—making the low-cost aspect of this text even more attractive. Volume 1 includes Parts One, Two, and Three, mechanics through thermodynamics. Volume 2 contains Parts Four, Five, and Six, electricity and magnetism along with optics and modern physics.

Instructor Supplements

NOTE: For convenience, all of the following instructor supplements (except the Instructor's Resource DVD) can be downloaded from the Instructor's Resource Area of MasteringPhysics® (www.masteringphysics.com) as well as from the Instructor's Resource Center on www.pearsonhighered.com/irc.

- The **Instructor's Solutions Manual** (ISBN 0-133-85713-1) contains solutions to all end-of-chapter exercises and problems, written in the Interpret/Develop/Evaluate/Assess (IDEA) problem-solving framework. The solutions are provided in PDF and editable Microsoft® Word formats for Mac and PC, with equations in MathType.
- The **Instructor's Resource DVD** (ISBN 0-133-85714-X) provides all the figures, photos, and tables from the text in JPEG format. All the problem-solving strategies, Tactics Boxes, key equations, and chapter summaries are provided in PDF and editable Microsoft® Word formats with equations in MathType. Each chapter also has a set of PowerPoint® lecture outlines and questions including the new GOT IT! Clickers. A comprehensive library of more than 220 applets from **ActivPhysics OnLine™**, a suite of over 70 PhET simulations, and 40 video tutor demonstrations are also included. Also, the complete Instructor's Solutions Manual is provided in both Word and PDF formats.
- MP **MasteringPhysics®** (www.masteringphysics.com) is the most advanced physics homework and tutorial system available. This online homework and tutoring system guides students through the toughest topics in physics with self-paced tutorials that provide individualized coaching. These assignable, in-depth tutorials are designed to coach students with hints and feedback specific to their individual errors. Instructors can also assign end-of-chapter problems from every chapter, including multiple-choice questions, section-specific exercises, and general problems. Quantitative problems can be assigned with numerical answers and randomized values (with sig fig feedback) or solutions. This third edition includes nearly 400 new problems written by the author explictly for use with MasteringPhysics.
- **Learning Catalytics** is a "bring your own device" student engagement, assessment, and classroom intelligence system that is based on cutting-edge research, innovation, and implementation of interactive teaching and peer instruction. With Learning Catalytics pre-lecture questions, you can see what students do and don't understand and adjust lectures accordingly.
- **Pearson eText** is available either automatically when MasteringPhysics® is packaged with new books or as a purchased upgrade online. Users can search for words or phrases, create notes, highlight text, bookmark sections, click on definitions to key terms, and launch PhET simulations and video tutor demonstrations as they read. Professors also have the ability to annotate the text for their course and hide chapters not covered in their syllabi.
- The **Test Bank** (ISBN 0-133-85715-8) contains more than 2000 multiple-choice, true-false, and conceptual questions in TestGen® and Microsoft Word® formats for Mac and PC users. More than half of the questions can be assigned with randomized numerical values.

Student Supplements

- **MasteringPhysics®** (www.masteringphysics.com) is the most advanced physics homework and tutorial system available. This online homework and tutoring system guides students through the most important topics in physics with self-paced tutorials that provide individualized coaching. These assignable, in-depth tutorials are designed to coach students with hints and feedback specific to their individual errors. Instructors can also assign end-of-chapter problems from every chapter including multiple-choice questions, section-specific exercises, and general problems. Quantitative problems can be assigned with numerical answers and randomized values (with sig fig feedback) or solutions.
- **Pearson eText** is available through MasteringPhysics®, either automatically when MasteringPhysics® is packaged with new books or as a purchased upgrade online. Allowing students access to the text wherever they have access to the Internet, Pearson eText comprises the full text with additional interactive features. Users can search for words or phrases, create notes, highlight text, bookmark sections, click on definitions to key terms, and launch PhET simulations and video tutor demonstrations as they read.

Acknowledgments

A project of this magnitude isn't the work of its author alone. First and foremost among those I thank for their contributions are the now several thousand students I've taught in calculus-based introductory physics courses at Middlebury College. Over the years your questions have taught me how to convey physics ideas in many different ways appropriate to your diverse learning styles. You've helped identify the "sticking points" that challenge introductory physics students, and you've showed me ways to help you avoid and "unlearn" the misconceptions that many students bring to introductory physics.

Thanks also to the numerous instructors and students from around the world who have contributed valuable suggestions for improvement of this text. I've heard you, and you'll find many of your ideas implemented in this third edition of *Essential University Physics*. And special thanks to my Middlebury physics colleagues who have taught from this text and who contribute valuable advice and insights on a regular basis: Jeff Dunham, Anne Goodsell, Noah Graham, Steve Ratcliff, and Susan Watson.

Experienced physics instructors thoroughly reviewed every chapter of this book, and reviewers' comments resulted in substantive changes—and sometimes in major rewrites—to the first drafts of the manuscript. We list all these reviewers below. But first, special thanks are due to several individuals who made exceptional contributions to the quality and in some cases the very existence of this book. First is Professor Jay Pasachoff of Williams College, whose willingness more than three decades ago to take a chance on an inexperienced coauthor has made writing introductory physics a large part of my professional career. Dr. Adam Black, physics editor and Ph.D. physicist at Pearson, had the vision to see promise in a new introductory text that would respond to the rising chorus of complaints about massive, encyclopedic, and expensive physics texts. Brad Patterson, developmental editor for the first edition, brought his graduate-level knowledge of physics to a role that made him a real collaborator. Brad is responsible for many of the book's innovative features, and it was a pleasure to work with him. John Murdzek and Matt Walker continued with Brad's excellent tradition of developmental editing on this third edition. We've gone to great lengths to make this book as error-free as possible, and much of the credit for that happy situation goes to Sen-Ben Liao, who solved every new and revised homework problem and updated the solutions manual.

I also wish to thank Nancy Whilton and Katie Conley at Pearson Education, and Haylee Schwenk at Lumina Datamatics, for their highly professional efforts in shepherding this book through its vigorous production schedule. Finally, as always, I thank my family, my colleagues, and my students for the patience they showed during the intensive process of writing and revising this book.

Reviewers

John R. Albright, *Purdue University–Calumet*
Rama Bansil, *Boston University*
Richard Barber, *Santa Clara University*
Linda S. Barton, *Rochester Institute of Technology*
Rasheed Bashirov, *Albertson College of Idaho*
Chris Berven, *University of Idaho*
David Bixler, *Angelo State University*
Ben Bromley, *University of Utah*
Charles Burkhardt, *St. Louis Community College*
Susan Cable, *Central Florida Community College*
George T. Carlson, Jr., *West Virginia Institute of Technology–West Virginia University*
Catherine Check, *Rock Valley College*
Norbert Chencinski, *College of Staten Island*
Carl Covatto, *Arizona State University*
David Donnelly, *Texas State University–San Marcos*
David G. Ellis, *University of Toledo*
Tim Farris, *Volunteer State Community College*
Paula Fekete, *Hunter College of The City University of New York*

Idan Ginsburg, *Harvard University*
James Goff, *Pima Community College*
Austin Hedeman, *University of California–Berkeley*
Andrew Hirsch, *Purdue University*
Mark Hollabaugh, *Normandale Community College*
Eric Hudson, *Pennsylvania State University*
Rex W. Joyner, *Indiana Institute of Technology*
Nikos Kalogeropoulos, *Borough of Manhattan Community College–The City University of New York*
Viken Kiledjian, *East Los Angeles College*
Kevin T. Kilty, *Laramie County Community College*
Duane Larson, *Bevill State Community College*
Kenneth W. McLaughlin, *Loras College*
Tom Marvin, *Southern Oregon University*
Perry S. Mason, *Lubbock Christian University*
Mark Masters, *Indiana University–Purdue University Fort Wayne*
Jonathan Mitschele, *Saint Joseph's College*
Gregor Novak, *United States Air Force Academy*
Richard Olenick, *University of Dallas*
Robert Philbin, *Trinidad State Junior College*
Russell Poch, *Howard Community College*
Steven Pollock, *Colorado University–Boulder*
Richard Price, *University of Texas at Brownsville*
James Rabchuk, *Western Illinois University*
George Schmiedeshoff, *Occidental College*
Natalia Semushkina, *Shippensburg University of Pennsylvania*
Anwar Shiekh, *Dine College*
David Slimmer, *Lander University*
Chris Sorensen, *Kansas State University*
Ronald G. Tabak, *Youngstown State University*
Gajendra Tulsian, *Daytona Beach Community College*
Brigita Urbanc, *Drexel University*
Henry Weigel, *Arapahoe Community College*
Arthur W. Wiggins, *Oakland Community College*
Fredy Zypman, *Yeshiva University*

Preface to the Student

Welcome to physics! Maybe you're taking introductory physics because you're majoring in a field of science or engineering that requires a semester or two of physics. Maybe you're premed, and you know that medical schools are increasingly interested in seeing calculus-based physics on your transcript. Perhaps you're really gung-ho and plan to major in physics. Or maybe you want to study physics further as a minor associated with related fields like math or chemistry or to complement a discipline like economics, environmental studies, or even music. Perhaps you had a great high-school physics course, and you're eager to continue. Maybe high-school physics was an academic disaster for you, and you're approaching this course with trepidation. Or perhaps this is your first experience with physics. Whatever your reason for taking introductory physics, welcome!

And whatever your reason, my goals for you are similar: I'd like to help you develop an understanding and appreciation of the physical universe at a deep and fundamental level; I'd like you to become aware of the broad range of natural and technological phenomena that physics can explain; and I'd like to help you strengthen your analytic and quantitative problem-solving skills. Even if you're studying physics only because it's a requirement, I want to help you engage the subject and come away with an appreciation for this fundamental science and its wide applicability. One of my greatest joys as a physics teacher is having students tell me after the course that they had taken it only because it was required, but found they really enjoyed their exposure to the ideas of physics.

Physics is fundamental. To understand physics is to understand how the world works, both in everyday life and on scales of time and space so small and so large as to defy intuition. For that reason I hope you'll find physics fascinating. But you'll also find it challenging. Learning physics will challenge you with the need for precise thinking and language; with subtle interpretations of even commonplace phenomena; and with the need for skillful application of mathematics. But there's also a simplicity to physics, a simplicity that results because there are in physics only a very few really basic principles to learn. Those succinct principles encompass a universe of natural phenomena and technological applications.

I've been teaching introductory physics for decades, and this book distills everything my students have taught me about the many different ways to approach physics; about the subtle misconceptions students often bring to physics; about the ideas and types of problems that present the greatest challenges; and about ways to make physics engaging, exciting, and relevant to your life and interests.

I have some specific advice for you that grows out of my long experience teaching introductory physics. Keeping this advice in mind will make physics easier (but not necessarily easy!), more interesting, and, I hope, more fun:

- *Read* each chapter thoroughly and carefully before you attempt to work any problem assignments. I've written this text with an informal, conversational style to make it engaging. It's not a reference work to be left alone until you need some specific piece of information; rather, it's an unfolding "story" of physics—its big ideas and their applications in quantitative problem solving. You may think physics is hard because it's mathematical, but in my long experience I've found that failure to *read* thoroughly is the biggest single reason for difficulties in introductory physics.
- *Look for the big ideas*. Physics isn't a hodgepodge of different phenomena, laws, and equations to memorize. Rather, it's a few big ideas from which flow myriad applications, examples, and special cases. In particular, don't think of physics as a jumble of equations that you choose among when solving a problem. Rather, identify those few big ideas and the equations that represent them, and try to see how seemingly distinct examples and special cases relate to the big ideas.
- *When working problems*, *re-read* the appropriate sections of the text, paying particular attention to the worked examples. Follow the IDEA strategy described in Chapter 1 and used in every subsequent worked example. Don't skimp on the final Assess step. Always ask: Does this answer make sense? How can I understand my answer in relation to the big principles of physics? How was this problem like others I've worked, or like examples in the text?
- *Don't confuse physics with math*. Mathematics is a tool, not an end in itself. Equations in physics aren't abstract math, but statements about the physical world. Be sure you understand each equation for what it says about physics, not just as an equality between mathematical terms.
- *Work with others*. Getting together informally in a room with a blackboard is a great way to explore physics, to clarify your ideas and help others clarify theirs, and to learn from your peers. I urge you to discuss physics problems together with your classmates, to contemplate together the "For Thought and Discussion" questions at the end of each chapter, and to engage one another in lively dialog as you grow your understanding of physics, the fundamental science.

Detailed Contents

Volume 1 contains Chapters 1–19
Volume 2 contains Chapters 20–39

PART FOUR

Electromagnetism 354

APPENDICES

ESSENTIAL UNIVERSITY PHYSICS

VOLUME 2 CHAPTERS 20–39

PART FOUR OVERVIEW

Electromagnetism

Electricity constitutes a significant portion of humankind's energy, as evidenced by this composite satellite image of Earth at night. Nearly all that electrical energy is produced by generators, devices that exploit an intimate relation between electricity and magnetism.

Electromagnetism is one of the fundamental forces, and it governs the behavior of matter from the atomic scale to the macroscopic world. Electromagnetic technology, from computer microchips to cell phones and on to large electric motors and generators, is essential to modern society. Even our bodies rely heavily on electromagnetism: Electric signals pace our heartbeat, electrochemical processes transmit nerve impulses, and the electric structure of cell membranes mediates the flow of materials into and out of the cell.

Four fundamental laws describe electricity and magnetism. Two deal separately with the two phenomena, while the others reveal profound connections that make electricity and magnetism aspects of a single phenomenon—electromagnetism. In this part you'll come to understand those fundamental laws, learn how electromagnetism determines the structure and behavior of nearly all matter, and explore the electromagnetic technologies that play so important a role in your life. Finally, you'll see how the laws of electromagnetism lead to electromagnetic waves and thus help us understand the nature of light.

Electric Charge, Force, and Field

What You Know

- You understand Newton's laws, so you know how to work with all kinds of forces.
- You know how to handle vector quantities.
- You're generally familiar with electricity and electronics on an everyday basis, and you probably know about electric charge and how like charges repel and opposite charges attract.

What You're Learning

- Electric charge is a fundamental property of matter, and here you'll learn how electric charge behaves and how two or more charges interact.
- You'll learn *Coulomb's law* for the *electric force* between two charges, written in vector form.
- You'll see how the *superposition principle* gives a simple prescription for combining electric forces from several charges.
- You'll review the *field concept* from Chapter 8 and learn how it applies to the *electric field*.
- You'll learn how to calculate the electric fields of distributions of multiple charges.
- You'll learn how matter behaves in electric fields and will learn the terms *conductor*, *insulator*, and *dielectric* as they apply to bulk matter.

How You'll Use It

- Electric charge and electric field are fundamental entities. You'll use these throughout your study of electromagnetism.
- A deeper understanding of electrical phenomena will also help you better appreciate everyday technology and use it more wisely and safely.

What's the fundamental criterion for initiating a lightning strike?

What holds your body together? What keeps a skyscraper standing? What holds your car on the road as you round a turn? What governs the electronic circuitry in your computer or smartphone, or provides the tension in your climbing rope? What enables a plant to make sugar from sunlight and simple chemicals? What underlies the awesome beauty of lightning? The answer, in all cases, is the **electric force**. With the exception of gravity, all the forces we've encountered in mechanics—including tension forces, normal forces, compression forces, and friction—are based on electric interactions; so are the forces responsible for all of chemistry and biology. The electric force, in turn, involves a fundamental property of matter—namely, electric charge.

20.1 Electric Charge

Electric charge is an intrinsic property of the electrons and protons that, along with uncharged neutrons, make up ordinary matter. What is electric charge? At the most fundamental level we don't know. We don't know what mass "really" is either, but we're familiar with it because we've spent our lives pushing objects around. Similarly, our knowledge of electric charge results from observing the behavior of charged objects.

Charge comes in two varieties, which Benjamin Franklin designated *positive* and *negative*. Those names are useful because the total charge on an object—the object's

net charge—is the algebraic sum of its constituent charges. Like charges repel, and opposites attract, a fact that constitutes a qualitative description of the electric force.

Quantities of Charge

All electrons carry the same charge, and all protons carry the same charge. The proton's charge has *exactly* the same magnitude as the electron's, but with opposite sign. Given that electrons and protons differ substantially in other properties—like mass—this electric relation is remarkable. Exercise 13 shows how dramatically different our world would be if there were even a slight difference between the magnitudes of the electron and proton charges.

The magnitude of the electron or proton charge is the **elementary charge** e. Electric charge is **quantized**; that is, it comes only in discrete amounts. In a famous experiment in 1909, the American physicist R. A. Millikan measured the charge on small oil drops and found it was always a multiple of a basic value we now know as the elementary charge.

Elementary particle theories show that the fundamental charge is actually $\frac{1}{3}e$. Such "fractional charges" reside on quarks, the building blocks of protons, neutrons, and many other particles. Quarks always join to produce particles with integer multiples of the full elementary charge, and it seems impossible to isolate individual quarks.

The SI unit of charge is the **coulomb** (C), named for the French physicist Charles Augustin de Coulomb (1736–1806). Since the late 19th century, the coulomb has been defined in terms of electric current. That will soon change, with the redefinition of SI units, to a definition based on assigning an exact value to the elementary charge e. Either way, it's convenient for our purposes to take $e \simeq 1.60\times10^{-19}$ C or to consider that 1 C is equivalent to about 6.24×10^{18} elementary charges.

Charge Conservation

Electric charge is a conserved quantity, meaning that the net charge in a closed region remains constant. Charged particles may be created or annihilated, but always in pairs of equal and opposite charge. The net charge always remains the same.

GOT IT? 20.1 The proton is a composite particle composed of three quarks, all of which are either *up quarks* (u; charge $+\frac{2}{3}e$) or *down quarks* (d; charge $-\frac{1}{3}e$). (More on quarks in Chapter 39.) Which of these quark combinations is the proton? (a) *udd*; (b) *uuu*; (c) *uud*; (d) *ddd*

20.2 Coulomb's Law

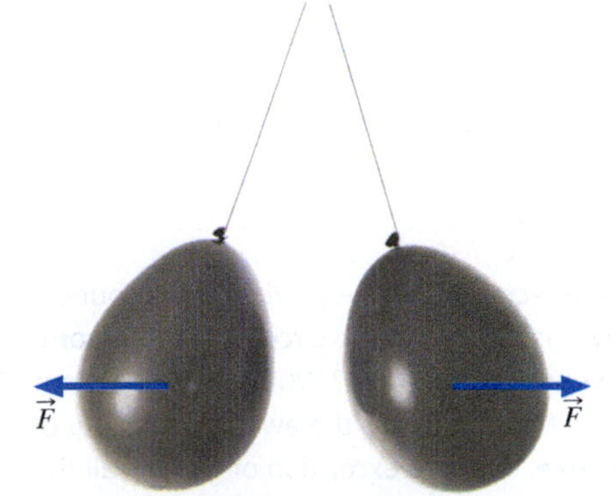

FIGURE 20.1 Two balloons carrying similar electric charges repel each other.

Rub a balloon; it gets charged and sticks to your clothing. Charge another balloon, and the two repel (Fig. 20.1). Socks cling to your clothes as they come from the dryer, and bits of Styrofoam cling annoyingly to your hands. Walk across a carpet, and you'll feel a shock when you touch a doorknob. All these are common examples where you're directly aware of electric charge.

Electricity would be unimportant if the only significant electric interactions were these obvious ones. In fact, the electric force dominates all interactions of everyday matter, from the motion of a car to the movement of a muscle. It's just that matter on a large scale is almost perfectly neutral, meaning it carries zero net charge. Therefore, electric effects aren't obvious. But at the molecular level, the electric nature of matter is immediately evident (Fig. 20.2).

PhET: Balloons and Static Electricity

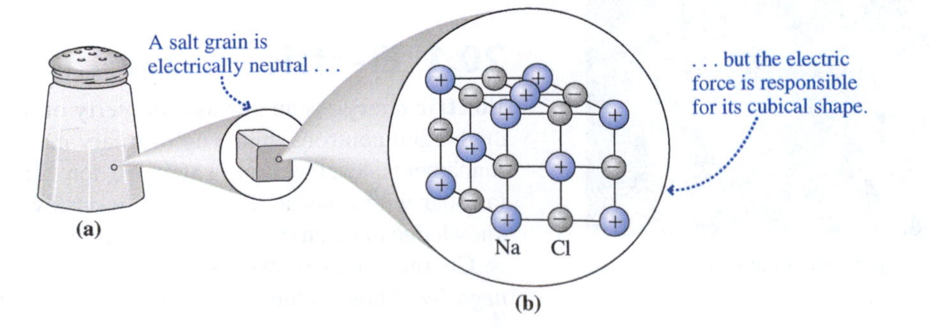

FIGURE 20.2 (a) A single salt grain is electrically neutral, so the electric force isn't obvious. (b) Actually, the electric force determines the structure of salt.

Attraction and repulsion of electric charges imply a force. Joseph Priestley and Charles Augustin de Coulomb investigated this force in the late 1700s. They found that the force between two charges acts along the line joining them, with the magnitude proportional to the product of the charges and inversely proportional to the square of the distance between them. **Coulomb's law** summarizes these results:

$$\vec{F}_{12} = \frac{kq_1q_2}{r^2}\hat{r} \quad \text{(Coulomb's law)} \tag{20.1}$$

where $\vec{F}_{12}$ is the force charge q_1 exerts on q_2 and r is the distance between the charges. In SI the proportionality constant k has the approximate value $9.0 \times 10^9\ \text{N}\cdot\text{m}^2/\text{C}^2$. Force is a vector, and $\hat{r}$ is a unit vector that helps determine its direction. Figure 20.3 shows that $\hat{r}$ lies on a line passing through the two charges and points in the direction *from* q_1 *toward* q_2. Reverse the roles of q_1 and q_2, and you'll see that $\vec{F}_{21}$ has the same magnitude as $\vec{F}_{12}$ but the opposite direction; thus Coulomb's law obeys Newton's third law. Figure 20.3 also shows that the force is in the same direction as the unit vector when the charges have the same sign, but opposite the unit vector when the charges have different signs. Thus Coulomb's law accounts for the fact that like charges repel and opposites attract.

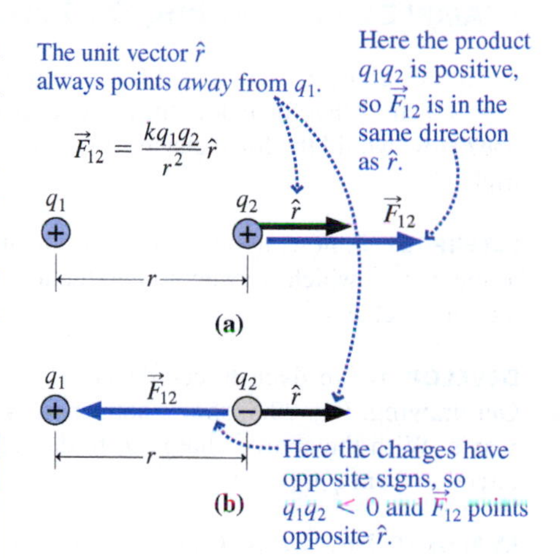

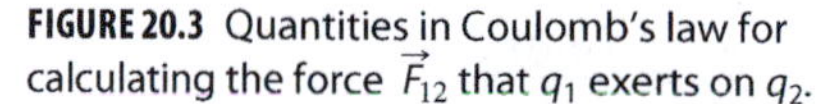

FIGURE 20.3 Quantities in Coulomb's law for calculating the force $\vec{F}_{12}$ that q_1 exerts on q_2.

PROBLEM-SOLVING STRATEGY 20.1 Coulomb's Law

The key to using Coulomb's law is to remember that force is a vector, and to realize that Coulomb's law in the form of Equation 20.1 gives both the magnitude and direction of the electric force. Dealing carefully with vector directions is especially important in situations with more than two charges.

INTERPRET First, make sure you're dealing with the electric force alone. Identify the charge or charges on which you want to calculate the force. Next, identify the charge or charges producing the force. These comprise the **source charge**.

DEVELOP Begin with a drawing that shows the charges, as in Fig. 20.4. If you're given charge coordinates, place the charges on the coordinate system; if not, choose a suitable coordinate system. For each source charge, determine the unit vector(s) in Equation 20.1. If the charges lie along or parallel to a coordinate axis, then the unit vector will be one of the unit vectors $\hat{\imath}$, $\hat{\jmath}$, or $\hat{k}$, perhaps with a minus sign. In Fig. 20.4, the force on q_3 due to q_1 is such a case. When the two charges don't lie on a coordinate axis, like q_1 and q_2 in Fig. 20.4, you can find the unit vector by noting that the displacement vector $\vec{r}_{12}$ points in the desired direction, from the source charge to the charge experiencing the force. Dividing $\vec{r}_{12}$ by its own magnitude then gives the unit vector in the direction of $\vec{r}_{12}$; that is, $\hat{r} = \vec{r}_{12}/r_{12}$.

EVALUATE For each source charge, determine the electric force using Equation 20.1,

$$\vec{F}_{12} = (kq_1q_2/r^2)\hat{r}$$

with $\hat{r}$ the unit vector you've just found.

ASSESS As always, assess your answer to see that it makes sense. Is the direction of the force you found consistent with the signs and placements of the charges giving rise to the force?

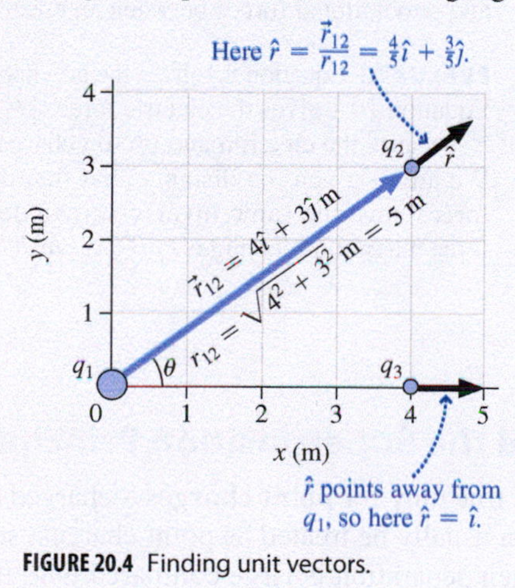

FIGURE 20.4 Finding unit vectors.

GOT IT? 20.2 Charge q_1 is located at $x = 1$ m, $y = 0$. What should you use for the unit vector $\hat{r}$ in Coulomb's law if you're calculating the force q_1 exerts on a charge q_2 located (1) at the origin and (2) at the point $x = 0, y = 1$ m? Explain why you can answer without knowing the sign of either charge.

EXAMPLE 20.1 Finding the Force: Two Charges

A 1.0-μC charge is at $x = 1.0$ cm, and a −1.5-μC charge is at $x = 3.0$ cm. What force does the positive charge exert on the negative one? How would the force change if the distance between the charges tripled?

INTERPRET Following our strategy, we identify the −1.5-μC charge as the one on which we want to find the force and the 1-μC charge as the source charge.

DEVELOP We're given the coordinates $x_1 = 1.0$ cm and $x_2 = 3.0$ cm. Our drawing, Fig. 20.5, shows both charges at their positions on the x-axis. With the source charge q_1 to the left, the unit vector in the direction from q_1 toward q_2 is $\hat{\imath}$.

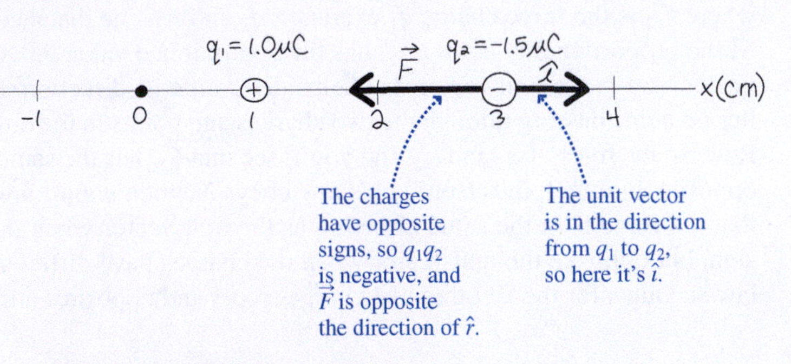

FIGURE 20.5 Sketch for Example 20.1.

EVALUATE Now we use Coulomb's law to evaluate the force:

$$\vec{F}_{12} = \frac{kq_1q_2}{r^2}\hat{r}$$

$$= \frac{(9.0\times10^9\ \text{N}\cdot\text{m}^2/\text{C}^2)(1.0\times10^{-6}\ \text{C})(-1.5\times10^{-6}\ \text{C})}{(0.020\ \text{m})^2}\hat{\imath}$$

$$= -34\hat{\imath}\ \text{N}$$

This force is for a separation of 2 cm; if that distance tripled, the force would drop by a factor of $1/3^2$, to $-3.8\hat{\imath}$ N.

ASSESS Make sense? Although the unit vector $\hat{\imath}$ points in the $+x$-direction, the charges have opposite signs and that makes the force direction opposite the unit vector, as shown in Fig. 20.5. In simpler terms, we've got two opposite charges, so they attract. That means the force exerted on a charge at $x = 3$ cm by an opposite charge at $x = 1$ cm had better be in the $-x$-direction. ■

CONCEPTUAL EXAMPLE 20.1 Gravity and the Electric Force

The electric force between elementary particles is far stronger than the gravitational force, yet gravity is much more obvious in everyday life. Why?

EVALUATE Gravity and the electric force obey similar inverse-square laws, and the magnitude of the force is proportional to the product of the masses or charges. There's a big difference, though: There's only one kind of mass, and gravity is always attractive, so large concentrations of mass—like a planet—result in strong gravitational forces. But charge comes in two varieties, and opposites attract, so large accumulations of matter tend to be electrically neutral, in which case large-scale electrical interactions aren't obvious.

ASSESS Ironically, it's the very strength of the electric force that makes it less obvious in everyday life. Opposite charges bind strongly, making bulk matter electrically neutral and its electrical interactions subtle.

MAKING THE CONNECTION Compare the magnitudes of the electric and gravitational forces between an electron and a proton.

EVALUATE Equation 8.1 gives the gravitational force: $F_g = Gm_em_p/r^2$. Equation 20.1 gives the electric force: $|F_E| = ke^2/r^2$, where we wrote e^2 because the electron and proton charges have the same magnitude. We aren't given the distance, but that doesn't matter because both forces have the same inverse-square dependence. The ratio of the force magnitudes is huge: $F_E/F_g = ke^2/Gm_em_p = 2.3\times10^{39}$!

Point Charges and the Superposition Principle

Coulomb's law is strictly true only for **point charges**—charged objects of negligible size. Electrons and protons can usually be treated as point charges; so, approximately, can any two charged objects if their separation is large compared with their size. But often we're interested in the electric effects of **charge distributions**—arrangements of charge spread over space. Charge distributions are present in molecules, memory cells in your computer, your heart, and thunderclouds. We need to combine the effects of two or more charges to find the electric effects of such charge distributions.

Figure 20.6 shows two charges q_1 and q_2 that constitute a simple charge distribution. We want to know the net force these exert on a third charge q_3. To find that net force, you might calculate the forces $\vec{F}_{13}$ and $\vec{F}_{23}$ from Equation 20.1, and then vectorially add them. And you'd be right: The force that q_1 exerts on q_3 is unaffected by the presence of q_2, and vice versa, so you can apply Coulomb's law separately to the pairs q_1q_3 and q_2q_3 and then combine the results. That may seem obvious, but nature needn't have been so simple.

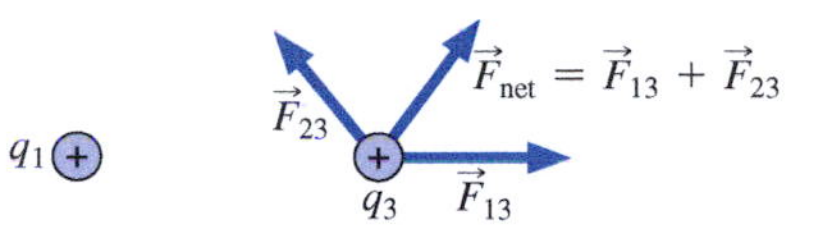

FIGURE 20.6 The superposition principle lets us add vectorially the forces from two or more charges.

The fact that electric forces add vectorially is called the **superposition principle**. Our confidence in this principle is ultimately based on experiments showing that electric and indeed electromagnetic phenomena behave according to the principle. With superposition we can solve relatively complicated problems by breaking them into simpler parts. If the superposition principle didn't hold, the mathematical description of electromagnetism would be far more complicated.

Although the force that one point charge exerts on another decreases with the inverse square of the distance between them, the same is not necessarily true of the force resulting from a charge distribution. The next example provides a case in point.

EXAMPLE 20.2 Finding the Force: Raindrops

Charged raindrops are ultimately responsible for lightning, producing substantial electric charge within specific regions of a thundercloud. Suppose two drops with equal charge q are on the x-axis at $x = \pm a$. Find the electric force on a third drop with charge Q at an arbitrary point on the y-axis.

INTERPRET Coulomb's law and the superposition principle apply, and we identify Q as the charge for which we want the force. The two charges q are the source charges.

DEVELOP Figure 20.7 is our drawing, showing the charges, the individual force vectors, and their sum. The drawing shows that the distance r in Coulomb's law is the hypotenuse $\sqrt{a^2 + y^2}$. It's clear from symmetry that the net force is in the y-direction, so we need to find only the y-components of the unit vectors. The y-components are clearly the same for each, and the drawing shows that they're given by $\hat{r}_y = y/\sqrt{a^2 + y^2}$.

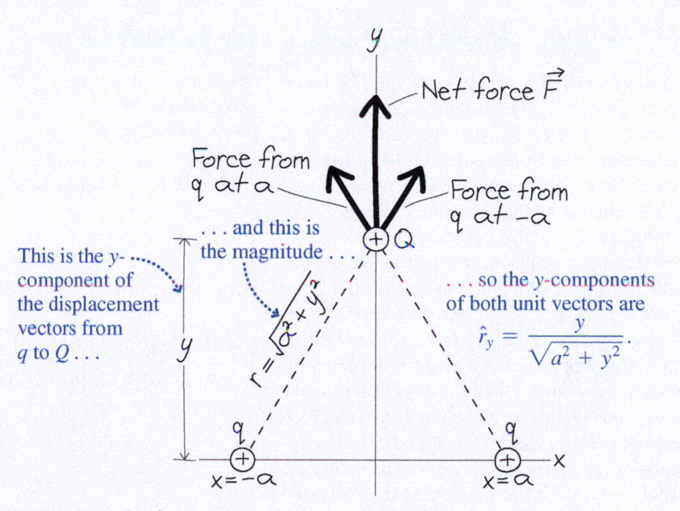

FIGURE 20.7 The force on Q is the vector sum of the forces from the individual charges.

EVALUATE From Coulomb's law, the y-component of the force from each q is $F_y = (kqQ/r^2)\hat{r}_y$, and the net force on Q becomes

$$\vec{F} = 2\left(\frac{kqQ}{a^2 + y^2}\right)\left(\frac{y}{\sqrt{a^2 + y^2}}\right)\hat{j} = \frac{2kqQy}{(a^2 + y^2)^{3/2}}\hat{j}$$

The factor of 2 comes from the two charges q, which contribute equally to the net force.

ASSESS Make sense? Evaluating $\vec{F}$ at $y = 0$ gives zero force. Here, midway between the two charges, Q experiences equal but opposite forces and the net force is zero. At large distances $y \gg a$, on the other hand, we can neglect a^2 compared with y^2, and the force becomes $\vec{F} = k(2q)Q\hat{j}/y^2$. This is just what we would expect from a single charge $2q$ a distance y from Q—showing that the system of two charges acts like a single charge $2q$ at distances that are large compared with the charge separation. In between our two extremes the behavior of force with distance is more complicated; in fact, its magnitude initially increases as Q moves away from the origin and then begins to decrease.

In drawing Fig. 20.7, we tacitly assumed that q and Q have the same signs. But our analysis holds even if they don't; then the product qQ is negative, and the forces actually point opposite the directions shown in Fig. 20.7. ■

20.3 The Electric Field

PhET: Charges and Fields

In Chapter 8 we defined the gravitational field at a point as the gravitational force per unit mass that an object at that point would experience. In that context, we can think of $\vec{g}$ as the *force per unit mass* that any object would experience due to Earth's gravity. So we can picture the gravitational field as a set of vectors giving the magnitude and direction of the gravitational force per unit mass at each point, as shown in Fig. 20.8*a* on the next page.

Right at this point the gravitational field is described by the vector $\vec{g}$. That means a mass m placed here would experience a gravitational force $m\vec{g}$.

The gravitational field is a continuous entity, so there are field vectors everywhere. We just can't draw them all.

(a)

Right at this point the electric field is described by the vector $\vec{E}_1$. That means a point charge q placed here would experience an electric force $q\vec{E}_1$.

Over here, farther from the charge producing the field, a point charge q would experience a weaker force $q\vec{E}_2$ in a different direction.

The electric field is a continuous entity, so there are field vectors everywhere. We just can't draw them all.

(b)

FIGURE 20.8 (a) Gravitational and (b) electric fields, here represented as sets of vectors.

We can do the same thing with the electric force, defining the **electric field** as the force per unit charge:

The electric field at any point is the force per unit charge that a charge would experience at that point. Mathematically,

$$\vec{E} = \frac{\vec{F}}{q} \quad \text{(electric field)} \qquad (20.2a)$$

The electric field exists at every point in space. When we represent the field by vectors, we can't draw one everywhere, but that doesn't mean there isn't a field at all points. Furthermore, we draw vectors as extended arrows, but each vector represents the field at only one point—namely, the tail end of the vector. Figure 20.8*b* illustrates this for the electric field of a point charge.

The field concept leads to a shift in our thinking about forces. Instead of the action-at-a-distance idea that Earth reaches across empty space to pull on the Moon, the field concept says that Earth creates a gravitational field and the Moon responds to the field at its location. Similarly, a charge creates an electric field throughout the space surrounding it. A second charge then responds to the field at its immediate location. Although the field reveals itself only through its effect on a charge, the field nevertheless exists at all points, whether or not charges are present. Right now you probably find the field concept a bit abstract, but as you advance in your study of electromagnetism you'll come to appreciate that fields are an essential feature of our universe, every bit as real as matter itself.

We can use Equation 20.2a as a prescription for measuring electric fields. Place a point charge at some location, measure the electric force it experiences, and divide by the charge to get the field. In practice, we need to be careful because the field generally arises from some distribution of source charges. If the charge we're using to probe the field—the **test charge**—is large, the field it creates may disturb the source charges, altering their configuration and thus the field they create. For that reason, it's important to use a very small test charge.

APPLICATION Electrophoresis

Electrophoresis is a widely used application of electric fields for separating molecules by size and molecular weight. It's especially useful in biochemistry and molecular biology for distinguishing larger molecules like proteins and DNA fragments. In the commonly used *gel electrophoresis*, molecules carrying electric charge move through a semisolid but permeable gel under the influence of a uniform electric field; the greater the charge, the greater the electric force. The gel exerts a retarding force that increases with increasing molecular size, with the result that each molecular species moves at a velocity that depends on its size and charge. After a given time, the electric field is switched off. The locations of the molecules then serve as indicators of their size, with the molecules that traveled farthest being the smallest. The photo shows a typical gel electrophoresis result. Here DNA fragments were introduced into the seven channels at the top of the gel and then moved downward; their final locations indicate molecular size. The smaller molecules—those with fewer nucleotide base pairs—end up farther down on the gel. The electric field is shown by the arrow; it needs to point upward because DNA fragments carry a negative charge.

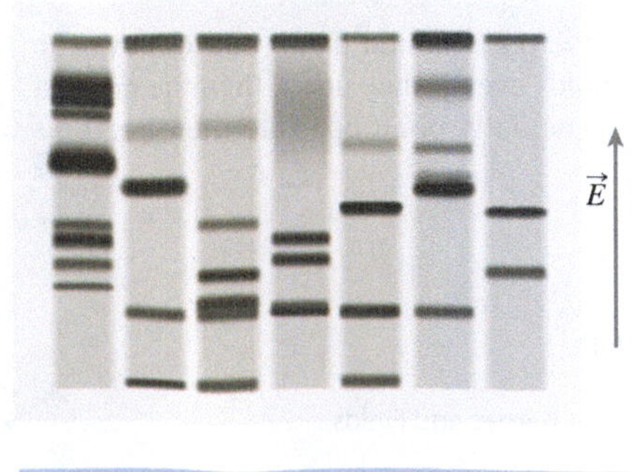

If we know the electric field $\vec{E}$ at a point, we can rearrange Equation 20.2a to find the force on any point charge q placed at that point:

$$\vec{F} = q\vec{E} \quad \text{(electric force and field)} \tag{20.2b}$$

If the charge q is positive, then this force is in the same direction as the field, but if q is negative, then the force is opposite to the field direction.

Equations 20.2 show that the units of electric field are newtons per coulomb. Fields of hundreds to thousands of N/C are commonplace, while fields of 3 MN/C will tear electrons from air molecules.

EXAMPLE 20.3 Force and Field: Inside a Lightning Storm

A charged raindrop carrying 10 μC experiences an electric force of 0.30 N in the $+x$-direction. What's the electric field at its location? What would the force be on a -5.0-μC drop at the same point?

INTERPRET In this problem we need to distinguish between electric force and electric field. The electric field exists with or without the charged raindrop present, and the electric force arises when the charged raindrop is in the electric field.

DEVELOP Knowing the electric force and the charge on the raindrop, we can use Equation 20.2a, $\vec{E} = \vec{F}/q$, to get the electric field. Once we know the field, we can use Equation 20.2b, $\vec{F} = q\vec{E}$, to calculate the force that would act if a different charge were at the same point.

EVALUATE Equation 20.2a gives the electric field:

$$\vec{E} = \frac{\vec{F}}{q} = \frac{0.30\hat{\imath}\ \text{N}}{10\ \mu\text{C}} = 30\hat{\imath}\ \text{kN/C}$$

Acting on a -5.0-μC charge, this field would result in a force

$$\vec{F} = q\vec{E} = (-5.0\ \mu\text{C})(30\hat{\imath}\ \text{kN/C}) = -0.15\hat{\imath}\ \text{N}$$

ASSESS Make sense? The force on the second charge is opposite the direction of the field because now we've got a negative charge in the same field.

✓TIP The Field Is Independent of the Test Charge

Does the electric field in this example point in the $-x$-direction when the charge is negative? No. The field is independent of the particular charge experiencing that field. Here the electric field points in the $+x$-direction *no matter what charge* you put in the field. For a positive charge, the force $q\vec{E}$ points in the *same* direction as the field; for a negative charge, $q < 0$, the force is *opposite* the field.

The Field of a Point Charge

Once we know the field of a charge distribution, we can calculate its effect on other charges. The simplest charge distribution is a single point charge. Coulomb's law gives the force on a test charge q_{test} located a distance r from a point charge q: $\vec{F} = (kqq_{\text{test}}/r^2)\hat{r}$, where $\hat{r}$ is a unit vector pointing *away* from q. The electric field arising from q is the force per unit charge, or

$$\vec{E} = \frac{\vec{F}}{q_{\text{test}}} = \frac{kq}{r^2}\hat{r} \quad \text{(field of a point charge)} \tag{20.3}$$

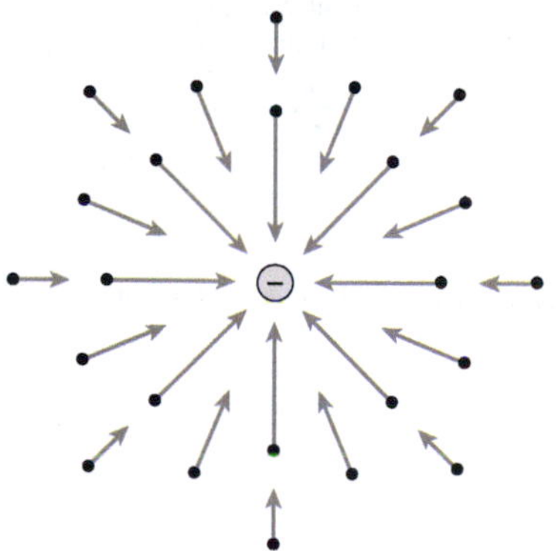

FIGURE 20.9 Field vectors for a negative point charge.

Since it's so closely related to Coulomb's law for the electric force, we also refer to Equation 20.3 as Coulomb's law. The equation contains no reference to the test charge q_{test} because the field of q exists independently of any other charge. Since $\hat{r}$ always points *away* from q, the direction of $\vec{E}$ is radially outward if q is positive and radially inward if q is negative. Figure 20.9 shows some field vectors for a negative point charge, analogous to those of the positive point charge in Fig. 20.8*b*.

GOT IT? 20.3 A positive point charge is located at the origin of an $x-y$ coordinate system, and an electron is placed at a location where the electric field due to the point charge is given by $\vec{E} = E_0(\hat{\imath} + \hat{\jmath})$, where E_0 is positive. Is the direction of the force on the electron (a) toward the origin, (b) away from the origin, (c) parallel to the x-axis, or (d) impossible to determine without knowing the coordinates of the electron's position?

MP

PhET: Charges and Fields

20.4 Fields of Charge Distributions

Since the electric force obeys the superposition principle, so does the electric field. That means the field of a charge distribution is the vector sum of the fields of the individual point charges making up the distribution:

$$\vec{E} = \vec{E}_1 + \vec{E}_2 + \vec{E}_3 + \cdots = \sum_i \vec{E}_i = \sum_i \frac{kq_i}{r_i^2}\hat{r}_i \qquad (20.4)$$

Here the $\vec{E}_i$'s are the fields of the point charges q_i located at distances r_i from the point where we're evaluating the field—called, appropriately, the **field point**. The $\hat{r}_i$'s are unit vectors pointing *from* each point charge *toward* the field point. In principle, Equation 20.4 gives the electric field of *any* charge distribution. In practice, the process of summing the individual field vectors is often complicated unless the charge distribution contains relatively few charges arranged in a symmetric way.

Finding electric fields using Equation 20.4 involves the same strategy we introduced for finding the electric force; the only difference is that there's no charge to experience the force. The first step then involves identifying the field point. We still need to find the appropriate unit vectors and form the vector sum in Equation 20.4.

EXAMPLE 20.4 Finding the Field: Two Protons

Two protons are 3.6 nm apart. Find the electric field at a point between them, 1.2 nm from one of the protons. Then find the force on an electron at this point.

INTERPRET We follow our electric-force strategy, except that instead of identifying the charge experiencing the force, we identify the field point as being 1.2 nm from one proton. The source charges are the two protons; they produce the field we're interested in.

DEVELOP Let's have the protons define the x-axis, as drawn in Fig. 20.10. Then the unit vector $\hat{r}_1$ from the left-hand proton toward the field point (which we've marked P) is $+\hat{\imath}$, while $\hat{r}_2$ from the right-hand proton toward P is $-\hat{\imath}$.

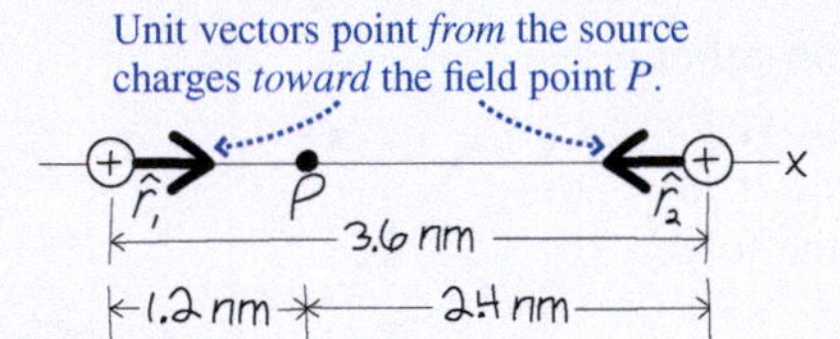

FIGURE 20.10 Finding the electric field at P.

EVALUATE We now evaluate the field at P using Equation 20.4:

$$\vec{E} = \vec{E}_1 + \vec{E}_2 = \frac{ke}{r_1^2}\hat{\imath} + \frac{ke}{r_2^2}(-\hat{\imath}) = ke\left(\frac{1}{r_1^2} - \frac{1}{r_2^2}\right)\hat{\imath}$$

We wrote e for q here because the protons' charge is the elementary charge.

Using $e = 1.6\times10^{-19}$ C, $r_1 = 1.2$ nm, and $r_2 = 2.4$ nm gives $\vec{E} = 750\hat{\imath}$ MN/C. An electron at P will therefore experience a force $\vec{F} = qE = -eE = -0.12\hat{\imath}$ nN.

ASSESS Make sense? The field points in the positive x-direction, reflecting the fact that P is closer to the left-hand proton with its stronger field at P. The force on the electron, on the other hand, is in the $-x$-direction; that's because the electron is negative (we used $q = -e$ for its charge), so the force it experiences is opposite the field. That field of almost 1 GN/C sounds huge—but that's not unusual at the microscopic scale, where we're close to individual elementary particles. ■

Sometimes we're interested in finding not the electric field but a point or points where the field has a particular value—often zero. Conceptual Example 20.2 explores such a case.

CONCEPTUAL EXAMPLE 20.2 Zero Field, Zero Force

A positive charge $+2Q$ is located at the origin, and a negative charge $-Q$ is at $x = a$. In which region of the x-axis is there a point where the force on a test charge—and therefore the electric field—is zero?

INTERPRET We're asked to locate qualitatively a point where the field is zero. Our sketch of the situation, Fig. 20.11, shows that the two charges divide the x-axis into three regions: (1) to the left of $2Q$ ($x < 0$), (2) between the charges ($0 < x < a$), and (3) to the right of $-Q$ ($x > a$). We need to determine which region could include a point where the electric force on a test charge is zero.

EVALUATE Consider what would happen to a positive test charge placed in each of these three regions. Anywhere in region (1), the test charge is closer to the charge with greater magnitude ($2Q$). That charge dominates throughout region (1), where our test charge would experience a repulsive force (to the left). The electric field, then, can't be zero in region (1). Between the two charges, the repulsive force from $2Q$ on a positive test charge points to the right; so does the attractive force from $-Q$. The field, therefore, can't be zero in region (2). That leaves region (3). Could the field be zero here? Put a positive test charge very close to $-Q$, and it experiences an attractive force toward the left. But far away, the distance between $2Q$ and $-Q$ becomes negligible. The fields of both charges drop off as the inverse square of the distance, so at large distances the field of the stronger charge will dominate. Therefore there *is* a point somewhere to the right of $-Q$ where the force on a test charge, and therefore the electric field, will be zero.

ASSESS This answer is consistent with our insight from Example 20.2 that when we get far from a charge distribution it begins to resemble a point charge with the net charge of the distribution. Here that net charge is $2Q - Q = +Q$, so at large distances we should indeed have a field pointing away from the charge distribution—and that's to the right in region (3). Although we considered a positive test charge, you'll reach the same conclusion with a negative test charge.

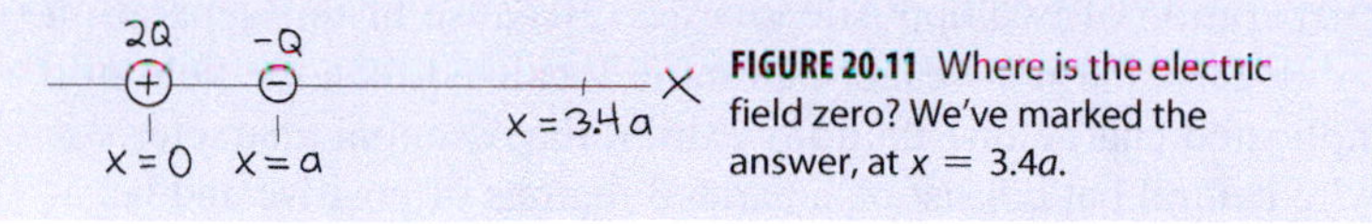

FIGURE 20.11 Where is the electric field zero? We've marked the answer, at $x = 3.4a$.

MAKING THE CONNECTION Find an expression for the position where the electric field in this example is zero.

EVALUATE In Fig. 20.11 we've taken the origin at $2Q$, so at any position x in region (3) we're a distance x from $2Q$ and a distance $x-a$ from $-Q$. Since we're to the right of both charges, the unit vector in Equation 20.3 for the point-charge field—a vector that always points away from the point charge—becomes $+\hat{\imath}$ for both charges. Applying Equation 20.3, $\vec{E} = (kq/r^2)\hat{r}$, for the fields of the two charges and summing gives

$$\vec{E} = \frac{k(2Q)}{x^2}\hat{\imath} + \frac{k(-Q)}{(x-a)^2}\hat{\imath}$$

If we set this expression to zero, we can cancel k, Q, and $\hat{\imath}$; inverting both sides of the remaining equation gives $x^2/2 = (x-a)^2$. Finally, taking the square root and solving for x gives the answer: $x = a\sqrt{2}/(\sqrt{2}-1) \simeq 3.4a$. As a check, note that this point does indeed lie to the right of $x = a$. We've marked this point in Fig. 20.11.

The Electric Dipole

One of the most important charge distributions is the **electric dipole**, consisting of two point charges of equal magnitude but opposite sign. Many molecules are essentially dipoles, so understanding the dipole helps explain molecular behavior (Fig. 20.12). During contraction the heart muscle becomes essentially a dipole, and physicians performing electrocardiography are measuring, among other things, the strength and orientation of that dipole. Technological devices, including radio and TV antennas, often use the dipole configuration.

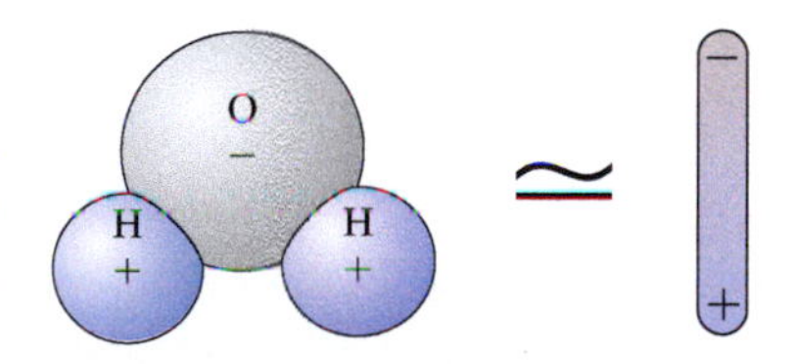

FIGURE 20.12 A water molecule behaves like an electric dipole. Its net charge is zero, but regions of positive and negative charge are separated.

EXAMPLE 20.5 The Electric Dipole: Modeling a Molecule

A molecule may be modeled approximately as a positive charge q at $x = a$ and a negative charge $-q$ at $x = -a$. Evaluate the electric field on the y-axis, and find an approximate expression valid at large distances ($|y| \gg a$).

INTERPRET Here's another example where we'll use our strategy in applying Equation 20.4 to calculate the field of a charge distribution. We identify the field point as being anywhere on the y-axis and the source charges as being $\pm q$.

DEVELOP Figure 20.13 is our drawing. The individual unit vectors point from the two charges toward the field point, but the *negative* charge contributes a field *opposite* its unit vector; we've indicated the individual fields in Fig. 20.13. Here symmetry makes the y-components cancel, giving a net field in the $-x$-direction. So we need only the x-components of the unit vectors, which Fig. 20.13 shows are $\hat{r}_{x-} = a/r$ for the negative charge at $-a$ and $\hat{r}_{x+} = -a/r$ for the positive charge at a.

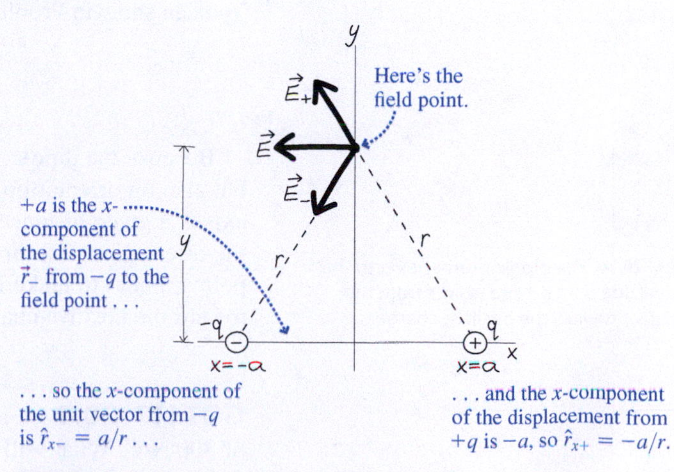

FIGURE 20.13 Finding the field of an electric dipole.

(continued)

EVALUATE We then evaluate the field using Equation 20.4:

$$\vec{E} = \frac{k(-q)}{r^2}\left(\frac{a}{r}\right)\hat{\imath} + \frac{kq}{r^2}\left(-\frac{a}{r}\right)\hat{\imath} = -\frac{2kqa}{(a^2 + y^2)^{3/2}}\hat{\imath}$$

where in the last step we used $r = \sqrt{a^2 + y^2}$. For $|y| \gg a$ we can neglect a^2 compared with y^2, giving

$$\vec{E} \simeq -\frac{2kqa}{|y|^3}\hat{\imath} \qquad (|y| \gg a)$$

ASSESS Make sense? The dipole has no net charge, so at large distances its field can't have the inverse-square drop-off of a point-charge field. Instead the dipole field falls faster, here as $1/|y|^3$. Note that we were careful to put absolute value signs on y^3; that way, our result applies for both positive and negative values of y.

✓TIP Approximations

Making approximations requires care. Here we're basically asking for the field when y is so large that a is negligible *compared with* y. So we neglect a^2 compared with y^2 when the two are summed, but we *don't* neglect a when it appears in the numerator, where it isn't being directly compared with y.

Example 20.5 shows that the dipole field at large distances decreases as the inverse *cube* of distance. Physically, that's because the dipole has zero *net* charge. Its field arises entirely from the slight separation of two opposite charges. Because of this separation the dipole field isn't exactly zero, but it's weaker and more localized than the field of a point charge. Many complicated charge distributions exhibit the essential characteristic of a dipole—that is, they're neutral but consist of separated regions of positive and negative charge—and at large distances such distributions all have essentially the same field configuration.

At large distances the dipole's physical characteristics q and a enter the equation for the electric field only through the product qa. We could double q and halve a, and the dipole's electric field would remain unchanged. At large distances, therefore, a dipole's electric properties are characterized completely by its **electric dipole moment** p, defined as the product of the charge q and the separation d between the two charges making up the dipole:

$$p = qd \quad \text{(dipole moment)} \qquad (20.5)$$

In Example 20.5 the charge separation was $d = 2a$, so there the dipole moment was $p = 2aq$. In terms of the dipole moment, the field in Example 20.5 can then be written

$$\vec{E} = -\frac{kp}{|y|^3}\hat{\imath} \quad \left(\begin{array}{c}\text{dipole field for } |y| \gg a,\\ \text{on perpendicular bisector}\end{array}\right) \qquad (20.6a)$$

You can show in Problem 52 that the field on the dipole axis is given by

$$\vec{E} = \frac{2kp}{|x|^3}\hat{\imath} \quad \left(\begin{array}{c}\text{dipole field}\\ \text{for } |x| \gg a, \text{ on axis}\end{array}\right) \qquad (20.6b)$$

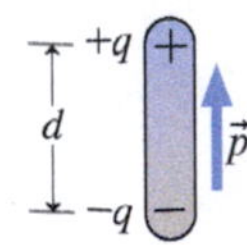

FIGURE 20.14 The dipole moment vector has magnitude $p = qd$ and points from the negative toward the positive charge.

Because the dipole isn't spherically symmetric, its field depends not only on distance but also on orientation; for instance, Equations 20.6 show that the field along the dipole axis at a given distance is twice as strong as along the bisector. So it's important to know the orientation of a dipole in space, and therefore we generalize our definition of the dipole moment to make it a vector of magnitude $p = qd$ in the direction from the negative toward the positive charge (Fig. 20.14).

GOT IT? 20.4 Far from a charge distribution, you measure an electric field strength of 800 N/C. What will the field strength be if you double your distance from the charge distribution, if the distribution consists of (1) a point charge or (2) a dipole?

Continuous Charge Distributions

Charge distribution

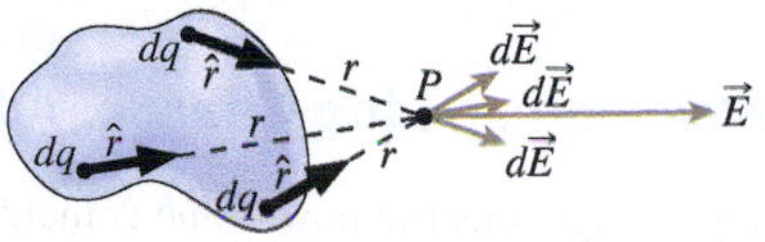

FIGURE 20.15 The electric field at P is the vector sum of the fields $d\vec{E}$ arising from the individual charge elements dq, each calculated using the appropriate distance r and unit vector $\hat{r}$.

Although any charge distribution ultimately consists of pointlike electrons and protons, it would be impossible to sum all the field vectors from the 10^{23} or so particles in a typical piece of matter. Instead, it's convenient to make the approximation that charge is spread continuously over the distribution. If the charge distribution extends throughout a volume, we describe it in terms of the **volume charge density** ρ, with units of C/m^3. For charge distributions spread over surfaces or lines, the corresponding quantities are the **surface charge density** σ (C/m^2) and the **line charge density** λ (C/m).

To calculate the field of a continuous charge distribution, we divide the charged region into very many small charge elements dq, each small enough that it's essentially a point charge. Each dq then produces an electric field $d\vec{E}$ given by Equation 20.3: $d\vec{E} = (k\,dq/r^2)\hat{r}$. We then form the vector sum of all the $d\vec{E}$'s (Fig. 20.15). In the limit of infinitely many infinitesimally small dq's and their corresponding $d\vec{E}$'s, that sum becomes an integral and we have

$$\vec{E} = \int d\vec{E} = \int \frac{k\,dq}{r^2}\hat{r} \quad \left(\begin{array}{c}\text{field of a continuous}\\ \text{charge distribution}\end{array}\right) \qquad (20.7)$$

The limits of this integral include the entire charge distribution.

Calculating the field of a continuous charge distribution involves the same strategy we've already used: We identify the field point and the source charges—although now the source is a continuous charge distribution. Summing the individual field contributions now presents us with an integral, and that means writing the unit vectors $\hat{r}$ and distances r in terms of coordinates over which we can integrate. Setting up the integral involves the same strategy we outlined in Chapter 9 to find the center of mass of a continuous distribution of matter, and used again in Chapter 10 to find rotational inertias.

EXAMPLE 20.6 Evaluating the Field: A Charged Ring

A ring of radius a carries a charge Q distributed evenly over the ring. Find an expression for the electric field at any point on the axis of the ring.

INTERPRET We identify the field point as lying anywhere on the ring's axis, and the source charge as the entire ring.

DEVELOP Let's take the x-axis to coincide with the ring axis, with the center of the ring at $x = 0$ (Fig. 20.16). The figure shows that the y-components of the field contributions from pairs of charge elements on opposite sides of the ring cancel; therefore, the net field points in the $+x$-direction (for $x > 0$) and we need only the x-components of the unit vectors. Those are the same for all unit vectors—namely, $\hat{r}_x = x/r$.

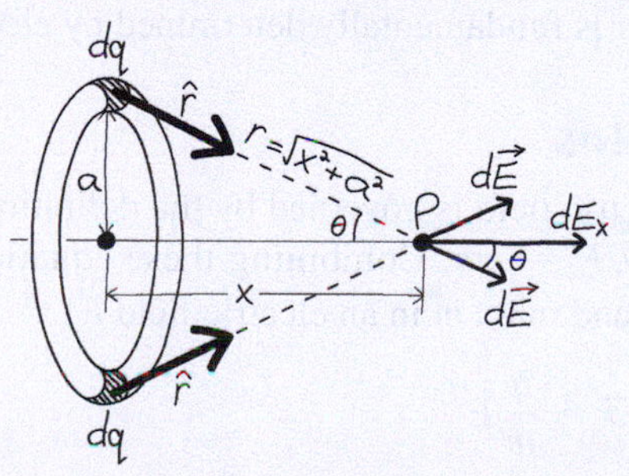

FIGURE 20.16 The electric field of a charged ring points along the ring axis, since field components perpendicular to the axis cancel in pairs.

EVALUATE We're now ready to set up the integral in Equation 20.7. Here each charge element contributes the same amount $dE_x = (k\,dq/r^2)\hat{r}_x = (k\,dq/r^2)(x/r)$ to the field. Figure 20.16 shows that $r = \sqrt{x^2 + a^2} = (x^2 + a^2)^{1/2}$, so the integral becomes

$$E = \int_{\text{ring}} dE_x = \int_{\text{ring}} \frac{kx\,dq}{(x^2 + a^2)^{3/2}} = \frac{kx}{(x^2 + a^2)^{3/2}} \int_{\text{ring}} dq$$

The last step follows because we have a fixed field point P, so its coordinate x is a constant for the integration. But the remaining integral is just the sum of all the charge elements on the ring—namely, the total charge Q. So our result becomes

$$E = \frac{kQx}{(x^2 + a^2)^{3/2}} \qquad \text{(on-axis field, charged ring)}$$

This is the magnitude; the direction is along the x-axis, away from the ring if Q is positive and toward it if Q is negative.

ASSESS Make sense? At $x = 0$ the field is zero. A charge placed at the ring center is pulled (or pushed) equally in all directions—no net force, so no electric field. But for $x \gg a$, we get $E = kQ/x^2$—just what we expect for a point charge Q. As always, a finite-size charge distribution looks like a point charge at large distances. ■

EXAMPLE 20.7 Line Charge: A Power Line's Field

A long, straight electric power line coincides with the x-axis and carries a uniform line charge density λ (unit: C/m). Find the electric field on the y-axis using the approximation that the wire is infinitely long.

INTERPRET We identify the field point as being a distance y from the wire, and the source charge as the whole wire.

DEVELOP Figure 20.17 is our drawing, showing a coordinate system with the field point P along the y-axis. We divide the wire into small charge elements dq and note that field contributions from two such elements dq on opposite sides of the y-axis contribute fields $d\vec{E}$ whose x-components cancel. Then we need only the y-component of each unit vector, and Figure 20.17 shows that's $\hat{r}_y = y/r$, where $r = \sqrt{x^2 + y^2}$.

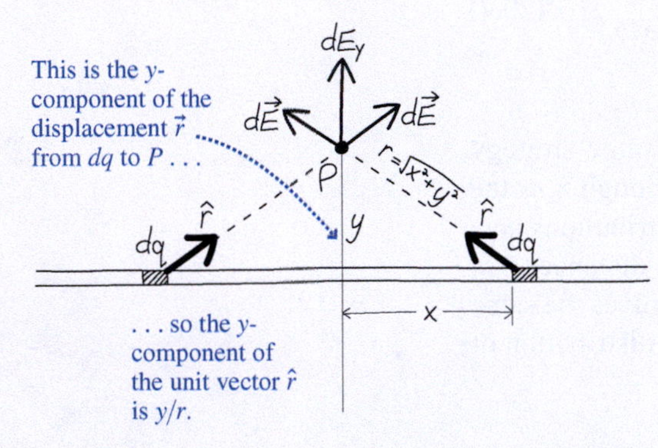

FIGURE 20.17 The field of a charged line is the vector sum of the fields $d\vec{E}$ from all the individual charge elements dq along the line.

EVALUATE We're now ready to set up the integral in Equation 20.7. As described in Chapter 9's integral strategy, we need to relate dq to a geometric variable so we can do the integral. Here our wire has charge density λ C/m, so if a charge element has length dx, then its charge is $dq = \lambda\, dx$. Putting all this together gives the y-component of the field from an arbitrary dq anywhere on the wire:

$$dE_y = \frac{k\,dq}{r^2}\hat{r}_y = \frac{k\lambda\,dx}{r^2}\frac{y}{r} = \frac{k\lambda y}{(x^2 + y^2)^{3/2}}dx$$

where we used $r = \sqrt{x^2 + y^2}$. Since the x-components cancel, we can sum—that is, integrate—the y-components to get the net field:

$$E = E_y = \int_{-\infty}^{+\infty} \frac{k\lambda y\,dx}{(x^2 + y^2)^{3/2}} = k\lambda y \int_{-\infty}^{+\infty} \frac{dx}{(x^2 + y^2)^{3/2}}$$

$$= k\lambda y\left[\frac{x}{y^2\sqrt{x^2 + y^2}}\right]_{-\infty}^{+\infty} = k\lambda y\left[\frac{1}{y^2} - \left(-\frac{1}{y^2}\right)\right] = \frac{2k\lambda}{y}$$

Here we used the integral table in Appendix A and applied the limits $x = \pm\infty$. Our result is the field's magnitude; the direction is away from the line for positive λ and toward the line for negative λ.

ASSESS Make sense? For an infinite line there's nothing to favor one direction along the line over another, so the only way the field can point is radially, away from or toward the line (Fig. 20.18). And because the line is infinite, it never resembles a point no matter how far away we are. As a result the field falls more slowly than the field of a point charge—in this case, as $1/y$. If we let r designate the radial distance from the line rather than the diagonal in Fig. 20.17, then the field decreases as $1/r$. An infinite line is impossible, but our result holds approximately for finite lines of charge as long as we're much closer to the line than its length, and not near an end. Far from a *finite* line, on the other hand, its field will resemble that of a point charge. You can explore the finite charged line in Problem 72. ■

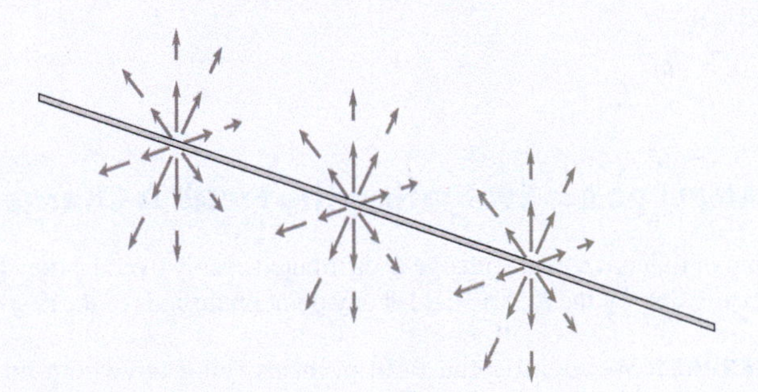

FIGURE 20.18 Field vectors for an infinite line of positive charge point radially outward, with magnitude decreasing inversely with distance.

MP®

PhET: Electric Field Hockey

20.5 Matter in Electric Fields

Electric fields give rise to forces on charged particles. Because matter consists of such particles, much of the behavior of matter is fundamentally determined by electric fields.

Point Charges in Electric Fields

The motion of a single charge in an electric field is governed by the definition of the electric field, $\vec{F} = q\vec{E}$, and Newton's law, $\vec{F} = m\vec{a}$. Combining these equations gives the acceleration of a particle with charge q and mass m in an electric field $\vec{E}$:

$$\vec{a} = \frac{q}{m}\vec{E} \tag{20.8}$$

This equation shows that it's the charge-to-mass ratio, q/m, that determines a particle's response to an electric field. Electrons, nearly 2000 times less massive than protons but carrying the same charge, are readily accelerated by electric fields. Many practical devices, from X-ray machines to fluorescent lights, use electrons accelerated in electric fields.

When the electric field is uniform, problems involving the motion of charged particles reduce to the constant-acceleration problems of Chapter 2. An ink-jet printer is one application; a pair of oppositely charged plates creates a uniform field that "steers" charged ink droplets to the right place on the page (Fig. 20.19).

When the field isn't uniform, it's generally more difficult to calculate particle trajectories. An important exception is a particle moving perpendicular to a field that points radially. Under appropriate conditions, the result is uniform circular motion (see Section 5.3), as shown in the next example.

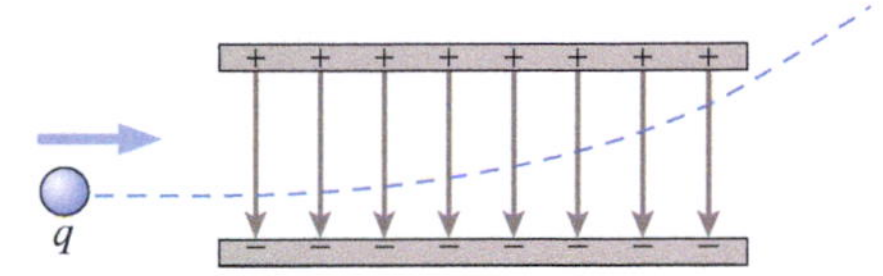

FIGURE 20.19 A pair of parallel charged plates creates a uniform electric field that deflects a charged particle. Can you tell the sign of the charge q?

EXAMPLE 20.8 Particle Motion: An Electrostatic Analyzer

Two oppositely charged curved metal plates establish an electric field given by $E = E_0(b/r)$, where E_0 and b are constants with the units of electric field and length, respectively. The field points toward the center of curvature, and r is the distance from the center. Find an expression for the speed v with which a proton entering vertically from below in Fig. 20.20 will leave the device moving horizontally.

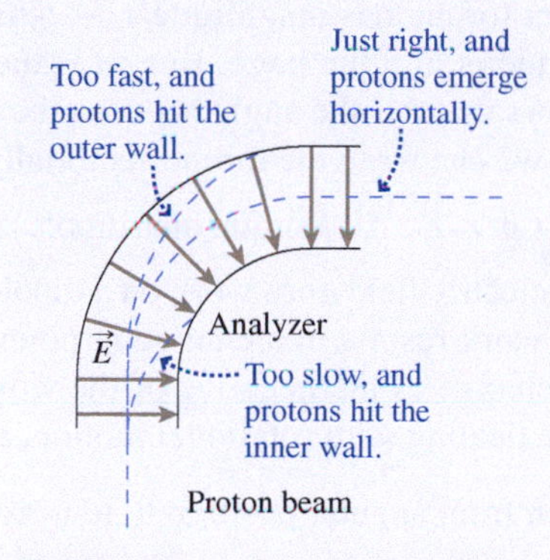

FIGURE 20.20 An electrostatic analyzer.

INTERPRET This problem is about charged-particle motion in an electric field that points radially. We're asked for the condition that will have a proton exiting the field region moving horizontally. Figure 20.20 shows that this requires its trajectory to be a circular arc.

DEVELOP Equation 20.8, $\vec{a} = (q/m)\vec{E}$, determines the acceleration of a charged particle in an electric field. Here we want uniform circular motion, so our plan is to write this equation with the given field and the acceleration v^2/r that we know applies in circular motion. Then we'll solve for v.

EVALUATE Under these conditions, Equation 20.8 becomes

$$a = \frac{v^2}{r} = \frac{eE}{m} = \frac{e}{m}E_0\frac{b}{r}$$

We then solve to get $v = \sqrt{eE_0b/m}$.

ASSESS Make sense? Strengthen the field by increasing E_0 or b, and the electric force becomes greater. For a given speed, that would result in more bending of the trajectory; to maintain the desired trajectory, we must therefore increase the speed. Note that the radius r canceled from our equations, showing that it doesn't matter where the protons enter the device. That's because the $1/r$ decrease in field strength matches the $1/r$ dependence of the acceleration. This device is called an electrostatic analyzer because it can sort charged particles by speed and charge-to-mass ratio. Spacecraft use such analyzers to characterize charged particles in interplanetary space. ■

GOT IT? 20.5 An electron, a proton, a deuteron (a neutron combined with a proton), a helium-3 nucleus (2 protons, 1 neutron), a helium-4 nucleus (2 protons, 2 neutrons), a carbon-13 nucleus (6 protons, 7 neutrons), and an oxygen-16 nucleus (8 protons, 8 neutrons) all find themselves in the same electric field. Rank in order their accelerations from lowest to highest under the assumption (only approximately correct) that the neutron and proton have the same mass and that the mass of a composite particle is the sum of the masses of its constituent neutrons and protons. Note any that have the same acceleration.

Dipoles in Electric Fields

Earlier in this chapter we calculated the field of an electric dipole, which consists of two opposite charges of equal magnitude. Here we study a dipole's response to electric fields. Since the dipole provides a model for many molecules, our results help explain molecular behavior.

Figure 20.21 shows a dipole with charges $\pm q$ separated a distance d, located in a uniform electric field. The dipole moment vector $\vec{p}$ has magnitude qd and points from the negative to the positive charge (recall Fig. 20.14). Since the field is uniform, it's the same at both ends of the dipole. Since the dipole charges are equal in magnitude but opposite in sign, they experience equal but opposite forces $\pm q\vec{E}$—and therefore there's no net force on the dipole.

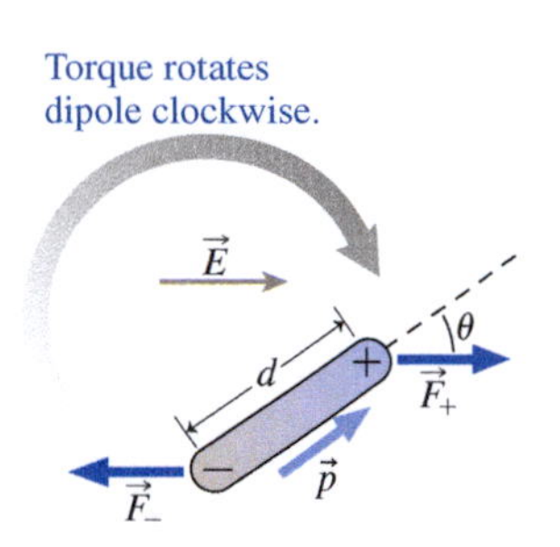

FIGURE 20.21 A dipole in a uniform electric field experiences a torque, but no net force.

However, Fig. 20.21 shows that the dipole does experience a torque that tends to align it with the field. In Chapter 11 we described torque as the cross product of the position vector with the force: $\vec{\tau} = \vec{r} \times \vec{F}$, where the magnitude of the torque vector is $rF\sin\theta$ and its direction is given by the right-hand rule. Figure 20.21 thus shows that the torque about the center of the dipole due to the force on the positive charge has magnitude $\tau_+ = rF\sin\theta = (\frac{1}{2}d)(qE)\sin\theta$. The torque associated with the negative charge has the same magnitude, and both torques are in the same direction since both tend to rotate the dipole clockwise. Thus the net torque has magnitude $\tau = qdE\sin\theta$; applying the right-hand rule shows that this torque is into the page. But qd is the magnitude of the dipole moment $\vec{p}$, and Fig. 20.21 shows that θ is the angle between the dipole moment vector and the electric field $\vec{E}$; therefore, we can write the torque vectorially as

$$\vec{\tau} = \vec{p} \times \vec{E} \qquad \text{(torque on a dipole)} \tag{20.9}$$

Because of this torque, the electric field does work on a dipole as it rotates. The electric force is conservative, so that work results in a change in potential energy. In Chapter 7 we defined potential-energy change as the negative of the work done by a conservative force: $\Delta U = -W$. Here we're dealing with rotational motion, and Equation 10.19 shows that the work done in a rotation from angular position θ_1 to θ_2 is given by $W = \int_{\theta_1}^{\theta_2} \tau\, d\theta$. Figure 20.21 shows that we're taking $\theta = 0$ when the dipole is aligned with the field. The figure also shows that the direction of increasing θ is counterclockwise or, in terms of rotational vectors, out of the page. The torque, in contrast, is clockwise or, vectorially, into the page. Thus the sign of the torque is opposite the angular change, so we need to write $\tau = -pE\sin\theta$ in the integral for the work. Let's now consider a dipole that's initially perpendicular to the field, so $\theta_1 = \pi/2$. Then the work done by the electric force as the dipole rotates to an arbitrary angle θ becomes

$$W = \int_{\pi/2}^{\theta} \tau\, d\theta = \int_{\pi/2}^{\theta} (-pE\sin\theta)\, d\theta = -pE[-\cos\theta]_{\pi/2}^{\theta} = pE\cos\theta$$

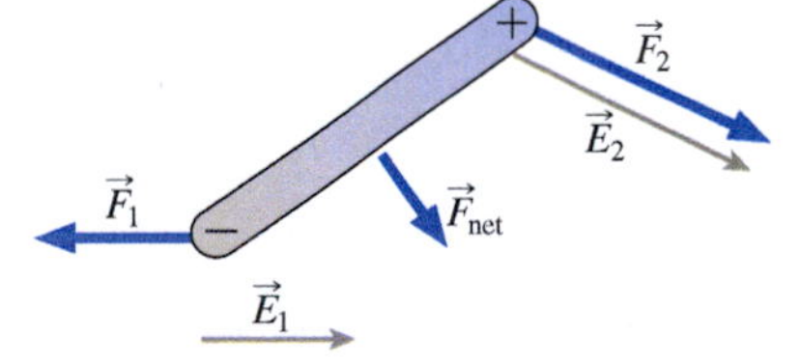

FIGURE 20.22 When the electric field differs in magnitude or direction at the two ends of the dipole, the dipole experiences a nonzero net force as well as a torque.

The potential-energy change is the negative of this work, and we note that $pE\cos\theta$ can be expressed as the dot product $\vec{p}\cdot\vec{E}$, so we can write the potential energy as

$$U = -\vec{p}\cdot\vec{E} \tag{20.10}$$

where $U = 0$ corresponds to the dipole at right angles to the field.

When the electric field isn't uniform, the charges at opposite ends of the dipole experience forces that differ in magnitude and/or aren't exactly opposite in direction. Then the dipole experiences a net force as well as a torque (Fig. 20.22). An important instance of this effect is the force on a dipole in the field of another dipole (Fig. 20.23). Because the dipole field falls off rapidly with distance and because the dipole responding to the field has closely spaced charges of equal magnitude but opposite sign, the dipole–dipole force is quite weak and falls extremely rapidly with distance. This weak force, which Fig. 20.23 shows to be attractive, is partly responsible for the van der Waals interaction between gas molecules that we mentioned in Chapter 17.

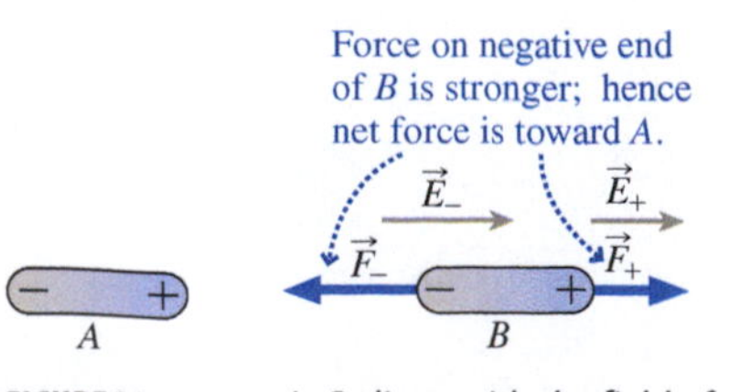

FIGURE 20.23 Dipole B aligns with the field of dipole A and then experiences a net force toward A.

Conductors, Insulators, and Dielectrics

Bulk matter contains vast numbers of point charges—namely, electrons and protons. In some matter—notably metals, ionic solutions, and ionized gases—individual charges are free to move throughout the material. In these **conductors**, the application of an electric

APPLICATION Microwave Cooking and Liquid Crystals

The torque on dipoles in electric fields forms the basis of two widespread contemporary technologies: the microwave oven and the liquid-crystal display (LCD).

A microwave oven works by generating an electric field whose direction changes several billion times per second. Water molecules, whose dipole moment is much greater than most others, attempt to align with the field. But the field is changing, so the molecules swing rapidly back and forth. As they jostle against each other, the energy they gain from the field is dissipated as heat that cooks the food.

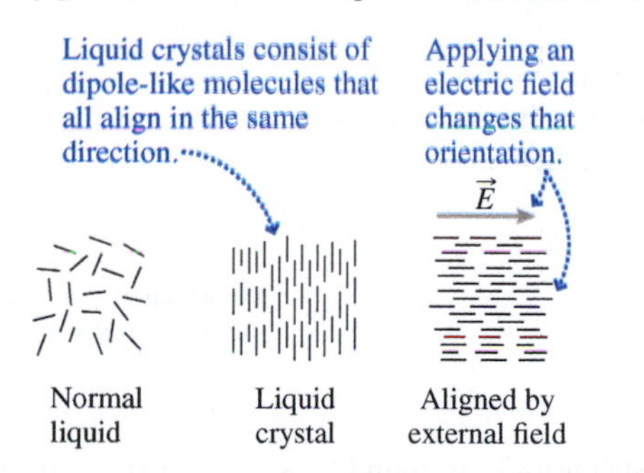

Computer displays, TVs, digital cameras, cell phones, watches, and many other devices display visual images using liquid crystals. These unique materials combine the fluidity of a liquid with the order of a solid. The liquid crystal consists of long molecules whose chemical structure results in a dipole-like charge separation. In response to each others' electric fields, the molecules tend to align. As the figure shows, an external electric field can rotate the liquid-crystal dipoles, altering the material's optical properties. With optical components we'll study in Chapter 29, different sections of a liquid-crystal display can then be made to appear visible or invisible. Liquid-crystal displays consume very little power, but they generate no light of their own and therefore most have a built-in light source. This photo of an iPhone shows its high-resolution display; also shown is a microphoto of the liquid crystals.

field results in the ordered motion of electric charge that we call **electric current**. We'll consider conductors and current in later chapters.

Materials in which charge is not free to move are **insulators**, since they don't support electric current. Insulators, however, still contain charges—it's just that their charges are bound into neutral molecules. Some molecules, like water, have intrinsic dipole moments and therefore rotate in response to an applied electric field. Even if they don't have dipole moments, molecules may respond to an electric field by stretching and acquiring **induced dipole moments** (Fig. 20.24). In either case, the application of an electric field results in the alignment of molecular dipoles with the field (Fig. 20.25). The fields of the dipoles, pointing from their positive to their negative charges, then reduce the applied electric field within the material. We'll explore the consequences of this effect further in Chapter 23. Materials in which molecules either have intrinsic dipole moments or acquire induced moments are called **dielectrics**.

Video Tutor Demo | **Charged Rod and Aluminum Can**

If the electric field applied to a dielectric becomes too great, individual charges are ripped free, and the material then acts like a conductor. Such **dielectric breakdown** can cause severe damage in materials and in electric equipment (Fig. 20.26). On a larger scale, lightning results from dielectric breakdown in air.

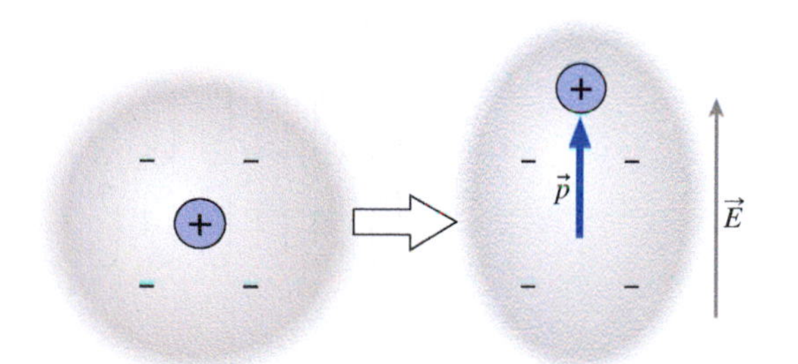

FIGURE 20.24 A molecule stretches in response to an electric field, acquiring a dipole moment.

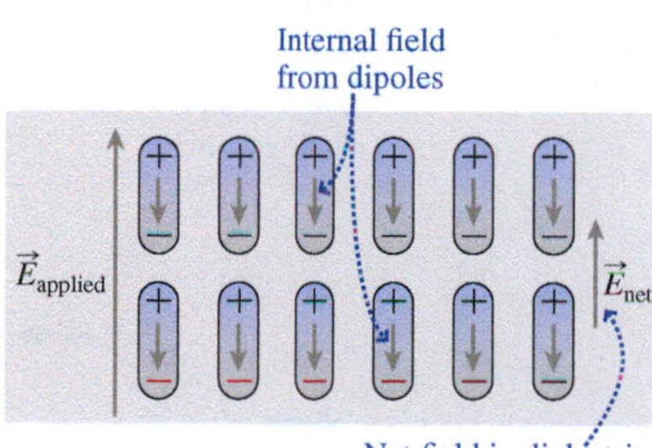

FIGURE 20.25 Alignment of molecular dipoles in a dielectric reduces the electric field within the dielectric.

FIGURE 20.26 Dielectric breakdown in a solid piece of Plexiglas produced this striking fractal pattern that marks permanent changes in the material.

CHAPTER 20 SUMMARY

Big Idea

This chapter introduces several big ideas. First is **electric charge**, a fundamental property of matter that comes in positive and negative forms. Like charges repel and opposites attract; this is the **electric force**. It's convenient to define the **electric field** as the force per unit charge that a charge would experience if placed in the vicinity of other charges. Both force and field obey the **superposition principle**, meaning that the effects of several charges add vectorially.

Key Concepts and Equations

Coulomb's law describes the electric force between point charges:

$$\vec{F}_{12} = \frac{kq_1q_2}{r^2}\hat{r}$$

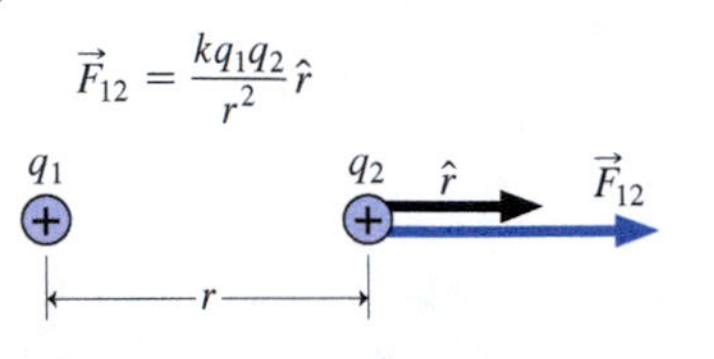

The electric field is the force per unit charge, $\vec{E} = \vec{F}/q$, and therefore the force a given charge q experiences in a field is $\vec{F} = q\vec{E}$.

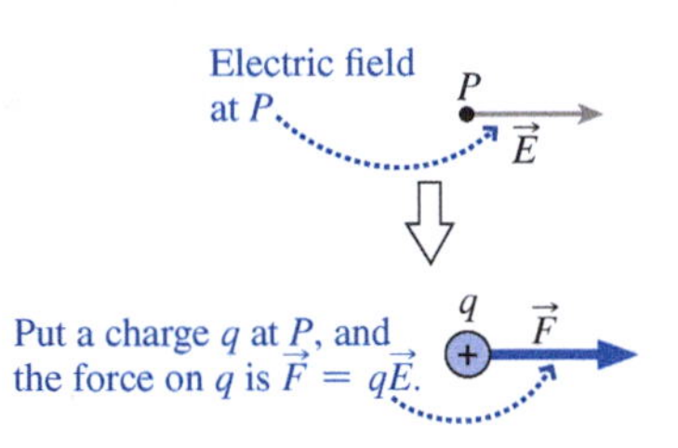

The field of a point charge follows from Coulomb's law:

$$\vec{E} = \frac{kq}{r^2}\hat{r}$$

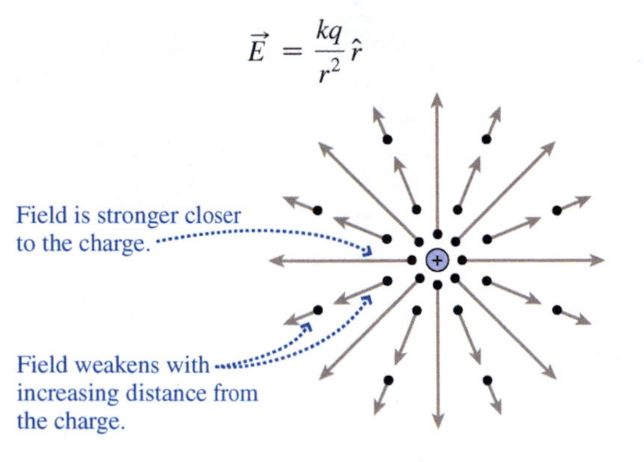

Fields of charge distributions are found by summing fields of individual point charges, or by integrating in the case of continuously distributed charge:

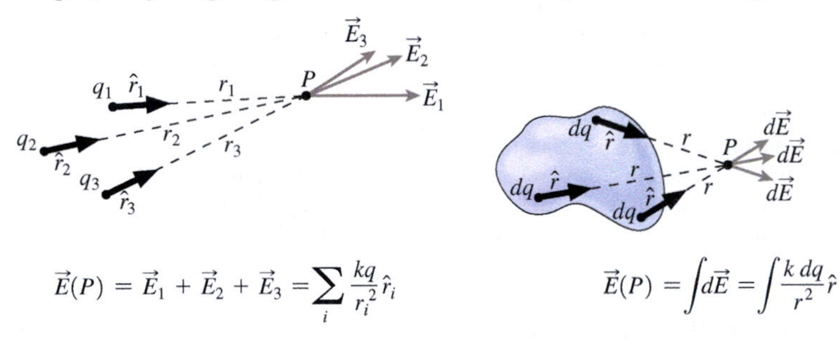

$$\vec{E}(P) = \vec{E}_1 + \vec{E}_2 + \vec{E}_3 = \sum_i \frac{kq}{r_i^2}\hat{r}_i \qquad \vec{E}(P) = \int d\vec{E} = \int \frac{k\,dq}{r^2}\hat{r}$$

Applications

A **dipole** consists of equal but opposite charges $\pm q$ a distance d apart. For distances large compared with d, the dipole field drops as $1/r^3$, and the dipole is completely characterized by its **dipole moment** $p = qd$.

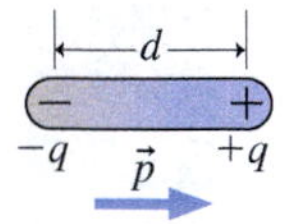

The field of an infinite line drops as $1/r$: $E = 2k\lambda/r$, with λ the charge per unit length. This is a good approximation to the field near an elongated structure like a wire.

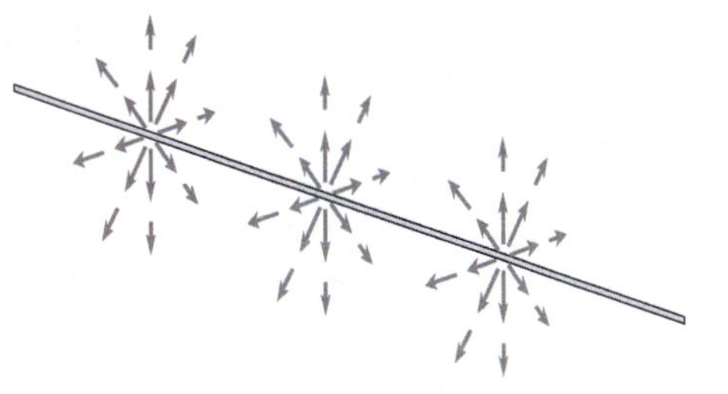

Point charges respond to electric fields with acceleration proportional to the charge-to-mass ratio q/m.

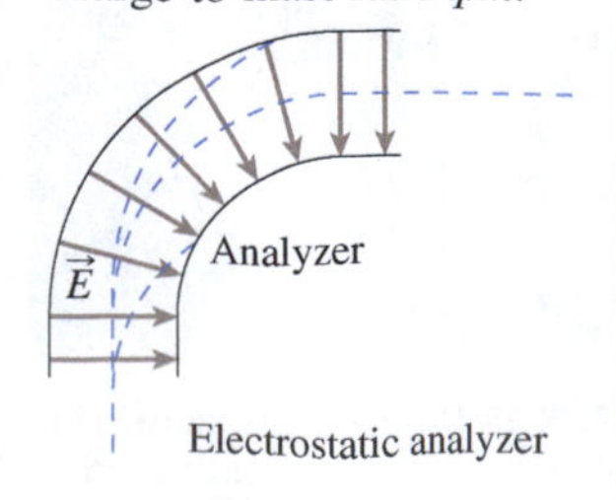

A dipole in an electric field experiences a torque that tends to align it with the field: $\vec{\tau} = \vec{p} \times \vec{E}$.

Torque rotates dipole clockwise.

$\vec{E}$ $\vec{F}_+$ $\vec{F}_-$ $\vec{p}$

If the field is nonuniform, there's also a net force on the dipole.

Dielectrics are insulating materials whose molecules act like electric dipoles.

MP® *For homework assigned on MasteringPhysics, go to www.masteringphysics.com*

BIO *Biology and/or medicine-related problems* **DATA** *Data problems* **ENV** *Environmental problems* **CH** *Challenge problems* **COMP** *Computer problems*

For Thought and Discussion

1. Conceptual Example 20.1 shows that the gravitational force between an electron and a proton is about 10^{-40} times weaker than the electric force between them. Since matter consists largely of electrons and protons, why is the gravitational force important at all?
2. A free neutron is unstable and soon decays to other particles, one of them a proton. Must there be others? If so, what electric properties must it or they have?
3. Where in Fig. 20.5 could you put a third charge so it would experience no net force? Would it be in stable or unstable equilibrium?
4. Why should the test charge used to measure an electric field be small?
5. Equation 20.3 gives the electric field of a point charge. Does the direction of (a) $\hat{r}$ or (b) $\vec{E}$ depend on whether the charge is positive or negative?
6. Is the electric force on a charged particle always in the direction of the field? Explain.
7. Why does a dipole, which has no net charge, produce an electric field?
8. The ring in Example 20.6 carries total charge Q, and the point P is the same distance $r = \sqrt{x^2 + a^2}$ from all parts of the ring. So why isn't the electric field of the ring just kQ/r^2?
9. A spherical balloon is initially uncharged. If you spread positive charge uniformly over the balloon's surface, would it expand or contract? What would happen if you spread negative charge instead?
10. Under what circumstances is the path of a charged particle a parabola? A circle?
11. Why should there be a force between two dipoles, which each have zero net charge?
12. Dipoles A and B are both located in the field of a point charge Q, as shown in Fig. 20.27. Does either experience a net torque? A net force? If each dipole is released from rest, describe qualitatively its subsequent motion.

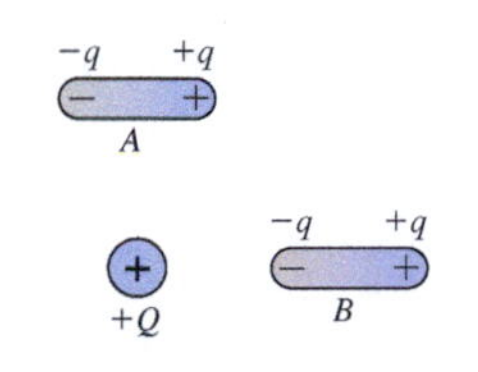

FIGURE 20.27 For Thought and Discussion 12

Exercises and Problems

Exercises

Section 20.1 Electric Charge

13. Suppose the electron and proton charges differed by one part in one billion. Estimate the net charge on your body, assuming it contains equal numbers of electrons and protons.
14. A typical lightning flash delivers about 25 C of negative charge from cloud to ground. How many electrons are involved?
15. Protons and neutrons are made from combinations of the two most common quarks, the u quark (charge $+\frac{2}{3}e$) and the d quark (charge $-\frac{1}{3}e$). How could three of these quarks combine to make (a) a proton and (b) a neutron?
16. Earth carries a net charge of about -5×10^5 C. How many more electrons are there than protons on Earth?
17. **BIO** As they fly, honeybees may acquire electric charges of about 180 pC. Electric forces between charged honeybees and spider webs can make the bees more vulnerable to capture by spiders. How many electrons would a honeybee have to lose to acquire a charge of +180 pC?

Section 20.2 Coulomb's Law

18. The electron and proton in a hydrogen atom are 52.9 pm apart. Find the magnitude of the electric force between them.
19. An electron at Earth's surface experiences a gravitational force $m_e g$. How far away can a proton be and still produce the same force on the electron? (Your answer should show why gravity is unimportant on the molecular scale!)
20. You break a piece of Styrofoam packing material, and it releases lots of little spheres whose electric charge makes them stick annoyingly to you. If two of the spheres carry equal charges and repel with a force of 21 mN when they're 15 mm apart, what's the magnitude of the charge on each?
21. A charge q is at the point $x = 1$ m, $y = 0$ m. Write expressions for the unit vectors you would use in Coulomb's law if you were finding the force that q exerts on other charges located at (a) $x = 1$ m, $y = 1$ m; (b) the origin; and (c) $x = 2$ m, $y = 3$ m. You're not given the sign of q. Why doesn't this matter?
22. A proton is at the origin and an electron is at the point $x = 0.41$ nm, $y = 0.36$ nm. Find the electric force on the proton.

Section 20.3 The Electric Field

23. An electron experiences an electric force of 0.61 nN. What's the field strength at its location?
24. Find the magnitude of the electric force on a 2.0-μC charge in a 100-N/C electric field.
25. A 68-nC charge experiences a 150-mN force in a certain electric field. Find (a) the field strength and (b) the force that a 35-μC charge would experience in the same field.
26. **BIO** The electric field inside a cell membrane is 8.0 MN/C. What's the force on a singly charged ion in this field?
27. A −1.0-μC charge experiences a $10\hat{\imath}$-N electric force in a certain electric field. What force would a proton experience in the same field?
28. The electron in a hydrogen atom is 52.9 pm from the proton. At this distance, what's the strength of the electric field due to the proton?

Section 20.4 Fields of Charge Distributions

29. In Fig. 20.28, point P is midway between the two charges. Find the electric field in the plane of the page (a) 5.0 cm to the left of P, (b) 5.0 cm directly above P, and (c) at P.

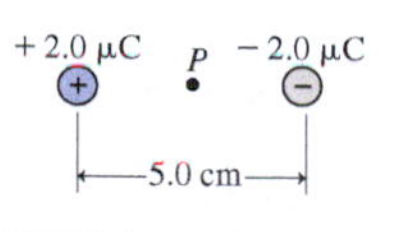

FIGURE 20.28 Exercise 29

30. The water molecule's dipole moment is 6.2×10^{-30} C·m. What would be the separation distance if the molecule consisted of charges $\pm e$? (The effective charge is actually less because H and O atoms share the electrons.)
31. The electric field 22 cm from a long wire carrying a uniform line charge density is 1.9 kN/C. What's the field strength 38 cm from the wire?
32. Find the line charge density on a long wire if the electric field 45 cm from the wire has magnitude 260 kN/C and points toward the wire.
33. Find the magnitude of the electric field due to a charged ring of radius a and total charge Q on the ring axis at distance a from the ring's center.

Section 20.5 Matter in Electric Fields

34. In his famous 1909 experiment that demonstrated quantization of electric charge, R. A. Millikan suspended small oil drops in an electric field. With field strength 20 MN/C, what mass drop can be suspended when the drop carries 10 elementary charges?
35. How strong an electric field is needed to accelerate electrons in an X-ray tube from rest to one-tenth the speed of light in a distance of 5.0 cm?
36. A proton moving to the right at 3.8×10^5 m/s enters a region where a 56-kN/C electric field points to the left. (a) How far will the proton get before it momentarily stops? (b) Describe its subsequent motion.
37. An electrostatic analyzer like that of Example 20.8 has $b = 7.5$ cm. What value of E_0 will enable the device to select protons moving at 84 m/s?

Problems

38. A 2-g ping-pong ball rubbed against a wool jacket acquires a net positive charge of 1 μC. Estimate the fraction of the ball's electrons that have been removed.
39. Two charges, one whose magnitude is twice as large as the other's, are located 12.5 cm apart and experience an attractive force of 143 N. (a) What's the magnitude of the larger charge? (b) Can you determine the sign of the larger charge?
40. A proton is on the x-axis at $x = 1.6$ nm. An electron is on the y-axis at $y = 0.85$ nm. Find the net force the two exert on a helium nucleus (charge $+2e$) at the origin.
41. A 9.5-μC charge is at $x = 15$ cm, $y = 5.0$ cm and a -3.2-μC charge is at $x = 4.4$ cm, $y = 11$ cm. Find the force on the negative charge.
42. A charge $3q$ is at the origin, and a charge $-2q$ is on the positive x-axis at $x = a$. Where would you place a third charge so it would experience no net electric force?
43. A negative charge $-q$ lies midway between two positive charges $+Q$. What must Q be such that the electric force on all three charges is zero?
44. In Fig. 20.29, take $q_1 = 68$ μC, $q_2 = -34$ μC, and $q_3 = 15$ μC. Find the electric force on q_3.

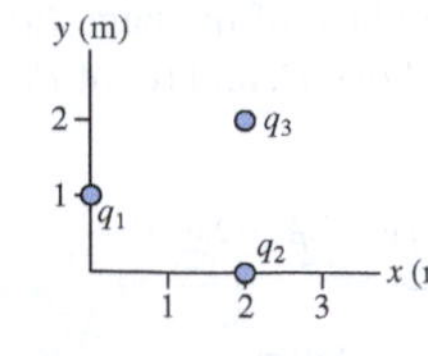

FIGURE 20.29 Problems 44 and 45

45. In Fig. 20.29, take $q_1 = 25$ μC and $q_2 = 20$ μC. If the force on q_1 points in the $-x$-direction, find (a) q_3 and (b) the magnitude of the force on q_1.
46. DATA BIO DNA fragments introduced into an electrophoresis apparatus (see Application, page 360) generally carry negative charges equivalent to two extra electrons per base pair of nucleotides in the fragment. The table below shows the forces on several DNA fragments in an electrophoresis apparatus, as a function of the number of base pairs. Plot these data, establish a best-fit line, and use the resulting slope to determine the strength of the electric field in the electrophoresis apparatus.

Base pairs	400	800	1200	2000	3000	5000
Force (pN)	0.235	0.472	0.724	1.15	1.65	2.87

47. A 65-μC point charge is at the origin. Find the electric field at the points (a) $x = 50$ cm, $y = 0$ cm; (b) $x = 50$ cm, $y = 50$ cm; and (c) $x = 25$ cm, $y = -75$ cm.
48. A 1.0-μC charge and a 2.0-μC charge are 10 cm apart. Find a point where the electric field is zero.
49. A proton is at the origin and an ion is at $x = 5.0$ nm. If the electric field is zero at $x = -5.0$ nm, what's the ion's charge?
50. (a) Find an expression for the electric field on the y-axis due to the two charges q in Fig. 20.7. (b) At what point is the field on the y-axis a maximum?
51. A dipole lies on the y-axis and consists of an electron at $y = 0.60$ nm and a proton at $y = -0.60$ nm. Find the electric field (a) midway between the two charges; (b) at the point $x = 2.0$ nm, $y = 0$ nm; and (c) at the point $x = -20$ nm, $y = 0$ nm.
52. Show that the field on the x-axis for the dipole of Example 20.5 is given by Equation 20.6b, for $|x| \gg a$.
53. You're 1.44 m from a charge distribution that is well under 1 cm in size. You measure an electric field strength of 296 N/C due to this distribution. You then move to a distance of 2.16 m from the distribution, where you measure a field strength of 87.7 N/C. What's the net charge of the distribution? *Hint*: Don't try to calculate the charge. Determine instead how the field decreases with distance, and from that infer the charge.
54. Three identical charges q form an equilateral triangle of side a, with two charges on the x-axis and one on the positive y-axis. (a) Find an expression for the electric field at points on the y-axis above the uppermost charge. (b) Show that your result reduces to the field of a point charge $3q$ for $y \gg a$.
55. Two identical small metal spheres initially carry charges q_1 and q_2. When they're 1.0 m apart, they experience a 2.5-N attractive force. Then they're brought together so charge moves from one to the other until they have the same net charge. They're again placed 1.0 m apart, and now they repel with a 2.5-N force. What were the original charges q_1 and q_2?
56. COMP Two 38.0-μC charges are attached to opposite ends of a spring with spring constant $k = 145$ N/m and equilibrium length 52.6 cm. By how much does the spring stretch? *Hint*: You'll need to use a computer or advanced calculator to solve the cubic equation that arises in this problem.
57. A thin rod lies on the x-axis between $x = 0$ and $x = L$ and carries total charge Q distributed uniformly over its length. Show that the electric field strength for $x > L$ is given by $E = kQ/[x(x - L)]$.
58. An electron is moving in a circular path around a long, uniformly charged wire carrying 2.5 nC/m. What's the electron's speed?
59. Find the line charge density on a long wire if a 6.8-μg particle carrying 2.1 nC describes a circular orbit about the wire with speed 280 m/s.
60. A dipole with dipole moment 1.5 nC·m is oriented at 30° to a 4.0-MN/C electric field. Find (a) the magnitude of the torque on the dipole and (b) the work required to rotate the dipole until it's antiparallel to the field.

61. You have a job examining patent applications. You're presented with the device in Fig. 20.30, which its inventor claims will separate isotopes of a particular element. Atoms are first stripped completely of their electrons, then accelerated from rest through an electric field chosen to give the desired isotope exactly the right speed to pass through the electrostatic analyzer (see Example 20.8). Will the device work?

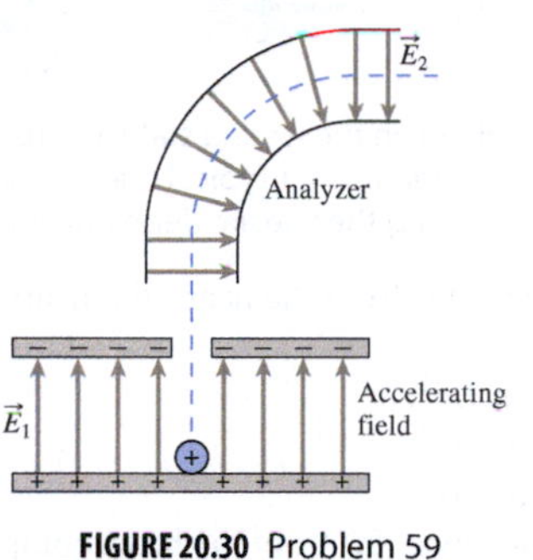

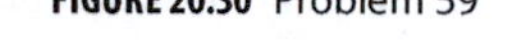

FIGURE 20.30 Problem 59

62. **BIO** A 5.0-μm strand of DNA carries charge $+e$ per nm of length. Treating it as a charged line, what's the electric field strength 25 nm from the DNA, not near either end?

63. A molecule has its dipole moment aligned with a 1.2-kN/C electric field. If it takes 3.1×10^{-27} J to reverse the molecule's orientation, what's its dipole moment?

64. **CH** Two identical dipoles, each of charge q and separation a, are a distance x apart, as shown in Fig. 20.31. (a) By considering forces between pairs of charges in the different dipoles, calculate the force between the dipoles and show that, in the limit $a \ll x$, it has magnitude $6kp^2/x^4$, where $p = qa$ is the dipole moment. (b) Is the force attractive or repulsive?

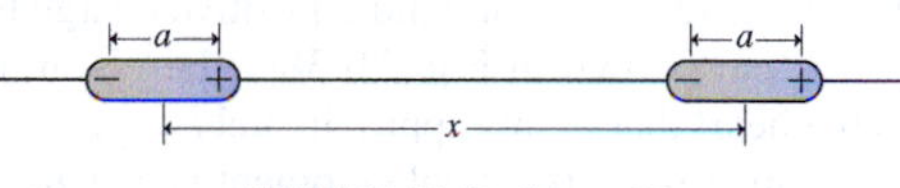

FIGURE 20.31 Problem 64

65. **CH** A dipole with charges $\pm q$ and separation $2a$ is located a distance x from a point charge $+Q$, oriented as shown in Fig. 20.32. Find expressions for the magnitude of (a) the net torque and (b) the net force on the dipole, both in the limit $x \gg a$. (c) What's the direction of the net force?

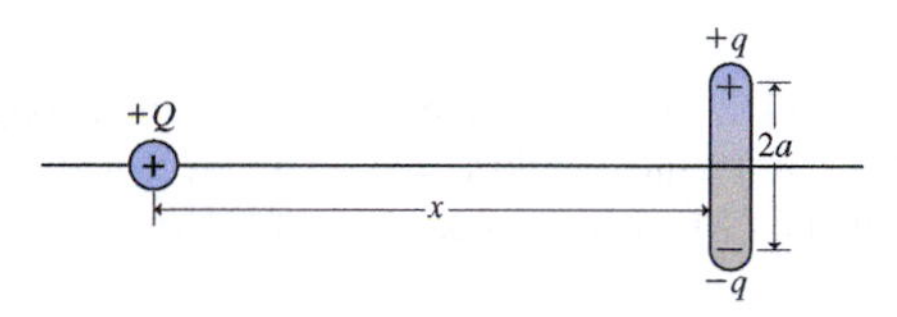

FIGURE 20.32 Problem 65

66. An electron is at the origin, and an ion with charge $+5e$ is at $x = 10$ nm. Find a point where the electric field is zero.

67. You're taking physical chemistry, and your professor is discussing molecular dipole moments. Water, he says, has a dipole moment of "1.85 debyes," while carbon monoxide's dipole moment is only "0.12 debye." Your physics professor wants these moments expressed in SI. She tells you that the atomic separation in these two covalent compounds is about the same, and asks what that indicates about the way shared charge is distributed. What do you tell her?

68. The electric field on the axis of a uniformly charged ring has magnitude 380 kN/C at a point 5.0 cm from the ring center. The magnitude 15 cm from the center is 160 kN/C; in both cases the field points away from the ring. Find (a) the ring's radius and (b) its charge.

69. **CH** An *electric quadrupole* consists of two oppositely directed dipoles in close proximity. (a) Calculate the field of the quadrupole shown in Fig. 20.33 for points to the right of $x = a$ and (b) show that for $x \gg a$ the quadrupole field falls off as $1/x^4$.

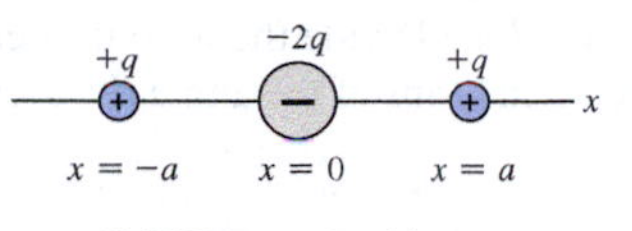

FIGURE 20.33 Problem 69

70. You measure the electric field on a dipole's axis, at a distance from the dipole that's large compared with the separation of the two charges in the dipole. If you maintain that distance but move to a point located at 45° from the dipole axis, by what factor will the magnitude of the electric field change? Will the field increase or decrease?

71. A straight wire 10 m long carries 25 μC distributed uniformly over its length. (a) What's the line charge density on the wire? Find the electric field strength (b) 15 cm from the wire axis, not near either end, and (c) 350 m from the wire. Make suitable approximations in both cases.

72. **CH** Figure 20.34 shows a thin rod of length L carrying charge Q distributed uniformly over its length. (a) What's the line charge density on the rod? (b) Modify the calculation of Example 20.7 to find an expression for the electric field at a point P a distance y along the perpendicular bisector. (c) Show that your result reduces the the field of a point charge when $|y| \gg L$.

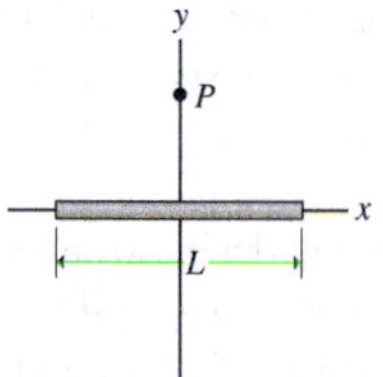

FIGURE 20.34 Problem 72

73. **CH** Figure 20.35 shows a thin, uniformly charged disk of radius R. Imagine the disk divided into rings of varying radii r, as suggested in the figure. (a) Show that the area of such a ring is very nearly $2\pi r\,dr$. (b) If the disk carries surface charge density σ, use the result of part (a) to write an expression for the charge dq on an infinitesimal ring. (c) Use the result of (b) along with the result of Example 20.6 to write the infinitesimal electric field dE of this ring at a point on the disk axis, taken to be the positive x-axis. (d) Integrate over all such rings to show that the net electric field on the axis has magnitude

$$E = 2\pi k\sigma\left(1 - \frac{x}{\sqrt{x^2 + R^2}}\right)$$

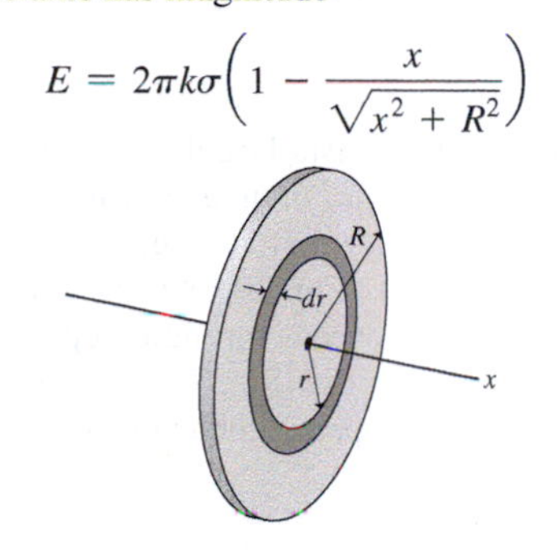

FIGURE 20.35 Problem 73

74. **CH** Use the result of Problem 73 to show that the field of an *infinite*, uniformly charged flat sheet is $2\pi k\sigma$, where σ is the surface charge density. (This result is independent of distance from the sheet.)

75. Use the binomial theorem to show that, for $x \gg R$, the result of Problem 73 reduces to the field of a point charge whose total charge is the charge density times the disk area.

76. **CH** A semicircular loop of radius a carries positive charge Q distributed uniformly. Find the electric field at the loop's center (point P in Fig. 20.36). (*Hint:* Divide the loop into charge elements dq as shown, write dq in terms of the angle $d\theta$, then integrate over θ.)

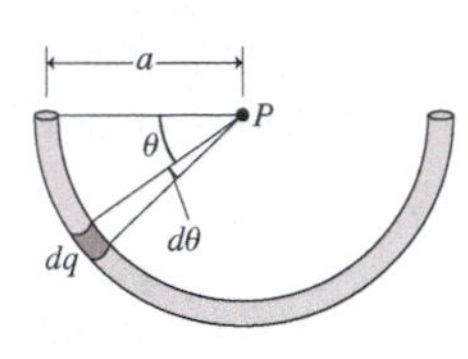

FIGURE 20.36 Problem 76

77. In Example 20.2, find the position on the y-axis where Q will experience the greatest force.

78. **CH** A thin rod carries charge Q distributed uniformly over its length L and is situated on the x-axis between $x = \pm L/2$. (a) Find the electric field at an arbitrary point (x, y). (You'll have to do separate integrals for the x- and y-components.) (b) Show that your result reduces to that of Problem 72 when $x = 0$ and to that of Problem 57 when $y = 0$ and $x > L/2$.

79. **CH** A thin rod extends along the x-axis from $x = 0$ to $x = L$ and carries line charge density $\lambda = \lambda_0(x/L)^2$, where λ_0 is a constant. Find the electric field at $x = -L$.

80. **CH** A rod of length $2L$ lies on the x-axis, centered at the origin, and carries line charge density $\lambda = \lambda_0(x/L)$, where λ_0 is a constant. (a) Find an expression for the electric field strength at points on the x-axis for $x > L$. (b) Show that for $x \gg L$ your result has the $1/x^3$ dependence of a dipole field, and determine the dipole moment of the rod.

81. You're working on the design of an ink-jet printer. Ink drops of mass m, speed v, and charge q will enter a region of uniform electric field E between two charged plates (Fig. 20.37). The drops enter midway between the plates, and the electric field deflects them toward the correct place on the page. Find an expression for the maximum electric field for which drops can still get through without hitting either plate.

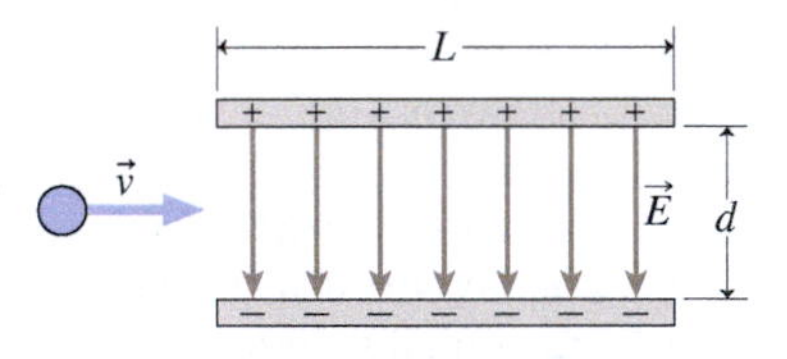

FIGURE 20.37 Problem 81

Passage Problems

BIO The human heart consists largely of elongated muscle cells, some 100 μm long and 15 μm in diameter. In its resting state, a cell contains two concentric layers of charge, which confine the electric field to the cell membrane (Fig. 20.38*a*). When the heart contracts, a wave of depolarization sweeps through, depleting charge and giving each cell a dipole moment (Fig. 20.38*b*). As a result, the entire organ acts like an electric dipole, producing an external field, which is indirectly detected by electrocardiography. Although the direction of the heart's dipole moment varies, Fig. 20.38*c* is typical. In answering the questions that follow, consider the heart in isolation—don't concern yourself with the effect of surrounding tissues on its electric field.

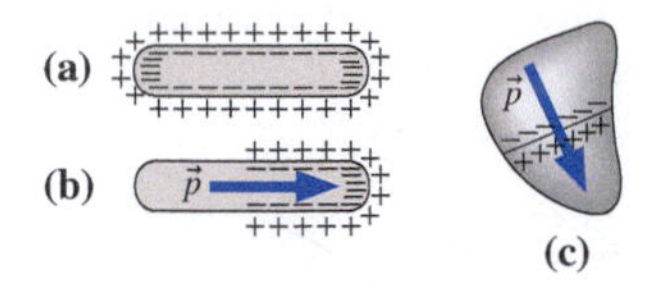

FIGURE 20.38 Heart cells (a) in the resting state and (b) partially depolarized, resulting in a dipole moment $\vec{p}$. (c) Typical orientation of the heart's dipole moment vector. Cells along the line are depolarizing.

82. At a distance r, far from the heart, the heart's electric field
    a. falls off as $1/r$.
    b. falls off as $1/r^2$.
    c. falls off as $1/r^3$.
    d. falls off as $1/r^4$.

83. At a given distance, far from the heart compared with its size, the electric field
    a. is weaker along an extension of the line shown in Fig. 20.38*c* than on a perpendicular line.
    b. is stronger along a an extension of the line shown in Fig. 20.38*c* than on a perpendicular line.
    c. has the same value at positions perpendicular and parallel to the line in Fig. 20.38*c*.

84. The difference between Figs. 20.38*a* and 20.38*b* that results in an external electric field in one case but not the other is that
    a. there's no net charge in Fig. 20.38*a* but there is a net charge in Fig. 20.38*b*.
    b. the total charge is greater in Fig. 20.38*a*.
    c. the charge is distributed in Fig. 20.38*b* so there's more negative charge to the left and more positive charge to the right.

85. At the instant shown in Fig. 20.38*c*, there's an electric field within the heart that points approximately
    a. in the direction of the dipole moment vector $\vec{p}$.
    b. opposite the dipole moment vector $\vec{p}$.
    c. perpendicular to the dipole moment vector $\vec{p}$.

Answers to Chapter Questions

Answer to Chapter Opening Question

The electric field in the atmosphere must be strong enough to rip electrons from air molecules, making the air an electrical conductor. This typically happens when E exceeds about 3 MV/m.

Answers to GOT IT? Questions

20.1 (c) *uud* because its net charge is $+e$ (*udd* is the neutron)

20.2 (1) $-\hat{\imath}$; (2) $-\frac{\sqrt{2}}{2}\hat{\imath} + \frac{\sqrt{2}}{2}\hat{\jmath}$; The unit vector always points away from the source charge regardless of the sign.

20.3 (a)

20.4 (1) drops by $1/2^2$, to 200 N/C; (2) drops by $1/2^3$, to 100 N/C

20.5 carbon-13, (oxygen-16, helium-4, deuteron—all the same), helium-3, proton, electron

21 Gauss's Law

What You Know

- You understand how *Coulomb's law* describes the electric force between two point charges.
- You know that the *superposition principle* allows you to sum vectorially the forces from multiple point charges.
- You're familiar with the concept of *electric field* as a way of representing the force per unit charge produced by an individual charge or charge distribution.
- You've calculated forces and fields associated with several simple charge distributions.

What You're Learning

- You'll learn how to describe electric fields visually using *electric field lines*.
- You'll see how counting the field lines emerging from closed surfaces leads to a profound but simple statement about the behavior of the electric field.
- You'll learn the concept of *electric flux* and use it to develop *Gauss's law*, one of the four fundamental laws of electromagnetism that expresses, in a very different way, the content of Coulomb's law.
- You'll see how Gauss's law makes it straightforward to calculate electric fields of symmetric charge distributions.
- You'll learn the implications of Gauss's law for *electrostatic equilibrium* in electrical conductors.

How You'll Use It

- Gauss's law will continue to provide a practical approach to finding electric fields in upcoming chapters.
- You'll use the flux concept again with magnetic fields.
- You'll eventually join Gauss's law with the other three laws of electromagnetism to give a complete description of electromagnetic phenomena.

Huge sparks jump to the operator's cage in the Hall of Electricity at the Boston Museum of Science, but the operator is unharmed. Why?

In this chapter we introduce an elegant way of describing electric fields that makes it much easier to calculate the fields of certain charge distributions. In the process we'll formulate one of the four fundamental laws of electromagnetism—a statement that embodies Coulomb's law but that gives deeper insights into the electric field.

21.1 Electric Field Lines

We can visualize electric fields by drawing **electric field lines**, continuous lines whose direction is everywhere the same as that of the electric field. To draw a field line, determine the field direction at some point. Move a small distance in the direction of the field, and evaluate the field direction at the new point. Extending the process in both directions from your starting point traces out an electric field line. You'll find that field lines begin on positive charges and either end on negative charges or extend to infinity. Drawing many field lines gives a picture of the overall field.

To explore the field of a positive point charge, as shown in Fig. 21.1*a*, start near the charge. The field points radially outward, so move a little way outward. The field is still radial. Repeat the process, and you'll trace a straight line that extends indefinitely.

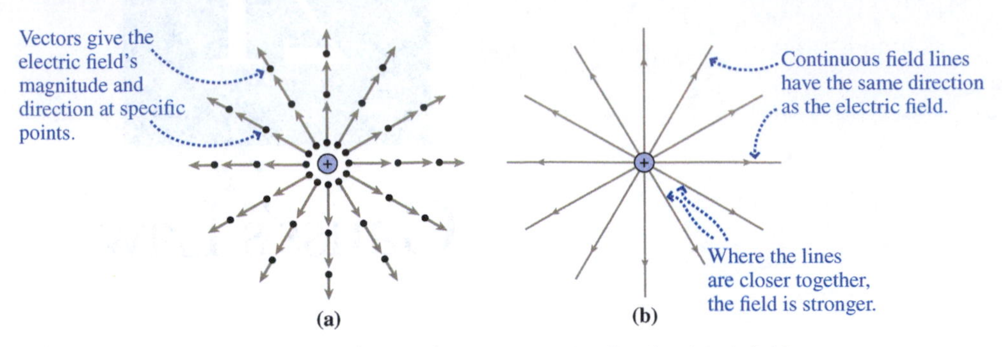

FIGURE 21.1 Vectors (a) and field lines (b) provide two ways to visualize the electric field.

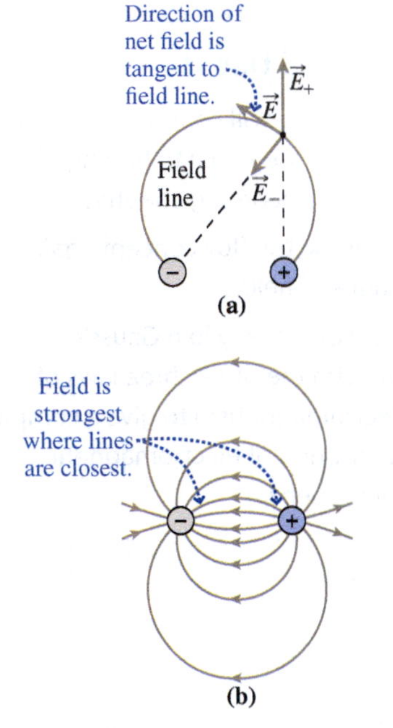

FIGURE 21.2 Field of an electric dipole. (a) At each point, the field-line direction is that of the *net* electric field, $\vec{E} = \vec{E}_+ + \vec{E}_-$. (b) Tracing many field lines shows the overall dipole field.

So the field lines of a positive point charge are straight lines that begin on the charge and extend radially to infinity (Fig. 21.1*b*).

In Fig. 21.1*b* the field lines spread apart as they extend farther from the point charge. Coulomb's law shows that the field weakens farther from the charge, so in Fig. 21.1*b* the field is stronger where the lines are closer and weaker where they're farther apart. This is generally true, and lets us infer the field's relative magnitude as well as direction from field-line pictures.

To trace the field lines of a charge distribution, follow the net field—the vector sum of the field contributions from all charges in the distribution. Usually the field direction varies, so the line is curved. Figure 21.2*a* shows the details for one field line of a dipole. Figure 21.2*b* shows a number of dipole field lines; here you can see that the field is strongest near the individual charges and in the region between them. The electric field exists everywhere, so there are really infinitely many field lines. We can't draw them all, but in order to make field-line pictures somewhat precise, we associate a fixed number of field lines with a charge of a given magnitude. In Fig. 21.3, for example, eight field lines correspond to a charge of magnitude q. Study the figure, and you'll see how all the fields are consistent with this convention.

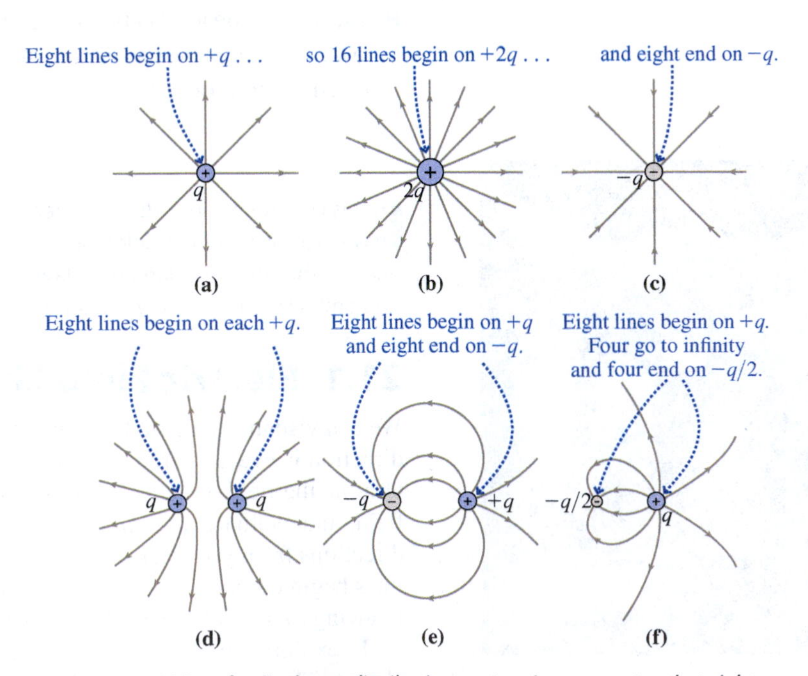

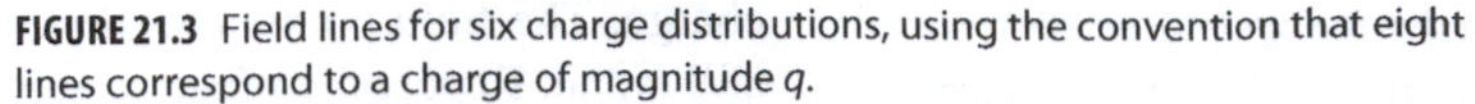
FIGURE 21.3 Field lines for six charge distributions, using the convention that eight lines correspond to a charge of magnitude q.

GOT IT? 21.1 Which figure represents the electric field of a charge distribution consisting of $+2q$ and $-q/2$, using the convention that eight lines correspond to a charge of magnitude q?

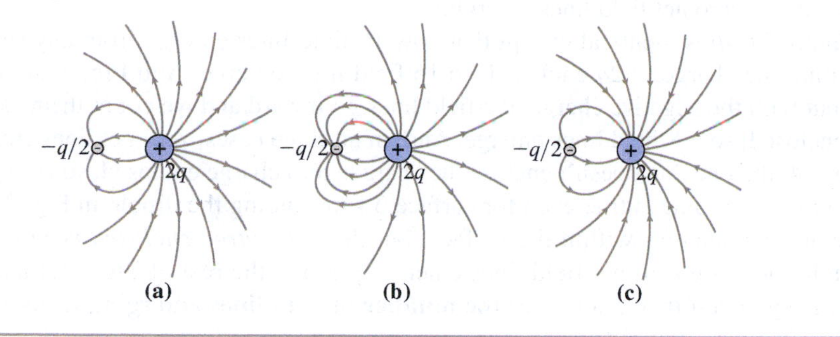

21.2 Electric Field and Electric Flux

In Fig. 21.4 we've surrounded each charge distribution from Fig. 21.3 with several surfaces. Each surface is closed, so it's impossible to get from inside to outside without crossing the surface. (Figure 21.4 shows only the two-dimensional cross section of each surface.) How many field lines emerge from within each surface?

In Fig. 21.4*a* the answer for surfaces 1 and 2 is eight. For surface 3 one field line crosses twice going out and once going in; if we count the inward-going line as negative, then there's still a net of eight lines going out. *Any* closed surface surrounding $+q$ will have eight lines emerging from it, for the simple reason that eight lines begin on the charge and extend to infinity; to get there they all have to cross the closed surface.

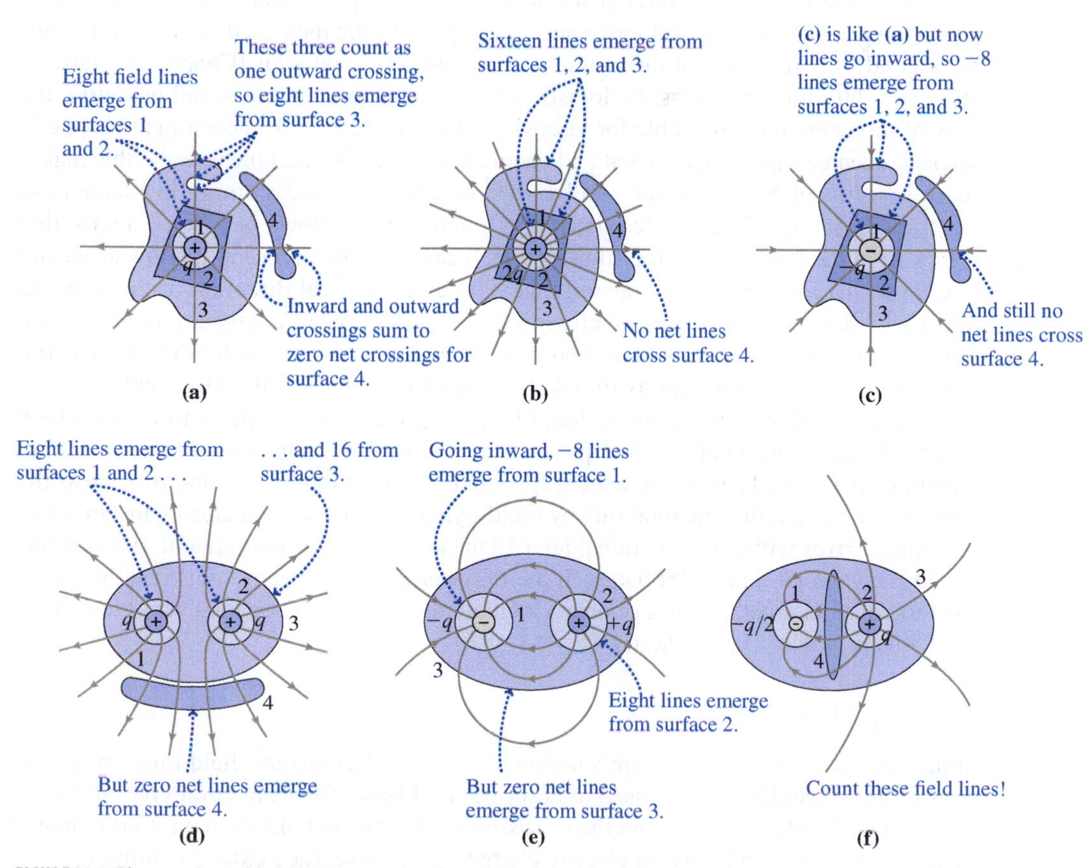

FIGURE 21.4 The number of field lines emerging from a closed surface depends only on the net charge enclosed.

What about surface 4? Two lines cross going outward and two inward, for a net of zero lines emerging. What's different is that surface 4 doesn't enclose the charge. You can convince yourself that *any* surface not enclosing the charge will have as many lines going in as out, for zero net field lines emerging.

Figure 21.4*b* is identical except that now 16 field lines emerge from any surface surrounding the charge: $+2q$ enclosed, so 16 field lines emerge. And Fig. 21.4*c* is similar, too, but with the negative charge, the field lines go inward and we count them as negative: $-q$ enclosed, so -8 field lines emerge. And, in all three cases, zero net lines emerge from surface 4, the one that doesn't enclose the charge: zero charge enclosed, so zero field lines emerging. The same is true even for surface 3 surrounding the dipole in Fig. 21.4*e*; here there are two charges within the surface, but the *net charge enclosed* is zero, and sure enough, there are zero net field lines emerging. Study the rest of Fig. 21.4 and you can convince yourself that in all cases **the number of field lines emerging from any closed surface is proportional to the net charge enclosed**.

This statement is very general. It doesn't matter what shape the surface is or whether the enclosed charge is a single point charge or a lot of charges carrying the same net charge. Nor does it matter how the charges are arranged, as long as they're *enclosed* by the surface in question. The presence of charges *outside* the surface doesn't alter our conclusion about the number of field lines emerging—although it may alter the shapes of the field lines. We'll now rephrase our statement mathematically to obtain one of the four fundamental laws of electromagnetism. As we proceed, remember that the mathematics just reflects what's clear from Fig. 21.4: The number of field lines emerging from a closed surface depends only on the net charge enclosed.

The Flux Concept

There are many situations in physics and engineering where we characterize flows, such as the flow of a fluid like blood through a vein or water in a pipe or river (Chapter 15—also of interest to physiologists, hydrologists, and civil engineers and measured, in SI, in kilograms per second); the flow of heat through insulation and other materials (Chapter 16—also of interest to building engineers, biologists, and many others, and measured in watts); the flow of momentum (responsible for viscosity, Chapter 15; also of interest in atmospheric science, oceanography, mechanical and chemical engineering and measured—this makes sense in terms of Newton's law—in $kg \cdot m/s^2$ or newtons); and, coming in Chapter 24, electric current as a flow of electric charge. Each of these flows is a **flux**—a term that comes from the Latin *fluxus*, for "flow." In each case we can represent the flow as we did in Chapter 15 for fluids—by drawing lines that give the local direction of the flow. As Fig. 15.11 showed, those lines are closer where the flow speed is higher and vice versa. Those last two sentences should remind you of Section 21.1's recipe for drawing electric field lines, and how the proximity of field lines describes the strength of the field.

We can describe flows in more detail by giving the flow rate per unit area—which we could equally well call the flux per unit area. For a fluid, that would be measured in kilograms per second per square meter ($kg/s \cdot m^2$). If that quantity were uniform across the flow, then we could find the total flux by multiplying the flux per unit area by the area. For example, a river with cross-sectional area 25 m^2 and a flux per unit area of 100 $kg/s \cdot m^2$ would carry a total flux of 2500 kg/s. If the flow per unit area weren't uniform, then we'd have to consider small patches of area, calculate the flux over each, and sum the results. We'll quantify that process shortly.

Electric Flux

In the case of electric fields, there's nothing "flowing." But electric field lines are analogous to the streamlines we've used to describe fluid flow. Their direction gives the local direction of the electric field, and their proximity reflects the field strength. Furthermore, field lines begin or end only on electric charges; otherwise, they extend continuously to infinity. This property of electric fields—which you'll soon see is intimately related to Coulomb's law—is analogous to the conservation of matter in fluid flows. So we use the term **electric flux** to describe the electrical analog of the flux of a fluid. In the simplest case of a uniform electric field of magnitude E perpendicular to an area A, the flux of the electric field is $\Phi = EA$, in analogy with our finding the flux in a river by multiplying the flux per unit area by the river's cross-sectional area. Here we use Φ, the capital Greek phi, as our symbol for electric flux.

What good is electric flux? Figure 21.5 gives the answer, showing that electric flux carries the same information as the "number of field lines crossing a surface" that was so important in our discussion of Fig. 21.4. Consider first the flux through a flat surface. Comparison of Figs. 21.5*a* and *b* shows that the number of field lines crossing a given surface is larger when the field is stronger—a fact that's precisely captured in our formula $\Phi = EA$. Comparison of Figs. 21.5*b* and *c* shows further that, for a given field strength, the number of field lines crossing a given surface is larger when the surface is larger. Again, $\Phi = EA$ captures this situation. Finally, Fig. 21.5*d* shows that the number of field lines is reduced when the surface is tilted relative to the field. You can also see this effect in Fig. 21.6, which shows a water-flow analogy. Specifically, the flux is reduced by a factor $\cos\theta$, where θ is the angle between the electric field $\vec{E}$ and a vector $\vec{A}$ that's normal to the surface. So our flux expression generalizes to $\Phi = EA\cos\theta$. If we define the normal vector $\vec{A}$ as having magnitude equal to the surface area A, then we can write the electric flux through a flat surface in a uniform electric field using the dot product:

$$\Phi = \vec{E} \cdot \vec{A} \tag{21.1}$$

where the dot product, defined in Chapter 6, is the product of the two vector magnitudes with the cosine of the angle between them. Since the units of $\vec{E}$ are N/C, flux is measured in $\text{N}\cdot\text{m}^2/\text{C}$.

The surfaces in Fig. 21.5 are *open* surfaces, meaning it's possible to get from one side to the other without passing through the surface. For open surfaces there's an ambiguity in the sign of Φ, since we could have taken $\vec{A}$ in either of the two directions along the perpendicular to the surface. But for *closed* surfaces, we unambiguously define the direction of $\vec{A}$ as the direction of the outward-pointing normal to the surface.

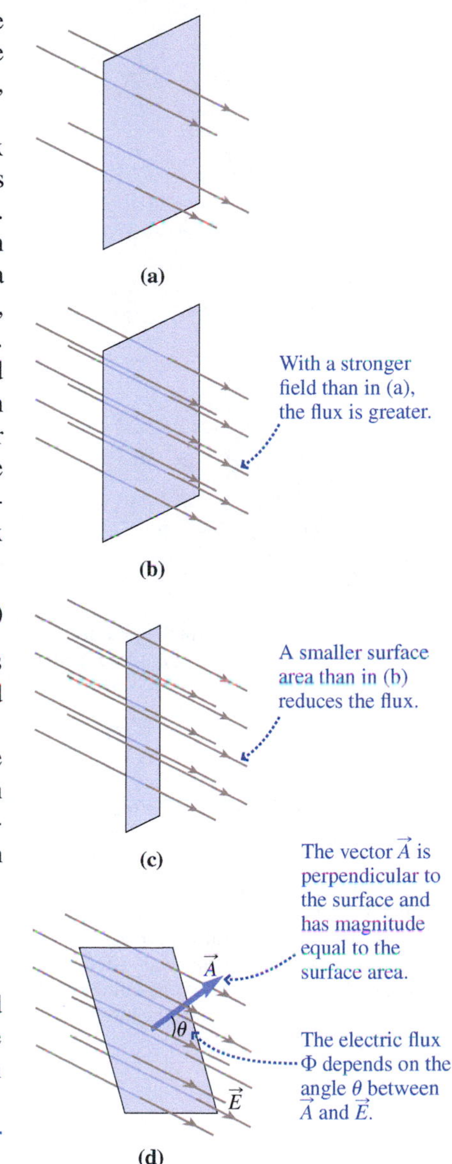

FIGURE 21.5 Electric flux through flat surfaces.

✓TIP The Flux Is Not the Field

The flux Φ and field $\vec{E}$ are related but distinct quantities. The field is a vector defined at each point in space. The flux is a scalar and a global property, depending on how the field behaves over an extended surface rather than at a single point; it's a quantification of the number of field lines crossing a surface.

What if a surface is curved and/or the field varies with position? Then we divide the surface into patches, each small enough that it's essentially flat and that the field is essentially uniform over each (Fig. 21.7). If a patch has area dA, then Equation 21.1 gives the flux through it: $d\Phi = \vec{E} \cdot d\vec{A}$, where the vector $d\vec{A}$ is normal to the patch. The total flux through the surface is then the sum over all the patches. If we make the patches arbitrarily small, that sum becomes an integral, and the flux is

$$\Phi = \int_{\text{surface}} \vec{E} \cdot d\vec{A} \tag{21.2}$$

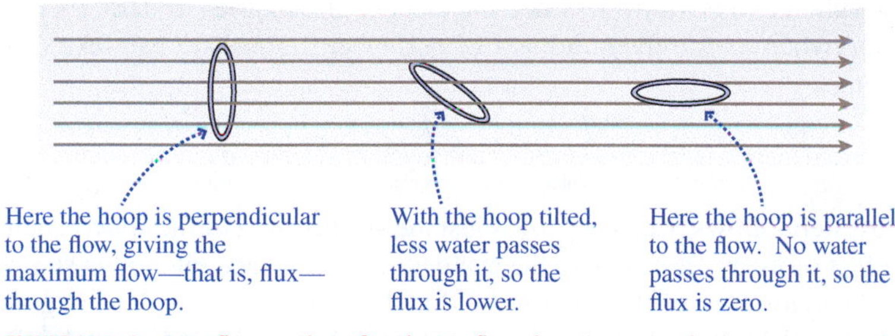

FIGURE 21.6 A water-flow analogy for electric flux, showing a circular hoop immersed in a flow of water.

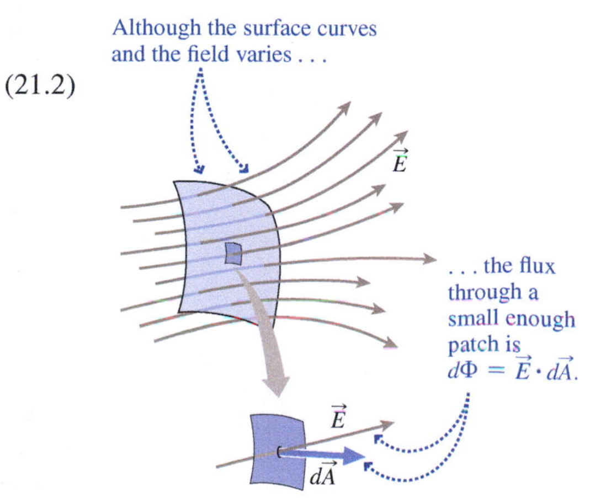

FIGURE 21.7 Finding the flux through a small area dA, so small it's essentially flat.

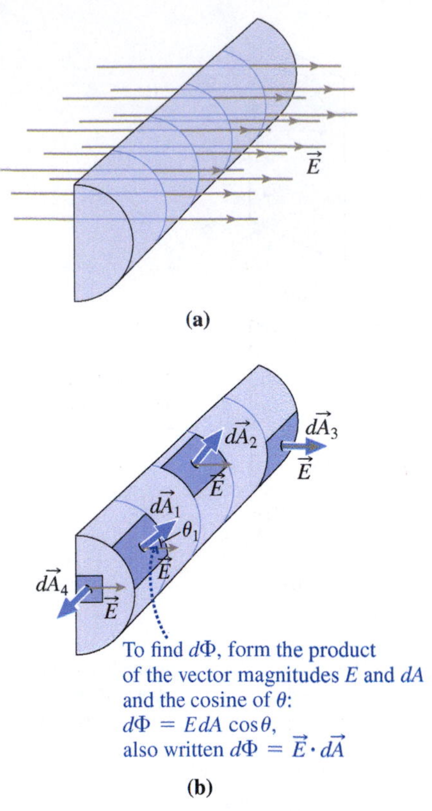

FIGURE 21.8 Meaning of the surface integral for electric flux, here for the case of a surface that isn't flat. (a) The surface is a half-cylinder in a uniform electric field $\vec{E}$. (b) The total flux through the surface is the sum of the fluxes $d\Phi = \vec{E} \cdot d\vec{A}$ over all the little patches constituting the surface, four of which are shown. In the limit that the patches become infinitesimally small, that sum becomes the integral in Equation 21.2.

The limits of the integral range over the entire surface, picking up contributions from all the patches $d\vec{A}$. The result is a **surface integral**—in general, the integral of a vector field over a surface, with the orientation of field and surface taken into account. Tactics Box 21.1 shows the meaning of such a surface integral. Although a surface integral can be difficult to evaluate, we'll find it most useful in situations where its evaluation is almost trivial.

TACTICS 21.1 Surface Integrals

The math in Equation 21.2 looks complicated, with two vectors and a dot product inside the integral and limits that specify an entire surface. Figure 21.8 helps decipher this math. The figure shows a surface, which could be arbitrary but in this case is half a cylinder including its semicircular end caps. This surface is immersed in an electric field $\vec{E}$ (Fig. 21.8a), which in this case is uniform but need not be. Figure 21.8b shows several small patches of the surface area. Each patch is described by a vector $d\vec{A}$ with magnitude equal to the area of the corresponding small patch and direction given by the normal to that patch, similar to the patch we showed in Fig. 21.7. At each patch we consider the direction of the normal vector $d\vec{A}$ relative to the electric field $\vec{E}$ at that patch, and we form the quantity $E\,dA\cos\theta$ or, written more compactly, $\vec{E} \cdot d\vec{A}$. What is this quantity? It's the flux, $d\Phi$, through the small patch. Although it's formed from two vectors, $d\Phi$ itself is a scalar. So to find the total flux through our surface, all we have to do is add up all those little $d\Phi$s. And that—in the limit that the $d\Phi$s become infinitesimally small and infinitely many—is the meaning of the surface integral in Equation 21.2. As you work through this chapter, you'll see many instances where we evaluate such surface integrals. We'll be able to choose surfaces that make the actual evaluation straightforward—but to appreciate that evaluation you'll need to understand the meaning of the surface integral as we've outlined it here.

GOT IT? 21.2 The figure shows a cube of side s in a uniform electric field $\vec{E}$. (1) What's the flux through each of the three cube faces A, B, and C with the cube oriented as in (a)? (2) Repeat for the orientation in (b), with the cube rotated 45°.

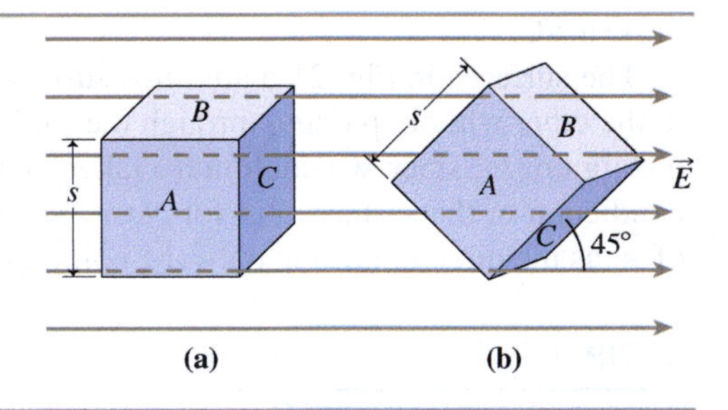

21.3 Gauss's Law

We've seen that the number of field lines emerging from a closed surface is proportional to the charge enclosed. Now that we've developed electric flux to express more rigorously the notion "number of field lines," we can state that **the electric flux through any closed surface is proportional to the net charge enclosed by that surface**. Writing the same thing mathematically gives $\Phi = \oint \vec{E} \cdot d\vec{A} \propto q_{\text{enclosed}}$, where the circle indicates that the integral is over a *closed* surface.

To evaluate the proportionality between flux and charge, consider a positive point charge q and a spherical surface of radius r centered on the charge (Fig. 21.9). The flux through this surface is given by Equation 21.2:

$$\Phi = \oint \vec{E} \cdot d\vec{A} = \oint E\,dA\cos\theta$$

But Fig. 21.9 shows that the surface normal $d\vec{A}$ and the electric field $\vec{E}$ are parallel at any point on the sphere, so $\cos\theta = 1$. Since the electric field varies as $1/r^2$, its magnitude is the same everywhere at the fixed radius r of our sphere. Thus, we can take E outside the integral, giving

$$\Phi = \oint_{\text{sphere}} E\,dA = E\oint_{\text{sphere}} dA = E(4\pi r^2)$$

where the last step follows because $\oint dA$ is just the surface area of the sphere, namely $4\pi r^2$. (Here's the first of many cases where the evaluation of a surface integral is straightforward because of the symmetry of the electric field and of the surface we're integrating over.) Now, the electric field of a point charge is given by Equation 20.3: $E = kq/r^2$. So we have

$\Phi = E(4\pi r^2) = (kq/r^2)(4\pi r^2) = 4\pi kq$. Since the point charge q is the only charge inside our spherical surface, the proportionality constant between flux and enclosed charge is $4\pi k$.

Before proceeding, we introduce the so-called permittivity constant ϵ_0, defined as $\epsilon_0 = 1/4\pi k$, where k is the Coulomb constant. The value of ϵ_0 is approximately $8.85 \times 10^{-12}\ \text{C}^2/\text{N}\cdot\text{m}^2$. There's no physics here, just a new constant that conveys the same information as k. That there are two redundant constants is a historical artifact, and we switch now from k to ϵ_0 because doing so makes subsequent formulas simpler. In terms of ϵ_0, the proportionality $4\pi k$ between flux and enclosed charge becomes $1/\epsilon_0$. So our statement that the flux through any closed surface is proportional to the net charge enclosed becomes

$$\oint \vec{E} \cdot d\vec{A} = \frac{q_{\text{enclosed}}}{\epsilon_0} \quad \text{(Gauss's law)} \qquad (21.3)$$

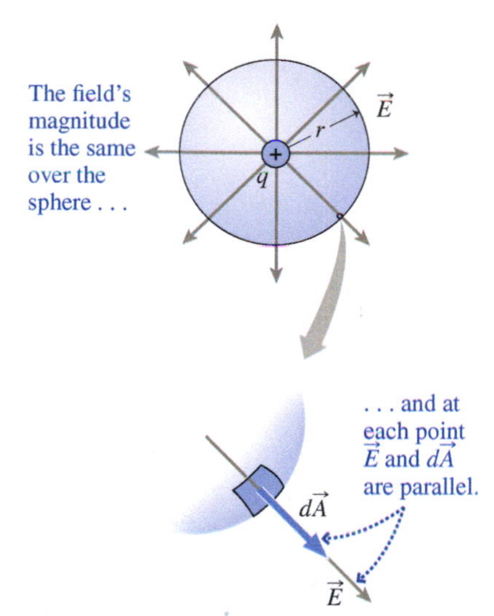

FIGURE 21.9 The electric field of a point charge, shown with a spherical surface centered on the charge.

Here the integral is taken over *any closed surface*, and q_{enclosed} is the charge enclosed *by that surface*.

Equation 21.3 is **Gauss's law**, one of four fundamental relations that govern the behavior of electromagnetic fields throughout the universe. Whether you journey into a star in some remote galaxy, down among the strands of a DNA molecule, or into the microprocessor chip at the heart of your computer, you'll find that the flux of the electric field through any closed surface depends only on the enclosed charge. In nearly 200 years of experiments, no electric field has ever been found to violate Gauss's law.

Gauss's law, though clothed in the mathematical finery of a surface integral, is just a more rigorous way of saying what's clear in Fig. 21.4: The number of field lines emerging from a closed surface is proportional to the net charge enclosed.

Gauss and Coulomb

Gauss's law and Coulomb's law look completely different, but they're closely related. Figure 21.10 shows that their relationship involves the inverse-square law. Gauss's law tells us that the flux through the two surfaces in the figure is the same and is equal to q/ϵ_0. But why? Because, as our arguments leading to Gauss's law show, the flux through a spherical surface of radius r centered on a point charge is the product of the surface area $4\pi r^2$ and the electric field E at the surface. But Coulomb's law says that the electric field drops off as $1/r^2$. As r increases, the surface area grows as r^2, but the $1/r^2$ decrease in field strength just compensates, giving a constant value for the flux. If the inverse-square law (e.g., Coulomb's law) didn't hold, then the flux wouldn't be constant and Gauss's law wouldn't hold either.

It's also the inverse-square law that makes electric field-line pictures useful for visualizing fields. Field lines begin or end only on charges; otherwise, they go off to infinity. As the field lines of a point charge spread in three dimensions, the number crossing any spherical surface (or any closed surface) remains the same. But larger spheres have larger surface areas, in proportion to r^2—and that means the density of field lines drops as $1/r^2$, accurately reflecting the field strength. Once again, the inverse-square law (Coulomb) and the relation between flux and enclosed charge (Gauss) are intimately connected. Incidentally, field-line pictures printed in a book or drawn on a blackboard generally can't be quantitatively correct because they don't show the spreading of field lines in all three dimensions.

We've been talking here only about isolated point charges, but Gauss's law applies to *all* electric fields, no matter how complicated the charge distributions that produce them. That's because the superposition principle allows us to add vectorially the electric fields described individually by Coulomb's law for the point-charge field. So our argument leading to Gauss's law still applies when the field $\vec{E}$ is a superposition of point-charge fields.

For static charge distributions, Gauss's and Coulomb's laws are completely equivalent. But with moving charges only Gauss's law remains exact. So Gauss's law is more fundamental, and we count it among the four basic laws of electromagnetism.

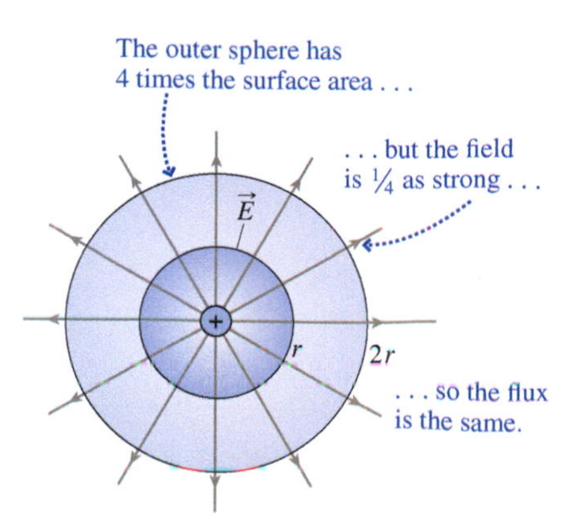

FIGURE 21.10 Gauss's law follows from the inverse-square aspect of Coulomb's law.

GOT IT? 21.3 A spherical surface surrounds an isolated positive charge, as shown. (1) If a second charge is placed outside the surface, which of the following will be true of the total flux through the surface? (a) It doesn't change; (b) it increases; (c) it decreases; (d) it increases or decreases depending on the sign of the second charge. (2) Repeat for the electric field on the surface at the point between the charges.

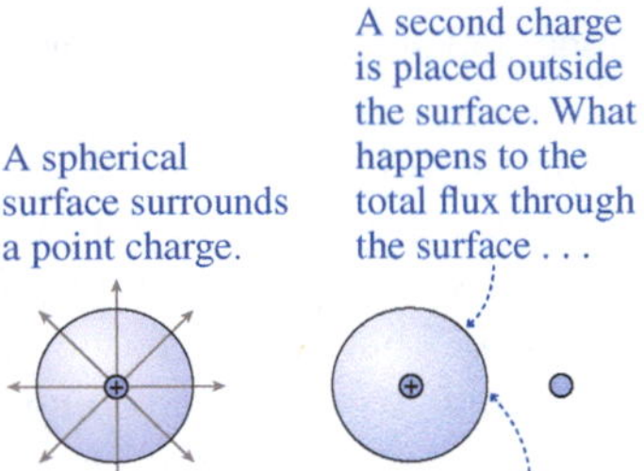

21.4 Using Gauss's Law

Gauss's law is a universal statement about electric fields; it's true for *any* surface enclosing *any* charge distribution. For charge distributions with sufficient symmetry—symmetric about a point, a line, or a plane—Gauss's law also provides a powerful alternative to Coulomb's law that makes electric-field calculations much easier. For such distributions it's possible to evaluate the flux integral on the left-hand side of Gauss's law (Equation 21.3) without actually knowing the field. We can then solve for E in terms of the enclosed charge. We begin with a general strategy for applying Gauss's law to symmetric charge distributions, followed by examples of the three symmetries.

PROBLEM-SOLVING STRATEGY 21.1 Gauss's Law

INTERPRET Check that your charge distribution has sufficient symmetry to use Gauss's law to find the electric field, and identify the symmetry. Is it spherical, line, or plane? If the charge distribution doesn't exhibit one of these symmetries, then Gauss's law—though always true—won't help you find the field.

DEVELOP Draw a diagram showing the charge distribution, and use the symmetry to infer the direction of the electric field. Then draw an appropriate **Gaussian surface**—an imaginary, closed surface that will let you evaluate the flux integral in Gauss's law. The field should have constant magnitude over the surface and should be perpendicular to the surface. Sketch some field lines; the symmetry should indicate their direction. With line and plane symmetry, there may be parts of the surface where the field is perpendicular and parts where it's parallel; we'll show in examples how to handle these situations. If you can't find a suitable Gaussian surface, there probably isn't sufficient symmetry to use Gauss's law to calculate the field.

EVALUATE

- Evaluate the flux $\Phi = \oint \vec{E} \cdot d\vec{A}$ over your Gaussian surface. Since you've found a surface to which the field is perpendicular, $\vec{E}$ and the surface normal vector $d\vec{A}$ are parallel, so $\cos\theta = 1$ and the dot product becomes $E\,dA$. With the field strength E constant over the surface, it can come out of the integral, leaving $\oint dA$. And that's the surface area A. So the flux will be EA. By carefully choosing your Gaussian surface, you've turned that messy integral in Equation 21.2 into a mere multiplication!
  If $\vec{E}$ is parallel to some parts of the area—as happens in line and plane symmetry—then $\vec{E} \perp d\vec{A}$, so $\vec{E} \cdot d\vec{A} = 0$ and there's no contribution to the flux from those areas.
- Evaluate the *enclosed* charge. This may or may not be the same as the total charge, depending on whether the position at which you're evaluating the field lies outside or inside the charge distribution.
- Evaluate the field E by invoking Gauss's law, equating the flux to $q_{\text{enclosed}}/\epsilon_0$, and solving for E. This is the field magnitude; the direction should be evident from symmetry.

ASSESS Does your answer make sense? Does the field behave as you would expect given what you know of simpler charge distributions—point charges, line charges, or charged sheets, depending on the symmetry?

You'll quickly get the hang of this strategy because Gauss's law is useful for finding the field only in the three common symmetries. That means you need to evaluate the integral for the flux just once for a given symmetry.

Spherical Symmetry

A charge distribution has spherical symmetry when the charge density depends only on the radial distance r from the center of the distribution—also called the *point of symmetry*. A point charge is one example, as are the three charge distributions we'll now treat in Examples 21.1 through 21.3. In fact, *any* spherical charge distribution has spherical symmetry, provided the charge density varies solely with distance from the center. The only electric field consistent with spherical symmetry is a field that points in the radial direction, either away from or toward the point of symmetry. We'll start with the simplest case, a charged spherical shell.

EXAMPLE 21.1 Gauss's Law: A Hollow Spherical Shell

A thin, hollow spherical shell of radius R has total charge Q distributed uniformly over its surface. Find the electric field (a) outside and (b) inside the shell.

INTERPRET This is a spherically symmetric charge distribution, so we can use Gauss's law to find the field.

DEVELOP With spherical symmetry, the field lines have to point radially outward for a positive charge and inward for a negative one. We've sketched the charged shell and some field lines outside it in Fig. 21.11. With spherical symmetry, the appropriate Gaussian surfaces are also spheres centered on the center of the shell. Since we're asked for the field both outside and inside the shell, we've sketched one such surface outside the shell and one inside.

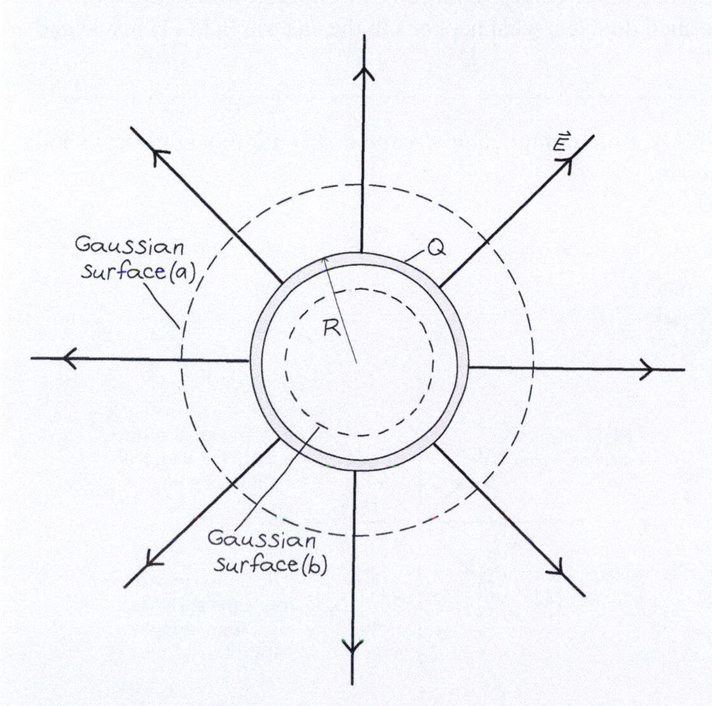

FIGURE 21.11 Finding the field of a charged shell.

EVALUATE

- We first evaluate the flux integral $\Phi = \oint \vec{E} \cdot d\vec{A}$ that appears on the left-hand side of Gauss's law. With our choice of a spherical Gaussian surface, the field is everywhere perpendicular to the surface, or parallel to the normal vector $d\vec{A}$ (recall Fig. 21.9). So $\cos\theta = 1$, and the dot product becomes just $E\,dA$. With our Gaussian surface being a sphere centered on the point of symmetry, the field magnitude E is the same all over the Gaussian surface. So E comes outside the integral, leaving $\Phi = E\oint dA$. The remaining integral is the sum of all the little areas dA over the spherical Gaussian surface—and that's the surface area of a sphere, $A = 4\pi r^2$. The flux then becomes $\Phi = 4\pi r^2 E$. Note that this result doesn't depend on whether the surface is outside or inside the shell or on the particular value of its radius r; it follows from the symmetry and our choice of a spherical Gaussian surface alone. So it holds for both Gaussian surfaces shown in Fig. 21.11, outside and inside the shell.
- Now we evaluate the *enclosed* charge. That *is* different for the two surfaces, so we'll deal first with (a), the field outside the shell. For this we consider the Gaussian surface marked (a); it's outside the shell, so it encloses the entire charge Q. Substituting our expression for the flux, $\Phi = 4\pi r^2 E$, into Gauss's law, $\Phi = \oint \vec{E} \cdot d\vec{A} = q_{\text{enclosed}}/\epsilon_0$, gives $4\pi r^2 E = Q/\epsilon_0$. We can then solve for E to get

$$E = \frac{Q}{4\pi\epsilon_0 r^2} \quad \text{(field outside any spherical charge distribution)} \tag{21.4}$$

- Now let's consider the interior of the shell. With all the charge on the shell's surface, there's no charge inside. So Gaussian surface (b) doesn't enclose any charge. But, as we argued above, the flux is still $4\pi r^2 E$, so Gauss's law becomes $4\pi r^2 E = 0$. Solving for E gives $E = 0$. We could have drawn Gaussian surface (b) anywhere inside the shell, including right up to its surface, so there's no electric field anywhere inside the shell.

ASSESS

- Consider first the field outside the shell (Equation 21.4). Since $1/4\pi\epsilon_0$ is the Coulomb constant k, this field can be written $E = kQ/r^2$—precisely the field of a point charge! In fact, our argument leading to this result didn't depend on the charge being on a spherical shell; as long as a charge distribution is spherically symmetric and we're evaluating the field *outside* the distribution, we would find the same result. That's why we've made the result a numbered equation and labeled it "field outside any spherical charge distribution." Outside *any* spherical charge distribution, then, its field is *exactly* the same as that of a point charge with

(*continued*)

the same total charge, located at the center of symmetry. Incidentally, this result also holds for gravity because it, too, obeys an inverse-square law. That's why we can treat planets as though they were point masses located at their centers.

- Now the interior field: How can it be exactly zero everywhere inside the shell? Figure 21.12 shows that this remarkable result is a consequence of the inverse-square law. ■

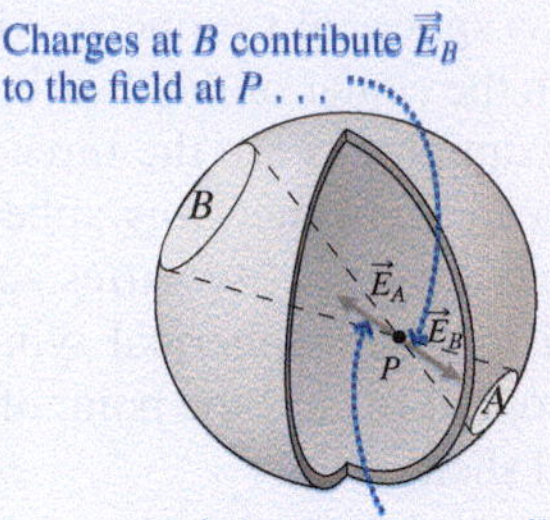

FIGURE 21.12 At any point P inside a charged shell, the field from the relatively few nearby charges at A is exactly canceled by the field from the more numerous but more distant charges at B. The result is zero field everywhere inside the shell.

✓TIP Symmetry Matters!

We used the fact that there's no charge inside an empty spherical shell to conclude that the field inside the shell is zero. But be careful: That conclusion follows only when there's enough symmetry, as Fig. 21.13 shows. Here, the Gaussian surface encloses zero net charge, so the flux through the surface is zero. But the electric field both on and inside the surface isn't zero.

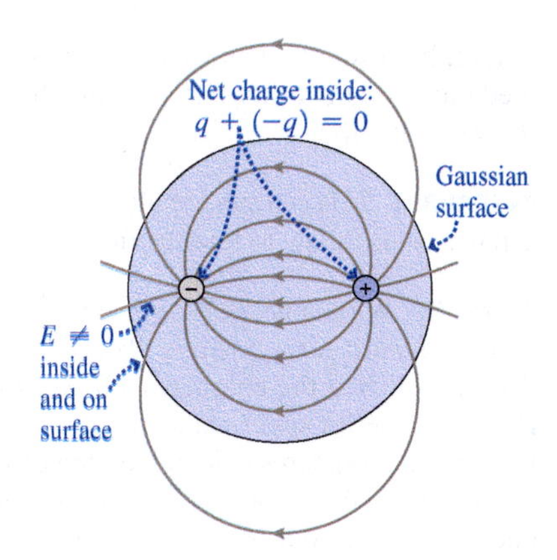

FIGURE 21.13 A spherical Gaussian surface surrounds a dipole. We have $q_{enclosed} = 0$, but $E \neq 0$ inside the surface because the dipole isn't spherically symmetric.

GOT IT? 21.4 A spherical shell carries charge Q distributed uniformly over its surface. If the charge on the shell doubles, what happens to the electric field (1) inside and (2) outside the shell?

Now let's consider a slightly more complicated charge distribution: a spherical shell with a point charge at the center.

EXAMPLE 21.2 Gauss's Law: A Point Charge within a Shell

A positive point charge $+q$ is at the center of a spherical shell of radius R carrying charge $-2q$, distributed uniformly over its surface. Find expressions for the field strength inside and outside the shell.

INTERPRET This problem is about a charge distribution with spherical symmetry.

DEVELOP We want to find the fields both inside and outside the distribution, so we show two spherical Gaussian surfaces in our sketch, Fig. 21.14. We've also drawn some field lines, which will help in assessing our answer.

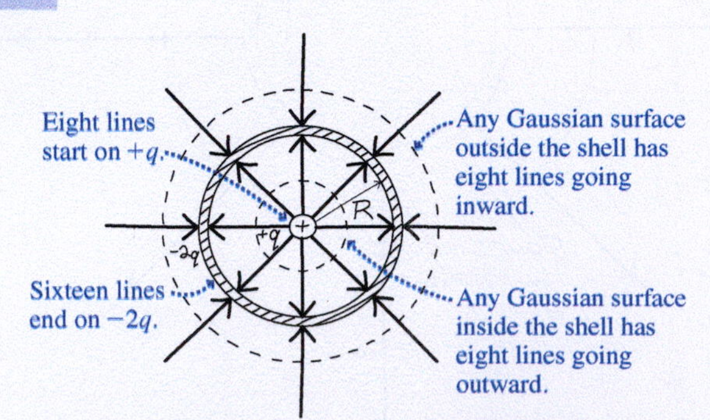

FIGURE 21.14 A shell carrying charge $-2q$ surrounds a point charge $+q$. Note that the number of field lines beginning on the point charge and the number ending on the shell correctly reflect their respective charges.

EVALUATE

- We already know that the flux through a spherical Gaussian surface is $\Phi = 4\pi r^2 E$ when we have spherical symmetry, so that's the flux for both surfaces.

- The outer surface encloses the charge $+q$ at the center and $-2q$ on the shell, so the net charge enclosed is $q_{\text{enclosed}} = -q$ for $r > R$. The inner surface encloses only the point charge $+q$, so $q_{\text{enclosed}} = +q$ for $r < R$.
- Now we find the field by equating the flux $\Phi = 4\pi r^2 E$ to $q_{\text{enclosed}}/\epsilon_0$. The result for $r > R$ is $E = -q/4\pi\epsilon_0 r^2$, where the minus sign appears because the enclosed charge is $-q$. For $r < R$ with enclosed charge $+q$, the result is $E = q/4\pi\epsilon_0 r^2$.

ASSESS Make sense? We've seen that the field outside a spherically symmetric charge distribution is that of an equivalent point charge at the center. Here the net charge is $-q$, and our result for $r > R$ is indeed the field of a point charge $-q$. Another way to see this is to apply superposition: The field outside due to the shell alone is that of a point charge $-2q$; that adds to the field of the central point charge $+q$ to give, again, the field of a point charge $-q$ in the region outside the shell. In Example 21.1 we found that the shell produces no field in its interior, so here superposition leaves us with the field of the central point charge alone, just as our result shows. Sketching the field lines, as we've done in Fig. 21.14, also shows how our results make sense. ■

What if we have charge distributed throughout a volume? As long as the distribution is spherically symmetric, we can find the electric field using Gauss's law just as we did in Examples 21.1 and 21.2. The only difference comes when we evaluate the enclosed charge. Example 21.3 considers this situation.

EXAMPLE 21.3 Gauss's Law: A Uniformly Charged Sphere

A charge Q is distributed uniformly throughout a sphere of radius R. Find the electric field at all points, (a) outside and (b) inside the sphere.

INTERPRET The charge distribution has spherical symmetry, so we can use Gauss's law to find the field.

DEVELOP Figure 21.15 is a sketch of the spherical charge distribution. As you've already seen, the field in any case of spherical symmetry must be radial, so we've sketched some field lines on the figure. As in Examples 21.1 and 21.2, Gaussian surfaces appropriate to spherical symmetry are themselves spheres. Since we're asked for the field both outside and inside the charge distribution, we've drawn two such Gaussian surfaces, one outside and one inside the distribution.

EVALUATE

- We already know the answer to (a), the field outside the charge distribution. That's because, as we found in Example 21.1, the field outside *any* spherically symmetric charge distribution is exactly that of a point charge located at the center. So the field outside the charged sphere has magnitude $E_{\text{out}} = Q/4\pi\epsilon_0 r^2$ and points radially outward for positive Q. You could also recover this result by considering the flux through surface (a) in Fig. 21.15 and equating it to $q_{\text{enclosed}}/\epsilon_0$—which would essentially repeat the calculation for the field outside the spherical shell in Example 21.1.
- For (b), the field inside the charged sphere, we need to apply Gauss's law to surface (b) in Fig. 21.15. In Example 21.1 we found that the left-hand side of Gauss's law—the flux—is given by $4\pi r^2 E$ whenever we have spherical symmetry and are using a spherical Gaussian surface. However, we still need to find the charge enclosed by surface (b). Since surface (b) lies *within* the sphere of charge, it doesn't enclose the entire charge Q. We can find just how much charge it encloses by comparing its volume with that of the entire sphere. Surface (b) has radius r, so its volume is $\frac{4}{3}\pi r^3$. The charged sphere has radius R, so its volume is $\frac{4}{3}\pi R^3$. So our Gaussian surface, (b), encloses a fraction r^3/R^3 of the total volume of the charged sphere. Because charge is distributed *uniformly* throughout the sphere, this is also the fraction of the total charge enclosed by our Gaussian surface. So q_{enclosed} in Gauss's law becomes Qr^3/R^3 for a Gaussian surface of radius r that's inside the charged sphere. We can now write Gauss's law for this situation, equating the flux, $4\pi r^2 E$, to $q_{\text{enclosed}}/\epsilon_0$: $4\pi r^2 E = Qr^3/\epsilon_0 R^3$. Solving for the field magnitude E gives

$$E = \frac{Qr}{4\pi\epsilon_0 R^3} \quad \text{(field inside uniformly charged sphere)} \quad (21.5)$$

ASSESS: Make sense? The field inside the sphere increases linearly with distance r from the center. This reflects two opposite effects: First, as we move outward from the center, the enclosed charge grows in proportion to the volume enclosed—that is, as r^3. But the distance from the center grows as well, and that causes a $1/r^2$ decrease in the field. The combined effect is a field that grows linearly from the center to the surface of the sphere, as shown in Equation 21.5. But once we're outside the sphere, the enclosed charge no longer grows, so we see only the $1/r^2$ decrease of the point-charge field. Figure 21.16 plots the resulting field magnitude E both inside and outside the sphere. As the figure shows, and as you can convince yourself by comparing Equations 21.4 and 21.5, our results for the exterior and interior fields agree at the sphere's surface, where $r = R$. ■

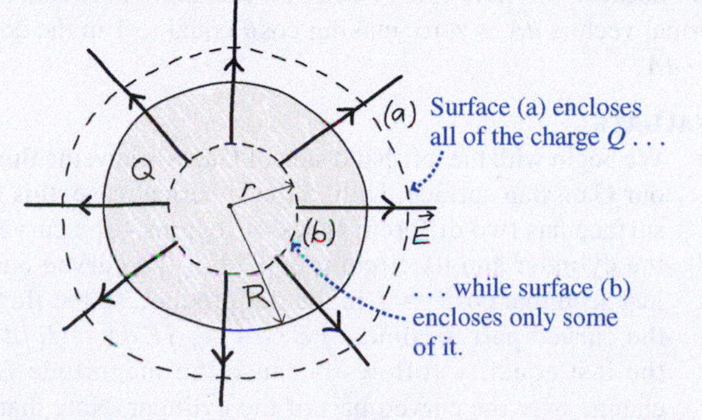

FIGURE 21.15 Finding the field of a uniformly charged sphere. The field also exists within the inner Gaussian surface, but we haven't shown it.

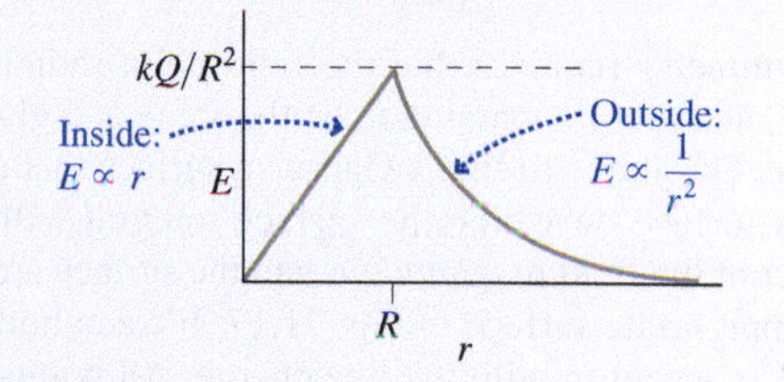

FIGURE 21.16 Field Strength versus radial distance for a uniformly charged sphere of radius R.

Before moving on, consider the three examples of Gauss's law in spherical symmetry that you've just seen. In each example we equated the left-hand side of Gauss's law to the right-hand side, and solved for the electric field magnitude E. We only had to find the left-hand side once: In every case of spherical symmetry with a spherical Gaussian surface, the left-hand side of Gauss's law is $\Phi = \oint \vec{E} \cdot d\vec{A} = 4\pi r^2 E$. We discovered this in Example 21.1, and you never need to do that calculation again as long as you've got spherical symmetry. So we used $4\pi r^2 E$ for the left-hand side of Gauss's law in both Examples 21.2 and 21.3. Example 21.1 showed us that the field *outside* any spherically symmetric charge distribution is *exactly* that of a point charge located at the center. Again, that's an exact result—not just an approximation that holds at large distances. It's true right up to the surface of the charge distribution, as Figure 21.16 shows graphically.

Example 21.1 also showed us that the field everywhere *inside* a spherical shell of charge is zero. You could think of any spherically symmetric charge distribution as made up of concentric shells—and that means there's never a contribution to the electric field from charge that's further from the center than the position where you're evaluating the field. In that light, here's another take on the most difficult of our three examples, Example 21.3 for the field within a spherical volume of charge: At any position inside that volume, there's no contribution from the charge that's further out. And, as Example 21.1 also showed, the charge interior to that position acts like a point charge. Work out that charge—essentially the calculation leading to q_{enclosed} in Example 21.3—and you'll get Equation 21.5 for the field inside a uniformly charged sphere. So all spherically symmetric charge distributions are similar, in that we apply Gauss's law in exactly the same way. The only difference is in finding the enclosed charge. Problems 69 and 71 provide additional examples involving charge distributions that are spherically symmetric but not uniform.

We'll now consider the two other symmetries where Gauss's law provides a quick route to finding the electric field. As you work through the examples for these symmetries, consider how they're similar to the spherical examples you've just seen. The application of Gauss's law works just the same, but now there's a different expression for the flux on the left-hand side, and calculations of enclosed charge are different.

Line Symmetry

A charge distribution has line symmetry when its charge density depends only on the perpendicular distance r from a line, called the *symmetry axis*. Symmetry then requires that the field point radially and that the field magnitude depend only on distance from the axis. It also requires the charge distribution to be infinitely long, so there's no variation parallel to the line. That's impossible, but nevertheless the infinite line is a reasonable approximation to elongated structures like wires. The next two examples explore the application of Gauss's law to line symmetry.

EXAMPLE 21.4 Gauss's Law: An Infinite Line of Charge

Use Gauss's law to find the electric field of an infinite line charge carrying charge density λ in coulombs per meter.

INTERPRET An infinite line has line symmetry, so we can apply Gauss's law to find the electric field.

DEVELOP Symmetry requires that the field point radially from the line of charge, and that its magnitude be the same at a given distance r from the line. Our job is to find a Gaussian surface that exploits this symmetry—a surface on which the surface integral will turn into a simple product of the field magnitude E and the surface area A. We've sketched an appropriate surface in Fig. 21.17: It's a cylinder of radius r and length L, concentric with the line charge. All points on the cylinder's curved surface are the same distance r from the line charge, so the field magnitude E is the same over the curved part of the surface. And because the field must be radial, the angle between $\vec{E}$ and the normal vectors $d\vec{A}$ is zero, making $\cos\theta$ equal to 1 in the dot product $\vec{E} \cdot d\vec{A}$.

EVALUATE

- We begin with the left-hand side of Gauss's law, the flux through our Gaussian surface. Unlike the spherical case, this Gaussian surface has two different types of regions—the curved part of the cylinder and its circular ends. For the curved part, we've just seen that $\cos\theta = 1$ in the dot product, so the flux through the curved part becomes $\int \vec{E} \cdot d\vec{A} = \int E\,dA = E\int dA$, where the last equality follows because the magnitude E doesn't change over the curved part of the cylinder. Note that we took

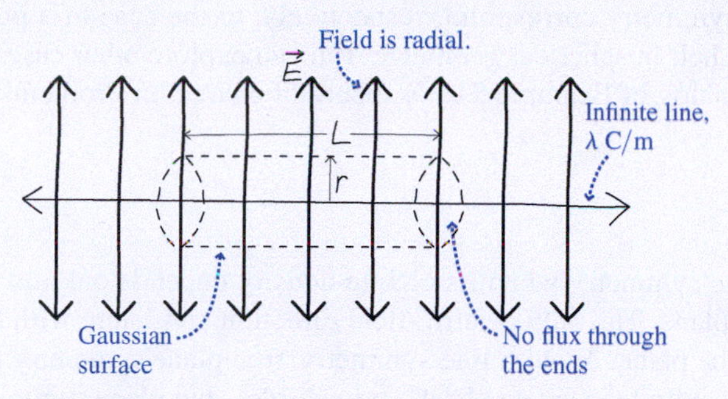

FIGURE 21.17 A cylindrical Gaussian surface surrounds a line charge. The sketch shows a slice in the plane of the page; the field extends radially outward from the line into and out of the page as well.

the circle off the integral sign because we're just considering the curved part, which alone doesn't constitute a *closed* surface. The remaining integral is just the area of the curved part of the cylinder. Imagine unrolling this curved surface and flattening it out; you'd get a rectangle of length L and width equal to the circumference $2\pi r$, so the flux through the curved part becomes $\Phi = 2\pi rLE$. What about the ends? No field lines emerge from them, so here the flux is zero; mathematically, $\vec{E}\cdot d\vec{A} = 0$ on the ends because the field $\vec{E}$ and the vectors $d\vec{A}$ normal to the ends are perpendicular. The left-hand side of Gauss's law, then, is $2\pi rLE$. This result depends only on the symmetry, so it applies in all situations involving line symmetry.

- Next we need the enclosed charge. The line carries λ C/m and our Gaussian cylinder encloses L meters of the line, so $q_{\text{enclosed}} = \lambda L$.
- Finally, we invoke Gauss's law by equating the flux Φ to $q_{\text{enclosed}}/\epsilon_0$. The result, $2\pi rLE = \lambda L/\epsilon_0$, solves to give

$$E = \frac{\lambda}{2\pi\epsilon_0 r} \quad \text{(field of a line charge)} \qquad (21.6)$$

ASSESS Make sense? We worked this same problem in Example 20.7 in a much more difficult calculation involving a complicated integral. Since $1/2\pi\epsilon_0 = 2k$, our result here is the same. But Gauss's law makes the problem much easier!

Although Example 21.4 involved an infinitesimally thin line of charge, you can see from Fig. 21.18 that our result must hold *outside* any charge with line symmetry. And, as we argued in Example 20.7, it's a good approximation for the field of any long, cylindrical structure as long as we're not too near its ends.

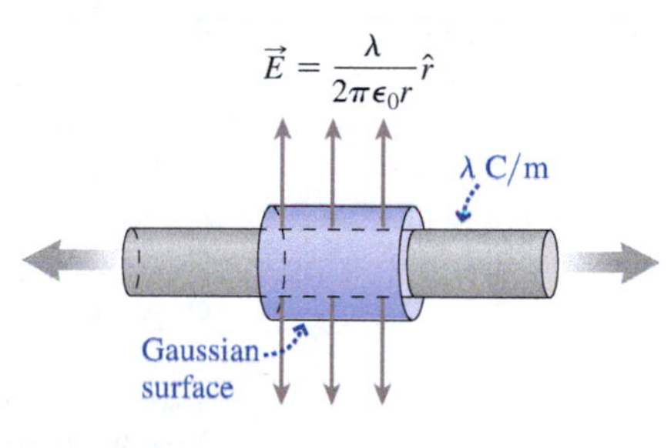

FIGURE 21.18 The arguments of Example 21.4 apply outside *any* cylindrical charge distribution.

EXAMPLE 21.5 Gauss's Law: A Hollow Pipe

A thin-walled pipe 3.0 m long and 2.0 cm in radius carries a net charge $q = 5.7\ \mu\text{C}$ distributed uniformly over its surface. Find the electric field both 1.0 cm and 3.0 cm from the pipe axis, far from either end.

INTERPRET Although the pipe has finite length, both distances are small compared with that length, so we can approximate the pipe as an infinitely long structure with line symmetry.

DEVELOP With line symmetry the appropriate Gaussian surface is a cylinder coaxial with the pipe. We've drawn two such cylinders in Fig. 21.19, one for each radius where we're asked to evaluate the field.

EVALUATE

- We showed in Example 21.4 that the flux integral in line symmetry gives $\Phi = 2\pi rLE$.

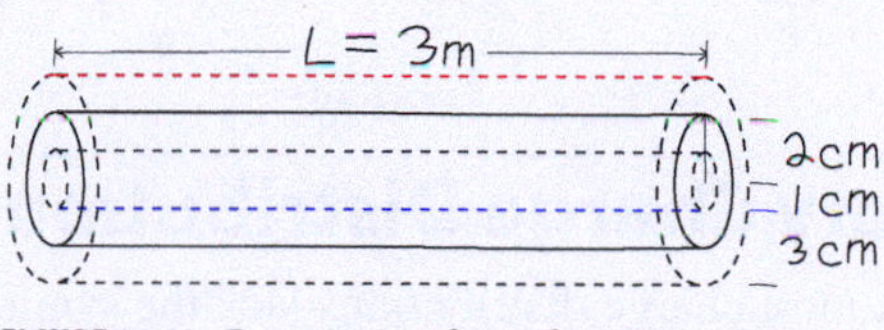

FIGURE 21.19 Gaussian surfaces for Example 21.5.

- Next we need the enclosed charge. At 3 cm we're outside the pipe, so the Gaussian surface with this radius encloses all the charge: $q_{\text{enclosed}} = 5.7\ \mu\text{C}$. The pipe is hollow, so at 1 cm the enclosed charge is zero.
- Equating the flux to $q_{\text{enclosed}}/\epsilon_0$ and solving for E give

$$E = \frac{q_{\text{enclosed}}}{2\pi\epsilon_0 rL} = \frac{5.7\ \mu\text{C}}{(2\pi\epsilon_0)(3.0\times10^{-2}\ \text{m})(3.0\ \text{m})} = 1.1\ \text{MN/C}$$

for the field at 3 cm and $E = 0$ for the field inside the pipe.

ASSESS Make sense? *Inside* the pipe, there's no field, and for the same reason as inside a uniformly charged hollow sphere—namely, that fields from near and far parts of the pipe cancel due, ultimately, to the inverse-square law. Again, be careful: That result follows because of the symmetry, although we'll soon see that with *conducting* pipes and shells there's no interior field even without that symmetry. We argued earlier that the field *outside* any line-symmetric distribution should be given by Equation 21.6, $E = \lambda/2\pi\epsilon_0 r$. In our result, the quantity q_{enclosed}/L is the line charge density λ, so our result is indeed consistent with that equation.

Our two examples of line symmetry correspond, respectively, to the case of a point charge and a hollow spherical shell in spherical geometry. You can explore other cases of line symmetry, including the analog of Example 21.3's sphere of charge, in Problems 56 and 73.

Plane Symmetry

A charge distribution has plane symmetry when its charge density depends only on the perpendicular distance from a plane. The only electric-field direction consistent with this symmetry is perpendicular to the plane. As with line symmetry, true plane symmetry implies a charge distribution that's infinite in extent. That's impossible—but plane symmetry remains a good approximation when charge is spread uniformly over large, flat surfaces or slabs. The next example applies Gauss's law to plane symmetry.

EXAMPLE 21.6 Gauss's Law: A Sheet of Charge

An infinite sheet of charge carries uniform surface charge density σ in coulombs per square meter. Find the resulting electric field.

INTERPRET Since the sheet is infinite, we have plane symmetry with the sheet itself as the symmetry plane.

DEVELOP We need a Gaussian surface chosen so the flux integral $\oint \vec{E} \cdot d\vec{A}$ reduces to a simple multiplication, at least over some parts of the area; over any remaining parts it should be zero. Because the electric field must be perpendicular to our charged sheet, *any* surface with ends parallel to the sheet will have $\cos\theta$ equal to 1 in the dot product. Also, the symmetry dictates that E can't change as we move in a direction parallel to the sheet, since we'd remain the same distance from it. Furthermore, any surface whose sides are perpendicular to the charged sheet will have zero flux through the sides. So we're free to choose for our Gaussian surface any surface with ends parallel to the sheet and sides perpendicular. To ensure that E has the same value over both ends of our surface, we'll have it extend equal distances on either side of the charged sheet. A simple surface that meets these criteria is a cylinder that straddles the charged sheet, as shown in Fig. 21.20.

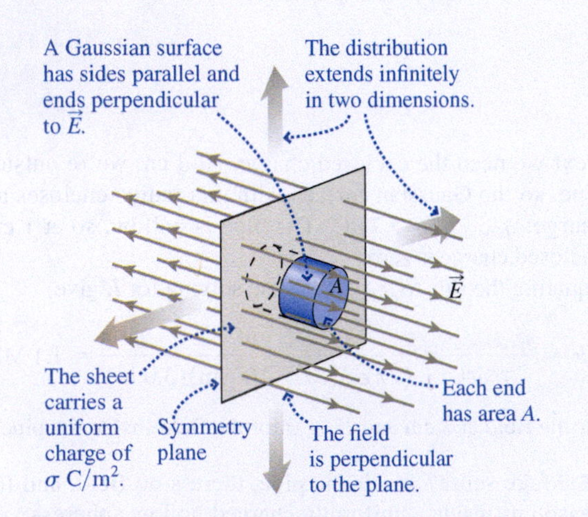

FIGURE 21.20 An infinite sheet of charge, with a Gaussian surface straddling the sheet.

EVALUATE

- First we evaluate the flux through our Gaussian cylinder. The field is parallel to the sides, so there's no flux contribution here. Then the total flux is through the ends, each of which has area A. Since the field is uniform over each end and perpendicular to the ends, the flux through each end is EA. So the total flux through both ends is $\Phi = 2EA$. This result is independent of the details of the charge distribution, so it holds in all cases of plane symmetry.
- Next we need the enclosed charge. The Gaussian surface encloses an area of the sheet equal to the area A of its ends. With surface charge density σ on the sheet, the enclosed charge is $q_{\text{enclosed}} = \sigma A$.
- Now we apply Gauss's law, equating the flux to $q_{\text{enclosed}}/\epsilon_0$. Thus $2EA = \sigma A/\epsilon_0$, so

$$E = \frac{\sigma}{2\epsilon_0} \quad \text{(field of a charged sheet)} \qquad (21.7)$$

The direction of this field on either side of the sheet is outward from the sheet if it's positively charged and inward if negative.

ASSESS Make sense? With an infinite plane, symmetry requires that the field lines be perpendicular to the plane. So they don't spread out, and that means the field doesn't vary with distance—as Equation 21.7 shows because it doesn't involve the distance from the sheet. Although our result is exact only for a truly infinite sheet, it's a good approximation near any large, flat, uniformly charged surface as long as we're not close to an edge. ■

21.5 Fields of Arbitrary Charge Distributions

Although Gauss's law is always true, most charge distributions lack the symmetry needed to apply Gauss's law to find the field. The alternative, Coulomb's law, is hard to use in all but the simplest cases. But we can often learn a lot by considering the distributions whose fields we calculated here and in Chapter 20. Figure 21.21 summarizes four of these

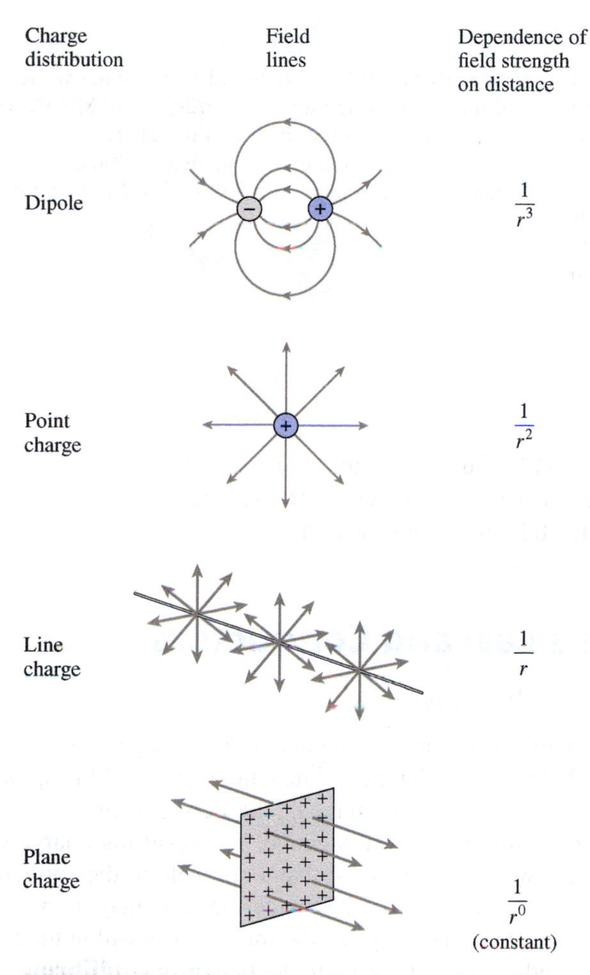

FIGURE 21.21 Fields of a dipole, a point charge, a charged line, and a charged plane.

fields. For the last three, note the simple relation between the number of dimensions and the behavior of the field. The plane has two dimensions, and its field doesn't decrease with distance. The line has one dimension, and its field decreases as $1/r$. The point has no dimensions, and its field falls as $1/r^2$. In a sense, the dipole continues this progression. It consists of opposite point charges whose effects nearly cancel; no wonder its field decreases faster still, as $1/r^3$. In fact, there's a hierarchy of charge distributions whose fields decrease ever faster, as dipoles nearly cancel dipoles, and so on. Scientists and engineers use this hierarchy in modeling charged structures ranging from molecules to radio antennas.

CONCEPTUAL EXAMPLE 21.1 A Charged Disk

Sketch some electric field lines for a uniformly charged disk, starting at the disk and extending out to several disk diameters.

EVALUATE When we're near the disk and not close to its edge, the disk looks like a large, flat, charged plane. Its field is essentially the uniform field of an infinite plane charge, so we draw straight field lines emanating perpendicular to the disk. Far from *any* finite distribution carrying a nonzero net charge, the field approximates that of a point charge, so farther away we draw field lines going radially outward. Field lines begin only on charges—in this case on the charged disk—so we have to connect close-in and far-out lines. We don't know exactly how the field looks in the intermediate region neither close to nor far from the disk, so we connect them as best we can. Figure 21.22 is the result.

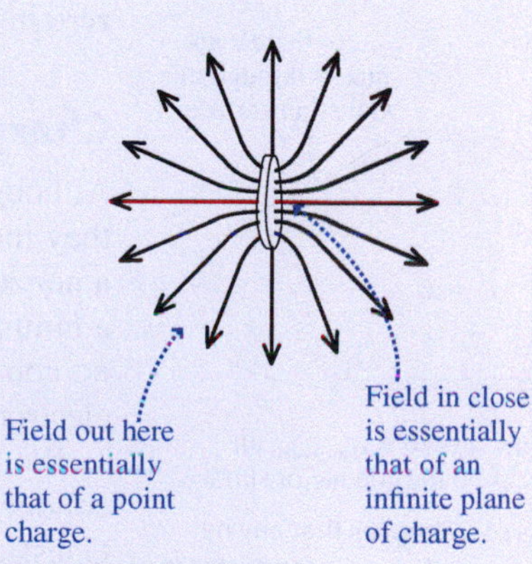

FIGURE 21.22 The field of a charged disk.

(continued)

ASSESS Our sketch is a good approximation to the field of a charged disk. And it obeys Gauss's law because the same number of field lines—namely, all 16 lines we chose to draw—cross any closed surface surrounding the entire disk.

MAKING THE CONNECTION Suppose the disk is 1.0 cm in diameter and carries charge 20 nC spread uniformly over its surface. Find the electric field strength (a) 1.0 mm from the disk surface and (b) 1.0 m from the disk.

EVALUATE (a) Close to the disk, assuming we're not near the edge, Equation 21.7 applies: $E = \sigma/2\epsilon_0 = 14$ MN/C, where we used the total charge and the disk area to get the surface charge density σ. (b) At 1 meter, the disk is so small it looks essentially like a point charge, so Equation 20.3 applies: $E = kq/r^2 = 180$ N/C.

GOT IT? 21.5 (1) If you're close to a *finite* line of charge (and not near its ends), does its field vary as (a) $1/r^3$, (b) $1/r^2$, or (c) $1/r$? (2) Repeat for the case when you're far from the line (i.e., much farther than its length).

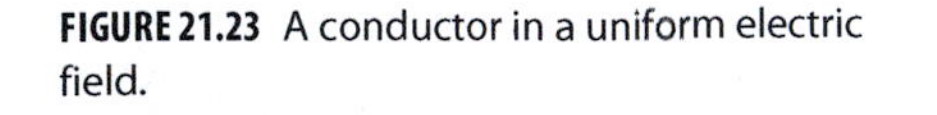

FIGURE 21.23 A conductor in a uniform electric field.

21.6 Gauss's Law and Conductors

Electrostatic Equilibrium

We've defined conductors as materials that contain free charges, like the free electrons in metals. Figure 21.23 shows what happens when an electric field is applied to a conductor. Free charges respond to the electric force $q\vec{E}$ by moving—in the direction of the field if they're positive, opposite the field if negative. The resulting charge separation gives rise to an electric field within the material that's opposite to the applied field. As more charge moves, this internal field becomes stronger until its magnitude eventually equals that of the applied field. Once that happens, the free charges within the conductor experience zero net force, and the conductor is in **electrostatic equilibrium**. Although individual charges continue to move about in random thermal motion, there's no longer any net charge motion. Once equilibrium is reached, the internal and applied fields are equal but opposite, and therefore:

The electric field is zero inside a conductor in electrostatic equilibrium.

It could not be otherwise: Since a conductor contains free charges, the presence of any internal electric field would result in bulk charge motion, and we wouldn't have equilibrium. This result doesn't depend on the size or shape of the conductor, the magnitude or direction of the applied field, or even the nature of the material as long as it's a conductor. This is a macroscopic view; it considers only average fields within the material. At the atomic and molecular level, there are still strong electric fields near individual electrons and positive ions. But the *average* field, taken over larger distances, is zero inside a conductor in electrostatic equilibrium.

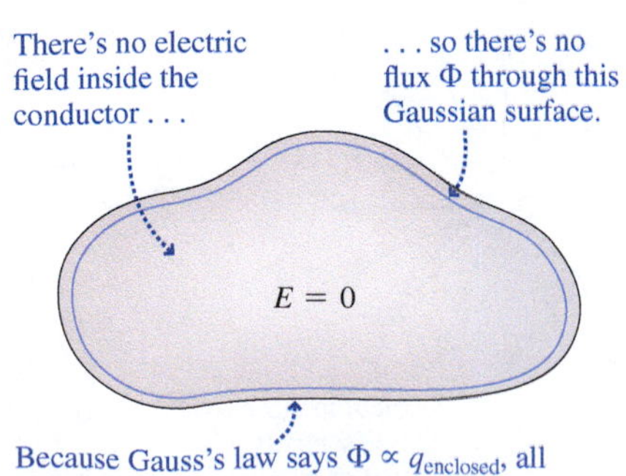

FIGURE 21.24 Gauss's law implies that any net charge resides on the surface of a conductor in electrostatic equilibrium.

Charged Conductors

Although they contain free charges, conductors are normally electrically neutral because they include equal numbers of electrons and protons. But suppose we give a conductor a nonzero net charge, for example, by injecting excess electrons into its interior. There's a mutual repulsion among the electrons and, because these are *excess* electrons, there's no compensating attraction from positive charges. We might expect, therefore, that the electrons will move as far apart as possible—namely, to the surface of the conductor.

We now use Gauss's law to prove that excess charge *must* be at the surface of a conductor in electrostatic equilibrium. Figure 21.24 shows a conducting material with a Gaussian surface drawn just below the material surface. In equilibrium there's no electric field inside the conductor, and thus the field is zero everywhere on the Gaussian surface.

The flux, $\oint \vec{E} \cdot d\vec{A}$, through the Gaussian surface is therefore also zero. But Gauss's law says that the flux through a closed surface is proportional to the net charge enclosed, and therefore the net charge inside our Gaussian surface must be zero. This is true no matter where the Gaussian surface is as long as it's *inside* the conductor. We can move it arbitrarily close to the conductor surface and it still encloses no net charge. If there is a net charge on the conductor, it lies outside the Gaussian surface, and therefore we conclude: **If a conductor in electrostatic equilibrium carries a net charge, that charge must reside on the conductor surface.**

EXAMPLE 21.7 Gauss's Law: A Hollow Conductor

An irregularly shaped conductor has a hollow cavity. The conductor itself carries a net charge of 1 μC, and there's a 2-μC point charge inside the cavity. Find the net charge on the cavity wall and on the outer surface of the conductor, assuming electrostatic equilibrium.

INTERPRET This problem involves a conductor in electrostatic equilibrium, which means (1) there's no electric field inside the conducting material and (2) the net charge resides on the conductor surface—which in this case includes both the inner and outer surfaces.

DEVELOP We sketch the situation in Fig. 21.25. Our plan is to apply Gauss's law to find the charges. We consider a Gaussian surface inside the conductor and enclosing the cavity, as shown.

EVALUATE Since there's no electric field inside the conductor, the flux through the Gaussian surface is zero, and therefore the net charge enclosed is also zero. But there's that +2-μC point charge in the cavity. For the Gaussian surface to enclose zero net charge, there must be −2 μC somewhere else—and the only place it can be is on the cavity wall. However, the entire conductor carries +1 μC. With −2 μC on the inside wall, that leaves +3 μC on the outer surface.

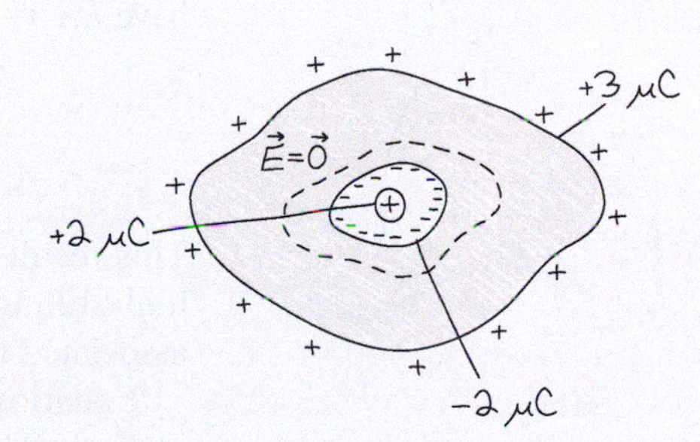

FIGURE 21.25 The Gaussian surface encloses zero net charge, so there must be −2 μC on the cavity wall.

ASSESS Make sense? Yes: This distribution of charge is the only one that's consistent with both Gauss's law and the requirement $E = 0$ inside a conductor in electrostatic equilibrium. As another check, think about what this charge distribution must look like from far away—namely, a point charge with net charge of 3 μC. Since the fields of the cavity wall and the inner point charge don't penetrate the conductor, the only field lines that reach out beyond the conductor are those from the charge on its outer surface. So that charge must be 3 μC, as we've found. ■

GOT IT? 21.6 A conductor carries a net charge $+Q$. There's a cavity inside the conductor that contains a point charge $-Q$. In electrostatic equilibrium, is the charge on the outer surface of the conductor (a) $-2Q$, (b) $-Q$, (c) 0, (d) Q, or (e) $2Q$?

Experimental Tests of Gauss's Law

That net charge moves to a conductor surface provides a sensitive test of Gauss's law and thus—through the arguments of Section 21.3 relating Gauss's and Coulomb's laws—a test of the inverse-square law for the electric force. Figure 21.26 shows a charged conducting ball touched to the inside of a hollow, initially neutral conductor. As required by Gauss's law, charge flows to the outer surface of the hollow conductor, leaving the ball uncharged. Measuring zero charge on the ball confirms Gauss's law, and thus the inverse-square law; such experiments show that the exponent in $1/r^2$ is indeed 2 to some 16 decimal places!

Video Tutor Demo | **Electroscope in Conducting Shell**

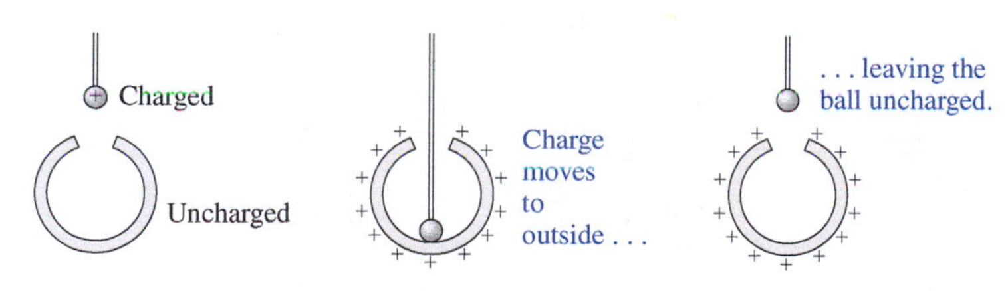

FIGURE 21.26 Experimental test of Gauss's law.

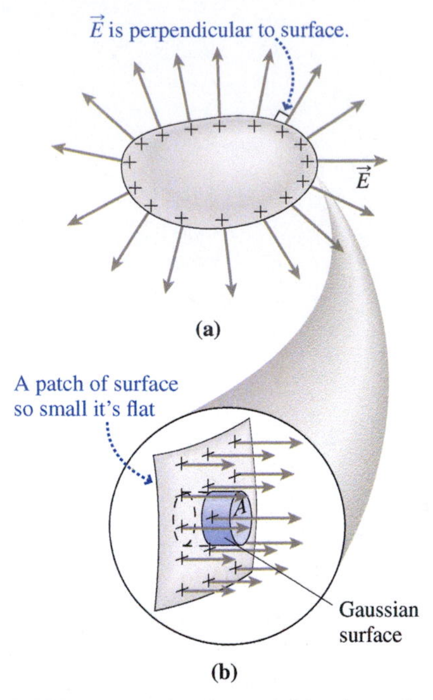

FIGURE 21.27 (a) The electric field at the surface of a charged conductor is perpendicular to the conductor surface. (b) A Gaussian surface straddles the conductor surface.

The Field at a Conductor Surface

There can't be an electric field *within* a conductor in electrostatic equilibrium, but there *may* be a field right at the conductor surface (Fig. 21.27*a*). Such a field must be perpendicular to the surface; otherwise, charge would move along the surface in response to the field's parallel component, and we wouldn't have equilibrium.

We can compute the strength of this surface field by considering a small Gaussian surface that straddles the conductor surface, as shown in Fig. 21.27*b*. There's no flux through the sides, and because the field is zero inside the conductor, there's no flux through the inner end either. So the only flux is through the outer end, with area *A*. Since the end is perpendicular to the field, the flux is *EA*. The Gaussian surface encloses charge σA, where σ is the surface charge density (C/m^2). Gauss's law equates the flux with $q_{enclosed}/\epsilon_0$, so we have $EA = \sigma A/\epsilon_0$, or

$$E = \frac{\sigma}{\epsilon_0} \quad \text{(field at conductor surface)} \tag{21.8}$$

This result shows that large fields develop where the charge density on a conductor is high. Engineers who design electrical devices must avoid high charge densities whose associated fields lead to sparks, arcing, and breakdown of electric insulation.

Equation 21.8 gives a field that depends only on the local charge density. Does that mean the field at a conductor surface arises only from the local charge? No! As always, the field is the vector sum of contributions from all charges. Remarkably, Gauss's law requires that charges on a conductor arrange themselves in such a way that the field at any point on the conductor surface depends only on the surface charge density right at that point—even though that field arises from *all* the charges on the surface (as well as from charges elsewhere if there are any)!

Consider a thin, flat, isolated, conducting sheet that has charge density σ on one of its two faces (Fig. 21.28*a*). Equation 21.8 shows that the field at the surface of this plate is σ/ϵ_0. But if the plate is large and flat, we can approximate it as an infinite sheet of charge—for which we found earlier (Equation 21.7) that the field should be $\sigma/2\epsilon_0$. Is there a contradiction here? No! If the plate is isolated, then symmetry requires that the charge spread itself evenly over *both* faces. If one face has charge density σ, so must the other—so we really have *two* charged sheets, each with density σ (Fig 21.28*b*). Each gives a field of magnitude $\sigma/2\epsilon_0$, and *outside* the conductor those fields superpose to give the net field σ/ϵ_0 (Fig. 21.28*c*). *Inside* the conductor their directions are opposite, and the result is zero field inside the conductor. Applying Equation 21.8 skips these details. But because Equation 21.8 was derived on the assumption that the field inside the conductor is zero, it "knows" about charges everywhere on the conductor—and in this case that means on the second face.

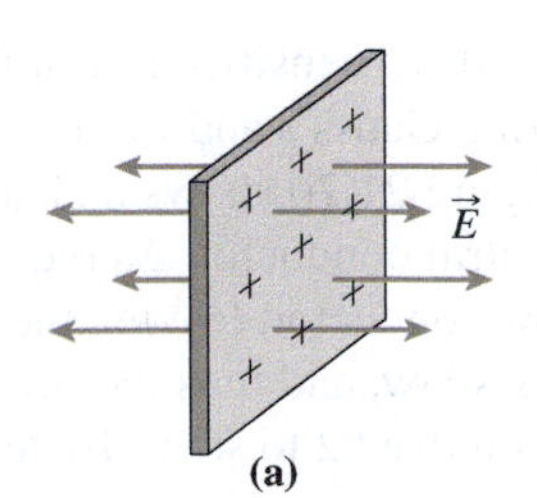

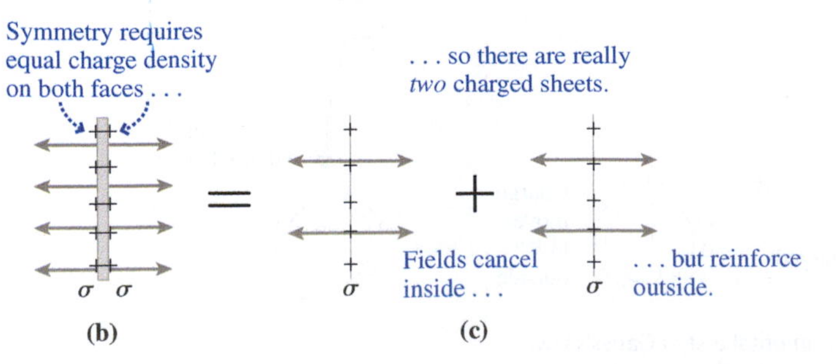

FIGURE 21.28 (a) An isolated, charged conducting plate. Its field points outward from both faces. (b) Edge-on view of the plate. (c) The field anywhere is the sum of the fields of the two faces, each treated as a single charged sheet.

APPLICATION Shielding and Lightning Safety

We've seen that charge moves to the outside of a conductor surface, leaving the interior free of charge and electric field—even if the interior is hollow. This is the basis of electric shielding, in which a conducting enclosure keeps out external electric fields. A common example is the *coaxial cable* that delivers TV signals from your cable company; coaxial cables also connect electronic instruments in scientific and medical research. A coaxial cable consists of an inner wire surrounded by a cylindrical conducting shield in which charge moves to block external electric fields that could cause interference. In another application of shielding, researchers doing experiments with very weak electric signals often construct entire rooms with conducting walls to minimize interference.

Shielding is also the reason a car is a relatively safe place in a thunderstorm. A lightning strike dumps charge on the car's metal body, but the charge distributes itself on the outside so as to prevent any electric fields from developing inside the car (see photo). That, in turn, prevents harmful currents from flowing through the occupants. The operator's cage in this chapter's opening photo has the same effect, harmlessly deflecting charge from the artificial lightning and keeping the interior free of electric fields.

Strictly speaking, charge resides on the outside of a conductor only in equilibrium. But electrons in metals respond so quickly that equilibrium results almost instantaneously—meaning that metallic shielding is effective even against the rapidly varying electric fields of high-frequency radio, TV, and microwave signals.

Equation 21.8 also applies to a pair of oppositely charged conducting plates (Fig. 21.29); the result, for the field between the plates, is σ/ϵ_0, where σ is the surface charge density on either plate. Why not $2\sigma/\epsilon_0$? Again, Equation 21.8 gives the field at a conductor surface—and it takes into account other charges that may be present. Here each plate's charge attracts the other's opposite charge, all to the *inner* face. Each plate is thus a single charge layer, giving a field $\sigma/2\epsilon_0$, and between the plates the fields sum to Equation 21.8's result, σ/ϵ_0. Beyond the plates the fields sum to zero—a result that also follows from Equation 21.8 because now there's zero surface charge on the outer faces.

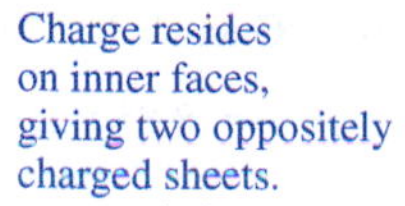

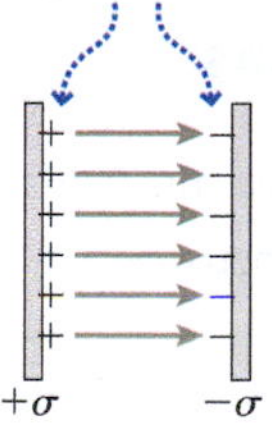

FIGURE 21.29 Edge-on view of two parallel conducting plates carrying opposite charges.

CHAPTER 21 SUMMARY

Big Idea

The big idea here is **Gauss's law**—a universal statement about electric fields that's closely related to Coulomb's inverse-square law but expressed in terms of the global behavior of the field over any closed surface. Using the **electric field-line** picture, Gauss's law says that the number of field lines emerging from a closed surface depends only on the net charge enclosed; more rigorously, it says that the **electric flux** through the surface is proportional to the enclosed charge.

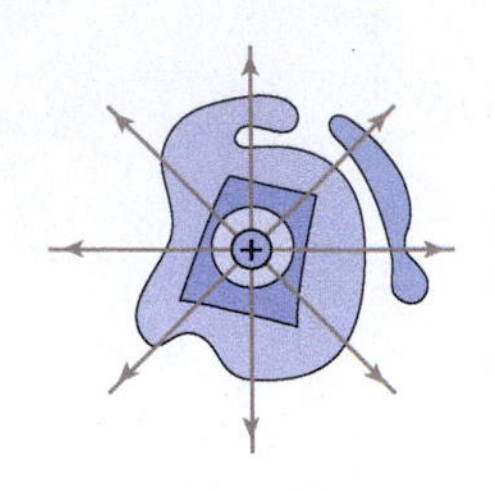

Point charge q

The number of lines through a surface depends on the net charge enclosed.

Point charges $+q$, $-q/2$

Key Concepts and Equations

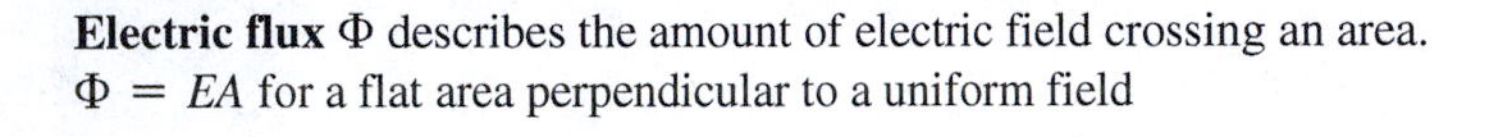

Electric flux Φ describes the amount of electric field crossing an area.
$\Phi = EA$ for a flat area perpendicular to a uniform field

$$\Phi = \int \vec{E} \cdot d\vec{A}, \text{ in general}$$

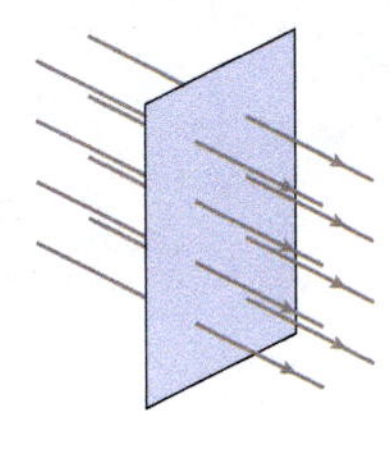

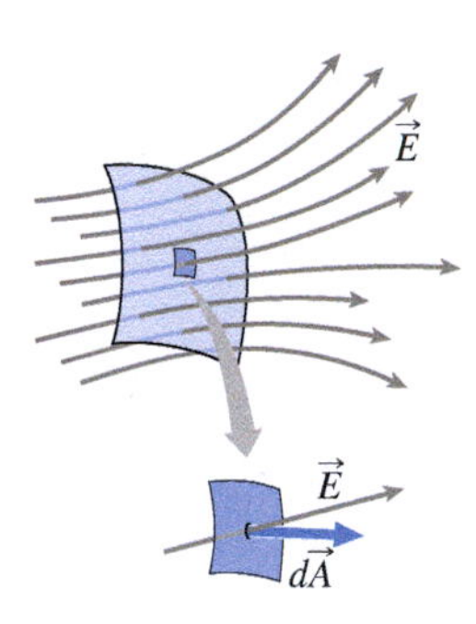

In terms of flux, Gauss's law reads $\oint \vec{E} \cdot d\vec{A} = q_{\text{enclosed}}/\epsilon_0$.

Here $\epsilon_0 = 1/4\pi k = 8.85 \times 10^{-12}\ \text{C}^2/\text{N}\cdot\text{m}^2$ is another way of expressing the Coulomb constant $k = 9.0 \times 10^9\ \text{N}\cdot\text{m}^2/\text{C}^2$.

Applications

Gauss's law gives the fields of symmetric charge distributions:

Spherical symmetry:

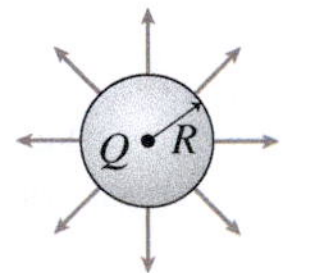

Outside: $E = \dfrac{Q}{4\pi\epsilon_0 r^2} = \dfrac{kQ}{r^2}$

Inside uniformly charged sphere:

$$E = \frac{kQr}{R^3}$$

Inside hollow sphere: $E = 0$

Line symmetry:

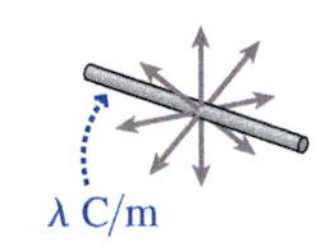

Outside: $E = \dfrac{\lambda}{2\pi\epsilon_0 r}$

Inside hollow pipe:
$E = 0$

Plane symmetry:

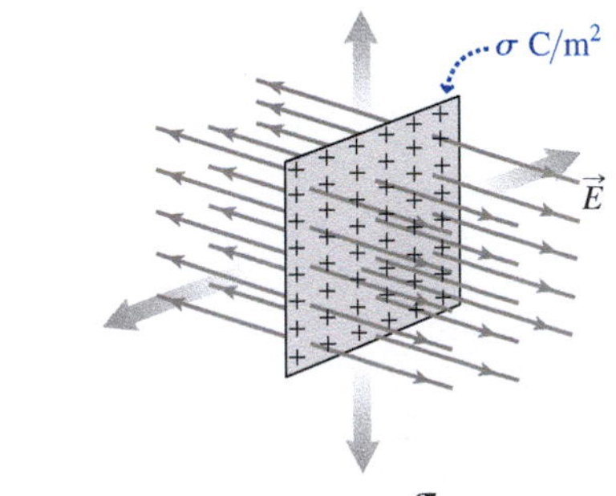

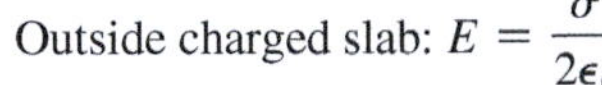

Outside charged slab: $E = \dfrac{\sigma}{2\epsilon_0}$

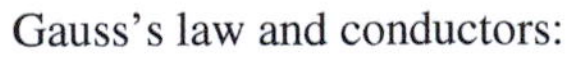

Gauss's law and conductors:

- The field is zero inside a conductor in electrostatic equilibrium.
- Any net charge resides on a conductor's surface.
- The field at the surface is perpendicular to the surface and has magnitude σ/ϵ_0.

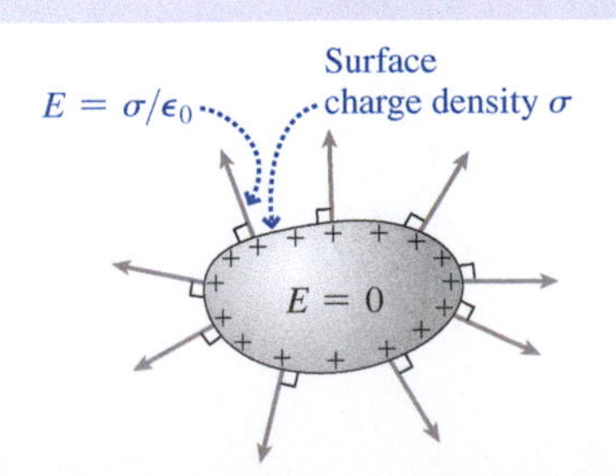

For homework assigned on MasteringPhysics, go to www.masteringphysics.com

BIO *Biology and/or medicine-related problems* **DATA** *Data problems* **ENV** *Environmental problems* **CH** *Challenge problems* **COMP** *Computer problems*

For Thought and Discussion

1. Can electric field lines ever cross? Why or why not?
2. The electric flux through a closed surface is zero. Must the electric field be zero on that surface? If not, give an example.
3. If the flux of the gravitational field through a closed surface is zero, what can you conclude about the region interior to the surface?
4. Under what conditions can the electric flux through a surface be written as EA, where A is the surface area?
5. Eight field lines emerge from a closed surface surrounding an isolated point charge. Would the number of field lines change if a second identical charge were brought to a point just *outside* the surface? If not, would anything change? Explain.
6. If a charged particle were released from rest on a curved field line, would its subsequent motion follow the field line? Explain.
7. In Gauss's law, $\oint \vec{E} \cdot d\vec{A} = q/\epsilon_0$, does the field $\vec{E}$ necessarily arise only from charges within the closed surface?
8. In a certain region the electric field points to the right and its magnitude increases as you move to the right, as shown in Fig. 21.30. Does the region contain net positive charge, net negative charge, or zero net charge?

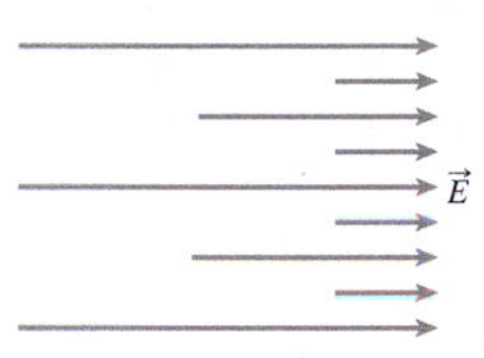

FIGURE 21.30 For Thought and Discussion 8. Left ends mark the beginnings of the field lines, which extend indefinitely to the right.

9. A point charge is located a fixed distance outside of a uniformly charged sphere. If the sphere shrinks in size without losing any charge, what happens to the force on the point charge?
10. The field of an infinite charged line decreases as $1/r$. Why isn't this a violation of the inverse-square law?
11. Why can't you use Gauss's law to determine the field of a uniformly charged cube? Why couldn't you use a cubical Gaussian surface?
12. You're sitting inside an uncharged, hollow spherical shell. Suddenly someone dumps a billion coulombs of charge on the shell, distributed uniformly. What happens to the electric field at your location?
13. Does Gauss's law apply to a spherical Gaussian surface not centered on a point charge, as shown in Fig. 21.31? Would this be a useful surface to use in calculating the electric field?

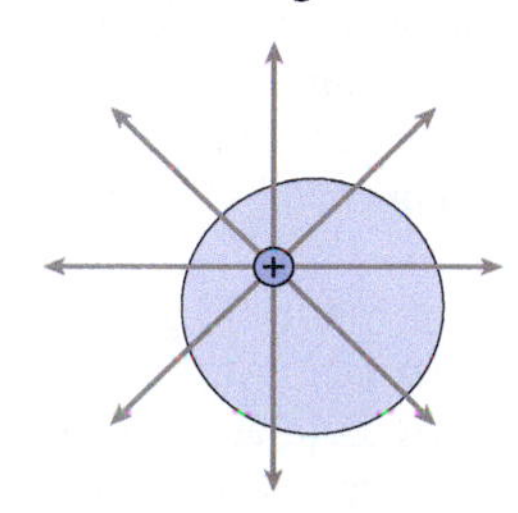

FIGURE 21.31 For Thought and Discussion 13

14. An insulating sphere carries charge spread uniformly throughout its volume. A conducting sphere has the same radius and net charge, but of course the charge is spread over its surface only (assume uniformly in this case). Compare the electric fields outside these two charge distributions.
15. Why must the electric field be zero inside a conductor in electrostatic equilibrium?
16. The electric field of a flat sheet of charge is $\sigma/2\epsilon_0$. Yet the field of a flat conducting sheet—even a thin one, like a piece of aluminum foil—is σ/ϵ_0. Explain this apparent discrepancy.

Exercises and Problems

Exercises

Section 21.1 Electric Field Lines

17. In Fig. 21.32, the magnitude of the middle charge is 3 μC. What's the net charge shown?

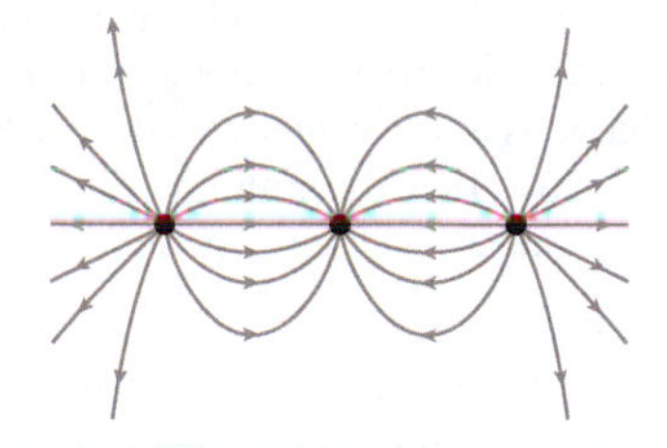

FIGURE 21.32 Exercise 17

18. Charges $+2q$ and $-q$ are near each other. Sketch some field lines for this charge distribution, using eight lines for a charge of magnitude q.
19. The net charge shown in Fig. 21.33 is $+Q$. Identify each of the charges A, B, and C shown.

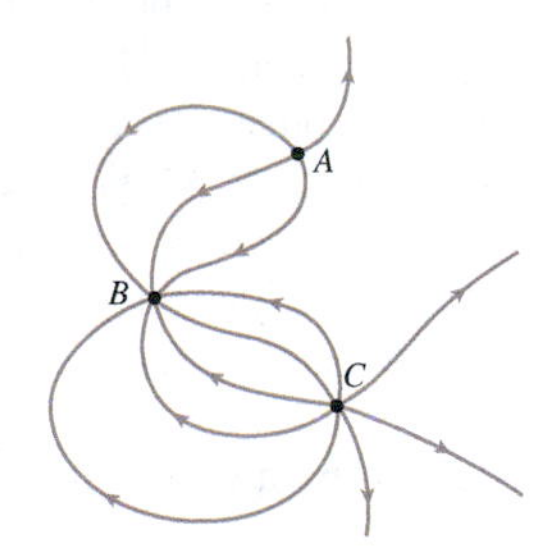

FIGURE 21.33 Exercise 19

Section 21.2 Electric Field and Electric Flux

20. A flat surface with area 2.0 m^2 is in a uniform 850-N/C electric field. Find the electric flux through the surface when it's (a) at right angles to the field, (b) at 45° to the field, and (c) parallel to the field.
21. What's the electric field strength in a region where the flux through a 1.0 cm × 1.0 cm flat surface is 65 N·m^2/C, if the field is uniform and the surface is at right angles to the field?
22. A flat surface with area 0.14 m^2 lies in the x-y plane, in a uniform electric field $\vec{E} = 5.1\hat{\imath} + 2.1\hat{\jmath} + 3.5\hat{k}$ kN/C. Find the flux through the surface.
23. The electric field on the surface of a 10-cm-diameter sphere is perpendicular to the sphere and has magnitude 47 kN/C. What's the electric flux through the sphere?
24. In the figure with GOT IT? 21.2, take $E = 1.75$ kN/C and $s = 125$ cm. Find the flux through faces B and C of cubes (a) and (b).

25. In Fig. 21.8, take the half-cylinder's radius and length to be 3.4 cm and 15 cm, respectively. If the electric field has magnitude 6.8 kN/C, find the flux through the half-cylinder. *Hint:* You don't need to do an integral! Why not?

Section 21.3 Gauss's Law

26. A sock comes out of the dryer with a trillion (10^{12}) excess electrons. What's the electric flux through a surface surrounding the sock?
27. What's the electric flux through the closed surfaces marked (a), (b), (c), and (d) in Fig. 21.34?

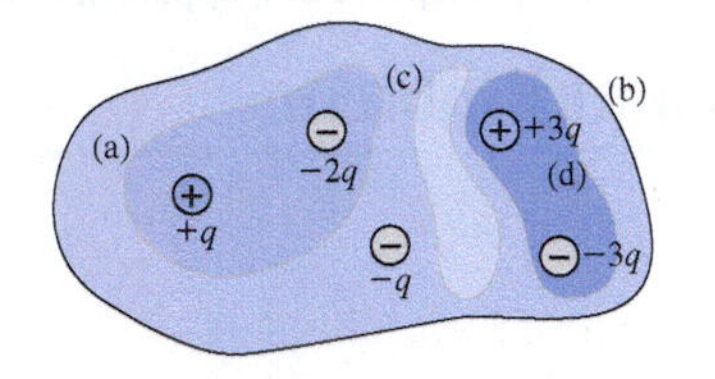

FIGURE 21.34 Exercise 27

28. A 6.8-μC charge and a −4.7-μC charge are inside an uncharged sphere. What's the electric flux through the sphere?
29. A 2.6-μC charge is at the center of a cube 7.5 cm on each side. What's the electric flux through one face of the cube? (*Hint:* Think about symmetry, and don't do an integral.)

Section 21.4 Using Gauss's Law

30. The electric field at the surface of a 5.0-cm-radius uniformly charged sphere is 90 kN/C. What's the field strength 10 cm from the surface?
31. A solid sphere 25 cm in radius carries 14 μC, distributed uniformly throughout its volume. Find the electric field strength (a) 15 cm, (b) 25 cm, and (c) 50 cm from its center.
32. A 15-nC point charge is at the center of a thin spherical shell of radius 10 cm, carrying −22 nC of charge distributed uniformly over its surface. Find the magnitude and direction of the electric field (a) 2.2 cm, (b) 5.6 cm, and (c) 14 cm from the point charge.
33. The electric field strength outside a charge distribution and 18 cm from its center has magnitude 55 kN/C. At 23 cm the field strength is 43 kN/C. Does the distribution have spherical or line symmetry?
34. An electron close to a large, flat sheet of charge is repelled from the sheet with a 1.8-pN force. Find the surface charge density on the sheet.
35. Find the field produced by a uniformly charged sheet carrying 87 pC/cm^2.
36. What surface charge density on an infinite sheet will produce a 1.4-kN/C electric field?

Section 21.5 Fields of Arbitrary Charge Distributions

37. A rod 50 cm long and 1.0 cm in radius carries a 2.0-μC charge distributed uniformly over its length. Find the approximate magnitude of the electric field (a) 4.0 mm from the rod surface, not near either end, and (b) 23 m from the rod.
38. What's the approximate field strength 1 cm above a sheet of paper carrying uniform surface charge density $\sigma = 45$ nC/m^2?
39. The disk in Fig. 21.22 has area 0.14 m^2 and is uniformly charged to 5.0 μC. Find the approximate field strength (a) 1 mm from the disk, not near the edge, and (b) 2.5 m from the disk.

Section 21.6 Gauss's Law and Conductors

40. What is the electric field strength just outside the surface of a conducting sphere carrying surface charge density 1.4 μC/m^2?
41. A net charge of 5.0 μC is applied on one side of a solid metal sphere 2.0 cm in diameter. Once electrostatic equilibrium is reached, and assuming no other conductors or charges nearby, what are (a) the volume charge density inside the sphere and (b) the surface charge density on the sphere?
42. A positive point charge q lies at the center of a spherical conducting shell carrying net charge $\frac{3}{2}q$. Sketch the field lines both inside and outside the shell, using eight field lines to represent a charge of magnitude q.
43. A total charge of 18 μC is applied to a thin, square metal plate 75 cm on a side. Find the electric field strength near the plate's surface.

Problems

44. What's the flux through the hemispherical open surface of radius R in a uniform field of magnitude E shown in Fig. 21.35? (*Hint:* Don't do a messy integral!)

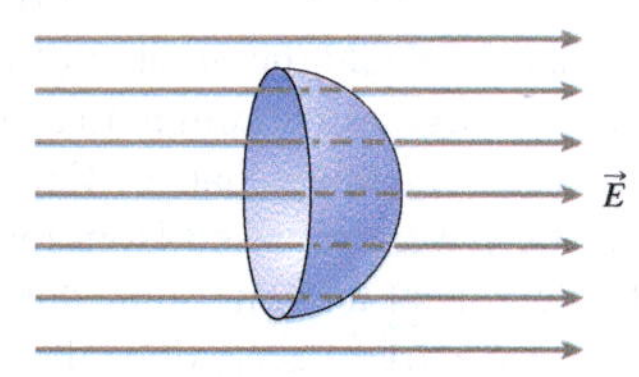

FIGURE 21.35 Problem 44

45. An electric field is given by $\vec{E} = E_0(y/a)\hat{k}$, where E_0 and a are constants. Find the flux through the square in the x-y plane bounded by the points $(0, 0)$, $(0, a)$, (a, a), $(a, 0)$.
46. The electric field in a certain region is given by $\vec{E} = ax\hat{\imath}$, where $a = 40$ N/C·m and x is in meters. Find the volume charge density in the region. (*Hint:* Apply Gauss's law to a cube 1 m on a side.)
47. **BIO** A study shows that mammalian red blood cells (RBCs) carry electric charge resulting from 4.4 million (for rabbit cells) to 15 million (for human cells) excess electrons spread over their surfaces. Approximating rabbit and human RBCs as spheres with radii 30 μm and 36 μm, respectively, find the electric field strengths at the cells' surfaces.
48. Positive charge is spread uniformly over the surface of a spherical balloon 70 cm in radius, resulting in an electric field of 26 kN/C at the balloon's surface. Find the field strength (a) 50 cm from the balloon's center and (b) 190 cm from the center. (c) What's the net charge on the balloon?
49. A solid sphere 2.0 cm in radius carries a uniform volume charge density. The electric field 1.0 cm from the sphere's center has magnitude 39 kN/C. (a) At what other distance does the field have this magnitude? (b) What's the net charge on the sphere?
50. A point charge of $-2Q$ is at the center of a spherical shell of radius R carrying charge Q spread uniformly over its surface. Find the electric field at (a) $r = \frac{1}{2}R$ and (b) $r = 2R$. (c) How would your answers change if the charge on the shell were doubled?
51. A friend is working on a biology experiment and needs to create an electric field of magnitude 430 N/C at 10 cm from the central portion of a large nonconducting square plate 4.5 m on each side. She needs to know how much charge to put on the plate. What do you tell her?
52. A spherical shell of radius 15 cm carries 4.8 μC distributed uniformly over its surface. At the center of the shell is a point charge. If the electric field at the sphere's surface is 750 kN/C and points outward, what are (a) the point charge and (b) the field just inside the shell?
53. A spherical shell 30 cm in diameter carries 85 μC distributed uniformly over its surface. A 1.0-μC point charge is located at the shell's center. Find the electric field strength (a) 5.0 cm from

the center and (b) 45 cm from the center. (c) How would your answers change if the charge on the shell were doubled?

54. A thick, spherical shell of inner radius a and outer radius b carries a uniform volume charge density ρ. Find an expression for the electric field strength in the region $a < r < b$, and show that your result is consistent with Equation 21.5 when $a = 0$.
CH

55. A long, thin wire carrying 5.6 nC/m runs down the center of a long, thin-walled, pipe with radius 1.0 cm carrying −4.2 nC/m spread uniformly over its surface. Find the electric field (a) 0.50 cm from the wire and (b) 1.5 cm from the wire.

56. An infinitely long rod of radius R carries a uniform volume charge density ρ. Show that the electric field strengths outside and inside the rod are given, respectively, by $E = \rho R^2/2\epsilon_0 r$ and $E = \rho r/2\epsilon_0$, where r is the distance from the rod axis. (Although an infinite rod is an impossibility, your answer is a good approximation for the field of a finite rod whose length is much greater than its diameter.)

57. A long, solid rod 4.5 cm in radius carries a uniform volume charge density. If the electric field strength at the surface of the rod (not near either end) is 16 kN/C, what's the volume charge density?

58. If you "painted" positive charge on the floor, what surface charge density would be necessary to suspend a 15 μC, 5.0-g particle above the floor?

59. A charged slab extends infinitely in two dimensions and has thickness d in the third dimension, as shown in Fig. 21.36. The slab carries a uniform volume charge density ρ. Find expressions for the electric field (a) inside and (b) outside the slab, as functions of the distance x from the center plane. (Although the infinite slab is impossible, your answer is a good approximation to the field of a finite slab whose width is much greater than its thickness.)

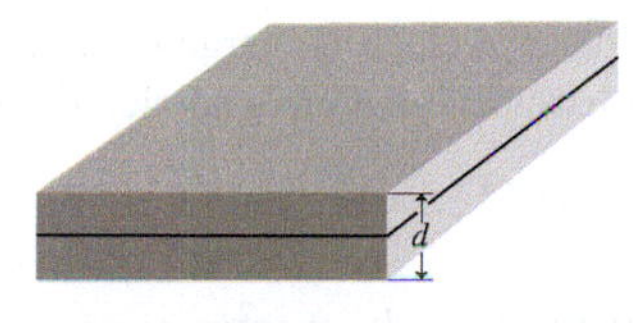

FIGURE 21.36 Problems 59 and 75

60. A solid sphere 10 cm in radius carries a 40-μC charge distributed uniformly throughout its volume. It's surrounded by a concentric shell 20 cm in radius, also uniformly charged with 40 μC. Find the electric field (a) 5.0 cm, (b) 15 cm, and (c) 30 cm from the center.

61. A nonconducting square plate 75 cm on a side carries a uniform surface charge density. The electric field strength 1 cm from the plate, not near an edge, is 45 kN/C. What's the approximate field strength 15 m from the plate?

62. A 250-nC point charge is placed at the center of an uncharged spherical conducting shell 20 cm in radius. Find (a) the surface charge density on the outer surface of the shell and (b) the electric field strength at the shell's outer surface.

63. An irregular conductor containing an irregular, empty cavity carries a net charge Q. (a) Show that the electric field inside the cavity must be zero. (b) If you put a point charge inside the cavity, what value must it have in order to make the charge density on the outer surface of the conductor everywhere zero?

64. You measure the electric field strength at points directly above the center of a square plate carrying charge spread uniformly over its surface. The data are tabulated in the next column, with x the perpendicular distance from the center of the plate. Use the data to determine (a) the total charge on the plate and (b) the plate's size. *Hint:* You'll need to consider separately data taken close to the plate and also far away. For the latter, plot E versus a quantity that should yield a straight line.
DATA

x (cm)	0.01	0.02	1.2	6.0	12.0	24.0
E (N/C)	5870	5860	4840	1960	754	221

x (cm)	48.0	72.0	96.0	120	240
E (N/C)	57.6	26.7	16.1	8.45	2.34

65. A point charge $-q$ is at the center of a spherical shell carrying charge $+2q$. That shell, in turn, is concentric with a larger shell carrying $-\frac{3}{2}q$. Draw a cross section of this structure, and sketch the electric field lines using the convention that eight lines correspond to a charge of magnitude q.

66. A point charge q is at the center of a spherical shell of radius R carrying charge $2q$ spread uniformly over its surface. Write expressions for the electric field strength at (a) $\frac{1}{2}R$ and (b) $2R$.

67. The volume charge density inside a solid sphere of radius a is $\rho = \rho_0 r/a$, where ρ_0 is a constant. Find (a) the total charge and (b) the electric field strength within the sphere, as a function of distance r from the center.
CH

68. Figure 21.37 shows a rectangular box with sides $2a$ and length L surrounding a line carrying uniform line charge density λ. The line passes directly through the center of the box faces. Integrate the field of the line charge over strips of width dx as shown to find the electric flux through one face of the box. Multiply by 4 to get the total flux, and show that your result is consistent with Gauss's law.
CH

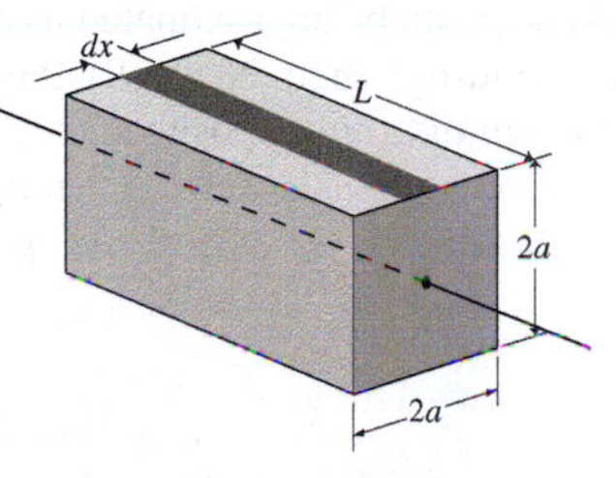

FIGURE 21.37 Problem 68

69. The charge density within a charged sphere of radius R is given by $\rho = \rho_0 - ar^2$, where ρ_0 and a are constants and r is the distance from the center. Find an expression for a such that the electric field outside the sphere is zero.
CH

70. Calculate the electric fields in Example 21.2 directly, using the superposition principle and integration. Consider the shell to be composed of charge elements that are coaxial rings, whose axes pass through the field point, which is a distance r from the center. (*Hint:* Consult Example 20.6. You'll have to evaluate the cases $r < R$ and $r > R$ separately.)
CH

71. A solid sphere of radius R carries a nonuniform volume charge density $\rho = \rho_0 e^{r/R}$, where ρ_0 is a constant and r is the distance from the center. Find an expression for the electric field strength at the sphere's surface.
CH

72. Problem 76 of Chapter 13 explored what happened to a person falling into a hole extending all the way through Earth's center and out the other side, assuming that $g(r) = g_0(r/R_E)$ for points inside Earth ($r < R_E$). Prove this assumption, treating Earth as a uniform sphere and using the gravitational version of Gauss's law: $\oint \vec{g} \cdot d\vec{A} = -4\pi G M_{\text{enclosed}}$.

73. An infinitely long solid cylinder of radius R carries a nonuniform charge density given by $\rho = \rho_0(r/R)$, where ρ_0 is a constant and r is the distance from the cylinder's axis. Find an expression for the magnitude of the electric field as a function of position r within the cylinder.
CH

74. CH A solid sphere of radius R carries a uniform volume charge density ρ. A hole of radius $R/2$ occupies a region from the center to the edge of the sphere, as shown in Fig. 21.38. Show that the electric field everywhere in the hole points horizontally and has magnitude $\rho R/6\epsilon_0$. *Hint:* Treat the hole as a superposition of two charged spheres with opposite charges.

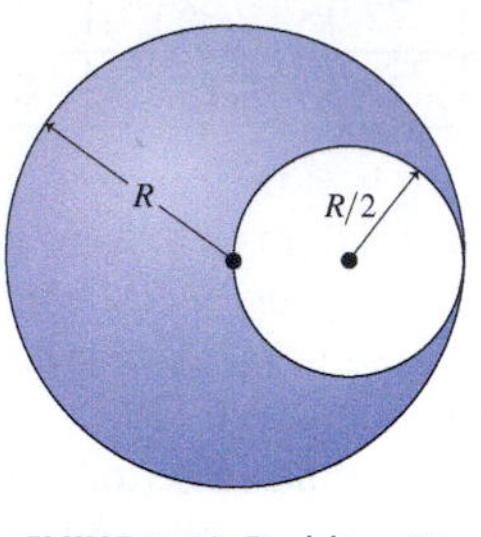

FIGURE 21.38 Problem 74

75. CH Repeat Problem 59 for the case where the charge density in the slab is given by $\rho = \rho_0|x/d|$, where ρ_0 is a constant.

Passage Problems

Coaxial cables are widely used with audio-visual technology, electronic instrumentation, and radio broadcasting, because they minimize interference with or from signals traveling on the cable. Coaxial cables consist of a wire inner conductor surrounded by a thin cylindrical conducting shield, usually of braided copper (Fig. 21.39). Flexible insulation separates the conductors. A straight length of coaxial cable can be approximated as an infinitely long wire surrounded by a cylindrical shell. Normally the two conductors carry charges of equal magnitude but opposite sign. (Charge actually varies with time and position as signals travel down the cable, but for these problems consider the charge to be fixed and spread uniformly.)

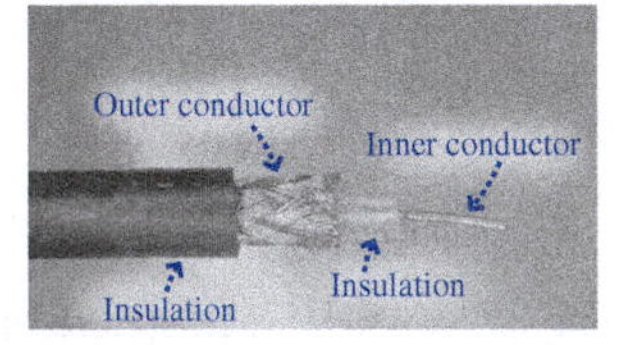

FIGURE 21.39 A coaxial cable (Passage Problems 76–79)

76. For a coaxial cable in electrostatic equilibrium carrying equal but opposite charges on its two conductors, there's a nonzero electric field
   a. only in the space between the wire and shield.
   b. in the space between wire and shield, and outside the shield.
   c. inside the metal conducting wire and shield, as well as between the wires and outside the shield.
   d. only outside the shield.

77. A coaxial cable carries equal but opposite charges on its two conductors. In electrostatic equilibrium, charge on the shield
   a. lies entirely on its outer surface.
   b. is divided evenly between inner and outer surfaces.
   c. lies entirely on its inner surface.
   d. distributes itself differently depending on the magnitude of the charge.

78. How does the electric field *between* the conductors in a coaxial cable in electrostatic equilibrium depend on the radial distance r from the cable's axis?
   a. it's constant
   b. as $1/r$
   c. as $1/r^2$
   d. as $1/r^3$

79. A coaxial cable in electrostatic equilibrium carries charge $-Q$ on its inner conductor and $+Q$ on its shield. If the charge on the shield *only* is doubled,
   a. the magnitude of the electric field between the conductors will double.
   b. the magnitude of the electric field outside the shield will double.
   c. the magnitude of the electric field at the outer surface of the shield will become twice the magnitude of the field at the shield's inner surface.
   d. the magnitude of the electric field at the outer surface of the shield will equal the magnitude of the field at the shield's inner surface.

Answers to Chapter Questions

Answer to Chapter Opening Question

Gauss's law requires that electric charge remain on the outside of the metal cage, arranging itself so there's no electric field within the cage.

Answers to GOT IT? Questions

21.1 (a)
21.2 (1) $\Phi_A = 0, \Phi_B = 0, \Phi_C = s^2E$;
(2) $\Phi_A = 0, \Phi_B = \Phi_C = s^2E\cos 45° = s^2E/\sqrt{2}$
21.3 (1) (a) flux doesn't change; (2) (d) field increases if charges are opposite, decreases if same
21.4 (1) field stays zero; (2) field (kQ/r^2) doubles
21.5 (1) (c); (2) (b)
21.6 (c)

22 Electric Potential

What You Know

- You understand electric fields and how to calculate them using Coulomb's law or Gauss's law.
- From Chapter 6, you know the definition of *work* as a line integral of force dotted with displacement.
- From Chapter 7, you understand what it means for a force to be *conservative*.
- You've probably heard the term *volt*, but you probably don't know its precise meaning.

What You're Learning

- You'll learn about *electric potential difference*, which is the work per unit charge involved in moving charges through electric fields.
- You'll learn the precise meaning of the term *volt*, as the unit of potential difference.
- You'll learn how to calculate potential differences by evaluating the line integral that defines potential difference.
- You'll see how to express potential differences in the field of a point charge.
- You'll learn to use superposition to calculate potential differences associated with distributions of point charges as well as continuous charge distributions.
- You'll explore the relation between potential difference and electric field, including the concept of *equipotentials*, and how to calculate field from potential.
- You'll learn how charge distributes itself on conductors.

How You'll Use It

- The concept of electric potential difference is central to the analysis of many electrical situations, including electric circuits.
- Potential difference is a scalar, making it generally easier to calculate than the electric field. Once you know potential difference as a function of position, you can find the electric field by differentiating.

Like gravity, the electric force is conservative. That means the work done in moving a charge against an electric force results in stored potential energy. It's convenient to consider the energy per unit charge, a measure that defines the concept of electric potential. Here we'll see how potential provides a simpler approach to calculating electric fields and also helps characterize everyday devices like batteries.

This parasailer landed on a 138,000-volt power line. Why wasn't he electrocuted?

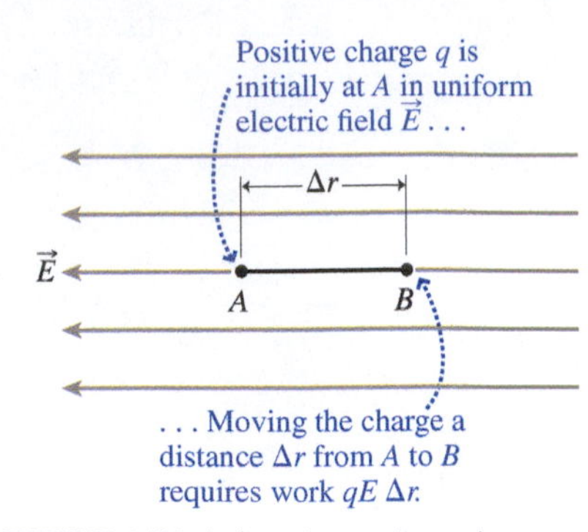

FIGURE 22.1 Work done in moving a charge q against an electric field $\vec{E}$.

22.1 Electric Potential Difference

In Chapter 7 we defined the potential-energy difference ΔU_{AB} as the negative of the work W_{AB} done by a conservative force $\vec{F}$ on an object moved from point A to point B (Equation 7.2):

$$\Delta U_{AB} = U_B - U_A = -W_{AB} = -\int_A^B \vec{F} \cdot d\vec{r}$$

where $d\vec{r}$ is an element of the path from A to B, and ΔU_{AB} is independent of the path taken from A to B. When the force doesn't vary, we can calculate the work more easily using Equation 6.5, $W = \vec{F} \cdot \Delta\vec{r}$.

Consider a positive charge q moved between points A and B a distance Δr apart in a uniform electric field $\vec{E}$, as shown in Fig. 22.1. Since the field is uniform, a constant electric force $\vec{F} = q\vec{E}$ acts on the charge, so we use Equation 6.5 to evaluate the work done by the field and the resulting potential-energy change:

$$\Delta U_{AB} = -W_{AB} = -q\vec{E} \cdot \Delta\vec{r} = -qE\,\Delta r \cos 180° = qE\,\Delta r$$

where the factor $\cos 180° = -1$ appears because $\vec{E}$ and $\Delta\vec{r}$ have opposite directions. Make sense? Pushing a positive charge from A to B against the electric field is like pushing a car up a hill: Potential energy increases in both cases. Let go of the charge, and the field accelerates it back, just as gravity would accelerate the car back down the hill.

Had we moved a charge $2q$ in Fig. 22.1, the potential-energy change ΔU would have been twice as great; a charge $\frac{1}{2}q$ would have cut ΔU in half. Since ΔU is proportional to charge, it's convenient to consider the *potential-energy change per unit charge* involved in moving a charge between two points. Mathematically, we write $\vec{F} = q\vec{E}$ in our general expression for ΔU_{AB} and divide by q. The result defines the **electric potential difference** ΔV:

Table 22.1 Force and Field, Potential Energy and Electric Potential

Quantity	Symbol/Equation	Units
Force	$\vec{F}$	N
Electric field	$\vec{E} = \dfrac{\vec{F}}{q}$	N/C or V/m
Potential-energy difference	$\Delta U = -\int_A^B \vec{F} \cdot d\vec{r}$	J
Electric potential difference	$\Delta V = \dfrac{\Delta U}{q}$ or $\Delta V = -\int_A^B \vec{E} \cdot d\vec{r}$	J/C or V

The electric potential difference from point A to point B is the potential-energy change per unit charge in moving a charge from A to B:

$$\Delta V_{AB} = \frac{\Delta U_{AB}}{q} = -\int_A^B \vec{E} \cdot d\vec{r} \quad \text{(electric potential difference)} \qquad (22.1a)$$

Here Δ and the subscripts AB show explicitly that we're talking about a *change* or *difference* from one point to another. We'll sometimes use just the symbol V for potential difference, in cases where the starting point A is understood. Note that potential difference, although computed from vectors, is itself a scalar quantity.

The switch from potential energy to electric potential is analogous to Chapter 20's introduction of the electric field as the electric force *per unit charge*; similarly, the electric potential difference is the change in potential energy *per unit charge*. The reason is the same: We want to express electric properties in terms that don't involve specific charges. Table 22.1 summarizes the relations among force and field, potential energy and electric potential.

In the special case of a uniform field, Equation 22.1a reduces to

$$\Delta V_{AB} = -\vec{E} \cdot \Delta\vec{r} \quad \text{(uniform field)} \qquad (22.1b)$$

where $\Delta\vec{r}$ is a vector from A to B. Figure 22.1 shows the special case when the field $\vec{E}$ and path $\Delta\vec{r}$ are in opposite directions; here, Equation 22.1b gives $\Delta V_{AB} = E\,\Delta r$.

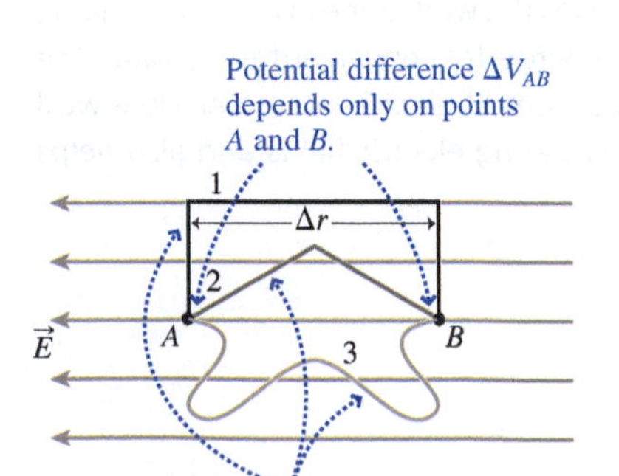

FIGURE 22.2 Potential difference is path independent.

Potential difference can be positive or negative, depending on whether the path goes against the field or with it. Moving a positive charge through a positive potential difference is like going uphill: Potential energy increases. Moving a positive charge through a negative potential difference is like going downhill: Potential energy decreases. The converse is true for a negative charge; even though the potential difference remains the same, the force is opposite and so the potential energy reverses sign.

We emphasize that potential difference is a property *of two points*; it doesn't depend on the path between those points. In Fig. 22.1, considering a straight path from A to B made the calculation of potential difference easy, but we would have found the same result—albeit with much more effort—using any path (Fig. 22.2).

GOT IT? 22.1 What would happen to the potential difference V_{AB} in Fig. 22.1 if (1) the electric field strength were doubled, (2) the distance Δr were doubled, (3) the points were moved so the path lay at right angles to the field, or (4) the positions of A and B were interchanged?

The Volt and the Electronvolt

Potential difference measures work or energy per unit charge, so its units are joules per coulomb, as shown in Table 22.1. Potential difference is important enough that this unit has a special name, the **volt** (V). To say that a car has a 12-V battery, for example, means the battery does 12 J of work on every coulomb of charge that moves between its terminals. Multiplying the first equality in Equation 22.1a by q shows that the change in potential energy of a charge q as it moves through a potential difference ΔV is $\Delta U = q\,\Delta V$. If a charge q "falls" freely through a potential difference ΔV, it therefore gains kinetic energy given by $|q\,\Delta V|$.

We often use the term **voltage** to mean potential difference, especially in electric circuits. The two are subtly different, however, when changing magnetic fields are present; more on this in Chapter 27. Table 22.2 lists some typical potential differences in technological and natural systems.

Table 22.2 Typical Potential Differences

Between human arm and leg due to heart's electrical activity	1 mV
Across biological cell membrane	80 mV
Between terminals of flashlight battery	1.5 V
Car battery	12 V
Electric outlet (depends on country)	100–240 V
Taser© (pulsed)	1200 V
Between long-distance electric transmission line and ground	365 kV
Between base of thunderstorm cloud and ground	100 MV

✓TIP Potential Difference Involves Two Points

Potential difference is the energy per unit charge involved in moving *between those points*. This is ultimately a practical matter; if you forget it, you won't be able to hook up a voltmeter properly or connect jumper cables safely to your car battery! This chapter's opening photo provides a dramatic illustration of this point.

Sometimes we say "the potential (or voltage) at point P." This is *always* a shorthand way of talking, and we *must* have in mind some other point. What we mean is the potential difference going from that other point to P. We'll consider choices for that "other point"—the so-called zero of potential—in the next section.

In molecular, atomic, and nuclear systems it's often convenient to measure energy in **electronvolts** (eV), defined as **the energy gained by a particle carrying one elementary charge when it moves through a potential difference of 1 volt**. Since one elementary charge is 1.6×10^{-19} C, 1 eV is 1.6×10^{-19} J. Energy in eV is particularly easy to calculate when charge is given in units of the elementary charge e; then, with ΔV in volts, $q\,\Delta V$ gives the energy in eV. However, the eV is *not* an SI unit and should be converted to joules before calculating other quantities, like velocity.

GOT IT? 22.2 (1) A proton (charge e), (2) an alpha particle (charge $2e$), and (3) a singly ionized oxygen atom each move through a 10-V potential difference. What's the work in eV done on each?

EXAMPLE 22.1 Potential Difference, Work, and Energy: X Rays

In an X-ray tube, a uniform electric field of 300 kN/C extends over a distance of 10 cm, from an electron source to a target; the field points from the target to the source. Find the potential difference between source and target and the energy gained by an electron as it accelerates from source to target (where its abrupt deceleration produces X rays). Express the energy in both electronvolts and joules.

INTERPRET This problem requires first calculating the potential difference from the field and then the energy from the potential difference.

DEVELOP Figure 22.3 is our drawing with point A the source and point B the target. Equation 22.1b, $\Delta V_{AB} = -\vec{E}\cdot\Delta\vec{r}$, determines the

(continued)

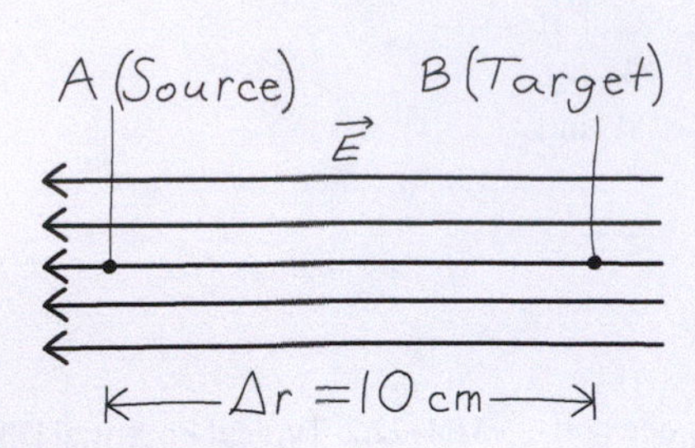

FIGURE 22.3 Sketch for Example 22.1.

potential difference in this uniform field. Given the potential difference, or energy per unit charge, we can find the energy gain from the magnitude of the product $q\,\Delta V$.

EVALUATE With the field and path in opposite directions, $\cos\theta = -1$ in the dot product, so Equation 22.1b gives

$$\Delta V_{AB} = E\,\Delta r = (300\ \text{kN/C})(0.10\ \text{m}) = 30\ \text{kV}$$

Although this difference is positive, a *negative* electron moves "downhill" from source to target, and thus gains kinetic energy as work gets done on it. With the charge measured in elementary charges, the product $|q\,\Delta V|$ gives this energy directly when it's expressed in electronvolts: (1 elementary charge e)(30 kV) = 30 keV. With 1.6×10^{-19} J/eV, this is 4.8 fJ.

ASSESS Make sense? An electronvolt is a lot smaller than a joule, so the SI answer (in fJ = 10^{-15} J) is numerically much smaller.

EXAMPLE 22.2 Potential of a Charged Sheet

An isolated, infinite charged sheet carries uniform surface charge density σ. Find an expression for the potential difference from the sheet to a point a perpendicular distance x from the sheet.

INTERPRET This is a question about calculating the potential difference from the field.

DEVELOP The result of Example 21.6 gives the field of a charged sheet: It's uniform, with magnitude $E = \sigma/2\epsilon_0$ and direction perpendicular to the sheet. We've drawn the sheet and a few of its field lines in Fig. 22.4. Since the field is uniform, Equation 21.1b, $\Delta V_{AB} = -\vec{E}\cdot\Delta\vec{r}$, determines the potential difference.

EVALUATE Moving away from the sheet means going in the direction of the field (assuming positive σ), so $\cos\theta = 1$ in the dot product, and we evaluate to get

$$V_{0x} = -Ex = -\frac{\sigma x}{2\epsilon_0}$$

Here we've used x for the displacement Δr and V_{0x} for the potential difference because we're measuring from the sheet ($x = 0$) to the point x.

ASSESS Make sense? Our result shows that the potential difference in a *uniform* field varies *linearly* with distance. Moving a positive charge away from the sheet is like going "downhill," in this case with a constant slope. Give the sheet a negative charge ($\sigma < 0$) and the potential difference changes sign; now moving a positive charge away from the sheet is going "uphill." (And moving a negative charge away is "downhill"—since like charges repel.)

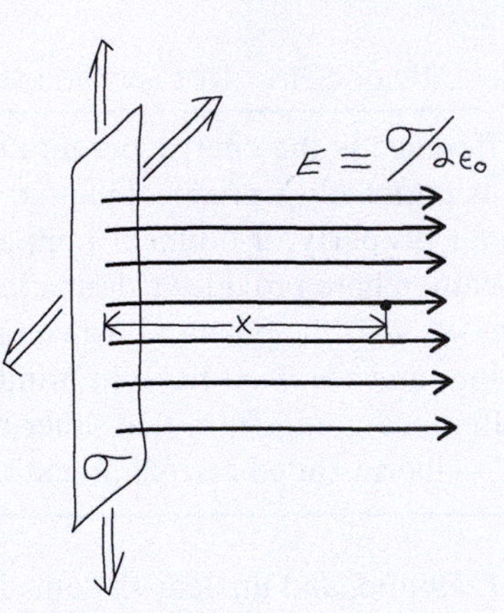

FIGURE 22.4 Sketch for Example 22.2. The field also extends to the left from the sheet, but we haven't drawn that.

Curved Paths and Nonuniform Fields

If the electric field isn't uniform or the path isn't straight, then we need to use the integral in Equation 22.1a to find the potential difference. With that equation we're dividing the path into segments $d\vec{r}$, each so short that it's essentially straight with a uniform field over its length (Fig. 22.5). Then Equation 22.1b gives the potential difference $dV = -\vec{E}\cdot d\vec{r}$ across the segment, and in integrating we're summing infinitely many infinitesimal dV values to get the potential difference between two points A and B. We'll see some examples in the next section.

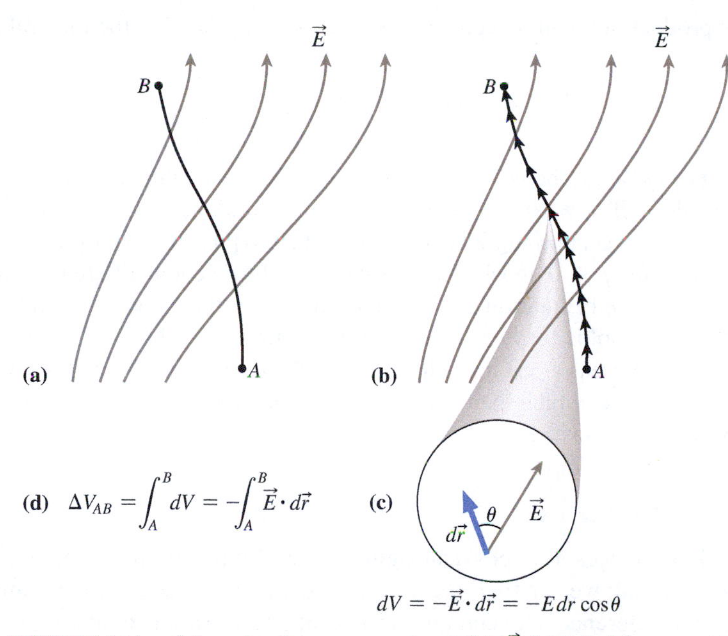

FIGURE 22.5 (a) A path from A to B traverses an electric field $\vec{E}$. (b) To find the potential difference ΔV_{AB}, we first divide the path into infinitesimal segments $d\vec{r}$. (c) The potential difference dV across one such segment is $dV = -\vec{E}\cdot d\vec{r}$. (d) Adding—that is, integrating—all the dVs along the path gives the expression for ΔV_{AB}.

GOT IT? 22.3 The figure shows three straight paths AB of the same length, each in a different electric field. The field at A is the same in each. Rank the potential differences ΔV_{AB}, from highest to lowest.

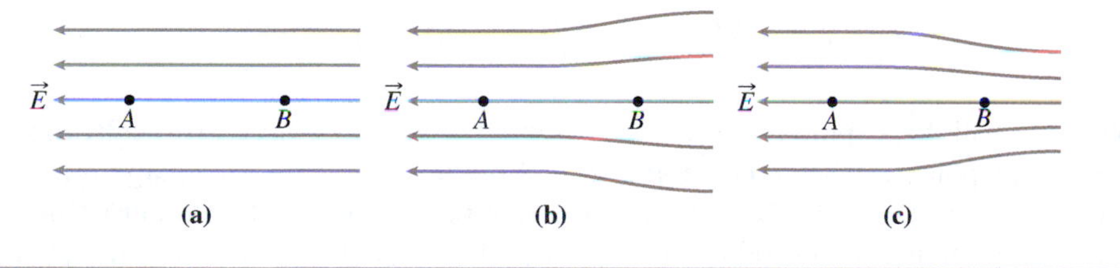

22.2 Calculating Potential Difference

Here we use Equation 22.1a to calculate the potential differences for several charge distributions. Most important is the point charge, which then provides an easy way to find potential differences for more complicated charge distributions.

The Potential of a Point Charge

Equation 20.3 gives the electric field of a point charge: $\vec{E} = (kq/r^2)\hat{r}$. Let's find the potential difference between two points A and B at distances r_A and r_B from a positive point charge, as shown in Fig. 22.6. We can't just multiply the distance $r_B - r_A$ by E because the field varies with position. Instead we integrate, using Equation 22.1a:

$$\Delta V_{AB} = -\int_{r_A}^{r_B} \vec{E}\cdot d\vec{r} = -\int_{r_A}^{r_B} \frac{kq}{r^2}\hat{r}\cdot d\vec{r}$$

As we move from A to B, the path elements are increments dr in the radial direction, so we write $d\vec{r} = \hat{r}\,dr$. Then the potential difference is

$$\Delta V_{AB} = -\int_{r_A}^{r_B} \frac{kq}{r^2}\hat{r}\cdot\hat{r}\,dr = -kq\int_{r_A}^{r_B} r^{-2}\,dr$$

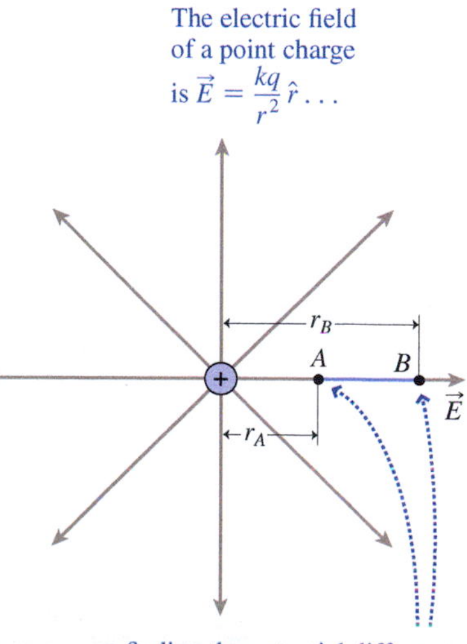

FIGURE 22.6 Potential difference in the field of a point charge.

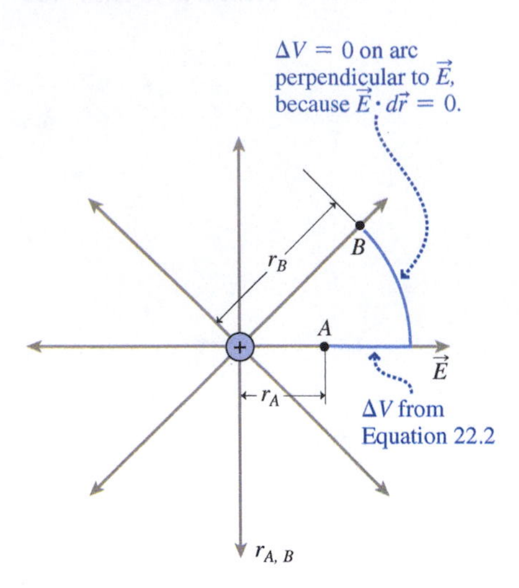

FIGURE 22.7 Potential difference is path independent, so ΔV_{AB} here still follows from Equation 22.2.

since the dot product of the unit vector $\hat{r}$ with itself is 1. Evaluating the integral gives

$$\Delta V_{AB} = -kq\left[-\frac{1}{r}\right]_{r_A}^{r_B} = kq\left(\frac{1}{r_B} - \frac{1}{r_A}\right) \tag{22.2}$$

Make sense? For $r_B > r_A$, the potential difference is negative, showing that a positive test charge at r_A would "fall" toward r_B. Going the other way would require an external force to do work on a positive charge, pushing it "uphill" against the repulsive force of the charge q. Our result holds as well for $q < 0$, in which case the sign of the potential difference changes.

Although we derived Equation 22.2 for two points on the same radial line, Fig. 22.7 shows that the result holds for *any* two points in the field of a charge q. It doesn't matter which point is at the greater distance either; if $r_B < r_A$, Equation 22.2 still gives the correct potential difference, which then becomes positive, showing that we would have to do work to move a positive test charge *toward* a positive q.

The Zero of Potential

Only potential differences have physical significance. But it's often convenient to define a point of zero potential; we can then speak of the potential V at some other point P, meaning the potential difference between our zero point and P. In this context the expression ΔV_{AB} can be written $V(B) - V(A)$. The choice for the zero of potential is usually based on mathematical or physical convenience. In electric power systems, Earth, called "ground," is usually taken as the zero of potential; in automobile electric systems, the car's metal frame is a convenient zero point.

When we deal with isolated charges, it's convenient to take the zero of potential at infinity. Then $r_A \to \infty$ in Equation 22.2 and $1/r_A$ becomes zero. We'll omit the subscript on r_B because it can be any point; then Equation 22.2 becomes

$$V_{\infty r} = V(r) = \frac{kq}{r} \quad \text{(point-charge potential)} \tag{22.3}$$

When we call this expression $V(r)$ "the potential of a point charge," we really mean that $V(r)$ is the potential difference going from a point very far from a charge q to a point a distance r from the charge—an interpretation that's consistent with our definition of potential difference as depending on *two* points. Because the field outside any spherically symmetric charge distribution is that of a point charge, Equation 22.3 also gives the potential outside a spherically symmetric charge distribution.

Does it bother you that potential difference can be finite over an infinite distance? The reason lies in the inverse-square dependence of the field, which drops so rapidly that the work done in moving a charge from infinity to the vicinity of a point charge remains finite. We found an analogous result in Chapter 8, where it took only a finite amount of energy to escape completely from a planet's gravitational attraction. As long as a charge distribution is finite in size—so its field at large distances falls at least as fast as $1/r^2$—it makes sense to take the zero of potential at infinity.

GOT IT? 22.4 You measure a potential difference of 50 V between two points a distance 10 cm apart along a line extending radially outward from a point charge. If you move closer to the charge and measure the potential difference over another 10-cm interval on the same line, will it be (a) greater, (b) less, or (c) the same?

EXAMPLE 22.3 Potential and Work: At the Science Museum

The Hall of Electricity at the Boston Museum of Science contains a large Van de Graaff generator, a device that builds up charge on a metal sphere (see Chapter 21's opening photo). The sphere has radius $R = 2.30$ m and develops a charge $Q = 640\ \mu\text{C}$. Treating this as a single isolated sphere, find (a) the potential at its surface, (b) the work needed to bring a proton from infinity to the sphere's surface, and (c) the potential difference between the sphere's surface and a point $2R$ from its center.

INTERPRET This problem is about potential differences in the field of a spherically symmetric charge distribution. In Chapter 21 we found that the field outside such a distribution is identical to that of a point charge. The term "potential" is meaningless unless we're talking about two points, so here, with a point-charge field, we interpret the question as asking us to take the zero of potential at infinity.

DEVELOP Because the field outside the spherical charge distribution is the same as that of a point charge, Equation 22.3, $V(r) = kQ/r$, determines the potential for $r \geq R$. We've sketched this $1/r$ potential curve in Fig. 22.8. Because the zero of potential is at infinity, we can multiply the potential at the surface by the proton's charge to get the work required to bring a proton from infinity. Finally, we can evaluate the potential difference $\Delta V_{R\,2R}$ from the potentials at R and $2R$.

EVALUATE (a) Equation 22.3 gives

$$V(R) = \frac{kQ}{R} = 2.50\ \text{MV}$$

using Q and R given for the museum's device. (b) This 2.5-MV result is the potential difference between infinity and the sphere's surface. Then the work needed to move a proton—1 elementary charge e—from infinity is 2.5 MeV or 4.0×10^{-13} J. (c) We find the potential difference from the surface to $2R$ by subtracting the potentials at the two points:

$$\Delta V_{R\,2R} = V(2R) - V(R) = \frac{kQ}{2R} - \frac{kQ}{R} = -\frac{kQ}{2R} = -1.25\ \text{MV}$$

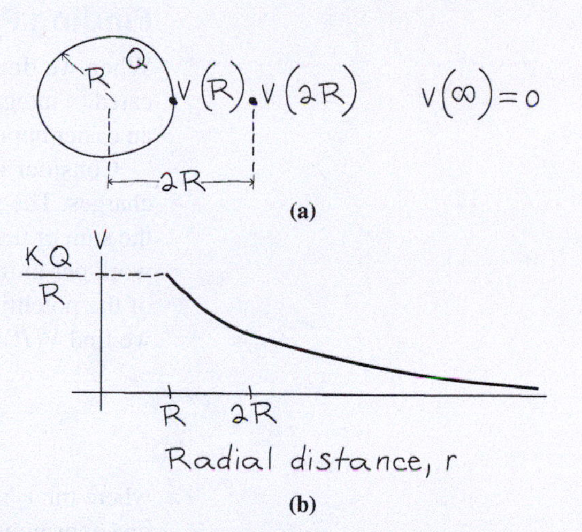

FIGURE 22.8 Sketch for Example 22.3.

ASSESS Make sense? The potential difference $\Delta V_{R\,2R}$ is negative because we're moving away from the positively charged sphere. Our result also shows that fully half the potential difference between the sphere and infinity occurs within one radius of the sphere's surface—a consequence of the rapid $1/r^2$ decrease in the field. ■

EXAMPLE 22.4 Potential Difference: A High-Voltage Power Line

A long, straight power-line wire has radius 1.0 cm and carries line charge density $\lambda = 2.6\ \mu\text{C/m}$. Assuming no other charges are present, what's the potential difference between the wire and the ground, 22 m below?

INTERPRET We can interpret the long, straight wire as essentially an infinitely long charge distribution with line symmetry.

DEVELOP In Chapter 21 we found that the field outside any line-symmetric distribution is that of a line charge, $\vec{E} = (\lambda/2\pi\epsilon_0 r)\hat{r}$, so this equation determines the power line's field. We haven't been given any explicit expression for potential differences in this field, so because the field varies with position, our plan is to apply Equation 22.1a, $\Delta V = -\int \vec{E} \cdot d\vec{r}$. We've drawn the situation in Fig. 22.9.

EVALUATE We evaluate the integral in Equation 22.1a over a straight path perpendicular to the wire, from its surface at r_A to the ground at r_B:

$$\begin{aligned}\Delta V_{AB} &= -\int_{r_A}^{r_B} \vec{E} \cdot d\vec{r} = -\int_{r_A}^{r_B} \frac{\lambda}{2\pi\epsilon_0 r}\hat{r} \cdot \hat{r}\, dr \\ &= -\frac{\lambda}{2\pi\epsilon_0}\int_{r_A}^{r_B} \frac{dr}{r} = -\frac{\lambda}{2\pi\epsilon_0} \ln r \Big|_{r_A}^{r_B} \\ &= \frac{\lambda}{2\pi\epsilon_0} \ln\left(\frac{r_A}{r_B}\right)\end{aligned} \qquad (22.4)$$

where the last step follows because $\ln x - \ln y = \ln(x/y)$ and $\ln(x/y) = -\ln(y/x)$. The numbers of this example give $\Delta V = -360\ \text{kV}$, a value typical of long-distance electric power transmission lines.

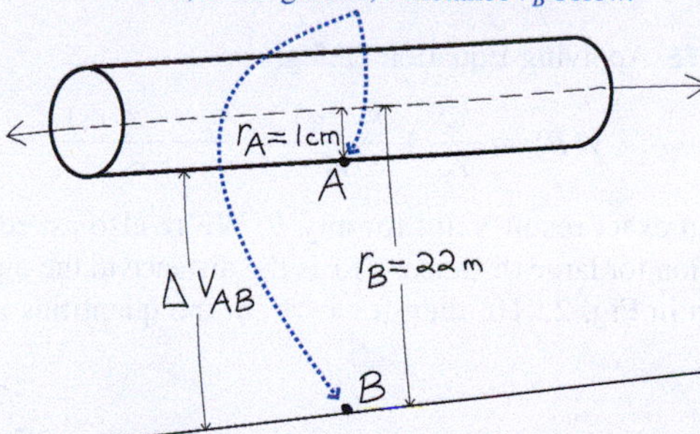

FIGURE 22.9 A long, straight power line approximated as an infinite charged rod whose field is that of a line charge.

ASSESS Make sense? Our result is negative because the path AB goes away from a positive charge. (Mathematically, $r_A < r_B$ so the logarithm is negative.) The symbolic form of our answer shows that we can't let r_B go to infinity. Physically, that's because we're assuming the charge distribution has infinite extent, so it never resembles a point charge no matter how far away we get; mathematically, it's because the field falls off more slowly than a point-charge field—namely, as $1/r$. In practice, our answer here should be modified to account for the presence of other wires and of charges drawn to the ground surface. ■

Finding Potential Differences Using Superposition

When we don't know the field of a charge distribution, or when the field is too complicated to integrate easily, we can find the potential using superposition. This often provides an easier approach to calculating the field, as we'll see in Sec. 22.3.

Consider a charge q brought from infinity to a point P in the vicinity of some other charges. The superposition principle states that the electric field of a charge distribution is the sum of the fields of the individual charges that make up the distribution. Therefore, the work per unit charge—that is, the potential difference—between infinity and P is the sum of the potential differences associated with the individual point charges. Mathematically, we find $V(P)$ by summing Equation 22.3 over the individual point charges q_i:

$$V(P) = \sum_i \frac{kq_i}{r_i} \qquad (22.5)$$

where the r_i's are the distances from each of the charges to the point P. Equation 22.5 has one enormous advantage over its counterpart for the electric field, Equation 20.4. Electric potential is a *scalar*, so the sum in Equation 22.5 is a scalar sum, with no angles, vector components, or unit vectors.

EXAMPLE 22.5 Discrete Charges: The Dipole Potential

An electric dipole consists of point charges $\pm q$ a distance $2a$ apart. Find the potential at an arbitrary point P, and an approximation to the potential for the case where the distance to P is large compared with the charge separation.

INTERPRET We have two point charges, so this problem is based on the point-charge potential, and therefore we'll take the zero of potential at infinity.

DEVELOP Figure 22.10 is our drawing, showing the distances from the two charges to a point P. Our plan is to apply superposition, summing the potentials of the individual point charges at P as determined in Equation 22.5, $V(P) = \Sigma(kq/r)$.

EVALUATE Applying Equation 22.5 gives

$$V(P) = \frac{kq}{r_1} + \frac{k(-q)}{r_2} = \frac{kq(r_2 - r_1)}{r_1 r_2}$$

This is an exact result valid for any P. We're also asked for an approximation for large distances. If r is the distance to the dipole center, as shown in Fig. 22.10, then for $r \gg a$, the quantities r_1, r_2, and r are nearly the same and the term $r_1 r_2$ is very nearly r^2. We have to be a little more careful with the term $r_1 - r_2$ because here we're comparing nearly equal quantities. Figure 22.10 shows that this term—the difference between the distances from the two charges to P—is approximately $2a\cos\theta$. So the dipole potential for $r \gg a$ becomes

$$V(r, \theta) = \frac{k(2aq)\cos\theta}{r^2} = \frac{kp\cos\theta}{r^2} \quad \text{(dipole potential)} \qquad (22.6)$$

with $p = 2aq$ being the dipole moment.

ASSESS Make sense? The dipole *potential* drops as $1/r^2$; earlier, we found the dipole *field* dropping as $1/r^3$. The difference of one power in r occurs because the potential results from integrating the field over distance. The same is true for the point charge, whose *field* drops as $1/r^2$ while its *potential* drops as $1/r$. Note also that Equation 22.6 gives $V = 0$ when $\theta = 90°$. There, on the dipole's perpendicular bisector, a charge brought from infinity is always moving at right angles to the dipole field (recall Fig. 21.2), so no work is involved (Fig. 22.11). ■

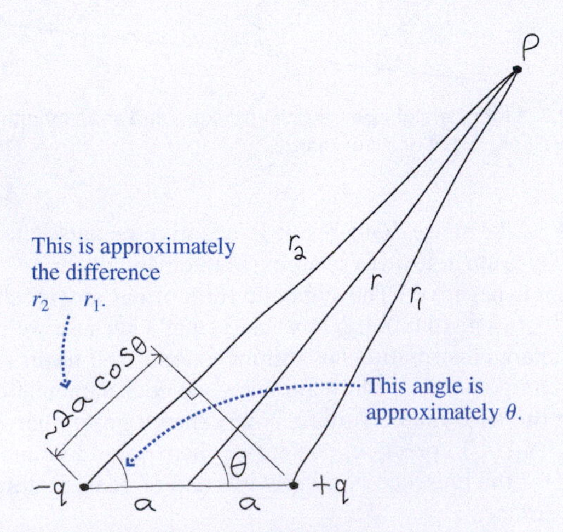

FIGURE 22.10 Finding the dipole potential.

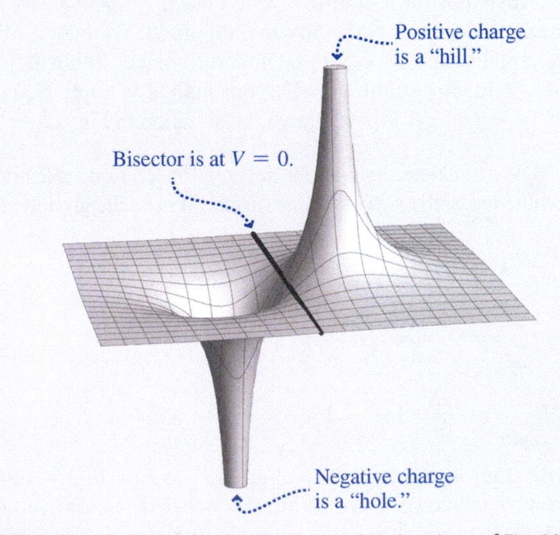

FIGURE 22.11 3-D plot of the dipole potential in the plane of Fig. 22.10.

GOT IT? 22.5 The figure shows three paths from infinity to a point P on a dipole's perpendicular bisector. Compare the work done in moving a charge to P on each of the paths.

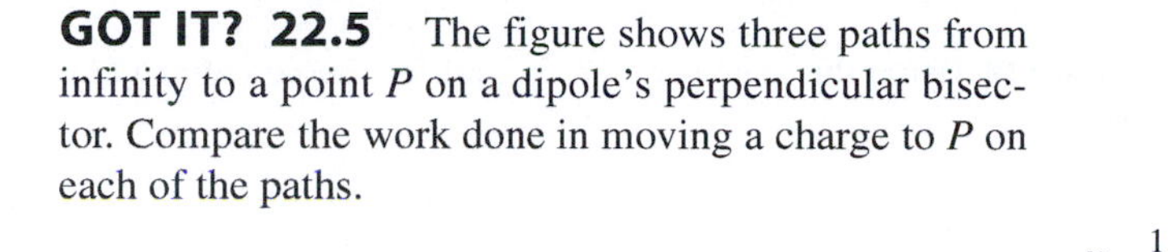

Continuous Charge Distributions

We can calculate the potential of a continuous charge distribution by considering it to be made up of infinitely many infinitesimal charge elements dq. Each acts like a point charge and therefore contributes to the potential at some point P an amount dV given by $dV = k\,dq/r$, where we take the zero of potential at infinity. The potential at P is the sum—that is, the integral—of the contributions dV from all the charge elements:

$$V = \int dV = \int \frac{k\,dq}{r} \quad \left(\begin{array}{c}\text{potential of a continuous}\\ \text{charge distribution}\end{array}\right) \tag{22.7}$$

where the integration is over the entire charge distribution. Example 22.6 provides a simple application of Equation 22.7 where the integration is straightforward, while Example 22.7 is more challenging.

EXAMPLE 22.6 Potential of a Continuous Distribution: A Charged Ring

A total charge Q is distributed uniformly around a thin ring of radius a. Find the potential on the ring's axis.

INTERPRET We interpret the ring as a continuous charge distribution.

DEVELOP Equation 22.7, $V = \int k\,dq/r$, gives the potential for continuous charge distributions. Figure 22.12 is our drawing, showing an x-axis coincident with the ring axis, with $x = 0$ at the ring center. Charge elements dq in this case are small segments of the ring, and

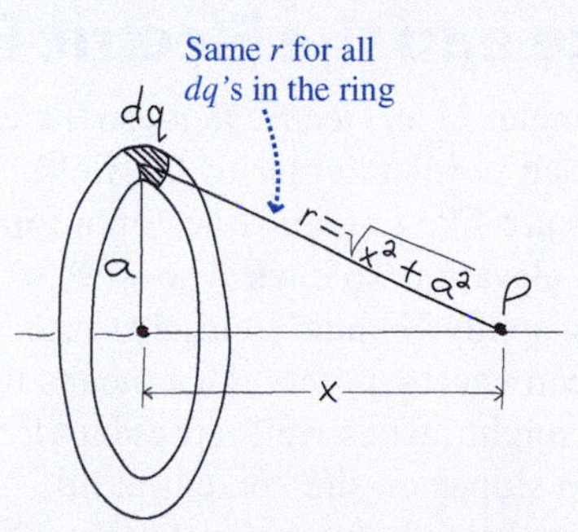

FIGURE 22.12 A charged ring.

Fig. 22.12 shows that the distance $r = \sqrt{x^2 + a^2}$ is the same for all charge elements.

EVALUATE Equation 22.7 becomes

$$V(x) = \int \frac{k\,dq}{r} = \frac{k}{r}\int dq = \frac{kQ}{r} = \frac{kQ}{\sqrt{x^2 + a^2}} \tag{22.8}$$

The integration here simplified because r is the same for all charge elements within the ring and so comes outside the integral; the remaining integral, $\int dq$, is the total charge Q.

ASSESS Make sense? At large distances ($x \gg a$), a^2 is negligible and our result becomes $V(x) = kQ/x$. This is the potential of a point charge Q—just as we'd expect when we're so far from the ring that its size isn't significant. At the ring's center, on the other hand, we have $V(0) = kQ/a$. Here we're a distance a from all parts of the ring, and since potential is a *scalar*, the different directions don't matter. The result is therefore the same as being a distance a from a point charge Q. ■

EXAMPLE 22.7 Potential of a Continuous Distribution: A Charged Disk

A charged disk of radius a carries a charge Q distributed uniformly over its surface. Find the potential at a point P on the disk axis, a distance x from the disk.

INTERPRET This problem, too, involves a continuous charge distribution.

DEVELOP Equation 22.7, $V = \int k\,dq/r$, determines the potential. But now all parts of the charge distribution aren't the same distance from P, so we have to set up the integral using the procedure outlined in Chapter 9's Tactics 9.1 and used most recently in Chapter 20 when we

(continued)

calculated the fields of continuous charge distributions. We've drawn the disk and its axis in Fig. 22.13. The preceding example suggests that we divide the disk into ring-shaped charge elements, as we've drawn in the figure. Each ring contributes a potential dV given by Equation 22.8: $dV = k\,dq/\sqrt{x^2 + r^2}$, where r is the ring radius. We get the potential of the entire disk by integrating over all the rings that make up the disk:

$$V(x) = \int_{\text{disk}} dV = \int_{r=0}^{r=a} \frac{k\,dq}{\sqrt{x^2 + r^2}}$$

Before we can evaluate this integral, we need to relate the charge element dq and the geometric variable r. Here the relation involves area: The ratio of ring area to disk area is the same as the ratio of dq to the total charge Q. "Unwinding" a ring gives a rectangle of length $2\pi r$ and width dr, so the ring area is $2\pi r\,dr$. The disk area is πa^2, so $dq/Q = 2\pi r\,dr/\pi a^2$, giving $dq = (2Q/a^2)r\,dr$.

EVALUATE Using this result in our integral for the potential $V(x)$ gives

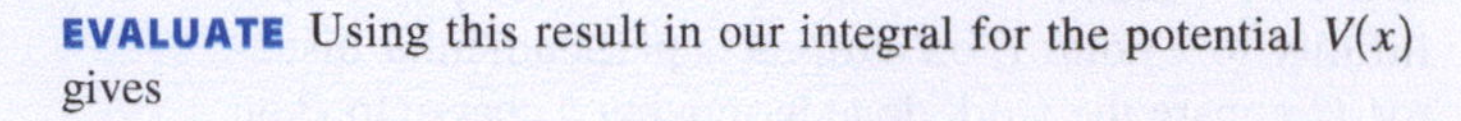

$$V(x) = \int_0^a \frac{2kQ}{a^2} \frac{r\,dr}{\sqrt{x^2 + r^2}} = \frac{kQ}{a^2} \int_0^a \frac{2r\,dr}{\sqrt{x^2 + r^2}}$$

Now, $2r\,dr = d(r^2) = d(x^2 + r^2)$ since x is a constant with respect to the integration. The integral therefore has the form $\int u^{-1/2}\,du$, where $u = x^2 + r^2$, and the result is $2u^{1/2}$ or

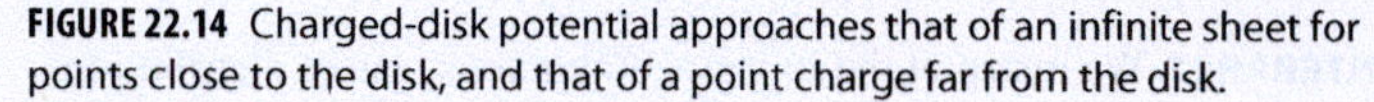

$$V(x) = \frac{2kQ}{a^2}\sqrt{x^2 + r^2}\,\Big|_{r=0}^{r=a} = \frac{2kQ}{a^2}\left(\sqrt{x^2 + a^2} - |x|\right)$$

ASSESS Make sense? Figure 22.14 shows that it does. Close to the disk, the potential resembles that of an infinite sheet, changing linearly with distance (recall Example 22.2); far away, it has the $1/r$ behavior of a point-charge potential. It's only at intermediate distances—on the order of the disk radius a—that we really need our exact expression. ■

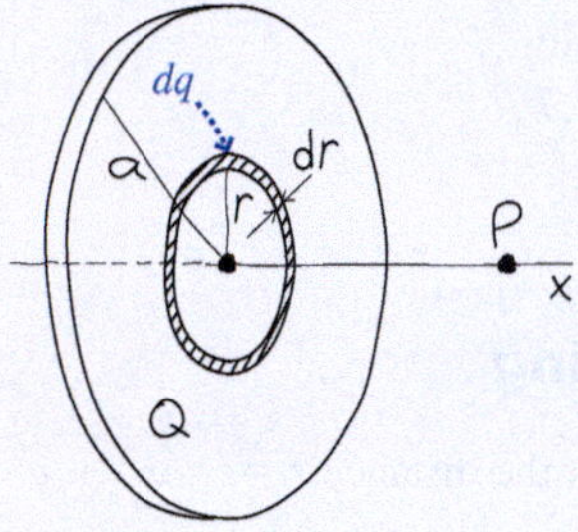

FIGURE 22.13 A charged disk, showing a ring-shaped charge element dq of radius r and width dr.

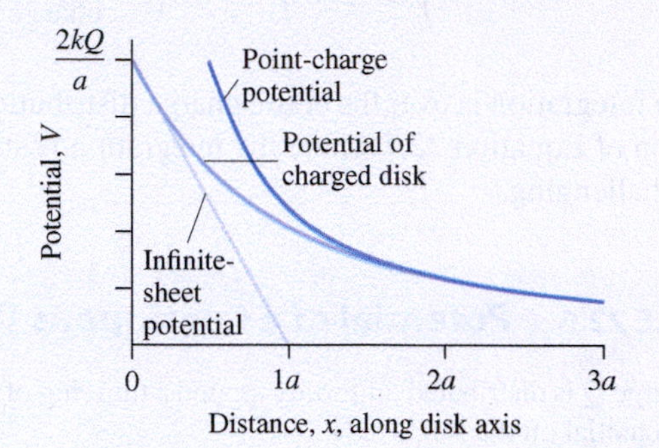

FIGURE 22.14 Charged-disk potential approaches that of an infinite sheet for points close to the disk, and that of a point charge far from the disk.

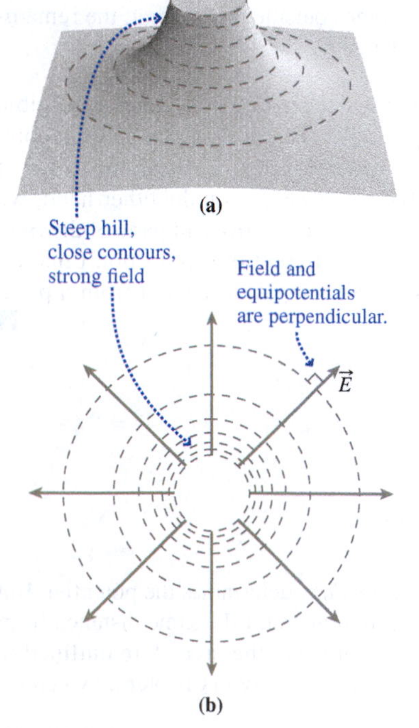

FIGURE 22.15 A flat-topped hill (a) and its contour map (b) represent equipotentials (dashed curves) for a charged spherical shell, in a plane through the shell's center.

22.3 Potential Difference and the Electric Field

It takes no work to move a charge at right angles to an electric field, so there's no potential difference between two points on a surface perpendicular to the field. Such surfaces are called **equipotentials**. Equipotentials are like contour lines on a topographic map (Fig. 22.15). A contour is a line of constant elevation, so it takes no work to move along it. Where contours are close, elevation changes rapidly. Similarly, closely spaced equipotentials indicate a large potential difference between nearby points. That means there must be a strong electric field present. Figure 22.15 might just as well represent electric potential, with closely spaced equipotentials—steep slopes on the "potential hill"—representing strong electric fields. Similarly, the equipotentials for a dipole (Fig. 22.16; see also Fig. 22.11) describe the steep "hill" of the positive charge and the "hole" of the negative charge that we showed in Fig. 22.11. There is one difference, though: Equipotentials are surfaces in three dimensions, and when we draw them as contour lines, we're showing only the surfaces' intersections with a plane.

GOT IT? 22.6 The figure shows cross sections through two equipotential surfaces. In both diagrams the potential difference between adjacent equipotentials is the same. Which could represent the field of a point charge? Explain.

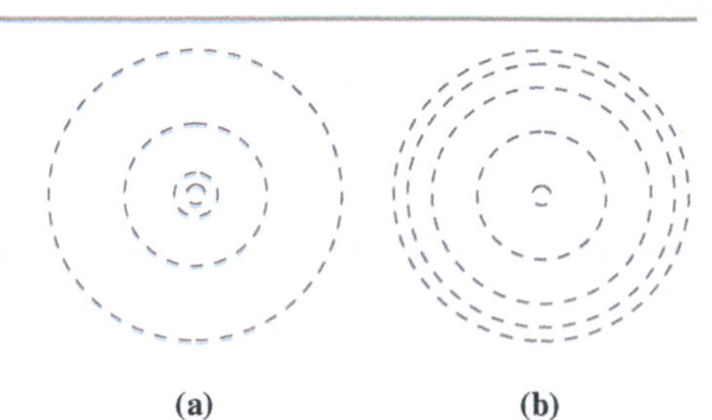

Calculating Field from Potential

Given electric field lines, we can construct equipotentials. Conversely, given equipotentials, we can reconstruct the field by sketching field lines at right angles to the equipotentials. Specifying the potential at each point thus conveys all the information needed to determine the field.

We can quantify the relation between potential and field by considering the potential difference dV between two nearby points. Suppose they're separated by a small displacement dx in the x-direction. Then Equation 22.1b becomes $dV = -E_x\,dx$, where we handled the dot product by considering only the component of $\vec{E}$ along the displacement. Dividing through by dx shows that we can write the electric-field component in the x-direction as $E_x = -dV/dx$. We can write similar expressions for the y- and z-components. When a function depends on more than one variable, as the potential generally does, we write derivatives with the partial derivative symbol ∂ instead of d to indicate the rate of change with respect to only one variable. Thus we have $E_x = -\partial V/\partial x$, $E_y = -\partial V/\partial y$, and $E_z = -\partial V/\partial z$. Putting together these three components lets us write the entire electric-field vector:

$$\vec{E} = -\left(\frac{\partial V}{\partial x}\hat{\imath} + \frac{\partial V}{\partial y}\hat{\jmath} + \frac{\partial V}{\partial z}\hat{k}\right) \quad (22.9)$$

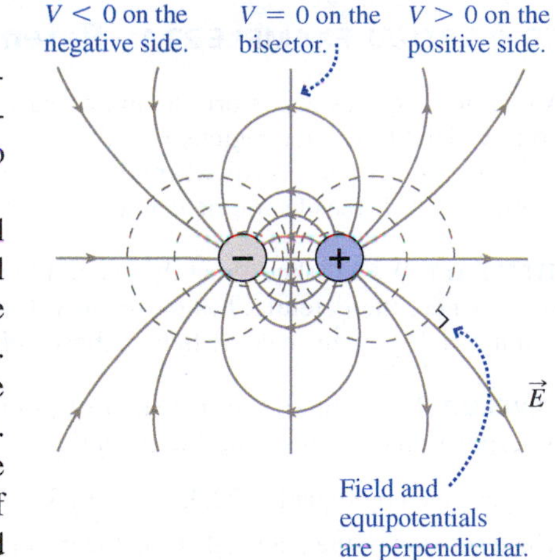

FIGURE 22.16 Equipotentials (dashed curves) and field lines for a dipole, in a plane containing the dipole. You should convince yourself that the equipotentials provide a contour map of Fig. 22.11.

PhET: Calculus Grapher
PhET: Charges and Fields

Equation 22.9 confirms that the electric field is strong where the potential changes rapidly. The minus sign here is the same as in Equation 22.1: It says that if we move in the direction of *increasing* potential, then we're moving *against* the electric field. Equation 22.9 also shows that the units of electric field, N/C, can be written equivalently as volts per meter—a unit widely used in both science and engineering.

Because potential is a scalar, it's often easier to calculate the potential and then use Equation 22.9 to get the field. The next example shows how.

EXAMPLE 22.8 Field from Potential: A Charged Disk

Use the result of Example 22.7 to find the electric field on the axis of a charged disk.

INTERPRET Example 22.7 gives the potential of a charged disk, so this problem is about calculating electric field from potential.

DEVELOP Example 22.7 gives the potential on the axis of a charged disk: $V(x) = (2kQ/a^2)\left(\sqrt{x^2 + a^2} - |x|\right)$. Equation 22.9 shows that we can get the electric field by differentiating the potential with respect to all three coordinates. But here the potential depends only on x, so $\vec{E}$ has only an x-component—a fact that should also be evident from symmetry. Our plan is to apply Equation 22.9, differentiating $V(x)$ to get the field component E_x.

EVALUATE We apply Equation 22.9 to $V(x)$ to get

$$E_x = -\frac{dV}{dx} = -\frac{d}{dx}\left[\frac{2kQ}{a^2}\left(\sqrt{x^2 + a^2} - |x|\right)\right]$$
$$= \frac{2kQ}{a^2}\left(\pm 1 - \frac{x}{\sqrt{x^2 + a^2}}\right)$$

where the $+/-$ signs apply for $x > 0$ and $x < 0$, respectively. Because V depends only on x, we wrote the total derivative dV/dx rather than the partial derivative.

ASSESS Make sense? In Fig. 22.14 we showed that the field of a disk ought to look like that of a charged sheet close up and like a point charge far away. For $|x| \ll a$, our result gives $|E_x| = 2kQ/a^2$. With $k = 1/4\pi\epsilon_0$ and $Q/\pi a^2 = \sigma$, this is indeed the field $E = \sigma/2\epsilon_0$ of a charged sheet. You can show in Problem 68 that the case $|x| \gg a$ reduces to the field of a point charge Q, as expected. ■

✓TIP Field and Potential

Note that the *values* of field and potential at a single point aren't related; rather, as Equation 22.9 shows, the field is determined by the *rate of change* of potential. Field and potential are related in the same way as acceleration and velocity; their *values* are independent, but one is the rate of change of the other—although with a negative sign in the case of field and potential. In particular, the field can be zero where the potential isn't, and vice versa. Conceptual Example 22.1 explores this situation.

CONCEPTUAL EXAMPLE 22.1 Potential and Field

A charge $+2Q$ lies at the origin, and a charge $-Q$ lies at $x = +a$. (a) In which of the three regions $x < 0$, $0 < x < a$, and $x > a$ could there be a point on the x-axis where the potential is zero? Take $V = 0$ at infinity. (b) Is the electric field at these points also zero?

INTERPRET We're asked about the general locations of points on the x-axis where the potential has the same value ($V = 0$) as it does at infinity, and about the electric field at those points.

DEVELOP The potential of a system of point charges is given by superposition—by summing the potentials for the individual point charges given by Equation 22.3, $V = \frac{kq}{r}$. So we want to locate points where the sum of the potentials from the positive charge $+2Q$ and the negative charge $-Q$ could sum to zero. In this respect this example is similar to Conceptual Example 20.2, except that there we were dealing with the electric field—a vector quantity—and here we're dealing with potential, which is a scalar.

EVALUATE As in Conceptual Example 20.2, all points with $x < 0$ are closer to the charge $2Q$, so its potential dominates and is always positive. So we can't have $V = 0$ in this region. Between the charges, where $0 < x < a$, we could be close to either charge depending on which end of the interval we're near. Near the positive charge, the potential must be positive; near the negative charge, it must be negative—and so there's a point somewhere in between where $V = 0$. (Because the charges have unequal magnitudes, that point won't be right in the middle. Can you tell whether it will be closer to $+2Q$ or to $-Q$?) Note that this middle situation is different from what we found for the electric field in Conceptual Example 20.2; there, the electric fields from the two charges are pointed in the same direction in the region between the charges, so we couldn't have $E = 0$ in that region. Finally, for $x > a$ but still very close to $-Q$, the potential is dominated by the negative charge and so must be negative. But farther away, the charge distribution begins to look like a single charge with net charge $2Q-Q = Q$, so the potential is positive at large distances. So there's a point in between where the potential is zero. Figure 22.17 is a plot of the potential on the x-axis, showing the two points we've identified. Now, Equation 22.9 shows that the electric field depends on the rate of change of potential, and that's zero where the curve in Fig. 22.17 is at a maximum or minimum. Since the maxima and minima don't occur where $V = 0$, we conclude that the electric field is *not* zero where the potential is zero.

ASSESS The two points we've located are actually the intersections of a $V = 0$ equipotential that surrounds the negative charge. Figure 22.18 shows a 3-D plot of the potential in the $x-y$ plane, with that $V = 0$ equipotential marked. You can see that nearly all points in the diagram lie above the $V = 0$ equipotential, indicating positive potential. The exception are those points in the "hole" created by the negative charge.

MAKING THE CONNECTION Find the exact locations of the two points on the x-axis that this example identified as having $V = 0$.

EVALUATE Between the charges, the potential is given by $V = \frac{k(2Q)}{x} + \frac{k(-Q)}{a - x}$, where we chose the signs in the second denominator to ensure that it's *positive* since the r in $V = kq/r$ is always a positive *distance*. Setting $V = 0$ in this expression and solving for x gives $x = \frac{2}{3}a$. To the right of both charges, we have $V = \frac{k(2Q)}{x} + \frac{k(-Q)}{x - a}$, where again we chose the second denominator to be positive—now $x-a$ since we're in the region $x > a$. Setting $V = 0$ and solving now gives $x = 2a$.

ASSESS Our answers are both in the expected regions, and the fact that the $V = 0$ point between the charges is closer to the negative charge reflects the greater influence of the larger-magnitude positive charge.

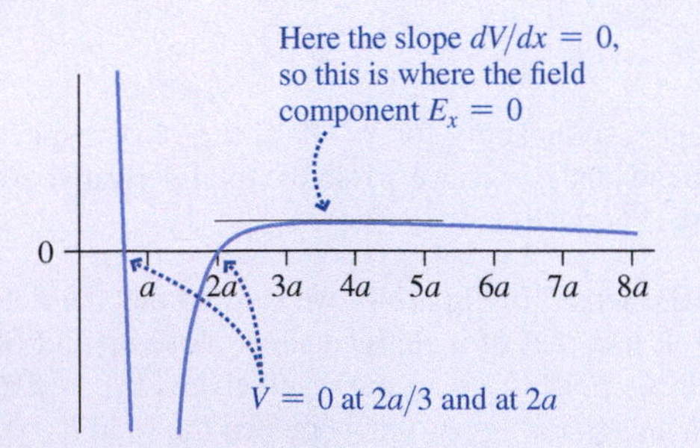

FIGURE 22.17 Potential V on the x-axis, showing points where $V = 0$. The potential goes to $\pm\infty$ at the locations of the positive charge ($x = 0$) and negative charge ($x = a$). Note that $E_x = 0$ not where the potential is zero but where $dV/dx = 0$. This point occurs at $x = 3.4a$, as found in Making the Connection for Conceptual Example 20.2.

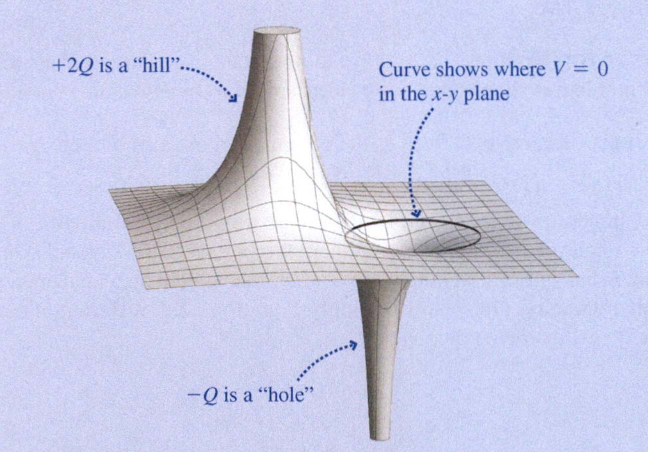

FIGURE 22.18 Three-dimensional plot of the potential V in the $x-y$ plane, showing the $V = 0$ equipotential in the $x-y$ plane. Where are the $V = 0$ points from Fig. 22.17 on this 3-D plot?

22.4 Charged Conductors

There's no electric field inside a conducting material in electrostatic equilibrium, and at the conductor surface there's no field component parallel to the surface. Therefore, it takes no work to move a test charge on or inside a conductor—and that means **a conductor in**

electrostatic equilibrium is an equipotential. We can exploit this fact to gain insight into the electric field at the surface of a conductor.

Consider an isolated, spherical conductor of radius R carrying charge Q. Charge is distributed uniformly over its surface, so the field outside the sphere is that of a point charge. Then the potential at its surface is $V(R) = kQ/R$, as we found in Example 22.3. Now consider two widely separated spheres of different sizes. If we connect them by a thin conducting wire (Fig. 22.19), charge will move through the wire until both spheres are at the same potential. But since the spheres are widely separated, each still has an essentially spherical charge distribution, so $V(R) = kQ/R$ gives each sphere's potential. Because the spheres have the same potential, $kQ_1/R_1 = kQ_2/R_2$. We can write each charge as the surface area multiplied by the surface charge density: $Q = 4\pi R^2\sigma$. Substituting for the Q's in the above equation and solving for the ratio of surface charge densities then gives

$$\frac{\sigma_1}{\sigma_2} = \frac{R_2}{R_1}$$

Thus the *smaller* sphere has the *higher* surface charge density. Since the electric field at a conductor surface has magnitude $E = \sigma/\varepsilon_0$, the field must also be stronger at the smaller sphere. Conceptual Example 22.2 explores this situation further.

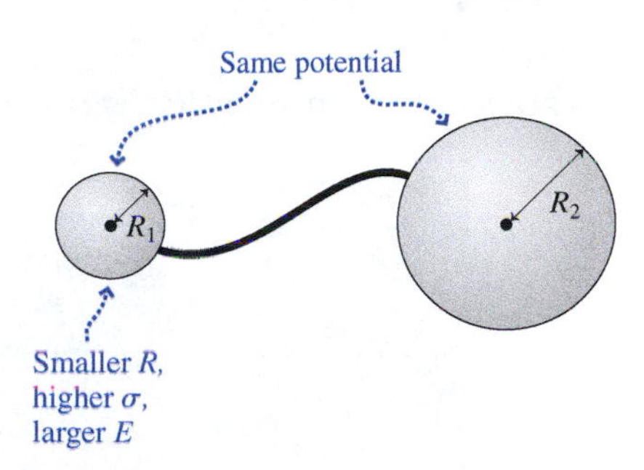

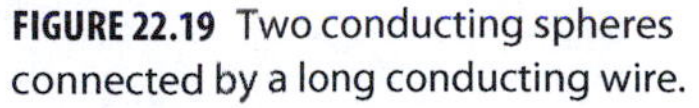
FIGURE 22.19 Two conducting spheres connected by a long conducting wire.

Video Tutor Demo | **Charged Conductor with Teardrop Shape**

CONCEPTUAL EXAMPLE 22.2 An Irregular Conductor

Sketch some equipotentials and electric field lines for an isolated egg-shaped conductor.

EVALUATE Where the conductor surface curves sharply, it's like the small sphere of Fig. 22.19. It therefore has higher surface charge density and a stronger electric field, which means more field lines emerge where the surface curves sharply. Since the field is perpendicular at the conductor surface, equipotentials just above the surface have essentially the same shape as the surface. Far from the charged conductor, on the other hand, its field resembles that of a point charge, with radial field lines and circular equipotentials. Figure 22.20 gives an approximate picture of the field and equipotentials based on these considerations.

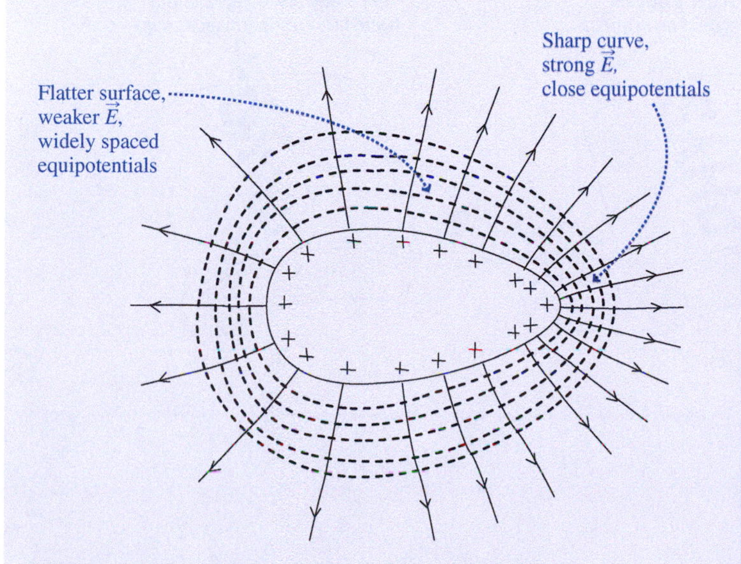

FIGURE 22.20 Equipotentials and field of a charged conductor.

ASSESS Our analysis here applies only to an *isolated* conductor. Figure 22.21 shows how the presence of nearby charges alters the charge distribution on a conductor.

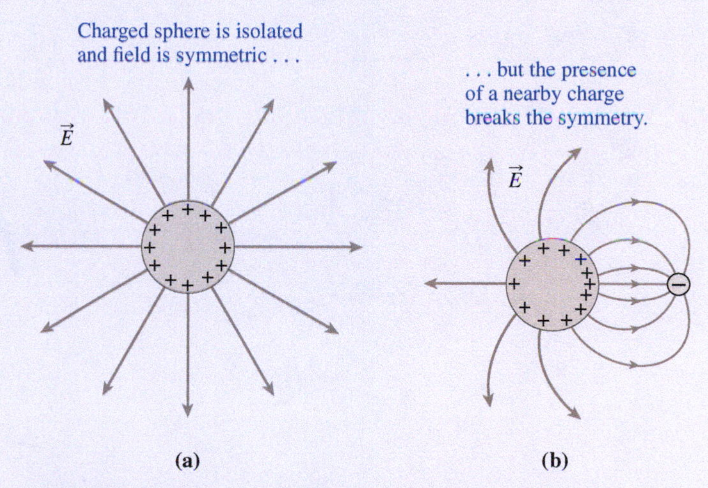

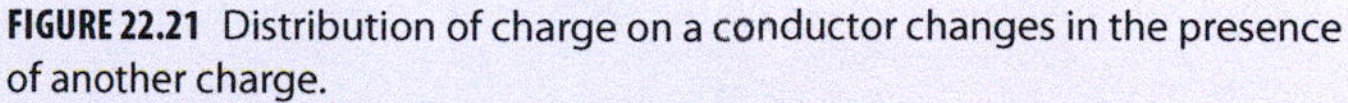
FIGURE 22.21 Distribution of charge on a conductor changes in the presence of another charge.

MAKING THE CONNECTION The potential difference between the conductor in Fig. 22.20 and the outermost equipotential shown is 70 V. Determine approximate values for the strongest and weakest electric fields in the region shown in the figure, assuming it's drawn at actual size.

EVALUATE Electric field is the rate of change of potential. At the tip, where the field is strongest, the outermost equipotential is about 7 mm from the conductor. So the field here is approximately (70 V)/(7 mm); that's 10 V/mm or 10 kV/m. At its most distant, the outer equipotential is about 12 mm from the conductor, giving a field of (70 V)/(12 mm) or just under 6 kV/m.

APPLICATION Corona Discharge, Pollution Control, and Xerography

The large electric fields that develop at sharply curved conductors can cause serious problems in electric equipment. Fields stronger than 3 MN/C strip electrons from air molecules, making air a conductor. The result is a blue glow, called corona discharge, resulting from electrons recombining with atoms. Corona discharge causes energy loss from high-voltage transmission lines, so engineers try to avoid sharp edges on conducting structures. The photo shows corona discharge leaking current across a power-line insulator.

Corona discharge can also be useful. Pollution-control devices called **electrostatic precipitators** use a thin wire at a high negative potential to produce a strong field that ionizes gas molecules. Ions adhere to pollutant particles, which are then attracted to positively charged collecting plates. Such devices remove up to 99% of particulate pollutants from power plants and factories, and their use under the Clean Air Act has substantially improved air quality in the United States.

You use corona discharge whenever you make a photocopy or print with a laser printer. The process here is *xerography*—literally, "dry writing"—and it starts with the uniform charging of a special photoconductive drum by corona discharge from a thin wire maintained at about 5 kV relative to the drum. The photoconductive material is a good insulator in the dark, but incident light dislodges electrons, which neutralize the charge on illuminated portions of the material. A laser beam scans the photoconductive drum, "writing" the image to be copied or printed by leaving darker areas charged and rendering lighter areas neutral. Next, a dusting of fine particles called *toner* is spread over the drum, adhering, via the electric force, only where the drum is charged. The toner is then transferred to a sheet of paper, which is heated to melt the particles into the paper, making a permanent copy. The diagram outlines this process.

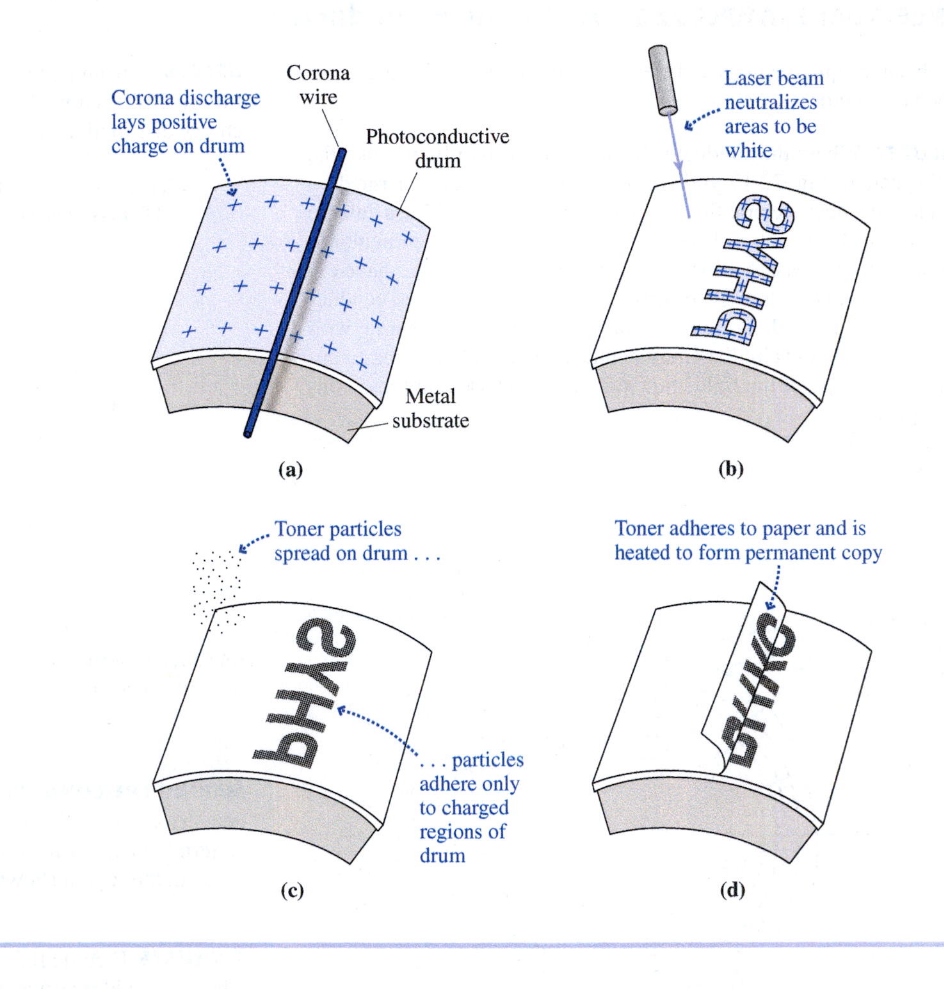

CHAPTER 22 SUMMARY

Big Idea

The big idea here is **electric potential difference**—a measure of the energy per unit charge involved in moving charge between two points in an electric field. Because the electric field is conservative, potential difference is path independent and thus depends only on the two points in question.

Key Concepts and Equations

Electric potential difference between points A and B is the negative of the line integral of the electric field over any path from A to B:

$$\Delta V_{AB} = \frac{\Delta U_{AB}}{q} = -\int_A^B \vec{E}\cdot d\vec{r}$$

When a charge "falls" through a potential difference ΔV, it gains energy $q\,\Delta V$.

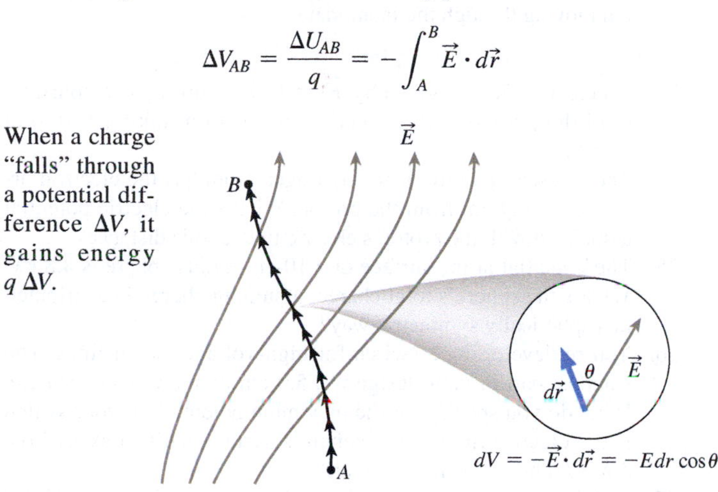

In a uniform field, the potential difference becomes

$$\Delta V_{AB} = -\vec{E}\cdot\Delta\vec{r}$$

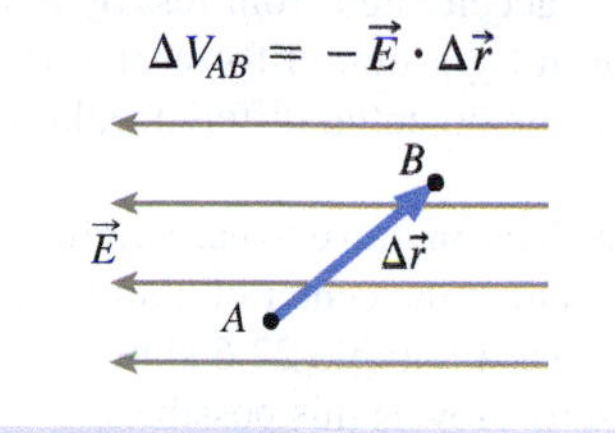

The potential in the field of a point charge is $V(r) = kq/r$, where the zero of potential is taken at infinity and r is the distance from the point charge.

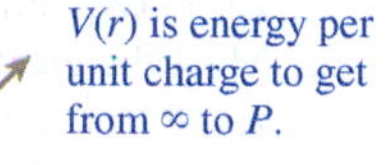

Potentials of charge distributions follow by summing or integrating the fields of pointlike charge elements:

$$V = \sum \frac{kq_i}{r_i} \quad \text{(discrete charges)}$$

$$V = \int \frac{k\,dq}{r} \quad \text{(continuous charge distribution)}$$

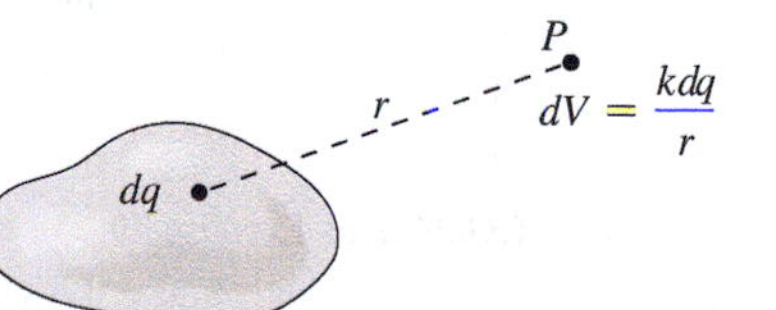

Equipotentials are surfaces of constant potential and are perpendicular to the electric field. Where equipotentials are close, the field is strong. The field component in a given direction depends on the rate at which potential changes with position; thus,

$$E_x = -\frac{dV}{dx}$$

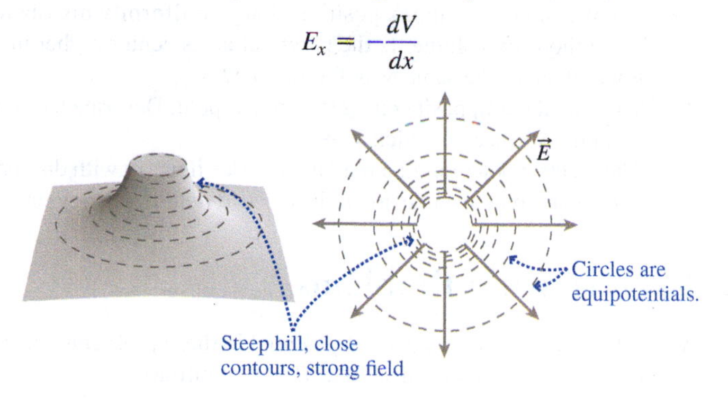

Applications

The dipole potential is

$$V = \frac{kp\cos\theta}{r^2}$$

where $p = qd$ is the dipole moment and the angle θ is measured from the dipole axis.

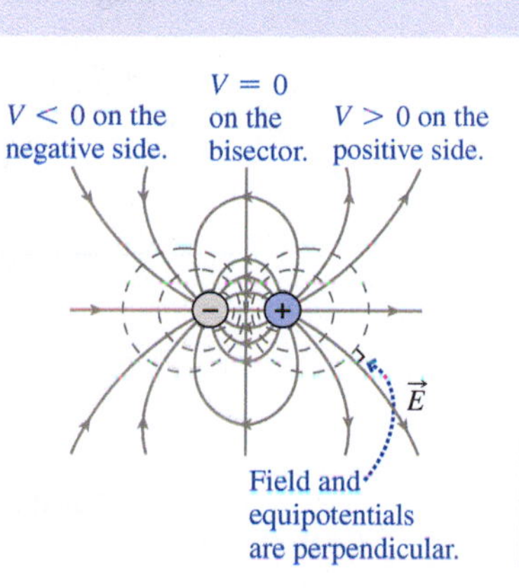

In charged conductors, the charge density is generally highest, and the field strongest, where a conductor curves sharply.

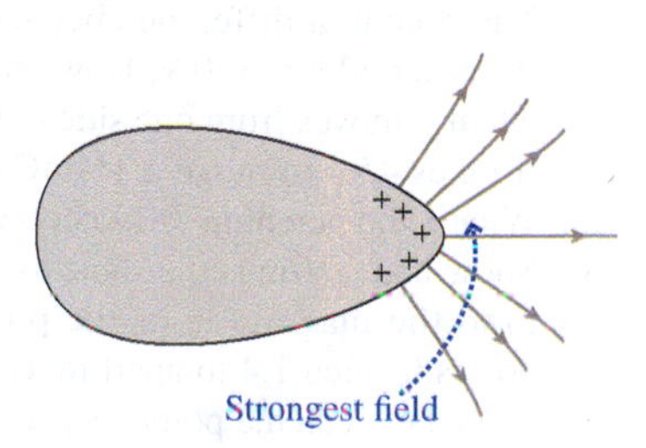

MP For homework assigned on MasteringPhysics, go to www.masteringphysics.com

BIO *Biology and/or medicine-related problems* **DATA** *Data problems* **ENV** *Environmental problems* **CH** *Challenge problems* **COMP** *Computer problems*

For Thought and Discussion

1. Why can a bird perch on a high-voltage power line without getting electrocuted?
2. One proton is accelerated from rest by a uniform electric field, another proton by a nonuniform electric field. If they move through the same potential difference, how do their final speeds compare?
3. Would a free electron move toward higher or lower potential?
4. The electric field at the center of a uniformly charged ring is obviously zero, yet Example 22.6 shows that the potential at the center isn't zero. How is this possible?
5. Must the potential be zero at any point where the electric field is zero? Explain.
6. Must the electric field be zero at any point where the potential is zero? Explain.
7. The potential is constant throughout an entire volume. What must be true of the electric field within that volume?
8. In considering the potential of an infinite flat sheet, why isn't it useful to take the zero of potential at infinity?
9. "Cherry picker" trucks for working on power lines often carry electrocution hazard signs. Explain how this hazard arises and why it might be more of a danger to someone on the ground than to a worker on the truck.
10. Can equipotential surfaces intersect? Explain.
11. Is the potential at the center of a hollow, uniformly charged spherical shell higher than, lower than, or the same as at the surface?
12. A solid sphere contains positive charge uniformly distributed throughout its volume. Is the potential at its center higher than, lower than, or the same as at the surface?
13. Two equal but opposite charges form a dipole. Describe the equipotential surface on which $V = 0$.
14. The electric potential in a region increases linearly with distance. What can you conclude about the electric field in this region?

Exercises and Problems

Note: If you're asked for values of potential in these problems, and the zero of potential isn't specified, take $V = 0$ at infinity.

Exercises

Section 22.1 Electric Potential Difference

15. How much work does it take to move a 50-μC charge against a 12-V potential difference?
16. The potential difference between the two sides of an ordinary electric outlet is 120 V. How much energy does an electron gain when it moves from one side to the other?
17. It takes 45 J to move a 15-mC charge from point A to point B. What's the potential difference ΔV_{AB}?
18. Show that 1 V/m is the same as 1 N/C.
19. Find the magnitude of the potential difference between two points located 1.4 m apart in a uniform 650-N/C electric field, if a line between the points is parallel to the field.
20. A charge of 3.1 C moves from the positive to the negative terminal of a 9.0-V battery. How much energy does the battery impart to the charge?
21. A proton, an alpha particle (a bare helium nucleus), and a singly ionized helium atom are accelerated through a 100-V potential difference. How much energy does each gain?
22. **BIO** The potential difference across a typical cell membrane is about 80 mV. How much work is done on a singly ionized potassium ion moving through the membrane?

Section 22.2 Calculating Potential Difference

23. An electric field is given by $\vec{E} = E_0\,\hat{j}$, where E_0 is a constant. Find the potential as a function of position, taking $V = 0$ at $y = 0$.
24. The classical picture of the hydrogen atom has the electron orbiting 0.0529 nm from the proton. What's the electric potential associated with the proton's electric field at this distance?
25. The potential at the surface of a 10-cm-radius sphere is 4.8 kV. What's the sphere's total charge, assuming charge is distributed in a spherically symmetric way?
26. You're developing a switch for high-voltage power lines. The smallest part in your design is a 5.0-cm-diameter metal sphere. What do you specify for the maximum potential on your switch if the electric field at the sphere's surface isn't to exceed the 3-MV/m breakdown field of air?
27. A 3.5-cm-diameter isolated metal sphere carries 0.86 μC. (a) Find the potential at the sphere's surface. (b) If a proton were released from rest at the surface, what would be its speed far from the sphere?

Section 22.3 Potential Difference and the Electric Field

28. In a uniform electric field, equipotential planes that differ by 5.00 V are 2.54 cm apart. What's the field strength?
29. Figure 22.22 shows a plot of potential versus position along the x-axis. Make a plot of the x-component of the electric field for this situation.

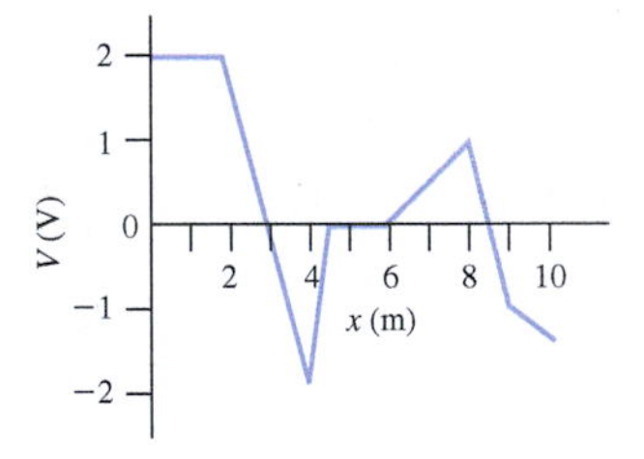

FIGURE 22.22 Exercise 29

30. Figure 22.23 shows some equipotentials in the x–y plane. (a) In what region is the electric field strongest? What are (b) the direction and (c) the magnitude of the field in this region?

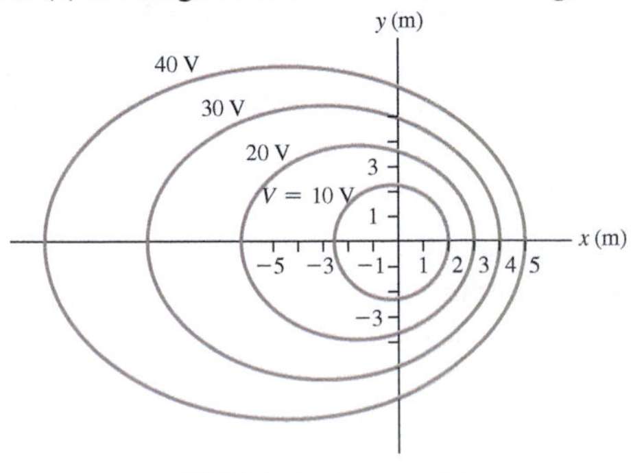

FIGURE 22.23 Exercise 30

31. The electric potential in a region is given by $V = 2xy - 3zx + 5y^2$, with V in volts and the coordinates in meters. Find (a) the potential and (b) the components of the electric field at the point $x = 1$ m, $y = 1$ m, $z = 1$ m.

Section 22.4 Charged Conductors

32. Dielectric breakdown of air occurs at fields of 3 MV/m. Find (a) the maximum potential (measured from infinity) for the sphere of Example 22.3 before dielectric breakdown occurs at the sphere's surface and (b) the charge on the sphere at this potential.
33. You're an automotive engineer working on the ignition system for a new engine. Its spark plugs have center electrodes made from 2.0-mm-diameter wire. The electrode ends gradually wear to a hemispherical shape, so they behave approximately like charged spheres. Your job is to specify the minimum potential that ensures these plugs will spark in air, neglecting the presence of the second electrode.
34. A large metal sphere has three times the diameter of a smaller sphere and carries three times the charge. Both spheres are isolated, so their surface charge densities are uniform. Compare (a) the potentials (relative to infinity) and (b) the electric field strengths at their surfaces.

Problems

35. Two points A and B lie 15 cm apart in a uniform electric field, with the path AB parallel to the field. If the potential difference ΔV_{AB} is 840 V, what's the field strength?
36. **BIO** The electric field within a cell membrane is approximately 8.0 MV/m and is essentially uniform. If the membrane is 10 nm thick, what's the potential difference across the membrane?
37. What's the potential difference between the terminals of a battery that can impart 7.2×10^{-19} J to each electron that moves between the terminals?
38. What's the charge on an ion that gains 1.6×10^{-15} J when it moves through a potential difference of 2500 V?
39. Two flat metal plates are a distance d apart, where d is small compared with the plate size. If the plates carry surface charge densities $\pm\sigma$, show that the magnitude of the potential difference between them is $V = \sigma d/\varepsilon_0$.
40. An electron passes point A moving at 6.5 Mm/s. At point B it comes to a stop. Find the potential difference ΔV_{AB}.
41. A 5.0-g object carries 3.8 μC. It acquires speed v when accelerated from rest through a potential difference V. If a 2.0-g object acquires twice the speed under the same circumstances, what's its charge?
42. Points A and B lie 32.0 cm apart on a line extending radially from a point charge Q, and the potentials at these points are $V_A = 362$ V and $V_B = 146$ V. Find Q and the distance r between point A and the charge.
43. A sphere of radius R carries negative charge of magnitude Q, distributed in a spherically symmetric way. Find an expression for the escape speed for a proton at the sphere's surface—that is, the speed that would enable the proton to escape to arbitrarily large distances starting at the sphere's surface.
44. **BIO** Proton-beam therapy can be preferable to X rays for cancer treatment (although much more expensive) because protons deliver most of their energy to the tumor, with less damage to healthy tissue. A cyclotron used to accelerate protons for cancer treatment repeatedly passes the protons through a 15-kV potential difference. (a) How many passes are needed to bring the protons' kinetic energy to 1.2×10^{-11} J? (b) What's that energy in eV?
45. A thin spherical shell has radius R and total charge Q distributed uniformly over its surface. Find the potential at its center.
46. A solid sphere of radius R carries charge Q distributed uniformly throughout its volume. Find the potential difference from the sphere's surface to its center. (*Hint*: Consult Example 21.1.)
47. Find the potential as a function of position in the electric field $\vec{E} = ax\hat{\imath}$, where a is a constant and where you're taking $V = 0$ at $x = 0$.
48. Your radio station needs a new coaxial cable to connect the transmitter and antenna. One possible cable consists of a 2.0-mm-diameter inner conductor and an outer conductor with diameter 1.6 cm and negligible thickness (Fig. 22.24); the maximum safe potential difference between the conductors is 2 kV. In your application, the conductors carry charge densities ±62 nC/m. Will this cable work for you?

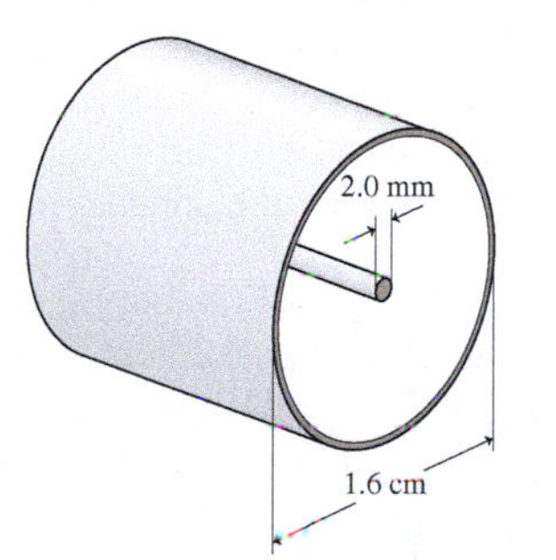

FIGURE 22.24 Problem 48

49. The potential difference between the surface of a 3.0-cm-diameter power line and a point 1.0 m distant is 3.9 kV. Find the line charge density on the power line.
50. Three equal charges q form an equilateral triangle of side a. Find the potential, relative to infinity, at the center of the triangle.
51. A charge $+Q$ lies at the origin and $-3Q$ at $x = a$. Find two points on the x-axis where $V = 0$.
52. Two identical charges q lie on the x-axis at $\pm a$. (a) Find an expression for the potential at all points in the x–y plane. (b) Show that your result reduces to the potential of a point charge for distances large compared with a.
53. A dipole of moment $p = 2.9$ nC·m consists of two charges separated by far less than 10 cm. Find the potential 10 cm from the dipole (a) on its axis, (b) at 45° to its axis, and (c) on its perpendicular bisector.
54. A thin plastic rod 20 cm long carries 3.2 nC distributed uniformly over its length. (a) If the rod is bent into a ring, find the potential at its center. (b) If it's bent into a semicircle, find the potential at the center of the semicircle.
55. A thin ring of radius R carries charge $3Q$ distributed uniformly over three-fourths of its circumference, and $-Q$ over the rest. Find the potential at the ring's center.
56. The potential at the center of a uniformly charged ring is 45 kV, and 15 cm along the ring axis the potential is 33 kV. Find the ring's radius and total charge.
57. The annulus shown in Fig. 22.25 carries a uniform surface charge density σ. Find an expression for the potential at an arbitrary point P on its axis.

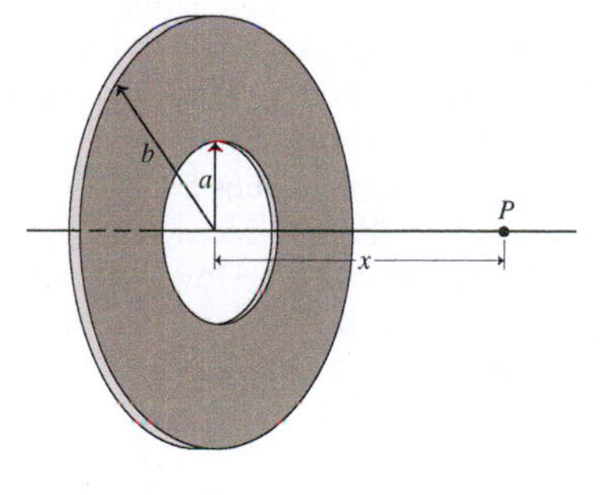

FIGURE 22.25 Problem 57

58. The potential in a region is given by $V = axy$, where a is a constant. (a) Determine the electric field in the region. (b) Sketch some equipotentials and field lines.
59. Use Equation 22.6 to calculate the electric field on the perpendicular bisector of a dipole, and show that your result is equivalent to Equation 20.6a.
60. Use the result of Example 22.6 to determine the on-axis field of a charged ring, and verify that your answer agrees with the result of Example 20.6.
61. The electric potential in a region is given by $V = -V_0(r/R)$, where V_0 and R are constants and r is the radial distance from the origin. Find expressions for the magnitude and direction of the electric field in this region.
62. Two metal spheres each 1.0 cm in radius are far apart. One sphere carries 38 nC, the other -10 nC. (a) What's the potential on each? (b) If the spheres are connected by a thin wire, what will be the potential on each once equilibrium is reached? (c) How much charge moves between the spheres in order to achieve equilibrium?
63. Two 5.0-cm-diameter conducting spheres are 8.0 m apart, and each carries 0.12 μC. Determine (a) the potential on each sphere, (b) the field strength at the surface of each sphere, (c) the potential midway between the spheres, and (d) the potential difference between the spheres.
64. A 2.0-cm-radius metal sphere carries 75 nC and is surrounded by a concentric spherical conducting shell of radius 10 cm carrying -75 nC. (a) Find the potential difference between shell and sphere. (b) How would your answer change if the shell's charge were $+150$ nC?
65. **CH** A sphere of radius R carries a nonuniform but spherically symmetric volume charge density that results in an electric field in the sphere given by $\vec{E} = E_0(r/R)^2\hat{r}$, where E_0 is a constant. Find the potential difference from the sphere's surface to its center.
66. The potential as a function of position in a region is given by $V(x) = 3x - 2x^2 - x^3$, with x in meters and V in volts. Find (a) all points on the x-axis where $V = 0$, (b) an expression for the electric field, and (c) all points on the x-axis where $E = 0$.
67. A conducting sphere 5.0 cm in radius carries 60 nC. It's surrounded by a concentric spherical conducting shell of radius 15 cm carrying -60 nC. (a) Find the potential at the sphere's surface, taking $V = 0$ at infinity. (b) Repeat for the case when the shell carries $+60$ nC.
68. **CH** Show that the result of Example 22.8 approaches the field of a point charge for $x \gg a$. (*Hint*: You'll need to apply the binomial approximation from Appendix A to the expression $1/\sqrt{x^2 + a^2}$.)
69. **CH** The potential on the axis of a uniformly charged disk at 5.0 cm from the disk center is 150 V; the potential 10 cm from the disk center is 110 V. Find the disk radius and its total charge.
70. A uranium nucleus (mass 238 u, charge $92e$) decays, emitting an alpha particle (mass 4 u, charge $2e$) and leaving a thorium nucleus (mass 234 u, charge $90e$). At the instant the alpha particle leaves the nucleus, the centers of the two are 7.4 fm apart and essentially at rest. Treating each particle as a spherical charge distribution, find their speeds when they're a great distance apart.
71. **BIO** The Taser, an ostensibly nonlethal weapon used by police to subdue unruly suspects, shoots two conducting darts into the victim's body. Thin wires connect the darts back to the weapon, and once the darts are embedded the weapon applies a 1200-V potential across them and delivers short pulses of electric charge to the darts. Each pulse carries 100 μC of charge. How much energy does the weapon need to supply to each charge pulse as it moves through the body from one dart to the other?
72. **CH** Using the dipole potential at points far from a dipole (given by Equation 22.6 in Example 22.5), show that the electric field at an arbitrary position can be written

$$\vec{E} = \frac{kp}{r^3}\left[\left(3\cos^2\theta - 1\right)\hat{\imath} + 3\sin\theta\cos\theta\,\hat{\jmath}\right]$$

where the x-axis coincides with the dipole axis.
73. **DATA** Measurements of the potential at points on the axis of a charged disk are given in the two tables below, one for measurements made close to the disk and the other for measurements made far away. In both tables x is the coordinate measured along the disk axis with the origin at the disk center, and the zero of potential is taken at infinity. (a) For each set of data, determine a quantity that, when you plot potential against it, should yield a straight line. Make your plots, establish a best-fit line, and determine its slope. Use your slopes to find (b) the total charge on the disk and (c) the disk radius. (*Hint*: Consult Example 22.7.)

Table 1

x (mm)	2.0	4.0	6.0	8.0	10.0
V (V)	900	876	843	820	797

Table 2

x (cm)	20	30	40	60	100
V (V)	165	118	80	58	30

74. **CH** Find an equation describing the $V = 0$ equipotential in the x–y plane for the situation of Conceptual Example 22.1. That is, find a relation between x and y that holds on this equipotential.
75. **CH** A thin rod of length L carries charge Q distributed uniformly over its length. (a) Show that the potential in the plane that perpendicularly bisects the rod is given by

$$V(r) = \frac{2kQ}{L}\ln\left[\frac{L}{2r} + \sqrt{1 + \frac{L^2}{4r^2}}\right]$$

where r is the perpendicular distance from the rod center and where the zero of potential is taken at infinity. (b) Show that this expression reduces to an expected result when $r \gg L$. (*Hint*: See Appendix A for a series expansion of the logarithm.)
76. **CH** For the rod of the preceding problem, (a) find an expression for the magnitude of the electric field in the perpendicular bisecting plane as a function of the distance r from the rod center. (b) Show that your expression reduces to an expected result when $r \gg L$. (c) What's the direction of the field?
77. **CH** A disk of radius a carries nonuniform surface charge density $\sigma = \sigma_0(r/a)$, where σ_0 is a constant. (a) Find the potential at an arbitrary point x on the disk axis, where $x = 0$ is the disk center. (b) Use the result of (a) to find the electric field on the disk axis, and (c) show that the field reduces to an expected form for $x \gg a$.
78. **CH** An open-ended cylinder of radius a and length $2a$ carries charge q spread uniformly over its surface. Find the potential at the center of the cylinder. (*Hint*: Treat the cylinder as a stack of charged rings, and integrate.)
79. **CH** A line charge extends along the x-axis from $-L/2$ to $L/2$. Its line charge density is $\lambda = \lambda_0(x/L)^2$, where λ_0 is a constant. Find an expression for the potential on the x-axis for $x > L/2$. Check that your expression reduces to an expected result for $x \gg L$.
80. **CH** Repeat Problem 79 for the charge distribution $\lambda = \lambda_0 x/L$. (*Hint*: What does this charge distribution resemble at large distances?)

81. COMP You're sizing a new electric transmission line, and you can save money with thinner wire. The potential difference between the line and the ground, 60 m below, is 115 kV. The field at the wire surface cannot exceed 25% of the 3-MV/m breakdown field in air. Neglecting charges in the ground itself, what minimum wire diameter do you specify? (*Hint*: You'll have to do a numerical calculation.)

Passage Problems

BIO Standard electrocardiography measures time-dependent potential differences between multiple points on the body, giving cardiologists multiple perspectives on the heart's electrical activity. In contrast, Fig. 22.26 is a "snapshot" showing a more detailed picture at an instant of time. The lines are equipotentials on the surface of a human torso, associated with the heart's electrical activity. Relative to the line marked $V = 0$, the potential is negative to the upper left (black) and positive to the lower right (color).

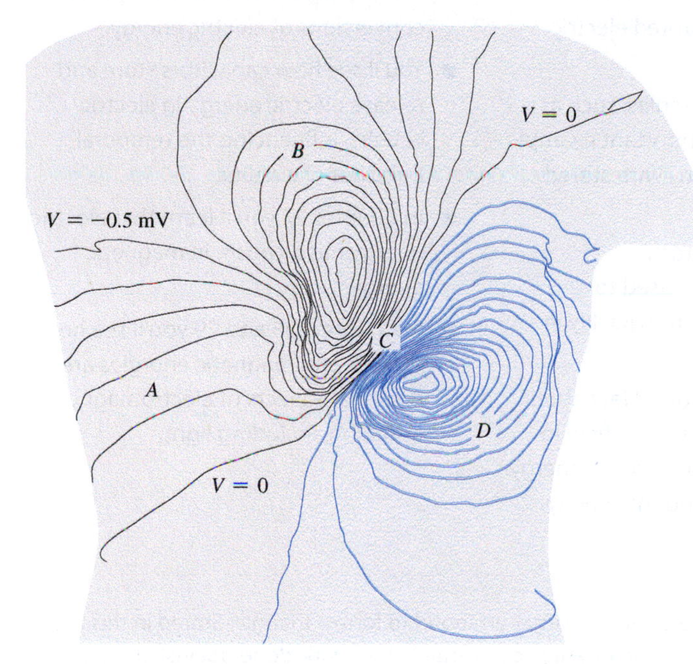

FIGURE 22.26 Equipotentials on a human torso (Passage Problems 82–85)

82. From the equipotentials, you can infer that the heart's electrical structure resembles that of a
    a. uniform charged sheet.
    b. dipole.
    c. point charge.
    d. uniformly charged sphere.
83. The electric field in the vicinity of the heart points approximately
    a. from upper left to lower right.
    b. from lower left to upper right.
    c. from upper right to lower left.
    d. from lower right to upper left.
84. The electric field is strongest in the region marked
    a. *A*.
    b. *B*.
    c. *C*.
    d. *D*.
85. The electric field in region *A* is approximately
    a. 20 μN/C.
    b. 2 mN/C.
    c. 20 mN/C.
    d. 2 kN/C.

Answers to Chapter Questions

Answer to Chapter Opening Question

138,000 volts is a measure of electric potential difference—the energy per unit charge involved in moving electric charge between two points. Luckily, the parasailer is in contact with only one wire, so he doesn't experience that lethal potential difference.

Answers to GOT IT? Questions

22.1 (1) doubles; (2) doubles; (3) becomes zero; (4) reverses sign
22.2 (1) 10 eV; (2) 20 eV; (3) 10 eV
22.3 (c) > (a) > (b)
22.4 (a) because the field is stronger
22.5 They're all equal to zero, because the potential anywhere on the perpendicular bisector of a dipole is zero, and they're all the same, because potential difference is path independent.
22.6 (a), because the equipotentials are closer nearer the center, indicating a stronger field. In (b) the field actually gets stronger farther from the center.

Electrostatic Energy and Capacitors

What You Know

- You understand electric potential as work per charge involved in moving charge between two points.
- You can calculate potential differences in simple electric field configurations.
- You can find the electric field given potential as a function of position.

What You're Learning

- You'll see that it takes energy to assemble distributions of electric charge, resulting in stored electric energy.
- You'll learn how molecules, such as fuels, represent an important example of charge distributions with stored electric energy.
- You'll explore capacitors, technological devices used to store and release energy quickly in electronic circuits.
- You'll discover a profound fact about electric fields: Every electric field in the universe represents stored energy. You'll learn how to find that energy.

How You'll Use It

- In Chapter 25 you'll learn about electric circuits and how they involve conversions of electric energy.
- You'll see how capacitors store and release electric energy in electric circuits, influencing the temporal behavior of circuits.
- In Chapter 27 you'll learn that electric energy has a cousin in magnetic energy.
- In Chapters 28 and 29 you'll see how electric and magnetic energies are essential aspects of electromagnetic radiation, including light.

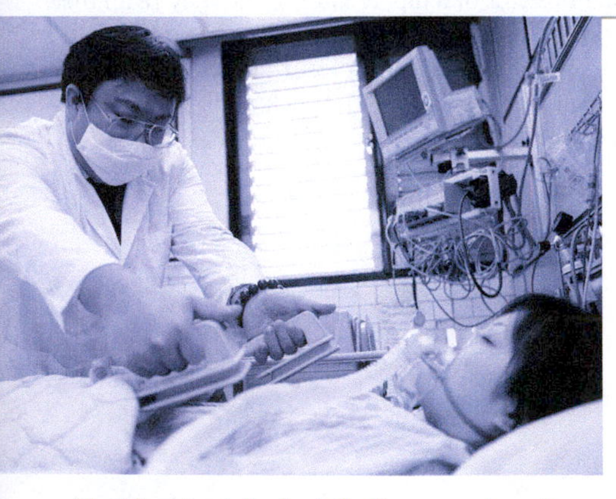

The lifesaving jolt of a defibrillator requires a large amount of energy delivered in a short time. Where does that energy come from?

Figure 23.1 shows three positive charges arranged to form a triangle. Stored in this charge distribution is **electrostatic energy** representing the work done against the repulsive electric forces as the charges were brought into proximity. Although this example may seem trivial, its implications are not. Energy storage in configurations of electric charge is a vital aspect of natural and technological systems. The energy of chemical reactions—including metabolizing food and burning fuels—is ultimately electric energy released in the rearrangement of molecular charge distributions. Energy storage using charged conductors is essential in technologies ranging from computer memories to cameras to high-powered lasers.

23.1 Electrostatic Energy

Let's find the energy stored in the configuration of Fig. 23.1, assuming we start with widely separated charges and bring them in sequentially. It takes no work to bring in charge q_1, since there's initially no electric field. But with q_1 in place, bringing in q_2 means doing work against q_1's electric field. In Chapter 22 we found that the potential of a point charge q is $V = kq/r$. So the potential V_1, which is due to q_1 and is evaluated at the eventual location of q_2, is kq_1/a, where a is the side of the triangle. That potential is the energy per unit charge; given q_2's charge, the work needed to bring in q_2 is $W_2 = q_2V_1 = kq_1q_2/a$. Then we bring in q_3, which experiences the electric fields of both q_1 and q_2. This requires work done against both fields; following our reasoning for q_2, that work is $W_3 = kq_1q_3/a + kq_2q_3/a$. The denominator is the same in both terms because the charges form an equilateral triangle. So the total work done to assemble this charge distribution is

$$W_2 + W_3 = \frac{kq_1q_2}{a} + \frac{kq_1q_3}{a} + \frac{kq_2q_3}{a}.$$

Because the electric field is conservative, this work becomes the stored electrostatic energy, U.

Although we considered the three charges in Fig. 23.1 to be positive, our expression for work holds no matter what the signs. That means electrostatic energy can be positive or negative, depending on the sign of the work done in assembling a charge distribution. If it's negative, then it takes work to separate the charges. Although we considered assembling our charges in the order 1, 2, 3, the expression for work would have been the same no matter what the order—showing that electrostatic energy is a property of a charge distribution, independent of how it's assembled. Figure 23.1 is a simple metaphor for a molecule. Water, for example, consists of a negatively charged oxygen atom and two positively charged hydrogen atoms. The electrostatic energy is negative and represents the energy it would take to dissociate the molecule; equivalently, it's the energy released when the water forms from individual atoms.

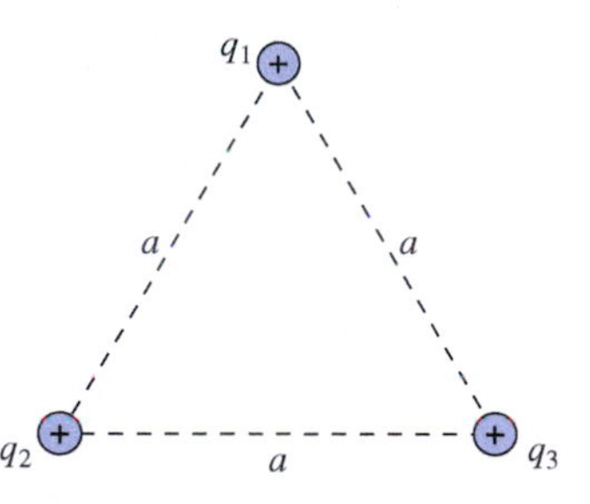

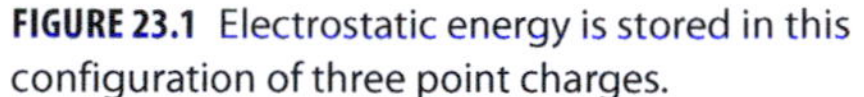

FIGURE 23.1 Electrostatic energy is stored in this configuration of three point charges.

GOT IT? 23.1 Three positive charges and one negative charge, all with the same charge magnitude Q, are assembled into a square, as shown. Is the stored electrostatic energy (a) positive, (b) negative, or (c) zero?

23.2 Capacitors

In technological applications, we often store energy in **capacitors**—pairs of electrical conductors that carry equal but opposite charges. Although capacitors come in many configurations, it's easiest to analyze the **parallel-plate capacitor** consisting of two closely spaced conducting plates (Fig. 23.2*a*). Understanding this device not only is technologically valuable, but will also give us deep insights into the electric field and electrostatic energy.

Initially both capacitor plates are electrically neutral. We charge the capacitor by transferring charge between the plates, building up positive charge on one plate and equal negative charge on the other. In practice, we accomplish this by connecting the capacitor to a battery, but here it's easier to imagine grabbing charge from one plate and physically moving it to the other. Charge on the plates produces an electric field between them, as shown in Fig. 23.2*b*. With closely spaced plates that field is essentially uniform in the region between the plates, except right near the edges. Outside, the field is so small as to be negligible. So we can approximate the parallel-plate capacitor as having a uniform field confined entirely to the region between its plates.

In Chapter 21 we showed that the electric field at the surface of a conductor is $E = \sigma/\epsilon_0$, with σ the charge per unit area. Here we've got charge spread uniformly over the capacitor plates, so if there's charge Q on a plate, then $\sigma = Q/A$, and the uniform field between the plates is $E = Q/\epsilon_0 A$. (If you think this should be doubled because there are two plates, reread the discussion around Figs. 21.28 and 21.29 to see why not.) In this uniform field, the potential difference between the plates is the product of the field and the plate separation: $V = Ed = Qd/\epsilon_0 A$.

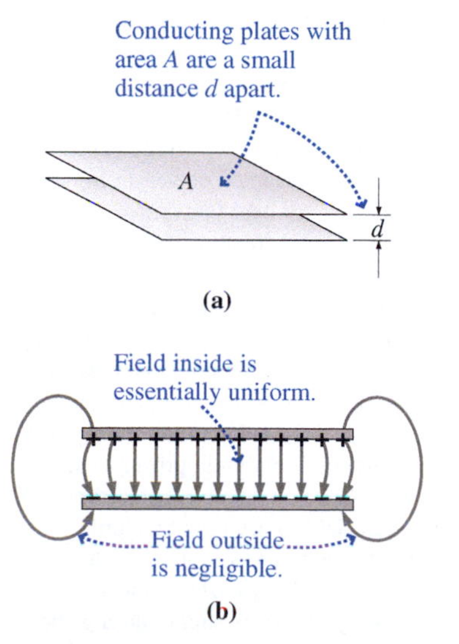

FIGURE 23.2 (a) A parallel-plate capacitor consists of closely spaced conducting plates with area A and spacing d. (b) Edge-on view, showing the electric field.

Capacitance

We can rewrite our expression for the potential difference between the plates of the capacitor in the form $Q = (\epsilon_0 A/d)V$. We added the parentheses to emphasize two things: First, charge is linearly proportional to potential difference and, second, the proportionality factor depends only on the constant ϵ_0 and on the geometry—here the plate area and spacing—of the two charged conductors. This factor gives the ratio of charge to potential difference, which defines the **capacitance** of a configuration of two conductors:

$$C = \frac{Q}{V} \quad \text{(capacitance)} \qquad (23.1)$$

Capacitance depends on the physical arrangement of the conductors, and it's a constant for a given capacitor. Our expression $Q = (\epsilon_0 A/d)V$ shows that the capacitance of a parallel-plate capacitor is

$$C = \frac{\epsilon_0 A}{d} \quad \text{(parallel-plate capacitor)} \qquad (23.2)$$

Problems 60 and 61 explore capacitance for other configurations.

Equation 23.1 shows that the units of capacitance are coulombs/volt. This unit has its own name, the **farad** (F), in honor of the 19th-century scientist Michael Faraday. One farad is a large capacitance; practical capacitors are often measured in μF (10^{-6} F) or pF (10^{-12} F). Incidentally, Equation 23.2 shows that the units of ϵ_0 may be expressed as F/m.

GOT IT? 23.2 If I give you a 5-gallon bucket, you know how much water it can hold. If I give you a 5-μF capacitor, do you know how much charge it can hold? Explain.

APPLICATION Uninterruptable Power

Computers, web servers, data centers, medical electronics, telecommunications, and a host of other vital electronic technologies are subject to failure or severe disruption in the event of even the briefest power outages. Traditionally, backup batteries have been used to provide sensitive equipment with uninterrupted power during outages. But batteries are heavy, expensive, relatively short-lived, and difficult to maintain. They can't easily deliver large amounts of power, and it takes time to switch to backup batteries. Increasingly, responsibility for maintaining uninterrupted power during short outages falls to ultracapacitors—large-capacitance capacitors capable of storing substantial energy and releasing it quickly to maintain power to sensitive equipment. Capacitor-based uninterruptable power supplies (UPSs) are now available with outputs ranging from a few watts to kilowatts and even megawatts. They ensure almost instantaneous continuity of energy supply during short outages, and for longer outages they provide a "bridge" to longer-running backup sources like fuel cells or generators.

Energy Storage in Capacitors

Imagine moving a small charge dQ from the negative to the positive plate of a capacitor when there's a potential difference V between the plates. Since potential difference is work per unit charge, this takes work $dW = V\,dQ$. The additional charge increases the electric field in the capacitor, resulting in an increase dV in the potential difference. Equation 23.1 shows that the increases dQ and dV are related by $dQ = C\,dV$. So the work involved in moving the charge dQ between the plates becomes $dW = V\,dQ = CV\,dV$.

If we start with the capacitor uncharged and then begin transferring charge between the plates, we'll need to do increasing amounts of work because the electric field and potential difference increase continuously with the charge we've already transferred. The total work involved will be the sum of all the dW values. Here the potential difference increases continuously, so that sum becomes an integral:

$$W = \int dW = \int_0^V CV\,dV = \tfrac{1}{2}CV^2$$

where the last step follows because the integral has the familiar form $\int x\,dx = \frac{1}{2}x^2$. The work we do in charging the capacitor is stored as potential energy U, so

$$U = \tfrac{1}{2}CV^2 \quad \text{(energy in a capacitor)} \qquad (23.3)$$

We can measure potential difference V directly, with a voltmeter, so it's more useful to express the energy in terms of voltage rather than charge.

✓**TIP** Charged but Neutral

A "charged capacitor" means a capacitor with one plate positive and the other negative; overall, the capacitor remains neutral (Fig. 23.3). The charge Q refers to the *magnitude* of the charge on either plate—*not* to the capacitor's net charge, which is zero.

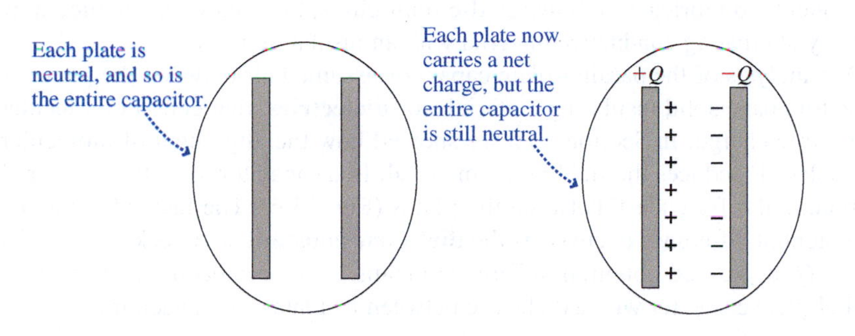

FIGURE 23.3 The net charge on the entire capacitor is zero, whether it's uncharged (left) or charged (right).

EXAMPLE 23.1 Capacitance, Charge, and Energy: A Parallel-Plate Capacitor

A capacitor consists of two circular metal plates of radius $R = 12$ cm, separated by $d = 5.0$ mm. (a) Find its capacitance. Find (b) the charge on the plates and (c) the stored energy when the capacitor is connected to a 12-V battery.

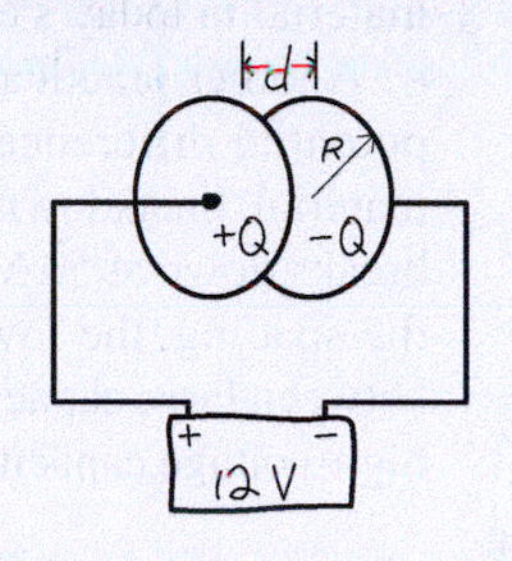

FIGURE 23.4 Sketch for Example 23.1.

INTERPRET Because the plates' area is much larger than their separation, we can treat the field between them as uniform. So we identify the configuration as a parallel-plate capacitor.

DEVELOP We've sketched the capacitor in Fig. 23.4. Equation 23.2, $C = \epsilon_0 A/d$, determines the capacitance for part (a) from the separation distance and plate area ($A = \pi R^2$). For parts (b) and (c) the 12-V battery maintains a 12-V potential difference across the capacitor. Knowing that voltage and the capacitance, we can find the capacitor's charge from Equation 23.1, $C = Q/V$, and the stored energy from Equation 23.3, $U = \frac{1}{2}CV^2$.

EVALUATE We first solve part (a) for the capacitance:

$$C = \frac{\epsilon_0 A}{d} = \frac{\epsilon_0 \pi R^2}{d} = 80 \text{ pF}$$

For part (b) the definition of capacitance then gives

$$Q = CV = (80 \text{ pF})(12 \text{ V}) = 960 \text{ pC}$$

or just under 1 nC. Then (c) the stored energy is

$$U = \tfrac{1}{2}CV^2 = \tfrac{1}{2}(80 \text{ pF})(12 \text{ V})^2 = 5760 \text{ pJ}$$

or about 5.8 nJ.

ASSESS Make sense? At 80 pF, this is a pretty small capacitor, so no wonder the charge and energy are measured in nano-units (nC and nJ). ■

23.3 Using Capacitors

Capacitors are essential in modern technology. They range from the billions of 25-fF (10^{-15} F) capacitors that store individual bits of information in your computer's memory, to millifarad-range capacitors that smooth 60-Hz AC power to provide steady current to your stereo, to so-called *ultracapacitors* measuring hundreds of farads that store electric energy for short bursts of power in systems ranging from power tools to buses to hybrid cars and subway trains. Figure 23.5 shows some typical capacitors used in electronic equipment.

Practical Capacitors

Equation 23.2 shows that the way to achieve a large capacitance is with large plate area and small spacing. That's true in general, whether or not a capacitor has parallel-plate

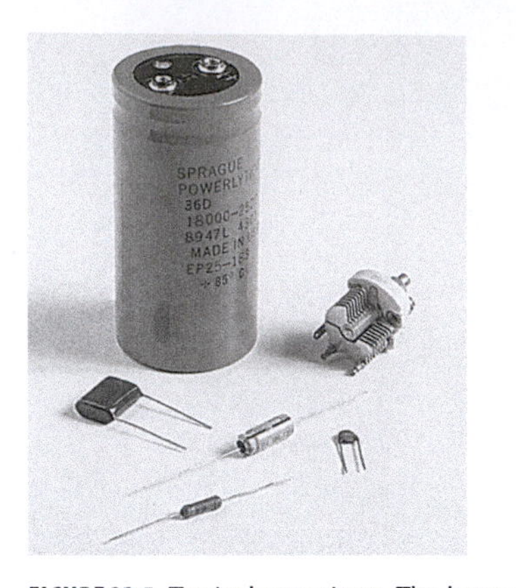

FIGURE 23.5 Typical capacitors. The large unit is an 18-mF electrolytic capacitor. At top right is an air-insulated variable capacitor in which one set of plates rotates to change the capacitance. The smaller capacitors range from 43 pF to 10 μF.

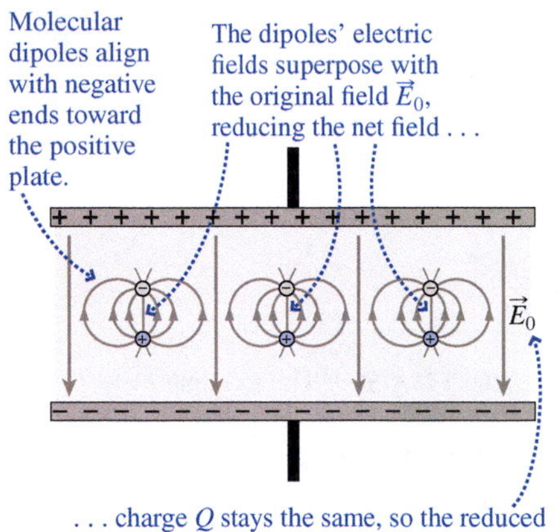

FIGURE 23.6 A capacitor with a dielectric.

geometry. Inexpensive capacitors are often made from two long strips of aluminum foil separated by thin plastic insulation. This foil "sandwich" is rolled into a compact cylinder, wires are attached, and the whole thing is dipped in a protective coating. Very large capacitances are achieved with electrolytic capacitors, in which a thin insulating layer develops chemically under the influence of the applied voltage. Capacitors are among the hardest components to fabricate on integrated-circuit chips, but small-capacitance units can be made by alternating conductive material with an insulating layer.

Our analysis of the parallel-plate capacitor assumed air between the plates. But most capacitors have solid insulating materials, or **dielectrics**, that contain molecular dipoles but no free charge. In Section 20.5 we showed how the alignment of molecular dipoles in a dielectric reduces the field in the material. In a capacitor, the effect is to reduce also the potential difference V between the plates (Fig. 23.6). The factor by which the field and potential difference decrease is the **dielectric constant**, κ (Greek kappa). For a given charge Q, decreased potential difference means a larger capacitance $C = Q/V$. Thus a parallel-plate capacitor with a dielectric between its plates has capacitance

$$C = \kappa \frac{\epsilon_0 A}{d} \quad \text{(parallel-plate capacitor with dielectric)} \qquad (23.4)$$

Most materials have dielectric constants between about 2 and 10; see Table 23.1. Some tantalum compounds have much higher values of κ, making this rare element a crucial material in today's electronic age.

Another practical consideration is a capacitor's **working voltage**, the maximum safe potential difference, beyond which there's a risk of dielectric breakdown. For a given material, breakdown occurs at a fixed electric field; for air it's 3 MV/m, while polyethylene breaks down at 50 MV/m. In a parallel-plate capacitor the field is $E = V/d$, so the smaller the spacing, the lower the allowed voltage before breakdown. Thus there's a trade-off between large capacitance (small d) and high working voltage (large d). Large-capacitance, high-voltage capacitors are expensive!

Table 23.1 Properties of Some Common Dielectrics

Dielectric Material	Dielectric Constant	Breakdown Field (MV/m)
Air	1.0006	3
Aluminum oxide	8.4	670
Glass (Pyrex)	5.6	14
Paper	3.5	14
Plexiglas	3.4	40
Polyethylene	2.3	50
Polystyrene	2.6	25
Quartz	3.8	8
Tantalum oxide	26	500
Teflon	2.1	60
Water	80	depends on time and purity

EXAMPLE 23.2 Finding Charge and Energy: Which Capacitor?

A 100-μF capacitor has a working voltage of 20 V, while a 1.0-μF capacitor is rated at 300 V. Which can store more charge? More energy?

INTERPRET This problem involves the charge and energy stored in capacitors, now constrained by the working voltage.

DEVELOP Equation 23.1, in the form $Q = CV$, determines the charge, and Equation 23.3, $U = \frac{1}{2}CV^2$, determines the stored energy. Setting V equal to the working voltage will give the maximum charge and energy.

EVALUATE For the charges on the two capacitors, we get from Equation 23.1,

$$Q_{100\mu\text{F}} = CV = (100\,\mu\text{F})(20\,\text{V}) = 2.0\,\text{mC}$$

and, similarly, $Q_{1\mu F} = 0.30$ mC. The energies follow from Equation 23.3:

$$U_{100\mu F} = \tfrac{1}{2}CV^2 = \tfrac{1}{2}(100\ \mu F)(20\ V)^2 = 20\ mJ$$

and, similarly, $U_{1\mu F} = \frac{1}{2}(1.0\ \mu F)(300\ V)^2 = 45$ mJ. So the 100-μF capacitor stores more charge, but the 1-μF capacitor stores more energy.

ASSESS Make sense? The larger capacitor holds more *charge*, despite its lower working voltage. But the *energy* in depends on V^2, so the smaller capacitor wins out because of its much higher working voltage.

GOT IT? 23.3 You need to replace a capacitor with one that can store more energy. Which will give you greater energy increase: (a) a capacitor with twice the capacitance and the same working voltage as the old one or (b) a capacitor with the same capacitance but twice the working voltage?

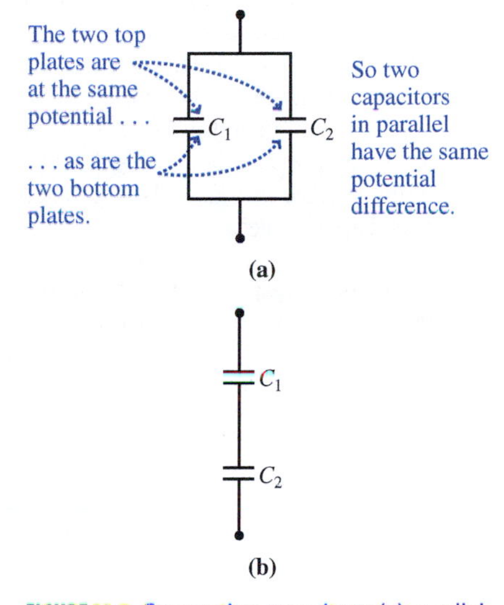

FIGURE 23.7 Connecting capacitors: (a) parallel and (b) series.  is the standard circuit symbol for a capacitor.

Connecting Capacitors: Parallel

Connecting capacitors together lets us achieve capacitance or working voltage that might not be available in a single capacitor. There are two simple ways to connect capacitors and other electronic components: in **parallel** and in **series** (Fig. 23.7).

With capacitors in parallel, a conducting wire connects the top plates of each capacitor and another connects the bottom plates. Therefore, both top plates are at the same potential, and so are both bottom plates. That means **two capacitors in parallel have the same potential difference between their plates**. We'll find that's always true for electric components in parallel. We want the equivalent capacitance of the parallel combination, meaning the ratio of the total charge on both capacitors to their common voltage V. Given the definition $C = Q/V$, we can write $Q_1 = C_1V$ and $Q_2 = C_2V$. So the total charge is $Q = Q_1 + Q_2 = C_1V + C_2V$. The equivalent capacitance is then $C = Q/V$ or $C = C_1 + C_2$. So capacitors in parallel add, a result that generalizes to any number of capacitors:

$$C = C_1 + C_2 + C_3 + \cdots \quad \text{(parallel capacitors)} \tag{23.5}$$

Video Tutor Demo | **Discharge Speed for Series and Parallel Capacitors**

Connecting Capacitors: Series

Figure 23.8 is a closer look at the series combination of Fig. 23.7*b*, showing what happens if we put charge $+Q$ on the top plate of C_1 and charge $-Q$ on the lower plate of C_2. Each of these charged plates pulls the opposite charge to the other plate of the individual capacitors—and that means **two capacitors in series carry the same charge**. But now the voltages can be different; they're given by Equation 23.1 as $V_1 = Q/C_1$ and $V_2 = Q/C_2$, where Q is the common charge. Since the electric fields in the two capacitors point the same way, the voltage across the series combination is $V = V_1 + V_2 = Q/C_1 + Q/C_2$. Dividing through by Q gives V/Q, which is the inverse of the equivalent capacitance Q/V. Thus

$$\frac{1}{C} = \frac{1}{C_1} + \frac{1}{C_2}$$

More generally, capacitors in series add reciprocally:

$$\frac{1}{C} = \frac{1}{C_1} + \frac{1}{C_2} + \frac{1}{C_3} + \cdots \quad \text{(series capacitors)} \tag{23.6a}$$

With two capacitors it's straightforward to invert Equation 23.6a to get

$$C = \frac{C_1C_2}{C_1 + C_2} \tag{23.6b}$$

Either way, the combined capacitance is less than any of the individual capacitances.

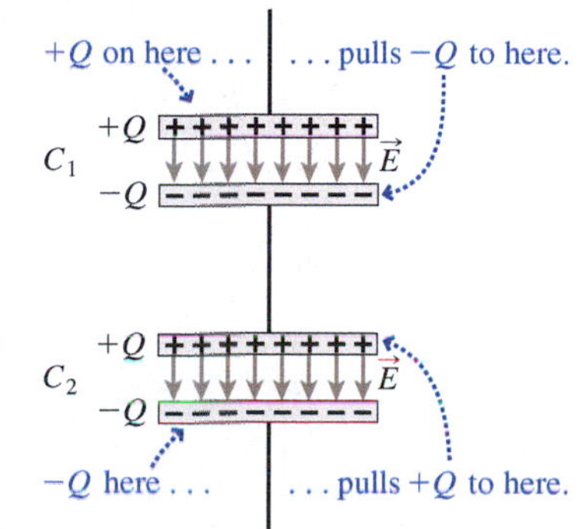

FIGURE 23.8 Capacitors in series carry the same charge.

CONCEPTUAL EXAMPLE 23.1 Parallel and Series Capacitors

Using parallel-plate capacitors, explain why capacitance should increase with capacitors in parallel and decrease with capacitors in series. What happens to the working voltage in each case?

EVALUATE Equation 23.2 shows that capacitance increases with increasing plate area and decreases with increasing plate separation. Figure 23.7*a* shows that two capacitors in parallel have greater plate area, with no change in spacing, so the combined capacitance increases. In contrast, the series combination in Figure 23.7*b* effectively increases the plate separation because it's the sum of the individual separations, so the capacitance goes down.

What about working voltage? In Fig. 23.7*a*, the parallel capacitors have the same voltage, so the working voltage of the combination is that of whichever capacitor has the lower working voltage. But in Fig. 23.7*b*, each series capacitor gets less than the total voltage, so the working voltage increases. How much it increases depends on the ratio of the capacitances.

ASSESS Series and parallel combinations let us build arbitrary capacitances and working voltages from standard capacitors available commercially. You might wonder about the wire connecting the series capacitors in Fig. 23.7*b*: Does it also affect the separation? No, because it doesn't separate charge, which is free to move along the conducting wire.

MAKING THE CONNECTION You've got two 10-μF capacitors rated at 15 V. What are the capacitances and working voltages of their parallel and series combinations?

EVALUATE Applying Equation 23.5 to equal capacitors shows that the capacitance doubles with two capacitors in parallel. So the parallel combination has $C = 20\,\mu\text{F}$, and its working voltage is still 15 V because each capacitor gets the full voltage. Apply Equation 23.6b to equal capacitances and you'll see that the series capacitance is half that of either capacitor, in this case 5 μF. Since the individual capacitances are equal, each must get half the applied voltage, giving the combination a working voltage of 30 V.

GOT IT? 23.4 You have two identical capacitors with capacitance C. How would you connect them to get equivalent capacitances (1) $2C$ and (2) $\frac{1}{2}C$? (3) Which combination would have the higher working voltage?

EXAMPLE 23.3 Equivalent Capacitance: Connecting Capacitors

Find the equivalent capacitance of the combination shown in Fig. 23.9*a*. If the maximum voltage to be applied between points A and B is 100 V, what should be the working voltage of C_1?

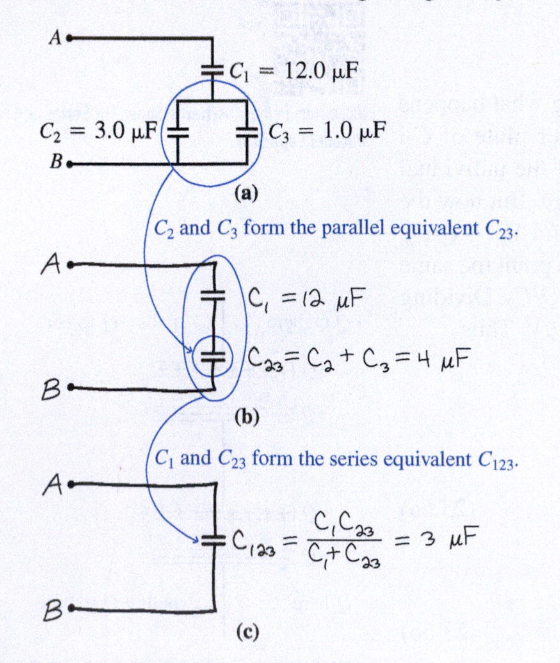

FIGURE 23.9 Finding the equivalent capacitance.

INTERPRET This problem is about an electric circuit—in this case, an assemblage of three capacitors.

DEVELOP To handle such circuit problems, we find combinations of series and parallel components, and then simplify the circuit by treating each combination as a single component. Here all components are capacitors, and each time we compute an equivalent capacitance for two capacitors, we'll redraw the circuit with the new equivalent capacitance. We begin by noting that C_2 and C_3 are in parallel, so the equivalent capacitance is given by Equation 23.5: $C_{23} = C_2 + C_3 = 4.0\,\mu\text{F}$. In Fig. 23.9*b* we've redrawn the original circuit showing this combination of the two individual capacitors. Next we see that C_1 is in series with C_{23}, so their equivalent capacitance follows from Equation 23.6b:

$$C_{123} = \frac{C_1 C_{23}}{C_1 + C_{23}} = \frac{(12\,\mu\text{F})(4.0\,\mu\text{F})}{12\,\mu\text{F} + 4.0\,\mu\text{F}} = 3.0\,\mu\text{F}$$

We've redrawn the circuit again with this equivalent capacitance (Fig. 23.9*c*).

Now we want the working voltage of C_1. Our plan is to go backward from the simple circuit of Fig. 23.9*c* until we have enough information to find the voltage on C_1. With the voltage V_{AB} across A and B known, we can calculate the charge on C_{123} using Equation 23.1: $Q_{123} = C_{123}V_{AB}$. But C_{123} is the series combination of C_1 and C_{23}, and we know that series capacitors carry the same charge—and that's the charge of their equivalent capacitance. So $Q_1 = Q_{123}$, and we can apply Equation 23.1 again, this time to C_1 alone, to get $V_1 = Q_1/C_1$.

EVALUATE With $V_{AB} = 100$ V across the combination C_{123}, the corresponding charge is $Q_{123} = C_{123}V_{AB} = (3.0\ \mu\text{F})(100\ \text{V}) = 300\ \mu\text{C}$. Because $Q_1 = Q_{123}$, the charge on C_1 is also 300 μC. We then substitute this into $V_1 = Q_1/C_1$ to get $V_1 = (300\ \mu\text{C})/(12.0\ \mu\text{F}) = 25$ V, the minimum working voltage for C_1.

ASSESS Make sense? Since C_1 is in series with C_{23}, it doesn't "feel" the full 100 V applied across *AB*, so its working voltage can be lower. And because its capacitance is *larger*, its share of the voltage is *smaller*, thanks to the relation $V = Q/C$ and the fact that series capacitors carry the *same charge*.

✓TIP Series and Parallel

Parallel components have their ends connected directly together; series components are connected in such a way that if you move through one component, the only place you can go is into the next. In Fig. 23.9*a*, C_2 and C_3 are definitely in parallel. But C_1 isn't in series with either of the other single capacitors because after C_1, the circuit splits and you could go into either C_2 or C_3. Equations 23.5 and 23.6 apply *only* to true parallel and series combinations. C_1 *is* in series with the combination C_{23}, so we could apply Equation 23.6b in analyzing Fig. 23.9*b*.

APPLICATION Bursts of Power

As San Francisco's BART trains decelerate, their kinetic energy is stored as electric energy in an ultracapacitor. The stored energy is then used to accelerate the train. This system saves BART some 320 megawatt-hours of energy each year.

Capacitors are excellent devices for short-term storage of electric energy because they can deliver their stored energy very quickly—much faster than a battery that might contain a lot more total energy.

When you use a flash camera, you have to wait a few seconds before the flash is ready to fire again. That's because the flash requires power—energy per time—far greater than the camera's battery could supply. So the battery gradually charges a capacitor, whose energy is then dumped abruptly to power the brief flash. It takes a while to recharge the capacitor before it's ready again. Much the same thing happens in a defibrillator, which delivers several hundred joules to restore a heart's normal beating. Again, the energy is stored in capacitors, which discharge in milliseconds. On a much larger scale, whole rooms full of capacitors store the energy that drives nanosecond laser pulses pouring millions of joules into tiny targets in experiments aimed at making nuclear fusion a viable energy source. And increasingly, ultracapacitors supply extra energy for bursts of power in machinery from amusement park rides to mass-transit trains to hybrid cars.

23.4 Energy in the Electric Field

What's the difference between a charged and an uncharged capacitor? Not the total charge, which is zero, but the arrangement of charge. And with the charge arrangement comes stored energy. Where, exactly, is this energy? We can ask the same question for the triangular charge distribution we assembled in Fig. 23.1. The individual charges didn't change, but their arrangement did. With the new arrangement came energy, but where is that energy?

What's changed in both cases is the electric field. There's no electric field in the uncharged capacitor, but once charged, there's a field between the plates. The triangular distribution started with three isolated point-charge fields and ended with a more complex field. So where's the stored energy? It's in the electric field. In fact, *every* electric field represents stored energy. Rearrange the charges to their original state—by discharging the capacitor or letting the three point charges fly apart—and you get back that energy. Because electric forces govern much of the behavior of everyday matter, many seemingly different forms of energy are actually electric. Burn gasoline or metabolize food, and you're rearranging the charge distributions we call molecules into new configurations whose electric fields contain less energy.

Since electric fields can vary with position, we specify the **energy density**, or energy stored per unit volume. For a capacitor, we can use Equation 23.1 in the form $V = Q/C$ to write the stored energy $U = \frac{1}{2}CV^2$ as $U = Q^2/2C$. For a parallel-plate capacitor,

Equation 23.2 gives $C = \epsilon_0 A/d$, so the stored energy becomes $U = Q^2 d/2\epsilon_0 A$. This energy is associated with the uniform electric field inside the capacitor, where it occupies a volume Ad. So the energy density is $U/Ad = Q^2/2\epsilon_0 A^2$. We can rewrite this in terms of the field, which we found to be $E = Q/\epsilon_0 A$. Then $Q = \epsilon_0 AE$, which we put in our expression $Q^2/2\epsilon_0 A^2$ for the energy density to get

$$u_E = \tfrac{1}{2}\epsilon_0 E^2 \quad \text{(electric energy density)} \qquad (23.7)$$

Although we derived Equation 23.7 for the uniform field of a parallel-plate capacitor, it is in fact universal. Anywhere there's an electric field, there's also stored energy with density, in J/m^3, given by $\frac{1}{2}\epsilon_0 E^2$. That's the deep significance of Equation 23.7: *Every electric field represents stored energy.* The energy that drives much of the physical universe, from everyday events here on Earth to happenings in distant galaxies, results from the release of energy stored in electric fields.

EXAMPLE 23.4 Electric Energy: A Thunderstorm

Typical electric fields in thunderstorms average around 10^5 V/m. Consider a cylindrical thundercloud with height 10 km and diameter 20 km, and assume a uniform electric field of 1×10^5 V/m. Find the electric energy contained in this cloud.

INTERPRET This problem is about stored electric energy.

DEVELOP Since the field and hence the energy density are uniform, our plan is to find the energy density and then multiply it by the cloud's cylindrical volume to calculate the total electric energy. We'll use Equation 23.7, $u_E = \frac{1}{2}\epsilon_0 E^2$, for the energy density.

EVALUATE The energy density is

$$u_E = \tfrac{1}{2}\epsilon_0 E^2 = \tfrac{1}{2}\epsilon_0 (1\times10^5\ \text{V/m})^2 = 4.4\times10^{-2}\ \text{J/m}^3$$

The cylindrical cloud has volume

$$V = \pi r^2 h = \pi(10\ \text{km})(10\ \text{km})^2 = 3.1\times10^{12}\ \text{m}^3$$

Multiplying energy density by volume gives the total stored energy:

$$U = u_E V = (4.4\times10^{-2}\ \text{J/m}^3)(3.1\times10^{12}\ \text{m}^3) = 140\ \text{GJ}$$

ASSESS Make sense? A gallon of gasoline contains about 0.1 GJ (see Appendix C), so the thundercloud stores the energy equivalent of about 1400 gallons of gasoline. That's not a whole lot for such a vast volume, showing that the energy density of macroscopic electric fields can't compare with the electric energy density locked into the molecular structure of a fuel. You'll never see cars running on the energy stored in atmospheric electric fields! ■

When the electric field is uniform, as in our thundercloud, the total energy is the product of energy density and volume. But when the field changes with position, we need calculus. Consider a small volume dV, so small that the electric field is essentially uniform over this volume. The stored energy is then $dU = u_E\,dV = \frac{1}{2}\epsilon_0 E^2\,dV$. The total energy in the field is the sum—here the integral—of all the dU values:

$$U = \tfrac{1}{2}\epsilon_0 \int E^2\,dV \qquad (23.8)$$

Because Equation 23.8 gives the energy stored in an electric field, it also represents the work done in assembling the charge distribution resulting in that field. The next example illustrates this point.

EXAMPLE 23.5 Work and Energy: A Shrinking Sphere

A sphere of radius R_1 carries charge Q distributed uniformly over its surface. How much work does it take to compress the sphere to a smaller radius R_2?

INTERPRET This problem asks for the work done in rearranging a charge distribution, which we know is equal to the change in stored electric energy. Here we start with a charged sphere already assembled, and rearrange the charge by shrinking the sphere to a smaller radius.

DEVELOP We have spherical symmetry, so the field and thus the stored energy outside the original radius R_1 don't change. Therefore, we need to find the energy stored in the new field created when the sphere shrinks. Figure 23.10 is our sketch of the situation before and

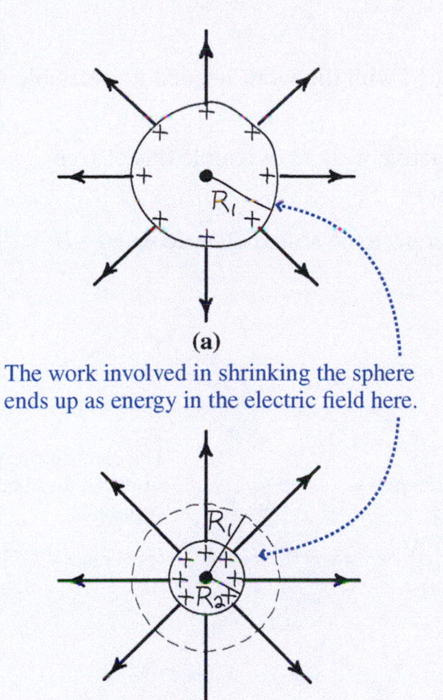

FIGURE 23.10 (a) A charged sphere and its electric field. (b) Shrinking the sphere creates field and energy in the region $R_2 < r < R_1$.

after the sphere shrinks. Here the field varies with position, so Equation 23.8,

$$U = \tfrac{1}{2}\epsilon_0 \int E^2\, dV$$

gives the stored energy. Our plan is to evaluate the field in the region $R_2 < r < R_1$ and use the result in Equation 23.8. Given the spherical

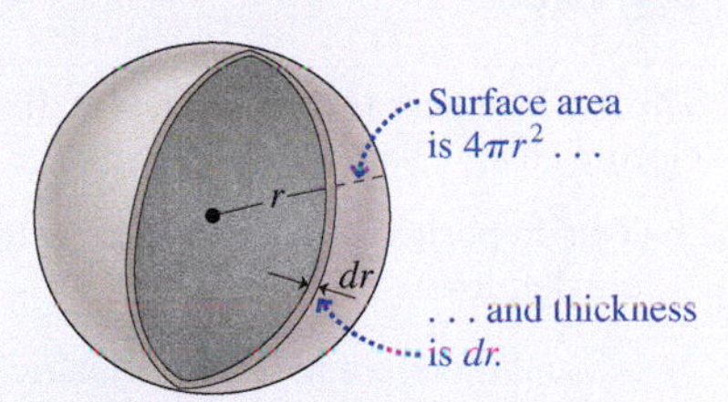

FIGURE 23.11 A thin spherical shell has volume $dV = 4\pi r^2\, dr$.

symmetry, the new field is a point-charge field: $E = kQ/r^2$. To use Equation 23.8 we need an appropriate volume element dV. With spherical symmetry, Fig. 23.11 shows that we can use a thin spherical shell of volume $dV = 4\pi r^2\, dr$. Then Equation 23.8 becomes

$$U = \tfrac{1}{2}\epsilon_0 \int E^2\, dV = \tfrac{1}{2}\epsilon_0 \int_{R_2}^{R_1} \left(\frac{kQ}{r^2}\right)^2 4\pi r^2\, dr = \frac{kQ^2}{2}\int_{R_2}^{R_1} r^{-2}\, dr$$

where we substituted $1/4\pi k$ for ϵ_0.

EVALUATE The integral is $\int r^{-2}\, dr = \dfrac{r^{-1}}{-1} = -\dfrac{1}{r}$, so

$$U = \frac{kQ^2}{2}\left(-\frac{1}{r}\right)\Bigg|_{R_2}^{R_1} = \frac{kQ^2}{2}\left(\frac{1}{R_2} - \frac{1}{R_1}\right)$$

ASSESS Make sense? Here $R_2 < R_1$, so the stored energy is positive and indicates that this much work had to be done to shrink the sphere. That makes sense, because the entire sphere carries charge of the same sign, and shrinking it moves that charge closer together, against the repulsive electric force. Letting R_1 go to infinity gives the work needed to assemble a spherical surface charge distribution. Putting $R_2 = 0$ makes the work and therefore the stored energy infinite—suggesting that the notion of a point charge is an impossible idealization. ■

GOT IT? 23.5 You're at a point P a distance a from a point charge $+q$. You then place a point charge $-q$ a distance a on the opposite side of P as shown. What happens to (1) the electric field strength and (2) the electric energy density at P? (3) Does the total electric energy $U = \int u_E\, dV$ of the entire field increase, decrease, or remain the same?

a — $+q$ — P

$+q$ — a — P — a — $-q$

CHAPTER 23 SUMMARY

Big Idea

The big idea here is that *all* electric fields represent stored energy. This energy is associated with the work needed to assemble a distribution of electric charge, and may be negative or positive.

You do positive work to assemble this charge distribution . . .

. . . and therefore the stored electric energy U is positive.

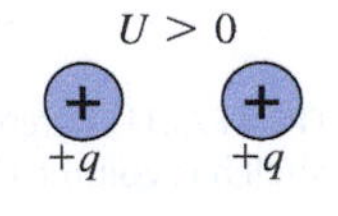

You do negative work to assemble this charge distribution . . .

. . . and therefore the stored electric energy U is negative.

$U < 0$

$+q$ $-q$

Key Concepts and Equations

The **energy density** in an electric field E is $u_E = \frac{1}{2}\epsilon_0 E^2$.

Integrating over volume gives the total electric energy U stored in the field:

$$U = \int u_E \, dV.$$

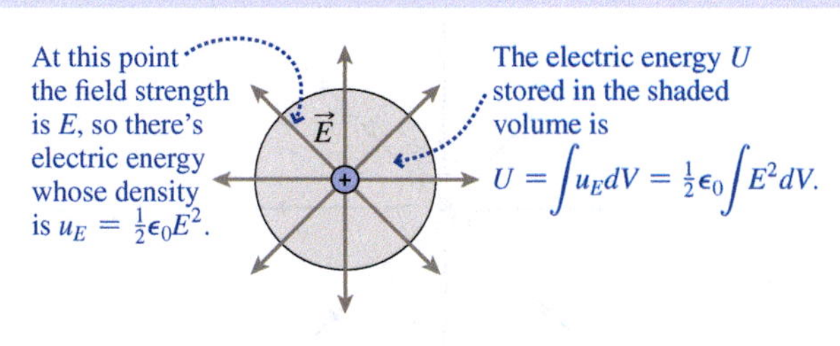

Applications

A **capacitor** is a pair of insulated conductors used to store electric energy. **Capacitance** is the ratio of charge to potential difference:

$$C = Q/V$$

For a parallel-plate capacitor:

$$C = \epsilon_0 A/d$$

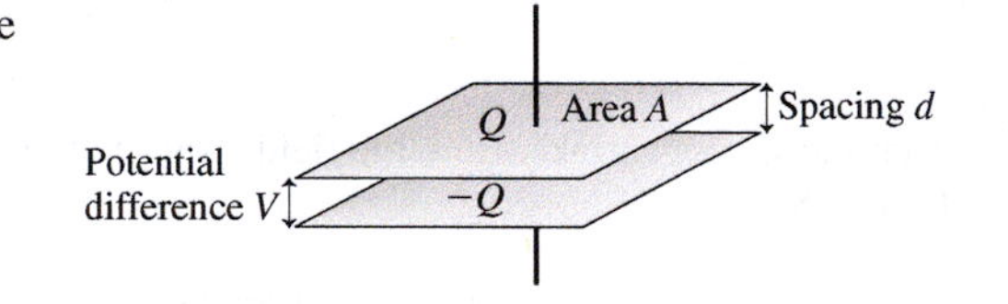

The energy stored in a capacitor with voltage V between its plates is $U = \frac{1}{2}CV^2$.

Capacitors in parallel add: $C = C_1 + C_2$.

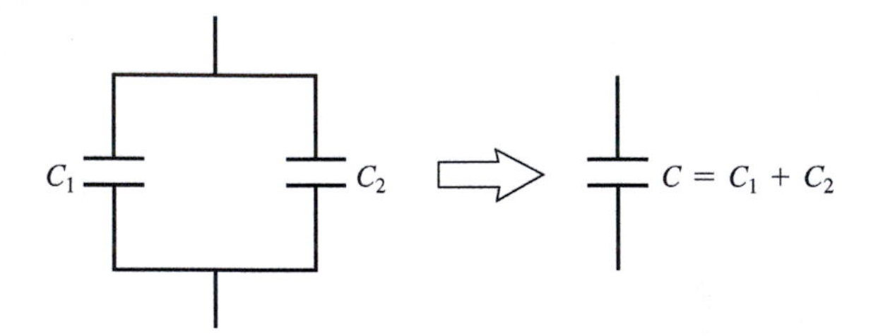

Capacitors in parallel have the same voltage.

Capacitors in series add reciprocally: $\frac{1}{C} = \frac{1}{C_1} + \frac{1}{C_2}$.

C_1

C_2

$$\frac{1}{C} = \frac{1}{C_1} + \frac{1}{C_2}$$

Capacitors in series have the same charge.

Complicated circuits are analyzed by breaking them into parallel and series combinations:

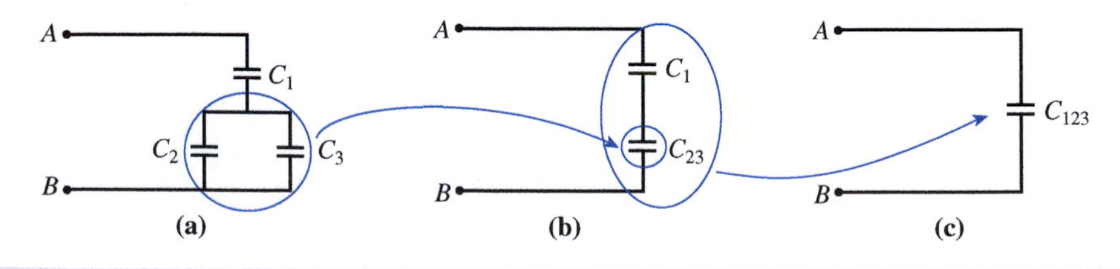

A **dielectric** between capacitor plates increases the capacitance, as determined by the **dielectric constant** κ of the material: $C \rightarrow \kappa C_0$.

MP For homework assigned on MasteringPhysics, go to www.masteringphysics.com

BIO *Biology and/or medicine-related problems* **DATA** *Data problems* **ENV** *Environmental problems* **CH** *Challenge problems* **COMP** *Computer problems*

For Thought and Discussion

1. Two positive point charges are infinitely far apart. Is it possible, using a finite amount of work, to move them until they're a small distance d apart?
2. How does the energy density at a certain distance from a negative point charge compare with the energy density at the same distance from a positive point charge of equal magnitude?
3. A dipole consists of two equal but opposite charges. Is the total energy stored in the dipole's electric field zero? Why or why not?
4. Charge is spread over the surface of a balloon, which is then allowed to expand. What happens to the energy of the electric field?
5. Does the superposition principle hold for electric-field energy densities? That is, if you double the field strength at some point, do you double the energy density as well?
6. A student argues that the total energy associated with the electric field of a charged sphere must be infinite because its field extends throughout an infinite volume. Critique this argument.
7. A capacitor is said to carry a charge Q. What's the net charge on the entire capacitor?
8. Does the capacitance describe the maximum amount of charge a capacitor can hold, in the same way that a bucket's capacity describes the maximum amount of water it can hold? Explain.
9. Is a force needed to hold the plates of a charged capacitor in place? Explain.
10. A solid conducting slab is inserted between the plates of a capacitor, not touching either plate. Does the capacitance increase, decrease, or remain the same?
11. Two capacitors contain equal amounts of energy, yet one has twice the capacitance. How do their voltages compare?
12. A parallel-plate capacitor is connected to a battery that imposes a potential difference V between its plates. If a dielectric slab is inserted between the plates, what happens to (a) the potential difference, (b) the capacitance C, and (c) the capacitor charge Q?

Exercises and Problems

Exercises

Section 23.1 Electrostatic Energy

13. Four 75-μC charges, initially far apart, are brought onto a line where they're spaced at 5.0-cm intervals. How much work does it take to assemble this charge distribution?
14. Three point charges $+q$ and a fourth, $-\frac{1}{2}q$, are assembled to form a square of side a. Find an expression for the electrostatic energy of this charge distribution.
15. Repeat Exercise 14 for the case when the fourth charge is $-q$.
16. If the three particles in Fig. 23.1 have identical charge q and mass m, and if they're released from their positions on the triangle, what speed v will they have when they're far away?
17. A crude model of the water molecule has a negatively charged oxygen atom and two protons, as shown in Fig. 23.12. Calculate the electrostatic energy of this configuration, which is therefore the magnitude of the energy released in forming this molecule. (*Note*: Your answer is an overestimate because electrons are actually "shared" among the three atoms, spending more time near the oxygen.)

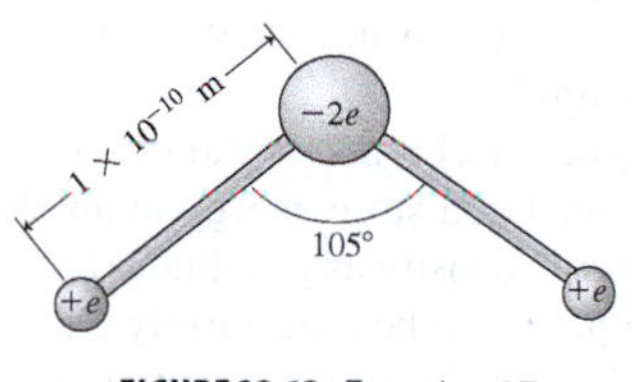

FIGURE 23.12 Exercise 17

Section 23.2 Capacitors

18. A capacitor consists of square conducting plates 25 cm on a side and 5.0 mm apart, carrying charges ±1.1 μC. Find (a) the electric field, (b) the potential difference between the plates, and (c) the stored energy.
19. An uncharged capacitor has parallel plates 5.0 cm on a side, spaced 1.2 mm apart. (a) How much work is required to transfer 7.2 μC from one plate to the other? (b) How much work is required to transfer an additional 7.2 μC?
20. (a) How much charge must be transferred between the initially uncharged plates of the capacitor in Exercise 19 in order to store 15 mJ of energy? (b) What will be the resulting potential difference between the plates?
21. A capacitor's plates hold 1.3 μC when charged to 60 V. What's its capacitance?
22. Show that the units of ϵ_0 may be written as F/m.
23. Find the capacitance of a parallel-plate capacitor with circular plates 20 cm in radius separated by 1.5 mm.
24. A parallel-plate capacitor with 1.1-mm plate spacing has ±2.3 μC on its plates when charged to 150 V. What's the plate area?
25. The power supply in a stereo receiver contains a 2500-μF capacitor charged to 35 V. How much energy does it store?
26. Find the capacitance of a capacitor that stores 350 μJ when the potential difference across its plates is 100 V.

Section 23.3 Using Capacitors

27. You have a 1.0-μF and a 2.0-μF capacitor. What capacitances can you get by connecting them in series or in parallel?
28. Two capacitors are connected in series and the combination is charged to 100 V. If the voltage across each capacitor is 50 V, how do their capacitances compare?
29. (a) Find the equivalent capacitance of the combination shown in Fig. 23.13. Find (b) the charge and (c) the voltage on each capacitor when a 12.0-V battery is connected across the combination.

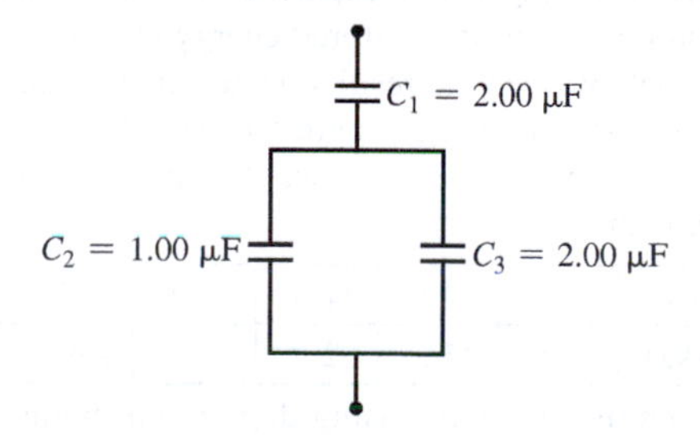

FIGURE 23.13 Exercise 29

30. You're given three capacitors: 1.0 μF, 2.0 μF, and 3.0 μF. Find (a) the maximum, (b) the minimum, and (c) two intermediate capacitances you could achieve using combinations of all three capacitors.

Section 23.4 Energy in the Electric Field

31. The energy density in a uniform electric field is 3.0 J/m^3. What's the field strength?
32. A car battery stores about 4 MJ of energy. If this energy were used to create a uniform 30-kV/m electric field, what volume would it occupy?
33. Air undergoes dielectric breakdown at a field strength of 3 MV/m. Could you store energy in an electric field in air with the same energy density as gasoline? (*Hint*: See Appendix C.)
34. Consider a proton to be a uniformly charged sphere 1 fm in radius. Find the electric energy density at the proton's surface.

Problems

35. A charge Q_0 is at the origin. A second charge, $Q_x = 2Q_0$, is brought from infinity to the point $x = a, y = 0$. Then a third charge Q_y is brought from infinity to $x = 0, y = a$. If it takes twice as much work to bring in Q_y as it did Q_x, what's Q_y in terms of Q_0?
36. A conducting sphere of radius a is surrounded by a concentric spherical shell of radius b. Both are initially uncharged. How much work does it take to transfer charge from one to the other until they carry charges $\pm Q$?
37. Two closely spaced square conducting plates measure 10 cm on a side. The electric-field energy density between them is 4.5 kJ/m^3. What's the charge on the plates?
38. **BIO** The potential difference across a cell membrane is 65 mV. On the outside are 1.5×10^6 singly ionized potassium atoms. Assuming an equal negative charge on the inside, find the membrane's capacitance.
39. Which can store more energy: a 1.0-μF capacitor rated at 250 V or a 470-pF capacitor rated at 3 kV?
40. A 0.01-μF, 300-V capacitor costs 25¢; a 0.1-μF, 100-V capacitor costs 35¢; and a 30-μF, 5-V capacitor costs 88¢. (a) Which can store the most charge? (b) Which can store the most energy? (c) Which is the most cost-effective energy-storage device, measured in J/¢?
41. **BIO** A medical defibrillator stores 950 J in a 100-μF capacitor. (a) What is the voltage across the capacitor? (b) If the capacitor discharges 300 J of its stored energy in 2.5 ms, what's the power delivered during this time?
42. A camera requires 5.0 J of energy for a flash lasting 1.0 ms. (a) What power does the flashtube use *while it's flashing*? (b) If the flashtube operates at 200 V, what size capacitor is needed to supply the flash energy? (c) If the flashtube is fired once every 10 s, what's its *average* power consumption?
43. Engineers testing an ultracapacitor (see Application on page 420) measure the capacitor's stored energy at different voltages. The table below gives the results. Determine a quantity that, when you plot stored energy against it, should give a straight line. Make your plot, establish a best-fit line, and use its slope to determine the capacitance.

Voltage (V)	12.2	20.1	31.8	37.9	45.7	50.2	56.0
Energy (kJ)	9.25	27.2	62.5	94	139	158	203

44. Your company's purchasing department bought lots of cheap 2.0-μF, 50-V capacitors. Your budget is maxed out and they won't let you buy additional capacitors for a circuit you're designing. You need 2.0-μF, 100-V capacitors and 0.5-μF, 200-V capacitors. How will you combine the available capacitors to make these?
45. What's the equivalent capacitance measured between A and B in Fig. 23.14?

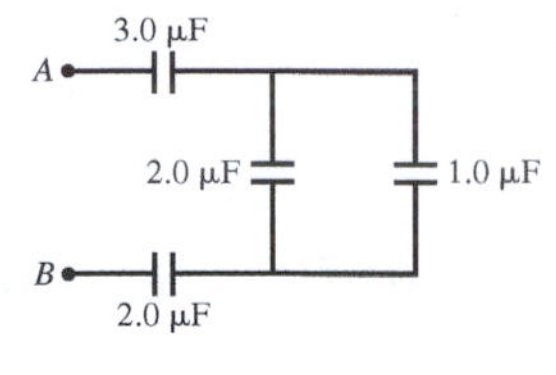

FIGURE 23.14 Problems 45 and 46

46. In Fig. 23.14, find the energy stored in the 1-μF capacitor when a 50-V battery is connected between A and B.
47. Capacitors C_1 and C_2 are in series, with voltage V across the combination. Show that the voltages across the individual capacitors are $V_1 = C_2V/(C_1 + C_2)$ and $V_2 = C_1V/(C_1 + C_2)$.
48. You're evaluating a new hire in your company's engineering department. Together you're working on a circuit where a 0.1-μF, 50-V capacitor is in series with a 0.2-μF, 200-V capacitor. The new engineer claims you can safely put 250 V across the combination. What do you say?
49. A parallel-plate capacitor has plates with area 50 cm^2 separated by 25 μm of polyethylene. Find its (a) capacitance and (b) working voltage.
50. A 470-pF capacitor consists of two 15-cm-radius circular plates, insulated with polystyrene. Find (a) the thickness of the polystyrene and (b) the capacitor's working voltage.
51. **BIO** The first accurate estimate of cell membrane thickness used a capacitive technique, which determined the capacitance per unit area of cell membrane in a macroscopic suspension of cells; the result was about 1 μF/cm^2. Assuming a dielectric constant of about 3 for the membrane, find the membrane's thickness. (*Note:* Your answer is the thickness of the bipolar lipid layer alone, and is lower by a factor of about 3 than values based on X-ray techniques.)
52. Your company is still stuck with those 2-μF capacitors from Problem 44. They turn out to be so cheap that their capacitances are all too low, ranging from 1.7 μF to 1.9 μF. A colleague suggests you put variable "trimmer" capacitors in parallel with the cheap capacitors and adjust the combination to precisely 2.00 μF. The available trimmers have variable capacitance from 25 nF to 350 nF. Will they work?
53. A cubical region 1.0 m on a side is located between $x = 0$ and $x = 1$ m. The region contains an electric field whose magnitude varies with x but is independent of y and z: $E = E_0(x/x_0)$, where $E_0 = 24$ kV/m and $x_0 = 6.0$ m. Find the total energy in the region.
54. **CH** A sphere of radius R contains charge Q spread uniformly throughout its volume. Find an expression for the electrostatic energy contained within the sphere itself. (*Hint*: Consult Example 21.3.)
55. A sphere of radius R carries total charge Q distributed uniformly over its surface. Show that the energy stored in its electric field is $U = kQ^2/2R$.
56. A uranium-235 nucleus has diameter 6.6 fm and contains 92 protons and 143 neutrons. Assuming that charge is distributed uniformly throughout the nucleus, use the results of Problems 54 and 55 to calculate the total electrostatic energy of this configuration.
57. Two widely separated 4.0-mm-diameter water drops each carry 15 nC. Assuming all charge resides on the drops' surfaces, find the change in electrostatic potential energy if they're brought together to form a single spherical drop.
58. A 2.1-mm-diameter wire carries a uniform line charge density $\lambda = 28$ μC/m. Find the energy in a region 1.0 m long within one wire diameter of the wire surface.

59. A typical lightning flash transfers 30 C across a potential difference of 30 MV. Assuming such flashes occur every 5 s in the thunderstorm of Example 23.4, roughly how long would the storm last if its electric energy were not replenished?

60. **CH** A capacitor consists of two long concentric metal cylinders (Fig. 23.15). Find an expression for its capacitance in terms of the dimensions shown.

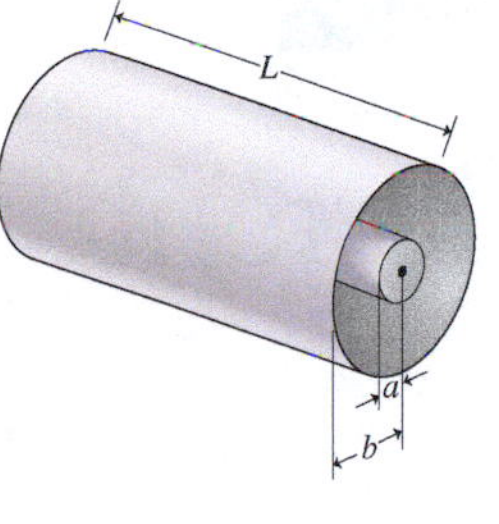

FIGURE 23.15 Problem 60

61. **CH** A capacitor consists of a conducting sphere of radius a surrounded by a concentric conducting shell of radius b. Show that its capacitance is $C = ab/k(b - a)$.

62. **CH** Show that the result of Problem 61 reduces to that of a parallel-plate capacitor when the separation $b - a$ is much less than the radius a.

63. **CH** A solid sphere contains a uniform volume charge density. What fraction of the total electrostatic energy of this configuration is contained *within* the sphere?

64. **CH** An air-insulated parallel-plate capacitor of capacitance C_0 is charged to voltage V_0 and then disconnected from the charging battery. A slab with dielectric constant κ and thickness equal to the capacitor spacing is then inserted halfway into the capacitor (Fig. 23.16). Determine (a) the new capacitance, (b) the stored energy, and (c) the force on the slab in terms of C_0, V_0, κ, and the plate length L.

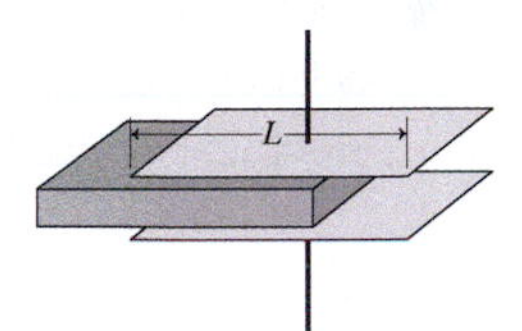

FIGURE 23.16 Problems 64 and 65

65. Repeat parts (b) and (c) of Problem 64, now assuming the battery remains connected while the slab is inserted.

66. A transmission line consists of two parallel wires, of radius a and separation b, carrying uniform line charge densities $\pm\lambda$, respectively. With $a \ll b$, their electric field is the superposition of the fields from two long straight lines of charge. Find the capacitance per unit length for this transmission line.

67. **CH** An infinitely long rod of radius R carries uniform volume charge density ρ. Find an expression for the electrostatic energy per unit length contained *within* the rod. (*Hint*: See Problem 21.56.)

68. **CH** (a) Write the electrostatic potential energy of a pair of oppositely charged, closely spaced parallel plates as a function of their separation x, their area A, and the charge magnitude Q. (b) Differentiate with respect to x to find the magnitude of the attractive force between the plates. Why isn't the force equal to the charge on one plate times the electric field between the plates?

69. An unknown capacitor C is connected in series with a 3.0-μF capacitor; this pair is placed in parallel with a 1.0-μF capacitor, and the entire combination is put in series with a 2.0-μF capacitor. (a) Make a circuit diagram of this network. (b) When a potential difference of 100 V is applied across the open ends of the network, the total energy stored in all the capacitors is 5.8 mJ. Find C.

Passage Problems

Nuclear fusion could provide humankind with limitless energy, making a gallon of seawater the energy equivalent of 300 gallons of gasoline. The National Ignition Facility (NIF) at Lawrence Livermore National Laboratory was designed for the "ignition" of nuclear fusion by bombarding a tiny deuterium-tritium pellet with energy from 192 converging laser beams. The NIF lasers deliver 2 MJ of energy in about 20 ns; Fig. 23.17 shows the target chamber where the laser beams converge. The energy is stored in capacitors that, because of conversion inefficiencies, have to store some 400 MJ. (*Note:* NIF is more complicated than described here, and the numbers and technical descriptions are only approximate.)

FIGURE 23.17 The NIF target chamber, shown during installation (Passage Problems 70–73)

70. What total capacitance is required if the capacitor system is charged to 20 kV?
   a. 100 μF
   b. 200 μF
   c. 1 F
   d. 2 F

71. If it were technically and economically feasible to double the voltage, how would the required capacitance change?
   a. drop to one-quarter its original value
   b. drop to one-half its original value
   c. would not change
   d. would double

72. While they're firing, the average power delivered by the laser beams is
   a. 100 KW.
   b. 100 MW.
   c. 100 GW.
   d. 100 TW.

73. Among the capacitors that store energy at NIF are 1200 300-μF units charged to about 20 kV. The energy stored in each capacitor is about
   a. 3 J.
   b. 20 kJ.
   c. 60 kJ.
   d. 400 MJ.

Answers to Chapter Questions

Answer to Chapter Opening Question

The energy is stored in the electric field of a pair of charged conductors—a capacitor—and dumped quickly to the defibrillator when it's needed.

Answers to GOT IT? Questions

23.1 (c)

23.2 No, because the charge the capacitor holds depends on the voltage across it (although there may be practical limits on the maximum voltage; see Section 23.3).

23.3 (b) because U depends on V^2

23.4 (1) parallel; (2) series; (3) The working voltage of the series combination is twice that of the parallel combination, which is the same as that of the individual capacitors.

23.5 (1) $E(P)$ doubles; (2) $u_E(P)$ quadruples; (3) U decreases, since the charges are attracted and therefore you do negative work to bring in the negative charge.

Electric Current

What You Know

- You understand electric charge and electric fields, and how charges respond to fields.
- You understand electric potential difference as a measure of the work per unit charge in moving charges through electric fields.
- You understand the distinction between energy and power.

What You're Learning

- You'll learn that the flow of charge constitutes *electric current* and how current relates to microscopic properties of a material in which current flows.
- You'll learn to distinguish between *current* and *current density*, with current density providing a more detailed look at electric current.
- You'll see why many materials, especially metals, exhibit a linear relation between electric field and current density—a relation known as *Ohm's law*.
- You'll explore the mechanisms of electrical conduction in different materials, including metals, ionic solutions, plasmas, semiconductors, and superconductors.
- You'll learn about *electrical resistance* and how it follows from basic material properties and geometry.
- You'll learn how to calculate electric power.
- You'll learn about electrical safety.

How You'll Use It

- In Chapter 25 you'll use the concepts of electric current and potential difference to analyze electric circuits.
- In Chapter 26 you'll see how electric current is a source of magnetism.
- In Chapter 28 you'll learn about the *alternating current* used in electric power systems.
- Understanding electric current will help you appreciate everyday electrical devices and use them safely.

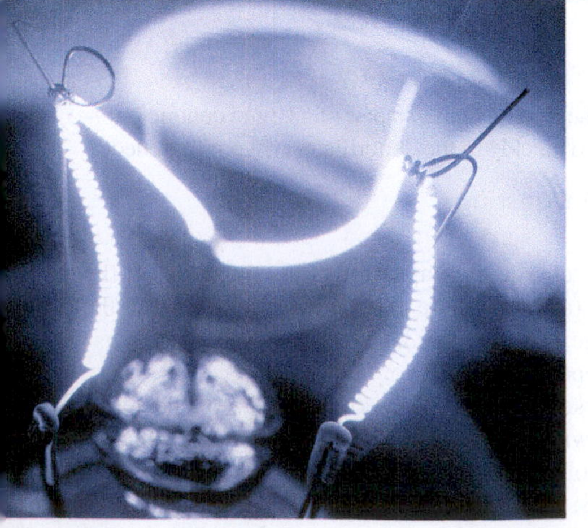

We now move beyond electrostatic equilibrium and consider situations in which charges are moving. The flow of charge constitutes **electric current**, and it occurs in materials containing free charges—that is, in conductors.

Electric current is essential in many technological and natural processes. Currents in lightbulbs, toasters, and stoves produce light and heat. Currents in electric motors run refrigerators, hybrid cars, and subway trains. In computers, currents process and move data. In your body, they regulate heartbeat and control muscles. Currents in Earth's liquid outer core generate the planet's magnetism, protecting us from cosmic radiation. And currents in the Sun are responsible for giant eruptions that can spew high-energy particles toward Earth.

How does electric current heat this lightbulb filament? Where does the energy come from?

24.1 Electric Current

Quantitatively, current is the net rate of charge crossing an area. Its units are coulombs per second, which is given the name **ampere** (A) after the French physicist André Marie Ampère (1775–1836). In electronics and biomedical applications, currents are small enough that milliamperes (mA) and microamperes (μA) are widely used. When current I is steady or a time average will do, we write

$$I = \frac{\Delta Q}{\Delta t} \quad \text{(steady current)} \tag{24.1a}$$

where ΔQ is the charge crossing the given area in time Δt. For time-varying currents we take the limit of small time intervals:

$$I = \frac{dQ}{dt} \quad \text{(instantaneous current)} \tag{24.1b}$$

Current is in the direction in which *positive* charge flows. If the moving charge is negative, as with electrons in a metal, then the current is opposite the charge motion.

A current may consist of one kind of moving charge, or both. If it's both, then the net current is the sum of the currents carried by positive and negative charges (Fig. 24.1*a*). That's why the bulk motion of a neutral object—even though it contains lots of positive and negative charge—doesn't constitute a current (Fig. 24.1*b*).

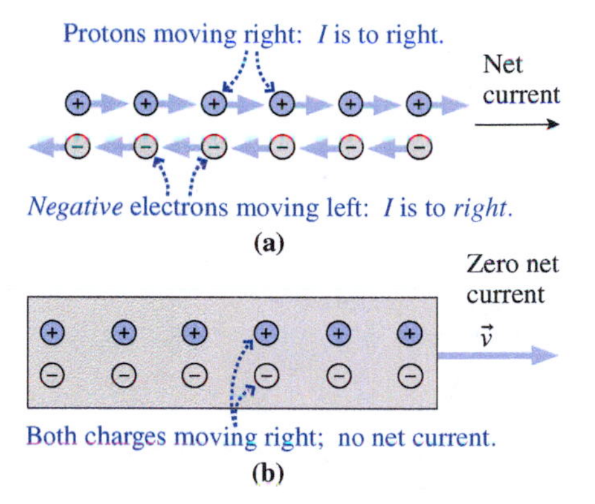

FIGURE 24.1 The net current is the sum of the currents carried by both positive and negative charges.

GOT IT? 24.1 Which of the following represents a nonzero current? What's its direction? (a) a beam of electrons moves from left to right; (b) a beam of protons moves upward; (c) in a solution, positive ions move left and negative ions move right; (d) blood, carrying positive and negative ions at the same speed, moves up through a vein; (e) a metal car with no net charge speeds westward

---

Current: A Microscopic Look

Current depends on the speed of the charge carriers, their density, and their charge. In some cases, like a beam of electrons in vacuum, "speed" here means the actual speed of the charges. But in typical conductors, charges are moving about at high speed with random thermal velocities that don't result in a net flow of charge. When a current is present, there's an additional and usually very small **drift velocity** superposed on the charges' random motion, and it's this drift velocity that determines the current. We'll see this in more detail when we consider metallic conductors.

Figure 24.2 shows a conductor that contains n charges per unit volume, each with charge q and drift speed v_d. We want to express the current in terms of these microscopic properties and macroscopic properties like length and area. With A the conductor's cross-sectional area, a length L of conductor has volume AL and contains nAL individual charges for a total charge $\Delta Q = nALq$. Moving at v_d, this charge takes time $\Delta t = L/v_d$ to pass a given point. Then the current is

$$I = \frac{\Delta Q}{\Delta t} = \frac{nALq}{L/v_d} = nAqv_d \tag{24.2}$$

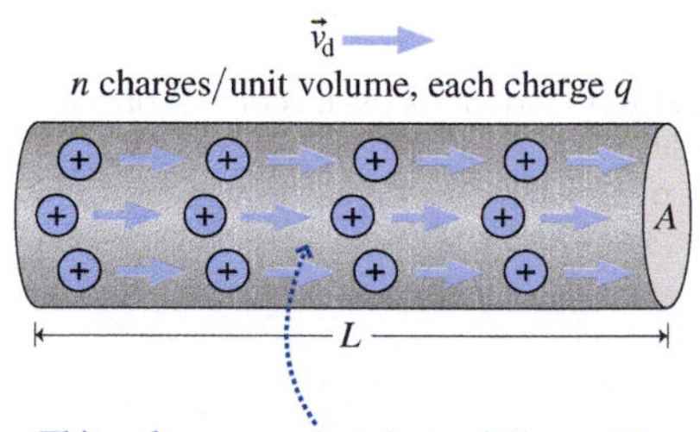

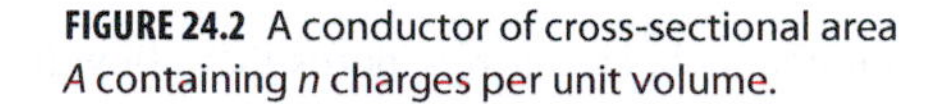

FIGURE 24.2 A conductor of cross-sectional area A containing n charges per unit volume.

EXAMPLE 24.1 Finding Current: A Copper Wire

A 5.0-A current flows in a copper wire with cross-sectional area 1.0 mm^2, carried by electrons with number density $n = 1.1 \times 10^{29}$ m^{-3}. Find the electrons' drift speed.

INTERPRET We're given microscopic parameters, so this is a problem about the relation between current and the parameters n, q, and v_d.

(continued)

DEVELOP Figure 24.3 is our sketch. Equation 24.2, $I = nAqv_d$, relates current to the macroscopic parameters, so our plan is to solve for v_d.

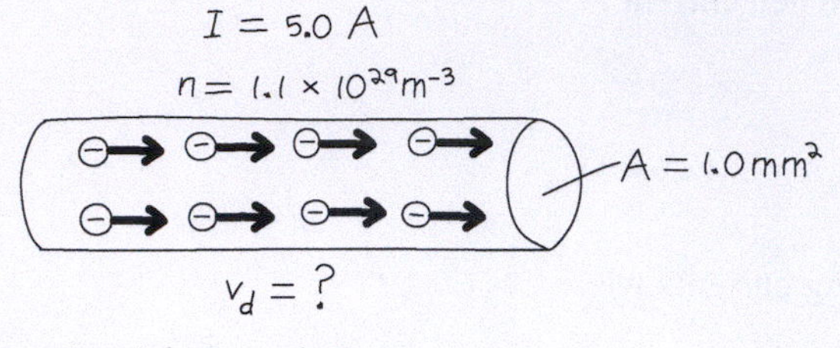

FIGURE 24.3 Sketch for Example 24.1.

EVALUATE Solving, we get

$$v_d = \frac{I}{nAq}$$

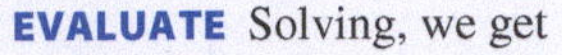

$$= \frac{5.0\text{ A}}{(1.1\times10^{29}\text{ m}^{-3})(10^{-6}\text{ m}^2)(1.6\times10^{-19}\text{ C})} = 0.28\text{ mm/s}$$

ASSESS Make sense? Our answer seems awfully small. After all, when you flip a light switch, the light comes on immediately, not several thousand seconds later as our answer might imply. But the answer is right. Electrons in the wire all get their "marching orders" from the electric field, and that's established almost instantaneously. As a result electrons throughout the wire start moving almost simultaneously. That's why the light comes on immediately. ■

✓TIP Drift Speed and Signal Speed

Example 24.1 points to an important distinction between the drift speeds of electrons and speed of electric signals. The former is typically about 1 mm/s, but the latter is close to the speed of light. When you connect a wire across a battery, for example, an electric field develops, starting at both battery terminals and moving along the wire at nearly the speed of light. As soon as electrons experience the field, they start to move. So there's almost no time delay before the start of the current.

Current Density

Currents aren't always confined to wires. Currents in the Earth, in chemical solutions, in your body, and in ionized gases flow in ill-defined paths, and their magnitude and direction may vary with position. We characterize such diffuse currents in terms of **current density**, $\vec{J}$, a vector whose direction at each point is that of the local current and whose magnitude is the current per unit area. Dividing Equation 24.2 by area and using the drift velocity vector $\vec{v}_d$ instead of speed v_d gives the current density:

$$\vec{J} = nq\vec{v}_d \quad \text{(current density)} \tag{24.3}$$

When the current density is uniform, as in a wire, the total current is just the product of the current density and the wire's cross-sectional area. A: $I = JA$. But when the current density varies, it's necessary to integrate. And because current may vary in magnitude or direction, and the area itself may not be flat, that integral becomes a surface integral similar to the one we introduced in Chapter 21 for electric flux. In fact, as Fig. 24.4 shows, the current is the flux of the current density: $I = \oint \vec{J}\cdot d\vec{A}$. You can explore the case of a nonuniform current density in Problem 63.

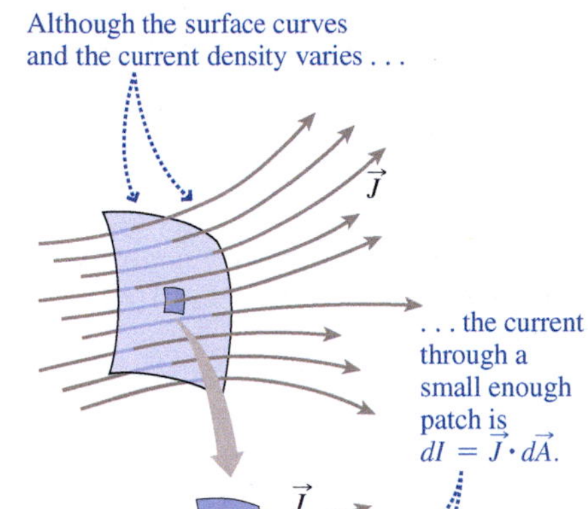

FIGURE 24.4 Current I is the flux of the current density $\vec{J}$, and finding the total current in the case of nonuniform $\vec{J}$ requires integration. Compare with Fig. 21.6.

EXAMPLE 24.2 **Current and Current Density: Through the Cell Membrane**

Ion channels are narrow pores that allow ions to pass through cell membranes (Fig. 24.5). A particular channel has a circular cross section 0.15 nm in radius; it opens for 1 ms and passes 1.1×10^4 singly ionized potassium ions. Find both the current and the current density in the channel.

INTERPRET This problem describes a flow of individual ions and asks for two distinct but related quantities: current and current density.

DEVELOP Current is the *rate* of charge passing through a given area, here the opening of an ion channel. Equation 24.1a, $I = \Delta Q/\Delta t$,

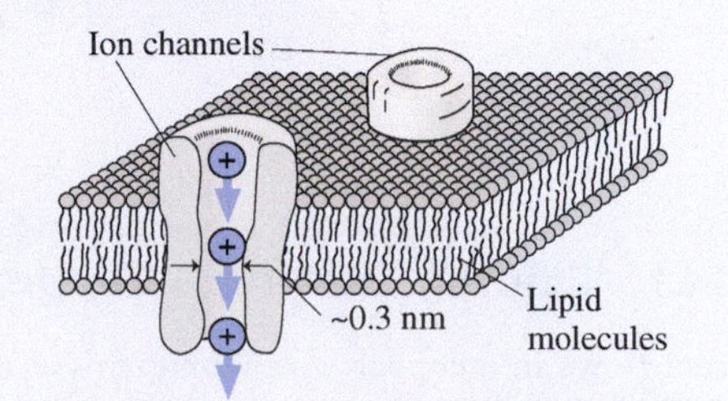

FIGURE 24.5 Diagram of a cell membrane, showing ions passing through an ion channel.

determines the current. Current density, however, is *current per unit area*, which we can compute from $J = I/A$.

EVALUATE With each ion carrying charge e, a total charge $\Delta Q = 1.1\times10^4 e = 1.8\times10^{-15}$ C flows through the channel in $\Delta t = 10^{-3}$ s, giving a current $I = \Delta Q/\Delta t = 1.8$ pA. For current density we then find

$$J = \frac{I}{A} = \frac{1.8\times10^{-12}\ \text{A}}{\pi(0.15\times10^{-9}\ \text{m})^2} = 2.5\times10^7\ \text{A/m}^2$$

ASSESS Make sense? How can something so tiny as a cell have a current density of 25 million amperes per square meter? No problem: Current density measures current *per unit area.* The ion channel is so small that the total current—1.8 picoamperes—is tiny. But that channel is impressive in its own right; its 25 MA/m^2 is about four times the maximum safe current density in typical household wiring. ■

24.2 Conduction Mechanisms

MP

PhET: Conductivity

Electric fields exert forces on charges, so it's the presence of electric fields in conductors that results in electric current. Fields in conductors? Yes. With moving charge we no longer have electrostatic equilibrium, so the field inside a conductor need not be zero. Newton's law suggests that an electric field should *accelerate* free charges in a conductor, resulting in an ever-increasing current. But in most conductors charges collide, usually with ions, and lose energy they've gained from the field. These collisions provide an effective force that counters the electric force, and the end result is that it takes an electric field to sustain a steady current. In most materials the field and current are in the same direction, and we can therefore express the relation between the two as

$$\vec{J} = \sigma\vec{E} \qquad \text{(Ohm's law, microscopic version)} \qquad (24.4a)$$

where the quantity σ is the material's **conductivity**.

Ohm's Law: A Microscopic View

For many common conductors, including metals, conductivity σ is independent of electric field. Such materials are called **ohmic**, and for them Equation 24.4a states that current density and electric field are linearly proportional. In **nonohmic** materials, conductivity does depend on field, and thus the relationship between $\vec{J}$ and $\vec{E}$ isn't linear.

You may be familiar with the *macroscopic* version of **Ohm's law**, which relates electric current and voltage in a piece of conducting material. Equation 24.4a is the *microscopic* version of Ohm's law, describing the relation between electric field and current density *at each point* within a conductor. The macroscopic version is helpful in analyzing electric circuits, and we'll derive it in the next section. But the microscopic version is important in biophysics, geophysics, astrophysics, semiconductor engineering, and other areas where electric fields vary with position and we want to know what's going on at each point.

Conductivity σ tells how large a current density will result from a given electric field; it's a measure of how easily charges in a material can move. A perfect conductor would have $\sigma = \infty$; a perfect insulator, $\sigma = 0$. A related quantity is **resistivity**, ρ, defined as the inverse of conductivity: $\rho = 1/\sigma$. Then Equation 24.4a can be written

$$\vec{J} = \frac{\vec{E}}{\rho} \qquad (24.4b)$$

Resistivity tells how hard it is for charge to move; the higher a material's resistivity, the stronger the electric field needed to produce a given current density. You may be familiar with electrical *resistance*, and you'll soon see how resistance and resistivity are related.

Equation 24.4b shows that the units of resistivity are V·m/A. One V/A is given the name **ohm**, Ω, after the German physicist Georg Ohm (1789–1854), who explored the relation between voltage and current. Thus the SI units of resistivity are Ω·m; reciprocally, those of conductivity are $(\Omega\cdot\text{m})^{-1}$. Conductivity and resistivity range widely, spanning some 24 orders of magnitude. Table 24.1 lists the resistivities of some typical materials. Measurement of electrical resistivity provides information on the composition of materials in fields from medicine to geophysics.

Table 24.1 Resistivities

Material	Resistivity (Ω·m)
Metallic conductors (20°C)	
Aluminum	2.65×10^{-8}
Copper	1.68×10^{-8}
Gold	2.24×10^{-8}
Iron	9.71×10^{-8}
Mercury	9.84×10^{-7}
Silver	1.59×10^{-8}
Ionic solutions (in water, 18°C)	
1-molar $CuSO_4$	3.9×10^{-4}
1-molar HCl	1.7×10^{-2}
1-molar NaCl	1.4×10^{-4}
H_2O	2.6×10^5
Blood, human	0.70
Seawater (typical)	0.22
Semiconductors	
Germanium	0.5
Silicon	3×10^3
Insulators	
Ceramics	$10^{11}-10^{14}$
Glass	$10^{10}-10^{14}$
Polystyrene	$10^{15}-10^{17}$
Rubber	$10^{13}-10^{16}$
Wood (dry)	10^8-10^{14}

EXAMPLE 24.3 Finding the Electric Field: Household Wiring

A 1.8-mm-diameter copper wire carries 15 A to a household appliance. Find the magnitude of the electric field in the wire.

INTERPRET This problem asks us to calculate the electric field within a conductor carrying an electric current.

DEVELOP Equation 24.4b, $\vec{J} = \vec{E}/\rho$, relates electric field and current density. Here we're given the total current I and the wire diameter, so we can write the current density as $J = I/A$, with A the wire's cross-sectional area. We also need the resistivity of copper from Table 24.1. Then we can solve Equation 24.4b for the electric field.

EVALUATE Solving for the field magnitude, we have

$$E = J\rho = \frac{I\rho}{A} = \frac{(15\text{ A})(1.68\times10^{-8}\ \Omega\cdot\text{m})}{\pi(0.90\times10^{-3}\text{ m})^2} = 99\text{ mV/m}$$

ASSESS Make sense? This number is a lot smaller than the electric fields we discussed in electrostatic situations. Because copper is such a good conductor, a weak field can drive a substantial current. In well-engineered circuits, the field inside conducting wires is often so small as to be negligible, even when the current is large. ■

GOT IT? 24.2 Two wires carry the same current I. Wire A has a larger diameter, a higher density of current-carrying electrons, and a lower resistivity than wire B. Rank in order, from smaller to larger, (1) the current densities, (2) the electric fields, and (3) the drift speeds in the two wires.

Conduction in Metals

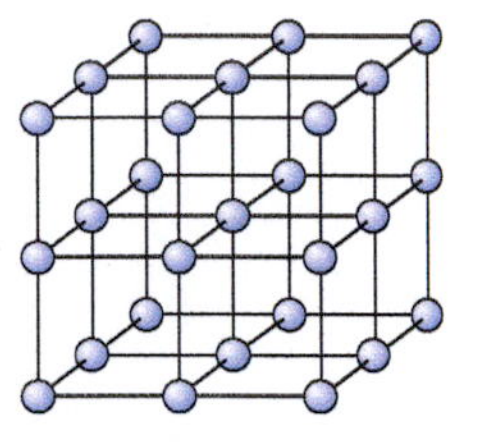

FIGURE 24.6 Atoms of a metal form a regular crystal lattice.

Metals are good conductors because they contain abundant free electrons, which respond readily to electric fields. Each atom in a metal typically contributes one or more electrons to this "sea" of free electrons. The remaining ions form a regular crystal lattice (Fig. 24.6). Electrons move through the lattice at about 10^6 m/s, colliding frequently with ions and bouncing off in random directions. In the absence of an electric field, there's no net flow of electrons in any particular direction, and so no current.

We'll now consider what happens when an electric field is applied to a metal, and we'll show why metals obey Ohm's law. However, our explanation is necessarily incomplete because a full description of metallic conduction involves quantum mechanics.

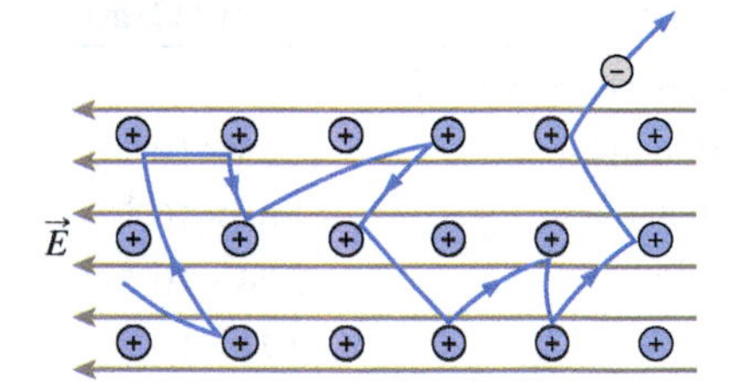

FIGURE 24.7 An electron's path in a metal is almost completely random, but in the presence of an electric field there's a slight drift antiparallel to the field.

An electric field accelerates negative electrons in the direction opposite the field. But like a car in stop-and-go traffic, the electron soon gives up the energy and speed it gained from the field. For the car, that happens at the next stoplight; for the electron, it's at the next collision with an ion, where it rebounds in a random direction (Fig. 24.7). Like the car, the electron thus acquires an average velocity that's proportional to the acceleration it experiences between collisions—that is, proportional to the electric field. There's one difference, though, between the electron and the car: The electron has also a high random thermal velocity, so the average velocity is a tiny effect superposed on the electron's random thermal motion. That average velocity is the drift velocity, $\vec{v}_d$. All electrons share this common drift velocity, so their motion constitutes a current proportional to v_d.

The drift velocity depends on two things: the electrons' acceleration and the rate at which they undergo collisions. The electric field provides the acceleration, so v_d is proportional to E. The collision rate depends on how fast the electrons are moving, and here's the important point: Because thermal motions are so fast, the additional drift velocity makes essentially no difference in the collision rate, so the latter is constant. Therefore, the drift velocity and hence the current are proportional to the electric field—and that makes Equation 24.4a a linear relationship between current density and field. That's why metals are ohmic.

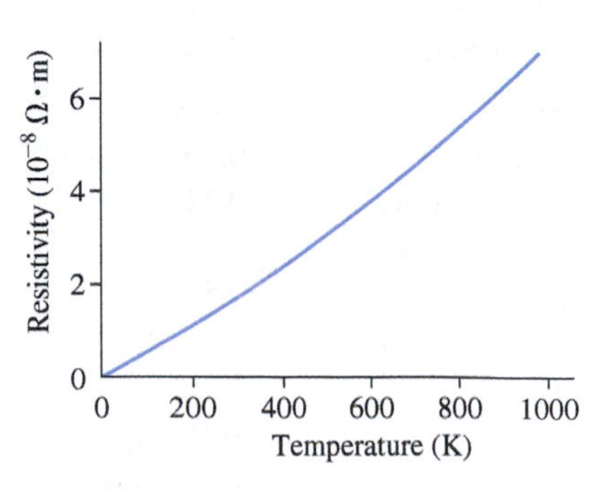

FIGURE 24.8 Resistivity of copper has a nearly linear dependence on temperature, in contrast to the classical prediction of a dependence on $\sqrt{T}$.

Although a metal's conductivity is independent of the applied field, it does depend on temperature T. That's because the thermal speed and hence the collision rate increase with temperature, decreasing conductivity and increasing resistivity. Classical physics gives thermal speed proportional to $\sqrt{T}$, as we saw in Section 17.1, so we might expect resistivity to depend similarly on temperature. Experiment, however, shows that resistivity is nearly linear with temperature (Fig. 24.8)—a result that can be explained using quantum mechanics.

Although the current associated with random thermal motions averages to zero, at any given instant short-term fluctuations can result in more electrons moving in a particular direction. The result is a very small current whose direction and magnitude fluctuate randomly. This **thermal noise** can overwhelm currents of interest in sensitive electronic equipment. Circuits like the amplifiers in radio telescopes are often cooled to decrease thermal noise.

Ionic Solutions

Liquid solutions contain positive and negative ions that respond to an electric field by moving in opposite directions, resulting in a net current. Conductivity is limited by collisions between ions and neutral atoms and, as Table 24.1 suggests, ionic solutions are poorer conductors than metals. Ionic conduction is essential to life, as the transport of ions through cell membranes in Example 24.2 suggests. Electric eels use ionic conduction to sense and kill their prey. Batteries and fuel cells use ionic conduction, which also plays a role in the corrosion of metals. And an ionic solution—sweat—increases your vulnerability to electric shock.

Plasmas

Plasma is ionized gas that conducts because it contains free electrons and ions. It takes substantial energy to ionize atoms, so plasmas usually exist only at high temperatures. Plasmas are rare on Earth; they're in fluorescent lamps, plasma TVs, neon signs, the ionosphere, flames, and lightning flashes. Yet much of the universe's ordinary matter is in the plasma state; stars, in particular, are mostly plasma.

The electric properties of plasma make it so different from ordinary gas that plasma is often called "the fourth state of matter." Some plasmas—like the Sun's corona—are so diffuse and therefore collisions so rare as to make them far better conductors than metals. These "collisionless" plasmas can sustain large currents with minimal electric fields.

Semiconductors

MP

PhET: Semiconductors

Even in insulators, random thermal motions dislodge a few electrons, giving these materials very modest conductivity. In a few materials—notably the element silicon—this effect is significant at room temperature. Such materials have conductivities between those of insulators and conductors, as reflected by their placement in Table 24.1, so they're called **semiconductors**. Semiconductors make possible the microelectronic technology so pervasive in modern civilization. Here we give a qualitative description of semiconductors based on classical physics; we'll revisit semiconductors from a quantum-mechanical viewpoint in Chapter 37.

A dislodged electron leaves behind a "hole," into which an adjacent electron, nudged by the electric field, can "fall" (Fig. 24.9). The result is a movement of holes in the direction

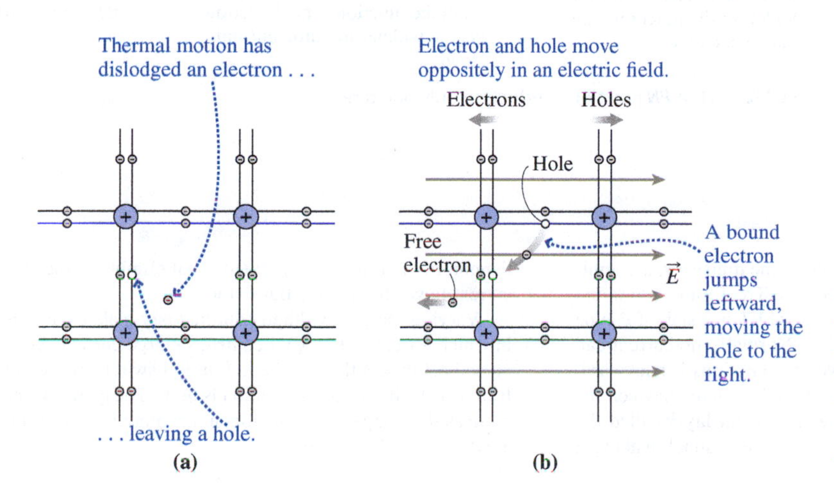

FIGURE 24.9 Structure of a silicon crystal, showing each atom bound to each of its neighbors by two shared electrons. (a) Thermal motion dislodges electrons, creating electron-hole pairs. (b) An electric field drives electrons and holes in opposite directions, creating an electric current.

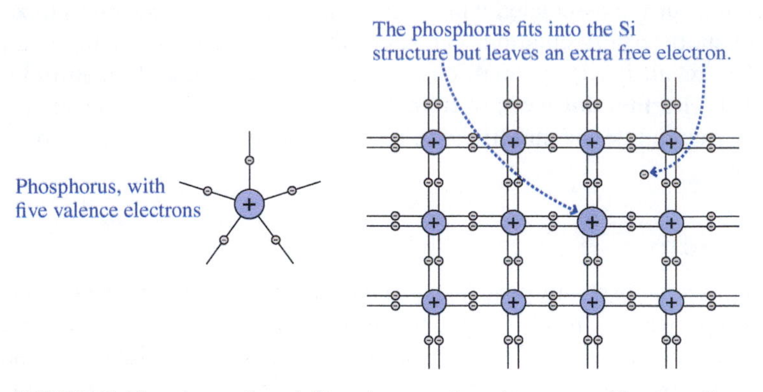

FIGURE 24.10 Phosphorus-doped silicon has extra free electrons, making it an *N*-type semiconductor.

of the field. Thus holes act as positive charges, so a pure semiconductor contains equal numbers of negative charge carriers (electrons) and positive carriers (holes).

The key to semiconductor technology lies in **doping** with impurities that greatly alter a semiconductor's intrinsic conductivity. Figure 24.10 shows how a single phosphorus atom, with five valence electrons, fits into silicon's crystal structure but leaves a free electron. It doesn't take much phosphorus for these extra electrons to constitute the vast majority of charge carriers. Since the charge carriers are negative, the material is called an ***N*-type semiconductor**. Doping with trivalent atoms like boron, in contrast, leaves extra holes and makes a ***P*-type** semiconductor.

The essential element of nearly every semiconductor device is the ***PN* junction**. Electrons and holes diffuse across such a junction and recombine, depleting the junction region of charge carriers and making it a poor conductor. Applying a voltage from the *P* to the *N* region—but not the other way—lets charge flow through the junction. So the *PN* junction conducts in one direction but not the other (Fig. 24.11). The wide range of semiconductor devices in use today results largely from carefully engineered combinations of *PN* junctions.

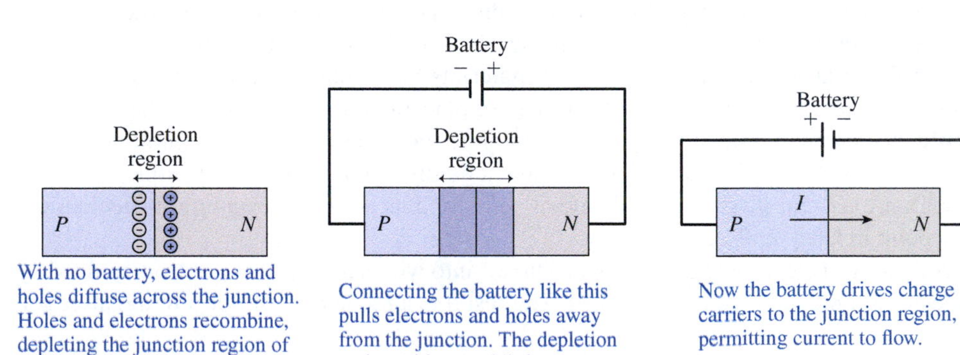

FIGURE 24.11 A *PN* junction conducts in only one direction.

APPLICATION The Transistor

Few inventions have revolutionized society as much as the transistor, the semiconductor device at the root of all modern electronics. The figure shows one type, the field-effect transistor, or FET. This particular FET is a slab of *P*-type semiconductor with two embedded *N*-type regions. Normally no current can flow through the transistor because one of its two *PN* junctions is backward, as shown in part (a) of the figure. But atop the so-called *channel* between the *N*-type regions is a thin insulating layer, and over it a metal layer called the *gate*. Make the gate positive, and it pulls electrons into the channel, making it temporarily *N*-type, as in part (b). That eliminates the *PN* junctions, and now current flows through the transistor.

Varying the gate voltage continuously makes the transistor an *amplifier*, in which a weak gate signal controls a large current. Swinging between fully on and off makes the transistor a digital switch, providing the binary 1 and 0 from which all digital information is built. Today, transistors by the billions are fabricated on single chips of silicon, making the powerful microprocessors that are the "brains" of computers.

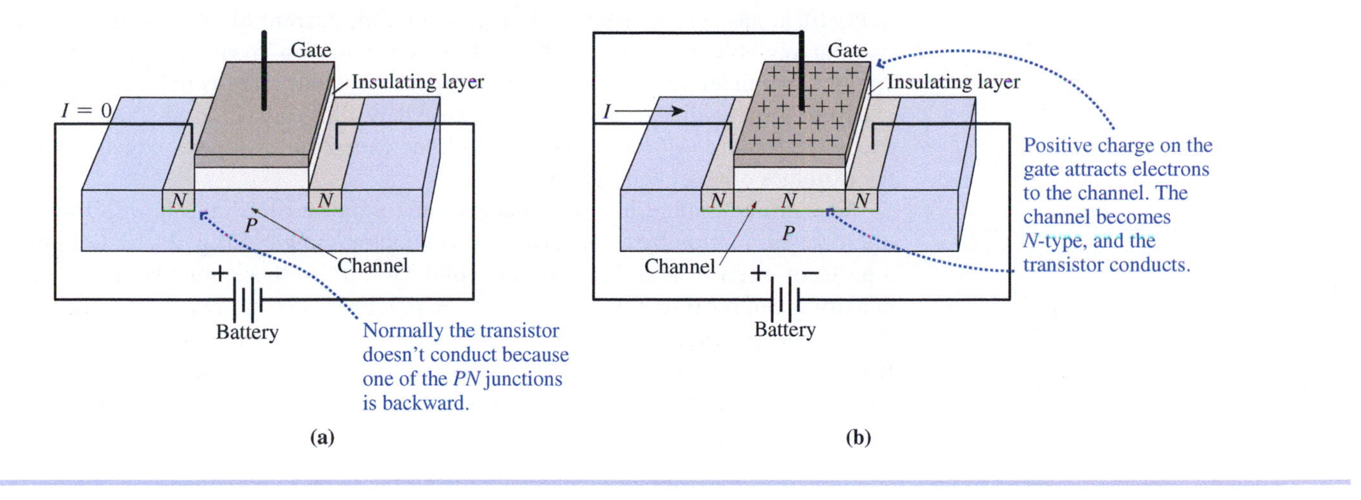

Superconductors

In 1911 the Dutch physicist H. Kamerlingh Onnes found that the resistivity of mercury dropped to zero at a temperature of 4.2 K. Today we know thousands of substances that become **superconductors** at sufficiently low temperatures. Currents in superconductors persist for years without measurable decrease, suggesting that the resistivity is truly zero (Fig. 24.12). For decades the known superconductors were metals and metal alloys that required cooling with liquid helium. Then, in 1986, physicists at IBM's Zurich laboratory stunned the scientific community with the discovery of ceramic materials that become superconducting at around 100 K—high enough to cool with inexpensive liquid nitrogen. The search for higher-temperature superconductors continues, with the highest temperature reported now over 200 K, and tantalizing hints of possible superconducting behavior at much higher temperatures. Development of a room-temperature superconductor could revolutionize much of electrical technology.

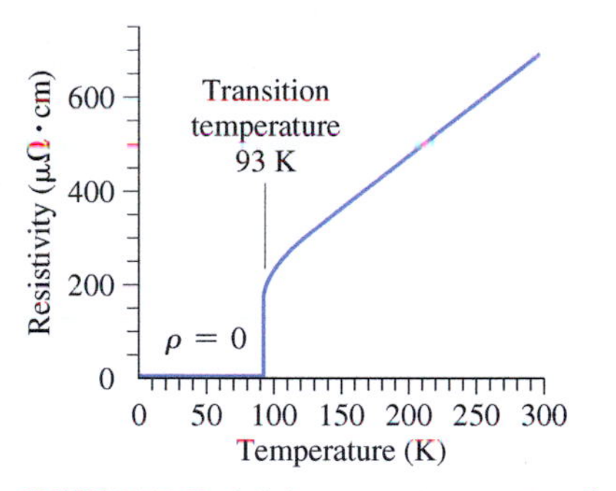

FIGURE 24.12 Resistivity versus temperature for a thin film of yttrium–barium–copper-oxide superconductor.

Superconductors offer loss-free flow of electric power. Today, superconductors are widely used in high-strength electromagnets, including those in MRI scanners; in filters that distinguish individual channels in cell-phone communications; in devices that measure weak magnetic fields in biomedical, geophysical, and other applications; for electric-power transmission in high-density urban applications; and in motors for ship propulsion. Expect more applications in the near future.

24.3 Resistance and Ohm's Law

PhET: Resistance in a Wire
PhET: Ohm's Law

How much current does it take to run this hair dryer? Do I risk a fatal shock if I touch this wire? How long an extension cord can I use with this electric saw? How long will it take to recharge my cell phone? Is the wiring in my house safe? All these questions are ultimately about the electric current flowing in wires, bodies, and batteries. The answer in each case depends on two things: the voltage V applied across the object and the **resistance** R that the object offers to the flow of electric current.

The macroscopic version of Ohm's law relates voltage, current, and resistance. Ohm's law states that the current through an object is proportional to the voltage across it and inversely proportional to the object's resistance:

$$I = \frac{V}{R} \quad \text{(Ohm's law, macroscopic version)} \tag{24.5}$$

Ohm's law shows that a given voltage can push more current through a lower resistance. It's worth noting two extreme cases: An **open circuit** is a nonconducting gap with infinite resistance. No matter what the voltage is across an open circuit, Equation 24.5 shows that no current can flow. A switch in its "off" position is an open circuit. A **short circuit**, in contrast, has zero resistance. In a short circuit, current of any magnitude is possible without any voltage or electric field. A switch in its "on" position approximates a short circuit. An unintentional short circuit is dangerous; short circuits in household wiring, for example, are a leading cause of fires because they allow large currents to flow, resulting in excessive heating. All real situations, with the exception of superconductors, lie between the extremes of short and open circuits.

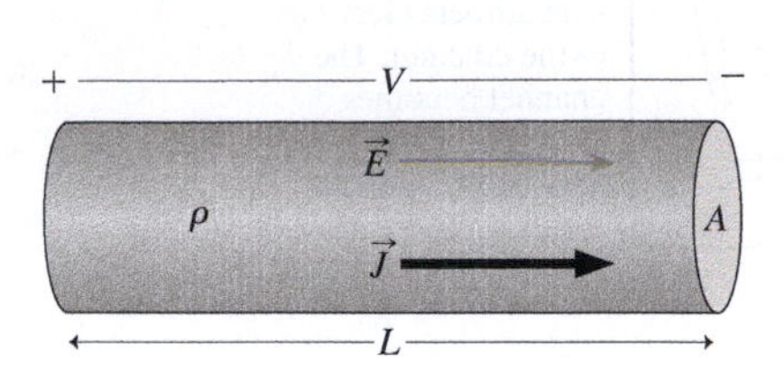

FIGURE 24.13 A cylinder of conducting material with resistivity ρ.

We can understand how the macroscopic version of Ohm's law follows from the microscopic version by considering the conductor shown in Fig. 24.13. Suppose there's a uniform electric field $\vec{E}$ within the conductor. Then there must be a uniform current density given by Equation 24.4b: $\vec{J} = \vec{E}/\rho$, where ρ is the resistivity of the material. Then the total current is $I = JA = EA/\rho$, where A is the conductor's cross-sectional area. If the conductor has length L, then the potential difference between its ends is $V = EL$, since the electric field is uniform. Solving to get $E = V/L$ and using the result in our expression for I gives

$$I = \frac{VA}{L\rho} = \frac{V}{\rho L/A}$$

Comparison with the macroscopic Ohm's law, Equation 24.5, lets us identify the resistance with the term $\rho L/A$. Thus resistance depends on the resistivity ρ and the geometry—length and area—of the particular piece of material:

$$R = \frac{\rho L}{A} \qquad (24.6)$$

We derived this expression for a conductor of uniform cross section; although Ohm's law still holds for a nonuniform conductor, integration is required to calculate the resistance in that case (see Problems 67 and 68). Equations 24.5 and 24.6 both show that the units of resistance are ohms (Ω).

We emphasize that Ohm's law is not fundamental; rather, it's an empirical law that describes electrical conduction in some materials. Table 24.2 summarizes the relation between microscopic and macroscopic quantities in Ohm's law.

Table 24.2 Microscopic and Macroscopic Quantities and Ohm's Law

Microscopic	Macroscopic	Relation
Electric field, $\vec{E}$	Voltage, V	$\vec{E}$ is defined at each point in a material; V is the integral of $\vec{E}$ over a path. In a uniform field, $V = EL$.
Current density, $\vec{J}$	Current, I	$\vec{J}$ is defined at each point in a material; I is the flux—the surface integral—of $\vec{J}$ over an area. With uniform current density, $I = JA$.
Resistivity, ρ	Resistance, R	ρ is a property of a given material; R is a property of a particular piece of material. In a piece with uniform cross section, $R = \rho L/A$.
Ohm's law $\vec{J} = \frac{\vec{E}}{\rho}$	Ohm's law $I = \frac{V}{R}$	Microscopic version relates current density to electric field at a point in a material. Macroscopic version relates current through to voltage across a given piece of material.

EXAMPLE 24.4 Resistance and Ohm's Law: Starting Your Car

A copper wire 0.50 cm in diameter and 70 cm long connects your car's battery to the starter motor. What's the wire's resistance? If the starter motor draws a current of 170 A, what's the potential difference across the wire?

INTERPRET This problem involves Ohm's law, and we identify the wire as the object in which we want to relate current, voltage, and resistance.

DEVELOP Figure 24.14 shows the wire. Equation 24.6, $R = \rho L/A$, determines the resistance, so our plan is first to use the resistivity of copper from Table 24.1 in Equation 24.6 and then to use the resulting resistance in Ohm's law (Equation 24.5), $I = V/R$, to find the potential difference.

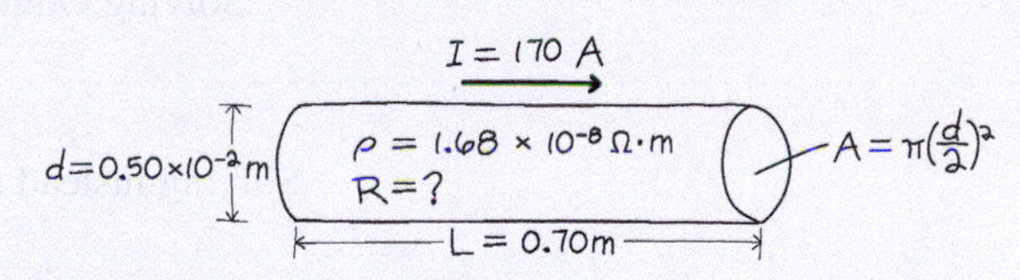

FIGURE 24.14 Sketch for Example 24.4.

EVALUATE Table 24.1 gives $\rho = 1.68\times10^{-8}\ \Omega\cdot\text{m}$ for copper, so for the resistance we get

$$R = \frac{\rho L}{A} = \frac{(1.68\times10^{-8}\ \Omega\cdot\text{m})(0.70\text{ m})}{\pi(0.25\times10^{-2}\text{ m})^2} = 0.60\text{ m}\Omega$$

Then Ohm's law gives the voltage: $V = IR = (170\text{ A})(0.60\text{ m}\Omega) = 0.10\text{ V}$.

ASSESS Make sense? These numbers seem awfully small. They should be! A wire carrying a large amount of current needs to have a very low resistance so the voltage across the wire remains low. We want that 12-V potential difference from the battery to appear across the starter motor, not the connecting wires. A thinner, higher-resistance wire would mean lower voltage across the starter and a significant reduction in current. ■

A **resistor** is a piece of conductor made to have a specific resistance. Heating elements in electric stoves, hair dryers, irons, space heaters, and the like are all essentially resistors; so are the filaments of incandescent lightbulbs. In all these cases the resistance—ultimately resulting from collisions between conduction electrons and lattice ions—provides a means of turning electric energy into heat. Resistors also set appropriate values of current and voltage in electronic circuits; for this purpose, they're made in a wide range of resistances. Resistors are rated not only by their resistance but also by the maximum power they can dissipate without overheating.

GOT IT? 24.3 The figure shows three pieces of wire. (a) and (b) are made from the same material, while (c) is made from a material with twice the resistivity. (a) and (c) have twice the diameter of (b), while (b) is twice as long as the others. (1) Which has the highest resistance? (2) If the same voltage is applied across each, which will pass the largest current?

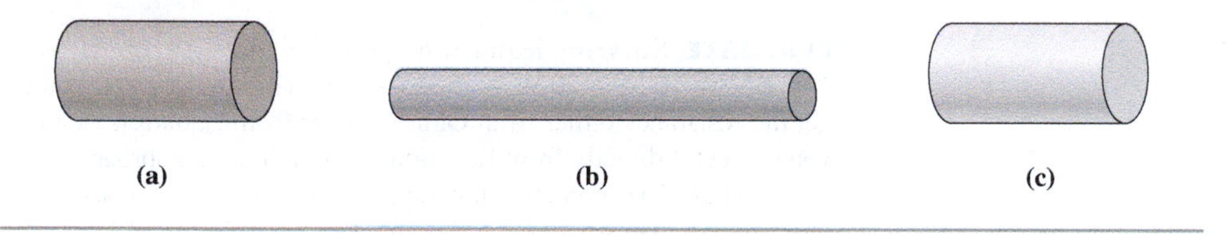

24.4 Electric Power

Impose a potential difference V across a resistor, and a current I flows through it. The quantity V is the energy gained per unit charge as charge "falls" through the potential difference. In a resistor, that energy is dissipated through collisions, heating the material. So V is also the energy per unit charge going into heating. Meanwhile, the current I is the rate at which charge flows through the resistor. Then the energy per unit time—that is, the power dissipated in heating the resistor—is the product of the energy per unit charge and the rate at which charge moves through the conductor:

Video Tutor Demo | **Resistance in Copper and Nichrome**

$$P = IV \quad \text{(electric power)} \tag{24.7}$$

Although we developed Equation 24.7 for power dissipated as heat in a resistor, it holds any time electrical energy is being converted to some other form. If we measure 5 V across an electric motor and 2 A through the motor, we can conclude that the motor is converting electrical to mechanical energy at the rate of 10 W (actually less because some of the power goes into heating).

Solving Ohm's law for V and putting the result in Equation 24.7 give

$$P = I^2R \tag{24.8a}$$

Solving instead for I gives

$$P = \frac{V^2}{R} \tag{24.8b}$$

These are useful forms when we know the resistance and either the voltage or the current.

✓TIP What's Constant?

Equation 24.8a seems to imply that power increases with increasing resistance, while Equation 24.8b seems to suggest the opposite. Both implications are correct—*if* I in Equation 24.8a and V in Equation 24.8b are constants. But there's no contradiction because I and V can't both be constant while the resistance R—the ratio of V to I—changes. In most cases we work with sources of constant voltage, and then the power dissipated is inversely proportional to the resistance.

CONCEPTUAL EXAMPLE 24.1 Power Transmission

Long-distance power transmission lines operate at very high voltages—often hundreds of kilovolts. Why?

EVALUATE Equation 24.7, $P = IV$, shows that we can get the same electric power from low voltage V and high current I, or vice versa. But Equation 24.8a shows that power loss in a transmission line increases as the *square* of the current. So use of high voltage and low current minimizes transmission losses (Fig. 24.15).

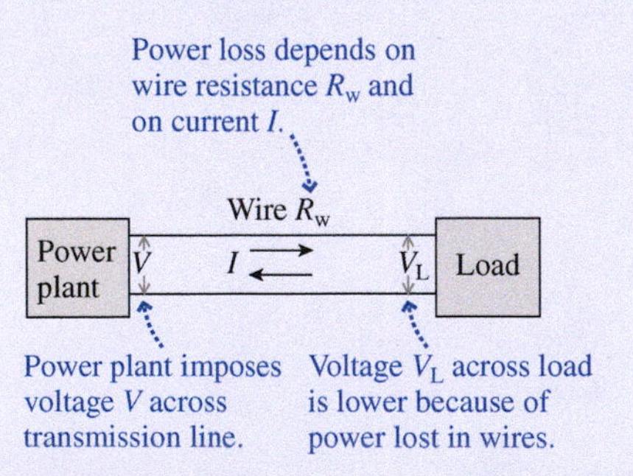

FIGURE 24.15 High voltage and low current minimize losses in power transmission.

ASSESS As a user, you don't encounter these high voltages. That's because transformers "step down" the voltage before it reaches the end user (you'll learn about transformers in Chapter 28). The lower voltages are safer and easier to handle, although even standard 120-V household power is far from "safe."

MAKING THE CONNECTION What's the current in a 120-V, 100-W incandescent lightbulb? What's the bulb's resistance?

EVALUATE Solving Equation 24.7 for the current I gives $I = P/V = 100\text{ W}/120\text{ W} = 0.833\text{ A}$. Knowing the current, you can get the resistance either from Ohm's law or from Equation 24.8a. Or you can get it directly from Equation 24.8b. All three approaches give $R = 144\ \Omega$. The filament temperature is 3000 K, so this resistance is much higher than what you'd measure with the bulb off.

GOT IT? 24.4 You put a variable resistance across a battery that maintains a fixed voltage across its terminals. If you lower the resistance, does the power supplied by the battery (a) increase or (b) decrease?

24.5 Electrical Safety

Whether you're in a lab hooking up electronic equipment, or in a hospital connecting instrumentation to a patient, or on a job designing electric devices, or simply at home plugging in appliances and tools, you should be concerned with electrical safety.

Everyone knows enough to be wary of "high voltage." People with a little more sophistication say, "It isn't the voltage but the current that kills." In fact, both points of view are partially correct. Current through the body is dangerous, but as with any resistor it takes voltage to drive that current.

Table 24.3 shows typical effects of electric currents entering the body through skin contact. A primary danger is disturbance of the electric signals that pace heartbeat; this is reflected in the lethal zone of 100–200 mA at which the heart goes into fibrillation—uncontrolled spasms of the cardiac muscle. With electric signals applied internally, much lower currents can be lethal. Doctors performing cardiac catheterization worry about currents at the microampere level.

Table 24.3 Effects of Externally Applied Current on Humans

Current Range	Effect
0.5–2 mA	Threshold of sensation
10–15 mA	Involuntary muscle contractions; can't let go
15–100 mA	Severe shock; muscle control lost; breathing difficult
100–200 mA	Fibrillation of heart; death within minutes
>200 mA	Cardiac arrest; breathing stops; severe burns

APPLICATION **The Taser©**

The Taser is a so-called electroshock weapon used by police to subdue unruly suspects. The Taser uses compressed nitrogen gas to fire two barbed darts into the victim with sufficient force and dart length to penetrate clothing and make contact with the victim's body. Very fine wires connect the darts back to the weapon, as far as 9 m away, and thus form a complete circuit once the darts are lodged in the body. In the most common police model, a potential difference of 1200 V is applied across the darts in pulses that last 100 μs. The weapon delivers 19 such pulses each second. The voltage, pulse shape, and pulse duration are engineered to cause major skeletal muscles to contract involuntarily without affecting the heart muscle and without dangerously strong contractions that would occur at much higher pulse rates. Thus the Taser effectively immobilizes the victim without danger or permanent damage.

Law-enforcement officials claim the Taser saves lives by substituting for lethal bullets or crude weapons like clubs. Others point to a number of cases of death following Taser use—although only a handful of those deaths have been attributed to the Taser's electrical effects. Debates over Taser safety are muddied by the fact that many Taser victims are already impaired by drug abuse and that police restraint of suspects often involves violent means that sometimes result in death.

In Problem 50 you can explore the physiological effects of the Taser in the context of Table 24.3.

Above 200 mA, complete cardiac arrest may occur, breathing may stop, and burns may occur. Sometimes high currents are useful: Emergency defibrillators briefly apply a high enough current to stop the heart, which often restarts normal beating. The figures in Table 24.3 are rough averages and vary from person to person as well as with duration of the shock and whether alternating or direct current is involved. Very young children and people with heart conditions are at higher risk.

Under dry conditions, the typical human has a resistance of about $10^5\ \Omega$ between two points on unbroken skin. What voltages are dangerous to such a person? At $10^5\ \Omega$ it takes

$$V = IR = (0.1\ \text{A})(10^5\ \Omega) = 10{,}000\ \text{V}$$

to drive the fatal 100 mA. But a person who's wet or sweaty has a much lower resistance and may be electrocuted by 120-V household electricity or even lower.

To be dangerous, an electric circuit must have high voltage *and* be capable of driving sufficient current. For example, a car battery can deliver 300 A, but it can't electrocute you

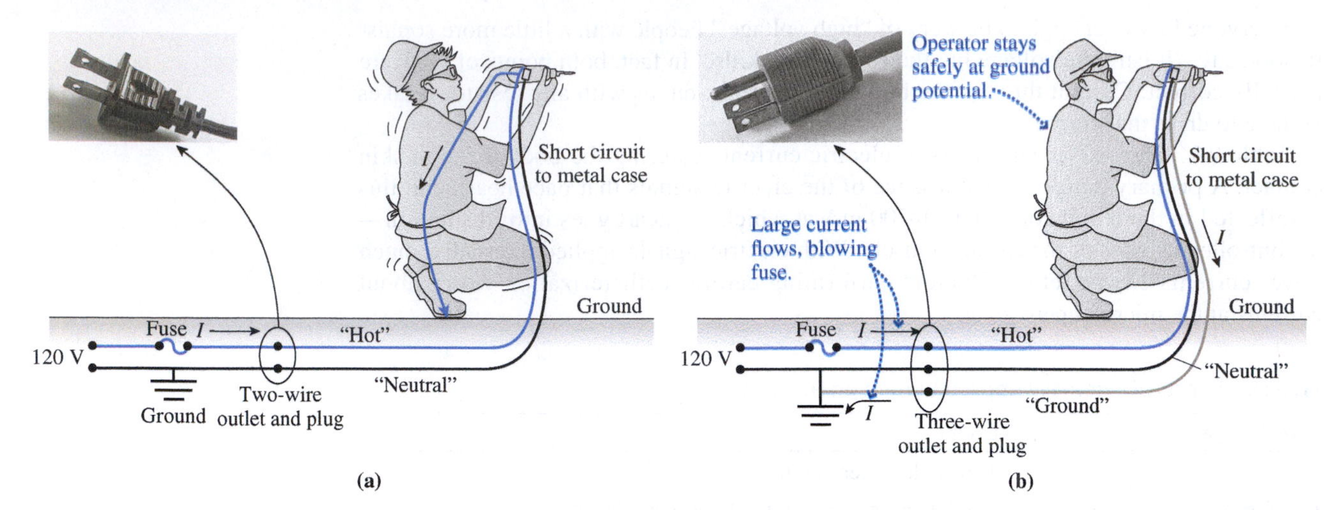

FIGURE 24.16 (a) A short circuit in an ungrounded tool could result in a lethal shock. (b) With a grounded tool, the fuse blows and the operator is safe.

because its 12 V won't drive much current through you. On the other hand, the 20,000 V that runs your car's spark plugs won't electrocute you either, since the high-voltage circuit can't deliver more than a few mA.

Because potential difference is a property of two points, receiving an electric shock requires that two parts of the body contact conductors at different potentials; this chapter's opening photo provides a dramatic example. In typical 120-V wiring used throughout North America, one of the two wires is connected physically to the ground. This ground connection prevents the wiring from reaching arbitrarily high potentials, as might otherwise happen in a thunderstorm or if a short circuit occurred in a power line. At the same time it means that an individual contacting the "hot" side of the circuit and any grounded conductor such as the ground, a water pipe, or a bathtub will receive a shock.

Many devices use three-wire cords to reduce shock hazard. Exposed metal parts connect directly to a third ground wire that normally carries no current. If something goes wrong and a "hot" wire accidentally short-circuits to the metal case, this wire provides a low-resistance path to ground (Fig. 24.16). A large current flows and blows the fuse or circuit breaker, shutting off the current. Even better are *ground fault circuit interrupters* used in kitchens, bathrooms, and other high-risk locations. These devices sense a slight imbalance in current on the two wires, and shut off the circuit on the assumption that the "missing" current is leaking to ground, perhaps through a person.

GOT IT? 24.5 Today's power tools are often cordless, powered by internal batteries. Are you completely safe from electric shock when using such a tool? Discuss.

CHAPTER 24 SUMMARY

Big Idea

The big idea here is **electric current**—the flow of electric charge—and its microscopic cousin, **current density**. With current we don't have electrostatic equilibrium, and there's usually an electric field in a current-carrying conductor. **Ohm's law** is an empirical statement—not a fundamental law of physics—that relates current and voltage, or current density and electric field.

Key Concepts and Equations

Quantitatively, **current** is defined as the rate of charge flow:

$$I = \frac{\Delta Q}{\Delta t}$$

Current density is the current per unit area. Its magnitude is

$$J = \frac{I}{A}$$

Charge ΔQ crosses this area in time Δt.

A

There are n charge carriers per unit volume, with charge q and drift velocity v_d.

Microscopically, current depends on the density of charge carriers, their charge, and the **drift velocity**:

$$I = nqAv_d$$

and

$$\vec{J} = nq\vec{v}_d$$

The microscopic version of **Ohm's law** relates electric field, current density, and **conductivity** σ (or its inverse, **resistivity** ρ):

$$\vec{J} = \sigma\vec{E}$$

The macroscopic version relates voltage, current, and resistance:

$$I = V/R$$

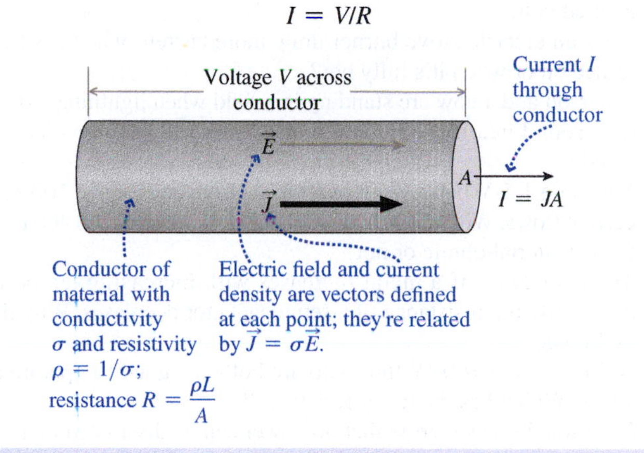

Electric power is the product of voltage and current:

$$P = IV$$

Using Ohm's law, this can also be written

$$P = I^2R$$

$$P = \frac{V^2}{R}$$

Applications

Different types of conductors have different conduction mechanisms. In **metals**, free electrons carry the current; in **ionic solutions**, both positive and negative ions are involved; in **plasmas**, the charge carriers are free electrons and ions; and in **semiconductors**, both electrons and positive holes carry current, with semiconductor conduction properties readily adjustable. **Superconductors** are materials that exhibit zero resistance at sufficiently low temperatures.

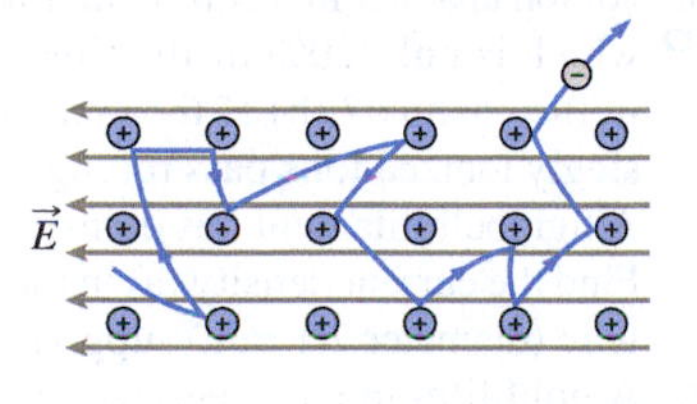

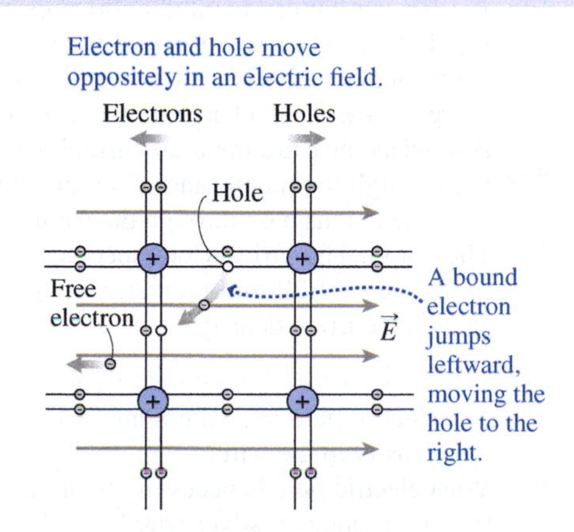

Electrical safety is a matter of avoiding currents high enough to cause biological harm, and that means avoiding voltages high enough to drive such currents.

MP For homework assigned on MasteringPhysics, go to www.masteringphysics.com

BIO *Biology and/or medicine-related problems* **DATA** *Data problems* **ENV** *Environmental problems* **CH** *Challenge problems* **COMP** *Computer problems*

For Thought and Discussion

1. Explain the difference between current and current density.
2. A constant electric field generally produces a constant drift velocity. How is this consistent with Newton's assertion that force results in acceleration, not velocity?
3. When caught in the open in a lightning storm, a person should crouch low with feet close together rather than lie flat on the ground. Why?
4. Good conductors of electricity are often good conductors of heat. Why might this be?
5. Why can current persist forever in a superconductor with no applied voltage?
6. Does an electric stove burner draw more current when it's first turned on or when it's fully hot?
7. A person and a cow are standing in a field when lightning strikes the ground nearby. Why is the cow more likely to be electrocuted?
8. You put a 1.5-V battery across a piece of material, and a 100-mA current flows. With a 9-V battery, the current increases to 400 mA. Is the material ohmic or not?
9. The resistance of a metal increases with increasing temperature, while the resistance of a semiconductor decreases. Why the difference?
10. A 50-W and a 100-W lightbulb are both designed to operate at 120 V. Which has the lower resistance?
11. Equation 24.8a suggests that no power can be dissipated in a superconductor because $R = 0$. But Equation 24.8b suggests the power should be infinite. Which is right, and why?
12. What's wrong with this news report: "A power-line worker was injured when 4000 volts passed through his body"?

Exercises and Problems

Exercises

Section 24.1 Electric Current

13. A wire carries 1.5 A. How many electrons pass through the wire in one second?
14. A 12-V car battery is rated at 80 ampere-hours, meaning it can supply 80 A of current for 1 hour before it becomes discharged. If you accidentally leave the headlights on until the battery discharges, how much charge moves through the lights?
15. **BIO** Biologists measure the total current due to potassium ions moving through the membrane of a rock crab neuron cell as 30 nA. How many ions pass through the membrane each second?
16. The National Electrical Code specifies a maximum current of 10 A in 16-gauge (1.29-mm-diameter) copper wire. What's the corresponding current density?

Section 24.2 Conduction Mechanisms

17. The electric field in an aluminum wire is 85 mV/m. Find the current density in the wire.
18. What electric field is necessary to drive a 7.5-A current through a 0.95-mm-diameter silver wire?
19. A cylindrical tube of seawater carries 350 mA of current. If the electric field in the water is 21 V/m, what's the tube's diameter?
20. A 1.0-cm-diameter rod carries a 50-A current when the electric field in the rod is 1.4 V/m. What's the resistivity of the rod material?
21. Use Table 24.1 to determine the conductivity of (a) copper and (b) seawater.

Section 24.3 Resistance and Ohm's Law

22. Find the resistance of a heating coil that draws 4.8 A when the voltage across it is 120 V.
23. What voltage does it take to drive 300 mA through a 1.2-kΩ resistance?
24. What's the current in a 47-kΩ resistor with 110 V across it?
25. The "third rail" that carries electric power to a subway train is an iron bar whose rectangular cross section measures 10 cm by 15 cm. Find the resistance of a 5.0-km length of this rail.
26. What current flows when a 45-V potential difference is imposed across a 1.8-kΩ resistor?
27. A uniform wire of resistance R is stretched until its length doubles. Assuming its density and resistivity remain constant, what's its new resistance?

Section 24.4 Electric Power

28. A car's starter motor draws 125 A with 11 V across its terminals. What's its power consumption?
29. A 4.5-W flashlight bulb draws 750 mA. (a) At what voltage does it operate? (b) What's its resistance?
30. A watch uses energy at the rate of 240 μW. What current does it draw from its 1.5-V battery?
31. A 35-Ω electric stove burner consumes 1.5 kW of power. At what voltage does it operate?
32. An incandescent lightbulb draws 0.50 A, while a compact fluorescent with the same light output draws 125 mA. Both operate on standard 120-V household power. How do their energy-consumption rates compare?

Section 24.5 Electrical Safety

33. Though rare, electrocution has been reported under wet conditions with voltages as low as 30 V. What resistance would be necessary for this voltage to drive a fatal current of 100 mA?
34. You touch a defective appliance while standing on the ground, and you feel the tingle of a 2.5-mA current. What's your resistance, assuming you're touching the "hot" side of the 120-V household wiring?
35. You have a typical resistance of 100 kΩ. (a) How much current could a 12-V car battery pass through you? (b) Would you feel this?

Problems

36. **BIO** An ion channel in a cell membrane carries 2.4 pA when it's open, which is only 20% of the time. (a) What's the average current in the channel? (b) If the channel opens for 1.0 ms, how many singly ionized ions pass through in this time?
37. A lightbulb filament has diameter 0.050 mm and carries 0.833 A. Find the current density (a) in the filament and (b) in the 12-gauge wire (diameter 2.1 mm) supplying current to the lightbulb.
38. A gold film in an integrated circuit measures 1.85 μm thick by 0.120 mm wide. It carries a current density of 0.482 MA/m^2. What's the total current?
39. A copper wire joins an aluminum wire whose diameter is twice that of the copper. The same current flows in both wires. The density of conduction electrons in copper is $1.1\times10^{29}\ m^{-3}$; in aluminum it's $2.1\times10^{29}\ m^{-3}$. Compare (a) the drift speeds and (b) the current densities in each wire.

40. In Fig. 24.17, a 100-mA current flows through a copper wire 0.10 mm in diameter, a salt solution in a 1.0-cm-diameter glass tube, and a vacuum tube where the current is carried by an electron beam 1.0 mm in diameter. The density of conduction electrons in copper is 1.1×10^{29} m^{-3}. The current in the solution is carried equally by positive and negative ions with charges $\pm 2e$; the density of each ion species is 6.1×10^{23} m^{-3}. The electron density in the beam is 2.2×10^{16} m^{-3}. Find the drift speed in each region.

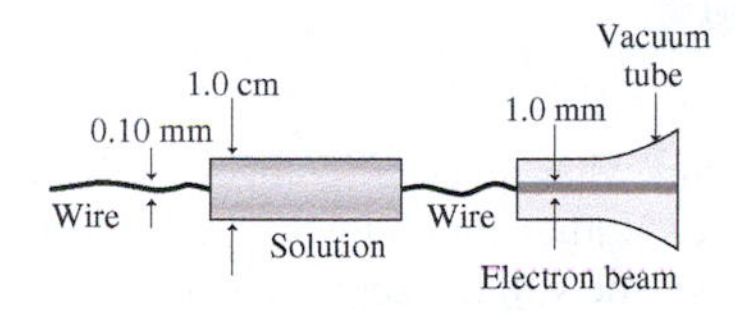

FIGURE 24.17 Problem 40

41. **BIO** In a study of proteins mediating cell membrane transport, biologists measure current versus time through the cell membranes of oocytes (nearly mature egg cells) taken from the African clawed frog, *Xenopus*. The measured current versus time is given approximately by $I = 60t + 200t^2 + 4.0t^3$ with t in seconds and I in nA. Find the total charge that flows through the cell membrane in the interval from $t = 0$ to $t = 5.0$ s.
42. There's a 2.5-V potential difference between opposite ends of a 6.0-m-long iron wire 1.0 mm in diameter. Assuming a uniform electric field in the wire, find (a) the current density and (b) the total current.
43. The maximum safe current in 12-gauge (2.1-mm-diameter) copper wire is 20 A. Find (a) the current density and (b) the electric field under these conditions.
44. Silver and iron wires of the same length and diameter carry the same current. How do the voltages across the two compare?
45. You have a cylindrical piece of material 2.4 cm long and 2.0 mm in diameter. When you attach a 9-V battery to its ends, a 2.6-mA current flows. Which material from Table 24.1 do you have?
46. How must the diameters of copper and aluminum wire be related if they're to have the same resistance per unit length?
47. You're writing the instruction manual for a power saw, and you have to specify the maximum permissible length for an extension cord made from 18-gauge copper wire (diameter 1.0 mm). The saw draws 7.0 A and needs a minimum of 115 V across its motor when the outlet supplies 120 V. What do you specify for the maximum length extension cord, given that they come in 25-foot increments?
48. **BIO** An implanted pacemaker supplies the heart with 72 pulses per minute, each pulse providing 6.0 V for 0.65 ms. The resistance of the heart muscle between the pacemaker's electrodes is 550 Ω. Find (a) the current that flows during a pulse, (b) the energy delivered in one pulse, and (c) the average power supplied by the pacemaker.
49. A solid rectangular iron bar measures 0.50 cm by 1.0 cm by 20 cm. Find the resistance between each of the three pairs of opposing faces, assuming the faces in question are equipotentials.
50. **BIO** Each pulse produced by the Taser described in the Application on page 449 typically delivers 100 μC of charge to the victim. Use this value, along with other quantities given in the Application, to find (a) the instantaneous current during a pulse, (b) the average current, and (c) the effective resistance of the victim between the points where the probes make contact.
51. The Nissan Leaf is an all-electric car powered by a 107-hp electric motor and a lithium-ion battery that stores 24 kWh and produces 394 V at its terminals when fully charged. The Leaf's battery can charge at the rate of 3.3 kW from a standard 120-V power outlet, at 6.6 kW from a 240-V outlet, and at 44 kW using a special 480-V charger. The Leaf's fuel economy is 3.38 miles per kWh, the equivalent of 114 miles per gallon in a gasoline-powered car. Find (a) the range of the Leaf, assuming the battery can be fully depleted, (b) the charging time for each mode, and (c) the current delivered by the fully charged battery when the motor is operating at full power.
52. **DATA** An electric heater is tested by immersing it in 0.500 kg of water and measuring the time Δt it takes to raise the water temperature by 10.0°C. The experiment is repeated for different currents I through the heater, and the results are tabulated below. Determine two quantities, based on heating time and current, which, when plotted, will give a straight line. Make your plot, determine a best-fit line, and use it to find the heater's resistance. *Hint*: You may need to consult Chapter 16.

I(A)	2.00	4.00	6.00	8.00	10.0
Δt (s)	422	112	44.3	28.2	16.9

53. Magnetic effects involving Jupiter's moon Io result in an effective voltage of about 400 kV, which drives current of some 5 MA between Io and Jupiter's polar regions. At Jupiter the current produces auroras analogous to those on Earth, as well as powerful bursts of radio waves that help radio astronomers analyze the Jovian current system. Estimate the total power associated with this current system, and compare with the 2-TW rate at which humankind consumes electrical energy.
54. At a particular point in a material with resistivity ρ the current density has magnitude J. Show that the power per unit volume dissipated at this point is $J^2\rho$.
55. A thermally insulated container of seawater carries a uniform current density of 75 mA/cm^2. How long does it take for its temperature to increase from 15°C to 20°C? Use the result of the preceding problem and any other information you might look up.
56. You're heading out for spring break, but your car won't start. Your friend says you might have corrosion at the battery terminals—a frequent cause of hard starting because of increased resistance. Having read Example 24.4, you know that the resistance between battery and starter should be around 1 mΩ. While your friend cranks the starter, you measure 4.2 V between the battery terminal and the wire carrying current to the starter motor. If the motor draws 125 A, is the resistance in its normal range?
57. Two cylindrical resistors are made from the same material and have the same length. When connected across the same battery, one dissipates twice as much power as the other. How do their diameters compare?
58. You're working on a new high-speed rail system. It uses 6000-horsepower electric locomotives, getting power from a single overhead wire with resistance 15 mΩ/km, at 25 kV potential relative to the track. Current returns through the track, whose resistance is negligible. Energy-efficiency standards call for no more than 3% power loss in the wire. How far from the power plant can the train go and still meet this standard?
59. A 100%-efficient electric motor is lifting a 15-N weight at 25 cm/s. How much current does it draw from a 6.0-V battery?
60. A power plant produces 1000 MW to supply a city 40 km away. Current flows from the power plant on a single wire with resistance 50 mΩ/km, through the city, and returns via the ground, which has negligible resistance. At the power plant the voltage between wire and ground is 115 kV. Find (a) the current in the wire and (b) the fraction of the power lost in transmission.
61. You're estimating costs for a new power line with your company's financial group. Engineering specifies a resistance per unit length of 50 mΩ/km. The costs of copper and aluminum wire are \$4.65/kg and \$2.30/kg and their densities are 8.9 g/cm^3 and 2.7 g/cm^3, respectively. Which material is more economical?

62. A 240-V electric motor is 90% efficient, meaning that 90% of the energy supplied to it ends up as mechanical work. If the motor lifts a 200-N weight at 3.1 m/s, how much current does it draw?
63. CH A metal bar has rectangular cross section 5.0 cm by 10 cm, as shown in Fig. 24.18. The bar has a nonuniform conductivity, and as a result the current density increases linearly from zero at the bottom to 0.10 A/cm^2 at the top. Find the total current in the bar.

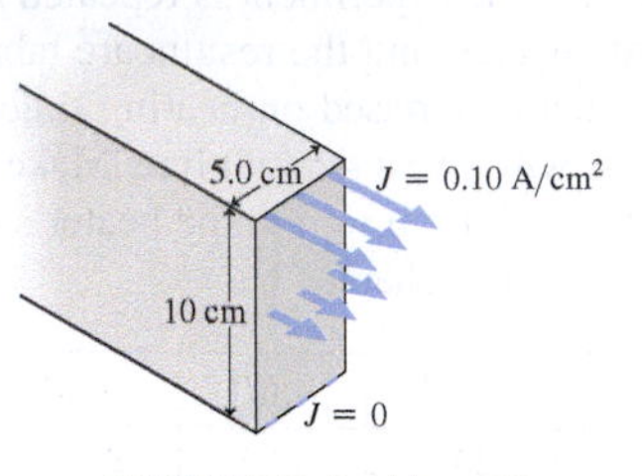

FIGURE 24.18 Problem 63

64. An immersion-type heating coil is connected to a 120-V outlet and immersed in a 250-mL cup of water initially at 10°C. The water comes to a boil in 85 s. Assuming no heat loss, and neglecting the heater's mass, find (a) the power and (b) the heater's resistance.
65. The resistivity of copper as a function of temperature is given approximately by $\rho = \rho_0[1 + \alpha(T - T_0)]$, where ρ_0 is Table 24.1's entry for 20°C, T_0 = 20°C, and $\alpha = 4.3\times10^{-3}\ ^\circ\text{C}^{-1}$. Find the temperature at which copper's resistivity is twice its room-temperature value.
66. Each atom in aluminum contributes about 3.5 conduction electrons. Find the drift speed in a 2.1-mm-diameter aluminum wire carrying 20 A.
67. CH A circular pan of radius b has a plastic bottom and metallic sidewall of height h. It's filled with a solution of resistivity ρ. A metal disk of radius a and height h is at the center, as shown in Fig. 24.19. The side and disk are essentially perfect conductors. Show that the resistance measured from side to disk is $R = \rho \ln (b/a)/2\pi h$.

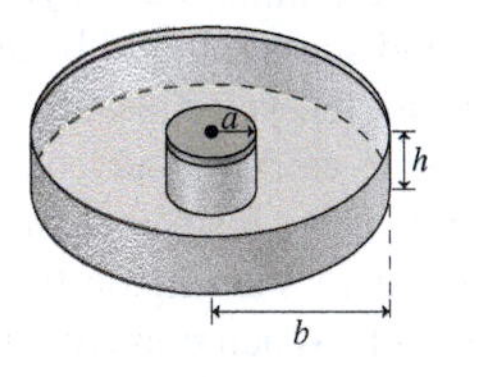

FIGURE 24.19 Problem 67

68. CH Figure 24.20 shows a truncated cone of material with resistivity ρ. Assume the equipotentials are planes parallel to the two faces, and integrate over slices of thickness dx like the one shown to find an expression for the total resistance between the faces.

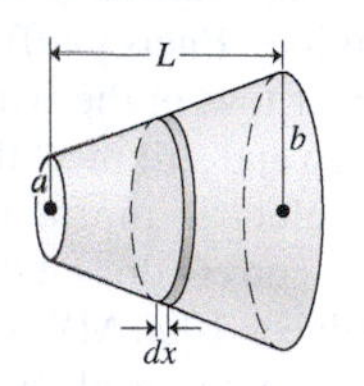

FIGURE 24.20 Problem 68

69. CH The current density in a particle beam with circular cross section of radius a points along the beam axis with a magnitude that decreases linearly from J_0 at the center ($r = 0$) to half that value at the edge ($r = a$). Find an expression for the total current in the beam.
70. CH A cylindrical resistor is 5.0 mm in diameter and 1.5 cm long. It's made of a composite material whose resistivity varies from one end to the other according to the equation $\rho = \rho_0(1 + x/L)e^{x/L}$, for $0 \le x \le L$, where $\rho_0 = 2.41\times10^{-3}\ \Omega\cdot\text{m}$. Find its resistance.
71. CH You work for an automobile manufacturer developing a new plug-in hybrid car. The car's mass is 1200 kg, and it uses a 360-V battery driving an electric motor that can handle a maximum current of 180 A. You're to specify the greatest slope the car can climb, maintaining 60 km/h, without its gasoline engine coming on to assist.

Passage Problems

A *brownout* occurs when an electric utility can't supply enough power to meet demand. Rather than cut off some customers completely, the utility reduces the voltage across its system. Brownouts are most likely on hot summer days, when heavy air-conditioning loads drive up demand for electricity. In a particular brownout, the utility reduces the voltage by 10%.

72. During the brownout, the current in conductors whose resistance is nearly independent of temperature
    a. decreases by approximately 10%.
    b. decreases by approximately 20%.
    c. decreases by approximately 5%.
    d. You can't tell without knowing the resistance.
73. Which of the following occurs in the conductors of the preceding problem during the brownout?
    a. Both the electric field and electron drift speed decrease.
    b. The electric field decreases but the electron drift speed doesn't.
    c. The current is carried by fewer electrons.
    d. The electrons undergo more frequent collisions.
74. During the brownout, the power dissipated in conductors whose resistance is nearly independent of temperature
    a. decreases by approximately 10%.
    b. decreases by approximately 20%.
    c. decreases by approximately 5%.
    d. You can't tell without knowing the resistance.
75. Metallic conductors like lightbulb filaments and electric stove burners have resistance that increases with increasing temperature. During the brownout, the current in such devices
    a. decreases by 10%.
    b. decreases by more than 10%.
    c. decreases by less than 10%.
    d. You can't tell without knowing more about how the resistance varies.

Answers to Chapter Questions

Answer to Chapter Opening Question

Collisions between electrons and the metal ions in the filament dissipate electric energy as heat. The energy results from the electrons' being accelerated by an electric field.

Answers to GOT IT? Questions

24.1 (a) current, right to left; (b) current, up; (c) current, left; (d), (e) no current
24.2 (1) $J_A < J_B$; (2) $E_A < E_B$; (3) $v_{dA} < v_{dB}$
24.3 (1) (b) it's twice as long as (c) but with one-fourth the area and half the resistivity; (2) (a) because it has the lowest resistance
24.4 (a)
24.5 No, you could still drill or cut into a live electric wire, putting metal parts of the tool at dangerous voltages.

25 Electric Circuits

What You Know

- You understand the concepts of voltage, current, and resistance.
- You know how Ohm's law describes a linear relationship among these three quantities that holds in many materials.
- You recognize several different mechanisms of conduction—in metals, ionic solutions, plasmas, semiconductors, and superconductors.
- You know how to calculate electric power.
- You understand the relation between charge and voltage in capacitors.

What You're Learning

- Here you'll learn how to "read" electric-circuit diagrams, identifying individual components and their interconnections.
- You'll see how series and parallel combinations allow you to analyze simple circuits.
- You'll learn how loop and node laws can help you analyze more complex circuits.
- You'll learn how to use electrical measuring instruments.
- You'll see how capacitors result in time-dependent behavior of circuits.

How You'll Use It

- You'll revisit circuit concepts in Chapter 28, when you study alternating current.
- Understanding circuits will help you to use and appreciate the many electrical and electronic devices in both your personal and professional lives.
- You'll encounter natural electric circuits in fields as diverse as cell biology, atmospheric science, oceanography, and astrophysics.

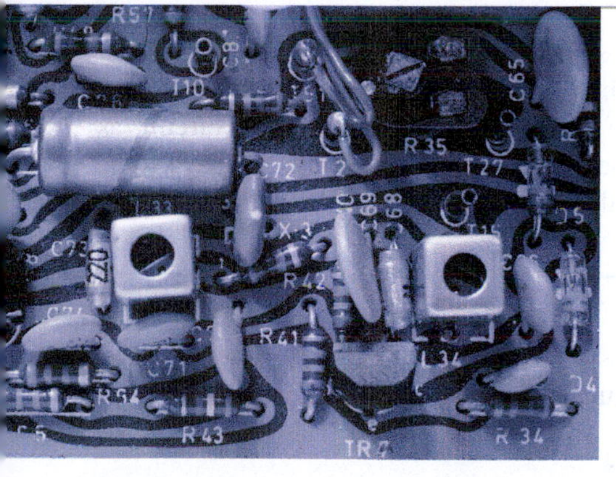

An electronic circuit board is a complex interconnection of electronic components. What two fundamental principles allow us to analyze even the most complex circuits?

An **electric circuit** is a collection of electrical components connected by conductors. Human-made circuits range from simple flashlights to computers. Electric circuits also exist in nature, including your own nervous system and Earth's atmospheric circuit in which thunderstorms are the batteries and the atmosphere a resistor. Understanding circuits will help you use effectively and safely the myriad electrical devices in your life and can even help you design new devices or troubleshoot old ones.

25.1 Circuits, Symbols, and Electromotive Force

We diagram circuits using standard symbols for circuit components and lines to represent wires (Fig. 25.1). We usually approximate wires as perfect conductors; then all points connected by a wire are at the same potential and are electrically equivalent. Realizing this will help you understand circuit diagrams.

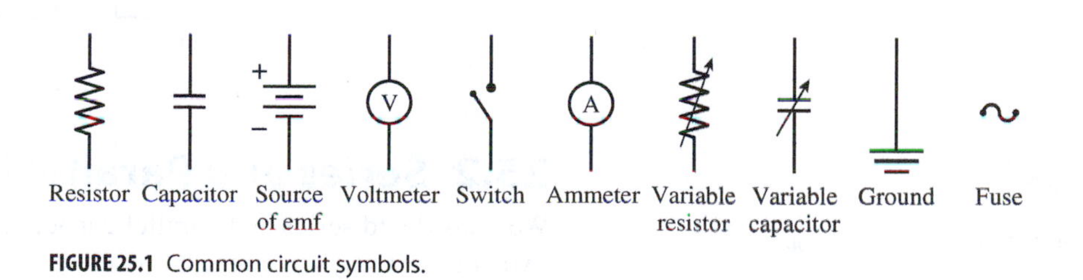

FIGURE 25.1 Common circuit symbols.

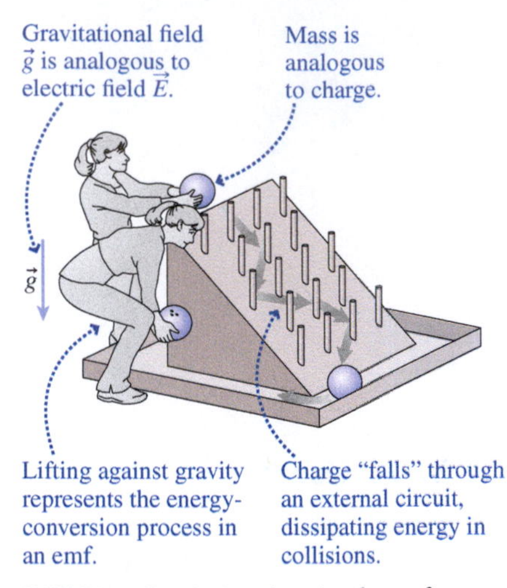

FIGURE 25.2 Gravitational analog for emf.

It takes an electric field to drive current through a conductor with nonzero resistance. But unless we actively maintain the field, charge will quickly move to establish electrostatic equilibrium, with no field inside the conductor and no current. So we need a device that can maintain a fixed potential difference and therefore an electric field in a current-carrying conductor. Such a device is called a source of **electromotive force**, or **emf**. (The name "force" here is inaccurate and is used only for historical reasons.) Most sources of emf have two **terminals** for connection to other circuit components. An emf converts some other form of energy to electrical energy by separating positive and negative charge to maintain a fixed potential difference between its terminals. The most familiar example is a battery, in which chemical reactions drive charge to the two terminals. Others include electric generators, which convert mechanical to electrical energy; photovoltaic cells, which use sunlight to separate charge; and cell membranes, which control ion flow into and out of the cell.

When a source of emf is connected to an external circuit, current flows through the circuit from the emf's positive terminal to the negative terminal. Energy-conversion processes in the emf then "lift" charge against the emf's internal electric field, maintaining a fixed potential difference across its terminals. The charge then "falls" through the external circuit, dissipating its energy in the circuit resistance. The result is a steady current, driven by the constant voltage across the emf. Figure 25.2 shows a gravitational analogy for an emf connected across an external circuit.

Quantitatively, emf is the work per unit charge involved in "lifting" charge against the electric field. Its units are therefore volts. An **ideal emf** maintains the same voltage across its terminals under all conditions. Real emfs have internal energy losses, and the terminal voltage may not equal the rated emf.

In Fig. 25.3 an ideal battery of emf $\mathcal{E}$ drives current through resistor R. We're assuming the wires connecting the battery and the resistor are perfect conductors, so the voltage across the resistor is equal to the battery's emf. Ohm's law then gives the resistor current: $I = \mathcal{E}/R$. Energetically, this circuit is analogous to Fig. 25.2: Charge gains $\mathcal{E}$ joules per coulomb as it's "lifted" against the electric field inside the battery, then dissipates that energy in heating the resistor.

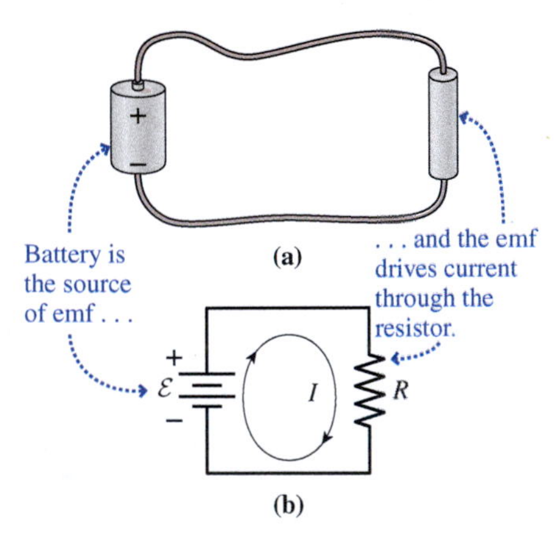

FIGURE 25.3 A circuit consisting of a battery and a resistor: (a) physical circuit; (b) schematic diagram.

✓TIP Don't Get Hung Up on Wires

We approximate wires as perfect conductors, so it takes no potential difference to drive current through a wire. Thus all points on the wire are at the same potential and are electrically equivalent. That means there are many ways to draw the same circuit; as long as two points are connected by a wire, that's all that matters. Real wires have some resistance, but if it's negligible compared with other resistances in the circuit, then we can approximate the wires as being ideal.

GOT IT? 25.1 The figure shows three circuits. Which are electrically equivalent?

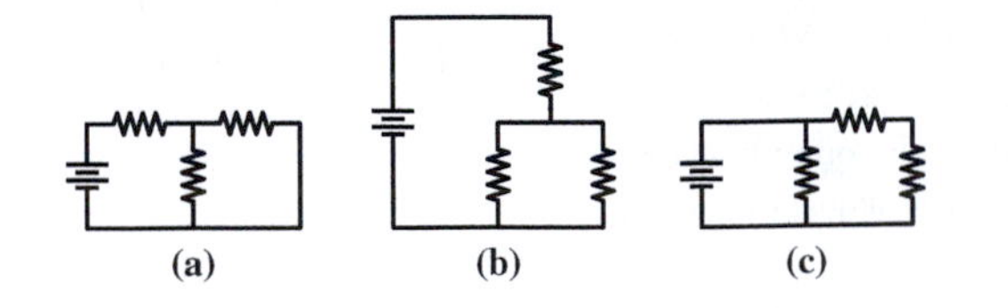

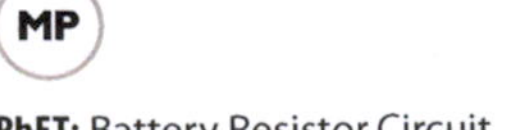

PhET: Battery Resistor Circuit

25.2 Series and Parallel Resistors

We considered series and parallel capacitors in Section 23.3. Series and parallel are the two simplest ways to connect *any* electric components. Two components are in series if the current flowing through one component has nowhere to go but through the other component. Two components are in parallel if they're connected together at each end. Here we'll consider series and parallel resistors.

Series Resistors

Figure 25.4 shows a circuit with two resistors in series. We'd like to know the current through and the voltage across each resistor. Neither is connected directly across the battery, so we can't argue that either resistor "sees" the battery emf. But the resistors are in series, and that means the only place for current to go after R_1 is through R_2. In a steady state, with no charge buildup in the circuit, that means the current through both resistors—and through the battery as well—must be the same. This is true whenever circuit components are in series:

The current through circuit components in series is the same.

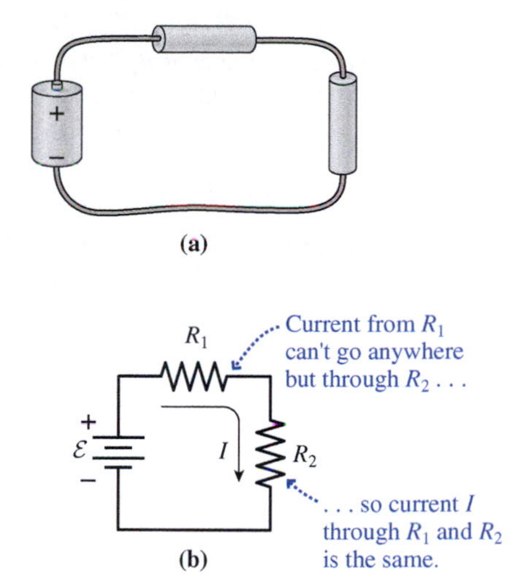

FIGURE 25.4 A battery and two resistors in series: (a) physical circuit; (b) schematic diagram.

If I is the current in Fig. 25.4, then by Ohm's law there must be a voltage $V_1 = IR_1$ across R_1 to drive the current through this resistor. Similarly, the voltage across R_2 is $V_2 = IR_2$. Thus, the voltage across the two resistors together is $V_1 + V_2 = IR_1 + IR_2$. But the battery is connected directly across this series combination, so the voltage across the two resistors together is the same as the battery emf $\mathcal{E}$. Therefore $IR_1 + IR_2 = \mathcal{E}$, or

$$I = \frac{\mathcal{E}}{R_1 + R_2}$$

Comparison with Ohm's law in the form $I = V/R$ shows that the two resistors in series behave like an equivalent resistance equal to the sum of their resistances. In an obvious generalization to more resistors in series, we have

MP PhET: Circuit Construction Kit (DC Only)

$$R_{\text{series}} = R_1 + R_2 + R_3 + \cdots \quad \text{(series resistors)} \tag{25.1}$$

In other words, resistors in series add.

Given the current, we can use Ohm's law in the form $V = IR$ to solve for the voltage across each resistor:

$$V_1 = \frac{R_1}{R_1 + R_2}\mathcal{E} \quad \text{and} \quad V_2 = \frac{R_2}{R_1 + R_2}\mathcal{E} \tag{25.2a, b}$$

These expressions show that the battery voltage divides between the two resistors in proportion to their resistance. For this reason a series combination of resistors is called a **voltage divider**.

✓TIP How Does the Battery Know?

How does the battery in Fig. 25.4 "know" how much current to supply? For a brief instant when the circuit is first connected, it doesn't. But in a very short time an electric field is established throughout the wires and resistors, and the circuit settles into a steady state, with the same current everywhere. Later, with circuits including capacitors, we'll analyze the approach to the steady state; for now, assume that the circuit reaches that state essentially instantaneously.

MP PhET: Signal Circuit

EXAMPLE 25.1 Series Resistors: Designing a Voltage Divider

A lightbulb with resistance 5.0 Ω is designed to operate at a current of 600 mA. To operate this lamp from a 12-V battery, what resistance should you put in series with it?

INTERPRET This problem is about a series circuit like Fig. 25.4, with the two resistors being the lightbulb and the unknown R_1.

DEVELOP We've sketched the circuit in Fig. 25.5, taking R_1 as the unknown and R_2 as the 5-Ω lightbulb. The same current flows through series resistors, so our plan is to find an expression for that current

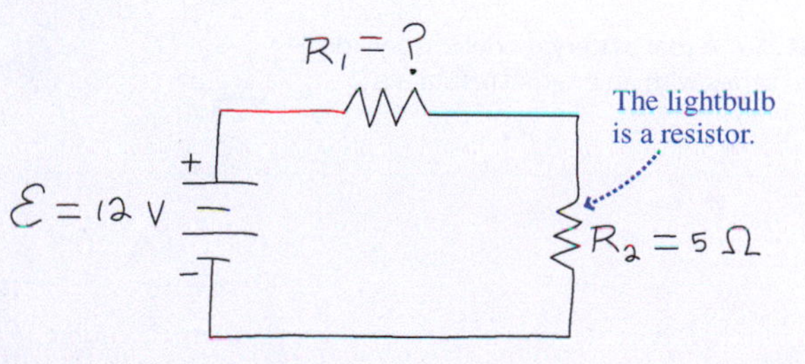

FIGURE 25.5 Sketch for Example 25.1.

(continued)

and then solve for the value of R_1 that will make the current 600 mA. Since resistors in series add, the current through both resistors follows from Ohm's law: $I = \mathcal{E}/(R_1 + R_2)$.

EVALUATE We solve for R_1 to get

$$R_1 = \frac{\mathcal{E} - IR_2}{I} = \frac{12\text{ V} - (0.60\text{ A})(5.0\,\Omega)}{0.60\text{ A}} = 15\,\Omega$$

ASSESS Make sense? The lightbulb's operating voltage is

$$V = IR_2 = (0.60\text{ A})(5.0\,\Omega) = 3.0\text{ V}$$

This is one-fourth of the battery voltage, so Equation 25.2b shows that the bulb's 5-Ω resistance should be one-fourth of the total. That makes the total 20 Ω, leaving 15 Ω for R_1. This isn't a very efficient way to run the bulb, since a lot more energy gets dissipated in R_1 than goes into lighting the bulb. Better to use a 3-V battery and no resistor. ■

GOT IT? 25.2 Rank from highest to lowest the voltages across the identical resistors R at the top of each circuit shown, and give the actual voltage for each. In (a) the second resistor has the same resistance R, and in (b) the gap is an open circuit (infinite resistance).

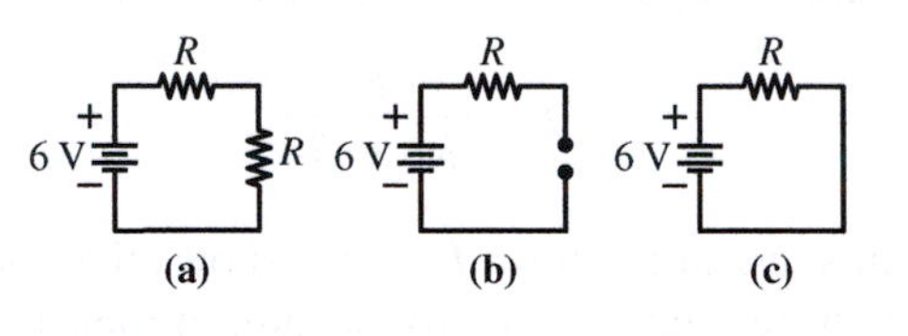

Real Batteries

FIGURE 25.6 Both batteries are rated at 1.5 V, but they have different internal resistances. Which do you think has the higher R_{int}?

What's the difference between the two 1.5-V batteries in Fig. 25.6? If they were ideal, both would maintain 1.5 V across their terminals no matter how much current was flowing. But these are real batteries. Chemical reaction rates limit the current, so it's not surprising that the larger battery can deliver more current.

We model a real battery as an ideal emf in series with an **internal resistance** (Fig. 25.7). Now, there's no such thing as an ideal emf; if there were, that's what battery companies would sell! And just one ideal emf could supply infinite power, thus solving all the world's energy problems. So a real battery isn't made by connecting an internal resistance to an ideal emf. Rather, the internal resistance is intrinsic to the battery, and there's no way to circumvent it. Some of it is actual resistance, but most represents the limited rate at which chemical reactions can separate charge. For a given battery voltage, lower internal resistance implies a more powerful battery—one that can deliver more current.

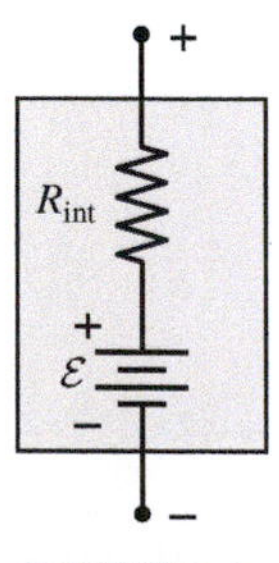

FIGURE 25.7 A real battery modeled as an ideal emf in series with an internal resistance.

Figure 25.8 shows that the internal resistance R_{int} is in series with the external load R_L to which the battery supplies power; the resulting circuit is a voltage divider. If R_{int} is small compared with R_L, Equation 25.2b shows that the voltage across the load will be very nearly the battery voltage. Then the battery is behaving nearly ideally because it has essentially $\mathcal{E}$ volts across its terminals. But if we lower R_L, more current flows and more voltage drops across R_{int}—and that leaves less voltage at the battery terminals and across the load. Even if we short-circuit the battery (not a good idea!), we won't get an infinite current; in fact, we'll get $I = \mathcal{E}/R_{int}$, which is the most current this battery can deliver.

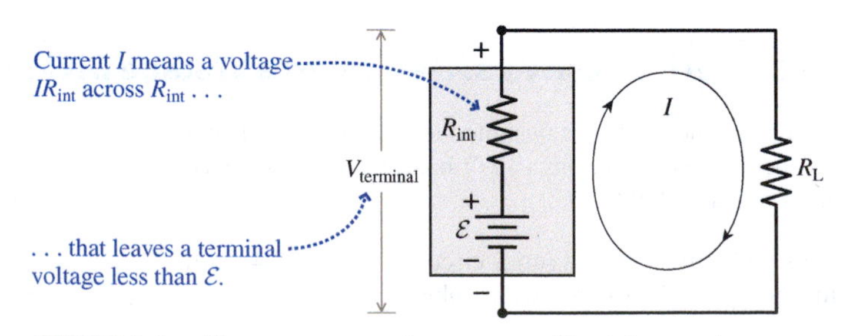

FIGURE 25.8 A real battery connected to an external load. Some voltage drops across the internal resistance, making the terminal voltage less than the battery's rated voltage.

EXAMPLE 25.2 Internal Resistance: Starting a Car

Your car has a 12-V battery with internal resistance 0.020 Ω. When the starter motor is cranking, it draws 125 A. What's the voltage across the battery terminals while starting?

INTERPRET This problem is about a real battery connected to a load, as in Fig. 25.8. We identify one resistor as the internal resistance and the load resistance as the starter motor.

DEVELOP Figure 25.9 is our sketch, showing the internal resistance in series with the load. The current is the same everywhere in a series circuit, so we can use Ohm's law to find the voltage across R_{int}. Subtracting that voltage from the battery's emf will then tell what's left across the load.

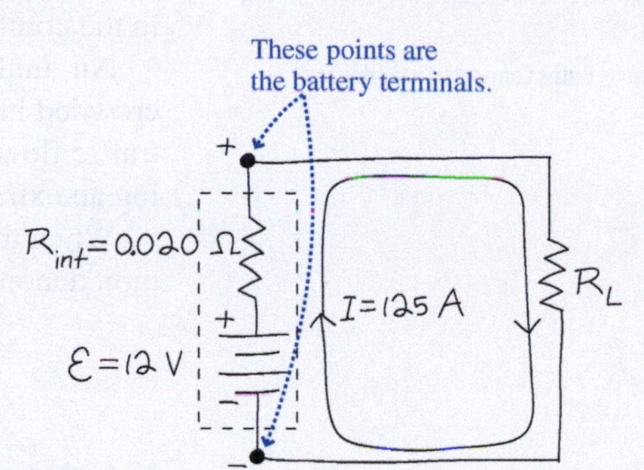

FIGURE 25.9 Sketch for Example 25.2.

EVALUATE For the internal resistance, Ohm's law gives

$$V_{\text{int}} = IR_{\text{int}} = (125\text{ A})(0.020\ \Omega) = 2.5\text{ V}$$

That leaves 12 V − 2.5 V or 9.5 V across the battery terminals.

ASSESS Make sense? That 9.5 V is substantially less than the battery's 12-V rating, so the battery is hardly behaving ideally. But the starter motor runs only briefly; most of the time the load on the battery—headlights, ignition system, electronics, and so on—draws far less current and so the battery behaves essentially like an ideal 12-V emf. A battery voltage of 9–11 V is typical during starting; much less than 9 V indicates a weak battery, a defective starter, or very cold weather. ■

Parallel Resistors

Figure 25.10 shows two resistors in parallel, connected across an ideal battery. Since the two resistors are connected at top and bottom by ideal wires, the voltage across each must be the same. We made this point in Chapter 23 when we discussed parallel capacitors, and it's worth repeating here:

> The voltage across circuit elements in parallel is the same.

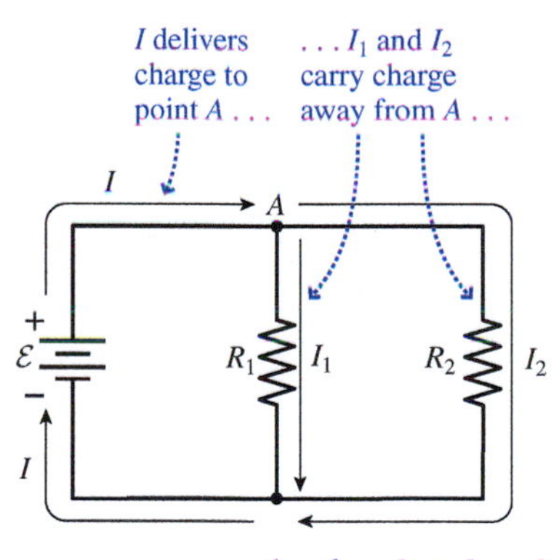

FIGURE 25.10 Parallel resistors connected across a battery.

The parallel resistors are connected directly across the battery, so their common voltage is the battery emf $\mathcal{E}$. Applying Ohm's law then gives the current through each resistor:

$$I_1 = \frac{\mathcal{E}}{R_1} \quad \text{and} \quad I_2 = \frac{\mathcal{E}}{R_2}$$

At point A in Fig. 25.10, a current I brings in charge from the battery, while the currents I_1 and I_2 carry charge away. Charge can't accumulate at this point (see Problem 67), so the incoming and outgoing currents must be equal: $I = I_1 + I_2$. Using our expressions for the two resistor currents gives

$$I = \frac{\mathcal{E}}{R_1} + \frac{\mathcal{E}}{R_2} = \mathcal{E}\left(\frac{1}{R_1} + \frac{1}{R_2}\right)$$

Comparison with Ohm's law in the form $I = V/R$ shows that the equivalent resistance of the parallel combination is given by

$$\frac{1}{R_{\text{parallel}}} = \frac{1}{R_1} + \frac{1}{R_2}$$

This result readily generalizes to more parallel resistors:

$$\frac{1}{R_{\text{parallel}}} = \frac{1}{R_1} + \frac{1}{R_2} + \frac{1}{R_3} + \cdots \quad \text{(parallel resistors)} \qquad (25.3\text{a})$$

Video Tutor Demo | **Bulbs Connected in Series and in Parallel**

In other words, resistors in parallel add reciprocally. Equation 25.3a shows that the resistance of a parallel combination is always lower than that of the lowest resistance in the combination. You should confirm this for yourself.

An analogy with highway traffic shows why this makes sense: Adding a lane to a crowded highway eases congestion (i.e., lowers the overall resistance), allowing a greater traffic flow (i.e., greater current). Putting one resistor in parallel with another is like adding an extra traffic lane.

When there are only two parallel resistors, we can rewrite Equation 25.3a using a common denominator to obtain

$$R_{\text{parallel}} = \frac{R_1 R_2}{R_1 + R_2} \tag{25.3b}$$

Note that *parallel* resistors combine in the same way as *series* capacitors, and vice versa.

GOT IT? 25.3 The figure shows all four possible combinations of three identical resistors. Rank them in order of highest to lowest resistance.

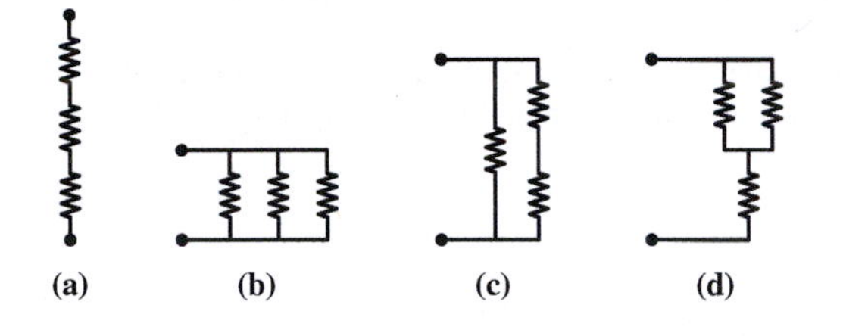

Analyzing Circuits

Many circuits contain series and parallel combinations. We analyze these circuits using the tactics outlined next, following the approach we used with series and parallel capacitors in Example 23.3.

TACTICS 25.1 Analyzing Circuits with Series and Parallel Components

1. Identify series and parallel combinations. Remember that components are in parallel *only* if they're connected directly together at each end. Components are in series *only* if current through one component has no place to go but through the next component. If you can't find at least one series or parallel combination, then you'll have to use the methods of Section 25.3.
2. Solve for the series and parallel equivalents using Equations 25.1 and 25.3 for resistors:

$$R_{\text{series}} = R_1 + R_2 + R_3 + \cdots \tag{25.1}$$

$$\frac{1}{R_{\text{parallel}}} = \frac{1}{R_1} + \frac{1}{R_2} + \frac{1}{R_3} + \cdots \tag{25.3a}$$

$$R_{\text{parallel}} = \frac{R_1 R_2}{R_1 + R_2} \tag{25.3b}$$

   If you're dealing with capacitors, use Equations 23.6 and 23.5, respectively.
3. Redraw the circuit, replacing series and parallel combinations with their one-component equivalents.
4. Repeat Steps 1–3, each time identifying series and parallel combinations and then reducing each to a single equivalent. Continue until either you've found the quantity you're asked for or the circuit consists of just an emf and one other component. You can then solve for the current in this component.
5. Work backward, replacing series and parallel equivalents with combinations of individual components. At each point apply Ohm's law, $I = V/R$, to find the currents through and/or the voltages across the individual components. As you work backward, remember that series components carry the same current as their series equivalent, and parallel components have the same voltage as their parallel equivalent. Continue until you're able to evaluate the quantity you're asked for.

EXAMPLE 25.3 Analyzing a Circuit: Series and Parallel Components

Find the current through the 2-Ω resistor in the circuit of Fig. 25.11*a*.

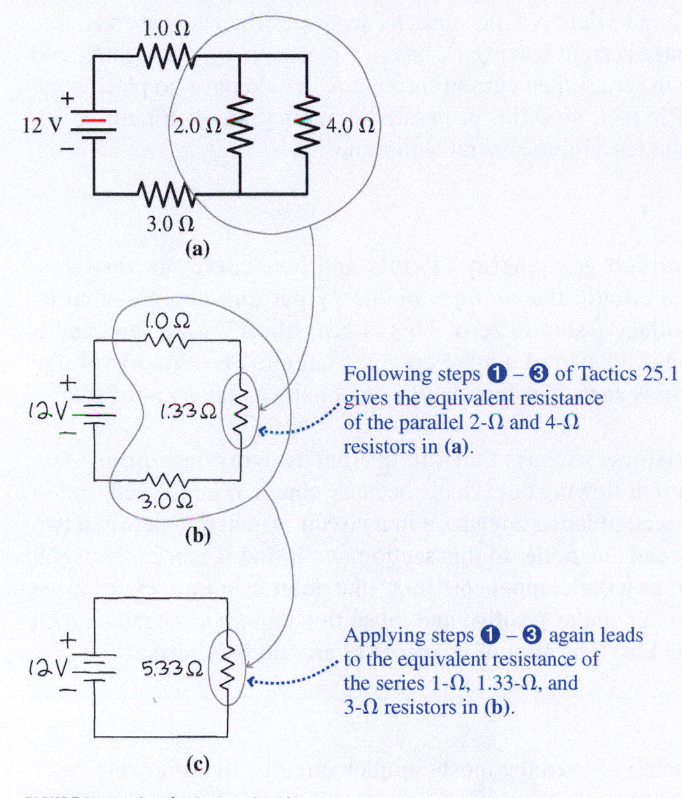

FIGURE 25.11 Analyzing a circuit.

INTERPRET This problem asks for the current in one resistor that's part of a more complex circuit. So it's about analyzing a circuit with series and parallel components.

DEVELOP We follow the steps in Tactics 25.1:

1. We identify the 2-Ω and 4-Ω resistors as being in parallel, and we find no other series or parallel resistor combinations. The 1-Ω resistor, for example, is *not* in series with either the 2-Ω or 4-Ω resistor because current leaving the 1-Ω resistor can take either of two paths.
2. We apply Equation 25.3b, $R_{\text{parallel}} = R_1R_2/(R_1 + R_2)$, to find the parallel combination: $(2\,\Omega)(4\,\Omega)/(2\,\Omega + 4\,\Omega) = 1.33\,\Omega$.
3. We redraw the circuit as Fig. 25.11*b*, replacing the two parallel resistors with their 1.33-Ω equivalent.
4. We repeat Steps 1–3 for the circuit in Fig. 25.11*b*, this time finding a series combination of three resistors. Applying Equation 25.1, $R_{\text{series}} = R_1 + R_2 + R_3$, gives 5.33 Ω for the equivalent resistance, and we redraw the circuit to get the simple circuit of Fig. 25.11*c*. Ohm's law, $I = V/R$, gives the current in the 5.33-Ω equivalent resistance: $I_{5.33\,\Omega} = (12\text{ V})/(5.33\,\Omega) = 2.25\text{ A}$.
5. Now we work backward, "unsimplifying" the circuit. That 5.33-Ω resistor is really the series combination in Fig. 25.11*b*; since the current through series components is the same, 2.25 A flows through each resistor—including the 1.33-Ω resistor that's really the parallel combination shown in Fig. 25.11*a*. We want the current in the 2-Ω member of that combination, and we could get that if we knew the voltage across it. But the voltage across parallel components is the same, and the same as the voltage across their parallel equivalent—in this case the 1.33-Ω resistance. We've found the current through that resistance, so Ohm's law gives the voltage: $V_{1.33\,\Omega} = I_{1.33\,\Omega}R_{1.33\,\Omega} = (2.25\text{ A})(1.33\,\Omega) = 3.0\text{ V}$.

EVALUATE Our last result is the voltage across each of the original parallel resistors, including the 2-Ω resistor whose current we want. So we're finally ready to compute our answer: $I_{2\Omega} = V_{2\Omega}/R_{2\Omega} = (3.0\text{ V})/(2.0\,\Omega) = 1.5\text{ A}$. Done!

ASSESS Make sense? A total of 2.25 A is flowing around the circuit; when it encounters the parallel combination, more should flow through the lower resistance, which is just what we found. Quantitatively, the current divides in inverse proportion to the parallel resistances, with 1.5 A through 2 Ω, and half as much, 0.75 A, through 4 Ω. Note how, in solving this problem, we used Ohm's law to find, alternately, voltage and then current in different resistances. ■

✓TIP Using Ohm's Law

Ohm's law relates the *voltage across a resistor* to the *current through that resistor*. It does *not* relate arbitrary voltages and currents anywhere in a circuit. Just because there's a 12-V battery in Fig. 25.11 doesn't mean there's 12 V across the 2-Ω resistor. And just because we found a total current of 2.25 A in Fig. 25.11*c* doesn't mean that's the current through the 2-Ω resistor.

GOT IT? 25.4 The figure shows a circuit with three identical lightbulbs and a battery. (1) Which, if any, of the bulbs is brightest? (2) What happens to each of the other two bulbs if you remove bulb C?

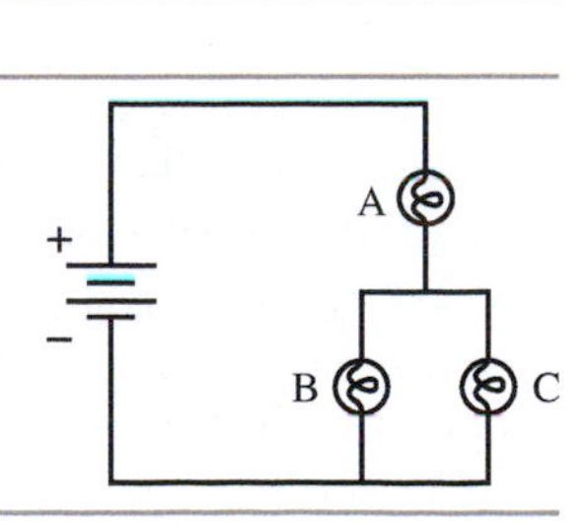

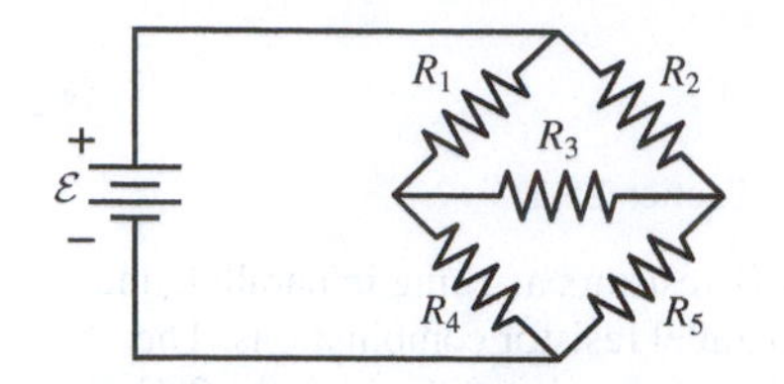

FIGURE 25.12 This circuit can't be analyzed using series and parallel combinations.

PhET: Circuit Construction Kit (DC Only)

25.3 Kirchhoff's Laws and Multiloop Circuits

Some circuits can't be simplified using series and parallel combinations. This happens when there's more than one emf, or when components are connected in complex ways. In Fig. 25.12, are R_1 and R_2 in parallel? No, because R_3 separates their lower ends. Are R_1 and R_4 in series? No, because current leaving R_1 has *two* places to go: through R_4 *and* through R_3. If R_1 and R_4 were in series, then current through R_1 would have no place to go except through R_4. There are, in fact, *no* series or parallel combinations in Figure 25.12. Analyzing circuits like this requires a more general technique.

Kirchhoff's Laws

Charges moving through a circuit gain energy at emfs and lose energy in resistors. If we go completely around a circuit, the *changes* in energy per unit charge—that is, increases and decreases in voltage—sum to zero. This is **Kirchhoff's loop law**, and it holds for any closed loop even if it's part of a more complex circuit: **The sum of voltage changes around a closed loop is zero**. The loop law is essentially a statement of energy conservation for circuits.

In analyzing parallel resistors, we saw that the current flowing into point *A* in Fig. 25.10 had to equal the current flowing out. That's because charge is conserved, and in a steady state charge can't be accumulating anywhere in a circuit. A junction between two or more circuit components is called a **node**. In this section, we'll find it especially useful to consider nodes where three or more components join, like point *A* in Fig. 25.10. If we count the currents flowing into a node as positive and those flowing out as negative, then we can state **Kirchhoff's node law**: **The sum of currents at any node is zero.**

Multiloop Circuits

Kirchhoff's laws allow us to analyze even the most complex circuits; the following strategy details the approach.

PROBLEM-SOLVING STRATEGY 25.1 Multiloop Circuits

INTERPRET

- Identify circuit loops and nodes. A loop is any complete closed path; a node is any point where three or more wires meet.
- Label the currents at each node, assigning a direction to each. The directions are arbitrary, and the actual direction may not be obvious.

DEVELOP

- For all but one node, write equations expressing Kirchhoff's node law: The sum of the currents at each node is zero. Take a current flowing into the node as positive, a current flowing out as negative.
- For as many independent loops as necessary, write equations expressing Kirchhoff's loop law: The sum of the voltage changes around a closed loop is zero. You can go either clockwise or counterclockwise, following these rules:
  - The voltage change going through a battery from the negative to the positive terminal is $+\mathcal{E}$; the voltage change from + to − is $-\mathcal{E}$.
  - For resistors traversed in the direction you've assigned to the current, the voltage change is $-IR$; for the opposite direction, it's $+IR$.
  - For other circuit components, use each component's characteristics to determine the voltage change.
- You don't need equations for all the nodes and loops because some are redundant, as you'll see in the next example.

EVALUATE Solve the equations to determine the unknown currents or other quantities.

ASSESS Assess your answer to see that it makes sense, paying particular attention to signs. A negative answer for a current means that the current actually flows opposite the direction you arbitrarily assigned.

EXAMPLE 25.4 Using Kirchhoff's Laws: A Multiloop Circuit

Find the current in resistor R_3 of Fig. 25.13*a*, following Problem-Solving Strategy 25.1.

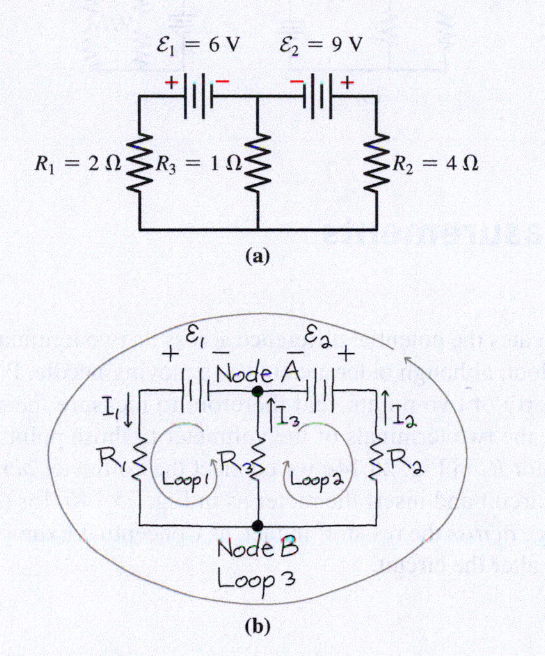

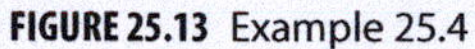

FIGURE 25.13 Example 25.4

INTERPRET In Fig. 25.13*b* we've redrawn the circuit with three loops and two nodes identified. We also labeled three currents at node A. The directions shown are completely arbitrary and may or may not be the actual directions of current flow. Because currents in series elements are the same, we can put the current labels anywhere on the series paths leading to the node. Our drawing shows that the same currents flow at node B; that's why one of the node equations is redundant. Note also that loop 3 comprises parts of loops 1 and 2, so equations for any two of these loops contain all the information we need. So one of the loop equations is redundant.

DEVELOP We need a current equation for one of the two nodes. Given the arbitrary directions we've assigned to the currents, the equation at node A is

$$-I_1 + I_2 + I_3 = 0 \qquad \text{(node A)}$$

To get the equation for loop 1, let's go counterclockwise around the loop, as shown. Starting at node A, we first encounter a positive voltage change $+\mathcal{E}_1$, then $-I_1R_1$, then $-I_3R_3$. So the loop 1 equation is $\mathcal{E}_1 - I_1R_1 - I_3R_3 = 0$. Here it's simplest if we substitute the numerical values shown in Fig. 25.13*a* and temporarily drop the units to avoid clutter. Then we have

$$6 - 2I_1 - I_3 = 0 \qquad \text{(loop 1)}$$

Loop 2 is similar except here we're going "backward" through R_2, so its term is positive:

$$9 + 4I_2 - I_3 = 0 \qquad \text{(loop 2)}$$

EVALUATE We want I_3, so we eliminate the other two currents. The node equation gives $I_1 = I_2 + I_3$; substituting in the loop 1 equation gives $6 - 2I_2 - 3I_3 = 0$ or $I_2 = \frac{1}{2}(6 - 3I_3)$. We use this result in the loop 2 equation to get $9 + 2(6 - 3I_3) - I_3 = 0$, or $21 - 7I_3 = 0$. Solving gives our answer: $I_3 = 3$ A.

ASSESS We assigned I_3 an upward direction through R_3, so our positive answer means that this is indeed the direction of the current in R_3. This makes sense because both batteries have their negative terminals at node A. If either battery had been reversed, however, the situation wouldn't have been so clear and we would have had to rely on the algebraic sign to determine the current direction. Even with the circuit as shown in Fig. 25.13, the directions of I_1 and I_2 depend on the relative strengths of the two batteries. With the values we're given, I_2 comes out -1.5 A, showing that the current actually flows downward in R_2. But if we reduce $\mathcal{E}_2$ to 2 V, I_2 becomes zero; lower still, and it flows upward and "backward" through $\mathcal{E}_2$. ■

APPLICATION The Cell Membrane

Many natural systems can be modeled as electric circuits. In 1952, Alan L. Hodgkin and Andrew F. Huxley developed a circuit model for the cell membrane; their work won them a share of the 1963 Nobel Prize for Physiology or Medicine. The figure shows a simplified version of the Hodgkin–Huxley model. The batteries $\mathcal{E}_K$, $\mathcal{E}_{Na}$, and $\mathcal{E}_L$ represent the electrochemical effects of potassium, sodium, and other ions, respectively; their emfs have values in the tens of millivolts. R_K, R_{Na}, and R_L are the resistances the cell membrane offers to each ionic species. The currents I_K, I_{Na}, and I_L represent ion flows across the membrane, and their values and signs follow from solving a multiloop-circuit problem similar to Example 25.4. The voltage V_M is the membrane potential between the inside and outside of the cell. The Hodgkin–Huxley model also contains a capacitor, which causes time-dependent behavior of the sort you'll see in Section 25.5.

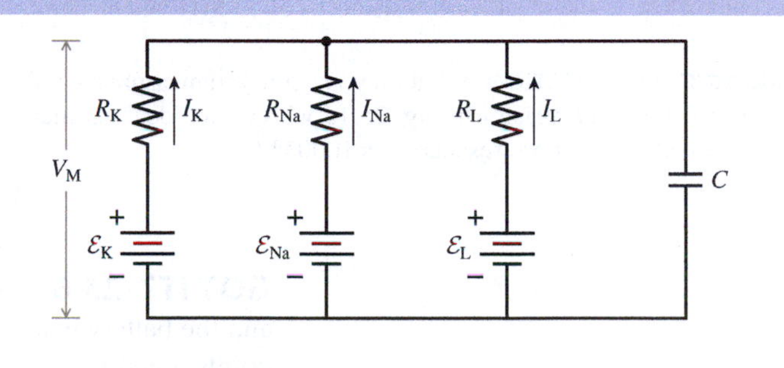

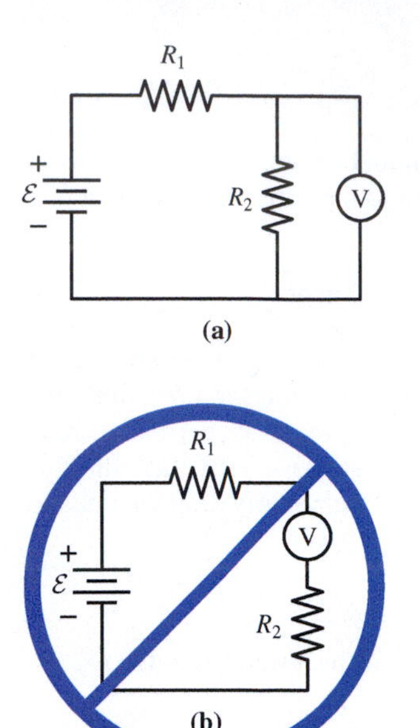

FIGURE 25.14 Correct (a) and incorrect (b) ways to connect a voltmeter for measuring the voltage across R_2.

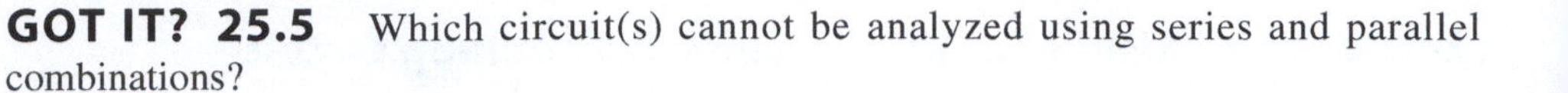

GOT IT? 25.5 Which circuit(s) cannot be analyzed using series and parallel combinations?

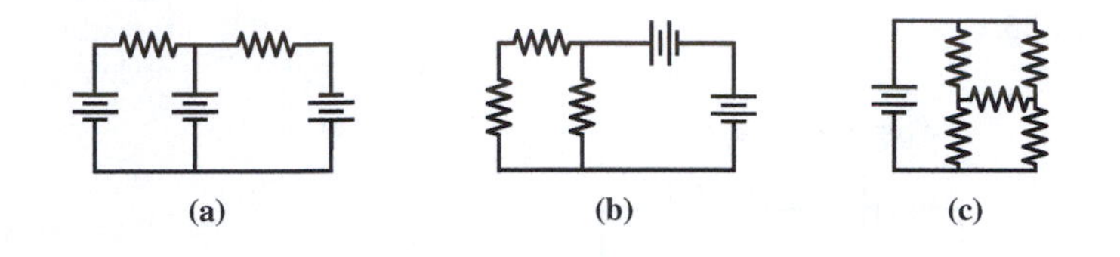

25.4 Electrical Measurements

Voltmeters

A **voltmeter** is a device that indicates the potential difference across its two terminals. The indicator is usually a digital readout, although older meters use a moving needle. Potential difference—voltage—is a property of two points, and therefore to measure the voltage between two points, we connect the two terminals of the voltmeter to those points. So to measure the voltage across resistor R_2 in Fig. 25.14*a* we connect the voltmeter *across* R_2, as shown. We do *not* break the circuit and insert the meter as in Fig. 25.14*b*, for then we wouldn't be measuring the voltage *across* the resistor; in fact, as Conceptual Example 25.1 makes clear, we would radically alter the circuit.

CONCEPTUAL EXAMPLE 25.1 Measuring Voltage

What should be the electrical resistance of an ideal voltmeter?

EVALUATE Before we attach the voltmeter in Fig. 25.14*a*, the battery "sees" the resistors R_1 and R_2 in series. When we connect a meter with resistance R_m, then we've got a parallel combination of R_1 and R_m where before we had just R_1. Since two parallel resistors have a lower resistance than either of the individual resistors, the overall circuit current increases—and so, therefore, does the voltage across R_1, which carries the total current. That in turn leaves the voltage across R_2 *lower* than before. Even if the meter is perfectly accurate, the voltage it reads will be lower than it was before we connected the meter.

How can we avoid this effect? By giving the meter a high resistance—ideally, infinite resistance, so the meter won't draw any current and therefore won't affect the circuit. The meter will therefore read the voltage that was there before we connected it.

ASSESS Truly infinite resistance is impossible—but as long as the meter's resistance is much larger than resistances in the circuit, the effect of finite meter resistance will be negligible. Modern digital meters come close to the ideal, with resistances of 10 MΩ and higher.

MAKING THE CONNECTION What do you get if you measure the voltage across the 40-Ω resistor in Fig. 25.15 with (a) an ideal voltmeter and (b) a voltmeter whose resistance is 1000 Ω?

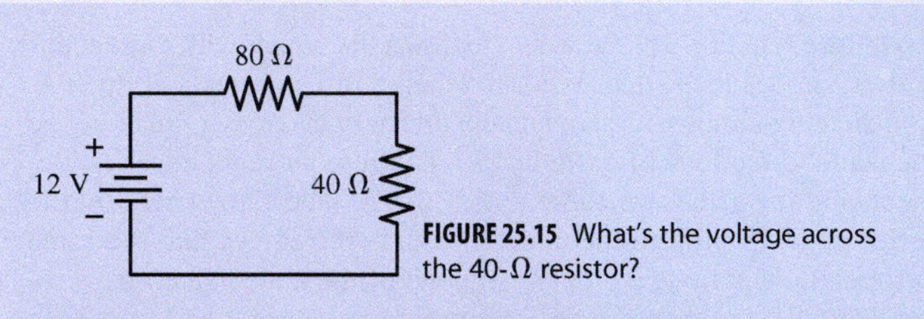

FIGURE 25.15 What's the voltage across the 40-Ω resistor?

EVALUATE (a) The circuit is a simple voltage divider, and Equation 25.2b shows that the voltage across the 40-Ω resistor is one-third of the battery voltage, or 4.00 V. With its infinite resistance, the ideal voltmeter doesn't alter the circuit, so it reads 4.00 V. (b) Connecting the voltmeter gives the circuit of Fig. 25.16; now Equation 25.3b gives 38.5 Ω for the parallel combination of the 1000-Ω meter and 40-Ω resistor. Applying Equation 25.2b to the resulting voltage divider gives 3.95 V, 2.5% lower than the ideal voltmeter.

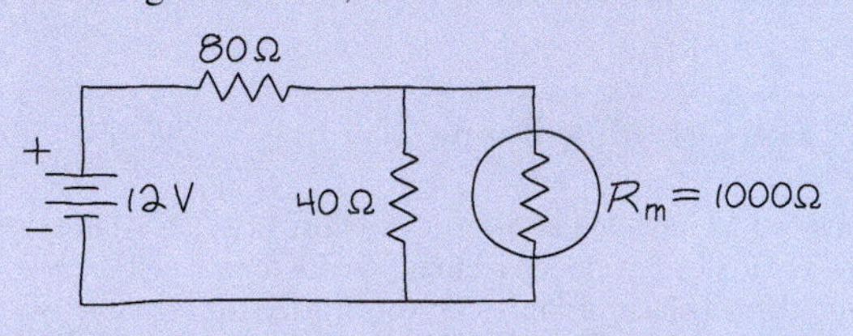

FIGURE 25.16 A nonideal voltmeter (R_m) alters the circuit.

GOT IT? 25.6 All resistors in the figure have the same value and the battery is ideal. If an ideal voltmeter is connected between points *A* and *B*, will it read (a) 10 V, (b) between 5 and 10 V, (c) 5 V, (d) between 0 and 5 V, or (e) 0 V?

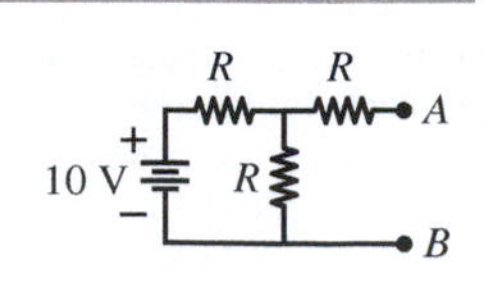

Ammeters

An **ammeter** measures the current flowing *through* itself. To measure the current through a circuit component, it's necessary to break the circuit and insert the ammeter in *series* with that component (Fig. 25.17*a*); only then will all the current also go through the meter. Connecting the ammeter across the resistor as in Fig. 25.17*b* is wrong because then the current through the resistor isn't going through the meter.

If the ammeter has any resistance, the total resistance of the circuit will increase with the meter connected in series. This in turn will decrease the current, giving an incorrect reading. So an ideal ammeter should have zero resistance. In practice, an ammeter's resistance should be much lower than typical resistances in the circuit being measured.

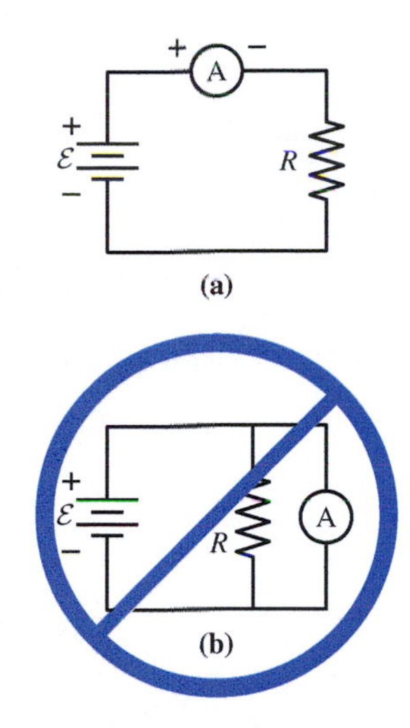

FIGURE 25.17 Correct (a) and incorrect (b) ways to connect an ammeter.

✓TIP Watch Your Language

A voltmeter measures potential difference *between* two points; hence, we connect it *across*—that is, in parallel with—the circuit element whose voltage we wish to measure. An ammeter measures the current *through* itself; hence, we connect it in *series* with the circuit element whose current we wish to measure. If you get used to voltages appearing *across* things and currents flowing *through* them, you'll have no trouble connecting meters. The ways to connect meters, and the words *across* for voltage and *through* for current, go back to the definitions of potential difference as a property of two points and of current as a flow.

Ohmmeters and Multimeters

Often we want the resistance of a particular component. Connecting a known voltage in series with an ammeter and the unknown resistance gives both current and voltage, letting us calculate the unknown resistance. A meter used for this purpose can be calibrated directly in ohms even though it's really measuring current; it's then an **ohmmeter**. The functions of voltmeter, ammeter, and ohmmeter are often combined in a single instrument called a **multimeter**.

25.5 Capacitors in Circuits

So far we've considered only circuits with steady current. A flashlight is a good example: Turn it on and current starts almost instantaneously, then flows steadily until you turn the flashlight off.

Capacitors alter this picture, causing circuit quantities to change more slowly. Recall that a capacitor is a pair of insulated conductors with charge and voltage related by $Q = CV$, where Q is the magnitude of the charge on either conductor, V is the potential difference between them, and C is the capacitance. Because charge and voltage are proportional in a capacitor, a change in voltage requires a change in charge. Charge changes when current flows through the wires connecting the capacitor to the rest of a circuit, and the magnitude of the current gives the rate at which capacitor charge increases or decreases. Since the current in any real circuit is finite, the charge on the capacitor cannot change instantaneously. But capacitor voltage is proportional to charge, so:

The voltage across a capacitor cannot change instantaneously.

This statement is the key to understanding circuits with capacitors. It says that the voltage on a capacitor can't jump abruptly from one value to another; mathematically, capacitor voltage V_C must be a continuous function of time, its derivative always finite. Just how rapidly the voltage can change depends on the capacitance and other circuit quantities, as you'll now see.

We consider an ***RC* circuit**, one that includes a resistor and capacitor. *RC* circuits are ubiquitous, appearing everywhere from microbiological structures to stereo amplifiers to giant energy-storage systems. We examine separately the two cases in which the capacitor is (1) charging and (2) discharging.

The *RC* Circuit: Charging

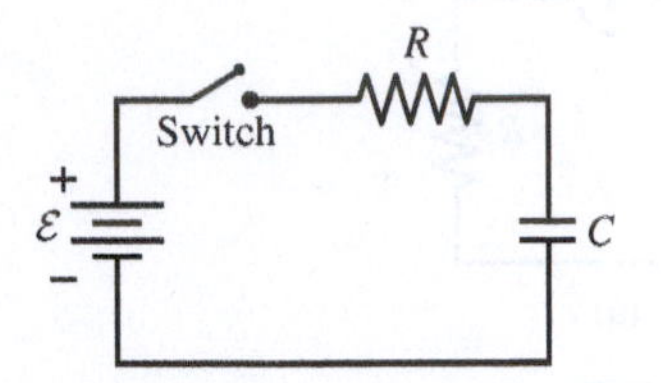

FIGURE 25.18 An *RC* circuit. The switch is closed at time $t = 0$.

In Fig. 25.18 the capacitor is initially uncharged, so the voltage across it is zero. Closing the switch connects the left side of the resistor to the battery, bringing its potential immediately to $\mathcal{E}$ volts (we're taking $V = 0$ at the negative battery terminal). The right end of the resistor remains at the same voltage as the upper capacitor plate—and that's still zero because the voltage across the capacitor can't change instantaneously.

So there are $\mathcal{E}$ volts across the resistor, and therefore a current $I = \mathcal{E}/R$ through it. This current delivers positive charge to the upper capacitor plate. At the same time, positive charge leaves the lower plate and flows back to the battery, making the lower plate negative. (As always with metallic conductors, it's actually negative electrons that are moving. But the effect is the same: a current flowing clockwise around the circuit in Fig. 25.18, resulting in the upper plate of the capacitor becoming more positive and the lower plate more negative.)

As charge accumulates on the plates, the capacitor voltage increases proportionately. But the capacitor and resistor voltages sum to the battery voltage $\mathcal{E}$, so as the capacitor voltage increases, the resistor voltage drops. By Ohm's law, the resistor current $I = V/R$ drops as well. This in turn decreases the *rate* at which charge accumulates in the capacitor. The capacitor voltage continues to increase as charge accumulates, but at an ever-slower rate.

Eventually the capacitor voltage approaches the battery voltage, and the resistor voltage and current tend to zero; so, therefore, does the rate at which charge accumulates on the capacitor. The whole system approaches a final state in which the capacitor is charged to the full battery voltage and the current in the circuit is zero. Figure 25.19 summarizes the interplay among current, charge, and voltage.

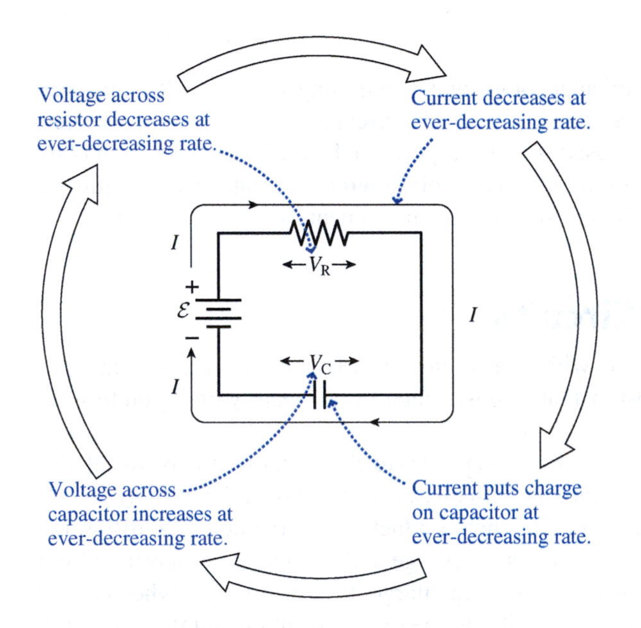

FIGURE 25.19 Interrelationships among quantities in a charging *RC* circuit.

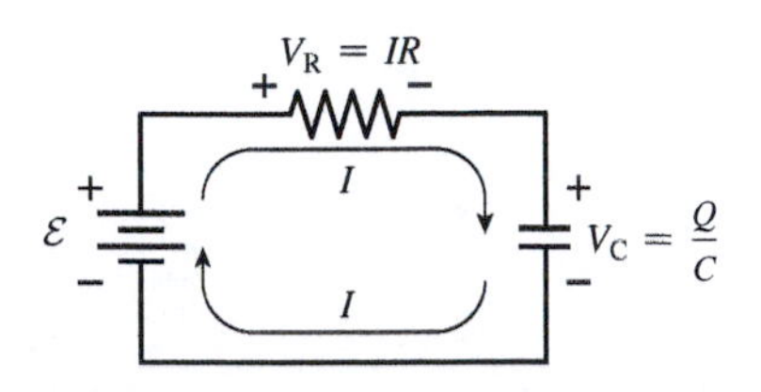

FIGURE 25.20 Voltage changes in a charging *RC* circuit.

We can analyze the circuit of Fig. 25.18 quantitatively using the loop law. Going clockwise around the loop, we first encounter a voltage increase $\mathcal{E}$ across the battery, then a drop IR across the resistor, then a drop V_C from the upper to lower capacitor plate (Fig. 25.20). But $V_C = Q/C$, so the loop equation becomes

$$\mathcal{E} - IR - \frac{Q}{C} = 0$$

This equation contains two unknowns, I and Q, but they're related because the current is the rate at which charge is accumulating on the capacitor: $I = dQ/dt$. To use this relation, we take the time derivative of the loop equation:

$$-R\frac{dI}{dt} - \frac{1}{C}\frac{dQ}{dt} = 0$$

The battery voltage $\mathcal{E}$ doesn't appear here because it's constant, so its derivative is zero. Using $I = dQ/dt$ and rearranging the equation gives

$$\frac{dI}{dt} = -\frac{I}{RC} \tag{25.4}$$

This equation shows that the rate of change of current is proportional to the current itself. Equations like this arise whenever a quantity changes at a rate proportional to the quantity itself. Population growth, the increase of money in a bank account, and the decay of a radioactive element are all described by similar equations.

Like the equation for simple harmonic motion in Chapter 13, Equation 25.4 is a *differential equation* because the unknown quantity I occurs in a derivative. The solution to a differential equation isn't a single number but a function expressing the relation between the unknown quantity—in this case current—and the independent variable—in this case time. We can solve this particular differential equation by multiplying both sides by dt/I in order to collect all terms involving I on one side of the equation. This gives

$$\frac{dI}{I} = -\frac{dt}{RC}$$

We can then integrate both sides, noting that RC is constant:

$$\int_{I_0}^{I} \frac{dI}{I} = -\frac{1}{RC}\int_0^t dt$$

where $I_0 = \mathcal{E}/R$ is the initial current at the time $t = 0$ just after the switch is closed and where the integration runs to an arbitrary time t, when the current has the value I. The integral on the left is the natural logarithm of I, and on the right it's just t. Then we have

$$\ln\left(\frac{I}{I_0}\right) = -\frac{t}{RC}$$

where we used $\ln I - \ln I_0 = \ln(I/I_0)$. To get an equation for I we exponentiate both sides, recalling that $e^{\ln x} = x$. This gives $I/I_0 = e^{-t/RC}$, or, since $I_0 = \mathcal{E}/R$,

$$I = \frac{\mathcal{E}}{R}e^{-t/RC} \tag{25.5}$$

Thus the current in the circuit decreases exponentially with time, in agreement with our qualitative analysis. The capacitor voltage is $V_C = \mathcal{E} - V_R$, or, since $V_R = IR = \mathcal{E}e^{-t/RC}$,

$$V_C = \mathcal{E}(1 - e^{-t/RC}) \quad (RC \text{ circuit, charging}) \tag{25.6}$$

Equation 25.6 shows that the capacitor voltage starts at zero and rises, with its rate of rise ever slowing as it gradually approaches the battery voltage $\mathcal{E}$—just as we reasoned in our qualitative analysis. Figure 25.21 plots capacitor voltage and current using the equations we've just derived.

When is the capacitor fully charged? Never, according to our equations! But the rate at which it approaches full charge is determined by the so-called **time constant**, RC—a characteristic time for changes to occur in a circuit containing a capacitor. Equation 25.6 shows that in one time constant, the voltage rises to $\mathcal{E}(1 - 1/e)$, or to about two-thirds of the battery voltage. A practical rule of thumb says that in five time constants ($t = 5RC$) a capacitor is 99% charged (see Exercise 33). The RC time constant clarifies our statement that the voltage across a capacitor can't change instantaneously. We can now say that the voltage can't change appreciably in times small compared with the time constant. On the other hand, after many time constants, we'll find essentially no current flowing to the capacitor. We've shown quantitatively the role of the time constant RC by marking the time in units of RC on Fig. 25.21.

Resistors and capacitors are available in a wide range of values, so practical values for RC span many orders of magnitude. RC circuits with time constants from microseconds to hours are widely used in electronic devices to control the rates at which electrical quantities vary. For example, circuits with RC many times the sixtieth-of-a-second period of standard AC power produce steady, direct-current power for audio and video equipment.

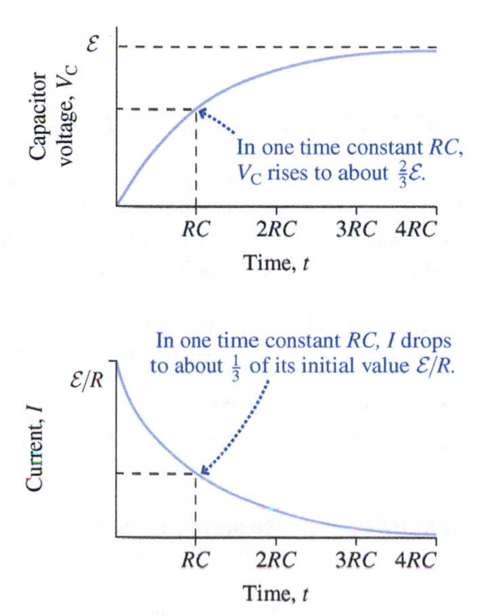

FIGURE 25.21 Time dependence of capacitor voltage and circuit current in a charging RC circuit. The approximate values $\frac{2}{3}$ and $\frac{1}{3}$ are actually $1 - 1/e$ and $1/e$, respectively.

Equalizers in audio systems are variable resistances in *RC* circuits; changing the resistance changes the time constant and therefore the way the circuit handles rapidly changing audio signals. Sometimes, though, the time constant can be a nuisance. Capacitance in audio systems can limit high-frequency response, decreasing the quality of music reproduction. With computer speeds in the GHz range—meaning basic operations occur billions of times a second—even tiny *RC* time constants associated with the resistance of wires and the capacitance of adjacent conductors can cause trouble.

The *RC* Circuit: Discharging

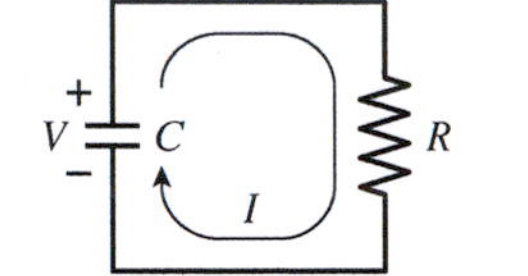

FIGURE 25.22 A discharging *RC* circuit.

Suppose we connect a charged capacitor across a resistor, as shown in Fig. 25.22. If the capacitor voltage is initially V_0, then when the circuit is connected, this voltage will drive a current $I_0 = V_0/R$ through the resistor. This current transfers charge from the positive to the negative capacitor plate, lowering the charge on the capacitor. Since capacitor charge and voltage are proportional, the capacitor voltage drops, too. So, therefore, does the current and therefore the rate at which the capacitor discharges. We expect both the voltage and current in this circuit to decay toward zero. In terms of energy, that happens because the energy stored in the capacitor's electric field is gradually dissipated as heat in the resistor.

The loop equation for Fig. 25.22 is particularly simple; going clockwise, we have $Q/C - IR = 0$, where the two terms are the voltage changes across the capacitor and resistor, respectively. Since we've indicated positive current in Fig. 25.22 in the direction that would *reduce* the capacitor charge Q, the rate of change dQ/dt and the current have opposite signs: $I = -dQ/dt$. Differentiating our loop equation and substituting this expression for I gives $dI/dt = -I/RC$. This is identical to Equation 25.4; the solution is therefore Equation 25.5, but with $I_0 = V_0/R$ instead of $\mathcal{E}/R$:

$$I = \frac{V_0}{R}e^{-t/RC} \qquad (25.7)$$

In this circuit the capacitor and resistor voltages are the same, since the two are in parallel (in this simplest of circuits, they're also in series). Because the resistor voltage and the current are proportional, the voltage across the capacitor and resistor is

$$V = V_0 e^{-t/RC} \quad (RC \text{ circuit, discharging}) \qquad (25.8)$$

Equations 25.7 and 25.8 show that the capacitor discharges with the same characteristic time constant *RC* that governs its charging.

EXAMPLE 25.5 Charging Capacitors: A Camera Flash

A camera flash gets its energy from a 150-μF capacitor and requires 170 V to fire. If the capacitor is charged by a 200-V source through an 18-kΩ resistor, how long must the photographer wait between flashes? Assume the capacitor is fully discharged with each flash.

INTERPRET This is a problem about a charging capacitor, and we want to find the time to reach a given voltage.

DEVELOP Equation 25.6, $V_C = \mathcal{E}(1 - e^{-t/RC})$, gives the voltage across a charging capacitor, so our plan is to solve this equation for the time t.

EVALUATE First we solve for the exponential term that contains the time:

$$e^{-t/RC} = 1 - \frac{V_C}{\mathcal{E}}$$

Then we take the natural logarithm of both sides, recalling that $\ln e^x = x$, so

$$-\frac{t}{RC} = \ln\left(1 - \frac{V_C}{\mathcal{E}}\right)$$

Solving for t and setting $V_C = 170$ V, $\mathcal{E} = 200$ V, $R = 18$ kΩ, and $C = 150$ μF gives

$$t = -RC\ln\left(1 - \frac{V_C}{\mathcal{E}}\right) = 5.1 \text{ s}$$

ASSESS The time constant here is $RC = 2.7$ s, and 170 V is well over two-thirds of the 200-V source. Therefore, we expect a charging time longer than one time constant. Our 5.1-s answer is nearly $2RC$. Problem 70 explores energy and power in this circuit. ■

RC Circuits: Long- and Short-Term Behavior

It's not always necessary to solve exponential equations in analyzing *RC* circuits. If we're interested only in times much shorter than the time constant, then it's enough to remember that the voltage across a capacitor can't change instantaneously. And after many time constants, a capacitor has essentially reached its final voltage, and no current is flowing to it. These conditions are sufficient to analyze circuits on short and long time scales.

TACTICS 25.2 Analyzing Long- and Short-Term Behavior of *RC* Circuits

Short-Term Behavior
For times much shorter than the time constant *RC*, capacitor voltage remains essentially unchanged. Therefore, you can replace the capacitor with a short circuit if it's uncharged or, if it's charged, with a battery whose emf is the capacitor's initial voltage. Then solve the circuit using the techniques of Section 25.2 or 25.3.

Long-Term Behavior
For times much longer than *RC*, no current is flowing to a capacitor. Therefore, you can replace the capacitor with an open circuit, and again solve using earlier techniques.

EXAMPLE 25.6 An *RC* Circuit: Long and Short Times

The capacitor in Fig. 25.23*a* is initially uncharged. Find the current through R_1 (a) the instant the switch is closed and (b) a long time after the switch is closed.

INTERPRET We interpret "the instant the switch is closed" to mean a time much shorter than the time constant *RC*, and "a long time" to mean a time much longer than *RC*. Then this is a problem involving the long- and short-term behavior of a circuit containing a capacitor.

DEVELOP We follow Tactics 25.2 and first redraw the circuit with the capacitor replaced by a short circuit (Fig. 25.23*b*). Solving this circuit will give the current in R_1 right after the switch is closed. For the long-term behavior we redraw the circuit with the capacitor an open circuit (Fig. 25.23*c*).

EVALUATE There can't be any voltage across a short circuit—a perfect conductor—so there's no voltage across R_2 in Fig. 25.23*b*. Thus for part (a) the entire battery voltage appears across R_1, giving a current $I = \mathcal{E}/R_1$. In Fig. 25.23*c* we have two resistors in series and the current in both is $I = \mathcal{E}/(R_1 + R_2)$, our answer to part (b).

ASSESS The current through R_1 starts out at $\mathcal{E}/R_1$ and gradually drops to $\mathcal{E}/(R_1 + R_2)$. That makes sense because the uncharged capacitor initially "shorts out" R_2, making it irrelevant. But as the capacitor charges, current starts flowing through R_2 and its presence is "felt." Without solving more complicated equations, we can't describe the intermediate behavior of the circuit, but getting the short- and long-term behavior is straightforward. ■

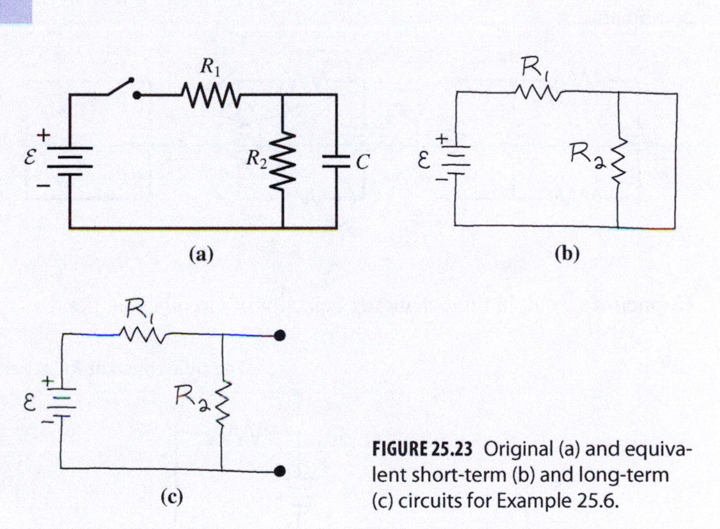

FIGURE 25.23 Original (a) and equivalent short-term (b) and long-term (c) circuits for Example 25.6.

GOT IT? 25.7 A capacitor is charged to 12 V and then connected between points *A* and *B* in the figure, with its positive plate at *A*. What's the current through the 2-kΩ resistor (1) immediately after the capacitor is connected and (2) a long time after it's connected?

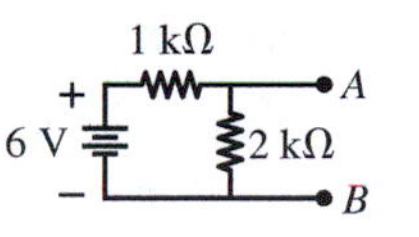

CHAPTER 25 SUMMARY

Big Idea

The big idea here is the **electric circuit**—an interconnection of electric components that usually includes one or more sources of electric energy, such as batteries.

Key Concepts and Equations

A source of **emf** or, simply, an *emf*, is a battery or other device that imparts energy to electric charge flowing through it. The value of the emf, $\mathcal{E}$, is the energy imparted per unit charge, measured in volts. An ideal emf maintains a fixed potential difference (voltage) across its terminals.

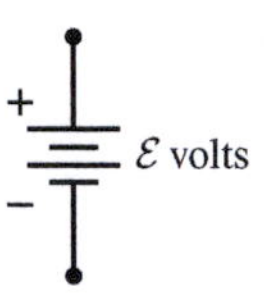

Resistors in series add:

$$R_{\text{series}} = R_1 + R_2 + R_3 + \cdots$$

Resistors in parallel add reciprocally:

$$\frac{1}{R_{\text{parallel}}} = \frac{1}{R_1} + \frac{1}{R_2} + \frac{1}{R_3} + \cdots$$

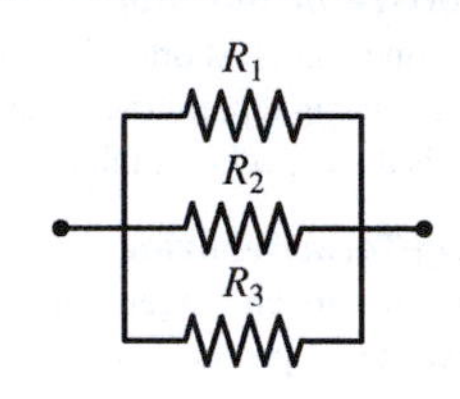

Simple circuits are analyzed by evaluating series and parallel combinations.

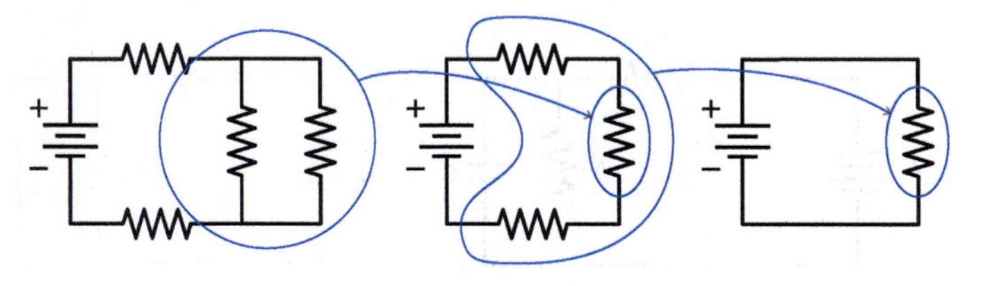

To analyze more complicated circuits, use Kirchhoff's node and loop laws.

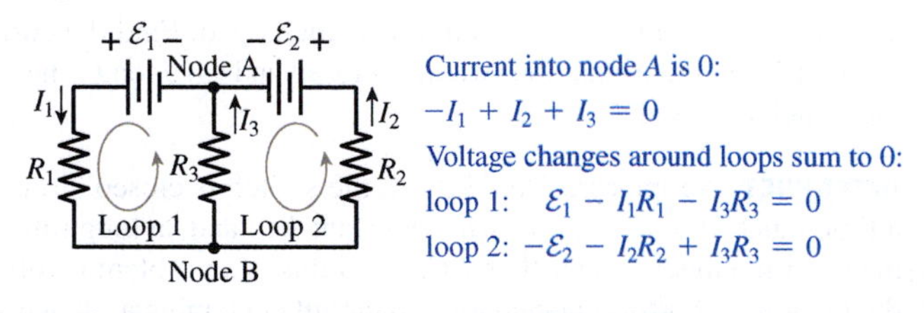

Capacitors result in time-changing behavior of circuit quantities.

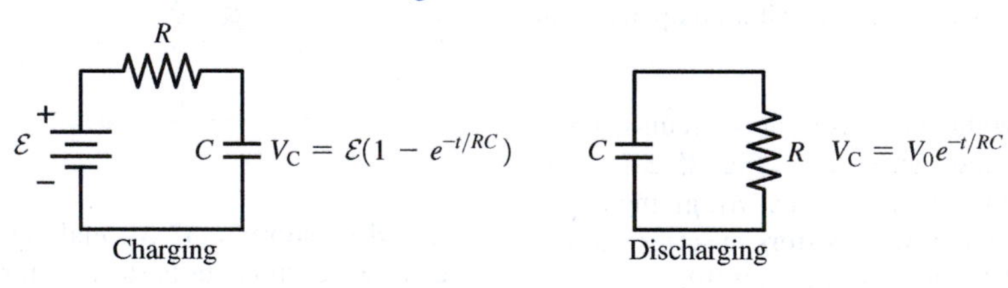

Applications

Batteries and other electric-energy sources have **internal resistance**. When they supply current, their terminal voltage is therefore less than their rated voltage $\mathcal{E}$.

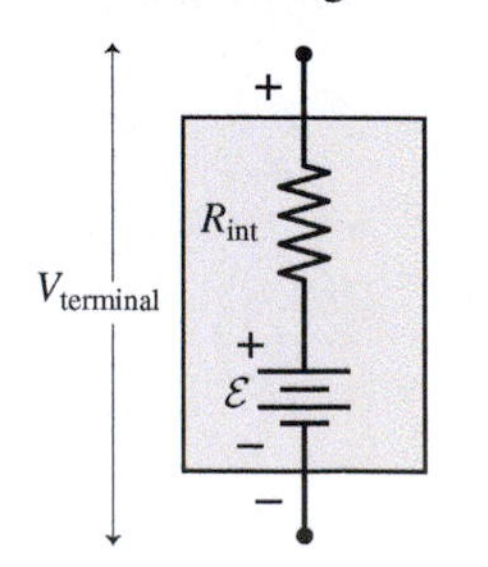

A **voltmeter** measures the voltage *across* its two terminals. It goes in parallel with the component whose voltage you want to measure.

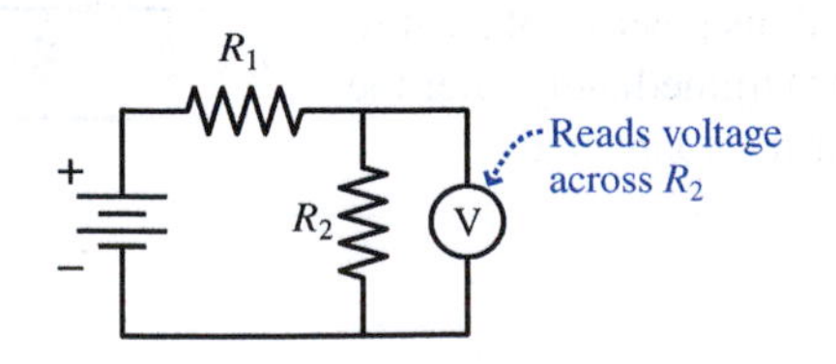

An ideal voltmeter has infinite resistance.

An **ammeter** measures the current *through* itself. It goes in series with the component whose current you want to measure.

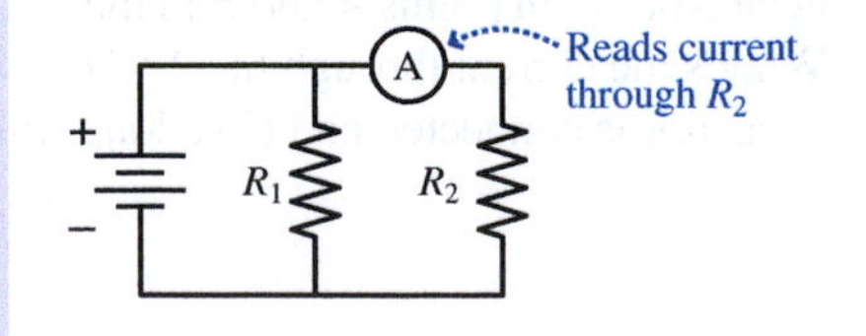

An ideal ammeter has zero resistance.

MP For homework assigned on MasteringPhysics, go to www.masteringphysics.com

BIO *Biology and/or medicine-related problems* **DATA** *Data problems* **ENV** *Environmental problems* **CH** *Challenge problems* **COMP** *Computer problems*

For Thought and Discussion

1. Are household electrical outlets connected in series or parallel? How do you know?
2. All the resistors in Fig. 25.24 have the same resistance. In which circuits does the battery supply the same current?

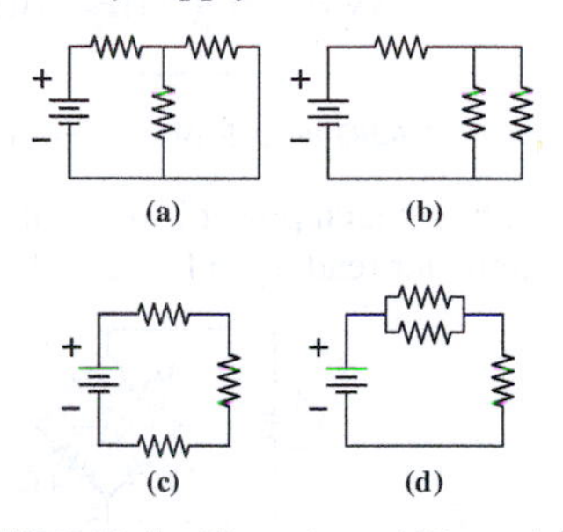

FIGURE 25.24 For Thought and Discussion 2

3. Can the voltage across a battery's terminals differ from the battery's rated voltage? Explain.
4. Can the voltage across a battery's terminals be higher than the battery's rated voltage? Explain.
5. In some cities, streetlights are wired in such a way that when one burns out, they all go out. Are the lights in series or parallel?
6. When the switch in Fig. 25.25 is open, what's the voltage across the resistor? Across the switch?

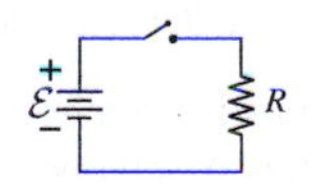

FIGURE 25.25 For Thought and Discussion 6

7. Two identical resistors in series dissipate equal power. How can this be, when electric charge loses energy in flowing through the first resistor?
8. When a large electric load such as a washing machine or oven comes on, lights throughout a house often dim. Why?
9. How would you connect a pair of equal resistors across an ideal battery in order to get the greatest power dissipation?
10. You have a battery whose voltage and internal resistance are unknown. Using an ideal voltmeter and an ideal ammeter, how would you determine each of these characteristics?
11. A student who's confused about voltage and current hooks a nearly ideal ammeter across a car battery. What happens?
12. A student who's confused about voltage and current tries to measure the voltage across a lighted lightbulb by inserting a voltmeter in series with the bulb. What happens to the bulb? Explain.

Exercises and Problems

Note: Work all circuit problems to two significant figures even if component values are given with one significant figure.

Exercises

Section 25.1 Circuits, Symbols, and Electromotive Force

13. Sketch a circuit diagram for a circuit that includes a resistor R_1 connected to the positive terminal of a battery, a pair of parallel resistors R_2 and R_3 connected to the lower-voltage end of R_1 and then returned to the battery's negative terminal, and a capacitor across R_2.
14. Sketch a diagram for a circuit consisting of two batteries, a resistor, and a capacitor, all in series. Does the circuit description allow you any flexibility?
15. Resistors R_1 and R_2 are in series, and the series combination is in parallel with R_3. This parallel combination is connected across a battery. Draw a diagram of this circuit.
16. What's the emf of a battery that delivers 27 J of energy as it moves 3.0 C between its terminals?
17. A 1.5-V battery stores 4.5 kJ of energy. How long can it light a flashlight bulb that draws 0.60 A?
18. If you accidentally leave your car headlights (current 5 A) on for an hour, how much energy drains from the car's 12-V battery?

Section 25.2 Series and Parallel Circuits

19. A 47-kΩ resistor and a 39-kΩ resistor are in parallel, and the pair is in series with a 22-kΩ resistor. What's the resistance of the combination?
20. What resistance should you place in parallel with a 56-kΩ resistor to make an equivalent resistance of 45 kΩ?
21. A defective starter motor draws 300 A from a car's 12-V battery, dropping the battery terminal voltage to 6 V. A good starter should draw only 100 A. What will the battery terminal voltage be with a good starter?
22. Find the internal resistance of the battery in Exercise 21.
23. When a 9-V battery is temporarily short-circuited, a 200-mA current flows. What's the battery's internal resistance?
24. You have a 1.0-Ω, a 2.0-Ω, and a 3.0-Ω resistor. What equivalent resistances can you form using all three?

Section 25.3 Kirchhoff's Laws and Multiloop Circuits

25. Find all three currents in the circuit of Fig. 25.13, but now with $\mathcal{E}_2 = 1.0$ V.
26. What's the current through the 3-Ω resistor in Fig. 25.26? (*Hint:* This is trivial. Can you see why?)

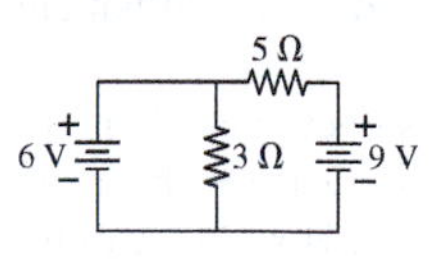

FIGURE 25.26 Exercise 26

27. Find I_2 in Example 25.4 for the case $\mathcal{E}_2 = 2.0$ V.

Section 25.4 Electrical Measurements

28. A voltmeter with 200-kΩ resistance is used to measure the voltage across the 10-kΩ resistor in Fig. 25.27. By what percentage is the measurement in error because of the finite meter resistance?

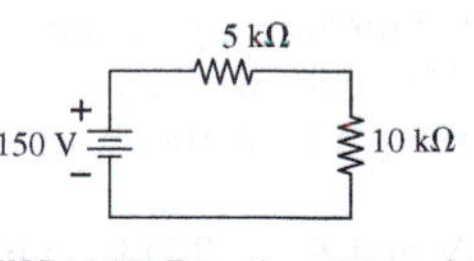

FIGURE 25.27 Exercises 28 and 29

29. An ammeter with 100-Ω resistance is inserted in the circuit of Fig. 25.27. By what percentage is the measured current in error because of the nonzero meter resistance?
30. A new mechanic foolishly connects an ammeter with 0.1-Ω resistance directly across a 12-V car battery with internal

resistance 0.01 Ω. What's the power dissipation in the meter? (No wonder it gets destroyed!)

Section 25.5 Capacitors in Circuits

31. Show that the quantity RC has the units of time (seconds).
32. If capacitance is in μF, what will be the units of the time constant RC when resistance is in (a) Ω, (b) kΩ, and (c) MΩ? (Your answers eliminate the need for tedious power-of-10 conversions.)
33. Show that a capacitor is charged to approximately 99% of the applied voltage in five time constants ($5RC$).
34. An uncharged 10-μF capacitor and a 470-kΩ resistor are in series, and 250 V is applied across the combination. How long does it take the capacitor voltage to reach 200 V?
35. Find an expression for the voltage across the capacitor in Example 25.6 when it's fully charged.

Problems

36. In Fig. 25.28, all resistors have the same value, R. What will be the resistance measured (a) between A and B or (b) between A and C?

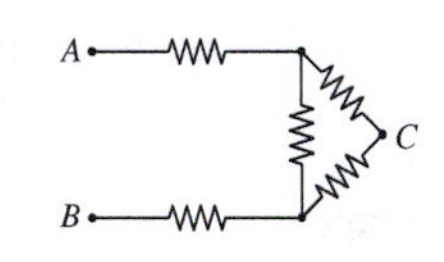

FIGURE 25.28 Problems 36 and 37

37. In Fig. 25.28, take all resistors to be 1 kΩ. Find the current in the vertical resistor when a 6.0-V battery is connected between A and B.
38. Three 1.5-V batteries, with internal resistances 0.01 Ω, 0.1 Ω, and 1 Ω, each have 1-Ω resistors connected across their terminals. What's the voltage between each battery's terminals, to three significant figures?
39. A partially discharged car battery can be modeled as a 9-V emf in series with a 0.08-Ω internal resistance. Jumper cables connect this battery to a fully charged battery, modeled as a 12-V emf in series with a 0.02-Ω internal resistance. The cables connect + to + and − to −. What current flows through the discharged battery?
40. Your company is overstocked on 50-Ω, $\frac{1}{2}$-W resistors. Your project requires 50-Ω resistors that can be safely connected across a 12-V power source. How many of the available resistors will you need, and how will you connect them?
41. A 6.0-V battery has internal resistance 2.5 Ω. If the battery is short-circuited, what's the rate of energy dissipation in its internal resistance?
42. How many 100-W, 120-V lightbulbs can be connected in parallel before they trip a 20-A circuit breaker?
43. **BIO** You company is designing a battery-based backup power source, and your job is to assess its safety. You know that under damp or sweaty conditions, the resistance between two points of unbroken skin on the human body can be as low as 500 Ω. Your product uses a 72-V battery whose internal resistance is 100 Ω. Is it capable of passing a fatal 100 mA (Table 24.3) through a damp human body?
44. Take $\mathcal{E} = 12$ V and $R_1 = 270$ Ω in Fig. 25.4. (a) What's the resistance R_2 if there's 4.5 V across it? (b) What will be the power dissipation in R_2?
45. In Fig. 25.29, R_1 is a variable resistor and the other two resistors have equal resistances R. (a) Find an expression for the voltage across R_1, and (b) sketch a graph of this voltage as R_1 varies from 0 to $10R$.

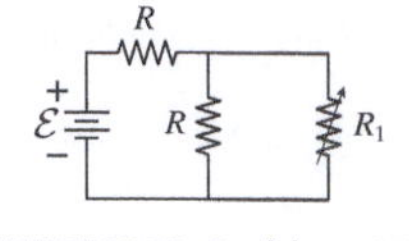

FIGURE 25.29 Problem 45

46. In the circuit of Fig. 25.30, find (a) the current supplied by the battery and (b) the current through the 6-Ω resistor.

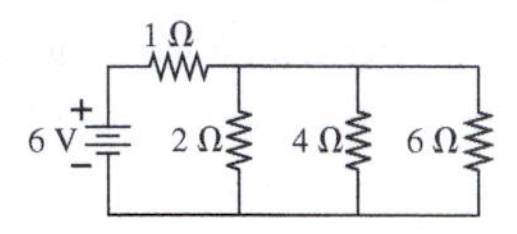

FIGURE 25.30 Problems 46 and 47

47. In Fig. 25.30, how much power is dissipated in the 4-Ω resistor?
48. What's the ammeter reading in Fig. 25.31?

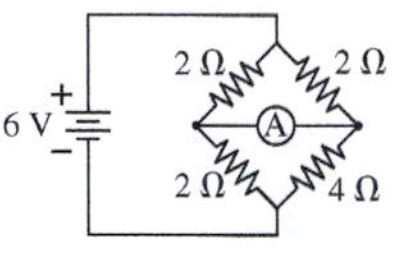

FIGURE 25.31 Problem 48

49. In Fig. 25.32, find the equivalent resistance measured between A and B.

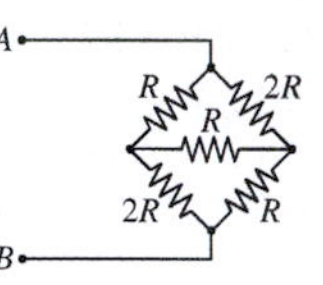

FIGURE 25.32 Problem 49

50. Find all three currents in the circuit of Fig. 25.13 with the values given, but with battery $\mathcal{E}_2$ reversed.
51. The voltage across the 30-kΩ resistor in Fig. 25.33 is measured with (a) a 50-kΩ voltmeter, (b) a 250-kΩ voltmeter, and (c) a 10-MΩ digital meter. What does each read, to two significant figures?

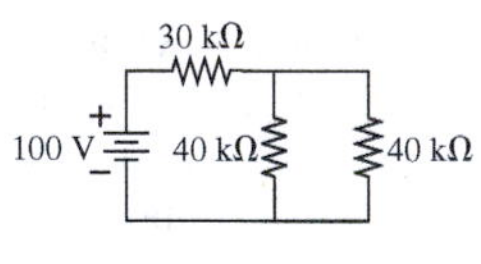

FIGURE 25.33 Problem 51

52. In Fig. 25.34, what are the meter readings when an ideal (a) voltmeter or (b) ammeter is connected between A and B?

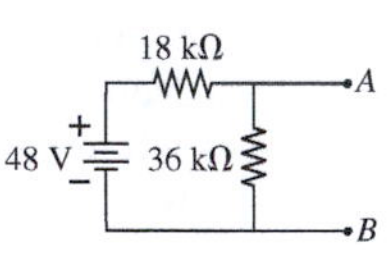

FIGURE 25.34 Problem 52

53. A resistor draws 1.00 A from an ideal 12.0-V battery. (a) If an ammeter with 0.10-Ω resistance is inserted in the circuit, what will it read? (b) If this current is used to calculate the resistance, by what percent will the result be in error?
54. The voltage across a charging capacitor in an RC circuit rises to $1 - 1/e$ of the battery voltage in 5.0 ms. (a) How long will it take to reach $1 - 1/e^3$ of the battery voltage? (b) If the capacitor is charging through a 22-kΩ resistor, what's the capacitance?
55. **BIO** You're designing an external defibrillator that discharges a capacitor through the patient's body, providing a pulse that stops

ventricular fibrillation. Specifications call for a capacitor storing 250 J of energy; when discharged through a body with 40-Ω transthoracic resistance, the capacitor voltage is to drop to half its initial value in 10 ms. Determine the capacitance (to the nearest 10 μF) and initial capacitor voltage (to the nearest 100 V) that meet these specs.

56. A capacitor used to provide steady voltages in the power supply of a stereo amplifier charges rapidly to 35 V every 1/60 second. It must then hold that voltage to within 1.0 V for the next 1/60 s while it discharges through the amplifier. If the amplifier draws 1.2 A from the 35-V supply, (a) what's its effective resistance, and (b) what capacitance is needed?

57. A capacitor is charged until it holds 5.0 J of energy, then connected across a 10-kΩ resistor. In 8.6 ms, the resistor dissipates 2.0 J. Find the capacitance.

58. In Fig. 25.35 the 2.0-μF capacitor is charged to 150 V, while the 1.0-μF capacitor is initially uncharged. Switch S is then closed. Find the total energy dissipated in the resistor as the circuit comes to equilibrium. (*Hint:* Think about charge conservation.)

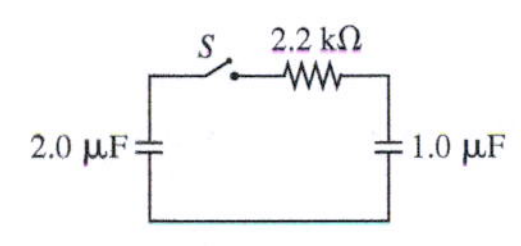

FIGURE 25.35 Problem 58

59. For the circuit of Example 25.6, take $\mathcal{E} = 100$ V, $R_1 = 4.0$ kΩ, and $R_2 = 6.0$ kΩ, and assume the capacitor is initially uncharged. Find the capacitor voltage and the currents in both resistors (a) just after the switch is closed, and (b) a long time after the switch is closed. Long after the switch is closed it's re-opened. What are V_C, I_1, and I_2 (c) just after this switch opening, and (d) a long time later?

60. In Fig. 25.36, the switch is initially open and both capacitors are initially uncharged. All resistors have the same value R. Find expressions for the current in R_2 (a) just after the switch is closed, and (b) a long time after the switch is closed.

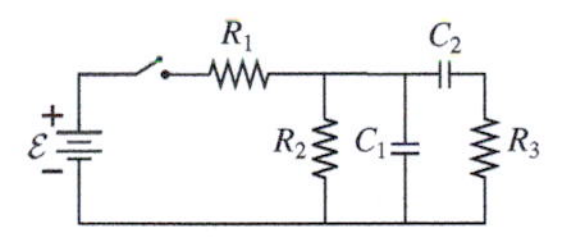

FIGURE 25.36 Problem 60

61. A battery's voltage is measured using a voltmeter with resistance 10.00 kΩ; the result is 4.982 V. A 15.00-kΩ meter gives 4.993 V. Find (a) the battery's voltage and (b) its internal resistance.

62. An ammeter with resistance 1.42 Ω is connected momentarily across a battery, and the meter reads 9.78 A. When the measurement is repeated with a 2.11-Ω meter, the reading is 7.46 A. Find (a) the battery voltage and (b) its internal resistance.

63. In Fig. 25.37, take $\mathcal{E}_1 = 12.0$ V, $\mathcal{E}_2 = 6.00$ V, $\mathcal{E}_3 = 3.00$ V, $R_1 = 1.00$ Ω, $R_2 = 2.00$ Ω, and $R_3 = 4.00$ Ω. Find the current in R_2 and give its direction.

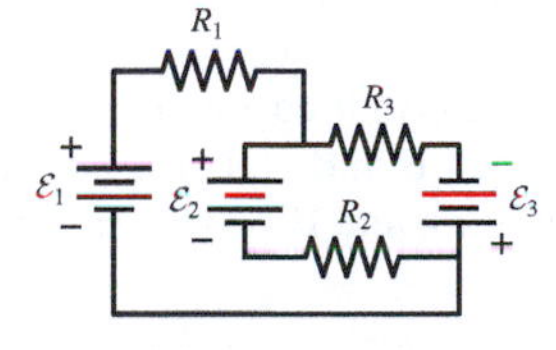

FIGURE 25.37 Problems 63 and 64

64. With all values except $\mathcal{E}_2$ as given in the preceding problem, (a) find $\mathcal{E}_2$ such that there is no current in this battery. (b) What are the currents in R_1 and R_3 under these conditions?

65. **DATA** The voltage on a charged capacitor is monitored with a voltmeter whose resistance is 1.00 MΩ. The table below gives the meter reading as a function of time. Determine a function of the voltage which, when you plot it against time, should give a straight line. Make your plot, establish a best-fit line, and use it to determine the capacitance.

Time (s)	0	1	2	3	4
Voltage (V)	15.0	10.3	6.36	3.78	2.43

66. Find the resistance needed in an RC circuit to bring a 20-μf capacitor from zero charge to 45% charge in 140 ms.

67. Suppose the currents into and out of a circuit node differ by 1 μA. If the node consists of a small metal sphere with diameter 1 mm, how long would it take for the electric field around the node to reach the 3-MV/m breakdown field in air?

68. **CH** Show that a battery delivers the most power when the load resistance across its terminals is equal to its internal resistance. (This is not the way to treat a battery, but it's the basis for load matching in amplifiers; see Problem 69.)

69. You're writing the instruction manual for a stereo amplifier with a maximum output of 100 W. The amplifier can be modeled as an emf in series with an 8-Ω resistance. What should you specify for the loudspeaker resistance to be used with the amplifier? How much power can the amplifier deliver to a speaker with half the optimum resistance?

70. **CH** Show that only half the total energy drawn from a battery in charging an RC circuit ends up stored in the capacitor. (*Hint:* What happens to the rest? You'll need to integrate.)

71. Find the equivalent resistance between A and B for the circuits in Fig. 25.38.

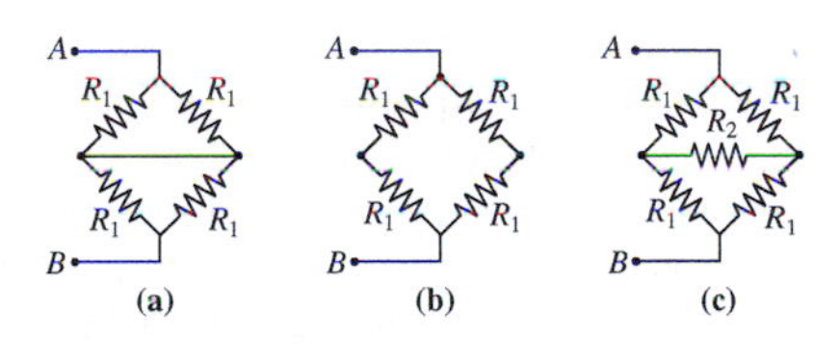

FIGURE 25.38 Problem 71

72. A 270-Ω resistor is connected across a battery and a 31-mA current flows. When the resistor is replaced with a 120-Ω resistor, the current increases to 63-mA. Find (a) the battery's voltage and (b) its internal resistance.

73. **CH** Obtain an expression for the rate of increase (dV/dt) of the voltage across a charging capacitor in an RC circuit. Evaluate your result at time $t = 0$, and show that if the capacitor continued charging steadily at this rate, it would reach full charge in exactly one time constant.

74. **CH** The circuit in Fig. 25.39 extends forever to the right, and all the resistors have the same value R. Show that the equivalent resistance measured across the two terminals at left is $R(1 + \sqrt{5})/2$. (*Hint:* You don't need to sum an infinite series.)

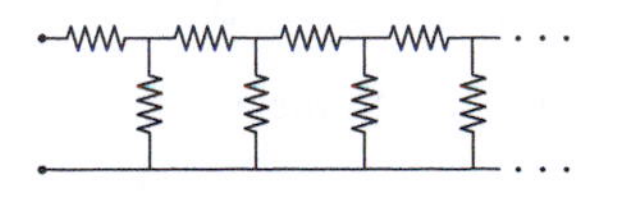

FIGURE 25.39 Problem 74

75. Figure 25.40 on the next page shows the voltage across a capacitor that's charging through a 4700-Ω resistor in the circuit of Fig. 25.18. Use the graph to determine (a) the battery voltage, (b) the time constant, and (c) the capacitance.

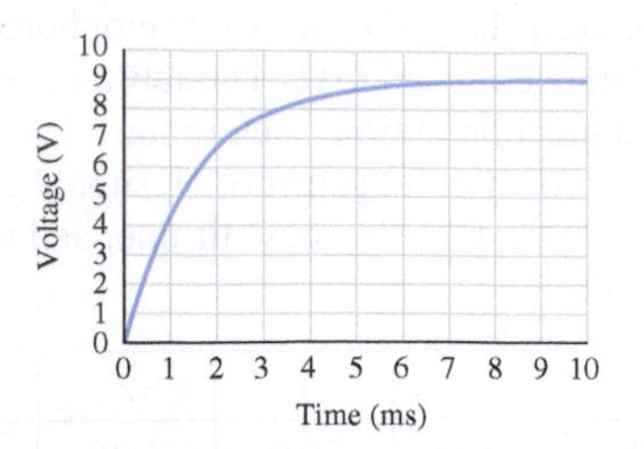

FIGURE 25.40 Problem 75

76. **BIO** Figure 25.41 shows a portion of a circuit used to model muscle cells and neurons. All resistors have the same value $R = 1.5\ \text{M}\Omega$, and the emfs are $\mathcal{E}_1 = 75\ \text{mV}$, $\mathcal{E}_2 = 45\ \text{mV}$, and $\mathcal{E}_3 = 20\ \text{mV}$. Find the current through $\mathcal{E}_3$, including its direction.

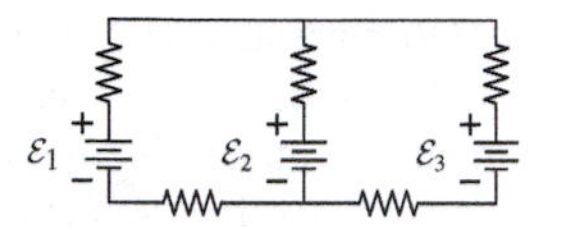

FIGURE 25.41 Problems 76 and 77

77. **BIO** An electrochemical impulse traveling along the cell modeled in Fig. 25.41 changes the value of $\mathcal{E}_3$ so now it supplies a 40-nA upward current. Assuming the rest of the circuit remains as described in Problem 76, what's the new value of $\mathcal{E}_3$?
78. **CH** A parallel-plate capacitor has plates of area $10\ \text{cm}^2$ separated by a 0.10-mm layer of glass insulation with resistivity $\rho = 1.2\times10^{13}\ \Omega\cdot\text{m}$ and dielectric constant $\kappa = 5.6$. Because of the finite resistivity, charge leaks through the insulation. (a) How can such a leaky capacitor be represented in a circuit diagram? (b) Find the time constant for this capacitor to discharge through its insulation, and show that it depends only on the properties of the insulating material and not on its dimensions.
79. Write the node and loop equations for the circuit in Fig. 25.23*a* (Example 25.6), and find the time constant.
80. **CH** In Problem 60, take $C_1 = C_2 = C$, and find the current through R_2 as a function of time. (*Hint:* Use the node and loop laws to get a differential equation for the current, and use the initial conditions on current and its derivative to evaluate the constants of integration.)
81. You're about to purchase a battery. Normally, batteries are rated in ampere-hours—the total charge they can deliver. Your application calls for a 5-A·h battery. But the 6-V battery you see while shopping online is rated at 50 watt-hours. Will it work?
82. In the circuit of Fig. 25.42 the switch is initially open and the capacitor is uncharged. Find expressions for the current I supplied by the battery (a) just after the switch is closed and (b) a long time after the switch is closed.

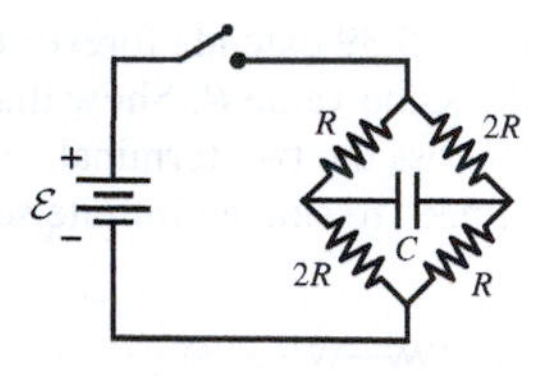

FIGURE 25.42 Problem 82

Passage Problems

BIO *Stray voltage* is a serious problem on dairy farms, often resulting from corroded wiring or poor wiring practices. These conditions can produce several volts between the ground and metal watering bowls, feed troughs, or milking equipment. Cows feel shocks that make them nervous, reducing milk output and sometimes leading to mammary gland infections. As a result, farmers can face serious financial losses. Figure 25.43 shows a typical stray-voltage situation, with the source of stray voltage modeled as a 6-V emf in series with a 1-kΩ resistance.

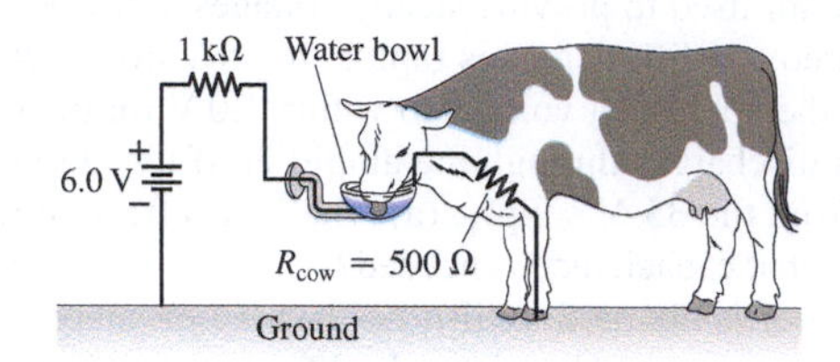

FIGURE 25.43 Stray voltage can bankrupt a dairy farm (Passage Problems 83–86)

83. The current through the 500-Ω cow will be
    a. 3 mA.
    b. 4 mA.
    c. 6 mA.
    d. 12 mA.
84. The voltage across the cow shown is
    a. 2 V.
    b. 4 V.
    c. 6 V.
    d. nearly 0 V.
85. In an effort to diagnose the problem, a farmer connects an ideal voltmeter between the water bowl and ground, with the cow absent. The voltmeter reading is
    a. 2 V.
    b. 4 V.
    c. 6 V.
    d. none of the above.
86. To explore the problem further, a farmer connects an ideal ammeter between the water bowl and ground, with the cow absent. The ammeter reading is
    a. 4 mA.
    b. 6 mA.
    c. 12 mA.
    d. infinite.

Answers to Chapter Questions

Answer to Chapter Opening Question

Conservation of charge and conservation of energy, as expressed by the node and loop laws, respectively.

Answers to GOT IT? Questions

25.1 (a) and (b)
25.2 (c) 6 V > (a) 3 V > (b) 0 V
25.3 $R_a > R_d > R_c > R_b$
25.4 (1) A is brightest because it carries more current; after A the current splits between B and C. (2) A and B become equally bright, with A dimming and B brightening relative to when C was in the circuit.
25.5 (a) and (c)
25.6 (c)
25.7 (1) 6 mA; (2) 2 mA

Magnetism: Force and Field

What You Know

- You understand the electric field and how it produces a force $\vec{F} = q\vec{E}$ on a charged particle.
- You can interpret and evaluate vector cross products; if not, you should review the introduction of the cross product in Chapter 11.
- You know about electric current.
- You understand the concept of flux, particularly for the electric field.

What You're Learning

- This chapter introduces the *magnetic field*, analogous to the electric field but in some ways more complex.
- You'll learn that magnetism is fundamentally a phenomenon involving *moving* electric charge.
- You'll see how the magnetic force on a charged particle depends on its charge, its velocity, and the magnetic field.
- You'll learn how magnetic fields originate from moving electric charges.
- You'll learn that there's no magnetic analog of electric charge, at least not in our everyday universe, and you'll see this expressed in *Gauss's law for magnetism*.
- You'll learn how *Ampère's law* describes the magnetic field in terms of electric current.
- You'll learn about three forms of magnetism that occur in matter.

How You'll Use It

- Magnetic phenomena are fundamental aspects of physical reality, important in both natural and technological systems, so your understanding of magnetism will be important in both your everyday life and your professional life.
- The remaining chapters of Part 4 will reveal an intimate connection between electricity and magnetism that will lead ultimately to an understanding of light.

People are fascinated with magnets and the mysterious, invisible force they produce. Magnetism plays essential roles in technology and the natural universe. We use magnetism for everything from holding notes on refrigerators to storing computer data to propelling high-speed trains. Earth's magnetism protects us from dangerous solar radiation, which itself originates in violent magnetic storms on the Sun. Without magnetism we wouldn't even see, for light itself results from an interaction between magnetism and electricity. In fact, magnetism and electricity are intimately related, and you'll soon see them as inseparable aspects of the same underlying phenomenon.

This ultraviolet image shows delicate loops of million-kelvin ionized gas—plasma—in the Sun's atmosphere. What force shapes the gas into such intricate structures, and why don't we see similar things in Earth's atmosphere?

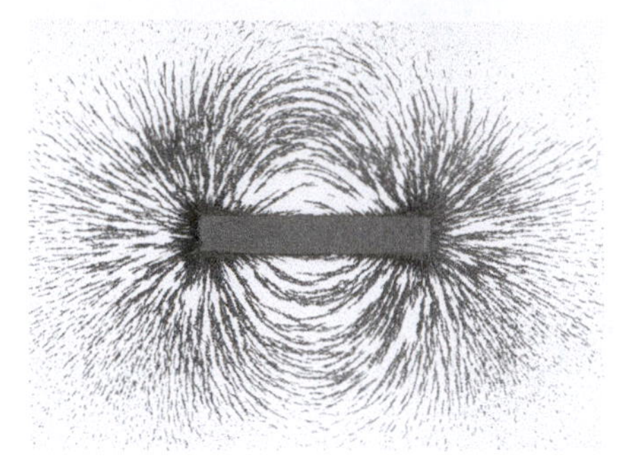

FIGURE 26.1 Iron filings align with the magnetic field, tracing out the field of a bar magnet.

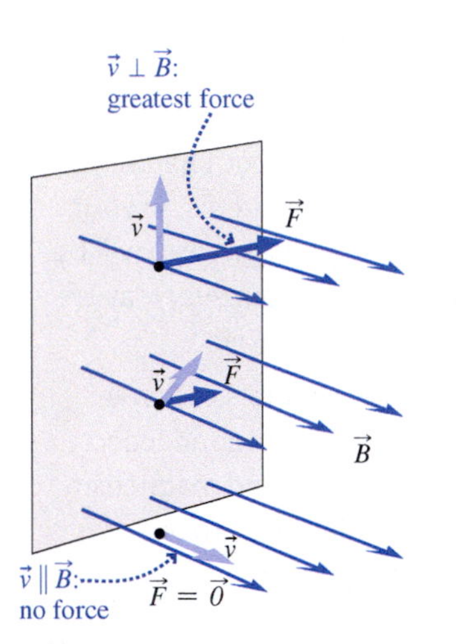

FIGURE 26.2
The magnetic force on a charged particle is perpendicular to both the particle's velocity $\vec{v}$ and the magnetic field $\vec{B}$.

MP

PhET: Magnet and Compass

Video Tutor Demo | **Magnet and Electron Beam**

26.1 What Is Magnetism?

You know from experience that magnets exert forces on each other and on certain materials, like iron. As we did with the gravitational and electric forces, it's convenient to describe this interaction in terms of a **magnetic field** (symbol $\vec{B}$). One magnet produces a magnetic field, and another responds to the field in its vicinity. We use field lines to picture the field, and we can trace those lines using small iron filings that align with the field (Fig. 26.1). In our illustrations, we'll use color for magnetic field lines to distinguish them from electric fields.

But the magnetism you're most familiar with is only one manifestation of a much more fundamental and universal phenomenon that's intimately linked with electricity. Here we'll go straight to the essence of magnetism; later, we'll see how familiar magnets fit into the big picture.

In Chapter 20 we introduced electric charge, a fundamental property of matter, and described its interactions using the concept of electric field. Magnetism, too, is based in *electric* charge. One crucial point both distinguishes and relates electricity and magnetism:

> The phenomena of magnetism involve *moving* electric charge.

In particular, *moving* electric charge is the source of magnetic fields, and *moving* electric charge is what responds to magnetic fields.

26.2 Magnetic Force and Field

In Chapter 20 we defined the electric field $\vec{E}$ with the equation $\vec{F}_E = q\vec{E}$, where $\vec{F}_E$ is the electric force on a charge q. Now consider a region where there's no electric field, but where there is a magnetic field. You could confirm the presence of the field and determine its direction with a compass, which is just a small magnet free to pivot into alignment with the field. Or, more fundamentally, you could explore the behavior of an electric point charge q in this field. If the charge is at rest, nothing happens. But if it's moving, it experiences a **magnetic force** as shown in Fig. 26.2. Experiment shows that:

1. The magnetic force is always at right angles to both the velocity $\vec{v}$ of the charge and the magnetic field $\vec{B}$.
2. The magnitude of the force is proportional to the product of the charge q, its speed v, and the magnetic field strength B.
3. The force is greatest when the charge moves at right angles to the field and is zero for motion parallel to the field. In general, the force is proportional to $\sin\theta$, where θ is the angle between the velocity $\vec{v}$ and the field $\vec{B}$.

Putting these facts together lets us write the magnetic force compactly using the vector cross product introduced in Chapter 11:

$$\vec{F}_B = q\vec{v} \times \vec{B} \qquad \text{(magnetic force)} \qquad (26.1)$$

Recall that the cross product $\vec{v} \times \vec{B}$ is a vector of magnitude $vB\sin\theta$, so the magnitude of the magnetic force is

$$|\vec{F}_B| = |q|vB\sin\theta$$

The direction of $\vec{v} \times \vec{B}$ is given by the right-hand rule (Fig. 26.3), and Equation 26.1 shows that the magnetic force has that same direction as $\vec{v} \times \vec{B}$ for positive q and the opposite direction for negative q.

Equation 26.1 shows that the units of magnetic field are N·s/(C·m), a unit given the name **tesla** (T) after the Serbian-American inventor Nikola Tesla (1856–1943). One tesla is a strong field, and a smaller unit called the gauss (G), equal to 10^{-4} T, is often used. Earth's magnetic field is a little less than 1 G, while the field of a refrigerator magnet is about 100 G. The fields used in magnetic resonance imaging (MRI) may be as strong as

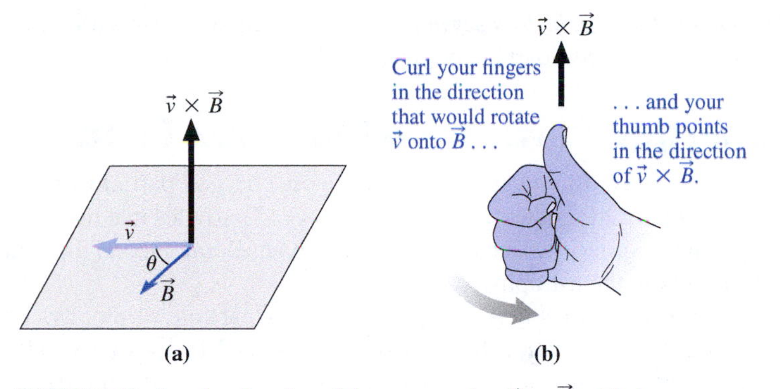

FIGURE 26.3 Finding the direction of the cross product $\vec{v} \times \vec{B}$ with the right-hand rule.

several tesla, while the incredibly dense, rapidly rotating collapsed stars called magnetars have fields up to 10^{11} T.

GOT IT? 26.1 The figure shows a proton in a magnetic field. (1) For which of the three proton velocities shown will the magnetic force be greatest? (2) What will be the direction of the force in all three cases?

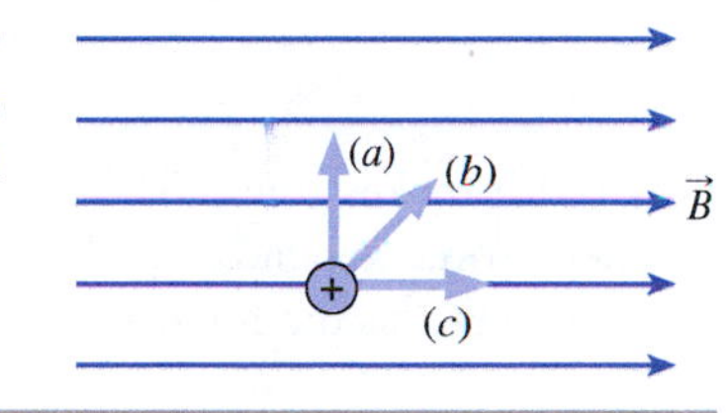

EXAMPLE 26.1 Finding the Magnetic Force: Steering Protons

Figure 26.4 shows three protons entering a 0.10-T magnetic field. All three are moving at 2.0 Mm/s. Find the magnetic force on each.

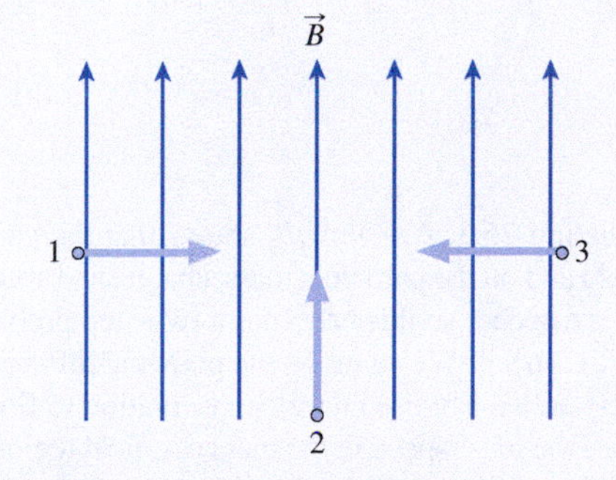

FIGURE 26.4 What's the magnetic force on each proton?

INTERPRET This problem is about the magnetic force on moving charged particles with the same speed but different directions of motion.

DEVELOP Equation 26.1, $\vec{F}_B = q\vec{v} \times \vec{B}$, gives the magnetic force, so we'll apply it to each of the particles.

EVALUATE Proton 2 is moving parallel to the field, so $\vec{v}_2 \times \vec{B} = 0$ and it experiences no magnetic force. Protons 1 and 3 are moving at right angles to the field, so $\sin\theta = 1$, and the magnitude of the force on each is $F_B = qvB \sin\theta = qvB$. Using the proton charge $q = e = 1.6\times10^{-19}$ C and the given values for B and v yields $F = 32$ fN. Since protons are positive, the direction of $\vec{F}_B$ is the same as that of $\vec{v} \times \vec{B}$; applying the right-hand rule shows that the direction is out of the page for proton 1 and into the page for proton 3.

ASSESS Our answer 32 fN (32×10^{-15} N) is a tiny force, but that's not surprising given the proton's tiny charge. Note that the magnetic field alone doesn't determine the force; in this example identical particles experience different forces because they're moving in different directions relative to the field. ■

Although electricity and magnetism are related, the electric and magnetic forces are distinct. Both may be present simultaneously, in which case a charged particle experiences both an electric force $\vec{F}_E = q\vec{E}$ and a magnetic force $\vec{F}_B$ given by Equation 26.1. The result is an **electromagnetic force**:

$$\vec{F} = q\vec{E} + q\vec{v} \times \vec{B} \quad \text{(electromagnetic force)} \qquad (26.2)$$

Because the magnetic force depends on velocity but the electric force doesn't, it's possible to use perpendicular electric and magnetic fields to select particles of a particular velocity

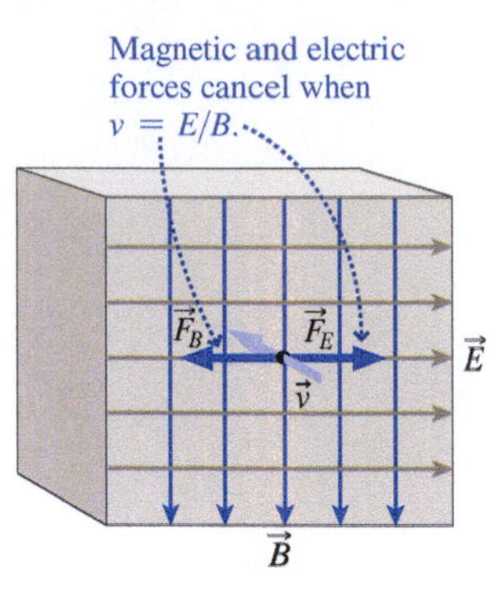

FIGURE 26.5 A velocity selector. The electric and magnetic forces cancel when $qE = qvB$, so only particles with speed $v = E/B$ pass through undeflected. The velocity $\vec{v}$ points into the page.

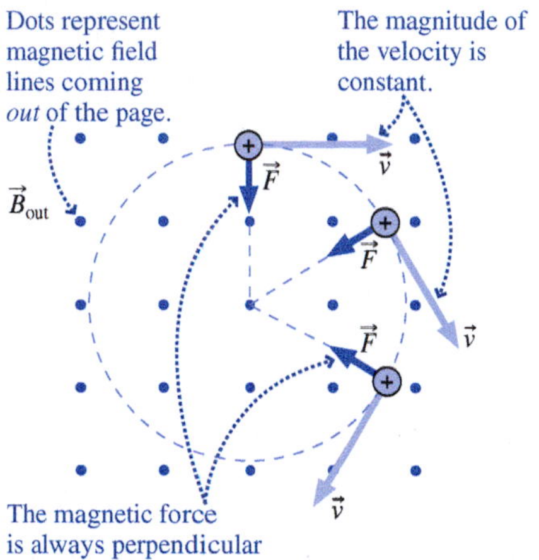

FIGURE 26.6 A charged particle moving at right angles to a uniform magnetic field describes circular motion.

(Fig. 26.5). Such *velocity selectors* serve to prepare particle beams with uniform velocity as well as to analyze charged-particle populations in interplanetary space.

26.3 Charged Particles in Magnetic Fields

Following Newton's law, the magnetic force deflects charged particles from their otherwise straight-line paths. Magnetic forces "steer" charged particles in a host of practical devices ranging from microwave ovens to giant particle accelerators, and they shape particle trajectories throughout the astrophysical universe.

The magnetic force always acts at right angles to a particle's velocity. **Therefore, it changes the direction of motion but not the speed, and it does no work.** In the special case of a particle moving at right angles to a uniform field, the magnetic force has a constant magnitude and the result, as Fig. 26.6 shows, is uniform circular motion. With $\vec{v}$ perpendicular to $\vec{B}$, the magnetic force of Equation 26.1 has magnitude qvB. This force provides the acceleration v^2/r that characterizes circular motion with radius r. Then Newton's law, $F = ma$, reads $qvB = mv^2/r$. We can solve for the radius of the particle's circular path to get

$$r = \frac{mv}{qB} \tag{26.3}$$

This result makes sense: The greater the particle's momentum mv, the harder it is for the magnetic force to bend its path and the larger the radius. On the other hand, a larger charge or field increases the force and makes a tighter orbit.

GOT IT? 26.2 A uniform magnetic field points out of this page. Will an electron that's moving in the plane of the page circle (a) clockwise or (b) counterclockwise as viewed from above the page?

EXAMPLE 26.2 Magnetic Deflection: A Mass Spectrometer

A mass spectrometer separates ions according to their ratio of charge to mass. Such devices are widely used in science and engineering to analyze unknown mixtures and to separate isotopes of chemical elements. Figure 26.7 shows ions of charge q and mass m first being accelerated from rest through a potential difference V and then entering a region of uniform magnetic field B pointing out of the page. Only the magnetic force acts on the ions in this region, so they undergo circular motion and, after half an orbit, land on a detector. Find an expression for the horizontal distance x from the entrance slit to the point where an ion lands on the detector.

INTERPRET This problem is about charged particles undergoing circular motion in a uniform magnetic field. The distance we're asked for is the diameter of the particles' circular path.

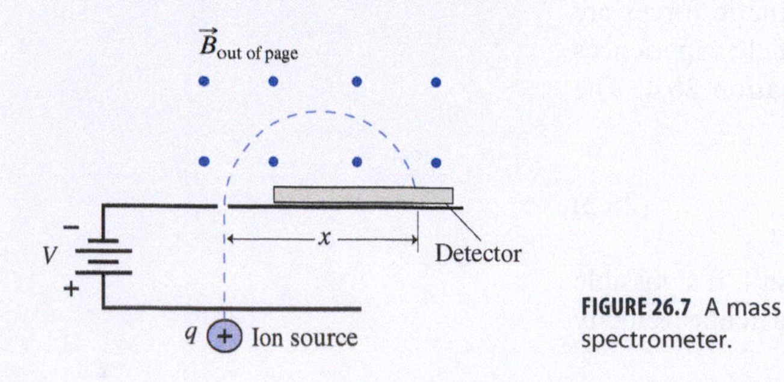

FIGURE 26.7 A mass spectrometer.

DEVELOP Equation 26.3, $r = mv/qB$, shows that the path radius depends on the field and on the particle's mass, charge, and speed. We know everything but the speed, so this becomes a two-step problem in which we'll first find the speed. We're given the potential difference—energy per unit charge—so we can use energy conservation to find the kinetic energy and hence the ions' speed in the magnetic-field region. Then we'll use Equation 26.3 to find the radius of the ions' circular path.

EVALUATE A charge q gains kinetic energy qV in "falling" through a potential difference V, so an ion's kinetic energy once it enters the magnetic field is $\frac{1}{2}mv^2 = qV$. Solving for v gives $v = \sqrt{2qV/m}$. Our answer, the path diameter x, is then twice the radius given in Equation 26.3:

$$x = 2r = \frac{2mv}{qB} = \frac{2m\sqrt{2qV/m}}{qB} = \frac{2}{B}\sqrt{\frac{2mV}{q}}$$

ASSESS Make sense? The greater the mass or speed—which increases with the accelerating voltage V—the harder it is to deflect the ion and the larger the diameter of its semicircular path. The larger the field or charge, the larger the force and the smaller the semicircle. Note that for a fixed voltage and magnetic field, this device sorts ions by their charge-to-mass ratio q/m. ■

The Cyclotron Frequency

What's the period of a particle's circular orbit in a uniform magnetic field? The orbit circumference is $2\pi r$, so the period is $T = 2\pi r/v$. Using Equation 26.3 for the radius gives

$$T = \frac{2\pi r}{v} = \frac{2\pi}{v}\frac{mv}{qB} = \frac{2\pi m}{qB}$$

Remarkably, the period is independent of the particle's speed and orbital radius. Equation 26.3 shows why: The higher the speed v, the larger the radius r and hence the circumference. So a faster particle describes a larger circle and ends up taking the same amount of time to go around.

Equivalently, we can describe the particle's circular motion in terms of its frequency f, in revolutions per second, which is just the inverse of the period:

$$f = \frac{qB}{2\pi m} \quad \text{(cyclotron frequency)} \tag{26.4}$$

This quantity is the **cyclotron frequency**. Because it depends only on the field and the charge-to-mass ratio, cyclotron motion provides astrophysicists with a direct measure of magnetic fields in distant objects. Conversely, a fixed magnetic field guarantees a specific cyclotron frequency regardless of the particles' speeds. Microwave ovens exploit this fact, with their microwaves generated by electrons circling 2.4 billion times per second in a special tube called a *magnetron*.

Particle Trajectories in Three Dimensions

When a charged particle moves in an arbitrary direction, we consider velocity components both perpendicular and parallel to the magnetic field. Our previous analysis applies to the perpendicular component, giving circular motion in a plane perpendicular to the field. And because there's no magnetic force with velocity parallel to the field, the parallel component is unaffected. The result, in a uniform field, is a spiral path along the field direction (Fig. 26.8).

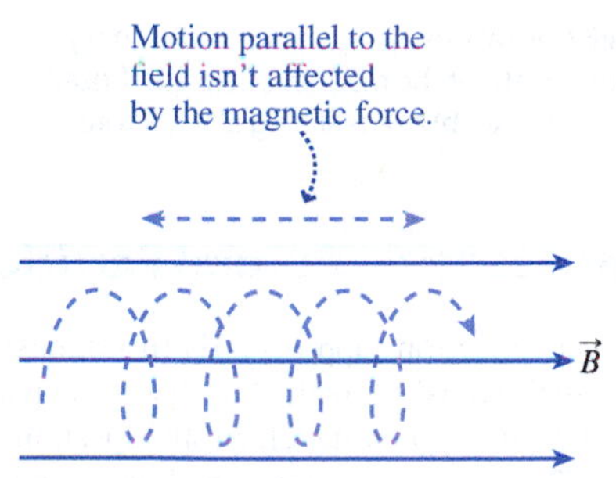

FIGURE 26.8 A particle in a uniform magnetic field describes a spiral path.

APPLICATION The Cyclotron

Physicists use high-energy particles to probe the structure of matter; engineers and physicians need high-energy particle beams in manufacturing, diagnostic, and therapeutic procedures. The easiest way to produce such beams is to accelerate ions through a potential difference, but the difficulties of handling high voltages make that impractical for all but the lowest energies. One of the earliest and most successful devices to circumvent this problem is the **cyclotron**, whose essential parts are shown in the figure. The device consists of an evacuated chamber between the poles of a magnet. Ions are produced at the center and undergo circular motion in the magnetic field.

Also in the chamber are two hollow conducting structures shaped like the letter D. A modest potential difference is applied across these "dees," and it alternates polarity at the cyclotron frequency. As ions circle around inside the cyclotron, they gain energy from the strong electric field associated with the potential difference at the gap. Inside the hollow conducting dee there's no electric field, so here the particles follow circular paths in the magnetic field. Halfway around they again encounter the dee gap. Because the potential is changing polarity in step with the particles' cyclotron motion, they again gain energy as they cross the gap. They move faster and in ever-larger circles,

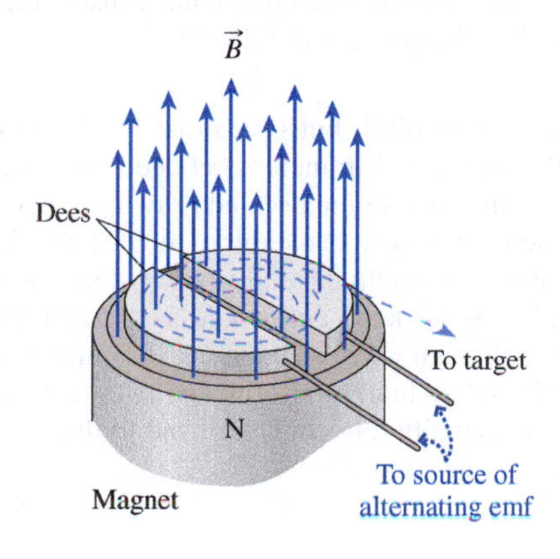

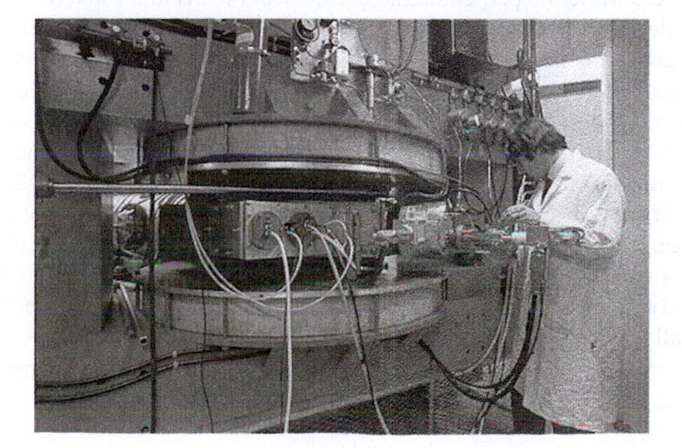

(continued)

but always with the same orbital period. When they approach the edge of the machine, an electric field deflects the ions and they emerge as a high-energy beam.

Cyclotrons produce ions with energies of millions of electronvolts. This is high enough to cause nuclear reactions, and many medically useful radioactive isotopes are made using cyclotrons. In particular, the diagnostic procedure called PET (positron emission tomography) relies on cyclotron-produced radioisotopes. Because these isotopes are short-lived, hospitals doing PET scans generally have cyclotrons on site. The photo shows a hospital-based cyclotron used in the development of PET technology. At higher energies the theory of relativity alters our conclusion that the cyclotron frequency is independent of energy, and the cyclotron becomes useless. An alternative design is the **synchrotron**, in which both the magnetic field and the frequency vary to account for increasing particle energy.

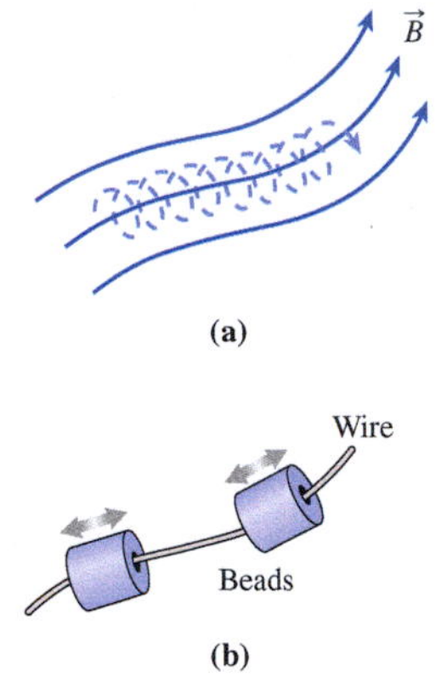

FIGURE 26.9 (a) Charged particles undergoing spiral motion about the magnetic field are "frozen" to the field like (b) beads sliding along a wire.

The absence of magnetic force in the field direction means it's easy to move charged particles along the field. But try to push a charged particle at right angles to the field, and it goes into circular motion; push harder and the circle only gets bigger. As a result, charged particles are effectively "frozen" to the field lines and move along the field like beads strung on a wire (Fig. 26.9). Nonuniform fields and particle collisions make this "freezing" less than perfect, but in many cases particle density is low enough that the "frozen" assumption is an excellent approximation. The coronal loops in this chapter's opening photo are a beautiful example of charged particles "frozen" to the solar magnetic field. Similarly, high-energy particles from the Sun get trapped on Earth's magnetic field lines; where the field intersects the atmosphere, particles collide with atmospheric nitrogen and oxygen to produce the spectacular displays we call the aurora. You can explore auroral particles quantitatively in Example 26.3. Here on Earth, trapping of charged particles on magnetic field lines enables researchers to confine plasmas at temperatures of 100 MK in attempts to harness the energy of nuclear fusion.

EXAMPLE 26.3 Auroral Particles

Energetic electrons approach Earth's atmosphere along magnetic field lines; their collisions with atmospheric atoms generate auroral displays. The field strength at the electrons' location is 47 nT. A given electron has 0.72 fJ of kinetic energy, and its velocity makes an angle $\theta = 60°$ with the magnetic field. Find the radius of its spiral trajectory.

INTERPRET This problem involves a charged particle undergoing three-dimensional motion in a magnetic field. The trajectory is a spiral, as suggested in Figs. 26.8 and 26.9. Because the magnetic field has no effect on motion parallel to the field, we anticipate needing to consider separately the parallel and perpendicular components of the proton's motion. Then we can think of the spiral trajectory as made up of uniform circular motion perpendicular to the field and uniform motion along the field.

EVALUATE The magnetic force on the electron is given by Equation 26.1, $\vec{F} = -e\vec{v} \times \vec{B}$, where we've written $-e$ for the electron charge. We break the electron's velocity into two components: $v_{\|}$ along the magnetic field and $v_{\perp}$ at right angles to the field. Equation 26.1 shows that the parallel component does not contribute to the magnetic force, while the perpendicular component results in a force of magnitude $F = ev_{\perp}B$. The right-hand rule shows that the force points toward the center of the electron's spiral path, and it's this force that keeps the electron in circular motion perpendicular to the field. The speed of this circular motion is $v_{\perp}$, so we can write Newton's second law as $ev_{\perp}B = mv_{\perp}^2/r$, where r is the radius of the circle and thus of the spiral trajectory. We've sketched the situation in Fig. 26.10, where it's evident that $v_{\perp} = v\sin\theta$. Solving our Newton's law equation for the unknown radius gives $r = mv_{\perp}/eB$—the same result we got in Equation 26.3, but now for three-dimensional motion where only the perpendicular velocity interacts with the magnetic field. Finally, we can get the electron's speed v from its given kinetic energy $K = \frac{1}{2}mv^2$: $v = \sqrt{2K/m}$, and thus $v_{\perp} = \sqrt{2K/m}\sin\theta$. Using this result in our expression for the radius r yields our answer:

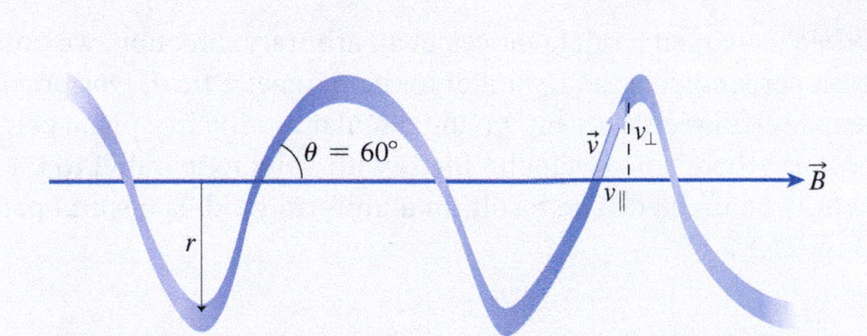

FIGURE 26.10 Our sketch of the electron's trajectory. We've marked the 60° angle on one turn of the spiral and on another turn we've shown the breakdown of the velocity vector into parallel and perpendicular components; the same angle applies on both turns.

$$r = \frac{m\sqrt{2K/m}\,\sin\theta}{eB} = \frac{\sqrt{2Km}\,\sin\theta}{eB} =$$

$$\frac{\sqrt{2(0.72\times10^{-15}\,\text{J})(9.11\times10^{-31}\,\text{kg})}\,\sin 60°}{(1.6\times10^{-19}\,\text{C})(47\times10^{-9}\,\text{T})} = 4.2\,\text{km}$$

where we found the electron mass in the table inside the front cover, along with the SI prefix giving 1 fJ = 10^{-15} J.

ASSESS: That's a big spiral, but given that Earth's magnetic field lines extend thousands of kilometers between the polar regions, it would be barely discernable on a scaled drawing. Energetic protons also spiral in the field, as you can explore in GOT IT? 26.3. Incidentally, a space physicist would have given that electron energy in eV, expressing 0.72 fJ as 4.5 keV, and would have called that 60° angle the *pitch angle*. Although we treated the field as uniform in this example, Earth's field isn't actually uniform, and over large distances that results in particles actually reflecting back and forth between the north and south polar regions.

GOT IT? 26.3 A proton of the same energy as the electron in Example 26.3 spirals in the same magnetic field. Compared with the electron, the proton (a) has a larger spiral and spirals in the same direction; (b) has a smaller spiral and spirals in the same direction; (c) has a smaller spiral and spirals in the opposite direction; (d) has a larger spiral and spirals in the opposite direction.

26.4 The Magnetic Force on a Current

An electric current consists of charges in motion, so a current in a magnetic field should experience a magnetic force. Figure 26.11 shows a straight wire in a magnetic field $\vec{B}$. Charges in the wire are moving about with thermal motions, but because these are random, the magnetic force on all the charges averages to zero. But if there's a current I in the wire, then the charges share a common drift velocity $\vec{v}_d$, and thus each experiences a magnetic force given by Equation 26.1: $\vec{F}_q = q\vec{v}_d \times \vec{B}$. If the wire has cross-sectional area A and contains n charges per unit volume, then the force on all the charge carriers in a length l of wire is $\vec{F} = nAlq\vec{v}_d \times \vec{B}$. But $nAqv_d$ is the current, I, as we found in Chapter 24. If we define a vector $\vec{l}$ whose magnitude is the wire length l and whose direction is along the current, then we can write

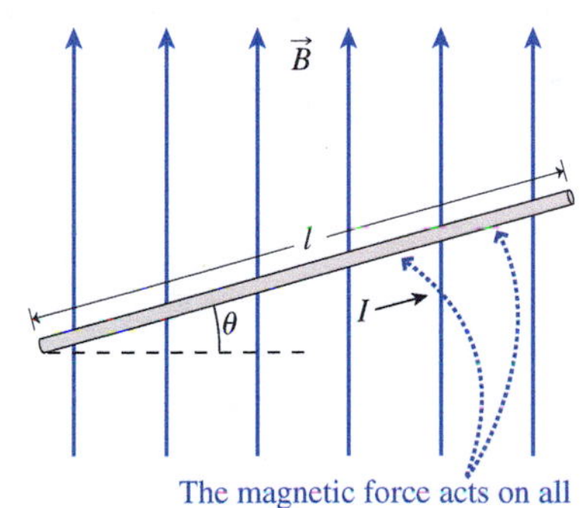

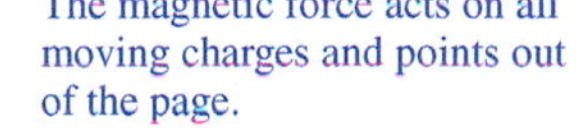

FIGURE 26.11 A straight wire carrying current I through a uniform magnetic field.

$$\vec{F} = I\vec{l} \times \vec{B} \quad \text{(magnetic force on a current)} \tag{26.5}$$

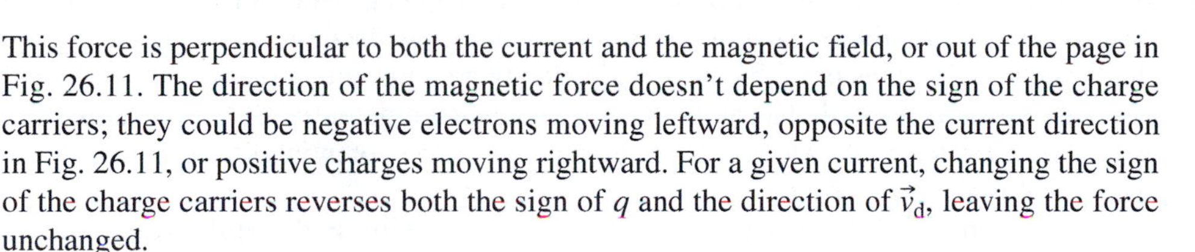

This force is perpendicular to both the current and the magnetic field, or out of the page in Fig. 26.11. The direction of the magnetic force doesn't depend on the sign of the charge carriers; they could be negative electrons moving leftward, opposite the current direction in Fig. 26.11, or positive charges moving rightward. For a given current, changing the sign of the charge carriers reverses both the sign of q and the direction of $\vec{v}_d$, leaving the force unchanged.

Equation 26.5 gives the net force on the charge carriers in the wire. In a physical wire, the magnetic force deflects charge carriers to one side of the wire, producing a charge separation and an electric field that results in a force on the rest of the wire (Fig. 26.12); this electric force is also what confines electrons within the wire. Although its origin is not entirely magnetic, we loosely call the force in Equation 26.5 "the magnetic force on a wire." The magnetic force on a current-carrying wire is the basis for many practical devices, including loudspeakers and the electric motors that start cars and run refrigerators, disk drives, subway trains, pumps, food processors, power tools, and myriad other instruments of modern society.

Equation 26.5 holds for straight wires in uniform magnetic fields. In other cases we apply Equation 26.5 to very short segments of a wire that's curved or in a nonuniform field, and we integrate to find the net force. Problem 57 explores this situation.

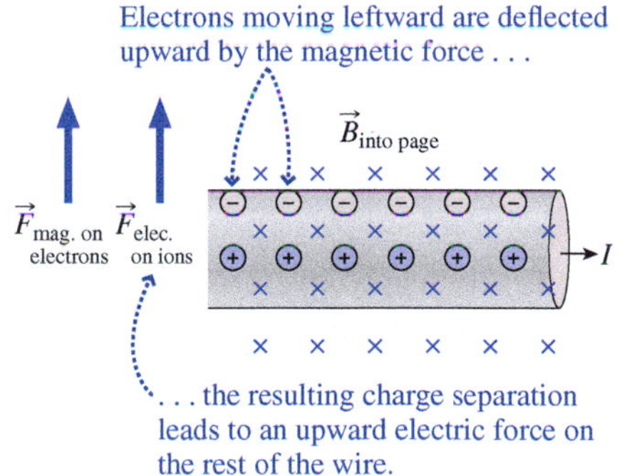

FIGURE 26.12 Origin of the magnetic force on a current-carrying wire.

GOT IT? 26.4 The figure shows a flexible wire passing through a magnetic field that points out of the page. The wire is deflected upward, as shown. Is the current flowing (a) to the left or (b) to the right?

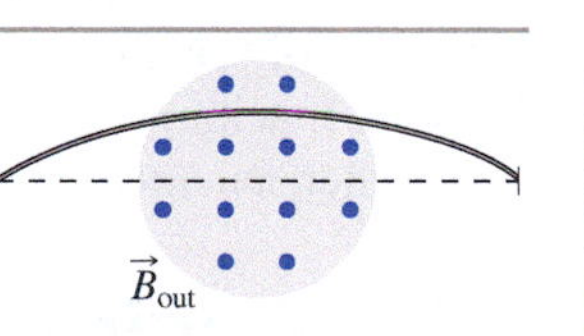

Video Tutor Demo | **Current-Carrying Wire in Magnetic Field**

CONCEPTUAL EXAMPLE 26.1 **Magnetic Force: A Power Line**

A power line runs along Earth's equator, where the magnetic field points horizontally from south to north; the line carries current flowing from west to east. What's the direction of the magnetic force on the power line?

EVALUATE We've sketched the situation in Fig. 26.13. Using the right-hand rule with the current eastward and magnetic field northward shows that the force is vertically upward.

(continued)

ASSESS As always, the force is at right angles to both the current and the magnetic field. As you'll see in "Making the Connection," below, this force is pretty feeble compared with the power line's weight.

MAKING THE CONNECTION Earth's equatorial field strength is 30 μT, and the power line carries 500 A. What's the magnetic force on a kilometer of the line?

EVALUATE Equation 26.5 gives

$$F = |I\vec{l} \times \vec{B}| = IlB \sin 90° = (500\text{ A})(1.0\text{ km})(30\text{ μT})(1) = 15\text{ N}$$

That's far less than the line's weight, which is on the order of 10 kN.

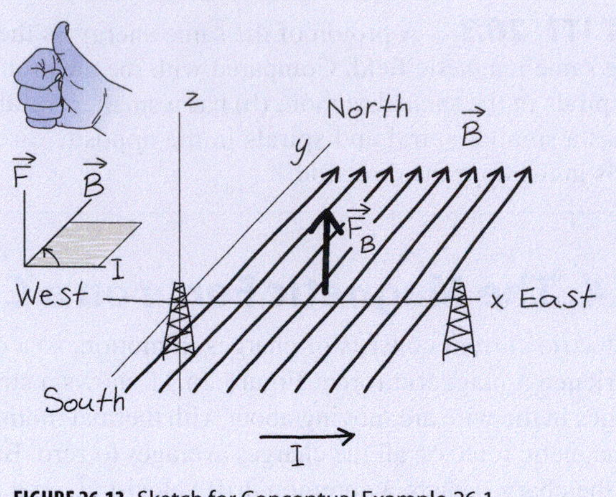

FIGURE 26.13 Sketch for Conceptual Example 26.1.

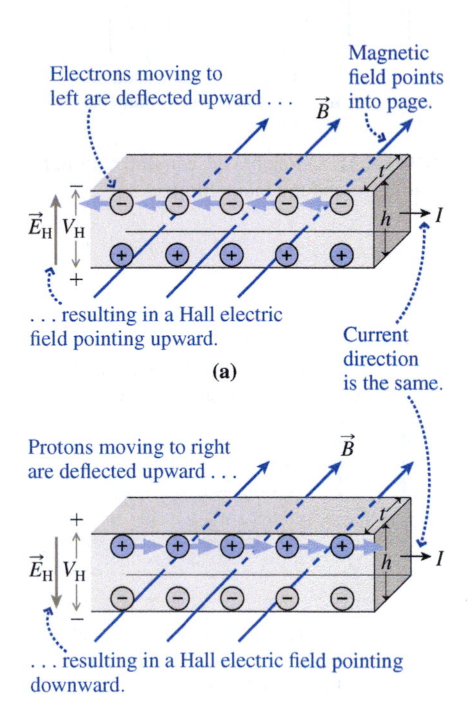

FIGURE 26.14 The Hall electric field $\vec{E}_H$ and Hall potential V_H arise from the magnetic deflection of charge carriers. In both (a) and (b) the current is to the right, carried in (a) by negative charge moving to the left and in (b) by positive charge moving to the right.

The Hall Effect

We noted earlier that the direction of the magnetic force depends on the direction of the current, not on the sign of the charge carriers. However, there's a subtle difference between two conductors with the same current yet different charge carriers. In Fig. 26.14, moving charge carriers of either sign are deflected to the upper surface of the conductor. Again, that's because the signs of both charge and velocity are opposite in the two cases. In both cases the result is a small electric field and its associated potential difference *across* the wire. The *direction* of the electric field and the *sign* of the potential difference depend on the sign of the charge carriers.

The separation of charges across a current-carrying wire is the **Hall effect**, and the potential difference is the **Hall potential**. In a steady state, the magnetic force on the charge carriers is just balanced by the electric force associated with charge separation, giving $qE = qv_dB$, or simply $E = v_dB$. In the rectangular conductor of Fig. 26.14, the electric field is uniform and the Hall potential is then $V_H = Eh = v_dBh$. Using $I = nqAv_d$ and solving for v_d, we can then write $V_H = IBh/nAq$. Since $A = ht$, with t the conductor thickness in the field direction (see Fig. 26.14), this becomes

$$V_H = \frac{IB}{nqt} \quad \text{(Hall potential)} \qquad (26.6)$$

The quantity $1/nq$ is the **Hall coefficient**. Measuring the Hall coefficient gives information on the nature and density of the charge carriers. Alternatively, measuring V_H in a material of known Hall coefficient carrying a known current gives a direct measure of the magnetic field strength.

Today, Hall-effect sensors are used not only to measure magnetic fields but also in a variety of practical applications including motion sensors using magnets mounted on rotating machinery such as car and bicycle wheels, as contact-free switches that change state when a magnet approaches, and as compasses in smartphones.

26.5 Origin of the Magnetic Field

Electric charges respond to electric fields, and electric charges produce electric fields. So it is with magnetism. We've just explored how *moving* electric charges respond to magnetic fields; we'll now see how *moving* electric charges produce magnetic fields. The first inkling of a relation between electricity and magnetism came in 1820 when the Danish scientist Hans Christian Oersted discovered that a compass needle is deflected by an electric current. A month after Oersted's discovery became known in Paris, the French scientists Jean Baptiste Biot (rhymes with "Leo") and Félix Savart (rhymes with "bazaar") had experimentally determined the form of the force arising from a steady current.

The Biot–Savart Law

The Biot–Savart law gives the contribution $d\vec{B}$ to the magnetic field at a point P due to a small element of current, in much the way that Coulomb's law gives the electric field $d\vec{E}$ due to a charge element dq. Figure 26.15 shows a wire carrying a steady current I and the contribution $d\vec{B}$ to the field at P from a small length dl of the wire. The current element $I\,dl$ is the source of the field; it plays the same role as the charge dq in Coulomb's law. The magnetic field decreases with the inverse square of the distance, just as in Coulomb's law for the electric field.

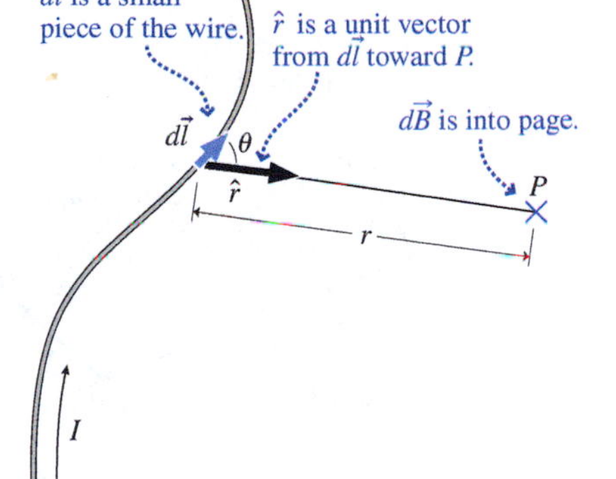

FIGURE 26.15 The Biot-Savart law gives the magnetic field $d\vec{B}$ at the point P arising from the current I flowing along the infinitesimal vector $d\vec{l}$.

There are important differences between the Coulomb and Biot–Savart laws. Charge—the source of electric field—is a scalar quantity. But *moving* charge—the source of magnetic field—has direction. The Biot–Savart law accounts for that direction by defining a vector $d\vec{l}$ along the current; then the field contribution $d\vec{B}$ from the source element $Id\vec{l}$ depends on the sine of the angle between $d\vec{l}$ and the unit vector $\hat{r}$ from the source toward the point where we're evaluating the field. Mathematically, all this is summarized in a compact vector equation:

$$d\vec{B} = \frac{\mu_0}{4\pi}\frac{Id\vec{l} \times \hat{r}}{r^2} \quad \text{(Biot–Savart law)} \tag{26.7}$$

where μ_0 is the **permeability constant**, whose exact value is $4\pi \times 10^{-7}$ N/A^2 (equivalent and often-used units are T·m/A).

Besides the more complicated directionality evidenced by the cross product in the Biot–Savart law, there's another distinction between the Coulomb and Biot–Savart laws. Both describe fields of localized structures—namely, point charges and current elements. It makes sense to talk about an isolated point charge. But an isolated current element is impossible in the steady state; any steady current must flow in a complete circuit. So a Biot–Savart calculation necessarily involves the fields produced by current elements around an entire circuit. The magnetic field obeys the superposition principle, so the net field at any point is the vector sum, or integral, of the field contributions of all the individual current elements:

$$\vec{B} = \int d\vec{B} = \frac{\mu_0}{4\pi}\int \frac{Id\vec{l} \times \hat{r}}{r^2} \quad \text{(Biot–Savart law, integrated)} \tag{26.8}$$

The field given in Equation 26.8 depends on the details of the current distribution, but the directionality associated with the cross product means that, quite generally, magnetic field lines encircle the current that is their source (Fig. 26.16). The next two examples use the Biot–Savart law; later we'll find a simpler way to calculate magnetic fields for some current distributions.

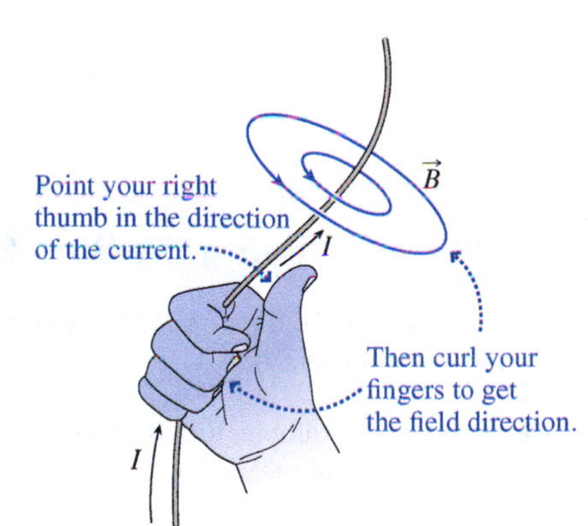

FIGURE 26.16 Magnetic field lines generally encircle a current, with direction given by the right-hand rule.

EXAMPLE 26.4 Using the Biot–Savart Law: The Field of a Current Loop

Find the magnetic field at an arbitrary point P on the axis of a circular loop of radius a carrying current I.

INTERPRET This is a problem involving the magnetic field produced by a specified current distribution.

DEVELOP Figure 26.17*a* shows the current loop with the point P a distance x along the axis. The Biot–Savart law determines the field at P, and we've identified the vectors $d\vec{l}$ and $\hat{r}$ that appear in the law. As Fig. 26.17*b* shows, the individual field components perpendicular to the axis cancel, giving a net field that's along the axis. So our plan is to find an expression for the x-components of the field contributions $d\vec{B}$, and then integrate to get the net field.

EVALUATE Figure 26.17*a* shows that the x-component of any $d\vec{B}$ is $dB_x = dB\cos\theta$, where $\cos\theta = a/r = a/\sqrt{x^2 + a^2}$. The figure also shows that $d\vec{l}$ and $\hat{r}$ are perpendicular; since $\hat{r}$ is a unit vector, the product $d\vec{l} \times \hat{r}$ has magnitude dl. Then the term $d\vec{l} \times \hat{r}/r^2$ in the Biot–Savart law has magnitude $dl/(x^2 + a^2)$, and we have

$$B = \int dB_x = \frac{\mu_0 I}{4\pi}\int_{\text{loop}} \frac{dl}{x^2 + a^2}\frac{a}{\sqrt{x^2 + a^2}}$$
$$= \frac{\mu_0 Ia}{4\pi(x^2 + a^2)^{3/2}}\int_{\text{loop}} dl$$

(continued)

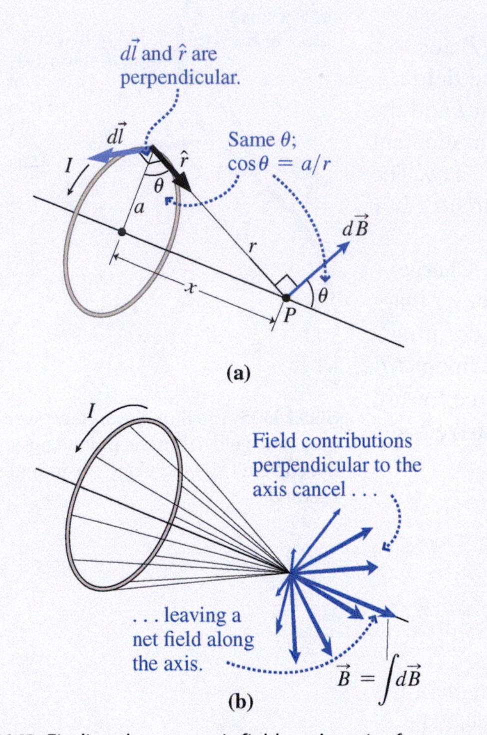

FIGURE 26.17 Finding the magnetic field on the axis of a current loop.

where the integral reduces to a simple form because the distance x is the same for all points on the loop. The remaining integral is the sum of infinitesimal lengths around the loop, or the loop circumference $2\pi a$. So we have

$$B = \frac{\mu_0 I a^2}{2(x^2 + a^2)^{3/2}} \tag{26.9}$$

The direction of the field, as suggested in Fig. 26.17*b*, is along the axis.

ASSESS The field is strongest right at the loop center ($x = 0$) because here we're closest to the loop and so the contributions from all segments of the loop are greatest. The field decreases as we move away from the loop. In general the field is a complicated function of distance, but for large distances ($x \gg a$) it falls off as $1/x^3$. That should remind you of the field we found for an electric dipole in Chapter 20. We'll have more to say about this dipole-like behavior in Section 26.6. ■

EXAMPLE 26.5 Using the Biot–Savart Law: The Field of a Straight Wire

Find the magnetic field produced by an infinitely long straight wire carrying steady current I.

INTERPRET This example, too, is about the field produced by a specified current distribution.

DEVELOP Figure 26.18 is our drawing of the wire on a coordinate system with the x-axis along the wire. Since the wire is infinite, the field magnitude must be the same at all points equidistant from the wire. We show one such point P, a distance y from the wire. We also show an infinitesimal segment $d\vec{l}$ of the wire and the unit vector $\hat{r}$ toward the field point. Our plan is to calculate the field contributions $d\vec{B}$ from all such current elements, and then integrate to find the field $\vec{B}$.

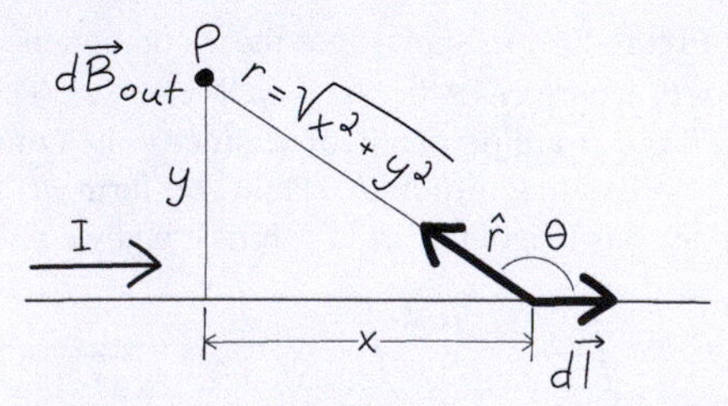

FIGURE 26.18 Calculating the magnetic field at P due to an infinite straight wire carrying current I along the x-axis.

EVALUATE Both $d\vec{l}$ and $\hat{r}$ lie in the plane of the page, so at P the vector $d\vec{l} \times \hat{r}$ in the Biot–Savart law is out of the page. This is true for any segment of the wire. Therefore, we can sum the magnitudes of the contributions $d\vec{B}$ to find the magnitude of the net field, and we know its direction at P will be out of the page. With $\hat{r}$ a unit vector, $|d\vec{l} \times \hat{r}| = dl \sin\theta$, where the triangle in Fig. 26.18 shows that $\sin\theta = y/r = y/\sqrt{x^2 + y^2}$. Then the Biot–Savart law gives a field contribution of magnitude

$$dB = \frac{\mu_0 I}{4\pi}\frac{|d\vec{l} \times \hat{r}|}{r^2} = \frac{\mu_0 I}{4\pi}\frac{dl \sin\theta}{r^2} = \frac{\mu_0 I}{4\pi}\frac{y\, dl}{(x^2 + y^2)^{3/2}}$$

Since the segment $d\vec{l}$ lies along the x-axis, $dl = dx$. Also, y is a constant here, so the net field becomes

$$B = \int dB = \frac{\mu_0 I y}{4\pi}\int_{-\infty}^{\infty}\frac{dx}{(x^2 + y^2)^{3/2}}$$

where we chose the limits to include the entire infinite wire. The integral is a standard one, given in the integral tables of Appendix A; the result is

$$B = \frac{\mu_0 I}{2\pi y} \tag{26.10}$$

ASSESS This result for the *magnetic* field of a long current-carrying wire should remind you of our earlier finding for the *electric* field of a line charge; both decrease as the inverse of the distance from the line. But where the electric field of a line charge points radially outward, the magnetic field of a line current encircles the current, as shown in Fig. 26.19. Of course, an infinite line current is impossible, but our result is a good approximation to the fields of finite wires if we're close compared with the wire's length. ■

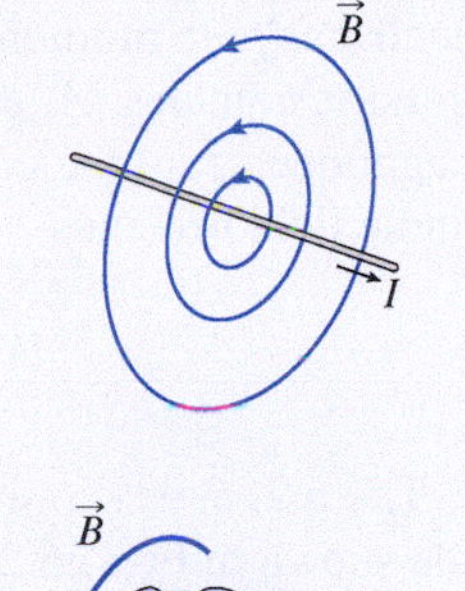

FIGURE 26.19 Magnetic field lines encircle a straight wire, with their direction given by the right-hand rule.

The Magnetic Force Between Conductors

In Section 26.4 we found the force on a current-carrying wire in a magnetic field: $\vec{F} = I\vec{l} \times \vec{B}$. Now you've seen that a straight wire produces a magnetic field. That means current-carrying wires exert magnetic forces on each other. Figure 26.20 shows the situation for two parallel wires carrying currents in the same direction. The wires are a distance d apart, so the field of wire 1 at the location of wire 2 follows from Equation 26.10: $B_1 = \mu_0 I_1/2\pi d$. The field is perpendicular to wire 2, so the force on a length l of wire 2 is

$$F_2 = I_2 l B_1 = \frac{\mu_0 I_1 I_2 l}{2\pi d} \quad \text{(magnetic force between two wires)} \qquad (26.11)$$

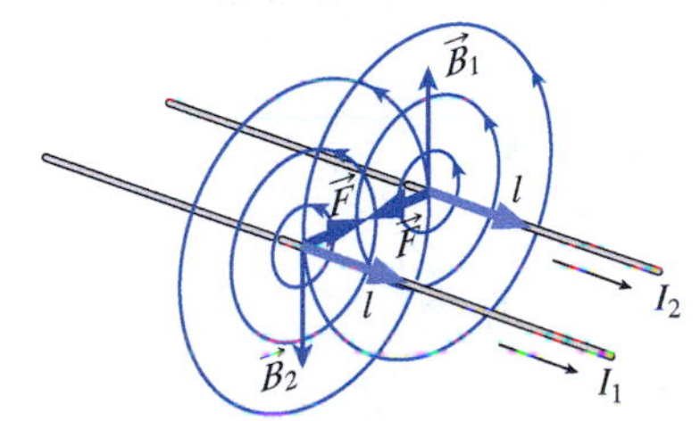

FIGURE 26.20 The magnetic force between parallel currents in the same direction is attractive.

Figure 26.20 shows that the direction of this force is toward wire 1, so the parallel currents *attract*. Analyzing the force on wire 1 from wire 2 amounts to interchanging the subscripts 1 and 2, giving an attractive force of the same magnitude. Reversing one of the currents would change the signs of both forces, showing that antiparallel currents *repel*.

The force between nearby conductors can be quite large, so engineers who design high-strength electromagnets must provide enough physical support to withstand the magnetic force (Problem 85 considers this situation). The hum you often hear around electrical equipment comes from the mechanical vibration of nearby conductors in transformers and other devices, as they respond to the changing force associated with 60-Hz alternating current.

GOT IT? 26.5 A flexible wire is wound into a flat spiral as shown in the figure. (1) If a current flows in the direction shown, will the coil (a) tighten or (b) become looser? (2) Does your answer depend on the current direction? Note: The current enters and leaves the coil through wires (not shown) at each end, perpendicular to the page.

26.6 Magnetic Dipoles

The current loop of Example 26.5 shows the essential characteristic of all steady-state currents—namely, a closed loop with current everywhere the same. Equation 26.9 gives the field on the loop axis: $B = \mu_0 I a^2/2(x^2 + a^2)^{3/2}$. For $x \gg a$ we can ignore a^2 compared with x^2 in the denominator, giving $B \simeq \mu_0 I a^2/2x^3$. Multiply both sides by 2π to get $B \simeq 2\mu_0 IA/4\pi x^3$, where A is the loop area. Compare this result with the field on the axis of an *electric* dipole, Equation 20.6b: Both show the inverse-cube dependence of the dipole field, and both involve fundamental constants from the Coulomb and Biot–Savart

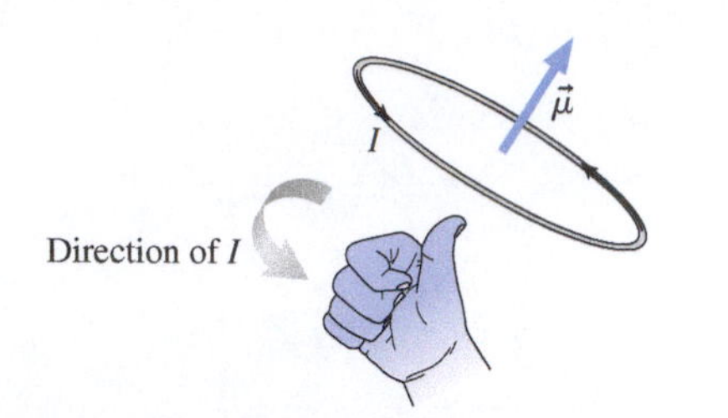

FIGURE 26.21 Finding the direction of a current loop's magnetic dipole moment.

laws that relate fields and their sources. Where the electric-field expression contains the electric dipole moment p, the product of charge and separation, the magnetic-field expression contains IA, the product of the loop current and loop area. We identify IA as the magnitude, μ, of the current loop's **magnetic dipole moment**. Then the on-axis magnetic dipole field becomes

$$B = \frac{\mu_0 \mu}{2\pi x^3} \quad \text{(on-axis field, magnetic dipole)} \tag{26.12}$$

The magnetic dipole moment is a vector whose direction follows from the right-hand rule shown in Fig. 26.21. If we describe the loop by a vector of magnitude A whose direction is perpendicular to the loop as shown in Fig. 26.21, then we can write the magnetic dipole moment as $\vec{\mu} = I\vec{A}$. Practical current loops often have multiple turns; since each carries the same current, an N-turn loop has effective current NI, so its dipole moment becomes

$$\vec{\mu} = NI\vec{A} \quad \left(\begin{array}{c}\text{magnetic dipole moment,}\\ N\text{-turn current loop}\end{array}\right) \tag{26.13}$$

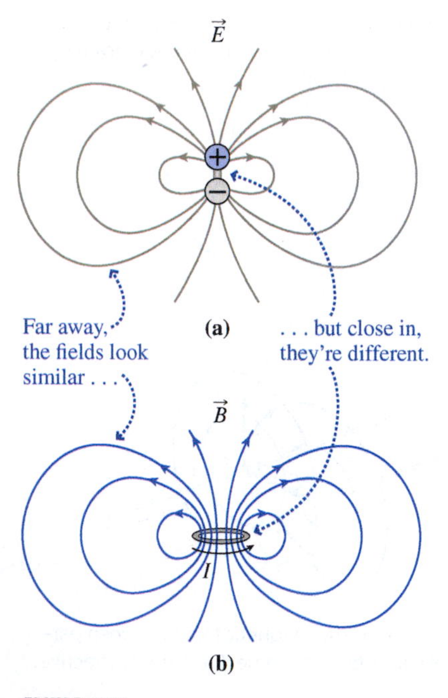

FIGURE 26.22 (a) The electric field of an electric dipole and (b) the magnetic field of a current loop. Far from their sources, both have the shape and the $1/r^3$ dependence of the dipole field.

Although we've found the magnetic field for a current loop only on the loop axis, a more elaborate calculation shows that the magnetic field anywhere far from the loop has exactly the same configuration as the electric field far from an electric dipole. And although we developed the magnetic dipole moment for a circular loop, Equation 26.13 in fact gives the dipole moment of *any* closed loop of current. We conclude that **any current loop constitutes a magnetic dipole**, and that far from the loop, its field will be that of a dipole. Electric and magnetic dipoles are analogous: Both have the same field configuration and mathematical form far from their sources (Fig. 26.22), and both are characterized by their respective dipole moments. But their fields aren't the same. One is an electric field, its origin in static electric charge; the other is a magnetic field, its origin in *moving* electric charge—specifically, charge moving in a closed loop. And the similarity in field configurations holds only at large distances; as Fig. 26.22 shows, the fields near electric and magnetic dipoles are very different, reflecting the different structures that give rise to each.

Current loops are ubiquitous, and so are dipole magnetic fields. Multiple turns of current-carrying wire produce the strong magnetic fields of electromagnets, and superconducting loops provide the fields in MRI scanners. At the atomic level, orbiting and spinning electrons constitute miniature magnetic dipoles. Even planets and stars have magnetic dipole fields.

APPLICATION Magnetic Fields of Earth and Sun

Many astrophysical objects have magnetic fields resulting from the interaction of conducting fluids with the objects' rotation. Earth's field arises in its liquid-iron outer core, where convective flows work with Earth's rotation to produce electric currents. The figure shows that Earth's field approximates that of a dipole; the magnitude of the dipole moment is approximately $\mu = 8.0\times10^{22}\ \text{A}\cdot\text{m}^2$. The direction of the dipole moment vector differs from that of Earth's rotation axis, which accounts for the difference between magnetic and true north. Earth's field reverses roughly every million years, and geologists track seafloor spreading from the resulting magnetization in rocks. Farther out, Earth's magnetic field traps high-energy particles and thus protects us from dangerous radiation. You can see from the figure that magnetic field lines concentrate toward the polar regions, which is why energetic particles tend to enter Earth's atmosphere near the poles, making the aurora a high-latitude phenomenon (recall Example 26.3).

The Sun's gaseous nature makes its magnetic field much more dynamic, and magnetism is the dominant force in its hot, electrically conducting atmosphere. The Sun's field reverses approximately every 11 years, coinciding with the rise and fall of sunspots—regions of intense magnetic field that are often sources of violent outbursts.

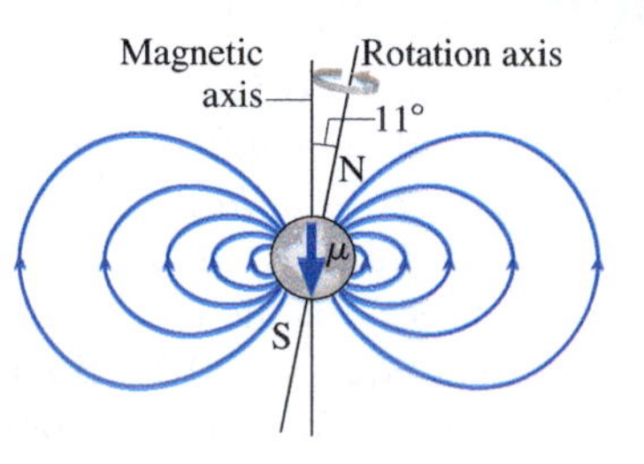

Dipoles and Monopoles

Atoms, molecules, and radio antennas are among the many structures that behave as *electric* dipoles. In all these, separation of positive and negative electric charge gives rise to the dipole. Magnetism is different. No one has ever found an isolated magnetic north or south pole analogous to an electric charge. Electromagnetic theory doesn't rule out such **magnetic monopoles**, and indeed some theories suggest that monopoles might have formed in the Big Bang. But they've never been found. All magnetic fields we've ever seen come from moving *electric* charge. As you'll see in Section 26.7, that includes the fields of permanent magnets. Because steady currents form closed loops, the simplest magnetic entity is the dipole.

Electric field lines generally begin or end on electric charges. But there aren't any "magnetic charges"—magnetic monopoles. Magnetic field lines don't begin or end, but form closed loops encircling the moving electric charges that are their sources. In Chapter 21 we developed Gauss's law to quantify the statement that the number of electric field lines emerging from any closed surface depends only on the charge enclosed. Because there's no magnetic charge, the net number of magnetic field lines and therefore the **magnetic flux** $\oint \vec{B} \cdot d\vec{A}$ emerging from any closed surface is always zero. Thus **Gauss's law for magnetism** is

$$\oint \vec{B} \cdot d\vec{A} = 0 \qquad \text{(Gauss's law for magnetism)} \qquad (26.14)$$

Like Gauss's law for electricity, Equation 26.14 is one of the four fundamental laws that govern all electromagnetic phenomena in the universe. We'll meet the remaining two laws shortly. Although Gauss's law for magnetism has zero on its right side, it's not devoid of content; rather, it says that all magnetic fields are configured so that their field lines have no beginnings or endings.

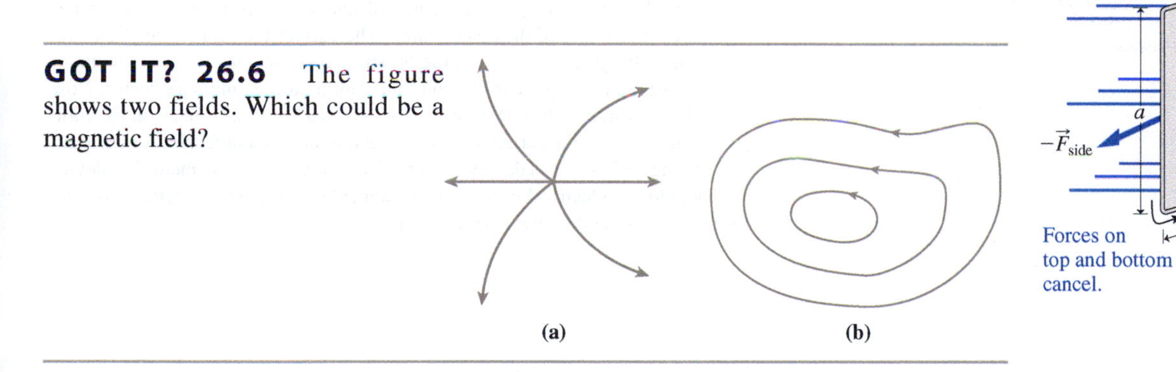

GOT IT? 26.6 The figure shows two fields. Which could be a magnetic field?

The Torque on a Magnetic Dipole

In Section 20.5 we found that an electric dipole $\vec{p}$ in a uniform electric field $\vec{E}$ experiences a torque $\vec{\tau} = \vec{p} \times \vec{E}$; in a nonuniform field there's a net force as well. The same is true for a magnetic dipole in a magnetic field, as you can see by considering the rectangular current loop in a uniform field shown in Fig. 26.23*a*. Current flowing along the top and bottom of the loop results in upward and downward forces of equal magnitude, and neither a net force nor a net torque is associated with these forces. Currents flowing along the vertical sides also result in equal but opposite forces. However, as Fig. 26.23*b* shows, these forces result in a net torque about a vertical axis through the center of the loop. The vertical sides have length a and the currents are perpendicular to the horizontal magnetic field, so the force on each has magnitude $F_{side} = IaB$. The vertical sides are half the loop width b from the axis, so the torque due to each is $\tau_{side} = \frac{1}{2}bF_{side} \sin\theta = \frac{1}{2}bIaB \sin\theta$. Torques on the two sides are in the same direction (out of the page in Fig. 26.23*b*), so the net torque is $\tau = IabB \sin\theta = IAB \sin\theta$, with A the loop area. We've already identified IA as the magnitude of the loop's magnetic dipole moment $\vec{\mu}$ and, given the direction of $\vec{\mu}$ as

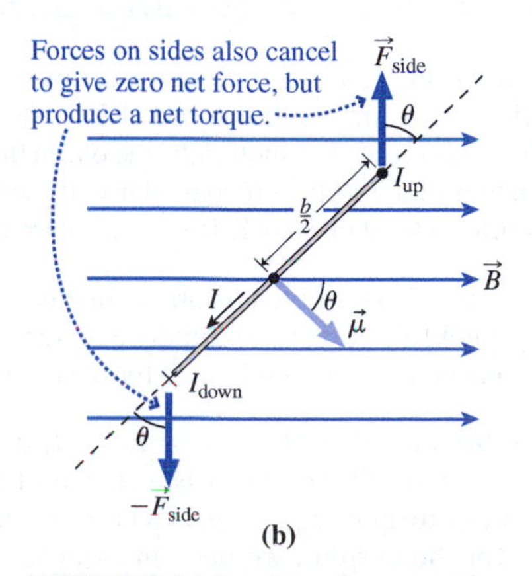

FIGURE 26.23 (a) A rectangular current loop in a uniform magnetic field. (b) Top view of the loop, showing that magnetic forces on the vertical sides result in a net torque.

shown in Figs. 26.21 and 26.23*b*, we can incorporate the directionality and the factor $\sin\theta$ into a cross product:

$$\vec{\tau} = \vec{\mu} \times \vec{B} \qquad \text{(torque on a magnetic dipole)} \qquad (26.15)$$

analogous to the torque on an electric dipole.

The magnetic torque of Equation 26.15 causes magnetic dipoles—current loops—to align with their dipole moment vectors along the magnetic field. It takes work to rotate a dipole out of alignment with the field, and in analogy with Equation 20.11 the associated potential energy is

$$U = -\vec{\mu}\cdot\vec{B} \qquad (26.16)$$

In a nonuniform field, a dipole also experiences a net force. That's why the nonuniform field near the poles of a bar magnet attracts magnetic materials that, as we'll see in the next section, contain magnetic dipoles.

The torque on a magnetic dipole is important in many technologies, including electric motors and MRI imaging. Some satellites use the torque produced by Earth's magnetic field to orient themselves in space; with electricity generated from solar panels powering current loops, there's no fuel to run out.

APPLICATION Electric Motors

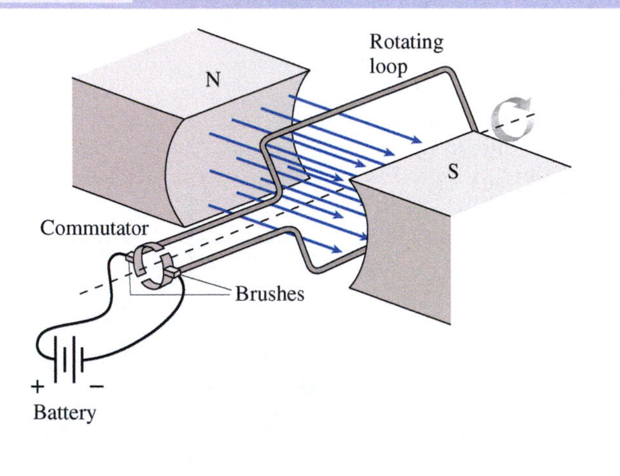

Electric motors are so much a part of our lives that we hardly think of them. Yet refrigerators, disk drives, subway trains, vacuum cleaners, power tools, food processors, fans, washing machines, water pumps, hybrid cars, and most industrial machinery would be impossible without electric motors.

At the heart of every electric motor is a current loop in a magnetic field. But instead of a steady current, the loop carries a current that reverses to keep the loop always spinning. In direct-current (DC) motors, this is achieved through the electrical contacts that provide current to the loop. The figure shows how current flows to the loop through a pair of stationary *brushes* that contact rotating conductors called the *commutator*. The current loop rotates to align with the field, but just as it does so, the brushes cross the gaps in the commutator and reverse the loop's current and therefore its dipole moment vector. Now the loop swings another 180° to its new "desired" position, but again the commutator reverses the current and so the loop rotates continuously. A rigid shaft spinning with the coils delivers mechanical energy. Thus the motor is a device that converts electrical energy to mechanical energy; the magnetic field is an intermediary in this energy conversion.

EXAMPLE 26.6 Torque on a Current Loop: Designing a Hybrid-Car Motor

Toyota's Prius gas–electric hybrid car uses a 60-kW electric motor that develops a maximum torque of 207 N·m. Suppose you want to produce this torque in a motor like the one in the preceding Application, consisting of a 700-turn rectangular coil measuring 30 cm by 20 cm in a uniform field of 50 mT. How much current does the motor need?

INTERPRET This problem is about an electric motor, which according to the Application is essentially a current loop in a magnetic field. We're given the torque and asked for the current.

DEVELOP Equation 26.15, $\vec{\tau} = \vec{\mu} \times \vec{B}$, determines the torque on a current loop. Figure 26.24 is a sketch of the loop at the point of maximum torque, $\tau_{max} = \mu B$, which occurs when $\sin\theta = 1$. To solve for the current, we need the magnetic dipole moment from Equation 26.13, $\mu = NIA$. Then $\tau_{max} = NIAB$.

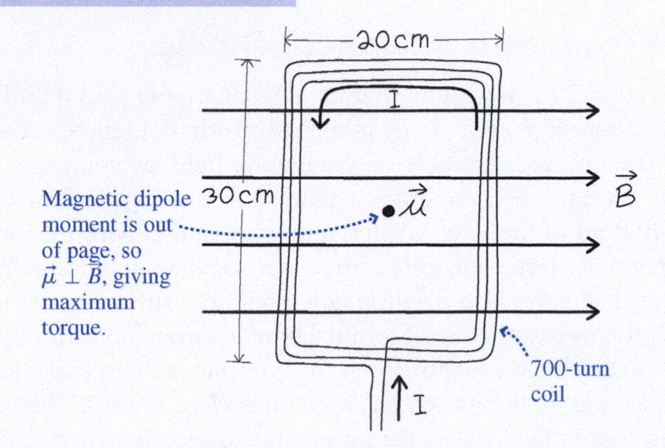

FIGURE 26.24 Loop for the motor of Example 26.6, shown in the position of maximum torque.

EVALUATE Solving for I using the maximum torque and the loop dimensions gives

$$I = \frac{\tau_{max}}{NAB} = \frac{207\ \text{N}\cdot\text{m}}{(700)(0.30\ \text{m})(0.20\ \text{m})(0.050\ \text{T})} = 99\ \text{A}$$

ASSESS That's a large current, but propelling a car is a big job. The actual Prius motor operates at 650 V, so its 60-kW power requires current $I = P/V = 92$ A, close to our answer. ■

26.7 Magnetic Matter

MP® PhET: Magnets and Electromagnets

So far, we've said remarkably little about magnets. That's because magnetism is fundamentally about moving electric charge. Magnets and magnetic matter are just a minor manifestation of this universal phenomenon.

The magnetism of everyday magnets and of magnetic materials like iron results from atomic-scale current loops. An electron orbiting a nucleus constitutes a simple current loop and therefore has a magnetic dipole moment (Fig. 26.25). More importantly, an electron possesses an intrinsic magnetic dipole moment associated with a quantum-mechanical angular momentum called *spin*. Interactions among these magnetic moments determine the magnetic properties of atoms and of bulk matter. The details necessarily involve quantum mechanics; here we give a qualitative overview of magnetism in matter, which manifests itself in three distinct forms.

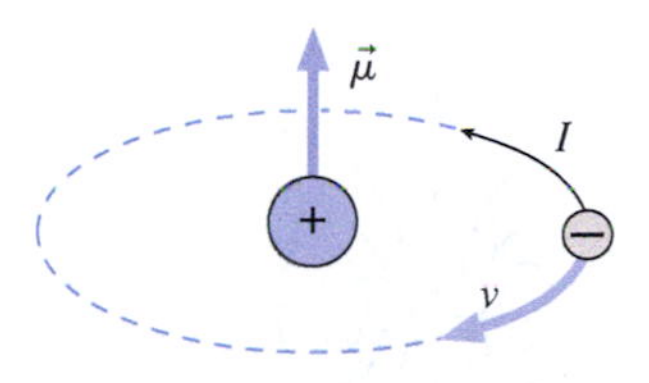

FIGURE 26.25 In the classical model of the atom, the circling electron constitutes a miniature current loop. The current is opposite the motion because the electron is negative. Drawing is only suggestive; the electron's intrinsic magnetic dipole moment is usually more important than that resulting from its orbital motion.

Ferromagnetism

The magnetism you're familiar with is **ferromagnetism**, which is limited to a few substances, including iron, nickel, cobalt, and some alloys and compounds. Strong interactions among atomic magnetic moments result in **magnetic domains**, regions that contain 10^{17}–10^{21} atoms whose magnetic moments are all aligned in the same direction. Normally the magnetic moments of different domains point in random directions, so there's no net magnetic moment. But when an external magnetic field is applied, the domains all align and the material acquires a net magnetic moment. If the field is nonuniform, the material then experiences a net force, which is why ferromagnetic materials are attracted to magnets.

So-called **hard** ferromagnetic materials retain their magnetism even after the applied field is removed; the result is a permanent magnet. A bar magnet, for example, has its internal magnetic moments aligned along its long dimension. You can think of its field as arising from currents circulating around the surface of the magnet (Fig. 26.26)—currents that ultimately result from the superposition of individual atomic current loops. Computer

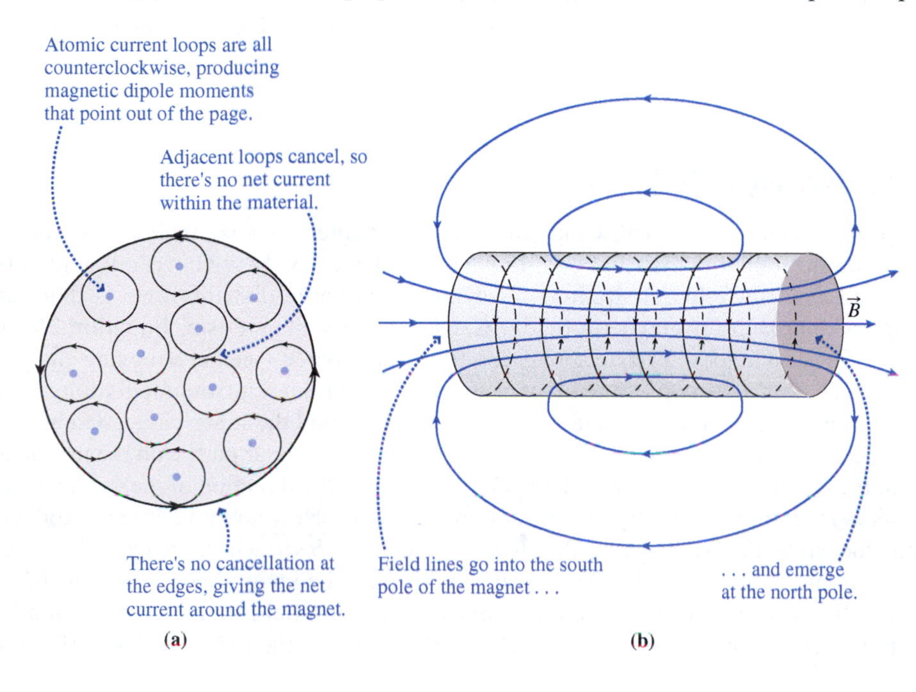

FIGURE 26.26 (a) Cross section of a bar magnet, showing atomic current loops all aligned the same way and making a net current around the magnet. (b) Side view showing the field that results from this *magnetization current*.

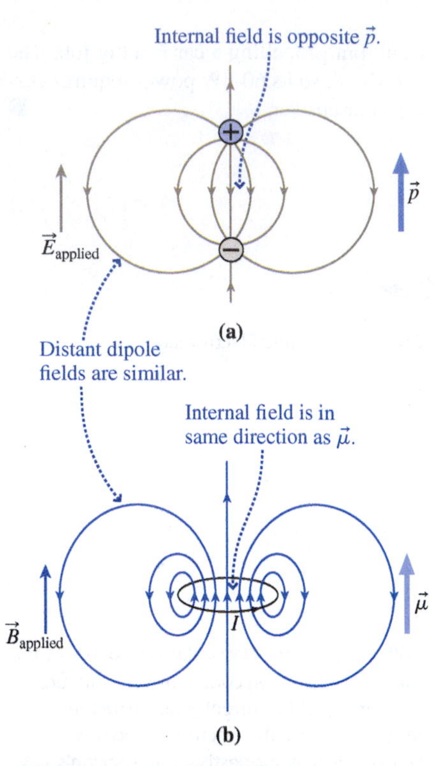

FIGURE 26.27 Internal fields of electric and magnetic dipoles have opposite directions. (a) Electric dipoles reduce an applied electric field; (b) magnetic dipoles increase an applied magnetic field.

disks and credit card strips use hard ferromagnetic materials that retain information as patterns of permanent magnetization. **Soft** ferromagnetic materials, in contrast, don't hold magnetization. They're used where magnetization must be turned on and off rapidly, as in the "heads" that write information to computer disks. Ferromagnetism disappears at the so-called **Curie temperature**, as random thermal motions disrupt the organized alignment of magnetic dipoles; for iron this phase transition occurs at 1043 K.

Paramagnetism

Many substances that aren't ferromagnetic nevertheless consist of atoms or molecules that have permanent magnetic dipole moments. There's no strong interaction among the individual dipoles, so these **paramagnetic** materials respond only weakly to external magnetic fields. Paramagnetic effects are generally significant only at very low temperatures.

Diamagnetism

Materials without intrinsic magnetic moments can have moments induced by changes in an applied magnetic field. Whereas ferromagnetic and paramagnetic materials are attracted to magnets, these **diamagnetic** materials are repelled. We'll explore the origins of diamagnetism in Chapter 27.

Magnetic Permeability and Susceptibility

We found in Chapter 20 that the alignment of molecular electric dipoles reduces the electric field in a material. In paramagnetic and ferromagnetic materials, alignment of magnetic dipoles causes an *increase* in the field. Figure 26.27 shows that this difference occurs because the magnetic field within a current loop points in the *same* direction as the loop's magnetic dipole moment, whereas the internal field of an electric dipole is *opposite* the dipole moment. Ferromagnetic behavior is further complicated because it depends on the material's past history, which is what makes permanent magnets possible. Coils for electromagnets and computer disk "heads" are wound on ferromagnetic cores to provide a much stronger magnetic field than the coil current alone could produce.

GOT IT? 26.7 Which of the following best describes the phenomenon responsible for ordinary magnets? (a) high concentrations of magnetic monopoles; (b) collective alignment of atomic magnetic dipoles; (c) electric currents due to free charges circulating in magnetic materials; (d) separation of positive and negative electric charges to the magnetic poles

---

26.8 Ampère's Law

Computing electric fields with Coulomb's law in Chapter 20 was cumbersome for all but the simplest charge distributions. In Chapter 21 we saw how Gauss's law greatly simplified electric-field calculations for symmetric charge distributions. Is there an analogous approach for magnetic fields? Gauss's law for magnetism, Equation 26.14, won't do because it doesn't relate a magnetic field to its source—namely, moving charge.

Figure 26.28 shows two of the circular magnetic field lines surrounding a long wire carrying a current I out of the page. Imagine moving around the inner circle, and as you go a little way, take the product of a small length dl of the circular path with the magnetic field in the direction you're going. Here you're moving in the direction of the field, so that product is $B\,dl$; more generally, it's the dot product $\vec{B}\cdot d\vec{l}$. Now add up all these products around the circle. Formally, the result is the line integral $\oint \vec{B}\cdot d\vec{l}$, where the circle indicates that we're integrating around a closed path. In this case the integral becomes just $\oint B\,dl$ because $\vec{B}$ and $d\vec{l}$ are in the same direction. But here the field magnitude is given by Equation 26.10: $B = \mu_0 I/2\pi r$, where we've replaced y with the radius r. Since r has the constant value r_1 on the inner circle in Fig. 26.28, the integral becomes $(\mu_0 I/2\pi r_1)\oint dl$. Now $\oint dl$ is the total length of the circular path, or its circumference $2\pi r_1$. So the value of $\oint \vec{B}\cdot d\vec{l}$ is $\mu_0 I$. If you try the same thing for the outer circle in Fig. 26.28, r_2 replaces r_1, but the result is the same: $\oint \vec{B}\cdot d\vec{l} = \mu_0 I$, independent of the radius.

We get the same result even if the path doesn't coincide with a field line, as Fig. 26.29 suggests. On the radial segments of the path shown, $\vec{B}\cdot d\vec{l} = 0$ and there's no contribution to the integral. On segment AB, the field is stronger than if we had stayed on CD, but the segment is proportionately shorter and the integral remains unchanged. We could approximate any arbitrary path as a sequence of radial segments and circular arcs, showing that the value of $\oint \vec{B}\cdot d\vec{l}$ is independent of path as long as the path surrounds the current I. The value of that integral is simply $\mu_0 I$. Magnetic fields obey the superposition principle, so this result must be true for any current distribution, not just a single line current. That is, the line integral $\oint \vec{B}\cdot d\vec{l}$ around *any* closed path is directly proportional to the current encircled by that path. This result is **Ampère's law**, a universal statement about current and magnetic field:

$$\oint \vec{B}\cdot d\vec{l} = \mu_0 I_{\text{encircled}} \quad \text{(Ampère's law, steady currents)} \tag{26.17}$$

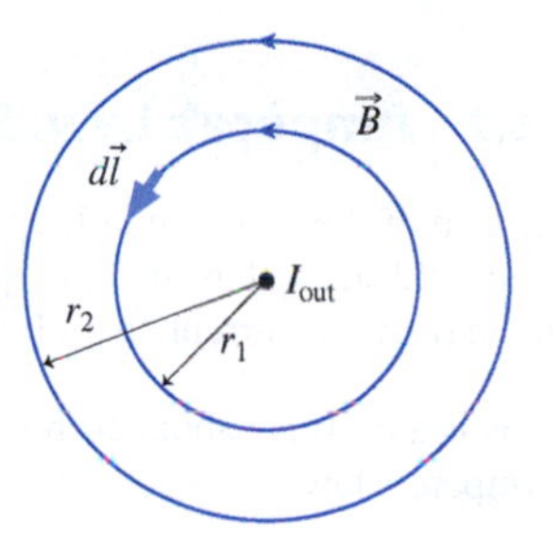

FIGURE 26.28 Two magnetic field lines surrounding a wire carrying current out of the page.

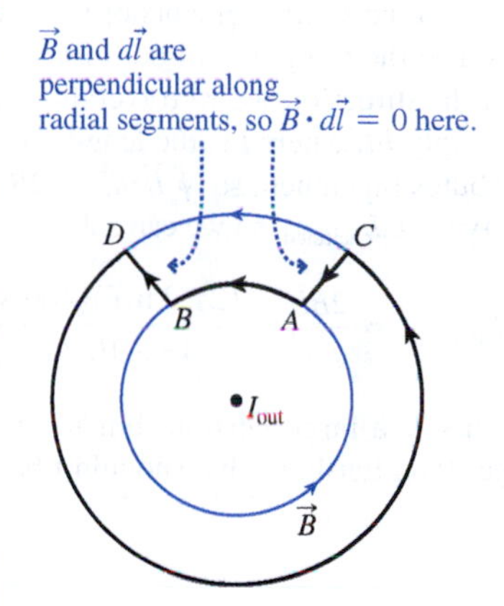

FIGURE 26.29 A closed loop that does not coincide with a field line. The line integral $\oint \vec{B}\cdot d\vec{l}$ around this loop has the same value $\mu_0 I$ that it has around a circular loop.

Ampère's law is another of the four fundamental laws of electromagnetism, although in the form of Equation 26.17 it's limited to steady currents; it also provides a decent approximation for slowly varying currents. In Chapter 29 we'll generalize Ampère's law to remove the restriction to steady currents.

Ampère's law relates the magnetic field to its source—namely, moving charge in the form of electric current—as does the Biot–Savart law. In fact, the laws of Ampère and Biot–Savart are related in the same way as Gauss's and Coulomb's laws. Coulomb and Biot–Savart show how fields arise from pointlike sources—charge elements dq and current elements $I\,d\vec{l}$. Gauss and Ampère are global descriptions, telling how the field must behave over a geometric structure (a closed surface for Gauss, a closed loop for Ampère) in relation to the source (charge or current) enclosed or encircled by that structure. In both cases the field $\vec{E}$ or $\vec{B}$ that appears in the integral is the *net* field arising from *all* sources, not just the enclosed charge or encircled currents.

Ampère's law, like Gauss's, is a truly universal statement. It holds in the electromagnetic devices we build, in atomic and molecular systems, and in distant astrophysical objects. Find a path around which $\oint \vec{B}\cdot d\vec{l}$ isn't zero, and you've found a region where electric current must be flowing (Fig. 26.30). Although it's difficult to show mathematically, the Biot–Savart law follows from Ampère's law just as Coulomb's law follows from Gauss's.

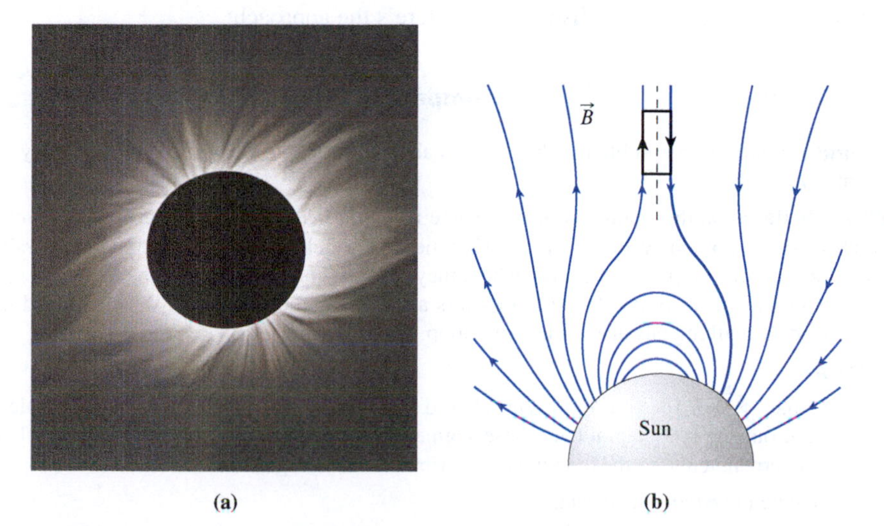

FIGURE 26.30 (a) Coronal streamers in the Sun's atmosphere contain oppositely directed magnetic fields. (b) A model calculation of the magnetic field in a single streamer. Since $\oint \vec{B}\cdot d\vec{l}$ is clearly nonzero around the loop shown, there must be an encircled current.

EXAMPLE 26.7 Ampère's Law: Solar Currents

The long dimension of the rectangular loop in Fig. 26.30*b* is 400 Mm, and the magnetic field strength near the loop has a constant magnitude of 2 mT. Estimate the total current encircled by the rectangle.

INTERPRET This is a problem about currents encircled by a loop, so it must involve Ampère's law.

DEVELOP Figure 26.30*b* provides our drawing. Ampère's law (Equation 26.17) equates $\oint \vec{B} \cdot d\vec{l}$ to $\mu_0 I_{\text{encircled}}$, so we want to evaluate the integral around the loop shown and then solve for $I_{\text{encircled}}$.

EVALUATE On the short segments of the path, $\vec{B} \perp d\vec{l}$, so there's no contribution to the integral. On both long segments, $\vec{B}$ is constant and lies in the direction we're traversing the path, so here $\int \vec{B} \cdot d\vec{l}$ becomes simply Bl, where l is the length of the path segments. Each side contributes this much, so $\oint \vec{B} \cdot d\vec{l} = 2Bl$. Ampère's law equates this quantity to $\mu_0 I_{\text{encircled}}$, so we can solve to get

$$I_{\text{encircled}} = \frac{2Bl}{\mu_0} = \frac{(2)(2\text{ mT})(400\text{ Mm})}{4\pi \times 10^{-7}\text{ N/A}^2} = 10^{12}\text{ A}$$

ASSESS This is a large current, but we're dealing with a region much larger than Earth, so that shouldn't be too surprising. You can get the direction of the current from the right-hand rule: Curl your fingers around the loop in the direction that gives positive $\oint \vec{B} \cdot d\vec{l}$, and your thumb points in the direction of the current. Here that's into the page. In three dimensions this current actually flows around the Sun in approximately the equatorial plane. Note that our result depends crucially on the field reversing across the equatorial plane. In a truly uniform field, one side of the loop would have contributed Bl to the line integral, the other $-Bl$. That would make the integral zero and imply no encircled current. This shows that we can have a uniform field in a current-free region, but not a field that reverses sign—at least not abruptly as in this solar example.

✓TIP Ampèrian Loops

The loop used with Ampère's law is truly arbitrary. It needn't coincide with a field line. In this example, the rectangular loop coincided with the field over its long sides but not along its ends. The loop used with Ampère's law is called an **Ampèrian loop**. Don't confuse Ampèrian loops with field lines; they might coincide, but they don't have to.

GOT IT? 26.8 The figure shows three parallel wires carrying current of the same magnitude I, but in one of them the current direction is opposite that of the other two. If $\oint \vec{B} \cdot d\vec{l} \neq 0$ around loop 2, (1) what's $\oint \vec{B} \cdot d\vec{l}$ around loop 1, and (2) which current is the opposite one?

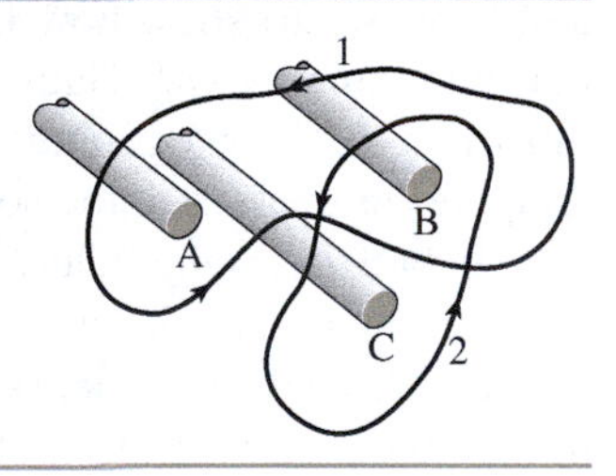

Using Ampère's Law

For charge distributions with sufficient symmetry, we used Gauss's law to solve for the electric field in a simple and elegant way. We can do the same with Ampère's law for sufficiently symmetric current distributions. Here's the approach:

PROBLEM-SOLVING STRATEGY 26.1 Ampère's Law

INTEPRET Interpret the problem to be sure it's about magnetic field and current. Identify the symmetry.

DEVELOP Based on the symmetry, sketch some field lines. Then find an Ampèrian loop over which you'll be able to evaluate $\oint \vec{B} \cdot d\vec{l}$. That means the field should be constant and parallel to the loop over all or part of the loop; where they're not parallel they should be perpendicular. Like a gaussian surface, the Ampèrian loop is a purely mathematical construct; it need not correspond to anything physical. Draw your loop.

EVALUATE

- Evaluate $\oint \vec{B} \cdot d\vec{l}$ for your loop. This should be straightforward because you'll be able to take B outside the integral over those segments where it's constant, and segments where $\vec{B}$ is perpendicular to the loop won't contribute to the integral.
- Evaluate the *encircled* current.
- Equate your result for $\oint \vec{B} \cdot d\vec{l}$ to $\mu_0 I_{\text{encircled}}$, and solve for B. Symmetry should give you the direction of $\vec{B}$.

ASSESS Does your answer make sense in terms of what you know about the fields of simple charge and current distributions?

EXAMPLE 26.8 Using Ampère's Law: Outside and Inside a Wire

A long, straight wire of radius R carries a current I distributed uniformly over its cross section. Find the magnetic field (a) outside and (b) inside the wire.

INTERPRET We follow our strategy and identify this as a situation with line symmetry. Therefore, the field depends only on the radial distance from the wire's central axis.

DEVELOP Magnetic field lines encircle their source, so the only field lines consistent with the symmetry are concentric circles. We've sketched some of these circular field lines in Fig. 26.31. The field is tangent to the field lines and, by symmetry, has the same magnitude B everywhere on a field line. So the field lines themselves make good Ampèrian loops.

EVALUATE

- The field is everywhere parallel to a circular Ampèrian loop, and its magnitude is constant on the loop, so for a loop of radius r, $\oint \vec{B} \cdot d\vec{l}$ becomes $B \oint dl$, or just $2\pi r B$, because $\oint dl$ is the circumference $2\pi r$. This is true both outside and inside the wire.

We'll first answer (a):

- For any loop *outside* the wire, the encircled current is the total current I.
- Equating our expression for $\oint \vec{B} \cdot d\vec{l}$ to μ_0 times the encircled current gives $2\pi r B = \mu_0 I$, so

$$B = \frac{\mu_0 I}{2\pi r} \quad \left(\begin{array}{c}\text{field outside any current} \\ \text{distribution with line symmetry}\end{array}\right) \qquad (26.18)$$

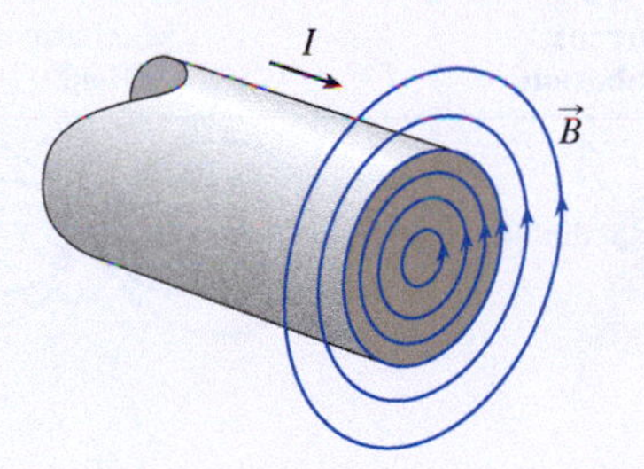

FIGURE 26.31 Cross section of a long cylindrical wire. Any field line can serve as an Ampèrian loop. Inside the wire, the loop's radius r is less than the wire's radius R; outside, $r > R$.

Now on to (b):

- *Inside* the wire, a circular Ampèrian loop encloses only some of the current. How much? With the current uniformly distributed over the wire's cross section, there's a uniform current density $J = I/A = I/\pi R^2$. The encircled current is the current density times the area πr^2 within our loop, so $I_{\text{encircled}} = I(r^2/R^2)$.
- Then we have $2\pi r B = \mu_0 I(r^2/R^2)$, which gives

$$B = \frac{\mu_0 I r}{2\pi R^2} \quad \left(\begin{array}{c}\text{field inside a uniform current} \\ \text{distribution with line symmetry}\end{array}\right) \qquad (26.19)$$

In both cases application of the right-hand rule shows that the field circles counterclockwise, as shown in Fig. 26.31.

ASSESS Equation 26.18 is identical to our result for the line current of Example 26.5, and shows that the field outside *any* current distribution with line symmetry is the same as that of a line current at the symmetry axis. We found the same thing for the electric fields outside cylindrical charge distributions, including the $1/r$ decrease with distance from the axis. Inside the wire, meanwhile, the field increases linearly with distance r from the cylinder axis. This makes sense because, as r increases, we encircle more current in proportion to r^2—while at the same time the field decreases as $1/r$. You found a similar result for a uniformly charged cylinder if you worked Problem 56 in Chapter 21. Now, the electric and magnetic fields of cylindrical distributions look very different—$\vec{E}$ is radial, while $\vec{B}$ forms circles—but the dependence on distance is the same in both cases.

✓TIP Symmetry Is Crucial

Our use of Ampère's law to derive the field of a long wire depends crucially on symmetry. We can't arbitrarily pull B outside the integral unless we know—as we do here from symmetry—that it's constant in magnitude and in direction relative to our Ampèrian loop. ■

EXAMPLE 26.9 Ampère's Law: A Current Sheet

An infinite flat sheet carries current out of this page. The current is distributed uniformly along the sheet, with current per unit width given by J_s. Find the magnetic field of this sheet.

INTERPRET We follow our strategy, identifying the current distribution as having plane symmetry. Then the only thing the field might depend on is the distance from the current-carrying sheet.

DEVELOP The only field lines consistent with the symmetry are straight lines parallel to the plane; we've drawn the current and some field lines in Fig. 26.32. The situation is similar to Example 26.7, and a suitable Ampèrian loop is a rectangle with sides along the field lines and perpendicular edges; we've sketched one such rectangle of width l.

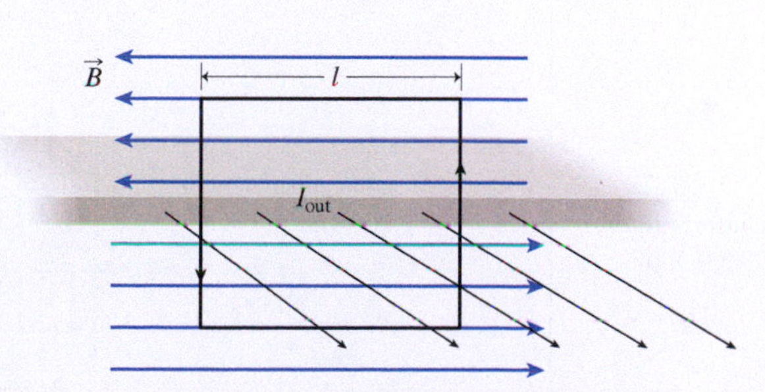

FIGURE 26.32 A current sheet extends infinitely to the left and right, as well as in and out of the page. Field lines and a rectangular Ampèrian loop are shown.

(continued)

We drew the field in opposite directions on either side of the sheet; it had better be that way if, as we discussed for the oppositely directed solar magnetic fields in Example 26.7, $\oint \vec{B}\cdot d\vec{l}$ is to be nonzero. And it has to be nonzero because we know that our Ampèrian loop encircles current.

EVALUATE

- We evaluate $\oint \vec{B}\cdot d\vec{l}$ just as in Example 26.7, getting $2Bl$.
- The sheet carries current J_s per unit width, so our rectangle of width l encircles a current $I_{\text{encircled}} = J_s l$.
- Equating our expression for $\oint \vec{B}\cdot d\vec{l}$ to $\mu_0 I_{\text{encircled}}$ gives $2Bl = \mu_0 J_s l$, or

$$B = \tfrac{1}{2}\mu_0 J_s \quad \text{(field of an infinite current sheet)} \qquad (26.20)$$

ASSESS Make sense? Like the electric field of an infinite plane charge, the magnetic field of an infinite current sheet doesn't depend on distance from the sheet. Now, there's no such thing as a truly infinite sheet, so our result is an approximation valid near a finite sheet but not close to its edges. As Fig. 26.33 shows, the lines of a finite sheet wrap around the ends to form closed loops, and far from the loop the field begins to resemble that of a wire. But close in, Equation 26.20 holds. ■

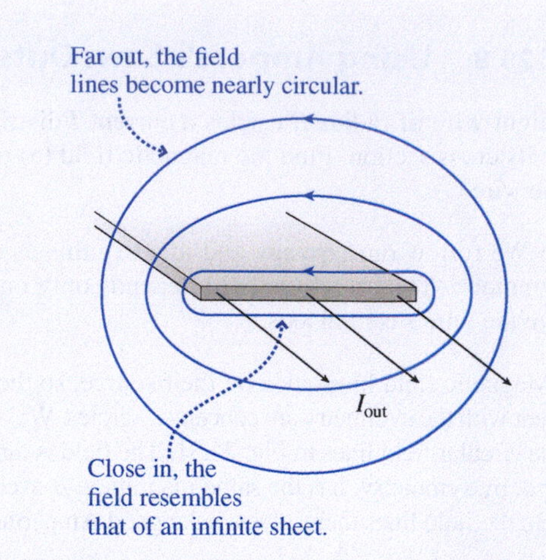

FIGURE 26.33 Field of a finite-width current sheet.

Fields of Simple Current Distributions

We've just used Ampère's law to calculate the fields of two symmetric current distributions, and we compared them with analogous results for electric fields. Table 26.1 summarizes these and other analogies. Although the magnetic and electric fields may look different, they

Table 26.1 Fields of Some Simple Charge and Current Distributions

Field Dependence on Distance[a]	Charge Distribution	Electric Field	Current Distribution	Magnetic Field
$\frac{1}{r^3}$	Electric dipole		Magnetic dipole	
$\frac{1}{r^2}$	Point charge or spherically symmetric		Impossible for steady current	
$\frac{1}{r}$	Charge distribution with line symmetry		Current distribution with line symmetry	
Uniform field; no variation	Infinite flat sheet of charge		Current sheet	

[a]For field *outside* distribution

exhibit the same general relationships between geometry and the way the fields decrease with distance. Real distributions are more complicated, but may often be approximated by these simple cases. Far from *any* current loop, for example, its field approximates that of a dipole. Very near *any* wire, its field is essentially that of a long, straight wire. Very near *any* flat sheet of current, the field is essentially that of Example 26.9.

Solenoids

We found in Chapter 23 that there's an essentially uniform electric field inside a parallel-plate capacitor. Here we explore a current configuration that produces an analogously uniform magnetic field.

Figure 26.34*a* shows a single current loop and its magnetic field. Add a few turns to form an extended coil, and the field isn't much different (Fig. 26.34*b*); more turns (Fig. 26.34*c*), and the region of strongest field is increasingly confined within the coil. With a very long coil (Fig. 26.34*d*), the field is strong and uniform deep within the coil and very weak outside. The limit of an infinitely long, tightly wound coil would produce a uniform field within and no field outside.

A tightly wound coil is a **solenoid**. For a long solenoid—much longer than its diameter—we can use Ampère's law to find the magnetic field inside the solenoid. Figure 26.35 shows a cross section through a solenoid, with a rectangular Ampèrian loop of width l. Since the field is zero outside, the only contribution to $\oint \vec{B} \cdot d\vec{l}$ is from the interior segment parallel to the field, and with a uniform field that gives Bl. If the solenoid carries current I and consists of n turns of wire per unit length, then Fig. 26.35 shows that our Ampèrian loop encircles a total current nlI. So Ampère's law reads $Bl = \mu_0 nlI$, or

$$B = \mu_0 nI \quad \text{(solenoid field)} \tag{26.21}$$

Since the rectangle's vertical dimension never entered the calculation, the field has this same magnitude everywhere inside the solenoid. Although Fig. 26.34 depicts circular coils, Equation 26.21 holds for a solenoid of any cross section.

Solenoids produce uniform magnetic fields in a variety of applications, including the long cylindrical "tunnel" of an MRI scanner. Because the field becomes nonuniform at

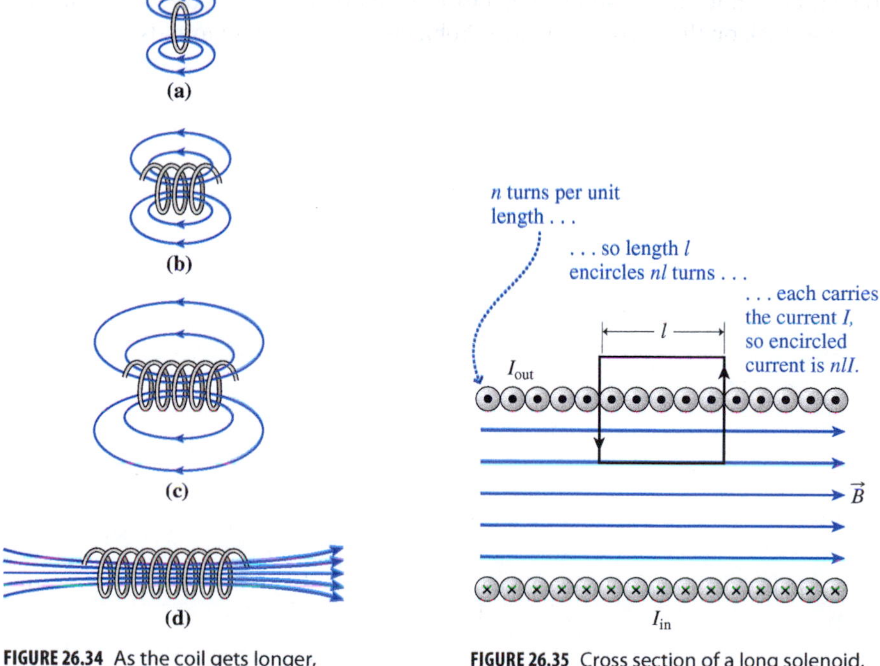

FIGURE 26.34 As the coil gets longer, the interior field stays nearly constant but the exterior field weakens as the field lines spread ever farther apart.

FIGURE 26.35 Cross section of a long solenoid, showing a rectangular Ampèrian loop straddling the region where solenoid coils emerge from the plane of the page.

the ends of a solenoid, ferromagnetic materials are attracted into the coil. Small solenoids can thus produce straight-line motion of an iron plunger. One application is the solenoid on a car starter, which engages the starter motor's gear with the gasoline engine. Solenoid-operated valves are widely used in controlling fluid flows; the valves that admit water to your washing machine and dishwasher are solenoid valves.

EXAMPLE 26.10 A Solenoid: The Current in an MRI Scanner

The solenoid used in an MRI scanner is 2.4 m long and 95 cm in diameter. It's wound from superconducting wire 2.0 mm in diameter, with adjacent turns separated by an insulating layer of negligible thickness. Find the current that will produce a 1.5-T magnetic field inside the solenoid.

INTERPRET This is a problem about a solenoid, which involves relating current and field.

DEVELOP Equation 26.21, $B = \mu_0 nI$, provides the relation we need. To use it we need n, the number of turns per unit length. Figure 26.36 shows how we find n from the wire diameter. Knowing n, we can use Equation 26.21 to find the current.

EVALUATE Figure 26.36 shows that $n = 500$ turns per meter. So now we can solve Equation 26.21 to get

$$I = \frac{B}{\mu_0 n} = \frac{1.5\ \text{T}}{(4\pi \times 10^{-7}\ \text{N/A}^2)(500\ \text{m}^{-1})} = 2.4\ \text{kA}$$

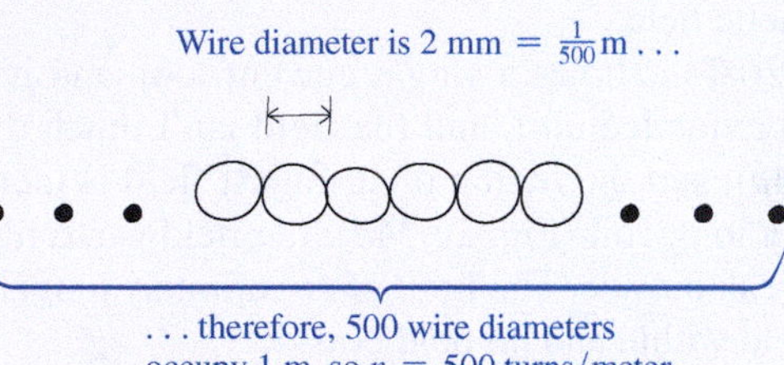

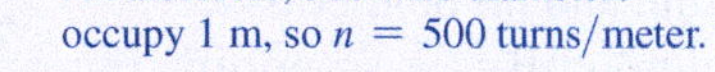

FIGURE 26.36 Finding n.

ASSESS That's a large current, but it's readily handled by the niobium–titanium superconductor in the MRI scanner. Notice that more turns per unit length would reduce the current demand; that's because each turn carries the same current I, so more turns increase the encircled current and thus the field for a given total current. Our answer here is only approximate; with its 2.4-m length and a diameter of nearly 1 m, our solenoid barely approaches the limit of a "long" solenoid. ■

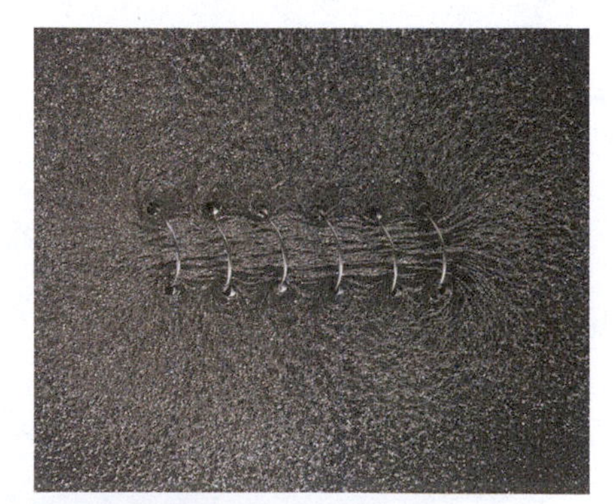

FIGURE 26.37 Iron filings trace the magnetic field of a loosely wound solenoid. Compare with the field of a bar magnet shown in Fig. 26.1.

With its current flowing around an essentially cylindrical surface, the solenoid might remind you of the bar magnet in Fig. 26.26. There, atomic current loops produce a magnetization current flowing around the cylindrical magnet. Indeed, a solenoid and a bar magnet are very similar, and they produce similar magnetic fields (Fig. 26.37). Wrap a solenoid around on itself and you've got a **toroid**—a donut-shaped coil whose circular field lines close back on themselves. Passage Problems 88–91 explore toroids.

CHAPTER 26 SUMMARY

Big Idea

The big new idea here is magnetism—an interaction that fundamentally involves *moving electric charge*. Moving charge produces magnetic fields, and moving charges respond to magnetic fields by experiencing a magnetic force.

Key Concepts and Equations

The **magnetic force** on a charge q moving with velocity $\vec{v}$ in a magnetic field $\vec{B}$ is

$$\vec{F} = q\vec{v} \times \vec{B} \qquad \text{(magnetic force)}$$

The force acts at right angles to both $\vec{v}$ and $\vec{B}$, and therefore it does no work.

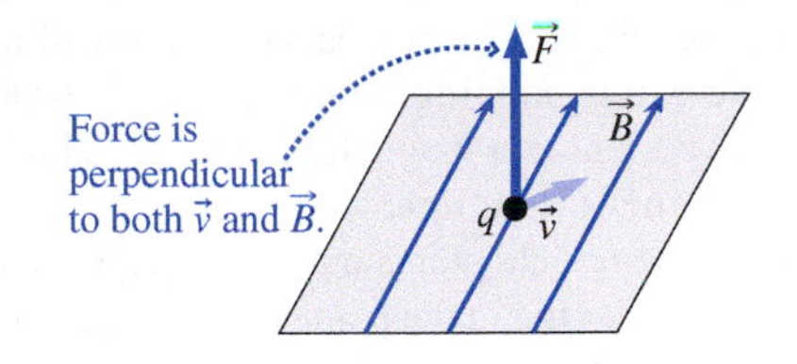

The **Biot–Savart law** describes the magnetic field $d\vec{B}$ arising from a small element of steady current:

$$d\vec{B} = \frac{\mu_0}{4\pi}\frac{I d\vec{l} \times \hat{r}}{r^2}$$

Here μ_0 is the **permeability constant**, with value $4\pi \times 10^{-7}$ N/A^2.

$d\vec{l}$ is a small piece of the wire.
$\hat{r}$ is a unit vector from $d\vec{l}$ toward P.
$d\vec{B}$ is into page.

Ampère's law provides a more global description of how magnetic fields arise from currents, relating the line integral around any closed loop to the encircled current:

$$\oint \vec{B} \cdot d\vec{l} = \mu_0 I_{\text{encircled}}$$

Ampère's law in this form applies only to steady currents.

Gauss's law for magnetism expresses the fact that there are no magnetic monopoles—magnetic analogs of electric charge—and that magnetic field lines therefore do not begin or end:

$$\oint \vec{B} \cdot d\vec{A} = 0$$

Static electric fields, in contrast, always begin or end on electric charges.

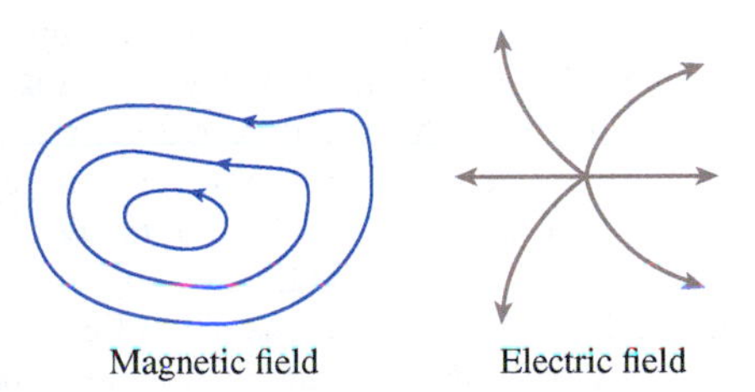
Magnetic field Electric field

Applications

A charged particle moving perpendicular to a uniform magnetic field undergoes circular motion with the **cyclotron frequency** $f = qB/2\pi m$. More generally, charged particles in magnetic fields follow spiral paths, "trapped" on the field lines.

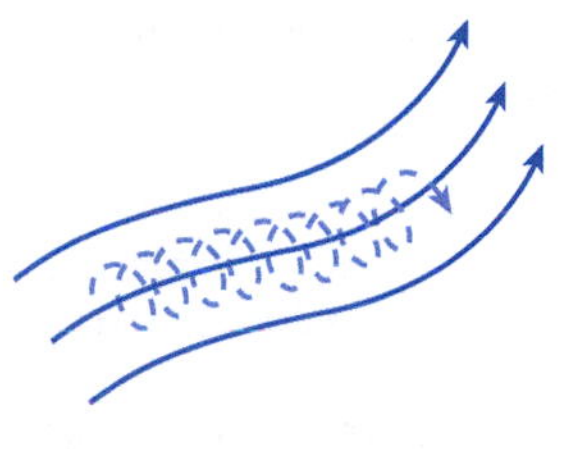

The magnetic force on a straight wire of length l carrying current I in a uniform magnetic field is $\vec{F} = I\vec{l} \times \vec{B}$. Parallel wires a distance d apart experience forces from each other's magnetic field: $F = \dfrac{\mu_0 I_1 I_2 l}{2\pi d}$. The force is attractive for currents in the same direction, repulsive for currents in opposite directions.

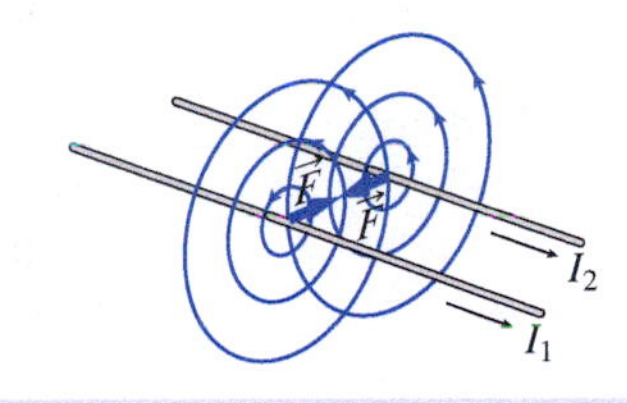

A current loop gives rise to a magnetic field that, at distances large compared with the loop's size, is a dipole field. The loop's magnetic dipole moment has magnitude $\mu = IA$, with A the loop area, and the loop responds to an external magnetic field by experiencing the torque typical of a dipole: $\vec{\tau} = \vec{\mu} \times \vec{B}$.

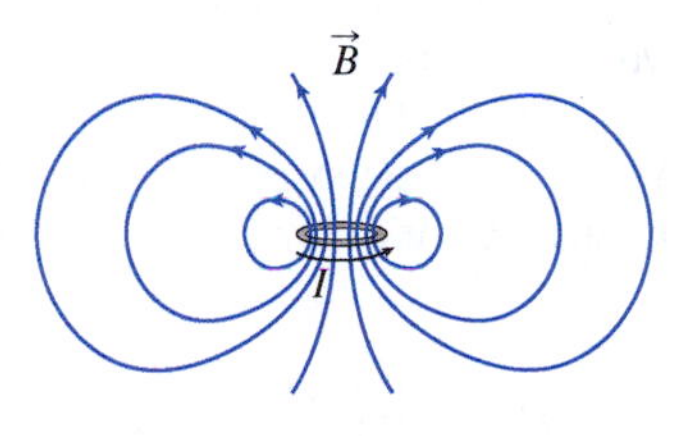

Fields of simple current distributions:

Line current: $B = \dfrac{\mu_0 I}{2\pi r}$ Current sheet: $B = \frac{1}{2}\mu_0 J_s$ Solenoid: $B = \mu_0 n I$

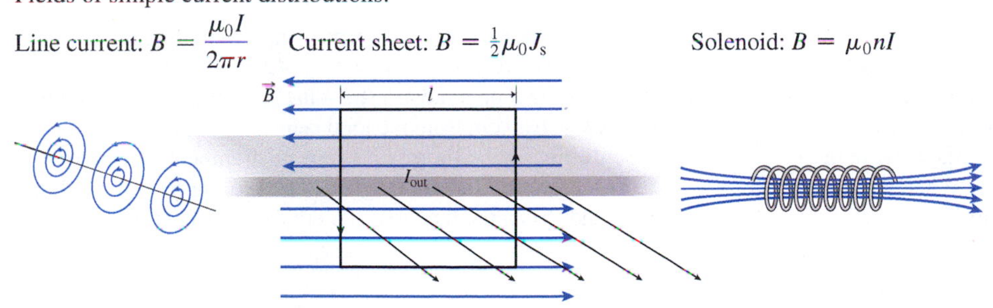

Magnetism in matter arises from the interactions of atomic-scale current loops. **Ferromagnetic** materials have strong interactions and exhibit the bulk magnetism associated with permanent magnets and with magnetic materials like iron. **Paramagnetism** and **diamagnetism** are weaker manifestations of magnetism in matter.

MP For homework assigned on MasteringPhysics, go to www.masteringphysics.com

BIO *Biology and/or medicine-related problems* **DATA** *Data problems* **ENV** *Environmental problems* **CH** *Challenge problems* **COMP** *Computer problems*

For Thought and Discussion

1. A charged particle moves through a region containing only a magnetic field. Under what condition will it experience no force?
2. An electron moving with velocity $\vec{v}$ through a magnetic field $\vec{B}$ experiences a magnetic force $\vec{F}$. Which of the vectors $\vec{F}, \vec{v}$, and $\vec{B}$ must be at right angles?
3. A magnetic field points out of this page. Will a positively charged particle moving in the plane of the page circle clockwise or counterclockwise as viewed from above?
4. Do particles in a cyclotron gain energy from the electric field, the magnetic field, or both? Explain.
5. An electron and a proton moving at the same speed enter a region containing a uniform magnetic field. Which is deflected more from its original path?
6. Two identical particles carrying equal charge are moving in opposite directions, perpendicular to a uniform magnetic field, when they collide elastically head-on. Describe their subsequent motion.
7. In what two senses does a current loop behave like a magnetic dipole?
8. The Biot–Savart law shows that the magnetic field of a current element decreases as $1/r^2$. Could you put together a complete circuit whose field exhibits this decrease? Why or why not?
9. Do currents in the same direction attract or repel? Explain.
10. If a current is passed through an unstretched spring, will the spring contract or expand? Explain.
11. Figure 26.38 shows some magnetic field lines associated with two parallel wires carrying equal currents perpendicular to the page. Are the currents in the same or opposite directions? How can you tell? *Note*: The only currents in Fig. 26.38 are those in the two wires.

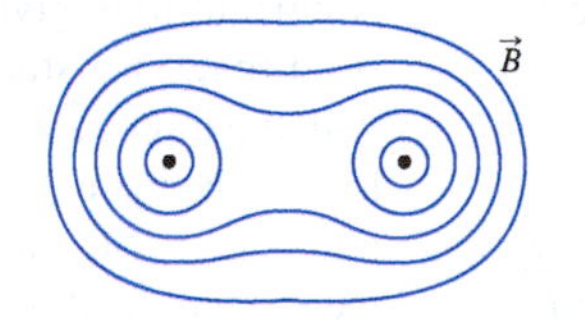

FIGURE 26.38 For Thought and Discussion 11

12. Why is a piece of iron attracted into a solenoid?
13. Would there be a magnetic force on a piece of iron deep inside a long solenoid? Explain.
14. An unmagnetized piece of iron has no net magnetic dipole moment, yet it's attracted to either pole of a bar magnet. Why?

Exercises and Problems

Exercises

Section 26.2 Magnetic Force and Field

15. Find (a) the minimum magnetic field needed to exert a 5.4-fN force on an electron moving at 21 Mm/s and (b) the field strength required if the field were at 45° to the electron's velocity.
16. An electron moving at right angles to a 0.10-T magnetic field experiences an acceleration of 6.0×10^{15} m/s^2. (a) What's its speed? (b) By how much does its speed change in 1 ns?
17. Find the magnitude of the magnetic force on a proton moving at 2.5×10^5 m/s (a) perpendicular; (b) at 30°; (c) parallel to a 0.50-T magnetic field.
18. The magnitude of Earth's magnetic field is about 0.5 gauss near Earth's surface. What's the maximum possible magnetic force on an electron with kinetic energy of 1 keV? Compare with the gravitational force on the electron.
19. A velocity selector uses a 60-mT magnetic field perpendicular to a 24-kN/C electric field. At what speed will charged particles pass through the selector undeflected?

Section 26.3 Charged Particles in Magnetic Fields

20. Find the radius of the path described by a proton moving at 15 km/s in a plane perpendicular to a 400-G magnetic field.
21. How long does it take an electron to complete a circular orbit perpendicular to a 1.0-G magnetic field?
22. Radio astronomers detect electromagnetic radiation at a frequency of 42 MHz from an interstellar gas cloud. If the radiation results from electrons spiraling in a magnetic field, what's the field strength?
23. In a microwave oven, electrons describe circular motion in a magnetic field within a special tube called a *magnetron*; as you'll learn in Chapter 29, the electrons' motion results in the production of micowaves. (a) If the electrons circle at a frequency of 2.45 GHz, what's the magnetic field strength? (b) If the magnetron can accommodate electron orbits with maximum diameter 2.72 mm, what's the electrons' energy in eV?
24. Two protons, moving in a plane perpendicular to a uniform 500-G magnetic field, undergo an elastic head-on collision. How much time elapses before they collide again?

Section 26.4 The Magnetic Force on a Current

25. Find the magnitude of the force on a 65.5-cm-long wire carrying 12.0 A at right angles to a 475-G magnetic field.
26. A wire carrying 15 A makes a 25° angle with a uniform magnetic field. The magnetic force per unit length of wire is 0.31 N/m. Find (a) the magnetic field strength and (b) the maximum force per unit length that could be achieved by reorienting the wire.
27. You're on a team performing a high-magnetic-field experiment. A conducting bar carrying 4.1 kA will pass through a 1.3-m-long region containing a 12-T magnetic field, making a 60° angle with the field. A colleague proposes resting the bar on wooden blocks. You argue that it will have to be clamped in place, and to back up your argument you claim that the magnetic force will exceed 10,000 pounds. Are you right?
28. A wire with mass per unit length 75 g/m runs horizontally at right angles to a horizontal magnetic field. A 6.2-A current in the wire results in its being suspended against gravity. What's the magnetic field strength?

Section 26.5 Origin of the Magnetic Field

29. A wire carries 6.71 A. You form it into a single-turn circular loop and measure a magnetic field of 42.8 μT at the loop center. (a) What's the loop's radius? (b) What's the field strength on the loop axis at 10.0 cm from the loop center?
30. A single-turn wire loop is 2.0 cm in diameter and carries a 650-mA current. Find the magnetic field strength (a) at the loop center and (b) on the loop axis, 20 cm from the center.

31. A 2.2-m-long wire carrying 3.5 A is wound into a tight coil 5.0 cm in diameter. Find the magnetic field at its center.
32. What's the current in a long wire if the magnetic field strength 1.2 cm from the wire's axis is 67 μT?
33. In standard household wiring, parallel wires about 1 cm apart carry currents of about 15 A. What's the force per unit length between these wires?

Section 26.6 Magnetic Dipoles

34. Earth's magnetic dipole moment is 8.0×10^{22} A·m^2. Find the magnetic field strength at Earth's magnetic poles.
35. A single-turn square wire loop 18.0 cm on a side carries a 1.25-A current. (a) What's the loop's magnetic dipole moment? (b) What's the magnitude of the torque the loop experiences when it's in a 2.12-T magnetic field with the loop's dipole moment vector at 65.0° to the field?
36. An electric motor contains a 250-turn circular coil 6.2 cm in diameter. If it develops a maximum torque of 1.2 N·m at a current of 3.3 A, what's the magnetic field strength?

Section 26.8 Ampère's Law

37. The line integral of the magnetic field on a closed path surrounding a wire has the value 8.8 μT·m. Find the current in the wire.
38. The magnetic field shown in Fig. 26.39 has uniform magnitude 75 μT, but its direction reverses abruptly. Find the current encircled by the rectangular loop shown.

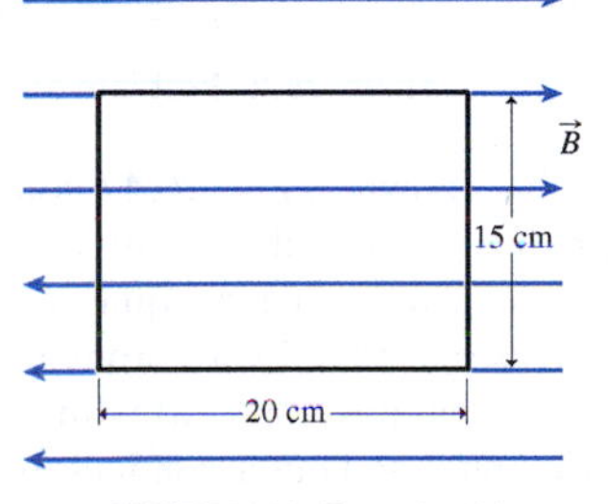

FIGURE 26.39 Exercise 38

39. Number 12 gauge wire, commonly used in household wiring, is 2.053 mm in diameter and can safely carry currents of up to 20.0 A. For a wire carrying this maximum current, find the magnetic field strength (a) 0.150 mm from the wire's axis, (b) at the wire's surface, and (c) 0.375 mm beyond the wire's surface.
40. Show that Equations 26.18 and 26.19 give the same results when evaluated at the wire's surface.
41. A superconducting solenoid has 3300 turns per meter and carries 4.1 kA. Find the magnetic field strength in the solenoid.

Problems

42. A particle carrying a 50-μC charge moves with velocity $\vec{v} = 5.0\hat{\imath} + 3.2\hat{k}$ m/s through a magnetic field given by $\vec{B} = 9.4\,\hat{\imath} + 6.7\,\hat{\jmath}$ T. (a) Find the magnetic force on the particle. (b) Form the dot products $\vec{F} \cdot \vec{v}$ and $\vec{F} \cdot \vec{B}$ to show explicitly that the force is perpendicular to both $\vec{v}$ and $\vec{B}$.
43. Jupiter has the strongest magnetic field in our solar system, about 14 G at its poles. Approximating the field as that of a dipole, find Jupiter's magnetic dipole moment. (*Hint:* Consult Appendix E.)
44. A proton moving with velocity $\vec{v}_1 = 3.6 \times 10^4\,\hat{\jmath}$ m/s experiences a magnetic force of $7.4 \times 10^{-16}\,\hat{\imath}$ N. A second proton moving on the x-axis experiences a magnetic force of $2.8 \times 10^{-16}\,\hat{\jmath}$ N. Find the magnitude and direction of the magnetic field (assumed uniform), and the velocity of the second proton.
45. A simplified model of Earth's magnetic field has it originating in a single current loop at the outer edge of the planet's liquid core (radius 3000 km). What current would give the 62-μT field measured at the north magnetic pole?
46. A beam of electrons moving in the x-direction at 8.7 Mm/s enters a region where a uniform 180-G magnetic field points in the y-direction. The boundary of the field region is perpendicular to the beam. How far into the field region does the beam penetrate?
47. Show that the orbital radius of a charged particle moving at right angles to a magnetic field B can be written $r = \sqrt{2Km}/qB$, where K is the kinetic energy in joules, m the particle's mass, and q its charge.
48. A 90-cm-diameter cyclotron with a 2.0-T magnetic field is used to accelerate deuterium nuclei (one proton plus one neutron). (a) At what frequency should the dee voltage be alternated? (b) What's the maximum kinetic energy of the deuterons? (c) If the magnitude of the potential difference between the dees is 1500 V, how many orbits do the deuterons complete before reaching maximum energy?
49. An electron is moving in a uniform 0.25-T magnetic field; its velocity components parallel and perpendicular to the field are both 3.1 Mm/s. (a) What's the radius of the electron's spiral path? (b) How far does it move along the field direction in the time it takes to complete a full orbit about the field?
50. A wire of negligible resistance is bent into a rectangle as in Fig. 26.40, and a battery and resistor are connected as shown. The right-hand side of the circuit extends into a region containing a uniform 38-mT magnetic field pointing into the page. Find the magnitude and direction of the net force on the circuit.

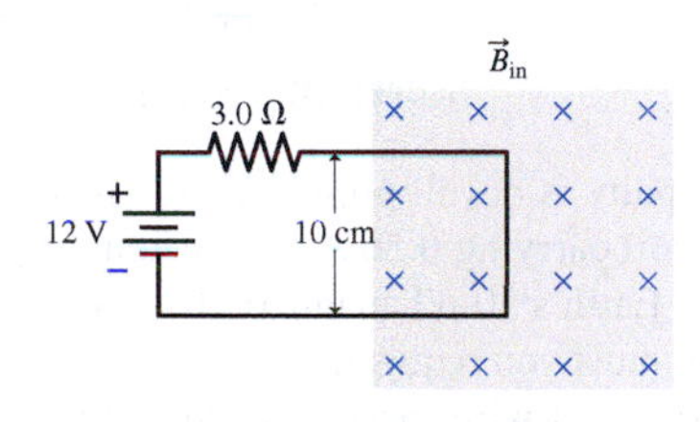

FIGURE 26.40 Problem 50

51. **BIO** You're designing a prosthetic ankle that includes a miniature electric motor containing a 150-turn circular coil 15 mm in diameter. The motor needs to develop a maximum torque of 3.1 mN·m. The strongest magnets available that will fit in the prosthesis produce a 220-mT field. What current do you need in your motor's coil?
52. A 20-cm-long conducting rod with mass 18 g is suspended by wires of negligible mass (Fig. 26.41). A uniform magnetic field of 0.15 T points horizontally into the page, as shown. An external circuit supplies current between the supports A and B. (a) What's the minimum current necessary to move the bar to the upper position, so it's supported against gravity? (b) What direction should the current flow?

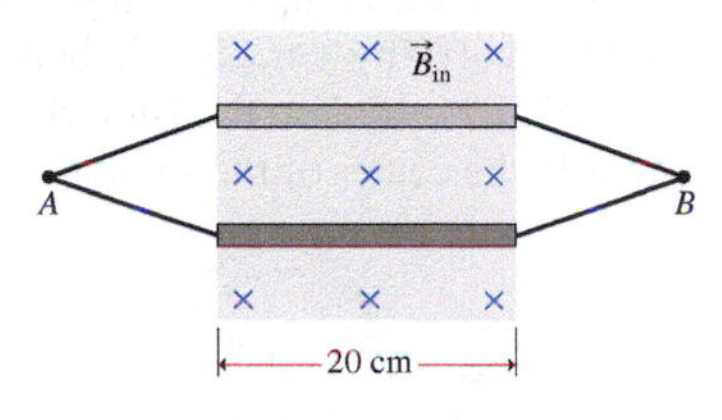

FIGURE 26.41 Problem 52

53. A rectangular copper strip measures 1.0 mm in the direction of a uniform 2.4-T magnetic field. When the strip carries a 6.8-A current perpendicular to the field, a 1.2-μV Hall potential develops across the strip. Find the number density of free electrons in the copper.

54. A single-turn wire loop 10 cm in diameter carries a 12-A current. It experiences a 0.015 N·m torque when the normal to the loop plane makes a 25° angle with a uniform magnetic field. Find the magnetic field strength.

55. A simple electric motor consists of a 220-turn coil, 4.2 cm in diameter, mounted between the poles of a magnet that produces a 95-mT field. When a 15-A current flows in the coil, what are (a) the coil's magnetic dipole moment and (b) the motor's maximum torque?

56. **BIO** Nuclear magnetic resonance (NMR) is a technique for analyzing chemical structures and also the basis of magnetic resonance imaging used for medical diagnosis. NMR relies on sensitive measurements of the energy needed to flip atomic nuclei by 180° in a given magnetic field. In an apparatus with a 9.4-T magnetic field, what energy is needed to flip a proton ($\mu = 1.41\times10^{-26}\ \text{A}\cdot\text{m}^2$) from parallel to antiparallel to the field?

57. A wire carrying 1.5 A passes through a 48-mT magnetic field. The wire is perpendicular to the field and makes a quarter-circle turn of radius 21 cm in the field region, as shown in Fig. 26.42. Find the magnitude and direction of the magnetic force on the curved section of wire.

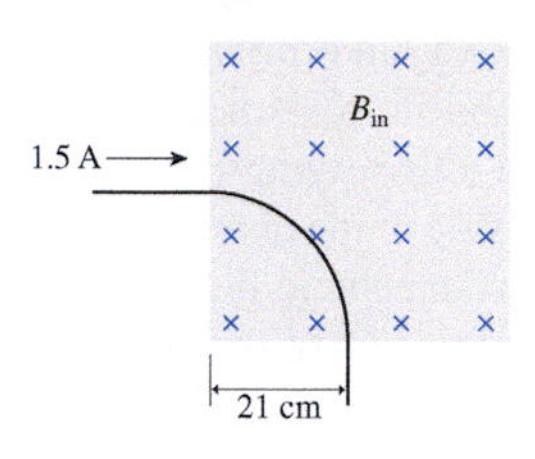

FIGURE 26.42 Problem 57

58. Your company is developing a device incorporating a 20-cm-diameter coil carrying 0.50 A that, when properly oriented, will just cancel Earth's 50-μT magnetic field at the coil's center. How much wire must you requisition for each coil?

59. A single piece of wire carrying current I is bent so it includes a circular loop of radius a, as shown in Fig. 26.43. Find an expression for the magnetic field at the loop center.

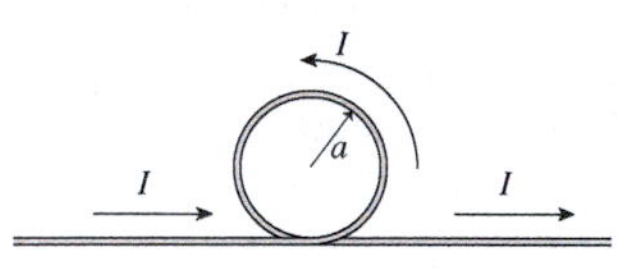

FIGURE 26.43 Problem 59

60. You and a friend get lost while hiking, so your friend pulls out a magnetic compass to get re-oriented. However, you're standing right under a power line carrying 1.5 kA toward magnetic north; it's 10 m above the compass. The horizontal component of Earth's magnetic field at your latitude points northward and has magnitude 0.24 G. Will the compass help you find your way?

61. **CH** Part of a long wire carrying current I is bent into a semicircle of radius a, as in Fig. 26.44. Use the Biot–Savart law to find the magnetic field at P, the center of the semicircle.

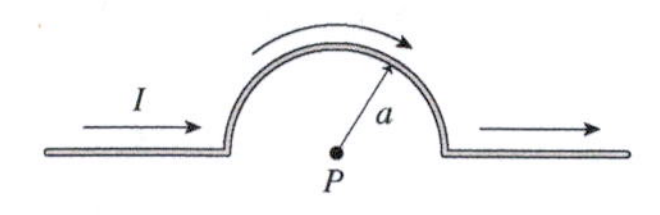

FIGURE 26.44 Problem 61

62. Three parallel wires of length l each carry current I in the same direction. They're positioned at the vertices of an equilateral triangle of side a, and oriented perpendicular to the triangle. Find an expression for the magnitude of the force on each wire.

63. A long, straight wire carries a 25-A current. A 10-cm by 15-cm rectangular wire loop carrying 850 mA is 3.0 cm from the wire, as shown in Fig. 26.45. Find the magnitude and direction of the net magnetic force on the loop.

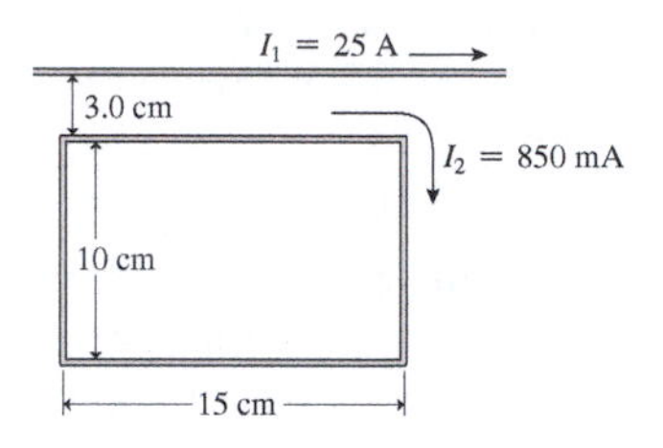

FIGURE 26.45 Problem 63

64. A long conducting rod of radius R carries a nonuniform current density $J = J_0 r/R$, where J_0 is a constant and r is the radial distance from the rod's axis. Find expressions for the magnetic field strength (a) inside and (b) outside the rod.

65. A long, hollow conducting pipe of radius R carries a uniform current I along the pipe, as shown in Fig. 26.46. Use Ampère's law to find the magnetic field strength (a) inside and (b) outside the pipe.

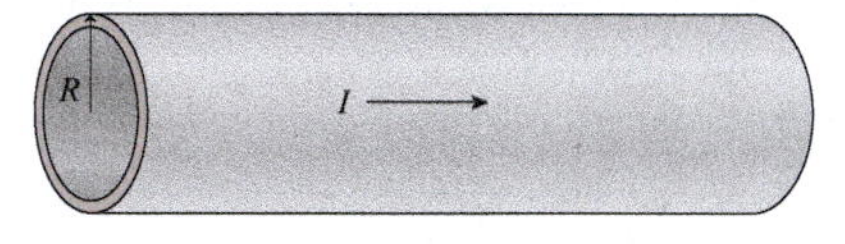

FIGURE 26.46 Problem 65

66. The coaxial cable shown in Fig. 26.47 consists of a solid inner conductor of radius a and a hollow outer conductor of inner radius b and thickness c. The two carry equal but opposite currents I, uniformly distributed. Find expressions for the magnetic field as a function of radial position r (a) within the inner conductor, (b) between the inner and outer conductors, and (c) beyond the outer conductor.

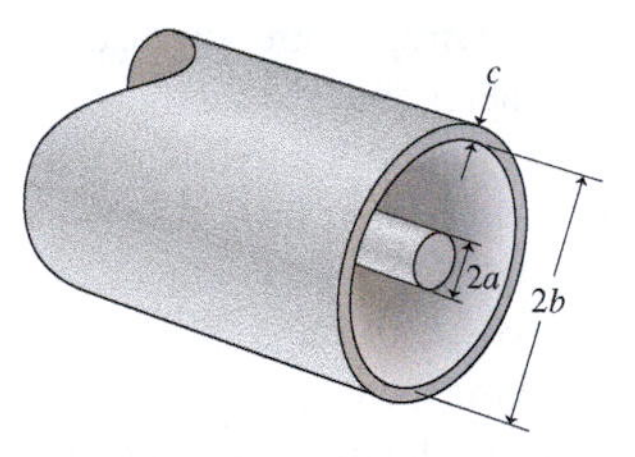

FIGURE 26.47 Problems 66 and 71

67. A solenoid used in a plasma physics experiment is 10 cm in diameter, is 1.0 m long, and carries a 35-A current to produce a 100-mT magnetic field. (a) How many turns are in the solenoid? (b) If the solenoid resistance is 2.7 Ω, how much power does it dissipate?

68. You have 10 m of 0.50-mm-diameter copper wire and a battery capable of passing 15 A through the wire. What magnetic field strengths could you obtain (a) inside a 2.0-cm-diameter solenoid wound with the wire as closely spaced as possible and (b) at the center of a single circular loop made from the wire?

69. **CH** Derive Equation 26.21 for the solenoid field by considering the solenoid to be made of infinitesimal current loops. Use Equation 26.9 for the loop fields, and integrate over all loops.

70. The largest lightning strikes have peak currents of around 250 kA, flowing in essentially cylindrical channels of ionized air. How far from such a flash would the resulting magnetic field be equal to Earth's magnetic field strength, about 50 μT?

71. A coaxial cable (see Fig. 26.47) consists of a 1.0-mm-diameter inner conductor and a 0.20-mm-thick outer conductor with interior diameter 1.0 cm. A 100-mA current flows down the inner conductor and back along the outer conductor. Find the magnetic field strength (a) 0.10 mm, (b) 5.0 mm, and (c) 2.0 cm from the cable axis.

72. **DATA** Indium antimonide (InSb) is a semiconductor commonly used in Hall-effect devices because of its relatively large Hall coefficient. A magnetic-field sensor is made from a 50-μm-thick strip of InSb, with Hall coefficient 228 cm^3/C. The table below shows the Hall potential as a function of current when the sensor is oriented with its current perpendicular to the unknown magnetic field. Plot the Hall potential against a quantity that should give a straight line, determine a best-fit line, and from it find the magnetic field strength.

I (mA)	10.0	20.0	30.0	40.0	50.0
V_H (mV)	0.393	0.750	1.24	1.56	1.97

73. Suppose the current sheet in Example 26.9 is actually a slab with non-negligible thickness d and that the current is distributed uniformly throughout its volume. Find an expression for the magnetic field inside the slab as a function of the perpendicular distance x from the center plane of the slab. Show that your result agrees with that of Example 26.9 at the surface of the slab.

74. A circular wire loop of radius 15 cm and negligible thickness carries a 2.0-A current. Use suitable approximations to find the magnetic field of this loop (a) in the loop plane, 1.0 mm outside the loop, and (b) on the loop axis, 3.0 m from the loop center.

75. A long, flat conducting bar of width w carries a total current I distributed uniformly, as shown in Fig. 26.48. Use approximations to write expressions for the magnetic field strength (a) near the conductor surface ($r \ll w$) but not near its edges and (b) far from the conductor ($r \gg w$).

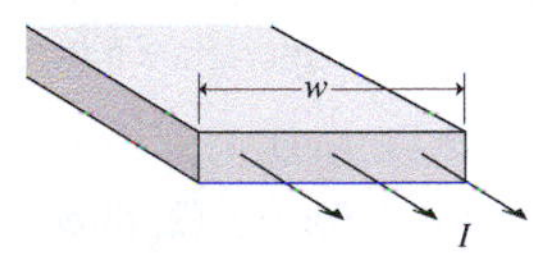

FIGURE 26.48 Problem 75

76. A long, hollow conducting pipe of radius R and length l carries a uniform current I flowing around the pipe (Fig. 26.49). Find expressions for the magnetic field (a) inside and (b) outside the pipe. (*Hint:* What configuration does this resemble?)

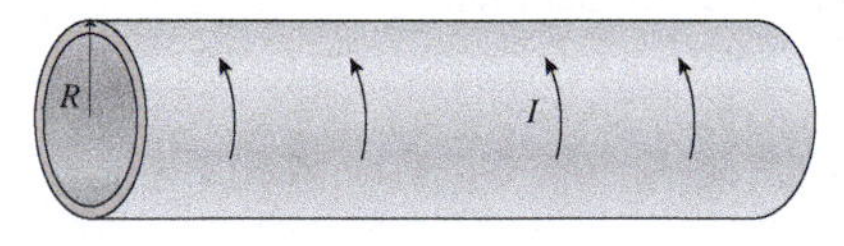

FIGURE 26.49 Problem 76

77. **CH** A solid conducting wire of radius R runs parallel to the z-axis and carries a current density given by $\vec{J} = J_0(1 - r/R)\hat{k}$, where J_0 is a constant and r is the distance from the wire axis. Find expressions for (a) the total current in the wire and (b) the magnetic field for $r > R$ and (c) $r < R$.

78. A disk of radius a carries uniform surface charge density σ and rotates with angular speed ω about the disk axis. Show that the magnetic field at the disk's center is $\frac{1}{2}\mu_0\sigma\omega a$.

79. You're developing a system to orient an orbiting telescope. The system uses three perpendicular coils, with torques developed in Earth's magnetic field when current passes through them. Weight limitations restrict you to a length l of wire for each coil. A colleague argues you'll get the greatest dipole moment and therefore the most torque with a multi-turn coil. You say a 1-turn coil is best. Who's right?

80. The structure shown in Fig. 26.50 is made from conducting rods. The upper horizontal rod (mass 22 g, length 95 cm) is free to slide vertically on the uprights while maintaining electrical contact. A battery connected across the insulating gap at the bottom of the left-hand upright drives 66 A through the structure. At what height h will the upper wire be in equilibrium?

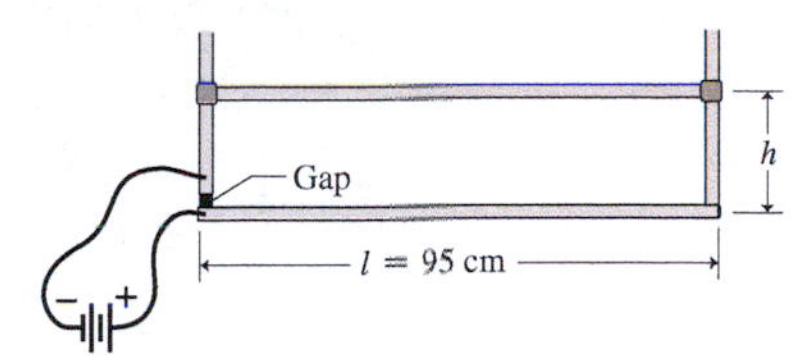

FIGURE 26.50 Problem 80

81. **CH** A long, flat conducting ribbon of width w is parallel to a long, straight wire; its near edge is a distance a from the wire (Fig. 26.51). Wire and ribbon carry the same current I; it's distributed uniformly over the ribbon. Use integration to show that the force per unit length between the two has magnitude $\dfrac{\mu_0 I^2}{2\pi w}\ln\left(\dfrac{a + w}{a}\right)$.

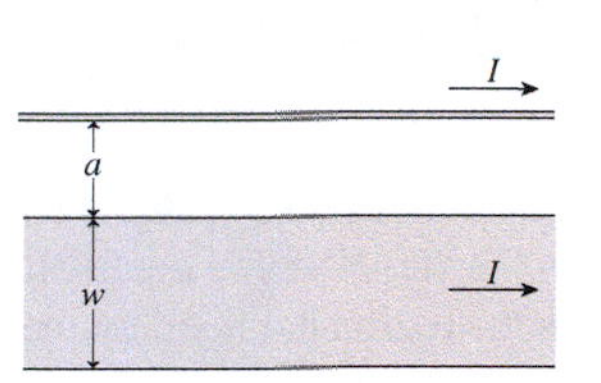

FIGURE 26.51 Problem 81

82. Find an expression for the magnetic field at the center of a square loop of side a carrying current I.

83. **CH** Repeat the calculation in Problem 69 for a solenoid of finite length l and cross-sectional radius a to find the magnetic field strength at the center of the solenoid's axis.

84. **CH** A magnetic dipole $\vec{\mu} = \mu\hat{\imath}$ is on the axis of a circular current loop of radius a oriented as shown in Fig. 26.17a, a distance x from the center. Differentiate Equation 26.16 to find the force on the dipole, and evaluate its magnitude for $x = a$. Is the force attractive or repulsive?

85. You're an engineer at a nuclear power plant, and one of your colleagues has drawn up plans to reroute the conductors carrying current from the plant's electric generator. Your colleague wants to carry this current on two parallel conducting rods 30 cm apart; each rod carries 15 kA with the currents flowing in opposite directions. The proposal calls for clamping the conductors in place every meter, with clamps capable of withstanding a maximum force of 100 N. Is the clamp design adequate?

86. **CH** Derive Equation 26.20 by considering the current sheet to be made of infinitely many infinitesimal line currents.

87. **BIO** Your roommate is sold on "magnet therapy," a sham treatment using small bar magnets attached to the body. You skeptically ask your roommate how this is supposed to work. He mumbles something about the Hall effect speeding blood flow. In reply, you estimate the Hall potential associated with typical blood parameters in the 100-G field of a bar magnet: red blood cells carrying 2-pC charge in a 12-cm/s flow through a 3.0-mm-diameter blood vessel containing 5 billion red blood cells per mL. To show that the Hall potential is negligible, you compare your estimate with the tens of mV typical of bioelectric activity. How do the two values compare?

Passage Problems

A *toroid* is a solenoid-like coil bent into a circle (Fig. 26.52*a*). Toroids are the configuration of choice in magnetic-confinement nuclear fusion experiments, which, if successful, could provide us with an almost unlimited energy source using deuterium fuel extracted from seawater. The ITER consortium, an international collaboration, is building a large toroidal fusion experiment in France; it's expected to be the first fusion device to produce energy on a large scale. Figure 26.52*b* shows a cross section of a toroid, with current emerging from the page at the inner edge and descending at the outer edge. The black circle is an Ampèrian loop.

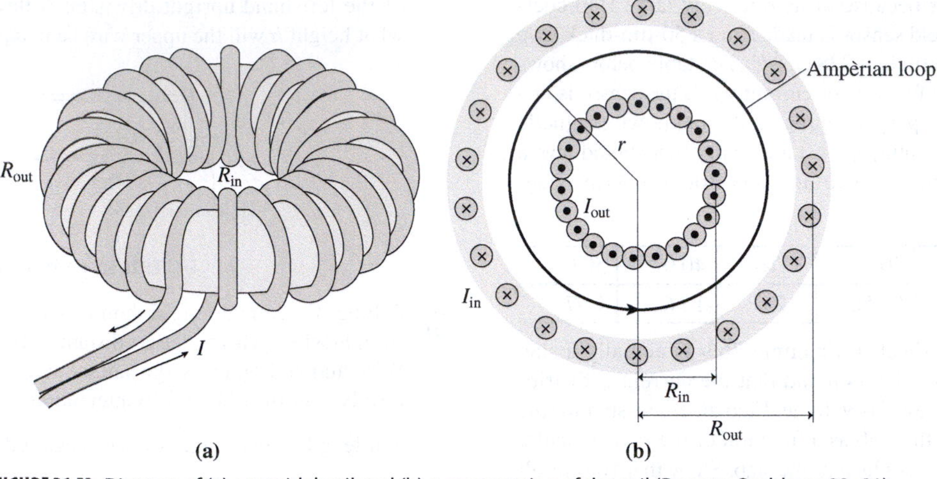

FIGURE 26.52 Diagram of (a) a toroidal coil and (b) a cross section of the coil (Passage Problems 88–91)

88. The magnetic field associated with the toroid is nonzero
    a. only within the "hole" in the donut-shaped coil.
    b. only within the region bounded by the coils.
    c. only outside the coils.
    d. everywhere.
89. In Fig. 26.52*b*, the magnetic field lines must be
    a. straight, and pointing into the page.
    b. straight, and pointing out of the page.
    c. straight, and pointing radially.
    d. circular.
90. Doubling the total number of turns N in the toroid, without changing its size or the current, will
    a. double the magnetic field.
    b. quadruple the magnetic field.
    c. halve the magnetic field.
    d. not change the magnetic field.
91. The toroid has inner radius R_{in} and outer radius R_{out}, while r is the radial coordinate measured from the center. The toroid is made from wire wound into a total of N turns, and carries current I. Which of the following is the correct formula for the magnetic field within the coils?
    a. $B = \mu_0 NI$
    b. $B = \mu_0 NI/2\pi R_{in}$
    c. $B = \mu_0 NI/2\pi R_{out}$
    d. $B = \mu_0 NI/2\pi r$

Answers to Chapter Questions

Answer to Chapter Opening Question

Magnetic force shapes the structure of the solar atmosphere. Magnetism is fundamentally an interaction involving moving electric charge, and the hot, ionized gas of the solar atmosphere contains free charge that responds to magnetism. Earth's cooler atmosphere consists of neutral molecules that don't experience a magnetic force.

Answers to GOT IT? Questions

26.1 (1) greatest for (*a*), 0 for (*c*); (2) direction for (*a*) and (*b*) is into the page
26.2 (b); it's a *negative* charge
26.3 (d)
26.4 (a)
26.5 (1) (a), because adjacent currents are in the same direction (2) Changing the current direction doesn't matter because the currents are still parallel.
26.6 (b), because the field lines form closed loops
26.7 (b)
26.8 (1) 0; (2) current A

27 Electromagnetic Induction

What You Know

- You understand the concepts of electric and magnetic fields.
- You know that electric fields originate in electric charge and magnetic fields in moving electric charge.
- You can determine the forces that both types of field exert on electric charges.
- You're familiar with the fields of simple charge and current distributions.
- You understand the concept of flux and how to evaluate surface integrals for electric flux.
- You know that the solenoid is a device that produces a uniform magnetic field.

What You're Learning

- Here you'll learn about *electromagnetic induction*—a deep connection between electric and magnetic fields whose existence enables a host of modern technologies and helped lead Einstein to the theory of relativity.
- You'll learn to use *Faraday's law* to calculate induced emfs and electric currents that result from *changing magnetic flux*.
- You'll see how conservation of energy manifests itself in electromagnetic induction.
- You'll learn that magnetic fields store energy, just as you learned in Chapter 23 that every electric field in the universe represents stored energy.
- You'll see how circuit elements called *inductors* use induction to store and release magnetic energy in circuits.
- You'll come to recognize that *changing magnetic fields* join electric charge as sources of electric fields.

How You'll Use It

- Myriad technologies, ranging from credit card swiping to electric power generation, make use of electromagnetic induction—so the concepts you learn here in Chapter 27 are behind many devices you use every day.
- In Chapter 28 you'll explore alternating-current circuits, and you'll see how capacitors and inductors are complementary devices that reflect an underlying complementarity between electricity and magnetism.
- In Chapter 29 you'll see how electromagnetic induction is crucial to the existence of electromagnetic waves, including light.

The electric and magnetic fields we've encountered so far originated in electric charge, either stationary or moving. We recognized a link between electricity and magnetism that lies in their common involvement with electric charge. In the remainder of our study, we'll explore a more intimate relation between electricity and magnetism, in which the fields themselves interact directly. This relation is the basis for new electromagnetic technologies, reveals the nature of light, and points toward the theory of relativity.

In 1989, a high-energy outburst from the Sun disrupted power grids in northeastern North America, and blacked out the entire Canadian province of Quebec. How was the energy stored at the Sun, and how did it result in power failures on Earth? How was the electricity generated when the power grid behaved normally?

27.1 Induced Currents

In 1831, the English scientist Michael Faraday and the American Joseph Henry independently found that electric currents arose in circuits subjected to changing magnetic fields. Here are four experiments that illustrate this phenomenon:

MP®

PhET: Faraday's Law

1. Move a bar magnet in the presence of a circuit consisting of a wire coil and an ammeter (Fig. 27.1). There's no battery or other obvious source of emf. As long as you hold the magnet stationary, there's no current. But move the magnet, and the ammeter registers a current—which we call an **induced current**. Move the magnet faster, and the induced current increases. Reverse the direction of motion, and the induced current reverses.

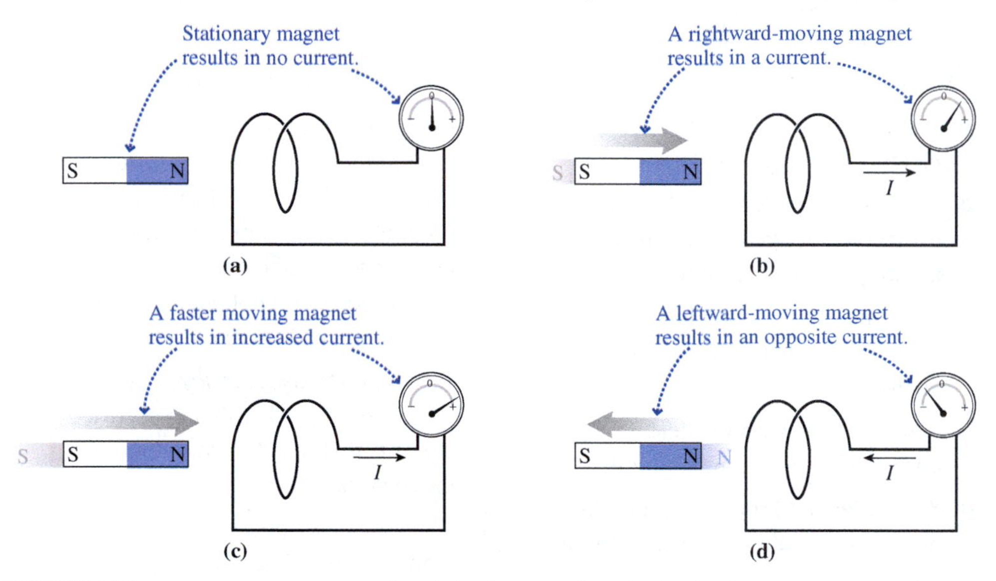

FIGURE 27.1 When a magnet moves near a closed circuit, current flows in the circuit.

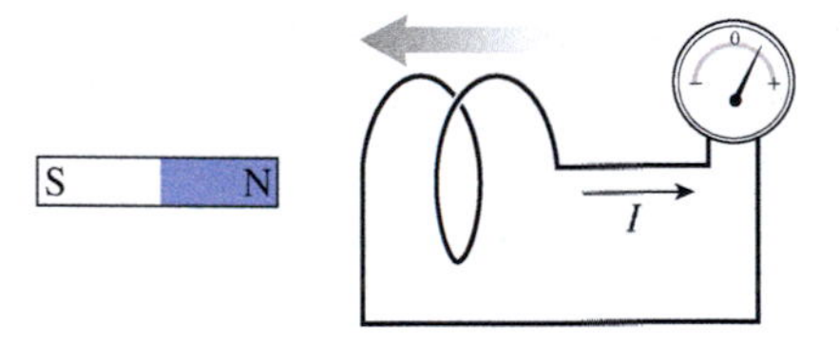

FIGURE 27.2 Moving the coil instead of the magnet gives the same result, as in Fig. 27.1*b*.

2. Move a coil near a stationary magnet, and a similar induced current results (Fig. 27.2). So the effect is the same whether it's the magnet that moves, or the coil. All that matters is the relative motion between magnet and coil.
3. Replace the bar magnet with a second coil, this one carrying a steady current from a battery (Fig. 27.3). The new coil creates a magnetic field like that of a bar magnet and, not surprisingly given the results of experiments 1 and 2, an induced current arises in the original coil when the two coils move relative to one another.

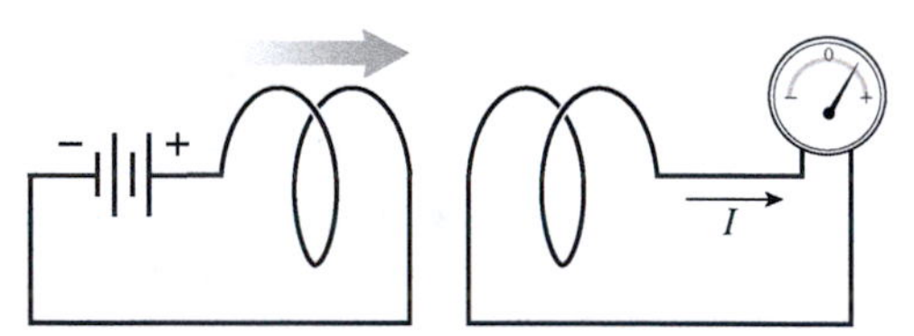

FIGURE 27.3 An induced current also results when a current-carrying circuit replaces the magnet.

4. Hold both coils stationary, and there's no induced current (Fig. 27.4). But now open the switch connecting the battery to the left-hand coil. The current in the left-hand coil drops quickly to zero, and during that brief interval the ammeter registers a current in the right-hand coil. Then the induced current ceases as the current in the left-hand coil remains at zero. Now close the switch again; as current briefly rises in the left-hand coil, the ammeter registers an induced current in the right-hand coil—and its direction is opposite what it was when you opened the switch. Once the current in the left coil reaches a steady value, the induced current in the right coil again ceases.

The common feature in these experiments is a *changing magnetic field*. It doesn't matter whether the field changes because a magnet moves, or a circuit moves, or because the current giving rise to the field changes. In each case, an induced current appears in a circuit

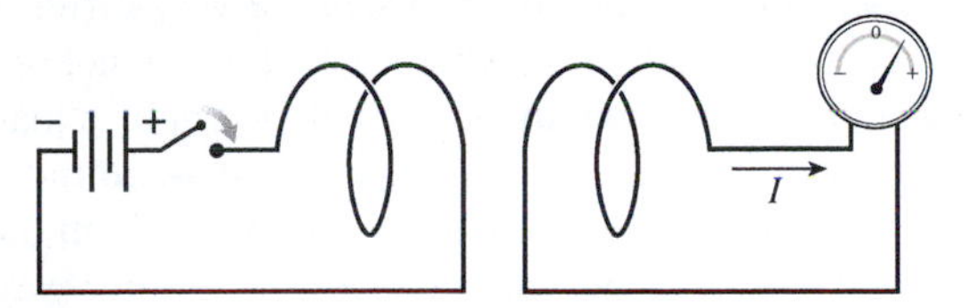

FIGURE 27.4 A current is also induced when the current in an adjacent circuit changes.

subjected to a changing magnetic field. Here's a new phenomenon—**electromagnetic induction**—whereby electrical effects arise from *changing* magnetic fields.

27.2 Faraday's Law

It takes a force acting on charged particles to drive electric current. In the circuits we've studied so far, that force was provided by devices like batteries, and we described a battery's effect by its emf—the energy per unit charge that it provides. With induced currents, there's no battery, but there still must be an emf. This **induced emf** isn't necessarily localized, as with a battery, but may be spread throughout the conductors making up the circuit. We'll now explore further just how emfs arise in the experiments of Section 27.1.

Motional EMF and Changing Fields

In experiment (1), described in Fig. 27.1, we moved a bar magnet toward a stationary coil and got an induced current. As you've just seen, that implies the presence of an induced emf in the coil. In experiment (2), described in Fig. 27.2, we held the magnet stationary and moved the coil. Again, we got an induced emf that drove an induced current. All that mattered was the *relative motion* of the magnet and the coil.

Although they produce the same effect—an induced current—the physical interpretation of the two experiments is very different. You can actually explain experiment (2) using what you already know of the magnetic force. Here the conducting coil is moving through the magnetic field of the stationary magnet, so free charges in the coil experience magnetic forces. We'll analyze a simpler version of this situation in Section 27.3, where you'll see that those forces do, in fact, lead to the induced current that's observed. We use the term **motional emf** to describe an induced emf that arises as a result of a conductor moving through a magnetic field.

Although it gives the same result, experiment (1) is very different. Here the coil is stationary, so its free charges don't experience magnetic forces. Yet there's an induced current, which must be caused by an induced emf. Here the emf arises not from any motion of the coil but from a *changing magnetic field*. This is a truly new phenomenon. In experiment (1) the change is a strengthening of the field at the coil as the magnet approaches, but it could equally well be a change in magnetic field that's not associated with any motion whatsoever; experiment (4) is a case in point, where the magnetic field changes because the current in an adjacent coil increases or decreases. You'll gain deeper insights into the remarkable phenomenon of electromagnetic induction as you study the remaining chapters of Part 4.

So there are two ways to produce induced emfs: through the motion of a conductor or by changing a magnetic field. We'll next explore how to describe induction quantitatively, but first it's worth a closer look at our observation that, in comparing moving-magnet and moving-coil experiments, *only the relative motion matters*. That fact was known since the first induction experiments of the early 19th century, but it took Albert Einstein to recognize its deep significance. Indeed, the second and third sentences of Einstein's 1905 paper introducing the special theory of relativity read, "Take, for example, the reciprocal electrodynamic action of a magnet and a conductor. The observable phenomenon here depends only on the relative motion of the conductor and the magnet, whereas the customary view draws a sharp distinction between the two cases in which either the one or the other of these bodies is in motion." The first of these sentences recognizes what we've been saying: all that matters is the relative motion of the coil and magnet. The "sharp distinction" in

Einstein's second sentence is the distinction between the very different physical descriptions of the two experiments, depending on whether it's the magnet or the coil that moves. Einstein went on to develop his relativity theory, which downplays that "sharp distinction" by saying that descriptions of the magnet–coil experiment—and indeed any descriptions of physical reality—are equally valid as long as they're made from the viewpoint of inertial reference frames. (You learned about inertial reference frames way back in Chapter 4, and you'll explore relativity theory in Chapter 33, where Section 33.8 will provide additional insights into the relation between magnetism and electricity.)

Magnetic Flux

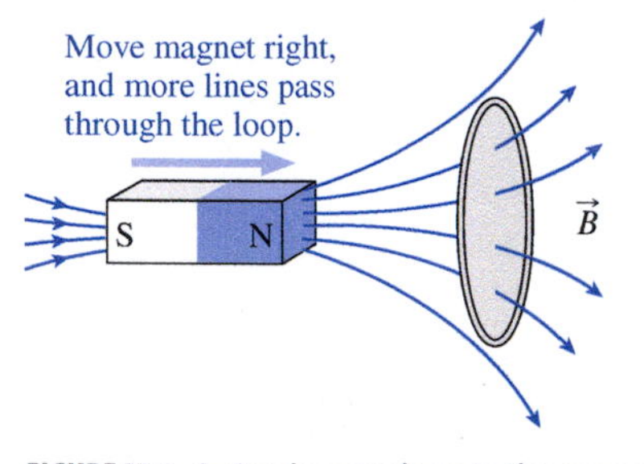

FIGURE 27.5 A circular wire loop in the magnetic field of a bar magnet. As the magnet moves closer, the flux through the loop increases.

To describe electromagnetic induction quantitatively we need to use the concept of *magnetic flux*. We introduced magnetic flux in Chapter 26 when we formulated Gauss's law for magnetism, the statement that the magnetic flux through any *closed* surface is zero.

Here we're interested in the flux through *open* surfaces, which need not be zero (Fig. 27.5). Like the electric flux defined in Chapter 21, magnetic flux is the integral of the magnetic field over a surface:

$$\Phi_B = \int \vec{B} \cdot d\vec{A} \quad \text{(magnetic flux)} \qquad (27.1a)$$

With electromagnetic induction, we're interested in the flux through a surface bounded by a circuit. For a loop like the one in Fig. 27.5, that surface can be the circular disk whose circumference is the loop. More generally, it can be *any* surface bounded by the loop.

Since we're dealing here with an *open* surface, there's an ambiguity in the direction of the area vector $d\vec{A}$ as we discussed when we introduced electric flux in Chapter 21. We won't resolve the ambiguity at this point, but will wait until Section 27.3, which introduces the all-important connection between electromagnetic induction and conservation of energy.

For a flat surface in a uniform magnetic field, Equation 27.1a reduces to

$$\Phi_B = \vec{B} \cdot \vec{A} = BA\cos\theta \quad \text{(magnetic flux, uniform field and flat area)} \qquad (27.1b)$$

where θ is the angle between the field and the normal to the area. When the field and area are perpendicular, as in the next example, Equation 27.1b reduces further to $\Phi_B = BA$. Magnetic flux has the units of field times area, or $\mathrm{T \cdot m^2}$. Unlike electric flux, magnetic flux is also given its own named unit, with $1\ \mathrm{T \cdot m^2}$ being a **weber** (Wb). Exercise 27.14 shows that the units of magnetic flux can also be expressed as $\mathrm{V \cdot s}$.

EXAMPLE 27.1 Magnetic Flux: A Solenoid

A solenoid of circular cross section has radius R, consists of n turns per unit length, and carries current I. Find the magnetic flux through each turn of the solenoid.

INTERPRET The solenoid creates a uniform magnetic field, and we're asked for the flux of this field through an area bounded by one turn of the solenoid.

DEVELOP The solenoid field is perpendicular to the turns of wire that make up the solenoid, as we've drawn in Fig. 27.6. So we have a uniform field at right angles to a flat area, and the flux becomes $\Phi_B = BA$. Equation 26.20 gives the solenoid field, $B = \mu_0 nI$, and the area is that of a circle, πR^2.

EVALUATE $\Phi_B = BA = \mu_0 nI\pi R^2$

ASSESS The flux increases with any factor that increases either the field or the area, so this result makes sense. We're being a little loose in thinking of a single turn of the solenoid as a closed loop, but if the solenoid is tightly wound, then this is an excellent approximation. What we've found is the flux through one turn; if the solenoid has N turns, then the flux through the entire solenoid is N times our result. ■

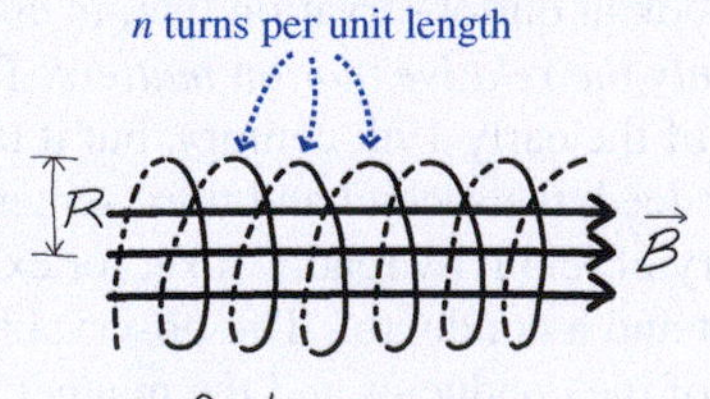

FIGURE 27.6 Sketch for Example 27.1.

EXAMPLE 27.2 Magnetic Flux: A Nonuniform Field

A long, straight wire carries current I. A rectangular wire loop of dimensions l by w lies in a plane containing the wire, with its closest edge a distance a from the wire and its dimension l parallel to the wire. Find the magnetic flux through the loop.

INTERPRET The long, straight wire gives rise to a magnetic field, and we're asked for the flux of this field through an adjacent rectangular area.

DEVELOP Figure 27.7 shows the situation. Field lines encircle the long wire, and at the rectangular loop they're pointing into the page, perpendicular to the loop area. Thus $\vec{B}\cdot d\vec{A}$ in Equation 27.1a becomes just $B\,dA$. Equation 26.17 gives the field strength: $B = \mu_0 I/2\pi r$. Since this field varies with distance from the wire, we have to integrate. We divide the rectangle into thin strips of width dr and area $dA = l\,dr$. Knowing B and dA, we can integrate over all such strips.

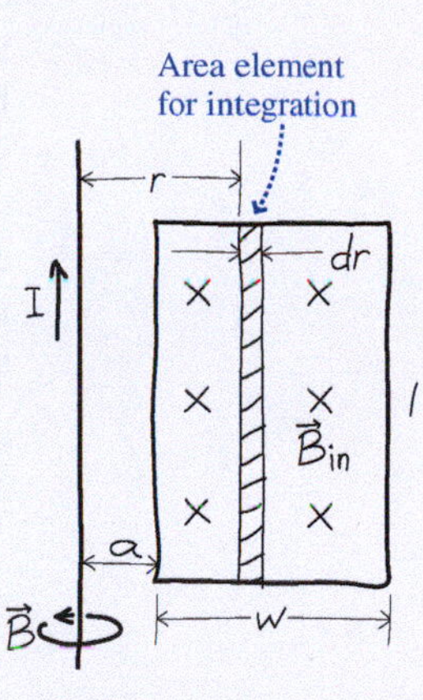

FIGURE 27.7 A rectangular loop in the magnetic field of a long wire.

EVALUATE We have

$$\Phi_B = \int B\,dA = \int_a^{a+w} \frac{\mu_0 I}{2\pi r}\,l\,dr = \frac{\mu_0 Il}{2\pi}\int_a^{a+w}\frac{dr}{r}$$

The integral is the natural logarithm, so

$$\Phi_B = \frac{\mu_0 Il}{2\pi}\ln r\Big|_a^{a+w} = \frac{\mu_0 Il}{2\pi}\ln\left(\frac{a+w}{a}\right)$$

ASSESS This result is directly proportional to the current, which determines the field strength, and to the loop length l. But it isn't directly proportional to the width w because the field strength falls off, and increasing w would expand the loop into regions of weaker field, contributing less to the overall flux.

Flux and Induced EMF

Although motional emfs and emfs induced by changing fields seem to be different phenomena, Faraday showed that both can be described in terms of *changing magnetic flux*. The result is a preliminary statement of **Faraday's law of electromagnetic induction**, another of the four basic laws of electromagnetism:

> The induced emf in a circuit is proportional to the rate of change of magnetic flux through any surface bounded by that circuit.

This statement is a special case of Faraday's law that describes electromagnetic induction specifically in circuits; later we'll present a more general form that applies even when no circuit is present. The induced emf tends to oppose the change in flux—a crucial point to which we'll devote all of Section 27.3—and so in SI the proportionality between emf and rate of change of flux is -1. Thus Faraday's law is

$$\mathcal{E} = -\frac{d\Phi_B}{dt} \quad \text{(Faraday's law)} \qquad (27.2)$$

where $\mathcal{E}$ is the induced emf in a circuit and Φ_B is the magnetic flux through any surface bounded by that circuit.

Faraday's law relates the induced emf to the *change* in flux. It isn't magnetic field or flux that causes an induced emf—it's the *change* in flux. The flux in a uniform field is given by Equation 27.1b, $\Phi_B = \vec{B}\cdot\vec{A} = BA\cos\theta$, which shows that we can change flux by changing the field strength B, the area A, or the angle θ describing the orientation between area and field.

A changing field strength can result from relative motion of a conductor and a magnet or other system that produces a field, or from a change in the current that produces a magnetic field. A changing orientation can result from a change in the orientation of a conducting system or of a nearby magnet. For an induced emf to occur as a result of a change in area, however, that change *must* involve the physical motion of conductors; the induced emf in such cases is always, therefore, motional emf. You'll see such a case in Example 27.4, and we'll analyze a similar situation in detail in Section 27.3. There are some unusual situations where it appears that a circuit area is changing but where the conductors either aren't moving or where their motion doesn't correspond to the area change. In general, analysis of true motional emf leads to an expression for the emf that involves the product of the magnetic field, the velocity of the moving conductors, and an appropriate length—as you'll find in Example 27.4.

PROBLEM-SOLVING STRATEGY 27.1 Faraday's Law and Induced emf

INTERPRET Make sure the problem involves a circuit in which current flows because of a changing magnetic flux. Identify the circuit and the cause of the changing flux. Possibilities include:

- A changing magnetic field, caused either by relative motion between the circuit and a magnet or by a changing current in an adjacent circuit. Alternatively, the problem may simply state that a magnetic field is changing at some specified rate, without giving the cause.
- A changing area, caused by the circuit expanding or contracting in the presence of a magnetic field.
- A changing orientation of the circuit relative to the field, causing a change in $\cos\theta$.

DEVELOP Find an appropriate expression for the magnetic flux through your circuit. If the field varies with position, you'll have to set up the integral in Equation 27.1a: $\Phi_B = \int \vec{B}\cdot d\vec{A}$; if not, you can use the simpler expression of Equation 27.1b: $\Phi_B = \vec{B}\cdot\vec{A} = BA\cos\theta$. Since the flux is changing, your expression for flux should either have an explicit time dependence or contain a quantity whose rate of change you're given.

EVALUATE Differentiate the flux with respect to time. Faraday's law, $\mathcal{E} = -d\Phi_B/dt$, then gives the induced emf. If you're asked about the circuit current, you can find it using Ohm's law: $I = \mathcal{E}/R$, with R the circuit resistance.

ASSESS Does your answer make sense? Does the induced emf or current increase with an increased rate of whatever quantity is changing? Do the induced effects vanish if you set the rate of change to zero?

MP

PhET: Faraday's Electromagnetic Lab

EXAMPLE 27.3 Induced Current: A Changing Magnetic Field

A wire loop of radius 10 cm has resistance 2.0 Ω. The plane of the loop is perpendicular to a uniform magnetic field $\vec{B}$ that's increasing at 0.10 T/s. Find the magnitude of the induced current in the loop.

INTERPRET We apply our problem-solving strategy, noting that this is a problem about induction in a circular loop, with the flux change caused by a changing magnetic field.

DEVELOP Figure 27.8 shows the loop with a field pointing into the page. With the field uniform and perpendicular to the loop area, we have $\Phi_B = BA = B\pi r^2$. We're given the rate of change dB/dt, so we can evaluate the derivative $d\Phi_B/dt$.

EVALUATE The rate of change of flux is

$$\frac{d\Phi_B}{dt} = \frac{d}{dt}(B\pi r^2)$$

Since the radius isn't changing, this becomes

$$\frac{d\Phi_B}{dt} = \pi r^2 \frac{dB}{dt}$$

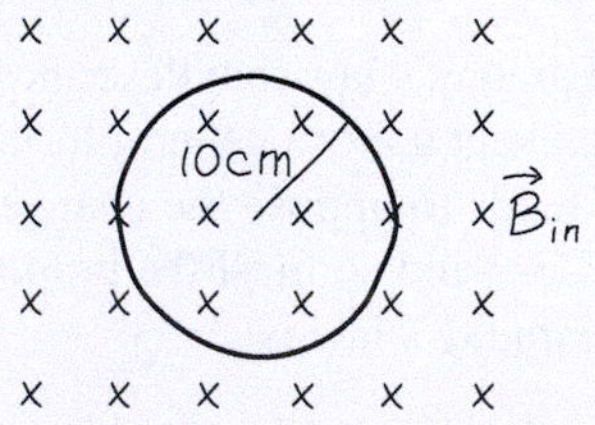

FIGURE 27.8 A circular conducting loop in a plane perpendicular to a uniform magnetic field.

We're given $dB/dt = 0.10$ T/s and $r = 10$ cm. So with $\mathcal{E} = -d\Phi_B/dt$, the magnitude of the induced emf then has the value $\pi r^2 dB/dt = 3.14$ mV. Ohm's law then gives the current: $I = \mathcal{E}/R = 3.14\ \text{mV}/2.0\ \Omega = 1.6$ mA.

ASSESS Make sense? The induced emf and hence the current scale directly with the value of dB/dt, confirming that the changing magnetic field is indeed the cause of the induced effects. Does it bother you that we took $\vec{B}$ as uniform even though it's changing, thus avoiding the integral of Equation 27.1a? The field is indeed changing, but that change is in *time*, not *space*, and the integral for the flux is over space. So at each instant the field is uniform, and we can dispense with the integral.

EXAMPLE 27.4 Induced Current: A Changing Area

Two parallel conducting rails a distance l apart are connected at one end by a resistance R. A conducting bar completes the circuit, joining the two rails electrically but free to slide along them. The whole circuit is perpendicular to a uniform magnetic field $\vec{B}$, as shown in Fig. 27.9. Find the current when the bar is pulled to the right with constant speed v.

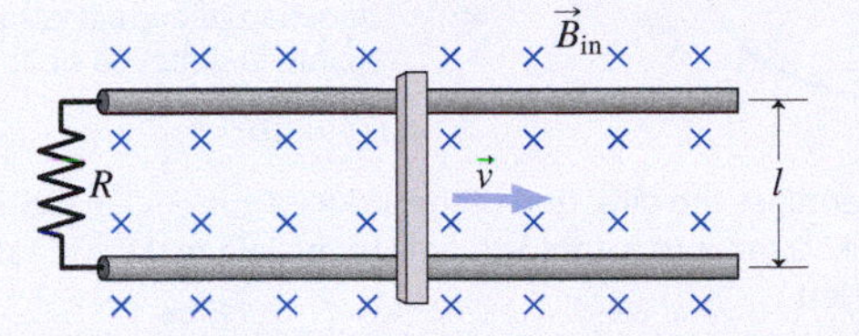

FIGURE 27.9 Pulling the bar to the right increases the circuit area, increasing the magnetic flux and inducing an emf that drives a current.

INTERPRET Here the circuit is formed by the rails, the resistance, and the conducting bar. The circuit area increases as the bar slides along the rails, so we've got a case of induction caused by a changing magnetic flux resulting from a changing area.

DEVELOP In this case of a uniform field perpendicular to the circuit, the flux is the product $\Phi_B = BA$. We can express this flux in terms of the changing position x of the sliding bar; since we're given the bar's speed, we'll be able to evaluate the rate of change of flux. If we take $x = 0$ at the left end of the rails, then the circuit area is $A = lx$, so the flux is $\Phi_B = BA = Blx$.

EVALUATE Differentiating the flux with respect to time gives

$$\frac{d\Phi_B}{dt} = Bl\frac{dx}{dt} = Blv$$

since dx/dt is the bar's velocity v. Faraday's law says that Blv is the magnitude of the induced emf $\mathcal{E}$, so the current in the circuit becomes

$$I = \frac{\mathcal{E}}{R} = \frac{Blv}{R}$$

ASSESS Make sense? Yes: The faster the bar moves, the greater the rate of change of flux, and so the greater the induced emf and current. ■

Example 27.4 provides a clear case of motional emf. Note that the induced emf $\mathcal{E} = Blv$ is, as we suggested earlier, expressed as a product of magnetic field strength, velocity, and a length—in this case the length of the bar. Although we worked this example using Faraday's law and changing magnetic flux, the physical mechanism behind motional emf always begins with the magnetic force on free charges in a conductor. This results in charge separation and therefore in an electric field that is the ultimate driver of the current in the case of motional emf. In Section 27.3 we'll explore a similar case in more detail, and we'll show how consideration of the forces on a moving conductor confirms conservation of energy in electromagnetic induction.

✓TIP It's the Change That Matters

You may wonder why, in problems like the preceding two examples, you're not given values for the magnetic field itself or for the location of the sliding bar—quantities that determine the magnetic flux. But the flux itself doesn't matter, only its *rate of change*. And in both cases the rate followed from the given information: in one case the rate of change of the field and in the other the speed of the bar.

Examples 27.3 and 27.4 take care of two ways to change magnetic flux. The third—changing orientation—is at the heart of an important electromagnetic technology, and we'll do an example in the next section.

27.3 Induction and Energy

Move a bar magnet toward a wire loop, as in Fig. 27.10. An induced current flows, dissipating energy as it heats the loop. Where did that energy come from? It came from work you did in moving the magnet.

Normally it doesn't take work to move with constant speed. But the induced current makes the loop a magnetic dipole whose field, as Fig. 27.10 shows, *opposes* the field of the approaching magnet. You have to do positive work to overcome the resulting repulsive force. It had better be this way! Otherwise, you'd get something for nothing, heating the loop without any source of energy.

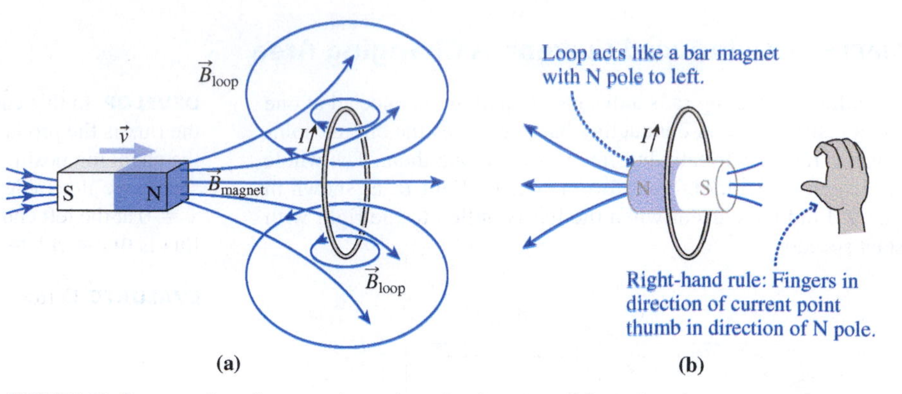

FIGURE 27.10 Conservation of energy determines the direction of the induced current. (a) Fields of bar magnet and loop. (b) The loop acts like a magnet with north pole to the left, making it hard to move the rectangular bar magnet at the left.

You can always find the direction of induced emfs and currents by asking: What direction of induced current will make it hard to move the magnet? The answer for Fig. 27.10 is a current that makes the loop a magnet with its north pole on the left, to repel the approaching bar magnet. By the right-hand rule, that gives the current direction shown: into the page at the top of the loop and out at the bottom. If, on the other hand, you move the magnet away from the loop, then the current flows in the opposite direction, putting the loop's south pole on the left and attracting the magnet, making it hard to pull the magnet away (Fig. 27.11).

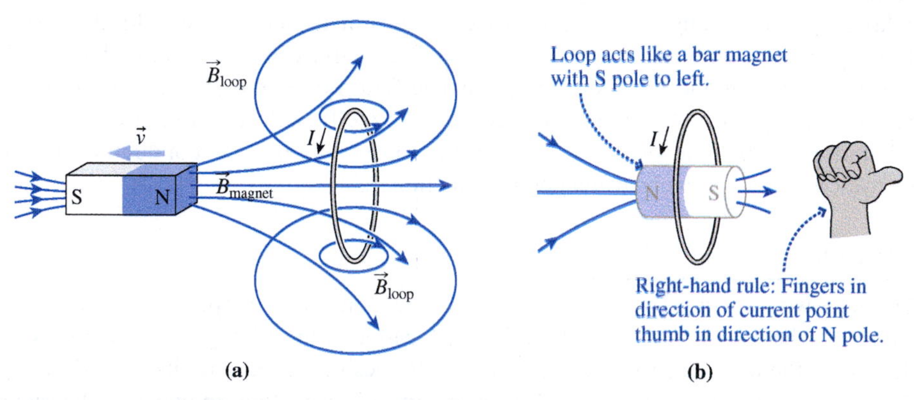

FIGURE 27.11 Now the direction of the induced current puts the loop's south pole to the left, making it hard to pull the magnet away.

This discussion is ultimately about energy conservation in the context of electromagnetic induction. **Lenz's law** summarizes what we've found:

> The direction of an induced emf or current is such that the magnetic field created by the induced current opposes the change in magnetic flux that created the current.

Mathematically, Lenz's law is contained in the minus sign that appears in Faraday's law, provided careful attention is paid to the direction of the area vector $d\vec{A}$ that appears in Equation 27.1a for magnetic flux. So it's possible to determine the direction of the induced effects from algebra alone. But we'll stick with the easier and more physically meaningful approach of using Faraday's law to find the magnitude of the induced emf and then reason out the direction using energy conservation.

GOT IT? 27.1 You push a bar magnet toward a loop, with the north pole toward the loop as in Fig. 27.10. If you keep pushing the magnet straight through the loop, (1) what will be the direction of the current as you pull it out the other side? (2) Will you need to do work, or will work be done on you?

Motional EMF and Lenz's Law

We've noted that motional emf is a special case of induction that we can explain in terms of the magnetic forces on charge carriers. Here we'll explore a case of motional emf to show explicitly that Lenz's law requires energy conservation.

Consider a square conducting loop of side l and resistance R pulled with constant speed v out of a uniform magnetic field $\vec{B}$ (Fig. 27.12). The magnetic flux through the loop is changing, so there's an induced emf that drives a current. Energy is dissipated as heat and so, as we've just argued, the agent pulling the loop must do work. We'll now demonstrate quantitatively that energy is conserved by showing that the rate of heating in the loop is exactly equal to the rate at which the agent pulling the loop does work.

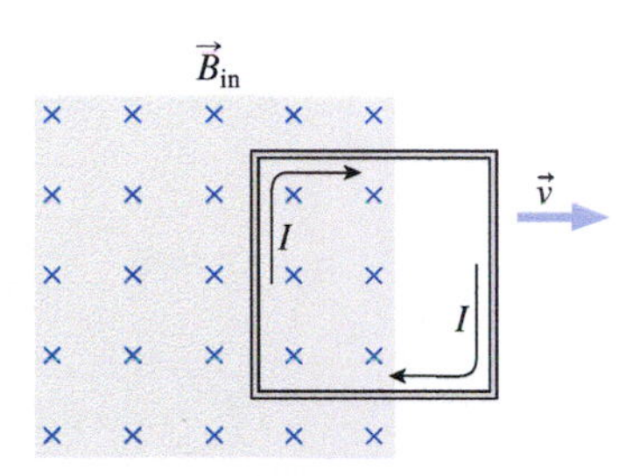

FIGURE 27.12 A conducting loop being pulled out of a magnetic field.

Pulling the conducting loop to the right moves its free electrons through the magnetic field; initially, that motion is also to the right. Then the magnetic force $q\vec{v} \times \vec{B}$ on the electrons is downward in Fig. 27.12 (opposite $\vec{v} \times \vec{B}$ because the electrons are negative). The result is an accumulation of negative charge near the bottom of the left-hand side of the square and therefore of positive charge near the top. As with a battery, the separated charge drives a current around the circuit, here in the clockwise direction. Although we say, loosely, that the magnetic force on the moving conductor is what produces the current, it's actually more complicated than that: The magnetic force results in separated charge, which gives rise to electric forces that drive the current. Furthermore, electric forces are what keep the electrons confined to the wire; otherwise they would exhibit circular motion as we showed in Section 26.3. Once the current is flowing, the magnetic force on the upward current causes a Hall-effect separation of charge, and the resulting electric field provides the force that opposes the applied force that's pulling the loop. These observations are consistent with our statement in Chapter 26 that magnetic forces themselves can do no work because they're always perpendicular to charge velocities, implying that electric forces must be involved as well.

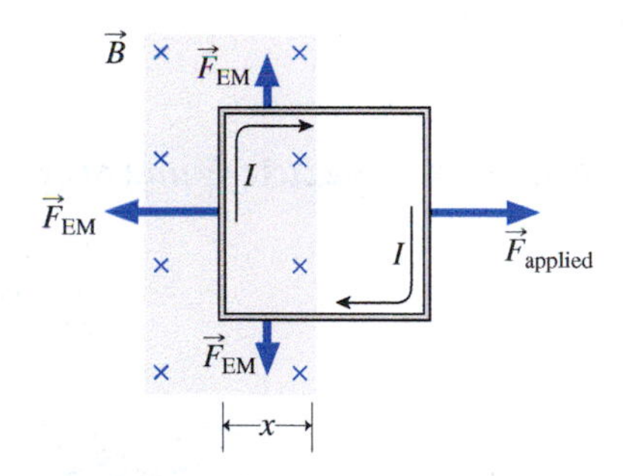

FIGURE 27.13 Forces on the loop. The subscript EM indicates that the forces on the conducting loop result from both electric and magnetic fields.

You saw in Chapter 26 that the magnetic force on a current-carrying conductor of length l is $\vec{F} = I\vec{l} \times \vec{B}$—although, again, "magnetic force" here is shorthand for a combination of magnetic and electric effects as we described in Section 26.4. Applying this expression to our conducting loop shows that there's no force on the right-hand side, where $\vec{B} = \vec{0}$, and that oppositely directed forces on the top and bottom of the loop cancel (Fig. 27.13). So the net force on the loop is that on the left side alone. The magnitude of this force is IlB, and the right-hand rule shows that it points to the left. This leftward force cancels the rightward-applied force, giving the zero net force that Newton's law requires for the loop to move with constant velocity.

We could equally well determine the current direction from magnetic-flux considerations. As the loop leaves the field, the flux decreases. The direction of the resulting induced current is such as to oppose this decrease. Therefore, the magnetic field of the induced current points *into* the page, as the induced current tries to maintain the flux. By the right-hand rule, a field within the loop and into the page requires that the induced current flow clockwise.

To calculate the current, we first find the induced emf. With the field perpendicular to the loop, and uniform in the region where it's nonzero, the magnetic flux is the product of the magnetic-field strength and the loop area that lies within the field: $\Phi_B = Blx$. Here x is the distance between the left edge of the loop and the right edge of the magnetic-field region. The magnetic field remains constant, but as the loop moves, the distance x decreases at the rate $dx/dt = -v$, where the minus indicates a decrease. Then the rate of change of flux is

$$\frac{d\Phi_B}{dt} = \frac{d(Blx)}{dt} = Bl\frac{dx}{dt} = -Blv$$

so Faraday's law gives

$$\mathcal{E} = -\frac{d\Phi_B}{dt} = Blv$$

This induced emf drives a current I around the loop, where $I = \mathcal{E}/R = Blv/R$. The rate of energy dissipation in the loop is the product of the emf and the current (Equation 24.7):

$$P = I\mathcal{E} = \frac{Blv}{R}Blv = \frac{B^2l^2v^2}{R} \qquad \text{(electric power dissipated in loop)}$$

We've found that the force on the loop resulting from its motion through the magnetic field has magnitude $F = IlB$; since the loop is moving with constant velocity, this is also the magnitude of the applied force. Equation 6.19 gives $P = \vec{F}\cdot\vec{v}$ for the power supplied. Here, with $\vec{F}$ and $\vec{v}$ in the same direction, we have

$$P = Fv = IlBv = \frac{Blv}{R}lBv = \frac{B^2l^2v^2}{R} \qquad \text{(mechanical power supplied to pull loop)}$$

the same as our expression for the power dissipated in the loop. Thus, all the work done by the agent pulling the loop ends up heating the resistor, showing explicitly that energy is indeed conserved.

GOT IT? 27.2 When the loop in Fig. 27.12 first enters the field, coming in from the left, will the loop current be (a) clockwise or (b) counterclockwise?

MP®

PhET: Generator

Electromagnetic induction is the principle behind many important technologies, from credit cards to electric-power generation. Induction also gives us the flexibility to transform voltage levels in electric-power systems, and to provide wireless charging systems for devices ranging from electric cars to toothbrushes.

APPLICATION Electric Generators

Rotation of loop changes the magnetic flux, inducing an emf.

N

S

Rotating slip rings

Stationary brushes

Rotating conducting loop

Electric load

Probably the most important technological application of induction is the electric generator. Humanity uses electrical energy at the phenomenal rate of about 2 TW, which is 2×10^{12} W and roughly equal to the power output of 20 billion human bodies. Virtually all this power comes from generators. A **generator** is just a system of conducting loops in a magnetic field, as shown in the figure. Mechanical energy rotates the conductors, resulting in a changing magnetic flux and therefore an induced emf. Current flows through the generator and on to whatever electrical loads are connected to it. Because the changing flux results from a change in the orientation of the loop relative to the field—that is, a change in θ in the expression $\Phi_B = BA\cos\theta$—a generator such as the one shown here produces an alternating emf that varies sinusoidally with time.

Any source of mechanical energy can power the generator, but the most common is steam from burning fossil fuels or from nuclear fission. Electrical energy is also generated from kinetic energy of water or wind. A small electric generator, driven by the car's engine, is used to recharge your car's battery.

Lenz's law, the conservation of energy in electromagnetic induction, is very much applicable to electric generators. Were it not for Lenz's law, generators would turn on their own and happily supply electricity without coal, oil, or uranium! The voluminous quantities of fuel consumed by power plants are dramatic testimony to the minus sign on the right-hand side of Equation 27.2.

Turn a hand-cranked or pedal-driven generator, and you can literally feel Lenz's law. Without any electrical load, turning the generator is easy. Switch on increasingly heavy loads, and the generator gets harder to turn. Most people find they can just sustain a 100-W lightbulb with a hand generator. Think about this next time you leave a light on!

If the diagram here reminds you of the motor in the Application in Chapter 26, that's no coincidence. Motors and generators are similar devices, just run in opposite ways. A motor converts electrical energy to mechanical energy; a generator converts mechanical energy to electrical energy. Often the same physical device serves both purposes. In a hybrid car, for example, an electric motor takes energy from a battery to provide propulsion. When the car brakes, the wheels turn the motor, which then acts as a generator and puts the car's energy back into the battery instead of dissipating it as heat. Such **regenerative braking** is one of the hybrid's several means of achieving greater energy efficiency.

GOT IT? 27.3 If you lower the electrical resistance connected across a generator while turning the generator at a constant speed, will the generator get (a) easier or (b) harder to turn?

EXAMPLE 27.5 Induction: Designing a Generator

An electric generator consists of a 100-turn circular coil 50 cm in diameter. It's rotated at $f = 60$ rev/s to produce standard 60-Hz alternating current like that used throughout North America. Find the magnetic-field strength needed for a peak output voltage of 170 V (which is the actual peak in standard 120-V household wiring).

INTERPRET Here we have a conducting coil rotating in a fixed magnetic field, so this is an induction problem where changing flux results from a changing orientation.

DEVELOP We sketched the coil and magnetic field in Fig. 27.14. With a uniform field and flat, circular area, the flux through one turn of the coil is given by Equation 27.1b, $\Phi_{1\text{ turn}} = \vec{B}\cdot\vec{A} = BA\cos\theta = B\pi r^2\cos\theta$. The angle θ changes as the loop rotates, and with it the flux. We need to express the total flux as a function of time so we can evaluate its derivative and thus the emf. Because the loop rotates with constant angular speed $\omega = 2\pi f$, its angular position is $\theta = 2\pi ft$. Then the flux through each turn is $B\pi r^2\cos(2\pi ft)$, and the total flux through all $N = 100$ turns is $NB\pi r^2\cos(2\pi ft)$.

EVALUATE Faraday's law equates the induced emf with the rate of change of this flux:

$$\mathcal{E} = -\frac{d\Phi_B}{dt} = -NB\pi r^2\frac{d}{dt}[\cos(2\pi ft)] = -NB\pi r^2[-2\pi f\sin(2\pi ft)]$$

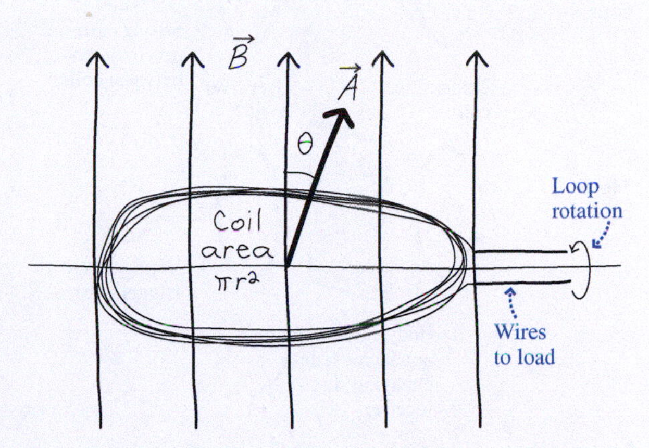

FIGURE 27.14 Coil in the generator of Example 27.5; at this instant the normal to the coil makes an angle θ with the magnetic field.

The emf has its peak value when the sine is 1, so $\mathcal{E}_{\text{peak}} = 2\pi^2 r^2 NBf$. We want this value to be 170 V; using $r = 25$ cm, $N = 100$ turns, and $f = 60$ rev/s then gives $B = 23$ mT.

ASSESS This value is about 200 G, typical of the field strength near the poles of a permanent magnet. Note that you don't need a value for time t to find the peak emf; when a quantity varies sinusoidally, its peak occurs when the sine or cosine function is 1, so the peak value is the magnitude of whatever quantity multiplies the sine or cosine.

Electromagnetic induction is also the basis of magnetic recording, once the dominant means of storing audio, video, and computer information but now more common in credit cards and similar applications. The magnetic strip on your credit card is a ferromagnetic material that stores information in regions of differing magnetization. Swiping your card induces current in a wire coil, which extracts the stored information as an electrical signal (Fig. 27.15). Early computer disks worked on the same principle, although in today's disks the magnetic field of the spinning disk causes changes in electrical resistance in the "head" that reads the disk information.

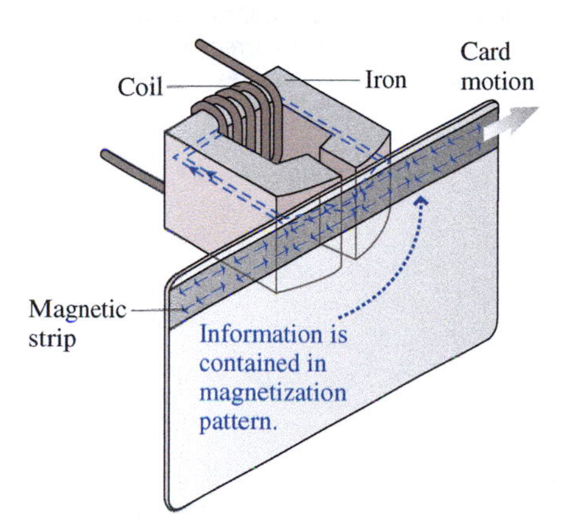

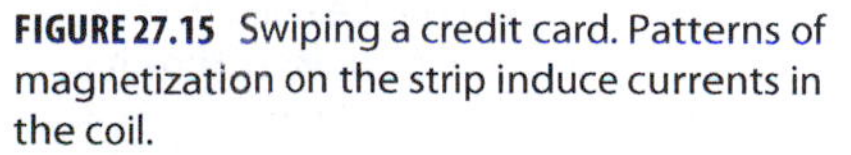

FIGURE 27.15 Swiping a credit card. Patterns of magnetization on the strip induce currents in the coil.

Eddy Currents

Induced currents aren't limited to conducting loops and circuits. They also occur in solid conductors subject to changing magnetic flux. The resistance of a solid conductor is low, which can result in large induced currents and significant power dissipation. That can make it hard to move a conducting material into or out of a magnetic field, as it's subject to a changing flux. The result is a kind of magnetic friction that saps energy. On the other hand, the effect can be useful in providing an alternative to friction brakes. Rapidly rotating saw blades or train wheels, for example, can be stopped quickly by turning on a nearby electromagnet; the resulting eddy currents quickly dissipate the rotational kinetic energy. The mechanical resistance you feel in exercise machines like elliptical trainers or stationary cycles results from a magnet positioned near the machine's rotating parts. And eddy currents are guardians of our security, as the next Application shows.

Video Tutor Demo | **Eddy Currents in Different Metals**

APPLICATION Metal Detectors

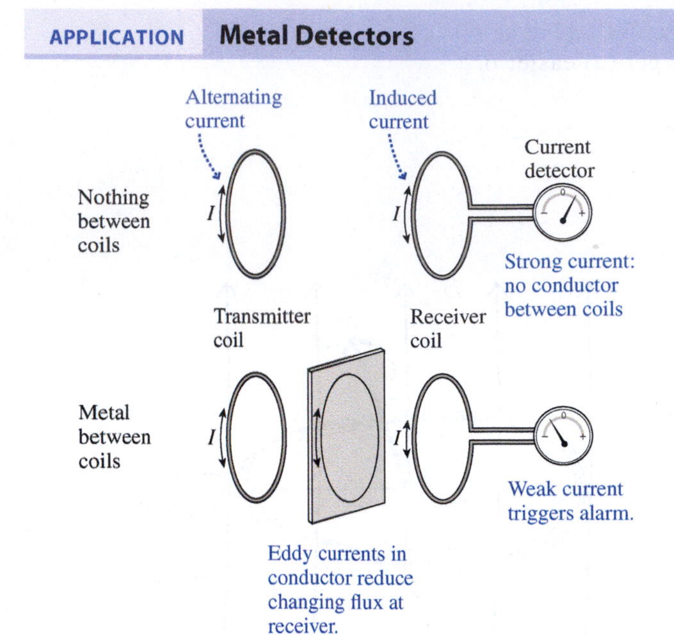

Metal detectors used in airports and other security checkpoints rely on eddy currents. In one type of detector, shown in the figure, an alternating current in one coil—the transmitter—produces a changing magnetic field that induces a current in a second coil, the receiver. A detector, basically an electronic ammeter, monitors the receiver current. Eddy currents are induced in any conducting material that comes between the two coils, and the direction of the induced currents is, as always, such as to reduce the changing flux. The superposition of the transmitter's changing flux with the changing flux from the eddy currents therefore reduces the changing flux at the receiver, dropping the receiver current and triggering an alarm. Other detectors have a single coil, using a short pulse of current to induce eddy currents and then "listening" for the currents induced back in the coil. Either way, you can thank Faraday's law if you've ever been stopped while going through a metal detector!

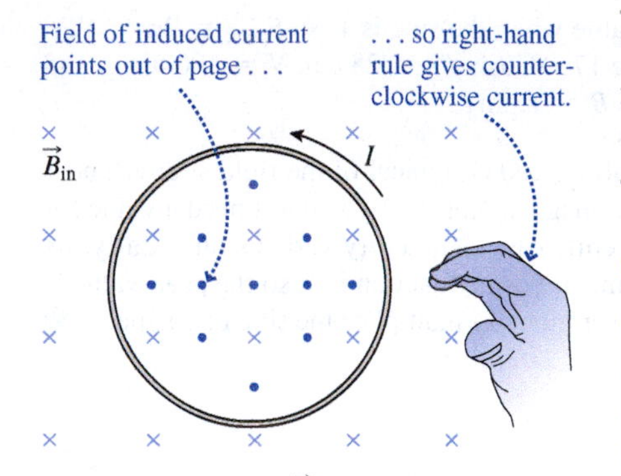

FIGURE 27.16 The field $\vec{B}$ is into the page and increasing; the induced current is counterclockwise, so its field opposes the increase.

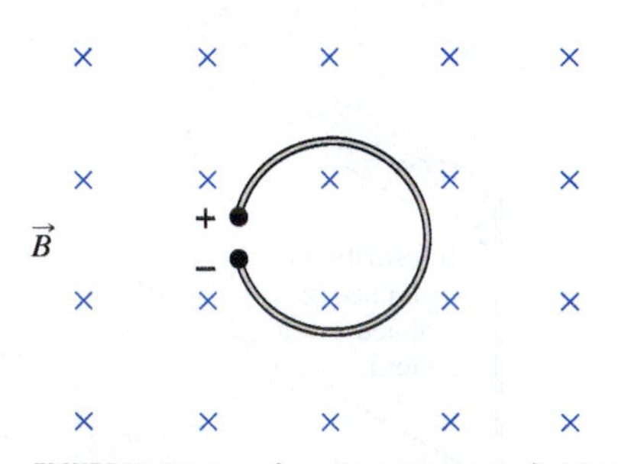

FIGURE 27.17 In a changing magnetic field, the induced emf results in charge buildup at the gap of an open circuit. The polarity shown results when the field is increasing.

GOT IT? 27.4 A copper penny falls on a path that takes it between the poles of a magnet. Does it hit the ground going (a) faster than, (b) slower than, or (c) at the same speed as if the magnet weren't present?

Closed and Open Circuits

Figure 27.16 shows a closed, conducting loop in a magnetic field that points into the page. Suppose the field is increasing in strength; then in order to oppose this change, the induced current must be in such a direction as to oppose the increase. Here that means the field in the interior of the loop needs to come out of the page, and by the right-hand rule that means the induced current is counterclockwise. It's not that the induced field always opposes the inducing field—rather, it opposes the *change*. If the field in Fig. 27.16 had been decreasing, then the induced current would have "tried" to reinforce it by flowing clockwise to make additional field into the page.

What if we have an open circuit, like the conducting loop with a small gap shown in Fig. 27.17? Then there's no induced current whose effects can oppose a change in the field. But we can imagine what *would* happen if the circuit were completed; as in Fig. 27.16, current would flow counterclockwise. Open the gap, and that means positive charge accumulates at the upper end of the gap and negative charge at the bottom. Charge buildup continues until the potential difference at the gap opposes the induced emf's tendency to move charge. The result is a steady state in which the gap voltage equals the induced emf.

GOT IT? 27.5 A long wire carries a current I as shown. What's the direction of the current in the circular conducting loop when I is (1) increasing and (2) decreasing?

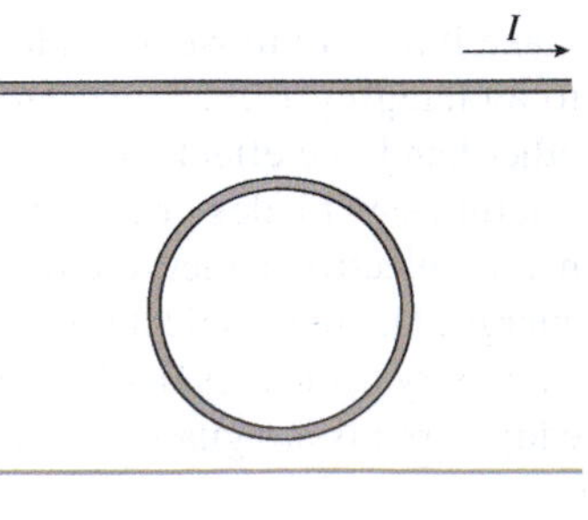

27.4 Inductance

There are many ways to change magnetic flux and thus induce emfs and currents. We can move a magnet, or move or rotate the circuit. Or, as in Fig. 27.4 or Fig. 27.18, we can change the magnetic flux by changing the current in a circuit and therefore the magnetic field it produces. In that case we speak of the **inductance** of a circuit or circuits.

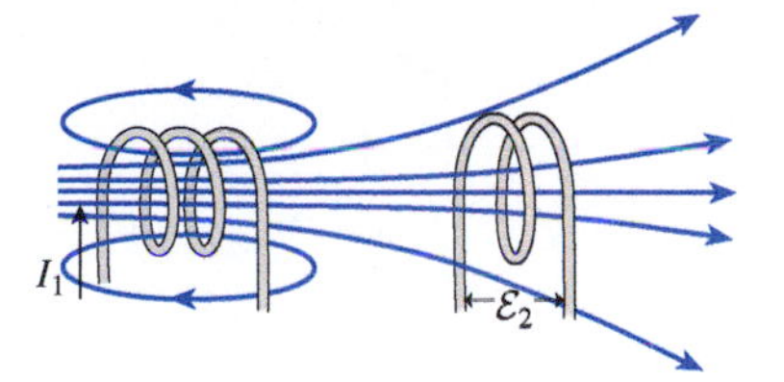

FIGURE 27.18 Mutual inductance. A changing current in either coil induces an emf in the other coil.

Mutual Inductance

Figure 27.18 shows two coils in proximity. If we send a changing current through the left-hand coil, there results a changing magnetic field, which gives rise to a changing magnetic flux in the right-hand coil. There's then an induced emf in the right-hand coil and, if it's connected in a complete circuit, an induced current as well.

The two coils in Fig. 27.18 have **mutual inductance**, meaning a changing current in one coil produces a changing flux at the other coil, thereby inducing an emf. Just how strong this effect is depends on the construction and orientation of the coils; for maximum inductance they should be arranged so that most of the flux from each coil goes through the other. Often coils are wound on iron cores to concentrate the flux and increase the mutual inductance.

Mutual inductance is the basis of transformers, which change voltage levels in alternating-current circuits; more on that in Chapter 28. Your car's ignition coil uses mutual inductance to produce the tens of kilovolts needed to fire the spark plugs and ignite the gasoline–air mixture in the engine. Current to the coil is interrupted at just the right instant, producing a rapid change in magnetic flux and inducing the emf that drives the spark. More mundane is an electric toothbrush, whose batteries charge without electrical connection to a power source. Instead, a small coil in the device is placed in proximity with a coil in the charging base, and alternating current in the base coil transfers energy via mutual inductance to provide the charging current.

Self-Inductance

Inductance isn't limited to two-coil systems. Magnetic flux from current in a single coil or circuit passes through that circuit itself (Fig. 27.19). If the current changes, so does the flux—and that induces an emf. As always, the induced emf opposes the *change* that produces it. Suppose, for example, that the current in Fig. 27.19 is increasing. Then the induced emf will be in the direction that opposes the current increase—clockwise, or opposite the current in Fig. 27.19. The induced emf therefore makes it harder to increase the current. On the other hand, if the current in Fig. 27.19 is decreasing, then the induced emf will try to drive additional current to counter the decrease; the induced emf is therefore in the same direction as the current. Either way, the induced emf makes it hard to change the current in a circuit.

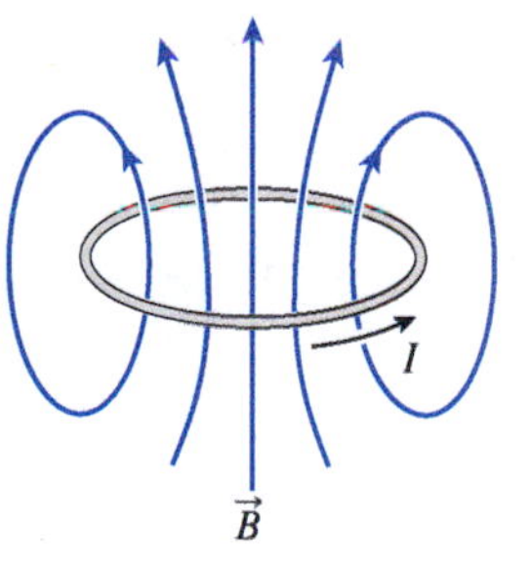

FIGURE 27.19 Magnetic flux from a current loop passes through the loop itself; a change in the current induces an emf that opposes the change.

This property whereby a circuit's own magnetic field opposes changes in the circuit current is called **self-inductance**. All circuits have self-inductance, but it's most important in circuits that encircle a great deal of their own magnetic flux, or when currents change rapidly. A simple piece of wire has little impact on the 60-Hz alternating current used for electric power. But in TVs and computers, where currents change billions of times per second, even the slightest self-inductance can have deleterious effects.

An **inductor** is designed specifically to exhibit self-inductance. Inductors have many uses in electric circuits, including establishing the frequencies of radio transmitters and helping "steer" high- and low-frequency signals to the tweeters and woofers of loudspeaker systems. We'll explore some of these uses in the next chapter. A typical inductor consists of a wire coil, sometimes wound on an iron core to promote flux concentration. Ideally, the only electrical property of an inductor is its inductance, but real inductors have resistance as well.

As long as the current in an inductor is steady, the magnetic flux is constant, so there's no induced emf and the inductor behaves, electrically, just like a wire. But when the current changes, the changing magnetic flux induces an emf that opposes the change in current. The more rapidly the current changes, the greater the rate of change of flux and so the greater the emf. The induced emf depends also on how much of its own magnetic flux the inductor encircles; consequently, we define self-inductance, L, as the ratio of magnetic flux through the inductor to current in the inductor:

$$L = \frac{\Phi_B}{I} \quad \text{(self-inductance)} \tag{27.3}$$

Equation 27.3 shows that the units of self-inductance are $\text{T}\cdot\text{m}^2/\text{A}$. This unit is given the name **henry** (H) in honor of the American scientist Joseph Henry (1797–1878). Inductances in common electronic circuits usually range from microhenrys (μH) up to several henrys.

Inductance is a constant determined by the physical design of an inductor. In principle we can calculate the inductance of any inductor, but in practice that's difficult unless the geometry is particularly simple.

EXAMPLE 27.6 Calculating Inductance: A Solenoid

A long solenoid of cross-sectional area A and length l has n turns per unit length. Find its self-inductance.

INTERPRET We're asked for self-inductance, which Equation 27.3 shows is the ratio of magnetic flux through the solenoid to current in the solenoid.

DEVELOP We'll assume a current in the solenoid and find the resulting magnetic flux. Then we can take their ratio to get the self-inductance. We need the magnetic field of a solenoid, which follows from Equation 26.20: $B = \mu_0 nI$. The field is uniform and perpendicular to the solenoid coils, as we showed in our drawing for Example 27.1, so the flux through each turn follows from Equation 27.1b: $\Phi_{1\text{ turn}} = BA$.

EVALUATE With n turns per unit length, the solenoid contains a total of nl turns, so the flux through all the turns is

$$\Phi_B = nlBA = (nl)(\mu_0 nI)A = \mu_0 n^2 AlI$$

Equation 27.3 gives the self-inductance as the ratio of flux to current, so

$$L = \frac{\Phi_B}{I} = \mu_0 n^2 Al \quad \text{(inductance of solenoid)} \tag{27.4}$$

ASSESS Make sense? As the area increases, so does the flux and therefore the inductance. As the length increases, so does the number of turns, so again the total flux increases. And as the number of turns per unit length increases, two things happen. First, the magnetic field of Equation 26.20 increases, increasing the flux BA through each turn. Second, the total number of turns increases, again increasing the total flux. That's why the inductance, L, depends on *n squared*. ■

The induced emf in an inductor always acts to oppose the *change* in current through the inductor. That change generally results from events happening in the rest of the circuit in which the inductor is connected, such as closing a switch, changing a resistance, or connecting a battery or other source of emf. If the inductor current is increasing, that means the inductor develops an emf that "pushes back" against current flowing in from the external circuit. In that case the inductor emf is called a **back emf**, and here you can think of the inductor as acting like a battery that's connected backward to oppose incoming current (Fig. 27.20*a*). If the inductor current is decreasing, on the other hand, the inductor emf acts to help keep the current flowing, as in Fig. 27.20*b*.

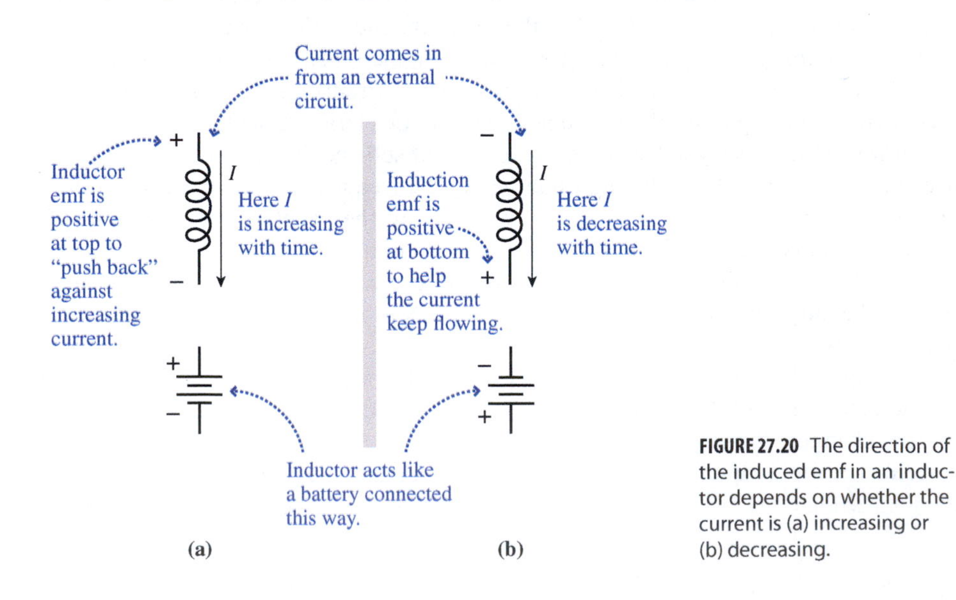

FIGURE 27.20 The direction of the induced emf in an inductor depends on whether the current is (a) increasing or (b) decreasing.

Quantitatively, the inductor emf follows from Faraday's law, which relates emf to the rate of change of magnetic flux: $\mathcal{E} = -d\Phi_B/dt$. If we differentiate Equation 27.3, the definition of self-inductance, we get

$$\frac{d\Phi_B}{dt} = L\frac{dI}{dt}$$

Substituting this expression for $d\Phi_B/dt$ in Faraday's law then gives

$$\mathcal{E}_L = -L\frac{dI}{dt} \quad \text{(inductor emf)} \tag{27.5}$$

This equation gives the emf $\mathcal{E}_L$ induced in an inductor L when the inductor current is changing at the rate dI/dt. The minus sign again tells us that the emf opposes the *change* in current. When the current isn't changing, $dI/dt = 0$ and there's no induced emf. In this case, the inductor acts like a piece of wire. But when the current changes, the inductor produces an emf whose magnitude depends on the rate of change of current, dI/dt. You can get the direction of that emf by thinking about which way the emf has to go to oppose the change in current, as described in Fig. 27.20. You can also get it formally from Equation 27.5, as described in Fig. 27.21. We'll use this approach in the next section, when we consider circuits containing inductors.

The dependence of the induced emf on dI/dt in Equation 27.5 isn't just mathematics! Rapid switching of inductive devices such as solenoid valves or electric motors results in induced emfs that can destroy delicate electronic components. And people have been electrocuted opening switches in circuits containing large inductors.

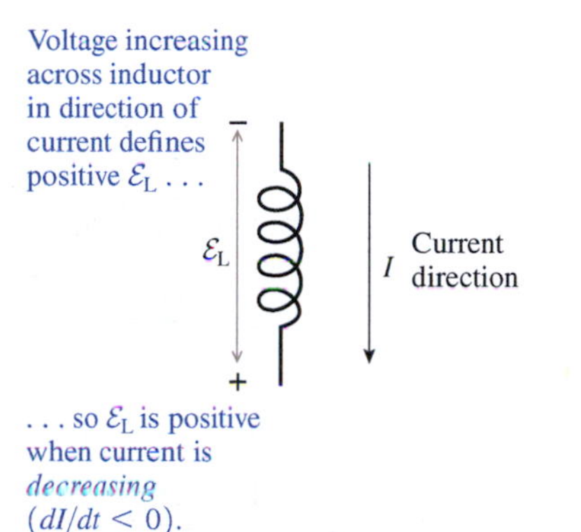

FIGURE 27.21 Sign conventions for the inductor emf of Equation 27.5. The coil is the circuit symbol for an inductor.

GOT IT? 27.6 Current flows from left to right through the inductor shown. A voltmeter connected across the inductor gives a constant reading, and shows that the left end of the inductor is positive. Is the current in the inductor (a) increasing, (b) decreasing, or (c) steady? Why?

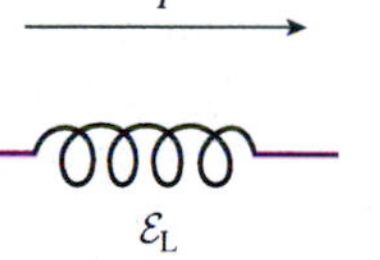

EXAMPLE 27.7 Back EMF: A Dangerous Inductor

A 5.0-A current is flowing in a 2.0-H inductor. The current is then reduced steadily to zero over 1.0 ms. Find the magnitude and direction of the inductor emf during this time.

INTERPRET There's an emf in the inductor because the current is changing; we want the magnitude and direction of that emf.

DEVELOP Figure 27.22 shows the situation, complete with an external circuit that's the source of the decreasing current. Equation 27.5, $\mathcal{E}_L = -L(dI/dt)$, gives the inductor emf in terms of the rate of change of current. Since the current changes steadily, the latter is just the change in current (-5.0 A) divided by the time involved.

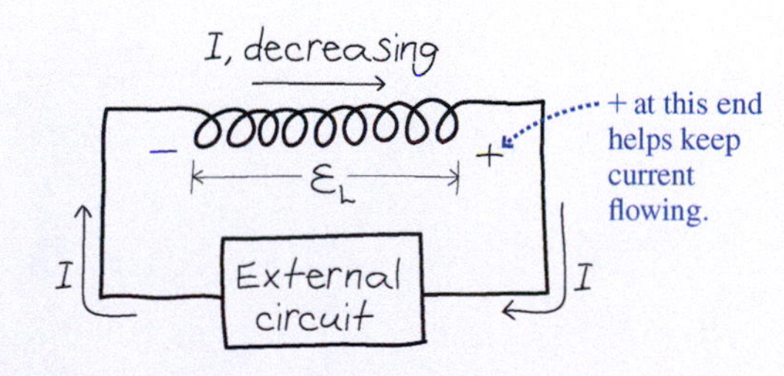

FIGURE 27.22 Sketch for Example 27.7.

EVALUATE

$$\mathcal{E}_L = -L\frac{dI}{dt} = -(2.0\ \text{H})\left(\frac{-5.0\ \text{A}}{1.0\ \text{ms}}\right) = 10{,}000\ \text{V}$$

That this answer is positive tells us that the emf increase across the inductor is in the same direction as the current, as we've indicated in Fig. 27.22.

ASSESS That's a potentially lethal voltage! Our answer is unrelated to the battery or whatever is supplying the inductor current. One could have a 6-V battery and still be electrocuted opening a circuit with a large inductance. Note that the direction we deduced is consistent with Lenz's law; here the inductor emf opposes the decrease in current, and that means it provides an emf in the direction that would keep current flowing in the external circuit—just the opposite of the situation in GOT IT? 27.6. ■

Inductors in Circuits

In Chapter 25 you saw that the voltage across a capacitor can't change instantaneously. We can make an analogous statement for inductors. Because the inductor emf depends on the rate of change of current and because an infinite emf is impossible, **the current through an inductor can't change instantaneously**. Much of your understanding of capacitors applies to inductors if you interchange the words "voltage" and "current."

Figure 27.23 shows a circuit with a battery, switch, resistor, and inductor. With the switch open there's no current (Fig. 27.23*a*). Close the switch, and the current at that instant is still zero because the inductor current can't change instantaneously. With no current, there's no voltage across the resistor, so the inductor must be producing a back emf equal in magnitude to the battery emf (Fig. 27.23*b*). Although at this instant there's zero current in the inductor, the nonzero emf $\mathcal{E}_L = -L(dI/dt)$ shows that the *rate of change* of current, dI/dt, isn't zero.

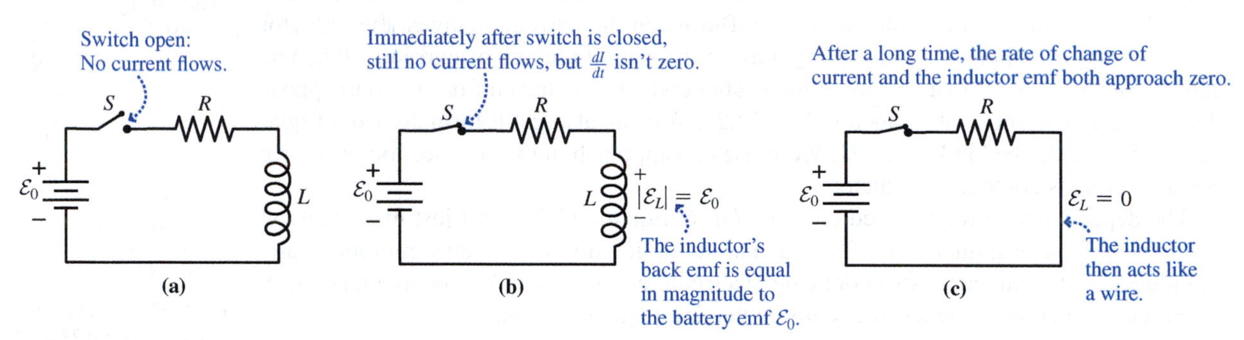

FIGURE 27.23 An *RL* circuit at three times.

So the inductor current rises from zero, and with it the resistor current and therefore the resistor voltage *IR*. The battery emf $\mathcal{E}_0$ is constant, so as *IR* increases, the magnitude of the inductor emf drops. Equation 27.5 shows that the rate of change of current drops as well. Eventually the whole circuit reaches a steady state in which dI/dt and therefore the inductor emf are both zero (Fig. 27.23*c*). At this point the inductor acts like a wire, and the resistor determines the current: $I = \mathcal{E}_0/R$. Figure 27.24 summarizes this analysis of the *RL* circuit.

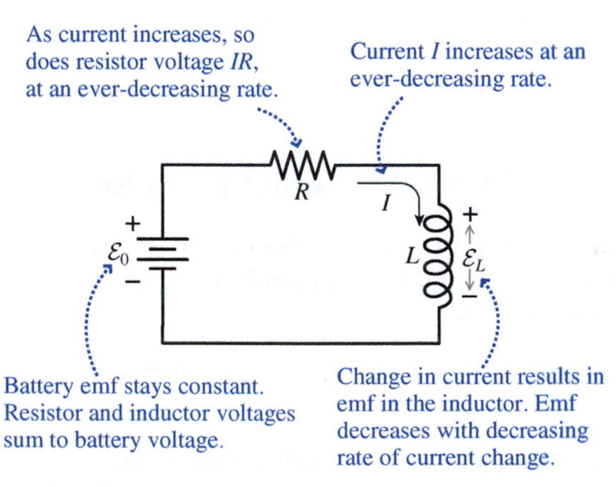

FIGURE 27.24 Interrelationships among circuit quantities as current builds up in an *RL* circuit. Compare with Fig. 25.19 for a charging capacitor.

We can analyze the circuit quantitatively using the loop law. Going clockwise, we encounter a voltage increase $\mathcal{E}_0$ at the battery, a decrease $-IR$ at the resistor, and a change $\mathcal{E}_L$ at the inductor. This change is actually a decrease, but we'll let Equation 27.5 take care of the signs. Then the loop law reads $\mathcal{E}_0 - IR + \mathcal{E}_L = 0$. The battery emf is constant, so if we differentiate this equation, we get

$$\frac{d\mathcal{E}_L}{dt} = R\frac{dI}{dt}$$

But Equation 27.5 gives $dI/dt = -\mathcal{E}_L/L$, so

$$\frac{d\mathcal{E}_L}{dt} = -R\frac{\mathcal{E}_L}{L}$$

This looks like Equation 25.4 for the *RC* circuit, but with $\mathcal{E}_L$ in place of current I, L in place of C, and $1/R$ in place of R. So the solution is that of Equation 25.4 with the appropriate substitutions:

$$\mathcal{E}_L = -\mathcal{E}_0 e^{-Rt/L} \tag{27.6}$$

This shows that the inductor emf decays exponentially from its initial value $-\mathcal{E}_0$ (negative because the inductor emf opposes the battery emf) to zero. Using the undifferentiated loop equation, we can now solve for the current:

$$I = \frac{\mathcal{E}_0 + \mathcal{E}_L}{R} = \frac{\mathcal{E}_0}{R}(1 - e^{-Rt/L}) \tag{27.7}$$

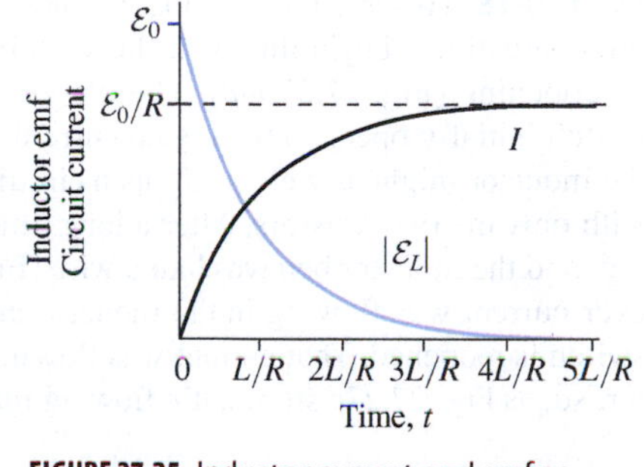

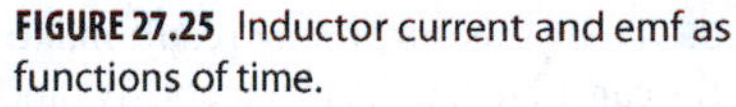
FIGURE 27.25 Inductor current and emf as functions of time.

With a capacitor, we characterized time-changing quantities with the capacitive time constant *RC*. Here the **inductive time constant** is L/R. In contrast to the capacitor case, the inductive time constant depends *inversely* on resistance. That's because a lower resistance means a higher steady-state current, which therefore requires a longer time to build up. Significant changes in current can't occur on time scales much shorter than L/R. Wait many time constants, and the circuit approaches a steady state with $\mathcal{E}_L = 0$. Figure 27.25 summarizes the time-dependent behavior of circuit quantities in an *RL* circuit.

EXAMPLE 27.8 The Inductive Time Constant: Firing Up an Electromagnet

A large electromagnet used for lifting scrap iron has self-inductance $L = 56$ H. It's connected to a constant 440-V power source; the total resistance of the circuit is 2.8 Ω. Find the time it takes for the current to reach 75% of its final value.

INTERPRET This is a problem about the buildup of current in an *RL* circuit.

DEVELOP Equation 27.7, $I = (\mathcal{E}_0/R)(1 - e^{-Rt/L})$, determines the current; here $\mathcal{E}_0/R$ is the final current, and we want to solve for the time t when I is 75% of this final value. That is, we want $0.75 = 1 - e^{-Rt/L}$.

EVALUATE Rearranging, we have $e^{-Rt/L} = 0.25$; then taking natural logs of both sides and using $\ln e^x = x$ gives $-Rt/L = \ln(0.25)$, or

$$t = -\frac{L}{R}\ln(0.25) = -\frac{56\text{ H}}{2.8\ \Omega}\ln(0.25) = 28\text{ s}$$

ASSESS This is a little more than one time constant ($L/R = 20$ s)—not surprising because we found with capacitors that we reach approximately two-thirds of the full charge in one time constant. Analogously, with inductors, we reach about two-thirds of the final current in one time constant. ■

Figure 27.26 shows a circuit with a two-way switch. Throw the switch to A, and current builds up as we just described. Throw it to B, and current continues to flow through the inductor and resistor because the inductor current can't change instantaneously. We won't go through the math, but it's straightforward to show that the current decays exponentially with the same time constant L/R:

$$I = I_0 e^{-Rt/L} \tag{27.8}$$

This is analogous to our result for the discharging capacitor.

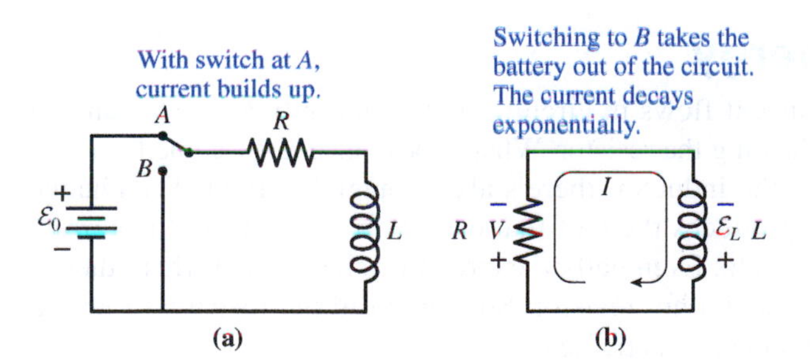

FIGURE 27.26 Buildup and decay of current in an *RL* circuit.

As with capacitors, it's not necessary to use exponential equations to analyze the short- and long-term behavior of circuits with inductors. All you need to remember is that for short times inductor current can't change instantaneously, and for long times inductors produce no emfs and therefore act like wires. The next example explores this situation.

CONCEPTUAL EXAMPLE 27.1 Inductors: Short Times, Long Times

The switch in Fig. 27.27*a* is initially open. It's then closed and, a long time later, reopened. What's the direction of the current in R_2 after the switch is reopened?

EVALUATE To see what's happening here, we sketch the circuit in three situations, beginning with the switch closing and ending with it reopening (Fig. 27.27*b–d*). There's no inductor current with the switch initially open, so there's no current right after it closes. Then the inductor might as well be an open circuit, so we drew Fig. 27.27*b* with only the two resistors. After a long time the current stops changing, and the inductor behaves like a wire (Fig. 27.27*c*). Finally, whatever current was flowing in the inductor continues to flow after the switch is reopened. That current was flowing downward in the inductor, so, as Fig. 27.27*d* shows, it's flowing *upward* through R_2.

ASSESS Does this surprising result make sense? Yes: Current in an inductor can't change instantaneously, and once the switch opens the current has nowhere to go but upward through R_2. The resistor has no say in the matter; as Making the Connection shows, its current is determined entirely by the battery voltage and R_1. If the switch stays open, the current in Fig. 27.27*d* decays exponentially as the resistor dissipates the energy that was stored in the inductor.

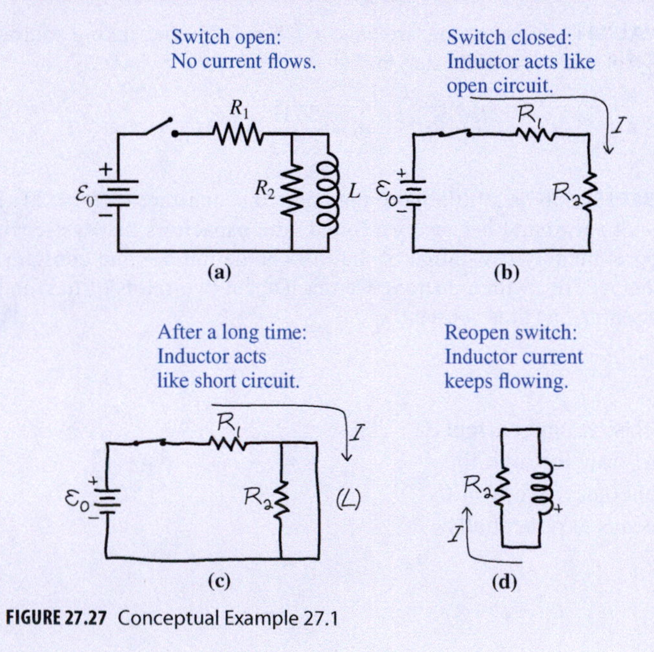

FIGURE 27.27 Conceptual Example 27.1

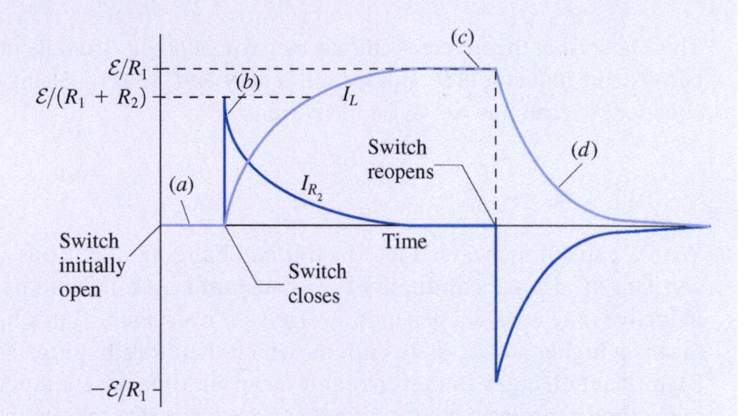

FIGURE 27.28 Currents in R_2 and L for Conceptual Example 27.1.

Figure 27.28 shows currents in the inductor and R_2 as functions of time. If R_2 weren't in the circuit, the voltage would rise dangerously high as the inductor tries to keep the current flowing. Resistors are often wired in parallel with large inductors to alleviate this danger.

MAKING THE CONNECTION Verify that the current in R_2 just after the switch is reopened has the value indicated in Fig. 27.28.

EVALUATE Just *before* the switch is reopened, Fig. 27.27*c* shows that the current through the inductor is $I_L = \mathcal{E}_0/R_1$; R_2 is irrelevant here because it's short-circuited by the inductor. Just *after* the switch opens, the current continues flowing, now going upward through R_2 as we reasoned above. So the current in R_2 is $-\mathcal{E}_0/R_1$, with the minus sign designating the upward direction according to the sign conventions in Fig. 27.28.

27.5 Magnetic Energy

In Figs. 27.26*b* and 27.27*d*, current flows in circuits containing only a resistor and an inductor. Energy is dissipated, heating the resistor. Where does this energy come from?

Because there's a current in the inductor, there's also a magnetic field. The change in that magnetic field is what produces the emf that drives the current. As the current decreases, so does the magnetic field. Eventually the circuit reaches a state where there's no current, no magnetic field—and a hot resistor. So where did the resistor's thermal energy come from? It came from the magnetic field.

Like the electric field, the magnetic field contains stored energy. Our decaying *RL* circuit is analogous to a discharging *RC* circuit, in which the electric field between the capacitor plates disappears as thermal energy appears in the resistor. As in the electric case, magnetic energy isn't limited to circuits: *Any* magnetic field contains energy. Release of magnetic energy drives a number of practical devices and also powers violent events throughout the universe; an important example that directly impacts us here on Earth is described in the Application on page 516.

Magnetic Energy in an Inductor

We can find the stored energy by reconsidering the buildup of current in the inductor. Earlier we wrote the loop law for the circuit of Fig. 27.23; if we multiply that equation by the current I, we get $I\mathcal{E}_0 - I^2R + I\mathcal{E}_L = 0$ or, using Equation 27.5 for $\mathcal{E}_L$,

$$I\mathcal{E}_0 - I^2R - LI\frac{dI}{dt} = 0$$

The three terms here have the units of voltage times current, or power. The first shows that the battery supplies energy *to* the circuit at the rate $I\mathcal{E}_0$. The second, $-I^2R$, is the rate of energy dissipation in the resistor; the negative sign means the resistor takes energy *from* the circuit. The current is increasing $(dI/dt > 0)$, so the third term is also negative; it describes energy the inductor takes from the circuit. But the inductor doesn't dissipate this energy; rather, it stores the energy in its magnetic field. The rate at which the inductor stores energy is thus

$$P = LI\frac{dI}{dt}$$

Suppose we increase the current in an inductor by some small amount dI over a small time interval dt. Since the power is the rate of energy storage, the energy dU stored during this time is

$$dU = P\,dt = LI\frac{dI}{dt}dt = LI\,dI.$$

We find the total energy stored in bringing the inductor current from zero to some final value I by summing—that is, integrating—all the dU values:

$$U = \int dU = \int P\,dt = \int_0^I LI\,dI = \tfrac{1}{2}LI^2\Big|_0^I$$

Evaluating at the two limits then gives the stored energy:

$$U = \tfrac{1}{2}LI^2 \qquad \text{(energy stored in inductor)} \qquad (27.9)$$

This much energy is therefore released when the magnetic field decays.

EXAMPLE 27.9 Magnetic Energy: An MRI Disaster

Superconducting electromagnets like the solenoids in MRI scanners store a lot of magnetic energy. Loss of coolant can be dangerous because the current is suddenly left without its zero-resistance path and quickly decays. The result is an explosive release of magnetic energy. A particular MRI solenoid carries 2.4 kA and has a 0.53-H inductance. When it loses superconductivity, its resistance goes abruptly to 31 mΩ. Find (a) the stored magnetic energy and (b) the rate of energy release at the instant superconductivity is lost.

INTERPRET We're asked first for the total stored energy and then for the power dissipated in the resistance of the coils at the instant they cease to be superconducting.

DEVELOP Equation 27.9, $U = \frac{1}{2}LI^2$, determines the stored energy, while $P = I^2R$ determines the resistor power. Just before the coolant loss, the MRI solenoid carries 2.4 kA; since current can't change instantaneously in an inductor, this current remains momentarily

(continued)

unchanged. Therefore, we have everything we need to find the power dissipation.

EVALUATE (a) Equation 27.8 gives

$$U = \tfrac{1}{2}LI^2 = (\tfrac{1}{2})(0.53\text{ H})(2.4\text{ kA})^2 = 1.5\text{ MJ}$$

while for (b) we have

$$P = I^2R = (2.4\text{ kA})^2(31\text{ m}\Omega) = 0.18\text{ MW}$$

ASSESS This is substantial power, equivalent to 1800 100-W lightbulbs burning in the space of this roughly human-size device! You can show in Problem 57 that it takes some 20 s before 90% of the energy has dissipated. To prevent explosive energy release, superconducting wires generally incorporate copper or silver to carry the current in the event of coolant loss that quenches the superconductivity. ■

APPLICATION **Solar Outbursts and Geomagnetic Storms**

This chapter's opening image depicts a power failure that blacked out the entire Canadian province of Quebec. That blackout was no accident but resulted from electrical failures caused, ultimately, by a massive outburst of particles from the Sun. Solar physicists—including your author—are still exploring the mechanisms of such outbursts, but it's clear that they involve the sudden release of energy stored in the magnetic field that permeates the Sun's atmosphere, or *corona*. Thus the energy described by Equation 27.10 becomes the kinetic energy of solar particles, mostly electrons and protons. The resulting *coronal mass ejections* (CMEs) propagate outward through the solar system at hundreds of kilometers per second and, if they happened to be aimed toward Earth, reach our planet in several days; the image shows a large CME erupting from the Sun. The particle burst and associated shock wave pummels Earth's magnetic field, and some particles become trapped on the field lines; there they cause auroras, as described in Section 26.3. The impact also compresses Earth's magnetic field, resulting in a rapidly changing field that extends down to Earth's surface. And time-changing magnetic fields, as Faraday's law shows, give rise to induced currents in any conductors that might be present. On Earth, such conductors include power lines and related electrical equipment, as well as Earth itself. Current surges in these conductors can damage critical components of the electric power grid, resulting in a cascade of failures that, in extreme cases, causes massive blackouts like the Quebec event. Even buried cables aren't immune, as induced currents surge through the solid Earth—although in Quebec, poorly conductive rock underlying the province reduced ground currents and actually exacerbated current surges in the aboveground power grid. Solar-induced magnetic disturbances at Earth are called *geomagnetic storms*, and in addition to power systems they can damage or interfere with satellites and communications. As humankind depends more on electrical, electronic, and space systems, we worry about massive geomagnetic storms that could plunge far greater areas than Quebec into power, communications, and information blackouts. Indeed, a record-strength geomagnetic storm in 1859 caused auroras that were seen as far south as Hawaii and Cuba and resulted in worldwide failures of newly developed electric telegraph systems. A similar event today could cost the world some $3 trillion in damages.

Magnetic-Energy Density

In Example 27.6 we found the inductance of a solenoid with length l and cross-sectional area A: $L = \mu_0 n^2 Al$. Equation 27.9 then gives the magnetic energy stored in the solenoid:

$$U = \tfrac{1}{2}LI^2 = \tfrac{1}{2}\mu_0 n^2 AlI^2 = \frac{1}{2\mu_0}(\mu_0 nI)^2 Al = \frac{B^2}{2\mu_0}Al$$

where we recognized the quantity $\mu_0 nI$ as B, the magnetic field in the solenoid (Equation 26.20). The quantity Al is the volume containing this field, so the energy per unit volume—the **magnetic-energy density**—is

$$u_B = \frac{B^2}{2\mu_0} \quad \text{(magnetic-energy density)} \qquad (27.10)$$

Although we derived this expression for the field of a solenoid, it is, in fact, a universal expression for the local magnetic-energy density. Wherever there's a magnetic field, there's stored energy.

Equation 27.10 is similar to Equation 23.7 for the energy density in an electric field: $u_E = \tfrac{1}{2}\epsilon_0 E^2$. Each energy density is proportional to the *square* of the field strength, and each contains the appropriate constant, μ_0 or ϵ_0. That the constant appears in the numerator in one case and in the denominator in the other is merely a consequence of the way SI units are defined.

GOT IT? 27.7 If you keep the current in a solenoid constant while doubling both its overall length and the number of turns per unit length, will the magnetic energy in the solenoid (a) double, (b) quadruple, (c) increase by a factor of 8, (d) increase by a factor of 16, or (e) decrease by a factor of $1/\sqrt{2}$?

27.6 Induced Electric Fields

So far we've been talking about induction in terms of emfs and circuits. But what, really, is emf? In the case of a battery, it results from chemical reactions that separate charge. With motional emf (Section 27.3), magnetic forces on a moving conductor act to separate charge. But what causes the emf in a conducting loop subject to a changing magnetic field? There's no motion, yet there must be a force on the free charges in the conductor. The only force we know that acts on stationary charges is the electric force, which results from electric fields. Therefore, there must be an electric field—an **induced electric field**—in the conducting loop. This field has the same effect on charges, exerting a force $q\vec{E}$, as did the electric fields we considered earlier. But the induced field originates not in electric charge but in *changing magnetic field.*

An induced electric field results whenever a magnetic field changes with time—whether or not an electric circuit is present. If there is a circuit, then the field drives induced currents. But the induced field, not the current, is fundamental. A single, stationary electron in a changing magnetic field experiences an *electric* force—clear evidence for the existence of the induced electric field.

We wrote Faraday's law as Equation 27.2, giving the relation between induced emf and changing magnetic flux. But the induced electric field is more fundamental, and emf simply means the work per unit charge gained as charge goes around a circuit—or for that matter any closed loop. Thus we can write $\mathcal{E} = \oint \vec{E} \cdot d\vec{l}$, and Faraday's law becomes

$$\oint \vec{E} \cdot d\vec{l} = -\frac{d\Phi_B}{dt} \quad \text{(Faraday's law)} \qquad (27.11)$$

Here we're using $d\vec{l}$ for the infinitesimal vector along the integration loop, to be consistent with the notation introduced in Chapter 26 for Ampère's law. Faraday's law in the form of Equation 27.11 is a universal statement about electric fields and changing magnetic flux. The line integral on the left-hand side is over *any* closed loop, which need not coincide with a circuit or conductor. The flux on the right-hand side is the surface integral of the magnetic field over any open surface bounded by the loop on the left-hand side.

Faraday's law tells us that there's another source of electric fields besides electric charge—namely, changing magnetic field:

A changing magnetic field creates an electric field.

This direct interaction between fields is the basis for many practical devices and, as we'll see in Chapter 29, is essential to the existence of light.

Faraday's law is similar to Ampère's law (Equation 26.16). On the left side, both involve the line integral of a field, $\vec{E}$ for Faraday and $\vec{B}$ for Ampère. On the right is a source of that field, changing magnetic field for $\vec{E}$ and moving electric charge—current—for $\vec{B}$. Both fields *encircle* their sources. That means the configuration of an induced electric field is very different from that of an electric field originating in charge. Field lines of an induced electric field have no beginnings or ends; they generally form closed loops encircling regions of changing magnetic field (Fig. 27.29).

When a changing magnetic field has sufficient symmetry, we can evaluate the induced electric field in the same way we did the magnetic field of a symmetric current distribution.

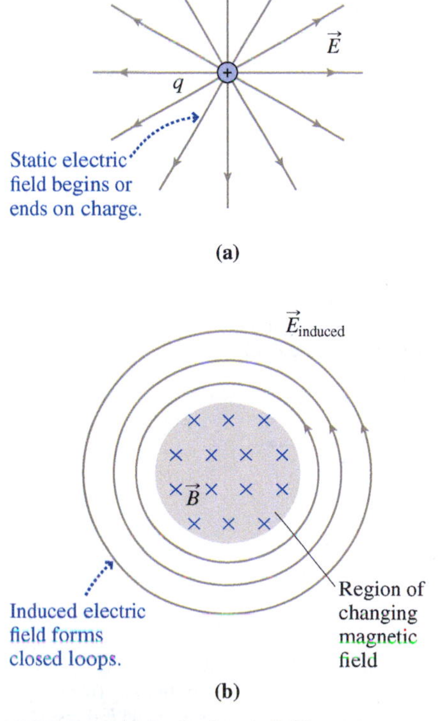

FIGURE 27.29 (a) Static electric fields originate in charges and look very different from (b) induced fields that result from changing magnetic fields.

EXAMPLE 27.10 Finding the Induced Electric Field: A Solenoid

A long solenoid has circular cross section of radius R. The solenoid current is increasing, and as a result so is the magnetic field in the solenoid. The field strength is given by $B = bt$, where b is a constant. Find the induced electric field outside the solenoid, a distance r from the axis.

INTERPRET Here's a problem about a changing magnetic field producing an electric field—that is, about Faraday's law. We'll follow the steps in Chapter 26's strategy for Ampère's law, modifying as appropriate to Faraday's law. We begin by identifying the symmetry, which here is line symmetry.

DEVELOP Symmetry requires that encircling electric field lines be circular. We've drawn some field lines in Fig. 27.30 and marked one of them as a loop for the integration in Faraday's law. We chose a loop coinciding with a field line because symmetry requires that the field strength E be constant over a circle concentric with the symmetry axis.

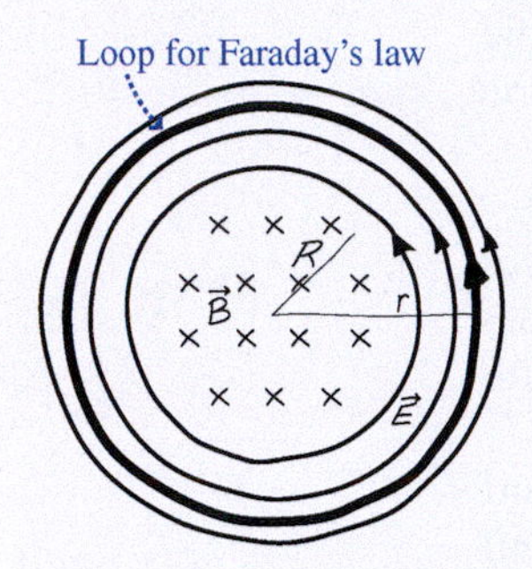

FIGURE 27.30 Cross section of a solenoid whose magnetic field points into the page and is increasing. Field lines of the induced electric field are circles concentric with the solenoid axis.

EVALUATE The situation on the left-hand side of Faraday's law is just as in Example 26.8, except with $\vec{E}$ instead of $\vec{B}$. So $\oint \vec{E} \cdot d\vec{l}$ evaluates to $2\pi rE$. Instead of current on the right-hand side, we have changing magnetic flux $-d\Phi_B/dt$. Here the loop is outside the solenoid, so it encircles the entire flux, which is $\Phi_B = BA = bt\pi R^2$. Then the rate of change of flux is $d\Phi_B/dt = \pi R^2 b$. As usual, we'll use Faraday's law for the magnitude of the induced effect, and then invoke energy conservation for the direction. Equating $2\pi rE$ to the rate of change of flux $\pi R^2 b$ then gives the magnitude of the induced electric field:

$$E = \frac{R^2 b}{2r}$$

What about the direction? If a current were flowing as a result of the induced electric field, its direction would be such as to *oppose* the increase in the solenoid's magnetic field. Since the solenoid's field points *into* the page, any induced current would have to produce a field pointing *out of* the page. Applying the right-hand rule by curling your fingers in the counterclockwise direction of the field arrows in Fig. 27.30 shows that a counterclockwise current would produce such an outward field. Since an induced current would be driven by the electric field, the counterclockwise direction we chose for the electric field lines in Fig. 27.30 is correct.

ASSESS The $1/r$ dependence here shouldn't surprise you; we found the same dependence for other fields, both electric and magnetic, resulting from sources with line symmetry. Note how we used a hypothetical induced current here to reason out the direction of the induced electric field. Even though there isn't any induced current in this case, you can still think about what *would* happen if there *were* such a current, and Lenz's law will then lead you to the direction of the induced electric field. Calculating the electric field *inside* the solenoid would be similar to this example, but a given field line would encircle only part of the magnetic flux; Exercise 33 covers that situation.

Conservative and Nonconservative Electric Fields

Static electric fields—those beginning and ending on stationary charge distributions—are *conservative*, meaning that the work required to move a charge between two points is path independent. A consequence is that it takes no work to move around a closed path in an electrostatic field; mathematically, we express this by writing

$$\oint \vec{E} \cdot d\vec{l} = 0 \qquad \text{(electrostatic field)}$$

In contrast, induced electric fields generally form closed loops, and here Faraday's law shows that the line integral of the electric field around a closed path is decidedly not zero. That means the induced electric field does work on a charge moved around a *closed* path and that the work done in moving between two points cannot be independent of the path taken (Fig. 27.31). The induced electric field, therefore, is not conservative.

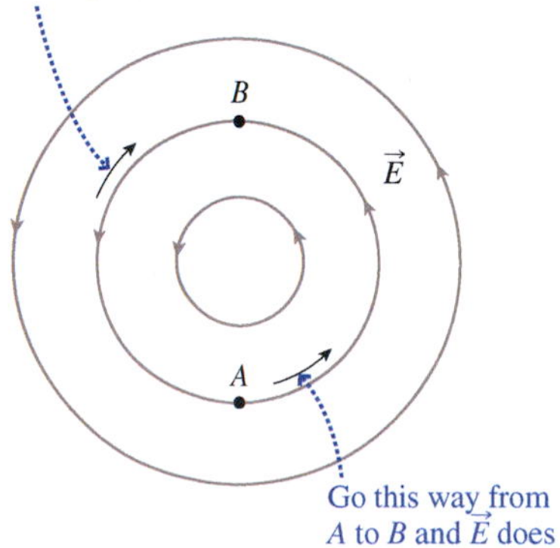

FIGURE 27.31 The work done to move charge in an induced electric field isn't path independent, so the induced field isn't conservative.

GOT IT? 27.8 The figure shows three resistors in series surrounding an infinitely long solenoid with a changing magnetic field; the resulting induced electric field drives a current counterclockwise, as shown. Two identical voltmeters are shown connected to the *same* points A and B. What does each read? Explain any apparent contradiction. *Hint:* This is a challenging question!

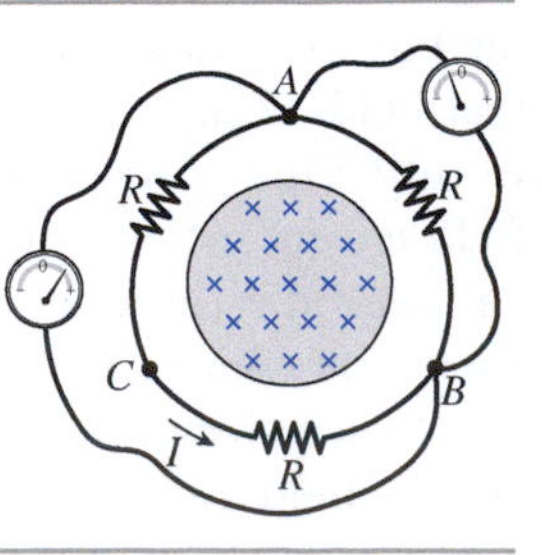

Diamagnetism

We introduced diamagnetism in Chapter 26 but couldn't explain it there because it involves induced electric fields. Figure 27.32 shows a highly simplified model representing two atomic electrons with equal but opposite magnetic moments. Although a proper treatment of diamagnetism requires quantum mechanics, this model shows qualitatively how diamagnetism arises.

The dipole moments in Fig. 27.32*a* cancel, so the associated atom has no magnetic dipole moment. But what happens when a magnetic field is applied, pointing into the page (Fig. 27.32*b*), perhaps by moving the north pole of a bar magnet toward the page? The changing magnetic field results in an electric field that alters the electrons' speeds. In order to oppose the imposition of the magnetic field, the electron on the right speeds up. Its dipole moment, which points out of the page, increases and opposes the bar magnet's field. Meanwhile the left-hand electron's dipole moment decreases. Now the atom has a net dipole moment pointing out of the page, opposing the incoming magnet and resulting in the repulsive force that characterizes diamagnetism.

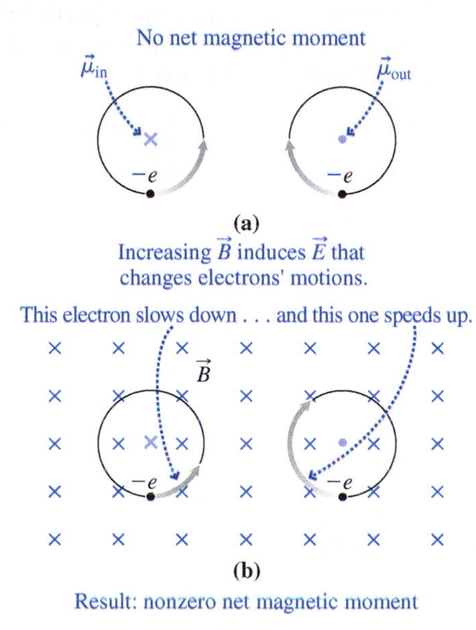

FIGURE 27.32 A simple model for diamagnetism.

A superconductor is perfectly diamagnetic, meaning that the magnetic field resulting from induced currents completely cancels any applied field. Since these induced currents persist in the zero-resistance superconductor, the material completely excludes magnetic fields from its interior, a phenomenon known as the Meissner effect (Fig. 27.33). The repulsive force associated with the magnetic moments of a permanent magnet and a nearby superconductor results in the widely publicized phenomenon of magnetic levitation (Fig. 27.34).

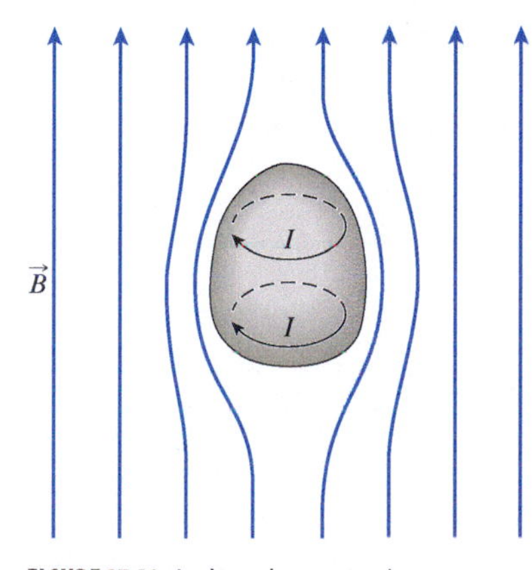

FIGURE 27.33 Induced currents in a superconductor completely cancel an applied magnetic field.

FIGURE 27.34 A small magnet levitates above a wafer of high-temperature superconductor in a bath of liquid nitrogen.

CHAPTER 27 SUMMARY

Big Idea

The big idea here is **electromagnetic induction**, a phenomenon in which **a changing magnetic field produces an electric field**. Applied to circuits, induction results in induced emfs that drive induced currents.

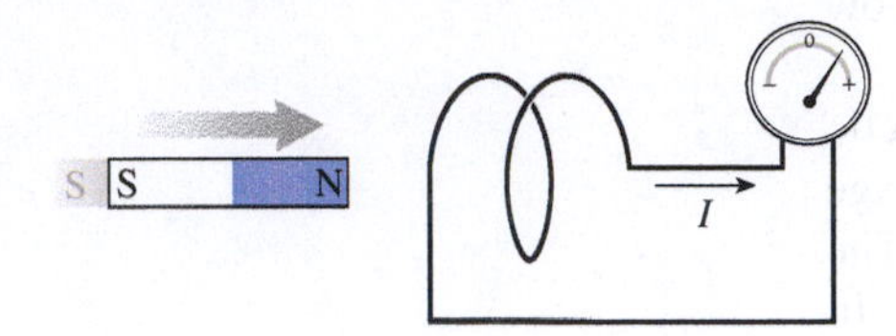

Here a moving magnet produces the changing magnetic field.

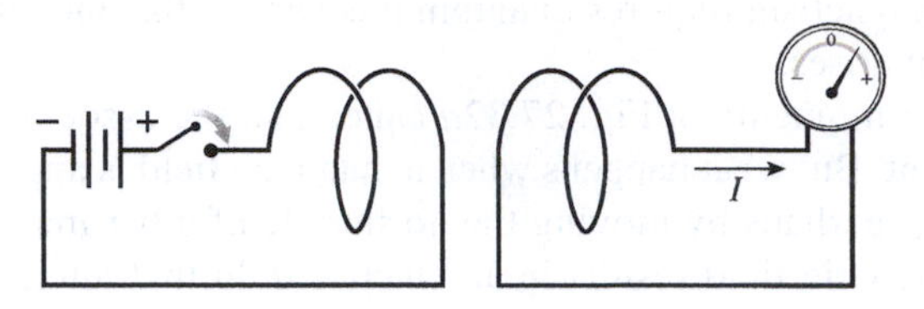

Here a change in current produces the changing magnetic field.

Key Concepts and Equations

Faraday's law describes induction quantitatively, relating the line integral of the induced electric field to changing magnetic flux:

$$\oint \vec{E} \cdot d\vec{l} = -d\Phi_B/dt$$

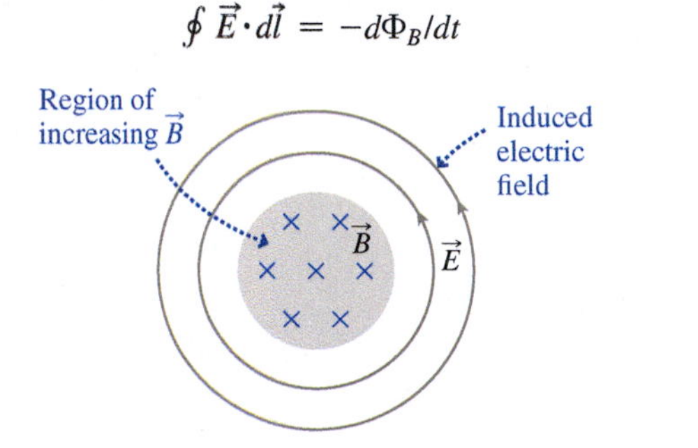

In conductors, Faraday's law gives the induced emf: $\mathcal{E} = -d\Phi_B/dt$.

Lenz's law shows that electromagnetic induction is consistent with conservation of energy, which requires that induced effects act to oppose the changes that give rise to them.

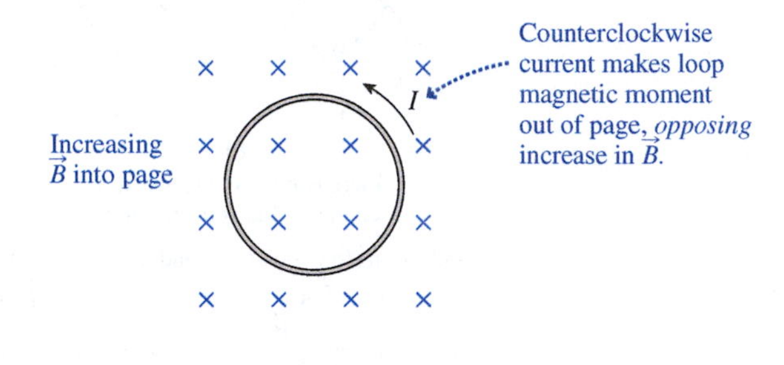

Magnetic fields contain stored energy, as do electric fields. The **magnetic-energy density** is

$$u_B = \frac{B^2}{2\mu_0}$$

Applications

Electric generators convert mechanical energy to electrical energy by moving conductors in magnetic fields to induce emfs that drive currents.

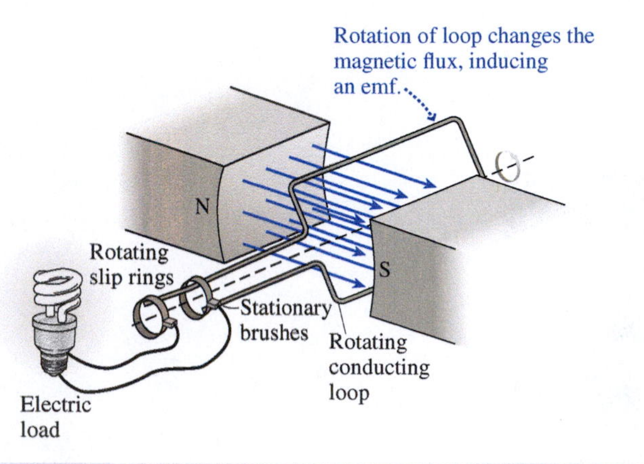

Inductors are wire coils that encircle their own magnetic flux, giving **self-inductance** $L = \Phi_B/I$. An inductor opposes changes in current, producing an emf given by $\mathcal{E} = -L(dI/dt)$. Circuit quantities in a simple *RL* circuit change with **inductive time constant** L/R.

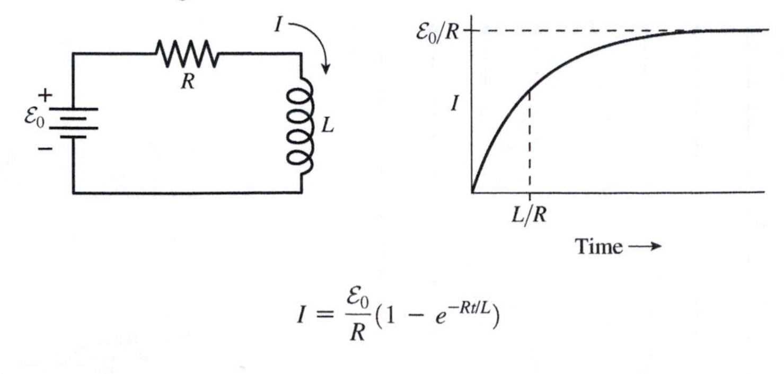

$$I = \frac{\mathcal{E}_0}{R}(1 - e^{-Rt/L})$$

Diamagnetism occurs when electromagnetic induction results in atoms acquiring net magnetic moments; the result is a repulsive interaction.

BIO *Biology and/or medicine-related problems* **DATA** *Data problems* **ENV** *Environmental problems* **CH** *Challenge problems* **COMP** *Computer problems*

For Thought and Discussion

1. In Fig. 27.35, a bar magnet moves toward a conducting ring. What's the direction of the induced current in the ring?

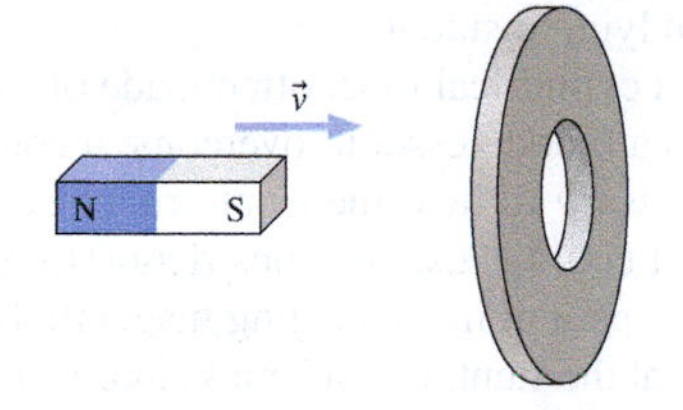

FIGURE 27.35 For Thought and Discussion 1

2. Figure 27.36 shows two concentric conducting loops, the outer connected to a battery and a switch. The switch is initially open. It's then closed, left closed for a while, and then reopened. Describe the currents in the inner loop during the entire procedure.

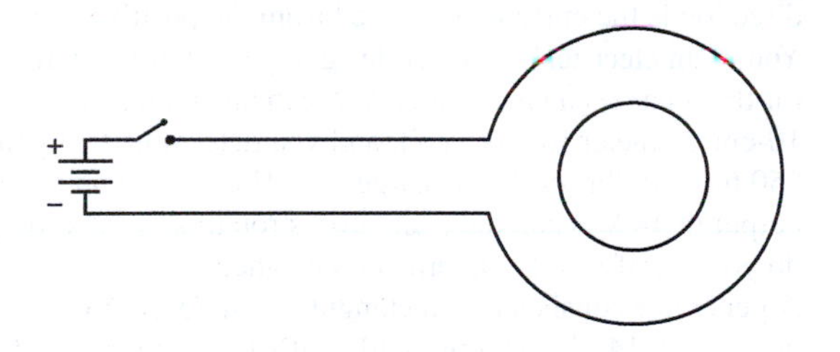

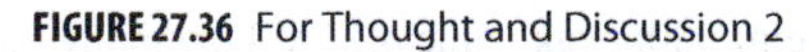

FIGURE 27.36 For Thought and Discussion 2

3. Fluctuations in Earth's magnetic field due to changing solar activity can wreak havoc with communications, even those using underground cables. How is this possible?
4. Chapter 26 stated that a static magnetic field cannot change the energy of a charged particle. Is this true of a changing magnetic field? Discuss.
5. Can an induced electric field exist in the absence of a conductor?
6. A car battery has a 12-V emf, yet energy from the battery provides the 30,000-V spark that ignites the gasoline. How is this possible?
7. You have a fixed length of wire to wind into an inductor. Will you get more inductance if you wind a short coil with large diameter, or a long coil with small diameter?
8. In a popular demonstration of induced emf, a lightbulb is connected across a large inductor in an *RL* circuit, as shown in Fig. 27.37. When the switch is opened, the bulb flashes brightly and may even burn out. Why?

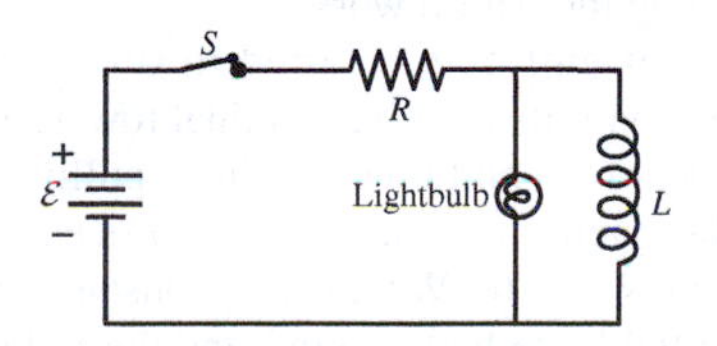

FIGURE 27.37 For Thought and Discussion 8

9. List some similarities and differences between inductors and capacitors.
10. A 1-H inductor carries 10 A, and a 10-H inductor carries 1 A. Which contains more stored energy?
11. It takes work to push two bar magnets together with like poles facing. Where does this energy go?
12. A small magnet is dropped into each of two hollow vertical tubes of equal length, one made of copper and one of aluminum. Does it take longer for the magnet to fall through the aluminum tube or the copper tube, or does it take the same amount of time for each? (*Hint*: Consult Table 24.1.)
13. Figures 27.1*b* and 27.2 actually describe the same situation, just from the viewpoints of two different inertial reference frames. In Fig. 27.2, in the reference frame of the magnet, you can think of the induced current as arising from the magnetic force on the electrons in the coil (motional emf). From the coil's reference frame (Fig. 27.1*b*), how would you describe the origin of the induced current? (This comparison played an important role in Einstein's thinking about relativity, and the phrase "the reciprocal electrodynamic action of a magnet and a conductor" appears in the second sentence of Einstein's 1905 paper introducing the special theory of relativity; more in Chapter 33.)

Exercises and Problems

Exercises

Sections 27.2 Faraday's Law and 27.3 Induction and Energy

14. Show that the volt is the SI unit for the rate of change of magnetic flux, making Faraday's law dimensionally correct. Your result also shows why the unit of flux itself can be expressed as V·s.
15. Find the magnetic flux through a 5.0-cm-diameter circular loop oriented with the loop normal at 36° to a uniform 75-mT magnetic field.
16. A circular wire loop 45 cm in diameter has resistance 120 Ω and lies in a horizontal plane. A uniform magnetic field points vertically downward, and in 25 ms it increases linearly from 5.0 mT to 55 mT. Find the magnetic flux through the loop at (a) the beginning and (b) the end of the 25-ms period. (c) What's the loop current during this time? (d) Which way does this current flow?
17. A conducting loop of area 240 cm^2 and resistance 12 Ω is perpendicular to a spatially uniform magnetic field and carries a 320-mA induced current. At what rate is the magnetic field changing?
18. The magnetic field inside a 23-cm-diameter solenoid is increasing at 2.4 T/s. How many turns should a coil wrapped around the outside of the solenoid have so that the emf induced in the coil is 15 V?

Section 27.4 Inductance

19. Find the self-inductance of a 1500-turn solenoid 55 cm long and 4.0 cm in diameter.
20. The current in an inductor is changing at 110 A/s and the inductor emf is 45 V. What's the self-inductance?
21. A 1.9-A current is flowing in a 22-H inductor. A switch opens, interrupting the current in 1.0 ms. Find the induced emf in the inductor.
22. Your little sister is building a radio from scratch. Plans call for a 450-μH inductor wound on a cardboard tube. She brings you the tube from a toilet-paper roll (12 cm long, 4.0 cm diameter), and asks how many turns she should wind on the full length of the tube. Your answer?
23. What inductance should you put in series with a 150-Ω resistor to give a time constant of 2.2 ms?
24. The current in a series *RL* circuit increases to 20% of its final value in 3.1 μs. If $L = 1.8$ mH, what's the resistance?

Section 27.5 Magnetic Energy

25. How much energy is stored in a 5.0-H inductor carrying 35 A?
26. What's the current in a 24-mH inductor storing 75 μJ of energy?
27. A 220-mH inductor carries 350 mA. How much energy must be supplied to the inductor in raising the current to 850 mA?
28. A 1250-turn solenoid 23.2 cm long and 1.58 cm in diameter carries 165 mA. How much magnetic energy does it contain?
29. Show that the quantity $B^2/2\mu_0$ has the units of energy density.
30. The world's strongest magnet that can produce a sustained field is a 45-T device at the National High Magnetic Field Laboratory in Florida. What's the corresponding magnetic-energy density?
31. Find the magnetic-field strength in a region where the magnetic-energy density is 7.8 J/cm^3.

Section 27.6 Induced Electric Fields

32. The induced electric field 12 cm from the axis of a 10-cm-radius solenoid is 45 V/m. Find the rate of change of the solenoid's magnetic field.
33. Find an expression for the electric-field strength *inside* the solenoid of Example 27.10, a distance r from the axis.

Problems

34. A conducting loop of area A and resistance R lies at right angles to a spatially uniform magnetic field. At time $t = 0$, the magnetic field and loop current are both zero. Subsequently, the current increases according to $I = bt^2$, where b is a constant with units A/s^2. Find an expression for the magnetic-field strength as a function of time.
35. A conducting loop with area 0.15 m^2 and resistance 6.0 Ω lies in the x–y plane. A spatially uniform magnetic field points in the z-direction. The field varies with time according to $B_z = at^2 - b$, where $a = 2.0\ \text{T/s}^2$ and $b = 8.0$ T. Find the loop current (a) at $t = 3.0$ s and (b) when $B_z = 0$.
36. A square wire loop of side l and resistance R is pulled with constant speed v from a region of no magnetic field until it's fully inside a region of constant, uniform magnetic field $\vec{B}$ perpendicular to the loop plane. The boundary of the field region is parallel to one side of the loop. Find an expression for the total work done by whatever is pulling the loop.
37. A 5-turn coil 1.0 cm in diameter is rotated at 10 rev/s about an axis perpendicular to a uniform magnetic field. A voltmeter connected to the coil through rotating contacts reads a peak value 360 μV. What's the magnetic-field strength?
38. A magnetic field is given by $\vec{B} = B_0(x/x_0)^2\hat{k}$, where B_0 and x_0 are constants. Find an expression for the magnetic flux through a square of side $2x_0$ that lies in the x–y plane with one corner at the origin and sides coinciding with the positive x- and y-axes.
39. A square wire loop 3.0 m on a side is perpendicular to a uniform 2.0-T magnetic field. A 6-V lightbulb is in series with the loop, as shown in Fig. 27.38. The magnetic field is reduced steadily to zero over time Δt. (a) Find Δt such that the bulb will shine at full brightness. (b) Which way will the loop current flow?

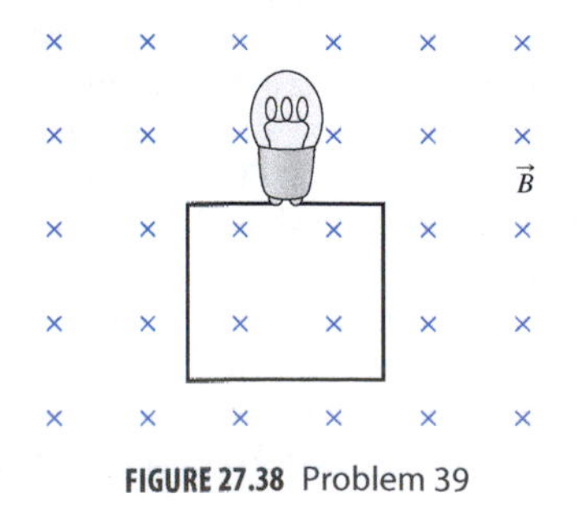

FIGURE 27.38 Problem 39

40. In Example 27.2 take $a = 1.0$ cm, $w = 3.5$ cm, and $l = 6.0$ cm. Suppose the rectangular loop is a conductor with resistance 50 mΩ, and the current I in the long wire is increasing at 25 A/s. Find the induced current in the loop. What's its direction?
41. A 2000-turn solenoid is 2.0 m long and 15 cm in diameter. The solenoid current is increasing at 1.0 kA/s. (a) Find the current in a 10-cm-diameter wire loop with resistance 5.0 Ω lying inside the solenoid and perpendicular to the solenoid axis. (b) Repeat for a loop whose diameter is 25 cm, so it now encircles the solenoid rather than lying inside it.
42. **BIO** A *stent* is a cylindrical tube, often made of metal mesh, that's inserted into a blood vessel to overcome a constriction. It's sometimes necessary to heat the stent after insertion to prevent cell growth that could cause the constriction to recur. One method is to place the patient in a changing magnetic field, so that induced currents heat the stent. Consider a stainless-steel stent 12 mm long by 4.5 mm in diameter, with total resistance 41 mΩ. Treating the stent as a wire loop in the optimum orientation, find the rate of change of magnetic field needed for a heating power of 250 mW.
43. A uniform magnetic field is given by $\vec{B} = bt\,\hat{k}$, where $b = 0.35$ T/s. Find the induced current in a conducting loop with area 240 cm^2 and resistance 0.20 Ω that lies in the x–y plane. In what direction is the current, as viewed from the positive z-axis?
44. You're an electrical engineer designing an alternator (the generator that charges a car's battery). Mechanical engineers specify a 10-cm-diameter rotating coil, and you determine that you can fit 250 turns in this coil. To charge a 12-V battery, you need a peak output of 14 V when the alternator is rotating at 1200 rpm. What do you specify for the alternator's magnetic field?
45. A generator consists of a rectangular coil 75 cm by 1.3 m, spinning in a 0.14-T magnetic field. If it's to produce a 60-Hz alternating emf with peak value 6.7 kV, how many turns must it have?
46. Figure 27.39 shows a pair of parallel conducting rails a distance l apart in a uniform magnetic field $\vec{B}$. A resistor R is connected across the rails, and a conducting bar of negligible resistance is being pulled along the rails with velocity $\vec{v}$ to the right. (a) What direction is the current in the resistor? (b) At what rate does the agent pulling the bar do work?

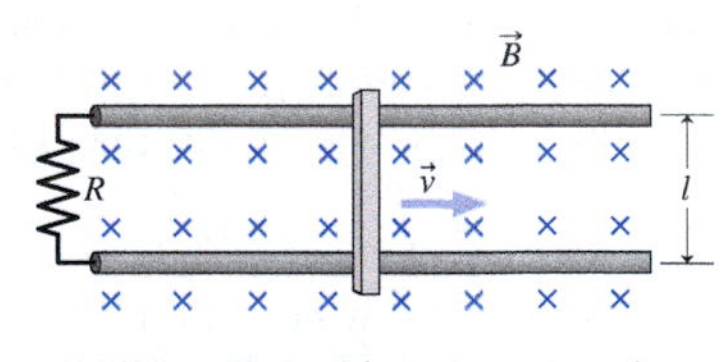

FIGURE 27.39 Problems 46–49 and 75

47. The resistor in Problem 46 is replaced by an ideal voltmeter. (a) To which rail should the positive meter terminal be connected if it's to indicate a positive voltage? (b) At what rate does the agent pulling the bar do work?
48. A battery of emf $\mathcal{E}$ is inserted in series with the resistor in Fig. 27.39, with its positive terminal toward the top rail. The bar is initially at rest, and now nothing's pulling it. (a) Describe the bar's subsequent motion. (b) The bar eventually reaches a constant speed. Why? (c) What is that constant speed, in terms of the magnetic field, the battery emf, and the rail spacing l? Does the resistance R affect the final speed? If not, what role does it play?
49. In Fig. 27.39, take $l = 10$ cm, $B = 0.50$ T, $R = 4.0$ Ω, and $v = 2.0$ m/s. Find (a) the current in the resistor, (b) the magnetic force on the bar, (c) the power dissipation in the resistor, and (d) the mechanical power supplied by the agent pulling the bar. Compare your answers to parts (c) and (d).
50. The magnetic field inside a solenoid of circular cross section is given by $\vec{B} = bt\,\hat{k}$, where $b = 2.1$ T/ms. At time $t = 0.40$ μs, a

proton is inside the solenoid at $x = 5.0$ cm, $y = z = 0$, and is moving with velocity $\vec{v} = 4.8\hat{\jmath}$ Mm/s. Find the electromagnetic force on the proton.

51. An electron is inside a solenoid, 28 cm from the axis. It experiences a 1.3-fN electric force. At what rate is the solenoid's magnetic field changing?
52. During lab, you're given a circular wire loop of resistance R and radius a with its plane perpendicular to a uniform magnetic field. You're supposed to increase the field strength from B_1 to B_2 and measure the total charge that moves around the loop. Your lab partner claims that the details of how you vary the field will make a difference in the total charge; your hunch is that it won't. By integrating the loop current over time, determine who's right.
53. A *flip coil* is used to measure magnetic fields. It's a small coil placed with its plane perpendicular to a magnetic field, and then flipped through 180°. The coil is connected to an instrument that measures the total charge Q that flows during this process. If the coil has N turns, area A, and resistance R, show that the field strength is $B = QR/2NA$.
54. The current in a series RL circuit rises to half its final value in 7.6 s. What's the time constant?
55. In a series RL circuit like Fig. 27.23*a*, $\mathcal{E}_0 = 45$ V, $R = 3.3\ \Omega$, and $L = 2.1$ H. If the current is 9.5 A, how long has the switch been closed?
56. In Fig. 27.23*a*, take $R = 2.5\ \text{k}\Omega$ and $\mathcal{E}_0 = 50$ V. When the switch is closed, the current through the inductor rises to 10 mA in 30 μs. Find (a) the inductance and (b) the current in the circuit after many time constants.
57. How long does it take to dissipate 90% of the magnetic energy in Example 27.9?
58. A series RL circuit like Fig. 27.23*a* has $\mathcal{E}_0 = 60$ V, $R = 22\ \Omega$, and $L = 1.5$ H. Find the rate of change of the current (a) immediately after the switch is closed and (b) 100 ms later.
59. You're a safety engineer reviewing plans for a university's new high-rise dorm. The elevator motors draw 20 A and behave electrically like 2.5-H inductors. You're concerned about dangerous voltages developing across the switch when a motor is turned off, and you recommend that a resistor be wired in parallel with each motor. (a) What should be the resistance in order to limit the emf to 100 V? (b) How much energy will the resistor dissipate?
60. In Fig. 27.26, take $\mathcal{E}_0 = 12$ V, $R = 2.7\ \Omega$, and $L = 20$ H. Initially the switch is in position B and there's no current anywhere. At $t = 0$ the switch is thrown to position A, and at $t = 10$ s it's returned to B. Find the inductor current at (a) $t = 5.0$ s and (b) $t = 15$ s.
61. In Fig. 27.40, take $\mathcal{E}_0 = 12$ V, $R_1 = 4.0\ \Omega$, $R_2 = 8.0\ \Omega$, and $R_3 = 2.0\ \Omega$. Find current I_2 (a) immediately after the switch is first closed and (b) a long time later. (c) After a long time, the switch is reopened. Now what's I_2?

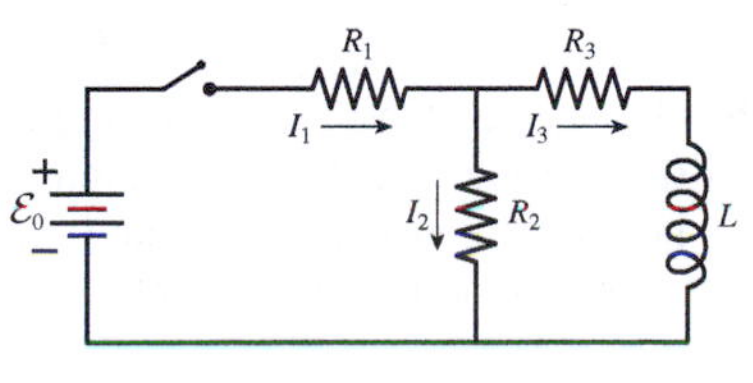

FIGURE 27.40 Problem 61

62. A battery, switch, resistor, and inductor are connected in series. When the switch is closed, the current rises to half its steady-state value in 1.0 ms. How long does it take for the magnetic energy in the inductor to rise to half its steady-state value?
63. When a nonideal 1.0-H inductor is short-circuited, its magnetic energy drops to one-fourth of its original value in 3.6 s. What is its resistance?
64. **BIO** Your hospital is installing a new MRI scanner using a 3.5-H superconducting solenoid carrying 1.8 kA. Copper is embedded in the coils to carry the current in the event of a quench (see Example 27.9). As safety officer, you're to specify (a) the maximum resistance that will limit power dissipation to 100 kW immediately after a loss of superconductivity and (b) the time it will take the power to drop to 50 kW. What specs do you give?
65. A neutron star's magnetic field is about 10^8 T. Consult Appendix C to compare the energy density in this field with that of (a) gasoline and (b) pure uranium-235 (mass density $19\times10^3\ \text{kg/m}^3$).
66. A single-turn loop of radius R carries current I. How does the magnetic-energy density at the loop center compare with that of a long solenoid of the same radius, carrying the same current, and consisting of n turns per unit length?
67. **CH** A wire of radius R carries current I distributed uniformly over its cross section. Find an expression for the total magnetic energy per unit length *within* the wire.
68. **CH** (a) Use Equation 27.8 to write an expression for the resistor's power dissipation as a function of time, and (b) integrate from $t = 0$ to $t = \infty$ to show that the total energy dissipated is equal to the energy initially stored in the inductor.
69. An electric field and a magnetic field have the same energy density. Find an expression for the ratio E/B and evaluate this ratio numerically. What are its units? Is your answer close to any of the fundamental constants listed inside the front cover?
70. A rectangular conducting loop of resistance R, mass m, and width w falls into a uniform magnetic field as shown in Fig. 27.41. (a) Explain why the loop eventually reaches a terminal speed. (b) Find an expression for the terminal speed.

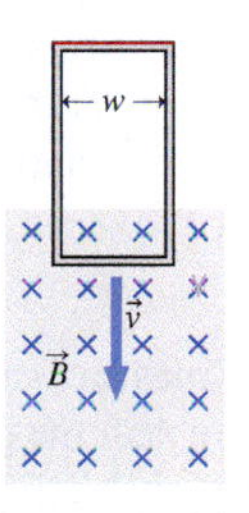

FIGURE 27.41 Problem 70

71. **CH** A conducting disk with radius a, thickness h, and resistivity ρ is inside a solenoid of circular cross section, its axis coinciding with the solenoid axis. The magnetic field in the solenoid is given by $B = bt$, where b is a constant. Find expressions for (a) the current density in the disk as a function of the distance r from the disk center and (b) the power dissipation in the entire disk. (*Hint:* Consider the disk as consisting of infinitesimal conducting loops.)
72. **CH** A long, straight coaxial cable consists of two thin, tubular conductors, the inner of radius a and the outer of radius b. Current I flows out along one conductor and back along the other. Show that the self-inductance per unit length of the cable is $\frac{\mu_0}{2\pi}\ln(b/a)$.
73. **DATA** The table below shows the current in a circuit like that of Fig. 27.26, where a current has been established with the switch in position A, and then it's thrown to position B at time $t = 0$. The resistance is $180\ \Omega$. Determine an appropriate function of current that, when plotted against time, should produce a straight line. Make your plot, determine a best-fit line, and use its slope to find the inductance in the circuit.

Time (ms)	0	20.0	40.0	60.0	80.0	100.0
Current (mA)	66.5	23.0	9.15	3.56	1.50	0.450

74. A circular wire loop of radius a and resistance R is pulled with constant speed v into a uniform magnetic field B. The loop is perpendicular to the field, and it begins entering the field at time $t = 0$. Find an expression for the current in the loop from $t = 0$ until the loop is fully immersed in the field.
CH

75. The bar in Problem 46 has mass m and is initially at rest. A constant force $\vec{F}$ to the right is applied to the bar. Formulate Newton's second law for the bar, and find its velocity as a function of time.
CH

76. Use the node and loop laws to determine the current in R_2 as a function of time after the switch is closed in Conceptual Example 27.1.
CH

77. (a) Find the magnetic-energy density as a function of radial distance for the coaxial cable of Problem 72, and integrate over the volume between the conductors to show that the total energy per unit length of the cable is given by $(\mu_0 I^2/4\pi)\ln(b/a)$. (b) Use the expression $U = \frac{1}{2}LI^2$ to find the inductance per unit length, and show that your result agrees with that of Problem 72.
CH

78. You and your roommate are headed to Cancún for spring break. Your roommate, who has had only high school physics, has read that an emf can be induced in the wings of an airplane and wonders whether this would give enough voltage to power a portable music player. What's your answer? (Assume that the wingspan of your 747 is 60 m, the plane is flying at 600 mph, and Earth's magnetic field is 0.3 G.)

79. One way to measure blood flow when blood vessels are exposed during surgery is to use an *electromagnetic flowmeter*. This device surrounds the blood vessel with an electromagnet, creating a magnetic field perpendicular to the blood flow. Since blood is a modest conductor, a motional emf develops across the blood vessel. Given vessel diameter d, magnetic field B, and voltage V measured across the vessel, show that the volume blood flow is given by $\pi d^2 V/4Bd$.
BIO

Passage Problems

Clever farmers with power lines crossing their land have been known to steal power by stringing wire near the power line and making use of the induced current. At least one such crime went to court and resulted in a conviction—despite the defense's claim that the defendant didn't touch the lines. Figure 27.42 shows a possible crime scene, with a rectangular wire loop mounted in a vertical plane beneath a power line. The power line carries a current of 10^4 A, alternating sinusoidally at 60 Hz.

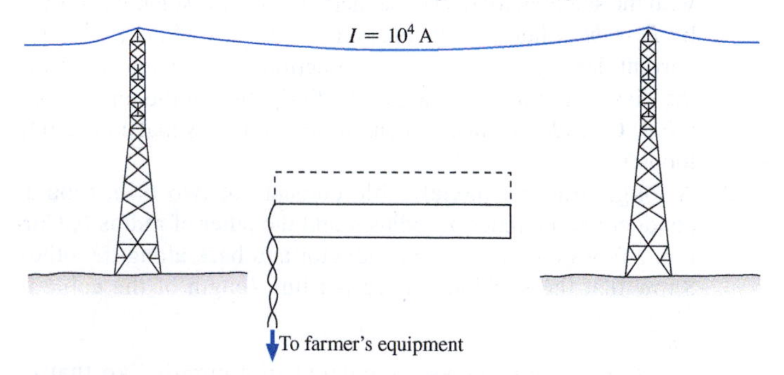

FIGURE 27.42 Crime scene for Passage Problems 80–83

80. If the loop were mounted in a horizontal rather than vertical plane at the same distance from the power line, the induced emf would
    a. increase slightly.
    b. decrease slightly.
    c. remain the same.
    d. become essentially zero.

81. If the loop's vertical dimension were doubled by extending it toward the power line (dashed line in Fig. 27.42), the induced emf would
    a. double.
    b. quadruple.
    c. more than double but not quadruple.
    d. increase but not quite double.

82. Suppose the same crime were committed in Europe, where the standard frequency is 50 Hz. Assuming everything else about the situation were the same, the induced emf would
    a. be greater.
    b. be less.
    c. be unchanged.
    d. depend on the nature of the energy source.

83. When this crime occurs,
    a. more fuel must be consumed at the power plant supplying the line.
    b. the power company does not suffer any economic damage.
    c. the power company can't determine that it's being robbed without an on-site inspection.
    d. there's no power left for customers further down the line.

Answers to Chapter Questions

Answer to Chapter Opening Question

The energy was stored in the Sun's magnetic field. The power grid was disrupted by surges of current induced by changes in Earth's magnetic field in response to the solar outburst. Normally, electricity is generated by moving electrical conductors in magnetic fields. All three of these answers involve electromagnetic induction.

Answers to GOT IT? Questions

27.1 (1) Opposite the direction shown in Fig. 27.10, but (2) you'll still have to do work.
27.2 (b) counterclockwise
27.3 (b) It gets harder to turn. Constant rate implies a fixed peak emf, so lowering the resistance increases the current and therefore the power.
27.4 (b)The penny hits with a slower speed because eddy currents dissipate some of its kinetic energy.
27.5 Changing current in the long wire produces an increasing magnetic field that points into the page at the loop. (1) The loop current opposes the increase in this field by producing a field in its interior that's out of the page. Therefore, the loop current is counterclockwise. (2) The loop current opposes the decrease in the into-the-page field, so it's clockwise.
27.6 (a) The current increases because the inductor emf is in such a direction as to oppose the current supplied by the external circuit.
27.7 (c)
27.8 Left-hand meter reads $2IR$, right-hand meter reads IR—even though they're electrically connected to the same points. There's no contradiction because the field isn't conservative, and electric potential therefore can't be defined unambiguously.

Alternating-Current Circuits

What You Know

- You understand the terms *amplitude*, *frequency*, and *phase* as they apply to oscillatory motion and waves.
- You've learned about three separate electrical components: resistors, capacitors, and inductors.
- You can analyze circuits containing batteries, resistors, and capacitors or batteries, resistors, and inductors.

What You're Learning

- You'll learn to characterize alternating voltages and currents (AC) in terms of amplitude, frequency, and phase.
- You'll learn to relate voltage and current in capacitors and inductors, including both amplitude and phase relations.
- You'll see how circuits containing both capacitors and inductors undergo electrical oscillations analogous to the simple harmonic motion of Chapter 13.
- You'll learn how transformers and power supplies make direct-current power from alternating-current sources.

How You'll Use It

- The electric power grid supplies you with AC power, so this chapter provides a basic understanding of everyday household electricity.
- Circuits with capacitors and inductors provide insight into a fundamental complementarity between electricity and magnetism, which will lead to your understanding of electromagnetic waves in Chapter 29.

Why do most power lines carry alternating current?

So far we've considered electric circuits energized by steady sources like batteries. But many circuits—from household power to audio and video signals to the "clock" that orchestrates events inside your computer—involve time-varying electrical quantities. Here we consider such **alternating-current** (AC) circuits.

28.1 Alternating Current

We saw in Chapter 27 how rotational motion in electric generators naturally leads to voltage and current that vary sinusoidally with time. Audio, video, and computer signals have more complicated time dependence, but, as we showed in Fig. 14.17, those signals can be analyzed as sums of sinusoidal terms. Studying circuits with sinusoidally varying electrical quantities therefore provides insights into all AC circuits.

A sinusoidal AC voltage or current is characterized by its amplitude, frequency, and phase constant—the same quantities we developed in Chapter 13 to describe simple harmonic motion. Amplitude is specified by the peak value (V_p, I_p) or the **root-mean-square (rms)** value (V_{rms}, I_{rms}). The rms is an average obtained by squaring the signal, taking the time average, and then taking the square root. For a sine wave, rms and peak values are related by

$$V_{rms} = \frac{V_p}{\sqrt{2}} \quad \text{and} \quad I_{rms} = \frac{I_p}{\sqrt{2}} \tag{28.1}$$

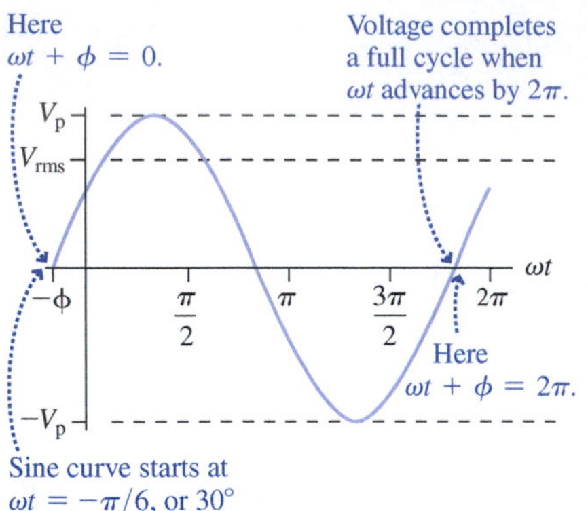

FIGURE 28.1 A sinusoidally varying AC voltage, showing peak and rms amplitudes and phase ϕ.

The 120 V of household wiring in North America, for example, is the rms value (see Fig. 28.1).

In practical situations we usually describe frequency f in cycles per second, or hertz (Hz). In mathematical analysis it's more convenient to use the angular frequency ω in radians per second or, equivalently, inverse seconds (s^{-1}). The relation between the two,

$$\omega = 2\pi f \tag{28.2}$$

is the same as for rotational and simple harmonic motion, and for the same reason: A full cycle contains 2π radians.

The phase constant ϕ of an AC signal tells when the sine curve crosses zero with positive slope (Fig. 28.1). A full mathematical description of an AC voltage or current then includes its amplitude (V_p, I_p), frequency (ω), and phase constant (ϕ):

$$V = V_p \sin(\omega t + \phi_V) \quad \text{and} \quad I = I_p \sin(\omega t + \phi_I) \tag{28.3}$$

Here we've labeled the phase constants with subscripts V and I to indicate that voltage and current—even in the same circuit element—need not have the same phase.

EXAMPLE 28.1 AC: Characterizing Household Voltage

Standard household wiring in North America supplies 120 V rms at 60 Hz. Express this mathematically in the form of Equation 28.3, assuming the voltage is rising through zero at time $t = 0$.

INTERPRET We're given an AC voltage in "practical" units, and we're asked to express it in the more mathematical form of Equation 28.3. We identify 120 V as the amplitude V_{rms}, 60 Hz as the frequency f, and the information about timing as describing the phase.

DEVELOP Equation 28.3, $V = V_p \sin(\omega t + \phi_V)$, contains the peak amplitude V_p and angular frequency ω. Equations 28.1, $V_{rms} = V_p/\sqrt{2}$, and 28.2, $\omega = 2\pi f$, determine these quantities from the values we're given.

EVALUATE Equation 28.1 gives $V_p = \sqrt{2}V_{rms} = (\sqrt{2})(120\text{ V}) = 170\text{ V}$, and Equation 28.2 gives $\omega = 2\pi f = (2\pi)(60\text{ Hz}) = 377\text{ s}^{-1}$. We don't have an equation for phase, but the fact that the sine curve rises through zero at $t = 0$ tells us that $\phi = 0$. So Equation 28.3's description of this AC voltage becomes $V = 170 \sin(377t)$ V.

ASSESS Make sense? Both the peak voltage and the angular frequency are numerically greater than their more familiar counterparts. That's because the rms voltage is an average, lower than the peak, and because the angular frequency measures radians per second rather than full cycles. Incidentally, wires entering your house actually carry 240 V rms, which is split into separate 120-V circuits except for major appliances like stoves, dryers, and water heaters; these operate at the full 240 V rms. In Europe and India, standard household voltage is 230 V rms at 50 Hz, while China, South Korea, and much of the rest of the world use 220-V, 50-Hz power. ■

GOT IT? 28.1 What are the peak voltage and angular frequency of the 220-V, 50-Hz AC power used in China and Korea? (a) 170 V, 20 ms; (b) 350 V, 377 s^{-1}; (c) 311 V, 314 s^{-1}; (d) 120 V, 50 ms

28.2 Circuit Elements in AC Circuits

MP

PhET: Circuit Construction Kit (AC + DC)

Here we examine separately the AC behavior of resistors, capacitors, and inductors so we can subsequently understand what happens when we combine these elements in AC circuits.

Resistors

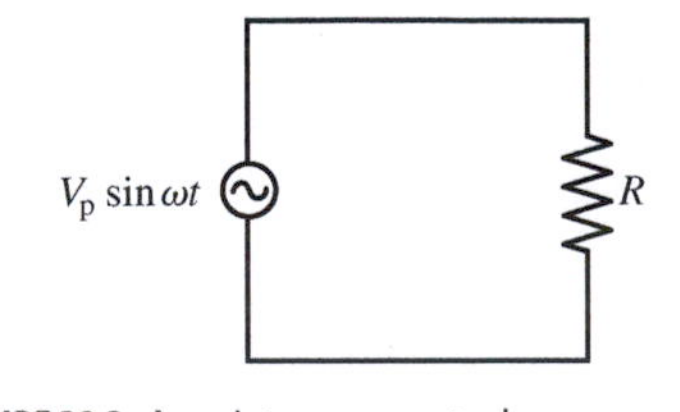

FIGURE 28.2 A resistor connected across an AC generator (symbol ⊙).

An ideal resistor is a device whose current and voltage are proportional: $I = V/R$. Figure 28.2 shows a resistor R connected across an AC generator, making the voltage across the resistor equal to the generator voltage. The generator voltage is described by Equation 28.3, where we take $\phi_V = 0$. Then the current is

$$I = \frac{V}{R} = \frac{V_p \sin \omega t}{R} = \frac{V_p}{R} \sin \omega t$$

The current has the same frequency as the voltage, and, since its phase constant is also zero, voltage and current are *in phase*—they peak at the same time. The peak current is the peak voltage divided by the resistance: $I_p = V_p/R$. Both voltage and current are sinusoidal, so their rms values are in the same ratio as their peak values; thus $I_{rms} = V_{rms}/R$.

Capacitors

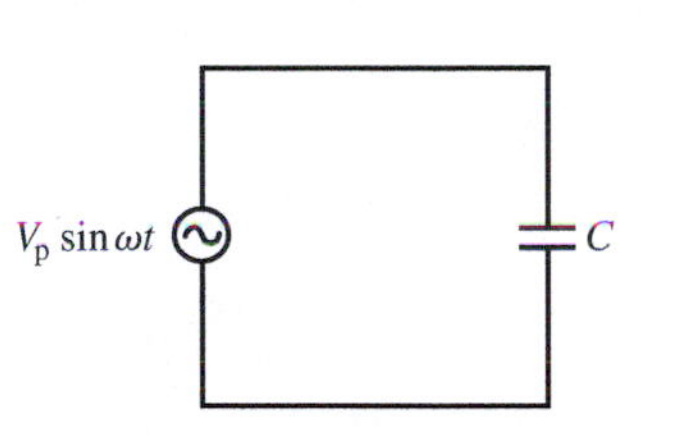

FIGURE 28.3 A capacitor connected across an AC generator.

Figure 28.3 shows a capacitor connected across an AC generator. In Chapter 23, we saw that voltage and charge are directly proportional in a capacitor: $q = CV$. Differentiating this relation gives

$$\frac{dq}{dt} = C\frac{dV}{dt}$$

But dq/dt is the current flowing to the capacitor plates (which we'll call the "capacitor current" even though charge doesn't actually flow through the space between the plates). So we have $I = C(dV/dt)$. The generator voltage $V_p \sin \omega t$ appears directly across the capacitor, so

$$\begin{aligned} I &= C\frac{d}{dt}(V_p \sin \omega t) \\ &= \omega C V_p \cos \omega t = \omega C V_p \sin\left(\omega t + \frac{\pi}{2}\right) \end{aligned} \quad (28.4)$$

Because the cosine curve is a sine curve shifted to the left by $\pi/2$ or 90°, Equation 28.4 tells us that **in a capacitor, current leads voltage by 90°** (Fig. 28.4).

The term $\omega C V_p$ multiplying the cosine in Equation 28.4 is the peak current, so $I_p = \omega C V_p$ or, in a form resembling Ohm's law,

$$I_p = \frac{V_p}{1/\omega C} = \frac{V_p}{X_C} \quad (28.5)$$

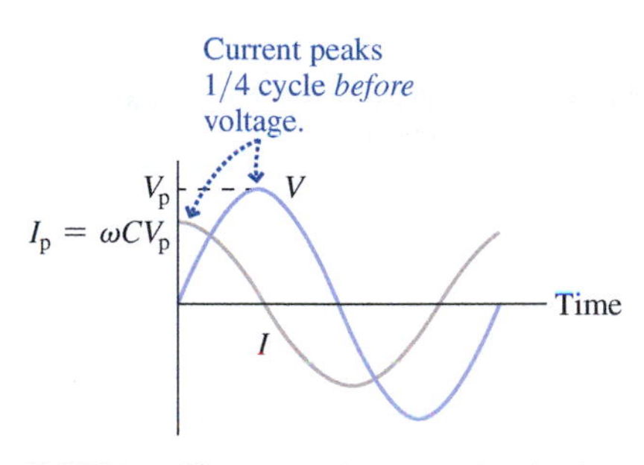

FIGURE 28.4 The current in a capacitor leads the voltage by one-fourth of a cycle, $\pi/2$ radians or 90°.

where we've defined $X_C = 1/\omega C$.

Equation 28.5 shows that the capacitor acts somewhat like a resistance $X_C = 1/\omega C$. But not quite! This "resistance" gives the relation between peak voltage and peak current, but it doesn't tell the whole story. The capacitor also introduces a phase difference between voltage and current. This phase difference reflects a fundamental physical difference between resistors and capacitors. A resistor dissipates electric energy as heat. A capacitor stores and releases electric energy. Over a complete cycle, the agent turning the generator in Fig. 28.3 does no net work, while the agent turning the generator with the resistive load of Fig. 28.2 continuously does work that gets dissipated as heat in the resistor. We give the quantity X_C in Equation 28.5 the name **capacitive reactance**. Like resistance, reactance is measured in ohms (Ω).

Does it make sense that X_C depends on frequency? Yes. As frequency goes to zero, X_C goes to infinity. At zero frequency nothing is changing; there's no charge moving on or off the plates, and the capacitor might as well be an open circuit. As frequency increases, larger currents flow to move charge on and off the plates in ever-shorter times, so the capacitor looks increasingly like a short circuit. To summarize, a capacitor at low frequencies acts like an open circuit, while at high frequencies it acts like a short circuit.

Why does the capacitor current *lead* the voltage? Because the capacitor voltage is proportional to its charge, and it takes current to move charge onto the capacitor plates. Therefore, current flows *before* the voltage changes significantly. We found this same relation in the *RC* circuit of Section 25.5, where closing the switch in the circuit of Fig. 25.18 resulted in an immediate current followed by a slow rise in the capacitor voltage.

Inductors

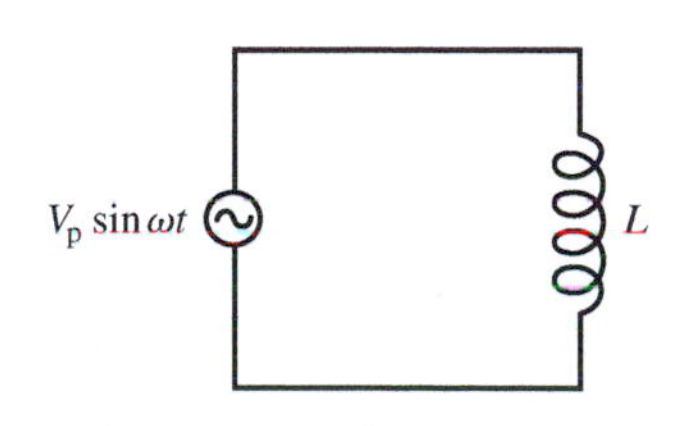

FIGURE 28.5 An inductor connected across an AC generator.

Figure 28.5 shows an inductor connected across an AC generator. The loop law for this circuit is $V_p \sin \omega t + \mathcal{E}_L = 0$. From Chapter 27 we know that the inductor emf is $\mathcal{E}_L = -L(dI/dt)$, so the loop law becomes

$$V_p \sin \omega t = L\frac{dI}{dt}$$

To obtain a relation involving the current I rather than its derivative, we integrate:

$$\int V_p \sin \omega t \, dt = \int L \frac{dI}{dt} dt = \int L\, dI = L \int dI = LI$$

The integral of sine is the negative cosine, so

$$-\frac{V_p}{\omega} \cos \omega t = LI$$

Here we've set the integration constants to zero because nonzero values would represent a DC emf and current that aren't in this circuit. Solving for I gives

$$I = -\frac{V_p}{\omega L} \cos \omega t = \frac{V_p}{\omega L} \sin\left(\omega t - \frac{\pi}{2}\right) \quad (28.6)$$

where the last step follows because $\sin(\alpha - \pi/2) = -\cos \alpha$ for any α.

Equation 28.6 shows that current in the inductor lags the voltage by $\pi/2$ or 90°. Equivalently, **the voltage across an inductor leads the inductor current by 90°** (Fig. 28.6). Equation 28.6 also shows that the peak current is

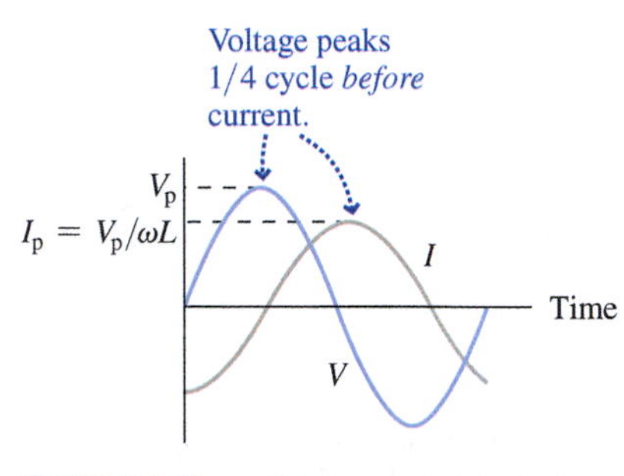

FIGURE 28.6 The voltage across an inductor leads the current by $\pi/2$ or 90°.

$$I_p = \frac{V_p}{\omega L} = \frac{V_p}{X_L} \quad (28.7)$$

Again, this equation resembles Ohm's law, with **inductive reactance** $X_L = \omega L$. As with the capacitor, no power is dissipated; instead, energy is alternately stored and released as the inductor's magnetic field builds and decays.

Does it make sense that inductive reactance increases with ω and L? Through its induced back emf, an inductor opposes changes in current. The greater the inductance, the greater the opposition. And the more rapidly the current is changing, the more vigorously the inductor opposes the change, so inductive reactance increases at high frequencies. At very high frequencies, an inductor looks like an open circuit. But at very low frequencies, it looks more and more like a short circuit, until with direct current (zero frequency), an inductor exhibits zero reactance because current isn't changing.

Why does the inductor voltage *lead* the current? Because a changing current in an inductor induces an emf. *Before* the current can build up significantly, there must first, therefore, be a voltage across the inductor.

Table 28.1 summarizes amplitude and phase relations in resistors, capacitors, and inductors.

Table 28.1 Amplitude and Phase Relations in Circuit Elements

Circuit Element	Peak Current versus Voltage	Phase Relation
Resistor	$I_p = \frac{V_p}{R}$	V and I in phase
Capacitor	$I_p = \frac{V_p}{X_C} = \frac{V_p}{1/\omega C}$	I leads V by 90°
Inductor	$I_p = \frac{V_p}{X_L} = \frac{V_p}{\omega L}$	V leads I by 90°

GOT IT? 28.2 A capacitor and inductor are connected across separate but identical electric generators, and the same current flows in each. If the frequency of the generators is doubled, how will the currents in the two components compare? (a) they will continue to carry equal currents; (b) the capacitor will carry twice as much current as the inductor; (c) the inductor will carry twice as much current as the capacitor; (d) the capacitor will carry four times as much current as the inductor; (e) the inductor will carry four times as much current as the capacitor

EXAMPLE 28.2 Inductors and Capacitors: Equal Currents?

A capacitor is connected across a 60-Hz, 120-V rms power line, and an rms current of 200 mA flows. (a) Find the capacitance. (b) What inductance, connected across the same power line, would result in the same current? (c) How would the phases of the inductor and capacitor currents compare?

INTERPRET We're being asked about the relation between AC voltage and current in capacitors and inductors. The idea here is that the voltage–current relation depends not only on the component values but also on frequency, and it involves phase as well as amplitude.

DEVELOP Equations 28.5, $I_{Cp} = V_{Cp}\omega C$, and 28.7, $I_{Lp} = V_{Lp}/\omega L$, relate the peak current and peak voltage in the two devices. Since rms and peak values are proportional, similar equations also relate rms current and voltage. The equations and associated phase relations also appear in Table 28.1.

EVALUATE (a) For the capacitor, we know the voltage and current. Equation 28.5 then gives $C = I_{Crms}/\omega V_{Crms} = 4.42\,\mu\text{F}$, where we used $I_{Crms} = 0.20$ A and $\omega = 2\pi f = 377\text{ s}^{-1}$ as we found in Example 28.1 for 60-Hz AC power. (b) For an inductor to pass the same current, it must have the same reactance; comparing Equations 28.5 and 28.7 shows that $\omega L = 1/\omega C$, or

$$L = \frac{1}{\omega^2 C} = \frac{1}{(377\text{ s}^{-1})^2(4.42\,\mu\text{F})} = 1.59\text{ H}$$

(c) Table 28.1 shows that the capacitor current *leads* the voltage by 90°, while the inductor current *lags* by 90°; therefore, the capacitor and inductor currents must be out of phase by 180° or π radians.

ASSESS Our expression for L shows that a larger capacitor would require a smaller inductor for the same current. That's because a larger capacitor has *lower* reactance and so passes more current at a given frequency. But an inductor is the opposite: A larger inductor has *higher* reactance. So at a fixed frequency, the inductance required for a given current scales inversely with the capacitance required for the same current. ■

Phasor Diagrams

Phasor diagrams summarize phase and amplitude relations in AC circuits. A **phasor** is an arrow whose length represents the amplitude of an AC voltage or current, rotating counterclockwise with the angular frequency ω of the AC quantity. The phasor's component on either axis represents the sinusoidally varying AC quantity. We'll use the vertical axis; others, especially electrical engineers, may use the horizontal.

Figure 28.7*a* shows phasors for current and voltage in a resistor. Since current and voltage are in phase in a resistor, the two phasors point in the same direction. For capacitors and inductors, current and voltage phasors are at right angles, indicating 90° phase differences (Fig. 28.7*b, c*). The phasor magnitudes are related by $V_p = I_pX$, with X being the appropriate reactance. As they rotate, the phasors' vertical components trace out current and voltage graphs like those of Figs. 28.4 and 28.6. You should convince yourself that the relations of Table 28.1 are correctly described by the phasor diagrams of Fig. 28.7.

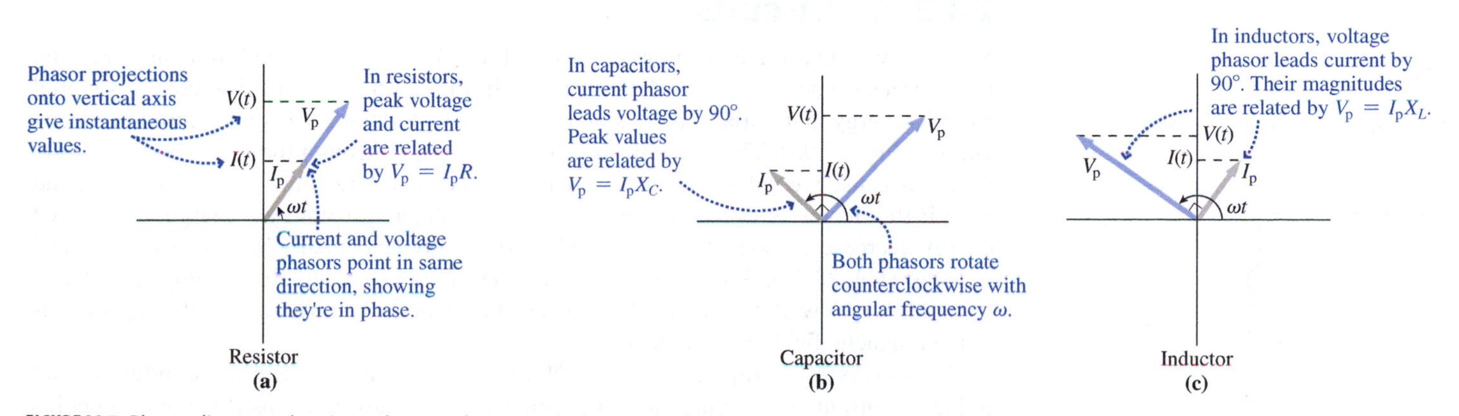

FIGURE 28.7 Phasor diagrams showing voltage and current in (a) a resistor, (b) a capacitor, and (c) an inductor.

Capacitors and Inductors: A Comparison

Capacitors and inductors are complementary. A capacitor opposes instantaneous changes in voltage; an inductor opposes instantaneous changes in current. In an *RC* circuit, voltage builds up across the capacitor. In an *RL* circuit, current builds up in the inductor. Similar curves describe capacitor voltage and inductor current over time. A capacitor stores electric energy $\frac{1}{2}CV^2$. An inductor stores magnetic energy $\frac{1}{2}LI^2$. A capacitor acts like an open circuit at low frequencies; an inductor like a short circuit at low frequencies. Each exhibits the opposite behavior at high frequencies. These comparisons reflect a deeper complementarity between electric and magnetic fields. Any verbal description of a capacitor applies to an inductor if we replace the words "capacitor" with "inductor," "electric" with "magnetic," and "voltage" with "current." Table 28.2 summarizes the complementary aspects of capacitors and inductors.

Table 28.2 Capacitors and Inductors

	Capacitor	Inductor
Defining relation	$C = \dfrac{q}{V}$	$L = \dfrac{\Phi_B}{I}$
Defining relation; differential form	$I = C\dfrac{dV}{dt}$	$\mathcal{E} = -L\dfrac{dI}{dt}$
Opposes changes in	Voltage	Current
Energy storage	In electric field $U = \frac{1}{2}CV^2$	In magnetic field $U = \frac{1}{2}LI^2$
Behavior in low-frequency limit	Open circuit	Short circuit
Behavior in high-frequency limit	Short circuit	Open circuit
Reactance	$X_C = 1/\omega C$	$X_L = \omega L$
Phase	Current leads voltage by 90°	Voltage leads current by 90°

APPLICATION Loudspeaker Systems

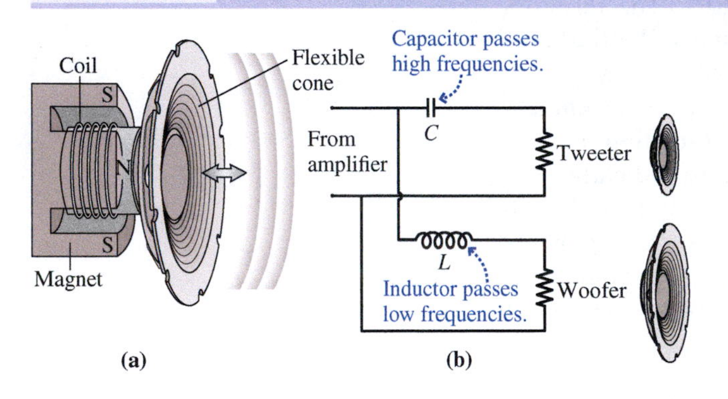

Loudspeakers convert electrical energy to sound, using the magnetic force on a coil that fits loosely around a permanent magnet. Part (a) of the figure shows that the coil is attached to a flexible cone. Cone and coil move back and forth as AC current corresponding to the audio signal flows in the coil, resulting in a time-varying magnetic force. The moving cone disturbs the air, producing sound waves.

Most loudspeaker systems include at least two separate units. A small *tweeter* produces high-frequency sound, while a larger, more massive *woofer* handles the low frequencies. A *crossover network* uses inductors and capacitors to "steer" the high- and low-frequency signals to the appropriate speakers. As the circuit diagram in part (b) shows, an inductor in series with the woofer blocks high frequencies but lets low frequencies pass unimpeded; a capacitor in series with the tweeter does the opposite. This circuit is an example of a *filter*, used in electronic systems to pass preferentially a range of frequencies.

28.3 *LC* Circuits

PhET: Circuit Construction Kit (AC + DC)

Suppose we charge a capacitor to some voltage V_p and corresponding charge q_p, and then connect it across an inductor, as shown in Fig. 28.8. The capacitor contains stored electric energy, but initially there's no current in the inductor and so no stored magnetic energy (Fig. 28.9*a*). The capacitor begins to discharge through the inductor, but slowly at first because the inductor opposes changes in current. Gradually the current rises, and with it the magnetic energy in the inductor. The capacitor voltage, charge, and stored energy decrease. At some time the initial energy is divided equally between capacitor and inductor (Fig. 28.9*b*). But the capacitor keeps discharging, eventually reaching zero charge (Fig. 28.9*c*). Now all the energy that was originally in the electric field of the capacitor is in the magnetic field of the inductor.

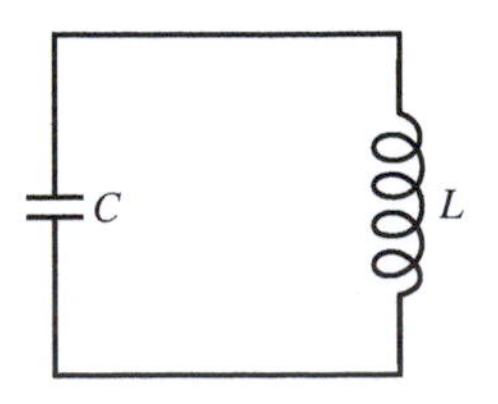

FIGURE 28.8 An *LC* circuit.

Does everything stop at this point? No, because there's current in the inductor, and inductor current can't change instantaneously. So the current keeps flowing and begins

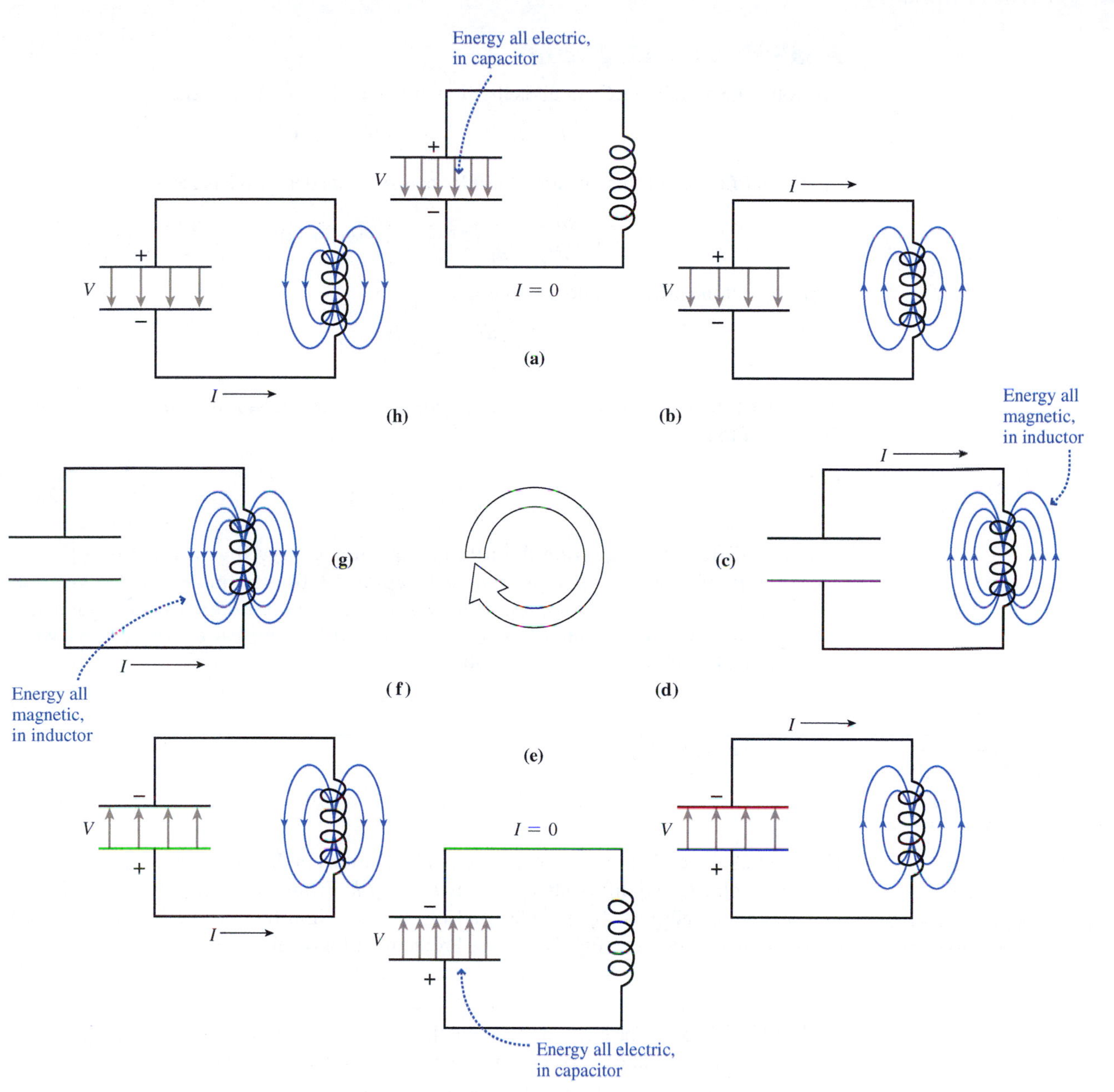

FIGURE 28.9 *LC* oscillations transfer energy between electric and magnetic fields.

piling positive charge on the bottom plate of the capacitor (Fig. 28.9*d*). Stored electric energy increases, and current and magnetic energy both decrease. Eventually the capacitor is fully charged but opposite its initial state (Fig. 28.9*e*). Again all the energy is in the capacitor. Now the capacitor begins to discharge, and the process repeats, with a counterclockwise current (Fig. 28.9*f*). All the energy is transferred to the inductor (Fig. 28.9*g*), and then back to the capacitor (Fig. 28.9*a* again). The circuit is now back to its initial state. Provided there's no energy loss, the oscillation repeats indefinitely.

This *LC* oscillation should remind you of the mass–spring system of Chapter 13. There, energy went back and forth between kinetic energy of the mass and potential energy of the spring. Here, energy goes between magnetic energy of the inductor and electric energy of the capacitor. The mass–spring system oscillates with frequency determined by the mass m and spring constant k. The *LC* circuit oscillates with frequency determined by the inductance L and capacitance C, as we'll show next. Figure 28.10 illustrates this analogy between the mass–spring system and the *LC* circuit.

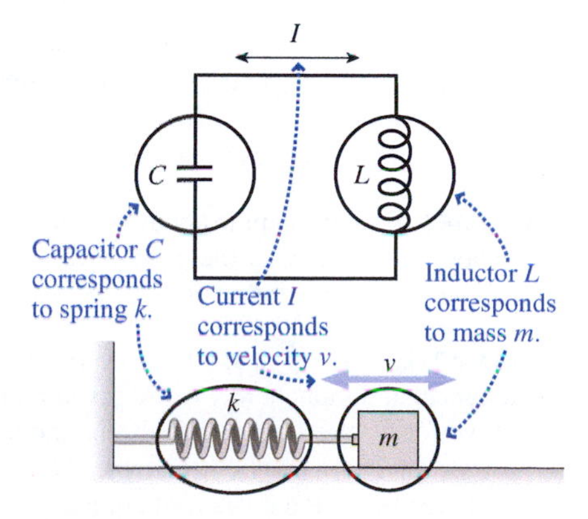

FIGURE 28.10 An *LC* circuit is the electrical analog of a mass–spring system.

Analyzing the *LC* Circuit

The total energy in the *LC* circuit is the sum of magnetic and electric energy:

$$U = U_B + U_E = \tfrac{1}{2}LI^2 + \tfrac{1}{2}CV^2$$

In an ideal *LC* circuit this quantity doesn't change, so its derivative is zero:

$$\frac{dU}{dt} = \frac{d}{dt}\left(\tfrac{1}{2}LI^2 + \tfrac{1}{2}CV^2\right) = 0$$

Carrying out the differentiation, we have

$$LI\frac{dI}{dt} + CV\frac{dV}{dt} = 0$$

We substitute $V = q/C$, $dV/dt = (1/C)dq/dt$, $I = dq/dt$, and $dI/dt = d^2q/dt^2$ and then divide by I to get

$$L\frac{d^2q}{dt^2} + \frac{1}{C}q = 0 \qquad (28.8)$$

This is a differential equation describing capacitor charge q as a function of time. We encountered a similar equation in Chapter 13 for the mass–spring system: $m(d^2x/dt^2) + kx = 0$. The solution there was a sinusoidal oscillation with angular frequency $\omega = \sqrt{k/m}$. In Equation 28.8, L replaces m and $1/C$ replaces k; therefore, the solution to Equation 28.8 is a sinusoidal oscillation:

$$q = q_p \cos \omega t \qquad (28.9)$$

with angular frequency

$$\omega = \frac{1}{\sqrt{LC}} \qquad (28.10)$$

Equation 28.9 readily provides other electrical quantities in the *LC* circuit. Using $q = CV$ gives the voltage, and differentiating gives $I = dq/dt$. From there you can get the electric and magnetic energies, $U_E = \tfrac{1}{2}CV^2$ and $U_B = \tfrac{1}{2}LI^2$. Sum them to verify that the total energy remains constant (Fig. 28.11; see Problem 62 for details).

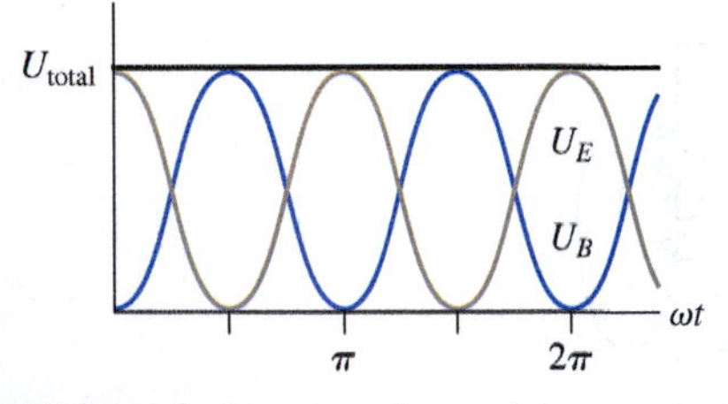

FIGURE 28.11 Electric and magnetic energies in an *LC* circuit sum to a constant total energy.

GOT IT? 28.3 You have an *LC* circuit that oscillates at a typical AM radio frequency of 1 MHz. You want to change the capacitor so it oscillates at a typical FM frequency, 100 MHz. (1) Should you make the capacitor (a) larger or (b) smaller? (2) By what factor?

EXAMPLE 28.3 An *LC* Circuit: Tuning a Piano

You wish to make an *LC* circuit oscillate at 440 Hz (A above middle C) to use in tuning pianos. You have a 25-mH inductor. (a) What value of capacitance should you use? (b) If you charge the capacitor to 5.0 V, what will be the peak current in the circuit?

INTERPRET This problem is about designing an *LC* circuit for a given frequency; in (b) we want the peak current—that is, the current when all the energy is in the inductor.

DEVELOP Equation 28.10, $\omega = 1/\sqrt{LC}$, relates frequency, capacitance, and inductance; we're given L and f. With $\omega = 2\pi f$, we can solve for C. We've recognized that the peak current comes when all the energy is the magnetic energy $\tfrac{1}{2}LI^2$ of the inductor. Given the initial capacitor voltage, we can equate this with the initial electric energy $\tfrac{1}{2}CV^2$ and solve for I.

EVALUATE (a) Equation 28.10 gives $C = 1/\omega^2 L = 1/4\pi^2 f^2 L = 5.23\ \mu\text{F}$, where $f = 440$ Hz and $L = 25$ mH. (b) Now that we know C, we equate the peak magnetic energy with the peak electric energy to get $\tfrac{1}{2}LI^2 = \tfrac{1}{2}CV^2$. Solving gives

$$I = \sqrt{\frac{C}{L}}\,V = \sqrt{\frac{5.23\ \mu\text{F}}{25\ \text{mH}}}\,(5.0\ \text{V}) = 72\ \text{mA}$$

ASSESS Our expression shows that the higher the initial voltage, the greater the current. That makes sense because a higher initial voltage means greater energy; therefore, a greater current is needed when this energy becomes all magnetic. A larger capacitance also raises the electric energy $\tfrac{1}{2}CV^2$, while a larger inductance lowers the current needed to achieve the same magnetic energy $\tfrac{1}{2}LI^2$. ■

Resistance in *LC* Circuits—Damping

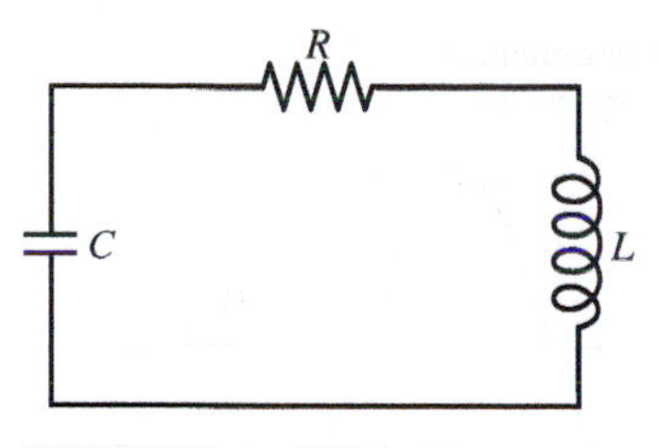

FIGURE 28.12 An *RLC* circuit.

Real inductors, capacitors, and wires have resistance (Fig. 28.12). If the resistance is low enough that only a small fraction of the energy is lost in each cycle, then the analysis in the preceding discussion applies. The circuit oscillates at a frequency given very nearly by Equation 28.10, but the oscillation amplitude slowly declines as energy is dissipated in the resistance.

We can analyze an *RLC* circuit by evaluating dU/dt as before, this time setting the result not to zero but to the rate of energy dissipation, $-I^2R$, where the minus sign indicates energy loss from the circuit:

$$\frac{dU}{dt} = \frac{d}{dt}\left(\tfrac{1}{2}LI^2 + \tfrac{1}{2}CV^2\right) = -I^2R$$

Making the same substitutions as before leads to

$$L\frac{d^2q}{dt^2} + R\frac{dq}{dt} + \frac{q}{C} = 0$$

This is mathematically identical to Equation 13.16 for damped harmonic motion, with L again replacing m, $1/C$ replacing k, and now R replacing the damping constant b. The solution follows by analogy with Equation 13.17, which is the solution to Equation 13.16:

$$q(t) = q_p e^{-Rt/2L} \cos \omega t \tag{28.11}$$

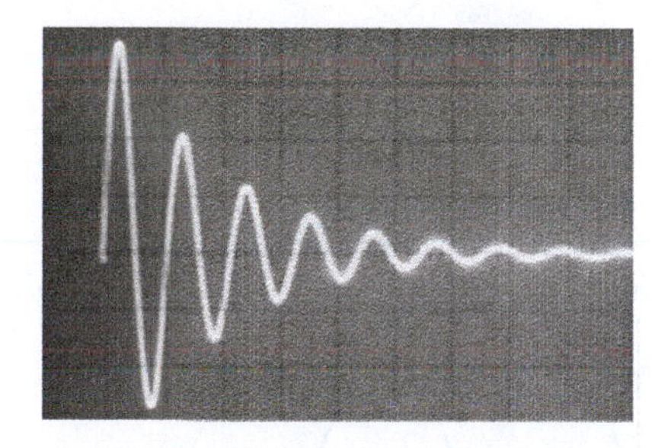

FIGURE 28.13 An oscilloscope displays the capacitor voltage in an *RLC* circuit.

Voltage and current behave similarly, with oscillation amplitude decaying exponentially with time constant $2L/R$ (Fig. 28.13).

As the resistance increases, oscillations decay more rapidly and the frequency of oscillation decreases. Eventually, when the time constant $2L/R$ equals the inverse of the frequency given in Equation 28.10, we have **critical damping**. Then all circuit quantities decay to zero without oscillation, just as we found for mechanical systems. In circuits designed to oscillate, like radio transmitters or TV tuners, engineers want to minimize damping. But in situations where oscillations would be a nuisance, it's important that circuits have enough resistance to suppress oscillation.

28.4 Driven *RLC* Circuits and Resonance

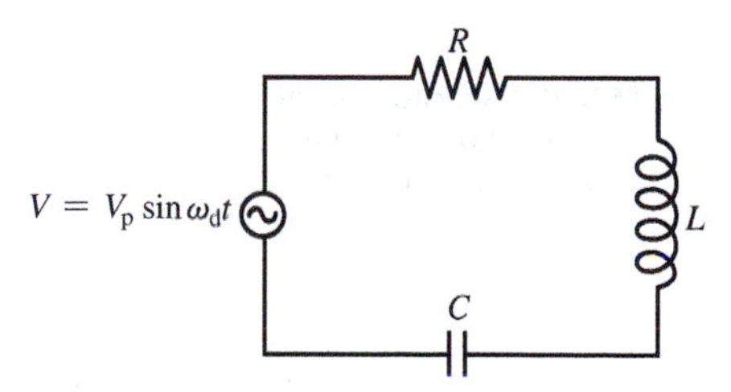

FIGURE 28.14 A series *RLC* circuit driven by an AC generator.

Figure 28.14 shows an *RLC* circuit connected across an AC generator. Adding the generator is like adding the external driving force on the mechanical oscillator that we considered in Section 13.7. We'll call the generator frequency ω_d, the driving frequency, just as we did in Chapter 13. Pursuing the mechanical analogy, we expect the driven *RLC* circuit to exhibit resonant behavior as we discussed in Section 13.7. Such electrical resonance is crucial to the operation of radio, TV, and other frequency-specific devices.

Resonance in the *RLC* Circuit

PhET: Circuit Construction Kit (AC + DC)

Suppose we vary the generator frequency ω_d in Fig. 28.14 while keeping the generator's peak voltage constant. At low frequencies the capacitor acts almost like an open circuit (its reactance $X_C = 1/\omega C$ is large), so little current flows. At high frequencies the inductor acts almost like an open circuit (its reactance $X_L = \omega L$ is large), so little current flows. At some intermediate frequency the current must be a maximum. We now show that this **resonant frequency** is the undamped natural frequency $\omega_0 = 1/\sqrt{LC}$.

Figure 28.14 is a series circuit, so the *same* current flows through all components. The voltage in a capacitor lags the current by 90°, while the voltage in an inductor leads by 90°. Since the same current flows in both series components, the inductor and capacitor voltages are therefore 180° out of phase and thus they tend to cancel (Fig. 28.15). But that cancellation is complete only when the two voltages have the same peak value. Since the current is the same in both components, comparison of Equations 28.5 for the capacitor, $I_p = V_p/X_C$, and 28.7 for the inductor, $I_p = V_p/X_L$, shows that the peak voltages are the

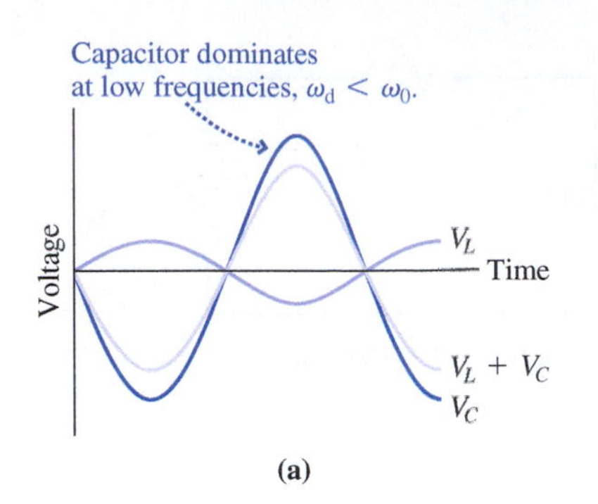

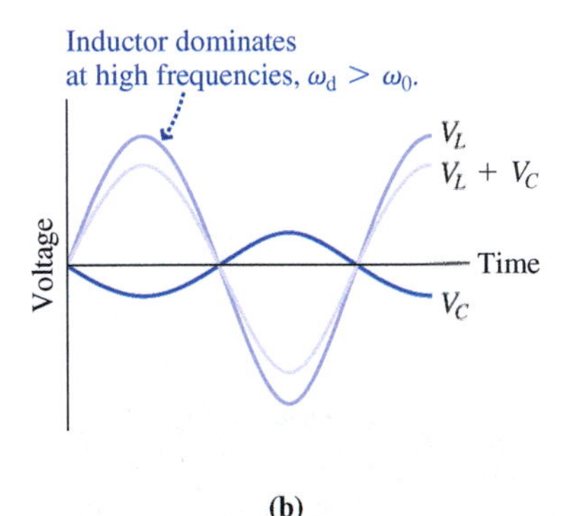

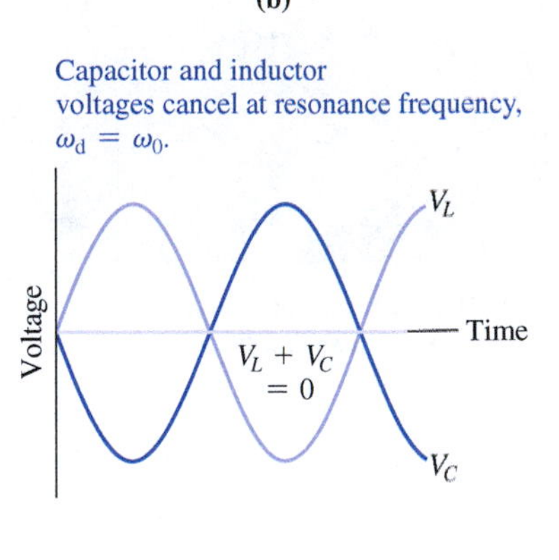

FIGURE 28.15 Capacitor and inductor voltages are 180° out of phase, but their relative magnitudes vary with frequency.

same when the capacitive reactance $X_C = 1/\omega C$ equals the inductive reactance $X_L = \omega L$. Equating the reactances gives

$$\omega_0 = \frac{1}{\sqrt{LC}} \quad \text{(resonant frequency)}$$

This is precisely the undamped natural frequency of Equation 28.10.

At resonance the capacitor and inductor voltages completely cancel. The voltage across the pair together is zero and—at the resonant frequency only—the pair might just as well be a wire. At resonance the resistance alone determines the circuit current. At any other frequency the capacitor and inductor voltages don't cancel, and the current is lower.

GOT IT? 28.4 You measure the capacitor and inductor voltages in a driven *RLC* circuit, and find 10 V for the rms capacitor voltage and 15 V for the rms inductor voltage. Is the driving frequency (a) above or (b) below resonance?

Frequency Response of the *RLC* Circuit

Here we'll use a phasor diagram (recall Section 28.2) to find the current in an *RLC* circuit as a function of the driving frequency (Fig. 28.16). Since the same current flows through all elements of this series circuit, a single phasor of length I_p represents the current. The resistor voltage is in phase with the current, so its phasor, V_{Rp}, is in the same direction as I_p. But the inductor voltage leads the current and the capacitor voltage lags, each by 90°, so their phasors, V_{Lp} and V_{Cp}, are perpendicular to the current phasor. At each instant the three voltages sum to give the generator voltage; Fig. 28.16 shows that this sum has magnitude $V_p = \sqrt{V_{Rp}^2 + (V_{Lp} - V_{Cp})^2}$. Expressing this in terms of the common current I_p and the resistance and reactances gives $V_p = \sqrt{I_p^2R^2 + (I_pX_L - I_pX_C)^2}$. Solving for I_p gives

$$I_p = \frac{V_p}{\sqrt{R^2 + (X_L - X_C)^2}} = \frac{V_p}{Z} \tag{28.12}$$

where we've defined the **impedance**, Z, as $Z = \sqrt{R^2 + (X_L - X_C)^2}$. Impedance is a generalization of resistance to include frequency-dependent effects of capacitance and inductance. Equation 28.12 is the corresponding generalization of Ohm's law. Impedance is lowest when $X_L = X_C$ or $\omega = 1/\sqrt{LC}$; then it's equal to the resistance alone. But Z becomes large at high frequencies, where $X_L = \omega L$ becomes large, and at low frequencies, where $X_C = 1/\omega C$ is large.

Figure 28.17 plots resonance curves from Equation 28.12, showing peak current versus frequency for three resistance values. With low resistance, the curve peaks sharply. Such a **high-*Q*** (for high-quality) circuit does a good job distinguishing its resonance frequency from nearby frequencies. High-*Q* circuits are important in applications such as radio, TV, and cell phones, where many signals occupy nearby frequencies. With higher resistance, the resonance curve broadens and the circuit responds to a range of frequencies; such a circuit has low *Q*. Problem 68 gives a rigorous definition of *Q*.

Equation 28.12 relates peak current and voltage in the *RLC* circuit, but it doesn't tell the whole story. As Fig. 28.16 shows, current and voltage are out of phase by the angle ϕ. Trigonometry gives $\tan\phi = (V_{Lp} - V_{Cp})/V_{Rp}$ or, since voltages are proportional to reactances and resistance,

$$\tan\phi = \frac{X_L - X_C}{R} = \frac{\omega L - 1/\omega C}{R} \tag{28.13}$$

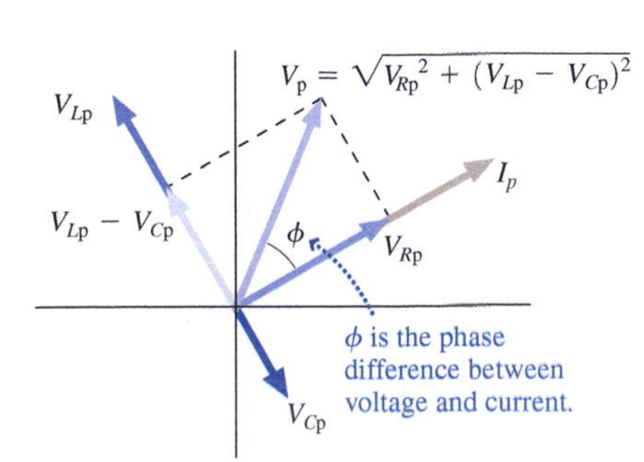

FIGURE 28.16 Phasor diagram for the driven *RLC* circuit, for the case $\omega > \omega_0$.

where $\phi = \phi_V - \phi_I$ is the phase difference between voltage and current. Positive ϕ means voltage leads current; negative ϕ means current leads voltage.

At resonance, $X_L = X_C$ and $\phi = 0$. Here capacitor and inductor voltages cancel, and the circuit behaves like a pure resistance. At low frequencies, capacitive reactance dominates; here ϕ is negative and the current leads the voltage. This is just what we expect in a capacitive circuit. The opposite is true at high frequencies, where the inductive reactance dominates. Figure 28.18 shows the phase difference as a function of frequency for three resistance values.

✓TIP Phase Matters

You can't analyze AC circuits by treating resistors, capacitors, and inductors all as "resistors" with resistances R, X_C, and X_L. That's because each component has a different phase relation between current and voltage. Phasor diagrams correctly account for these relations, which show up in the minus sign joining capacitive and inductive reactance, and in the Pythagorean addition of resistance and reactance in Fig. 28.16 and Equation 28.12.

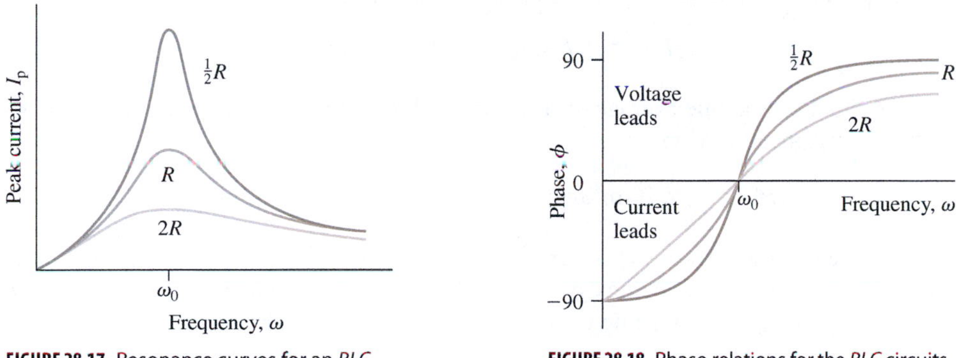

FIGURE 28.17 Resonance curves for an *RLC* circuit with three different resistances.

FIGURE 28.18 Phase relations for the *RLC* circuits whose resonance curves are shown in Fig. 28.17.

EXAMPLE 28.4 An *RLC* Circuit: Designing a Loudspeaker System

Current flows to the midrange speaker in a loudspeaker system through a 2.2-mH inductor in series with a capacitor. (a) What should the capacitance be so that a given voltage produces the greatest current at 1.0 kHz? (b) If the same voltage produces half this current at 618 Hz, what is the speaker's resistance? (c) If the peak output voltage of the amplifier is 24 V, what will the peak capacitor voltage be at 1 kHz?

INTERPRET This is a problem about the peak current and voltage in a series *RLC* circuit, where we identify the speaker as R and the amplifier as the generator in Fig. 28.14.

DEVELOP The peak current is at the resonant frequency of Equation 28.10, $\omega = 1/\sqrt{LC}$, so in (a) we can solve this equation for C. Equation 28.12 relates peak voltage and current to the component values and the frequency, so in (b) we can solve for R. In (c) we'll need to find the current at 1 kHz, and then use Equation 28.5, $I_p = V_p/X_C$, which relates peak voltage and current in a capacitor, to find V_{Cp}.

EVALUATE (a) We solve Equation 28.10 for C, using $\omega = 2\pi f$: $C = 1/[(2\pi f)^2 L]$; with $f = 1.0$ kHz and $L = 2.2$ mH, this gives $C = 11.5\ \mu\text{F}$. (b) Equation 28.12 shows that we'll have half the peak current when Z is twice the value $Z = R$ that it has at resonance. So we want $Z = \sqrt{R^2 + (X_L - X_C)^2} = 2R$ at the frequency $\omega_2 = (2\pi)(618\text{ Hz})$. Squaring and solving for R gives

$$R = \frac{1}{\sqrt{3}}\left|\omega_2 L - \frac{1}{\omega_2 C}\right| = 8.0\ \Omega$$

where we used $X_L = \omega_2 L$ and $X_C = 1/\omega_2 C$ for the reactances. For (c), note that the impedance at the 1-kHz resonant frequency is just R, so the peak current is $I_p = V_p/R$. Then Equation 28.5 gives the peak capacitor voltage:

$$V_{Cp} = I_p X_C = \left(\frac{V_p}{R}\right)\left(\frac{1}{\omega C}\right) = 43\text{ V}$$

where V_p is the 24-V peak voltage applied to the circuit and $\omega = 2\pi f$ with $f = 1$ kHz. (This answer is for the 1-kHz resonant frequency; Problem 71 shows that V_{Cp} is actually somewhat higher than 43 V at frequencies just below resonance.)

ASSESS Our 43-V answer for (c) is greater than the 24-V peak output of the amplifier, so how can it be right? Remember that there's another source of emf in the circuit—the inductor, whose emf depends on the rate of change of current. Although the capacitor and inductor voltages cancel at resonance, individually both can be higher than the applied voltage. In this low-Q circuit, the peak capacitor voltage isn't too much higher than the applied voltage, but in high-Q circuits like radio transmitters, capacitors may have to withstand voltages hundreds of times the applied voltage. Incidentally, that 8-Ω answer in (b) is typical of loudspeakers. ■

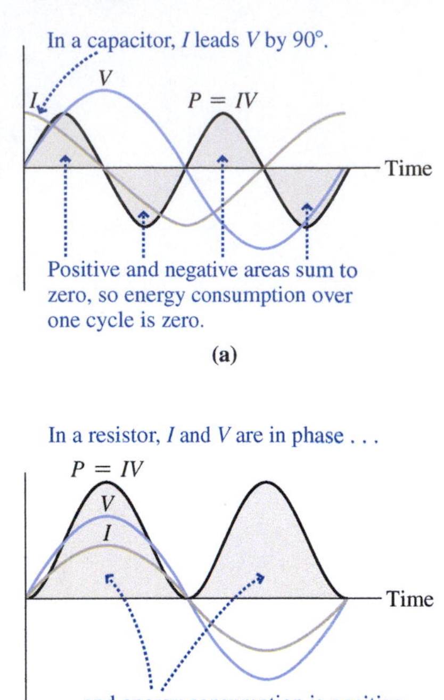

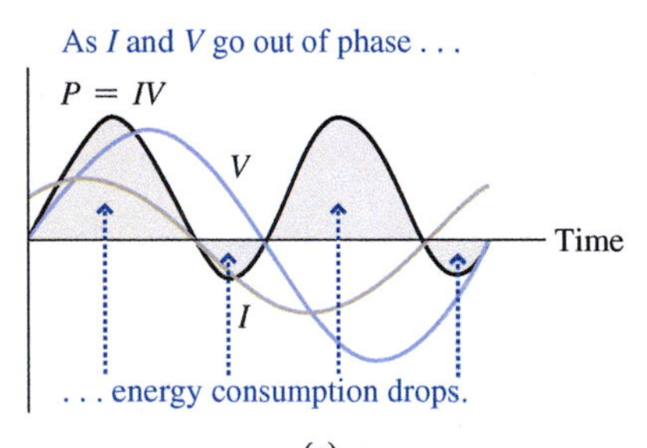

FIGURE 28.19 Energy consumption over one cycle is the area under the curve $P = IV$, with areas below the axis counted as negative.

28.5 Power in AC Circuits

Capacitors and inductors don't dissipate energy; rather, in an AC circuit they alternately store and release it. Therefore, the average power consumption over one cycle is zero in a purely reactive circuit—one containing only capacitance and/or inductance. You can see this mathematically in Fig. 28.19*a*, which shows current, voltage, and power—the product IV—over one cycle for a capacitor. Although the capacitor absorbs energy during part of the cycle—when the power is positive—it returns the same amount later (negative power), giving zero net power consumption over the cycle. That's because current and voltage are out of phase, so their product can be negative or positive at different times. In contrast, a resistor's V and I are in phase (Fig. 28.19*b*), so power is always positive and the resistor takes energy from the circuit. In a circuit containing resistance, capacitance, and inductance, the phase relation between current and voltage depends on the circuit details. Figure 28.19*c* shows a case where I and V are only slightly out of phase; the result is a net power consumption, but less than with a pure resistance.

We can develop a general expression for power in AC circuits by considering the time-average product of voltage and current with arbitrary phase difference ϕ:

$$\langle P \rangle = \langle [I_p \sin(\omega t - \phi)][V_p \sin \omega t] \rangle$$

where $\langle\ \rangle$ indicates a time average over one cycle. Expanding the current term using a trig identity (see Appendix A) gives

$$\langle P \rangle = I_p V_p \langle (\sin^2 \omega t)(\cos \phi) - (\sin \omega t)(\cos \omega t)(\sin \phi) \rangle$$

The average of $(\sin \omega t)(\cos \omega t)$ is zero, as we've just shown for two signals 90° out of phase. The quantity $\sin^2 \omega t$ swings symmetrically from 0 to 1, so its average is $\frac{1}{2}$. Then we have $\langle P \rangle = \frac{1}{2} I_p V_p \cos \phi$. Writing the peak values as $\sqrt{2}$ times the rms values gives

$$\langle P \rangle = \tfrac{1}{2}\sqrt{2}\, I_{rms} \sqrt{2}\, V_{rms} \cos \phi = I_{rms} V_{rms} \cos \phi \qquad (28.14)$$

This confirms our earlier graphical arguments. When the voltage and current are in phase, the average power is the product $I_{rms} V_{rms}$. But with current and voltage out of phase, the average power is lower; at 90° phase difference it's zero.

The factor $\cos \phi$ is the **power factor**. A purely resistive circuit has power factor 1, while a circuit with only inductance and capacitance has power factor 0. In general, the power factor depends on frequency; in the series *RLC* circuit, for example, it's 1 at resonance but lower at other frequencies.

CONCEPTUAL EXAMPLE 28.1 Managing the Power Factor

You're chief engineer for a power company. Should you strive for a high or a low power factor on your lines?

EVALUATE For a given power, Equation 28.14 shows that the product $I_{rms} V_{rms}$ will need to be higher as the power factor drops below 1. Your equipment operates at fixed voltage, so that means more current when $\cos \phi < 1$. Power lost in the lines is I^2R, and therefore you'll have greater transmission loss with a low power factor. Furthermore, you'll risk overloading your lines. So you're best served by keeping the power factor close to 1.

ASSESS Our answer helps explain some real-life power failures: The August 2003 blackout that affected 50 million people in the United States and Canada resulted in part from too low a power factor, resulting in an overloaded line that drooped from excessive heating, short-circuited to a tree, and triggered chain-reaction failures throughout the power grid.

MAKING THE CONNECTION Transmission losses on a well-managed electric grid average about 8% of the total power delivered. How does this figure change if the power factor drops from 1 to 0.71?

EVALUATE To get the same power down the line, Equation 28.14 shows that the current must increase by $1/\cos \phi = 1/0.71 = 1.4$. The transmission loss is I^2R, so the loss increases by a factor $1.4^2 = 2$. That will more than double the original 8% loss rate, because the line will need to carry still more power to overcome the loss, and the line will heat more, increasing its resistance.

GOT IT? 28.5 A resistor and capacitor are connected in series across an AC generator. If the capacitor is replaced with a second resistor whose resistance is equal to the capacitor's reactance, will the power supplied by the generator (a) increase, (b) decrease, or (c) stay the same?

28.6 Transformers and Power Supplies

A **transformer** is a pair of wire coils, often wound on an iron core to concentrate magnetic flux (Fig. 28.20). A changing current in the **primary** coil results in a changing magnetic flux through the **secondary**, and this induces an emf in the secondary. The induced emf, in turn, drives current in any circuit connected across the secondary. Thus the device transfers electric power between two circuits without direct electrical contact.

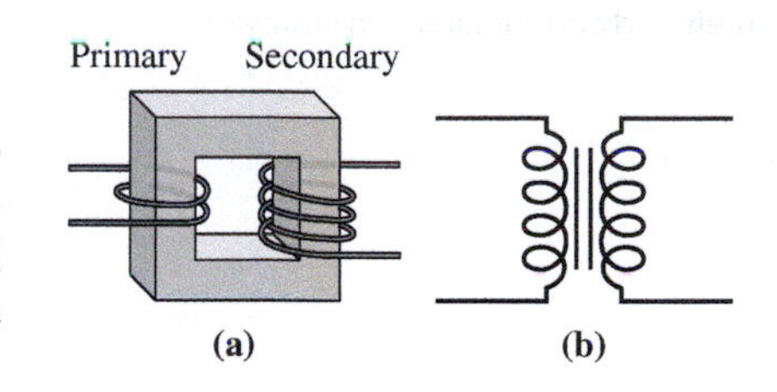

FIGURE 28.20 (a) A transformer consisting of two coils wound on an iron core. (b) Transformer circuit symbol.

The transformer in Fig. 28.20 is a **step-up transformer** because it has more turns in its secondary. Since each turn encircles the same changing magnetic flux, each gets the same induced emf and therefore the emf across the secondary is greater than across the primary. Interchanging primary and secondary in Fig. 28.20 would give a **step-down transformer**. In general, the ratio of the peak (or rms) secondary voltage V_2 to the peak (or rms) primary voltage V_1 is the same as the ratio of turns in the two coils:

$$V_2 = \frac{N_2}{N_1} V_1 \qquad (28.15)$$

Aren't we getting something for nothing with a step-up transformer? No. A step-up transformer increases voltage, but not power. An ideal transformer passes all the power supplied to its primary on to the secondary, so $I_1 V_1 = I_2 V_2$. If voltage goes up, current goes down, and vice versa. Real transformers have losses, but good engineering holds these to a few percent of the total power.

Transformers work only with AC because they use electromagnetic induction and therefore require *changing* current. One reason for the near-universal use of AC power is the ease of changing voltage levels (Fig. 28.21). Relatively low voltages are safer for the end user. But since power $P = IV$, using a higher voltage in long-distance transmission means lower current. Power dissipated in the conductors themselves is I^2R, so that in turn means less power lost in transmission. Transformers readily handle the voltage conversions in AC power systems. Changing the voltage from a DC source, in contrast, requires first interrupting the DC to produce a changing current; a car's ignition system is one example.

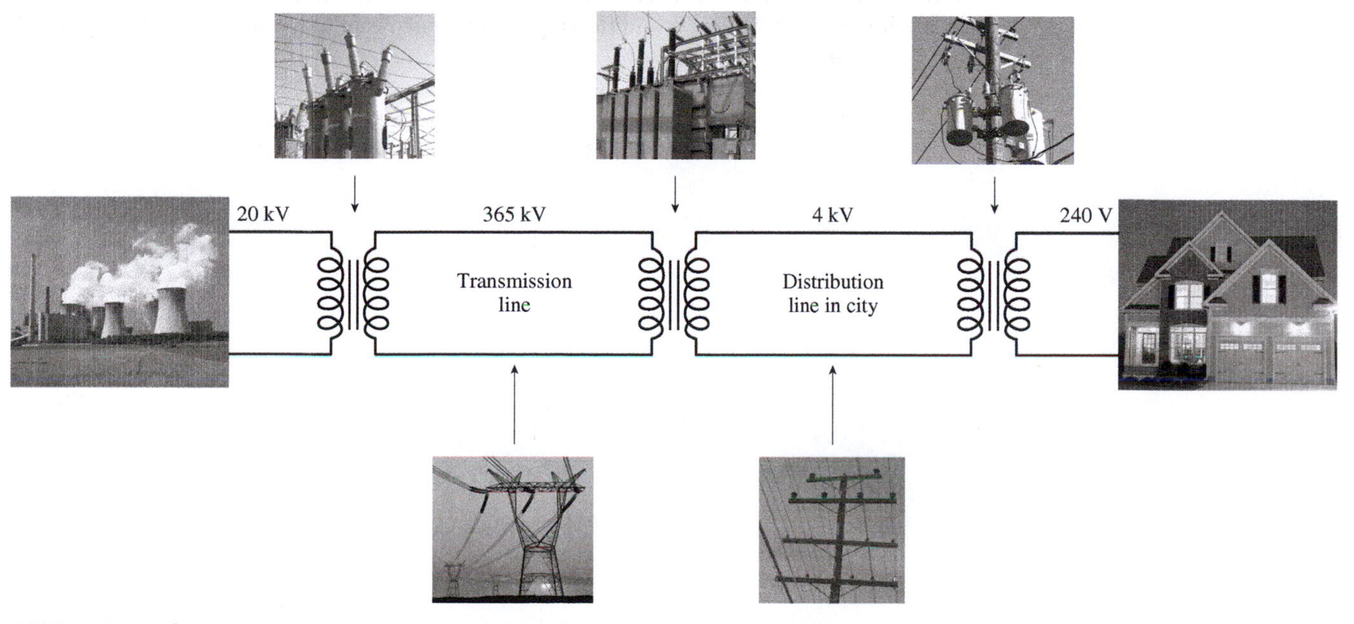

FIGURE 28.21 Transformers change voltage levels throughout the power distribution network.

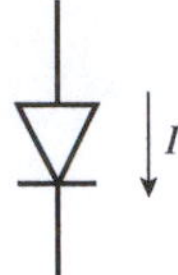

FIGURE 28.22 Circuit symbol for a diode, with preferred current direction indicated.

Direct-Current Power Supplies

Lightbulbs and heaters work equally well on AC or DC, but electronic equipment requires DC. In Chapter 24 we found that a junction between *P*- and *N*-type semiconductors passes current in one direction but not the other. A **diode** is a *PN* junction that serves as a "one-way valve" for electric current. An ideal diode acts like a short circuit in the preferred direction, and like an open circuit in the opposite direction (Fig. 28.22).

Figure 28.23*a* shows a DC power supply using a transformer, diode, and capacitor, delivering power to a load symbolized by the resistor *R*. The transformer steps the voltage to the desired level, while the diode passes current only in its preferred direction, "chopping off" the negative half of the AC cycle. The capacitor smoothes, or **filters**, the remaining half to produce nearly steady DC. Figure 28.23*b* shows how this works: As the AC voltage rises, the capacitor charges rapidly through the low resistance of the diode in its "on" state. But the diode "turns off" when the AC voltage drops, leaving only the resistor as a discharge path for the capacitor. If the *RC* time constant is long enough—much longer than the typical 1/60-s period of the 60-Hz AC cycle—then the capacitor voltage hardly drops before the next cycle again sends in a surge of charge. Large capacitors are expensive, so practical power supplies often use additional filtering and voltage regulation involving semiconductor devices.

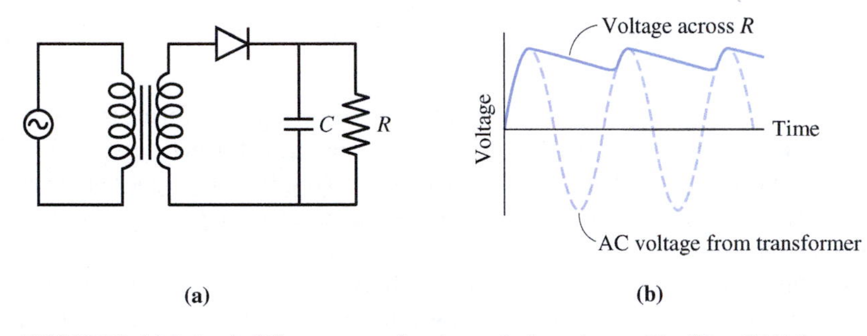

FIGURE 28.23 (a) A simple DC power supply using a diode and capacitive filter. (b) Voltage across *R* exhibits a variation called *ripple* as the capacitor discharges slightly between cycles. A practical power supply would use a larger capacitor, resulting in less ripple.

GOT IT? 28.6 A distribution line in a city supplies AC power at 7.2 kV to a transformer that steps it down to 240 V for an individual home. If the current in the transformer's primary is 1.5 A, the current flowing to the home is approximately (a) 3.0 A, (b) 1.5 A, (c) 240 A, or (d) 45 A.

CHAPTER 28 SUMMARY

Big Idea

The big idea here is **alternating current** (AC), which in its simplest form exhibits sinusoidal variation in current and voltage. Resistors respond to AC as to DC, with current directly proportional to voltage. In capacitors and inductors, the current–voltage relation depends on frequency, and the current and voltage are out of phase.

Key Concepts and Equations

An AC voltage or current is characterized by its amplitude (rms or peak value), its frequency, and phase:

$$V = V_p \sin(\omega t + \phi)$$

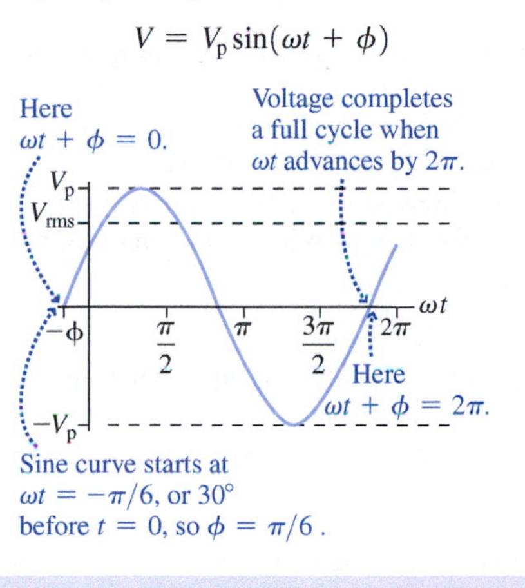

Reactance, X, characterizes the relation between peak (or rms) current and voltage in capacitors and inductors:

$$I_p = \frac{V_p}{X}$$

Capacitor: $X_C = 1/\omega C$

Inductor: $X_L = \omega L$

where $\omega = 2\pi f$ is the angular frequency of the AC voltage and current.

Phasors are arrows used to describe time-varying AC quantities. They rotate with angular velocity equal to the angular frequency ω, and the projection of the phasor on the vertical axis gives the instantaneous value of voltage or current.

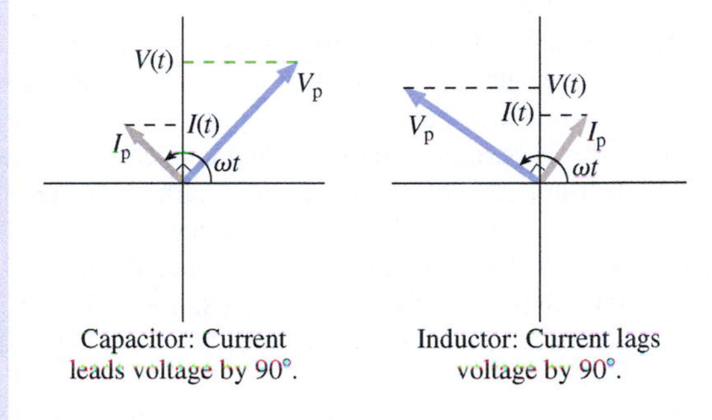

Applications

In an ***LC* circuit**, energy oscillates between electric and magnetic forms with frequency

$$\omega_0 = \frac{1}{\sqrt{LC}}$$

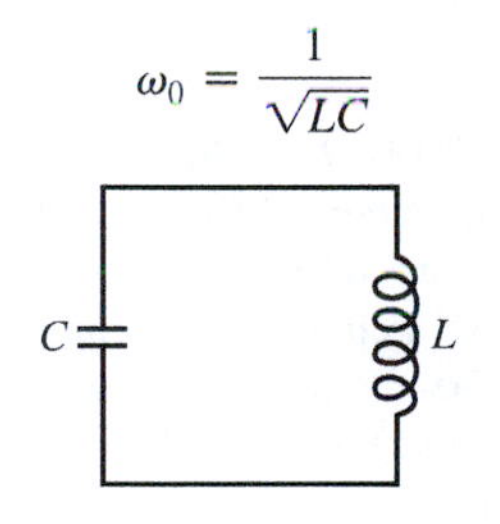

In a series ***RLC* circuit**, capacitor and inductor voltages cancel at the **resonant frequency**, ω_0. Here the circuit exhibits the minimum **impedance**, $Z = \sqrt{R^2 + (X_L - X_C)^2}$, and passes the maximum current. The phase difference between voltage and current is $\tan\phi = (X_L - X_C)/R$.

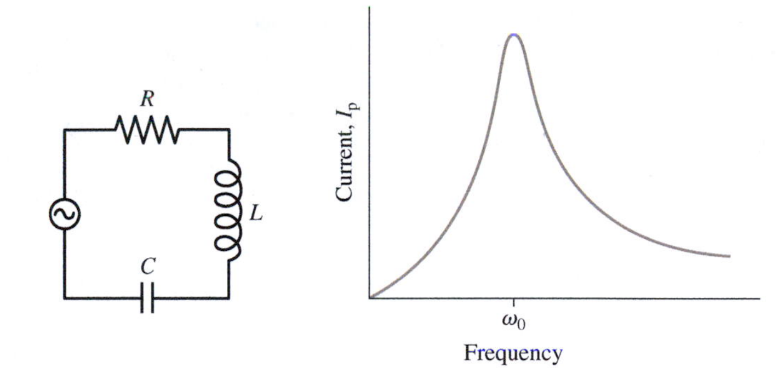

The average power in an AC circuit depends on the cosine of the phase difference, also called the **power factor**:

$$\langle P \rangle = I_{rms} V_{rms} \cos\phi$$

Transformers use electromagnetic induction to change voltage levels, transferring electric power between two circuits. **Diodes** and capacitive filters change AC to DC.

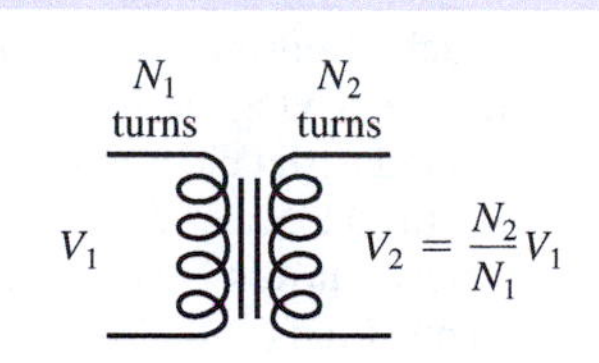

MP For homework assigned on MasteringPhysics, go to www.masteringphysics.com

BIO *Biology and/or medicine-related problems* **DATA** *Data problems* **ENV** *Environmental problems* **CH** *Challenge problems* **COMP** *Computer problems*

For Thought and Discussion

1. Two AC signals have the same amplitude but different frequencies. Are their rms amplitudes the same?
2. What's meant by the statement, "A capacitor acts like a DC open circuit"?
3. There's an insulating gap between capacitor plates, so how can current flow in an AC circuit containing a capacitor?
4. Why does it make sense that inductive reactance increases with frequency?
5. The same AC voltage appears across a capacitor and a resistor, and the same rms current flows in each. Is the power dissipation the same in each?
6. When a particular inductor and capacitor are connected across the same AC voltage, the current in the inductor is higher than in the capacitor. Is this true at all frequencies?
7. An inductor and capacitor are connected in series across an AC generator, and the voltage across the inductor is higher than across the capacitor. Is the generator frequency above or below resonance?
8. When the capacitor voltage in an undriven *LC* circuit reaches zero, why don't the oscillations stop?
9. Why is Equation 28.5 not a full description of the relation between voltage and current in a capacitor?
10. The applied voltage in a series *RLC* circuit lags the current. Is the frequency above or below resonance?
11. The voltage across two components in series is zero. Is it possible that the voltages across the individual components *aren't* zero? Give an example.
12. If you measure the rms voltages across the resistor, capacitor, and inductor in a series *RLC* circuit, will they add to the rms generator voltage?
13. A step-up transformer increases voltage, or energy per unit charge. Why doesn't this violate energy conservation?

Exercises and Problems

Exercises

Section 28.1 Alternating Current

14. Much of Europe uses AC power at 230 V rms and 50 Hz. Express this AC voltage in the form of Equation 28.3, taking $\phi_V = 0$.
15. An industrial electric motor runs at 208 V rms and 400 Hz. What are (a) the peak voltage and (b) the angular frequency?
16. An AC current is given by $I = 495\sin(9.43t)$, with I in mA and t in ms. Find (a) the rms current and (b) the frequency in Hz.
17. What are the phase constants for the signals in Fig. 28.24?

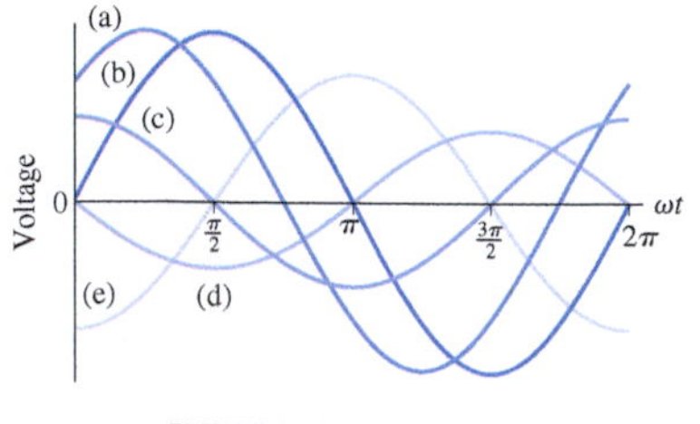

FIGURE 28.24 Exercise 17

Section 28.2 Circuit Elements in AC Circuits

18. Find the rms current in a 1.0-μF capacitor connected across 120-V rms, 60-Hz AC power.
19. A 470-Ω resistor, 10-μF capacitor, and 750-mH inductor are each connected across 6.3-V rms, 60-Hz AC power. Find the rms current in each.
20. Find the reactance of a 3.3-μF capacitor at (a) 60 Hz, (b) 1.0 kHz, and (c) 20 kHz.
21. A 15-μF capacitor carries 1.4 A rms. What's its minimum safe voltage rating if the frequency is (a) 60 Hz and (b) 1.0 kHz?
22. A capacitor and a 1.8-kΩ resistor pass the same current when connected across 60-Hz power. Find the capacitance.
23. A 50-mH inductor is connected across a 10-V rms AC generator, and a 2.0-mA rms current flows. What's the generator frequency?

Section 28.3 *LC* Circuits

24. Find the resonant frequency of an *LC* circuit consisting of a 0.22-μF capacitor and a 1.7-mH inductor.
25. An *LC* circuit with $C = 18$ mF undergoes oscillations with period 2.4 s. Find the inductance.
26. Your sister who's building the radio (Chapter 27 Problem 22) wants to use a variable capacitor with her toilet-paper-tube inductor to span the AM radio band (550–1600 kHz). What capacitance range do you suggest?
27. An *LC* circuit with a 20-μF capacitor oscillates with period 5.0 ms. The peak current is 25 mA. Find (a) the inductance and (b) the peak voltage.

Section 28.4 Driven *RLC* Circuits and Resonance

28. A series *RLC* circuit has $R = 75\,\text{k}\Omega$, $L = 20\,\text{mH}$, and resonates at 4.0 kHz. (a) What's the capacitance? (b) Find the circuit's impedance at resonance and (c) at 3.0 kHz.
29. Find the impedance at 10 kHz of a circuit consisting of a 1.5-kΩ resistor, 5.0-μF capacitor, and 50-mH inductor in series.
30. A series *RLC* circuit has $R = 18\,\text{k}\Omega$, $C = 14\,\mu\text{F}$, and $L = 0.20$ H. (a) At what frequency is its impedance lowest? (b) What's the impedance at this frequency?
31. If the peak voltage applied to produce the curves in Fig. 28.17 is 100 V, and if $R = 10\,\text{k}\Omega$, what are the peak currents at resonance for the three curves shown?

Section 28.5 Power in AC Circuits

Section 28.6 Transformers and Power Supplies

32. An electric drill draws 4.6 A rms at 120 V rms. If the current lags the voltage by 25°, what's the drill's power consumption?
33. A 40-W fluorescent lamp has power factor 0.85 and operates from the 120-V rms AC power line. How much current does it draw?
34. An electric water heater draws 20 A rms at 240 V rms and is purely resistive. An AC motor has the same current and voltage, but its inductance causes the voltage to lead the current by 20°. Find the power consumption in each device.
35. **BIO** For safety, medical equipment connected to patients is often powered by an *isolation transformer*, whose primary is connected to 120-V AC power and whose secondary delivers 120-V power. What's the turns ratio of such a transformer?
36. You're planning a semester in China, so you want to purchase a transformer to step the 220-V Chinese power down to 120 V to

power your stereo. (a) If the transformer's primary has 660 turns, how many should the secondary have? (b) You can save money with a transformer whose maximum primary current is 1.4 A. If your stereo draws 2.9 A, will this transformer work?

Problems

37. (a) A 2.2-H inductor is connected across 120-V rms, 60-Hz power. Find the rms inductor current. (b) Repeat if the same inductor is connected across the 230-V rms, 50-Hz power commonly used in Europe.
38. A 2.0-μF capacitor has 1.0-kΩ reactance. (a) What's the frequency of the applied voltage? (b) What inductance would give the same reactance at this frequency? (c) How would the reactances compare if the frequency were doubled?
39. Show that the unit of both capacitive and inductive reactance is the ohm.
40. **BIO** *Electroencephalography* (EEG) elucidates brain function by analyzing *brain waves*, AC voltages resulting from electrical activity in the brain. *Alpha waves* are brain waves with frequencies from 7.5 Hz to 12.5 Hz. A particular alpha wave has frequency 9.84 Hz and rms amplitude 31.8 μV. Express this voltage in the form of Equation 28.3, assuming zero phase constant.
41. At 15 kHz an inductor has 12 times the reactance of a capacitor. At what frequency will their reactances be equal?
42. A 0.75-H inductor is in series with a fluorescent lamp, and the combination is across 120-V rms, 60-Hz power. If the rms inductor voltage is 90 V, what's the rms lamp current?
43. A 2.2-nF capacitor and one of unknown capacitance are in parallel across a 10-V rms sine-wave generator. At 1.0 kHz, the generator supplies a total current of 3.4 mA rms. The generator frequency is then decreased until the rms current drops to 1.2 mA. Find (a) the unknown capacitance and (b) the lower frequency.
44. **BIO** Connections to the body for electrocardiography (ECG) and electroencephalography (EEG) are normally made with metal electrodes and conductive gels to ensure good electrical contact. An alternative is the *capacitively coupled noncontact electrode*, which uses a conductor near but not contacting the skin, to form a capacitor. Clothing can serve as the capacitor's insulation, eliminating skin contact. A particular EEG instrument calls for capacitive electrodes with maximum reactance 10 MΩ at a typical EEG beta wave frequency of 25 Hz. What's the minimum electrode capacitance?
45. The FM radio band covers the frequency range 88–108 MHz. If the variable capacitor in an FM receiver ranges from 10.9 pF to 16.4 pF, what inductor should be used to make an *LC* circuit whose resonant frequency spans the FM band?
46. An *LC* circuit includes a 0.025-μF capacitor and a 340-μH inductor. (a) If the peak capacitor voltage is 190 V, what's the peak inductor current? (b) How long after the voltage peak does the current peak occur?
47. One-eighth of a cycle after the capacitor in an *LC* circuit is fully charged, what are the following as fractions of their peak values: (a) capacitor charge, (b) energy in the capacitor, (c) inductor current, (d) energy in the inductor?
48. The 2420-μF capacitor in Fig. 28.25 is initially charged to 250 V. (a) Describe how you would manipulate switches *A* and *B* to transfer all the energy from the 2420-μF capacitor to the 605-μF capacitor. Include the times you would throw the switches. (b) What will be the voltage across the 605-μF capacitor once you've finished?

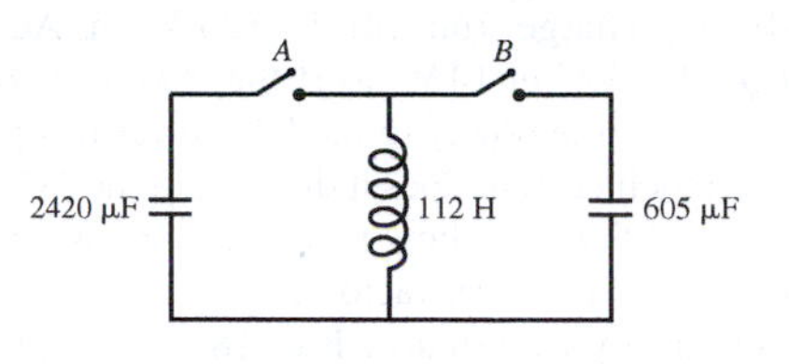

FIGURE 28.25 Problem 48

49. A damped *LC* circuit consists of a 0.15-μF capacitor and a 20-mH inductor with resistance 1.6 Ω. How many oscillation cycles will occur before the peak capacitor voltage drops to half its initial value?
50. A damped *RLC* circuit includes a 5.0-Ω resistor and a 100-mH inductor. If half the initial energy is lost after 15 cycles, what's the capacitance?
51. An *RLC* circuit includes a 1.5-H inductor and a 250-μF capacitor rated at 400 V. The circuit is connected across a sine-wave generator with $V_p = 32$ V. What minimum resistance will ensure that the capacitor voltage does not exceed its rated value when the circuit is at resonance?
52. **DATA** The table below shows the ratio of peak voltage to peak current—that is, the impedance *Z*—as a function of frequency for a series *RLC* circuit. Plot the data and use your graph to estimate (a) the resonant frequency and (b) the resistance *R*.

Frequency (Hz)	150	200	230	280	350	400
Impedance (Ω)	320	140	74	77	190	280

53. Figure 28.26 shows the phasor diagram for an *RLC* circuit. (a) Is the driving frequency above or below resonance? (b) Complete the diagram by adding the applied voltage phasor, and from your diagram determine the phase difference between applied voltage and current.

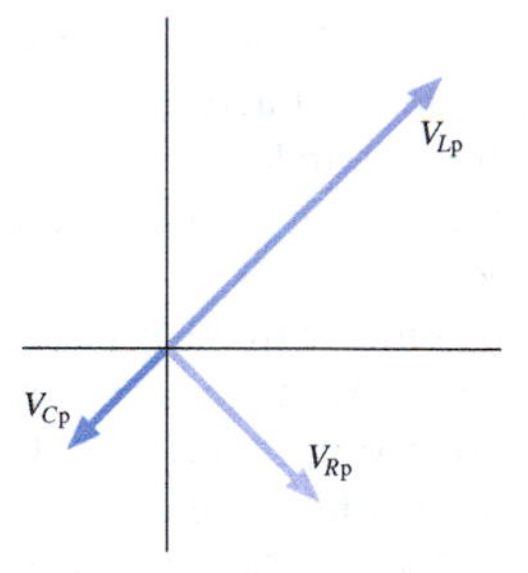

FIGURE 28.26 Problem 53

54. An AC voltage of fixed amplitude is applied across a series *RLC* circuit. The components are such that the current at half the resonant frequency is half the current at resonance. Show that the current at twice the resonant frequency is also half the current at resonance.
55. A series *RLC* circuit has resistance 127 Ω and impedance 344 Ω. (a) What's the power factor? (b) If the rms current is 225 mA, what's the power dissipation?
56. A series *RLC* circuit has power factor 0.764 and impedance 182 Ω at 442 Hz. (a) What's the resistance? (b) If the inductance is 25.0 mH, what's the resonant frequency?
57. You're Chief Financial Officer for a power company, and you consult your engineering department in an effort to minimize power-line losses. Your power plant produces 60-Hz power at 365 kV rms and 200 A rms, and delivers it via transmission lines with total resistance 100 Ω. You ask the engineers for the percentage of power that's lost. They reply that it depends on the power factor. What's the percentage loss for power factors of (a) 1.0 and (b) 0.60?

58. A car-battery charger runs off the 120-V rms AC power line and supplies 10-A DC at 14 V. (a) If the charger is 80% efficient in converting the line power to the DC power it supplies to the battery, how much current does it draw from the AC line? (b) If electricity costs 9.5¢/kWh, how much does it cost to run the charger for 10 hours if the power factor is 1?
59. A power supply like that of Fig. 28.23 is supposed to deliver 22-V DC at a maximum current of 150 mA. The transformer's peak output voltage can charge the capacitor to a full 22 V, and the primary is supplied with 60-Hz AC. What capacitance will ensure that the output voltage stays within 3% of the rated 22 V?
60. An *RLC* circuit includes a 3.3-μF capacitor and a 27-mH inductor. The capacitor is charged to 35 V, and the circuit begins oscillating. Ten full cycles later the capacitor voltage peaks at 28 V. Find the resistance.
61. A series *RLC* circuit with $R = 1.3\ \Omega, L = 27$ mH, and $C = 0.33\ \mu\text{F}$ is connected across a sine-wave generator. If the capacitor's peak voltage rating is 600 V, what's the maximum safe value for the generator's peak output voltage when it's tuned to resonance?
62. Differentiate Equation 28.9 to find the current in the *LC* circuit, and use $q = CV$ to find the voltage. From these, obtain the electric energy in the capacitor and the magnetic energy in the inductor, and sum to show that the total energy remains constant. (*Hint:* You'll need Equation 28.10 and a familiar trig identity.)
63. CH Find a second frequency where the current in the speaker of Example 28.4 has half its maximum value.
64. CH Two capacitors are connected in parallel across a 24-V rms, 7.5-kHz sine-wave generator, and the generator supplies a total rms current of 56 mA. With capacitors rewired in series, the rms current drops to 2.8 mA. What are the two capacitances?
65. CH A "black box" has two input connections and two output connections. With a 12-V rms, 60-Hz sine wave across the inputs, the output is a 6.0-V, 60-Hz sine wave leading the input voltage by 45°. Design a circuit that could be in the "black box."
66. CH A series *RLC* circuit with $R = 47\ \Omega, L = 250$ mH, and $C = 4.0\ \mu\text{F}$ is connected across a sine-wave generator whose peak output voltage is independent of frequency. Find the frequency range over which the peak current will exceed half its value at resonance.
67. CH A sine-wave generator with 20-V peak output is applied across a series *RLC* circuit. At the resonant frequency of 2.0 kHz, the peak current is 50 mA; at 1.0 kHz, it's 15 mA. Find *R*, *L*, and *C*.
68. CH For *RLC* circuits in which the resistance isn't too high, the *Q* factor may be defined as the ratio of the resonant frequency to the difference between the two frequencies where the power dissipated in the circuit is half the power dissipated at resonance. Using suitable approximations, show that this definition leads to $Q = \omega_0 L/R$, with ω_0 the resonant frequency.
69. CH A *triangle wave* swings linearly between voltages $-V_p$ and $+V_p$. Show that the rms voltage of a triangle wave is $V_p/\sqrt{3}$.
70. CH Substitute the expression for $q(t)$ in Equation 28.11 into the differential equation for an *LC* circuit with resistance, and find an expression for the angular frequency of the damped oscillations in terms of *R*, *L*, and *C*.
71. Although the maximum current flows in the speaker circuit of Example 28.4 at the 1-kHz resonant frequency, the peak voltage across the capacitor is a maximum at a somewhat lower frequency. Find that frequency and the corresponding peak voltage.
72. Your professor tells you about the days before digital computers when engineers used electric circuits to model mechanical systems. Suppose a 5.0-kg mass is connected to a spring with $k = 1.44$ kN/m. This is then modeled by an *LC* circuit with $L = 2.5$ H. What should *C* be in order for the *LC* circuit to have the same resonant frequency as the mass–spring system?

Passage Problems

A *filter* is a circuit designed to pass AC signals in some frequency range and to attenuate others. Common filters include *low-pass filters*, which allow low-frequency signals to pass but attenuate high frequencies; *high-pass filters*, which do the opposite; and *band-pass* filters, which pass a range of frequencies while attenuating signals with frequencies outside the band. Filters are widely used in electronics. Applications include tone and equalizer controls in audio equipment; filters to separate nearby frequencies at cell phone towers; and filters to eliminate unwanted electrical noise in biomedical instruments such as electrocardiographs. A simple design for an *RC* filter is shown in Fig. 28.27.

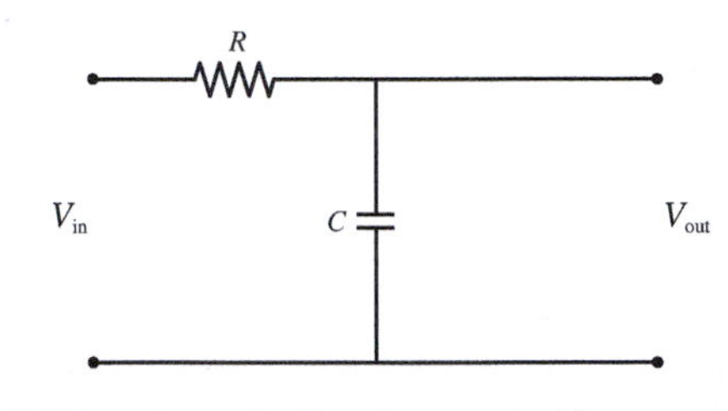

FIGURE 28.27 An *RC* filter (Passage Problems 73–76)

73. The circuit shown in Fig. 28.27 is
    a. a low-pass filter.
    b. a high-pass filter.
    c. a band-pass filter.
    d. impossible to tell without knowing the component values.
74. When the angular frequency ω of the input voltage V_{in} is such that the capacitor's reactance is equal to the resistance, the output voltage is
    a. $V_{in}/4$.
    b. $V_{in}/2$.
    c. $V_{in}/\sqrt{2}$.
    d. $2V_{in}$.
75. The circuit of Fig. 28.27
    a. exhibits resonance at frequency $\omega = 1/RC$.
    b. exhibits resonance at frequency $\omega = 1/\sqrt{RC}$.
    c. produces an output voltage whose frequency differs from that of the input.
    d. produces an output voltage whose phase differs from that of the input.
76. If you replace the capacitor in Fig. 28.27 with an inductor, the circuit
    a. continues to function as before.
    b. becomes the opposite kind of filter.
    c. produces zero output voltage because the inductor is a short circuit.
    d. produces an output voltage that exceeds the input voltage.

Answers to Chapter Questions

Answer to Chapter Opening Question

Alternating current is used because transformers, based on electromagnetic induction, can alter voltage levels to ensure both efficient long-distance power transmission and safety for end users.

Answers to GOT IT? Questions

28.1 (c)
28.2 (d)
28.3 (1) (b); (2) by a factor of 10^{-4}
28.4 (a) because the inductor's reactance must be greater
28.5 (b)
28.6 (d)

Maxwell's Equations and Electromagnetic Waves

What You Know

- You've seen four laws of electromagnetism: Gauss for electricity, Gauss for magnetism, Faraday's law, and Ampère's law.
- From Chapter 14, you understand the nature of mechanical waves.
- From Chapter 14, you know how to describe waves mathematically.

What You're Learning

- You'll see how Maxwell enhanced Ampère's law, completing the set of four equations describing all electromagnetic phenomena.
- You'll follow a physical and mathematical explanation of how the equations of electromagnetism explain the existence of *electromagnetic waves*.
- You'll learn the properties of electromagnetic waves, including the fact that their speed in vacuum is precisely the speed of light.
- You'll explore the electromagnetic spectrum, including visible light.
- You'll learn how electromagnetic waves are produced by accelerated charges.
- You'll see how electromagnetic waves carry energy and momentum.

How You'll Use It

- Electromagnetic waves, including light, are one of your primary means of interacting with the world around you.
- Electromagnetic waves are the basis of nearly all wireless communication systems, from radio to cell phones to WiFi and Bluetooth.
- Electromagnetic waves provide nearly all our knowledge of the universe beyond Earth.
- Electromagnetic waves are the basis of optics, which you'll explore in Chapters 30–32.

At this point we've introduced the four fundamental laws of electromagnetism—Gauss's law for electricity, Gauss's law for magnetism, Ampère's law, and Faraday's law—that govern the behavior of electric and magnetic fields throughout the universe. We've seen how these laws describe the electric and magnetic interactions that make matter act as it does, and we've explored practical electromagnetic devices. Here we extend the fundamental laws to their most general form and show how they predict the existence of electromagnetic waves. These include the visible light, radio, microwaves, X rays, ultraviolet, and infrared with which we see, communicate, cook our food, diagnose diseases, learn about the universe, and perform myriad other tasks from mundane to profound.

How does a conversation travel between cell phones?

29.1 The Four Laws of Electromagnetism

Table 29.1 summarizes the four laws as we introduced them in earlier chapters. Look at these laws and you'll notice some similarities. On the left-hand sides of the equations, the two laws of Gauss are identical except for the interchanging of $\vec{E}$ and $\vec{B}$; the same is true for Ampère's and Faraday's laws.

The right-hand sides are more different. Gauss's law for electricity involves charge, while Gauss's law for magnetism has zero on the right-hand side. Actually, though, these laws are similar. Since we have no experimental evidence for the existence of isolated magnetic charge, the magnetic charge on the right-hand side of Gauss's law for magnetism is zero. If and when magnetic monopoles are discovered, then the right-hand side of Gauss's law for magnetism would be nonzero for any surface enclosing net magnetic charge.

The right-hand sides of Ampère's and Faraday's laws are distinctly different. Ampère's law has current—the flow of electric charge—as a source of magnetic field. We can understand the absence of a similar term in Faraday's law because we've never observed a flow of magnetic monopoles. If we had such a flow, then we would expect this magnetic current to produce an electric field.

Two of the differences among the laws of electromagnetism would be resolved if we knew for sure that magnetic monopoles exist. That current theories of elementary particles suggest the existence of monopoles is a tantalizing hint that there may be a fuller symmetry between electric and magnetic phenomena.

Table 29.1 Four Laws of Electromagnetism (still incomplete)

Law	Mathematical Statement	What It Says
Gauss for $\vec{E}$	$\oint \vec{E}\cdot d\vec{A} = \dfrac{q}{\epsilon_0}$	How charges produce electric field; field lines begin and end on charges.
Gauss for $\vec{B}$	$\oint \vec{B}\cdot d\vec{A} = 0$	No magnetic charge; magnetic field lines don't begin or end.
Faraday	$\oint \vec{E}\cdot d\vec{l} = -\dfrac{d\Phi_B}{dt}$	Changing magnetic flux produces electric field.
Ampère (steady currents only)	$\oint \vec{B}\cdot d\vec{l} = \mu_0 I$	Electric current produces magnetic field.

29.2 Ambiguity in Ampère's Law

There's one difference that magnetic monopoles won't resolve. On the right-hand side of Faraday's law is the term $d\Phi_B/dt$ that describes changing magnetic flux as a source of electric field. There's no comparable term in Ampère's law. Are we missing something? Is it possible that a changing electric flux produces a magnetic field? So far, you haven't seen any experimental evidence for such a conjecture. It's suggested only by the sense that the near-symmetry between electricity and magnetism is not a coincidence. If a changing electric flux did produce a magnetic field, just as a changing magnetic flux produces an electric field, then we would expect a term $d\Phi_E/dt$ on the right-hand side of Ampère's law.

When we first stated Ampère's law in Chapter 26, we emphasized that it applied only to *steady* currents. Why that restriction? Figure 29.1 shows a situation in which current is *not* steady—namely, an *RC* circuit. Current in this circuit carries charge onto the capacitor plates. The current gradually decreases to zero as the capacitor charges. While it's flowing, the current should produce a magnetic field. Let's try to use Ampère's law to calculate that field.

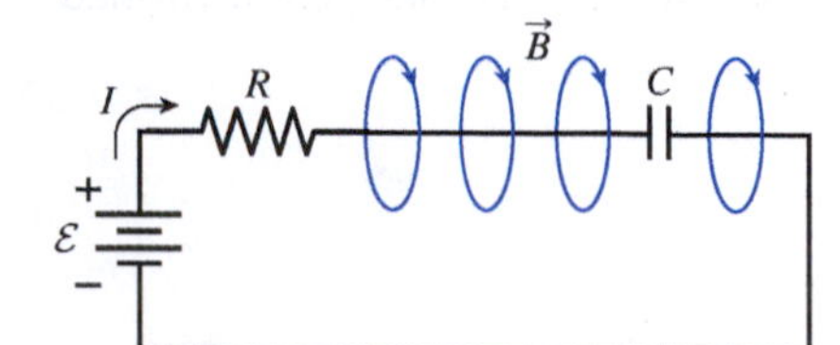

FIGURE 29.1 A charging *RC* circuit, showing some magnetic field lines surrounding the current-carrying wire.

Ampère's law says that the line integral of the magnetic field around any closed loop is proportional to the encircled current:

$$\oint \vec{B}\cdot d\vec{l} = \mu_0 I$$

The encircled current is the current through *any open surface* bounded by the loop. Figure 29.2 shows four such surfaces. The same current flows through surfaces 1, 2, and 4 because a current-carrying wire pierces each surface. But no current flows through surface 3 because it's in the gap between the capacitor plates. Charge flows onto the plates of the capacitor, but it doesn't flow *through* that gap. So for surfaces 1, 2, and 4 the right-hand side of Ampère's law is $\mu_0 I$, but for surface 3 it's zero. Thus Ampère's law is ambiguous in this case of a changing current.

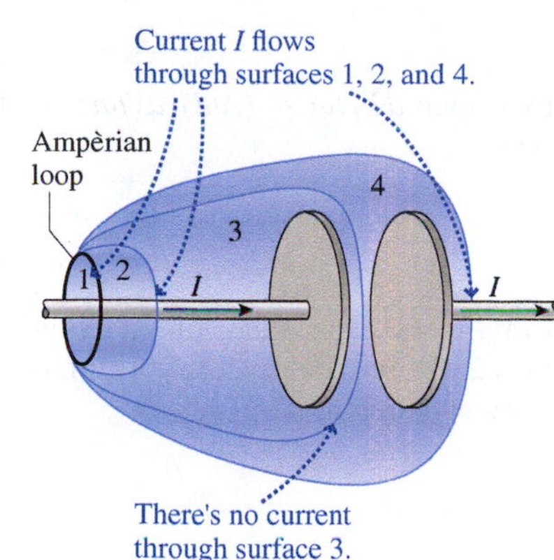

FIGURE 29.2 Four surfaces bounded by the same circular Ampèrian loop. Surface 1 is a flat, circular disk. The others are like soap bubbles in the process of being blown; they're open at the left end, so if current does pass through a surface, it does so at the right end only.

This ambiguity doesn't arise with steady currents. In an *RC* circuit the steady-state current is zero, and thus the right-hand side of Ampère's law is zero for *any* surface. It's only when currents are changing with time that Ampère's law becomes ambiguous. That's why the form of Ampère's law we've used until now is strictly valid only for steady currents.

Can we extend Ampère's law to cover unsteady currents without affecting its validity in the steady case? Symmetry between Ampère's and Faraday's laws suggests that a changing electric flux might produce a magnetic field. Between the plates of a charging capacitor is an electric field whose magnitude is increasing. That means there's a changing electric flux through surface 3 of Fig. 29.2.

It was the Scottish physicist James Clerk Maxwell who, in about 1860, suggested that a changing electric flux should give rise to a magnetic field. Since that time many experiments, including direct measurement of the magnetic field inside a charging capacitor, have confirmed Maxwell's remarkable insight. Maxwell quantified his idea by introducing a new term into Ampère's law, a term that describes changing electric flux:

$$\oint \vec{B} \cdot d\vec{l} = \mu_0 I + \mu_0 \epsilon_0 \frac{d\Phi_E}{dt} \quad \text{(Ampère's law with Maxwell's modification)} \qquad (29.1)$$

Now there's no ambiguity. The integral is taken around any loop, I is the current through *any* surface bounded by the loop, and Φ_E is the electric flux through that surface. With our charging capacitor, Equation 29.1 gives the same magnetic field no matter which surface we choose. For surfaces 1, 2, and 4 of Fig. 29.2, the current I makes all the contribution to the right-hand side of the equation. For surface 3, the right-hand side of Equation 29.1 comes entirely from the changing electric flux.

Although changing electric flux isn't the same thing as electric current, it has the same effect in producing a magnetic field. For this reason Maxwell called the term $\epsilon_0(d\Phi_E/dt)$ the **displacement current**. The word *displacement* has historical roots that don't provide much physical insight. But *current* is meaningful because displacement current is indistinguishable from real current in producing magnetic fields. Although we developed the idea of displacement current using the specific example of a charging capacitor, we emphasize that Ampère's law in its now complete form (Equation 29.1) is truly universal: *Any* changing electric flux results in a magnetic field. That fact will prove crucial in establishing the existence of electromagnetic waves.

GOT IT? 29.1 Would you expect to find a magnetic field between the capacitor plates in Fig. 29.2? Explain.

EXAMPLE 29.1 Displacement Current: A Capacitor

A parallel-plate capacitor with plate area A and spacing d is charging at the rate dV/dt. Show that the displacement current is equal to the current in the wires feeding the capacitor.

INTERPRET This is about a comparison between a familiar quantity—current—and a new quantity, namely, displacement current.

DEVELOP We're given the rate at which the capacitor voltage increases. Given that $q = CV$ for a capacitor, we can find the rate of charge buildup—and that's equal to the current I delivering charge to the capacitor. Equation 29.1 shows that the displacement current is $\epsilon_0(d\Phi_E/dt)$, so we'll need the rate of change of electric flux. A parallel-plate capacitor produces an essentially uniform field $E = V/d$. Since the field is uniform, the electric flux through a surface within the capacitor is simply the field strength times the plate area.

EVALUATE For the current, we differentiate the capacitor relation $q = CV$ to get $dq/dt = I = C\,dV/dt$. For the flux, we multiply the electric field by the plate area: $\Phi_E = EA = VA/d$. The rate of change

(continued)

of flux is then $d\Phi_E/dt = (A/d)(dV/dt)$, so the displacement current becomes

$$I_d = \epsilon_0 \frac{d\Phi_E}{dt} = \frac{\epsilon_0 A}{d}\frac{dV}{dt}$$

But $\epsilon_0 A/d$ is the capacitance of a parallel-plate capacitor (given by Equation 23.2), so the displacement current is $I_d = C\,dV/dt$, the same as the actual current I.

ASSESS Make sense? It had better be this way, or Ampère's law would still be ambiguous. For any surface pierced by the wire in Fig. 29.2, the only contribution to the right-hand side of Ampère's law is from the current I. For any surface between the capacitor plates, the only contribution is from the displacement current $I_d = \epsilon_0(d\Phi_E/dt)$. For Ampère's law to give the same magnetic field whichever surface we choose, I and I_d had better be the same. ■

29.3 Maxwell's Equations

It was Maxwell's genius to recognize that Ampère's law needed modifying to reflect the symmetry suggested by Faraday's law. To honor Maxwell, the four complete laws of electromagnetism are called **Maxwell's equations**. The complete set of equations, first published in 1864, governs the behavior of electric and magnetic fields everywhere. Table 29.2 summarizes Maxwell's equations.

Table 29.2 Maxwell's Equations

Law	Mathematical Statement	What It Says	Equation Number
Gauss for $\vec{E}$	$\oint \vec{E}\cdot d\vec{A} = \dfrac{q}{\epsilon_0}$	How charges produce electric field; field lines begin and end on charges.	(29.2)
Gauss for $\vec{B}$	$\oint \vec{B}\cdot d\vec{A} = 0$	No magnetic charge; magnetic field lines don't begin or end.	(29.3)
Faraday	$\oint \vec{E}\cdot d\vec{l} = -\dfrac{d\Phi_B}{dt}$	Changing magnetic flux produces electric field.	(29.4)
Ampère	$\oint \vec{B}\cdot d\vec{l} = \mu_0 I + \mu_0\epsilon_0 \dfrac{d\Phi_E}{dt}$	Electric current and changing electric flux produce magnetic field.	(29.5)

These four compact statements are all it takes to describe classical electromagnetic phenomena. Everything electric or magnetic that we've considered and will consider—from polar molecules to electric current; resistors, capacitors, inductors, and transistors; solar flares and cell membranes; electric generators and thunderstorms; computers, smartphones, and the northern lights—can be described using Maxwell's equations. And despite this wealth of phenomena, we have yet to discuss a most important manifestation of electromagnetism—namely, electromagnetic waves. We've put off waves until now because they depend crucially on Maxwell's extension of Ampère's law. It's easiest to understand electromagnetic waves when they propagate through empty space, so we'll first simplify Maxwell's equations for the case of a vacuum.

Maxwell's Equations in Vacuum

To express Maxwell's equations in vacuum, we simply remove all reference to matter—that is, to electric charge and current:

$$\oint \vec{E}\cdot d\vec{A} = 0 \quad (\text{Gauss}, \vec{E}) \qquad (29.6)$$

$$\oint \vec{B}\cdot d\vec{A} = 0 \quad (\text{Gauss}, \vec{B}) \qquad (29.7)$$

$$\oint \vec{E}\cdot d\vec{l} = -\frac{d\Phi_B}{dt} \quad (\text{Faraday}) \qquad (29.8)$$

$$\oint \vec{B}\cdot d\vec{l} = \mu_0\epsilon_0 \frac{d\Phi_E}{dt} \quad (\text{Ampère}) \qquad (29.9)$$

In vacuum the symmetry is complete, with electric and magnetic fields appearing on an equal footing. With charge and current absent, the only source of either field is a change in the other field—as shown by the time derivatives on the right-hand sides of Faraday's and Ampère's laws.

29.4 Electromagnetic Waves

Faraday's law shows that a changing magnetic field induces an electric field. Ampère's law shows that a changing electric field induces a magnetic field. Together, the two suggest the possibility of **electromagnetic waves**, in which each type of field continuously induces the other, resulting in an electromagnetic disturbance that propagates through space as a wave. We'll now confirm this suggestion with a rigorous demonstration, directly from Maxwell's equations, that electromagnetic waves are indeed possible. In the process we'll discover the properties of electromagnetic waves and come to a deep understanding of the nature of light.

A Plane Electromagnetic Wave

Here we describe the simplest type of electromagnetic wave—a plane wave in vacuum. A plane wave's properties don't vary in directions perpendicular to the wave propagation, so its wavefronts are infinite planes. A plane wave is an approximation to the more realistic case of a spherical wave expanding from a localized source, and it's a good approximation at distances from the wave source that are large compared with the wavelength. Light waves from the Sun, for example, or radio waves miles from the transmitter are essentially plane waves.

In vacuum, it turns out that the electric and magnetic fields of an electromagnetic wave are perpendicular. They're both also perpendicular to the direction of wave propagation—making the electromagnetic wave a transverse wave, as defined in Chapter 14. To be concrete, we'll take the x-direction to be the direction of propagation, the y-direction that of the electric field, and the z-direction that of the magnetic field (Fig. 29.3). We won't prove that a configuration like this is the only one possible for an electromagnetic wave (although in vacuum it is; see Problem 44). What we will do is prove that this configuration satisfies Maxwell's equations—thus showing that such electromagnetic waves are indeed possible. But first we need a mathematical description of our plane electromagnetic wave.

In Chapter 14 we described a sinusoidal wave propagating in the x-direction by a function of the form $A \sin(kx - \omega t)$, where A is the wave amplitude, k the wave num-

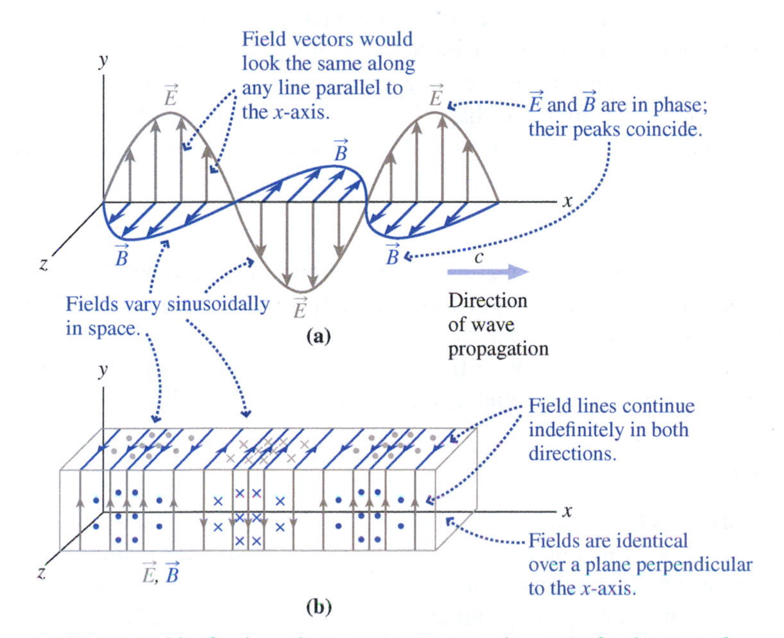

FIGURE 29.3 Fields of a plane electromagnetic wave, shown at a fixed instant of time. (a) Field vectors for points on the x-axis show sinusoidal variation in the fields. (b) A partial representation of the field lines in a rectangular slab. Lines on the facing surfaces of the slab are shown as arrows; lines going through the slab appear as dots or crosses. Spacing of the field lines reflects the sinusoidal variation shown in part (a).

ber, and ω the angular frequency. For mechanical waves, $A \sin(kx - \omega t)$ described some physical quantity such as the height of a water wave or the pressure variation in a sound wave. In an electromagnetic wave, the corresponding physical quantities are the electric and magnetic fields. It turns out that these two wave fields, though perpendicular, are in phase—meaning that their peaks and troughs coincide, as shown in Fig. 29.3*a*. Having chosen the y-direction for the electric field and z for the magnetic field, we can write the fields of our plane electromagnetic wave as

$$\vec{E}(x, t) = E_p \sin(kx - \omega t)\hat{j} \tag{29.10}$$

and

$$\vec{B}(x, t) = B_p \sin(kx - \omega t)\hat{k} \tag{29.11}$$

where the amplitudes E_p and B_p are constants and where $\hat{j}$ and $\hat{k}$ are unit vectors in the y- and z-directions. Figure 29.3*a* is a "snapshot" of some field vectors of this wave at points on the x-axis, shown at a fixed instant of time. That $\vec{E}$ and $\vec{B}$ are perpendicular is obvious from the figure, as is the fact that they're perpendicular to the propagation direction x. You can also see the sinusoidal variation, as the field vectors get alternately longer, then shorter, then reverse direction, and so on. And you can see that $\vec{E}$ and $\vec{B}$ are in phase because their peaks coincide. We emphasize that Fig. 29.3*a* shows field *vectors* for points on the x-axis only; the fields extend forever throughout space, and because this is a plane wave, a picture of field vectors along any line parallel to the x-axis would look the same.

We can also draw field *lines* for our wave, in contrast to the field vectors of Fig. 29.3*a*. We can't draw complete field lines because they extend forever in both directions. So in Fig. 29.3*b* we've shown the field lines only in a rectangular slab; that's enough to give a picture of what the fields look like everywhere. You should convince yourself that Figs. 29.3*a* and *b* show exactly the same thing—namely, a plane electromagnetic wave described by Equations 29.10 and 29.11. In one case we use field *vectors*, whose lengths are proportional to the field magnitudes, and in the other we use field *lines*, which extend forever and whose spacing indicates the field magnitudes.

We'll now show that the electric and magnetic fields pictured in Fig. 29.3 and described by Equations 29.10 and 29.11 satisfy Maxwell's equations. We've chosen a sinusoidal waveform for our wave fields because of its mathematical simplicity. But the superposition principle holds for electric and magnetic fields, and we know from Section 14.5 that we can represent *any* waveform by superposing sinusoids. So our proof that electromagnetic waves can exist holds for any wave shape. That means we can use electromagnetic waves to communicate the complex waveforms representing music, images, and computer data.

Gauss's Laws

In vacuum, Gauss's laws for electric and magnetic fields both have zero on the right-hand side, reflecting the absence of charge. That means the electric and magnetic flux through *any* closed surface must be zero, and therefore the field lines can't begin or end. With our plane wave, the field lines shown partially in Fig. 29.3*b* extend straight forever in both directions. So they don't begin or end, and therefore the fields satisfy Gauss's laws.

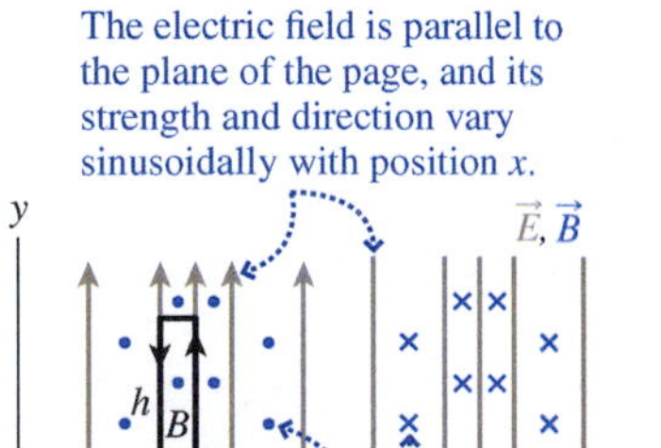

FIGURE 29.4 View of Fig. 29.3*b* in the x–y plane, with a rectangular loop for evaluating the line integral in Faraday's law.

Faraday's Law

To see that Faraday's law is satisfied, look directly toward the x–y plane in Fig. 29.3*b*. You see electric field lines going up and down and magnetic field lines coming straight in and out, as shown in Fig. 29.4. Consider the small rectangular loop of height h and infinitesimal width dx shown in the figure. Evaluating the line integral of the electric field $\vec{E}$ around this loop, we get no contribution from the short ends at right angles to the field. Going around counterclockwise, we get a contribution $-Eh$ as we go down the left side against the field direction. Then we get a positive contribution going up the right side. Because the field varies with position, the field on the right side of the loop is different from that on the left. Let the change be dE, so the field on the right side is $E + dE$, giving a contribution $(E+dE)h$ to the line integral. Then the line integral of $\vec{E}$ around the loop is

$$\oint \vec{E} \cdot d\vec{l} = -Eh + (E + dE)h = h\,dE$$

This nonzero line integral implies an induced electric field. Induced by what? By a changing magnetic flux through the loop. The electric field of the wave arises because of the changing magnetic field of the wave. The area of the loop is $h\,dx$, and the magnetic field $\vec{B}$ is at right angles to this area, so the magnetic flux through the loop is $\Phi_B = Bh\,dx$. The rate of change of flux through the loop is then

$$\frac{d\Phi_B}{dt} = h\,dx\frac{dB}{dt}$$

Faraday's law relates the line integral of the electric field to the rate of change of flux:

$$\oint \vec{E} \cdot d\vec{l} = -\frac{d\Phi_B}{dt}$$

or, using our expressions for the line integral and the rate of change of flux, $h\,dE = -h\,dx(dB/dt)$. Dividing through by $h\,dx$ gives $dE/dx = -dB/dt$. In deriving this equation, we considered changes in E with position at a fixed instant of time. Similarly, the change in B with respect to time is taken at a fixed position. That is, the derivatives are *partial derivatives*—rates of change with respect to one variable while another is held fixed. If you've studied partial derivatives in calculus, you know that the symbol ∂ designates a partial derivative. So our equation $dE/dx = -dB/dt$ should be written with partial derivatives:

$$\frac{\partial E}{\partial x} = -\frac{\partial B}{\partial t} \qquad (29.12)$$

This equation—which is Faraday's law applied to our electromagnetic wave—says that the rate at which the electric field changes with *position* depends on the rate at which the magnetic field changes with *time*.

Ampère's Law

Now look at Fig. 29.3*b* from above. You see the magnetic field lines in the x–z plane and electric field lines emerging perpendicular to the x–z plane (Fig. 29.5). Apply Ampère's law (Equation 29.9) to the rectangle shown. In the line integral there's no contribution from the short sides because they're perpendicular to the field. Going down the left side gives Bh. Going up the right, against the field, gives $-(B + dB)h$, where dB is the change in B across the rectangle. So the line integral in Ampère's law is

$$\oint \vec{B} \cdot d\vec{l} = Bh - (B + dB)h = -h\,dB$$

Again, the area of the rectangular loop is $h\,dx$, so the electric flux through the rectangle becomes $Eh\,dx$. Therefore the rate of change of electric flux is

$$\frac{d\Phi_E}{dt} = h\,dx\left(\frac{dE}{dt}\right)$$

Ampère's law relates the line integral of the magnetic field to this time derivative of the electric flux, giving $-h\,dB = \epsilon_0\mu_0 h\,dx(dE/dt)$. Dividing by $h\,dx$ and again using partial derivatives, we have

$$\frac{\partial B}{\partial x} = -\epsilon_0\mu_0\frac{\partial E}{\partial t} \qquad (29.13)$$

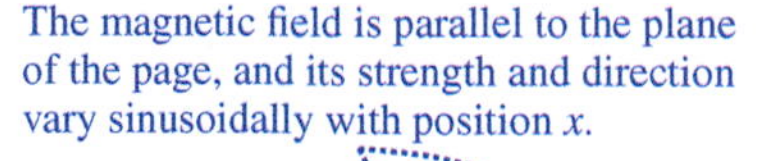

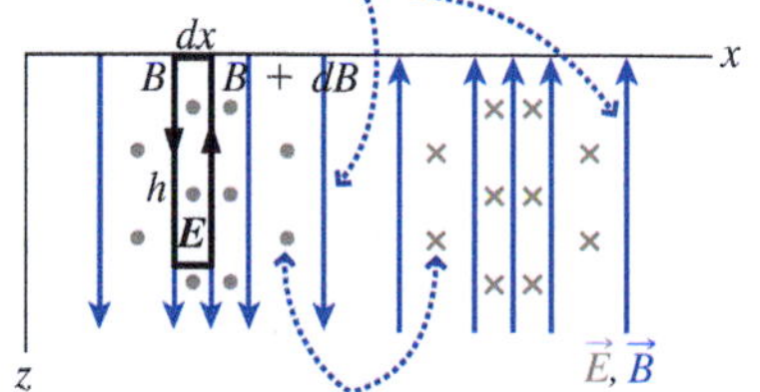

FIGURE 29.5 View of Fig. 29.3*b* in the x–z plane, with a rectangular loop for evaluating the line integral in Ampère's law.

This equation—which is Ampère's law applied to our electromagnetic wave—says that the rate at which the magnetic field changes with *position* depends on the rate at which the electric field changes with *time*. Note that this is the converse of what you saw with Equation 29.12—reflecting that omnipresent complementarity between electricity and magnetism.

Equations 29.12 and 29.13—derived from Faraday's and Ampère's laws—express fully the requirements that Maxwell's universal laws of electromagnetism pose on the field structure of Fig. 29.3. Each describes an induced field that arises from the changing of the other field. That other field, in turn, arises from the changing of the first field. Thus we have a self-perpetuating electromagnetic structure, whose fields exist and change without the need for charged matter. If Equations 29.10 and 29.11, which describe the fields in Fig. 29.3, can be made consistent with Equations 29.12 and 29.13, then we'll have shown that our electromagnetic wave satisfies Maxwell's equations and is thus a possible configuration of electric and magnetic fields. An alternative approach, which doesn't require the sinusoidal fields of Equations 29.10 and 29.11, is to show that Equations 29.12 and 29.13 lead to the wave equation that we introduced in Chapter 14. You can explore this approach in Problem 63.

Conditions on the Wave Fields

To see that Equation 29.12 is satisfied, we differentiate the electric field of Equation 29.10 with respect to x and the magnetic field of Equation 29.11 with respect to t:

$$\frac{\partial E}{\partial x} = \frac{\partial}{\partial x}\left[E_p \sin(kx - \omega t)\right] = kE_p \cos(kx - \omega t)$$

and

$$\frac{\partial B}{\partial t} = \frac{\partial}{\partial t}\left[B_p \sin(kx - \omega t)\right] = -\omega B_p \cos(kx - \omega t)$$

Putting these expressions in for the derivatives in Equation 29.12 gives

$$kE_p \cos(kx - \omega t) = -\left[-\omega B_p \cos(kx - \omega t)\right]$$

The cosine cancels, showing that the equation holds if

$$kE_p = \omega B_p \tag{29.14}$$

To see that Equation 29.13 is also satisfied, we differentiate the magnetic field given by Equation 29.11 with respect to x and the electric field of Equation 29.10 with respect to t:

$$\frac{\partial B}{\partial x} = kB_p \cos(kx - \omega t) \quad \text{and} \quad \frac{\partial E}{\partial t} = -\omega E_p \cos(kx - \omega t)$$

Using these expressions in Equation 29.13 then gives

$$kB_p \cos(kx - \omega t) = -\epsilon_0 \mu_0\left[-\omega E_p \cos(kx - \omega t)\right]$$

Again, the cosine cancels, so this equation is satisfied if

$$kB_p = \epsilon_0 \mu_0 \omega E_p \tag{29.15}$$

Our analysis has shown that electromagnetic waves whose form is given by Fig. 29.3 and Equations 29.10 and 29.11 can exist, provided that the amplitudes E_p and B_p, and the frequency ω and wave number k, are related by Equations 29.14 and 29.15. Physically, the existence of these waves is possible because a change in either field—electric or magnetic—induces the other field, giving rise to a self-perpetuating electromagnetic-field structure. Maxwell's theory thus leads to the prediction of an entirely new phenomenon—electromagnetic waves. We'll now explore some properties of these waves.

GOT IT? 29.2 Equations 29.12 and 29.13 will form the basis for our understanding of electromagnetic waves. From which two of the four Maxwell equations do they derive? (a) Gauss for electricity and Gauss for magnetism; (b) Gauss for electricity and Ampère; (c) Faraday and Ampère; (d) Gauss for magnetism and Faraday

29.5 Properties of Electromagnetic Waves

Wave Speed

In Chapter 14 we found that the speed of a sinusoidal wave is the ratio of the angular frequency to the wave number: speed $= \omega/k$. To determine the speed of our electromagnetic wave, we use Equation 29.14 to get $E_p = \omega B_p/k$, and then use this expression in Equation 29.15:

$$kB_p = \epsilon_0\mu_0\omega E_p = \frac{\epsilon_0\mu_0\omega^2 B_p}{k}$$

The amplitude B_p cancels, and we solve for the wave speed ω/k to get

$$\text{wave speed} = \frac{\omega}{k} = \frac{1}{\sqrt{\epsilon_0\mu_0}} \quad \text{(EM wave speed in vacuum)} \tag{29.16a}$$

This result shows that the speed of an electromagnetic wave in vacuum depends only on the electric and magnetic constants ϵ_0 and μ_0. All electromagnetic waves in vacuum, regardless of frequency or amplitude, share this speed. Although we derived this result for sinusoidal waves, the superposition principle ensures that it holds for any wave shape.

Using the known values of ϵ_0 and μ_0, let's evaluate the speed given in Equation 29.16a:

$$\frac{1}{\sqrt{\epsilon_0\mu_0}} = \frac{1}{\sqrt{(8.85\times10^{-12}\,\text{C}^2/\text{N}\cdot\text{m}^2)(4\pi\times10^{-7}\,\text{N/A}^2)}} = 3.00\times10^8\ \text{m/s}$$

But this is the speed of light! During the two centuries before Maxwell, scientists had measured light's speed with increasing accuracy. They had also recognized, thanks to Thomas Young's 1801 interference experiment, that light consists of waves. Then, in the 1860s, came Maxwell. Using a theory developed from laboratory experiments on electricity and magnetism, with no reference to optics or light, Maxwell showed how the interplay of electric and magnetic fields results in electromagnetic waves. The wave speed—calculated from the constants ϵ_0 and μ_0—was the known speed of light. Maxwell's conclusion was inescapable: *Light is an electromagnetic wave.*

Maxwell's identification of light as an electromagnetic phenomenon is a classic example of the unification of knowledge in science. With one simple calculation, Maxwell brought the entire science of optics under the umbrella of electromagnetism. Maxwell's work stands as a crowning intellectual triumph, one whose implications are still expanding our view of the universe.

Maxwell's discovery lets us recast Equation 29.16a in the form

$$\frac{\omega}{k} = c \quad \text{(EM wave speed in vacuum: the speed of light, } c\text{!)} \tag{29.16b}$$

where $c = 1/\sqrt{\epsilon_0\mu_0}$ is the speed of light. Because $\omega = 2\pi f$ and $k = 2\pi/\lambda$, we can rewrite Equation 29.16b in terms of the more familiar frequency f and wavelength λ as

$$f\lambda = c \quad \text{(frequency, wavelength, and the speed of light)} \tag{29.16c}$$

As we saw in Chapter 1, the SI definition of the meter gives c the exact value 299,792,458 m/s. Some non-SI values for c are approximately 186,000 miles per second, approximately 1 foot per nanosecond (see Exercise 20), and exactly 1 light-year per year.

Wave Amplitude

The amplitudes E_p and B_p dropped out of our analysis, showing that an electromagnetic wave's speed is independent of amplitude. But the field strengths E and B aren't independent. Using $\omega/k = c$, we can recast Equation 29.14 to show that

$$E = \frac{\omega}{k}B = cB \qquad (E, B \text{ relation in vacuum EM wave}) \qquad (29.17)$$

Here we dropped the "peak" subscript because E_p and B_p multiply identical cosine terms in our wave description, so Equation 29.14 applies whether or not we're at the peak field.

Phase, Orientation, and Waves in Matter

The wave of Fig. 29.3 and Equations 29.10 and 29.11 has $\vec{E}$ and $\vec{B}$ in phase in *time*—peaking at the same time—while they're perpendicular in *space* and also perpendicular to the propagation direction. Our derivation of the wave speed used these properties, so we've confirmed that electromagnetic waves in vacuum are transverse waves with $\vec{E}$ and $\vec{B}$ perpendicular and in phase. Specifically, the direction of propagation is that of the cross product $\vec{E} \times \vec{B}$. These geometrical properties also apply to electromagnetic waves in common materials like air and glass. The wave speed in these materials is lower than in vacuum, although for air the difference is minuscule. Electromagnetic waves in more complex materials can have very different properties and propagation speeds.

GOT IT? 29.3 At a particular point the electric field of an electromagnetic wave points in the $+y$-direction, while the magnetic field points in the $-z$-direction. Is the propagation direction (a) $+x$; (b) $-x$; (c) either $+x$ or $-x$ but you can't tell which; (d) $-y$; (e) $+z$; or (f) not along any of the coordinate axes?

EXAMPLE 29.2 Electromagnetic-Wave Properties: Laser Light

A laser beam with wavelength 633 nm is propagating through air in the $+z$-direction. Its electric field is parallel to the x-axis and has amplitude 6.0 kV/m. (a) Find the wave frequency, (b) the amplitude of the magnetic field, and (c) the direction of the magnetic field.

INTERPRET Light is an electromagnetic wave, so the laser beam shares the wave properties we've just discussed. As we noted, its speed in air is nearly the same as in vacuum. Here we're given the wavelength and peak electric field, E_p.

DEVELOP Equation 29.16c, $f\lambda = c$, relates wavelength and frequency, so we'll use that for (a). Equation 29.17, $E = cB$, relates E and B, so we can get (b) from the given value of E_p. For (c), we'll draw a reoriented version of Fig. 29.3*a* to help infer the direction of $\vec{B}$.

EVALUATE (a) Solving for the frequency gives

$$f = c/\lambda = (3.0\times10^8 \text{ m/s})/(633 \text{ nm}) = 4.7\times10^{14} \text{ Hz}$$

(b) Solving for the magnetic-field amplitude gives $B_p = E_p/c = 20\ \mu\text{T}$. (c) Figure 29.6 shows that with propagation in the z-direction and $\vec{E}$ along the x-axis, $\vec{B}$ must be parallel to the y-axis.

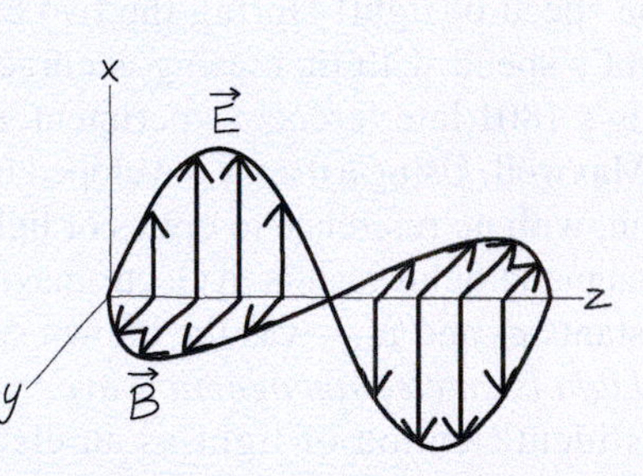

FIGURE 29.6 Reoriented version of Fig. 29.3*a* for Example 29.2. It's important that we still have a right-handed coordinate system, with the *x*-, *y*-, and *z*-axes in the same relation as in Fig. 29.3. Equivalently, the direction of $\vec{E} \times \vec{B}$ is the propagation direction.

ASSESS That 10^{14}-Hz frequency sounds huge, but light has such a short wavelength that its frequency is indeed high; more on this shortly. Notice both in Fig. 29.3 and in our reoriented wave of Fig. 29.6 that the vectors $\vec{E}$ and $\vec{B}$ and the propagation direction form a right-handed coordinate system, so any two of those vectors determine the direction of the third. ■

Polarization

Although $\vec{E}$ and $\vec{B}$ are necessarily perpendicular, their orientation is still arbitrary within a plane perpendicular to the propagation direction. **Polarization** specifies the direction of the electric field and thus determines the perpendicular magnetic-field direction as well (Fig. 29.7).

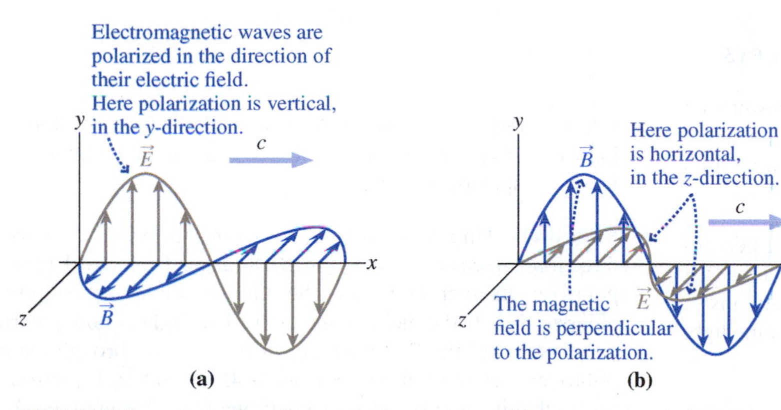

FIGURE 29.7 The polarization direction is the direction of the wave's electric field.

Electromagnetic waves used in radio and TV originate from antennas that give the waves a definite polarization. Most laser light is also polarized. In contrast, light from hot sources like the Sun or a lightbulb is unpolarized, consisting of a mix of waves with random field orientations. Unpolarized light becomes polarized when it reflects off surfaces or passes through substances whose structure has a preferred direction. Many crystals and synthetic materials such as Polaroid exhibit such a *transmission axis*. Light reflecting off roads and a car's hood becomes partially polarized in the horizontal direction, and Polaroid sunglasses, with their transmission axis vertical, block the resulting glare. The same thing happens with reflection off water, making Polaroid sunglasses especially useful for boaters.

A polarizing material passes unattenuated only the component of the wave field $\vec{E}$ along the transmission axis—namely, $E\cos\theta$, where θ is the angle between the field and the transmission axis. We'll show shortly that the intensity of an electromagnetic wave is proportional to the square of the field strength. As a result, a wave of intensity S_0 emerges from a polarizer with intensity given by the **law of Malus**:

$$S = S_0 \cos^2\theta \qquad (29.18)$$

FIGURE 29.8 Two pieces of polarizing material with their transmission axes at right angles. Where they overlap, no light gets through.

Thus electromagnetic waves are blocked completely by a polarizer with its transmission axis oriented perpendicular to the waves' polarization (Fig. 29.8).

Measuring polarization tells us about sources of electromagnetic waves and about materials through which they propagate. Many astrophysical processes produce polarized waves; their polarization gives clues to mechanisms operating in the cosmos. Geologists pass polarized light through thin sections of rock to reveal the rocks' composition, and engineers use polarization to locate stresses in mechanical structures. Polarization is essential in many technologies, including the ubiquitous liquid crystal displays (LCDs) in our cell phones, cameras, computers, and TVs (Fig. 29.9).

Video Tutor Demo | **Parallel-Wire Polarizer for Microwaves**

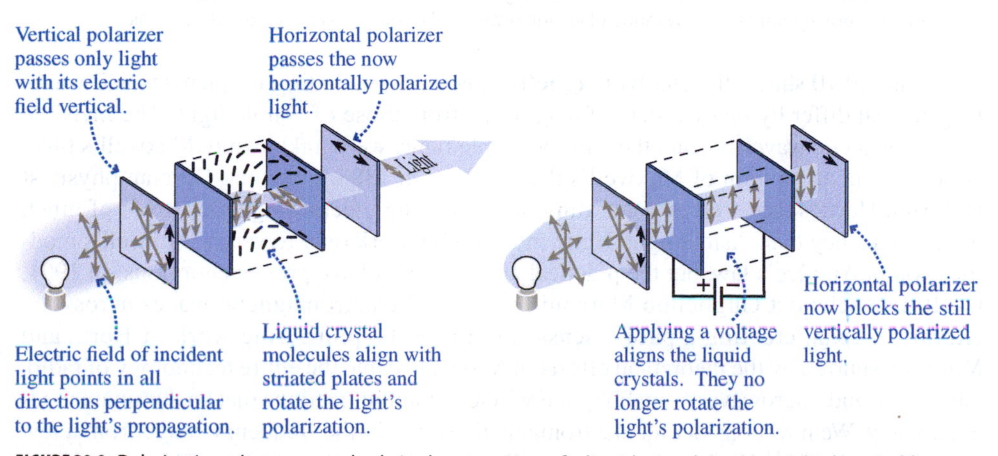

FIGURE 29.9 Polarization plays a central role in the operation of a liquid crystal display. Multiple units like the one shown—millions in a TV or computer screen—produce the individual pixels on an LCD.

CONCEPTUAL EXAMPLE 29.1 Crossed Polarizers

Unpolarized light shines on a pair of polarizers with their transmission axes perpendicular, so no light gets through the combination. What happens when a third polarizer is sandwiched in between, with its transmission axis at 45° to the others?

EVALUATE The middle polarizer's transmission axis isn't perpendicular to the first one's, so some of the light coming through the first polarizer gets through the middle one. That light's polarization isn't perpendicular to the last polarizer's transmission axis, so some light gets all the way through the combination.

ASSESS This result may seem surprising: If the two outer polarizers are perpendicular, how can a third polarizer change the situation? But it does. No pair of adjacent polarizers is perpendicular, so each pair transmits some light. Inserting the third polarizer lets light through where none came through before.

MAKING THE CONNECTION How does the intensity of light emerging from this polarizer "sandwich" compare with the intensity of the incident unpolarized light?

EVALUATE Unpolarized light is a random mix of polarization directions, so $\cos^2\theta$ in Equation 29.18 ranges from 0 to 1 for the first polarizer. Its average is $\frac{1}{2}$, so the intensity emerging from the first polarizer is half the incident intensity. This light is now polarized in the direction of the first polarizer; it then passes through the middle polarizer, oriented at 45°. Since $\cos 45° = 1/\sqrt{2}$, Equation 29.18 shows that its intensity is cut in half again. Light emerging from the middle polarizer then passes through the last one, oriented at 45° to the light's new polarization, so its intensity is halved yet again. The effect of three reductions by one-half each is that light emerges from the "sandwich" with one-eighth its incident intensity.

29.6 The Electromagnetic Spectrum

Although Equations 29.16 relate an electromagnetic wave's frequency and wavelength, one of these quantities remains arbitrary. That means electromagnetic waves can have any frequency or, equivalently, any wavelength. Visible light occupies a wavelength range from about 400 nm to 700 nm, corresponding to frequencies of 7.5×10^{14} Hz to 4.3×10^{14} Hz. The different wavelengths or frequencies correspond to different colors, with red at the long-wavelength, low-frequency end of the visible region and violet at the short-wavelength, high-frequency end (see the enlargement in Fig. 29.10).

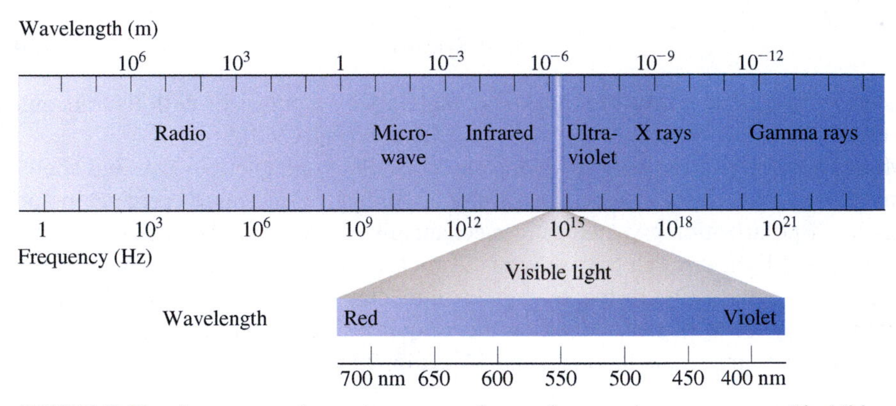

FIGURE 29.10 The electromagnetic spectrum ranges from radio waves to gamma rays, with visible light occupying only a narrow range of wavelengths and frequencies on a logarithmic scale.

Figure 29.10 shows the **electromagnetic spectrum**, including frequencies and wavelengths that differ by many orders of magnitude from those of visible light. The invisible electromagnetic waves beyond the narrow visible range were unknown in Maxwell's time. A brilliant confirmation of Maxwell's theory came in 1888, when the German physicist Heinrich Hertz succeeded in generating and detecting electromagnetic waves of much lower frequency than visible light. Hertz intended his work only to verify Maxwell's modification of Ampère's law, but the practical consequences have proven enormous. In 1901, the Italian scientist Guglielmo Marconi transmitted electromagnetic waves across the Atlantic Ocean, creating a public sensation. From the pioneering work of Hertz and Marconi, spurred by the theoretical efforts of Maxwell, came the entire technology of radio, television, and microwaves—enabling the wireless communications that so dominate modern society. We now consider all electromagnetic waves in the frequency range from a few Hz to about 3×10^{11} Hz as radio waves, with AM radio at about 1 MHz, FM at 100 MHz, television in patches of the spectrum from about 50 MHz to 1 GHz, and microwaves for WiFi, radar, cooking, cell phones, and satellite communications at 1 GHz and above.

Between radio waves and visible light lies the infrared frequency range. Electromagnetic waves in this region are emitted by warm objects, even when they're not hot enough to glow visibly. For this reason, infrared cameras are used to determine subtle body-temperature differences in medical diagnosis, to examine buildings for heat loss, and to study the birth of stars in clouds of interstellar gas and dust.

Beyond visible light are the ultraviolet rays responsible for sunburn, then the highly penetrating X rays, and finally the gamma rays whose primary terrestrial source is radioactive decay. All these phenomena, from radio to gamma rays, are fundamentally the same: All are electromagnetic waves, differing only in frequency and wavelength. All travel with speed c in vacuum, and all consist of electric and magnetic fields produced from each other through the induction processes described by Faraday's and Ampère's laws. Naming the different types of electromagnetic waves is just a convenience; there are no gaps in the continuous range of frequencies and wavelengths. Practical differences arise because waves of different wavelengths interact differently with matter; in particular, shorter wavelengths tend to be generated and absorbed most efficiently by smaller systems.

Earth's atmosphere is transparent to visible light and to most radio frequencies. But it's opaque to most infrared, the higher-frequency ultraviolet, X rays, and all but the highest-frequency gamma rays. Earth's surface would be hazardous to life if ultraviolet weren't absorbed by ozone gas high in the atmosphere, and our planet would be a lot cooler if water vapor, carbon dioxide, and other gases didn't absorb outgoing infrared. But that same infrared absorption is at the heart of our worries about global climate change, because we humans are increasing the levels of infrared-trapping gases. And, as the Application shows, our knowledge of the universe beyond Earth would be severely limited if we could only observe electromagnetic waves that make it through Earth's atmosphere.

GOT IT? 29.4 Figure 29.10 shows that the frequency of gamma rays is (a) about 50% greater than that of visible light, (b) about one one-millionth that of visible light, or (c) more than a million times that of visible light.

APPLICATION Windows on the Universe

For centuries humankind's only information about the universe beyond Earth—except for the occasional meteorite—came from visible light. Optical astronomy gave us a good picture of objects like the Sun and other stars, but visible light alone can't reveal the richness of our vast universe. The discovery of radio waves from outer space led to the development of radio astronomy in the 1930s, complementing optical telescopes by receiving electromagnetic waves of much lower frequency than visible light. But, as the figure shows, much of the rest of the electromagnetic spectrum is blocked by Earth's atmosphere. So a truly comprehensive picture of the universe had to await the development of space flight in the late 20th century. Today we have space-based observatories that gather electromagnetic waves in the infrared, ultraviolet, X-ray, and even gamma-ray regions of the spectrum. We also observe from space in visible light, eliminating atmospheric distortion. Together, astronomical observations across the full electromagnetic spectrum let us observe phenomena from low-energy processes like the birth of stars to high-energy events such as matter plunging into black holes. The photos show two extremes: a ground-based radio telescope array and a spacecraft that detects mysterious gamma-ray bursts from the distant universe.

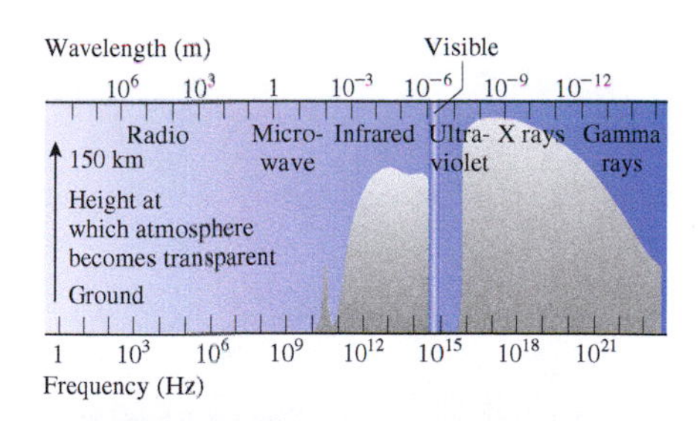

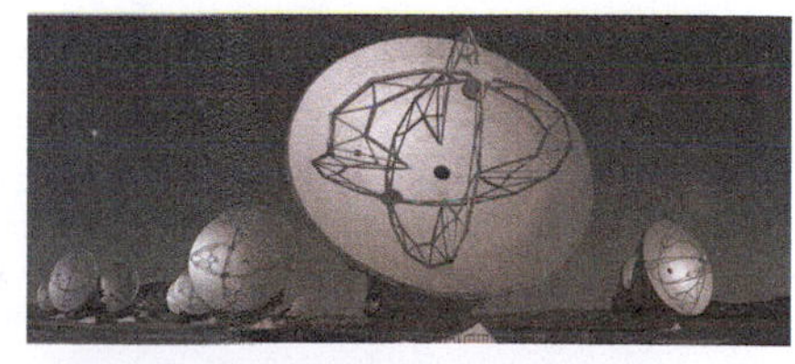

29.7 Producing Electromagnetic Waves

All it takes to produce an electromagnetic wave is a changing electric or magnetic field. Once there's a change in one field, induction provides the other field, and together the changing fields continuously regenerate one another. The wave is on its way! Ultimately, changing fields of both types result when we alter the motion of electric charge. Therefore, **accelerated charge is the source of electromagnetic waves**.

In a radio transmitter, the accelerated charges are electrons moving back and forth in an antenna, driven by alternating voltage from an *LC* circuit (Fig. 29.11). In an X-ray tube, high-energy electrons decelerate rapidly as they slam into a target; their deceleration is the source of the electromagnetic waves, now in the X-ray region of the spectrum. In the magnetron tube of a microwave oven, electrons circle in a magnetic field; their centripetal acceleration is the source of the microwaves that cook your food. And the altered movement of electrons in atoms—although described accurately only by quantum mechanics—is the source of most visible light. If the motion of the accelerated charges is periodic, then the wave frequency is that of the motion; more generally, systems are most efficient at producing (and receiving) electromagnetic waves whose wavelength is comparable to the size of the system. That's why TV antennas are on the order of 1 m in size, while nuclei—some 10^{-15} m in diameter—produce gamma rays.

Calculation of electromagnetic waves from accelerated charges presents challenging but important problems for physicists and engineers. Figure 29.12 shows the field of an oscillating dipole—a configuration approximated by many systems from antennas to atoms and molecules. Note that the waves are strongest in the direction at right angles to

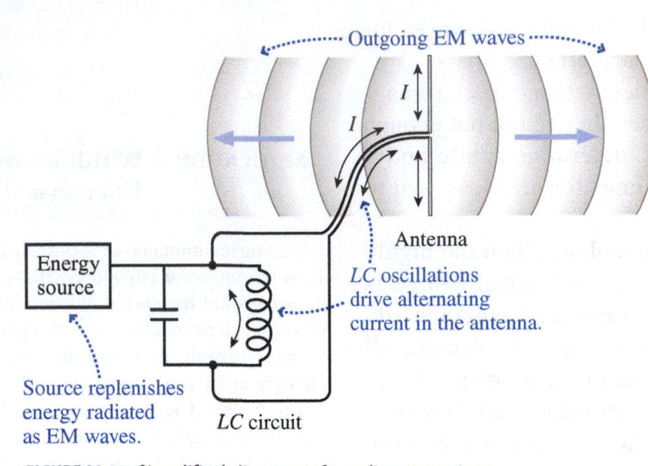

FIGURE 29.11 Simplified diagram of a radio transmitter.

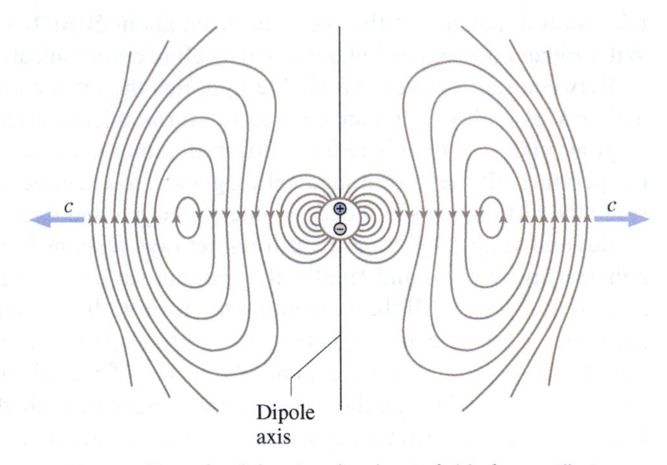

FIGURE 29.12 "Snapshot" showing the electric field of an oscillating electric dipole at an instant in time.

MP

PhET: Radio Waves & Electromagnetic Fields
PhET: Radiating Charge

the acceleration of the charge distribution and that there's no radiation in the direction of the acceleration. This accounts for, among other phenomena, the directionality of radio and TV antennas, which transmit and receive most effectively perpendicular to the long direction of the antenna.

The field shown in Fig. 29.12 seems to bear little resemblance to the plane-wave fields of Fig. 29.3 that we used to demonstrate the possibility of electromagnetic waves. We could produce true plane waves only with an infinite sheet of accelerated charge—an impossibility. But far from the source, the curved field lines in Fig. 29.12 would appear straight, and the wave would approximate a plane wave. So our plane-wave analysis is a valid approximation at great distances—typically many wavelengths—from a localized wave source. Closer to the source more complicated expressions for the wave fields apply, but these, too, satisfy Maxwell's equations.

GOT IT? 29.5 Molecular biologists and pharmaceutical companies are increasingly turning to *synchrotron radiation* for the study of biomolecules and the development of new drugs. This intense electromagnetic radiation comes from electrons circling at high speed in so-called storage rings that use magnetic fields to keep the electrons in circular paths. Which of the following is most essential for the generation of these electromagnetic waves? (a) the fact that the electrons move in circular paths and are thus accelerated; (b) the presence of a strong magnetic field; (c) the electrons' high speed

---

29.8 Energy and Momentum in Electromagnetic Waves

We showed in earlier chapters that electric and magnetic fields contain energy. Electromagnetic waves are combinations of electric and magnetic fields; as they propagate, they transport the energy contained in those fields.

Wave Intensity

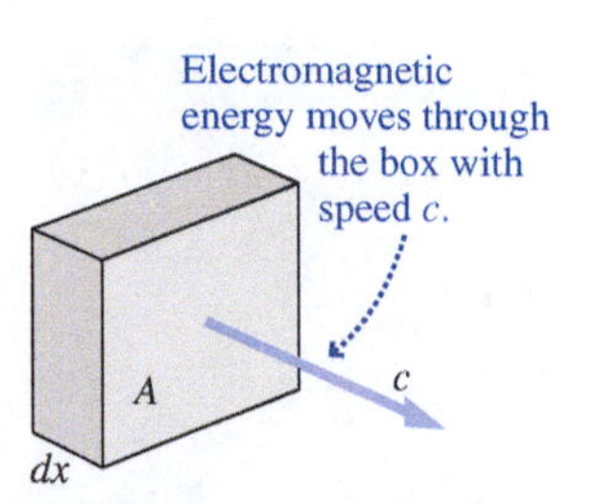

FIGURE 29.13 A box of length dx and cross-sectional area A at right angles to the propagation of an electromagnetic wave.

In Chapter 14 we defined wave intensity as the rate at which a wave transports energy across a unit area; its units are W/m^2. We can calculate the intensity S of a plane electromagnetic wave by considering a rectangular box of thickness dx and cross-sectional area A with its face perpendicular to the wave propagation (Fig. 29.13). Within this box are wave fields $\vec{E}$ and $\vec{B}$ whose energy densities are given by Equations 23.7 and 26.9: $u_E = \frac{1}{2}\epsilon_0 E^2$ and $u_B = B^2/2\mu_0$. If dx is small enough that E and B don't vary significantly, the total energy in the box is the sum of the electric and magnetic energy densities multiplied by the box volume $A\,dx$:

$$dU = (u_E + u_B)A\,dx = \frac{1}{2}\left(\epsilon_0 E^2 + \frac{B^2}{\mu_0}\right)A\,dx$$

This energy moves with speed c, so all the energy moves out of the box in a time $dt = dx/c$. The rate at which energy moves through the cross-sectional area A is then

$$\frac{dU}{dt} = \frac{1}{2}\left(\epsilon_0 E^2 + \frac{B^2}{\mu_0}\right)\frac{A\,dx}{dx/c} = \frac{c}{2}\left(\epsilon_0 E^2 + \frac{B^2}{\mu_0}\right)A$$

So the intensity S, or rate of energy flow per unit area, is

$$S = \frac{c}{2}\left(\epsilon_0 E^2 + \frac{B^2}{\mu_0}\right)$$

We can recast this equation in simpler form using $E = cB$ and $B = E/c$ for an electromagnetic wave. Replacing one of the E's in E^2 with cB and one of the B's in B^2 with E/c, we have

$$S = \frac{c}{2}\left(\epsilon_0 cEB + \frac{EB}{\mu_0 c}\right) = \frac{1}{2\mu_0}(\epsilon_0 \mu_0 c^2 + 1)EB$$

But $c = 1/\sqrt{\epsilon_0 \mu_0}$, so $\epsilon_0 \mu_0 c^2 = 1$, giving

$$S = \frac{EB}{\mu_0} \qquad (29.19a)$$

Although we derived Equation 29.19a for an electromagnetic wave, it is in fact a special case of the more general result that nonparallel electric and magnetic fields entail a flow of electromagnetic energy. In general, the rate of energy flow per unit area is given by

$$\vec{S} = \frac{\vec{E} \times \vec{B}}{\mu_0} \quad \text{(Poynting vector)} \qquad (29.19b)$$

Here the vector $\vec{S}$ gives the direction of the energy flow as well as its magnitude. For an electromagnetic wave in vacuum, with $\vec{E}$ and $\vec{B}$ at right angles, Equation 29.19b reduces to Equation 29.19a, with the direction of energy flow the same as the direction of wave travel. The vector intensity $\vec{S}$ is called the **Poynting vector** after the English physicist J. H. Poynting, who suggested it in 1884. Problem 60 explores an important application of the Poynting vector to fields that don't constitute an electromagnetic wave.

In an electromagnetic wave the fields oscillate, and so, therefore, does the intensity. We're usually not interested in this rapid oscillation. For example, an engineer designing a solar collector doesn't care that sunlight intensity oscillates at about 10^{14} Hz. What she really wants is the *average* intensity, $\bar{S}$. Because the instantaneous intensity given by Equation 29.19a contains a product of sinusoidally varying terms, which are in phase, the average intensity is just half the peak intensity:

$$\bar{S} = \frac{\overline{EB}}{\mu_0} = \frac{E_p B_p}{2\mu_0} \quad \text{(average intensity)} \qquad (29.20a)$$

Typical values for $\bar{S}$ in visible light range from a few W/m^2 in the faint light of a candle to many MW/m^2 in the most intense laser beams.

We wrote Equation 29.20a in terms of both the electric and magnetic fields, but we can use the wave condition $E = cB$ to eliminate either field in terms of the other:

$$\bar{S} = \frac{E_p^{\,2}}{2\mu_0 c} \quad \text{and} \quad \bar{S} = \frac{cB_p^{\,2}}{2\mu_0} \qquad (29.20\text{b, c})$$

GOT IT? 29.6 Lasers 1 and 2 emit light of the same color, and the electric field in the beam from laser 1 is twice as strong as the field in laser 2's beam. How do their (1) magnetic fields, (2) intensities, and (3) wavelengths compare?

EXAMPLE 29.3 Fields and Power: Solar Energy

The average intensity of noontime sunlight on a clear day is about $1\,\mathrm{kW/m^2}$. (a) What are the peak electric and magnetic fields in sunlight? (b) At this intensity, what area of 40% efficient solar collectors would you need to replace a 4.8-kW water heater?

INTERPRET In (a) we're asked for the peak electric and magnetic fields, and we identify $1\,\mathrm{kW/m^2}$ as the average intensity, $\bar{S}$. In (b) we want solar collectors to replace an electric heater whose power we're given.

DEVELOP For (a), Equations 29.20b, $\bar{S} = E_p^2/2\mu_0 c$, and 29.20c, $\bar{S} = cB_p^2/2\mu_0$, relate peak fields and average intensity. We could solve both of these or, more easily, solve one of them and then use Equation 29.17, $E = cB$, to get the other field. In (b) we'll need to use the 40% efficiency to get the effective power per square meter of solar collector, and then find the area needed to replace the 4.8-kW power of the electric heater.

EVALUATE (a) We solve Equation 29.20c for B_p:

$$B_p = \sqrt{2\,\mu_0 \bar{S}/c} = 2.9\,\mu\mathrm{T}.$$

Then Equation 29.17 gives $E = cB = 0.87\,\mathrm{kV/m}$. (b) At 40% efficiency, each square meter of solar collector in 1-kW/m^2 sunlight produces 0.40 kW. So to get 4.8 kW, we need $(4.8\,\mathrm{kW})/(0.40\,\mathrm{kW/m^2}) = 12\,\mathrm{m^2}$ of collector area.

ASSESS The fields we've calculated are relatively modest, showing that even bright sunlight doesn't entail large electric and magnetic fields. That 12-m^2 collector area is also modest—much smaller than the $100\,\mathrm{m^2}$ of a typical house roof—showing that water heating is a practical use of solar technology. Indeed, your author's house, in cloudy Vermont, gets 95% of its summertime hot water from just $9\,\mathrm{m^2}$ of solar collectors. ■

Waves from Localized Sources

When electromagnetic waves originate in a localized source such as an atom, a radio transmitting antenna, a lightbulb, or a star, the wavefronts aren't planes but expanding spheres (recall Fig. 14.13). As the waves expand, their energy is spread over ever-larger spheres—whose area increases as the square of the distance from the source. Therefore, as we found for mechanical waves in Chapter 14, the power per unit area—also called the *intensity* and designated S for electromagnetic waves—decreases as the inverse square of the distance:

$$S = \frac{P}{4\pi r^2} \tag{29.21}$$

Video Tutor Demo | **Point of Equal Brightness between Two Light Sources**

Here S and P can be either peak or average intensity and power, and r is the distance from the source. The intensity decreases not because electromagnetic waves "weaken" and lose energy but because their energy gets spread ever more thinly.

Because the intensity of an electromagnetic wave is proportional to the *square* of the field strengths (Equations 29.20), Equation 29.21 shows that the *fields* of a spherical wave decrease as $1/r$. Contrast that with the $1/r^2$ decrease in the electric field of a stationary point charge, and you can see why the wave fields associated with an accelerated charge dominate in all but the immediate vicinity of the charge.

EXAMPLE 29.4 Electromagnetic-Wave Intensity: Cell-Phone Reception

A cell phone's typical average radiated power is about 0.6 W. If the receiver at a cell tower can handle signals with peak electric fields as weak as 1.2 mV/m, what's the maximum distance from phone to tower?

INTERPRET We're asked to find the distance from the 0.6-W cell phone to a cell tower on the condition that the electric field of the cell phone's electromagnetic wave is no weaker than 1.2 mV/m when measured at the tower.

DEVELOP Assuming the 0.6-W signal spreads in all directions, Equation 29.21, $\bar{S} = \bar{P}/4\pi r^2$, gives the average intensity at a distance r from the phone. (Here the bars indicate we're dealing with average quantities.) Using this expression for $\bar{S}$ in Equation 29.20b gives

$\overline{P}/4\pi r^2 = E_p^2/2\mu_0 c$. Our plan is to solve for the distance r that gives $E_p = 1.2$ mV/m.

EVALUATE Solving for r gives $r = \sqrt{2\mu_0 c P/4\pi E_p^2} = 5$ km, using $P = 0.6$ W and $E_p = 1.2$ mV/m.

ASSESS This answer is about 3 miles, a bit more than the cell radius discussed in the Application below. That's enough to provide a margin of safety, ensuring reliable communications for all phones within the cell.

APPLICATION Cell Phones

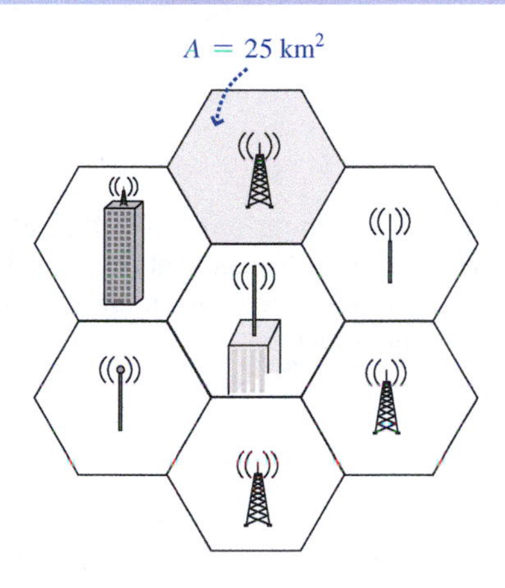

Your cell phone contains a tiny, low-power radio transmitter whose signal intensity decreases as the inverse square of the distance from the phone. The cell-phone network consists of antennas and associated circuits that receive and transmit signals from and to individual phones. Because of the phones' low power, antennas need to be closely spaced so a phone is rarely out of range. The figure shows a typical urban cell-phone network consisting of multiple cells—hence the term *cell* phone—each with an antenna mounted on a tower or building. Cells are typically hexagonal regions about 25 km^2 in area; approximating them as circles gives a radius of about 2.8 km—roughly the maximum distance between a phone and an antenna. As you move through an urban area, the network automatically "hands off" your phone to the nearest cell tower. Cell phones transmit on one frequency and receive on another, allowing two-way communications with both parties able to talk at once. The system uses thousands of frequency channels, and thus a single cell tower can handle many simultaneous calls. Cell towers are more widely spaced in rural regions, and phones automatically boost their power to compensate.

Momentum and Radiation Pressure

Moving objects carry not only energy but also momentum. So do electromagnetic waves. Maxwell showed that wave energy U and momentum p are related by $p = U/c$. The wave intensity $\overline{S}$ is the average rate at which the wave carries energy per unit area, and therefore the wave carries momentum per unit area at the rate $\overline{S}/c$. An object that absorbs the wave energy (like a black object exposed to sunlight) absorbs this momentum as well. Newton's law in its general form $F = dp/dt$ shows that the object then experiences a force. Since $\overline{S}/c$ is the rate of momentum absorption per unit area, the result is a **radiation pressure**:

MP

PhET: Optical Tweezers and Applications

$$P_{\text{rad}} = \frac{\overline{S}}{c} \quad \text{(radiation pressure)} \qquad (29.22)$$

Radiation pressure doubles if an object reflects electromagnetic waves, just as bouncing a basketball off the backboard changes the ball's momentum by $2mv$ and therefore delivers momentum $2mv$ to the backboard.

The pressure of ordinary light is tiny and difficult to measure, but high-power lasers can actually levitate small particles. Light pressure has even been suggested for spacecraft propulsion (see Passage Problems 71–74). The idea that electromagnetic waves carry momentum played a crucial role in Einstein's development of his equation $E = mc^2$. Today, biologists exploit the transfer of momentum from electromagnetic waves to matter, trapping and manipulating viruses, DNA strands, and other microbiological structures with laser-based *optical tweezers*. You can explore optical tweezers in the PhET simulation referenced here.

CHAPTER 29 SUMMARY

Big Idea

The big idea here—and one of the biggest ideas in physics—is that electric and magnetic fields together form self-regenerating structures that propagate through space as **electromagnetic waves**. What makes these waves possible is that changing magnetic fields induce electric fields (Faraday's law), and changing electric fields induce magnetic fields (Ampère's law, with Maxwell's modification). Electromagnetic (EM) waves in vacuum consist of electric and magnetic fields perpendicular to each other and to the direction of wave propagation, and in phase.

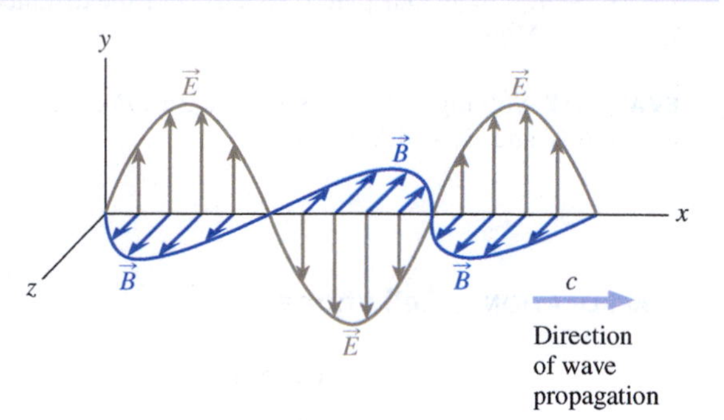

Key Concepts and Equations

Maxwell's equations describe completely the behavior of electric and magnetic fields in classical physics:

Law	Mathematical Statement	What It Says
Gauss for $\vec{E}$	$\oint \vec{E}\cdot d\vec{A} = \dfrac{q}{\epsilon_0}$	How charges produce electric fields; field lines begin and end on charges.
Gauss for $\vec{B}$	$\oint \vec{B}\cdot d\vec{A} = 0$	No magnetic charge; magnetic field lines don't begin or end.
Faraday	$\oint \vec{E}\cdot d\vec{l} = -\dfrac{d\Phi_B}{dt}$	Changing magnetic flux produces electric fields.
Ampère	$\oint \vec{B}\cdot d\vec{l} = \mu_0 I + \mu_0\epsilon_0\dfrac{d\Phi_E}{dt}$	Electric current and changing electric flux produce magnetic fields.

Maxwell's equations show that electromagnetic waves are possible and that their speed in vacuum, the speed of light c, is related to the electric and magnetic constants ϵ_0 and μ_0:

$$c = \frac{1}{\sqrt{\epsilon_0\mu_0}}$$

The value of c is very nearly 3.00×10^8 m/s. Its exact value, used in defining the meter, is 299,792,458 m/s.

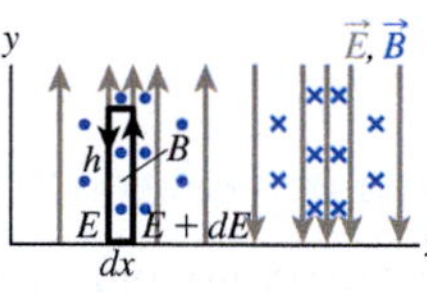

In vacuum, the electric and magnetic fields of a wave are related by

$$E = cB$$

The wave's frequency and wavelength are related by

$$f\lambda = c$$

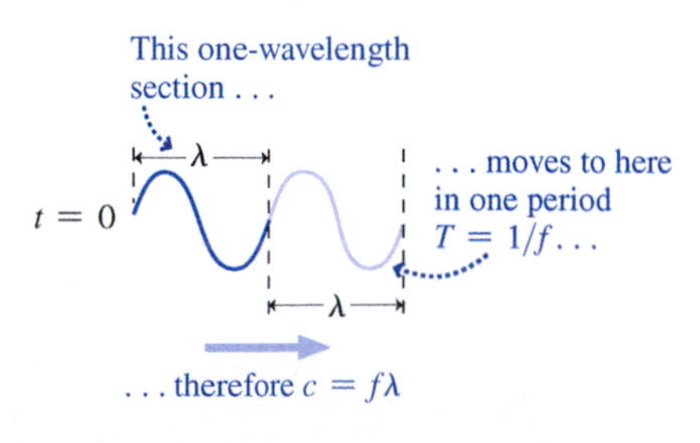

EM waves can have any wavelength; the whole range constitutes the **electromagnetic spectrum**.

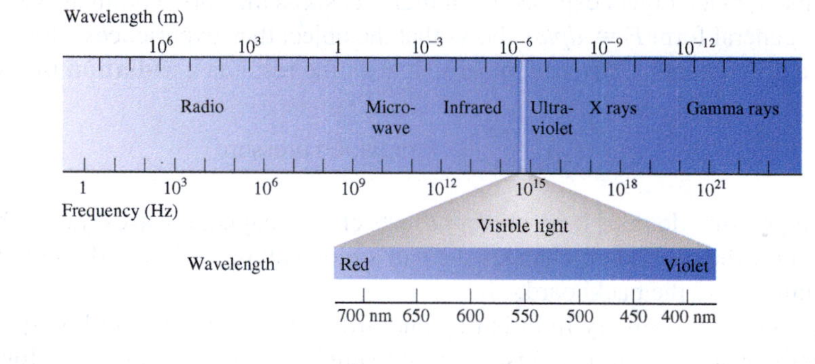

Applications

Polarization describes the direction of an EM wave's electric field and is a property widely used in scientific research and in technological devices including the ubiquitous liquid crystal displays. When polarized light of intensity S_0 is incident on a polarizer with its transmission axis at angle θ to the polarization, the light emerges with intensity

$$S = S_0\cos^2\theta$$

EM waves carry both energy and momentum. The **Poynting vector**

$$\vec{S} = \frac{\vec{E}\times\vec{B}}{\mu_0}$$

describes the rate of energy flow per unit area, while the momentum flow results in a **radiation pressure**:

$$P_{\text{rad}} = \frac{\bar{S}}{c}$$

MP For homework assigned on MasteringPhysics, go to www.masteringphysics.com

BIO *Biology and/or medicine-related problems* **DATA** *Data problems* **ENV** *Environmental problems* **CH** *Challenge problems* **COMP** *Computer problems*

For Thought and Discussion

1. Why is Maxwell's modification of Ampère's law essential to the existence of electromagnetic waves?
2. The presence of magnetic monopoles would require a modification of Gauss's law for magnetism. Which other Maxwell equation would need modification?
3. Is there displacement current in an electromagnetic wave? Is there ordinary conduction current?
4. List some similarities and differences between electromagnetic waves and sound waves.
5. The speed of an electromagnetic wave is given by $c = \lambda f$. How does the speed depend on frequency? On wavelength?
6. When astronomers observe a supernova explosion in a distant galaxy, they see a sudden, simultaneous rise in visible light and other forms of electromagnetic radiation. How is this evidence that the speed of light is independent of frequency?
7. Turning a TV antenna so its rods point vertically may change the quality of your TV reception. Why?
8. The Sun emits about half of its electromagnetic-wave energy in the visible region of the spectrum. Where do you think it emits most of the remainder?
9. An *LC* circuit is made entirely from superconducting materials, yet its oscillations eventually damp out. Why?
10. If you double the field strength in an electromagnetic wave, what happens to the intensity?
11. The intensity of light drops as the inverse square of the distance from the source. Does this mean that electromagnetic energy is lost? Explain.
12. Electromagnetic waves don't readily penetrate metals. Why not?

Exercises and Problems

Exercises

Section 29.2 Ambiguity in Ampère's Law

13. A uniform electric field is increasing at 1.5 (V/m)/μs. Find the displacement current through a 1-cm^2 area perpendicular to the field.
14. A parallel-plate capacitor has square plates 10 cm on a side and 0.50 cm apart. The voltage across the plates is increasing at 220 V/ms. What's the displacement current in the capacitor?

Section 29.4 Electromagnetic Waves

15. The fields of an electromagnetic wave are $\vec{E} = E_p \sin(kz + \omega t)\hat{\jmath}$ and $\vec{B} = B_p \sin(kz + \omega t)\hat{\imath}$. Give a unit vector in the wave's propagation direction.
16. A radio wave's electric field is given by the expression $\vec{E} = E \sin(kz - \omega t) \times (\hat{\imath} + \hat{\jmath})$. (a) Find the peak electric field. (b) Give a unit vector in the direction of the magnetic field at a place and time where $\sin(kz - \omega t)$ is positive.

Section 29.5 Properties of Electromagnetic Waves

17. A *light-minute* is the distance light travels in 1 minute. Show that the Sun is about 8 light-minutes from Earth.
18. Your intercontinental telephone call is carried by electromagnetic waves routed via a satellite in geostationary orbit at 36,000 km altitude. Approximately how long does it take before your voice is heard at the other end?
19. An airplane's radar altimeter works by bouncing radio waves off the ground and measuring the round-trip travel time. If that time is 74.7 μs, what should the pilot report to the passengers as the current altitude?
20. Roughly how long does it take light to travel 1 foot?
21. If you speak via radio from Earth to an astronaut on the Moon, how long is it before you can receive a reply?
22. What are the wavelengths of (a) a 100-MHz FM radio wave, (b) a 5.0-GHz WiFi signal, (c) a 600-THz light wave, and (d) a 1.0-EHz X ray?
23. A 60-Hz power line emits electromagnetic radiation. What's the wavelength?
24. Microwave ovens for consumers' use operate at 2.45 GHz. What's the distance between wave crests in such a microwave?
25. An electromagnetic wave is propagating in the z-direction. What's its polarization direction if its magnetic field is in the y-direction?
26. Polarized light is incident on a sheet of polarizing material, and only 20% of the light gets through. Find the angle between the electric field and the material's transmission axis.
27. Vertically polarized light passes through a polarizer with its axis at 70° to the vertical. What fraction of the incident intensity emerges from the polarizer?

Section 29.8 Energy and Momentum in Electromagnetic Waves

28. A typical laboratory electric field is 1500 V/m. Find the average intensity of an electromagnetic wave with this value for its peak field.
29. What would be the average intensity of a laser beam so strong that its electric field produced dielectric breakdown of air (which requires $E_p = 3$ MV/m)?
30. Estimate the peak electric field inside a 1.1-kW microwave oven under the simplifying approximation that the microwaves propagate as a plane wave through the oven's 750-cm^2 cross-sectional area.
31. Your new radio says it can pick up signals with peak electric fields as weak as 450 μV/m. Will it work if you take it to your remote cabin, where the intensity of your favorite radio station is 0.35 nW/m^2?
32. A laser pointer delivers 0.10-mW average power in a beam 0.90 mm in diameter. Find (a) the average intensity, (b) the peak electric field, and (c) the peak magnetic field.
33. Your university radio station has a 5.0-kW radio transmitter that broadcasts uniformly in all directions; listeners within 15 km have reliable reception. You want to increase the power to double that range. What should be the new power?

Problems

34. A parallel-plate capacitor has circular plates with radius 50.0 cm and spacing 1.0 mm. A uniform electric field between the plates is changing at the rate of 1.0 MV/m·s. Find the magnetic field between the plates (a) on the symmetry axis, (b) 15 cm from the axis, and (c) 150 cm from the axis.
35. You're engineering a new cell phone, and you'd like to incorporate the antenna entirely within the phone, which is 9 cm long when closed. The antenna is to be a quarter-wavelength long—a common design for vertically oriented antennas. If the cell-phone frequency is 2.4 GHz, will the antenna fit?

36. An electric field points into the page and occupies a circular region of radius 1.0 m, as shown in Fig. 29.14. There are no electric charges in the region, but there is a magnetic field forming closed loops pointing clockwise, as shown. The magnetic-field strength 50 cm from the center of the region is 2.0 μT. (a) What's the rate of change of the electric field? (b) Is the electric field increasing or decreasing?

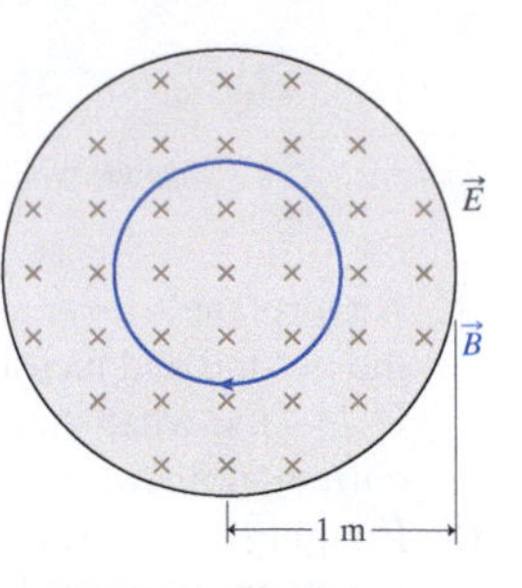

FIGURE 29.14 Problem 36

37. **BIO** The medical profession divides the ultraviolet region of the electromagnetic spectrum into three bands: UVA (320 nm–420 nm), UVB (290 nm–320 nm), and UVC (100 nm–290 nm). UVA and UVB promote skin cancer and premature skin aging; UVB also causes sunburn, but helpfully fosters production of vitamin D. Ozone in Earth's atmosphere blocks most of the more dangerous UVC. Find the frequency range associated with UVB radiation.
38. Dielectric breakdown in air occurs when the electric field is approximately 3 MV/m. What would be the peak magnetic field in an electromagnetic wave with this peak electric field?
39. A radio receiver can detect signals with electric fields as weak as 320 μV/m. Find the corresponding magnetic field.
40. A polarizer blocks 75% of a polarized light beam. What's the angle between the beam's polarization and the polarizer's axis?
41. An electro-optic modulator is a device that switches a laser beam rapidly from off to on by switching the polarization direction through 90° when a voltage is applied. But a brownout results in only enough voltage for a 72° rotation. What fraction of the light is transmitted during the brownout when the beam is supposed to be fully on?
42. Unpolarized light of intensity S_0 passes first through a polarizer with its axis vertical and then through one with its axis at 35° to the vertical. Find the intensity after the second polarizer.
43. Vertically polarized light passes through two polarizers, the first at 60° to the vertical and the second at 90° to the vertical. What fraction of the light gets through?
44. **CH** Show that it's impossible for an electromagnetic wave in vacuum to have a time-varying component of its electric field in the direction of its magnetic field. (*Hint:* Assume $\vec{E}$ does have such a component, and show that you can't satisfy both Gauss's and Faraday's laws.)
45. **BIO** High microwave intensities can cause biological damage through heating of tissue; a particular concern is cataract formation. The U.S. Food and Drug Administration limits microwave radiation near the door of a microwave oven to 5.0 mW/m^2. The window in a particular oven door measures 40 cm by 17 cm and is covered with a metal screen to block microwaves. Assuming power leaks uniformly through the window area, what percent of the oven's 900-W microwave power can leak without exceeding the FDA standards?
46. Use the fact that sunlight intensity at Earth's orbit is 1364 W/m^2 to calculate the Sun's total power output.
47. A quasar 10 billion light-years from Earth appears the same brightness as a star 50,000 light-years away. How do the power outputs of quasar and star compare?
48. **BIO** Lasers are classified according to the eye-damage danger they pose. Class 2 lasers, including many laser pointers, produce visible light with no greater than 1 mW total power. They're relatively safe because the eye's blink reflex limits exposure time to 250 ms. Find (a) the intensity of a 1-mW class 2 laser with beam diameter 1.0 mm, (b) the total energy delivered before the blink reflex shuts the eye, and (c) the peak electric field in the laser beam.
49. At 1.5 km from a radio transmitter, the peak electric field is 350 mV/m. Assuming the transmitter broadcasts equally in all directions, find (a) the transmitted power and (b) the peak electric field 10 km from the transmitter.
50. Find the peak electric and magnetic fields 1.5 m from a 60-W lightbulb that radiates equally in all directions.
51. A typical fluorescent lamp is a little more than 1 m long and a few cm in diameter. How do you expect the light intensity to vary with distance (a) near the lamp but not near either end and (b) far from the lamp?
52. A camera flash delivers 2.5 kW of light power for 1.0 ms. Find (a) the total energy and (b) the total momentum carried by the flash.
53. A laser produces an average power of 7.0 W in a 1.0-mm-diameter beam. Find (a) the average intensity and (b) the peak electric field of the laser light.
54. A 180-W/cm^2 laser beam shines on a light-absorbing surface. What's the radiation pressure on the surface?
55. A 65-kg astronaut is floating in empty space. If she shines a 1.0-W flashlight in a fixed direction, how long will it take her to accelerate to 10 m/s?
56. A *photon rocket* emits a beam of light instead of hot gas. How powerful a beam would be needed to equal the 40-MN thrust of NASA's new Space Launch System (SLS)? Compare your answer with humanity's total electric power-generating capability, about 2 TW.
57. A white dwarf star is approximately the size of Earth but radiates about as much power as the Sun. Estimate the radiation pressure on a light-absorbing object at the white dwarf's surface.
58. Use appropriate data from Appendix E to calculate the radiation pressure on a light-absorbing object at the Sun's surface.
59. A radar system produces pulses consisting of exactly 100 full cycles of a sinusoidal 72.5-GHz electromagnetic wave. The average power while the transmitter is on is 66.0 MW, and the waves are confined to a beam 22.4 cm in diameter. Find (a) the peak electric field, (b) the wavelength, (c) the total energy in one pulse, and (d) the total momentum in one pulse. (e) If the transmitter produces 945 pulses each second, what's its average power output?
60. **CH** A cylindrical resistor of length L, radius a, and resistance R carries current I. Calculate the electric and magnetic fields at the surface of the resistor, assuming the electric field is uniform over the surface. Calculate the Poynting vector and show that it points into the resistor. Calculate the flux of the Poynting vector (that is, $\int \vec{S} \cdot d\vec{A}$) over the resistor's surface to get the rate of electromagnetic energy flow into the resistor, and show that the result is I^2R. Your result shows that the energy heating the resistor comes from the fields surrounding it. These fields are sustained by the source of electric energy that drives the current.
61. In a stack of polarizing sheets, each sheet has its transmission axis rotated 14° with respect to the preceding sheet. If the stack passes 37% of the incident unpolarized light, how many sheets does it contain?
62. **CH** You're an astronomer studying the origin of the solar system, and you're evaluating a hypothesis that sufficiently small particles were blown out of the solar system by the force of sunlight. To see how small such particles must be, compare the force of sunlight with the force of solar gravity, and solve for the particle radius at which the two are equal. Assume spherical particles with density 2 g/cm^3. (*Note:* Distance from the Sun doesn't matter. Why not?)
63. **CH** Differentiate Equation 29.12 with respect to x and Equation 29.13 with respect to t. Then, using the fact that mixed derivatives are equal (e.g., $\frac{\partial}{\partial t}\left(\frac{\partial B}{dx}\right) = \frac{\partial}{\partial x}\left(\frac{\partial B}{dt}\right)$), combine the resulting

equations and show that the result is the wave equation (Equation 14.5) for waves with speed $c = 1/\sqrt{\epsilon_0\mu_0}$.

64. Maxwell's equations in a dielectric resemble those in vacuum (Equations 29.6–29.9) but with ϵ_0 replaced by $\kappa\epsilon_0$, where κ is the dielectric constant introduced in Chapter 23. Show that the speed of electromagnetic waves in a dielectric is $c/\sqrt{\kappa}$.
65. A friend buys a used pickup truck that comes with a CB radio. However, the antenna is broken off, and your friend asks you to help make one out of a steel rod that he will affix to the rear bumper. You know that the CB channel frequency is 27.3 MHz and that the antenna must be a quarter-wavelength long. How long should you make the rod?
66. Your roommate's father is CEO of a coal company, so your roommate is understandably skeptical of alternative energy proposals. He claims that there's no future for solar energy, because the power in sunlight is insufficient to meet humankind's energy demand. Is he right? To find out, compare the solar power incident on Earth with our human energy consumption rate of about 16 TW.
67. The *Voyager 1* spacecraft is now beyond the outer reaches of our solar system, but earthbound scientists still receive data from the spacecraft's 20-W radio transmitter. *Voyager* is expected to continue transmitting until about 2025, when it will be some 25 billion km from Earth. What's the diameter of a dish antenna that will receive 10^{-20} W of power from *Voyager* at this time?
68. Your friend who works for the college radio station must make electric-field measurements for a report to be filed with the station's application for license renewal. The measurement is made 4.6 km from the antenna, where your friend measures the electric field at 380 V/m. The station is allowed to broadcast at no more than 55-kW power. Assuming power spreads equally in all directions, is the station in compliance with its license?
69. The National Ignition Facility at Lawrence Livermore National Laboratory initiates nuclear fusion by converging 192 laser beams on a deuterium–tritium target. Each beam has a square cross section 38 cm on a side, and each beam delivers 10.0 kJ of energy in 20.0 ns. Find (a) the peak electric field and (b) the peak magnetic field in each laser beam. (c) Find the combined power of all 192 laser beams while they're firing, and compare with humankind's energy consumption rate of about 16 TW.
70. **DATA** The table below shows the intensity of the radio signal received at Earth from a spacecraft on its way to the outer solar system, as a function of its distance from Earth. Distances are in astronomical units (AU, with 1 AU being the mean Earth–Sun distance; see Appendix E). Determine a quantity that, when you plot $\bar{S}$ versus that quantity, should give a straight line. Make your plot, establish a best-fit line, and use it to determine the spacecraft's transmitter power.

Distance (AU)	1.56	1.81	2.14	2.78	3.17	4.25
Intensity, $\bar{S}$ (10^{-23} W/m^2)	22.5	17.8	11.6	7.10	5.63	3.01

Passage Problems

Proposals have been made to "sail" spacecraft to the outer solar system using the pressure of sunlight, or even to propel interstellar spacecraft with high-powered, Earth-based lasers. Sailing spacecraft would need no fuel—a great advantage because fuel constitutes much of the initial weight of any space mission. NASA's Sunjammer, launched in 2015, is the first sailing spacecraft deployed beyond Earth orbit. Figure 29.15 shows what a similar solar sail might look like in space.

FIGURE 29.15 Artist's conception of a solar sailing spacecraft. The sail is hundreds of square meters in area, but is less than 10 μm thick (Passage Problems 71–74).

71. If a sunlight-powered sailing spacecraft accelerated at 1 m/s^2 in the vicinity of Earth's orbit, what would be its acceleration at Mars, about 1.5 times as far from the Sun as Earth?
    a. about 0.25 m/s^2
    b. a little less than 0.5 m/s^2
    c. a little more than 0.5 m/s^2
    d. about 0.66 m/s^2
72. One spacecraft has a sail that absorbs all light incident on it; the other has a perfectly reflective sail. How do their accelerations compare in light with the same intensity?
    a. The absorptive sail gives twice the acceleration.
    b. The reflective sail gives twice the acceleration.
    c. The absorptive sail gives greater acceleration, but not twice as much.
    d. The reflective sail gives greater acceleration, but not twice as much.
73. A sail capable of propelling a spacecraft to the outer solar system must be able to overcome the Sun's gravity. Suppose a spacecraft is designed so the force of sunlight on its sail is 20 times that of solar gravity in the vicinity of Earth's orbit. If the spacecraft reaches Jupiter, some 5 times as far from the Sun as Earth,
    a. the sail force will still exceed solar gravity, now by a factor of 4.
    b. the sail force will be slightly less than solar gravity.
    c. the sail force will now be 25 times solar gravity.
    d. the sail force will still be 20 times solar gravity.
74. The intensity of sunlight at Earth's orbit is about 1.4 kW/m^2. A 100-kg sailing spacecraft with 1-km^2 sail area would experience an acceleration of about
    a. 5 mm/s^2.
    b. 5 cm/s^2.
    c. 5 m/s^2.
    d. 5 km/s^2.

Answers to Chapter Questions

Answer to Chapter Opening Question

Electromagnetic waves, comprising changing electric and magnetic fields, carry not only cell-phone conversations but also TV shows, the energy of sunlight, and signals from physical processes in the farthest reaches of the universe.

Answers to GOT IT? Questions

29.1 Yes; the displacement current $\epsilon_0\, d\Phi_E/dt$ has the same effect as the real current I in producing magnetic fields.
29.2 (c)
29.3 (b)
29.4 (c)
29.5 (a)
29.6 (1) $B_1 = 2B_2$; (2) $S_1 = 4S_2$; (3) $\lambda_1 = \lambda_2$

PART FOUR SUMMARY

Electromagnetism

Electromagnetism is a fundamental force of nature. The strong attraction between positive and negative charge makes most bulk matter electrically neutral, and hides from us the essential role electricity and magnetism play in the structure of matter.

Electromagnetic interactions are best described in terms of **electric fields** and **magnetic fields**. Electric charges create electric fields, and electric charges respond to the fields of other charges.

Moving electric charges create magnetic fields, and moving electric charges respond to magnetic fields. Both electric and magnetic fields store energy.

A changing magnetic field creates an electric field, and vice versa. Together, changing fields combine to make **electromagnetic waves**—self-replicating structures that propagate through empty space at the speed of light, c. Light itself is an electromagnetic wave.

Maxwell's equations are the four fundamental laws of electromagnetism:

Law	Mathematical Statement	What It Says
Gauss for $\vec{E}$	$\oint \vec{E}\cdot d\vec{A} = \dfrac{q}{\epsilon_0}$	How charges produce electric field; field lines begin and end on charges.
Gauss for $\vec{B}$	$\oint \vec{B}\cdot d\vec{A} = 0$	No magnetic charge; magnetic field lines don't begin or end.
Faraday	$\oint \vec{E}\cdot d\vec{l} = -\dfrac{d\Phi_B}{dt}$	Changing magnetic flux produces electric field.
Ampère	$\oint \vec{B}\cdot d\vec{l} = \mu_0 I + \mu_0\epsilon_0\dfrac{d\Phi_E}{dt}$	Electric current and changing electric flux produce magnetic field.

Coulomb's law and the **Biot–Savart law** provide alternatives to Gauss's and Ampère's laws for determining electric and magnetic fields of pointlike elements of charge and moving charge, respectively:

$$\vec{E} = \frac{kq}{r^2}\hat{r}$$

$$d\vec{B} = \frac{\mu_0}{4\pi}\frac{I\,d\vec{l}\times\hat{r}}{r^2}$$

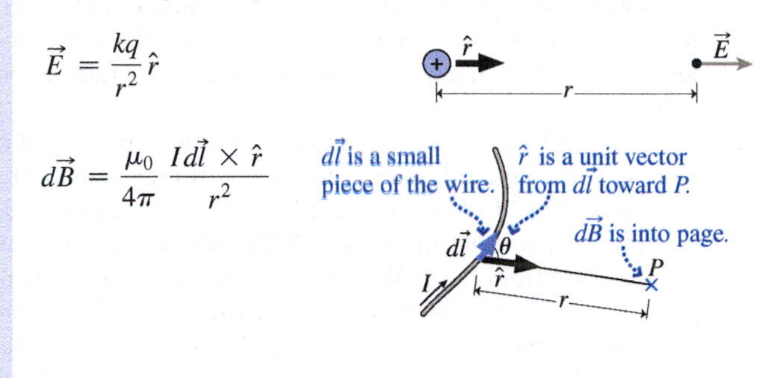

The **electromagnetic force** on a charged particle consists of the **electric force** and the **magnetic force**. Both are proportional to the charge and to the appropriate field; the magnetic force depends also on the particle's velocity $\vec{v}$:

$$\vec{F}_{EM} = \vec{F}_E + \vec{F}_B = q\vec{E} + q\vec{v}\times\vec{B}$$

The **electric potential difference** describes the work per unit charge needed to move charge between two points in an electric field; its units are N/C or **volts** (V):

$$V_{AB} = -\int_A^B \vec{E}\cdot d\vec{l}$$

Electric current is a flow of electric charge:

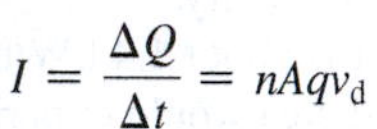

$$I = \frac{\Delta Q}{\Delta t} = nAqv_d$$

$\vec{v}_d$

n charges/unit volume, each charge q

A

L

In ohmic materials, **Ohm's law** relates voltage, current, and resistance: $I = V/R$.

Electric circuits are interconnections of electric components, including batteries, resistors, and others. They can often be analyzed by considering **series** and **parallel** combinations.

R_1, R_2 ⇨ $R_1 + R_2$ Series

R_1 R_2 ⇨ $\dfrac{1}{R} = \dfrac{1}{R_1} + \dfrac{1}{R_2}$ Parallel

Electromagnetic induction, described by Faraday's law, is the basis of electric generators and a host of other electromagnetic technologies and natural phenomena.

Electromagnetic waves result from changing electric and magnetic fields. EM waves include light, and all EM waves propagate in vacuum at the speed of light, $c = 1/\sqrt{\mu_0\epsilon_0}$.

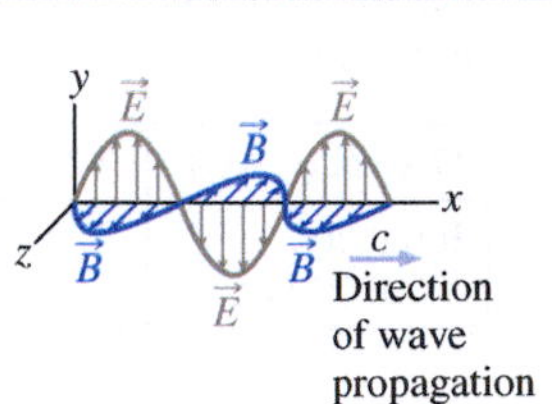

Part Four Challenge Problem

A wire of length L and resistance R forms a rectangular loop twice as long as it is wide. It's mounted on a nonconducting horizontal axle parallel to its longer dimension, as shown in the figure. A uniform magnetic field $\vec{B}$ points into the page. A long string of negligible mass is wrapped many times around a drum of radius a attached to the axle, and a mass m is attached to the string. When the mass is released, it falls and eventually reaches a speed that, averaged over one cycle of the loop's rotation, is constant. Find an expression for that average speed.

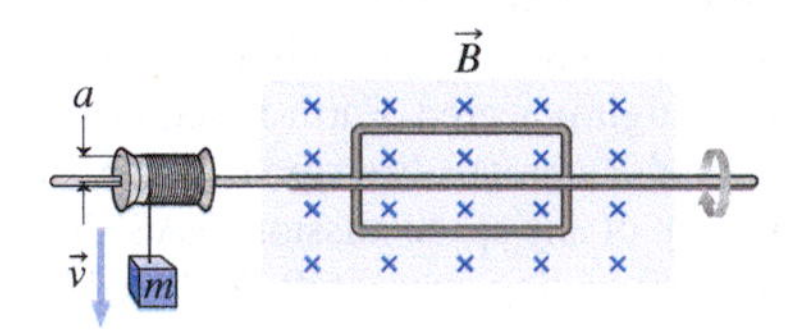

PART FIVE OVERVIEW

Optics

Drops of dew act as miniature optical systems, with light refracting through the drops to form myriad images of the background flowers.

Imagine a world without light. We see because light reflects off objects, and our eyes form images because light refracts in our corneas and lenses. When our built-in optical systems aren't perfect, we correct them with additional lenses or we use lasers to reshape the cornea. Microscopes and telescopes extend the range of our vision. The phenomenon of interference makes possible some of the most precise measurements and is behind the operation of everyday technologies like CDs and DVDs. Light signals carry e-mail, web pages, telephone conversations, and computer data through the optical fibers that form the world's communications networks. Although the behavior of light is ultimately grounded in Maxwell's equations of electromagnetism, we can learn much about light from the simpler perspective of optics. The next three chapters explore the behavior of light, images and optical instruments, and phenomena associated with the wave nature of light.

Reflection and Refraction

What You Know

- Although this chapter's material is fundamentally based on the electromagnetic waves of Chapter 29, you don't need that level of understanding to follow this introduction to optics.
- You have considerable real-world experience with optical processes—especially reflection in mirrors.
- You will need your knowledge of trigonometry.

What You're Learning

- You'll learn how to describe *reflection* at plane surfaces like mirrors and configurations of multiple mirrors.
- You'll learn about *refraction* at flat interfaces between different transparent materials.
- You'll see how refraction results from differences in the speed of light in different materials.
- You'll use *Snell's law* to describe refraction quantitatively.
- You'll see how *total internal reflection* results when light is incident on an interface from a medium of greater optical density to one of lesser density, as in water to air.
- You'll explore how wavelength-dependent refraction results in *dispersion*, a phenomenon that helps explain the rainbow.

How You'll Use It

- In Chapter 31 you'll use your knowledge of reflection to see how images form in both flat and curved mirrors.
- You'll use your knowledge of refraction to study image formation by lenses and in optical instruments such as microscopes and telescopes.
- If you pursue a career in any scientific field, you'll probably use *spectroscopy*, a technique that exploits dispersion to analyze light and infer chemical composition, molecular structure, and other useful information.
- Understanding reflection and refraction will enhance your appreciation of many natural phenomena, including rainbows and other atmospheric optical effects.

Maxwell's brilliant work shows that the phenomena of **optics**—the behavior of light—are manifestations of electromagnetism. Except in the atomic realm, where quantum physics reigns, all optical phenomena are understandable in terms of electromagnetic-wave fields described by Maxwell's equations. But when objects with which light or any other wave interacts are much larger than the wavelength, light generally travels in straight lines called **rays**. **Geometrical optics** describes the behavior of light in this approximation. Here we'll use geometrical optics to explore the behavior of light at interfaces between different materials. In Chapter 31 we'll see how geometrical optics explains lenses, the human eye, and many optical instruments.

What process causes the upside-down image above the shark, and what's this got to do with the Internet?

30.1 Reflection

Some materials, notably metals, **reflect** nearly all the light incident on them. It's no coincidence that these materials are also good electrical conductors. The oscillating electric field of a light wave drives a metal's free electrons into oscillatory motion, which, in turn, produces electromagnetic waves. The net effect is to reradiate the wave back into the original medium. Other materials reflect only part of the incident light. Either way, reflection satisfies the same geometrical conditions: The incident ray, the reflected ray, and the normal to the interface between two materials all lie in the same plane. The **angle of reflection** θ_1' that the reflected ray makes with the normal is the same as the **angle of incidence** θ_1 made by the incident ray (Fig. 30.1*a*):

$$\theta_1' = \theta_1 \tag{30.1}$$

where the subscript 1 denotes the first medium.

In **specular reflection**, parallel rays reflect off a smooth surface and the entire beam is reflected without distortion (Fig. 30.1*b*). In contrast, a rough surface reflects individual rays in different directions (Fig. 30.1*c*)—even though each ray still obeys Equation 30.1. This is **diffuse reflection**. White paper is a diffuse reflector, while the aluminum coating of a mirror is an excellent specular reflector.

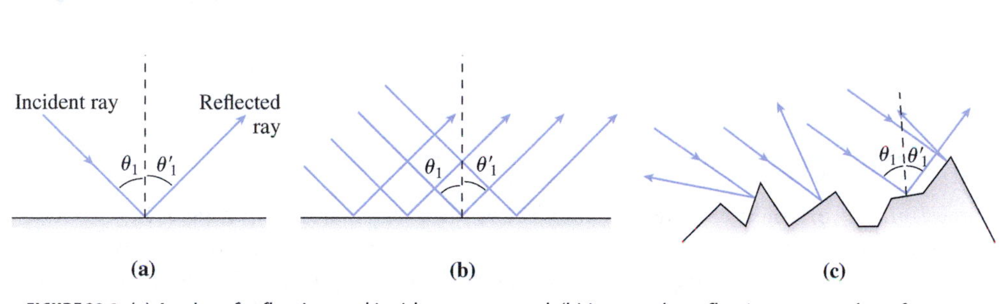

FIGURE 30.1 (a) Angles of reflection and incidence are equal. (b) In specular reflection, a smooth surface reflects a light beam undistorted. (c) A rough surface results in diffuse reflection.

APPLICATION Moon's Distance

Knowing the precise distance to the Moon and how it varies with time helps scientists confirm Einstein's general theory of relativity, a fundamental theory of physics that describes, among other things, black holes and the overall structure of the universe. In the late 1960s and early 1970s, *Apollo* astronauts left arrays of corner reflectors on the Moon, and they've been used ever since to reflect laser beams originating from Earth. Timing the beams' round-trip travel provides precise measures of the Moon's distance. Since 2006, a lunar-ranging experiment at New Mexico's Apache Point Observatory has been measuring the lunar distance to within about 1 millimeter! The photo shows the largest of the *Apollo* reflector arrays, placed by *Apollo 15* astronauts in 1971.

EXAMPLE 30.1 Reflection: The Corner Reflector

Two mirrors join at right angles. Show that any light ray incident in a plane perpendicular to both mirrors will return antiparallel to its incident direction.

INTERPRET We've sketched the situation in Fig. 30.2. We're asked to show that the lines representing incident and outgoing rays in the figure are parallel.

DEVELOP The outgoing ray will be antiparallel to the incident ray if the light turns through a total of 180°. Applying Equation 30.1 to each of the two reflections shows that the angles θ are the same, and so are the angles ϕ. Figure 30.2 shows that the first reflection turns the incident ray through an angle $180° - 2\theta$. Similarly, the second turns it through $180° - 2\phi$.

EVALUATE The pair of mirrors thus turns the ray through a total angle of

$$(180° - 2\theta) + (180° - 2\phi) = 360° - 2(\theta + \phi)$$

But Fig. 30.2 shows that θ and ϕ sum to 90°, so the total angle is $360° - 180° = 180°$—which is what we set out to prove.

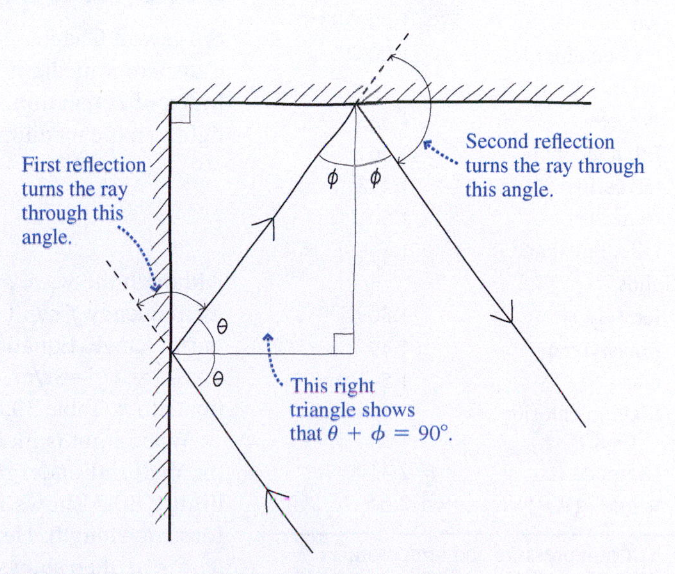

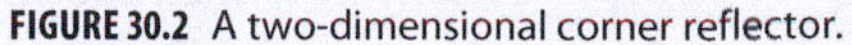

FIGURE 30.2 A two-dimensional corner reflector.

(continued)

ASSESS Here we've explored a remarkable device, a pair of perpendicular mirrors that returns a light ray in exactly the direction from which it came—provided the light is in the plane perpendicular to both mirrors. Add a third mirror at right angles to the other two, and you have a **corner reflector**—a device that returns a light ray in the direction from which it came, period. Corner reflectors, often made with prisms rather than mirrors, are widely used in optics. You can explore the corner reflector in Problem 58, while the Application shows how corner reflectors help verify one of the most fundamental theories of physics. ■

GOT IT? 30.1 The reflection of a London steeple shows in a wet pavement. Would you consider this to be (a) specular reflection, (b) diffuse reflection, or (c) somewhere in between?

Medium 1 | Medium 2

$\vec{v}_1$ | $\vec{v}_2$

λ_1 | λ_2

A | B

Observers at A and B see the same number of wave crests pass in a given time, so they both measure the same frequency f . . .

. . . therefore, because $v = f\lambda$, the wavelength λ must be shorter where the wave moves more slowly.

FIGURE 30.3 Wave frequency doesn't change as a wave goes from one medium to another, but wavelength does change.

Partial Reflection

Some light is reflected even at the interface with a transparent material. The detailed description of such partial reflection follows from Maxwell's equations, and is akin to the partial reflection of waves on strings described in Chapter 14. The least reflection occurs with normal incidence; for glass, about 4% of normally incident light is reflected. Reflection increases with larger incidence angles. Camera lenses, binoculars, solar photovoltaic cells, and other devices often have special antireflection coatings to reduce light loss. You'll see how these coatings work in Chapter 32.

Table 30.1 Indices of Refraction*

Substance	Index of Refraction, n
Gases	
Air	1.000293
Carbon dioxide	1.00045
Liquids	
Water	1.333
Ethyl alcohol	1.361
Glycerine	1.473
Benzene	1.501
Diiodomethane	1.738
Solids	
Ice (H_2O)	1.309
Polystyrene	1.49
Glass	1.5–1.9
Sodium chloride (NaCl)	1.544
Diamond (C)	2.419
Rutile (TiO_2)	2.62

*At 1 atm pressure and temperatures ranging from 0°C to 20°C, measured at a wavelength of 589 nm (the yellow line of sodium).

30.2 Refraction

We saw in Chapter 14 that wave speeds differ in different media. With light, the speed in transparent media is lower than in vacuum. We characterize a transparent medium by its **index of refraction**, defined as the ratio of the speed of light c in vacuum to the speed of light v in the medium:

$$n = \frac{c}{v} \quad \text{(index of refraction)} \qquad (30.2)$$

Although the wave speed changes when light enters a new medium, Fig. 30.3 shows that its frequency f can't change and therefore, since the wave speed is $v = f\lambda$, the wavelength must change. Equation 30.2 shows that the wavelength in a medium with refractive index n is $\lambda = v/f = c/nf$. Because c and f don't change, the wavelength is inversely proportional to n. Table 30.1 lists some refractive indices.

When light is incident at an angle on a transparent material, the light transmitted into the material undergoes **refraction**—a change in its propagation direction (Fig. 30.4). Figure 30.5 shows how refraction results from the change in wave speed and therefore wavelength. Here we assume the refractive index is higher in medium 2; our result $\lambda = c/nf$ then shows that the wavelength is shorter in medium 2. We've shaded two right triangles with a common hypotenuse and one side equal to the appropriate wavelength. The angles opposite these sides are the angles of incidence and refraction. In each case the

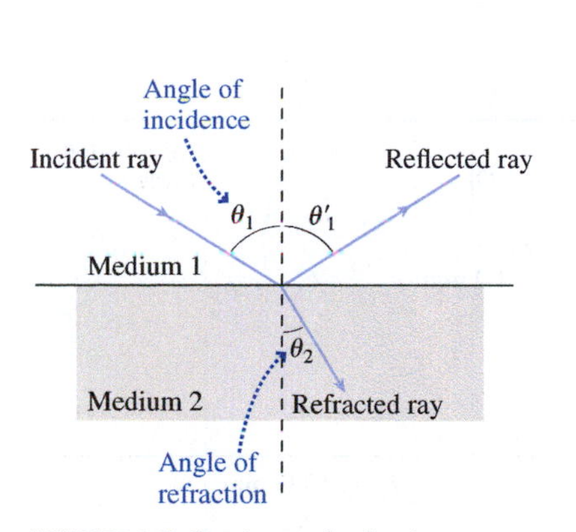

FIGURE 30.4 Refraction and reflection at an interface, here when medium 2 has the higher refractive index.

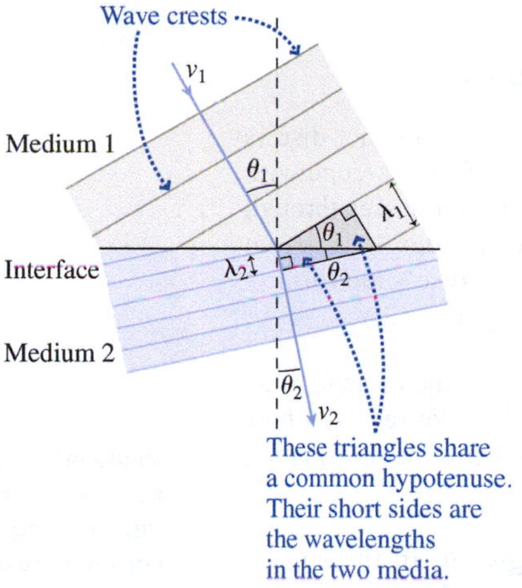

FIGURE 30.5 Refraction occurs because wave speed and wavelength differ in the two media.

hypotenuse is given by $\lambda/\sin\theta$. Equating expressions for this common hypotenuse gives $\lambda_1/\sin\theta_1 = \lambda_2/\sin\theta_2$. Since $\lambda = c/nf$ with f the same in both media, we get **Snell's law**:

$$n_1 \sin\theta_1 = n_2 \sin\theta_2 \qquad \text{(Snell's law)} \tag{30.3}$$

First developed geometrically in 1621 by van Roijen Snell of the Netherlands, and described analytically in the 1630s by René Descartes in France, Snell's law lets us predict what will happen at an interface given the refractive indices of the two media.

PhET: Bending Light (Intro)

Snell's law applies whether light goes from a medium of lower to higher refractive index or vice versa. When going from lower to higher index, the beam bends *toward* the normal; when going from higher to lower index, it bends *away* from the normal.

In some situations, including the human eye and Earth's atmosphere, the refractive index varies continuously with position, so light refracts continuously, following a curved path. You can explore two examples in Passage Problems 63–66.

EXAMPLE 30.2 Refraction: A Plane Slab

A light ray propagating in air strikes a glass slab of thickness d and refractive index n at incidence angle θ_1. Show that it emerges from the slab propagating parallel to its original direction.

INTERPRET This is a problem about refraction, which in this case occurs twice. We identify the two interfaces as first the air–glass interface where the light enters the glass and then the glass–air interface where the light exits the glass.

DEVELOP Figure 30.6 is a sketch showing the path of the light through the glass. There are two pairs of incidence and refractive angles, which we labeled θ_1, θ_2 and θ_3, θ_4. Our plan is to apply Snell's law at each interface and thus prove that $\theta_4 = \theta_1$. At the air–glass interface, we have $n_1 \sin\theta_1 = n_2 \sin\theta_2$, and at the glass–air interface, $n_3 \sin\theta_3 = n_4 \sin\theta_4$. Note that θ_1 and θ_4 are in the same medium (air), so we set $n_4 = n_1$. Similarly, $n_3 = n_2$ for θ_2 and θ_3 in glass. Then our equations become $n_1 \sin\theta_1 = n_2 \sin\theta_2$ and $n_2 \sin\theta_3 = n_1 \sin\theta_4$.

EVALUATE Taking $n_1 = 1$ for air and $n_2 = n$ for glass at the air–glass interface, we have $\sin\theta_2 = \sin\theta_1/n$. But at the glass–air interface, $n_1 = n$ and $n_2 = 1$, so here $\sin\theta_4 = n \sin\theta_3$. But the slab faces are parallel, so $\theta_3 = \theta_2$. Thus $\sin\theta_4 = n \sin\theta_2$. Using our expression

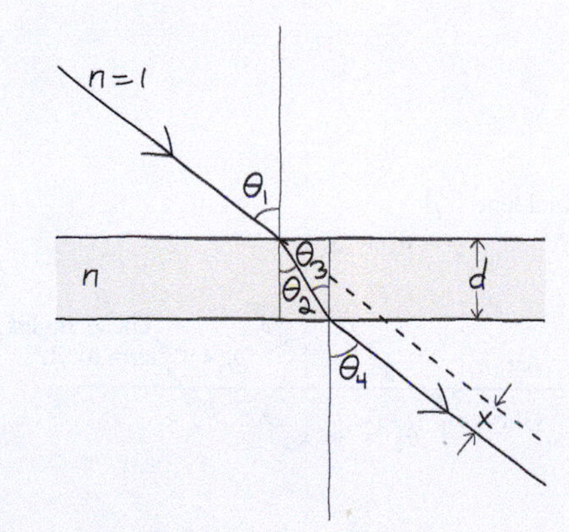

FIGURE 30.6 Light passing through a transparent slab.

for θ_2 from the first interface then gives

$$\sin\theta_4 = n\left(\frac{\sin\theta_1}{n}\right) = \sin\theta_1$$

showing that the incident and outgoing rays are indeed parallel.

ASSESS This result shows that light isn't deflected when it passes through a parallel-faced slab of transparent material. It is, however, displaced by the distance x shown in Fig. 30.6. You can find that displacement in Problem 54.

EXAMPLE 30.3 Refraction: CD Music

The laser beam that "reads" information from a compact disc is 0.737 mm wide when it strikes the disc, and it forms a cone with half-angle $\theta_1 = 27.0°$ as shown in Fig. 30.7. It then passes through a 1.20-mm-thick layer of plastic with refractive index 1.55 before reaching the reflective information layer near the disc's top surface. What's the beam diameter d at the information layer?

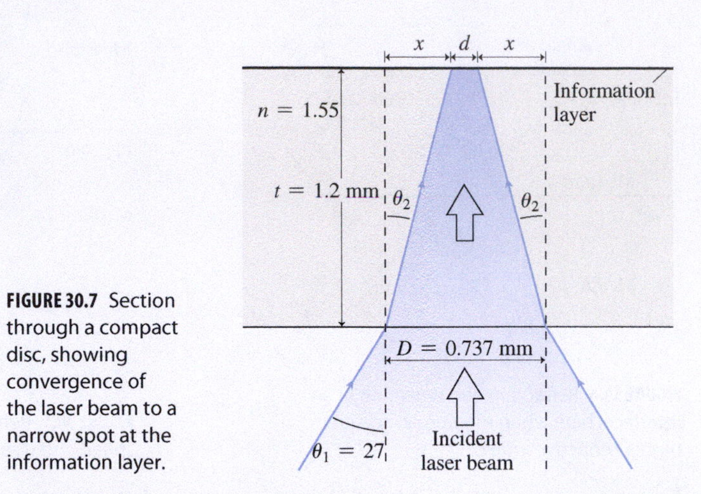

FIGURE 30.7 Section through a compact disc, showing convergence of the laser beam to a narrow spot at the information layer.

INTERPRET Rays defining the beam refract toward the normal, making a smaller convergence angle within the disc. We're asked how much this converging beam narrows when it reaches the information layer.

DEVELOP Snell's law will give us the angles θ_2 in the figure: $n_1 \sin\theta_1 = n_2 \sin\theta_2$. Given θ_2, we can use trigonometry to find the distance x marked in the drawing: $x = t\tan\theta_2$, with t the 1.2-mm thickness. The beam diameter d then follows from $d = D - 2x$, where D is the 0.737-mm beam diameter at the disc surface.

EVALUATE With $n_1 = 1$ and $n_2 = 1.55$, Snell's law gives

$$\theta_2 = \sin^{-1}(\sin\theta_1/n_2) = 17.03°$$

Therefore

$$d = D - 2x = D - 2t\tan\theta_2 = 1.80\,\mu\text{m}.$$

ASSESS This answer makes sense because d is just a bit larger than the "pits" cut into the CD to store information. Narrowing of the laser beam plays a crucial role in keeping CDs noise free. The tiniest dust speck would blot out information at the μm-scale information layer, but at the point where the beam enters the disc, it would take mm-size dust to cause problems. We'll explore CD and DVD technology further in Chapter 32. ■

GOT IT? 30.2 The figure shows the path of a light ray through three different media. Rank the media according to their refractive indices, in decreasing order.

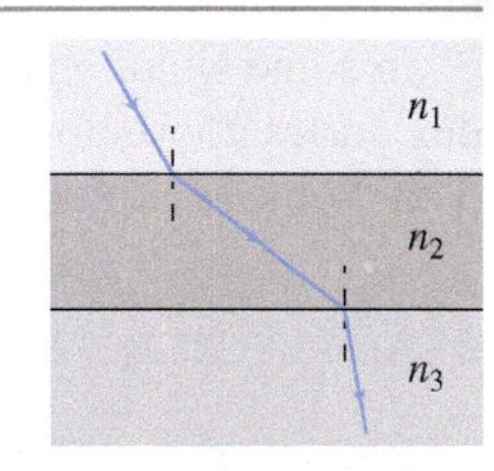

Refraction, Reflection, and Polarization

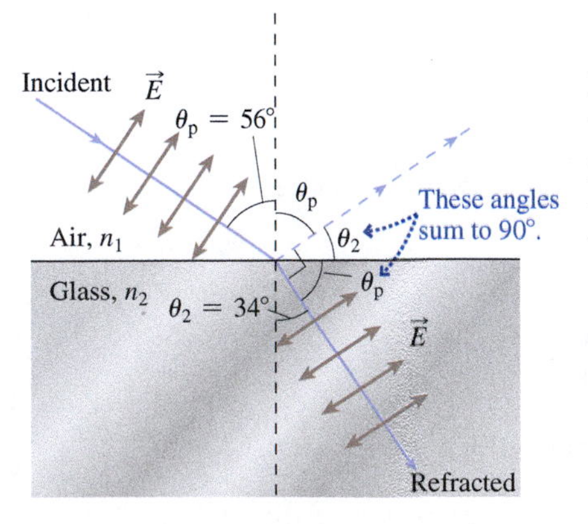

FIGURE 30.8 The polarizing incidence angle θ_p occurs when the angles of incidence and refraction sum to 90°.

Both reflection and refraction ultimately involve interactions between the incident light wave's electric field and charges in a material. The oscillation of molecular dipoles in response to the field gives rise to both refracted and reflected light. It's not surprising, therefore, that details depend on the direction of the electric field—that is, on polarization. When the field lies in the plane defined by the incident and reflected rays, there's a special angle of incidence at which no reflection occurs. This is the **Brewster angle**, or **polarizing angle**, and it occurs when the reflected ray would be perpendicular to the transmitted ray (Fig. 30.8). Then the molecular dipoles are oscillating along the direction a reflected ray would take, and, as we saw in Section 29.7, there's no electromagnetic radiation from an oscillating dipole along the oscillation direction.

Figure 30.8 shows that the polarizing incidence angle θ_p occurs when θ_p and the angle θ_2 of the reflected ray sum to 90°; equivalently, $\theta_2 = 90° - \theta_p$. Since $\sin\theta = \cos(90° - \theta)$, that means $\sin\theta_2 = \cos\theta_p$. Now, Snell's law gives $\sin\theta_2 = (n_1/n_2)\sin\theta_p$; substituting $\cos\theta_p$ for $\sin\theta_2$, this becomes $\cos\theta_p = (n_1/n_2)\sin\theta_p$. Multiplying both sides by (n_2/n_1) and dividing by $\cos\theta_p$ then gives

$$\tan\theta_p = \frac{n_2}{n_1} \quad \text{(polarizing angle)} \qquad (30.4)$$

For the air–glass interface shown in Fig. 30.8, θ_p is about 56°.

When unpolarized light is incident at the polarizing angle, only the component of the light's electric field that's perpendicular to the plane of Fig. 30.8 gets reflected, and the result is polarized light. Polarization using this effect is important in a number of technologies, including lasers. The window through which light emerges from a laser is usually cut at the polarizing angle, and as a result most laser light is intrinsically polarized. A similar polarizing phenomenon occurs for reflection from metals and other opaque surfaces, as well as from water. Figure 30.9 shows that polarizing sunglasses can reduce glare caused by such reflections.

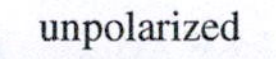

FIGURE 30.9 Light reflecting off horizontal surfaces is partially polarized for reasons similar to those suggested in Fig. 30.8. Polarizing sunglasses dramatically reduce the glare from such reflections.

30.3 Total Internal Reflection

Light propagating from a medium with a higher refractive index into one with a lower index is bent *away* from the normal, as shown for a glass–air interface in Fig. 30.10. In other words, the angle of refraction in this case is larger than the angle of incidence. So at some incidence angle, the angle of refraction becomes 90°. Then what?

As Fig. 30.10 shows, light incident at this **critical angle** or larger cannot escape from the glass. Instead, **total internal reflection** occurs, returning all the light to the medium with the larger refractive index. We can find the critical angle by setting $\theta_2 = 90°$ in Snell's law (Equation 30.3). The critical angle θ_c is then θ_1, and we have

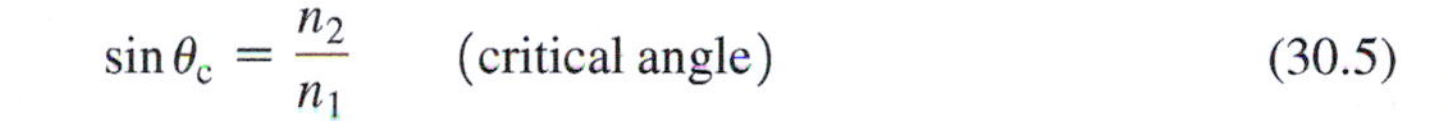

$$\sin\theta_c = \frac{n_2}{n_1} \quad \text{(critical angle)} \qquad (30.5)$$

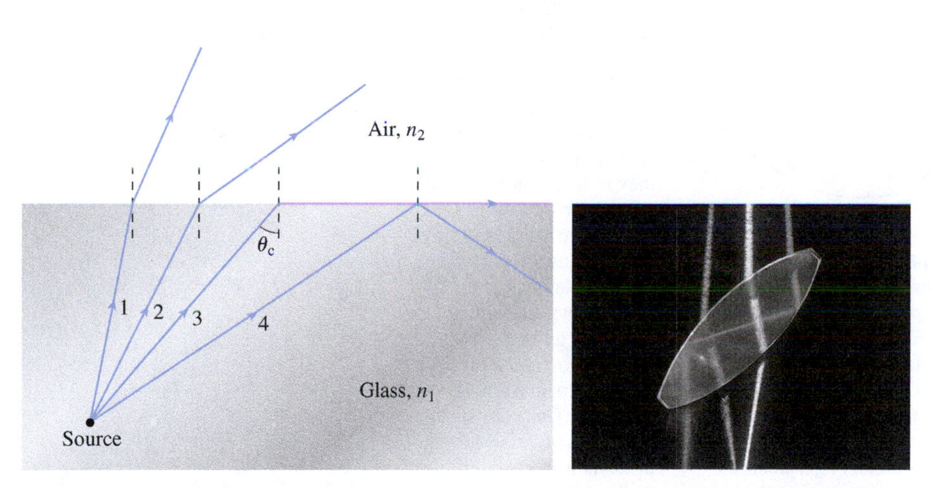

FIGURE 30.10 Light propagating in glass is refracted away from the normal at the glass–air interface. Ray 3, incident at the critical angle θ_c, just skims along the interface. At larger incidence angles (ray 4), the light undergoes total internal reflection. The rightmost beam in the photo (incident from above) undergoes two total internal reflections.

Total internal reflection makes uncoated glass an excellent reflector when it's oriented appropriately (Fig. 30.11). Binoculars owe their compact size to glass prisms that reflect light internally to provide a longer light path. For an underwater observer, the existence of the critical angle affects the view of the outside world, as the next example shows. Finally, total internal reflection is the basis of the optical fibers that carry signals over the global Internet, as the Application below describes.

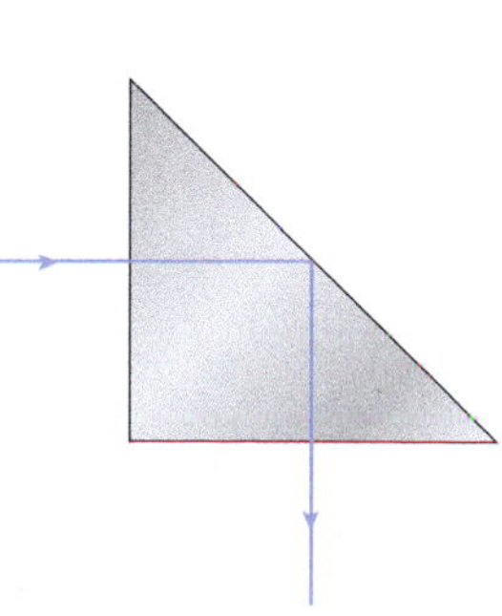

FIGURE 30.11 Light undergoes total internal reflection in a glass prism.

GOT IT? 30.3 The glass prism in Fig. 30.11 has $n = 1.5$ and is surrounded by air $(n = 1)$. What would happen to the incident light ray shown if the prism were immersed in water $(n = 1.333)$?

CONCEPTUAL EXAMPLE 30.1 Total Internal Reflection: A Watching Whale

Planeloads of whale watchers fly over the ocean. What does a submerged whale see when it looks up at them?

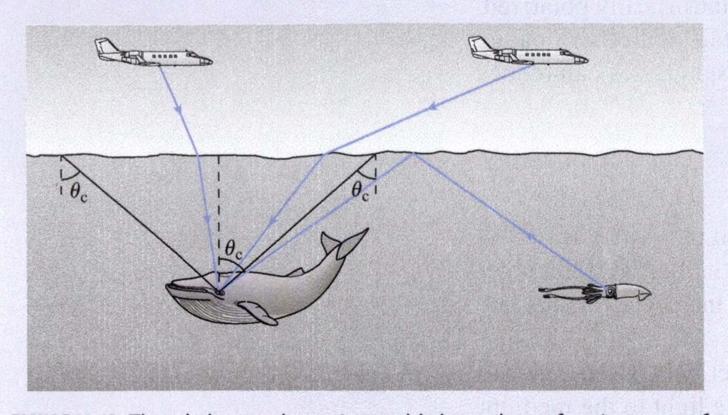

FIGURE 30.12 The whale sees the entire world above the surface in a cone of half-angle θ_c; beyond that, it sees reflections of objects below the surface.

EVALUATE The whale is underwater, so it's in a medium with a higher refractive index than air. Some of the light reaching the whale is from objects above the water surface, like the planes in Fig. 30.12. But the whale also sees objects below the surface, like the squid in Fig. 30.12, light from which is totally reflected at the water–air interface.

ASSESS The whale sees the entire above-surface world within a cone, as Fig. 30.12 shows. If you've ever looked upward from underwater, you've experienced the same phenomenon.

MAKING THE CONNECTION What's the half-angle of the cone in which the whale sees objects above the water surface?

ANSWER The cone's half-angle is the critical angle θ_c, as shown in Fig. 30.12. For water, Table 30.1 gives $n = 1.333$, so, by Equation 30.5, $\theta_c = \sin^{-1}(1/1.333) = 48.6°$.

APPLICATION Optical Fibers

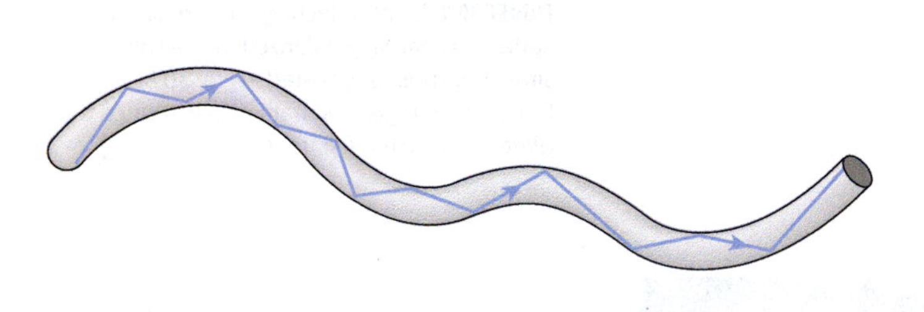

Refraction and total internal reflection are the basis for **optical fibers**, which carry much of the world's communications. Optical fibers provide the physical connectivity of the global Internet and handle information ranging from telephone and television to light signals within medical, astronomical, and industrial instruments.

A typical fiber consists of a glass core only 8 μm in diameter, surrounded by a so-called cladding consisting of glass with a lower refractive index. Total internal reflection at the core–cladding interface guides light along the fiber, as shown in the figure. The glass used in optical fibers is so clear that a 1-km-thick slab would be as transparent as an ordinary window pane. Today's fibers carry light produced by semiconductor lasers at infrared wavelengths of 850 nm, 1350 nm, or 1550 nm.

An optical fiber's main advantage over copper wire is its huge rate of information flow, called **bandwidth**. Communicating information—audio, video, or digital data—requires a range of frequencies, and the greater the rate of information transfer, the wider that range. With its frequency of around 10^{14} Hz, light can accommodate a much wider frequency range within a channel than can radio communication systems operating at frequencies on the order of 10^{10} Hz. A single optical fiber, for example, can carry tens of thousands of simultaneous telephone conversations. Fibers are also lighter and more rugged than copper cables, and they're less vulnerable than copper or open-air transmission to illicit tapping. And because they're made from insulators, optical fibers are less susceptible to electrical noise. The photo shows two cables that can carry information at the same rate. One consists of a few optical fibers while the other is a thick bundle of copper wires.

30.4 Dispersion

MP

PhET: Bending Light (Prism Break)

Refraction ultimately involves the interaction of electromagnetic-wave fields with atomic electrons. It's not surprising, therefore, that the electrons' behavior and consequently also the refractive index depend on frequency (or, equivalently, on wavelength or color). The result is **dispersion**—refraction of different colors through different angles. Figure 30.13 shows the wavelength dependence of the refractive index for a type of glass engineered to exhibit high dispersion. The classic example of dispersion is Newton's demonstration that white light is a mixture of all colors in the visible spectrum (Fig. 30.14). The rainbow is a beautiful natural manifestation of dispersion combined with internal reflection, as the Application on the next page describes.

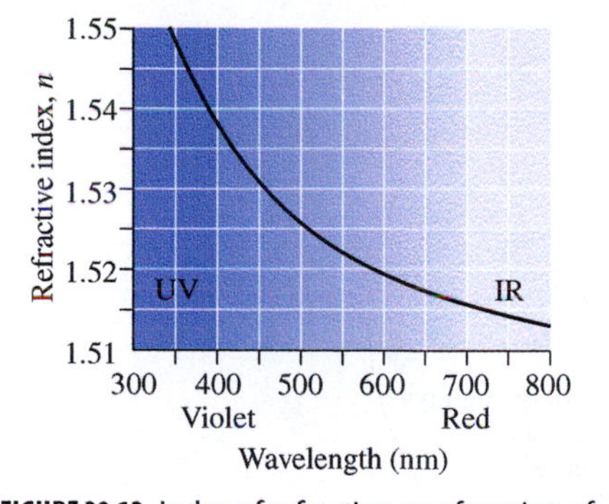

FIGURE 30.13 Index of refraction as a function of wavelength for high-dispersion crown glass.

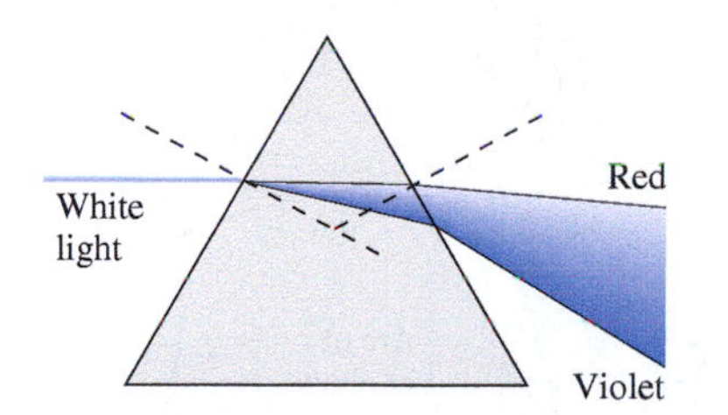

FIGURE 30.14 Dispersion separates the colors in white light, with shorter-wavelength violet experiencing the greatest refraction.

EXAMPLE 30.4 Dispersion: A Prism

White light strikes the prism in Fig. 30.15 normal to one surface. The prism is made of glass whose refractive index is plotted in Fig. 30.13. Find the angle between outgoing red and violet light, with wavelengths of 700 nm and 400 nm, respectively.

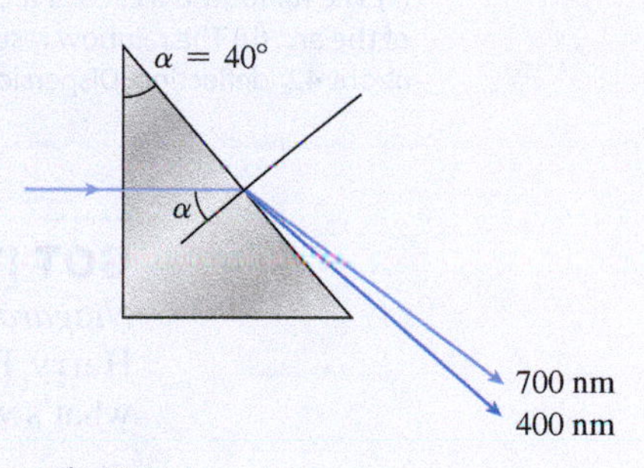

FIGURE 30.15 Example 30.4

INTERPRET This is a refraction problem, like those of Section 30.2, except that here we've got two different refractive indices for two different wavelengths. Since the incident beam is normal to the vertical face of the prism, there's no refraction at the first interface.

DEVELOP To assess the refraction at the second interface, from glass to air, we can use Snell's law, Equation 30.3, just as we did in Section 30.2. A look at Fig. 30.13 shows that $n_{400} = 1.538$ and $n_{700} = 1.516$. We'll have to apply Snell's law twice, once for each of these refractive indices. We'll also need the angle of incidence at the glass–air interface, which the geometry of Fig. 30.15 shows is equal to the 40° angle at the top of the prism. We've labeled both angles α.

EVALUATE We can get the angles of refraction—the angles that the outgoing beams make with the normal to the slanting face of the prism—by solving Snell's law for θ_2 and using the appropriate refractive index for n_1. We'll take $n_2 = 1$ for air. So we have

$$\theta_{400} = \sin^{-1}(n_{400}\sin\alpha) = \sin^{-1}[(1.538)(\sin 40°)] = 81.34°$$

$$\theta_{700} = \sin^{-1}(n_{700}\sin\alpha) = \sin^{-1}[(1.516)(\sin 40°)] = 77.02°$$

The angle between the two outgoing beams is, therefore, $\Delta\theta = \theta_{400} - \theta_{700} = 4.32°$.

ASSESS This is a pretty small angle, despite the fact that both beams undergo substantial refraction. Dispersion is generally a small effect, which is why it isn't always obvious. For more insight into geometrical optics, try reworking this example for the case of an isosceles prism, with $\alpha = 45°$ (see Question 8).

APPLICATION The Rainbow

Rainbows occur when sunlight strikes rain or other airborne water droplets. An observer standing between the Sun and the rain then sees a circular arc of colored bands. Part (a) of the figure shows that the center of that arc lies on the line joining the Sun to the observer's head. That means each observer sees a different rainbow! Furthermore, the rainbow's arc always subtends an angle of about 42°.

Isaac Newton provided the first full explanation of the rainbow, invoking both internal reflection and dispersion. Part (b) of the figure shows light passing through a spherical raindrop. Parallel rays striking the curved drop experience a range of incidence angles, giving a range of angles ϕ between incident and outgoing rays. As the figure shows, however, there's a maximum angle ϕ_{max} of about 42°, and more light returns at angles close to ϕ_{max} than at other angles. That's why the rainbow appears as a bright arc at an angle of about 42° to the direction of the Sun's rays. Problems 55 and 56 detail how to find ϕ_{max}.

The "bunching" of light rays near ϕ_{max} shows why a bright band appears, but why the different colors? The refractive index varies with wavelength, and so, therefore, does ϕ_{max}. Thus each color appears at a slightly different angle. For water, the refractive index ranges from $n_{red} = 1.330$ to $n_{violet} = 1.342$. Using these values with the results of Problems 55 and 56 yields $\phi_{red} = 42.53°$ and $\phi_{violet} = 40.78°$. Thus the rainbow appears as a band of colors subtending an angle of about 1.75°, with red at the top.

You'll occasionally see a fainter *secondary rainbow* above the primary arc. This results from two internal reflections, which causes the order of colors to be reversed. Problem 57 explores the secondary rainbow.

(continued)

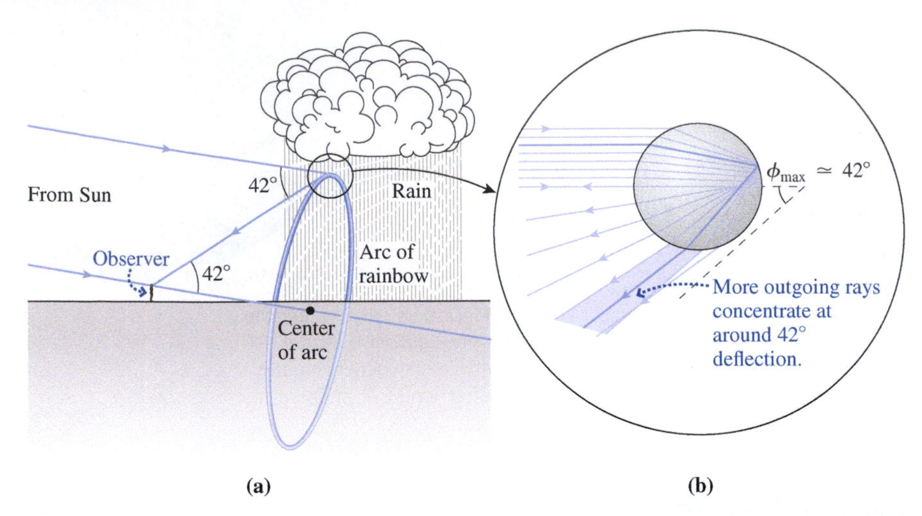

(a) The rainbow is a circular arc at 42° from the line connecting the Sun, the observer, and the center of the arc. (b) The rainbow results from total internal reflection in raindrops, concentrating light at about 42° deflection. Dispersion separates wavelengths slightly, resulting in the rainbow's colors.

GOT IT? 30.4 The painting shown is *Niagara*, by the English-born American artist Harry Fenn. From an optical standpoint, what's wrong with the painting?

Dispersion is the basis of **spectroscopy**, the analysis of light and other electromagnetic radiation in terms of its constituent wavelengths. Hot, dense objects emit a continuous range of wavelengths, while diffuse gases emit and absorb radiation at only a few specific wavelengths (Fig. 30.16). Such discrete spectra provide some of the strongest evidence for the nature of atoms, and today spectroscopy is a powerful tool throughout the sciences. Spectroscopy helps astronomers to determine the composition and motions of distant astrophysical objects, geologists to identify minerals, and chemists to study molecules. Although early spectroscopy used prisms, most modern instruments use instead diffraction gratings, which we'll describe in Chapter 32.

Dispersion can be a nuisance in optical systems. Glass lenses, for example, focus different colors at different points, resulting in distortion known as *chromatic aberration*. Dispersion in optical fibers—based not on wavelength but on different paths taken by rays reflecting at different angles—can degrade digital information. So-called single-mode fibers reduce this effect by passing only those rays that have a single specific reflection angle. On the other hand, dispersion of radio waves provides a crucial correction to the global positioning system (GPS). Ionization in the upper atmosphere introduces an uncertain but frequency-dependent variation in the travel time for radio waves from GPS satellites. It's this travel time that provides GPS location information. Sending waves at two different frequencies and comparing their travel times reveals the atmospheric conditions, and makes dual-frequency GPS receivers accurate to within a few centimeters.

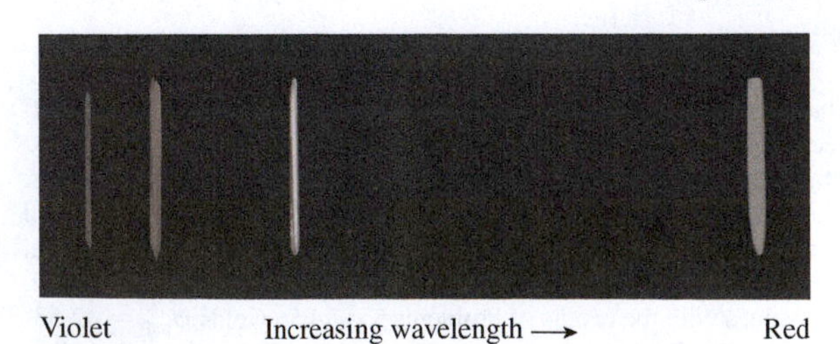

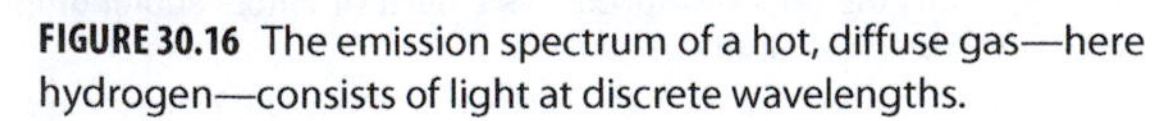

FIGURE 30.16 The emission spectrum of a hot, diffuse gas—here hydrogen—consists of light at discrete wavelengths.

CHAPTER 30 SUMMARY

Big Idea

The big idea here is that light can be considered to travel in straight **rays** when the objects with which it interacts are much larger than the wavelength. Under these conditions, light rays **reflect** and **refract** at interfaces between different materials.

Key Concepts and Equations

The **angle of incidence** and **angle of reflection** are equal:

$$\theta_1' = \theta_1$$

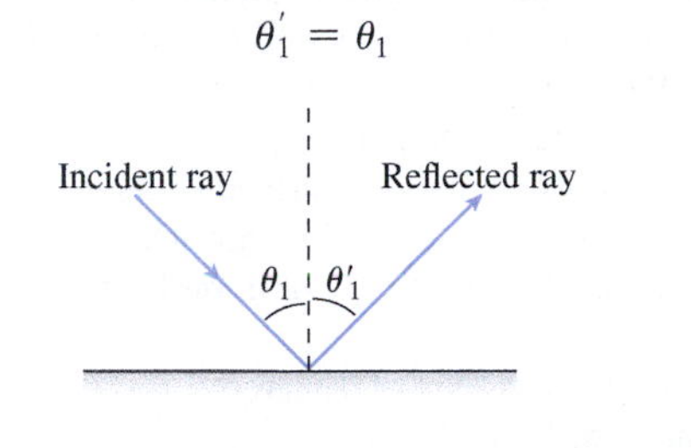

Snell's law relates the angle of incidence and angle of refraction:

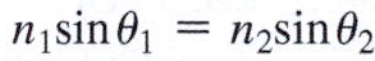

$$n_1 \sin\theta_1 = n_2 \sin\theta_2$$

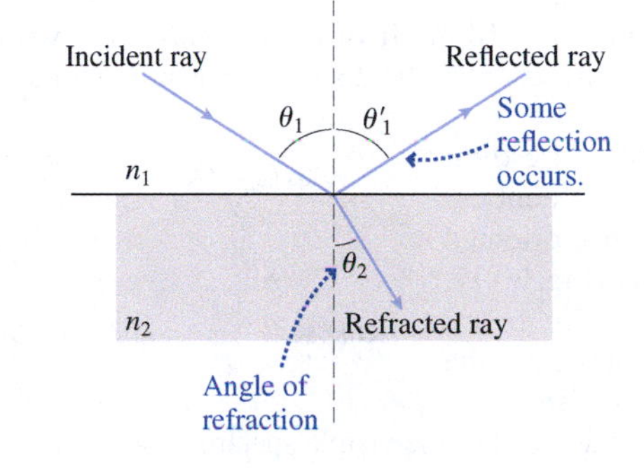

Applications

Total internal reflection results when light is incident at greater than the **critical angle**, θ_c, on an interface with a medium with lower refractive index n_2:

$$\sin\theta_c = \frac{n_2}{n_1}$$

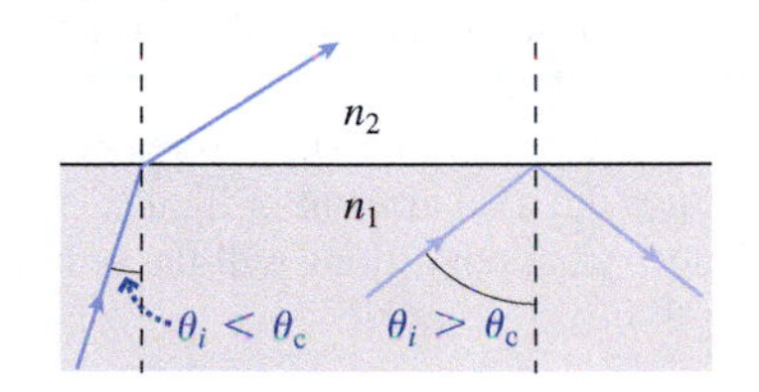

Light polarized in the plane of the incident and refracted rays undergoes no reflection at an interface; this special **polarizing angle**, θ_p, is given by

$$\tan\theta_p = \frac{n_2}{n_1}$$

For an air–glass interface, $\theta_p \simeq 56°$.

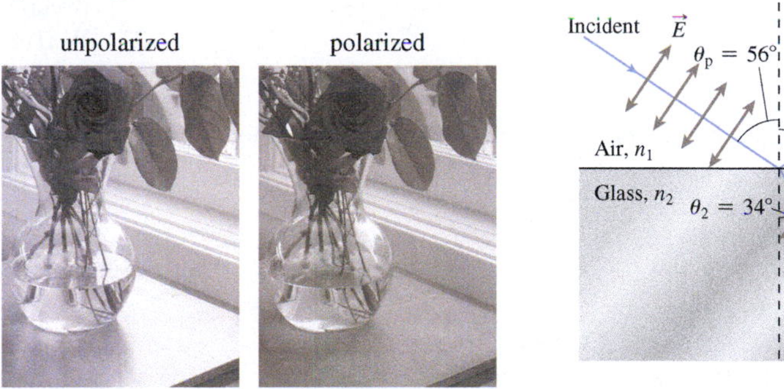

Dispersion results from the wavelength dependence of the refractive index and causes different colors to refract at different angles.

A combination of total internal reflection and dispersion in raindrops accounts for the rainbow.

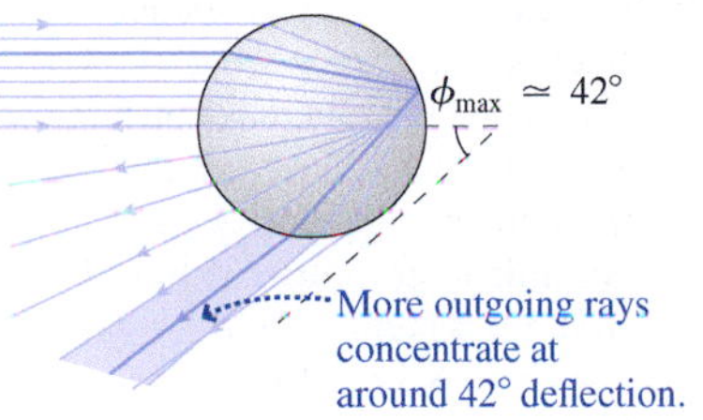

Total internal reflection guides signals in optical fibers.

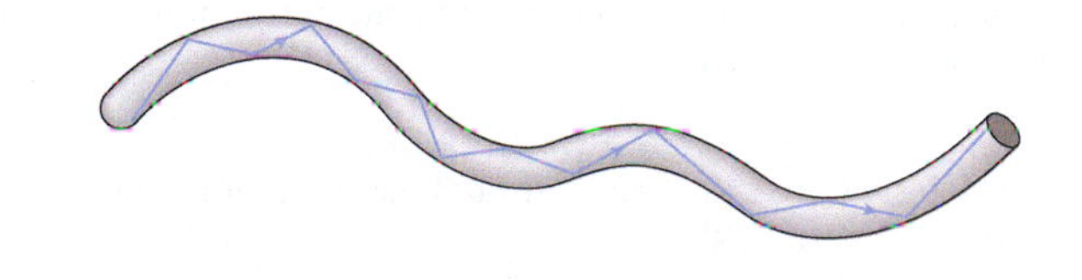

MP For homework assigned on MasteringPhysics, go to www.masteringphysics.com

BIO Biology and/or medicine-related problems **DATA** Data problems **ENV** Environmental problems **CH** Challenge problems **COMP** Computer problems

For Thought and Discussion

1. Why is it usually inappropriate to consider low-frequency sound waves as traveling in rays? Why is the ray approximation more appropriate for high-frequency sound and for light?
2. Why does a spoon appear bent when it's in a glass of water?
3. Why do a diamond and an identically shaped piece of glass sparkle differently?
4. White light goes from air through a glass slab with parallel surfaces. Will its colors be dispersed when it emerges from the glass?
5. You send white light through two identical glass prisms, oriented as shown in Fig. 30.17. Describe the beam that emerges from the right-hand prism.

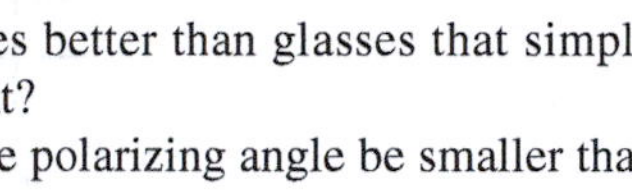

FIGURE 30.17 For Thought and Discussion 5

6. In glass, which end of the visible spectrum has the smallest critical angle for total internal reflection?
7. Why can't you walk to the end of the rainbow?
8. An attempt to rework Example 30.4 with an isosceles prism results in imaginary numbers for the two angles of refraction. What could this mean?
9. Why are polarizing sunglasses better than glasses that simply reduce the total amount of light?
10. Under what conditions will the polarizing angle be smaller than 45°?

Exercises and Problems

Exercises

Section 30.1 Reflection

11. Through what angle should you rotate a mirror so that a reflected ray rotates through 30°?
12. The mirrors in Fig. 30.18 make a 60° angle. A light ray enters parallel to the symmetry axis, as shown. (a) How many reflections does it make? (b) Where and in what direction does it exit the mirror system?

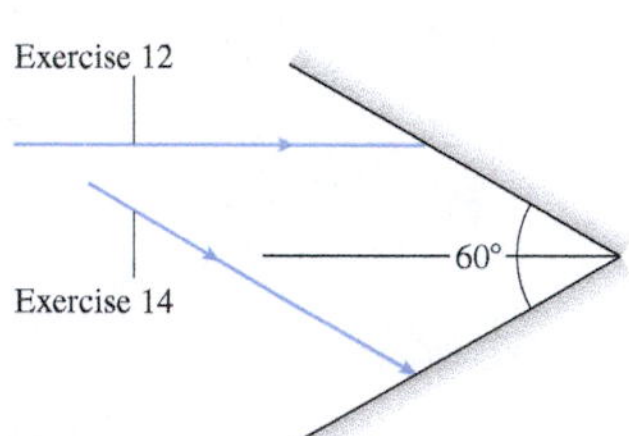

FIGURE 30.18 Exercises 12 and 14 and Problem 28

13. To what angular accuracy must two ostensibly perpendicular mirrors be aligned so that an incident ray returns within 1° of its incident direction?
14. If a light ray enters the mirror system of Fig. 30.18 propagating in the plane of the page and parallel to one mirror, through what angle will it be turned?

Section 30.2 Refraction

15. In which substance in Table 30.1 does the speed of light have the value 2.292×10^8 m/s?
16. Information in a compact disc is stored in "pits" whose depth is essentially one-fourth the wavelength of the laser light used to "read" the information. That wavelength is 780 nm in air, but the wavelength on which the pit depth is based is measured in the $n = 1.55$ plastic that makes up most of the disc. Find the pit depth.
17. Light is incident on an air–glass interface, and the refracted light in the glass makes a 40° angle with the normal to the interface. The glass has refractive index 1.52. Find the incidence angle.
18. A light ray propagates in a transparent material at 15° to the normal to the surface. It emerges into the surrounding air at 24° to the normal. Find the material's refractive index.
19. Light propagating in the glass ($n = 1.52$) wall of an aquarium tank strikes the wall's interior surface with incidence angle 12.4°. What's the angle of refraction in the water?
20. Find the polarizing angle for diamond when light is incident from air.
21. Find the refractive index of a material for which the polarizing angle in air is 62°.

Section 30.3 Total Internal Reflection

22. Find the critical angle for total internal reflection in (a) ice, (b) polystyrene, and (c) rutile, when the surrounding medium is air.
23. A drop of water is trapped in a block of ice. What's the critical angle for total internal reflection at the water–ice interface?
24. What is the critical angle for light propagating in glass with $n = 1.52$ when the glass is immersed in (a) water, (b) benzene, and (c) diiodomethane?
25. Total internal reflection occurs at an interface between plastic and air at incidence angles greater than 37°. Find the plastic's refractive index.

Section 30.4 Dispersion

26. Blue and red laser beams strike an air–glass interface with incidence angle 50°. If the glass has refractive indices of 1.680 for the blue light and 1.621 for the red, what will be the angle between the two beams in the glass?
27. White light propagating in air is incident at 45° on the equilateral prism of Fig. 30.19. Find the angular dispersion γ of the outgoing beam if the prism has refractive indices $n_{red} = 1.582$ and $n_{violet} = 1.633$.

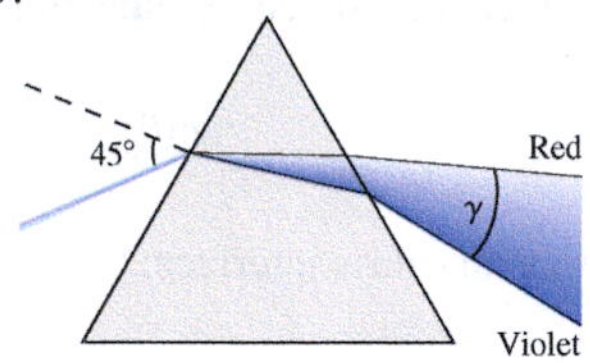

FIGURE 30.19 Exercise 27 (angles of dispersed rays aren't accurate)

Problems

28. Suppose the 60° angle in Fig. 30.18 is changed to 75°. A ray enters the mirror system parallel to the axis. (a) How many reflections does it make? (b) Through what angle is it turned when it exits the system?
29. **BIO** The refractive index of a human cornea is 1.40. If 550-nm light strikes a cornea at incidence angle 25°, find (a) the angle of refraction and (b) the wavelength in the cornea.
30. Two plane mirrors make an angle ϕ. A light ray enters the system and is reflected once off each mirror. Show that the ray is turned through an angle $360° - 2\phi$.
31. An unlabeled bottle of liquid has spilled, and you're trying to find out whether it's relatively harmless ethyl alcohol or toxic benzene. You submerge a glass block with $n = 1.52$ in the liquid, and shine a laser beam so it strikes the submerged glass with incidence angle 31.5°. You measure the angle of refraction in the glass at 27.9°. Which liquid is it? (See Table 30.1.)

32. A meter stick lies on the bottom of the rectangular tank in Fig. 30.20, with its zero mark at the tank's left edge. You look into the long dimension of the tank at a 45° angle, with your line of sight just grazing the top edge, as shown. What mark on the meter stick do you see when the tank is (a) empty, (b) half full of water, and (c) full of water?

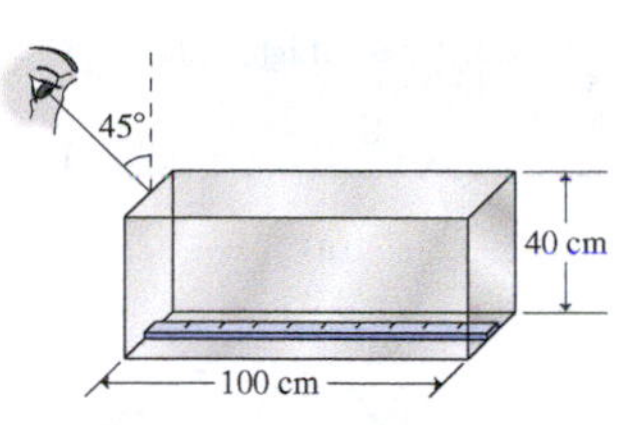

FIGURE 30.20 Problem 32

33. You look at the center of one face of a solid glass cube of glass, on a line of sight making a 55° angle with the normal to the cube face. What minimum refractive index of the glass will let you see through the cube's opposite face?
34. At the aquarium where you work, a fish has gone missing in a 10-m-deep, 11-m-diameter cylindrical tank. You shine a flashlight in from the top edge of the tank, hoping to see if the missing fish is on the bottom. What's the smallest angle your flashlight beam can make with the horizontal if it's to illuminate the bottom?
35. You're standing 2.3 m horizontally from the edge of a 4.5-m-deep lake, with your eyes 1.7 m above the water's surface. A diver holding a flashlight at the lake bottom shines the light so you can see it. If the light in the water makes a 42° angle with the vertical, at what horizontal distance is the diver from the edge of the lake?
36. You've dropped your car keys at night off the end of a dock into water 1.6 m deep. A flashlight held directly above the dock edge and 0.50 m above the water illuminates the keys when it's aimed at 40° to the vertical, as shown in Fig. 30.21. What's the horizontal distance x from the edge of the dock to the keys?

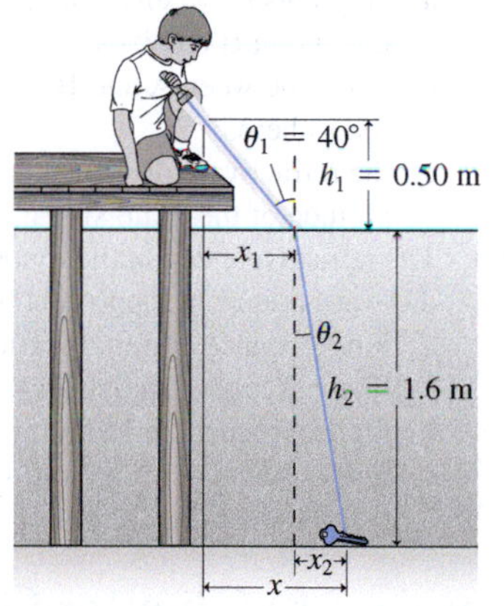

FIGURE 30.21 Problem 36

37. **BIO** Laser eye surgery uses ultraviolet light with wavelength 193 nm. What's the UV light's wavelength within the eye's lens, where $n = 1.39$?
38. The prism in Fig. 30.22 has $n = 1.52$ and $\alpha = 60°$ and is surrounded by air. A light beam is incident at $\theta_1 = 37°$. Find the angle δ through which the beam is deflected.

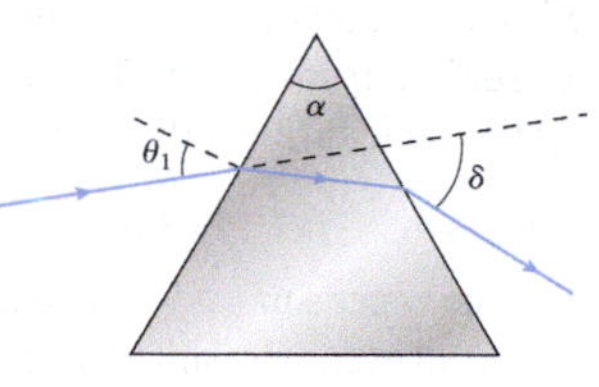

FIGURE 30.22 Problems 38 and 39

39. Repeat Problem 38 for the case $n = 1.75, \alpha = 40°$, and $\theta_1 = 25°$.
40. Find the minimum refractive index for the prism in Fig. 30.11 if total internal reflection occurs as shown when the prism is surrounded by air.
41. Where and in what direction would the main beam emerge if the prism in Fig. 30.11 were made of ice, surrounded by air?
42. Find the speed of light in a material for which the critical angle at an interface with air is 61°.
43. The prism of Fig. 30.11 has $n = 1.52$. When it's immersed in a liquid, a beam incident as shown in the figure ceases to undergo total reflection. What's the minimum value for the liquid's refractive index?
44. For the interface between air (refractive index 1) and a material with refractive index n, show that the critical angle and the polarizing angle are related by $\sin\theta_c = \cot\theta_p$.
45. A scuba diver sets off a camera flash at depth h in water with refractive index n. Show that light emerges from the water's surface through a circle of diameter $2h/\sqrt{n^2 - 1}$.
46. Suppose the red and blue beams of Exercise 26 are now propagating in the same direction *inside* the glass. For what range of incidence angles on the glass–air interface will one beam be totally reflected and the other not?
47. **BIO** In cataract surgery, ophthalmologists replace the eye's natural lens with a synthetic *intraocular lens*, or IOL. A particular IOL has refractive index 1.452. Find the angle of refraction for a light ray striking this lens with incidence angle 77.0°. The medium before the IOL is the eye's *aqueous humor*, a liquid with $n = 1.337$.
48. In a ruby laser, light is produced in a solid rod of ruby, with refractive index 1.77. The light emerges from the end of the rod into the surrounding air. At what angle should the rod end be cut so the emerging light is fully polarized?
49. Reconsider Example 30.4, now in a glass with $n_{700} = 1.482$ and $n_{400} = 1.615$. Determine what happens to the red and violet light with these respective wavelengths, and speculate on what happens to the entire incident beam of white light.
50. A cylindrical tank 2.4 m deep is full to the brim with water. Sunlight first hits part of the tank bottom when the rising Sun makes a 22° angle with the horizon. Find the tank's diameter.
51. For what diameter tank in Problem 50 will sunlight strike some part of the tank bottom whenever the Sun is above the horizon?
52. Light is incident from air on the flat wall of a polystyrene water tank. If the incidence angle is 40°, what is the angle of refraction in the water?
53. **BIO** You're an optometrist, mounting a projector at the back of your 4.2-m-long exam room, 2.6 m above the floor. It shines an eye-test pattern on a white wall opposite the projector. Patients will sit with their eyes 3.3 m from the wall and 1.4 m above the floor to view the pattern. At what height should you center the pattern on the wall?
54. Find an expression for the displacement x in Fig. 30.6, in terms of θ_1, d, and n.
55. **CH** Figure 30.23 shows light passing through a spherical raindrop, undergoing two refractions and total internal reflection, resulting in an angle ϕ between the incident and outgoing rays. Show that $\phi = 4\sin^{-1}(\sin\theta/n) - 2\theta$, where θ is the incidence angle.

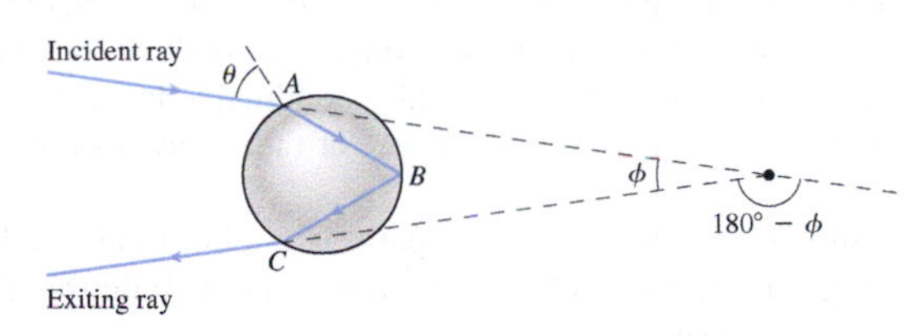

FIGURE 30.23 Problem 55

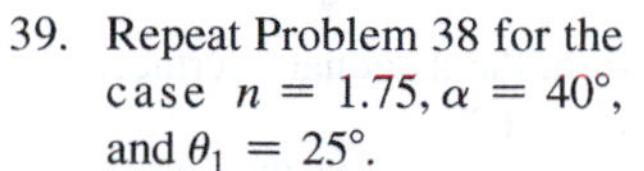

56. **CH** (a) Differentiate the result of Problem 55 to show that the maximum value of ϕ occurs when the incidence angle θ is given by $\cos^2\theta = \frac{1}{3}(n^2 - 1)$. (b) Use this result and that of Problem 55 to find the maximum ϕ in a raindrop with $n = 1.333$. This is the angle at which the rainbow appears, as shown in the Application on pages 573–574.
57. **CH** Figure 30.24 shows the approximate path of a light ray that undergoes internal reflection twice in a

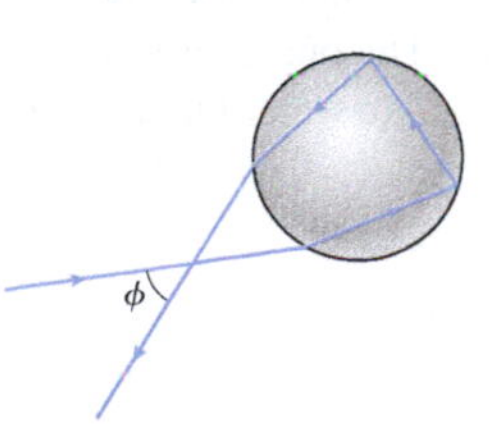

FIGURE 30.24 Problem 57

spherical water drop. Repeat Problems 55 and 56 for this case to find the angle at which the secondary rainbow occurs.

58. **CH** Show that a three-dimensional corner reflector (three mutually perpendicular mirrors, or a solid cube in which total internal reflection occurs) turns an incident light ray through 180°. (*Hint:* Let $\vec{q} = q_x\hat{\imath} + q_y\hat{\jmath} + q_z\hat{k}$ be a vector in the propagation direction. How does this vector get changed on reflection by a mirror in a plane defined by two of the coordinate axes?)
59. **CH** *Fermat's principle* states that a light ray's path is such that the time to traverse that path is an extremum (a minimum or a maximum) when compared with times for nearby paths. Show that Fermat's principle implies Snell's law by proving that a light ray going from point *A* in one medium to point *B* in a second medium will take the least time if it obeys Snell's law.
60. You're an automotive engineer charged with evaluating safety glass, which is made by bonding a layer of flexible plastic between two layers of glass, thus eliminating dangerous glass fragments during accidents. A new product uses glass with refractive index $n = 1.55$ and plastic with $n = 1.48$. You're asked to determine whether total internal reflection at the glass–plastic interface could cause problems with visibility. What do you conclude, and why?
61. **CH** A slab of transparent material has thickness d and refractive index n that varies across the material: $n(x) = n_1 + (n_2 - n_1)(x/d)^2$, where x is measured from one face of the slab. A light ray is incident normally on the slab. Find an expression for the time it takes to traverse the slab.
62. **DATA** For common materials like glass, the wavelength dependence of the refractive index at visual wavelengths is given approximately by $n(\lambda) = b + c/\lambda^2$, where b and c are constants. The table below gives values of λ and n for the crown glass of Example 30.4. Determine a quantity that, when you plot n against it, should give a straight line. Make your plot, establish a best-fit line, and use the line to determine the constants b and c.

λ (nm)	425	475	525	575	625	675
n	1.534	1.528	1.523	1.521	1.518	1.517

Passage Problems

Mirages occur when air's refractive index varies with position as a result of uneven heating. Under such conditions, light undergoes refraction continually and thus follows a curved path. Other examples where a varying refractive index is important include the eye's lens and Earth's *ionosphere*, an electrically conductive layer in the upper atmosphere, where the refractive index for radio waves varies with altitude.

63. Figure 30.25*a* depicts light's path over a hot road, producing a mirage. From the path shown, you can conclude that the air's refractive index
    a. increases from left to right.
    b. increases from right to left.
    c. increases upward.
    d. increases downward.
64. The observer in Fig. 30.25*a* sees a shimmering mirage that looks like water but actually results from sky light following the curved path. To the observer, the mirage appears to be at

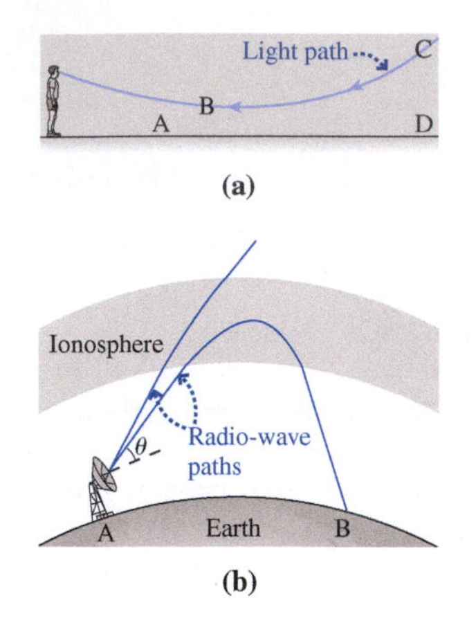

FIGURE 30.25 Passage Problems 63–66. (a) Light path in a mirage. (b) Long-distance radio communication via ionospheric refraction (not to scale).

   a. point A.
   b. point B.
   c. point C.
   d. point D.
65. Figure 30.25*b* shows how continuous refraction in the ionosphere enables long-distance radio communication. Waves launched at angles steeper than θ don't refract enough to return to Earth, so they propagate through the ionosphere and on to space. You can therefore conclude that
    a. all points between A and B receive stronger signals from A than point B receives.
    b. points between A and B can't receive signals from A via the ionosphere.
    c. the refractive index must become infinite at the maximum altitude of the radio signal.
66. The refractive index in the ionosphere is strongly dependent on radio-wave frequency, approaching 1 for high frequencies. Therefore,
    a. long-distance communication via the ionosphere is more likely at higher frequencies.
    b. higher frequencies won't penetrate as far into the ionosphere.
    c. higher frequencies are more appropriate for satellite-based communication.

Answers to Chapter Questions

Answer to Chapter Opening Question

The image results from total internal reflection of light striking the water surface from below. This is the same process that guides signals along the optical fibers that carry data over the Internet.

Answers to GOT IT? Questions

30.1 The water surface is uneven, causing different reflection angles, but the unevenness isn't so fine as to result in fully diffuse reflection.

30.2 $n_3 > n_1 > n_2$

30.3 It would emerge into the water from the diagonal interface, because the critical angle now becomes 63°.

30.4 Rainbows always subtend a half-angle of 42°. The diameter of the rainbow, therefore, subtends 84°. Fenn's entire painting covers about three times the diameter of the rainbows shown, or some 250°. That's too much to capture in a single scene, which can't exceed 180°.

Images and Optical Instruments

What You Know

- You understand reflection and refraction at plane surfaces.
- You have optical experience with your own eyes, and possibly with corrective lenses, and perhaps with instruments such as microscopes, telescopes, and magnifying glasses.

What You're Learning

- You'll learn the process of *ray tracing*, and use it to describe both graphically and algebraically how plane and curved mirrors form images.
- You'll learn to describe image formation with lenses, using both ray tracing diagrams and algebraic equations.
- With both mirrors and lenses, you'll learn to distinguish *real images* and *virtual images*.
- You'll learn to calculate the focal lengths of thin lenses.
- You'll explore the optics of the human eye, including vision correction.
- You'll learn about common optical instruments, including cameras, microscopes, and telescopes.

How You'll Use It

- If you go on in any field of science, engineering, or medicine, you'll almost certainly use optical instruments to explore small or distant systems, to diagnose and correct vision problems, to image internal regions of the human body, or to record visual images.
- If you do photography or videography, you'll be using optical instruments—ranging from the simple lens system in your smartphone's camera to elaborate telephoto lenses for wildlife viewing.

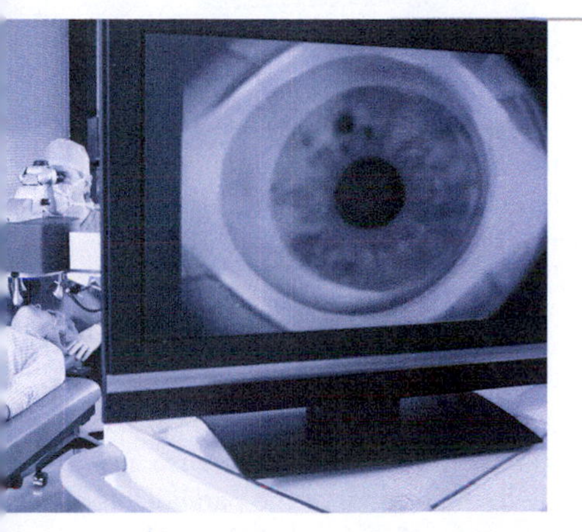

How does laser surgery provide permanent vision correction?

Reflection and refraction alter the direction of light propagation, according to the laws developed in Chapter 30. Microscopes, telescopes, cameras, contact lenses, scanners, and your own eyes use reflection or refraction to form **images** that provide visual representations of reality. Here we study image formation using geometrical optics—a valid approximation provided the objects we're imaging are much larger than the wavelength of light.

When you view an object through an optical system, light reflects or refracts, so it doesn't propagate in straight paths from the object. As a result you see an image that may differ in size, orientation, or apparent position from the actual object. In some cases light actually comes from the image to your eyes; the image is then a **real image**. In other cases light only *appears* to come from the image location; then the image is a **virtual image**.

31.1 Images with Mirrors

Plane Mirrors

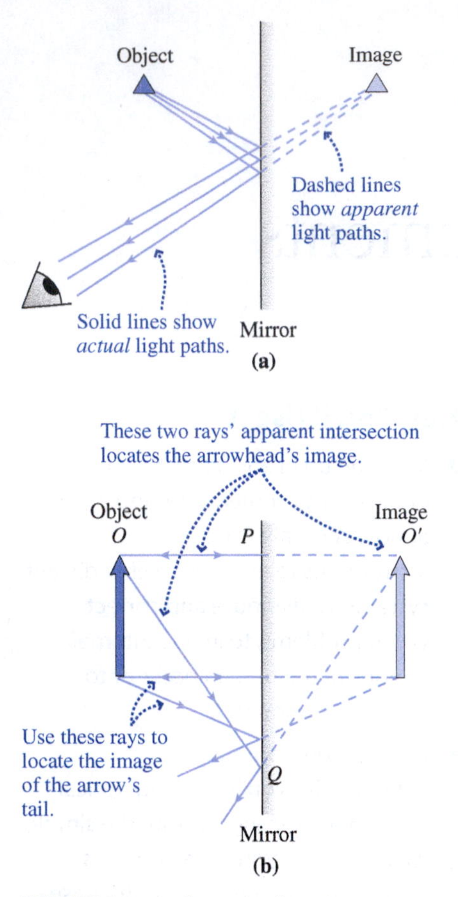

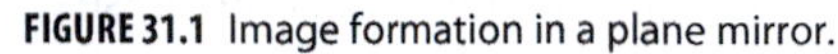

FIGURE 31.1 Image formation in a plane mirror.

In Fig. 31.1*a* we show three light rays that leave the tip of an arrowhead and reflect off a plane mirror to reach the observer's eye. The light looks to the observer like it's coming from a point behind the mirror, so that's the location of the arrowhead's image. The image is virtual because no light actually comes from behind the mirror—even though that's where the image is.

Since two lines define a point, we need only two rays to locate the arrowhead in Fig. 31.1*a*. We've repeated the image-location procedure in Fig. 31.1*b*, now using as one of the rays the ray that reflects normally. The same procedure locates the bottom of the arrow, and we could easily fill in to get the entire arrow; the resulting image is shown in Fig. 31.1*b*.

Because the angles of incidence and reflection are equal, angles OQP and $O'QP$ in Fig. 31.1*b* are equal. The right triangles OPQ and $O'PQ$ share a common side as well, so they're congruent. Thus the distances OP and $O'P$ are equal, so the image is located behind the mirror as far as the object is in front of it. Furthermore, rays from the top and bottom of the arrow and normal to the mirror are parallel, so the image is the same height as the actual arrow.

Images in plane mirrors preserve an object's length and upright orientation, but they reverse the object. When you look in the mirror, you face the mirror. So does your image—meaning the reversal is front to back, not left to right as you might think. This front-to-back reversal makes the image of your right hand look like a left hand (Fig. 31.2). Mathematically, the mirror reverses the coordinate axis perpendicular to the mirror plane. This alters handedness, rotation, and all other phenomena connected with the right-handed coordinate systems we've been using.

FIGURE 31.2 The palm of a right hand faces the mirror. So does the image's palm. That makes the image look like a left hand, but it's still the image of a *right* hand.

GOT IT? 31.1 You stand in front of a plane mirror whose top is at the same height as the top of your head. Approximately how far down must the mirror extend for you to see your full image? (a) to your chest; (b) to your waist; (c) to your knees; (d) to the floor

Curved Mirrors

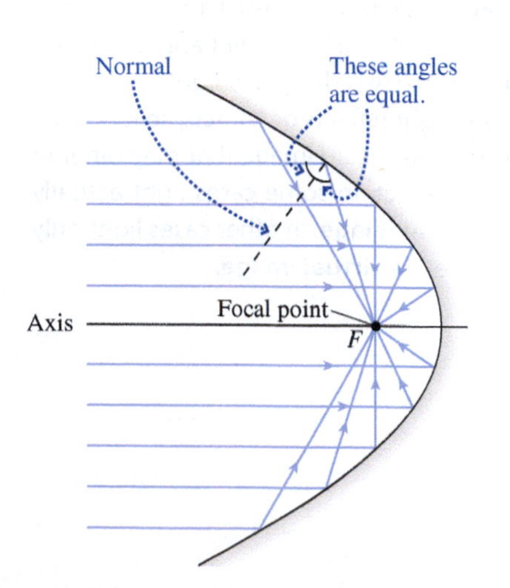

FIGURE 31.3 A parabolic mirror reflects rays parallel to its axis to a common focus.

In contrast to plane mirrors, curved mirrors form images that may be upright or inverted, virtual or real, large or small. The best curved mirrors are parabolic. That's because any line parallel to the parabola's axis makes the same angle to the normal of the parabola as does a second line drawn to a special point called the **focus** or **focal point** (Fig. 31.3). Because the angles of incidence and reflection are equal, this means a parabolic mirror reflects rays parallel to the mirror axis so they converge at the focus. This effect is used to concentrate light to high intensities or, conversely, to create a parallel beam from a point source of light at the focus.

Near the apex of the parabolic mirror in Fig. 31.3, you can't tell whether the shape is parabolic or spherical; a sphere closely approximates the parabola. Because a spherical surface is easier to form, many focusing mirrors are, in fact, spherical. The slight distortion this causes is called **spherical aberration**; a notable case is the Hubble Space Telescope mirror, which was ground to the wrong curve and had substantial aberration (Fig. 31.4). Normally spherical aberration is minimized by making the mirror only a tiny fraction of the entire sphere. In that case the **focal length**—the distance from the mirror's apex to its focal point—is much larger than the mirror, so most rays striking the mirror are nearly parallel to the mirror axis. It's only for such **paraxial rays** that the approximation of a parabola by a sphere results in accurate focusing. But for clarity our diagrams will often show mirrors with exaggerated curvature, and consequently not all rays will seem paraxial.

We can see how spherical mirrors form images by tracing two rays from each of several points on the object, as we did for plane mirrors. Some special rays simplify this process; their properties all follow from the law of reflection and the properties of a spherical mirror in the paraxial approximation.

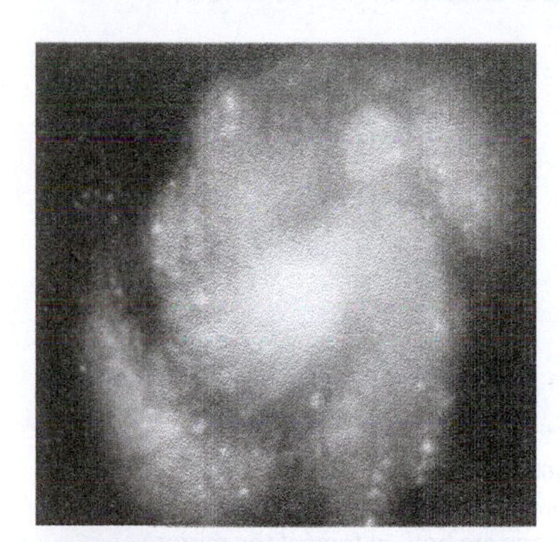

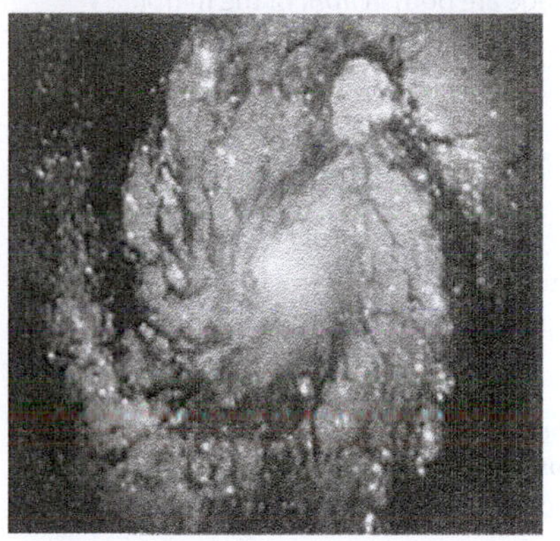

FIGURE 31.4 Incorrect curvature gave the Hubble Space Telescope mirror substantial spherical aberration. Astronauts later installed corrective optics. Images are of the same galaxy before and after the repair.

TACTICS 31.1 Ray Tracing with Mirrors

Figure 31.5 shows four special rays, any two of which suffice to locate an image:

1. Any ray parallel to the mirror axis reflects through the focal point F.
2. Conversely, any ray that passes through F reflects parallel to the axis.
3. Any ray that strikes the center of the mirror reflects symmetrically about the mirror axis.
4. Any ray through the center of curvature, C, strikes the mirror normal to the mirror surface and thus returns on itself.

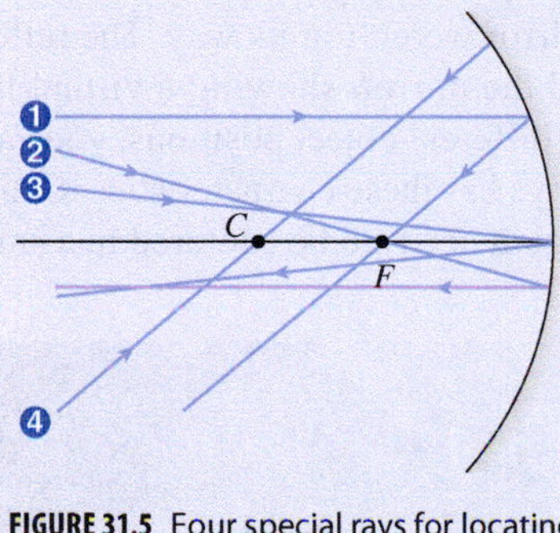

FIGURE 31.5 Four special rays for locating images in curved mirrors. Any two suffice to locate the image.

Figure 31.6 shows ray tracings, using our special rays 1 and 2 that go through the focal point, to find the image location in three cases. In each case symmetry ensures that the bottom of the image arrow is on the axis, so we haven't bothered to trace it. In Fig. 31.6*a* we see that an object beyond the mirror's center of curvature, C, forms a smaller, inverted image. Light actually emerges from this image, so it's a *real* image. If you looked from the

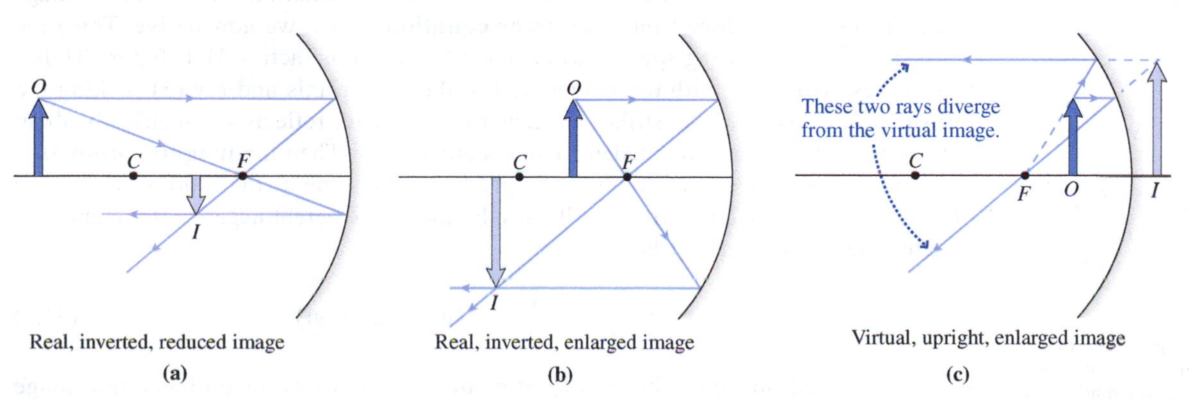

FIGURE 31.6 Image formation with a concave spherical mirror, using rays 1 and 2 described in Tactics 31.1. O denotes the object and I its image.

FIGURE 31.7 A bear meets its real image, formed by the concave mirror at the rear. Bear and image are both in *front* of the mirror.

left toward the mirror in Fig. 31.6*a* you would actually see the image in space in front of the mirror (Fig. 31.7).

As the object moves closer to the mirror, the real image grows; with the object between the center of curvature and the focus, the image is larger than the object and farther from the mirror (Fig. 31.6*b*). As the object moves toward the focus, the image grows larger and moves rapidly away from the mirror. With the object right at the focus, the rays emerge in a parallel beam and there's no image. Finally, rays from an object closer to the mirror than the focus diverge after reflection. To an observer they appear to come from a point behind the mirror. Thus there is a virtual image, in this case upright and enlarged (Fig. 31.6*c*).

GOT IT? 31.2 Where would you place an object so that its real image is the same size as the object, as in Fig. 31.7? (a) at the center of curvature C; (b) at the focal point F; (c) between F and the mirror; (d) between F and C; (e) beyond C

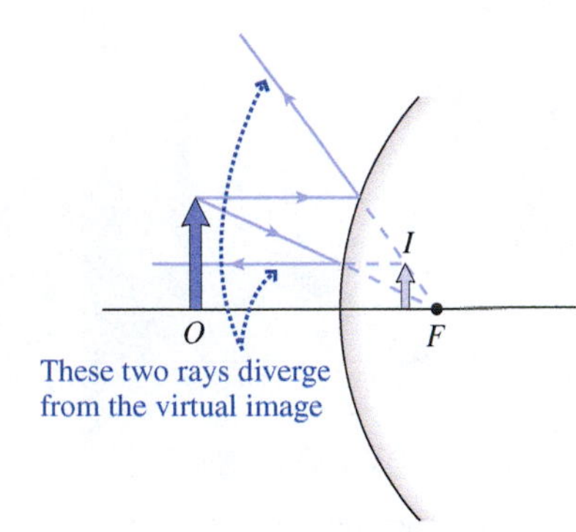

FIGURE 31.8 Image formation with a convex mirror. The image is always virtual, upright, and reduced in size.

Convex Mirrors

A convex mirror reflects on the outside of its spherical curvature, causing light to diverge and therefore to form only virtual images (Fig. 31.8). Although the focus has less obvious physical significance in this case, its location still controls the geometry of reflected rays. As Fig. 31.8 shows, we can still draw a ray parallel to the axis and another ray that would go through the focus if the mirror weren't in its way. The reflected rays appear to diverge from a common point behind the mirror, showing a virtual image that's upright and reduced in size. By considering different object positions, you can convince yourself that the image in a convex mirror always has these characteristics. Convex mirrors are widely used where an image of a broad region needs to be captured in a small space (Fig. 31.9).

FIGURE 31.9 A convex mirror gives a wide-angle view.

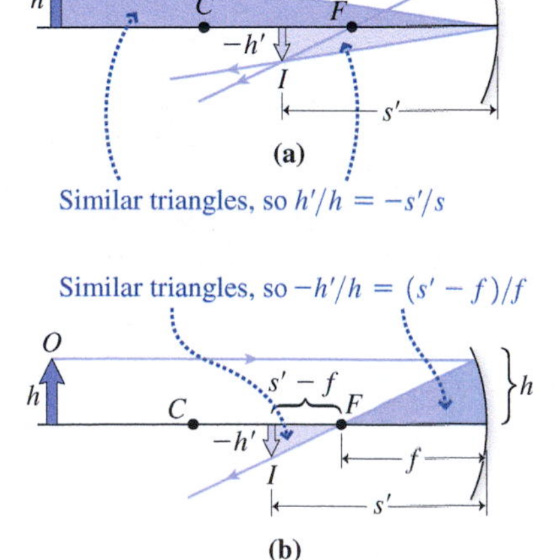

FIGURE 31.10 Finding the image *I* using rays 1 and 3 of Tactics 31.1. For an inverted image the height h' is negative, so we've marked the arrow length—a positive quantity—as $-h'$.

The Mirror Equation

Drawing ray diagrams gives an intuitive feel for image formation. More precise image locations and sizes follow from the **mirror equation**, which we now derive. This time we'll find the image using special rays (1) and (3) listed in Tactics 31.1. Figure 31.10*a* shows these two rays, with ray (1) parallel to the mirror axis and ray (3) striking the mirror's center. The ray that strikes the center of the mirror reflects symmetrically about the axis; therefore, the two shaded triangles are similar. Then the **magnification** M—the ratio of image height h' to object height h—is the same as the ratio of image and object distances from the mirror. We'll consider the image height negative if the image is inverted; then from Fig. 31.10*a* we have

$$M = \frac{h'}{h} = -\frac{s'}{s} \quad \text{(magnification)} \tag{31.1}$$

Here object and image are both in front of the mirror, so we take object and image distances s and s' as positive quantities; the negative sign in Equation 31.1 then shows that in this case the image is inverted. Also, for the object location of Fig. 31.10*a*, it's clear that $|M| < 1$, meaning the image is reduced rather than enlarged.

Figure 31.10*b* is the same as Fig. 31.10*a* except that now we show only the ray that reflects through the focus. We've also labeled the focal length f and shaded another pair of similar triangles. From these you can see that $-h'/h = (s' - f)/f$. Here we put the minus sign on h' because we've defined h' as negative for the inverted image, but comparing similar triangles requires a ratio of positive quantities. Equation 31.1 shows that the ratio h'/h is the magnification $M = -s'/s$. So we have $s'/s = (s' - f)/f$ or, dividing both sides by s' and doing a little algebra,

$$\frac{1}{s} + \frac{1}{s'} = \frac{1}{f} \quad \text{(mirror equation)} \qquad (31.2)$$

Although we derived the mirror equation using a real image, the equation applies to virtual images with the convention that a *negative* image distance s' means the image is *behind* the mirror. And we can handle *convex* mirrors as well by taking the focal length to be a *negative* quantity. Table 31.1 summarizes image formation with mirrors, including these sign conventions.

Table 31.1 Image Formation with Mirrors: Sign Conventions

Focal Length, f	Object Distance, s	Image Distance, s'	Type of Image	Ray Diagram
+ (concave)	+ (in front of mirror) $s > 2f$	+ (in front of mirror) $s' < 2f$	Real, inverted, reduced	O C F I
+ (concave)	+ (in front of mirror) $2f > s > f$	+ (in front of mirror) $s' > 2f$	Real, inverted, enlarged	O C F I
+ (concave)	+ (in front of mirror) $s < f$	− (behind mirror)	Virtual, upright, enlarged	C F O I
− (convex)	+ (in front of mirror)	− (behind mirror)	Virtual, upright, reduced	I O F

A diagram similar to Fig. 31.10 but using the ray through the center of curvature gives another useful fact about curved mirrors: The magnitude of the focal length is half the radius:

$$|f| = \frac{R}{2} \qquad (31.3)$$

You can prove this in Problem 75.

EXAMPLE 31.1 A Concave Mirror: The Hubble Space Telescope

During assembly of the Hubble Space Telescope, one technician stood 3.85 m in front of the telescope's concave mirror (Fig. 31.11). Given the telescope's 5.52-m focal length, find (a) the location and (b) the magnification of the technician's image.

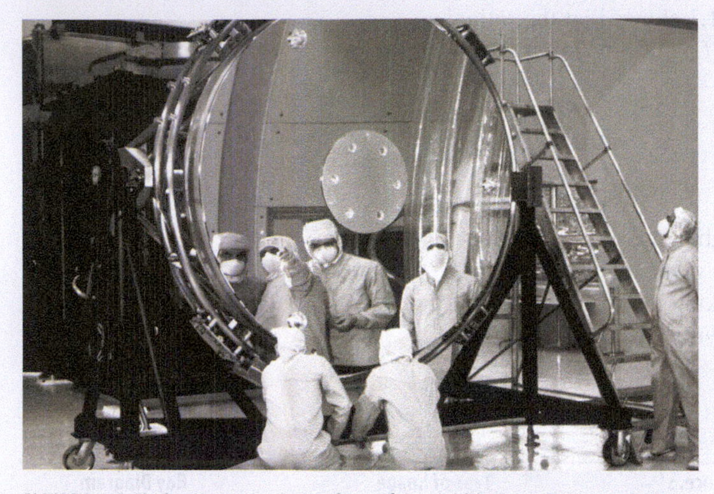

FIGURE 31.11 Technicians standing in front of the Hubble Space Telescope mirror.

INTERPRET This problem is about image formation in a concave mirror. We identify the technician as the object, the 5.52-m focal length as f, and the 3.85-m distance as the object distance s.

DEVELOP We've sketched the situation in Fig. 31.12. With the object closer than the focal length, our sketch resembles Fig. 31.6c, so we anticipate an enlarged, virtual image, as shown. For (a), we'll solve the mirror equation 31.2 for the image distance s' to get the image location. Then for (b), we can find the magnification from Equation 31.1, $M = -s'/s$.

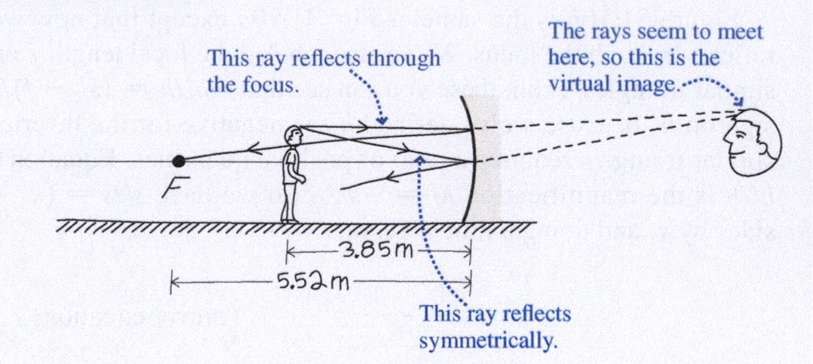

FIGURE 31.12 Sketch for Example 31.1, showing two rays that locate the virtual image of the technician's head.

EVALUATE (a) Solving Equation 31.2 for s' gives

$$s' = \frac{fs}{s - f} = \frac{(5.52\,\text{m})(3.85\,\text{m})}{3.85\,\text{m} - 5.52\,\text{m}} = -12.7\,\text{m}$$

(b) Using this result in Equation 31.1 gives the magnification:

$$M = -\frac{s'}{s} = -\frac{-12.7\,\text{m}}{3.85\,\text{m}} = 3.30$$

ASSESS The *negative* image distance confirms what our sketch anticipated: This is a virtual image, located behind the mirror. The negative distance cancels the minus sign in Equation 31.1, giving a positive threefold magnification. Thus the image is upright and enlarged, just as Fig. 31.11 shows. ■

EXAMPLE 31.2 A Convex Mirror: Jurassic Park

In the film *Jurassic Park*, horrified passengers watch in a car's convex side-view mirror as a *Tyrannosaurus rex* pursues them. Printed on the mirror is the warning "OBJECTS IN MIRROR ARE CLOSER THAN THEY SEEM." If the mirror's curvature radius is 12 m and the *T. rex* is actually 9.0 m from the mirror, by what factor does the dinosaur appear reduced in size?

INTERPRET This is about a convex mirror, governed by the same basic equations as the concave mirror in Example 31.1. We identify the 12-m length as the curvature radius R and the 9.0-m distance as the object distance s.

DEVELOP We've sketched the situation in Fig. 31.13. Since this resembles Fig. 31.8, we see that the image will indeed be reduced in size. Equation 31.1, $M = -s'/s$, gives the magnification we want, but to use it we need the image distance s'. In Example 31.1 we solved Equation 31.2 to get $s' = fs/(s - f)$. Although that mirror was concave and this one is convex, Equation 31.2 applies to both mirrors. So we can use $s' = fs/(s - f)$ in Equation 31.1 to get the magnification:

$$M = -\frac{s'}{s} = -\frac{fs/(s-f)}{s} = -\frac{f}{s-f}$$

We aren't given the focal length, but Equation 31.3, $|f| = R/2$, shows that its magnitude is half the 12-m radius. Since this is a *convex* mirror, Table 31.1 shows that the focal length is *negative*, so $f = -6.0\,\text{m}$.

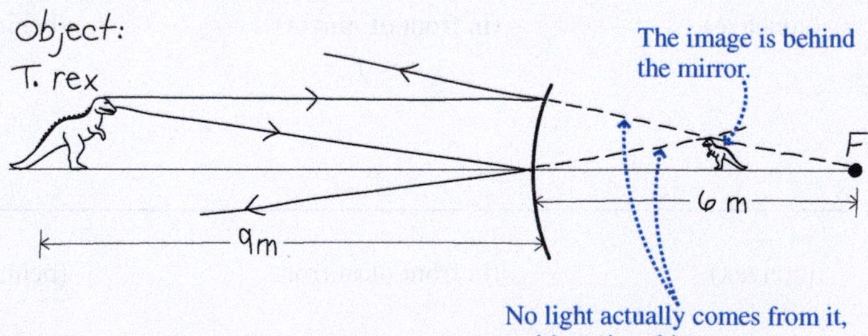

FIGURE 31.13 Sketch for Example 31.2, showing two rays that locate the image of *T. rex*. Mirror is not to scale.

EVALUATE Using our expression for M gives

$$M = -\frac{f}{s-f} = -\frac{(-6.0\,\text{m})}{9.0\,\text{m} - (-6.0\,\text{m})} = 0.40$$

where the answer comes out positive because f is negative.

ASSESS Our result shows that *T. rex* appears in the mirror at only 40% of its actual size; in other words, it looks farther away than it really is. ■

31.2 Images with Lenses

A **lens** is a piece of transparent material that uses refraction to form images. Like mirrors, lenses can be either concave or convex. But light goes *through* a lens, while it reflects off a mirror, so the roles of concave and convex are reversed. A convex lens focuses parallel rays to a **focal point** and is therefore a **converging lens** (Fig. 31.14). As we'll see, a convex lens can form real and virtual images, depending on the object location. A concave lens, in contrast, is a **diverging lens**; it refracts parallel rays so they appear to diverge from a common focus. Like a *convex* mirror, a *concave* lens forms only virtual images (Fig. 31.15).

PhET: Geometric Optics

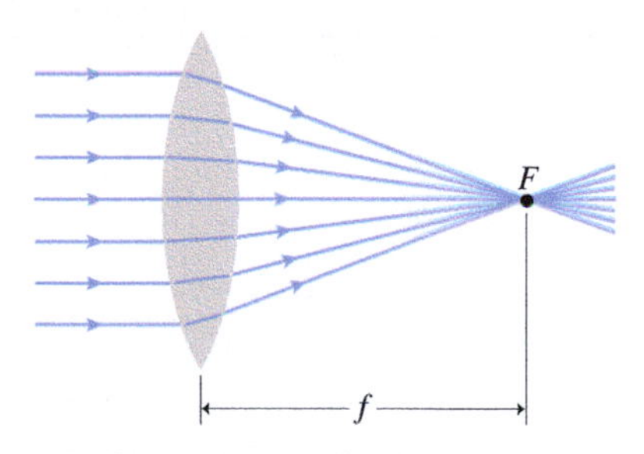

FIGURE 31.14 A convex lens brings parallel light rays to a focus at *F*.

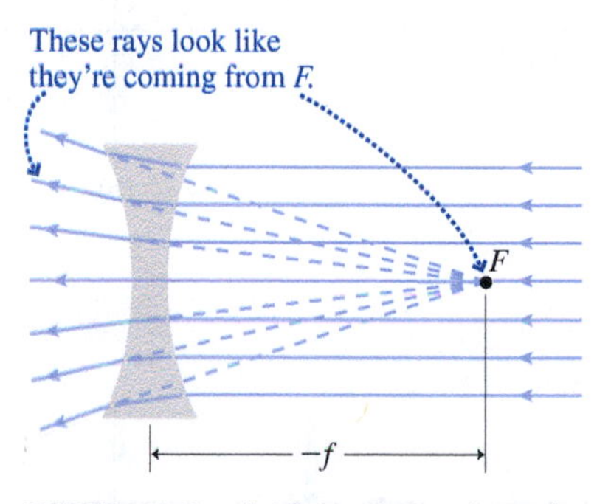

FIGURE 31.15 Parallel light passing through a concave lens diverges so it looks as though it's coming from a common focus.

We first explore a **thin lens**—one whose thickness is small compared with the curvature radii of its two surfaces. Although light refracts as it enters a lens and again as it exits, in the thin-lens approximation the two surfaces are so close that it suffices to consider that the light bends just once, as it crosses the center plane of the lens. Unlike a mirror, light can go either way through a lens; that means the lens has two focal points, one on either side. For a thin lens, the focal length proves to be the same in either direction, so it doesn't matter which way we orient the lens. We'll consider all lenses to be thin unless otherwise stated, and later we'll justify the thin-lens approximation mathematically.

Lens Images by Ray Tracing

As with mirrors, any two rays serve to locate the image formed by a lens. For lenses, two special rays simplify ray tracing.

TACTICS 31.2 Ray Tracing with Lenses

Figure 31.16 shows two special rays:

1. Any ray parallel to the lens axis refracts through the focal point.
2. Any ray through the center of the lens passes undeflected.

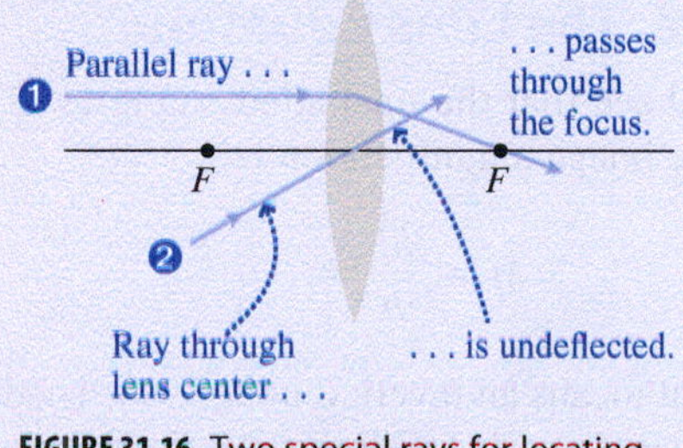

FIGURE 31.16 Two special rays for locating images formed with lenses.

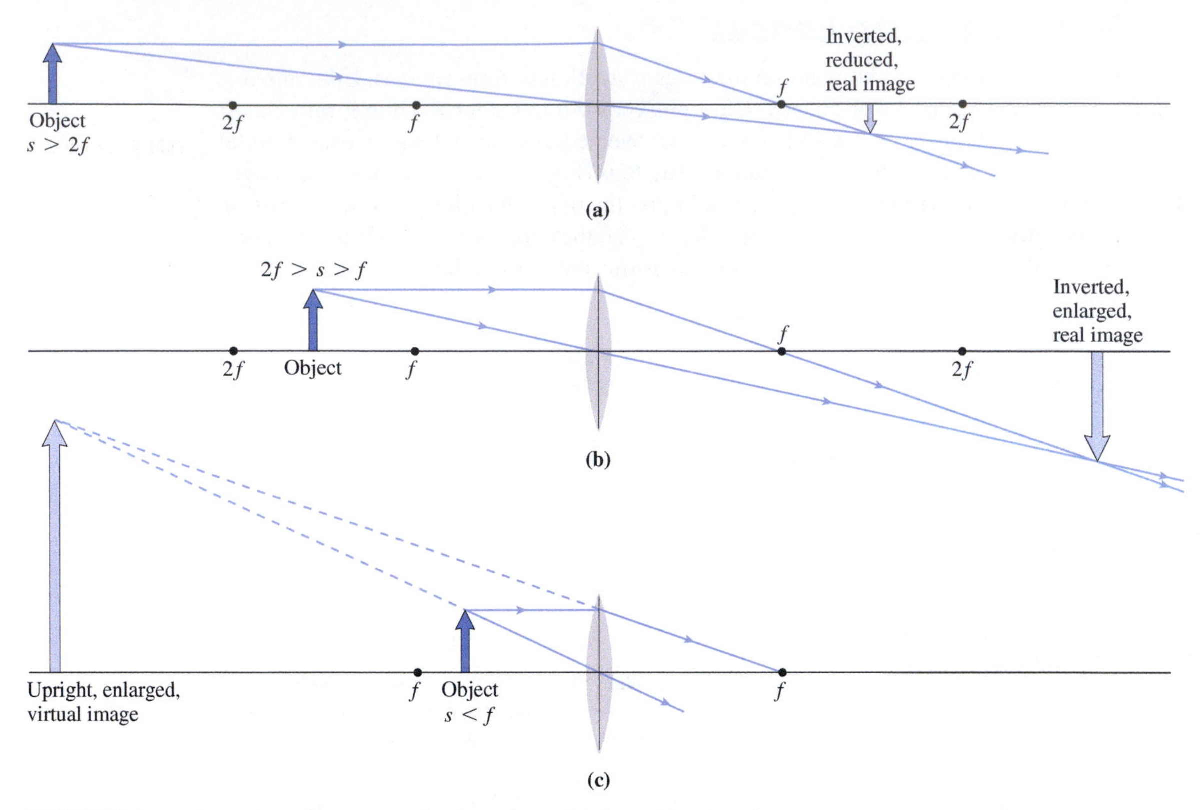

FIGURE 31.17 Image formation with a converging lens, shown for three object locations.

Figure 31.17 shows ray tracings for different object placements in relation to a converging lens. In Fig. 31.17*a* we see that an object farther out than two focal lengths produces a smaller, inverted, real image on the other side of the lens. Since light really emanates from this image, you could see it without looking through the lens. As the object moves toward the lens, the image moves away and grows. When the object is between one and two focal lengths from the lens, the image has moved beyond $2f$ and is enlarged (Fig. 31.17*b*). The image on a movie screen is formed in this way. Moving the object closer than the focal point produces an enlarged, virtual image that can be seen only by an observer looking *through* the lens (Fig. 31.17*c*).

Figure 31.18 shows ray tracings for a diverging lens. Like a convex mirror, this lens produces only virtual images that are upright and reduced in size; they're visible only through the lens. The basic geometry of Fig. 31.18 doesn't change even if the object moves within the focal length.

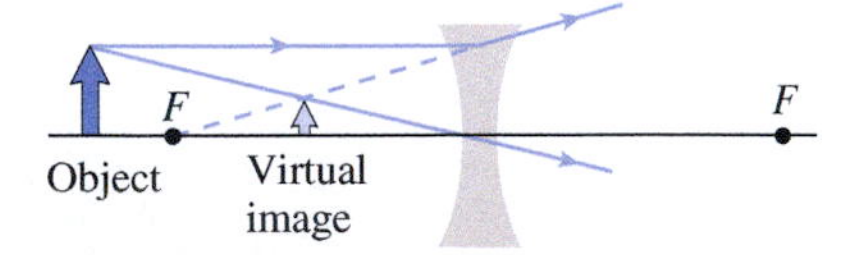

FIGURE 31.18 A diverging lens always forms a reduced, upright, virtual image, visible only through the lens.

Getting Quantitative: The Lens Equation

Study Fig. 31.19 and you'll see that the unshaded triangles *OAB* and *IDB* are similar. Therefore, as for mirrors, the image magnification is

$$M = \frac{h'}{h} = -\frac{s'}{s} \tag{31.4}$$

where again a negative height means an inverted image. The shaded triangles in Fig. 31.19 are also similar, so $-h'/(s' - f) = h/f$. Combining this result with Equation 31.4 and doing some algebra then gives

$$\frac{1}{s} + \frac{1}{s'} = \frac{1}{f} \quad \text{(lens equation)} \tag{31.5}$$

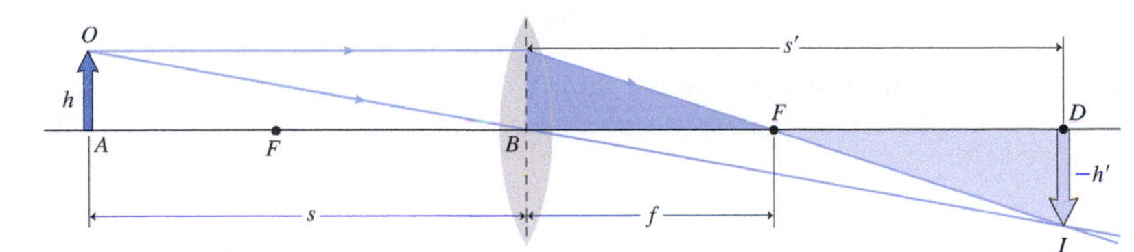

FIGURE 31.19 Ray diagram for deriving the lens equation. Triangles OAB and IDB are similar, as are the shaded triangles.

Video Tutor Demo | **Partially Covering a Lens**

which is identical to Equation 31.2, the mirror equation. Although we derived Equation 31.5 for the case of a real image, it holds for virtual images if we consider the image distance negative; in that case the image is on the same side of the lens as is the object. And it holds for diverging lenses if we consider the focal length negative. Table 31.2 summarizes image formation with lenses, including these sign conventions. Figure 31.20 describes graphically the sizes and types of images formed at different object distances.

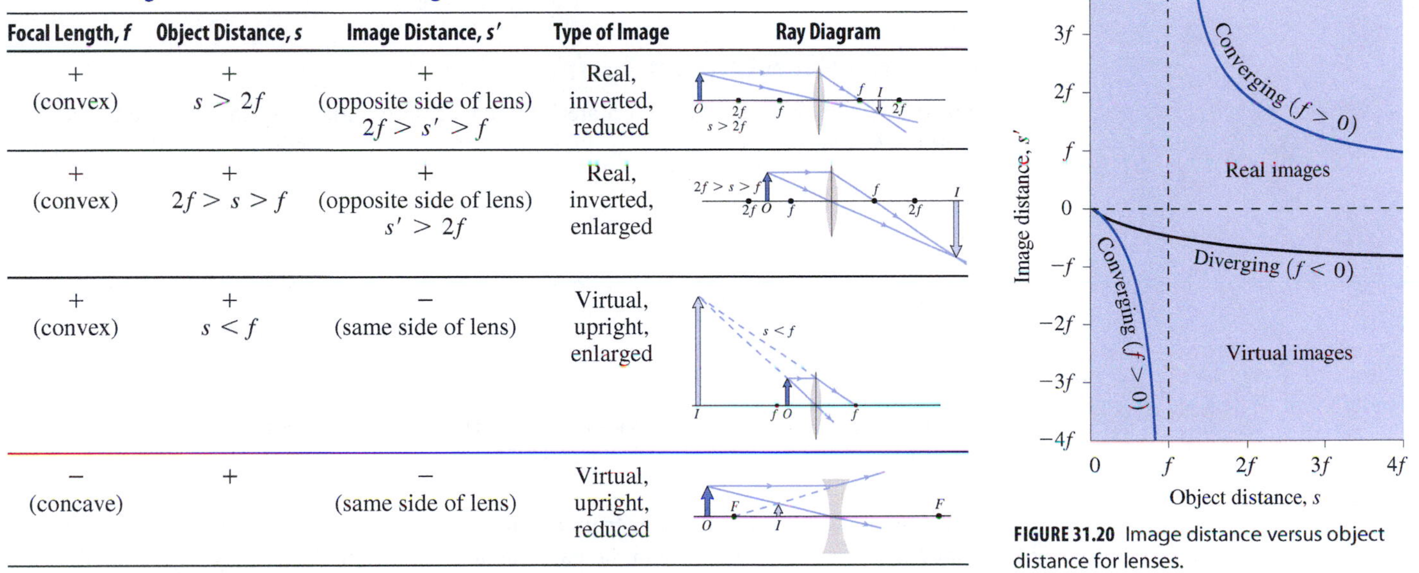

Table 31.2 Image Formation with Lenses: Sign Conventions

Focal Length, f	Object Distance, s	Image Distance, s'	Type of Image	Ray Diagram
+ (convex)	+ $s > 2f$	+ (opposite side of lens) $2f > s' > f$	Real, inverted, reduced	
+ (convex)	+ $2f > s > f$	+ (opposite side of lens) $s' > 2f$	Real, inverted, enlarged	
+ (convex)	+ $s < f$	− (same side of lens)	Virtual, upright, enlarged	
− (concave)	+	− (same side of lens)	Virtual, upright, reduced	

FIGURE 31.20 Image distance versus object distance for lenses.

EXAMPLE 31.3 Using the Lens Equation: Fine Print

You're using a magnifying glass (a converging lens) with 21-cm focal length to read a telephone book (Fig. 31.21). How far from the page should you hold the lens in order to see the print enlarged three times?

FIGURE 31.21 Using a converging lens as a magnifying glass (Example 31.3).

INTERPRET This is a problem involving image formation with a converging lens. The object is the phone book, so we identify the book-to-lens distance we're asked for as the object distance s. Since this is a converging (i.e., convex) lens, the focal length is positive: $f = +21$ cm. The factor-of-3 magnification we want is the quantity M.

DEVELOP The situation is like Fig. 31.17c, with an enlarged, upright, virtual image. We're given the focal length and magnification, but we don't know either the object distance s, which we're looking for, or the image distance s'. Equation 31.4, $M = -s'/s$, relates the two, so we can first use that equation to eliminate s' in terms of s and then solve the lens Equation 31.5, $1/s + 1/s' = 1/f$, for s.

EVALUATE With $M = 3$, Equation 31.4 gives $s' = -3s$. Then Equation 31.5 becomes

$$\frac{1}{s} - \frac{1}{3s} = \frac{2}{3s} = \frac{1}{f} = \frac{1}{21\,\text{cm}}$$

so $s = (2)(21\,\text{cm})/3 = 14\,\text{cm}$.

(continued)

ASSESS Our answer is less than the focal length, as Fig. 31.17*c* shows is required for a virtual image. Figure 31.21 confirms that the image is enlarged, upright, and virtual, and it appears farther away than the object. It's also on the same side of the lens as the object, which explains the negative image distance $s' = -42\,\text{cm}$.

GOT IT? 31.3 You look through a lens at this page and see the words enlarged and right-side up. Is the image you observe real or virtual? Is the lens concave or convex?

31.3 Refraction in Lenses: The Details

So far we've treated lenses as being arbitrarily thin and neglected details of the refraction process. Here we develop a more general description of refraction in lenses, of which our thin-lens approximation is a special case.

Refraction at a Curved Surface

Figure 31.22 shows a transparent material with refractive index n_2 and a curved surface of radius R. Outside the material is a medium with refractive index n_1. We'll now prove

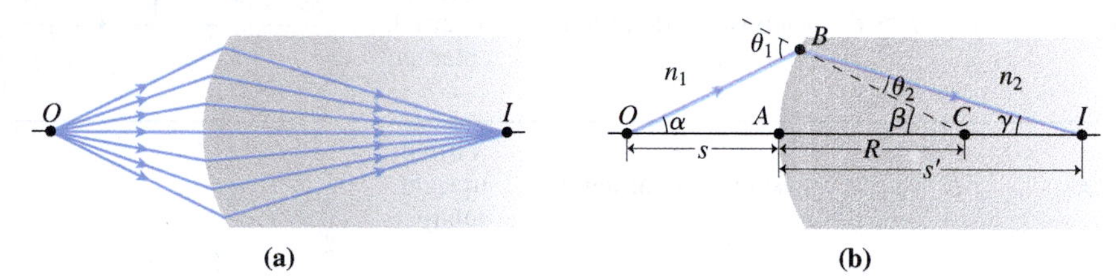

FIGURE 31.22 Refraction at an interface with a curved surface. All labeled angles are considered small, even though the drawing doesn't show them as such.

what Fig. 31.22*a* shows: that rays from a point object O are refracted to a common image point I. Our proof is valid only in the paraxial approximation that all rays make small angles with the optic axis. However, as with mirrors, our drawings won't always show these angles as being small.

Figure 31.22*b* shows a single ray. With all the labeled angles small, we can approximate $\sin x \simeq \tan x \simeq x$, with x in radians. Then Snell's law, $n_1 \sin\theta_1 = n_2 \sin\theta_2$, becomes $n_1\theta_1 = n_2\theta_2$. Triangles BCI and OBC give $\theta_2 = \beta - \gamma$ and $\theta_1 = \alpha + \beta$, so Snell's law becomes $n_1(\alpha + \beta) = n_2(\beta - \gamma)$. Furthermore, in the small-angle approximation the arc BA is so close to a straight line that we can write $\alpha \simeq \tan\alpha \simeq BA/s$, with $s = OA$ the object's distance from the refracting surface. Similarly, $\beta \simeq BA/R$ and $\gamma \simeq BA/s'$. Thus our expression of Snell's law becomes

$$n_1\left(\frac{BA}{s} + \frac{BA}{R}\right) = n_2\left(\frac{BA}{R} - \frac{BA}{s'}\right)$$

or, on canceling BA and rearranging,

$$\frac{n_1}{s} + \frac{n_2}{s'} = \frac{n_2 - n_1}{R} \qquad (31.6)$$

The angle α doesn't appear here, showing that this relation between object and image distances holds for *all* rays that satisfy the small-angle approximation. So Fig. 31.22*a* is correct: All such rays do indeed come to a common focus at I.

We derived Equation 31.6 for the case of a real image, but as usual it applies to virtual images if we take the image distance as negative. And it applies to concave surfaces if we take R to be negative. It even works for flat surfaces, with $R = \infty$.

EXAMPLE 31.4 Refraction at a Curved Surface: A Cylindrical Aquarium

An aquarium consists of a thin-walled plastic tube 70.0 cm in diameter. For a cat looking directly into the aquarium, what's the apparent distance to a fish 15.0 cm from the aquarium wall?

INTERPRET We interpret the cylindrical aquarium as a two-dimensional version of the spherical surface we analyzed with Fig. 31.22 and Equation 31.6. The plastic tube is thin, so we can neglect refraction within the plastic and consider that we have just a cylinder of water. The object is the fish, and since it's *inside* the water, the cylindrical edge of the aquarium is concave *toward* the object. Then the curvature radius is *negative*; we're given the diameter as 70.0 cm, so $R = -35.0$ cm. With the object in the water, $n_1 = 1.333$ from Table 30.1 and $n_2 = 1$ for air. The 15-cm distance from the edge to the fish is the object distance s.

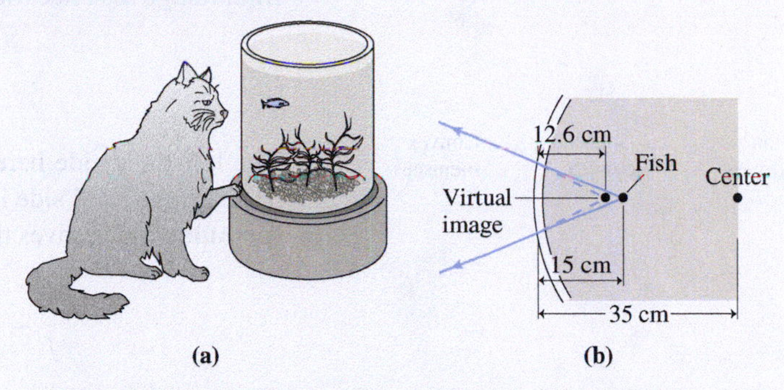

FIGURE 31.23 (a) A cylindrical aquarium. (b) Top view, showing the formation of a virtual image of a fish that's actually 15 cm from the edge.

DEVELOP Figure 31.23 shows the physical situation and a ray diagram viewed from above. Our plan is to solve Equation 31.6, $n_1/s + n_2/s' = (n_2 - n_1)/R$, for s' and evaluate using $R = -35.0$ cm, $n_1 = 1.333$, $n_2 = 1$, and $s = 15.0$ cm.

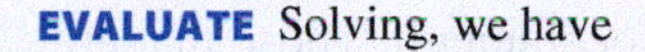

EVALUATE Solving, we have

$$s' = n_2\left(\frac{n_2 - n_1}{R} - \frac{n_1}{s}\right)^{-1} = -12.6 \text{ cm}$$

ASSESS Make sense? The fish is actually 15 cm from the edge, but refraction makes the image distance s' shorter. The same effect occurs when you look down into a swimming pool or lake: Objects on the bottom look closer, and you can find out how much by applying Equation 31.6 with $R = \infty$ (see Exercise 28). ■

Lenses, Thick and Thin

Figure 31.24 shows a lens of thickness t with refractive index n, surrounded by air with $n = 1$. Object O_1 lies a distance s_1 from the left-hand surface. This surface forms an image I_1, which we'll also call O_2 because it acts as an object for the right-hand surface. Refraction at that surface forms a second image I_2. We want to relate the original object O_1 and the final image I_2.

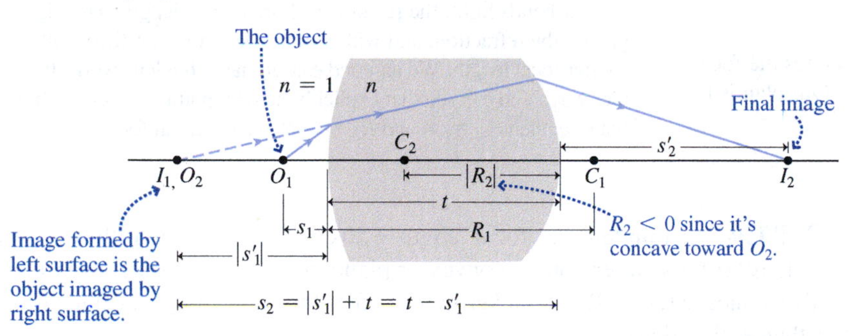

FIGURE 31.24 Analysis of a thick lens with different curvature radii. C_1 and C_2 are the centers of curvature of the left and right sides, respectively; t is the lens thickness.

At the left-hand surface, the quantities in Equation 31.6 become

$$\frac{1}{s_1} + \frac{n}{s_1'} = \frac{n-1}{R_1} \qquad \text{(left-hand surface)}$$

We've placed O_1 so close that I_1 is a *virtual* image, so s_1' is *negative*. Now we set up another instance of Equation 31.6, this time for the right-hand surface. Here I_1 is the object, whose distance s_2 is $t - s_1'$ because s_1' is negative. Also at the right-hand surface, $s' = s_2'$, $n_1 = n$, and $R = R_2$, where, for the case shown, R_2 is negative because the right-hand surface is concave toward the object (I_1, O_2) that's being imaged. So at the right-hand surface, Equation 31.6 reads

$$\frac{n}{t - s_1'} + \frac{1}{s_2'} = \frac{1-n}{R_2} \qquad \text{(right-hand surface)}$$

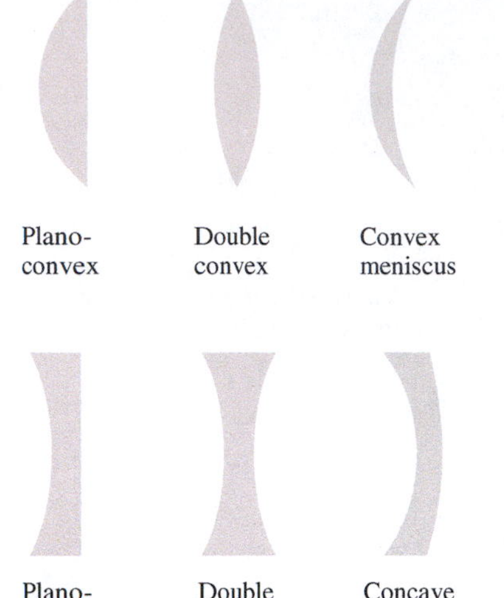

FIGURE 31.25 Common lens types.

Now we'll let the lens become arbitrarily thin, so $t \to 0$. Then we add the two equations; the intermediate-image term n/s'_1 cancels, leaving only the first object distance and final image distance. So we drop the subscripts 1 and 2, and the result is

$$\frac{1}{s} + \frac{1}{s'} = (n-1)\left(\frac{1}{R_1} - \frac{1}{R_2}\right)$$

The left-hand side here is identical to the left-hand side of Equation 31.5, and Equation 31.5's right-hand side is $1/f$. Equating the two right-hand sides results in the **lensmaker's formula**, which gives the focal length:

$$\frac{1}{f} = (n-1)\left(\frac{1}{R_1} - \frac{1}{R_2}\right) \quad \text{(lensmaker's formula)} \quad (31.7)$$

Again, the radii here can be positive or negative; in Fig. 31.24 R_1 is positive because the left-hand surface is convex toward the object, while R_2 is negative because the right-hand surface is concave toward its object, the intermediate image I_1. Although we derived Equation 31.7 for the case of a virtual intermediate image, the lensmaker's formula is a general result for the focal length of a thin lens.

Lenses come in a variety of shapes (Fig. 31.25). Those that are thicker in the center are converging lenses, for which Equation 31.7 gives a positive focal length. Those that are thinner in the center are diverging lenses, with negative f. These behaviors reverse if the medium surrounding the lens has a higher refractive index (see Problem 74).

EXAMPLE 31.5 The Lensmaker's Formula: A Plano-Convex Lens

Find an expression for the focal length of the plano-convex lens in Fig. 31.25, given refractive index n and radius R for the one curved surface.

INTERPRET This is a thin lens, such as we just analyzed in deriving the lensmaker's formula. With an object at the left of the lens, we identify $R_1 = R$ and $R_2 = \infty$ for the flat surface.

DEVELOP The lensmaker's formula, Equation 31.7, relates the focal length, the lens surface radii, and the refractive index. Our plan is to solve for f.

EVALUATE With $R_1 = R$ and $R_2 = \infty$, Equation 31.7 gives

$$f = \left[(n-1)\left(\frac{1}{R} - \frac{1}{\infty}\right)\right]^{-1} = \frac{R}{n-1}$$

ASSESS Make sense? The smaller R, the more curved the lens and the more it bends light; the result is a shorter focal length. The higher n, the greater the refraction, and with n in the denominator, the result is again a shorter focal length. We asserted earlier that a thin lens works the same either way. You can see that explicitly here by putting the object beyond the flat side; then $R_1 = \infty$ and $R_2 = -R$, but the result for f is unchanged. ■

GOT IT? 31.4 A thin lens has focal length +50 cm. Which of the following *must* be true of this lens? (a) it's either double convex or plano-convex; (b) it's a convex meniscus lens; (c) it's a concave lens; (d) it's thicker in the center than at the edges; (e) it's thinner in the center than at the edges

Lens Aberrations

Lenses exhibit several optical defects. We described **spherical aberration** in mirrors; this same defect occurs with spherical lenses (Fig. 31.26*a*). Our lens analysis required that all rays make small angles with the lens axis; if not, then they don't share a common focus, causing spherical aberration. Small angles occur naturally with distant objects, but not with objects close to the lens. Using only the central portion of the lens can eliminate those rays with larger angles (Fig. 31.26*b*), leading to sharper focus. That's why a camera focuses over a wider range when it's "stopped down," with an opaque iris covering the outer part of the lens. The trade-off is that less light is available.

We mentioned **chromatic aberration** in Chapter 30; it occurs because the refractive index varies with wavelength, causing different colors to focus at different points. High-quality optical systems minimize this effect by using composite lenses of materials with different refractive indices. **Astigmatism** occurs when a lens has different curvature radii in different directions. This is a common defect in the human eye, corrected with glasses or contact lenses that have compensating asymmetric curvature.

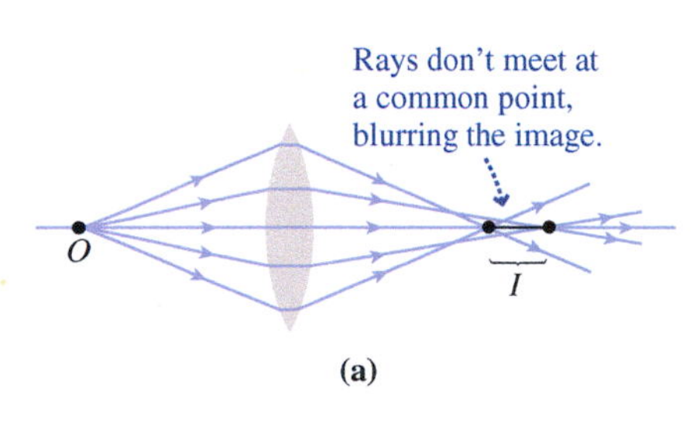

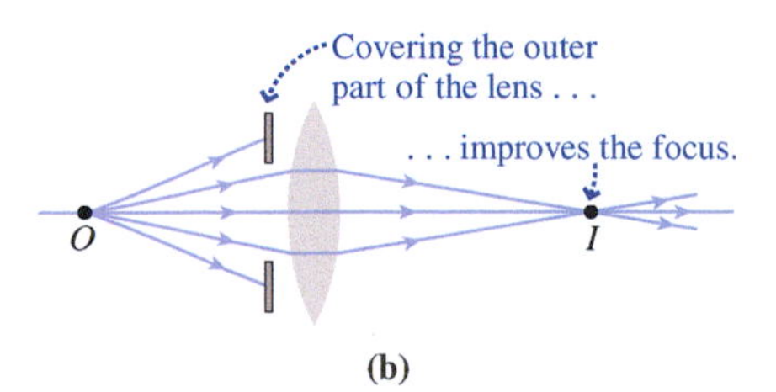

FIGURE 31.26 (a) Spherical aberration. (b) Using only the central portion of the lens minimizes aberration, but at the expense of a dimmer image.

31.4 Optical Instruments

Numerous optical instruments use lenses, mirrors, or both to form images. All but the simplest have more than one optical element, but the principles we've developed here still apply. We analyze such instruments by tracing light through the sequence of optical elements, using the image formed by one element as the object for the next.

The Eye

Our eyes are complex optical systems with several refracting surfaces and mechanisms to vary the focal length and amount of light admitted (Fig. 31.27). Light enters through the *cornea* and traverses a liquid called the *aqueous humor* before passing through the lens. It then traverses the *vitreous humor*, a liquid in the main body of the roughly 2.3-cm-diameter eyeball. Finally it strikes the *retina*, where special cells called rods and cones produce electrochemical signals that carry visual information to the brain.

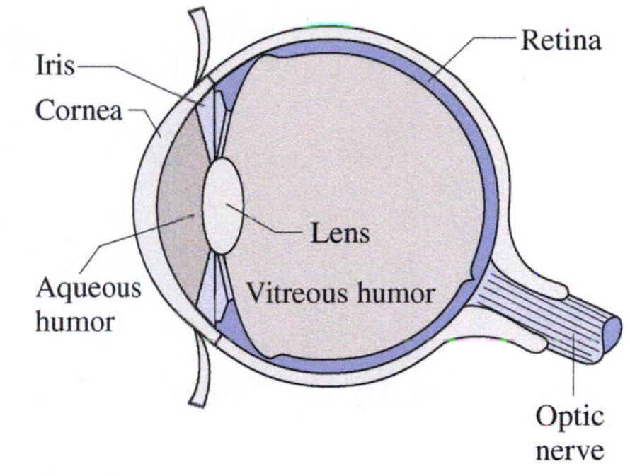

FIGURE 31.27 The human eye.

A properly functioning eye produces well-focused real images on the retina, with the cornea providing most of the refractive focusing. Muscles adjust the lens, changing its focal length to compensate for different object distances. Other muscles adjust the iris, resizing the pupil opening to adjust for different light levels.

In nearsighted (myopic) people, the image forms in front of the retina, causing distant objects to appear blurred (Fig. 31.28*a*). Diverging corrective lenses produce closer intermediate images that the myopic eye can then focus (Fig. 31.28*b*). In farsighted (hyperopic) people, the image of nearby objects would form behind the retina, and converging corrective lenses are used (Fig. 31.29). Even normal eyes can't focus much closer than the so-called **near point** at about 25 cm. This distance greatly increases with age, a condition called *presbyopia*.

Prescriptions for corrective lenses specify the corrective power, P, in **diopters**, which is the inverse of the focal length in meters. Thus a 1-diopter lens has $f = 1\,\text{m}$, while a 2-diopter lens has $f = 0.5\,\text{m}$ and is more powerful in that it refracts light more sharply. Like f itself, the sign of a lens's corrective power is positive or negative depending on whether the lens is converging or diverging.

It doesn't matter whether a corrective lens is several centimeters from the eye, as with glasses, or right on the cornea, as with contact lenses. Contact lenses can be thin because, as Equation 31.7 shows, it's the curvature radii and not the thickness that determine the focal length. A more radical approach to vision correction is laser surgery, described in the Application on the next page.

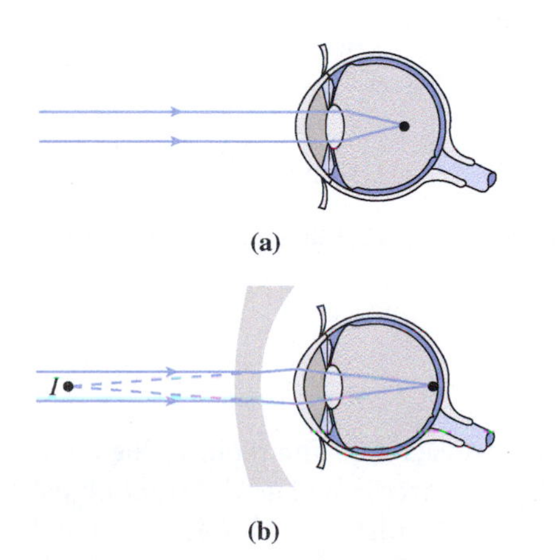

FIGURE 31.28 (a) A nearsighted eye focuses light from distant objects in front of the retina. (b) A diverging lens corrects the problem, creating a closer virtual image that the eye can focus.

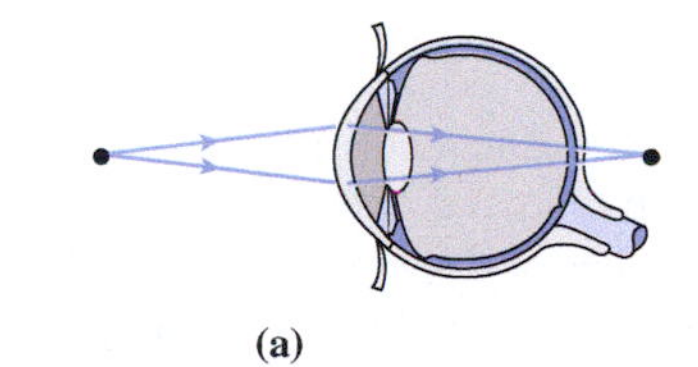

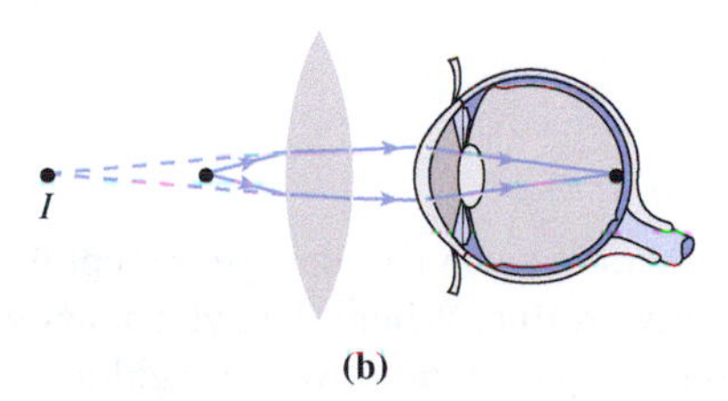

FIGURE 31.29 (a) A farsighted eye can't focus light from nearby objects. (b) A converging lens produces a more distant image that the eye can focus.

APPLICATION Laser Vision Correction

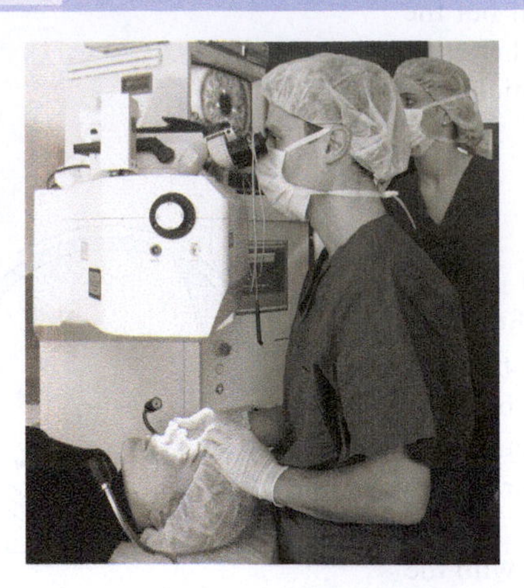

The cornea provides most of the eye's refractive power, with the adjustable lens compensating for different object distances. The popular LASIK procedure corrects vision by reshaping the cornea. In LASIK, the surgeon begins by mechanically cutting a flap of the outermost corneal layer. Then a precision laser beam breaks molecular bonds in the corneal tissue, vaporizing the material and reshaping the cornea according to a prescription customized for the individual eye. With a nearsighted eye, the laser thins the central cornea, making it less sharply curved and thus reducing its refractive power. This has the same effect as the corrective lens in Fig. 31.28*b*. It's harder, but still possible, to correct farsightedness with LASIK. This involves thinning a ring-shaped region around the central cornea, making the cornea more steeply curved and thus increasing its refractive power. The corrective lens in Fig. 31.29*b* accomplishes the same thing. The corneal reshaping doesn't have to be symmetric, so LASIK can also correct an asymmetric cornea that causes astigmatism. What it can't do is restore the ability to focus both near and far, since that's handled by the lens, which stiffens with age. So older LASIK patients still need reading glasses. It's possible to correct one eye for near vision and another for distance, but then the patient loses some of the depth perception that comes with binocular vision.

The laser used in vision correction is a precisely controllable *excimer laser*, which produces intense bursts of ultraviolet light. Each pulse removes only 0.25 μm of tissue—one four-thousandth of a millimeter. The laser is so precise that it can cut notches in a human hair! The surgeon determines the necessary corneal adjustments and feeds the information to a computer that controls the laser. Thanks to the laser's precision, most patients achieve nearly complete vision correction.

CONCEPTUAL EXAMPLE 31.1 Contact Lens Mix-Up

You and your roommate have gotten your boxes of disposable contact lenses mixed up. One box is marked "−1.75 diopter," the other "+2.5 diopter." You're farsighted and your roommate is nearsighted. Which lenses are yours?

EVALUATE Figure 31.29 shows that you need a converging lens to correct your farsightedness. From Table 31.2's sign conventions, that means a positive focal length and therefore a positive corrective power $P = 1/f$. So yours are the +2.5-diopter lenses.

ASSESS Your lenses don't actually look like the lens in Fig. 31.29*b*. Since your cornea is curved, they're more like the convex meniscus lens of Fig. 31.25. The important point is that they're thicker in the middle, which makes them converging lenses.

MAKING THE CONNECTION What's the focal length of your contact lenses?

EVALUATE The diopter measure is the inverse of the focal length in meters, so, conversely, $f = 1/P = (1/2.5)\text{ m} = 40\text{ cm}$.

EXAMPLE 31.6 The Power of Lenses: Lost Your Glasses!

You're on vacation and have lost your reading glasses; without them, your eyes can't focus closer than 70 cm. Fortunately, you can buy nonprescription reading glasses at the pharmacy, where they come in 0.25-diopter increments. Which glasses should you buy so you can focus at the standard 25-cm near point?

INTERPRET Your eyes can't focus closer than 70 cm, so this problem is asking for the power of a lens that will make an object at 25 cm appear as if it's at 70 cm. In other words, when the object distance s is 25 cm, the image distance s' should be −70 cm. We put the minus sign here because, as Fig. 31.29*b* shows, the image is on the same side of the lens as the object, so this is a *virtual* image.

DEVELOP Equation 31.5, $1/s + 1/s' = 1/f$, relates the inverse of the focal length to the object and image distances. But with the focal length in meters, $1/f$ is just the power, P, in diopters. So we can get the required power directly from Equation 31.5.

EVALUATE Applying Equation 31.5 gives

$$P = \frac{1}{f} = \frac{1}{s} + \frac{1}{s'} = \frac{1}{0.25\text{ m}} + \frac{1}{-0.70\text{ m}} = 2.57\text{ diopters}$$

ASSESS The closest available power is 2.5 diopters, so that's what you should buy. ■

Cameras

A camera is much like the eye, except that an electronic detector or film replaces the light-sensitive retina. Where the eye changes the lens shape to accommodate different object distances, a camera moves its rigid lens to change the image distance. Simple "point and shoot" cameras use infrared beams to determine the object distance, and then automatically adjust the lens position for optimum focus. The camera also adjusts the lens aperture and exposure time for ambient light conditions. Zoom lenses have moveable elements that alter the focal length for wide-angle to telephoto views.

Magnifiers and Microscopes

Examining very small objects requires bringing them closer than the 25-cm near point below which the human eye can't focus. We therefore use lenses to put enlarged images at greater distances, where we can focus. What matters is not the actual image size, but how much bigger an object looks to us—and that depends on how much of our field of view it occupies. **Angular magnification**, m, is the ratio of the angle an object subtends when seen through a lens to the angle subtended when it's at the 25-cm near point and viewed with the naked eye. Figure 31.30*a* shows that the latter angle, measured in radians, is approximately $\alpha = h/25\,\text{cm}$, where h is the object height. We get the most comfortable viewing with the eye close to the lens and the object just inside the focal length, forming a large and distant virtual image. With this geometry, Fig. 31.30*b* shows that the image angle is very nearly h/f. Then the angular magnification is

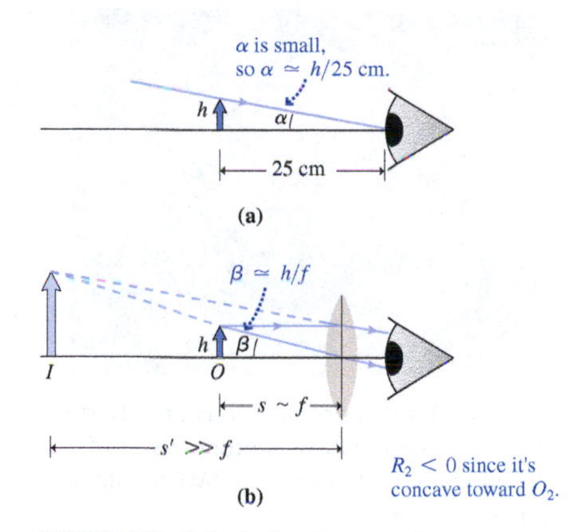

FIGURE 31.30 Calculating the angular magnification $m = \beta/\alpha$.

$$m = \frac{\beta}{\alpha} = \frac{h/f}{h/25\,\text{cm}} = \frac{25\,\text{cm}}{f} \quad \text{(simple magnifier)} \qquad (31.8)$$

Single lenses produce angular magnifications up to about four before aberrations compromise image quality. Greater magnification requires more than one lens. In a **compound microscope**, an **objective lens** of short focal length forms a magnified real image. This image is viewed through a second lens, the **eyepiece**, used as a simple magnifier (Fig. 31.31). The object being viewed is positioned just beyond the focus of the objective lens, and its image falls just inside the focal length of the eyepiece. If both focal lengths are small compared with the distance between the lenses, then the object distance for the objective lens is approximately the objective focal length f_o, and the resulting image distance is approximately the lens spacing L. The real image formed by the objective lens is larger than the object by the ratio of the image and object distances, or $M_o = -L/f_o$. The eyepiece makes the real image look larger still, by a factor of its angular magnification $m = 25\,\text{cm}/f_e$. So the overall magnification of the microscope is

$$M = M_o m_e = -\frac{L}{f_o}\left(\frac{25\,\text{cm}}{f_e}\right) \quad \text{(compound microscope)} \qquad (31.9)$$

where, as usual, the minus sign signifies an inverted image.

Optical microscopes work well as long as the approximation of geometrical optics holds—that is, when the object is much larger than the wavelength of light. Viewing smaller objects requires shorter wavelengths than those of visible light. In the electron microscope, those "waves" are electrons, whose wavelike nature we'll examine in Chapter 34.

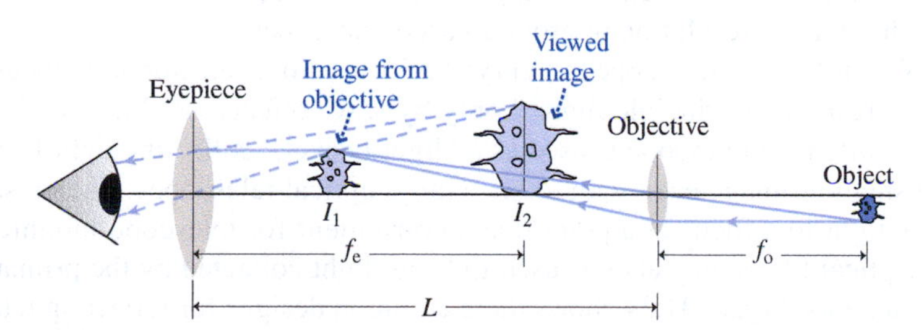

FIGURE 31.31 Image formation in a compound microscope. Figure is not to scale; L should be much greater than either focal length, and image I_1 should be very near the eyepiece's focus, resulting in greater magnification.

Telescopes

A telescope collects light from distant objects, either forming an image or supplying light to instruments for analysis. Modern astronomical instruments are invariably **reflectors**, whose main light-gathering element is a mirror. Small handheld telescopes, binoculars, and telephoto lenses are **refractors**, which use lenses to gather light.

A simple refractor consists of an objective lens that images distant objects at essentially its focal point, followed by an eyepiece to view this image (Fig. 31.32). The

FIGURE 31.33 Artist's conception of the Thirty Meter Telescope, whose mirror consists of 492 individual segments. The TMT should be operational by 2020.

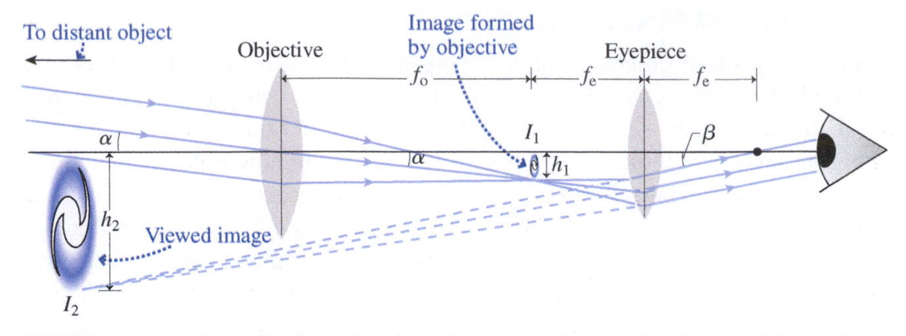

FIGURE 31.32 Image formation in a refracting telescope. A distant object is imaged first at the focus of the objective lens (image I_1). An eyepiece with its focus at nearly the same point then gives an enlarged virtual image (I_2). The angles α and β are given by $\alpha \simeq h_1/f_o$ and $\beta \simeq h_1/f_e$, leading to Equation 31.10.

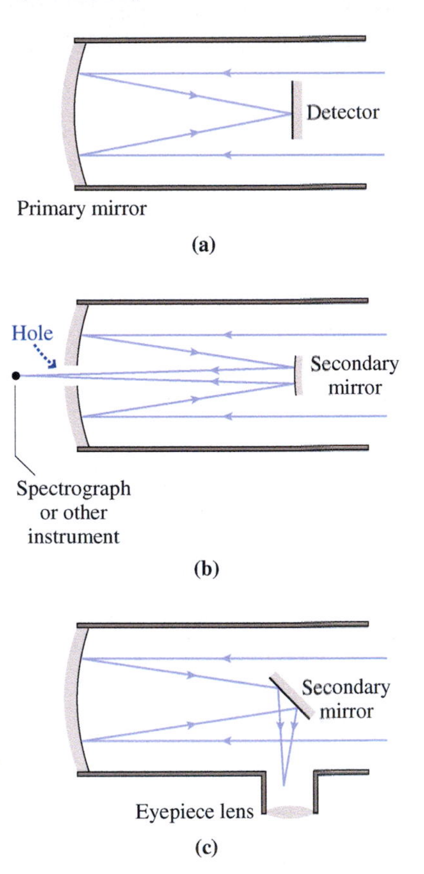

FIGURE 31.34 Reflecting telescopes. (a) A detector at the prime focus gives the best image quality. (b) The Cassegrain design is widely used in large telescopes. (c) The Newtonian design is used primarily in small telescopes.

focal points of objective and eyepiece are nearly coincident, so the real image at the objective's focus is then seen through the eyepiece as a greatly enlarged, virtual image. The angular magnification is the ratio of the angle β subtended by the final image to the angle α subtended by the actual object; Fig. 31.32 shows that this ratio is

$$m = \frac{\beta}{\alpha} = \frac{f_o}{f_e} \quad \text{(refracting telescope)} \tag{31.10}$$

Since a real image is inverted and a virtual image is upright, a two-lens refracting telescope gives an inverted image. This is fine for astronomical work, but telescopes designed for terrestrial use have an extra lens, a diverging eyepiece, or a set of reflecting prisms to produce an upright image.

Reflecting telescopes offer many advantages over refractors. Telescope mirrors have reflective coatings on their front surfaces, eliminating chromatic aberration because light doesn't pass through glass. Reflectors can be much larger since mirrors are supported across their entire back surfaces—unlike lenses, which must be supported at the edges. Whereas the largest refracting telescope ever built has a 1-m-diameter lens, today's largest reflectors boast diameters on the order of 10 m. Still larger telescopes are in the works, including the 24-m Giant Magellan Telescope, to be built in Chile, and the Thirty Meter Telescope (TMT), slated for construction atop Hawaii's Mauna Kea (Fig. 31.33). Ten-meter and larger class telescopes incorporate segmented mirrors whose shape can be adjusted under computer control for optimum focusing. With so-called *adaptive optics*, such systems may adjust rapidly enough to compensate for the atmospheric turbulence that has traditionally limited the resolution of ground-based telescopes.

The simplest reflecting telescope is a curved mirror with a detector at its focus. Superb image quality results, in principle limited only by wave effects we'll discuss in the next chapter. More often the telescope is used as a "light bucket," collecting light from distant sources too small to image even with today's large optical telescopes. Then a secondary mirror sends light to a focus at a point that's convenient for telescope-mounted instrumentation. Optical fibers may also be used to bring light collected by the primary mirror to fixed instruments. Figure 31.34 shows three common designs for reflecting telescopes.

Magnification is not a particularly important quantity in astronomical telescopes, which are used more for spectral and other analysis than for direct imaging. More important is the light-gathering power of the instrument, which is determined simply by the area of its objective lens or primary mirror. Each of the two 10-m Keck Telescopes, for instance, has 100 times the light-gathering power of the 1-m Yerkes refractor and more than 17 times the power of the 2.4-m Hubble Space Telescope. The Thirty Meter Telescope will further expand that light-gathering by nearly tenfold.

GOT IT? 31.5 If you look backward through a refracting telescope like that shown in Fig. 31.32, looking into the objective and with a tiny object very near the eyepiece, will the instrument function as a microscope? Explain.

CHAPTER 31 SUMMARY

Big Idea

The big idea here is how reflection and refraction form **images**. Images are **real** or **virtual** depending on whether or not light actually comes from the image location.

Key Concepts and Equations

A curved mirror or lens has a **focal point**, F, at which parallel light rays converge:

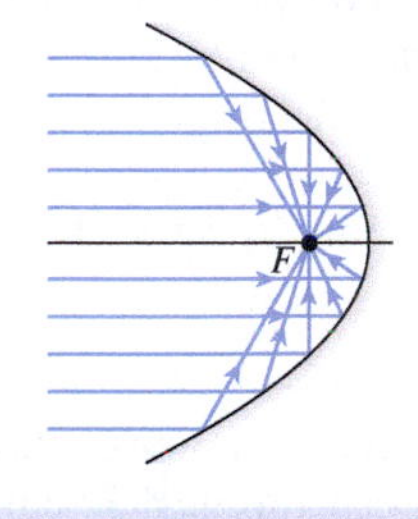

The same equation describes image formation with mirrors and lenses:

$$\frac{1}{s} + \frac{1}{s'} = \frac{1}{f}$$

The table summarizes the sign conventions for each term.

Value	Symbol	Condition	Sign
Object distance	s	Object on same side as *incoming* light rays	+
		Object on opposite side from *incoming* light rays	−
Image distance	s'	Image on same side as *outgoing* light rays	+
		Image on opposite side from *outgoing* light rays	−
Focal length	f	Focus on same side as *outgoing* light rays	+
		Focus on opposite side from *outgoing* light rays	−

Image formation with mirrors and lenses, shown by ray tracing:

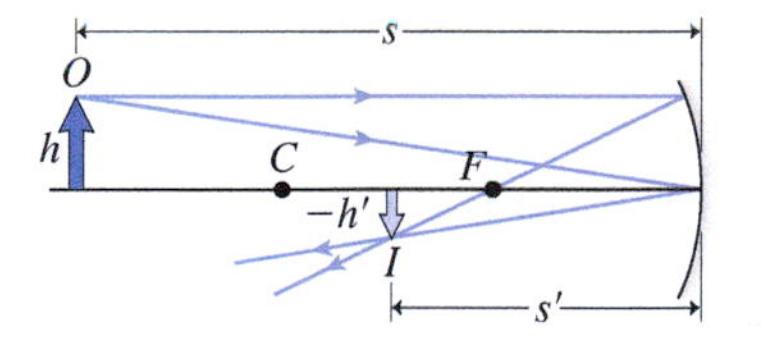

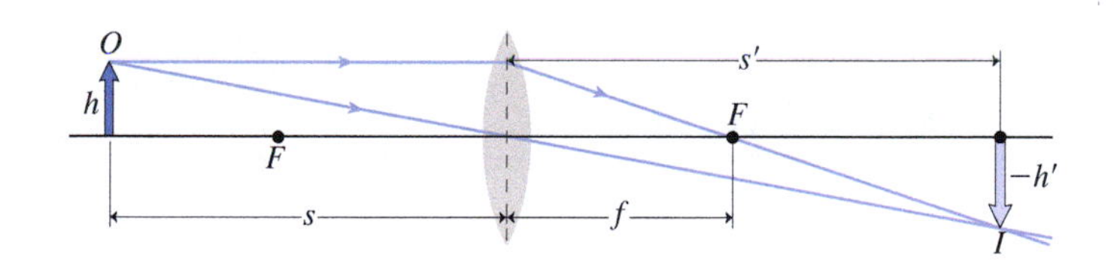

The **lensmaker's formula** gives the focal length f of a thin lens:

$$\frac{1}{f} = (n - 1)\left(\frac{1}{R_1} - \frac{1}{R_2}\right)$$

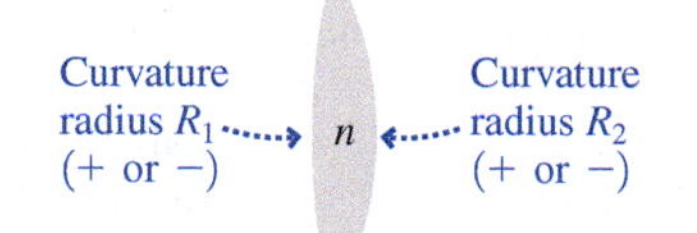

Applications

Compound microscope

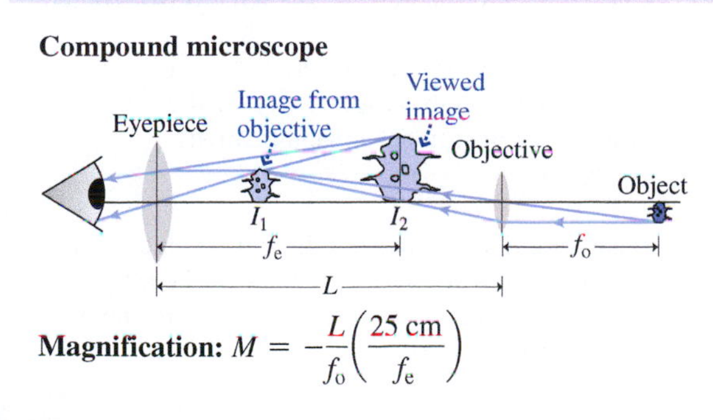

Magnification: $M = -\frac{L}{f_o}\left(\frac{25\text{ cm}}{f_e}\right)$

Refracting telescope

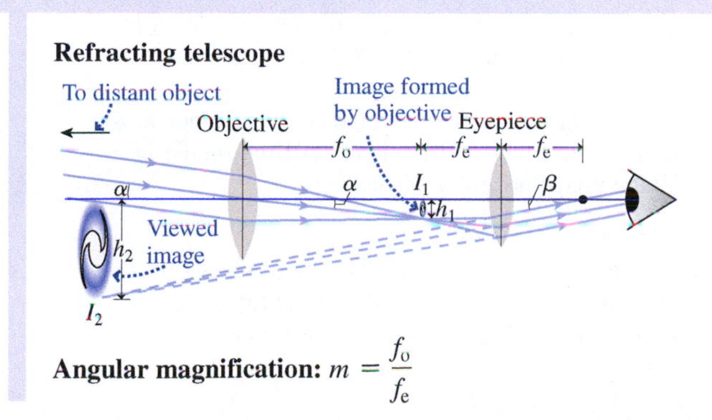

Angular magnification: $m = \frac{f_o}{f_e}$

MP® *For homework assigned on MasteringPhysics, go to www.masteringphysics.com*

BIO *Biology and/or medicine-related problems* **DATA** *Data problems* **ENV** *Environmental problems* **CH** *Challenge problems* **COMP** *Computer problems*

For Thought and Discussion

1. How can you see a virtual image, when it's not "really there"?
2. Under what circumstances will the image in a concave mirror be the same size as the object?
3. If you're handed a converging lens, what can you do to estimate its focal length quickly?
4. A diverging lens always makes a reduced image. Could you use such a lens to start a fire by focusing sunlight? Explain.
5. Is there any limit to the temperature you can achieve by focusing sunlight? (*Hint:* Think about the second law of thermodynamics.)
6. Can a concave mirror make a reduced real image? A reduced virtual image? An enlarged real image? An enlarged virtual image? Specify conditions for each possible image.
7. If you placed a screen at the location of a virtual image, would the image appear on the screen? Why or why not?
8. If you look into the bowl of a metal spoon, you see yourself upside down. Flip the spoon so you're looking at the back side, and now you're right-side up. Explain.
9. Is the image on a movie screen real or virtual? How do you know?
10. Does a fish in a spherical bowl appear larger or smaller than it actually is?
11. A block of ice contains a hollow, air-filled space in the shape of a double-convex lens. Describe the optical behavior of this space.
12. The refractive index of the human cornea is about 1.4. If you can see clearly in air, why can't you see clearly underwater? Why do goggles help?
13. Do you want a long or short focal length for a telescope's objective lens? What about a microscope's?
14. Give at least three reasons why reflecting telescopes are superior to refractors.

Exercises and Problems

Exercises

Section 31.1 Images with Mirrors

15. A shoe store uses small floor-level mirrors to let customers view prospective purchases. At what angle should such a mirror be inclined so that a person standing 50 cm from the mirror with eyes 140 cm off the floor can see her feet?
16. A candle is on the axis of a 15-cm-focal-length concave mirror, 36 cm from the mirror. (a) Where is its image? (b) How do the image and object sizes compare? (c) Is the image real or virtual?
17. An object is five focal lengths from a concave mirror. (a) How do the object and image heights compare? (b) Is the image upright or inverted?
18. A virtual image is located 40 cm behind a concave mirror with focal length 18 cm. (a) Where is the object? (b) By how much is the image magnified?
19. (a) Where on the axis of a concave mirror would you place an object to get a half-size image? (b) Where will the image be located? (c) Will the image be real or virtual?

Section 31.2 Images with Lenses

20. A lightbulb is 56 cm from a convex lens. Its image appears on a screen 31 cm from the lens, on the other side. Find (a) the lens's focal length and (b) how much the image is enlarged or reduced.
21. By what factor is the image magnified for an object 1.5 focal lengths from a converging lens? Is the image upright or inverted?
22. A lens with 50-cm focal length produces a real image the same size as the object. How far from the lens are image and object?
23. By holding a magnifying glass 25 cm from your desk lamp, you can focus an image of the lamp's bulb on a wall 1.6 m from the lamp. What's the focal length of your magnifying glass?
24. A real image is four times as far from a lens as is the object. What's the object distance, measured in focal lengths?
25. A magnifying glass enlarges print by 50% when it's 9.0 cm from a page. What's its focal length?

Section 31.3 Refraction in Lenses: The Details

26. You're writing specifications for a new line of magnifying glasses that have double-convex lenses with equal 32-cm curvature radii, made from glass with $n = 1.52$. What do you list for the focal length?
27. You're standing in a wading pool and your feet appear to be 30 cm below the surface. How deep is the pool?
28. The bottom of a swimming pool looks to be 1.5 m below the surface. Find the pool's actual depth.
29. A tiny insect is trapped 1.0 mm from the center of a spherical dewdrop 4.0 mm in diameter. As you look straight into the drop, what's the insect's apparent distance from the drop's surface?
30. You're underwater, looking through a spherical air bubble (Fig. 31.35). What's its actual diameter if it appears, along your line of sight, to be 1.5 cm in diameter?

FIGURE 31.35 Exercise 32

Section 31.4 Optical Instruments

31. **BIO** You have to hold a book 55 cm from your eyes for the print to be in focus. What power lens would correct your farsightedness?
32. What focal length should you specify if you want a magnifying glass with angular magnification 3.2?
33. **BIO** You're an optometrist helping a nearsighted patient who claims he can't see clearly beyond 80 cm. Prescribe a lens that will put the images of distant objects at 80 cm, giving your patient clear vision at all distances beyond the normal near point.
34. **BIO** A particular eye has a focal length of 2.0 cm instead of the 2.2 cm that would put a sharply focused image on the retina. (a) Is this eye nearsighted or farsighted? (b) What corrective lens is needed?
35. A compound microscope has objective and eyepiece focal lengths of 6.1 mm and 1.7 cm, respectively. If the lenses are 8.3 cm apart, what is the instrument's magnification?

Problems

36. (a) Find the focal length of a concave mirror if an object placed 38.4 cm in front of the mirror has a real image 55.7 cm from the mirror. (b) Where and what type will the image be if the object is moved to a point 16.0 cm from the mirror?

37. A 12-mm-high object is 10 cm from a concave mirror with focal length 17 cm. (a) Where is the image, (b) how high is it, and (c) what type is it?
38. Repeat Problem 37 for a convex mirror, assuming all numbers stay the same.
39. An object's image in a 27-cm-focal-length concave mirror is upright and magnified by a factor of 3. Where is the object?
40. You're asked to design a concave mirror that will produce a virtual image, enlarged 1.8 times, of an object 22 cm from the mirror. What do you specify for the mirror's curvature radius?
41. Viewed from Earth, the Moon subtends an angle of 0.52° in the sky. What will be the physical size of the Moon's image formed by either of the twin Keck telescopes, with 10-m-diameter mirrors and 17.5-m focal length?
42. At what two distances could you place an object from a 45-cm-focal-length concave mirror to get an image 1.5 times the object's size?
43. LCD projectors commonly used for computer and video projection create an image on a small LCD display (see Application on page 369). The display is mounted before a lens and illuminated from behind. In a projector using a 7.50-cm-focal-length convex lens, where should the LCD display be located so the projected image is focused on a screen 6.30 m from the lens?
44. An object 15 cm from a concave mirror has a virtual image magnified 2.5 times. What's the mirror's focal length?
45. How far from a page should you hold a lens with 32-cm focal length in order to see the print magnified 1.6 times?
46. A converging lens has focal length 4.0 cm. A 1.0-cm-high arrow is located 7.0 cm from the lens with its lowest point 5.0 mm above the lens axis. Make a full-scale ray-tracing diagram to locate both ends of the image. Confirm using the lens equation.
47. A lens has focal length $f = 35$ cm. Find the type and height of the image produced when a 2.2-cm-high object is placed at distances (a) $f + 10$ cm and (b) $f - 10$ cm.
48. How far apart are the object and image produced by a converging lens with 35-cm focal length when the object is (a) 40 cm and (b) 30 cm from the lens?
49. A candle and a screen are 70 cm apart. Find two points between candle and screen where you could put a convex lens with 17-cm focal length to give a sharp image of the candle on the screen.
50. **BIO** The cornea of the human eye has refractive index 1.38, while the eye's lens has a graduated index in the range 1.38 to 1.40; use 1.39 for this problem. For the aqueous humor between cornea and lens, $n = 1.34$. Find the angle through which light is deflected at the first surface of (a) the cornea and (b) the lens, if it's incident at 20° to the normal at each surface. Your result shows that the cornea is the dominant refractive element in the eye.
51. How far from a 25-cm-focal-length lens should you place an object to get an upright image magnified 1.8 times?
52. An object and its lens-produced real image are 2.4 m apart. If the lens has 55-cm focal length, what are the possible values for the object distance and magnification?
53. An object is 68 cm from a plano-convex lens whose curved side has curvature radius 26 cm. The refractive index of the lens is 1.62. Where is the image, and what type is it?
54. Use Equation 31.6 to show that an object at the center of a glass sphere will appear to be its actual distance—one radius—from the edge. Draw a ray diagram showing why this makes sense.
55. Rework Example 31.4 for a fish 15.0 cm from the *far* wall of the tank.
56. Consider the inverse of Example 31.4: You're inside a 70.0-cm-diameter hollow tube containing air, and the tip of your nose is 15.0 cm from the tube's wall. The tube is immersed in water, and a fish looks in. To the fish, what's the apparent distance from your nose to the tube wall?
57. Two specks of dirt are trapped in a crystal ball, one at the center and the other halfway to the surface. If you peer into the ball on a line joining the two specks, the outer one appears to be only one-third of the way to the other. Find the refractive index of the ball.
58. **BIO** A contact lens is in the shape of a convex meniscus (see Fig. 31.25). The inner surface is curved to fit the eye, with curvature radius 7.80 mm. The lens is made from plastic with refractive index $n = 1.56$. If it has a 44.4-cm focal length, what's the curvature radius of its outer surface?
59. For what refractive index would the focal length of a plano-convex lens be equal to the curvature radius of its one curved surface?
60. An object is 28 cm from a double-convex lens with $n = 1.5$ and curvature radii 35 cm and 55 cm. Where is the image, and what type is it?
61. **BIO** You're an optician who's been asked to design a new replacement lens for cataract patients. The lens must be 5.5 mm in diameter, with focal length 17 mm, and it can't be thicker than 0.8 mm. For the lens material, you have a choice of plastic with refractive index 1.49 or more expensive silicone with $n = 1.58$. Which material do you choose, and why?
62. A double-convex lens with equal 28.5-cm curvature radii is made from glass with refractive indices $n_{\text{red}} = 1.512$ and $n_{\text{violet}} = 1.547$. If a point source of white light is located on the lens axis at 75.0 cm from the lens, over what distance will its visible image be smeared?
63. An object placed 17.5 cm from a convex lens of glass with $n = 1.524$ forms a virtual image twice the object's size. If the lens is replaced with an identically shaped one made of diamond, (a) what type of image will appear and (b) what will be its magnification?
64. You're taking a photography class, working with a camera whose zoom lens covers the focal-length range 38 mm–110 mm. Your instructor asks you to compare the sizes of the images of a distant object when photographed at the two zoom extremes. Your answer?
65. A camera can normally focus as close as 60 cm, but it has provisions for mounting additional lenses just in front of the main lens to provide close-up capability. What type and power of auxiliary lens will allow the camera to focus as close as 20 cm?
66. A 300-power compound microscope has a 4.5-mm-focal-length objective lens. If the distance from objective to eyepiece is 10 cm, what should be the focal length of the eyepiece?
67. To the unaided eye, Jupiter has an angular diameter of 50 arcseconds. What will its angular size be when viewed through a 1-m-focal-length refracting telescope with a 40-mm-focal-length eyepiece?
68. A Cassegrain telescope like that shown in Fig. 31.34*b* has 1.0-m focal length, and the convex secondary mirror is located 0.85 m from the primary. What should be the focal length of the secondary in order to put the final image 0.12 m behind the front surface of the primary mirror?
69. You stand with your nose 6.0 cm from the surface of a reflecting ball, and your nose's image appears three-quarters full size. What's the ball's diameter?
70. **BIO** A contact lens prescription calls for +2.25-diopter lenses with inner curvature radius 8.6 mm to fit the patient's cornea. (a) If the lenses are plastic with $n = 1.56$, what should be the outer curvature radius? (b) With these lenses, the patient comfortably reads a newspaper 30 cm from her eyes. Where's the image as viewed through the lenses?
71. Show that placing a 1-diopter lens in front of a 2-diopter lens gives the equivalent of a single 3-diopter lens (i.e., the powers of closely spaced lenses add).

72. Derive an expression for the thickness t of a plano-convex lens with diameter d, focal length f, and refractive index n.
73. Show that identical objects placed equal distances on either side of the focal point of a concave mirror or converging lens produce images of equal size. Are the images of the same type?
74. CH Generalize the derivation of the lensmaker's formula (Equation 31.7) to show that a lens of refractive index n_{lens} in an external medium with index n_{ext} has focal length given by

$$\frac{1}{f} = \left(\frac{n_{lens}}{n_{ext}} - 1\right)\left(\frac{1}{R_1} - \frac{1}{R_2}\right)$$

75. Draw a diagram like Fig. 31.10, but showing a ray from the arrowhead through the center of curvature. Using the fact that this ray reflects back on itself, draw similar triangles with object and image as their vertical sides, and show that the center of curvature is twice as far from the mirror as the focal point—that is, $R = 2f$, with R the curvature radius.
76. Galileo's first telescope used the arrangement shown in Fig. 31.36, with a double-concave eyepiece slightly before the focus of the objective lens. Use ray tracing to show that this design gives an upright image, which makes the Galilean telescope useful in terrestrial observing.

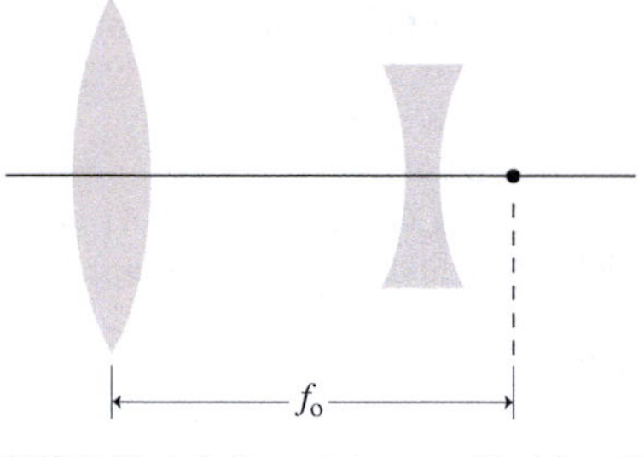

FIGURE 31.36 A Galilean telescope (Problem 76)

77. The maximum magnification of a simple magnifier occurs with the image at the 25-cm near point. Show that the angular magnification is $m = 1 + (25\,\text{cm}/f)$, where f is the focal length.
78. CH Chromatic aberration results from variation of the refractive index with wavelength. Starting with the lensmaker's formula, find an expression for the fractional change df/f in the focal length of a thin lens in terms of the change dn in refractive index.
79. CH For visible wavelengths, the refractive index of the polycarbonate plastic widely used in eyeglasses is given approximately by $n(\lambda) = b + c/\lambda^2$, where $b = 1.55$ and $c = 11{,}500\,\text{nm}^2$. (a) Find an expression for the change in refractive index dn corresponding to a small wavelength change $d\lambda$. (b) Use the results of part (a) and of Problem 78 to determine the variation df in focal length for a +2.25-diopter polycarbonate lens, over a wavelength range of 10.0 nm centered on 589 nm.
80. DATA The table below shows measurements of magnification versus object distance for a lens. Determine a quantity that, when you plot object distance against it, should give a straight line. Make your plot, establish a best-fit line, and use your line to find the focal length of the lens.

Object distance, s (cm)	10.1	29.2	51.6	78.3	98.9
Magnification, M	1.31	4.77	−4.38	−1.27	−0.724

Passage Problems

The *speed* of a camera lens measures its ability to photograph in dim light. Speed is characterized by *f-ratio*, also called the *f-number*, defined as the ratio of focal length f to lens diameter d. Thus an $f/2.8$ lens, for example, has diameter $d = f/2.8$. The actual amount of light a lens admits depends on its area A, but the inverse-square law shows that the light intensity at the camera's imaging sensor is proportional to A/f^2. Most cameras have an adjustable iris that obscures part of the lens to change the f-ratio in response to available light. Point-and-shoot cameras adjust the f-ratio automatically, but serious photographers use their camera's manual f-ratio adjustment (Fig. 31.37). *Stopping down* is the photographer's term for reducing the lens area using the adjustable iris.

FIGURE 31.37 A 35-mm camera lens (Passage Problems 81–84). The numbers from 22 to 2.8 at the bottom are values for the f-ratio, f/d. Turning the ring with these numbers adjusts the iris that covers the outer part of the lens, thus changing the f-ratio.

81. Zooming your camera's lens for telephoto shots increases the focal length. With no change in the lens area, this will
    a. increase the f-ratio and increase the lens speed.
    b. decrease the f-ratio and decrease the lens speed.
    c. increase the f-ratio and decrease the lens speed.
    d. not change the f-ratio or the lens speed.
82. Increasing the f-ratio from 2.8 to 5.6
    a. decreases the light admitted by a factor of 2.
    b. decreases the light admitted by a factor of 4.
    c. increases the light admitted by a factor of 2.
    d. increases the light admitted by a factor of 4.
83. You're given two lenses with different diameters. Knowing nothing else, you can conclude that
    a. the larger lens is faster.
    b. the smaller lens has the shorter focal length.
    c. the smaller lens suffers less spherical aberration.
    d. none of the above
84. If a lens suffers from spherical aberration, stopping down will
    a. worsen the focus.
    b. improve the focus.
    c. not affect the focus.

Answers to Chapter Questions

Answer to Chapter Opening Question

High-intensity laser light reshapes the cornea, so that refracted light converges to produce sharp images.

Answers to GOT IT? Questions

31.1 (b)
31.2 (a)
31.3 Virtual image; convex lens
31.4 (d)
31.5 No; the positions of the lenses in relation to their focal points aren't correct for a microscope. You won't be able to image a small object placed very near the eyepiece. If you look at a more distant object, though, you'll see it reduced in size.

Interference and Diffraction

What You Know

- You understand the behavior of waves, especially wave interference.
- You know about electromagnetic waves and how they involve electric and magnetic fields.
- You know that light consists of electromagnetic waves.

What You're Learning

- You'll see how coherent light waves that travel on different paths undergo interference when they recombine.
- You'll learn to describe quantitatively the interference resulting from light passing through two slits.
- You'll extend your understanding of two-slit interference to systems with multiple slits.
- You'll learn the principles involved in grating spectroscopy and X-ray diffraction.
- You'll learn how interferometry enables exquisitely precise distance measurements.
- You'll learn about diffraction and how it puts fundamental limits on our ability to form perfect optical images.

How You'll Use It

- The knowledge you gain from Chapter 32 applies to the optical systems you'll encounter in your professional and personal life.
- Waves and wave interference will be important again in your study of quantum physics in Part 6, and the quantum nature of matter will help you understand how scientists use alternatives to visible light in imaging small systems.

Photo of the Khalifa Sports City in Qatar, taken by the GeoEye-1 satellite from 423 miles up. What's the fundamental limitation on our ability to image fine details in satellite photos?

The preceding chapters described the behavior of light using geometrical optics—an approximation that's valid when we're dealing with length scales much larger than the wavelength of light, so we can ignore light's wave nature. We now turn to **physical optics**, which treats optical phenomena for which the wave nature of light plays an essential role. Two related phenomena, interference and diffraction, are central in physical optics.

32.1 Coherence and Interference

In Chapter 14 we showed how wave displacements add to produce **constructive interference** or **destructive interference**. Electromagnetic waves, including light, are no exception: Since electric and magnetic fields obey the superposition principle, the net fields at any point are the vector sums of individual wave fields. That summation may increase (constructive interference) or decrease (destructive interference) the net field.

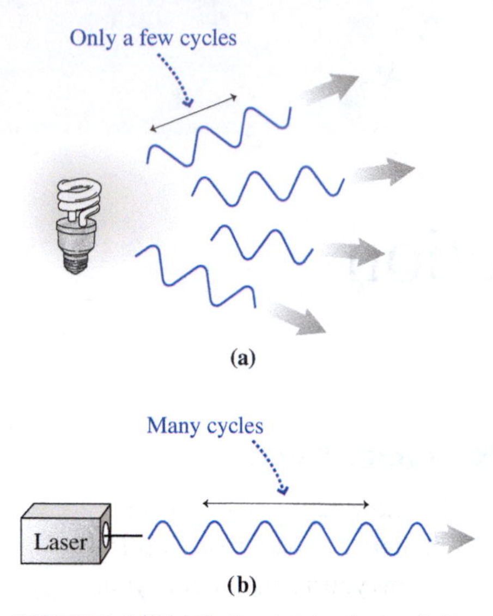

FIGURE 32.1 (a) Lightbulbs emit incoherent light consisting of short wavetrains with random phases. (b) Laser light consists of much longer wavetrains, making it more coherent.

Coherence

Any two or more waves will interfere at points where they overlap, but a steady interference pattern results only when the waves maintain the same frequency and phase relationship—in which case the light is said to be **coherent**. Light emerging from different sources, different parts of the same source, or at different times is unlikely to be coherent. Sending such light through two or more holes or slits, however, produces two versions of each individual wave, and those two can then interfere as they meet beyond the slits. That's how Thomas Young produced coherent light in his 1801 experiment that demonstrated the wave nature of light.

Today most interference experiments are done with lasers, and the intrinsic coherence of laser light is often cited as a reason. As Fig. 32.1 suggests, laser light does indeed maintain coherence longer than light from common sources such as lightbulbs or the Sun. But even with lasers, a steady interference pattern generally requires splitting the laser beam with a system of slits or other devices that produce two versions of the same light wave. You can see why by considering the **coherence length**—the distance over which light maintains its phase and frequency. For ordinary light sources, as Fig. 32.1*a* suggests, the coherence length is very short—it's on the order of 1 μm for sunlight. Laser light, in contrast, exhibits substantial coherence lengths—tens of centimeters for everyday lasers, and many kilometers with special designs. But given that the speed of light is 300 Mm/s, those lengths imply that even laser light retains its coherence for only a tiny fraction of a second. So laser-based interference experiments, like those using ordinary light sources, require that the light beam be split into two or more parts that are later recombined. What makes lasers particularly useful in interference experiments—and indeed throughout optics—is that their light is very nearly **monochromatic**, meaning it consists of a very narrow band of wavelengths.

Destructive and Constructive Interference

Consider light waves that originate together at a single source, travel two different paths, and then rejoin. Suppose one wave's path is exactly half a wavelength longer than the other. Then, when the waves recombine, they'll be out of phase by half a wavelength (Fig. 32.2*a*) and thus their superposition has smaller amplitude (zero, if the two interfering waves have exactly the same amplitude). If, on the other hand, the path lengths don't differ, or they differ by a full wavelength, then the two waves recombine in phase (Fig. 32.2*b*) and their superposition has larger amplitude. These two cases correspond, respectively, to destructive interference and constructive interference. It doesn't matter whether path lengths for the waves in Fig. 32.2*a* differ by half a wavelength, or $1\frac{1}{2}$ wavelengths, or $2\frac{1}{2}$ wavelengths; as long as the difference is an odd multiple of a half-wavelength, the waves recombine out of phase and destructive interference results. Thus:

> Destructive interference results when light paths differ by an odd-integer multiple of a half-wavelength.

Similarly, it doesn't matter whether the path lengths in Fig. 32.2*b* are the same, or differ by 1, 2, 3, or any other integer number of wavelengths. Thus:

> Constructive interference results when light paths differ by an integer multiple of the wavelength.

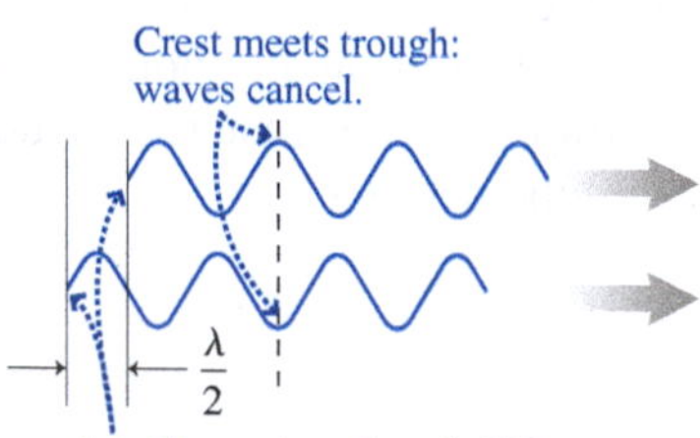

Crest meets crest: waves reinforce.

λ

A full-wavelength path difference results in constructive interference.

(b)

FIGURE 32.2 Two waves that start out in phase but travel different paths before rejoining.

There's one caveat to our statements: The path difference can't be greater than the coherence length; otherwise, the waves won't be coherent when they recombine. Laser light has the advantage here because of its greater coherence length.

Light paths don't have to differ by half or full multiples of the wavelength. In intermediate cases, interfering waves superpose to make a composite wave whose amplitude may be enhanced or diminished, depending on the relative phase.

APPLICATION **More CD Music**

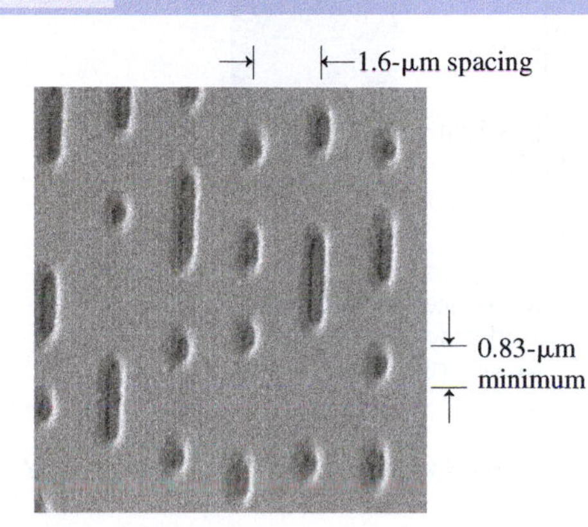

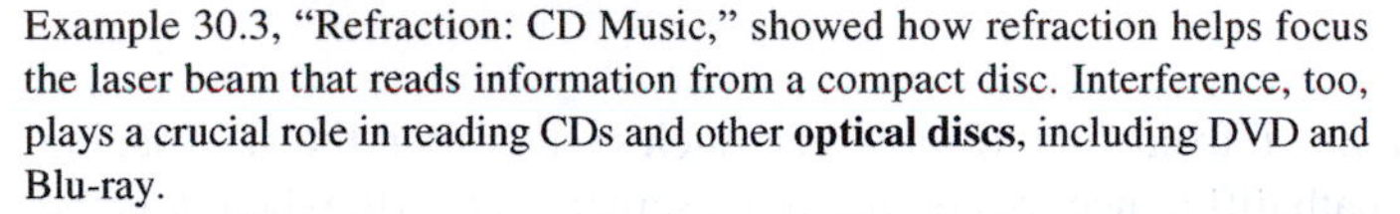

Example 30.3, "Refraction: CD Music," showed how refraction helps focus the laser beam that reads information from a compact disc. Interference, too, plays a crucial role in reading CDs and other **optical discs**, including DVD and Blu-ray.

Information on an optical disc is stored digitally in a sequence of pits stamped into a reflective metallic information layer, as shown in the photo. The pits' depth is very nearly one-quarter wavelength of the laser light used to read the disc. From the transparent underside of the disc, each pit appears as an elevated bump. Since the bumps stick down one-quarter wavelength, light reflecting off a bump follows a round-trip path that's shorter by half a wavelength than that of light reflecting off the undisturbed information layer (see the figure below). The laser beam is wider than the pit, so the reflected beam includes light both from the undisturbed disc and from the bump. The two interfere destructively, making the reflected beam less intense when a bump is present. As the disc spins, the result is a pattern of fluctuating light intensity conveying the information associated with the pattern of pits. A photodetector then converts that pattern to electrical signals that ultimately drive loudspeakers, headphones, or a video display.

You'll see later in the chapter how principles of physical optics determine how much information can be stored on an optical disc.

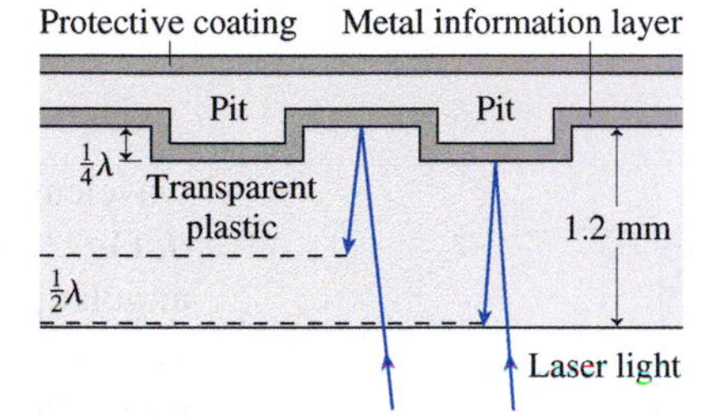

GOT IT? 32.1 Laser light is split into two beams, one of which is sent on a path that takes it through a slab of glass, where light's speed is about two-thirds of what it is in air. The other beam travels an identical-length path, entirely in air. Which of the following describes the interference that occurs when the beams are recombined? (a) It's constructive; (b) it's destructive; (c) it could be constructive, destructive, or in between, depending on the thickness of the glass slab; (d) there is no interference.

32.2 Double-Slit Interference

In Chapter 14 we looked briefly at interference patterns produced by a pair of coherent sources. Such a pair can be made by passing light through two narrow slits. In 1801, Thomas Young used this approach in a historic experiment that confirmed the wave nature of light. Young first admitted sunlight to his laboratory through a hole small enough to ensure coherence of the incoming light. The light then passed through a pair of narrow, closely spaced slits, after which it illuminated a screen. Each slit acts as a source of cylindrical wavefronts that interfere in the region between slits and screen (Fig. 32.3*a*). Constructive and destructive interference produce **interference fringes**—alternating bright and dark bands (Fig. 32.3*b*).

PhET: Wave Interference: Light

Bright fringes represent constructive interference, and therefore they occur where the difference in the path length for light traveling from the two slits is a multiple of the

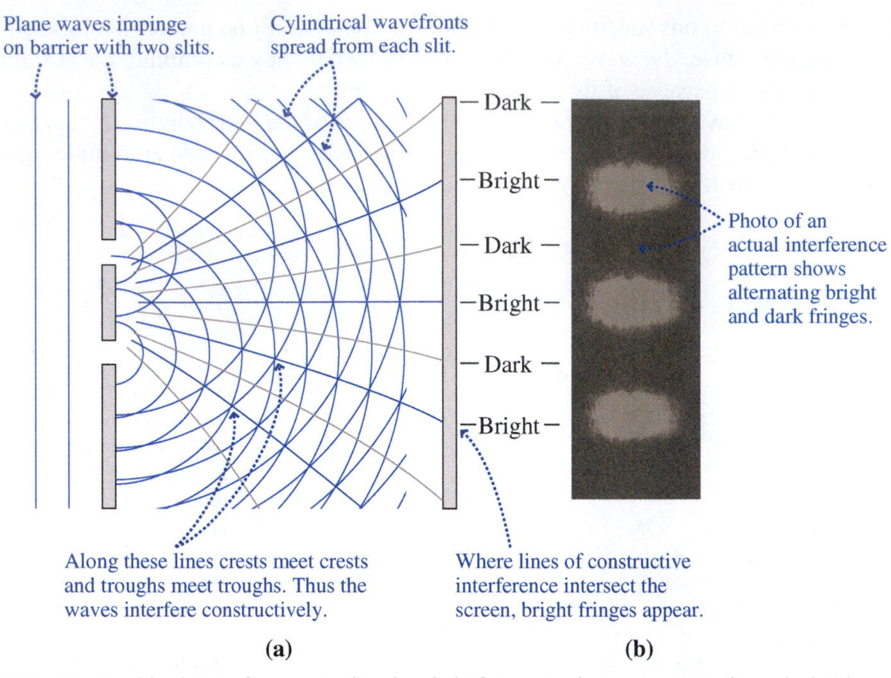

FIGURE 32.3 Double-slit interference results when light from a single source passes through closely spaced slits.

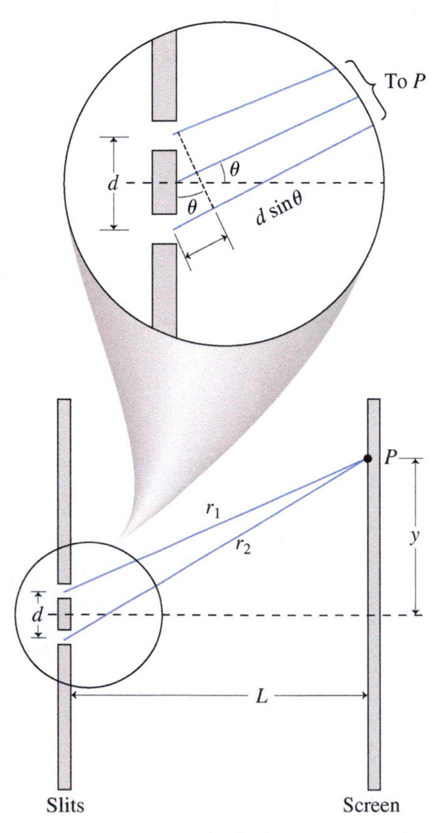

FIGURE 32.4 Geometry for finding locations of the interference fringes. In the blowup you can see that for $L \gg d$, the paths to P are nearly parallel and differ by $d \sin\theta$.

wavelength. When the distance L from slits to screen is much greater than the slit spacing d, Fig. 32.4 shows that the path difference to a point on the screen is $d \sin\theta$, where θ is the angular position of a point on the screen measured from an axis perpendicular to slits and screen. So our criterion for constructive interference—that this difference be an integer number of wavelengths—becomes

$$d \sin\theta = m\lambda \qquad (\text{bright fringes}, m = 0, 1, 2, \ldots) \tag{32.1a}$$

The integer m is the **order** of the fringe, with the central bright fringe being the zeroth-order fringe and with higher-order fringes on either side.

Waves interfere destructively when their path lengths differ by an odd-integer multiple of a half-wavelength:

$$d \sin\theta = \left(m + \tfrac{1}{2}\right)\lambda \qquad (\text{dark fringes}, m = 0, 1, 2, \ldots) \tag{32.1b}$$

where m is any integer.

In a typical double-slit experiment, L may be on the order of 1m, d a fraction of 1mm, and λ the sub-μm wavelength of visible light. Then we have the additional condition that $\lambda \ll d$. This makes the fringes very closely spaced on the screen, so the angle θ in Fig. 32.4 is small even for large orders m. Then $\sin\theta \simeq \tan\theta = y/L$, and a fringe's position y on the screen, measured from the central maximum, becomes

$$y_{\text{bright}} = m\frac{\lambda L}{d} \quad \text{and} \quad y_{\text{dark}} = \left(m + \tfrac{1}{2}\right)\frac{\lambda L}{d} \qquad \left(\begin{array}{c}\text{fringe position,}\\ \lambda \ll d\end{array}\right) \tag{32.2a, b}$$

GOT IT? 32.2 If you increase the slit separation in a two-slit system, do the interference fringes become (a) closer together or (b) farther apart?

EXAMPLE 32.1 Measuring Wavelength: Laser Light

Two slits 0.075 mm apart are located 1.5 m from a screen. Laser light shining through the slits produces an interference pattern whose third-order bright fringe is 3.8 cm from the screen center. Find the light's wavelength.

INTERPRET The concept behind this problem is two-slit interference. The phrase "third order" tells us we're dealing with the $m = 3$ bright fringe located at $y_{\text{bright}} = 3.8\,\text{cm}$.

DEVELOP Our plan is to use Equation 32.2a, $y_{\text{bright}} = m(\lambda L/d)$, and solve for λ. Since that equation requires $\lambda \ll d$, we'll then check to see whether our answer is consistent with this condition.

EVALUATE Solving, we have

$$\lambda = \frac{y_{\text{bright}}d}{mL} = \frac{(0.038\,\text{m})(0.075\times10^{-3}\,\text{m})}{(3)(1.5\,\text{m})} = 633\,\text{nm}$$

ASSESS This is indeed much less than the slit spacing of 0.075 mm or 75,000 nm. Our 633-nm result is in fact the wavelength of the red light from widely used helium–neon lasers. ■

Intensity in the Interference Pattern

We located the maxima and minima in two-slit interference using geometrical arguments alone. To find the actual intensity we need to superpose the interfering waves. You might think we could do that by adding the intensities of the two waves. But no! It's the electric and magnetic fields of the wave that obey the superposition principle, not the wave intensity. Intensity is proportional to the *square* of either field (recall Equation 29.20); if we added intensities we could never get the cancellation that occurs in destructive interference.

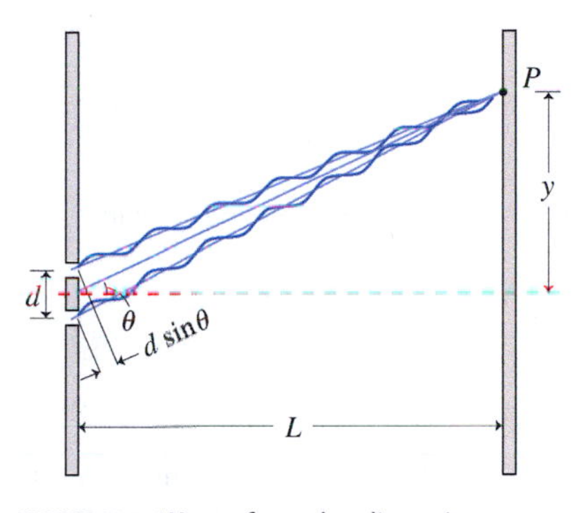

FIGURE 32.5 Waves from the slits arrive at P displaced by the path-length difference $d\sin\theta$. For $L \gg d$, $\sin\theta \simeq \tan\theta = y/L$.

Consider again a point P in the interference pattern (Fig. 32.5). In the approximation $d \ll L$, the difference in path lengths is so small that we can neglect any difference in the amplitudes of the two waves resulting from the falloff in intensity with distance. However, the difference in phase is crucial; it's what causes the interference. So we consider waves whose electric fields at P vary sinusoidally in time, with equal amplitude E_p but an explicit phase difference ϕ:

$$E_1 = E_p \sin\omega t \qquad \text{and} \qquad E_2 = E_p \sin(\omega t + \phi)$$

We aren't bothering with vectors because the two waves are polarized in the same direction and therefore their fields add algebraically. Then the net electric field at P is

$$E = E_1 + E_2 = E_p[\sin\omega t + \sin(\omega t + \phi)]$$

Appendix A gives the trig identity $\sin\alpha + \sin\beta = 2\sin[(\alpha+\beta)/2]\cos[(\alpha-\beta)/2]$, which, with $\alpha = \omega t$ and $\beta = \omega t + \phi$, gives

$$E = 2E_p \sin\left(\omega t + \frac{\phi}{2}\right)\cos\left(\frac{\phi}{2}\right)$$

where we also used $\cos(-x) = \cos x$. Thus, the electric field at P oscillates with the wave frequency ω, and its amplitude is $2E_p\cos(\phi/2)$. Since the phase difference ϕ depends on the difference in path lengths from the two slits, this amplitude varies across the screen, giving the interference pattern.

We've seen that the path-length difference is $d\sin\theta$, with d the slit spacing and θ the angle to P; under our approximation $d \ll L$, θ is small and $\sin\theta \simeq \tan\theta = y/L$, where y is the position of P as shown in Fig. 32.5. Then the path difference becomes yd/L. Now, that all-important phase difference ϕ is whatever fraction of a full cycle (2π radians) this path difference yd/L is of the wavelength λ; that is,

$$\phi = 2\pi\left(\frac{yd}{\lambda L}\right)$$

Then the amplitude $2E_p\cos(\phi/2)$ becomes $2E_p\cos(\pi yd/\lambda L)$. The average intensity follows from Equation 29.20b:

$$\bar{S} = \frac{[2E_p\cos(\pi yd/\lambda L)]^2}{2\mu_0 c} = 4\bar{S}_0\cos^2\left(\frac{\pi d}{\lambda L}y\right) \qquad (32.3)$$

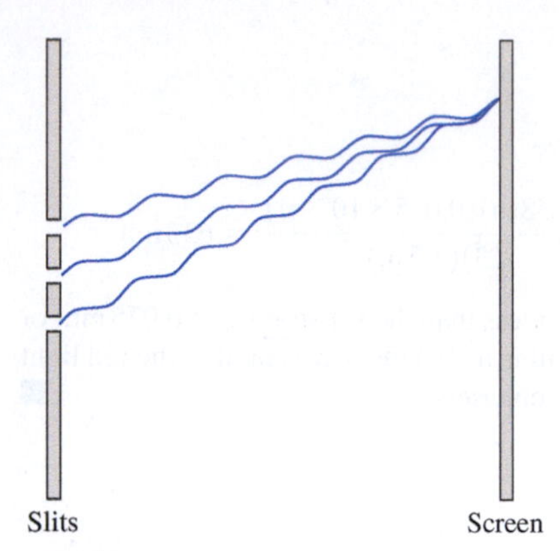

FIGURE 32.6 Waves from three evenly spaced slits interfere constructively when they reach the screen in phase.

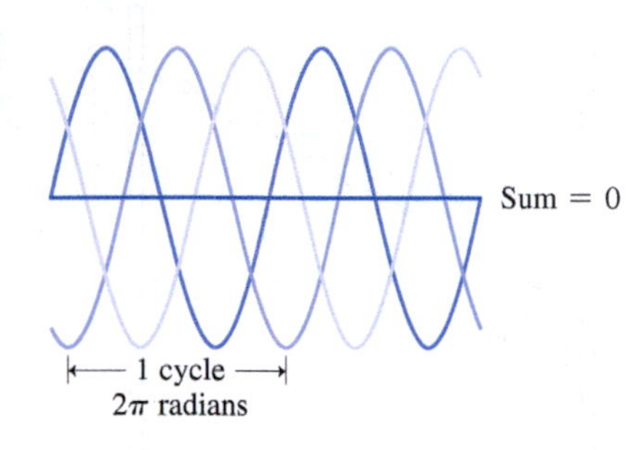

FIGURE 32.7 Waves from three slits must be out of phase by one-third of a cycle in order to interfere destructively.

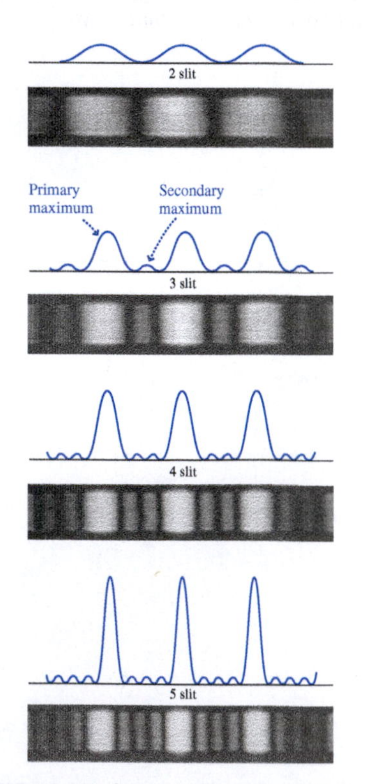

FIGURE 32.8 Interference patterns for multiple-slit systems with the same slit spacing. The bright fringes stay in the same place but become narrower and brighter as the number of slits increases. As the graphs show, the peak intensity scales as the *square* of the number of slits.

where $\bar{S}_0 = E_p^2/2\mu_0 c$ is the average intensity of either wave alone. Now, $\cos^2$ has its maximum value, 1, when its argument is a multiple of π. Thus Equation 32.3 gives maximum intensity when $yd/\lambda L$ is an integer m, or when $y = m\lambda L/d$. This is just the condition 32.2*a*, showing that our intensity calculation is fully consistent with the simpler geometrical analysis. But the intensity calculation tells more: It gives not only the fringe positions, but also the intensity variation in between.

32.3 Multiple-Slit Interference and Diffraction Gratings

Systems with multiple slits play a crucial role in optical instrumentation and in the analysis of materials. As we'll see, gratings with several thousand slits per centimeter make possible high-resolution spectroscopic analysis. At a much smaller scale the regularly spaced rows of atoms in a crystal act much like a multiple-slit system for X rays, and the resulting X-ray patterns reveal the crystal structure.

Figure 32.6 shows waves from three evenly spaced slits interfering at a screen. Maximum intensity requires that all three waves be in phase or, equivalently, travel paths differing by an integer number of wavelengths. Our criterion for the maximum in a two-slit pattern, $d \sin\theta = m\lambda$, ensures that waves from two adjacent slits will add constructively. Since the slits are evenly spaced with distance d between each pair, waves coming through a third slit will be in phase with the other two if this criterion is met. So the criterion for a maximum in an N-slit system is still Equation 32.1a:

$$d \sin\theta = m\lambda \quad \text{(maxima in multiple-slit interference, } m = 0, 1, 2, \ldots) \qquad (32.1a)$$

With more than two waves, however, the criterion for destructive interference is more complicated. Somehow all the waves need to sum to zero. Figure 32.7 shows that this happens for three waves when each is out of phase with the others by one-third of a cycle. Thus, the path-length difference $d \sin\theta$ must be either $\left(m + \frac{1}{3}\right)\lambda$ or $\left(m + \frac{2}{3}\right)\lambda$, where m is an integer. The case $\left(m + \frac{3}{3}\right)\lambda$ is excluded because then the path lengths differ by a full wavelength, giving constructive interference and thus a maximum in the interference pattern. More generally we can write

$$d \sin\theta = \frac{m}{N}\lambda \qquad (32.4)$$

for destructive interference in an N-slit system, where m is an integer *but not an integer multiple of N*.

Figure 32.8 shows interference patterns and intensity plots from some multiple-slit systems. Note that the bright, or *primary*, maxima are separated by several minima and fainter, or *secondary*, maxima. Why this complex pattern? Our analysis of the three-slit system shows two minima between every pair of primary maxima; for example, we considered the minima at $d \sin\theta = \left(m + \frac{1}{3}\right)\lambda$ and $d \sin\theta = \left(m + \frac{2}{3}\right)\lambda$, which lie between the maxima at $d \sin\theta = m\lambda$ and $d \sin\theta = (m + 1)\lambda$. More generally, Equation 32.4 shows that there are $N - 1$ minima between each pair of primary maxima given by Equation 32.1a. The secondary maxima that lie between these minima result from interference that is neither fully destructive nor fully constructive. The figure shows that the primary maxima become brighter and narrower as the number of slits increases, while the secondary maxima become relatively less bright. With a large number N of slits, then, we should expect a pattern of bright but narrow primary maxima, with broad, essentially dark regions in between.

Diffraction Gratings

A set of many closely spaced slits is called a **diffraction grating** and proves very useful in the spectroscopic analysis of light. Diffraction gratings are commonly several centimeters across and have several thousand slits—usually called lines—per cm. Gratings

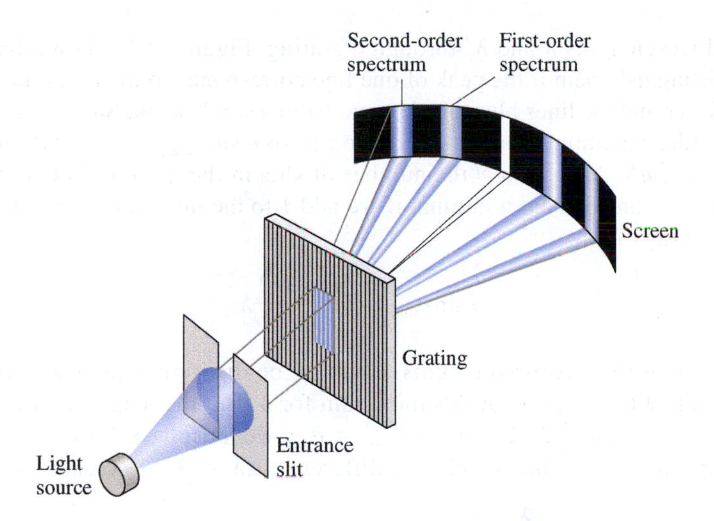

FIGURE 32.9 Essential elements of a grating spectrometer. An electronic detector normally replaces the screen.

like the slit systems we've been discussing are **transmission gratings**, since light passes through the slits. **Reflection gratings** produce similar interference effects by reflecting incident light.

We've seen that the maxima of the multiple-slit interference pattern are given by the same criterion, $d \sin\theta = m\lambda$, that applies to a two-slit system. For $m = 0$ this equation implies that all wavelengths peak together at the central maximum, but for larger values of m the angular position of the maximum depends on wavelength. Thus, a diffraction grating can be used in place of a prism to disperse light into its component wavelengths, and the integer m is therefore called the **order** of the dispersion. Figure 32.9 shows a spectrometer that works on this principle. Because the maxima in N-slit interference are very sharp for large N (recall Fig. 32.8), a grating with many slits diffracts individual wavelengths to very precise locations.

EXAMPLE 32.2 Finding the Separation: A Grating Spectrometer

Light from glowing hydrogen contains many discrete spectral lines, of which two are Hα (hydrogen-alpha) and Hβ (hydrogen-beta), with wavelengths of 656.3 nm and 486.1 nm, respectively. Find the first-order angular separation between these wavelengths in a spectrometer that uses a grating with 6000 slits per cm.

INTERPRET The concept behind the grating spectrometer is multiple-slit interference, so our job is to find the angles to which the grating sends the given wavelengths. "First-order" means we have $m = 1$.

DEVELOP Equation 32.1a, $d \sin\theta = m\lambda$, gives the location of the interference maxima as a function of wavelength λ, order m, and slit spacing d. We're given m and two values for λ, but we don't know d. However, we're told that there are 6000 slits per cm, so we can find d. Our plan is first to calculate d, then use Equation 32.1a to find the angular positions for the two wavelengths, and finally take their difference to get the angular separation.

EVALUATE With 6000 slits/cm, the spacing is $d = 1/6000\,\text{cm} = 1.667\,\mu\text{m}$. Applying Equation 32.1a with $m = 1$ gives

$$\theta_\alpha = \sin^{-1}\left(\frac{\lambda}{d}\right) = \sin^{-1}\left(\frac{0.6563\,\mu\text{m}}{1.667\,\mu\text{m}}\right) = 23.2^\circ$$

A similar calculation gives $\theta_\beta = 17.0^\circ$. Thus the angular separation is 6.2°.

ASSESS Our 6.2° result is certainly adequate to distinguish clearly these two wavelengths. For greater angular separation, or to separate closer wavelengths, we could look at the higher-order dispersion (see Exercise 20). ■

Resolving Power

The detailed shapes and wavelengths of spectral lines contain a wealth of information about the systems in which light originates. Studying these details requires a high dispersion in order to separate nearby spectral lines or to analyze the intensity-versus-wavelength profile of a single line. Suppose we pass light containing two spectral lines of

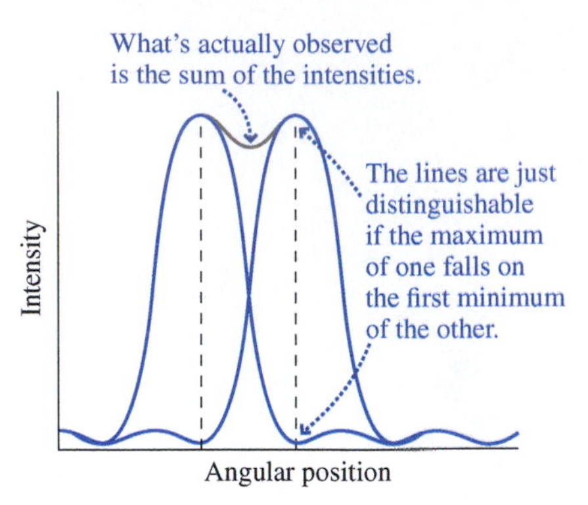

FIGURE 32.10 Intensity versus angular position for spectral lines with slightly different wavelengths, as dispersed with a grating.

nearly equal wavelengths λ and λ' through a grating. Figure 32.10 shows that we'll just be able to distinguish them if the peak of one line corresponds to the first minimum of the other; any closer and the lines blur together and form a single peak. Suppose wavelength λ has its mth-order maximum at angular position θ, so $d \sin\theta_{\text{max}} = m\lambda$. We can write this as $d \sin\theta_{\text{max}} = (mN/N)\lambda$, with N the number of slits in the grating. Equation 32.4 then shows that we get an adjacent minimum if we add 1 to the numerator mN. Thus the adjacent minimum satisfies

$$d \sin\theta_{\text{min}} = \frac{mN + 1}{N}\lambda$$

Our criterion that the two wavelengths λ and λ' be distinguishable is that the maximum for λ' fall at the location of this minimum for λ. But the maximum for λ' satisfies $d \sin\theta'_{\text{max}} = m\lambda' = (mN/N)\lambda'$, so for $\theta'_{\text{max}} = \theta_{\text{min}}$ we must have $(mN + 1)\lambda = mN\lambda'$. Expressing this in terms of the wavelength difference $\Delta\lambda = \lambda' - \lambda$ leads to

$$\frac{\lambda}{\Delta\lambda} = mN \qquad \text{(resolving power)} \qquad (32.5)$$

The quantity $\lambda/\Delta\lambda$ is the grating's **resolving power**, a measure of its ability to distinguish closely spaced wavelengths. The higher the resolving power, the smaller the wavelength difference $\Delta\lambda$ that we can distinguish. Equation 32.5 shows that the resolving power increases with the number of lines, N, on the grating and also with the order, m, of the spectrum we observe.

EXAMPLE 32.3 Resolving Power: "Seeing" a Double Star

A double-star system consists of a massive star essentially at rest, with a smaller companion in circular orbit. It's far too distant for the pair to appear as anything but a single point even to the largest telescopes. Yet astronomers can "see" the companion star through the Doppler shift in wavelengths of its spectral lines. The Hα spectral line from the stationary massive star is at $\lambda = 656.272$ nm; for the companion when it's moving away from Earth, the Hα line Doppler-shifts to 656.329 nm (corresponding to a speed of about 26 km/s). If a spectrometer has 5000 lines, what order spectrum will resolve the Hα lines from the two stars?

INTERPRET The concept here is resolution of distinct spectral lines using a grating spectrometer.

DEVELOP Equation 32.5, $\lambda/\Delta\lambda = mN$, determines the resolving power. We can solve for m to get the order: $m = \lambda/(N\,\Delta\lambda)$. We're given the two wavelengths, so we can readily find $\Delta\lambda$.

EVALUATE We have $\Delta\lambda = 656.329\text{ nm} - 656.272\text{ nm} = 0.057\text{ nm}$. Then

$$m = \frac{\lambda}{N\,\Delta\lambda} = \frac{656.272\text{ nm}}{(5000)(0.057\text{ nm})} = 2.3$$

ASSESS Since the order m must be an integer, we'll have to use the third-order spectrum. ■

X-Ray Diffraction

The wavelengths of X rays, on the order of 0.1 nm, are far too short for diffraction with gratings produced mechanically or photographically. Instead, **X-ray diffraction** occurs when X rays interact with the regularly spaced atoms in a crystal. At the microscopic level, reflection of an electromagnetic wave occurs when the wave's electric field sets electrons oscillating. The electrons re-radiate, producing the reflected beam. With X rays reflecting from a crystal, the regular atomic spacing results in interference that enhances the reflected radiation at certain angles. Figure 32.11a shows an X-ray beam interacting with the atoms in a crystal. In Fig. 32.11b we see that waves reflecting at one layer of atoms travel a distance $2d \sin\theta$ farther than those reflecting at the layer above, where θ is the angle between the incident beam and the atomic planes. Constructive interference occurs when this difference is an integer number of wavelengths:

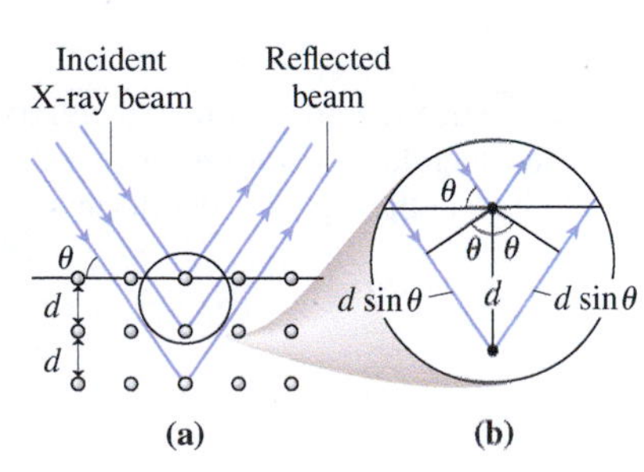

FIGURE 32.11 (a) X rays reflecting off the planes of atoms in a crystal. (b) Constructive interference enhances the outgoing beam when the extra distance $2d \sin\theta$ is an integer multiple of the X-ray wavelength.

$$2d \sin\theta = m\lambda \qquad \text{(Bragg condition, } m = 1, 2, 3, \ldots) \qquad (32.6)$$

This **Bragg condition** lets us use a crystal with known spacing as a diffraction grating for X rays. More important is the converse: Much of what we know about crystal structure comes from probing crystals with X rays and using the resulting patterns to deduce positions of their atoms. X-ray diffraction measurements by British scientist Rosalind Franklin in 1952 were crucial in establishing the structure of DNA. Today, geologists and materials scientists use X-ray diffraction routinely in studying the structure of rocks, metals, and other substances. Molecular biologists and pharmaceutical manufacturers use X-ray diffraction to analyze biomolecules and as an aid in designing new drugs. There's even an X-ray diffraction instrument on Mars, used by the *Curiosity* rover to analyze Martian soils and determine mineral abundances.

Other Gratings

Anything with regularly spaced structures can act as a diffraction grating for waves of suitable wavelength. The rainbow of colors you see on the underside of a CD or DVD results because adjacent pits of CD tracks (shown in the Application earlier in this chapter) act as a diffraction grating. Ocean waves can act as a diffraction grating for radio waves; oceanographers exploit such diffraction in radar surveys that yield the wavelength and amplitude of the ocean waves. Finally, sound waves in a solid set up refractive index variations that act as diffraction gratings for light; the Application describes some everyday uses of this phenomenon.

GOT IT? 32.3 If you increase the number of slits in a grating while keeping the spacing the same, does each of the following (a) increase, (b) decrease, or (c) remain unchanged? (1) The spacing between intensity maxima in the interference pattern, (2) the intensity of the maxima, and (3) the width of the maxima.

APPLICATION **Laser Printers, DVDs, and AOMs**

Last time you used a laser printer, made a PowerPoint presentation, or burned a DVD, chances are that an acousto-optic modulator (AOM) played a role. As the figure shows, these devices use a loudspeaker-like *transducer* to beam sound waves into a transparent crystal. The regular spacing of the acoustic wavefronts constitutes a diffraction grating, and laser light entering the crystal is diffracted at an angle determined by the wavefront spacing—that is, by the acoustic wavelength. Varying the acoustic frequency alters the acoustic wavelength, which, in turn, changes the diffraction angle of the light. Thus the AOM may be used to "steer" a laser beam to different locations. Varying the acoustic amplitude alters the amplitude of the diffracted beam; this allows the AOM to modulate the light beam's intensity or to switch it on and off altogether. AOMs are commonly used in laser printers, laser-based projection systems, and DVD burners for either steering or modulating a laser beam, or both. AOMs also find applications in optical communications systems, in switching high-power lasers, in trapping and manipulating biomolecules and other small particles, and in a host of other technologies.

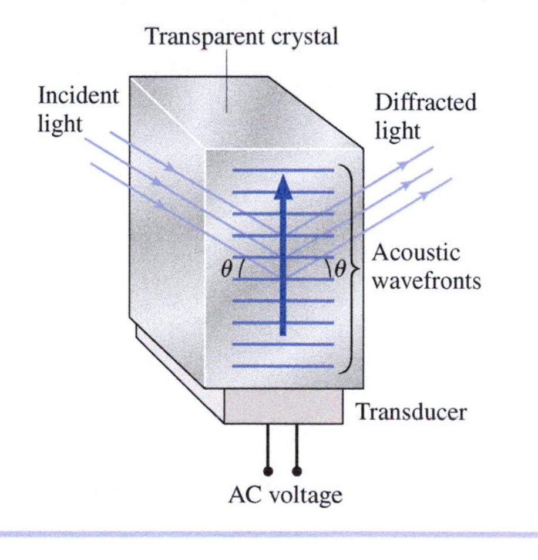

32.4 Interferometry

Passing light through multiple slits isn't the only way to produce interference. So will any process that separates light into several beams, sends them on different paths, and then rejoins them. Such processes are the basis of **interferometry**, an exquisitely sensitive technique for measuring small displacements, time intervals, and other quantities.

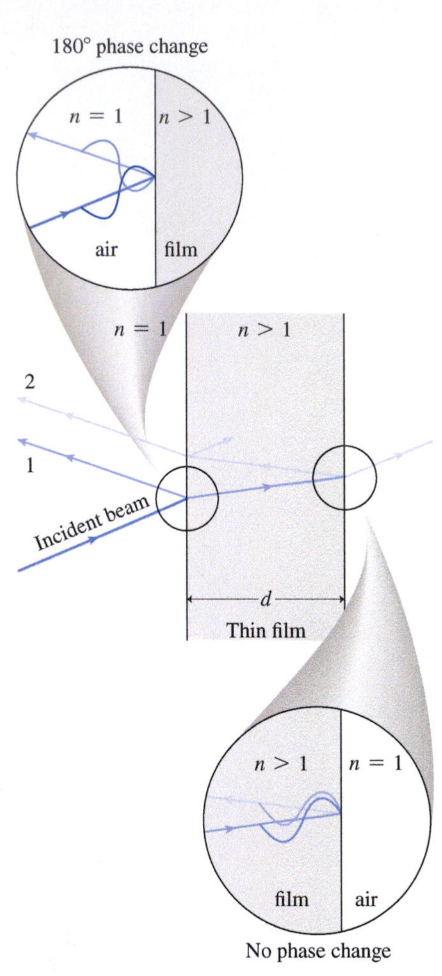

FIGURE 32.12 Reflection and refraction at a thin film of transparent material, showing a 180° phase change at the first interface and none at the second.

Thin Films

Light passing through thin, transparent films is partially reflected at both the front and back surfaces, and the resulting beams recombine to produce interference. In Section 14.6 we saw how waves on a string reflect where one string joins another with different properties; in particular, the reflected wave is inverted if the second string has greater mass per unit length. Light behaves analogously: It's reflected with a 180° phase change at the interface between a material with lower refractive index and one with higher refractive index. But it reflects without a phase change at an interface from higher to lower refractive index. For a thin film with refractive index n higher than its surroundings, Fig. 32.12 shows that there's a 180° phase change at the first interface and no change at the second.

If the film in Fig. 32.12 has thickness d, there's also a phase change due to the additional path length for beam 2. We'll consider the case of normal incidence, although for clarity the figure shows slightly oblique incidence. With normal incidence, the extra length is twice the film thickness, or $2d$. Because reflected beam 1 forms with a 180° phase change but beam 2 has no phase change, it takes another 180° phase shift—half a wavelength path difference—to put beams 1 and 2 back in phase to give constructive interference. That occurs if beam 2's extra path length $2d$ is half a wavelength, or $1\frac{1}{2}$ wavelengths, or any odd-integer multiple of one-half wavelength: $2d = \left(m + \frac{1}{2}\right)\lambda_n$, where $m = 0, 1, 2, 3, \ldots$, and where the subscript n indicates that this is the wavelength as measured in the material. In Chapter 30 we found that the wavelength in a material with refractive index n is reduced by a factor $1/n$ from its value in air or vacuum; thus $\lambda_n = \lambda/n$, and our condition for constructive interference becomes

$$2nd = \left(m + \tfrac{1}{2}\right)\lambda \qquad \text{(constructive interference, thin film)} \qquad (32.7)$$

Interference in thin layers is the basis of some very sensitive optical techniques. The shape of a lens, for example, can be measured to within a fraction of a wavelength of light using interference in a thin "film" consisting of the air between the lens and a flat glass plate (Fig. 32.13).

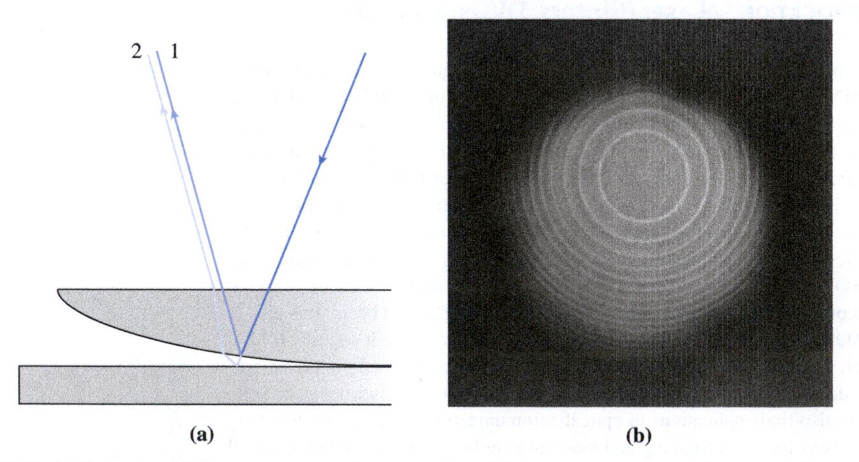

FIGURE 32.13 (a) A portion of a lens sitting on a flat glass plate. (b) Newton's rings arise from the difference in path lengths between rays like 1 and 2 and provide precise information about the lens shape.

CONCEPTUAL EXAMPLE 32.1 Interference: A Soap Film

Figure 32.14 shows a soap film in a circular ring with different colored bands that run horizontally across the film. Why do these bands occur?

EVALUATE The bands must be the result of interference, as described in Fig. 32.12. But why are they different colors? The soap film is vertical and consists mostly of water, so gravity makes it thicker at the bottom. Therefore, the wavelengths that undergo constructive interference vary with vertical position on the film. There are many bands, with colors repeating, because there are multiple orders of interference, with different values of m.

ASSESS The dark region at the top confirms our explanation: Here the film is so thin that no visible wavelength undergoes constructive interference, so it appears dark. The film is probably about to break!

MAKING THE CONNECTION A 20-cm-high soap film is 1 μm thick at the bottom, tapering to near zero thickness at the top. If it's illuminated with 650-nm laser light, how many bright bands appear?

EVALUATE Equation 32.7 gives the condition for constructive interference; the number of bands will be the number of interference orders m possible in this film. Solving for m at the bottom of the film gives $m = 2n\lambda/d - 1/2 = 3.6$, with $n = 1.333$ for water. Since m must be an integer, there's no bright band right at the bottom. The $m = 3$ band is higher up, and above it are the $m = 2$, $m = 1$, and $m = 0$ bands, for a total of four bright bands.

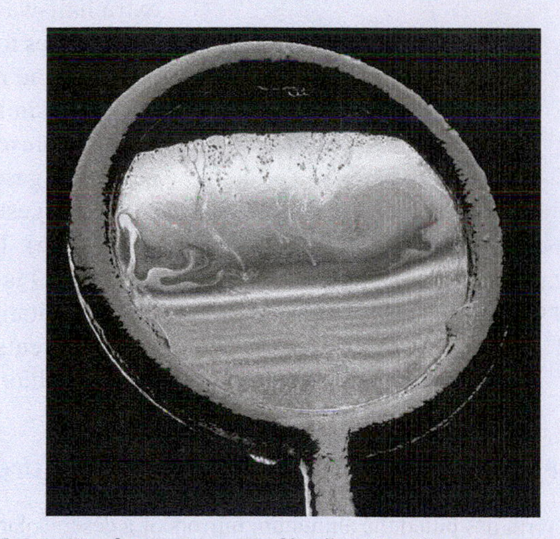

FIGURE 32.14 Interference in a soap film illuminated with white light.

In analyzing thin films, we've considered only the first reflection at each interface. Actually, multiple reflections occur within the film, producing ever-weaker rays. A fuller treatment involving Maxwell's equations shows that when a film of refractive index n_2 is sandwiched between materials with indices n_1 and n_3, complete cancellation of reflected rays in the incident medium occurs if the thickness is right and if $n_2 = \sqrt{n_1 n_3}$. This is the basis of the antireflection coatings, mentioned in Chapter 30, which ensure maximum light transfer in camera lenses, solar photovoltaic cells, and other applications.

GOT IT? 32.4 If you photographed the soap film in Fig. 32.14 with a camera that's sensitive to both visible and infrared light, would the dark region at the top appear (a) smaller or (b) larger?

The Michelson Interferometer

Several optical instruments use interference for precise measurement. Among the simplest and most important is the **Michelson interferometer**, invented by the American physicist Albert Michelson and used in the 1880s in a groundbreaking experiment that paved the way for the theory of relativity. We discuss this experiment in the next chapter; here we describe the interferometer, which is still used for precision measurements.

Figure 32.15 shows the basic Michelson interferometer. The key idea is that light from a monochromatic source is split into two beams by a half-silvered mirror called a **beam splitter**. The beam splitter is set at a 45° angle, so the reflected and transmitted beams travel perpendicular paths. Each then reflects off a flat mirror and returns to the beam splitter. The beam splitter again transmits and reflects half the light incident on it, with the result that some light from the originally separated beams is recombined. The recombined beams interfere, and the interference pattern is observed with a viewing lens; an example of the resulting pattern is shown at the bottom of Fig. 32.15.

If the path lengths for the two beams were exactly the same, they would recombine in phase and interfere constructively. In reality, the path lengths are never exactly the same, the mirrors are never exactly perpendicular, and the beams aren't perfectly parallel. But that's no problem: What happens is that different parts of the beams recombine with different phase differences, and the result is a pattern of light and dark interference fringes, as shown in Fig. 32.15. The distance between successive fringes corresponds to a path-length difference of one full wavelength.

Now suppose one mirror moves slightly. The path-length differences change and therefore the interference pattern shifts. A mere quarter-wavelength mirror movement adds an

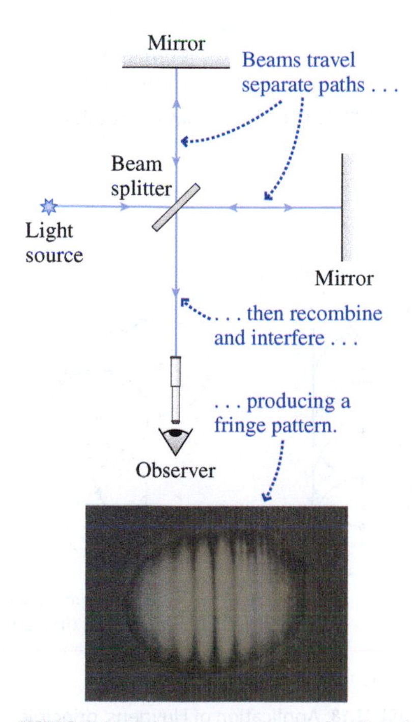

FIGURE 32.15 Schematic diagram of a Michelson interferometer, with a photo of the interference fringes.

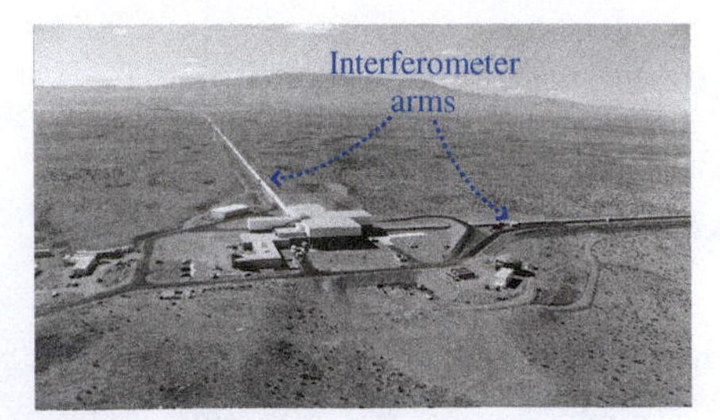

FIGURE 32.16 The LIGO instrument at Hanford, Washington, is an interferometer with 4-km arms. The light undergoes multiple reflections, giving an effective arm length of 300 km.

extra half-wavelength to the round-trip path. That results in a 180° phase shift, moving dark fringes to where light ones were. Shifts a fraction of this amount are readily detected, allowing the measurement of mirror displacements to within a small fraction of a wavelength. A similar shift occurs if a transparent material is placed in one path, retarding the beam because of its refractive index. This provides accurate measures of the refractive indices of gases, which are so close to 1 that less-sensitive techniques don't work.

The largest Michelson interferometers ever built are twin instruments with 4-km arms (Fig. 32.16). These comprise LIGO, the Laser Interferometer Gravitational Wave Observatory. LIGO is designed to detect mirror displacements on the order of 10^{-18} m resulting from gravitational waves—"ripples" in the structure of space and time caused by distant cosmic events. LIGO will eventually be dwarfed by a space-based interferometer with arms 5 million km long!

EXAMPLE 32.4 An Interferometric Measurement: Sandstorm!

A sandstorm has pitted the aluminum mirrors of a desert solar-energy installation, and engineers want to know the depths of the pits. They construct a Michelson interferometer with a sample from one of the pitted mirrors in place of one flat mirror. With 633-nm laser light, the interference pattern in Fig. 32.17 results. What is the approximate depth of the pit?

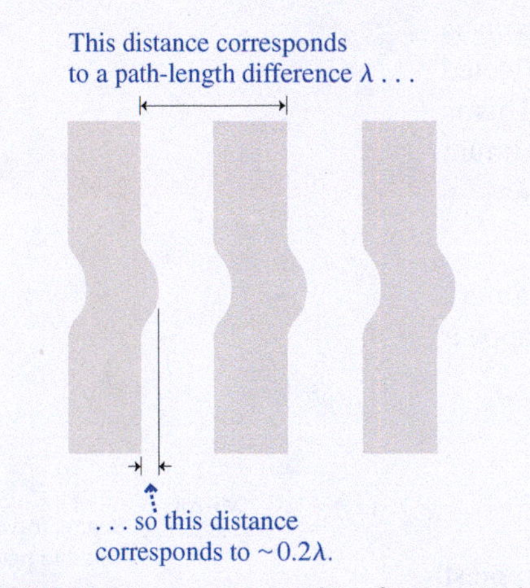

FIGURE 32.17 Fringe pattern resulting from a pitted mirror.

INTERPRET The concept here is interferometry—inferring distance from observations of light interference, in this case with the Michelson configuration of Fig. 32.15. The distortions of the interference fringes shown in Fig. 32.17 result because some of the light travels a little farther—namely, into the bottom of a pit and back out.

DEVELOP The full fringe spacing corresponds to a path difference of one wavelength, and Fig. 32.17 shows that the fringe distortion due to the pit gives a shift about one-fifth of the distance between fringes. We need to use this information to find the extra distance traveled by light reflecting from the bottom of the pit, and from that the pit depth. We're given the wavelength, so we can estimate the round trip, into and out of the pit, as approximately 0.2λ. The pit depth will be half this quantity.

EVALUATE The extra path length for the light reflecting off the pit is 0.2λ, so the pit depth is about 0.1λ. With $\lambda = 633$ nm, the pit depth is about 63 nm.

ASSESS Try measuring that with a meter stick! Interferometry provides an exquisitely sensitive measurement of small distances. ■

32.5 Huygens' Principle and Diffraction

The interference we've been studying in this chapter isn't the only optical phenomenon where the wave nature of light is important. There's also **diffraction**—the bending of light or other waves as they pass by objects. Interference and diffraction are closely related, and the double- and multiple-slit interference we've studied actually involves diffraction as well—hence the term *diffraction grating*.

Diffraction, like other optical phenomena, is ultimately governed by Maxwell's equations. But we can understand diffraction more readily using **Huygens' principle**, articulated in 1678 by the Dutch scientist Christian Huygens, who was the first to suggest that light might be a wave. Huygens' principle states:

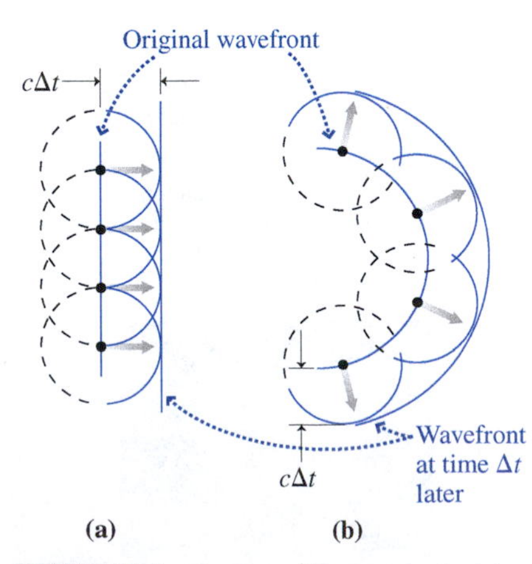

FIGURE 32.18 Application of Huygens' principle to (a) plane and (b) spherical waves. In each case, the wavefront acts like a set of point sources emitting circular waves that expand to produce a new wavefront.

> All points on a wavefront act as point sources of spherically propagating *wavelets* that travel at the speed of light appropriate to the medium. At a short time Δt later, the new wavefront is the unique surface tangent to all the forward-propagating wavelets.

Figure 32.18 shows how Huygens' principle accounts for the propagation of plane and spherical waves.

Diffraction

Figure 32.19 shows plane waves incident on an opaque barrier containing a hole. Since the waves are blocked by the barrier, Huygens' wavelets produced near each barrier edge cause the wavefronts to bend at the barrier. When the width of the hole is much greater than the wavelength, as in Fig. 32.19*a*, this diffraction is of little consequence, and the waves effectively propagate straight through the hole in a beam defined by the hole size. But when the hole size and wavelength are comparable, wavefronts emerging from the hole spread in a broad pattern (Fig. 32.19*b*). Thus diffraction, although it always occurs, is significant only on length scales comparable to or smaller than the wavelength. That's why we could ignore diffraction and assume that light always travels in straight lines when we considered optical systems with dimensions much larger than the wavelength of light.

Diffraction ultimately limits our ability to image small objects and to focus light precisely. The chapter opening photo and its associated question are one example. Next, we'll show the reason for this fundamental limitation on optical systems by examining the behavior of light as it passes through a single slit. The result will help you understand optical challenges ranging from telescopic imaging of distant astrophysical objects to the development of Blu-ray discs.

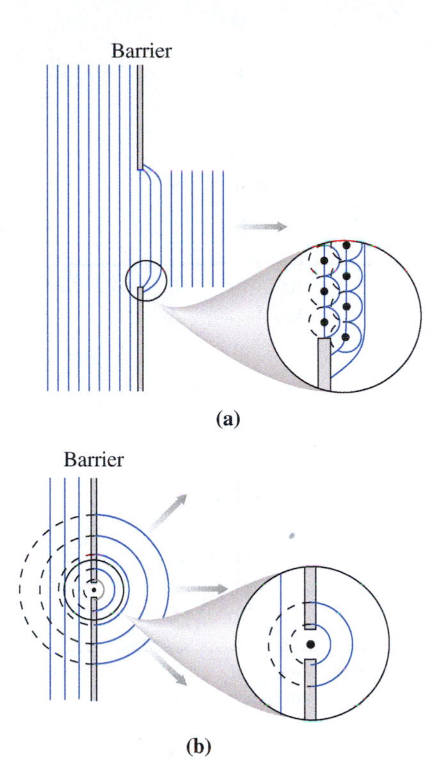

FIGURE 32.19 Plane waves incident on an opaque barrier with a hole. Diffraction is negligible for a hole large compared with the wavelength (a), but pronounced for a small hole (b).

Single-Slit Diffraction

In treating double-slit and multiple-slit interference, we assumed that plane waves passing through a slit emerged with circular wavefronts. According to Fig. 32.19*b*, that's true only if the slit width is small compared with the wavelength, so the slit can be treated as a single, localized source of new waves. When the slit width isn't small, Huygens' principle implies that we have to consider each point in the slit as a separate source—and then we can expect interference from waves originating at different points in the same slit. Thus a single wide slit is really like a multiple-slit system with infinitely many slits!

Figure 32.20*a* shows light incident on a slit of width a. Each point in the slit acts as a source of spherical wavelets propagating in all directions to the right of the slit. We focus on a particular direction described by the angle θ, and we'll look at interference of light from the five points shown. Figure 32.20*b* concentrates on the points from which rays 1, 2, and 3 originate and shows that the path lengths for rays 1 and 3 differ by $\frac{1}{2}a\sin\theta$. These two beams will interfere destructively if this distance is half the wavelength—that is, if $\frac{1}{2}a\sin\theta = \frac{1}{2}\lambda$ or $a\sin\theta = \lambda$. But if rays 1 and 3 interfere destructively, so do rays 3 and 5, which have the same geometry, and so do rays 2 and 4, for the same reason. In fact, a ray leaving *any* point in the lower half of the slit will interfere destructively with the point located a distance $a/2$ above it. Therefore, an observer viewing the slit system at the angle θ satisfying $a\sin\theta = \lambda$ will see no light.

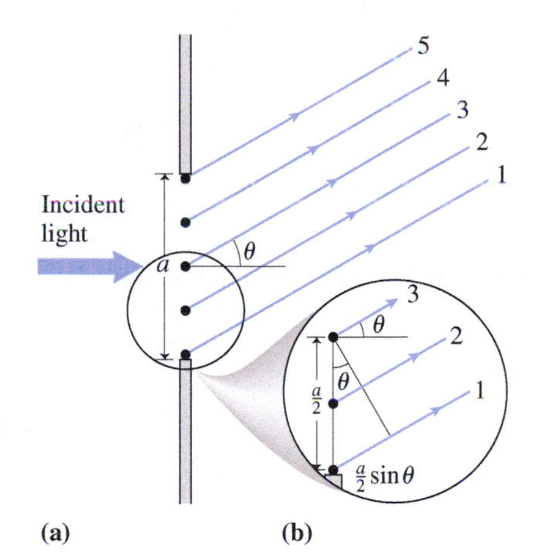

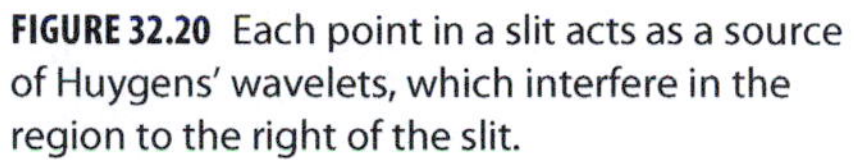

FIGURE 32.20 Each point in a slit acts as a source of Huygens' wavelets, which interfere in the region to the right of the slit.

Similarly, the sources for rays 1 and 2 are $a/4$ apart and will therefore interfere destructively if $\frac{1}{4}a\sin\theta = \frac{1}{2}\lambda$, or $a\sin\theta = 2\lambda$. But then so will rays 2 and 3, and rays 3 and 4; in fact, any ray from a point in the lower three-quarters of the slit will interfere destructively with a ray from the point $a/4$ above it, and therefore, an observer looking at an angle θ satisfying $a\sin\theta = 2\lambda$ will see no light.

We could equally well have divided the slit into six sections with seven evenly spaced points; we would then have found destructive interference if $\frac{1}{6}a\sin\theta = \frac{1}{2}\lambda$, or $a\sin\theta = 3\lambda$. We could continue this process for any number of points in the slit, and therefore, we conclude that destructive interference occurs for all angles θ satisfying

$$a\sin\theta = m\lambda \quad \text{(destructive interference, single-slit diffraction)} \qquad (32.8)$$

with m any nonzero integer and a the slit width. Note that the case $m = 0$ is excluded; it produces not destructive interference but a central maximum in which all waves are in phase.

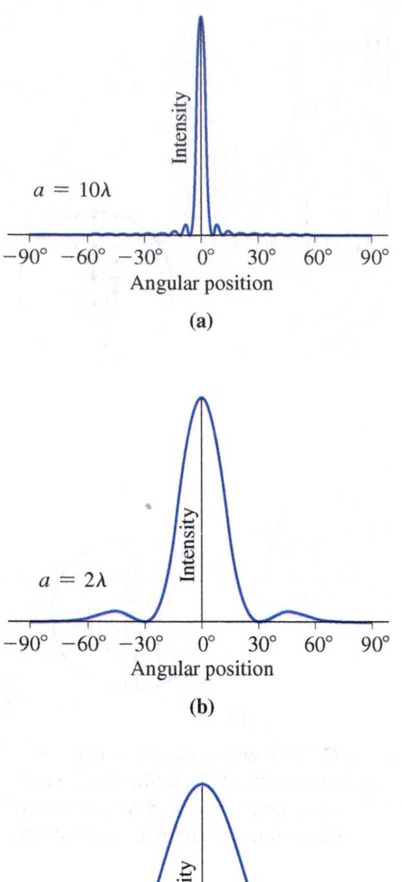

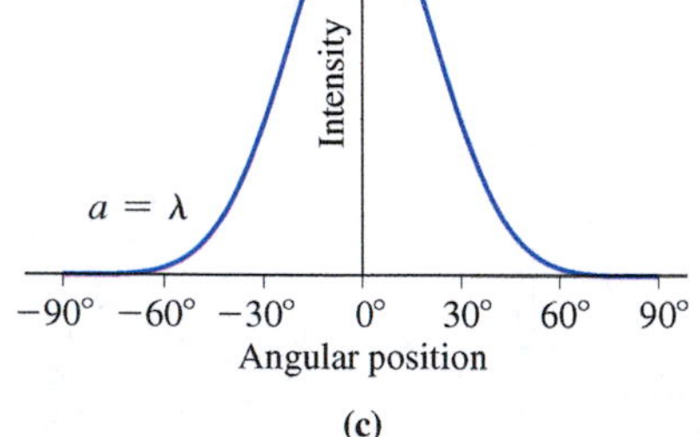

FIGURE 32.21 Intensity in single-slit diffraction, as a function of the angle θ from the centerline, for three values of slit width a.

✓TIP Interference and Diffraction

Equation 32.8 for the *minima* of a single-slit diffraction pattern looks just like Equation 32.1a for the *maxima* of a multiple-slit interference pattern, except that the slit width a replaces the slit spacing d. Why does the same equation give the minima in one case and the maxima in another? Because we're dealing with two distinct but related phenomena. In the multiple-slit case, each slit was so narrow that it could be considered a single source, neglecting the interference of waves originating within the same slit. In the single-slit case, the diffraction pattern occurs precisely because of the interference of waves from different points within the same slit.

Intensity in Single-Slit Diffraction

Geometry gave us the *positions* of the maxima in single-slit diffraction, just as it did for multiple-slit interference. But to get the *intensity* across the diffraction pattern, we'd have to superpose electric fields of the interfering waves, as we did in deriving Equation 32.3 for two-slit interference. But now we've got infinitely many fields to sum, corresponding to waves from every part of the slit. It's possible to do this using a calculus-like graphical technique involving the phasor concept introduced in Chapter 28. We won't go through the derivation; however, we'll motivate the result by noting that the diffraction pattern at any point occurs because of phase differences among waves originating at different parts of the slit. It's not surprising, therefore, that a key factor is the phase difference between waves from opposite ends of the slit. Applying the analysis of Fig. 32.20 to the entire slit width a gives a path-length difference $a\sin\theta$ for these waves. As usual, this path difference is to the wavelength as the phase difference ϕ is to a full cycle, 2π radians. Thus

$$\phi = \frac{2\pi}{\lambda}a\sin\theta \tag{32.9}$$

is the phase difference between rays from the ends of the slit to a point at angular position θ. The phasor-summing process relates the amplitude of the net electric field to this phase difference ϕ and shows that the field is proportional to $\sin(\phi/2)/(\phi/2)$. Used in Equation 29.20 to get the intensity from the electric field, this result gives the intensity as a function of angle in single-slit diffraction:

$$\bar{S} = \bar{S}_0\left[\frac{\sin(\phi/2)}{\phi/2}\right]^2 \tag{32.10}$$

Here $\bar{S}_0$ is the average intensity at the central maximum of the pattern ($\theta = \phi = 0$), and ϕ is given by Equation 32.9. At $\theta = \phi = 0$, Equation 32.10 appears to be indeterminate, but using the limit $\sin x/x \to 1$ as $x \to 0$ shows that the result is indeed $\bar{S}_0$. Problem 68 explores another approach to Equation 32.10.

Figure 32.21 plots Equation 32.10 for three values of the slit width a in relation to the wavelength λ. For wide slits—a large compared with λ—the central peak is narrow and the secondary peaks are much lower and also half as wide as the central peak. Here diffraction is negligible, and the beam essentially propagates through the slit in the ray approximation of geometrical optics. But as the slit narrows, the diffracted beam spreads until, with $a = \lambda$, it covers an angular width of some 120°.

The intensity given by Equation 32.10 will be zero when the numerator on the right-hand side is zero—that is, when the argument of the sine function is an integer multiple of π. That occurs when $\phi/2 = (\pi a/\lambda)\sin\theta = m\pi$, or when $a\sin\theta = m\lambda$. Thus, we recover our result of Equation 32.8 for the angular positions where destructive interference gives zero intensity.

Multiple Slits and Other Diffracting Systems

In treating multiple-slit systems in Section 32.2, we assumed the slits were so narrow compared with the wavelength that the central diffraction peak spread into the entire space

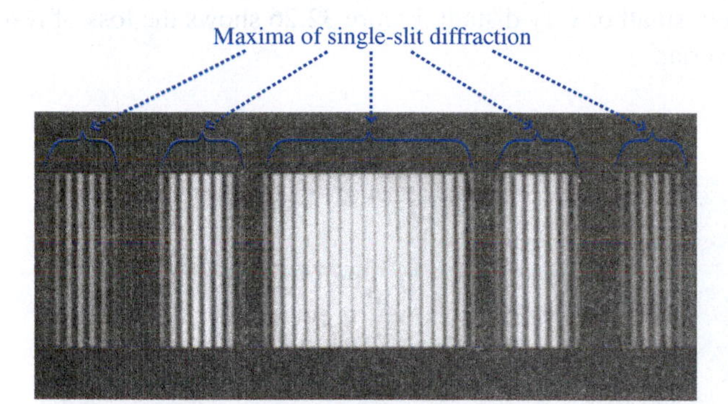

FIGURE 32.22 When the slit width is not negligible, a double-slit system produces the regular variations of double-slit interference within a single-slit diffraction pattern.

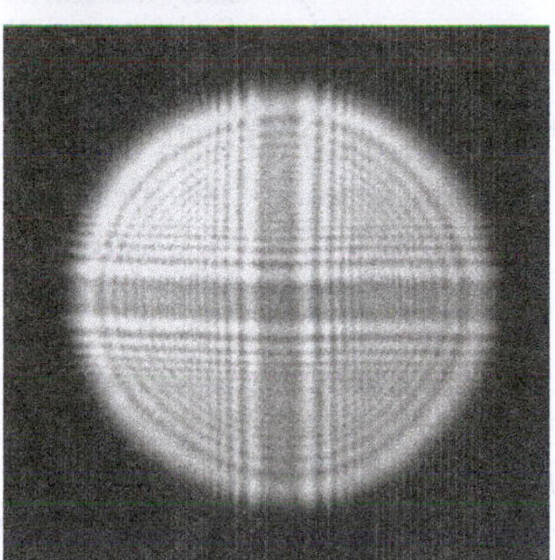

FIGURE 32.23 Diffraction patterns from light passing sharp edges. (a) Straight edge of an opaque barrier. (b) Circular aperture with crosshairs.

beyond the slit system. When the slit width isn't negligible, each slit produces a single-slit diffraction pattern. The result is a pattern that combines single-slit diffraction with multiple-slit interference (Fig. 32.22).

Diffraction occurs any time light passes a sharp, opaque edge like the edges of the slits we've been considering. Close examination of the shadow produced by a sharp edge shows parallel fringes resulting from interference of the diffracting wavefronts (Fig. 32.23*a*). More complex diffraction patterns result from objects of different shape (Fig. 32.23*b*). Such diffraction limits our ability to form sharp optical images, as we show in the next section.

GOT IT? 32.5 A classmate down the hall is playing obnoxiously loud music, and you both have your doors open. Why are you most annoyed by the music's thumping bass?

32.6 The Diffraction Limit

Diffraction imposes a fundamental limit on the ability of optical systems—telescopes, microscopes, cameras, and even eyes—to distinguish closely spaced objects. Consider two point sources of light illuminating a slit. The sources are so far from the slit that waves reaching the slit are essentially plane waves, but the different source positions mean the waves reach the slit at different angles. We assume the sources are incoherent, so they don't produce a regular interference pattern. Then light diffracting at the slit produces two single-slit diffraction patterns, one for each source. Because the sources are at different angular positions, the central maxima of these patterns don't coincide, as shown in Fig. 32.24.

If the angular separation between the sources is great enough, then the central maxima of the two diffraction patterns will be entirely distinct. In that case we can clearly distinguish the two sources (Fig. 32.24*a*). But as the sources get closer, the central maxima begin to overlap (Fig. 32.24*b*). They remain distinguishable as long as the total intensity pattern shows two peaks. Since the sources are incoherent, the total intensity is just the sum of the individual intensities. Figure 32.25 shows how that sum loses its two-peak structure as the diffraction patterns merge. In general, two peaks are barely distinguishable if the central maximum of one coincides with the first minimum of the other. This condition is called the **Rayleigh criterion**, and when it's met the two sources are just barely **resolved**.

Optical systems are analogous to the single slit we've just considered. Every system has an aperture of finite size through which light enters. That aperture may be an actual slit or hole, like the diaphragm that stops down a camera lens, or it may be the full size of a microscope lens or a telescope mirror. So all optical systems ultimately suffer loss of resolution if two sources—or two parts of the same object—have too small an angular separation. Thus, diffraction fundamentally limits our ability to probe the structure of

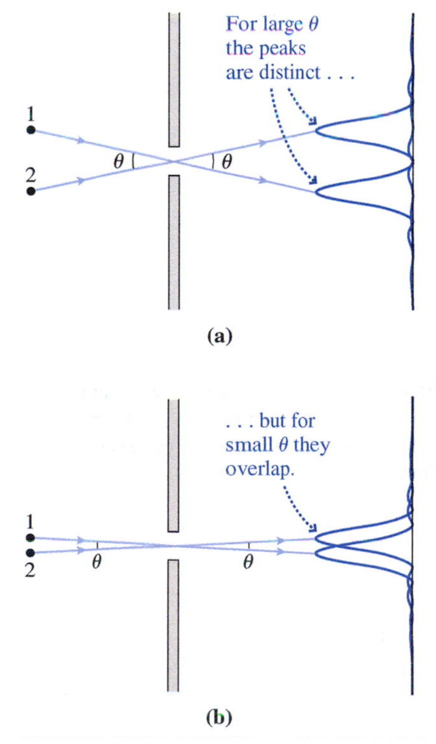

FIGURE 32.24 Two distant light sources at different angular positions produce diffraction patterns whose central peaks have the same angular separation θ as the sources.

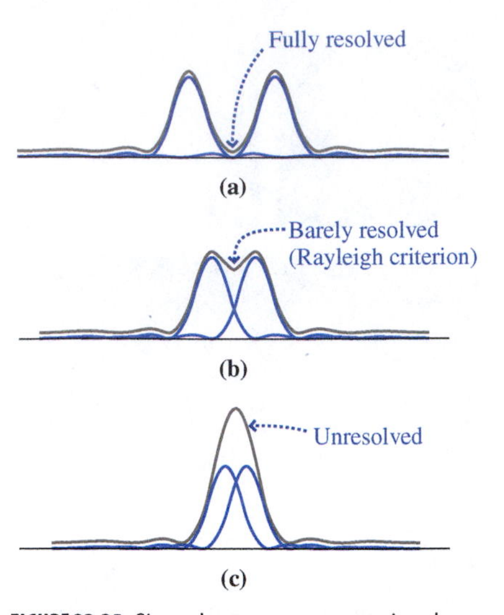

FIGURE 32.25 Since the two sources are incoherent, the total intensity is just the sum (gray curve) of the intensities of the two diffraction patterns.

objects that are either very small or very distant. Figure 32.26 shows the loss of resolution as diffraction patterns overlap.

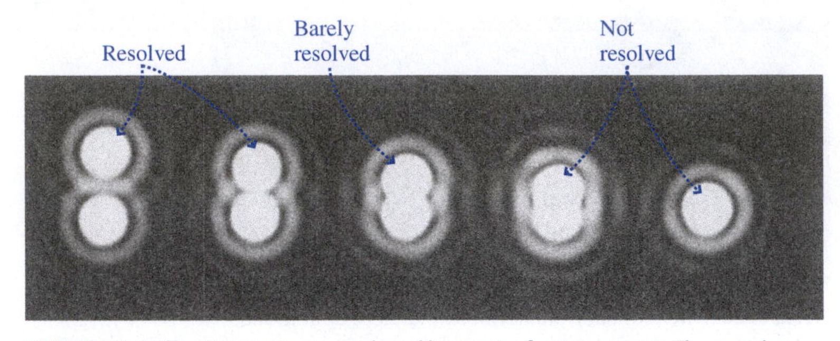

FIGURE 32.26 Diffraction patterns produced by a pair of point sources. The angular separation of the sources decreases until they can't be resolved.

Figure 32.24 shows that the angular separation between the diffraction peaks is equal to the angular separation between the sources themselves. Then the Rayleigh criterion is just met if the angular separation between the two sources is equal to the angular separation between a central peak and the first minimum. We found earlier that the first minimum in single-slit diffraction occurs at the angular position given by $\sin\theta = \lambda/a$, with a the slit width and with θ measured from the central peak. In most optical systems the wavelength is much less than the size of any apertures, so we can use the small-angle approximation $\sin\theta \simeq \theta$. Then the Rayleigh criterion—the condition that two sources be just resolvable—for single-slit diffraction becomes

$$\theta_{\text{min}} = \frac{\lambda}{a} \quad \text{(Rayleigh criterion, slit)} \tag{32.11a}$$

Most optical systems have circular apertures rather than slits. The diffraction pattern from such an aperture is a series of concentric rings (Fig. 32.27). Mathematical analysis shows that the angular position of the first ring and therefore the minimum resolvable source separation for a circular aperture is

$$\theta_{\text{min}} = \frac{1.22\lambda}{D} \quad \text{(Rayleigh criterion, circular aperture)} \tag{32.11b}$$

with D the aperture diameter.

FIGURE 32.27 3-D plot of intensity versus position in circular diffraction. The right-most image in Fig. 32.26 shows the corresponding diffraction pattern.

Equations 32.11 show that increasing the aperture size allows smaller angular differences to be resolved. In optical instrument design, that means larger mirrors or lenses. An alternative is to decrease the wavelength, which may or may not be an option depending on the source. In high-quality optical systems, diffraction is often the limiting factor preventing perfectly sharp image formation; such systems are said to be **diffraction limited**. For example, the diffraction limit sets a minimum size for objects resolvable with optical microscopes; that's why electron microscopes—with shorter effective wavelength—are used to image smaller biological structures. Large ground-based telescopes are an exception to the diffraction limit; their image quality is limited by atmospheric turbulence, although this can be reduced with adaptive optics. From their vantage points above Earth's atmosphere, space telescopes such as Hubble are truly diffraction limited. And although they look down through the atmosphere, Earth-imaging satellites are generally diffraction limited as well because the turbulent lower atmosphere is far from the satellites' optical systems and close to what they're imaging, namely, the ground. This greatly reduces atmospheric distortion.

Astronomers circumvent the diffraction limit by combining data from several telescopes to produce, in effect, a single instrument with aperture equal to the telescope separation. Radio astronomers achieve exquisite resolution by combining telescopes on different continents; for optical astronomy the technique is limited to smaller separations. You can explore astronomical interferometry further in the Passage Problems.

GOT IT? 32.6 You're a biologist trying to resolve details of structures within a cell, but they look fuzzy even at the highest power of your microscope. Which of the following might help: (a) substituting an eyepiece with shorter focal length, as suggested by Equation 31.10; (b) putting a red filter over the white light source used to illuminate the microscope slide; or (c) putting a blue filter over the white light source?

EXAMPLE 32.5 The Diffraction Limit: Asteroid Alert

An asteroid 20×10^6 km away appears on a collision course with Earth. What's the minimum size for the asteroid that could be resolved with the 2.4-m-diameter diffraction-limited Hubble Space Telescope, using 550-nm reflected sunlight?

INTERPRET This is a problem about the diffraction limit with a circular aperture. We're after the minimum physical size for the asteroid at a given distance. We identify $D = 2.4$ m as the aperture size and $\lambda = 550$ nm as the wavelength of the light.

DEVELOP Equation 32.11b, $\theta_{\text{min}} = 1.22\lambda/D$, determines the diffraction limit, expressed as the minimum angular size that can be resolved. So our plan is to express the unknown physical size l in terms of angular size θ and then apply Equation 32.11b.

EVALUATE If we consider the opposite ends of the asteroid to be like the two peaks in Fig. 32.24, then the angular size for small θ is just l/L, where L is the distance to the asteroid. Then Equation 32.11b becomes

$$\frac{l}{L} = \frac{1.22\lambda}{D}$$

or $l = 1.22\lambda L/D = 5.6$ km using the numbers given.

ASSESS An object this size poses a grave danger, being comparable to the asteroid whose impact caused the extinction of the dinosaurs. If astronomers see only a fuzzy blur, then they'll have to wait until the asteroid is closer to resolve its physical size and assess the danger. ■

APPLICATION Movies on Disc: CD to DVD to Blu-ray

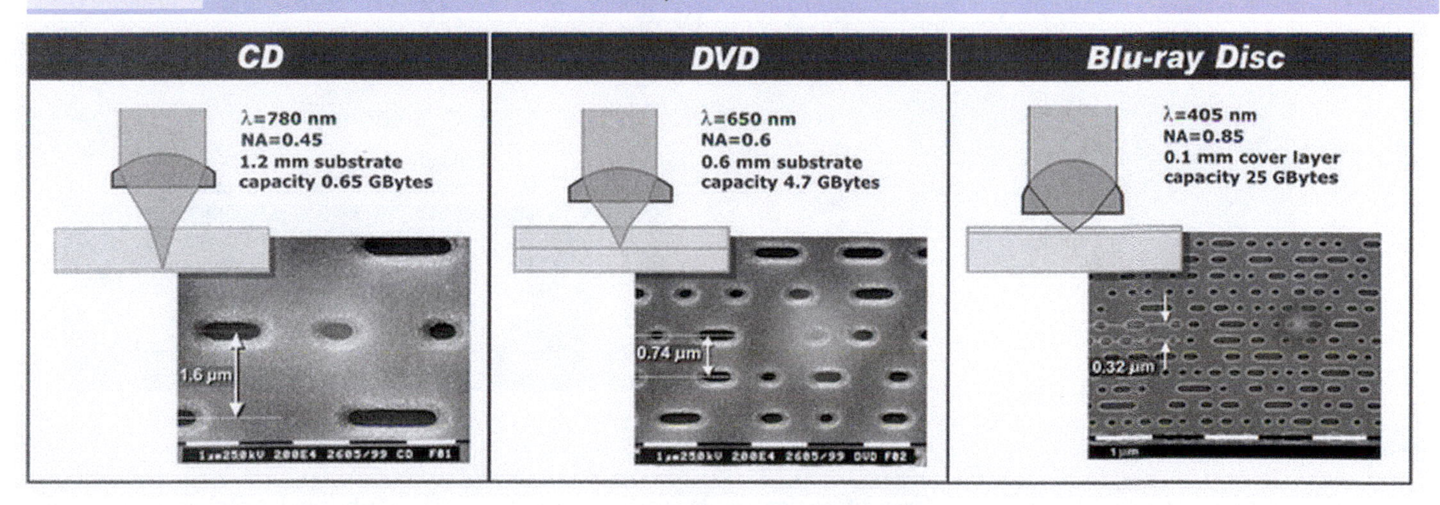

The Application earlier in this chapter described how a CD encodes information in pits 1.6 μm apart and as short as 0.83 μm. CDs are read with 780-nm infrared laser light. The pit size and spacing have to be large enough that diffraction effects at the laser's wavelength don't cause the CD player's optical system to confuse adjacent pits. The result is a maximum capacity of about 650 MB (megabytes; 1 byte is 8 binary bits, with a bit the fundamental piece of binary information represented by a digital 1 or 0). This translates into 74 minutes of audio.

CDs were developed in the 1980s, when inexpensive semiconductor lasers were available only in the infrared. By the 1990s inexpensive visible-light lasers became available, and that enabled the development of DVDs (for "digital video disc" or "digital versatile disc"). Read with 635-nm or 650-nm red light, DVDs can use smaller pit size and spacing because of the lower diffraction limit. That, coupled with a two-layer structure and more sophisticated data-compression schemes, gives standard DVDs a capacity of about 4.7 GB—enough for 2 or more hours of video, depending on quality.

Despite their large capacity, DVDs aren't adequate for today's high-definition TV (HDTV). But improvements in laser technology give us a 405-nm violet laser that enables high-definition video discs. Again, the shorter wavelength and hence lower diffraction limit, along with other improvements, allow much more information to fit on a disc. The resulting Blu-ray technology stores 25 GB on a single-layer disc. That corresponds to 4.5 hours of high-definition video, or 12 hours of standard video. The figure compares CDs, DVDs, and Blu-ray discs.

CHAPTER 32 SUMMARY

Big Idea

The big idea here is that—despite our use of the geometrical-optics approximation in the preceding chapters—light is indeed a wave and therefore exhibits the two related phenomena of **interference** and **diffraction**. These wave effects are important whenever light or any other wave interacts with objects whose size is comparable to or smaller than the wavelength.

Key Concepts and Equations

Constructive interference occurs when two waves combine in phase:

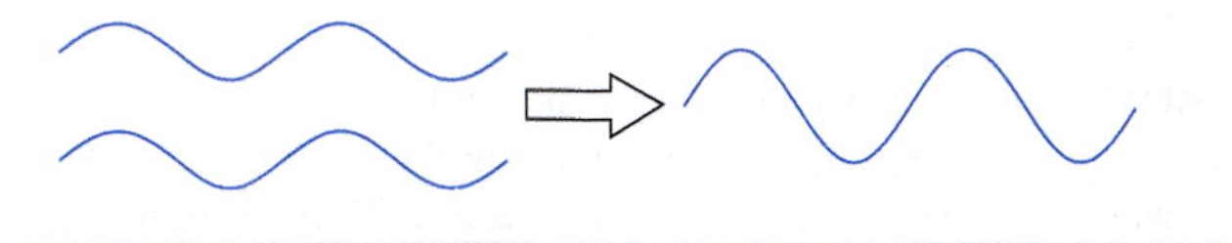

Destructive interference occurs when two waves combine 180° out of phase:

When light of wavelength λ passes through two or more narrow slits, the resulting interference shows maxima when

$$d\sin\theta = m\lambda$$

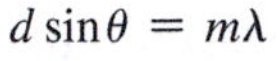

where m is an integer called the order. With multiple slits the maxima become stronger and narrower, but their position doesn't change.

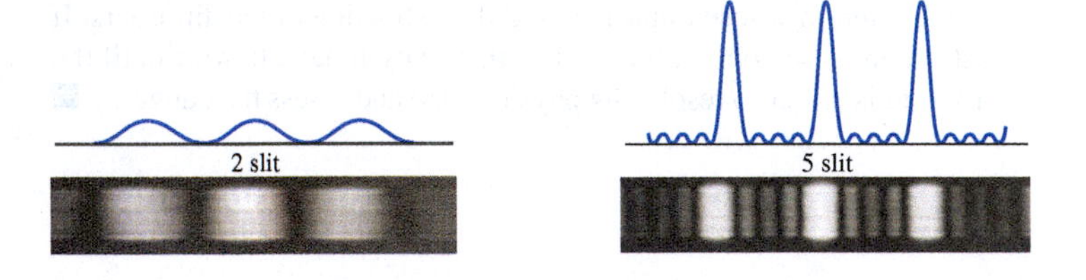

Diffraction occurs because, according to **Huygens' principle**, each point on a wavefront acts as a source of spherical waves, causing light to bend as it encounters sharp edges. Waves from different parts of a wavefront interfere to produce diffraction patterns.

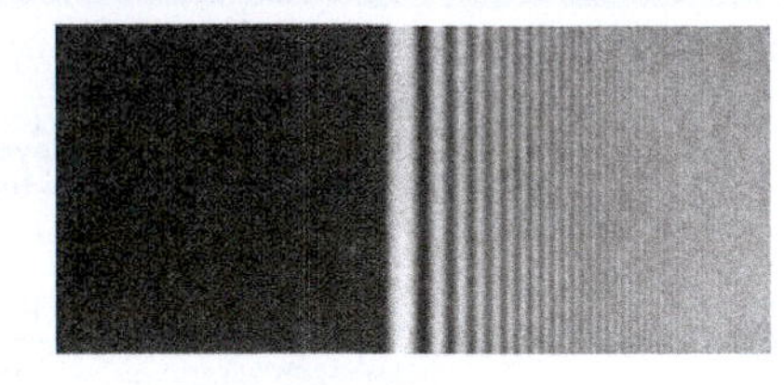

The **diffraction limit** is a fundamental restriction on our ability to image small or distant objects. For a circular aperture of diameter D, the **Rayleigh criterion** gives the minimum angular separation that can be resolved with light of wavelength λ:

$$\theta_{\min} = \frac{1.22\lambda}{D}$$

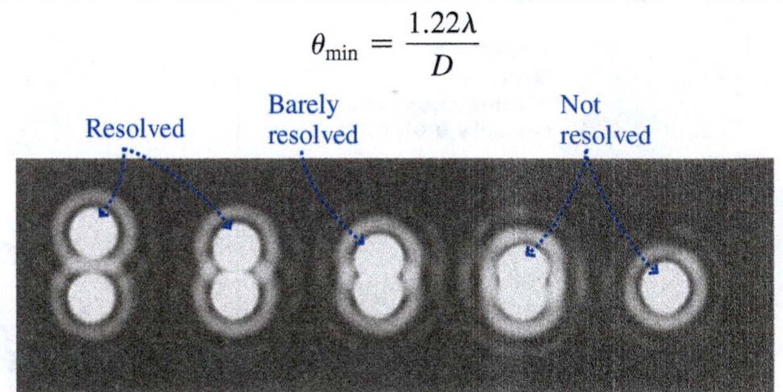

Applications

A **diffraction grating** consists of multiple slits or lines that result in constructive interference at different positions for different wavelengths. Diffraction gratings are used in spectrometers to disperse individual wavelengths. A grating's **resolving power**, the ratio of wavelength to the minimum resolvable difference in wavelengths, is given by

$$\frac{\lambda}{\Delta\lambda} = mN$$

where N is the number of lines in the grating and m is the order of the dispersion.

X-ray diffraction uses regularly positioned atoms as a grating, and is a powerful technique for analyzing crystal and molecular structure. Maximum intensity occurs when

$$2d\sin\theta = m\lambda$$

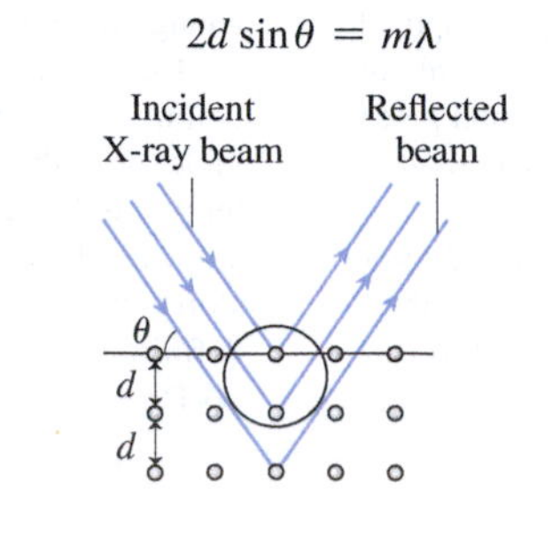

The **Michelson interferometer** splits light into two beams that travel on perpendicular paths. They recombine, and the resulting interference allows precision measurements.

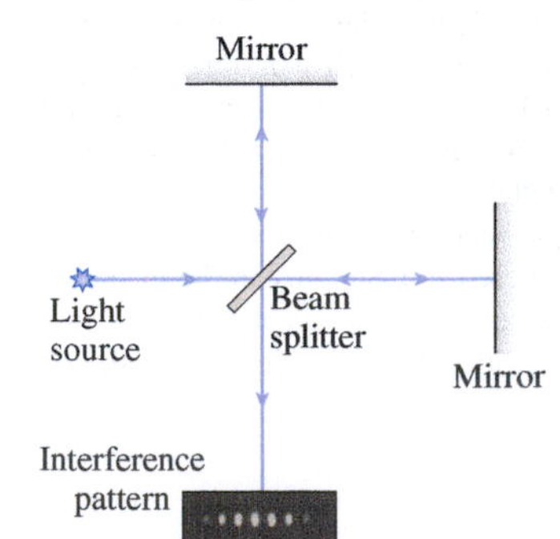

BIO *Biology and/or medicine-related problems* **DATA** *Data problems* **ENV** *Environmental problems* **CH** *Challenge problems* **COMP** *Computer problems*

For Thought and Discussion

1. A prism bends blue light more than red. Is the same true of a diffraction grating?
2. Why does an oil slick show colored bands?
3. Why does a soap bubble turn colorless just before it dries up and pops?
4. Why don't you see interference effects between the front and back of your eyeglasses?
5. You can hear around corners, but you can't see around corners. Why?
6. In deriving the intensity in double-slit interference, why can't you simply add the intensities from the two slits?
7. The primary maxima in multiple-slit interference are in the same angular positions as those in double-slit interference. Why, then, do diffraction gratings have thousands of slits instead of just two?
8. When the Moon passes in front of a star, the starlight intensity fluctuates before going to zero instead of dropping abruptly. Explain.
9. Sketch roughly the diffraction pattern you would expect for light passing through a square hole a few wavelengths wide.

Exercises and Problems

Exercises

Section 32.2 Double-Slit Interference

10. A double-slit system is used to measure the wavelength of light. The system has slit spacing $d = 15\,\mu\text{m}$ and slit-to-screen distance $L = 2.2\,\text{m}$. If the $m = 1$ maximum in the interference pattern occurs 7.1 cm from screen center, what's the wavelength?
11. A double-slit experiment with $d = 0.025\,\text{mm}$ and $L = 75\,\text{cm}$ uses 550-nm light. Find the spacing between adjacent bright fringes.
12. A double-slit experiment has slit spacing 0.12 mm. (a) What should be the slit-to-screen distance L if the bright fringes are to be 5.0 mm apart when the slits are illuminated with 633-nm laser light? (b) What will be the fringe spacing with 480-nm light?
13. The interference pattern from two slits separated by 0.37 mm has bright fringes with angular spacing 0.065°. Find the light's wavelength.
14. The 546-nm green line of gaseous mercury falls on a double-slit apparatus. If the fifth dark fringe is at 0.113° from the centerline, what's the slit separation?

Section 32.3 Multiple-Slit Interference and Diffraction Gratings

15. In a five-slit system, how many minima lie between the zeroth-order and first-order maxima?
16. In a three-slit system, the first minimum occurs at angular position 5°. Where is the next maximum?
17. A five-slit system with 7.5-μm slit spacing is illuminated with 633-nm light. Find the angular positions of (a) the first two maxima and (b) the third and sixth minima.
18. Green light at 520 nm is diffracted by a grating with 3000 lines/cm. Through what angle is the light diffracted in (a) first and (b) fifth order?
19. Light is incident normally on a grating with 10,000 lines/cm. Find the maximum order in which (a) 450-nm and (b) 650-nm light will be visible.
20. Find the second-order angular separation of the two wavelengths in Example 32.2.

Section 32.4 Interferometry

21. Find the minimum thickness of a soap film ($n = 1.333$) in which 550-nm light will undergo constructive interference.
22. Light of unknown wavelength shines on a precisely machined glass wedge with refractive index 1.52. The closest point to the apex of the wedge where reflection is enhanced occurs where the wedge is 98 nm thick. Find the wavelength.
23. Monochromatic light shines on a glass wedge with refractive index 1.65, and enhanced reflection occurs where the wedge is 450 nm thick. Find all possible values for the wavelength in the visible range.
24. White light shines on a 75.0-nm-thick sliver of fluorite ($n = 1.43$). What wavelength is most strongly reflected?
25. For the soap film described in Conceptual Example 32.1's "Making the Connection," what portion of the film will appear dark when it's illuminated with white light?

Section 32.5 Huygens' Principle and Diffraction

26. For what ratio of slit width to wavelength will the first minima of a single-slit diffraction pattern occur at $\pm 90°$?
27. Light with wavelength 633 nm is incident on a 2.50-μm-wide slit. Find the angular width of the central peak in the diffraction pattern, taken as the angular separation between the first minima.
28. You're inside a metal building that blocks radio waves, but you're trying to make a call with your cell phone, which broadcasts at a frequency of 950 MHz. Down the hall from you is a narrow window measuring 35 cm wide. What's the horizontal angular width of the beam (i.e., the angle between the first minima) from your phone as it emerges from the window?
29. Find the intensity as a fraction of the central peak intensity for the second secondary maximum in single-slit diffraction, assuming the peak lies midway between the second and third minima.

Section 32.6 The Diffraction Limit

30. Find the minimum angular separation resolvable with 633-nm laser light passing through a circular aperture of diameter 2.1 cm.
31. Find the minimum telescope aperture that could resolve an object with angular diameter 0.35 arcsecond, observed at 520-nm wavelength. (*Note:* 1 arcsec = 1/3600°.)
32. What's the longest wavelength of light you could use to resolve a structure with angular diameter 0.44 mrad, using a microscope with a 1.2-mm-diameter objective lens?
33. **BIO** In bright light, the human eye's pupil diameter is about 2 mm. If diffraction were the limiting factor, what's the eye's minimum angular resolution under these conditions, assuming 550-nm light?

Problems

34. Find the angular position of the second-order bright fringe in a double-slit system whose slit spacing is 1.5 μm for (a) red light at 640 nm, (b) yellow light at 580 nm, and (c) violet light at 410 nm.

35. A double-slit experiment has slit spacing 0.035 mm, slit-to-screen distance 1.5 m, and wavelength 490 nm. What's the phase difference between two waves arriving at a point 0.56 cm from the center line of the screen?
36. For a double-slit system with slit spacing 0.0525 mm and wavelength 633 nm, at what angular position is the path difference a quarter wavelength?
37. A screen 1.0 m wide is 2.0 m from a pair of slits illuminated by 633-nm laser light, with the screen's center on the centerline of the slits. Find the highest-order bright fringe that will appear on the screen if the slit spacing is (a) 0.10 mm and (b) 10 μm.
38. A tube of glowing gas emits light at 550 nm and 400 nm. In a double-slit apparatus, what's the lowest-order 550-nm bright fringe that will fall on a 400-nm dark fringe, and what are the fringes' corresponding orders?
39. On the screen of a multiple-slit system, the interference pattern shows bright maxima separated by 0.86° and seven minima between each bright maximum. (a) How many slits are there? (b) What's the slit separation if the incident light has wavelength 656.3 nm?
40. You're designing a spectrometer whose specifications call for a minimum of 5° separation between the red hydrogen-α line at 656 nm and the yellow sodium line at 589 nm when the two are observed in third order with a grating spectrometer. Available gratings have 2500 lines/cm, 3500 lines/cm, or 4500 lines/cm. What's the coarsest grating you can use?
41. For visible light with wavelengths from 400 nm to 700 nm, show that the first-order spectrum is the only one that doesn't overlap with the next higher order.
42. Find the total number of lines in a 2.5-cm-wide diffraction grating whose third-order spectrum puts the 656-nm hydrogen-α spectral line 37° from the central maximum.
43. What order is necessary to resolve 647.98-nm and 648.07-nm spectral lines using a 4500-line grating?
44. A thin film of toluene ($n = 1.49$) floats on water. Find the minimum film thickness if the most strongly reflected light has wavelength 460 nm.
45. NASA asks you to assess the feasibility of a single-mirror space-based optical telescope that could resolve an Earth-size planet 5 light-years away. What do you conclude?
46. In the second-order spectrum from a diffraction grating, yellow light at 588 nm overlaps violet light (wavelength range 390 nm–450 nm) diffracted in a different order. What's the exact wavelength of the violet light, and what's the order of its diffraction?
47. X-ray diffraction in potassium chloride (KCl) results in a first-order maximum when 97-pm-wavelength X rays graze the crystal plane at 8.5°. Find the spacing between crystal planes.
48. As a soap bubble with $n = 1.333$ evaporates and thins, reflected colors gradually disappear. What are (a) the bubble thickness just as the last vestige of color vanishes and (b) the last color seen?
49. An oil film with refractive index 1.25 floats on water. The film thickness varies from 0.80 μm to 2.1 μm. If 630-nm light is incident normally on the film, at how many locations will it undergo enhanced reflection?
50. **DATA** The table below lists the angular positions of the bright fringes that result when monochromatic laser light shines through a diffraction grating, as a function of order m. The spacing between lines of the grating is $d = 3.2\,\mu\text{m}$. Determine a quantity that, when plotted against m, should give a straight line. Plot the data, determine a best-fit line, and use the result to find the wavelength of the light.

Order, m	0	1	2	3	4	5
Angular position	0.0°	9.2°	22°	30°	48°	64°

51. Two perfectly flat glass plates are separated at one end by a sheet of paper 0.065 mm thick. 550-nm light illuminates the plates from above, as shown in Fig. 32.28. How many bright bands appear to an observer looking down on the plates?

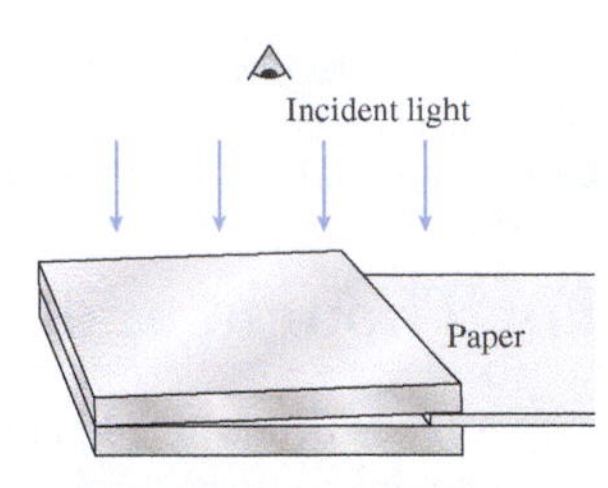

FIGURE 32.28 Problems 51, 52, and 64

52. An air wedge like that of Fig. 32.28 shows N bright bands when illuminated from above. Find an expression for the number of bands if the air is replaced by a liquid of refractive index n different from that of the glass.
53. A Michelson interferometer uses light from glowing hydrogen at 486.1 nm. As you move one mirror, 530 bright fringes pass a fixed point in the viewer. How far did the mirror move?
54. Find the wavelength of light used in a Michelson interferometer if 550 bright fringes go by a fixed point when the mirror moves 0.150 mm.
55. One arm of a Michelson interferometer is 42.5 cm long and is enclosed in a box that can be evacuated. The box initially contains air, which is gradually pumped out. In the process, 388 bright fringes pass a point in the viewer. If the interferometer uses light with wavelength 641.6 nm, what's the air's refractive index?
56. Your stereo is in a dead spot caused by direct reception from an FM radio station at 89.5 MHz interfering with the signal reflecting off a wall behind you. How much farther from the wall should you move so that the interference is fully constructive?
57. A proposed "star wars" antimissile laser is to focus infrared light with 2.8-μm wavelength to a 50-cm-diameter spot on a missile 2500 km distant. Find the minimum diameter for a concave mirror that can achieve this spot size, given the diffraction limit. (Your answer suggests one of many technical difficulties faced by antimissile defense systems.)
58. Suppose one of the 10-m-diameter Keck Telescopes in Hawaii is trained on San Francisco, 3400 km away. Assuming 550-nm light, and ignoring atmospheric distortion, would it be possible to read (a) newspaper headlines or (b) a billboard sign at this distance? (c) Repeat for the case of the Keck optical interferometer, formed from the two 10-m Keck Telescopes and several smaller ones, with a 50-m effective aperture.
59. A camera has an $f/1.4$ lens, meaning the ratio of focal length to lens diameter is 1.4. Find the smallest spot diameter (i.e., the diameter of the first diffraction minimum) to which this lens can focus parallel light with 580-nm wavelength.
60. The CIA wants your help identifying individual terrorists in a photo of a training camp taken from a spy satellite at 100-km altitude. You ask for details of the optical system used, but they're classified. However, they do tell you that the optics are diffraction limited and can resolve facial features as small as 5 cm. Assuming a typical optical wavelength of 550 nm, what do you conclude about the size of the mirror or lens in the satellite camera?
61. **BIO** While driving at night, your eyes' irises dilate to 3.1-mm diameter. If your vision were diffraction limited, what would be the greatest distance at which you could see as distinct the two headlights of an oncoming car, spaced 1.5 m apart? Take $\lambda = 550\,\text{nm}$.
62. Under the best conditions, atmospheric turbulence limits ground-based telescopes' resolution to about 1 arcsecond (1/3600 of a degree). For what apertures is this limitation more severe than that of diffraction at 550 nm? (Your answer shows why large ground-based telescopes don't generally produce better images than small ones, although they do gather more light.)

63. **BIO** You're a biologist studying rhinoviruses, which cause the common cold. These are among the smallest viruses, some 50 nm in diameter, and you can't image them with your optical microscope using visible light (average wavelength about 560 nm). A sales rep tries to sell you an expensive microscope using 280-nm ultraviolet light, saying it will resolve structures half the size that's resolvable with your current microscope. Is the rep correct? Will the new microscope resolve your rhinoviruses?

64. An air wedge like that of Fig. 32.28 displays 10,003 bright bands when illuminated from above. If the region between the plates is then evacuated, the number of bands drops to 10,000. Find the refractive index of the air.

65. A thin-walled glass tube of length L containing a gas of unknown refractive index is placed in one arm of a Michelson interferometer using light of wavelength λ. The tube is then evacuated. During the process, m bright fringes pass a fixed point in the viewer. Find an expression for the refractive index of the gas.

66. Light is incident on a diffraction grating at angle α to the normal. Show that the condition for maximum light intensity becomes $d(\sin\theta \pm \sin\alpha) = m\lambda$.

67. An arrangement known as Lloyd's mirror (Fig. 32.29) allows interference between direct and reflected beams from the same source. Find an expression for the separation of bright fringes on the screen, given the distances d and D and the light's wavelength λ.

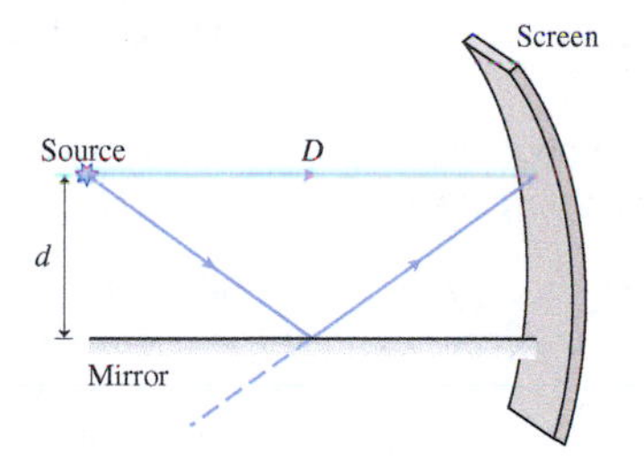

FIGURE 32.29 Lloyd's mirror (Problem 67)

68. **CH** The intensity of the single-slit diffraction pattern can be calculated by summing the amplitudes of infinitely many field amplitudes corresponding to waves from every infinitesimal part of the slit. (a) Referring to Fig. 32.20, show that the field from an element of slit width dy, a distance y from the bottom edge of the slit, is $dE = (E_p dy/a)\sin(\omega t + \phi(y))$, where $\phi(y) = (2\pi y/\lambda)\sin\theta$. (b) Integrate dE over the entire slit from $y = 0$ to $y = a$, and use trig identities from Appendix A, to find the total amplitude and from there show that the average intensity is given by Equation 32.10.

69. You're on an international panel charged with allocating "real estate" for communications satellites in geostationary orbit. You're to estimate how many satellites could fit in geostationary orbit without receivers on the ground picking up multiple signals. Satellites broadcast at 12 GHz and receiver dishes are 45 cm in diameter. Use the Raleigh criterion to estimate the number of satellites allowed in geostationary orbit if each receiver dish is to "see" just one satellite. Make the simplifying assumption that all dishes are located on the equator, and that each is pointed at the nearest satellite. (*Hint:* Consult Example 8.3).

70. **ENV** You're investigating an oil spill for your state environmental protection agency. There's a thin film of oil on water, and you know its refractive index is $n_{\text{oil}} = 1.38$. You shine white light vertically on the oil, and use a spectrometer to determine that the most strongly reflected wavelength is 580 nm. Assuming first-order thin-film interference, what do you report for the thickness of the oil slick?

Passage Problems

Even the nearest stars are so distant that a single diffraction-limited telescope capable of imaging Earth-size planets orbiting them would be hopelessly large (see Problem 45). Astronomers get around this limitation using *interferometry* to combine data from several telescopes, producing an instrument that acts like a single telescope with aperture equal to the distance between the individual telescopes (Fig. 32.30). The technological challenge is to combine the signals with their relative phase intact; for this reason, interferometry has been used successfully for decades in radio astronomy but only recently with optical telescopes.

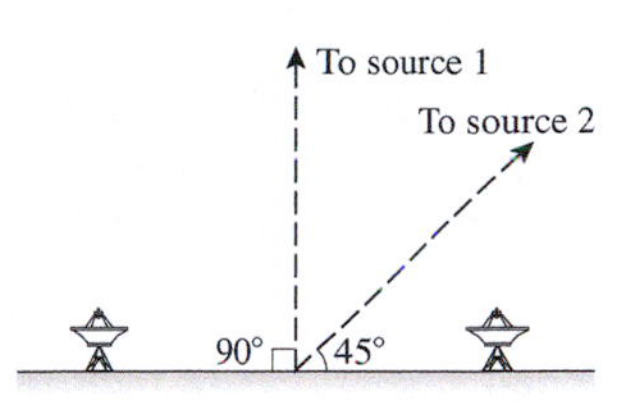

FIGURE 32.30 A two-dish interferometer used for radio astronomy (Passage Problems 71–74). Dashed lines show directions to sources in Problems 73 and 74.

71. If the separation of two telescopes comprising an interferometer is doubled, the angular separation between two sources just barely resolvable by the interferometer will
    a. not change.
    b. decrease by a factor of $1/\sqrt{2}$.
    c. halve.
    d. double.

72. If the separation of two telescopes comprising an interferometer is doubled, the instrument's light-collecting power will
    a. not change.
    b. increase by a factor of $\sqrt{2}$.
    c. double.
    d. quadruple.

73. If a point source is located directly above a two-telescope interferometer, on the perpendicular bisector of the line joining the telescopes (source 1 in Fig. 32.30), electromagnetic waves reaching the two will be
    a. in phase.
    b. out of phase by 45°.
    c. out of phase by 90°.
    d. you can't tell without further information

74. If a point source is located on a line at 45° to the line joining the two telescopes (source 2 in Fig. 32.30), electromagnetic waves reaching the two will be
    a. in phase.
    b. out of phase by 45°.
    c. out of phase by 90°.
    d. you can't tell without further information

Answers to Chapter Questions

Answer to Chapter Opening Question

The phenomenon of diffraction ultimately limits our ability to form images with any optical systems, including spy satellites and other telescopes, as well as microscopes.

Answers to GOT IT? Questions

32.1 (a)
32.2 (a)
32.3 (1) (c); (2) (a); (3) (b)
32.4 (b)
32.5 The lower frequency sound waves have longer wavelengths, and therefore they diffract more at both doors. So the bass gets into your room, but the higher frequencies don't.
32.6 (c)

PART FIVE SUMMARY Optics

Optics is the study of light and its behavior. **Geometrical optics** is an approximation that holds when the objects with which light interacts are much larger than its wavelength. In this case, light generally travels in straight lines called **rays**. **Physical optics**, in contrast, treats light explicitly as a wave. Physical optics explains a host of phenomena that ultimately involve the interference of light waves.

When light rays are incident on an interface between two materials, they generally undergo **reflection** and, for transparent materials, **refraction**. The angles of incidence and reflection are equal. **Snell's law** relates the angles of incidence and refraction:

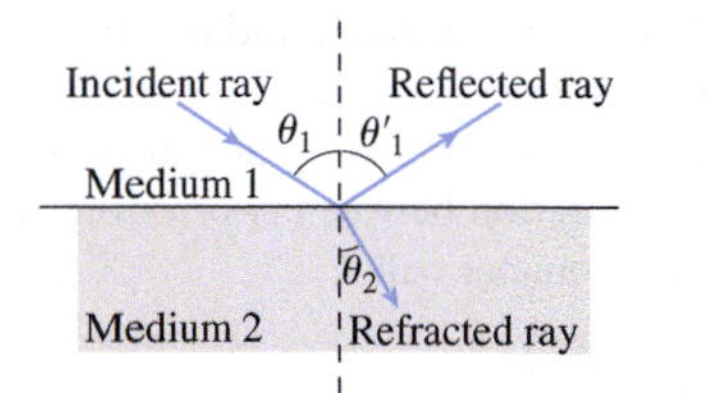

Reflection: $\theta'_1 = \theta_1$

Refraction (Snell's law):

$n_1 \sin\theta_1 = n_2 \sin\theta_2$

The **index of refraction** n relates light's speed in a medium to its speed c in vacuum:

$$n = \frac{c}{v}$$

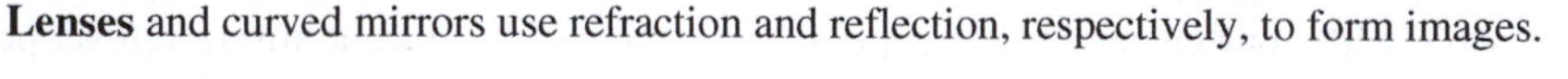

Lenses and curved mirrors use refraction and reflection, respectively, to form images.

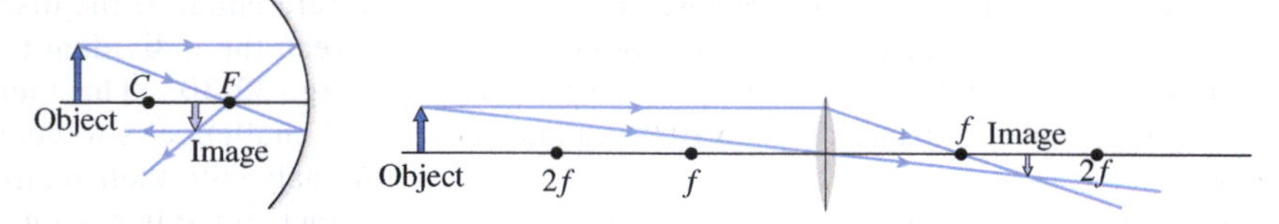

In both cases the object distance s, image distance s', and **focal length** f are related by $\frac{1}{s} + \frac{1}{s'} = \frac{1}{f}$.

With **real images**, shown in both figures above, light actually comes from the image. With **virtual images**, shown in both figures below, light only *appears* to come from the image:

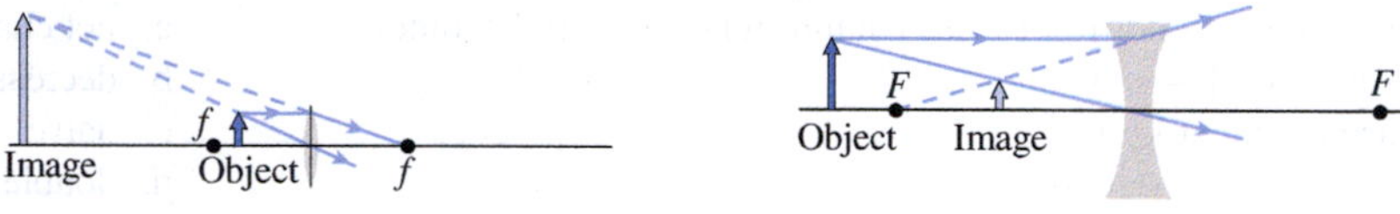

The wave nature of light becomes important when light interacts with objects comparable in size to its wavelength, or when light travels different paths and recombines to produce interference.

Destructive interference occurs when waves are out of step by an odd-integer multiple of a half-wavelength.

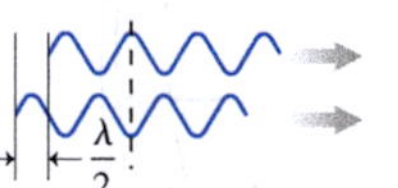

Constructive interference occurs when waves are out of step by an integer multiple of the wavelength.

A system consisting of two narrow slits produces a pattern of **interference fringes** resulting from alternating regions of constructive and destructive interference:

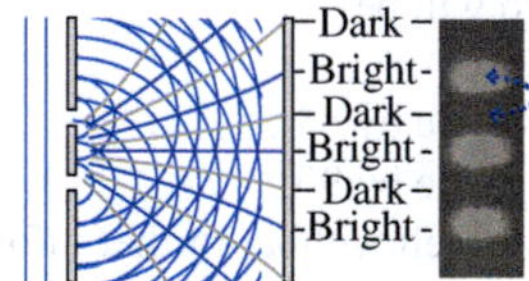

Photo showing alternating bright and dark fringes.

$d\sin\theta = m\lambda$ (bright fringes)

$d\sin\theta = (m + \frac{1}{2})\lambda$ (dark fringes)

With multiple slits the bright fringes become narrower and brighter:

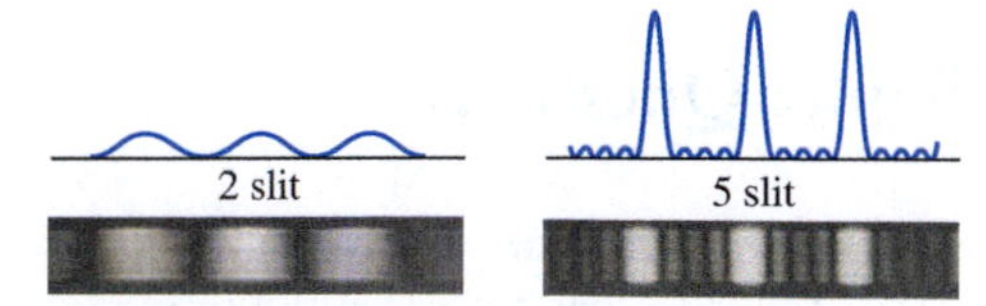

A multiple-slit system constitutes a **diffraction grating** and is used to separate different wavelengths in **spectroscopy**.

Huygens' principle explains the propagation of waves by stating that each part of a wavefront acts as a source of circular waves that spread out and interfere to propagate the wave. When light passes through small apertures or by sharp edges, Huygens' principle shows that the light **diffracts**, bending and producing interference fringes as waves from different points interfere.

Diffraction fundamentally limits our ability to resolve small objects or to see closely spaced but distant objects as separate.

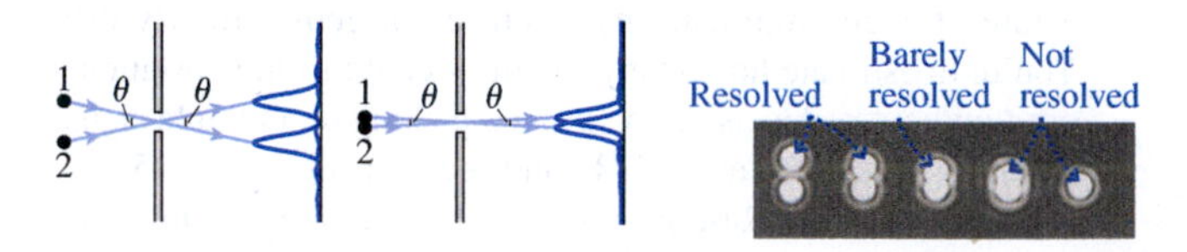

For a circular aperture of diameter d (such as a telescope with d being its mirror diameter), the **diffraction limit** gives the smallest angular separation that can be resolved at a given wavelength λ:

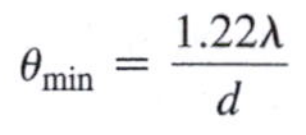

$$\theta_{\min} = \frac{1.22\lambda}{d}$$

Part Five Challenge Problem

A double-slit system consists of two slits each of width a, with separation d between the slit centers ($d > a$). Light of intensity S_0 and wavelength λ is incident on the system, perpendicular to the plane containing the slits. Find an expression for the outgoing intensity as a function of angular position θ, taking into account both the slit width and the separation. Plot your result for the case $d = 4a$, and compare with Fig. 32.22.

PART SIX

OVERVIEW

Modern Physics

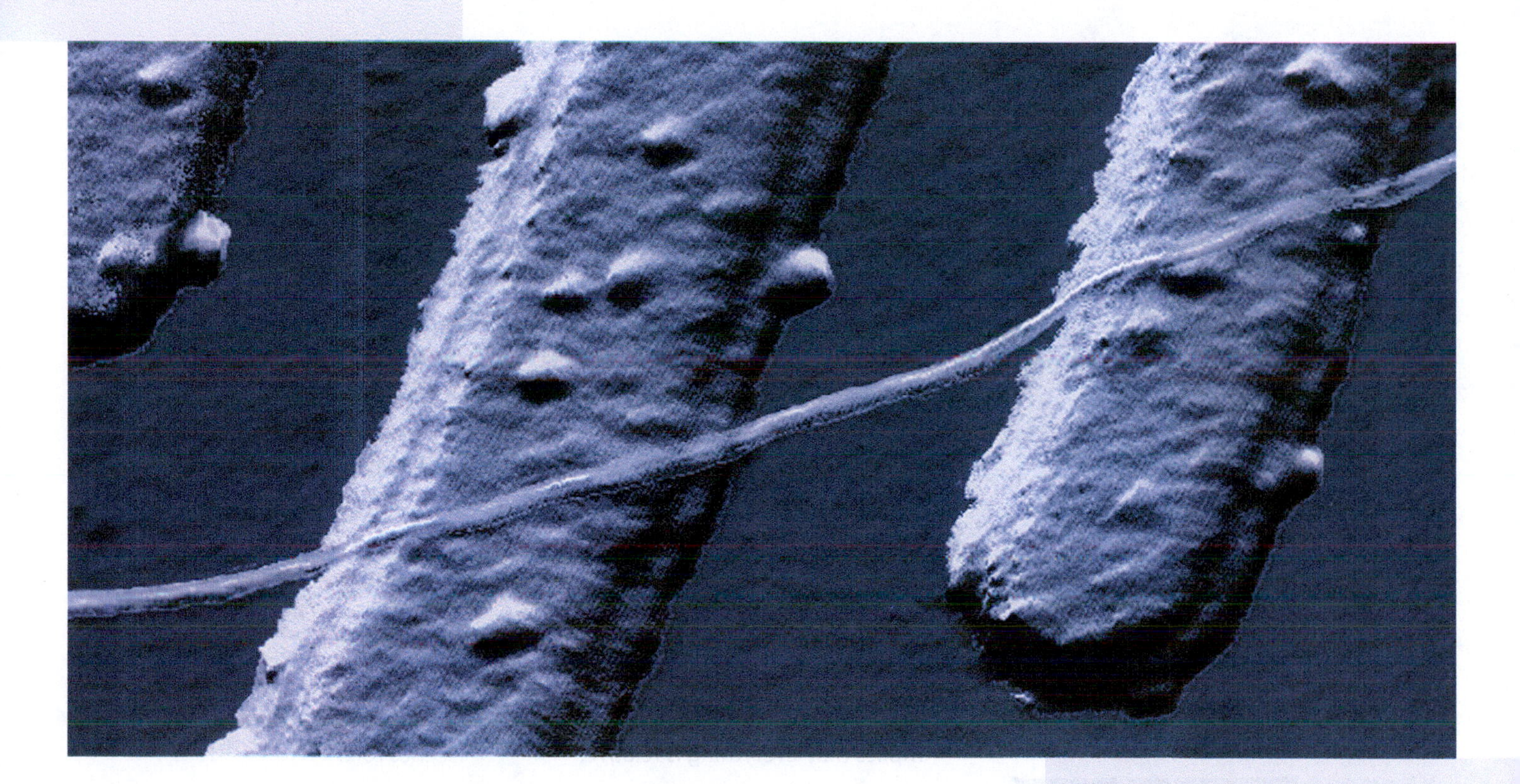

What are the fundamental particles of matter? What holds them together to make protons, neutrons, nuclei, atoms, molecules, and solids? Is nature fundamentally predictable, or does uncertainty rule in the microscopic world? At the other extreme, how big is the universe? How did it begin, and how will it end? All these are questions for relativity and quantum physics—collectively called "modern physics" because they were developed after the turn of the 20th century. In Part 6, we give a brief account of Einstein's theory of relativity, followed by a glimpse at quantum physics and its applications. We end with an overview of the latest developments in fundamental physics, from the nature of elementary particles to surprising new findings about the origin and composition of the universe.

The world's smallest electrical wire, a carbon nanotube, is only 10 atoms across. In this image made with an atomic force microscope, the nanotube wire runs across a backdrop of platinum electrodes. Our understanding of physics at the atomic and molecular level lets us construct an increasing variety of practical nanoscale devices.

Relativity

What You Know

- You have an intuitive feel for space and time.
- You understand concepts of motion in Newtonian physics, especially velocity.
- You know about relative motion and frames of reference.
- You understand the concept of inertial reference frames.
- You know that velocities add in a straightforward manner in the context of Newtonian physics.

What You're Learning

- You'll see how 19th-century understandings of physics led to quandaries that Einstein resolved with his theory of relativity.
- You'll learn the principle of special relativity and how it underlies all of special relativity's startling implications.
- You'll see how measures of space, time, and the simultaneity and even order of events depend on one's reference frame.
- You'll learn about the Lorentz transformations and will use them to find space and time coordinates of events in different reference frames.
- You'll learn to discredit the common misconception that Einstein's work implies that "everything is relative."
- You'll understand the relation between mass, energy, and momentum in relativity.
- You'll see how relativity connects electricity and magnetism.
- You'll have a brief, qualitative introduction to general relativity.

How You'll Use It

- With your understanding of relativity and its implications for space and time, you'll gain a richer appreciation of the universe you inhabit.
- If you go into in a field like high-energy particle physics, you'll need special relativity in your everyday work.
- If you study high-energy astrophysics or cosmology, you'll find general relativity at the basis of your work.
- If you go into medicine and order PET scans for your patients, you'll be using relativistic physics.
- Whenever you use your smartphone's location services, or navigate with GPS, relativity plays a behind-the-scenes role.

Behind Einstein's theories is a profoundly simple principle that can be stated in a single sentence. What is it?

Maxwell's electromagnetic theory was a crowning achievement of 19th-century physics, providing an understanding of the nature of light and enabling a host of practical technologies. At the same time, Maxwell's electromagnetism led to baffling questions and contradictions that shook the roots of physical understanding and even of common sense.

The theory of relativity resolved these contradictions. It radically altered our fundamental understanding of the physical world, and its influence spilled over into all areas of human thought. Relativity stands as a monument to human intellect and imagination, and it reveals a universe far richer than earlier physicists could conceive of. We'll approach relativity historically, building on your understanding of electromagnetism. That way you'll get a sense of the questions that electromagnetism posed to 19th-century physicists and of how Einstein's answer to these questions was at once profoundly bold and sweepingly simple.

33.1 Speed *c* Relative to What?

Maxwell's equations show that electromagnetic waves in vacuum propagate with speed c. Speed c relative to what? In Chapter 14 we found the speed of waves on a string; the speed in that case was relative to the string. Similarly, the 340-m/s speed of sound in air is relative to the air. If you move through the air, the speed of sound *relative to you* won't be the same. In these and other cases of mechanical waves, the wave speed is the speed relative to the medium in which the wave is a disturbance.

The Ether Concept

What about light? Nineteenth-century physicists, their worldview built on the highly successful mechanical paradigm of Newtonian physics, supposed that light waves were like mechanical waves and required a medium. They postulated a tenuous substance called the **ether** that permeated the entire universe, allowing light from distant stars to reach us. Electric and magnetic fields were stresses in the ether, and electromagnetic waves were propagating disturbances moving through the ether at speed c.

The ether had to have some unusual properties. It must offer no resistance to material bodies, or the planets would lose energy and spiral into the Sun. It must be very stiff, to account for the high speed of light. And it had to be more jelly-like than fluid, because fluids can't support *transverse* waves, and, as you saw in Chapter 29, electromagnetic waves are transverse. These properties make ether a rather improbable substance, but to 19th-century physicists the ether was essential in understanding electromagnetic waves.

The speed of light c follows from Maxwell's equations. But in the 19th-century view, light has speed c only for an observer at rest with respect to the ether. Therefore, Maxwell's equations could be correct only in the ether frame of reference. This put electromagnetism in a rather different position from mechanics. In mechanics, the concept of absolute motion is meaningless. You can eat your dinner, toss a ball, or do any mechanical experiment as well on an airplane moving steadily at 1000 km/h as you can when the plane is at rest on the ground. This is the principle of **Galilean relativity**, which states that the laws of mechanics are valid in all inertial reference frames—that is, frames of reference in uniform motion (Section 4.2). But the laws of electromagnetism seemed valid only in the ether's reference frame because only in this frame was the prediction of electromagnetic waves moving at speed c correct.

So for 19th-century physicists, the laws for one branch of physics (mechanics) seemed to work in all inertial frames, while those of another branch (electromagnetism) could not. Despite this dichotomy, physicists had great faith in mechanical models and in the ether concept, for without the ether the question "speed c relative to what?" seemed impossible to answer. Thus the late 19th century saw a flurry of experiments to detect the ether. Ultimately they failed, paving the way for the new worldview of relativity that, in Einstein's own words, "arose from necessity, from serious and deep contradictions in the old theory from which there seemed no escape."

33.2 Matter, Motion, and the Ether

It was natural for 19th-century physicists to ask about Earth's motion relative to the ether. If Earth is moving through the ether, then the speed of light should be different in different directions. On the other hand, Earth might be at rest relative to the ether. Because other planets, stars, and galaxies move with respect to Earth, it's hard to imagine that the ether is everywhere fixed with respect to Earth alone: This violates the Copernican view that Earth doesn't occupy a privileged spot in the universe. But maybe Earth drags with it the ether in its immediate vicinity. If this "ether drag" occurs, then the speed of light must be independent of direction, but if there's no ether drag, then the speed of light measured on Earth must depend on direction. Through observation and experiment, 19th-century physicists sought to resolve the question of Earth's motion through the ether.

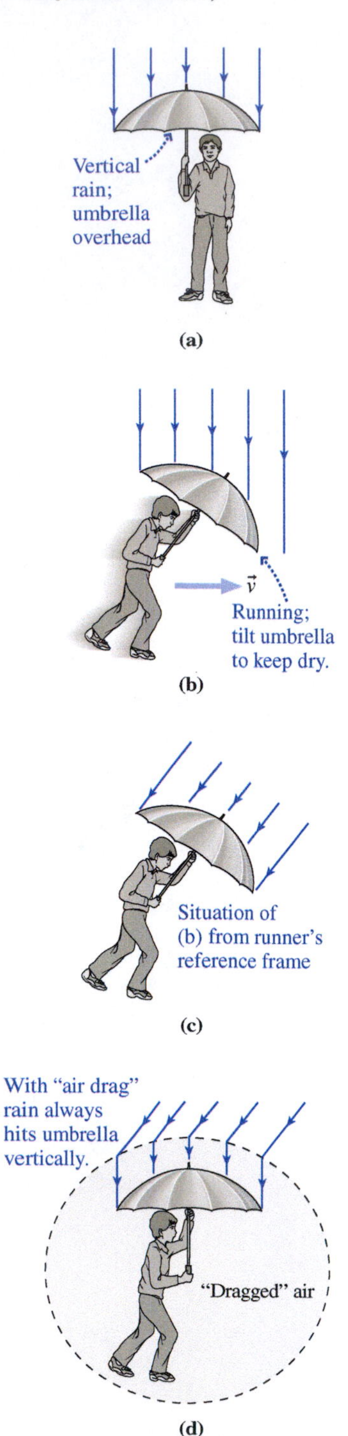

FIGURE 33.1 A rain/umbrella analogy for aberration of starlight.

Aberration of Starlight

Imagine standing in a rainstorm with rain falling vertically. To keep dry, you hold your umbrella with its shaft straight up, as shown in Fig. 33.1*a*. But if you run, as in Fig. 33.1*b*, you'll keep driest if you tilt your umbrella forward. Why? Because the direction of rainfall *relative to you* is at an angle, as shown in Fig. 33.1*c*. This assumes you don't drag with you a large volume of air. If such an "air drag" occurred, raindrops entering the region around you would be accelerated quickly in the horizontal direction by the air moving with you, so they would now fall vertically relative to you, as in Fig. 33.1*d*. No matter which way you ran, as long as you dragged air with you, you would point your umbrella vertically upward to stay dry.

This umbrella example is analogous to the observation of light from stars, with the rain being starlight and the umbrella a telescope. If Earth doesn't drag ether, then the direction from which starlight comes will depend on Earth's motion relative to the ether. But if "ether drag" occurs in analogy with Fig. 33.1*d*, then light from a particular star will always come from the same direction.

In fact we do observe a tiny change in the direction of starlight. As Earth swings around in its orbit, we must first point a telescope one way to see a particular star. Then, six months later, Earth's orbital motion is in exactly the opposite direction, and we must point the telescope in a slightly different direction. This phenomenon is called **aberration of starlight** and shows that *Earth does not drag the ether*.

The Michelson–Morley Experiment

If we reject the pre-Copernican notion that Earth alone is at rest relative to the ether, then aberration of starlight forces us to conclude that Earth moves through the ether. Furthermore, the relative velocity of the motion must change throughout the year as Earth orbits the Sun.

In 1881–1887, American scientists Albert A. Michelson and Edward W. Morley attempted to determine Earth's velocity relative to the ether. They used Michelson's interferometer (Fig. 33.2), whose operation we described in Chapter 32. Recall that the interferometer produces a pattern of interference fringes that shifts if the round-trip travel time for light on one of its two perpendicular arms changes. The interference pattern reflects, among other things, possible differences in travel times that arise from differences in the speed of light in different directions—differences that should result from Earth's motion through the ether. Rotating the apparatus through 90° would interchange the directions of the arms and should therefore shift the interference pattern.

Now suppose Earth moves at speed v relative to the ether. Then to an observer on Earth, there's an "ether wind" blowing past Earth. Suppose the Michelson–Morley apparatus is oriented with one light path parallel to the wind and the other perpendicular. Consider a light beam moving the distance L at right angles to the wind. The beam must be aimed slightly upwind so that it will actually move perpendicular to the wind. The

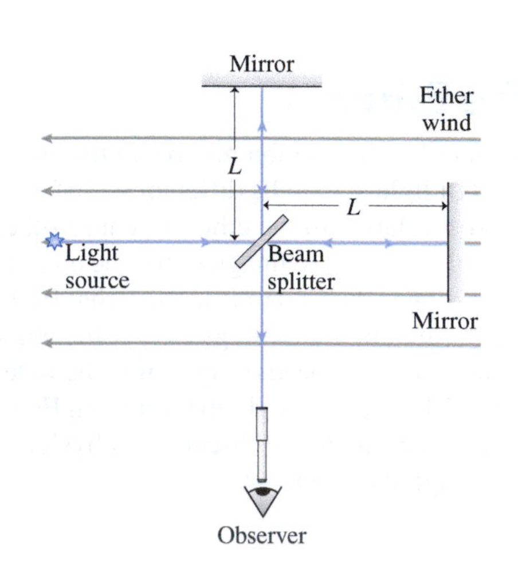

FIGURE 33.2 Simplified diagram of the Michelson–Morley experiment. Ether wind should result in a longer travel time for light on the horizontal arm.

light moves in this direction at speed c relative to the ether, but the ether wind sweeps it back so its path in the Michelson–Morley apparatus is at right angles to the wind. From Fig. 33.3, we see that its speed relative to the apparatus is $u = \sqrt{c^2 - v^2}$, so the round-trip travel time is

$$t_{\text{perpendicular}} = \frac{2L}{u} = \frac{2L}{\sqrt{c^2 - v^2}} \quad (33.1)$$

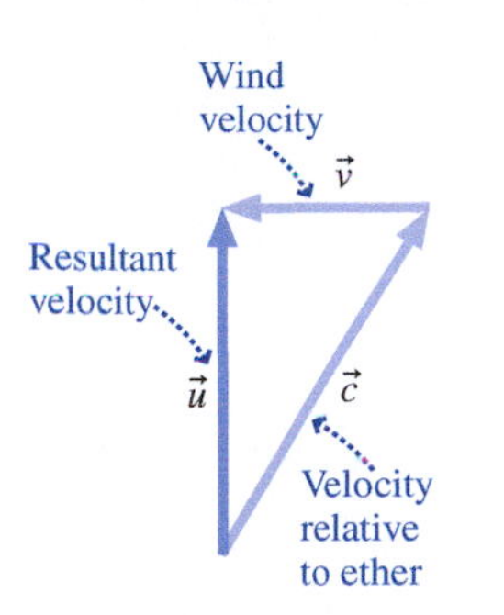

FIGURE 33.3 Vector diagram for light moving at right angles to an ether wind.

Light sent a distance L "upstream"—against the ether wind—travels at speed c relative to the ether but at speed $c - v$ relative to Earth. It therefore takes time $t_{\text{upstream}} = L/(c - v)$. Returning, the light moves at $c + v$ relative to Earth, so $t_{\text{downstream}} = L/(c + v)$. The round-trip time parallel to the ether wind is then

$$t_{\text{parallel}} = \frac{L}{c - v} + \frac{L}{c + v} = \frac{2cL}{c^2 - v^2} \quad (33.2)$$

The two round-trip travel times differ, with the trip parallel to the ether wind always taking longer (see Exercises 13, 14, and 27). Light on the parallel trip slows when it moves against the ether wind, then speeds up when it moves with the wind. But slowing always dominates because the light spends more time moving against the wind than with it.

The Michelson–Morley experiment of 1887 was sensitive enough to detect differences in the speed of light an order of magnitude smaller than Earth's orbital speed. The experiment was repeated with the apparatus oriented in different directions, and at different times throughout the year, and the same simple but striking result always emerged: There was never any difference in the travel times for the two light beams. In terms of the ether concept, the Michelson–Morley experiment showed that *Earth does not move relative to the ether.*

GOT IT? 33.1 Which sentence best describes the Michelson–Morley experiment? (a) The M–M experiment attempted to detect differences in the speed of light propagating in different directions relative to the ether wind. (b) The M–M experiment measured values for the speed of light in two mutually perpendicular directions. (c) The M–M experiment confirmed the aberration of starlight.

A Contradiction in Physics

Aberration of starlight shows that Earth doesn't drag ether with it. Earth must therefore move relative to the ether. But the Michelson–Morley experiment shows that it doesn't. This contradiction is a deep one, rooted in the fundamental laws of electromagnetism and in the analogy between mechanical waves and electromagnetic waves. The contradiction arises directly in trying to answer the simple question: With respect to what does light move at speed c?

Physicists at the end of the 19th century made ingenious attempts to resolve the dilemma of light and the ether, but their explanations either were inconsistent with experiment or lacked sound conceptual bases.

FIGURE 33.4 In 1905, when he formulated special relativity, Einstein was a 26-year-old father.

33.3 Special Relativity

In 1905, at the age of 26, Albert Einstein (Fig. 33.4) presented his **special theory of relativity**, which resolved the dilemma but altered the very foundation of physical thought. Einstein declared simply that the ether is a fiction. But then with respect to what does light move at speed c? With respect, Einstein declared, to anyone who cares to observe it. This statement is at once simple, radical, and conservative. Simple, because its meaning is clear and obvious. Anyone who measures the speed of light in vacuum will get the value $c = 3.0 \times 10^8$ m/s. Radical, because it alters our commonsense notions of space and time. Conservative, because it asserts for electromagnetism what had long been true in mechanics: that the laws of physics don't depend on the motion of the observer. Einstein summarized his new ideas in the **principle of relativity**, which is expressed in this simple sentence:

The laws of physics are the same in all inertial reference frames.

Recall that inertial frames are unaccelerated—that is, frames in which the laws of *mechanics* were already valid. Einstein's statement encompasses *all* laws of physics, including mechanics and electromagnetism. The prediction that electromagnetic waves move at speed c must, then, be a universal prediction that holds in *all* inertial reference frames. The *special* theory of relativity is special because it's valid only for the special case of inertial frames. Later we'll discuss the *general* theory of relativity, which removes this restriction.

Einstein's relativity explains the result of the Michelson–Morley experiment: No matter what Earth's speed is relative to anything, an observer on Earth should measure the same speed for light in all directions. But at the same time, relativity flagrantly violates our commonsense notions of space and time. We'll see just how in Sections 33.4–33.6.

33.4 Space and Time in Relativity

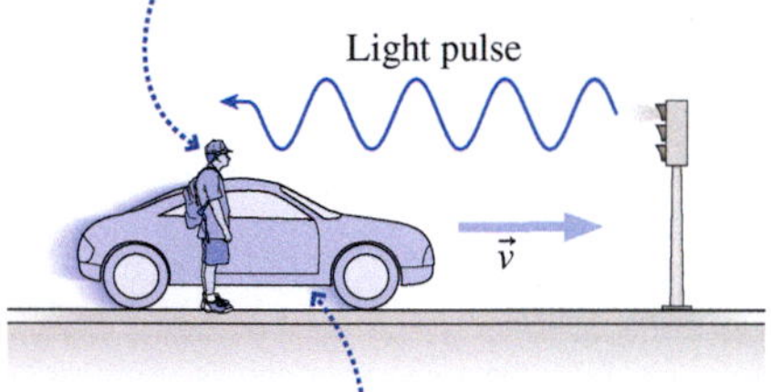

A standing pedestrian measures the speed of light as c.

Light pulse

$\vec{v}$

Although the car moves toward the light source at speed v, the driver also measures the speed of light as c.

FIGURE 33.5 Both driver and pedestrian measure the same speed c for light, even though they're in relative motion.

A pedestrian stands by the roadside as a car drives by (Fig. 33.5). Driver and pedestrian each measure the speed of light from a blinking traffic signal. The theory of relativity says they'll both get the same value, $c = 3.0 \times 10^8$ m/s, even though the car is moving toward the light source. How is that possible? Consider how each observer might make the measurement. Each has a meter stick and an accurate stopwatch. Suppose a light flash passes the front end of each meter stick just as they coincide. Each observer measures the time for the flash to traverse the 1-m stick and calculates *speed* = *distance*/*time*. They get the same answer—even though common sense suggests that the light should pass the far end of the "moving" meter stick sooner.

How can this be? Maybe the car's motion affects the driver's stopwatch, making it inaccurate. But no; this suggestion violates relativity's assertion that *all* uniformly moving reference frames are equally good vantage points for doing physics. There can't be anything special about the "moving" reference frame; in fact, it's meaningless to talk about the car as "moving" and the pedestrian "at rest." This is the point of relativity: The concept of absolute motion is meaningless.

The only way out, consistent with relativity, is to let go of absolute space and time. Our two observers' instruments are measuring different quantities that depend on their reference frame—namely, the distance and travel time for the light flash. Those quantities differ in just the right way to make the speed of light come out the same for both observers. This is certainly not what common sense tells us about space and time. But in relativity it's the laws of physics, not measures of space and time, that must be the same for all. Keep in mind the principle of relativity, and you'll see how the rest of special relativity's remarkable consequences follow.

Time Dilation

Figure 33.6a shows a "light box," consisting of a box of length L with a light source at one end and a mirror at the other. A light flash leaves the source, reflects off the mirror, and returns to the source. We want the time between two events: the emission of the flash and its return to the source. An **event** is an occurrence specified by giving its position and its time.

For concreteness, we'll imagine that the light box is in a spaceship moving past Earth at a uniform velocity. But don't think there's something special about space or spaceships. The whole point of relativity is that all inertial frames are equivalent places for doing physics,

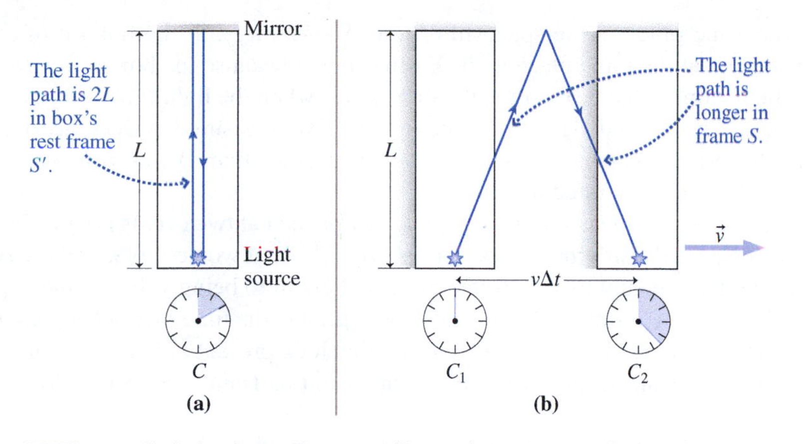

FIGURE 33.6 A "light box" to explore time dilation shown (a) in a reference frame S' at rest with respect to the box and (b) in a frame S where the box is moving to the right.

and our spaceship is just one inertial reference frame. We'll call that frame S'. There's also an accurate clock, C, in the spaceship, and C reads zero just as the light flash is emitted.

In Fig. 33.6*a* we consider the light-box experiment viewed in the spaceship's reference frame. Since the light box is at rest in this frame, the light travels a round-trip distance $2L$ from source to mirror and back, giving a round-trip travel time of $\Delta t' = 2L/c$. This is the time read on the spaceship's clock C.

Now consider the situation as viewed in Earth's reference frame, which we'll call S. In this frame, spaceship and light clock are moving to the right with speed v, as shown in Fig. 33.6*b*. Suppose there are two clocks in Earth's frame, positioned so the light box passes clock C_1 just as the flash is emitted, and passes C_2 just as the flash returns to the source. The clocks are synchronized, and C_1 reads zero just as the light flash is emitted. We want to know C_2's reading at the instant the light box passes it and the flash returns to its source; that will be the time, Δt, between the flash emission and return as measured in Earth's frame S.

Figure 33.6*b* shows that the box moves to the right a distance $v\,\Delta t$ in the time between emission and return of the light flash. Meanwhile the light takes a diagonal path up to the mirror of the moving box and then back down. The path length is twice the diagonal from source to mirror or, by the Pythagorean theorem, $2\sqrt{L^2 + (v\,\Delta t/2)^2}$. The time for light to go this distance is the distance divided by the speed of light, or $\Delta t = 2\sqrt{L^2 + (v\,\Delta t/2)^2}/c$. We explicitly used relativity here, assuming the speed of light remained c in Earth's frame. If we didn't believe relativity, we would have vectorially added light's velocity $\vec{c}$ and the box's velocity $\vec{v}$. But that would make the spaceship's frame the only one in which the speed of light was c—in violation of the relativity principle.

The unknown Δt appears on both sides of our expression; multiplying through by c and squaring gives

$$c^2(\Delta t)^2 = 4L^2 + v^2(\Delta t)^2$$

We then solve for $(\Delta t)^2$ to get

$$(\Delta t)^2 = \frac{4L^2}{c^2 - v^2} = \frac{4L^2}{c^2}\left(\frac{1}{1 - v^2/c^2}\right)$$

Taking the square root of both sides, and noting that $2L/c$ is just the time $\Delta t'$ measured in the frame S' at rest with respect to the box, we have $\Delta t = \Delta t'/\sqrt{1 - v^2/c^2}$ or

$$\Delta t' = \Delta t\sqrt{1 - v^2/c^2} \quad \text{(time dilation)} \qquad (33.3)$$

Equation 33.3 describes the phenomenon of **time dilation**, in which the time between two events is shortest in a frame of reference in which the two events occur at the same place. In our example, the events are the emission and return of the light flash, and they

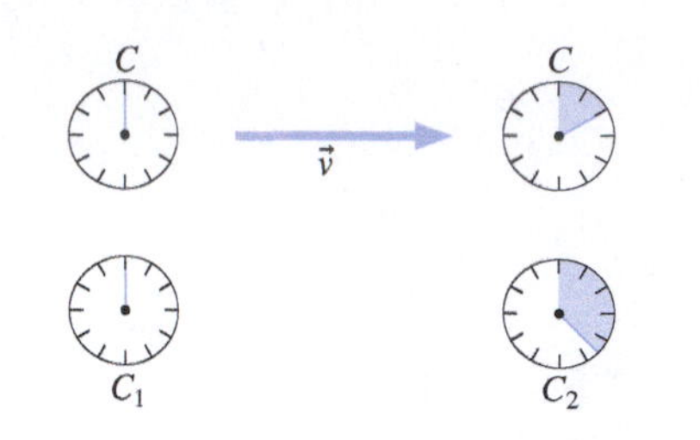

FIGURE 33.7 Clock C moves between clocks C_1 and C_2, which are at rest relative to each other and synchronized in their rest frame. C measures a shorter elapsed time.

occur at the same place in the spaceship frame S'—namely, at the bottom of the box. They don't occur at the same place in the Earth frame S because the box moves relative to Earth, so the bottom of the box isn't at the same place when the light flash returns as it was when the flash was emitted. Thus $\Delta t'$—the time interval measured in the spaceship frame S', where the events occur at the same place—is shorter than Δt, as you can see from Equation 33.3 and as illustrated in Fig. 33.7.

The shorter time $\Delta t'$ measured by a single clock present at two events is called a **proper time**. Here "proper" doesn't mean "correct" or "right"—that would violate relativity, since a time measurement in any inertial frame has equal claim to being valid. Rather, "proper" is used in the sense of "proprietary" in that proper time is the time that *belongs* to this one particular clock. In Earth frame S there's no single clock present at both the emission and return of the light flash, so the time measurement in this frame doesn't "belong" to any one clock.

Time dilation is sometimes characterized by saying "moving clocks run slow," but this statement violates the spirit of relativity because it suggests that some frames are "really" moving and others aren't. The whole point of relativity is that all inertial frames are equally good for describing physical reality, so none can claim to be "at rest" while others are "moving." What the statement "moving clocks run slow" is trying to convey is what we've just seen: The time interval between two events is shortest in a reference frame where the two occur at the same place. There's no significance whatever to our putting the light box in a "moving" spaceship and comparing ship time with Earth time. We could equally well have put the light box and its clock on Earth, and two separate clocks in the ship. Then Earth with its light box would be moving past the ship at speed v—so the Earth clock would measure the time $\Delta t'$ in Equation 33.3, and the two clocks in the ship frame would measure the longer time Δt. (That may sound like a contradiction, but it can't be because *there's nothing special about any inertial frame, including Earth's*. We'll return to this point shortly.)

We used a light box to illustrate time dilation. But time dilation isn't something that happens only when we use light to determine time intervals. It's something that happens to *time itself*. Take away the light box in Fig. 33.6, and the clocks will show the same discrepancy. Don't look for a physical mechanism that slows things down. All manifestations of time—the oscillations of the quartz crystal in a digital watch, the swing of a pendulum clock, the period of vibration of atoms in an atomic clock, biological rhythms, and human lifetimes—are affected in the same way.

EXAMPLE 33.1 Calculating Time Dilation: Star Trek

A spaceship leaves Earth on a one-way star trip that earthbound observers judge will take 25 years. If the ship travels at $0.95c$ relative to Earth, how long will the trip take as judged by observers in the ship?

INTERPRET This is a problem about time dilation. Since the events of departure from Earth and arrival at the star occur at the same place in the ship's reference frame, we identify the ship time as the proper time $\Delta t'$ and the Earth time as Δt in our discussion of time dilation.

DEVELOP Equation 33.3, $\Delta t' = \Delta t\sqrt{1 - v^2/c^2}$, relates the two times. We're given the Earth time Δt, so we can use this equation to find $\Delta t'$.

EVALUATE With $v = 0.95c$, $v/c = 0.95$ and the quantity v^2/c^2 in Equation 33.3 becomes 0.95^2. Then

$$\Delta t' = \Delta t\sqrt{1 - v^2/c^2} = (25\,\text{y})\sqrt{1 - 0.95^2} = 7.8\,\text{y}$$

ASSESS This time is considerably shorter than 25 years, confirming our statement that the time between events is shortest as measured in a reference frame where the events occur at the same place. We'll soon explore what happens if the ship turns around and returns to Earth. ■

You don't notice time dilation in everyday life because the factor v^2/c^2 is so small for even your fastest motion relative to Earth. Even in a jet airplane, the time difference amounts to a few milliseconds per century. This illustrates the important point that any results from relativity should agree with commonsense Newtonian physics when relative velocities are small compared with the speed of light. Since your intuition and common sense are built on experience at low relative velocities, it's not surprising that effects at high relative velocities seem counter to your common sense.

APPLICATION Mountains and Muons

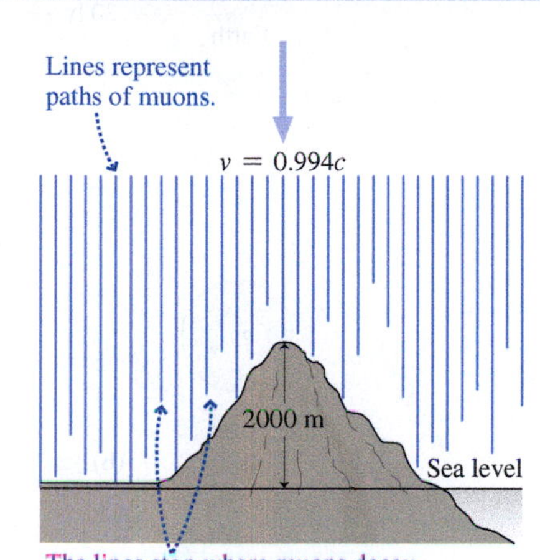

The lines stop where muons decay.

Time dilation is obvious in experiments with subatomic particles moving, relative to us, at speeds near c. In a classic experiment, the "clocks" are the lifetimes of particles called muons, which are created by the interaction of cosmic rays with Earth's upper atmosphere and subsequently decay. The experiment consists in counting the number of muons incident each hour on the top of Mt. Washington in New Hampshire, about 2000 m above sea level. The measurement is then repeated at sea level. The figure shows the situation in Earth's reference frame.

Using a detector that records only those muons moving at about $0.994c$ at the mountaintop altitude, the experiment shows that an average of about 560 muons with this speed are incident on the mountaintop each hour. If the mountain weren't there, the muons would travel from the mountaintop altitude to sea level in a time given by

$$\Delta t = \frac{2000\,\text{m}}{(0.994)(3.0\times 10^8\,\text{m/s})} = 6.7\,\mu\text{s}$$

The muon's decay rate is such that one should expect only about 25 of the original 560 muons to remain after a 6.7-μs interval, so that's approximately the number we might expect to detect each hour at sea level. However, that 6.7-μs interval is measured in Earth's reference frame—not the muons'. In the muons' frame, time dilation should reduce that interval to

$$\Delta t' = (6.7\,\mu\text{s})\sqrt{1 - 0.994^2} = 0.73\,\mu\text{s}$$

The muons' decay is determined by *their* measure of time, and their decay rate is such that we should expect 414 muons to survive for 0.73 μs.

So what happens? Observers count just over 400 muons per hour at sea level. This is no subtle effect. The difference between 25 and 414 is dramatic. At $0.994c$, the nonrelativistic description is hopelessly inadequate, and time dilation is obvious.

The Twin Paradox

Time dilation lets us travel into the future! The so-called twin paradox shows how. One twin boards a spaceship for a journey to a distant star; the other stays on Earth. There are clocks at both Earth and star, like clocks C_1 and C_2 in Fig. 33.7. There's a clock on the spaceship, like C in Fig. 33.7. Ship clock and Earth clock read the same time as the ship departs (Fig. 33.8*a*), but when it arrives at the star, time dilation means the ship clock reads less time than the star clock (Fig. 33.8*b*). Now the ship turns around and returns home. Again, the situation is just like Fig. 33.7 or our light box of Fig. 33.6, so less time elapses on the ship (Fig. 33.8*c*). The traveling twin arrives home younger than her earthbound brother! Depending on how far and how fast she goes, that age difference can be arbitrarily large. But this is a one-way trip to the future. If the traveling twin doesn't like what she finds in the future, there's no going back.

EXAMPLE 33.2 Time Dilation: The Twin Paradox

Earth and a star are 20 light years (ly) apart, measured in a frame at rest with respect to Earth and star. Twin A boards a spaceship, travels at $0.80c$ to the star, and then returns immediately to Earth at $0.80c$. Determine the round-trip travel times in Earth and ship reference frames.

INTERPRET This problem involves time dilation, here applied to the two separate legs of the round-trip journey. We identify the ship clock as C in Fig. 33.7, and the ship time as $\Delta t'$.

(continued)

DEVELOP Earth–star time for the one-way journey follows from *distance* = *speed* × *time*, so that's how we'll find Δt. Then we can apply Equation 33.3, $\Delta t' = \Delta t\sqrt{1 - v^2/c^2}$, to get the ship time $\Delta t'$. We'll double both to get the round-trip times.

EVALUATE At $0.80c$, the time to go 20 ly is $\Delta t = (20\text{ ly})/(0.80\text{ ly/y}) = 25\text{ y}$. Equation 33.3 then gives

$$\Delta t' = (25\text{y})\sqrt{1 - 0.80^2} = 15\text{y}$$

Doubling these values gives round-trip times of 50 years and 30 years for the Earth and ship, respectively.

ASSESS The traveling twin returns younger, by 20 years! We've marked the various times on the clocks in Fig. 33.8.

✓TIP Years, Light Years, and the Speed of Light

A light year (ly) is the distance light travels in one year. By definition, therefore, the speed of light is 1 ly/y. It's often easiest in relativity to work in units where the speed of light is 1, whether those units be light years and years, light seconds and seconds, or whatever.

■

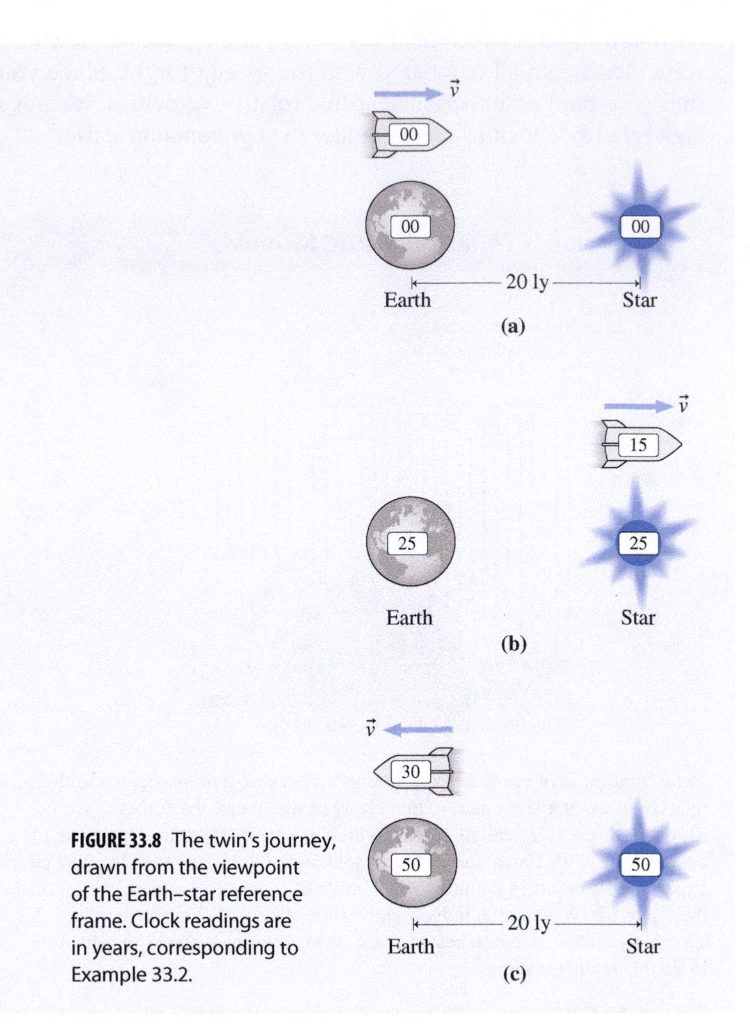

FIGURE 33.8 The twin's journey, drawn from the viewpoint of the Earth–star reference frame. Clock readings are in years, corresponding to Example 33.2.

Here's the seeming paradox: From the ship's viewpoint, it looks like Earth recedes, turns around, and returns. So why isn't the Earth twin younger? The answer lies in what's *special* about special relativity—namely, it applies only to reference frames in uniform motion. The traveling twin accelerates when turning around, so briefly she's in a noninertial reference frame. Although relativity precludes us from saying that one twin is moving and the other is not, we can say that one twin's motion *changes* and the other's doesn't. This is obvious to the traveling twin, who experiences forces associated with the turnaround. The earthbound twin, in contrast, doesn't feel anything unusual when the ship turns around. During the journey the ship occupies two different inertial frames, separated by the turnaround acceleration, while Earth remains in a single inertial frame. It's that asymmetry that resolves the paradox. The traveling twin really is younger!

What if the traveling twin didn't turn around? Then the situation would be symmetric, and each could argue that the other's clocks "run slow." But unless they get together again at the same place, there's no unambiguous way to compare their clock readings or ages. And they can't get together without at least one of them accelerating. As we'll soon see, clocks that are synchronized in one reference frame aren't synchronized in another—and that takes the seeming contradiction out of two observers each finding that the other's clocks "run slow."

GOT IT? 33.2 Triplets A and B board spaceships and head away from Earth in opposite directions, each traveling the same distance at the same speed before returning to Earth. Triplet C remains on Earth. Which expression describes the triplets' relative ages once they're reunited on Earth? (a) $A < B < C$; (b) $A = B > C$; (c) $A = B < C$; (d) $A = B = C$

Length Contraction

In Example 33.2 the 20-ly distance, 25-y time, and $0.8c$ speed are related through the expression $\Delta x = v\,\Delta t$, where Δx and Δt are measured in the Earth frame. But relativity tells us that this relation must hold in *all* inertial reference frames, so it's also valid in the spaceship frame except during the turnaround. From the ship's viewpoint, Earth and star are moving at $v = 0.80c$, and it takes $\Delta t' = 15\,\text{y}$ for the Earth–star system to pass the ship. Then $\Delta x' = v\,\Delta t' = (0.80\,\text{ly/y})(15\,\text{y}) = 12\,\text{ly}$ is the Earth–star distance as measured in the ship frame. Thus measures of space as well as time depend on one's reference frame. Equation 33.3 gives $\Delta t' = \Delta t\sqrt{1 - v^2/c^2}$, so we can write generally that

$$\Delta x' = v\,\Delta t' = v\,\Delta t\sqrt{1 - v^2/c^2}$$

or

$$\Delta x' = \Delta x\sqrt{1 - v^2/c^2} \quad \text{(length contraction)} \qquad (33.4)$$

for the distance between two objects measured in a reference frame in which they move with speed v. Here Δx is the distance in a reference frame at rest with respect to the two objects. Since $\sqrt{1 - v^2/c^2}$ is less than 1 for $v > 0$, Equation 33.4 shows that $\Delta x > \Delta x'$. Therefore, the distance is greatest in this so-called rest frame. The two points in question could be the ends of a single object, in which case Δx is the object's length. This phenomenon of **length contraction** is also called **Lorentz–Fitzgerald contraction**, after Dutch physicist H. A. Lorentz and Irish physicist George F. Fitzgerald, who independently proposed it as an ad hoc way of explaining the Michelson–Morley experiment. Only with Einstein's work did the contraction acquire a solid conceptual basis.

Length contraction shows that an object is longest in its own rest frame and is shorter to observers for whom it's moving. As with time dilation, don't go looking for a physical mechanism that squashes "moving" objects. That presupposes an absolute space with respect to which contraction occurs. Rather, it's space itself that's different for different observers. Accepting relativity means giving up notions of absolute space and time; length contraction and time dilation are necessary consequences.

FIGURE 33.9 The Stanford Linear Accelerator is 3.2 km (2 miles) long. But to electrons moving through it at $0.9999995c$, it's only 3.2 m long. Photo shows the accelerator passing under Interstate Highway 280 west of Palo Alto, California.

EXAMPLE 33.3 Length Contraction and Time Dilation: SLAC

At the Stanford Linear Accelerator Center (SLAC; see Fig. 33.9), subatomic particles are accelerated to high energies over a straight path whose length, in Earth's reference frame, is 3.2 km. For an electron traveling at $0.9999995c$, how long does the trip take as measured (a) in Earth's frame and (b) in the electrons' frame? (c) What's the length of the linear accelerator in the electrons' frame?

INTERPRET We're being asked about time dilation and length contraction. The Earth frame here is like the Earth frame of Example 33.2, with the ends of the accelerator replacing Earth and star, and $\Delta x = 3.2\,\text{km}$ in place of the 20-ly Earth–star separation. In (a) we're therefore being asked for Δt, in (b) for $\Delta t'$, and in (c) for $\Delta x'$.

DEVELOP As always $\Delta x = v\,\Delta t$ relates distance, time, and speed in a single reference frame. We're given Δx and v, so we'll first solve for Δt. Then we can use Equation 33.3, $\Delta t' = \Delta t\sqrt{1 - v^2/c^2}$, for $\Delta t'$, and Equation 33.4, $\Delta x' = \Delta x\sqrt{1 - v^2/c^2}$, for $\Delta x'$.

EVALUATE The electrons' speed is so close to c that it suffices to calculate the travel time with $v = c$. (a) We have $\Delta t = \Delta x/c = (3.2\,\text{km})/(3.0\times10^8\,\text{m/s}) = 11\,\mu\text{s}$. (b) Equation 33.3 gives $\Delta t' = \Delta t\sqrt{1 - v^2/c^2}$; with $v/c = 0.9999995$, the square root works out to be 10^{-3}, so $\Delta t' = 11\,\text{ns}$. (c) Equation 33.4 shows that the length shrinks by the same factor, to 3.2 m.

ASSESS In this case of extremely relativistic speed, the relativistic factor $\sqrt{1 - v^2/c^2}$ is tiny, and the effects of time dilation and length contraction are dramatic. Note that we could approximate v as c in finding Δt, but not in working with the relativistic factor, where even the slightest difference from c is crucial. As a check on our answer, note that $\Delta x' = v\,\Delta t'$, as required by the principle of relativity. ■

Equations 33.3 and 33.4 show that relativistic effects are significant only at high relative speeds, with v^2/c^2 comparable to 1. You've no experience of such speeds in your everyday life, so relativity seems counterintuitive. Had you grown up moving relative to your surroundings at speeds approaching c, the relativity of space and time would be as obvious as your commonsense notions seem now. For physicists working with high-energy particles or studying distant, rapidly moving galaxies, relativistic effects *are* obvious features of physical reality.

33.5 Simultaneity Is Relative

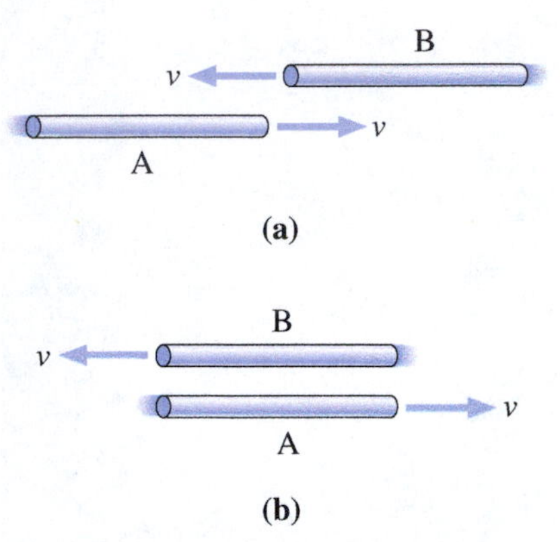

FIGURE 33.10 (a) In frame S, rods A and B have the same speed v and both are contracted by the same amount. (b) Their ends coincide at the same time.

One remarkable consequence of relativity is that simultaneity of events and sometimes even their time order depend on one's reference frame. Here we explore how this comes about.

Figure 33.10*a* shows two identical rods approaching each other in a reference frame S where they have the same speed v. Figure 33.10*b* shows that the right end of rod A passes the right end of rod B at the same instant that the left ends pass. Call the passing of the two right ends event E_1 and the passing of the left ends E_2. Events E_1 and E_2 are **simultaneous**, meaning they occur at the same time in S.

Now consider a reference frame S' in which rod A is at rest and is therefore longer than it was in frame S. Rod B, meanwhile, is moving toward rod A with a greater speed relative to S' than it had relative to S. Therefore, it's shorter than it was in S. Figure 33.11 shows that, as a result of their different lengths, the right ends of the two rods coincide before the left ends; in other words, event E_1 precedes E_2. Now look at the situation from a reference frame in which rod B is at rest, and you'll see that E_2 precedes E_1 (Fig. 33.12). So events that are simultaneous in one reference frame aren't simultaneous in another frame.

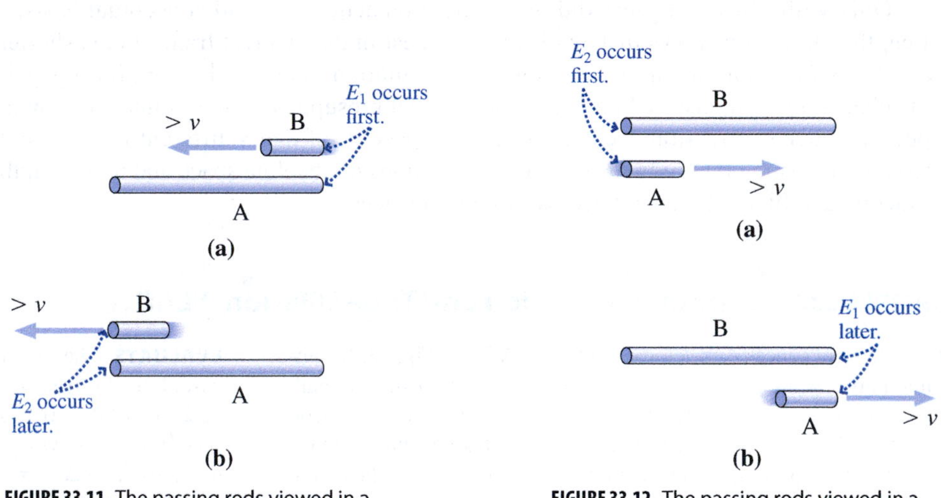

FIGURE 33.11 The passing rods viewed in a reference frame S' at rest with respect to rod A.

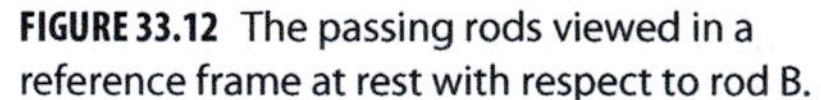

FIGURE 33.12 The passing rods viewed in a reference frame at rest with respect to rod B.

Isn't this just an illusion resulting from apparent length differences due to the rods' motion? Isn't the picture in frame S (Fig. 33.10) the "real" one? No! Relativity assures us that all inertial reference frames are equally valid for describing physical reality. The length differences and changes in the ordering of events aren't "apparent" and they aren't "illusions." They result from valid descriptions in different reference frames, and each has equal claim to reality. If you insist that one frame—say S—has more validity, then you're clinging to the 19th-century notion that there's one favored reference frame in which alone the laws of physics are valid.

But how can observers disagree about the *order* of events? After all, if one event causes another, we expect cause always to precede effect. But, as you'll soon see, the only events whose time order is different for different observers are those that are so far apart in space, and so close in time, that not even a light signal from one event could reach the location of the other event before it happened. There's no way for such events to influence each other, so they can't be causally related.

CONCEPTUAL EXAMPLE 33.1 **"Running Slow": A Contradiction?**

In Example 33.2, the outbound trip from Earth to star took 25 years in the Earth–star reference frame but only 15 years in the spaceship's frame. Thus, observers in the Earth–star frame can say that clocks on the ship "run slow." What do passengers on the ship say about clocks in the Earth–star frame?

EVALUATE During the outbound trip, the spaceship is in a perfectly good inertial reference frame. So the laws of physics are the same for the ship's passengers as they are for earthbound observers, and they can make exactly the same argument: They see Earth and star moving, and they conclude that clocks in the Earth–star frame must "run slow."

Were you inclined to give the seemingly more logical answer that, since the ship clocks "run slow," the Earth–star clocks "run fast"? If so, you haven't applied the principle of relativity: The laws of physics are the same in all inertial reference frames. There's nothing special about the spaceship's frame that would make it see Earth's clocks "running fast" when observers on Earth, in an exactly analogous situation, see the ship's clocks "running slow."

ASSESS How is this not a contradiction? The answer lies in the relativity of simultaneity. Earth and star clocks are synchronized in the Earth–star frame, but not in the ship's frame. From the ship's viewpoint, Earth–star clocks are "running slow" by the factor $\sqrt{1 - v^2/c^2} = 0.6$ that we found in Example 33.2. So the 15-year trip time in the ship frame takes, from the ship's perspective, only $(0.6)(15 \text{ years}) = 9$ years on the Earth–star clocks. The Earth clock read zero when the ship left Earth, so in the ship frame it reads 9 years when the ship reaches the star. Yet the star clock reads 25 years at that event. So, from the ship's viewpoint, Earth and star clocks are out of sync by 16 years. Figure 33.13 shows the situation from the ship's frame. Compare with Figs. 33.8*a* and *b*: Each observer thinks the other's clocks "run slow," yet there's no contradiction!

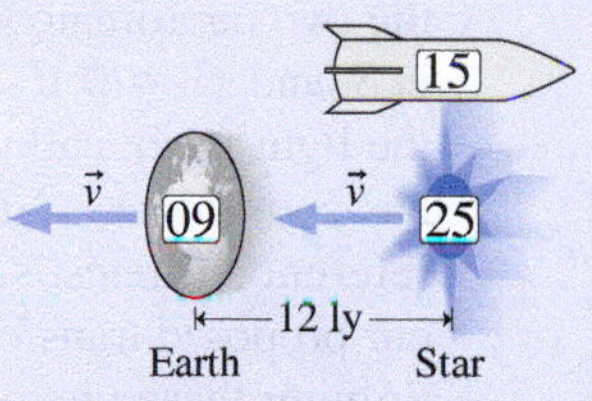

FIGURE 33.13 The situation in the ship's frame. Note that Earth, star, and the distance between them are contracted, while the ship appears longer than in Fig. 33.8.

This example shows a situation in which the space and time coordinates of an event (here the ship's arrival at the star) are different in the two reference frames considered. In Section 33.6, you'll see how the *Lorentz transformations* describe quantitatively the relationship between event coordinates in different reference frames.

MAKING THE CONNECTION For the star trek of Example 33.1, how do Earth and the star clock readings differ as judged in the spaceship's reference frame?

EVALUATE The ship sees Earth–star clocks "running slow" by the factor $\sqrt{1 - 0.95^2} = 0.312$. Given the 7.8-year time for the trip on the ship clock, observers on the ship judge that the elapsed time in the Earth–star frame is only $(7.8 \text{ years})(0.312) = 2.4 \text{ years}$. But we know the star clock reads 25 years when the ship arrives, so the Earth clock is behind by $25 \text{ years} - 2.4 \text{ years} = 22.6$ years as judged in the ship's reference frame.

GOT IT? 33.3 A comet plunges into the planet Jupiter. At the instant this happens, your physics class on Earth begins; in other words, the comet collision is simultaneous with your class beginning. A friend traveling from Earth toward Jupiter in a high-speed spaceship would say that (a) the comet collision occurs before your class begins; (b) the comet collision occurs after your class begins; (c) the comet collision and the beginning of your class are simultaneous.

33.6 The Lorentz Transformations

Events are determined by where (three spatial coordinates) and when (time) they occur. Our work with time dilation, length contraction, and the ordering of events suggests that these coordinates depend on one's frame of reference. Here we develop general expressions, called the **Lorentz transformations**, that relate the time and space coordinates of events in different reference frames.

Consider coordinate axes in a reference frame S and in another frame S' moving in the x-direction with speed v relative to S. The origins of the two systems coincide at time $t = t' = 0$. Given an event with coordinates x, y, z, t in S, what are its coordinates x', y', z', t' in S'? Were it not for relativity, we'd expect y, z, and t to remain unchanged, while the relative motion along x means a given x in S would correspond to $x' = x - vt$ in S' (Fig. 33.14).

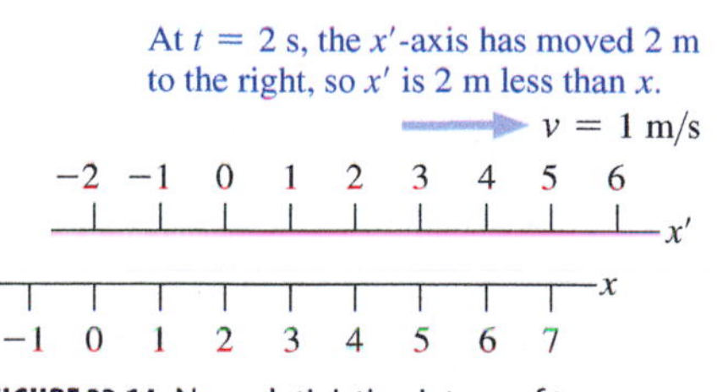

FIGURE 33.14 Nonrelativistic picture of two coordinate axes in relative motion, shown at time $t = 2$ s. In general, $x' = x - vt$.

Relativity should alter the transformations for both time and the spatial coordinate, x, along the direction of relative motion. However, our results must reduce to the nonrelativistic results $x' = x - vt$ and $t' = t$ in the limit $v \ll c$. A simple form with this property is $x' = \gamma(x - vt)$, where γ is a factor, still to be determined, that reduces to 1 as $v \rightarrow 0$. We could also transform the other way; the only difference is that the x-axis is moving in the negative direction relative to x'. Therefore, we should have a similar form, with the sign of v reversed: $x = \gamma(x' + vt')$. Now suppose a light flash goes off just as the origins of the two coordinate systems coincide. The coordinates of this event, E_1, are $x = 0, t = 0$ in S, and $x' = 0, t' = 0$ in S'. At some later time t, an observer at position x in S observes the light flash; call this event E_2. Since light travels with speed $c, x = ct$. In frame S', E_2 has coordinates x', t'. But relativity requires that light travel with speed c in *all* inertial reference frames, so we must have $x' = ct'$. Putting these expressions for x and x' in our proposed transformation equations gives $ct' = \gamma t(c - v)$ and $ct = \gamma t'(c + v)$. Multiplying these two equations yields $c^2 = \gamma^2(c - v)(c + v) = \gamma^2(c^2 - v^2)$. Therefore, $\gamma = 1/\sqrt{1 - v^2/c^2}$. Taking $v \rightarrow 0$ in this expression shows that $\gamma \rightarrow 1$ in the nonrelativistic limit, as required. So we have our transformation equations for x.

What about y, z, and t? The y- and z-axes are perpendicular to the direction of motion, so there's no length contraction and therefore $y' = y$ and $z' = z$. The fact of time dilation makes clear that measures of time differ in different reference frames, so it's not surprising that $t' \neq t$. You can derive the transformation equations for t from those for x (see Problem 44). The results, along with the equations we found for x, y, and z, are summarized in Table 33.1.

Table 33.1 The Lorentz Transformations

S to S'	S' to S	
$y' = y$	$y = y'$	
$z' = z$	$z = z'$	
$x' = \gamma(x - vt)$	$x = \gamma(x' + vt')$	where $\gamma = \dfrac{1}{\sqrt{1 - v^2/c^2}}$
$t' = \gamma(t - vx/c^2)$	$t = \gamma(t' + vx'/c^2)$	

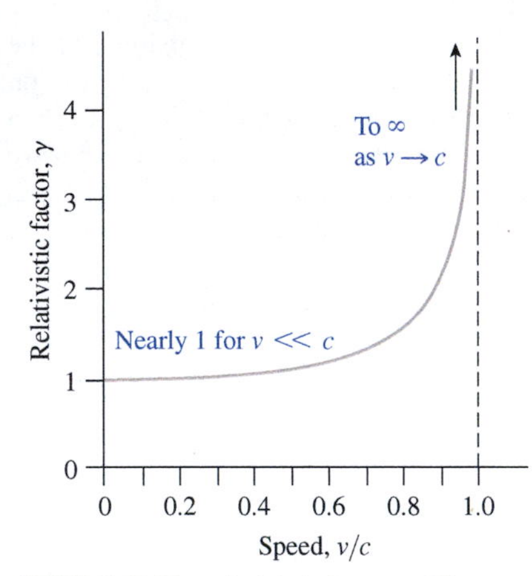

FIGURE 33.15 The relativistic factor γ differs significantly from 1 only as relative speeds approach c.

The relativistic factor $\gamma = 1/\sqrt{1 - v^2/c^2}$ appears throughout special relativity; in fact, you've already encountered it (or, rather, its inverse) in Equations 33.3 and 33.4, the expressions for time dilation and length contraction. Figure 33.15 shows that γ increases very slowly as the relative speed v increases; even at $v = \frac{1}{2}c, \gamma$ is only 1.15. That slow increase occurs because γ depends not on v in comparison with c but on the *square* of v relative to c^2. If v is much less than c, then v^2/c^2 is truly tiny. But at high relative speeds, γ shoots up rapidly, asymptotically approaching infinity as $v \rightarrow c$. That's why we don't notice relativistic effects except at high relative speeds. But they're there, and sensitive measurements can distinguish relativistic from Newtonian predictions even at relatively low speeds. The Michelson–Morley experiment is an example; it used Earth's orbital speed, only about $0.0001c$, and therefore needed the exquisite sensitivity of an interference-based technique.

PROBLEM-SOLVING STRATEGY 33.1 Lorentz Transformations

INTERPRET Make sure the problem involves space and time coordinates of events as measured in two different frames of reference. Identify the frames, which generally are those of specific objects introduced in the problem statement. Identify also the events and the particular coordinates you're interested in.

DEVELOP Establish coordinate systems in the two reference frames, arbitrarily designated S and S'. Choosing the direction of relative motion along the x-axis will let you use the Lorentz transformations as they appear in Table 33.1. Remember that time is also a coordinate, and take the two coordinate systems to coincide at time $t = t' = 0$. You'll make the math simpler by choosing the origin in space and time to occur at one of the events in the problem. Then determine any other event coordinates that are implicit in the problem statement.

EVALUATE Apply the appropriate Lorentz transformations from Table 33.1 to calculate the unknown coordinates.

ASSESS Ask whether your results make sense. If your calculated order of events differs in different frames, be sure you're dealing with events that are far enough apart in space and close enough in time that they can't be causally related.

EXAMPLE 33.4 Galactic Fireworks: Using the Lorentz Transformations

Our Milky Way and the Andromeda Galaxy are approximately at rest with respect to each other and are 2 million light years (Mly) apart. Supernova explosions occur simultaneously in both galaxies, as judged in the galaxies' reference frame. A spacecraft is traveling at $0.8c$ from the Milky Way toward Andromeda. Find the time between the supernova events as measured in the spacecraft's reference frame.

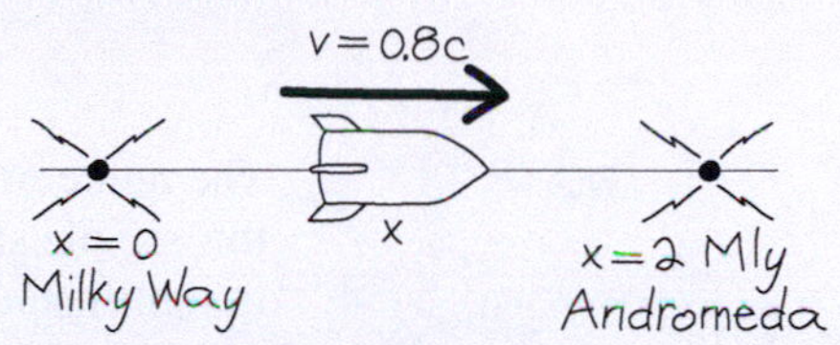

FIGURE 33.16 Our sketch for Example 33.4, drawn in frame S at time $t = 0$, when observers in S judge both supernova explosions to occur. This picture does not apply in frame S'!

INTERPRET We're given two distant events—both supernova explosions—that are simultaneous in a particular reference frame. We're asked to find the time between them in a different reference frame. So this problem is about using the Lorentz transformations for time coordinates.

DEVELOP Following our strategy, we establish coordinate system S, the galaxy reference frame. We take the origin at the Milky Way, with the x-axis extending through Andromeda (here we're treating the galaxies as points). We take $t = 0$ to be the time of the supernova explosions, which are simultaneous in S. Then the coordinates of the two supernovas in S are $x_{\text{MW}} = 0, t_{\text{MW}} = 0$ and $x_{\text{A}} = 2\,\text{Mly}, t_{\text{A}} = 0$. Here the subscripts designate not just the galaxies but the specific events of the supernova explosions. We show the situation at time $t = 0$ in Fig. 33.16. The other reference frame is that of the spaceship, which we designate S'; it's moving at speed $v = 0.8c$ along the x-axis relative to frame S. Taking the two coordinate systems to coincide at time $t = 0$ gives $t'_{\text{MW}} = 0$, and we're in the situation we used to derive the Lorentz transformations of Table 33.1. So our plan is to apply the transformation equation that gives t', namely, $t' = \gamma(t - vx/c^2)$.

EVALUATE We first evaluate the factor γ, finding that $\gamma = 1/\sqrt{1 - 0.8^2} = 5/3$. Then we apply the transformation equation for t' to get the time of the Andromeda supernova in the spacecraft's reference frame:

$$t'_{\text{A}} = \gamma\left(t_{\text{A}} - \frac{vx_{\text{A}}}{c^2}\right) = \left(\frac{5}{3}\right)\left(0 - \frac{(0.8\,\text{ly/y})(2\,\text{Mly})}{(1\,\text{ly/y})^2}\right) = -2.7\,\text{My}$$

ASSESS Since the Milky Way supernova goes off at $t'_{\text{MW}} = 0$, our *negative* answer means the Andromeda event occurs 2.7 million years *earlier* in the spacecraft's reference frame. Again, there's no problem with causality; since the distant events are simultaneous in *some* frame (the galaxy frame), they can't possibly be cause and effect. You can easily show that a spacecraft observer going the other way would judge the Andromeda supernova to occur 2.7 My later. Problems 39–42 explore a similar situation, including cases where two events occur far enough apart in time that they *could* be causally related. ■

Relativistic Velocity Addition

If you're in an airplane going 1000 km/h relative to the ground and you walk toward the front of the plane at 5 km/h, common sense suggests that you move at 1005 km/h relative to the ground. But measures of time and distance vary among frames of reference in relative motion. For this reason the velocity of an object with respect to one frame doesn't simply add to the relative velocity between frames to give the object's velocity with respect to another frame. In the airplane your speed with respect to the ground is actually a little less than 1005 km/h as you stroll down the aisle, although the difference is insignificant at such a low speed.

The correct expression for **relativistic velocity addition** follows from the Lorentz transformations. Consider a reference frame S and another frame S' moving in the positive x-direction with speed v relative to S. Let their origins coincide at time $t = t' = 0$, so the Lorentz transformations of Table 33.1 apply.

Suppose an object moves with velocity u' along the x'-axis in S'. We seek the velocity u of the object relative to the frame S. (We're using u and u' for the object because we've already used v for the relative velocity of the reference frames.)

In either frame, velocity is the ratio of change in position to change in time, or $u = \Delta x/\Delta t$. Designating the beginning of the interval Δt by the subscript 1 and the end by 2, we can use the Lorentz transformations to write

$$\Delta x = x_2 - x_1 = \gamma[(x'_2 - x'_1) + v(t'_2 - t'_1)] = \gamma(\Delta x' + v\,\Delta t')$$

and

$$\Delta t = t_2 - t_1 = \gamma[(t'_2 - t'_1) + v(x'_2 - x'_1)/c^2] = \gamma(\Delta t' + v\,\Delta x'/c^2)$$

Forming the ratio of these quantities, we have

$$\frac{\Delta x}{\Delta t} = \frac{\Delta x' + v\,\Delta t'}{\Delta t' + v\,\Delta x'/c^2} = \frac{(\Delta x'/\Delta t') + v}{1 + v(\Delta x'/\Delta t')/c^2}$$

But $\Delta x'/\Delta t'$ is the velocity u' of the object in frame S', and $\Delta x/\Delta t$ is the velocity u, so

$$u = \frac{u' + v}{1 + u'v/c^2} \quad \text{(relativistic velocity addition)} \tag{33.5a}$$

The numerator of this expression is just what we would expect from common sense. But this simple sum of two velocities is altered by the second term in the denominator, which is significant only when both the object's velocity u' and the relative velocity v are comparable with c. Solving Equation 33.5a for u' in terms of u, v, and c gives the inverse transformation:

$$u' = \frac{u - v}{1 - uv/c^2} \quad \text{(relativistic velocity addition)} \tag{33.5b}$$

EXAMPLE 33.5 Relativistic Velocity Addition: Collision Course

Two spacecraft approach Earth from opposite directions, each moving at $0.80c$ relative to Earth, as shown in Fig. 33.17*a*. How fast do the spacecraft move relative to each other?

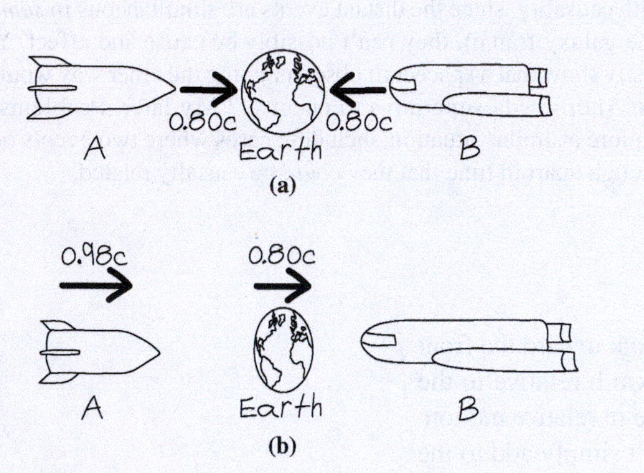

FIGURE 33.17 Sketch for Example 33.5 (a) in Earth's frame S' and (b) in spacecraft B's frame S.

INTERPRET The naïve answer, $1.6c$, isn't consistent with relativity. Instead, this is a problem involving relativistic velocity addition. We identify the Earth frame of reference as S' and ship B's frame as S. Ship A is moving at $u' = 0.8c$ relative to the Earth frame S', and we want to know its speed u relative to ship B's frame S. Ship B is also moving toward Earth at $0.8c$, or, equivalently, Earth is moving toward ship B at this speed, so we identify $v = 0.8c$ as the relative velocity between frames.

DEVELOP With all terms identified, our plan is to apply Equation 33.5a to find the velocity u of ship A relative to ship B.

EVALUATE $u = \dfrac{u' + v}{1 + u'v/c^2} = \dfrac{0.80c + 0.80c}{1 + (0.80c)(0.80c)/c^2} = \dfrac{1.6c}{1.64} = 0.98c$

ASSESS The relative speed is less than the $1.6c$ we get from a naïve addition and also less than the speed of light. This result is quite general: Equations 33.5 imply that as long as an object moves with speed $u < c$ relative to some frame, then its speed relative to any other frame is less than c. And if you set $u = c$ to describe a light beam, you'll find that Equations 33.5 give $u' = c$ as well—reaffirming the relativistic point that the speed of light is the same in all inertial reference frames. ■

GOT IT? 33.4 You're driving down the highway, and your speedometer reads exactly 30 km/h. A car passes you, going in the same direction at exactly 20 km/h relative to you. Does its speedometer—which measures the car's speed relative to the road—read (a) more or (b) less than 50 km/h?

Is Everything Relative?

You already know the answer: The laws of physics aren't relative—that's the fundamental principle of relativity. Neither is the speed of light, whose existence and value follow from laws of physics—specifically, Maxwell's equations. And there are a host of other **relativistic invariants**, independent of reference frame. One such invariant is the **spacetime interval**, a kind of four-dimensional "distance" between two events in space and time. The spacetime interval is given by an expression that looks like a modified Pythagorean theorem:

$$(\Delta s)^2 = c^2(\Delta t)^2 - [(\Delta x)^2 + (\Delta y)^2 + (\Delta z)^2] \tag{33.6}$$

where the Δ quantities are the differences between the time and space coordinates of the events. The invariance of Δs follows directly from the Lorentz transformations, as you can show in Problem 63.

The spacetime interval describes a relation between two events that's independent of reference frame. The invariance of the spacetime interval suggests that something absolute underlies the shifting sands of relativistic space and time. That something is **spacetime**—a four-dimensional framework linking space and time into a single continuum. The spacetime interval is the magnitude of a four-dimensional vector—a **4-vector**—whose components involve the three spatial distances Δx, Δy, Δz and the time Δt between two events. The individual space and time components differ in different reference frames, but they always conspire to give the same invariant interval. This is analogous to the vectors of ordinary two- and three-dimensional space, where the vector components depend on your choice of coordinate system. But the actual vector quantity—for example, a force—has a reality independent of your coordinate choices, and its magnitude doesn't depend on the coordinate system (Fig. 33.18). This analogy isn't perfect because of the negative sign in Equation 33.6's expression for the spacetime interval. That sign reflects the fact that the underlying geometry of spacetime isn't the Euclidean geometry you learned in high school.

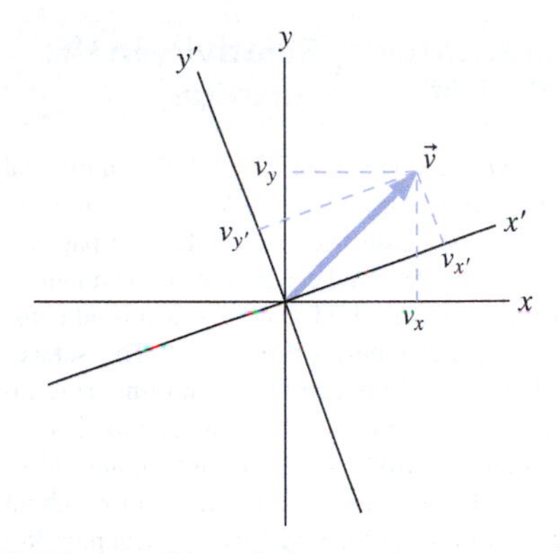

FIGURE 33.18 Although the x- and y-components of an ordinary vector depend on the choice of coordinate system, the magnitude of the vector does not.

Other 4-vectors play a role in more advanced treatments of relativity. These include a four-dimensional electric-current density, whose components involve charge density and the three components of ordinary current density; a four-dimensional *wave vector* that links frequency and wavelength and whose invariant magnitude yields the Doppler effect in its correct relativistic form; and a *4-potential* that yields both the electric and magnetic fields. A particularly important example is the *energy–momentum 4-vector*. Its invariant magnitude is famously related to mass, as you'll see in the next section.

33.7 Energy and Momentum in Relativity

Conservation of momentum and conservation of energy are cornerstones of Newtonian mechanics, where they hold in any inertial reference frame. But momentum and energy are functions of velocity, and we've just seen that relativity alters the Newtonian picture of how velocities transform from one reference frame to another. How, then, can momentum and energy be conserved in all reference frames?

Momentum

In Newtonian mechanics the momentum of a particle with mass m and velocity $\vec{u}$ is $m\vec{u}$. (Here we use $\vec{u}$ for particle velocities, reserving $\vec{v}$ for the relative velocity between reference frames.) But if a system's momentum—the sum of its individual particles' momenta $m\vec{u}$—is conserved in one frame of reference, then relativistic velocity addition suggests that it won't be conserved in another. But conservation of momentum *is* a fundamental law, one that transcends Newtonian physics. So momentum is conserved in relativity (and, indeed, in quantum physics as well). The problem here lies not with momentum conservation but with the Newtonian expression for momentum. The expression $m\vec{u}$ is actually an approximation valid only for speeds u much less than c. The measure of momentum valid at any speed is

$$\vec{p} = \frac{m\vec{u}}{\sqrt{1 - u^2/c^2}} = \gamma m\vec{u} \quad \text{(relativistic momentum)} \qquad (33.7)$$

where γ is the relativistic factor that we introduced with the Lorentz transformations. The momentum in Equation 33.7 is conserved in all reference frames, and at low velocities it reduces to the Newtonian expression $\vec{p} = m\vec{u}$.

As $u \to c$, the factor γ grows arbitrarily large, and so does the relativistic momentum (Fig. 33.19). Since force is the rate of change of momentum, that means a very large force is required to produce even the slightest change in the velocity of a rapidly moving particle. This helps answer a common question about relativity: Why is it impossible

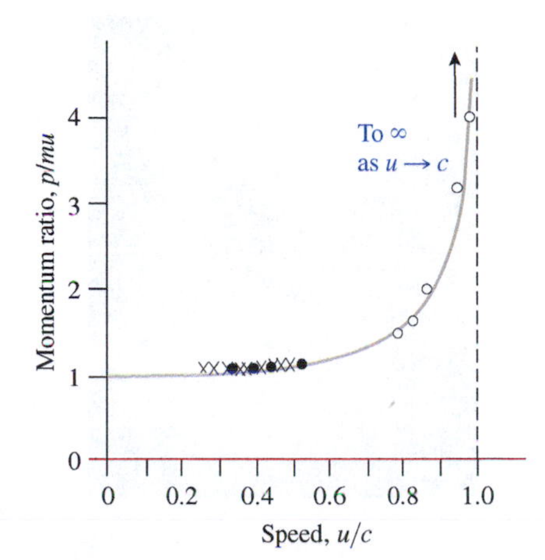

FIGURE 33.19 The ratio of relativistic momentum to Newtonian momentum mu. The curve follows Equation 33.7; crosses and circles mark experimental data. Compare with Figure 33.15.

APPLICATION PET SCANS **Relativity in the Hospital**

Positron emission tomography (PET) is a medical imaging technology based on electron–positron annihilation (see Example 33.6 on the next page for more on positrons and electron–positron annihilation). To produce a PET scan, a patient is administered a positron-emitting radioisotope. The isotopes used are short-lived, so there's no lingering radioactivity; common isotopes include oxygen-15 (2 minutes), carbon-11 (20 minutes), and nitrogen-13 (10 minutes). (You'll learn more about radioisotopes and their half-lives in Chapter 38.) These substances are combined chemically to replace ordinary oxygen, carbon, or nitrogen in such materials as water (H_2O), carbon dioxide (CO_2), ammonia (NH_3), or more complex biomolecules such as glucose. Given their short lives, the radioisotopes must be produced on site, using a cyclotron (see Chapter 26's Application on page 473).

Each radioactive nucleus decays by emitting a positron. In body tissue, the positron very soon meets an ordinary electron and the pair annihilate. Positron and electron have the same mass m, and the rest mass of both particles is converted completely to energy according to $E = mc^2$. In order to conserve momentum, this energy emerges in the form of two 511-keV gamma rays moving in opposite directions. The PET scanning technology uses multiple gamma-ray detectors arrayed around the patient. Electronic circuitry identifies pairs of gamma rays emitted simultaneously. Each pair determines a line, connecting two detectors, on which the emission took place; examining multiple pairs determines the point where the emission occurred. Since each positron travels a negligible distance before encountering an electron, the gamma-ray emission region is where the radioisotopes are located. By choosing radioactively "tagged" substances that concentrate in specific tissues, clinicians can image particular physiological processes. The image shows two pet scans of a human brain, with the different patterns corresponding to active word recognition.

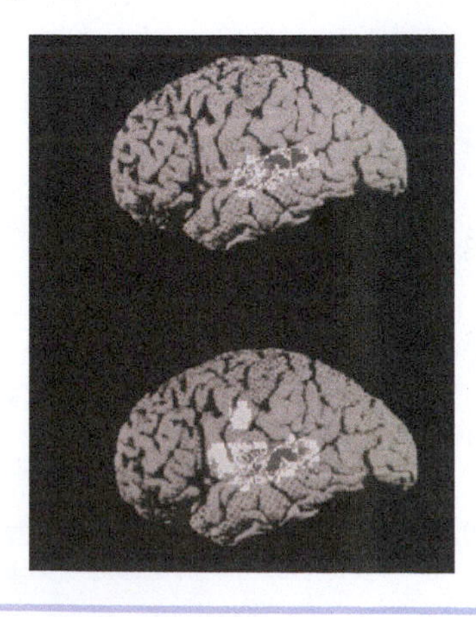

to accelerate an object to the speed of light? One answer is that the object's momentum would approach infinity, and no matter how close to c it was moving, it would still require infinite force to give it the last bit of speed.

Energy and Mass

The most widely known result of relativity is the equation $E = mc^2$. Here we develop a general expression for relativistic energy and show just what $E = mc^2$ means. In the process you'll gain new insights into energy, momentum, and mass.

In Chapter 6 we derived the work–energy theorem, and thus developed the concept of kinetic energy, by calculating the work needed to accelerate a mass m from rest to some final speed. We did that by integrating force over distance, using Newton's law $F = dp/dt$ to write force in terms of momentum. We'll do the same thing here, now using our relativistic expression for momentum. So we have

$$\frac{dp}{dt} = \frac{d}{dt}\left[\frac{mu}{\sqrt{1 - u^2/c^2}}\right] = \frac{m(du/dt)}{(1 - u^2/c^2)^{3/2}}$$

Then the kinetic energy gained as a particle accelerates from rest in our reference frame to a final speed u is

$$K = \int \frac{dp}{dt}dx = \int \frac{dp}{dt}u\,dt = \int \frac{m(du/dt)}{(1 - u^2/c^2)^{3/2}}u\,dt = \int_0^u \frac{mu}{(1 - u^2/c^2)^{3/2}}du$$

where we used $u = dx/dt$ to replace dx with $u\,dt$. The integral is readily evaluated using the fact that $u\,du = \frac{1}{2}d(u^2)$, giving

$$K = \frac{mc^2}{\sqrt{1 - u^2/c^2}} - mc^2 = \gamma mc^2 - mc^2 \qquad (33.8)$$

for a particle's kinetic energy in a reference frame where the particle has speed u. Once again, γ here is the relativistic factor $1/\sqrt{1 - u^2/c^2}$.

For speeds low compared with c, Equation 33.8 reduces to the Newtonian $K = \frac{1}{2}mu^2$, as you can show in Problem 61. But Equation 33.8 suggests, more generally, that kinetic energy is the difference of two energies—the velocity-dependent quantity γmc^2 and the term mc^2 that depends only on mass, not velocity. Pursuing this interpretation, we identify γmc^2 as the particle's **total energy** and mc^2 as its **rest energy**. Rearranging Equation 33.8 lets us write the total energy as the sum of the kinetic energy and the rest energy: $E = K + mc^2$ or, more simply

$$E = \gamma mc^2 = \frac{mc^2}{\sqrt{1 - u^2/c^2}} \qquad \text{(total energy)} \qquad (33.9)$$

What does all this mean? Put $u = 0$ in Equation 33.9 and you get $E = mc^2$—showing that the total energy of a stationary particle isn't zero but is directly proportional to its mass. Thus that particle has energy simply by virtue of having mass or, as Einstein first recognized, *mass and energy are equivalent*. The proportionality between mass and energy is a whopping big c^2—about 9×10^{16} J/kg in SI units.

Although we developed Equation 33.9 by considering kinetic energy, the mass–energy equivalence $E = mc^2$ is universal. Energy, like mass, exhibits inertia. A hot object is slightly harder to accelerate than an otherwise identical cold one because of the inertia of its thermal energy. A stretched spring is more massive than an otherwise identical unstretched one because of its extra potential energy. When a system loses energy, it loses mass as well.

To the public, $E = mc^2$ is synonymous with nuclear energy. The equation does describe mass changes in nuclear reactions, but it applies equally well to chemical reactions and all other energy conversions. Weigh a nuclear power plant and weigh it again a month later, and you'll find it weighs slightly less. Weigh a coal-burning power plant and all the coal and oxygen that go into it for a month, then weigh all the carbon dioxide and other combustion products, and you'll find a difference. If both plants produce the same amount of energy, the mass difference is the same for both. The distinction lies in the amount of mass released as energy in individual reactions. Fission of a single uranium nucleus involves about 50 million times as much energy, and therefore mass, as the reaction of a carbon atom with oxygen to make carbon dioxide. That's why a coal-burning power plant consumes many hundred-car trainloads of coal each week, while a nuclear plant needs only a few truckloads of uranium every year or so. Incidentally, neither process converts very much of the fuel mass to energy; if we could convert *all* the mass in a given object to energy, ordinary matter would be an almost limitless source of energy. Such conversion is in fact possible, but only in the annihilation of matter and antimatter.

EXAMPLE 33.6 Mass–Energy Equivalence: Annihilation

A positron is an antimatter particle with the same mass as the electron but the opposite electric charge. When an electron and positron meet, they annihilate and produce a pair of identical gamma rays (bundles of electromagnetic energy). Find the energy of each gamma ray.

INTERPRET This is a problem about mass–energy equivalence—in this case all the mass of the electron–positron pair changing to gamma-ray energy. We'll assume the pair has negligible kinetic energy K, so the total energy available is just their rest energy.

DEVELOP With $K = 0$, Equation 33.9 reduces to $E = mc^2$. Since there are two particles, each of mass m, and two identical gamma rays, each gets energy mc^2. So our plan is to evaluate this quantity. We'll need the electron mass m, which is given inside the front cover of this book.

EVALUATE We have $E = mc^2 = (9.11\times10^{-31}\,\mathrm{kg})(3.00\times10^{8}\,\mathrm{m/s})^2 = 82.0\,\mathrm{fJ}$.

ASSESS High-energy physicists usually work in electronvolts, and our 82-fJ answer is equivalent to 511 keV. Detection of 511-keV gamma rays from laboratory or astrophysical sources is a sure sign of electron–positron annihilation. The Application on the preceding page shows how the medical technique PET (positron emission tomography) uses these annihilation gamma rays to image processes occurring inside the body. ■

Given the fame of $E = mc^2$, it's easy to overlook the fact that the rest energy mc^2 is generally only part of a particle's total energy. For a particle moving at velocity u that's small compared with c, the total energy is only slightly greater than mc^2; the extra is very nearly the Newtonian kinetic energy, $\frac{1}{2}mu^2$. (Here "at rest" and "moving" are, as always, relative to some inertial reference frame.) But when a particle moves with nearly the speed of light, the relativistic factor $\gamma = 1/\sqrt{1 - u^2/c^2}$ becomes much greater than 1, and the total energy γmc^2 is many times the rest energy. Such a particle is termed **relativistic**.

EXAMPLE 33.7 Total Energy: A Relativistic Electron

An electron has total energy 2.50 MeV. Find (a) its kinetic energy and (b) its speed.

INTERPRET Here we're given the electron's total energy and asked how much of that is the kinetic energy associated with its speed, which we're also asked for.

DEVELOP Equation 33.9 gives total energy, which is the sum of kinetic energy K and rest energy mc^2. So our plan is to subtract the rest energy to find the kinetic energy. Equation 33.9 also expresses the total energy as γmc^2, so we can find $\gamma = 1/\sqrt{1 - u^2/c^2}$ and then solve for the speed u. We don't need to calculate mc^2 because we found it in Example 33.6: For the electron, $mc^2 = 511$ keV or 0.511 MeV.

EVALUATE (a) We have

$$K = E - mc^2 = 2.50\,\mathrm{MeV} - 0.511\,\mathrm{MeV} = 1.99\,\mathrm{MeV}$$

(b) Since $E = \gamma mc^2, \gamma = E/mc^2 = 2.50\,\mathrm{MeV}/0.511\,\mathrm{MeV} = 4.89$. But $\gamma = 1/\sqrt{1 - u^2/c^2}$, which we solve to get

$$u = c\sqrt{1 - 1/\gamma^2} = 2.94\times10^{8}\,\mathrm{m/s}$$

ASSESS Our answer for kinetic energy is considerably greater than the electron's rest energy, and our speed u is close to c, both confirming that this is a relativistic electron. ■

GOT IT? 33.5 The rest energy of a proton is 938 MeV. Without doing any calculations, quickly estimate the speed of a proton with total energy 1 TeV (10^{12} eV).

The Energy–Momentum Relation

In Newtonian physics the equations $p = mu$ and $K = \frac{1}{2}mu^2$ yield $p^2 = 2K/m$. Similarly, Problem 62 shows that in relativity we can combine the equations $p = \gamma mu$ and $E = \gamma mc^2$ to get

$$E^2 = p^2c^2 + (mc^2)^2 \qquad \text{(energy–momentum relation)} \qquad (33.10)$$

which involves E rather than K because in relativity the energy includes both kinetic and rest energies. For a particle at rest, $p = 0$ and Equation 33.10 shows that the total energy is just the rest energy. For highly relativistic particles, the rest energy is negligible and the total energy becomes very nearly $E = pc$. Some "particles"—like the photons that, in quantum physics, are "bundles" of electromagnetic energy—have no mass. These particles exist only in motion at the speed of light, and for them Equation 33.10 gives the exact relation $E = pc$.

Rearranging Equation 33.10 gives $(mc^2)^2 = E^2 - p^2c^2$. This should remind you of Equation 33.6 for the spacetime interval, whose square is the difference between the squares of the time component $c\,\Delta t$ and the spatial separation $\sqrt{(\Delta x)^2 + (\Delta y)^2 + (\Delta z)^2}$. Similarly, our rearranged Equation 33.10 gives the square of the rest energy mc^2 as the difference between the squares of the total energy and the magnitude of the momentum multiplied by c. You can therefore think of energy and momentum as the time and space components of a 4-vector. Your frame of reference determines how this energy–momentum 4-vector breaks out into time and space components—that is, into energy and momentum. In a particle's rest frame, for example, $p = 0$ and the vector has only a time component equal to the rest energy. But no matter what frame you're in, the magnitude of the 4-vector is the same, and it's equal to the rest energy mc^2. Therefore, mass—the rest energy divided by the constant c^2—is a relativistic invariant.

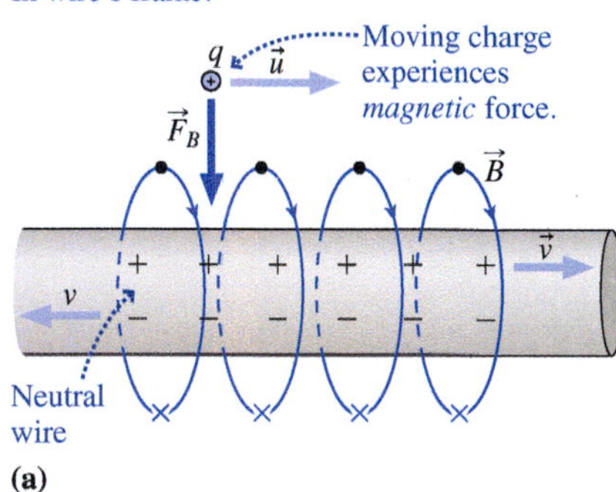

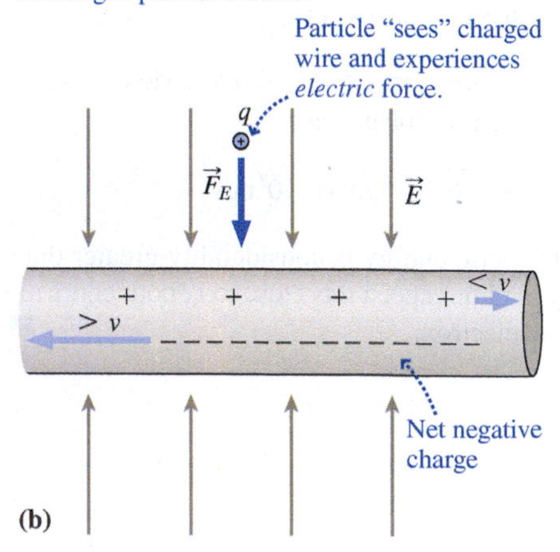

FIGURE 33.20 The force on a charged particle is magnetic or electric, depending on the reference frame.

33.8 Electromagnetism and Relativity

Historically, relativity arose from deep questions about the propagation of electromagnetic waves. We've seen that relativity alters concepts, such as space, time, energy, and momentum, that are fundamental to Newtonian physics. For that reason, Newtonian physics becomes an approximation valid at low speeds. But relativity is built on the premise that Maxwell's equations of electromagnetism are correct in all reference frames—including the prediction of electromagnetic waves propagating at speed c. Indeed, Einstein's 1905 paper wasn't titled "The Theory of Relativity," but "On the Electrodynamics of Moving Bodies"—showing how intimately related are electromagnetism and relativity. Maxwell's equations are relativistically correct and require no modification.

Although electric and magnetic fields in any frame of reference obey the same Maxwell equations, the fields themselves aren't invariant. Sit in the rest frame of a point charge, and you see a spherically symmetric point-charge field. Move relative to the charge, and you see a magnetic field as well, associated with the moving charge. The electric field you see is altered; it's no longer spherically symmetric. So electric and magnetic fields aren't absolutes; what one observer sees as a purely electric field another may see as a mix of electric and magnetic fields, and vice versa. You can think of the electric and magnetic fields as components of a more fundamental electromagnetic field; how that field breaks out into electric and magnetic fields depends on your frame of reference. Although the fields are different in different reference frames, there's an important electromagnetic quantity that's invariant—namely, electric charge.

We'll illustrate the deep relationship that relativity imposes on electricity and magnetism by considering the force on a positively charged particle moving relative to a current-carrying wire. For simplicity, assume the wire contains equal line-charge densities of positive and negative charge, moving in opposite directions with the same speed v relative to the wire (Fig. 33.20*a*). The resulting current produces a magnetic field that encircles the wire, and the charged particle moving to the right with velocity $\vec{u}$ as shown in Fig. 33.20*a* experiences a *magnetic* force $\vec{F}_B = q\vec{u} \times \vec{B}$ toward the wire. Because the positive and negative line-charge densities are equal, the wire is neutral, so there's no *electric* force.

Now look at the situation in the particle's reference frame. The positive charges in the wire have a lower speed relative to the particle than the negative charges, so in the particle's frame the distances between negative charges are contracted *more* than the distances between positive charges, as shown in Fig. 33.20*b*. But charge is invariant, so that means the charge per unit length is *greater* for the negative charges. In the frame of the charged particle, *the wire carries a net negative charge!* That results in an *electric* field pointing toward the wire, and therefore the charged particle experiences an *electric* force $\vec{F}_E = q\vec{E}$ toward the wire. There's still a magnetic field as well, but since the particle is at rest in its own frame, the magnetic force $q\vec{u} \times \vec{B}$ is zero.

We've given two quite different descriptions of the force on the charged particle. In the wire's reference frame, there's a purely magnetic force that we could determine knowing the magnetic force $q\vec{u} \times \vec{B}$ and how magnetic fields arise from currents. In contrast, describing the force in the particle's reference frame requires no knowledge of magnetism whatever. We need to know only the electric force $q\vec{E}$ and how electric fields arise from charges. This illustrates a profound point: Electricity and magnetism aren't independent phenomena that happen to be related. Rather, they're two aspects of a single phenomenon—electromagnetism. Given the relativity principle, it's impossible to have electricity without magnetism, or vice versa. Relativity provides the complete unification of electromagnetism that we've hinted at throughout our study of these phenomena. You can explore this unification further in Problem 74.

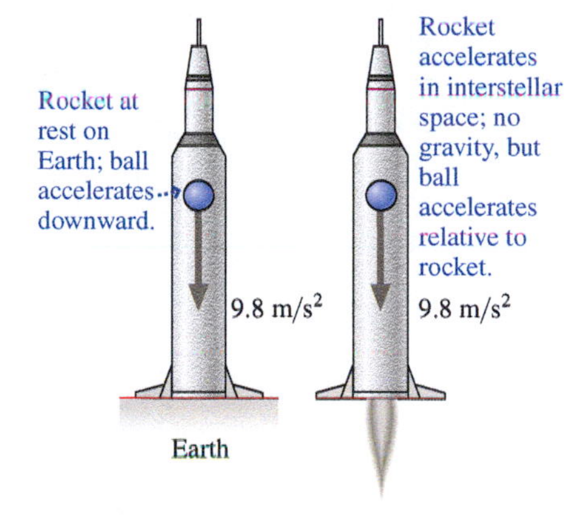

FIGURE 33.21 It's impossible to distinguish the effects of uniform gravitation from acceleration, which is why general relativity is about gravity.

33.9 General Relativity

The *special* theory of relativity is restricted to reference frames in uniform motion. Following special relativity, Einstein attempted to formulate a theory that would encompass observers in accelerated motion. But he recognized that it's impossible to distinguish the effects of uniform acceleration from those of a uniform gravitational field (Fig. 33.21). Consequently, Einstein's 1916 **general theory of relativity** became a theory of gravity. General relativity describes gravity as the geometrical curvature of four-dimensional spacetime. In this description, matter and energy curve spacetime in their vicinity, and objects moving through the curved spacetime follow the straightest possible paths—which aren't the straight lines of Euclidean geometry. Figure 33.22 shows a two-dimensional analogy for particles in curved spacetime.

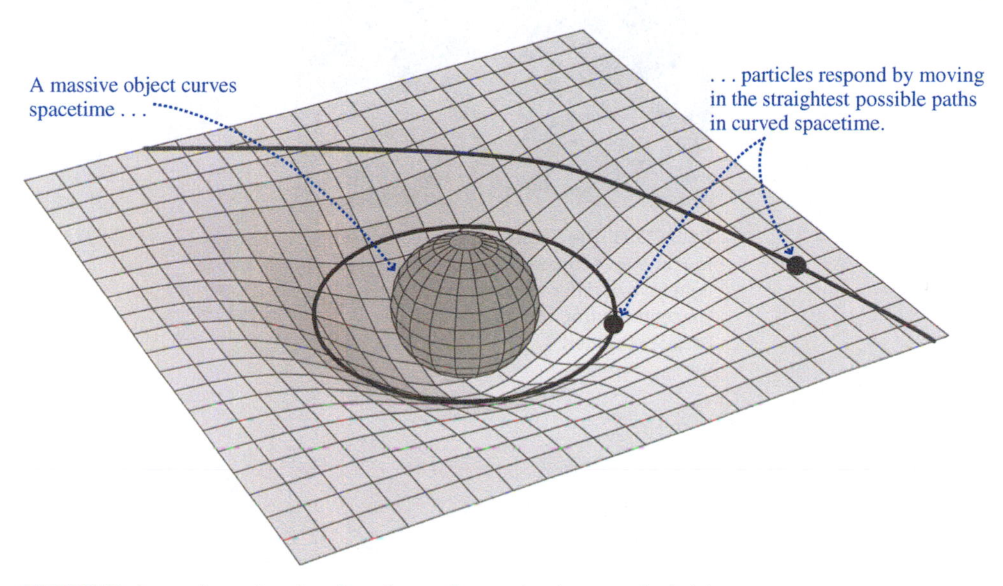

FIGURE 33.22 A two-dimensional analog of curved spacetime in general relativity.

General relativity also predicts the existence of *gravitational waves*—literally, ripples in the fabric of spacetime that propagate at the speed of light. These waves are produced by rapid motions of large masses such as collapsed stars in close orbits, or the merger of two neutron stars or black holes. For decades, physicists have been constructing ever more sensitive instruments in an attempt to detect these gravitational waves. They were finally rewarded in 2015, when LIGO, the Laser Interferometer Gravitational Wave Observatory, detected gravitational waves from the merging of two black holes in a galaxy 1.3 billion light-years from Earth. LIGO uses what are essentially a pair of Michelson interferometers (recall Figs. 32.15 and 33.2) located in Livingston, Louisiana and Hanford, Washington; these detect changes in interference patterns occurring in response to distortions of space and time by passing gravitational waves. The LIGO gravitational-wave detection came almost exactly 100 years after the publication of Einstein's general relativity, and confirmed the last unverified prediction of the theory.

General relativity's predictions differ substantially from those of Newton's theory of gravity only in regions of very strong gravitational fields—far stronger than those found anywhere in our solar system. For that reason general relativity has become a cornerstone of modern astrophysics, describing the physics of such bizarre objects as neutron stars, black holes, and so-called *gravitational lenses* that can produce multiple images of astrophysical objects (Fig. 33.23). General relativity also addresses cosmological questions of the origin and ultimate fate of the universe, as you'll see in Chapter 39. But we're not without terrestrial uses for general relativity; the Global Positioning System (GPS) would be off by several kilometers if the effects of general relativity on the GPS satellites' clocks weren't taken into account.

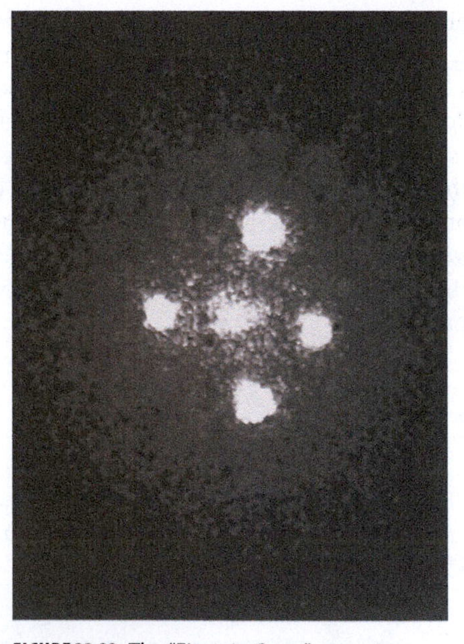

FIGURE 33.23 The "Einstein Cross" comprises four images of the same quasar, formed as light follows different paths in the curved spacetime surrounding a massive galaxy (visible at center).

CHAPTER 33 SUMMARY

Big Idea

The big idea here is a simple one: The laws of physics are the same for all observers and don't depend on one's state of motion. That's the **principle of relativity**. **Special relativity** is restricted to inertial reference frames; **general relativity** removes that restriction and in so doing becomes a theory of gravity. Both relativity theories radically alter our commonsense notions of space and time. In relativity, measures of space and time—but not the laws of physics—depend on one's reference frame.

Key Concepts and Equations

Among the laws of physics are Maxwell's equations, with their prediction of electromagnetic waves (e.g., light) propagating at the speed of light c. Therefore, the speed of light is the same in all inertial reference frames.

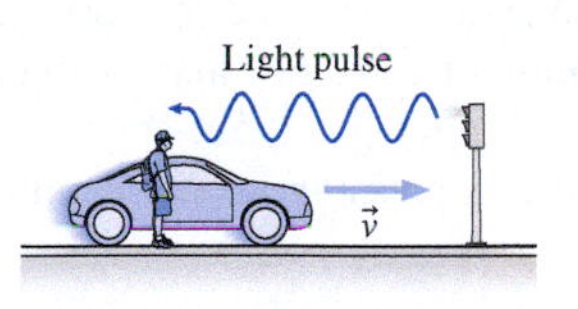

Invariance of c leads to **time dilation** and **length contraction**:

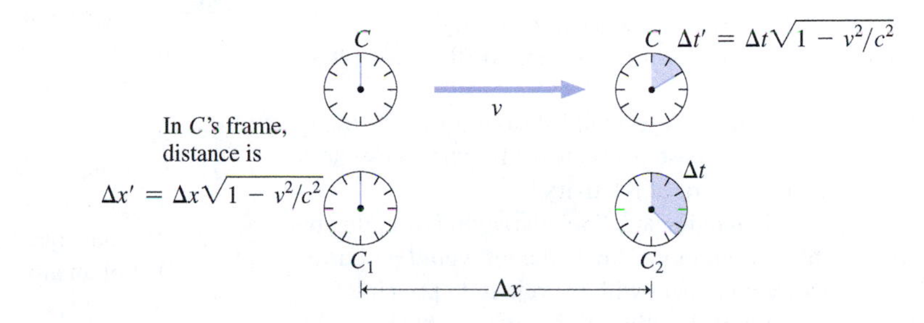

Additionally, events simultaneous in one reference frame may not be simultaneous in another frame.

Time dilation and length contraction are specific instances of the **Lorentz transformations** of space and time coordinates of events observed in different reference frames. The transformations here apply to relative motion in the x-direction.

S to S'	S' to S	
$y' = y$	$y = y'$	
$z' = z$	$z = z'$	
$x' = \gamma(x - vt)$	$x = \gamma(x' + vt')$	where $\gamma = \dfrac{1}{\sqrt{1 - v^2/c^2}}$
$t' = \gamma(t - vx/c^2)$	$t = \gamma(t' + vx'/c^2)$	

Underlying the changing measures of space and time is four-dimensional **spacetime**, in which exist **4-vectors** whose magnitude is independent of reference frame.

Invariant spacetime interval:

$$(\Delta s)^2 = c^2(\Delta t)^2 - [(\Delta x)^2 + (\Delta y)^2 + (\Delta z)^2]$$

Invariant particle mass:

$$(mc^2)^2 = E^2 - p^2c^2$$

Energy, momentum, and mass are closely related in relativity:

Momentum: $\vec{p} = \dfrac{m\vec{u}}{\sqrt{1 - u^2/c^2}} = \gamma m\vec{u}$

Energy: $E = \dfrac{mc^2}{\sqrt{1 - u^2/c^2}} = \gamma mc^2 = K + mc^2 = \sqrt{(pc)^2 + (mc^2)^2}$

Kinetic energy (K)   Rest energy (mc^2)

Applications

The **Michelson–Morley experiment** of 1887 failed to detect any motion of Earth relative to the **ether**, the medium in which 19th-century physicists believed light propagated with speed c. This result helped pave the way for relativity, with no ether and with c the same in all inertial reference frames.

Relativity shows that velocities don't simply add; rather,

$$u = \frac{u' + v}{1 + u'v/c^2}$$

where u' is an object's velocity relative to a reference frame S', u its velocity relative to frame S, and v the relative velocity between S and S'.

Einstein's equation $E = mc^2$ describes a universal interchangeability between matter and energy; contrary to common opinion, it isn't just about nuclear energy.

MP For homework assigned on MasteringPhysics, go to www.masteringphysics.com

BIO *Biology and/or medicine-related problems* **DATA** *Data problems* **ENV** *Environmental problems* **CH** *Challenge problems* **COMP** *Computer problems*

For Thought and Discussion

1. Why was the Michelson–Morley experiment a more sensitive test of motion through the ether than independent measurements of the speed of light in two perpendicular directions?
2. Why was it necessary to repeat the Michelson–Morley experiment throughout the year?
3. What's *special* about the *special* theory of relativity?
4. Does relativity require that the speed of sound be the same for all observers? Why or why not?
5. Time dilation is sometimes described by saying that "moving clocks run slow." In what sense is this true? In what sense does the statement violate the spirit of relativity?
6. If you're in a spaceship moving at $0.95c$ relative to Earth, do you perceive time to be passing more slowly than it would on Earth? Think! Is your answer consistent with the relativity principle?
7. The Andromeda Galaxy is 2 million light years from the Milky Way. Although nothing can go faster than light, it would still be possible to travel to Andromeda in much less than 2 million years. How?
8. Is matter converted to energy in a nuclear reactor? In a burning candle? In your body?
9. If you took your pulse while traveling in a high-speed spacecraft, would it be faster than, slower than, or the same as on Earth?
10. The rest energy of an electron is 511 keV. What's the approximate speed of an electron whose total energy is 1 GeV? (*Note:* No calculations needed!)
11. An atom in an excited state emits a burst of light. What happens to the atom's mass?
12. The quantity $\vec{E}\cdot\vec{B}$ is invariant. What does this say about how different observers will measure the angle between $\vec{E}$ and $\vec{B}$ in a light wave?

Exercises and Problems

Exercises

Section 33.2 Matter, Motion, and the Ether

13. An airplane makes a round trip between two points 1800 km apart, flying with airspeed 800 km/h. What's the round trip flying time (a) if there's no wind, (b) with wind at 130 km/h perpendicular to a line joining the two points, and (c) with wind at 130 km/h along a line joining the two points?
14. Consider a Michelson–Morley experiment with 11-m light paths perpendicular and parallel to the ether wind. What would be the difference in light travel times on the two paths if Earth moved relative to the ether at (a) its orbital speed (Appendix E); (b) $0.01c$; (c) $0.5c$; and (d) $0.99c$?

Section 33.4 Space and Time in Relativity

15. Two stars are 50 ly apart, measured in their common rest frame. How far apart are they to a spaceship moving between them at $0.75c$?
16. How long would it take a spacecraft traveling at $0.65c$ to get from Earth to Pluto according to clocks (a) on Earth and (b) on the spacecraft? Assume Earth and Pluto are on the same side of the Sun.
17. A spaceship passes by you at half the speed of light, and you determine that it's 35 m long. Find its length as measured in its rest frame.
18. An extraterrestrial spacecraft whizzes through the solar system at $0.80c$. How long does it take to go the 8.3 light-minute-distance from Earth to Sun (a) according to an observer on Earth and (b) according to an alien aboard the ship?
19. How fast would you have to move relative to a meter stick for it to be 99 cm long in your reference frame?
20. **BIO** A hospital's linear accelerator produces electron beams for cancer treatment. The accelerator is 1.6 m long and the electrons reach a speed of $0.98c$. How long is the accelerator in the electrons' reference frame?

Section 33.7 Energy and Momentum in Relativity

21. By what factor does an object's momentum change if you double its speed when its original speed is (a) 25 m/s and (b) 100 Mm/s?
22. At what speed will the momentum of a proton (mass 1 u) equal that of an alpha particle (mass 4 u) moving at $0.5c$?
23. At what speed will the Newtonian expression for momentum be in error by 1%?
24. A particle is moving at $0.90c$. If its speed increases by 10%, by what factor does its momentum increase?
25. Find (a) the total energy and (b) the kinetic energy of an electron moving at $0.97c$.
26. At what speed will the relativistic and Newtonian expressions for kinetic energy differ by 10%?

Problems

27. Show that the time of Equation 33.2 is longer than that of Equation 33.1 when $0 < v < c$.
28. You're designing a Michelson interferometer in which a speed-of-light difference of 100 m/s in two perpendicular directions is supposed to shift the interference pattern so a bright fringe of 550-nm light ends up where the adjacent dark fringe would be in the absence of a speed difference. How long should you make the interferometer's arms?
29. Earth and Sun are 8.3 light minutes apart, as measured in their rest frame. (a) What's the speed of a spacecraft that makes the trip in 5.0 min according to its on-board clocks? (b) What's the trip time as measured by clocks in the Earth–Sun reference frame?
30. You're the communications officer on a fast spaceship that takes 50 years in ship time to reach the Andromeda Galaxy, 2 million light years from Earth in the common rest frame of Earth and Andromeda. As soon as you reach Andromeda, your captain orders you to send a radio message to Earth announcing your arrival; he claims the message will reach Earth about a century after you left. You claim it will be much later when the message arrives. Who's right?
31. You wish to travel to a star N light years from Earth. How fast must you go if the one-way journey is to occupy N years of your life?
32. The nearest star beyond our solar system is about 4 light years away. If a spaceship can get to the star in 5 years, as measured on Earth, (a) how long would the ship's pilot judge the journey to take? (b) How far from Earth would the pilot find the star to be?
33. Twins A and B live on Earth. On their 20th birthday, twin B climbs into a spaceship and makes a round-trip journey at $0.95c$ to a star 30 light years distant, as measured in the Earth–star reference frame. What are their ages when twin B returns to Earth?

34. Radioactive oxygen-15 decays at such a rate that half the atoms in a given sample decay every 2 min. If a tube containing 1000 O-15 atoms is moved at $0.80c$ relative to Earth for 6.67 min according to clocks on Earth, how many atoms will be left at the end of that time?
35. Two distant galaxies are receding from Earth at $0.75c$ in opposite directions. How fast does an observer in one galaxy measure the other to be moving?
36. Two spaceships are racing. The "slower" one passes Earth at $0.70c$, and the "faster" one moves at $0.40c$ relative to the slower one. What's the faster ship's speed relative to Earth?
37. Use relativistic velocity addition to show that if an object moves at speed $v < c$ relative to some inertial reference frame, then its speed relative to any other inertial frame must also be less than c.
38. Earth and Sun are 8.33 light minutes apart. Event A occurs on Earth at time $t = 0$ and event B on the Sun at $t = 2.45$ min, as measured in the Earth–Sun frame. Find the time order and time difference between A and B for observers (a) moving on a line from Earth to Sun at $0.750c$, (b) moving on a line from Sun to Earth at $0.750c$, and (c) moving on a line from Earth to Sun at $0.294c$.
39. You're writing a galactic history involving two civilizations that evolve on opposite sides of a 1.0×10^5-ly-diameter galaxy. In the galaxy's reference frame, civilization B launched its first spacecraft 45,000 years after civilization A. You and your readers, from a more advanced civilization, are traveling through the galaxy at $0.99c$ on a line from A to B. Which civilization do you record as having first achieved interstellar travel, and how much in advance of the other?
40. Repeat Problem 39, now assuming that civilization B lags A by 1.2 million years in the galaxy's reference frame.
41. Could there be observers who would judge the two events in Problem 39 to be simultaneous? If so, how fast and in what direction must these observers be moving?
42. Could there be observers who would judge the two events in Problem 40 to be simultaneous? If so, how fast and in what direction must these observers be moving?
43. The *Curiosity* rover touched down on Mars when Earth and Mars were 14 light-minutes apart. At the instant of touchdown, clocks at Mission Control in Pasadena, California, read 10:31 PM. As judged by observers on a spacecraft heading along the Earth–Mars line at $0.35c$, did touchdown occur before or after the time the clocks in Pasadena read 10:31 PM, and by how much?
44. Derive the Lorentz transformations for time from the transformations for space.
45. In the light box of Fig. 33.6, let event A be the emission of the light flash and event B its return to the source. Assign suitable space and time coordinates to these events in the frame in which the box moves with speed v. Apply the Lorentz transformations to show that the time between the two events in the box frame is given by Equation 33.3.
46. You're a consultant for the director of a sci-fi movie. The film starts with two spaceships, each measuring 25 m long in its rest frame, approaching Earth in opposite directions with speeds shown in Fig. 33.24. The director wants to know how long to make ship B for scenes shot (a) in Earth's reference frame and (b) in ship A's frame. Your answers?

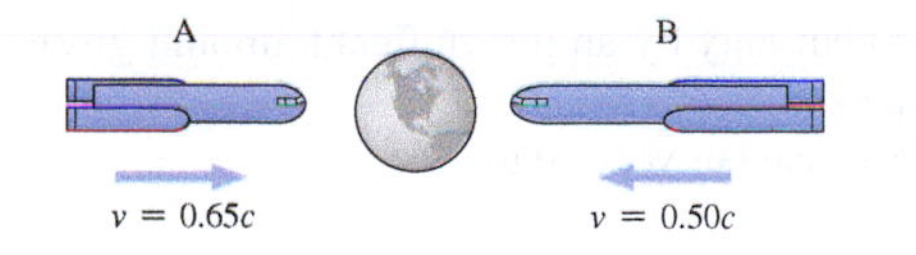

FIGURE 33.24 Problem 46; the drawing is in Earth's reference frame

47. How fast would you have to go to reach a star 240 light years away in an 85-year human lifetime?
48. An advanced civilization has developed a spaceship that goes, with respect to the galaxy, only 50 km/s slower than light. (a) According to the ship's crew, how long does it take to cross the galaxy's 100,000-ly diameter? (b) What's the galactic diameter measured in the ship's reference frame?
49. A spaceship travels at $0.80c$ from Earth to a star 10 light years distant, as measured in the Earth–star reference frame. Let event A be the ship's departure from Earth and event B its arrival at the star. (a) Find the distance and time between the two events in the Earth–star frame. (b) Repeat for the ship's frame. (*Hint:* The distance in the ship frame is the distance an observer has to move *with respect to that frame* to be at both events—not the same as the Lorentz-contracted distance between Earth and star.) (c) Compute the square of the spacetime interval in both frames to show explicitly that it's invariant.
50. Use Equation 33.6 to calculate the square of the spacetime interval between the events (a) of Problem 39 and (b) of Problem 40. Comment on the signs of your answers in relation to the possibility of a causal relationship between the events.
51. A light beam is emitted at event A and arrives at event B. Show that the spacetime interval between the two events is zero.
52. Compare the momentum changes needed to boost a spacecraft (a) from $0.10c$ to $0.20c$ and (b) from $0.80c$ to $0.90c$.
53. Event A occurs at $x = 0$ and $t = 0$ in reference frame S. Event B occurs at $x = 3.8$ light years and $t = 1.6$ years in S. Find (a) the distance and (b) the time between A and B in a frame moving at $0.80c$ along the x-axis of S.
54. When a particle's speed doubles, its momentum increases by a factor of 3. What was its original speed?
55. Find (a) the speed and (b) the momentum of a proton with kinetic energy 500 MeV.
56. The Large Hadron Collider accelerates protons to energies of 14 TeV. (a) Compare that proton energy with the kinetic energy of a 25-mg bug crawling at 2.0 mm/s. (b) Compare the proton momentum with that of the same bug.
57. A large city consumes electrical energy at the rate of 1 GW. If you converted all the rest mass in a 1-g raisin to electrical energy, for how long could it power the city?
58. In a nuclear-fusion reaction, two deuterium nuclei combine to make a helium nucleus plus a neutron, releasing 3.3 MeV of energy in the process. By how much do the combined masses of the helium nucleus and the neutron differ from the combined masses of the original deuterium nuclei?
59. Find the kinetic energy of an electron moving at (a) $0.0010c$, (b) $0.60c$, and (c) $0.99c$. Use suitable approximations where possible.
60. **CH** Find the speed of an electron with kinetic energy (a) 100 eV, (b) 100 keV, (c) 1 MeV, and (d) 1 GeV. Use suitable approximations where possible.
61. **CH** Use the binomial approximation (Appendix A) to show that Equation 33.8 reduces to the Newtonian expression for kinetic energy in the limit $u \ll c$.
62. **CH** Show that Equation 33.10 follows from the expressions for relativistic momentum and total energy.
63. **CH** Show from the Lorentz transformations that the spacetime interval of Equation 33.6 has the same value in all reference frames.
64. How fast would you have to travel to reach the Crab Nebula, 6500 light years from Earth, in 15 years? Give your answer to seven significant figures.
65. At what speed are a particle's kinetic and rest energies equal?

66. The highest-energy cosmic rays ever detected are protons with energies on the order of 300 EeV. (a) What's Earth's radius as measured in the reference frame of such a proton as it approaches Earth? (b) Compare the high-energy proton's total energy with that of a 143-g baseball moving at 100 km/h. (c) Compare the proton's momentum with that of the same baseball.
67. When an object's speed increases by 5%, its momentum increases by a factor of 5. What was its original speed?
68. **CH** Use the Lorentz transformations to show that if two events are separated in space and time so that a light signal leaving one event cannot reach the other, then there is an observer for whom the two events are simultaneous. Show that the converse is also true: If a light signal can get from one event to the other, then no observer will find them simultaneous.
69. **CH** A source emitting light with frequency f moves toward you at speed u. By considering both time dilation and the effect of wavefronts "piling up" as shown in Fig. 14.33, show that you measure a Doppler-shifted frequency given by

$$f' = f\sqrt{\frac{c+u}{c-u}}$$

Use the binomial approximation (Appendix A) to show that this result can be written in the form of Equation 14.15 when $u \ll c$.
70. **CH** Equation 33.5a transforms the velocity $\vec{u}$ of an object moving in the x-direction—the same direction as the relative velocity $\vec{v}$ of the two reference frames. Now suppose the object's velocity also has a component u_y perpendicular to the two frames' relative velocity $\vec{v}$. Find the transformation from u_y' to u_y.
71. **CH** Consider a relativistic particle of mass m moving along a straight line. Use Equation 33.7 to find an expression for the force on the particle, defined as $F = dp/dt$, in terms of its acceleration $a = du/dt$.
72. Find the speed of a particle whose relativistic kinetic energy is 50% greater than the Newtonian value calculated for the same speed.
73. **CH** It's the 24th century, and you're a curator at the Starfleet Museum of Ancient Technology. Archaeologists have unearthed a "TV tube," an ancient device for displaying moving images. Your job is to get it working. One reference says the device accelerated electrons, which then bombarded a screen to produce images; to the electrons, the tube was 57 cm long. You measure the tube and find it's 60 cm long. To get it working, you need to know the electrons' speed and the potential difference needed to accelerate them. The electron's rest energy is 511 keV. Your answers?
74. **CH** Consider a line of positive charge with line charge density λ as measured in a frame S at rest with respect to the charges. (a) Show that the electric field a distance r from this charged line has magnitude $E = \lambda/2\pi\epsilon_0 r$, and that there's no magnetic field (no relativity needed here). Now consider the situation in a frame S' moving at speed v parallel to the line of charge. (b) Show that the charge density measured in S' is given by $\lambda' = \gamma\lambda$, where $\gamma = 1/\sqrt{1 - v^2/c^2}$. (c) Use the result of (b) to find the electric field in S'. Since the charge is moving with respect to S', there's a current in S'. (d) Find an expression for this current and (e) for the magnetic field it produces. Determine the values of the quantities (f) $\vec{E}\cdot\vec{B}$ and (g) $E^2 - c^2B^2$ in both reference frames, and show that these quantities are invariant. Your result gives a hint at how electric and magnetic fields transform, and demonstrates one instance of the fact that $\vec{E}\cdot\vec{B}$ and $E^2 - c^2B^2$ are always invariant.
75. **DATA** The table below lists the total energy and corresponding momentum for a particle. They're measured in MeV and MeV/c, respectively—commonly used units in particle physics. Determine suitable functions of these quantities to plot such that the resulting plot should be a straight line. Plot your data, determine a best-fit line, and use it to find (a) the value of c and (b) the particle's mass. You may need to convert to SI before plotting. Can you identify the particle?

Total energy, E (MeV)	0.511	1.01	1.51	2.51	3.51	4.51	5.51
Momentum, p (MeV/c)	0	0.872	1.41	2.46	3.45	4.61	5.49

Passage Problems

You've been named captain of NASA's first interstellar mission since the Voyager robotic spacecraft. You board your spaceship, accelerate quickly to 0.8c, and cruise at constant speed toward Proxima Centauri, the closest star to our Sun. Proxima Centauri is 4 light-years distant as measured in the two stars' common rest frame. On the way, you conduct various medical experiments to determine the effects of a long space voyage on the human body.

76. Taking your pulse, you find
    a. it's significantly slower than when you're on Earth.
    b. it's the same as when you're on Earth.
    c. it's significantly faster than when you're on Earth.
77. How much do you age during your interstellar journey?
    a. 3 years
    b. just under 4 years
    c. just over 4 years
    d. 5 years
78. Back on Earth, Mission Control judges that your shipboard clocks run slow. What do you judge about clocks at Mission Control?
    a. They run fast.
    b. They keep time at the same rate as your clocks.
    c. They run slow.
    d. You can't tell anything about their clocks.
79. In your spaceship's reference frame, the distance from the Sun to Proxima Centauri is
    a. 2.4 light years.
    b. just under 4 light years.
    c. 4 light years.
    d. 5 light years.

Answers to Chapter Questions

Answer to Chapter Opening Question

The laws of physics are the same for all observers, regardless of their state of motion.

Answers to GOT IT? Questions

33.1 (a)
33.2 (c)
33.3 (a)
33.4 (b) but only by an insignificant amount given the low relative speed
33.5 less than but very nearly c

Particles and Waves

What You Know

- You understand concepts of energy and angular momentum.
- You've seen energy calculations for circular orbits with gravity.
- You've worked with the Stefan–Boltzmann law for the energy radiated by a hot object.
- You know that, according to Maxwell's equations, accelerated charge is the source of electromagnetic radiation, including light.
- You have a solid understanding of Newtonian physics and how it can be used make deterministic predictions of future motion.

What You're Learning

- You'll see how blackbody radiation, the photoelectric effect, the Compton effect, and atomic spectra all contradict the predictions of classical physics.
- You'll learn how the quantization hypotheses of Planck, Einstein, and Bohr overcame these classical contradictions.
- You'll see how *Planck's constant* sets the scale at which quantum effects become noticeable.
- You'll use the quantization condition $E = hf$ to determine photon energies.
- You'll learn to calculate photon energies and wavelengths associated with electron transitions in Bohr's atomic model.
- You'll learn about the wave–particle duality and how to calculate the wavelengths of matter particles.
- You'll learn the uncertainty principle and Bohr's complementarity principle, and how they're at the basis of our understanding of quantum physics.

How You'll Use It

- Chapter 34 presents an overview of early quantum ideas. In Chapter 35 you'll see a more focused and quantitative development of quantum physics.
- Subsequent chapters will apply quantum ideas to atoms, molecules, solids, and atomic nuclei.

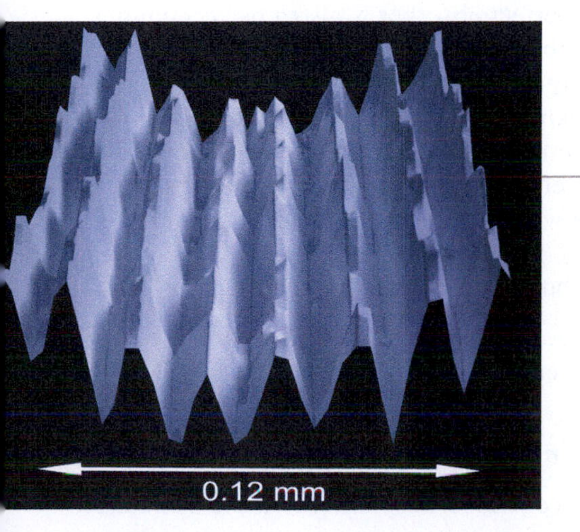

Ridges and valleys represent bright and dark fringes in the interference pattern produced by two beams of ultracold sodium atoms. What does this picture tell us about the nature of matter?

Newtonian mechanics and Maxwell's electromagnetism constitute the core of **classical physics**, providing a deep understanding of physical reality. Although these theories were firmly established by the middle of the 19th century, they remain central to the work of many contemporary scientists and engineers.

Nevertheless, at the end of the 19th century a few seemingly minor phenomena defied classical explanation. Most physicists felt that it was only a matter of time before these, too, came under the classical umbrella. But that was not to be. We've seen how questions about light led to a radical restructuring of our concepts of space and time. Other questions, especially those concerning matter at the atomic scale, brought about an even more radical transformation of physical thought.

This chapter explores some phenomena that led to quantum physics and recounts early attempts to explain them. The next chapter gives a fuller account of quantum theory, and subsequent chapters explore its application to atoms, molecules, nuclei, and quantum-based technologies.

34.1 Toward Quantum Theory

Are matter and energy continuously divisible? The essential difference between classical and quantum physics is that the former answers this question "in general, yes," while the latter says definitively "no." Most physical quantities are **quantized**, coming in only certain discrete values.

The idea that physical quantities might come in discrete "chunks" is not new. Some 2400 years ago, the Greek philosopher Democritus proposed that all matter consists of indivisible atoms. By the start of the 20th century, a more scientifically grounded atomic theory was widely accepted. J. J. Thomson's discovery of the electron in 1897 showed that atoms might be divisible after all, but at the same time it revealed a finer division of matter into discrete "chunks." Robert A. Millikan's 1909 oil-drop experiment showed that electric charge is similarly quantized. Discovery of the proton and later the neutron further solidified the notion that matter comprises fundamental building blocks with definite values for their physical properties.

Quantization of matter into particles with discrete properties is not incompatible with classical physics as long as those particles behave according to classical laws—in particular, that they move continuously through space and can have *any* amount of energy. Add electromagnetism to the picture and the classical viewpoint requires that the electric and magnetic fields be continuous, exerting forces on charged particles and changing, in a gradual and continuous way, the particles' energies.

The startling fact of quantum physics is that this classical behavior does not occur at the atomic scale; instead, energy itself is often quantized. Reconciling the implications of that fact with our commonsense notions of matter and motion has proved impossible; instead, the quantum world speaks a different language, one in which deeply ingrained ideas about causality and the solid reality of matter seem no longer to apply. Here we look at three distinct phenomena that force us to accept the idea that energy can be quantized.

34.2 Blackbody Radiation

Heat an object hot enough and it glows, emitting electromagnetic radiation in the form of light. As we saw in Section 16.3, the total power radiated is proportional to the fourth power of the temperature. There's also a change in wavelength with increasing temperature: The first visible glow is a dull red, changing with higher temperatures to orange and then yellow colors corresponding to ever-shorter wavelengths.

A perfect absorber of electromagnetic radiation is called a **blackbody** because it absorbs all light and thus appears black. A perfect absorber is also a perfect emitter, and when a blackbody is heated it emits electromagnetic radiation in a broad range of wavelengths; this is **blackbody radiation**. Many objects—such as the Sun or an electric-stove burner—behave approximately like blackbodies. An excellent approximation to a blackbody is a hollow piece of *any* material with a small hole. As Fig. 34.1 shows, any radiation entering the hole undergoes multiple reflections and is eventually absorbed. The hole, therefore, is a nearly perfect absorber, so when the material is heated, the radiation emerging is blackbody radiation.

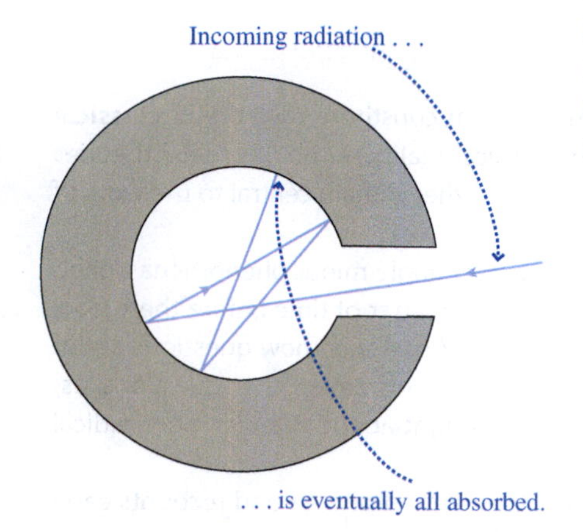

FIGURE 34.1 A cavity with a small hole absorbs nearly all incident radiation and hence is a near-perfect blackbody.

Experimental study of blackbody radiation shows three characteristic features:

1. The radiation covers a continuous range of wavelengths, with the total power radiated at all wavelengths combined given by the Stefan–Boltzmann law that we introduced in Chapter 16:

$$P_{\text{blackbody}} = \sigma A T^4 \tag{34.1}$$

   where A is the area of the radiating surface, T is its absolute (kelvin) temperature, and $\sigma = 5.67 \times 10^{-8}\ \text{W}/(\text{m}^2 \cdot \text{K}^4)$ is the Stefan–Boltzmann constant.
2. The radiation peaks at a wavelength that's inversely proportional to the temperature; this is known as **Wien's law**.
3. The distribution of wavelengths depends only on temperature, not on the material of which the blackbody is made.

A blackbody's **radiance** measures the radiated power as a function of wavelength. Because the blackbody emits a continuous spectrum, we have to express radiance as power per unit spectral interval. If we choose intervals in wavelength, then the relation implied in feature 2 above gives a peak radiance at wavelength λ_{peak} such that

$$\lambda_{\text{peak}}T = 2.898\,\text{mm}\cdot\text{K} \qquad \text{(Wien's law)} \tag{34.2a}$$

We emphasize, however, that the choice of fixed wavelength intervals is arbitrary. If we had chosen fixed frequency intervals, then the constant in Equation 34.2a would be different, and a plot of radiance versus wavelength would peak at a different wavelength (see Problem 78). So λ_{peak} in Equation 34.2a is not some absolute measure of the wavelength at which the blackbody emits the "most" radiation, but rather the wavelength of the maximum radiation if you choose to keep track of power in intervals of fixed wavelength. A more physically based quantity is the median wavelength, below and above which half the power is radiated; it's given by

$$\lambda_{\text{median}}T = 4.11\,\text{mm}\cdot\text{K} \tag{34.2b}$$

Whatever measure one chooses, though, the important point is that the peak wavelength is inversely proportional to temperature. In our subsequent discussion, we'll adopt a definition of radiance as the power emitted per unit area per unit wavelength interval; then Equation 34.2a describes the peak wavelength. Figure 34.2 plots this measure of blackbody radiance at three temperatures.

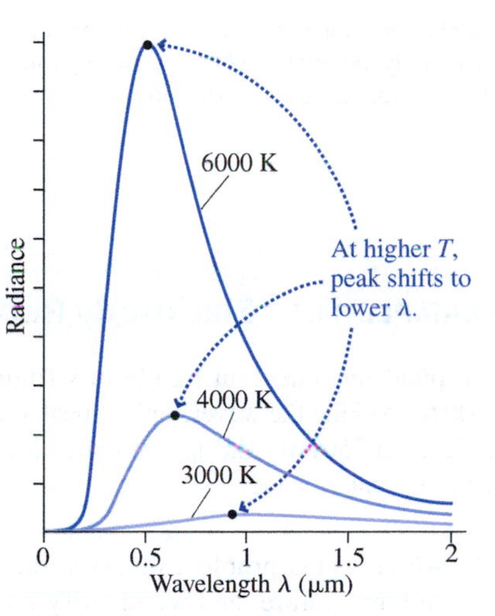

FIGURE 34.2 Blackbody radiance—energy per unit wavelength interval—as a function of wavelength.

Microscopically, blackbody radiation is associated with the thermal motions of atoms and molecules, so it's not surprising that the radiation increases with temperature. In the late 1800s, physicists tried to apply the laws of electromagnetism and statistical mechanics to explain the experimental observations of blackbody radiation. They met with some success in describing such aspects as the T^4 dependence of the total energy radiated, and the shifting of the radiation distribution toward shorter wavelengths with increasing temperature, but they could not reproduce the actual observed distribution at all wavelengths.

In 1900, the German physicist Max Planck formulated an equation that fit the observed radiance curves for blackbody radiation at all wavelengths:

$$R(\lambda, T) = \frac{2\pi hc^2}{\lambda^5(e^{hc/\lambda kT} - 1)} \tag{34.3}$$

Two familiar quantities here are Boltzmann's constant $k = 1.38\times10^{-23}$ J/K, introduced in Chapter 17, and the speed of light c. A new quantity is the constant h, whose value Planck chose in order to make the equation fit experimental data.

Planck first presented his law as a purely empirical equation describing blackbody experiments. Later he showed that his equation had a remarkable physical interpretation:

> The energy of a vibrating molecule is quantized, meaning it can have only certain discrete values. Specifically, if f is the vibration frequency, then the energy must be an integer multiple of the quantity hf:
>
> $$E = nhf, \qquad n = 0, 1, 2, 3, \ldots \tag{34.4}$$

where h is the constant Planck introduced in Equation 34.3. Today we know h as one of the fundamental constants of nature and call it **Planck's constant**. Its value is approximately 6.63×10^{-34} J·s, and it's because h is so small that quantum phenomena are usually obvious only in the atomic and molecular realm. Planck's quantization of the energy of vibrating molecules implies further that a molecule can absorb or emit energy only in discrete "bundles" of size hf, and that in doing so it jumps abruptly from one of its allowed energy levels to another (Fig. 34.3). (Later developments showed that Planck was correct about the size of the energy jumps but that the factor n in Equation 34.4 should actually be $n + \frac{1}{2}$.)

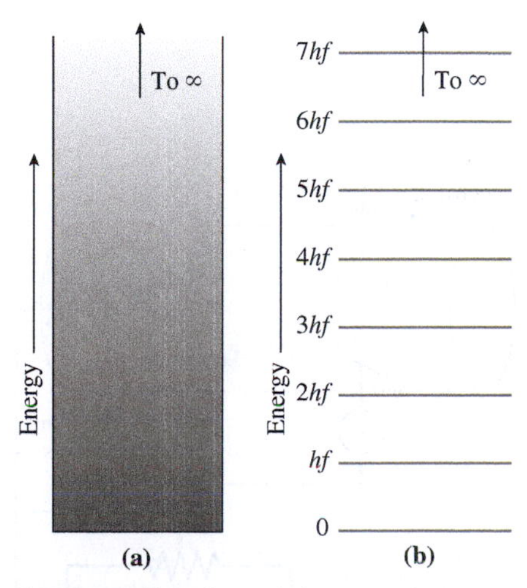

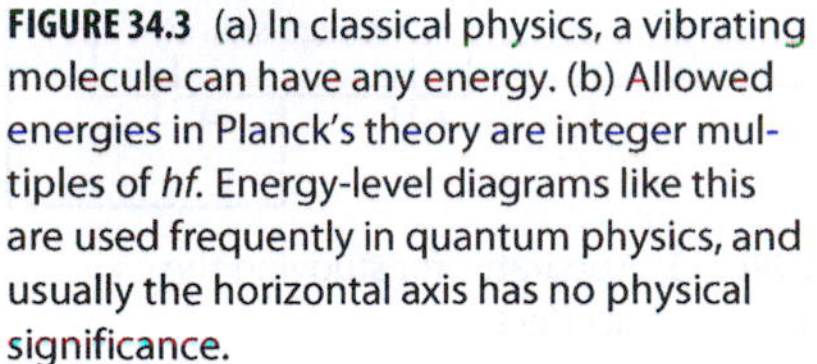

FIGURE 34.3 (a) In classical physics, a vibrating molecule can have any energy. (b) Allowed energies in Planck's theory are integer multiples of hf. Energy-level diagrams like this are used frequently in quantum physics, and usually the horizontal axis has no physical significance.

Planck himself was very conservative and reluctant to accept or elaborate on his theory's evident disagreement with classical physics; nevertheless, his revolutionary work won Planck the 1918 Nobel Prize. Other physicists subsequently emphasized the contrast

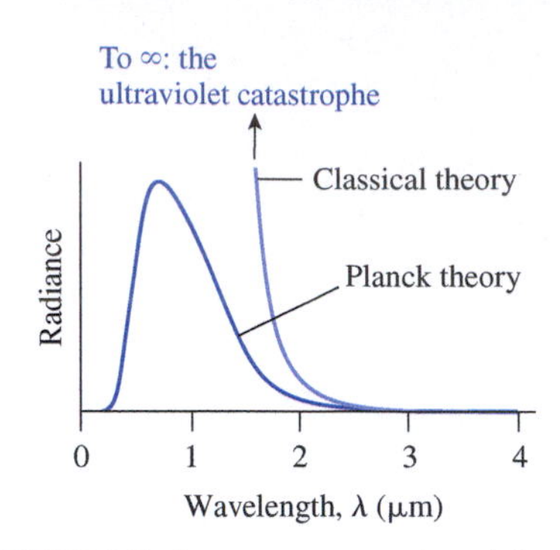

FIGURE 34.4 Radiance versus wavelength for blackbody radiation at 6000 K, showing also the incorrect classical prediction.

between Planck's work and the classical treatment of blackbody radiation. That earlier treatment, based on the assumption that energy is shared equally among all possible vibrational modes, had led to the **Rayleigh–Jeans law** for the radiance of a blackbody:

$$R(\lambda, T) = \frac{2\pi ckT}{\lambda^4} \tag{34.5}$$

Not only did the Rayleigh–Jeans law contradict experimental measurements, but it also led to the absurd conclusion that every object, at every nonzero temperature, should emit electromagnetic energy at an infinite rate, with that energy concentrated at the shortest wavelengths (Fig. 34.4). Since the shortest wavelength electromagnetic radiation known at the time was ultraviolet light, this phenomenon was called the **ultraviolet catastrophe**. In Planck's equation, the exponential term in the denominator grows rapidly with decreasing wavelength, diminishing the radiance and averting the ultraviolet catastrophe. Problems 72, 76, and 79 show that Planck's law reduces to the Rayleigh–Jeans law for longer wavelengths, and that it also leads to Wien's law and the Stefan–Boltzmann law.

EXAMPLE 34.1 Blackbody Radiation: Lightbulb Efficiency

A typical incandescent lightbulb's filament temperature is about 2900 K. (a) Find the wavelength of peak radiance, and (b) compare the radiance at 550 nm—the approximate center of the visible spectrum—with the peak radiance.

INTERPRET This problem involves radiation from an object at a known temperature, and we identify the filament as a blackbody. We're asked for both the peak wavelength and a comparison of radiances at two different wavelengths. We're implicitly adopting our definition of radiance as power emitted per unit area per unit wavelength, and our answers will reflect this choice.

DEVELOP Equation 34.2a gives the peak wavelength, and Equation 34.3 gives the radiance as a function of wavelength. So our plan for (a) is to solve Equation 34.2a, $\lambda_{\text{peak}}T = 2.898\,\text{mm}\cdot\text{K}$, for $T = 2900\,\text{K}$. For (b) we'll form a ratio of radiances from Equation 34.3, using the result of (a) and the given 550-nm visible wavelength.

EVALUATE (a) Equation 34.2a gives $\lambda = 2.898\,\text{mm}\cdot\text{K}/2900\,\text{K} = 1.0\,\mu\text{m}$. (b) To compare radiances, we form a ratio of the right-hand sides of Equation 34.3 evaluated at the wavelengths $\lambda_2 = 550\,\text{nm}$ and $\lambda_1 = 1.0\,\mu\text{m}$ or 1000 nm. The numerators cancel, giving

$$\frac{R(\lambda_2, T)}{R(\lambda_1, T)} = \frac{\lambda_1^5(e^{hc/\lambda_1 kT} - 1)}{\lambda_2^5(e^{hc/\lambda_2 kT} - 1)} = 0.34$$

ASSESS Our 1.0-μm answer for (a) lies in the infrared, suggesting that incandescent lightbulbs aren't very efficient at producing visible light; that's the reason they're being phased out. Our answer to (b) confirms this point, showing there's much less radiance in the visible than at the infrared peak. And remember that we've defined radiance as power per unit area per unit wavelength interval. If we adopt the more physical median wavelength given by $\lambda_{\text{median}}T = 4.11\,\text{mm}\cdot\text{K}$, we find $\lambda_{\text{median}} = 1.4\,\mu\text{m}$, well into the infrared. Since half the radiation occurs at wavelengths longer than this median, the bulb emits far more than half its energy as invisible infrared. ■

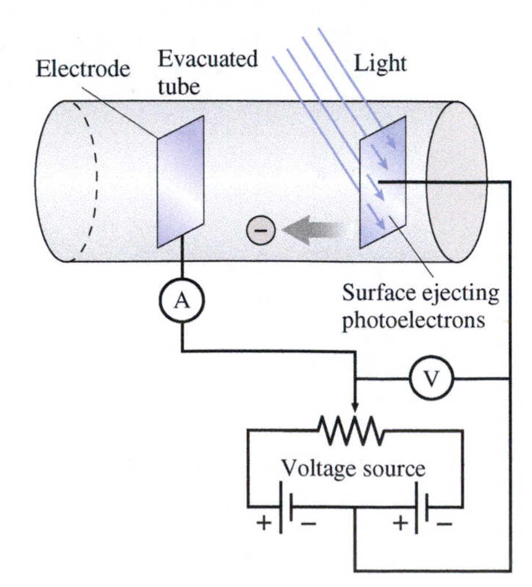

FIGURE 34.5 Apparatus for studying the photoelectric effect.

GOT IT? 34.1 Two identical blackbodies, A and B, are heated until A's temperature is twice B's. Compare (1) their total radiated power and (2) their wavelengths of peak radiance.

34.3 Photons

Planck showed that vibrating molecules could exchange energy with electromagnetic radiation only in quantized bundles of size *hf*. Is the radiation's energy similarly quantized?

The Photoelectric Effect

In 1887 Heinrich Hertz observed that metals emit electrons when struck by light. Observations of this **photoelectric effect** continued with experiments involving metal electrodes in evacuated glass containers (Fig. 34.5). Illuminating one electrode causes it to emit electrons. Making the second electrode positive attracts the electrons, and the resulting current measures the rate at which electrons are ejected. Make the second electrode sufficiently negative, on the other hand, and the electron energy isn't great enough to overcome the repulsive potential; then the current ceases. This so-called *stopping potential* provides a measure of the ejected electrons' maximum kinetic energy $K_{max} = eV_s$.

Classical physics suggests that the photoelectric effect should occur because an electron experiences a force in the oscillating electric field of a light wave. As the electron absorbs energy from the wave, the amplitude of its motion should grow until eventually it has enough energy to escape from the metal. Because the energy in a wave is spread throughout the entire wave, it might take a while for a single tiny electron to absorb enough energy. Increasing the light intensity should increase the electric field, resulting in the electron being ejected sooner and with more energy. Changing the wave frequency should have little effect.

The photoelectric effect does occur, but not in the way classical physics suggests. Figure 34.6 shows results from a photoelectric experiment, in the form of current versus voltage as read by the meters in Fig. 34.5. These results, along with observations made by varying the frequency of the incident light, show three major disagreements with the classical prediction:

1. Current begins immediately, showing that electrons are ejected immediately, even in dim light.
2. The maximum electron energy, as measured by the stopping potential V_s, is independent of the light intensity.
3. Below a certain cutoff frequency *no* electrons are emitted, no matter how intense the light. Above the cutoff frequency electrons are emitted with a maximum energy that increases in proportion to the light-wave frequency.

PhET: Photoelectric Effect

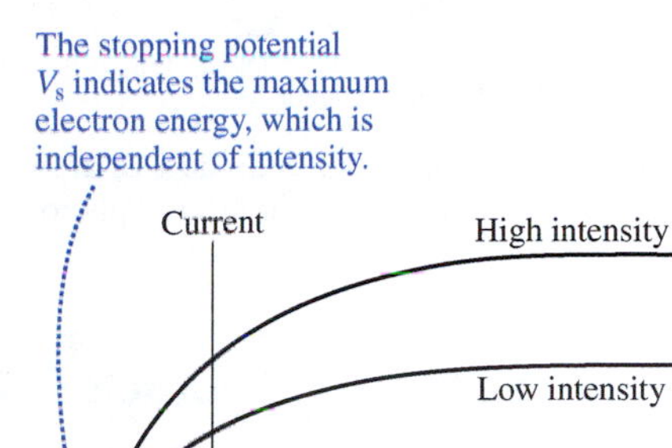

FIGURE 34.6 Current versus voltage for the photoelectric experiment of Fig. 34.5, shown for two light intensities at the same frequency.

In 1905, the same year he formulated the special theory of relativity, Albert Einstein proposed an explanation for the photoelectric effect. Einstein suggested that an electromagnetic wave's energy is concentrated in "bundles" called **quanta** or **photons**. Einstein applied to these photons the same energy-quantization condition that Planck had already proposed for molecular vibrations: that photons in light with frequency f have energy hf, where again h is Planck's constant:

$$E = hf \qquad \text{(photon energy)} \tag{34.6}$$

The more intense the light, the more photons—but the energy of each photon is unrelated to the light intensity.

Table 34.1 Work Functions

Element (Symbol)	ϕ (eV)
Silver (Ag)	4.26
Aluminum (Al)	4.28
Cesium (Cs)	2.14
Copper (Cu)	4.65
Potassium (K)	2.30
Sodium (Na)	2.75
Nickel (Ni)	5.15
Silicon (Si)	4.85

Einstein's idea explains all three nonclassical aspects of the photoelectric effect. Each material has a minimum energy—called the *work function*, ϕ—required to eject an electron. (Table 34.1 lists work functions for selected elements.) Since the energy in a photon of light with frequency f is hf, the photons in low-frequency light have less energy than the work function and are therefore unable to eject electrons—no matter how many photons there are. At the cutoff frequency, the photon energy equals the work function, and the photons have just enough energy to eject electrons. As the frequency increases still further, the electrons emerge with maximum kinetic energy K equal to the difference between the photon energy and the work function:

$$K_{max} = hf - \phi \tag{34.7}$$

Thus, the electrons' maximum kinetic energy depends only on the photon energy—that is, on the light frequency but not on its intensity (Fig. 34.7). Finally, the immediate ejection of electrons follows because an individual photon delivers its entire bundle of energy to an electron all at once. Einstein received the 1921 Nobel Prize primarily for his explanation of the photoelectric effect rather than for his more controversial relativity theories. In 1914 Millikan, who had earlier demonstrated the quantization of electric charge, carried out meticulous photoelectric experiments that confirmed Einstein's hypothesis and helped earn Millikan the 1923 Nobel Prize.

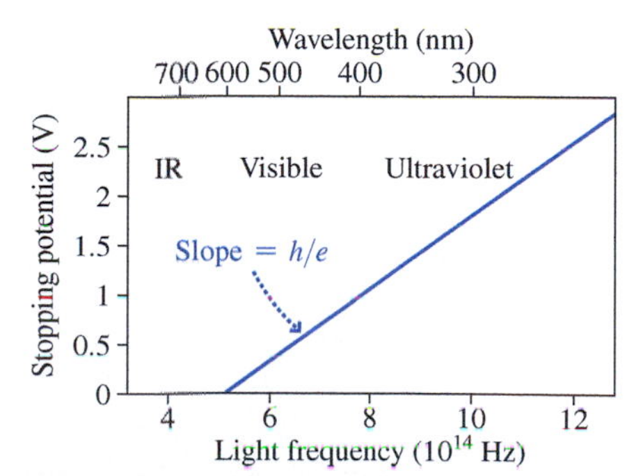

FIGURE 34.7 Results of a photoelectric experiment, showing stopping potential as a function of light frequency and wavelength. The stopping potential in volts is a direct measure of the electron energy in electronvolts.

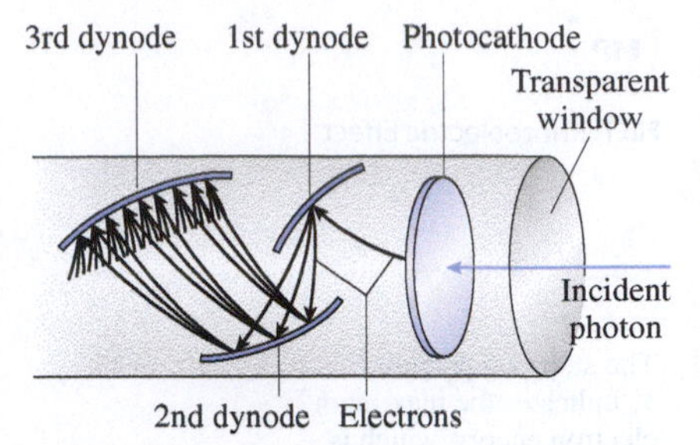

FIGURE 34.8 A photomultiplier produces a large pulse of electrons from a single incident photon.

GOT IT? 34.2 If you replot Fig. 34.7 for a material with a different work function, (1) will the slope of the line change? (2) Will the point at which it intersects the horizontal axis change?

Today, the photoelectric effect is used in extremely sensitive light detectors called **photomultipliers**. In these devices, one photon dislodges a single electron, which in turn liberates multiple electrons from an electrode called a *dynode*; a chain of dynodes then results in a cascade of as many as a billion electrons for each incident photon (Fig. 34.8).

EXAMPLE 34.2 The Photoelectric Effect: Designing a Photomultiplier

(a) Choose a suitable material from Table 34.1 for the light-sensitive surface in a photomultiplier that will respond to visible light at wavelengths of 575 nm and shorter. (b) Find the maximum kinetic energy of electrons ejected with the shortest-wavelength visible light, around 400 nm.

INTERPRET This problem is about the photoelectric effect. In (a) we're asked for a material in which 575-nm light can eject electrons. That means the work function can be no greater than the energy of a 575-nm photon. In (b) we need to find the excess electron energy, over the work function, for 400-nm light.

DEVELOP Equation 34.6, $E = hf$, relates the quantized energy of a photon to its frequency. Since $f\lambda = c$, we can rewrite Equation 34.6 as $E = hc/\lambda$. We'll use this to find the photon energy, and then we'll consult Table 34.1 for an appropriate material. Finally, we can use $f = c/\lambda$ in Equation 34.7, $K_{max} = hf - \phi$, to get the maximum kinetic energy of electrons when $\lambda = 400$ nm.

EVALUATE (a) At 575 nm, $E = hc/\lambda = 3.46\times10^{-19}$ J, or 2.16 eV, where 1 eV $= 1.6\times10^{-19}$ J. This energy must be enough for the electron to overcome the work function; the only material in Table 34.1 for which this is possible is cesium, with $\phi = 2.14$ eV. (b) At $\lambda = 400$ nm and $f = c/\lambda$, Equation 34.7 gives $K_{max} = hc/\lambda - \phi = 0.96$ eV Here we converted the SI value of hc/λ to electronvolts before subtracting ϕ.

ASSESS Make sense? The work function we've chosen is just under the 2.16-eV photon energy at 575 nm, so electrons ejected with photons of this wavelength have negligible kinetic energy. The 400-nm minimum visible wavelength corresponds to roughly 50% higher frequency and therefore energy, or roughly 3 eV. It takes about 2 eV to overcome the work function, leaving about 1 eV of electron kinetic energy.

✓TIP Working with Electronvolts

Recall that 1 electronvolt (eV) is the *energy* gained by an electron across a 1-V potential. So electronvolts are a unit of energy, but *not* the standard SI unit, which is the joule. We computed an energy $E = hc/\lambda$ in SI units and then converted to electronvolts using the factor 1 eV $= 1.6\times10^{-19}$ J. In general it's safest to work in SI units and then convert to eV as needed.

■

Waves or Particles?

In positing the existence of photons, Einstein gave the first inklings of the **wave–particle duality**—the seemingly dual nature of light, which acts in some situations like a wave and in others, as in the photoelectric effect, more like a localized particle. We now turn to another phenomenon that demonstrates light's particle-like aspect. Later we'll see how the wave–particle duality encompasses not only light but matter as well.

The Compton Effect

In 1923 the American physicist Arthur Holly Compton, at Washington University in St. Louis, did an experiment that dramatically confirmed the particle-like aspect of electromagnetic radiation. Although Compton's work came much later in the history of quantum theory than Einstein's, we include it here because it so strongly corroborates Einstein's photon hypothesis.

Compton was studying the interaction of X rays with electrons. Classically, an electron subject to an electromagnetic wave should undergo oscillatory motion, driven by the wave's oscillating electric field. Since accelerated charge is the source of electromagnetic waves, the electron should itself produce electromagnetic waves *of the same frequency as*

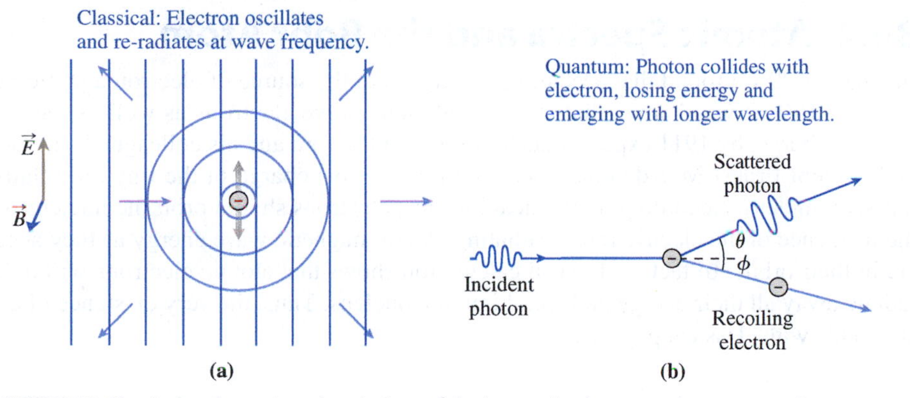

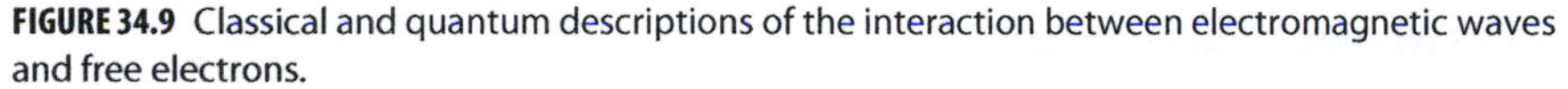

FIGURE 34.9 Classical and quantum descriptions of the interaction between electromagnetic waves and free electrons.

the incident waves (Fig. 34.9*a*). As we saw in Section 29.7, the electron should radiate in all directions, with maximum radiation perpendicular to its oscillatory motion.

Compton and his coworkers measured the intensity of scattered X rays as a function of wavelength for different scattering angles. Remarkably, they found the greatest concentration of scattered X rays at a wavelength *longer* than that of the incident radiation (Fig. 34.10). They interpreted their results as implying that particle-like photons had collided with electrons, losing energy to the electrons and therefore, since $E = hf$, emerging with lower frequency and correspondingly longer wavelength (Fig. 34.9*b*).

We can understand this **Compton effect** by treating the interaction as an elastic collision between the incident photon and a stationary electron. The photon moves at c, so it's necessary to use relativistic expressions for energy and momentum. You can work out the details in Problem 77; the result gives the **Compton shift**: $\Delta\lambda = \lambda - \lambda_0$—that is, the change from the photon's original wavelength λ_0:

$$\Delta\lambda = \frac{h}{mc}(1 - \cos\theta) \qquad \text{(Compton shift)} \qquad (34.8)$$

Figure 34.10 shows that this equation is in excellent agreement with experimental data.

The term h/mc in Equation 34.8 is the **Compton wavelength** of the electron and gives the wavelength shift for a photon scattering at $\theta = 90°$. Its value is $\lambda_C = h/mc = 0.00243\,\text{nm}$, or 2.43 pm. Equation 34.8 shows that the largest wavelength shift will be $2\lambda_C$, occurring at $\theta = 180°$. For the shift to be noticeable it should be a significant fraction of the incident wavelength, which therefore can't be too many times the Compton wavelength. For X rays, λ is in the range from approximately 10 pm to 10 nm, and therefore, detection of the Compton shift in X rays is already difficult. It would be totally impossible with visible light.

Today, Compton scattering with gamma rays is a widely used technique for studying the structure of matter. For example, abnormalities in human bone can be detected through Compton scattering of gamma rays emitted by a radioactive source embedded in bone. And the inverse Compton effect—the scattering of a rapidly moving electron off a photon—is a common process in high-energy astrophysical systems and is used in the laboratory to produce beams of gamma radiation.

The wavelength shift in Compton scattering admits no classical explanation. Coming after a decade of experimental and theoretical work that pointed increasingly to quantization as the essence of the atomic world, Compton's experimental results were for many physicists the convincing evidence for the reality of quanta.

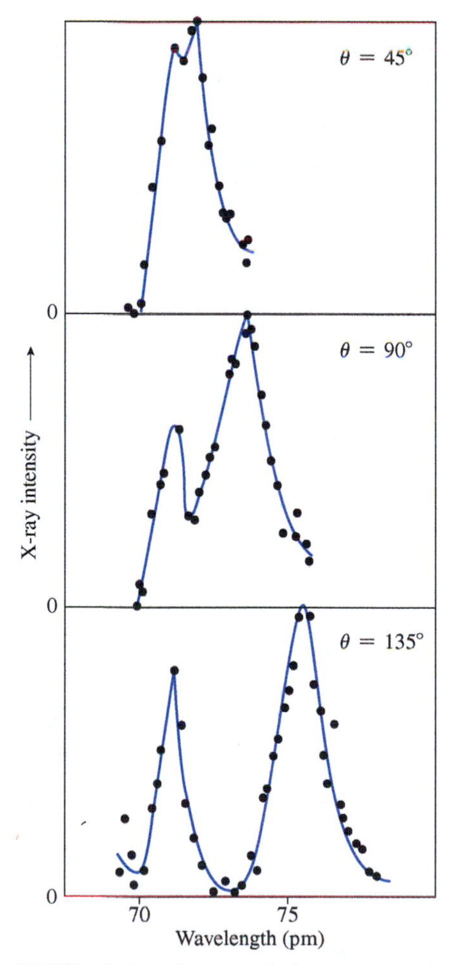

FIGURE 34.10 Compton's results for scattering of X rays with $\lambda = 71$ pm. Right-hand peak shows the wavelength shift of the Compton effect. The unshifted left-hand peak is from photons scattering off tightly bound atomic electrons, which don't absorb significant energy. Solid curves are theoretical predictions.

GOT IT? 34.3 Will the Compton wavelength shift be (a) greater or (b) less for photons of a given wavelength scattering off protons rather than electrons?

34.4 Atomic Spectra and the Bohr Atom

PhET: Neon Lights & Other Discharge Lamps

In Chapter 29 we found that accelerated charges are the source of electromagnetic radiation. By 1900 it was known that atoms contain negative electrons as well as regions of positive charge; by 1911 experiments by Ernest Rutherford and his colleague Hans Geiger and student Ernest Marsden had localized the positive charge in the tiny but relatively massive nucleus. According to classical physics, electrons should orbit the nucleus under the influence of the electric force, radiating electromagnetic wave energy as they accelerate in their orbits. In fact, a classical calculation shows that atomic electrons will quickly radiate away all their energy and spiral into the nucleus. Thus, the very existence of atoms is at odds with classical physics.

The Hydrogen Spectrum

A more subtle problem involving atoms dates to 1804, when William Wollaston noticed lines between some of the colors dispersed by a prism. Ten years later, the German optician Josef von Fraunhofer dispersed the solar spectrum sufficiently that he could see hundreds of narrow, dark lines against the otherwise continuous spectrum. Studies of light emitted by diffuse gases excited by electric discharges show similar **spectral lines**, these bright against an otherwise dark background (recall Fig. 30.16). Such **emission spectra** result when atoms emit light of discrete frequencies. **Absorption spectra**, in contrast, arise when atoms in a diffuse gas absorb discrete frequencies of light from a continuous source. We emphasize the word "diffuse": Discrete spectra generally arise only when a gas is sufficiently diffuse that light from one atom stands a strong chance of escaping the gas before it interacts with other atoms. In dense gases, multiple interactions result in the continuous spectrum of blackbody radiation.

Every element produces its own unique spectral lines, so analysis of spectra, even from the remote reaches of the cosmos, allows us to identify and characterize the material emitting the light. Spectral analysis led to the discovery of helium in the Sun's atmosphere before that element had been identified on Earth—hence the name, from the Greek word *helios* for Sun. Measuring the Doppler shift of spectral lines lets us "see" stars orbiting black holes in distant galaxies, and also gives direct evidence for the expansion of the universe. Back on Earth, the technique of atomic absorption spectroscopy uses spectral lines to determine the elemental composition of substances, helping identify pollutants or trace the flow of elements in biological samples.

In 1884, a Swiss schoolteacher named Johann Balmer realized that the wavelengths of the first four lines in the visible spectrum of hydrogen (see Fig. 30.16) were related by the equation

$$\frac{1}{\lambda} = R_{\mathrm{H}}\left(\frac{1}{2^2} - \frac{1}{n^2}\right)$$

where $n = 3, 4, 5, 6, \ldots$ and R_{H} is the **Rydberg constant for hydrogen**, with the approximate value $1.0968\times10^7\,\mathrm{m}^{-1}$. Other series of lines in the hydrogen spectrum were soon found, and Balmer's equation was generalized to

$$\frac{1}{\lambda} = R_{\mathrm{H}}\left(\frac{1}{n_2^2} - \frac{1}{n_1^2}\right) \qquad (34.9)$$

where $n_1 = n_2 + 1, n_2 + 2, \ldots$. The Balmer series of lines has $n_2 = 2$; the Lyman series, in the ultraviolet, has $n_2 = 1$; and the infrared Paschen series has $n_2 = 3$. There are in fact infinitely many such series, corresponding to $n_2 = 1, 2, 3, \ldots$.

But why should atoms emit discrete spectral lines? And why should the hydrogen lines form patterns with the simple regularity of Equation 34.9?

The Bohr Atom

In 1913, the great Danish physicist Niels Bohr proposed an atomic theory that accounted for the spectral lines of hydrogen. In the **Bohr atom** the electron moves in a circular orbit about the nucleus, held by the electric force. Classically, any orbital radius and correspondingly any energy and angular momentum should be possible. But Bohr quantized the atom, stating that the only possible orbits were those with angular momentum an integer multiple of Planck's constant divided by 2π. Angular momentum quantization implies energy quantization, which, as we'll show, leads to Equation 34.9 for the hydrogen spectral lines.

Bohr asserted that an electron in an allowed orbit does not radiate energy, in contradiction to the predictions of classical electromagnetism. But an electron can jump from one orbit to another, emitting or absorbing a photon whose energy is equal to the energy difference between the two orbital levels. We can therefore find the expected photon energies—and the corresponding wavelengths—if we know the allowed energy levels.

To find the quantized atomic energy levels in Bohr's model, consider a hydrogen atom consisting of a fixed proton and an electron in circular orbit. Treating the proton as fixed is a good approximation because its mass is nearly 2000 times the electron's. We consider only electron speeds much less than that of light, which is an excellent approximation in hydrogen.

In Example 11.1 we found that the angular momentum of a particle with mass m and speed v, moving in a circular path of radius r, is mvr. Thus, Bohr's quantization condition reads

$$mvr = n\hbar \qquad \text{(quantization, Bohr atom)} \qquad (34.10)$$

where $n = 1, 2, 3, \ldots$ and where we define $\hbar \equiv h/2\pi$ (read "h bar"). We need to relate the electron's angular momentum to its energy so we can find out what Equation 34.10 implies about energy quantization.

You studied circular orbits for the inverse-square force of gravity in Chapter 8, where you saw that kinetic and potential energies in a circular orbit are related by $K = -\frac{1}{2}U$, with the zero of potential energy taken at infinity. The total energy $K + U$ is therefore $\frac{1}{2}U$. These results hold for any $1/r^2$ force, including the electric force. In the electric case the potential energy U is the point-charge potential of the proton, ke/r, multiplied by the electron charge, $-e$. Then the total energy is $E = \frac{1}{2}U = -ke^2/2r$. The minus sign means the electron is *bound* to the proton, in that it would take energy to separate them. Solving this equation for r then gives

$$r = -\frac{ke^2}{2E} \qquad (34.11)$$

Since the kinetic energy is $K = -\frac{1}{2}U = -E$, we also have $\frac{1}{2}mv^2 = -E$ or $v = \sqrt{-2E/m}$. Using our expressions for r and v in the quantization condition 34.10 gives $m\sqrt{-2E/m}\,(-ke^2/2E) = n\hbar$. Solving for the energy E, we find

$$E = -\frac{k^2e^4m}{2\hbar^2n^2}$$

It's convenient to define the **Bohr radius**, a_0, as

$$a_0 = \frac{\hbar^2}{mke^2} = 0.0529\,\text{nm}$$

With this definition the energy becomes

$$E = -\frac{ke^2}{2a_0}\left(\frac{1}{n^2}\right) \qquad \text{(energy levels, Bohr atom)} \qquad (34.12a)$$

Equation 34.12a gives us the allowed energy levels under Bohr's quantization condition. Evaluating this expression for the case $n = 1$ gives $E_1 = -2.18\times10^{-19}\,\text{J} = -13.6\,\text{eV}$; it's then convenient to write Equation 39.12a numerically in the form

$$E = -\frac{13.6\,\text{eV}}{n^2} \qquad (34.12b)$$

where in both forms $n = 1, 2, 3, \ldots$. The lowest energy state, $n = 1$, is called the **ground state**; the others are **excited states**.

Now we have the allowed energy levels. What about spectra? When an electron jumps between energy levels, it emits or absorbs a photon whose energy hf is equal to the energy difference between the levels. So imagine an electron going from a higher level n_1 to a lower level n_2. The energy difference, according to Equation 34.12a, is

$$\Delta E = -\frac{ke^2}{2a_0}\left(\frac{1}{n_1^2} - \frac{1}{n_2^2}\right) = \frac{ke^2}{2a_0}\left(\frac{1}{n_2^2} - \frac{1}{n_1^2}\right)$$

and this is equal to the energy of the emitted photon. But the photon energy is $\Delta E = hf = hc/\lambda$, and therefore $1/\lambda = \Delta E/hc$ or, using our expression for ΔE,

$$\frac{1}{\lambda} = \frac{ke^2}{2a_0hc}\left(\frac{1}{n_2^2} - \frac{1}{n_1^2}\right)$$

This looks just like Equation 34.9 for the hydrogen spectral lines, except that $ke^2/2a_0hc$ replaces the Rydberg constant R_H. Evaluating this quantity gives $R_\infty = ke^2/2a_0hc = 1.0974\times10^7\,\text{m}^{-1}$, which is very close to the experimentally observed Rydberg constant for hydrogen. The small discrepancy results from our approximation that the proton is stationary. That approximation is equivalent to assuming an infinite proton mass; hence the subscript ∞ on this theoretically calculated Rydberg constant.

Bohr's theory of quantized angular momentum thus accounts brilliantly for the observed spectrum of hydrogen. We can understand the origin of the various spectral line series using Fig. 34.11, an **energy-level diagram** for the Bohr model of hydrogen. Allowed energy levels are shown as horizontal lines, and various possible transitions among levels as vertical arrows. Transitions with a common final state are grouped, and each group represents a different series of spectral lines.

Knowing the energy levels of Equation 34.12, we can also find the radii of the allowed electron orbits, as given by Equation 34.11:

$$r = -\frac{ke^2}{2E} = \left(\frac{ke^2}{2}\right)\left(\frac{2a_0n^2}{ke^2}\right) = n^2a_0 \qquad (34.13)$$

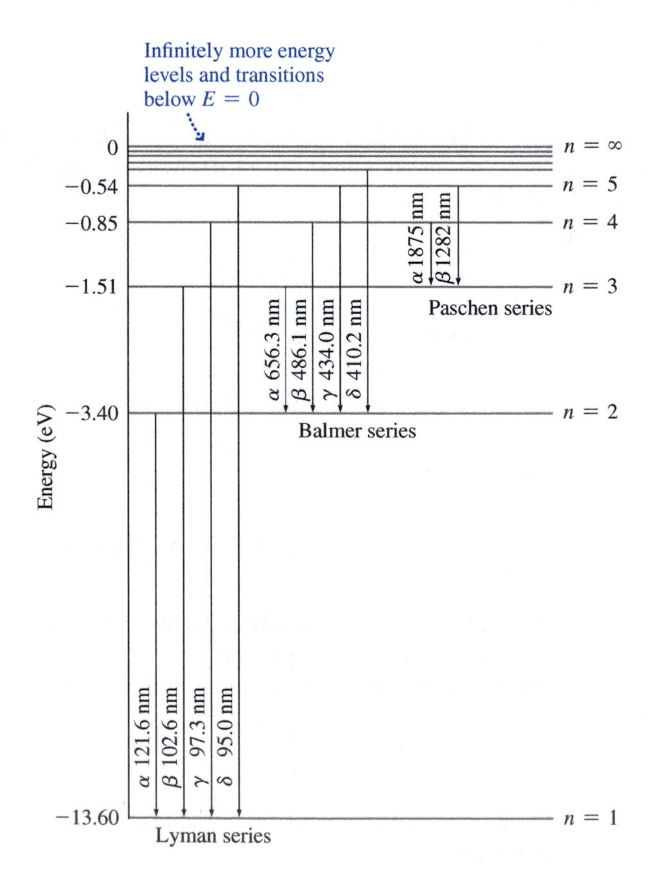

FIGURE 34.11 Energy-level diagram for the Bohr model of the hydrogen atom, showing transitions responsible for the first three series of spectral lines. Each series arises from jumps to a common final state.

Thus, the lowest energy orbit has a radius of one Bohr radius, with higher-energy orbits growing rapidly with increasing n. A hydrogen atom in its ground state—$n = 1$—has a diameter of two Bohr radii, or about 0.1 nm. As we'll see in Chapter 35, the Bohr model's precise electron orbits aren't compatible with the fully developed theory of quantum mechanics; nevertheless, Equation 34.13 does give the approximate size of atoms.

EXAMPLE 34.3 The Bohr Model: Big Atoms

Hydrogen atoms are normally in their ground state, with diameter approximately 0.1 nm. But in the diffuse gas of interstellar space, atoms exist in highly excited states with sizes approaching a fraction of a millimeter. Such **Rydberg atoms** can also be produced temporarily in the lab. Transitions among Rydberg states result in photons at radio wavelengths. One of the longest wavelengths observed corresponds to a transition from $n = 273$ to $n = 272$. (a) What's the diameter of a hydrogen atom in the $n = 273$ state? (b) At what wavelength should a radio telescope be set to observe this transition?

INTERPRET This is a problem about electron transitions in hydrogen atoms, albeit of unusual size. The Bohr model applies.

DEVELOP We'll use Equation 34.13, $r = n^2 a_0$, to find the atomic diameter, with $n = 273$. Equation 34.9, $1/\lambda = R_H(1/n_2^2 - 1/n_1^2)$, will give the transition wavelength with $n_1 = 273$ and $n_2 = 272$.

EVALUATE (a) The diameter is twice the radius, so Equation 34.13 gives $d = (2)(273^2)a_0 = 7.9\ \mu\text{m}$. (b) Inverting Equation 34.9 to get the wavelength gives

$$\lambda = \left[R_H\left(\frac{1}{272^2} - \frac{1}{273^2}\right)\right]^{-1} = 92\ \text{cm}$$

with $R_H = 1.097 \times 10^7\ \text{m}^{-1}$.

ASSESS Our atomic diameter is some 75,000 times that of ground-state hydrogen and about the size of a red blood cell! A wavelength of 92 cm corresponds to a frequency $f = c/\lambda$ of about 300 MHz, which happens to lie in a gap between VHF TV channel 13 and UHF channel 14. ■

Equation 34.12 shows, and Fig. 34.11 suggests, that there are infinitely many electron energy levels between the ground state at −13.6 eV and zero energy. It's possible to give an atomic electron enough energy to bring it above the $E = 0$ level, but then it's no longer bound to the proton. Removing an electron is **ionization**, and Equation 34.12b and Fig. 34.11 show that it takes 13.6 eV to ionize a hydrogen atom in its ground state. This quantity is the **ionization energy**.

GOT IT? 34.4 Figure 34.11 shows some of the transitions available to the electron in the Bohr model of hydrogen, but not all of them. The figure suggests that the shortest possible wavelength emitted in *any* electron transition in hydrogen is (a) arbitrarily small, (b) between 95 nm and 1282 nm, or (c) a little shorter than 95 nm.

Limitations of the Bohr Model

Bohr's theory proved astoundingly successful in explaining the hydrogen spectrum. It also explains the spectra of hydrogen-like ions—atoms with all but one of their electrons removed—with the appropriate change in the value of the nuclear charge. And it has some success in predicting the spectra of atoms such as lithium and sodium that have a single valence electron beyond a group of more tightly bound electrons. But it can't account for the spectra of more complicated atoms, even two-electron helium. And with hydrogen, there are subtle spectral details that the Bohr model doesn't address. Furthermore, like Planck's original quantum hypothesis, Bohr's quantization of atomic energy levels lacked a convincing theoretical basis. You'll see in the next chapter how the much more comprehensive theory of quantum mechanics overcomes these limitations.

34.5 Matter Waves

In classical physics, light is purely a wave phenomenon. Einstein's photons gave light a particle-like quality as well. In 1923, 10 years after Bohr's atomic theory, a French prince named Louis de Broglie (pronounced "de Broy") set forth a remarkable hypothesis in his doctoral thesis. If light has both wave-like and particle-like properties, he reasoned, why shouldn't matter also exhibit both properties?

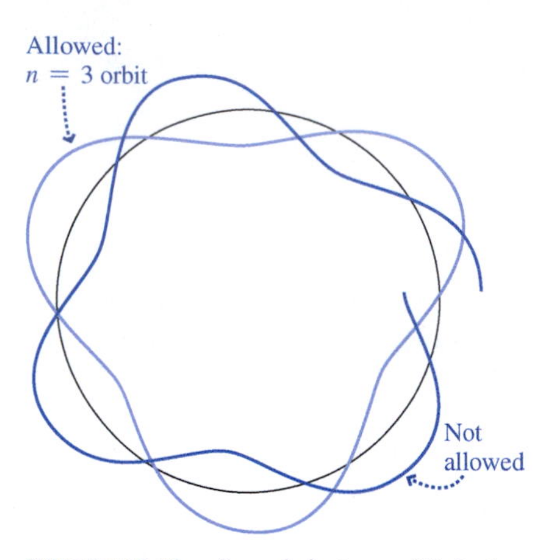

FIGURE 34.12 The allowed electron orbits in the Bohr atom are those that can fit an integral number of de Broglie wavelengths around the circular Bohr orbit.

We saw in Chapter 29 that light with energy E also carries momentum $p = E/c$. Combined with Equation 34.6, that means a photon of light with frequency f has momentum $p = hf/c$. Since $f\lambda = c$, the photon's momentum and wavelength are therefore related by

$$\lambda = \frac{h}{p} \quad \text{(de Broglie wavelength)} \tag{34.14}$$

De Broglie proposed that this same relation should hold for particles of matter; at nonrelativistic speed, for example, an electron should have associated with it a **de Broglie wavelength** given by h/mv.

De Broglie used his matter-wave hypothesis to explain why atomic electron orbits are quantized. He proposed that Bohr's allowed orbits were those in which standing waves could exist (Fig. 34.12), in much the same way that a violin string can support only certain frequencies of standing waves. Suppose that n full wavelengths of a de Broglie electron wave fit around the circumference of the electron's circular orbit. Then we must have $n\lambda = n(h/p) = n(h/mv) = 2\pi r$, with r the orbit radius. Multiplying both sides by $mv/2\pi$ then gives $mvr = nh/2\pi = n\hbar$, which is Bohr's quantization condition. Thus, de Broglie's hypothesis provides a natural explanation for the quantization of atomic energy levels.

CONCEPTUAL EXAMPLE 34.1 The de Broglie Wavelength: Large and Small

If matter has wave properties, why don't we observe baseballs, cars, and people undergoing quantum interference?

EVALUATE Planck's constant h is tiny, and the masses of macroscopic objects are large. That makes the de Broglie wavelength (Equation 34.14) of macroscopic objects minuscule if they have any velocity whatsoever. Since wave behavior is evident only when waves interact with systems comparable in size to the wavelength, the wave aspect of macroscopic objects isn't evident. Even with subatomic particles, wave behavior isn't obvious at high velocities (i.e., high values of momentum p in the denominator of Equation 34.14). In the atom, though, it's a different story, as Making the Connection shows.

ASSESS Couldn't we make a macroscopic object's wavelength large by making its momentum mv small? Yes—but at normal temperatures, thermal agitation always means a significant random velocity. Only at very low temperatures can macroscopic systems exhibit quantum interference.

MAKING THE CONNECTION Find the de Broglie wavelength of (a) a 150-g baseball pitched at 45 m/s and (b) an electron moving at 1 Mm/s. Compare your results with the sizes of home plate and an atom, respectively.

EVALUATE Given mass and speed, Equation 34.14 becomes $\lambda = h/mv$. This gives $\lambda_{\text{baseball}} \simeq 10^{-34}$ m, unimaginably smaller than home plate. But $\lambda_{\text{electron}} \simeq 0.7$ nm, several times the size of an atom. Therefore, wave effects dominate this electron's interactions with atoms.

APPLICATION The Electron Microscope

In Chapter 32 we found that light can't sharply image objects whose size is on the order of the wavelength or smaller—a factor that limits the resolving power of conventional microscopes. But Equation 34.14 shows that we can control the wavelength of electrons by adjusting their speed—and therefore we can achieve electron wavelengths much shorter than that of light. The **electron microscope** exploits this effect, providing resolutions down to about 1 nm and magnifications of 10^6.

Electron microscopes accelerate electron beams to energies of 50–100 keV, with corresponding wavelengths of about 0.005 nm. Magnetic fields act as focusing lenses, forming an image of whatever object is placed in the beam path. An electronic detector reads the image, which is then displayed on a screen.

Electron microscopes are indispensable tools in biology, chemistry, and materials science. A related device, the scanning electron microscope, produces dramatic three-dimensional images at magnifications of $10-10^5$, as shown in the photo of an ant carrying a microelectronic chip.

GOT IT? 34.5 De Broglie's matter-wave hypothesis explains the quantization of electron orbits in the Bohr atom most fundamentally in terms of (a) angular momentum, (b) wavelength, or (c) energy.

PhET: Quantum Wave Interference

Electron Diffraction and Matter-Wave Interference

In 1927, the American physicists Clinton Davisson and Lester Germer gave a convincing verification of de Broglie's matter-wave hypothesis. Davisson and Germer were studying the interaction of an electron beam with a nickel crystal, and they noticed regular intensity peaks reminiscent of X-ray diffraction. Shortly afterward, the Scottish physicist George Thomson observed electron diffraction directly, further evidence of the electron's wave nature (Fig. 34.13). Thomson was the son of J. J. Thomson, who had discovered the electron as a particle in 1897. Together their work captured the electron's wave–particle duality. Today, experiments with entire atoms and even larger clusters of matter exhibit wave interference—as shown in this chapter's opening photo.

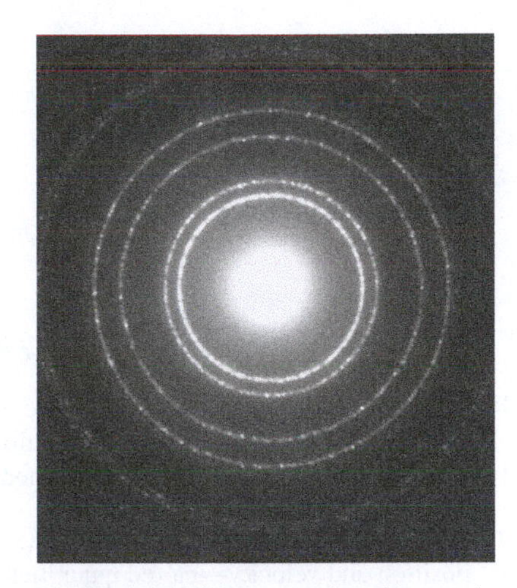

FIGURE 34.13 Diffraction produced by passing an electron beam through a circular aperture shows that electrons have a wave-like character.

34.6 The Uncertainty Principle

In classical physics it's possible, in principle, to know the exact position and velocity of a particle and therefore to predict with certainty its future behavior. But not so in quantum physics! In 1927, the German physicist Werner Heisenberg presented his **uncertainty principle**, which states that some pairs of quantities cannot be measured simultaneously with arbitrary precision. Position and momentum constitute one such pair; if we measure a particle's position to within an uncertainty Δx, then we can't simultaneously determine its momentum to an accuracy better than Δp, where

$$\Delta x\, \Delta p \geq \hbar \qquad \text{(uncertainty principle)} \tag{34.15}$$

Why this limitation? The fundamental reason is quantization. To measure some property of a system requires interacting with the system—for example, shining light on it. Interaction involves energy, and the interaction energy disturbs the system slightly. As a result, values inferred from the measurement are no longer quite right. In classical physics the energy can be arbitrarily small, resulting in a negligible disturbance. But in quantum theory the minimum energy is a single quantum, like a photon of light, and thus the disturbance can't be arbitrarily small.

So why not use lower-frequency light, whose photon energy hf is lower? Because lower frequency means longer wavelength and, as we found in Chapter 32, diffraction effects limit resolution at longer wavelengths. Heisenberg summarized this dilemma with the "thought experiment" illustrated in Fig. 34.14, which uses a single photon to observe an electron. A short-wavelength photon allows precise localization of the electron (Fig. 34.14*a*). But short wavelength means high frequency and thus high photon energy. The high-energy photon imparts considerable momentum to the electron, and thus the very act that fixes the electron's position degrades our knowledge of its momentum. We can decrease this disturbance with a lower-energy, longer-wavelength photon (Fig. 34.14*b*). But now diffraction precludes precisely determining the electron's position. So we can measure the electron's position accurately, at the expense of not knowing its momentum. Or we can measure its momentum accurately, but then we can't know its position. With a photon of intermediate wavelength we could measure both quantities, but neither precisely. The uncertainty principle, Equation 34.15, quantifies this trade-off.

The uncertainty principle is intimately connected with de Broglie's wave hypothesis. Suppose we pass an electron beam through a slit, as shown in Fig. 34.15 (next page). Then we know the electrons' vertical position to within the slit width. If the slit is much wider than the electrons' de Broglie wavelength, there's minimal diffraction. The electrons follow straight lines and we're quite sure of their vertical momentum, in this case zero (Fig. 34.15*a*). But with a wide slit we don't know much about the electrons' vertical position. Making the slit smaller gives a more precise position, but then diffraction spreads the beam, increasing the uncertainty in the electrons' vertical momentum (Fig. 34.15*b*). So the wave nature of matter ultimately imposes a trade-off: The more we know of a particle's position, the less we know of its momentum, and vice versa.

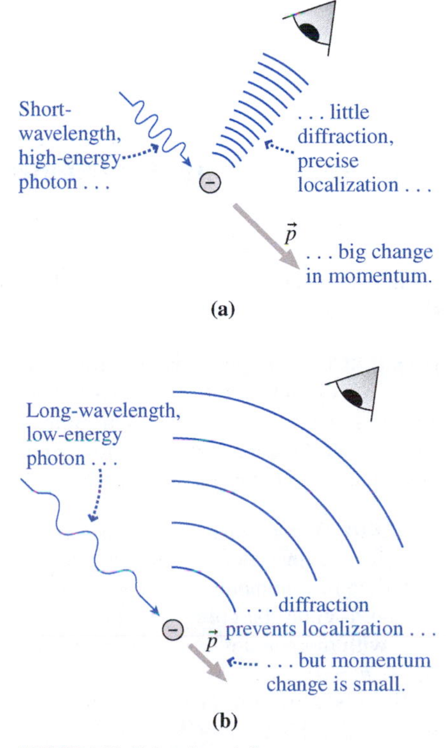

FIGURE 34.14 Heisenberg's "quantum microscope" thought experiment.

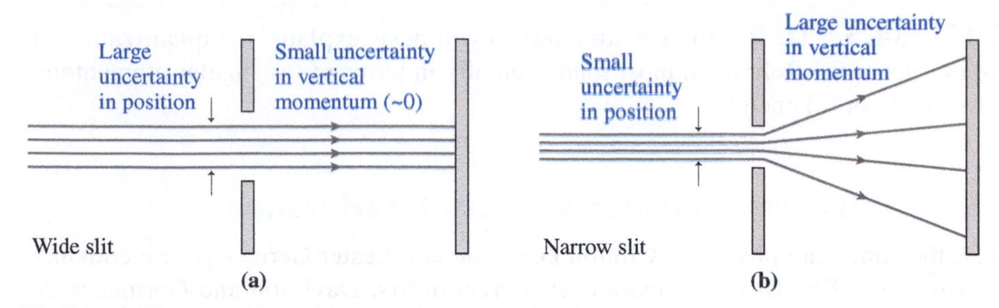

FIGURE 34.15 The wave nature of matter is intimately related to the uncertainty principle, as shown here for beams of electrons passing through wide and narrow slits. In (b), diffraction introduces the uncertainty in vertical momentum.

EXAMPLE 34.4 The Uncertainty Principle: Microelectronics

A beam of aluminum atoms is used to dope a semiconductor chip to set its electrical properties. If the atoms' velocity is known to within 0.2 m/s, how accurately can they be positioned?

INTERPRET This is a question about simultaneously knowing the atoms' position and velocity—paired quantities governed by the uncertainty principle.

DEVELOP We're given the velocity uncertainty Δv, from which we'll find the momentum uncertainty $\Delta p = m\,\Delta v$. Then we can use the uncertainty principle, Equation 34.15, $\Delta x\,\Delta p \geq \hbar$, to find the uncertainty Δx in position. To find the mass, we'll need aluminum's atomic weight from Appendix D, along with Appendix C's conversion from unified mass units (u) to kilograms.

EVALUATE We have

$$\Delta p = m\,\Delta v = (26.98\,\text{u})(1.66\times10^{-27}\,\text{kg/u})(0.20\,\text{m/s}) = 9\times10^{-27}\,\text{kg}\cdot\text{m/s}$$

Then Equation 34.15 gives the position uncertainty:

$$\Delta x = \frac{\hbar}{\Delta p} = 12\,\text{nm}$$

where, again, $\hbar = h/2\pi$.

ASSESS Our 12-nm answer is about 100 atomic diameters and shows that the uncertainty principle constrains our ability to fabricate very small microelectronic structures. ■

It sounds like the uncertainty principle only limits our knowledge. But in fact it proves useful in estimating the size and energies of atomic-scale systems, as the next example shows.

EXAMPLE 34.5 The Uncertainty Principle: Estimating Atomic and Nuclear Energies

Use the uncertainty principle to estimate the minimum energy possible for (a) an electron confined to a region of atomic dimensions, about 0.1 nm, and (b) a proton confined to a region of nuclear dimensions, about 1 fm.

INTERPRET We're given the uncertainty in position; that's the width of the region in which the particles are confined. The particles can't be at rest, or we'd know that their momentum was exactly zero—in violation of the uncertainty principle. So they must have a minimum momentum and therefore energy. We're asked to find that energy.

DEVELOP We need to find the minimum momentum consistent with the uncertainty principle, and from it the energy. Suppose a particle has momentum of magnitude p, but we don't know its direction. It could be going one way, with momentum p, or the other way, with momentum $-p$. Then the momentum itself is uncertain by $\Delta p = p - (-p) = 2p$. The uncertainty principle says $\Delta p \geq \hbar/\Delta x$, so there's a minimum magnitude for the momentum given by $p \geq \hbar/2\,\Delta x$. Using $p = mv$ and $K = \frac{1}{2}mv^2$ gives $K = p^2/2m$, and therefore the uncertainty principle requires

$$K \geq \frac{1}{2m}\left(\frac{\hbar}{2\,\Delta x}\right)^2$$

EVALUATE Evaluating this constraint for an electron with $\Delta x = 0.1$ nm and for a proton with $\Delta x = 1$ fm gives minimum energies of about 1 eV and 5 MeV, respectively.

ASSESS Energies in electronvolts are typical of atomic-scale systems, as we saw in Fig. 34.11. Our result shows that nuclear energies are some 5 million times greater—indicating the dramatic difference between chemical and nuclear energy sources. We'll have more to say about that difference in Chapter 38. ■

Energy–Time Uncertainty

A second pair of variables that defy simultaneous measurement are the energy of a system and the time it remains at that energy. The energy uncertainty ΔE is related to the time Δt through the inequality

$$\Delta E \, \Delta t \geq \hbar \quad (34.16)$$

One effect of energy–time uncertainty is to render atomic and nuclear energy levels inexact and therefore to broaden spectral lines. If an atom were forever in a fixed energy state, we could take infinitely long to measure its energy and therefore make ΔE arbitrarily small. But excited states of atoms have characteristic lifetimes (typically $\sim 10^{-8}$ s), which limit the measurement time and therefore set a minimum uncertainty in the energy level. Problem 70 and Passage Problems 84–87 explore energy–time uncertainty, as does the Application on this page.

GOT IT? 34.6 An object is moving in one dimension. If you know that its momentum has magnitude p, but you don't know which direction it's going in, then the uncertainty in its momentum is (a) 0, (b) p, or (c) $2p$.

Observers, Uncertainty, and Causality

The uncertainty principle moves the observer from a passive onlooker to an active participant in physical events. To observe is necessarily to disturb, and quantum theory is therefore concerned with the role of the observer and the process of measurement. The uncertainty principle is fundamentally a statement about what can and cannot be learned through measurement.

Position and momentum cannot be measured simultaneously with perfect accuracy. Surely, though, a particle has well-defined values of both, even though we can't know them? The answer seems to be no. The standard interpretation of quantum mechanics suggests that it makes no sense to talk about what can't be measured, and recent experiments have ruled out "hidden variables" that might be active at a lower level to guide the particle in a deterministic path. Its wave aspect makes a particle a "fuzzy" thing, and it really makes no sense to think of it as a tiny ball with definite momentum and position. For that reason it also makes no sense to think of the particle's future as being fully determined in the sense that Newton's laws determine the future path of, say, a baseball. We're left with uncertainty—or indeterminacy, as Heisenberg's word also translates—as a fundamental fact of our universe.

APPLICATION Femtosecond Photography of Chemical Reactions

Chemists and biochemists studying chemical reactions in detail need to image events that take place in times measured in femtoseconds (1 fs = 10^{-15} s). For this, they use ultrashort laser pulses that can "freeze" the action of electrons participating in chemical reactions. Because the temporal duration of the laser pulses is so short, energy–time uncertainty requires that the photons in the laser beam have considerable uncertainty in their energies. Since photon energy is $E = hf = hc/\lambda$, that translates into a broad spectrum of wavelengths—an advantage for the femtosecond laser technique because it allows a single laser pulse to probe multiple energy transitions in the molecule under study. The photo shows a femtosecond laser system constructed on an optical bench.

34.7 Complementarity

One of the most disturbing aspects of quantum theory is the wave–particle duality—the seeming contradiction that matter and light have both wave-like and particle-like properties. If this bothers you, you're in good company: Heisenberg himself expressed frustration in trying to understand the quantum world:

> I remember discussions with Bohr which went through many hours till very late at night and ended almost in despair; and when at the end of the discussion I went alone for a walk in the neighboring park I repeated to myself again and again the question: Can nature possibly be as absurd as it seems to us in these atomic experiments?*

Bohr dealt with the wave–particle duality through his **principle of complementarity**. The wave and particle pictures, he said, are complementary aspects of the same reality. If we do an experiment to measure a wave-like property—for example, the diffraction of electrons—then we find wave properties but not particle properties. If we do an experiment

*Werner Heisenberg, *Physics and Philosophy: The Revolution in Modern Science* (New York: Harper & Brothers, 1962).

to measure a particle-like property—for example, localizing an electron—then we won't find wave properties. The two measurements require different experiments, and we can't perform both simultaneously on the same entity. So we'll never catch wave and particle in an outright contradiction, and the answer to the question "Which is it, wave or particle?" has to be that it's both, and which you find depends on what experiment you choose to perform.

Bohr articulated a second principle that helps reconcile the seeming contradiction between classical and quantum physics. His **correspondence principle** states that the predictions of classical and quantum physics should agree in situations where the size of individual quanta is negligible. Taking $h \rightarrow 0$ in Planck's law, for example, gives the classical Rayleigh–Jeans law (see Problem 72). Or, for large n, the energies of adjacent atomic states in the Bohr model become so close that the levels appear essentially as a continuum of allowed energies—as expected in classical physics. Or consider a 1000-W radio beam; the photon energy hf is so low that the beam contains an enormous number of photons per unit beam length, and we can consider the energy distributed essentially continuously over the beam. But in a 1000-W X-ray beam, the photon energy is much higher and the number of photons correspondingly fewer; it's therefore difficult to avoid the fundamental fact of energy quantization. Visible light lies somewhere in between; we can often treat its energy as being continuously distributed, except when it interacts with systems as small as individual atoms.

CHAPTER 34 SUMMARY

Big Ideas

The big ideas here are at the heart of quantum physics—a radically different view of reality at the atomic scale. **Quantization** means that some physical quantities—often including energy—come only in discrete values. Another fundamental aspect of quantum reality is **wave–particle duality**, wherein light and matter exhibit both wave-like and particle-like aspects. Bohr's **complementarity principle** precludes these ever being in direct conflict. Finally, quantization and wave–particle duality lead to the **uncertainty principle**, which states that it's impossible to measure simultaneously and with arbitrary precision a particle's position and momentum.

Key Concepts and Equations

Planck's constant,

$$h = 6.63\times10^{-34}\ \text{J}\cdot\text{s}$$

sets the fundamental scale of quantization.

It's also expressed as "h bar":

$$\hbar = h/2\pi$$

The energy of electromagnetic radiation with frequency f is quantized in **photons** with energy

$$E = hf$$

Electron energies in the **Bohr model** of hydrogen are quantized according to

$$E = -\frac{ke^2}{2a_0}\left(\frac{1}{n^2}\right) \simeq -\frac{13.6\,\text{eV}}{n^2}$$

where n is an integer and $a_0 = 0.0529$ nm is the **Bohr radius**.

The **de Broglie wavelength** of a particle with momentum p is

$$\lambda = \frac{h}{p}$$

The **uncertainty principle** relates uncertainties in position and momentum:

$$\Delta x\,\Delta p \geq \hbar$$

Applications

A correct description of **blackbody radiation** requires Planck's quantization hypothesis. The peak radiance—energy radiated per unit wavelength interval—from a blackbody at temperature T occurs at a wavelength given by $\lambda T = 2.898\ \text{mm}\cdot\text{K}$.

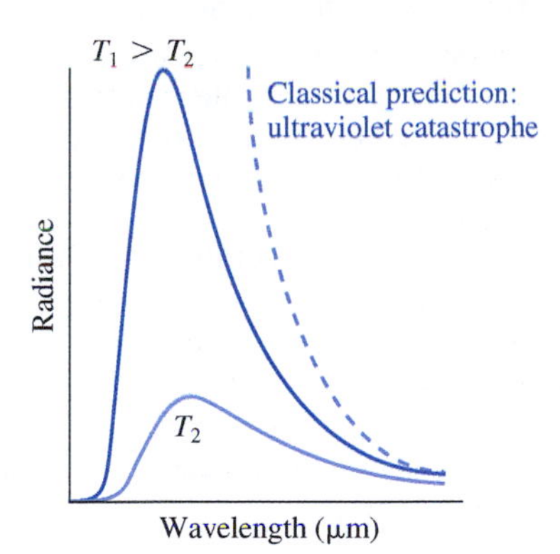

The **photoelectric effect** involves the ejection of electrons from a metal surface illuminated with electromagnetic waves. Explanation of the effect led Einstein to propose **photons** as the quanta of electromagnetic-wave energy.

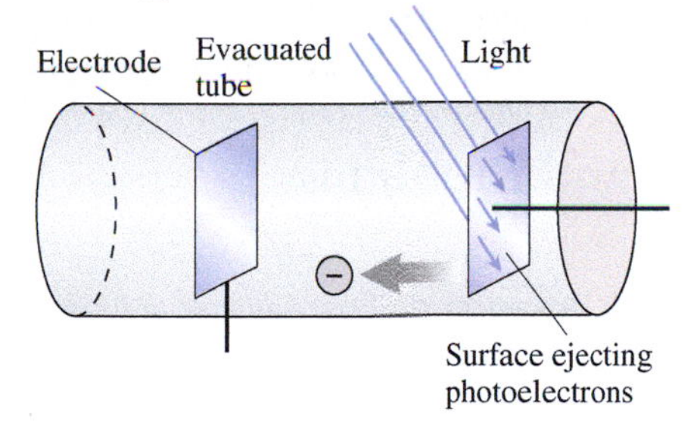

The **Compton effect** shows that photons interact with free electrons exactly like colliding particles, losing energy and emerging with longer wavelength.

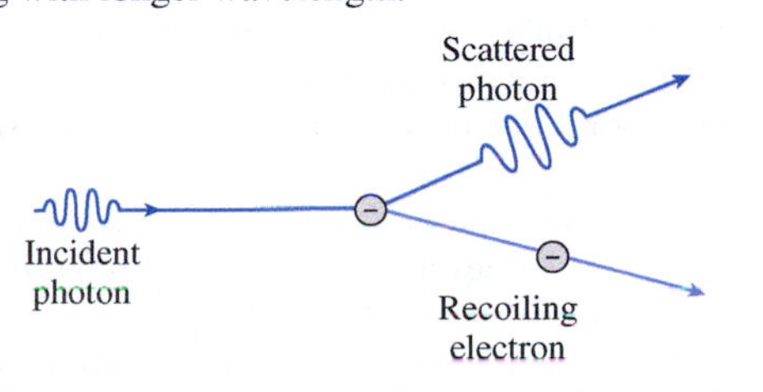

Quantization of atomic energy levels leads directly to **atomic spectra**. In the Bohr model of hydrogen, the spectral line produced in a transition from the n_1 to the n_2 energy level is given by

$$\frac{1}{\lambda} = R_H\left(\frac{1}{n_2^2} - \frac{1}{n_1^2}\right)$$

where $R_H = 1.0968\times10^7\ \text{m}^{-1}$.

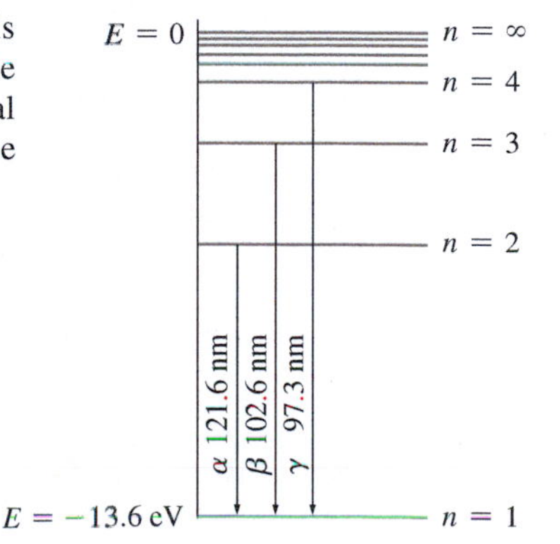

MP *For homework assigned on MasteringPhysics, go to www.masteringphysics.com*

BIO *Biology and/or medicine-related problems* **DATA** *Data problems* **ENV** *Environmental problems* **CH** *Challenge problems*

For Thought and Discussion

1. Why does classical physics predict that atoms should collapse?
2. Looking at the night sky, you see one star that appears red, another yellow, and another blue. Compare their temperatures.
3. Imagine an atom that, unlike hydrogen, had only three energy levels. If these levels were evenly spaced, how many spectral lines would result? How would their wavelengths compare?
4. What colors of visible light have the highest-energy photons?
5. Why is the immediate ejection of electrons in the photoelectric effect surprising from a classical viewpoint?
6. Suppose the Compton effect were significant at radio wavelengths. What problems might this present for radio and TV broadcasting?
7. How are the uncertainty principle and wave–particle duality related?
8. How many spectral lines are in the entire Balmer series?
9. Why are the lines of the Lyman series in the ultraviolet while some Balmer lines are in the visible?
10. Why does the photoelectric effect suggest that light has particle-like properties?
11. Energy–time uncertainty limits the precision with which we can know the mass of unstable particles (those that decay after a finite time). Why?
12. If you measure a particle's position with perfect accuracy, what do you know about its momentum?
13. How might our everyday experience be different if Planck's constant had the value 1 J·s?
14. Why are the energies given by Equations 34.12 negative?

Exercises and Problems

Exercises

Section 34.2 Blackbody Radiation

15. If you double a blackbody's temperature, by what factor does its radiated power increase?
16. The surface temperature of the star Rigel is 10^4 K. Find (a) the power radiated per square meter of its surface, (b) its λ_{peak}, and (c) its λ_{median}.
17. Find λ_{peak} and λ_{median} for Earth, considered a 288-K blackbody.
18. Spacecraft instruments measure the radiation from an asteroid, and the data show that the power per unit wavelength peaks at 40 μm. Assuming the asteroid is a blackbody, find its surface temperature.
19. The Sun approximates a blackbody at 5800 K. (a) Find the wavelength of peak radiance on the per-unit-wavelength basis implicit in Equation 34.2a. (b) Find the median wavelength, below which half the radiation is emitted (Equation 34.2b). Identify the spectral region of each.

Section 34.3 Photons

20. Find the energy in electronvolts of (a) a 1.0-MHz radio photon, (b) a 5.0×10^{14}-Hz optical photon, and (c) a 3.0×10^{18}-Hz X-ray photon.
21. **BIO** The human eye is sensitive to wavelengths from about 400 nm to 700 nm. What's the corresponding range of photon energies?
22. A microwave oven uses electromagnetic radiation at 2.4 GHz. (a) What's the energy of each microwave photon? (b) At what rate does a 900-W oven produce photons?
23. A red laser at 650 nm and a blue laser at 450 nm emit photons at the same rate. How do their total power outputs compare?
24. Find the maximum work function, in eV, for a surface to emit electrons when illuminated with 945-nm infrared light.

Section 34.4 Atomic Spectra and the Bohr Atom

25. Calculate the wavelengths of the first three lines in the Lyman series for hydrogen.
26. Which spectral line of the hydrogen Paschen series ($n_2 = 3$) has wavelength 1282 nm?
27. What's the maximum wavelength of light that can ionize hydrogen in its ground state? In what spectral region is this?
28. At what energy level does the Bohr hydrogen atom have diameter 5.18 nm?

Section 34.5 Matter Waves

29. Find the de Broglie wavelength of (a) Earth, orbiting the Sun at 30 km/s, and (b) an electron moving at 10 km/s.
30. How slowly must an electron be moving for its de Broglie wavelength to be 1 mm?
31. A proton and electron have the same de Broglie wavelength. How do their speeds compare, assuming $v \ll c$ for both?
32. Find the de Broglie wavelength of electrons with kinetic energies (a) 10 eV, (b) 1.0 keV, and (c) 10 keV.

Section 34.6 The Uncertainty Principle

33. A proton is confined to a space 1 fm wide (about the size of an atomic nucleus). What's the minimum uncertainty in its velocity?
34. Is it possible to determine an electron's velocity accurate to ±1 m/s while simultaneously finding its position to within ±1 μm? What about a proton?
35. A proton has velocity $v = (1500 \pm 0.25)$ m/s. What's the uncertainty in its position?
36. An electron is moving in the $+x$-direction with speed measured at 50 Mm/s, accurate to ±10%. What's the minimum uncertainty in its position?
37. Find the minimum energy for a neutron in a uranium nucleus whose diameter is 15 fm.

Problems

38. Find the power per unit area emitted by a 3000-K incandescent lamp filament in the wavelength interval from 500 nm to 502 nm.
39. Treating the Sun as a 5800-K blackbody, compare its UV radiance at 200 nm with its visible radiance at its 500-nm peak wavelength.
40. For a 2.0-kK blackbody, by what percentage is the Rayleigh–Jeans law in error at wavelengths of (a) 1.0 mm, (b) 10 μm, and (c) 1.0 μm?
41. The radiance of a blackbody peaks at 558 nm. (a) What's its temperature? (b) How does its radiance at 382 nm (violet light) compare with that at 694 nm (red light)?
42. (a) Find the Compton wavelength for a proton. (b) Find the energy in electronvolts of a gamma ray whose wavelength equals the proton's Compton wavelength.
43. Find the rate of photon production by (a) a radio antenna broadcasting 1.0 kW at 89.5 MHz, (b) a laser producing 1.0 mW of

633-nm light, and (c) an X-ray machine producing 0.10-nm X rays with total power 2.5 kW.

44. Electrons in a photoelectric experiment emerge from an aluminum surface with maximum kinetic energy 1.3 eV. Find the wavelength of the illuminating radiation.
45. (a) Find the cutoff frequency for the photoelectric effect in copper. (b) Find the maximum energy of the ejected electrons if the copper is illuminated with light of frequency 1.8×10^{15} Hz.
46. The stopping potential in a photoelectric experiment is 1.8 V when the illuminating radiation has wavelength 365 nm. Determine (a) the work function of the emitting surface and (b) the stopping potential for 280-nm radiation.
47. **BIO** Chlorophyll is a photosynthetic molecule common in green plants. On a per-unit-wavelength basis, its ability to absorb visible light has two peaks, at 430 nm and 662 nm. (a) Find the corresponding photon energies. (b) Use these peak wavelengths to explain why plants are green.
48. Find the initial wavelength of a photon that loses half its energy when it Compton-scatters from an electron and emerges at 90° to its initial direction.
49. When light shines on potassium, the photoelectrons' maximum speed is 4.2×10^5 m/s. Find the light's wavelength.
50. The maximum electron energy in a photoelectric experiment is 2.8 eV. When the wavelength of the illuminating radiation is increased by 50%, the maximum energy drops to 1.1 eV. Find (a) the work function of the emitting surface and (b) the original wavelength.
51. A 150-pm X-ray photon Compton-scatters off an electron and emerges at 135° to its original direction. Find (a) the wavelength of the scattered photon and (b) the electron's kinetic energy.
52. Find the kinetic energy of an initially stationary electron after a 0.10-nm X-ray photon scatters from it at 90°.
53. A photocathode ejects electrons with maximum energy 0.85 eV when illuminated with 430-nm blue light. Will it eject electrons when illuminated with 633-nm red light? If so, what will be the maximum electron energy?
54. A cosmic-ray particle interacts with an energy-measuring device for a mere 12 zs. What's the minimum uncertainty in the measured energy? Express your answer in joules and in eV.
55. An electron is known to be somewhere inside a carbon nanotube that's 370 nm long and 1.2 nm in diameter. Find the minimum uncertainties in the components of its velocity (a) along the tube and (b) perpendicular to the tube's long dimension.
56. Find the de Broglie wavelength of an electron that's been accelerated from rest through a 4.5-kV potential difference.
57. An experimental transistor uses a single electron trapped in a channel 6.6 nm wide. What's the minimum kinetic energy this electron could have, consistent with the uncertainty principle? Give your answer in joules and in eV.
58. (a) Find the highest possible energy for a photon emitted as the electron jumps between two adjacent energy levels in the Bohr hydrogen atom. (b) Which energy levels are involved?
59. Find (a) the wavelength and (b) the energy in electronvolts of the photon emitted when a Rydberg hydrogen atom drops from the $n = 180$ level to the $n = 179$ level.
60. The wavelengths of a spectral line series tend to a limit as $n_1 \rightarrow \infty$. Evaluate the series limit for (a) the Lyman series and (b) the Balmer series in hydrogen.
61. A Rydberg hydrogen atom makes a downward transition to the $n = 225$ state, emitting a 9.32-μeV photon. What was the original state?
62. A hydrogen atom is in its ground state when its electron absorbs a 48-eV photon. What's the energy of the resulting free electron?
63. How much energy does it take to ionize a hydrogen atom in its first excited state?
64. Ultraviolet light with wavelength 75 nm shines on hydrogen atoms in their ground states, ionizing some of the atoms. What's the energy of the electrons freed in this process?
65. Helium with one of its two electrons removed acts very much like hydrogen, and the Bohr model successfully describes it. Find (a) the radius of the ground-state electron orbit and (b) the photon energy emitted in a transition from the $n = 2$ to the $n = 1$ state in this singly ionized helium.
66. Through what potential difference should you accelerate an electron from rest so its de Broglie wavelength will be the size of a hydrogen atom, about 0.1 nm?
67. Find the minimum electron speed that would make an electron microscope superior to an optical microscope using 450-nm light.
68. **BIO** You're a cell biologist who wants to image microtubules that form the "skeletons" of living cells. The microtubules are 25 nm in diameter, and, as Chapter 32 shows, you need to image with waves whose wavelength is at least this small. You can use either an inexpensive electron microscope that accelerates electrons to kinetic energies of 40 keV, or a more expensive unit that produces 100-keV electrons. Will the less expensive microscope work?
69. An electron is trapped in a "quantum well" 23 nm wide. Find its minimum possible speed.
70. Typically, an atom remains in an excited state for about 10^{-8} s before it drops to a lower state, emitting a photon in the process. What's the uncertainty in the energy of this transition?
71. An electron is moving at 10^6 m/s and you wish to measure its energy to an accuracy of ±0.01%. What's the minimum time necessary for this measurement?
72. **CH** Use the series expansion for e^x (Appendix A) to show that Planck's law (Equation 34.3) reduces to the Rayleigh–Jeans law (Equation 34.5) when $\lambda \gg hc/kT$.
73. A photon's wavelength is equal to the Compton wavelength of a particle with mass m. Show that the photon's energy is equal to the particle's rest energy.
74. Show that the frequency range of the hydrogen spectral line series involving transitions ending at the nth level is $\Delta f = cR_H/(n+1)^2$.
75. **CH** A photon undergoes a 90° Compton scattering off a stationary electron, and the electron emerges with *total* energy $\gamma m_e c^2$, where γ is the relativistic factor introduced in Chapter 33. Find an expression for the initial photon energy.
76. **CH** Show that Wien's law (Equation 34.2a) follows from Planck's law (Equation 34.3). (*Hint:* Differentiate Planck's law with respect to wavelength.)
77. **CH** Consider an elastic collision between a photon with initial wavelength λ_0 moving in the x-direction and a stationary electron, as depicted in Fig. 34.9*b*. Use relativistic expressions for energy and momentum from Chapter 33 to show that conservation of energy and momentum yield the equations $hc/\lambda_0 + mc^2 = hc/\lambda + \gamma mc^2$, $h/\lambda_0 = (h/\lambda)\cos\theta + \gamma mu\cos\phi$, and $0 = (h/\lambda)\sin\theta - \gamma mu\sin\phi$, where λ is the post-collision photon wavelength and the angles θ and ϕ are as shown in Fig. 34.9*b*. Solve these equations to find the Compton shift (Equation 34.8).
78. **COMP** What would the constant in Equation 34.2a be if blackbody radiance were defined for fixed intervals of frequency rather than wavelength? (*Hint:* Use $\lambda = c/f$ to express the radiance as $R(f, T)$, then differentiate to find the maximum, and solve the resulting relation numerically. Express your answer in a form like Equations 34.2a and b.)

79. Integrate Equation 34.3 over all wavelengths to get the total power radiated per unit area. Show that your result is equivalent to Equation 34.1, with the Stefan–Boltzmann constant given by $\sigma = 2\pi^5 k^4/15c^2h^3$. (*Hint:* Use $hc/\lambda kT$ as the integration variable.)
CH

80. Perform a numerical integration of Equation 34.3 to the wavelength given by Equation 34.2b. Divide by the result of Problem 79, and thus verify that Equation 34.2b gives the wavelength above and below which a blackbody radiates half its energy.
COMP

81. Use the momentum conservation equations given in Problem 77 and Equation 34.8 for the Compton shift to show that the electron's recoil angle in Fig. 34.9*b* is given by $\tan\phi = \sin\theta/[(1 + \lambda_C/\lambda_0)(1 - \cos\theta)]$.
CH

82. Show that in the Bohr model, the frequency of a photon emitted in a transition between levels $n + 1$ and n, in the limit of large n, is equal to the electron's orbital frequency. (This is an example of Bohr's correspondence principle.)

83. The table below lists the stopping potential as a function of wavelength in a photoelectric effect experiment. Determine quantities to plot that should yield a straight line. Make your plot, establish a best-fit line, and use your line to determine (a) an experimental value for Planck's constant and (b) the work function of the material comprising the photocathode. (c) Use Table 34.1 to identify the material.
DATA

Wavelength, λ (nm)	225	275	325	375	425	475	525
Stopping potential, V (V)	3.25	2.17	1.52	0.962	0.646	0.312	0.065

Passage Problems

Particle physicists use the energy–time uncertainty relation to estimate the lifetimes of unstable particles produced in high-energy particle accelerators (Chapter 39). Some particles have lifetimes of 10^{-24} s and shorter—impossible to measure directly. However, physicists can measure particle masses, and they do so for many instances of the same particle to get a distribution of masses. By Einstein's $E = mc^2$, that corresponds to a distribution of energies (Fig. 34.16). Measuring the distribution's width at half its peak (see Fig. 34.16) gives an estimate of the energy uncertainty, and the corresponding Δt from inequality 34.16 provides the particle's lifetime.

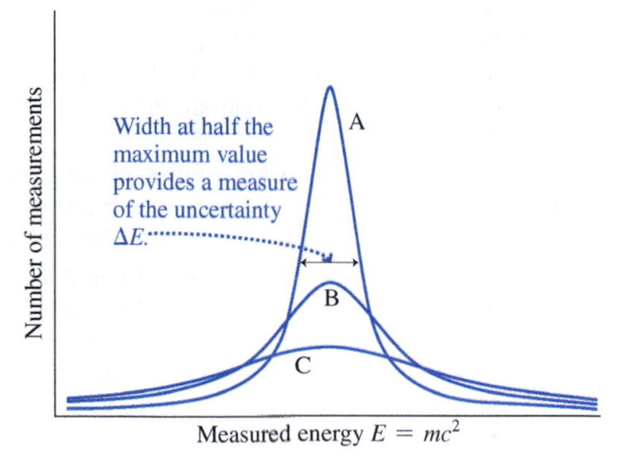

FIGURE 34.16 Mass distributions for high-energy particles (Passage Problems 84–87). The vertical axis gives the number of measurements that yield a given value on the horizontal axis

84. Which of the curves in Fig. 34.16 represents the particle with the shortest lifetime?
    a. A
    b. B
    c. C
    d. You can't tell from the graph.

85. An energy uncertainty of 1 MeV corresponds to a particle lifetime closest to
    a. 10^{-34} s.
    b. 10^{-21} s.
    c. 10^{-9} s.
    d. 1 μs.

86. The converse approach is used for particles with longer lifetimes: Direct measurement of the lifetime yields, through energy–time uncertainty, a range of expected values for particle energies or masses. The longer the lifetime,
    a. the wider the mass range and the narrower the energy range.
    b. the wider the mass and energy ranges.
    c. the narrower the mass range and the wider the energy range.
    d. the narrower the mass and energy ranges.

87. For a particle with lifetime 10^{-7} s, the corresponding mass range is closest to
    a. 10^{-44} u.
    b. 10^{-27} u.
    c. 10^{-17} u.
    d. 1 u.

Answers to Chapter Questions

Answer to Chapter Opening Question

That matter, like light, behaves as waves under some circumstances.

Answers to GOT IT? Questions

34.1 (1) A's power is 16 times greater; (2) A's peak wavelength is half that of B

34.2 (1) No; slope remains h/e; (2) yes; the horizontal intercept is the cutoff frequency, which depends on the work function of the material

34.3 (b) much less because the proton is ~2000 times more massive than the electron

34.4 (c)

34.5 (a)

34.6 (c)

Quantum Mechanics

What You Know

- You've some basic ideas of quantum physics—quantization, the uncertainty principle, the correspondence principle—but at this point they've got no firm basis.
- You understand Newtonian mechanics and can express the relationship between kinetic energy and momentum.
- You know about potential energy and potential-energy curves.

What You're Learning

- You'll learn to describe quantitatively the relationship between wave and particle descriptions of quantum systems.
- You'll learn how the *Schrödinger equation* describes the quantum-mechanical *wave function*.
- You'll see how the wave function is usually interpreted, and you'll come to appreciate why there's still controversy about the philosophical interpretation of quantum physics and its wave function.
- You'll learn to solve the Schrödinger equation for one-dimensional square-well potentials.
- You'll explore solutions to the Schrödinger equation for the quantum harmonic oscillator.
- You'll see how *quantum tunneling* allows particles to penetrate into regions that classical energy conservation forbids them from entering.
- You'll understand *quantum degeneracy* and other complexities that arise in two- and three-dimensional quantum systems.
- You'll see that special relativity provides the theoretical pass for *antimatter* and *spin*.

How You'll Use It

- Quantum mechanics, based on the Schrödinger equation, provides the basis for understanding atomic structure—as you'll see in Chapter 36.
- Quantum mechanics will provide a firm theoretical basis for your understanding of the periodic table of the elements and thus for all of chemistry.
- Applied to many-particle systems in Chapter 37, quantum mechanics will explain the behavior of molecules and solids.

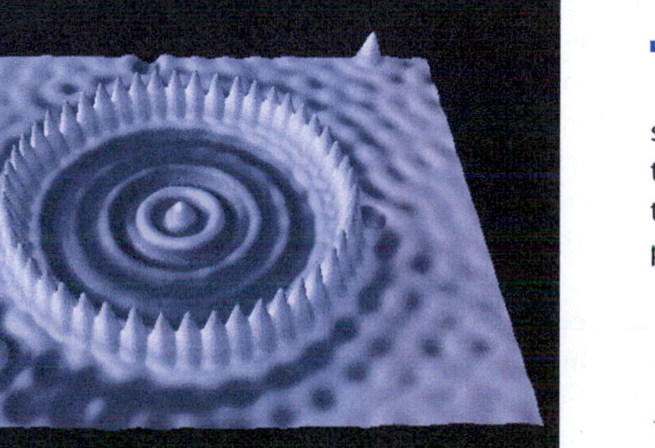

The ideas developed in the preceding chapter are at the core of the **old quantum theory**. The old quantum theory introduced the basic concepts of quantum physics and was successful in explaining a number of quantum phenomena—for example, blackbody radiation, the photoelectric effect, and the hydrogen spectrum. On the other hand, it couldn't treat even the simplest multielectron atoms, and it left some subtle spectral features unexplained. Furthermore, the old quantum theory was a hodgepodge of separate but loosely

This scanning-tunneling microscope image shows a "quantum corral" of 48 iron atoms on a copper surface. What unusual quantum phenomenon enables this type of microscopy?

related ideas, each developed to explain a particular phenomenon; it lacked coherence and clear guiding principles.

Is there a more coherent theory that predicts the behavior of systems at the atomic and subatomic scales, and that offers a satisfying description of how such systems really work? The answer is at once an emphatic yes and a disappointing no. Yes, because **quantum mechanics**, developed in the 1920s, predicts with remarkable precision the observed properties of atomic systems, including their energies, the wavelengths of spectral lines, and the lifetimes of excited atoms. No, because quantum mechanics doesn't give a satisfying visual *picture* of the atomic and subatomic worlds. The uncertainty principle and wave–particle duality are essential aspects of quantum mechanics. Any picture we formulate of electrons or photons whizzing around like miniature balls with precise positions and momenta is inappropriate. But quantum mechanics does provide a self-consistent description that lets us explore and predict the behavior of atoms, the organization of chemical elements, the physics of semiconductors and superconductors, the extraordinary behavior of matter at low temperature, the formation of white dwarf stars, the operation of lasers, and a host of other phenomena for which classical physics is at best inaccurate and at worst totally inadequate. In this chapter you'll explore the mathematical structure and physical interpretation of quantum mechanics. In Chapters 36 and 37 you'll apply quantum mechanics first to the atom and then to more complex systems that involve quantum-mechanical interactions among many atoms.

35.1 Particles, Waves, and Probability

Photons and Light Waves

In Maxwell's electromagnetic theory, we had a seemingly complete description of light as an electromagnetic wave. Now we find, through the photoelectric and Compton effects, that light sometimes manifests itself as particles. What's the connection between wave and particle descriptions?

In a photoelectric experiment, the rate at which electrons are ejected depends on the light's intensity. Since an electron is ejected when it absorbs a photon, we conclude that the number of photons in the incident light is proportional to light intensity. Now, the intensity of an electromagnetic wave depends on the square of the electric or magnetic field (Equations 29.20b, c). The fields, in turn, obey Maxwell's equations, so one aspect of a photoelectric experiment—namely, the rate of electron ejection—relates to Maxwell's description of light as an electromagnetic wave.

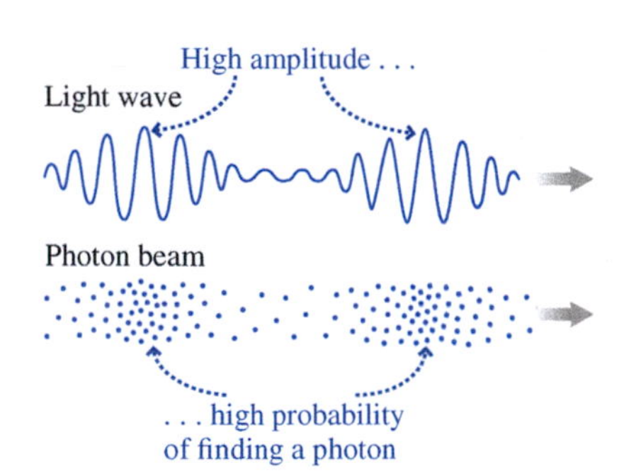

FIGURE 35.1 The probability of finding a photon is directly proportional to the intensity of the electromagnetic wave. The figure is only suggestive because we can't depict photons as localized particles.

We can quantify the relation between waves and photons, but only in a statistical sense. The ejection of individual electrons in a photoelectric experiment is quite random. The uncertainty principle prevents us from following a photon trajectory and predicting when and where an electron will be ejected. All we can say is that electrons are more likely to be ejected where the wave intensity is greater. Specifically, the probability that an electron will be ejected is directly proportional to the intensity of the incident electromagnetic waves—that is, to the square of the wave fields. More generally, the probability of finding a photon in a beam of electromagnetic waves is directly proportional to the wave intensity (Fig. 35.1).

In this quantum-mechanical description, the fields still evolve according to Maxwell's equations. For example, the fields of an electromagnetic wave undergoing double-slit interference develop regions of maximum and minimum wave intensity—the bright and dark bands of the interference pattern. But the wave fields determine only the *probability* that individual photons will be detected in the interference pattern. That's why a very short exposure or a low-intensity beam results not in a weak version of the interference pattern but in a seemingly arbitrary pattern. Only with large numbers of photons does the statistical pattern emerge (Fig. 35.2).

In quantum mechanics, then, the relation between the wave and particle aspects of light is this: As long as we don't try to detect the light, it propagates as a wave governed by Maxwell's equations. But when we detect the light, we do so through interactions involving individual photons. Those interactions are random events whose probability depends on the wave intensity—that is, on the square of the wave fields.

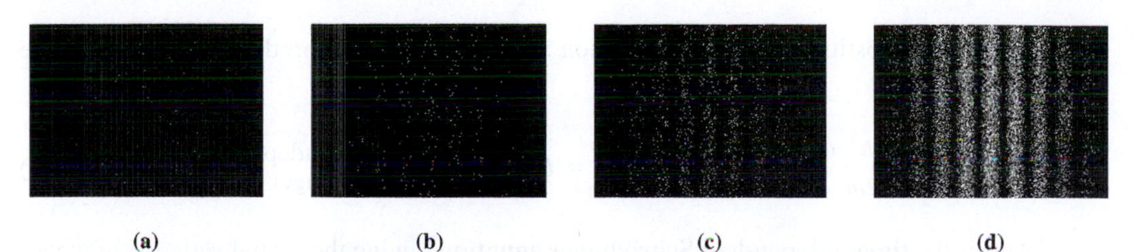

FIGURE 35.2 Development of a two-slit interference pattern from random photon events: (a) approximately 50 photons, (b) 250 photons, (c) 1000 photons, (d) 10,000 photons.

Electrons and Matter Waves

In Chapter 34 we introduced de Broglie's remarkable hypothesis that matter, as well as light, exhibits both wave and particle properties. The wave–particle duality puts matter and light on essentially the same footing, and the statistical interpretation is the same for each. Figure 35.3 shows a beam of particles and its associated de Broglie matter wave. Just as the probability of finding a photon is proportional to the wave intensity—that is, the *square* of the electromagnetic-field amplitude—so the probability of finding a particle is directly proportional to the square of the matter-wave amplitude. And as with light, the particle nature of matter manifests itself only when we try to detect a particle; leave it alone, and the particle's behavior is governed by its wave nature.

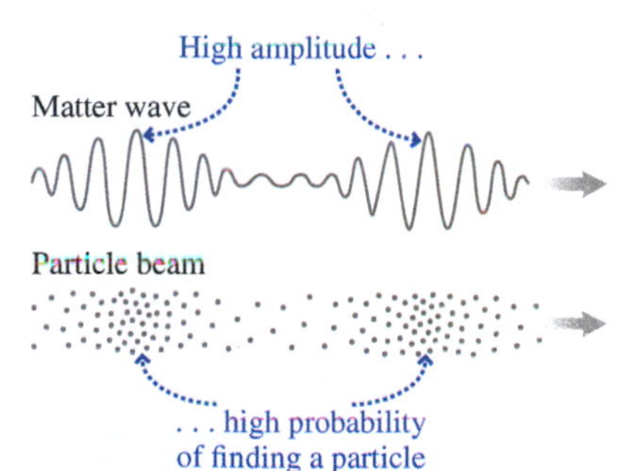

FIGURE 35.3 A beam of particles and its associated matter wave.

Maxwell's equations determine the behavior of light waves, but what equation describes matter waves? In 1926 the Austrian physicist Erwin Schrödinger answered this question with his **Schrödinger wave equation.** In the same year, Schrödinger showed that his wave theory was equivalent to a matrix-based theory that Heisenberg, Max Born, and Pascual Jordan had formulated in 1925. Heisenberg received the 1932 Nobel Prize in physics, and Schrödinger shared the 1933 Nobel Prize with Paul Dirac for their contributions to quantum theory.

GOT IT? 35.1 Focusing a particular laser beam results in a 10-fold increase in the electric field of the associated light wave. The probability of finding a photon at a point in the focused beam is increased by (a) a factor of 10, (b) a factor of $\sqrt{10}$, (c) a factor of 2, or (d) a factor of 100.

35.2 The Schrödinger Equation

The Schrödinger equation describes matter waves in terms of a **wave function,** ψ (Greek psi), which depends on both space and time. The solution of differential equations in two variables is beyond the mathematical level of this text, so here we'll consider only spatial variations, and for now we'll further restrict ourselves to one dimension.

We can understand the Schrödinger equation by considering a sinusoidal wave of the form $\psi(x) = A\sin kx$, where, as usual in describing waves, $k = 2\pi/\lambda$, with λ the wavelength. Differentiating this expression twice gives

$$\frac{d^2\psi(x)}{dx^2} = -Ak^2\sin kx = -k^2\psi(x)$$

But $k = 2\pi/\lambda$ and, for matter waves, de Broglie's hypothesis gives $\lambda = h/p$, with h Planck's constant and p the particle's momentum. Thus we can write k in terms of momentum as $k = 2\pi p/h = p/\hbar$. Now in classical physics a particle of mass m has kinetic energy K and momentum p related by $K = p^2/2m$. Furthermore, kinetic energy is the difference between the total energy E and the potential energy U; thus $E - U = p^2/2m$. Putting this all together, we can write the quantity k^2 in our differentiated wave expression as

$$k^2 = \frac{p^2}{\hbar^2} = \frac{2m(E - U)}{\hbar^2}$$

Make this substitution in our expression above for $d^2\psi/dx^2$ and do a little algebra; the result is

$$-\frac{\hbar^2}{2m}\frac{d^2\psi(x)}{dx^2} + U(x)\psi(x) = E\psi(x) \qquad \left(\begin{array}{c}\text{time-independent}\\ \text{Schrödinger equation}\end{array}\right) \qquad (35.1)$$

This is the **time-independent Schrödinger equation**, giving the spatial variation of matter waves in one dimension. A solution of the full time-dependent equation consists of a solution to Equation 35.1 multiplied by a sinusoidal oscillation with frequency $f = E/h$, where E is the particle energy. We developed the time-independent Schrödinger equation by merging de Broglie's matter-wave hypothesis $\lambda = h/p$ with the Newtonian relation $K = p^2/2m$; for that reason, we expect the equation to hold only for nonrelativistic particles.

The Schrödinger equation provides a description of physical reality in remarkable agreement with experiments. As we'll see, Schrödinger's equation goes a long way toward explaining the structure of atoms, their chemical properties, and indeed the entire science of chemistry. Furthermore, the Schrödinger description obeys the correspondence principle, agreeing with Newtonian mechanics for macroscopic systems where quantum effects are small.

The Meaning of ψ

What's the meaning of the wave function ψ? That's a deep question that physicists and philosophers continue to debate. In the standard interpretation, ψ is not an observable quantity. It manifests itself only in the statistical distributions of particle detections. More specifically, the probability per unit volume—also called the **probability density**—that we'll find a particle is given by ψ^2. For a particle confined to one dimension, the probability density becomes the probability per unit length, and we interpret ψ to mean that the probability $P(x)$ of finding the particle in a small interval dx at position x is

$$P(x) = \psi^2(x)\,dx \qquad \text{(probability and the wave function)} \qquad (35.2)$$

We can interpret Equation 35.2 in two ways. At face value, it gives the probability that a single experiment, with a detector at position x set up to find particles in an interval of width dx, will detect the particle (Fig. 35.4). Or, if we do many such experiments, the equation gives the fraction of the experiments in which we'll find a particle in our detector.

But *what is* ψ? How can it be unobservable yet govern the behavior of matter? There can't be a direct causal link between the wave function and individual particles, since ψ determines only the *probability* that a particle will behave in a certain way. Think about this! In quantum mechanics the outcome of an experiment isn't fully determined. The

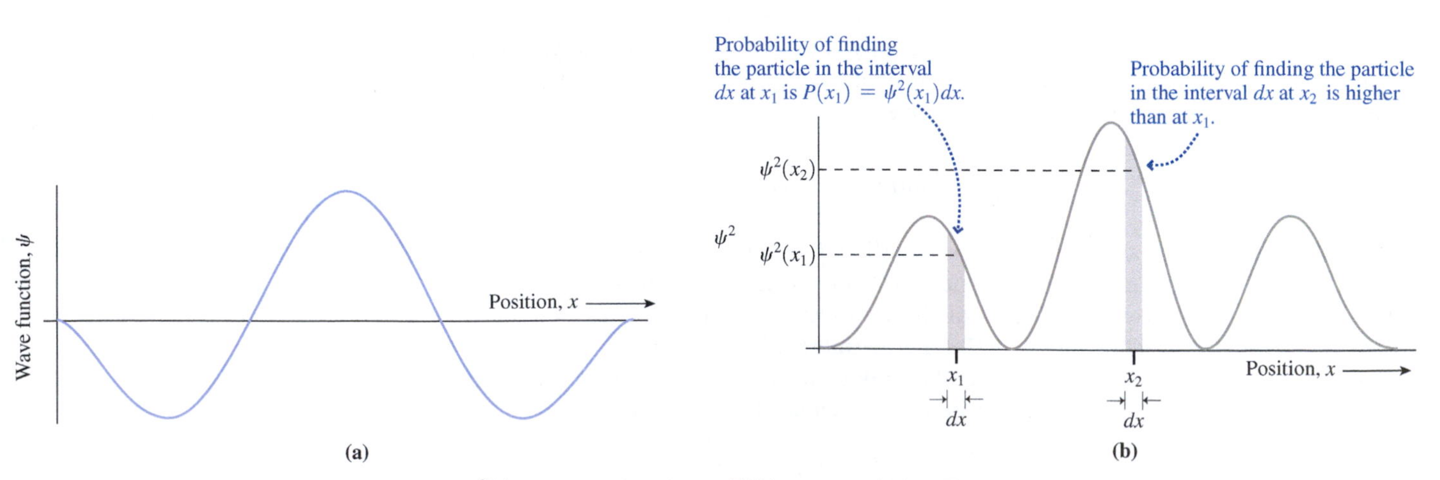

FIGURE 35.4 The meaning of the probability density $\psi^2(x)$. (a) A wave function and (b) its square, which gives the probability density.

Schrödinger equation describes only the probability of a given outcome. The quantum world is so different, according to the standard interpretation, that our macroscopic language, concepts, and pictorial models are simply inadequate. In particular, macroscopic causality gives way to microscopic indeterminacy in which quantum events are truly random; physical laws govern only the statistical pattern of events.

Are you bothered by the strange implications of quantum mechanics, with its description of a universe governed ultimately by probability? If so, you're in good company. Einstein himself never accepted the idea of a probabilistic universe, asserting, in a common paraphrase, that "God does not play dice." Einstein and Bohr frequently debated, and in a widely cited 1935 paper, Einstein and his colleagues Boris Podolsky and Nathan Rosen argued that quantum mechanics could not be a complete theory but required an underlying, deterministic physics governing so-called *hidden variables*, hidden from us by the uncertainty principle. But experiments done since the 1980s place severe constraints on such hidden-variable theories. Nevertheless, fascinating discussions on the interpretation of quantum mechanics continue to this day. Here, however, we'll take a more practical route, turning to the Schrödinger equation to see how it's used in analyzing quantum-mechanical systems.

GOT IT? 35.2 In Fig. 35.4b, are the probabilities of finding a particle in small regions of width dx given by (a) the areas of the shaded rectangles, (b) the values of ψ^2 in each region, or (c) the value of ψ in each region?

Normalization and Other Constraints on the Wave Function

In one dimension, the quantity $\psi^2 dx$ represents the probability of finding the particle in the interval dx. But the particle *must* be *somewhere*. Therefore, if we sum the probabilities of finding the particle in all such intervals dx, the result must be 1; there must be a 100% chance that we'll find the particle somewhere. Since the probability density may vary with position, that sum becomes an integral:

$$\int_{-\infty}^{+\infty} \psi^2 dx = 1 \qquad \text{(normalization condition)} \qquad (35.3)$$

Once we have a solution $\psi(x)$ to the Schrödinger equation 35.1, this **normalization condition** sets the overall amplitude of the function ψ.

The Schrödinger equation contains the second derivative of ψ. In order that this term be well defined, both ψ itself and its first derivative must be continuous. (An exception to the continuity condition on $d\psi/dx$—possible only in unrealistic example situations— occurs if the potential energy U becomes infinite.)

35.3 Particles and Potentials

The Infinite Square Well

We first solve the Schrödinger equation for a particularly simple system—a particle trapped in one dimension between two perfectly rigid walls. Although unrealistic in some respects, this system nevertheless is a surprisingly good approximation to some real quantum systems, including electronic devices and simple nuclei. More important, its analysis illustrates the general procedure for applying the Schrödinger equation and shows how energy quantization emerges from Schrödinger's theory.

PhET: Quantum Bound State: One Well

In classical physics, a particle trapped between rigid walls moves back and forth with constant speed. In the absence of friction or other losses, the particle's energy remains constant at its initial value. And in classical physics, that value can be anything.

We can describe the particle's situation using its potential-energy curve. Since the particle experiences no forces while it's between the walls, its potential energy U is constant

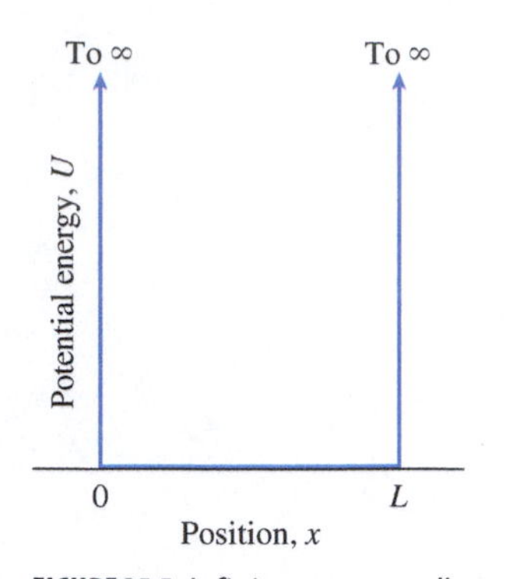

FIGURE 35.5 Infinite square-well potential-energy curve describes a particle constrained to move in one dimension between rigid walls separated by a distance L.

in this region, and we can fix the arbitrary zero of potential energy by setting $U = 0$. If the walls are perfectly rigid, then the particle can't penetrate them, no matter what its energy. This means that the potential energy becomes abruptly infinite at the walls. Then the potential-energy curve for our particle looks like Fig. 35.5; you can see from the figure why this curve is called an **infinite square well**. In this case the well extends from $x = 0$ to $x = L$.

We now consider the quantum-mechanical description of a particle in the infinite square well. The particle has a wave function whose time-independent part is given by the Schrödinger equation (Equation 35.1):

$$-\frac{\hbar^2}{2m}\frac{d^2\psi}{dx^2} + U(x)\psi = E\psi$$

where the potential energy $U(x)$ is that of the square well in Fig. 35.5:

$$U = 0 \text{ for } 0 < x < L$$
$$U = \infty \text{ for } x < 0 \text{ or } x > L$$

Since there's no chance that the particle can penetrate the rigid walls, the function ψ must be exactly zero in the region where $U = \infty$. All we need to calculate, then, is ψ inside the well, where $0 \le x \le L$. To ensure that the particle is confined to the well, our solution must satisfy so-called **boundary conditions**: $\psi = 0$ at $x = 0$ and at $x = L$.

Within the well, $U = 0$ and the Schrödinger equation becomes

$$-\frac{\hbar^2}{2m}\frac{d^2\psi}{dx^2} = E\psi \tag{35.4}$$

To find solutions, recall de Broglie's hypothesis that the allowed orbits in the Bohr atom are those for which standing waves just "fit" around the orbit. We have a similar situation with the infinite square well, in which the allowed solutions should be standing waves with nodes at the ends of the well—exactly analogous to standing waves on a string with both ends clamped that we discussed in Chapter 14. So we want a sinusoidal wave for $\psi(x)$, subject to the boundary conditions $\psi(0) = 0$ and $\psi(L) = 0$. The first condition is satisfied if we take a wave of the form $\psi = A \sin kx$, with A and k both constants. The second condition requires that $k = n\pi/L$, where n is any integer—a condition equivalent to saying that an integer number of half-wavelengths fit in the well. So we propose a solution of the form

$$\psi(x) = A \sin\left(\frac{n\pi x}{L}\right)$$

with the constant A still undetermined. This equation represents standing waves with nodes at the ends of the square well, but does it satisfy the Schrödinger equation? We can find out by substituting into Equation 35.4. We need not only ψ but also its second derivative; twice differentiating our proposed solution gives

$$\frac{d^2\psi}{dx^2} = -A\frac{n^2\pi^2}{L^2}\sin\left(\frac{n\pi x}{L}\right)$$

Substituting ψ and $d^2\psi/dx^2$ into Equation 35.4 gives

$$\left(-\frac{\hbar^2}{2m}\right)\left[-A\frac{n^2\pi^2}{L^2}\sin\left(\frac{n\pi x}{L}\right)\right] = EA\sin\left(\frac{n\pi x}{L}\right)$$

which reduces to

$$E = \frac{n^2\pi^2\hbar^2}{2mL^2} = \frac{n^2h^2}{8mL^2} \quad \text{(energy levels for an infinite square-well potential)} \tag{35.5}$$

Equation 35.5 says that our proposed solution can indeed satisfy the Schrödinger equation—provided the particle energy E has any one of the values given by Equation 35.5, with n an integer.

Our standing-wave solutions show how the quantization of energy arises naturally from the Schrödinger equation. Physically, the reason for quantization remains as de Broglie had postulated: Matter waves in a confined system must be standing waves with an integer number of half-wavelengths. Although de Broglie's hypothesis and the Schrödinger equation lead to exactly the same conclusion for the infinite square well, we'll see that with more complicated potential-energy functions only the Schrödinger equation can give us the full story.

The integer n that appears in Equation 35.5 is the **quantum number** for the particle in the square well. The physical state of a quantum-mechanical system is its **quantum state**. Here one quantum number suffices to specify the quantum state, which then tells us everything quantum mechanics has to say about the situation. As far as the Schrödinger equation is concerned, it looks like all integer values of n are allowed. The choice of negative or positive n has no physical significance, since ψ^2 has the same value with either sign of ψ; for this reason, negative n's are redundant. But $n = 0$ implies $\psi = 0$ everywhere, giving no chance of finding the particle anywhere. So we're left with positive integer values of n.

With only nonzero n's allowed, Equation 35.5 shows that the particle's energy is always positive; zero energy isn't allowed. The lowest possible energy is $E_1 = h^2/8mL^2$, obtained with $n = 1$. This is the **ground-state energy**; the corresponding wave function is the **ground-state wave function**. A nonzero ground-state energy is a common feature of quantum systems and one with no classical counterpart. Figure 35.6 is an energy-level diagram for the infinite square well.

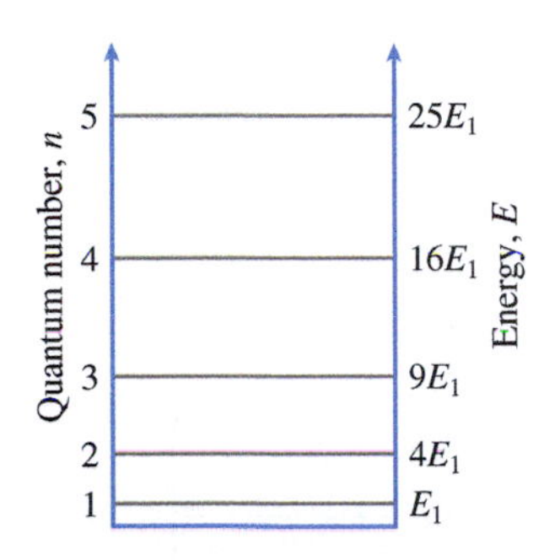

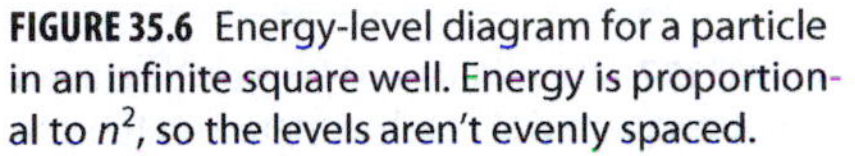
FIGURE 35.6 Energy-level diagram for a particle in an infinite square well. Energy is proportional to n^2, so the levels aren't evenly spaced.

CONCEPTUAL EXAMPLE 35.1 Ground-State Energy

Why can't the ground-state energy of the square well be zero?

EVALUATE Consider the uncertainty principle, $\Delta x\, \Delta p \geq \hbar$. If the ground-state energy were zero, then we would know precisely the particle's kinetic energy $p^2/2m$—zero—and therefore we would know that its momentum p was also zero. But we know that the particle is within the well, so the uncertainty in its position is at most the well width L. The product $\Delta p\, \Delta x$ would then be zero, in violation of the uncertainty principle.

ASSESS We used the uncertainty principle in the preceding chapter to estimate the minimum energies of confined particles. The ground-state energy for the square well is a specific instance of this so-called *zero-point energy*.

MAKING THE CONNECTION An electron drops from the $n = 2$ state to the ground state of a 0.75-nm-wide infinite square well, emitting a photon in the process. Find the photon's energy.

EVALUATE Equation 35.5 gives the square-well energies. Here the photon's energy is the difference between E_2 and the ground-state energy E_1: $\Delta E = 3.2\times10^{-19}$ J, or 2.0 eV.

GOT IT? 35.3 Electron A is confined to a square well 1 nm wide; electron B to a similar well only 1 pm wide. How do their ground-state energies compare? (a) $E_B = 10E_A$; (b) $E_B = 1000E_A$; (c) $E_B = 10^6E_A$; (d) $E_B = 10^{-3}E_A$

Normalization, Probabilities, and the Correspondence Principle

We still don't know the constant A in our solution for the infinite square well. We find this using the normalization condition 35.3: $\int_{-\infty}^{\infty}\psi^2 dx = 1$. Inside the well, $\psi = A\sin(n\pi x/L)$; outside, $\psi = 0$. So we can write the normalization condition as an integral over $0 < x < L$:

$$\int_0^L A^2 \sin^2\!\left(\frac{n\pi x}{L}\right) dx = 1$$

If we divided $\int_0^L \sin^2(n\pi x/L)\, dx$ by the well width L, we would have the average of sine squared over an integer number of half-cycles—or just $\frac{1}{2}$. So the integral of $\sin^2(n\pi x/L)$

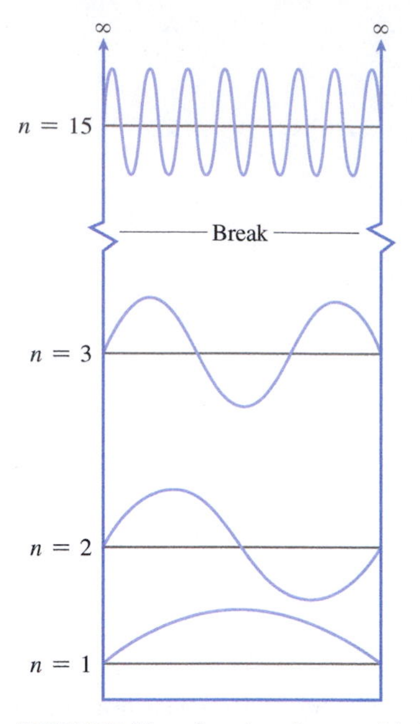

FIGURE 35.7 Wave functions for a particle in an infinite square well, each centered on the corresponding energy level.

from 0 to L is $\frac{1}{2}L$, and therefore $A^2(L/2) = 1$, or $A = \sqrt{2/L}$. The normalized wave function is then

$$\psi_n = \sqrt{\frac{2}{L}}\sin\left(\frac{n\pi x}{L}\right) \tag{35.6}$$

where the subscript n refers to the function associated with the nth quantum state. Figure 35.7 shows wave functions for the ground state and three excited states.

Where are we likely to find the particle? Classically, it would move back and forth at constant speed and therefore would be equally likely to be anywhere in the well. Quantum-mechanically, the probability of finding it at some position x is proportional to the probability density ψ^2 at that point. Figure 35.8 shows the probability densities given by squaring the wave functions of Fig. 35.7. For $n = 1$, we're clearly most likely to find the particle near the middle of the well—in marked contrast to the classical prediction of equal probability everywhere. For other low-n states there are obvious regions of high and low probability. But as the quantum number increases, the maxima and minima of the probability density get closer together. Any instrument we use to detect the electron has a finite resolution, and once the periodicity of the wave function drops below that resolution, we measure an average probability, which is essentially constant over the interval (Fig. 35.8).

This is a manifestation of Bohr's correspondence principle: For large quantum numbers n, the interval between adjacent energy levels becomes small compared with the energy itself, and a measurement of the electron's position gives results in agreement with classical physics. But classical physics is totally inadequate at low n, where the nonclassical zero-point energy and quantization are most evident.

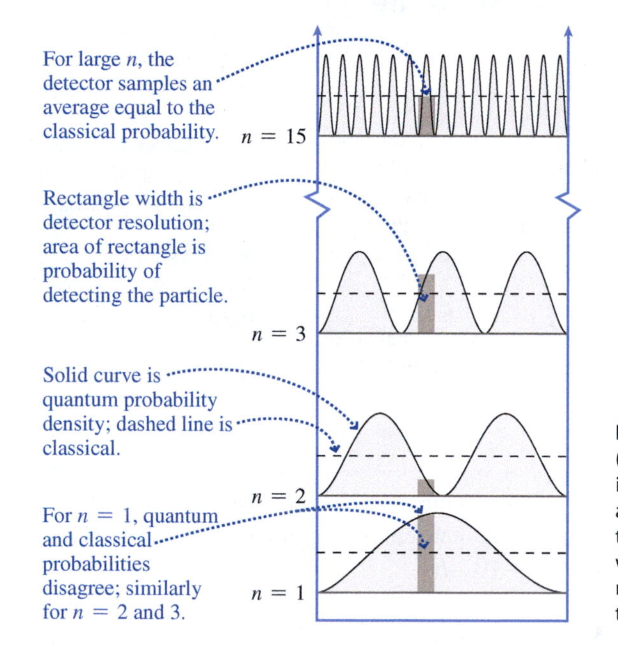

FIGURE 35.8 Classical (dashed) and quantum (solid) probability densities for a particle in an infinite square well. The shaded area under each curve is 1, indicating that the particle must be somewhere in the well. Width of the colored rectangle is the resolution of an instrument used to detect the particle.

EXAMPLE 35.1 Quantum Probability: The Square-Well Ground State

A particle is in the ground state of an infinite square well. Find the probability that it will be found in the left-hand quarter of the well.

INTERPRET This is a question about probability, and we know that the probability density is the square of the wave function. So our solution is going to involve ψ^2.

DEVELOP The ground-state wave function from Equation 35.6 is $\psi_1 = \sqrt{2/L}\sin(\pi x/L)$. We normalized the wave function so the integral of ψ_1^2 over the entire well is 1, showing that the particle is *somewhere* in the well. That is, the area under the entire plot of ψ_1^2 is 1. We sketched ψ_1^2 in Fig. 35.9, showing that the probability of finding the particle in some region is the area under the curve in that region. So to find the probability that the particle is in the left-hand quarter of the well, we'll evaluate $\int \psi^2 dx$ from 0 to $\frac{1}{4}L$.

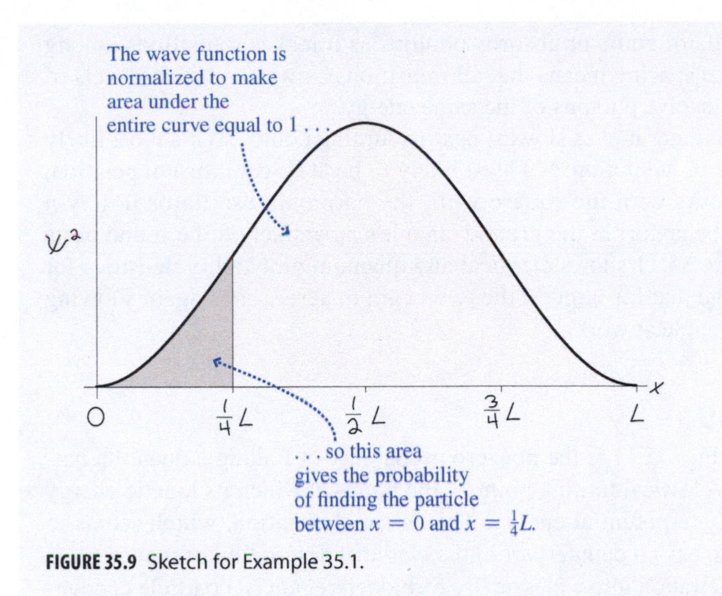

FIGURE 35.9 Sketch for Example 35.1.

EVALUATE The probability becomes

$$P = \frac{2}{L}\int_0^{L/4} \sin^2\left(\frac{\pi x}{L}\right) dx$$

We can integrate using the table at the end of Appendix A; the result is

$$P = \frac{2}{L}\left(\frac{x}{2} - \frac{\sin(2\pi x/L)}{4\pi/L}\Bigg|_0^{L/4}\right) = \frac{2}{L}\left(\frac{L}{8} - \frac{L}{4\pi}\right) = 0.091$$

ASSESS This is considerably lower than the probability $P = 0.25$ we would expect classically for finding the particle in any quarter of the well, and reflects the lower value of ψ^2 nearer the well ends. Problem 54 repeats the calculation of this example for arbitrary quantum numbers, showing that classical and quantum probabilities agree at large n.

GOT IT? 35.4 Which of the following would be a reasonable answer if Example 35.1 had asked for the probability that the particle would be found in the central quarter of the well: (a) 0.091, (b) 0.25, (c) 0.475, (d) 0.90?

The infinite square well gives insights into important quantum phenomena shared by more realistic systems such as atoms. These include quantized energy levels, nonzero ground-state energy, nonclassical probabilities, and agreement with classical physics at large quantum numbers. In Chapter 36 we'll apply the Schrödinger equation to atoms, where we'll find many of the same phenomena. First, though, we look at some other simple systems that exhibit additional quantum behaviors.

The Harmonic Oscillator

In Chapter 13 you studied simple harmonic motion, which occurs when a particle is subject to a restoring force that's directly proportional to the displacement from equilibrium. Such a *linear* restoring force implies a *quadratic* potential-energy function, and conversely, as you saw in Section 13.5, any system with a quadratic potential-energy function is a harmonic oscillator. That includes many systems at the atomic and molecular scale. Understanding the quantum-mechanical harmonic oscillator is therefore crucial in describing the behavior of matter on small scales.

MP

PhET: Quantum Bound State: One Well

A mass–spring system has potential energy $U = \frac{1}{2}kx^2$ and oscillates with angular frequency given by Equation 13.7a: $\omega = \sqrt{k/m}$. Combining these equations gives $U = \frac{1}{2}m\omega^2x^2$, providing a potential-energy function suitable for an electron or atom vibrating at the end of a molecular bond. Solving the Schrödinger equation for this potential requires advanced math techniques, and shows that normalizable wave functions exist only for discrete values of the energy E:

$$E_n = (n + \tfrac{1}{2})\hbar\omega \qquad (35.7)$$

where now $n = 0$ is the ground state. Figure 35.10 shows an energy-level diagram for the harmonic oscillator; note the even spacing implied by Equation 35.7. The additive factor $\frac{1}{2}$ in Equation 35.7 shows that Planck wasn't quite right in suggesting that the allowed harmonic-oscillator energies should be multiples of $hf(=\hbar\omega)$. Planck's spectral distribution (Equation 34.3) is nevertheless correct, but he did not foresee the existence of nonzero ground-state energy.

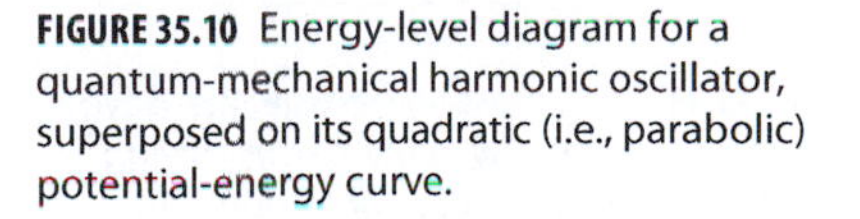

FIGURE 35.10 Energy-level diagram for a quantum-mechanical harmonic oscillator, superposed on its quadratic (i.e., parabolic) potential-energy curve.

The even spacing between the energy levels of the harmonic oscillator is in marked contrast to the situation in atoms (Fig. 34.11) or in the infinite square well (Fig. 35.6).

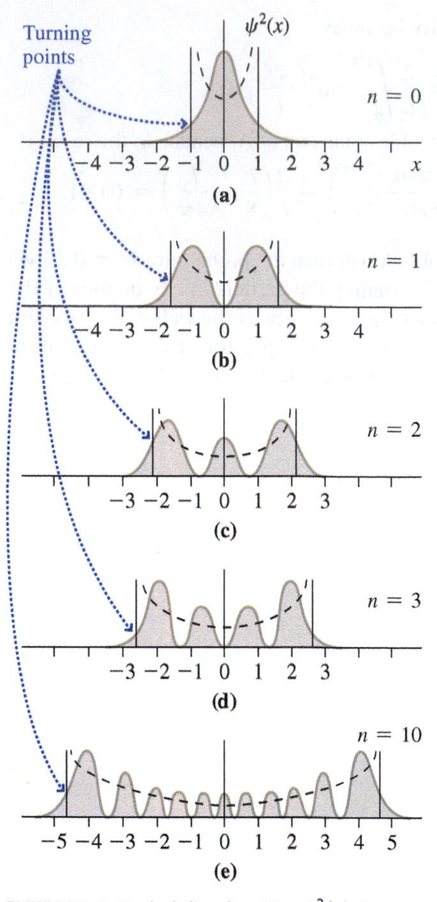

FIGURE 35.11 Probability densities $\psi^2(x)$ for some states of the harmonic oscillator. Dashed curves are classical predictions. Increasing spread in the classical turning points reflects the higher energy of the higher-n states.

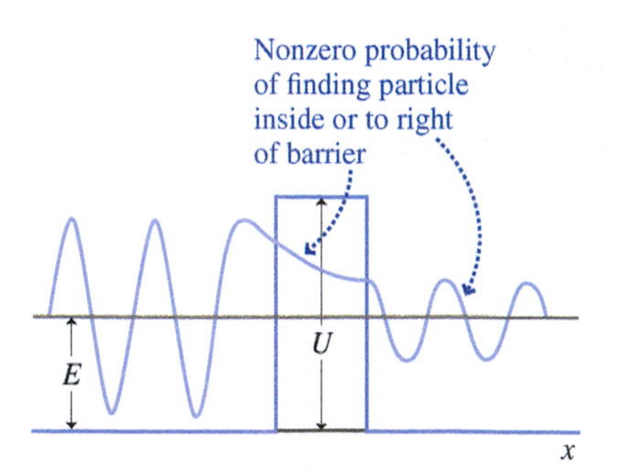

FIGURE 35.12 A potential barrier of height U, showing the wave function for a particle incident from the left with energy E lower than the barrier energy U.

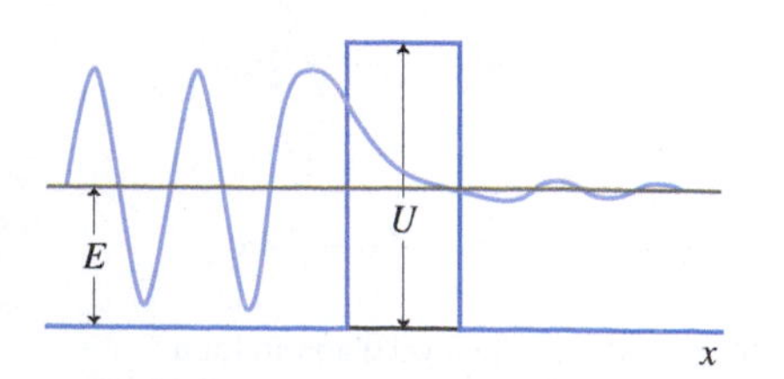

FIGURE 35.13 For a massive particle, the wave function drops rapidly in the barrier, giving negligible probability of penetration.

A quantum harmonic oscillator emits or absorbs photons as it makes transitions among adjacent levels, and the even spacing means that all transitions between adjacent levels of a pure harmonic oscillator involve photons of the same energy.

A classical harmonic oscillator moves slowest near its turning points, so it's most likely to be found at the extremes of its motion. It's least likely to be at its equilibrium position, where it's moving fastest. As with the square well, the harmonic oscillator in low-n states exhibits unclassical behavior; in the ground state it's *most* likely to be found at its equilibrium position! Figure 35.11 shows classical and quantum probability densities for the harmonic oscillator; note that for larger n the two begin to agree, once again showing Bohr's correspondence principle at work.

Quantum Tunneling

One remarkable feature of Fig. 35.11 is the nonzero probability of finding a quantum harmonic oscillator beyond its classical turning points—the points at which its kinetic energy has been converted entirely to potential energy. This unusual situation, which seems to violate energy conservation, has no counterpart in the classical description of matter.

Another example of penetration into a classically forbidden region is a particle encountering a potential barrier (Fig. 35.12). Examples of such barriers include electric potential differences associated with atomic nuclei, gaps between solid materials, and insulating layers in some semiconductor devices. Classically, a particle whose total energy is lower than the barrier energy is confined to one side of the barrier. If we solve the Schrödinger equation for this potential-energy curve, however, we find oscillatory solutions on either side of the barrier, joined according to the continuity conditions on ψ and $d\psi/dx$ by exponential functions within the barrier. Such a solution is shown superimposed on the barrier in Fig. 35.12. The probability density ψ^2 associated with this solution remains nonzero through the barrier and continues to give a nonzero probability of finding the particle on the far side—implying that a particle initially on one side of the barrier may later be found on the other side.

How likely is this phenomenon, called **quantum tunneling**? That depends on the relation of the particle energy E to the barrier energy U, and also on the width of the barrier. As you can show in Problem 49, the ψ function inside the barrier involves exponential functions of the form $e^{\pm\sqrt{2m(U-E)}x/\hbar}$. In general, these exponentials drop very rapidly across the barrier width unless the particle energy E is close to the barrier energy or the particle mass m is small. The probability that a particle will be found on the far side of the barrier is therefore very low when the mass m is large, so quantum tunneling is a microscopic phenomenon (Fig. 35.13).

It looks as if tunneling violates energy conservation. But we're saved by the uncertainty principle. If we catch the particle within the barrier, the uncertainty in its position is no greater than the barrier width. We know from Example 34.5 that this implies a minimum energy. A quantitative analysis shows that minimum to be such that we can no longer be sure the particle energy is lower than the barrier energy. If we don't try to detect a particle within the barrier, its penetration is a purely wave phenomenon to which our particulate energy considerations don't apply. Again we see the wave–particle duality at work: If we don't observe the particle, its behavior is governed by the associated waves and may result in most unparticle-like phenomena such as tunneling. If we do try to catch it in the act of such behavior, it ceases to be wave-like and the surprising phenomena cease.

Tunneling is important in a number of quantum-mechanical phenomena and technological devices. That the Sun shines—and therefore that we're alive—is a consequence of quantum tunneling of nuclei in the Sun's core. Classically, those nuclei don't have sufficient energy to get close enough to overcome their mutual electric repulsion. But they can tunnel through this "Coulomb barrier" and fuse to release the enormous energy that powers the Sun. An opposite process, alpha decay, occurs as alpha particles tunnel through a potential barrier that traps them inside large nuclei like uranium. Measurement of the alpha particles' energy shows it to be lower than the barrier energy, confirming that tunneling occurs. Tunneling is the basis of the scanning tunneling microscope (STM), a remarkable device that lets us image individual atoms (see Application, next page). Quantum tunneling moves electrons on and off the transistors that store information in flash memory—the memory used in your smartphone, tablet, camera, flash drive, and increasingly as replacements for hard disks in computers. Finally, tunneling of individual electrons is the basis of one proposed implementation of a new elementary-charge based definition of the coulomb, as described in Chapter 20.

MP®

PhET: Quantum Tunneling and Wave Packets

GOT IT? 35.5 A proton and an electron approach a barrier. Both have the same energy E, which is lower than the barrier potential U. Is (a) the proton or (b) the electron more likely to get through, or (c) are they equally likely to get through?

APPLICATION Scanning Tunneling Microscope

Developed in the 1980s by Heinrich Rohrer and Gerd Binnig of IBM Zurich Research Laboratory, the scanning tunneling microscope (STM) has become a vital tool for semiconductor engineers, biologists, chemists, and nanotechnologists. The STM works by quantum tunneling between an extraordinarily fine conducting tip and the surface under study. The photo shows a scanning electron microscope image of an STM tip, which may be only one atom wide. As in the barrier of Fig. 35.12, the electron wave function tapers off exponentially in the space outside the surface. Place a conducting tip near but not touching the surface, and there's a nonzero probability that electrons will tunnel through the gap to reach the tip, resulting in an electric current. The exponential falloff of the wave function means this **tunneling current** is extremely sensitive to the tip-to-surface gap, and therefore changes significantly with surface irregularities.

A practical STM scans the tip over the surface, and feedback devices move the tip to keep the tunneling current constant despite surface irregularities, as shown in the figure. Therefore, the tip traces out the surface topography, and this information is used to construct an image of the surface—even at the scale of individual atoms (see this chapter's opening photo).

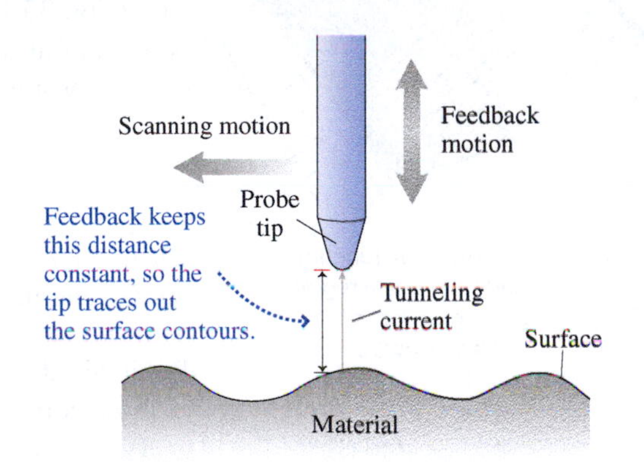

Finite Potential Wells

Both the infinite square well and the harmonic oscillator have potential wells of infinite depth. No matter what its energy a particle is bound in such a well; it can't escape to large distances. Its quantized energy states are therefore all **bound states**. Provided they aren't too shallow, wells of finite depth also exhibit quantized bound states whose wave functions resemble those of the infinite square well (Fig. 35.14), although they show a small but nonzero probability of tunneling into the classically forbidden region outside the well. Problems 58–60 explore quantitatively the energy levels in finite wells.

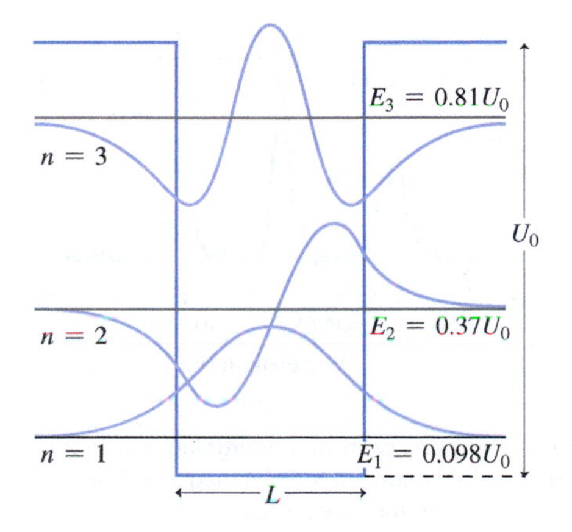

FIGURE 35.14 Bound-state wave functions for a finite square well, superposed on the associated energy levels. For this combination of well depth, well width, and particle mass there are only three bound states.

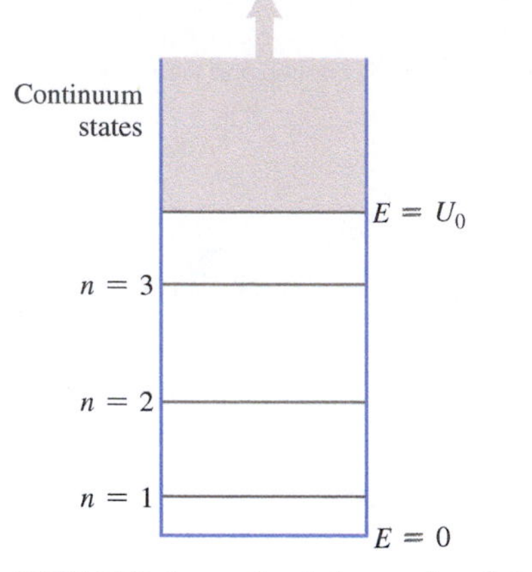

FIGURE 35.15 Energy-level diagram for a finite square well shows discrete bound states and a continuum of unbound states.

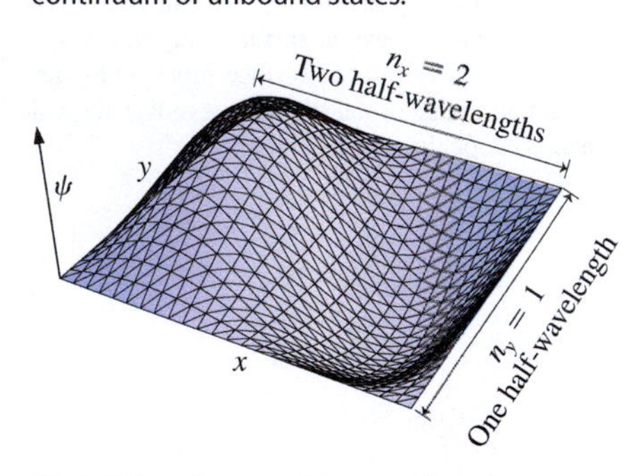

FIGURE 35.16 One possible wave function for a particle confined to a square region in two dimensions. This function is $\psi(x, y) = \sin(n_x\pi x/L)\sin(n_y\pi y/L)$, with $n_x = 2$ and $n_y = 1$.

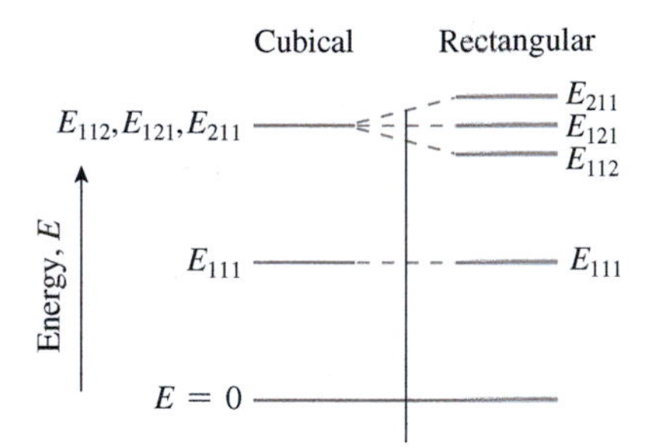

FIGURE 35.17 Energy-level diagrams showing the ground state and first excited state for a particle in a three-dimensional box. Making the sides different lengths removes the degeneracy.

Quantized bound states represent particles with energy lower than the well height. Particles with higher energy are free to move anywhere, and their wave functions are everywhere oscillatory. Furthermore, particles in these **unbound states** can have any energy whatsoever as long as it exceeds the well height; unbound energies aren't quantized. Rather, there's a **continuum** of allowed energies above the well top, in contrast to the discrete, quantized levels below (Fig. 35.15). We'll find both bound and unbound states again in the next chapter when we study the atom.

35.4 Quantum Mechanics in Three Dimensions

One-dimensional quantum systems show important features of the quantum world, like energy quantization and tunneling. But atoms and most other quantum systems are three-dimensional. The wave function then depends on all three spatial variables, and the Schrödinger equation reflects this complexity. You can explore the three-dimensional Schrödinger equation in Problem 51; here we just point out some new features of three-dimensional quantum systems.

A single quantum number n characterizes quantum states in one dimension. With the infinite square well, for example, an integer number of half-wavelengths can fit in the well, and n is that number. Each n is associated with a distinct energy level. In two or three dimensions, similar considerations lead to independent quantum numbers for each dimension (Fig. 35.16). For each set of quantum numbers there's an associated energy. For a particle of mass m confined to a cubical box of side L, for example, a generalization of the one-dimensional square well leads to the energy levels

$$E = \frac{h^2}{8mL^2}(n_x^2 + n_y^2 + n_z^2) \tag{35.8}$$

where the n's are the quantum numbers associated with each spatial dimension. As in one dimension, the allowed values for the n's are positive integers. Thus, the ground state has $n_x = n_y = n_z = 1$. But what's the first excited state? It could be $n_x = 2, n_y = n_z = 1$. But it could equally well be $n_x = n_y = 1, n_z = 2$, or $n_x = n_z = 1, n_y = 2$, since all three of these combinations give the same energy.

Two or more quantum states with the same energy are termed **degenerate**. The first excited state of a particle confined to a cubical box is threefold degenerate, meaning there are three distinct states with the same energy. Degeneracy is often associated with symmetry of the quantum-mechanical system. In the cubical box, the equal-length sides result in different combinations of quantum numbers with the same energy. Making the sides different would remove the degeneracy, splitting a single energy level into three (Fig. 35.17). The same thing happens in more realistic quantum systems. For example, imposing a magnetic field on an otherwise spherically symmetric atom breaks the symmetry and may split energy levels that were previously degenerate (Fig. 35.18). Detection of this splitting in optical spectra allows measurement of magnetic fields on the Sun and in other remote objects.

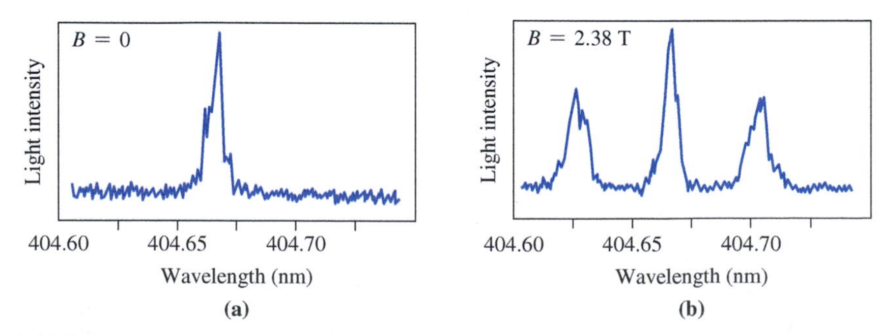

FIGURE 35.18 (a) Spectral line at 404.66 nm produced by mercury atoms undergoing transitions from $n = 7$ to $n = 6$. The upper level is actually threefold degenerate. (b) A magnetic field breaks the symmetry and removes the degeneracy, splitting the spectral line.

GOT IT? 35.6 Is the energy of the first excited state for a particle in a cubical box (a) twice, (b) four times, or (c) eight times that of the ground state?

35.5 Relativistic Quantum Mechanics

Like Newtonian physics, quantum mechanics based on the Schrödinger equation is not consistent with special relativity's requirement that the laws of physics be the same in all inertial reference frames. It's therefore an approximation valid for particle speeds v much lower than c. For most applications in atomic, molecular, and condensed-matter physics, $v \ll c$ so the Schrödinger equation applies. But when particle speeds are a significant fraction of c, the Schrödinger equation becomes inadequate and must be replaced with a relativistic wave equation. And even for slowly moving particles, the requirement of relativistic invariance leads to some surprising new phenomena.

The Dirac Equation and Antiparticles

In 1928 the English physicist Paul Dirac formulated a relativistic wave equation for electrons. In the process he encountered several unexpected mathematical requirements with deep physical significance.

Dirac replaced the Newtonian energy–momentum relation $K = p^2/2m$ with the relativistic expression $E^2 = (mc^2)^2 + p^2c^2$ that we saw in Chapter 33. But this expression implies two values for E, depending on which sign one chooses in taking the square root. Dirac argued that both roots are meaningful and that the negative root implies the existence of a particle identical in mass to the electron but carrying positive charge. The 1932 discovery of this **positron** vindicated Dirac's brilliant idea. Today we know that every elementary particle has a corresponding **antiparticle**, identical in mass but opposite in electric, magnetic, and other properties.

Einstein's energy–mass equivalence implies that **pair creation** of a particle–antiparticle pair is possible, given energy $2mc^2$ equivalent to the mass of the pair. The opposite process, annihilation, occurs as particle and antiparticle meet and disappear to form a pair of photons. (Recall Chapter 33's Application "PET Scans: Relativity in the Hospital," which described pair annihilation.) Although pair creation is rare today, it was commonplace in the hot, early universe, where thermal energy alone was high enough to create particle–antiparticle pairs. In those early times Einstein's mass–energy equivalence would have been obvious, and the number of particles in a closed volume wouldn't have remained constant.

Electron Spin

Another unexpected mathematical result of Dirac's work was that the wave function had to involve matrices. This, Dirac showed, implied physically that the electron must possess an intrinsic angular momentum—something physicists had already inferred from experiments, but without any theoretical grounding. This angular momentum, called **spin**, has enormous significance in quantum mechanics and particularly in atomic structure, as we'll see in the next chapter.

CHAPTER 35 SUMMARY

Big Idea

The big idea here is the description of particles in the quantum realm using **wave functions**, whose square relates to the probability of finding a particle. Thus the link between the most thorough description physics can provide—the wave function—and the behavior of an individual particle is only statistical. The **Schrödinger equation** gives the wave function for nonrelativistic particles and leads to energy quantization for confined particles.

Key Concepts and Equations

The **time-independent Schrödinger equation** gives the wave function ψ for a particle of mass m with total energy E and potential energy U:

$$-\frac{\hbar^2}{2m}\frac{d^2\psi}{dx^2} + U(x)\psi = E\psi$$

The square of the wave function is the **probability density**. In one dimension, the probability of finding the particle in some small interval dx at position x is

$$P(x) = \psi^2(x)\,dx$$

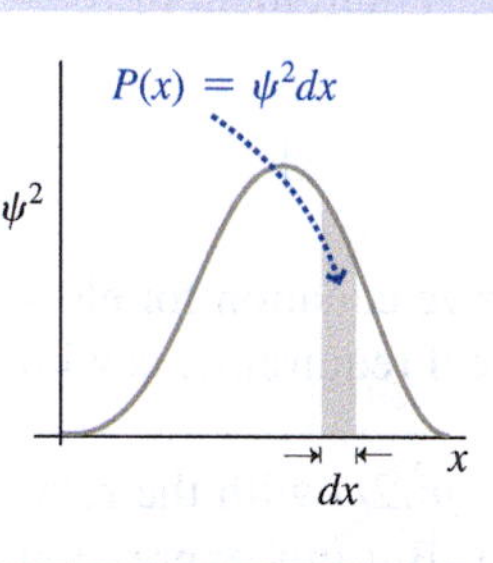

Normalization: A particle must be *somewhere*, so

$$\int_{-\infty}^{+\infty} \psi^2(x)dx = 1$$

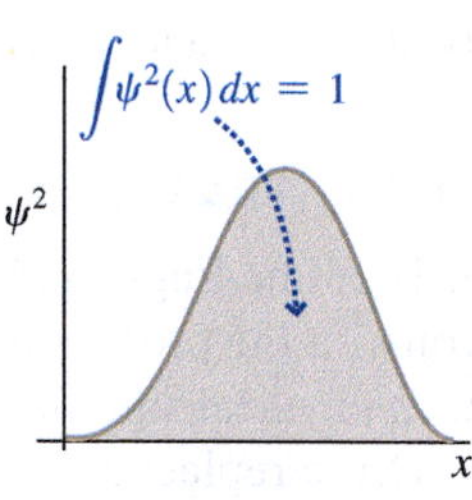

Applications

Infinite square well

Wave functions: $\psi_n = \sqrt{\frac{2}{L}}\sin\left(\frac{n\pi x}{L}\right)$

Energy levels: $E_n = \frac{n^2h^2}{8mL^2}$

3-D well: $E = \frac{h^2}{8mL^2}(n_x^2 + n_y^2 + n_z^2)$

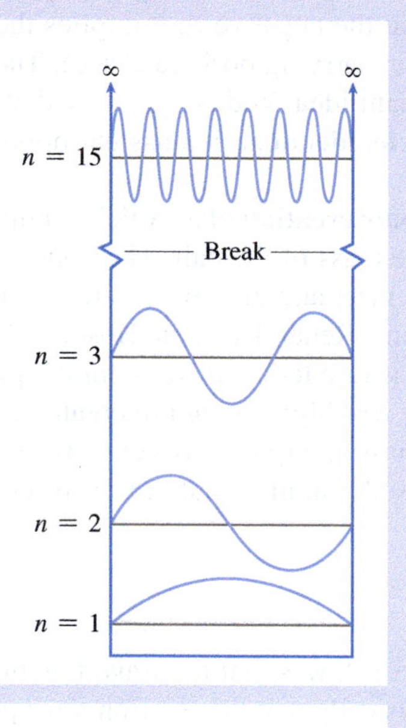

Harmonic oscillator

Energy levels: $E_n = (n + \frac{1}{2})\hbar\omega$

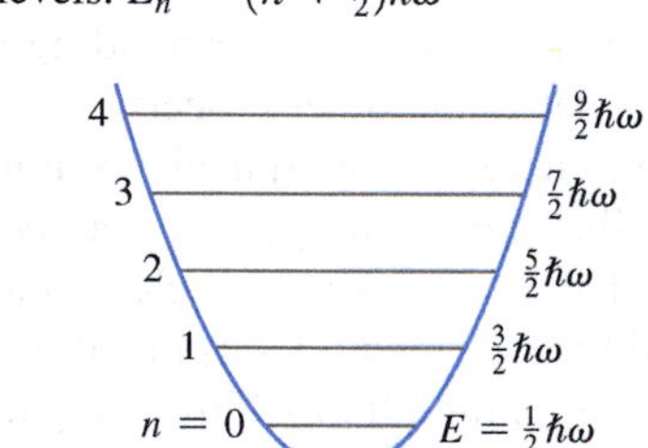

Finite well

Discrete bound states; continuum of unbound states

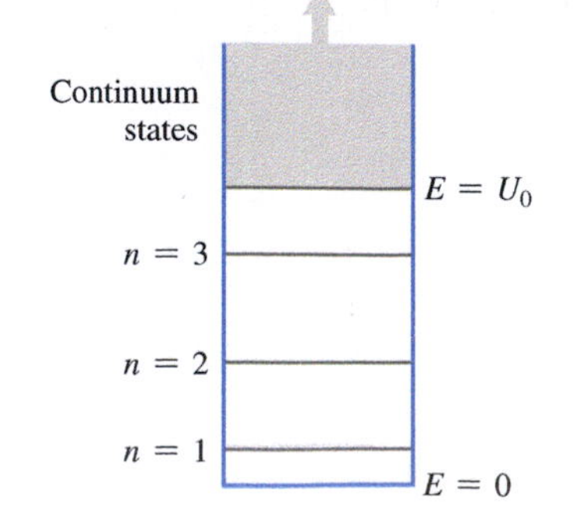

Quantum tunneling

Nonzero probability of finding a quantum particle in a region forbidden by classical energy conservation leads to the possibility of barrier penetration:

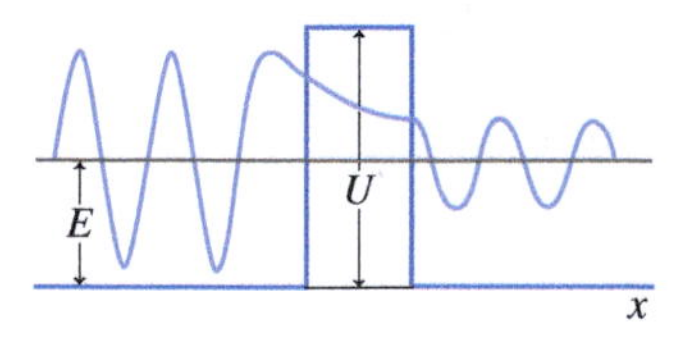

For Thought and Discussion

1. Explain qualitatively why a particle confined to a finite region cannot have zero energy.
2. Does quantum tunneling violate energy conservation? Explain.
3. Bohr's correspondence principle states that quantum and classical mechanics must agree in a certain limit. Give an example.
4. The ground-state wave function for a quantum harmonic oscillator has a single central peak. Why is this at odds with classical physics?
5. What's the essential difference between the energy-level structures of infinite and finite square wells?
6. In terms of de Broglie's matter-wave hypothesis, how does making the sides of a box different lengths remove the degeneracy associated with a particle confined to that box?
7. A particle is confined to a two-dimensional box whose sides are in the ratio 1:2. Are any of its energy levels degenerate? If so, give an example. If not, why not?
8. What did Einstein mean by his remark, loosely paraphrased, that "God does not play dice"?
9. Some philosophers argue that the strict determinism of classical physics is inconsistent with human free will, but that the indeterminacy of quantum mechanics does leave room for free will. Others claim that physics has no bearing on the question of free will. What do you think?
10. What fundamental principle of physics, when combined with quantum mechanics, provides a theoretical basis for the existence of antimatter?
11. Figure 35.19 shows an infinite square well with a step-like potential at the bottom. Sketch qualitatively what you think a wave function might look like for a particle whose energy is (a) less than the step height and (b) greater than the step height.

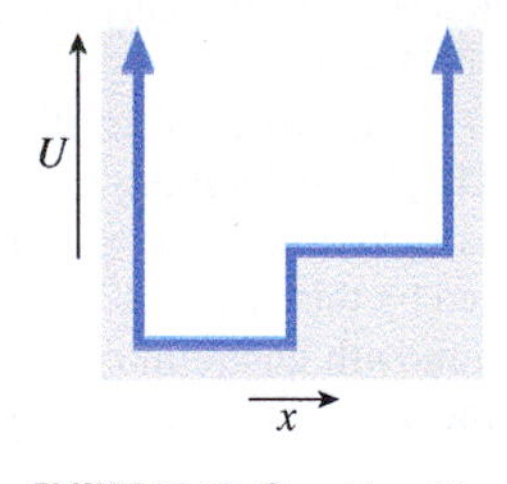

FIGURE 35.19 Question 11

Exercises and Problems

Exercises

Section 35.2 The Schrödinger Equation

12. What are the units of the wave function $\psi(x)$ in a one-dimensional situation?
13. A particle's wave function is $\psi = Ae^{-x^2/a^2}$, where A and a are constants. (a) Where is the particle most likely to be found? (b) Where is the probability per unit length half its maximum value?
14. The solution to the Schrödinger equation for a particular potential is $\psi = 0$ for $|x| > a$ and $\psi = A\sin(\pi x/a)$ for $-a \le x \le a$, where A and a are constants. In terms of a, what value of A is required to normalize ψ?

Section 35.3 Particles and Potentials

15. What's the quantum number for a particle in an infinite square well if the particle's energy is 25 times the ground-state energy?
16. A particle in an infinite square well makes a transition from a higher to a lower energy state; the corresponding energy decrease is 33 times the ground-state energy. Find the quantum numbers of the initial and final states.
17. Determine the ground-state energy for an electron in an infinite square well of width 10.0 nm.
18. Find the width of a square well in which a proton's first excited state has energy 1.5 keV.
19. A carbon nanotube traps an electron in a hollow cylindrical structure 0.48 nm in diameter. Approximating the nanotube as a one-dimensional infinite square well, find the energies in eV of (a) the ground state and (b) the first excited state.
20. One reason we don't notice quantum effects in everyday life is that Planck's constant h is so small. Treating yourself as a particle (mass 60 kg) in a room-sized one-dimensional infinite square well (width 2.6 m), how big would h have to be if your minimum possible energy corresponded to a speed of 1.0 m/s?
21. A particle is confined to a 1.0-nm-wide infinite square well. If the energy difference between the ground state and the first excited state is 1.13 eV, is the particle an electron or a proton?
22. A 3-g snail crawls at 0.5 mm/s between two rocks 15 cm apart. Treating this system as an infinite square well, determine the approximate quantum number. Does the correspondence principle permit the use of the classical approximation in this case?
23. An alpha particle (mass 4 u) is trapped in a uranium nucleus with diameter 15 fm. Treating the system as a one-dimensional square well, what would be the minimum energy for the alpha particle?
24. A quantum harmonic oscillator has ground-state energy 0.14 eV. What would be the system's classical oscillation frequency f?
25. Find the ground-state energy for a particle in a harmonic oscillator potential whose classical angular frequency ω is $1.0\times10^{17}\ \text{s}^{-1}$.
26. A harmonic oscillator emits a 1.1-eV photon as it undergoes a transition between adjacent states. Find its classical oscillation frequency f.
27. The ground-state energy of a harmonic oscillator is 4.0 eV. Find the energy separation between adjacent quantum states.
28. Your roommate is taking Newtonian physics, while you've moved on to quantum mechanics. He claims that QM can't be right, because he didn't see any evidence of quantized energy levels in a mass–spring harmonic oscillator experiment. You reply by calculating the spacing between energy levels of this system, which consists of a 1-g mass on a spring with $k = 80$ N/m. What is that spacing, and how does this help your argument?

Section 35.4 Quantum Mechanics in Three Dimensions

29. If all sides of a cubical box are doubled, what happens to the ground-state energy of a particle in that box?
30. A very crude model for an atomic nucleus is a cubical box 1 fm on a side. What would be the energy of a gamma ray emitted if a proton in such a nucleus made a transition from its first excited state to the ground state?

31. An electron is confined to a cubical box. For what box width will a transition from the first excited state to the ground state result in emission of a 950-nm infrared photon?

Problems

32. Find an expression for the normalization constant A for the wave function given by $\psi = 0$ for $|x| > b$ and $\psi = A(b^2 - x^2)$ for $-b \le x \le b$.
33. Suppose ψ_1 and ψ_2 are solutions of the Schrödinger equation for the same energy E. Show that the linear combination $a\psi_1 + b\psi_2$ is also a solution, where a and b are arbitrary constants.
34. An electron is trapped in an infinite square well 25 nm wide. Find the wavelengths of the photons emitted in these transitions: (a) $n = 2$ to $n = 1$; (b) $n = 20$ to $n = 19$; (c) $n = 100$ to $n = 1$.
35. An electron drops from the $n = 7$ to the $n = 6$ level of an infinite square well 1.5 nm wide. Find (a) the energy and (b) the wavelength of the photon emitted.
36. Show explicitly that the difference between adjacent energy levels in an infinite square well becomes arbitrarily small compared with the energy of the upper level, in the limit of large quantum number n.
37. An electron is in a narrow molecule 4.4 nm long, a situation that approximates a one-dimensional infinite square well. If the electron is in its ground state, what is the maximum wavelength of electromagnetic radiation that can cause a transition to an excited state?
38. The ground-state energy for an electron in infinite square well A is equal to the energy of the first excited state for an electron in well B. How do the wells' widths compare?
39. Electrons in an ensemble of 0.834-nm-wide square wells are all initially in the $n = 4$ state. (a) How many different wavelengths of spectral lines could be emitted as the electrons cascade to the ground state through all possible downward transitions? (b) Find those wavelengths. (c) What regions of the electromagnetic spectrum do these spectral lines encompass?
40. Sketch the probability density for the $n = 2$ state of an infinite square well extending from $x = 0$ to $x = L$, and determine where the particle is most likely to be found.
41. An infinite square well extends from $-L/2$ to $L/2$. (a) Find expressions for the normalized wave functions for a particle of mass m in this well, giving separate expressions for even and odd quantum numbers. (b) Find the corresponding energy levels.
42. A particle is in the ground state of an infinite square well. What's the probability of finding the particle in the left-hand third of the well?
43. A laser emits 1.96-eV photons. If this emission is due to electron transitions from the $n = 2$ to $n = 1$ states of an infinite square well, what's the well width?
44. What's the probability of finding a particle in the central 80% of an infinite square well, assuming it's in the ground state?
45. **BIO** Is quantization significant for macromolecules confined to biological cells? To find out, consider a protein of mass 250,000 u confined to a 10-μm-diameter cell. Treating this as a particle in a one-dimensional square well, find the energy difference between the ground state and the first excited state. Given that biochemical reactions typically involve energies on the order of 1 eV, what do you conclude about the role of quantization?
46. In your physical chemistry course, you model hydrogen chloride as a hydrogen atom on a spring; the other end of the spring is attached to a rigid wall (the massive chlorine atom). In order to determine the spring constant in your model, you measure the minimum photon energy that will promote HCl molecules to their first excited state. The result is 0.358 eV. What do you calculate for the effective k?
47. A particle detector has a resolution 15% of the width of an infinite square well. What's the probability that the detector will find a particle in the ground state of the square well if the detector is centered on (a) the midpoint of the well and (b) a point one-fourth of the way across the well?
48. Find the probability that a particle in an infinite square well is located in the central one-fourth of the well for the quantum states $n =$ (a) 1, (b) 2, (c) 5, and (d) 20. (e) What's the classical probability in this situation?
49. A particle of mass m is in a region where its total energy E is less than its potential energy U. Show that the Schrödinger equation has nonzero solutions of the form $Ae^{\pm\sqrt{2m(U-E)}x/\hbar}$. Such solutions describe the wave function in quantum tunneling, beyond the turning points in a quantum harmonic oscillator, or beyond the well edges in a finite potential well.
50. (a) Use Equation 35.8 to draw an energy-level diagram for the first six energy levels of a particle in a cubical box, in terms of $h^2/8mL^2$, and (b) give the degeneracy of each.
51. The generalization of the Schrödinger equation to three dimensions is

$$-\frac{\hbar^2}{2m}\left(\frac{\partial^2\psi}{\partial x^2} + \frac{\partial^2\psi}{\partial y^2} + \frac{\partial^2\psi}{\partial z^2}\right) + U(x, y, z)\psi = E\psi$$

(a) For a particle confined to the cubical region $0 \le x \le L, 0 \le y \le L, 0 \le z \le L$, show by direct substitution that the equation is satisfied by wave functions of the form $\psi(x, y, z) = A\sin(n_x\pi x/L)\sin(n_y\pi y/L)\sin(n_z\pi z/L)$, where the n's are integers and A is a constant. (b) In the process of working part (a), verify that the energies E are given by Equation 35.8.
52. A 9-W laser beam shines on an ensemble of 10^{24} electrons, each in the ground state of a one-dimensional infinite square well 0.72 nm wide. The photon energy is just high enough to raise an electron to its first excited state. How many electrons can be excited if the beam shines for 10 ms?
53. A large number of electrons are confined to infinite square wells 1.2 nm wide. They're undergoing transitions among all possible states. How many visible lines (400 nm to 700 nm) are in the spectrum emitted by this ensemble of square-well systems?
54. **CH** A particle is in the nth quantum state of an infinite square well. (a) Show that the probability of finding it in the left-hand quarter of the well is

$$P = \frac{1}{4} - \frac{\sin(n\pi/2)}{2n\pi}$$

(b) Show that for odd n, the probability approaches the classical value $\frac{1}{4}$ as $n \to \infty$.
55. **CH** (a) Using the potential energy $U = \frac{1}{2}m\omega^2x^2$ discussed on page 675, develop the Schrödinger equation for the harmonic oscillator. (b) Show by substitution that $\psi_0(x) = A_0e^{-\alpha^2x^2/2}$ satisfies your equation, where $\alpha^2 = m\omega/\hbar$ and the energy is given by Equation 35.7 with $n = 0$. (c) Find the normalization constant A_0. You then have the ground-state wave function for the harmonic oscillator.
56. You're trying to convince a friend that nuclear energy represents a much more concentrated energy source than fossil fuels, whose combustion involves rearranging atomic electrons. For a rough comparison, you calculate the ground-state energy of a proton

confined to 1-fm-diameter atomic nucleus and that of an electron confined to a 0.1-nm-diameter atom. Approximating each system as a one-dimensional infinite square well, what's the ratio of their ground-state energies?

57. **DATA** The table below lists the wavelengths emitted as electrons in identical square-well potentials drop from various states n to the ground state. Determine a quantity that, when you plot λ against it, should yield a straight line. Make your plot, establish a best-fit line, and use your line to determine the width of the square well.

Initial state, n	4	5	7	8	10
Wavelength, λ (nm)	1110	674	354	281	169

58. **CH** The next three problems solve the Schrödinger equation for finite square wells like that shown in Fig. 35.14. It's convenient to work in dimensionless forms of the particle energy E and well depth U_0, given respectively by $\epsilon = 2mL^2E/\hbar^2$ and $\mu = 2mL^2U_0/\hbar^2$. Assuming that $E < U_0$, or, equivalently, $\epsilon < \mu$, show by substitution that the following wave functions satisfy the Schrödinger equation in the regions indicated:

$$\psi_1 = A\sin(\sqrt{\epsilon}x/L), \quad 0 \le x \le L$$

$$\psi_2 = Be^{-\sqrt{\mu-\epsilon}x/L}, \quad x \ge L$$

where A and B are constants.

59. **CH** The wave functions of Problem 58, as well as their derivatives, need to be continuous at $x = L$ if these functions are to represent the quantum state of a particle in the finite square well. (a) Show that these conditions lead to two equations:

$$A\sin(\sqrt{\epsilon}) = Be^{-\sqrt{\mu-\epsilon}}$$

$$\sqrt{\epsilon}A\cos(\sqrt{\epsilon}) = -\sqrt{\mu-\epsilon}Be^{-\sqrt{\mu-\epsilon}}$$

(b) then show that these lead to the single equation

$$\tan(\sqrt{\epsilon}) = -\sqrt{\frac{\epsilon}{\mu-\epsilon}}$$

60. **COMP** Solve the final equation of Problem 59 to find all possible values of ϵ for (a) $\mu = 2$, (b) $\mu = 20$, and (c) $\mu = 50$. You'll need to use a numerical root-finding routine on a calculator or computer. The number of solutions may vary with μ, and it's possible that there are no solutions for some values of μ.

Passage Problems

BIO *Quantum dots*, or *qdots*, are nanoscale crystals of semiconductor material that trap electrons in a potential well closely resembling the three-dimensional square well discussed in Section 35.4. Physicists, materials scientists, and semiconductor engineers have been studying qdots for their potential to miniaturize electronic components. More recently, qdots have been used in biology and medicine to "tag" individual molecules, helping scientists follow cellular processes (Fig. 35.20). Qdots also facilitate high-resolution imaging within the cell, and they show promise for medical diagnostics and targeting tumors for the delivery of anticancer agents. In the biomedical context, qdots work as replacements for traditional fluorescent dyes. Illuminating qdots promotes their electrons to higher energy levels; as they drop back, they emit photons of precise wavelength. A dot's size and structure determine this wavelength.

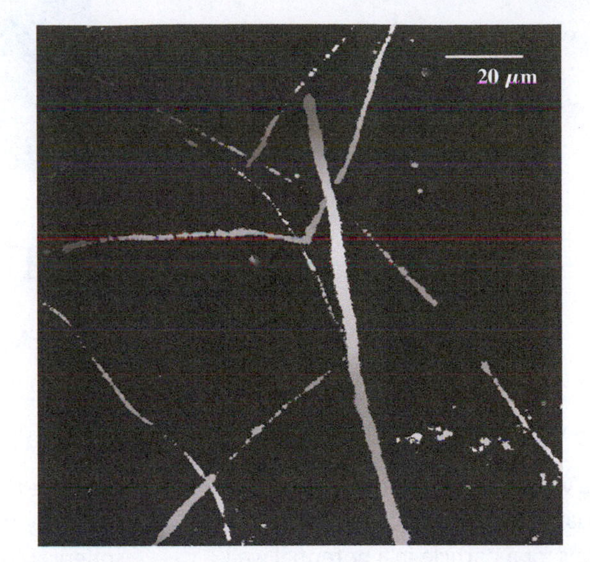

FIGURE 35.20 In this microscopic photo, motor protein molecules called dynein have been tagged with quantum dots, allowing their paths to be tracked (Passage Problems 61–64).

61. If a qdot's size is decreased, what happens to the wavelength of the photon emitted in a transition from the dot's first excited state to the ground state?
    a. The wavelength increases.
    b. The wavelength decreases.
    c. The wavelength is unchanged.
62. If the dot behaves as a perfectly cubical 3-D square well, the first excited state is
    a. nondegenerate.
    b. twofold degenerate.
    c. threefold degenerate.
    d. You can't tell without knowing the energy.
63. If the dot behaves as a perfectly cubical 3-D square well, the ground state is
    a. nondegenerate.
    b. twofold degenerate.
    c. threefold degenerate.
    d. You can't tell without knowing the energy.
64. If all three sides of a qdot are halved, its ground-state energy
    a. is halved.
    b. drops to one-fourth its original value.
    c. doubles.
    d. quadruples.

Answers to Chapter Questions

Answer to Chapter Opening Question

Quantum tunneling, the ability of particles to penetrate a barrier that classical physics says they don't have sufficient energy to overcome.

Answers to GOT IT? Questions

35.1 (d)
35.2 (a)
35.3 (c)
35.4 (c)
35.5 (b)
35.6 (a)

Atomic Physics

What You Know

- You've seen how the Schrödinger equation leads to quantized energy levels for a particle in a potential well.
- You understand the distinction between bound states and continuum states.
- You know how to calculate the probability of finding a quantum particle in a particular region, given its wave function.

What You're Learning

- You'll see the Schrödinger equation for the hydrogen atom expressed in spherical coordinates.
- You'll solve explicitly for the hydrogen ground states.
- You'll understand how excited states of hydrogen arise from the Schrödinger equation.
- You'll learn about quantized angular momentum in the hydrogen atom, including *space quantization*.
- You'll learn about *electron spin* and how it's quantized.
- You'll learn the *exclusion principle* and how it's responsible for the structure of multielectron atoms and, ultimately, the periodic table of the elements.
- You'll learn rules governing transitions among atomic energy levels and how they explain, among other things, the operation of lasers.

How You'll Use It

- If your future studies or career involve chemistry, you'll appreciate that chemistry is based, ultimately, on the quantum mechanical description of atoms. Increasingly, chemists determine complicated molecular structures with computer calculations based on the Schrödinger equation.
- If you use spectroscopy as an analysis tool in any area of science or engineering, you'll be exploiting the quantization of energy in atoms and molecules, and electron transitions among energy.
- Understanding the exclusion principle will show you why the universe is so rich with different forms of matter and why complex systems such as yourself can even exist.

How do the principles of quantum physics explain the different chemical elements?

In Chapter 35 we applied the Schrödinger equation to simplified quantum systems. Here we turn to the more realistic case of the atom, and explore how quantum mechanics explains atomic structure and the periodic table of the elements. We'll deal most thoroughly with the simplest atom, hydrogen, and we'll be more qualitative in describing multielectron atoms.

36.1 The Hydrogen Atom

Like a particle in a three-dimensional box, the electron in hydrogen is confined to a three-dimensional potential well. For the electron, the well results from the proton's electrostatic attraction. From Chapter 22 you know that the electric potential due to the proton, treated as a point charge e, is $V(r) = ke/r$, with r the distance to the proton and the zero of potential at infinity. Electric potential is energy per unit charge, so multiplying by the electron charge $-e$ gives the potential energy of the electron–proton system—that is, of the hydrogen atom:

$$U(r) = -\frac{ke^2}{r} \tag{36.1}$$

We'll approximate the massive proton as being at rest at the origin, so Equation 36.1 gives the electron's potential energy as a function of radial position r. We can therefore use Equation 36.1 as the potential energy in the Schrödinger equation for the hydrogen atom.

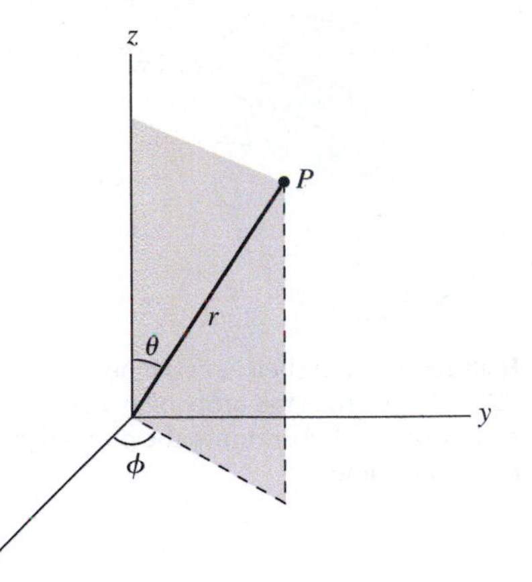

FIGURE 36.1 Spherical coordinates r, θ, ϕ provide an alternative to rectangular coordinates x, y, z.

The Schrödinger Equation in Spherical Coordinates

Because the electron's potential energy depends on radial distance r, it's best to work in spherical coordinates, where the position of a point is given by its distance r from the origin along with two angles θ and ϕ that specify its orientation (Fig. 36.1). Converting the Schrödinger equation to spherical coordinates is straightforward but tedious; the result is

$$-\frac{\hbar^2}{2mr^2}\left[\frac{\partial}{\partial r}\left(r^2\frac{\partial\psi}{\partial r}\right) + \frac{1}{\sin\theta}\frac{\partial}{\partial\theta}\left(\sin\theta\frac{\partial\psi}{\partial\theta}\right) + \frac{1}{\sin^2\theta}\frac{\partial^2\psi}{\partial\phi^2}\right] - \frac{ke^2}{r}\psi = E\psi \quad \left(\begin{array}{l}\text{Schrödinger equation,}\\ \text{spherical coordinates}\end{array}\right) \tag{36.2}$$

where we've used Equation 36.1 for the potential-energy function.

Although Equation 36.2 looks forbidding, it can be solved using advanced techniques. For total energy E less than zero, corresponding to bound states in hydrogen's potential well, most solutions become infinite at large r and therefore aren't normalizable. As a result, only certain values of the energy E give acceptable bound-state solutions. For total energy greater than zero, the electron is unbound and any energy proves possible, as with the finite square well.

The Hydrogen Ground State

In general, solutions to Equation 36.2 depend on all three variables r, θ, ϕ. But some solutions, including the ground state, are spherically symmetric—they depend only on r. Here we show that the ground state has the form of an exponential, and in the process derive the ground-state energy. Consider the function

$$\psi = Ae^{-r/a_0} \tag{36.3}$$

where A and a_0 are as yet undetermined constants, the latter with the units of length. For this spherically symmetric function, nothing depends on the angular variables θ and ϕ, so derivatives with respect to those variables are strictly zero. We're then dealing with a function of only one variable, so we can write total instead of partial derivatives. Equation 36.2 then becomes

$$-\frac{\hbar^2}{2mr^2}\frac{d}{dr}\left(r^2\frac{d\psi}{dr}\right) - \frac{ke^2}{r}\psi = E\psi \tag{36.4}$$

PhET: Quantum Bound State: One Well: 3D Coulomb

Substituting the proposed solution 36.3 for ψ gives

$$-\frac{\hbar^2}{2ma_0^2} + \frac{\hbar^2}{mra_0} - \frac{ke^2}{r} = E_1$$

(see Problem 42), where E_1 is the ground-state energy. This equation must be satisfied for all values of r, so the two r-dependent terms must cancel:

$$\frac{\hbar^2}{mra_0} = \frac{ke^2}{r}$$

or

$$a_0 = \frac{\hbar^2}{mke^2} = 5.29\times10^{-11}\,\text{m} = 0.0529\,\text{nm}$$

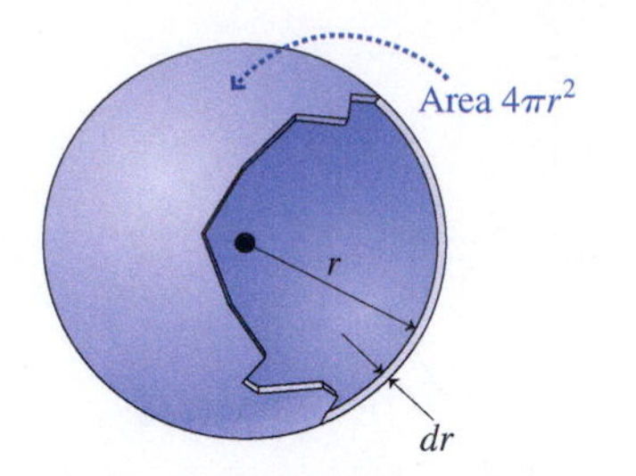

FIGURE 36.2 A thin shell has volume $dV = 4\pi r^2 dr$; thus the probability per unit radial distance is $4\pi r^2$ times the probability per unit volume.

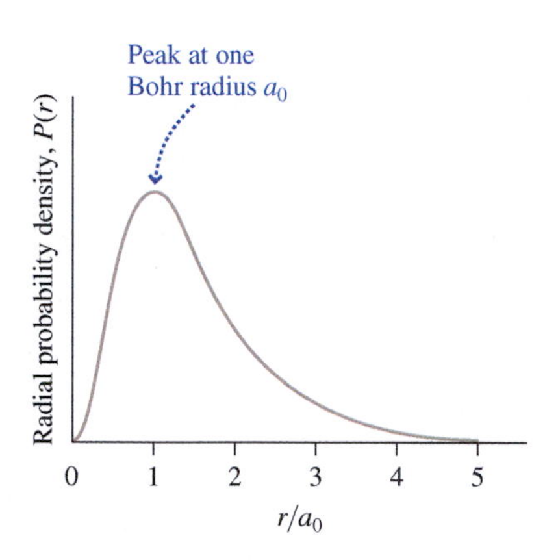

FIGURE 36.3 Radial probability density for the hydrogen ground state.

This is precisely the **Bohr radius** that we introduced in Section 34.4.

With the r terms gone, our expression for the ground-state energy becomes $E_1 = -\hbar^2/2ma_0^2 = -13.6\text{ eV}$, where the minus sign shows that the atom is a bound system. Thus Equation 36.3 is indeed a solution to the Schrödinger equation for hydrogen, with energy $E_1 = -13.6\text{ eV}$.

In deriving expressions for a_0 and E_1, we've shown how Schrödinger's theory gives two fundamental parameters of atomic physics: the Bohr radius and the hydrogen ground-state energy. Both agree with the values we found in Chapter 34 using the simpler Bohr model. But Bohr's theory still clings to the notion of classical orbits, with a_0 the ground-state orbital radius. Schrödinger's theory is truly quantum mechanical, representing the electron with its wave function ψ and associated probability distribution. The Bohr radius is no longer an actual orbital radius but instead determines atomic size only in a statistical sense.

The Radial Probability Distribution

Because the ground-state wave function falls off exponentially as e^{-r/a_0}, we're unlikely to find the electron at distances far greater than the Bohr radius. But where are we most likely to find it? Although ψ is greatest at $r = 0$, that's not the answer. In three dimensions the probability density ψ^2 is the probability *per unit volume* of finding the electron. In asking where we're most likely to find the electron, we want the probability *per unit radial distance*. Figure 36.2 shows a thin spherical shell with radius r and therefore area $4\pi r^2$. It has thickness dr, so its volume is $dV = 4\pi r^2 dr$. Then the probability of finding the electron in this shell is $\psi^2 dV = 4\pi r^2 \psi^2 dr$. The **radial probability density**, $P(r)$, is the probability per unit radius, or

$$P(r) = 4\pi r^2 \psi^2 \qquad \text{(radial probability density)} \qquad (36.5)$$

For the hydrogen ground state, we use Equation 36.3 for ψ to get $P_1 = 4\pi r^2 A^2 e^{-2r/a_0}$, where the subscript 1 designates the ground state. Figure 36.3 is a plot of this probability density, which peaks at $r = a_0$. Thus the single most likely place to find the electron in ground-state hydrogen is one Bohr radius from the proton.

EXAMPLE 36.1 The Hydrogen Atom: Normalization and the Probability Distribution

(a) Determine the normalization constant A in Equation 36.3. (b) Use the resulting wave function to find the probability that the electron in the hydrogen ground state will be found beyond the Bohr radius.

INTERPRET The wave function 36.3 contains an undetermined constant A. This problem is asking us to apply the normalization condition to find A and then use the concept of radial probability density to determine the probability of finding the electron beyond $r = a_0$.

DEVELOP The electron must be *somewhere* in the range $r = 0$ to $r = \infty$. Since $P(r)\,dr$ is the probability of finding the electron in a region of width dr, the normalization condition becomes $\int_0^\infty P(r)\,dr = 1$. So our plan is to evaluate this integral using the ground-state probability density $P_1 = 4\pi r^2 A^2 e^{-2r/a_0}$. We'll then solve for the unknown A. Then we can integrate again, this time from $r = a_0$ to $r = \infty$, to get the probability of finding the electron beyond a_0.

EVALUATE (a) Using the probability density P_1, the normalization condition becomes

$$\int_{r=0}^{r=\infty} 4\pi r^2 A^2 e^{-2r/a_0} dr = 1$$

We could evaluate using integration by parts; however, the result is in the integral table at the end of Appendix A. Replacing x by r and a by $-2/a_0$ in the table's expression for $\int x^2 e^{ax} dx$, we have

$$\int_0^\infty 4\pi A^2 r^2 e^{-2r/a_0} dr$$
$$= 4\pi A^2 \left\{ \frac{r^2 e^{-2r/a_0}}{(-2/a_0)} - \frac{2}{(-2/a_0)}\left[\frac{e^{-2r/a_0}}{(-2/a_0)^2}\left(-\frac{2}{a_0}r - 1\right)\right]\right\}\Bigg|_0^\infty = 1$$

The expression in curly brackets vanishes at $r = \infty$, and at $r = 0$ the exponentials are just 1, so we have $4\pi A^2\left[0 - \left(-\tfrac{1}{4}a_0^3\right)\right] = 1$, or $A = 1/\sqrt{\pi a_0^3}$.

(b) In part (a) we integrated from $r = 0$ to $r = \infty$ because we wanted the probability that the electron was *somewhere*. Here we want the probability that it's beyond $r = a_0$, so we change the lower limit on the integral from 0 to a_0. The result is

$$P(r > a_0) = \int_a^\infty 4\pi r^2 A^2 e^{-2r/a_0} dr$$
$$= 4\pi A^2 a_0^3\left(\tfrac{1}{2}e^{-2} + \tfrac{3}{4}e^{-2}\right) = 5\pi A^2 a_0^3 e^{-2}$$

With $A^2 = 1/\pi a_0^3$, this becomes $P(r > a_0) = 5e^{-2} \simeq 0.677$.

ASSESS Our result shows that about two-thirds of the time, the electron will be found beyond the Bohr radius. So although it's reasonable to say that the atom's radius is roughly the Bohr radius, both Fig. 36.3 and our result here show that there's no sharp cutoff that marks the "size" of the atom. ■

Excited States of Hydrogen

So far we've examined only the ground state of hydrogen. But Equation 36.2 admits many more normalizable solutions, corresponding to the excited states of hydrogen.

In general, each energy level is associated with one spherically symmetric wave function and a number of nonsymmetric ones. For historical reasons, the spherically symmetric states are called ***s* states**. The distinct energy levels are labeled by the quantum number n, called the **principal quantum number**. The ground state, for example, is the $1s$ state. The energy of the nth level, derivable from the Schrödinger equation, turns out to agree exactly with the earlier Bohr theory:

$$E_n = -\frac{1}{n^2}\frac{\hbar^2}{2ma_0^2} = \frac{E_1}{n^2} = \frac{-13.6\text{ eV}}{n^2} \quad \text{(hydrogen energy levels)} \tag{36.6}$$

The spherically symmetric state with energy E_2—that is, the $2s$ state—has wave function given by

$$\psi_{2s} = \frac{1}{4\sqrt{2\pi a_0^3}}\left(2 - \frac{r}{a_0}\right)e^{-r/2a_0} \tag{36.7}$$

By substituting this function into Equation 36.4, you can verify that the energy E_2 is given by Equation 36.6 (see Problem 66). The radial probability densities for the first three spherically symmetric states are plotted in Fig. 36.4; note that the excited states correspond to larger, more "smeared-out" atoms.

Although we're discussing hydrogen, our results generalize to any single-electron atom—that is, to an atom of atomic number Z ionized $Z - 1$ times. For such an atom the potential-energy function becomes $-kZe^2/r$, and our calculations go through as before except that the factor e^2 is replaced by Ze^2. Then the energy levels become

$$E_n = -\frac{Z^2}{n^2}\frac{\hbar^2}{2ma_0^2} = \frac{Z^2E_1}{n^2} = -\frac{(13.6\text{ eV})Z^2}{n^2} \tag{36.8}$$

reflecting the tighter binding of the more highly charged nucleus (see Fig. 36.5 and Problem 67).

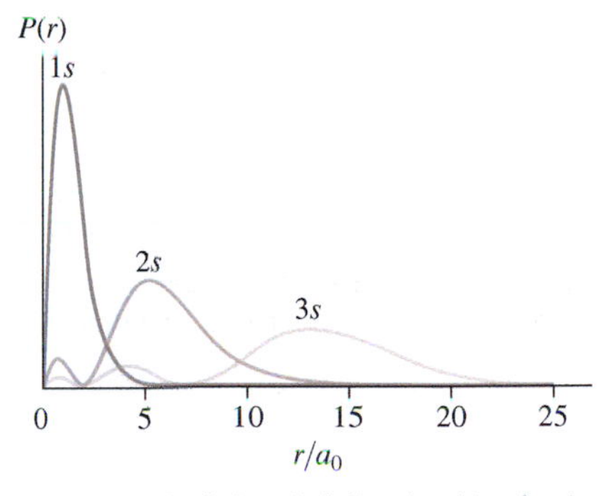

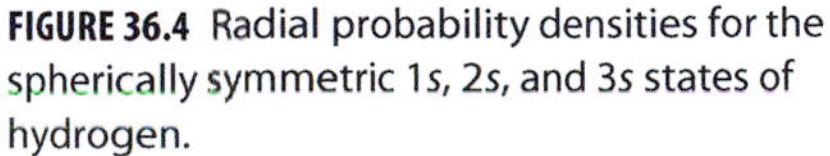

FIGURE 36.4 Radial probability densities for the spherically symmetric $1s$, $2s$, and $3s$ states of hydrogen.

GOT IT? 36.1 Which is the most appropriate estimate for the radial "size" of a hydrogen atom in its $2s$ state: (a) a_0, (b) $2a_0$, (c) $5a_0$, (d) $15a_0$?

Orbital Quantum Numbers and Angular Momentum

In the spherically symmetric s states, it turns out that the **orbital angular momentum** associated with the electron's motion is zero. This is at odds with Bohr's prediction that angular momentum should be an integer multiple of $\hbar$. And it makes clear that we can't be talking here about classical orbits, since motion in an elliptical or circular path entails angular momentum. But there are other solutions to the Schrödinger equation for hydrogen, solutions that aren't spherically symmetric and that have nonzero angular momentum.

For a given principal quantum number n, there are in fact n distinct solutions with different angular momenta. The **orbital quantum number** l distinguishes these states and ranges from 0 to $n - 1$. Thus the ground state ($n = 1$) corresponds to the single value $l = 0$. Higher energy levels, however, are degenerate, meaning there's more than one l value for each $n > 1$. The orbital quantum number determines the magnitude L of the electron's orbital angular momentum:

$$L = \sqrt{l(l+1)}\hbar \quad \text{(quantization of orbital angular momentum)} \tag{36.9}$$

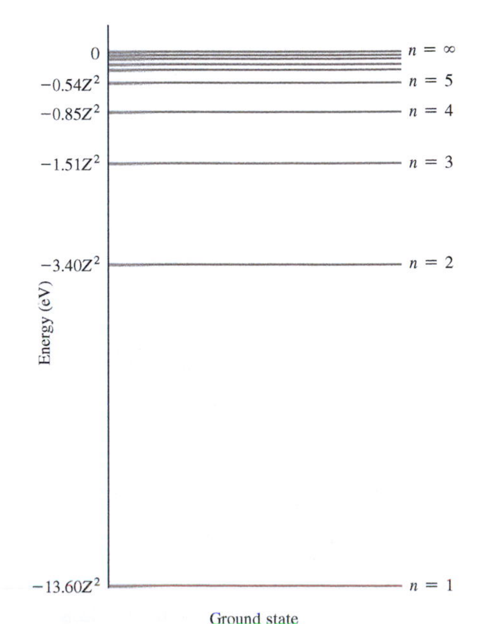

FIGURE 36.5 Energy-level diagram for a one-electron atom with atomic number Z. Energies scale as Z^2.

EXAMPLE 36.2 Orbital Angular Momentum: An Excited State

Find the possible values for the orbital angular momentum of an electron in the $n = 3$ state of hydrogen.

INTERPRET We're asked about the orbital angular momentum L, whose value follows from the orbital quantum number l. Thus we'll need the possible l values for $n = 3$.

DEVELOP For any n, there are n distinct l values, from 0 to $n - 1$. For $n = 3$, that means $l = 0$, 1, or 2. So our plan is to evaluate L using Equation 36.9, $L = \sqrt{l(l+1)}\hbar$, for these three l values.

EVALUATE With $l = 0$, Equation 36.9 gives $L = 0$; for $l = 1$, $L = \sqrt{2}\hbar$; and for $l = 2$, $L = \sqrt{6}\hbar$.

ASSESS $l = 0$ is the spherically symmetric 3s state, which we've seen has zero angular momentum. The higher-l states have increasing angular momentum. ■

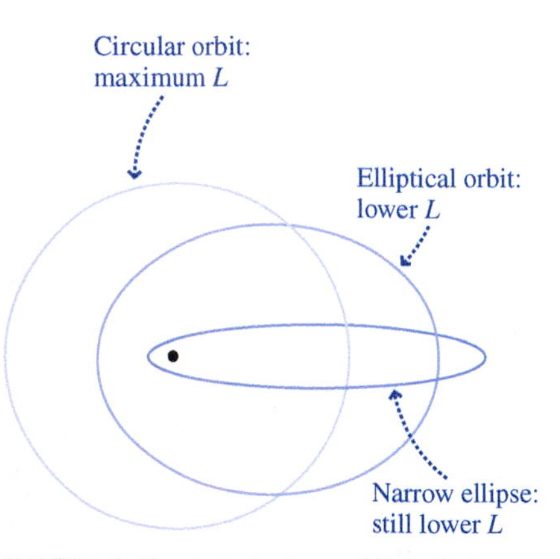

FIGURE 36.6 Classical electron orbits with the same energy but different angular momenta.

States with l values 0, 1, 2, 3, 4, 5, . . . are given the letter labels $s, p, d, f, g, h, \ldots$. These combine with the principal quantum number n to specify both the energy and angular momentum of a state. Thus, the ground state is 1s, and the $n = 2$ state with $l = 1$ is the 2p state. (The lowercase letters $s, p, d, \ldots$ are used in discussing individual electrons, while the corresponding capital letters denote orbital angular momentum states of an entire atom. For one-electron hydrogen, the two are the same.)

Quantization of orbital angular momentum is another nonclassical aspect of quantum mechanics. In classical physics, an electron of a given energy can have any angular momentum, up to the maximum of a circular orbit (Fig. 36.6). At high n, the number of l values is so large that we don't notice quantization—another manifestation of Bohr's correspondence principle. But for low n, the quantum-mechanical discreteness of both energy and angular momentum is clearly evident.

Space Quantization

Angular momentum is a vector, and the angular momentum vector is quantized not only in magnitude but also in direction—a phenomenon called **space quantization**. Space quantization of orbital angular momentum gives rise to a third quantum number, m_l. Space quantization becomes evident when an atom is in a magnetic field that establishes a preferred axis along which the angular momentum component can be measured; for this reason m_l is called the **orbital magnetic quantum number**.

Space quantization requires that the component L_z of orbital angular momentum along any chosen axis have only values given by

$$L_z = m_l\hbar \quad \text{(space quantization)} \tag{36.10}$$

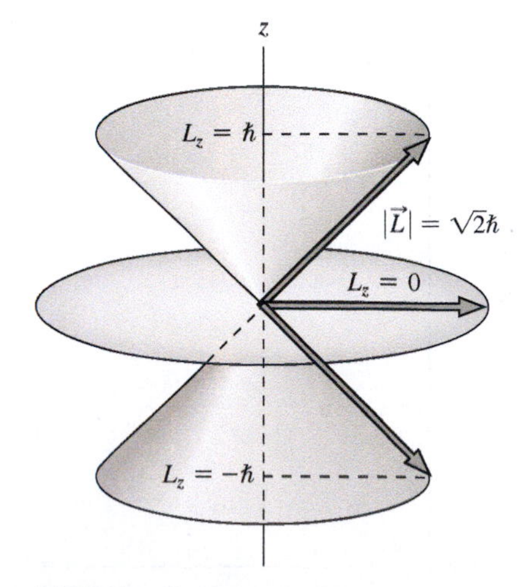

FIGURE 36.7 The three possible orientations for the angular momentum vector in the $l = 1$ state, where $L = \sqrt{2}\hbar$. Only the z-component is fixed; the x- and y-components are uncertain.

where m_l takes integer values from $-l$ to l. Thus an $l = 1$ state can have one of three possible m_l values: -1, 0, or $+1$, corresponding to angular momentum components $-\hbar$, 0, or $+\hbar$ along some axis. Since the magnitude of the angular momentum in an $l = 1$ state is $\sqrt{2}\hbar$ (see Example 36.2), none of these values corresponds to full alignment with the axis. Instead, we can think geometrically of the angular momentum vectors as being constrained to lie at angles $\cos^{-1}(L_z/L)$ to the axis; for $l = 1$ these angles are $\pm 45°$ and 90° (Fig. 36.7). Although the angle is useful for diagramming the angular momentum vector, we emphasize that the quantum numbers l and m_l tell everything there is to know about quantized orbital angular momentum. Quantum physicists, therefore, aren't usually concerned with the orientation of angular momentum vectors.

36.2 Electron Spin

Detailed observation of the hydrogen spectrum shows that spectral lines exhibit a fine splitting; where a lower-resolution spectrum shows one spectral line, at higher resolution there appears a closely spaced pair of lines. This splitting could not be explained using the three quantum numbers n, l, and m_l. In 1925 the Austrian physicist Wolfgang Pauli suggested that a fourth quantum number, capable of taking only two values, might be needed.

Soon Samuel Goudsmit and George Uhlenbeck realized that the spectral splitting could be explained if this fourth quantum number were associated with an intrinsic angular momentum, or **spin**, carried by the electron. Later, as we indicated in Chapter 35, Paul Dirac showed that electron spin follows from the requirement of relativistic invariance. Spin is an inherently quantum-mechanical property with no classical analog. Although it can be visualized crudely by imagining the electron to be a small sphere spinning about an axis, this classical picture is really inappropriate.

Spin angular momentum is quantized similarly to orbital angular momentum. But unlike the orbital quantum number l that takes a range of integer values, the electron spin quantum number s has only the single value $s = \frac{1}{2}$. The electron is therefore a **spin-$\frac{1}{2}$** particle. The magnitude of the spin angular momentum is related to the spin quantum number in the same way that the magnitude of orbital angular momentum is related to the orbital quantum number l:

$$S = \sqrt{s(s+1)}\hbar \quad \text{(quantization of spin angular momentum)} \quad (36.11)$$

Since s takes only the value $\frac{1}{2}$, the magnitude of the electron spin angular momentum is $S = \frac{\sqrt{3}}{2}\hbar$.

Spin angular momentum also exhibits space quantization. That is, the component of spin along a chosen axis takes only the values

$$S_z = m_s\hbar \quad (36.12)$$

where the quantum number m_s has the two possible values $-\frac{1}{2}$ and $+\frac{1}{2}$. Figure 36.8 shows space quantization of electron spin.

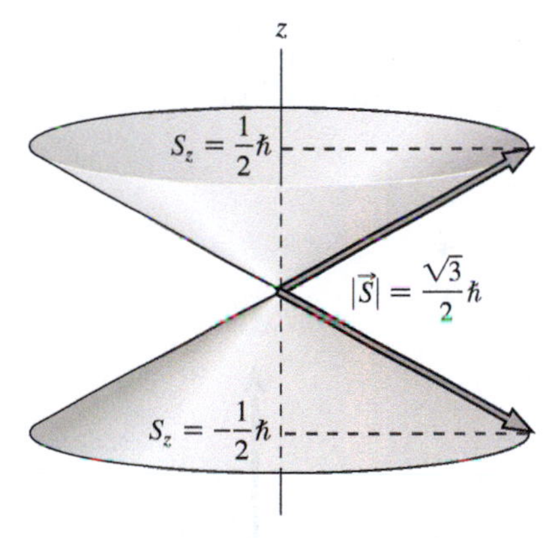

FIGURE 36.8 Space quantization of electron spin.

Magnetic Moment of the Electron

Together, the electron's spin and electric charge mean the electron behaves like a miniature current loop, with an intrinsic magnetic dipole moment. The dipole moment vector $\vec{M}$ associated with the spin angular momentum vector $\vec{S}$ is given by

$$\vec{M} = -\frac{e}{m}\vec{S} \quad (36.13)$$

with e/m the electron's charge-to-mass ratio (see Problem 73). Since the component of $\vec{S}$ on any axis can take only the values $\pm\frac{1}{2}\hbar$, the components of the magnetic moment can be only

$$M_z = \pm\frac{e\hbar}{2m} \quad (36.14)$$

The quantity $\mu_B = e\hbar/2m$ is a fundamental unit for measuring magnetic moments. It's called the **Bohr magneton**, and its value is approximately $9.27\times10^{-24}\,\text{A}\cdot\text{m}^2$.

The ratio of magnetic moment to spin angular momentum is twice what we would expect classically for a charged particle in circular motion. Like spin itself, the factor of 2 is a relativistic effect first explained by Dirac. Actually, the factor is not quite 2 but approximately 2.00232, a result that follows from the theory of quantum electrodynamics.

The Stern–Gerlach Experiment

In 1922, Otto Stern and Walther Gerlach at the University of Hamburg demonstrated the quantization of atomic angular momentum vectors. The **Stern–Gerlach experiment** used a nonuniform magnetic field to separate a beam of silver atoms according to the component of their angular momentum along the field direction. T. E. Phipps and J. B. Taylor repeated the experiment in 1927, giving unambiguous verification of quantized electron spin. They used hydrogen atoms in the ground state; as we've seen, this state has zero orbital angular momentum, so the only angular momentum effects are due to electron spin. Classically a beam of hydrogen should be spread into a continuous band corresponding to angular momentum components from $\frac{-\sqrt{3}}{2}\hbar$ to $\frac{+\sqrt{3}}{2}\hbar$. But in fact the beam always splits in two, corresponding to the two angular momentum components $\pm\frac{1}{2}\hbar$. Figure 36.9 shows the experiment.

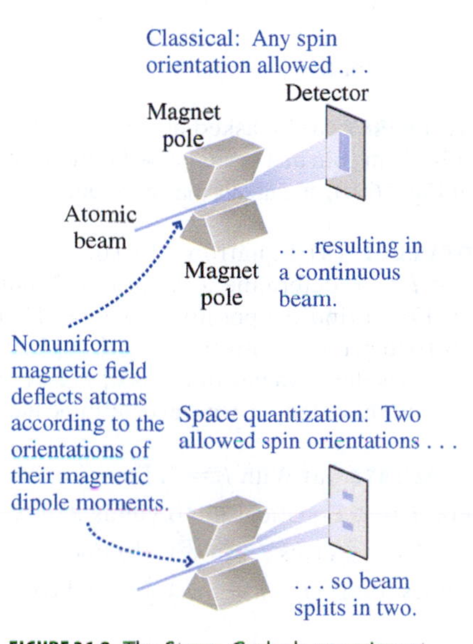

FIGURE 36.9 The Stern–Gerlach experiment.

GOT IT? 36.2 The nucleus of oxygen-17 has spin $\frac{5}{2}$. How many possible orientations are there for its spin angular momentum vector? (a) 2; (b) 5; (c) 6; (d) 7; (e) 17

Total Angular Momentum and Spin-Orbit Coupling

Orbital and spin angular momenta combine through the process of **spin-orbit coupling** to give an atom's total angular momentum, $\vec{J}$:

$$\vec{J} = \vec{L} + \vec{S} \tag{36.15}$$

The magnitude J is quantized similarly to orbital and spin angular momenta:

$$J = \sqrt{j(j+1)}\hbar \quad \text{(quantization of total angular momentum)} \tag{36.16}$$

For an atom with a single electron, the quantum number j takes the values

$$j = l \pm \tfrac{1}{2} \qquad \text{for } l \neq 0 \tag{36.17a}$$

$$j = \tfrac{1}{2} \qquad \text{for } l = 0 \tag{36.17b}$$

The state of an atom with total angular momentum J is specified by the principal quantum number, the capital letter designating the orbital angular momentum $(S, P, D, F, G, \ldots)$, and, as a subscript, the j value. Thus a hydrogen atom with $n = 3$, $l = 2$, and $j = \frac{3}{2}$ is designated $3D_{3/2}$.

Total angular momentum also exhibits space quantization, with the component of $\vec{J}$ on some axis given by

$$J_z = m_j\hbar \tag{36.18}$$

Here the quantum number m_j takes the values $(-j, -j+1, \ldots, j-1, j)$.

Derivation of these so-called **angular momentum coupling rules** is not easy, but we can understand them in terms of simple vector diagrams like those shown in Fig. 36.10.

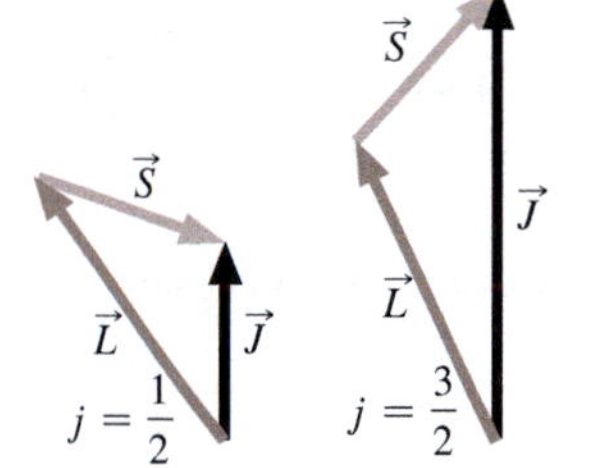

FIGURE 36.10 Spin-orbit coupling with $l = 1$, for which Equation 36.17a gives $j = \frac{1}{2}$ or $j = \frac{3}{2}$.

EXAMPLE 36.3 Spin-Orbit Coupling: Finding the Angular Momentum

(a) Find the possible magnitudes for the total angular momentum of hydrogen in the $l = 2$ state. (b) For each possible J, how many values are there for the component of $\vec{J}$ on a given axis?

INTERPRET We're asked about total angular momentum, which results from spin and orbital angular momentum contributions as shown in Fig. 36.10, and about the space quantization of $\vec{J}$.

DEVELOP (a) Equations 36.16, $J = \sqrt{j(j+1)}\hbar$, and 36.17a, $j = l \pm \frac{1}{2}$, determine J. With $l = 2$, our plan is to use Equation 36.17a to find the possible j values. Then we can apply Equation 36.16 to get each corresponding J. (b) Equation 36.18, $J_z = m_j\hbar$, determines the J_z values in terms of m_j, which ranges from $-j$ to j. So once we have the j's, we can determine the number of J_z values.

EVALUATE (a) With $l = 2$, Equation 36.17a gives $j = \frac{3}{2}$ or $j = \frac{5}{2}$. For $j = \frac{3}{2}$, Equation 36.16 yields $J = \sqrt{\frac{3}{2}(\frac{3}{2}+1)}\hbar = \frac{\sqrt{15}}{2}\hbar$; similarly, $j = \frac{5}{2}$ gives $J = \frac{\sqrt{35}}{2}\hbar$. (b) For $j = \frac{3}{2}$ there are four possible m_j values from $-j$ to j: $-\frac{3}{2}, -\frac{1}{2}, \frac{1}{2}, \frac{3}{2}$, and correspondingly four J_z values, given by Equation 36.18: $-\frac{3}{2}\hbar, -\frac{1}{2}\hbar, \frac{1}{2}\hbar, \frac{3}{2}\hbar$. Similar counting gives six values for $j = \frac{5}{2}$.

ASSESS Figure 36.11 shows how the spin and orbit angular momenta combine to give the two possible j values in this example. ■

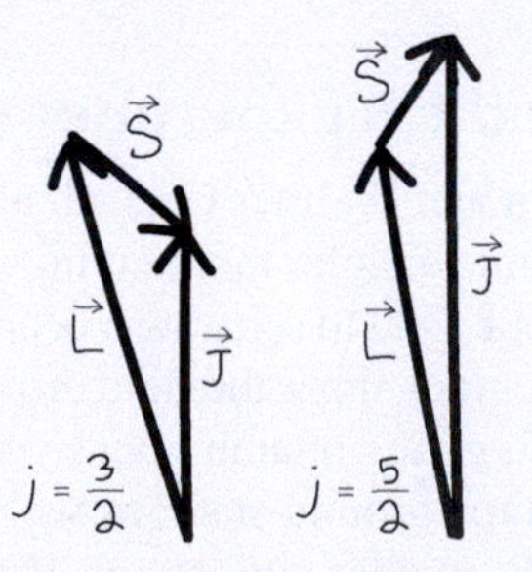

FIGURE 36.11 Vector diagrams for spin-orbit coupling with $l = 2$ (Example 36.3).

The two j values for a given l correspond to two distinct quantum states, and these states have slightly different energies. This energy difference is associated with the orientation of the electron's magnetic moment in a magnetic field that, in the electron's reference frame, results from the apparent motion of the positively charged nucleus around the electron—a field whose direction is that of the electron's orbital angular momentum $\vec{L}$. As Equation 36.13 shows, the electron's negative charge means its spin $\vec{S}$ and magnetic moment $\vec{M}$ have opposite directions. Because a magnetic dipole has the highest energy when it's oriented opposite the magnetic field, this means that the more nearly parallel alignment of $\vec{S}$ and $\vec{L}$ —corresponding to $j = l + \frac{1}{2}$—has the higher energy.

In hydrogen, the magnitude of the energy difference between the $j = \frac{1}{2}$ and $j = \frac{3}{2}$ states of the first excited level is only 5×10^{-5} eV, far smaller than the 10.2-eV separation between this level and the ground state. Because the $n = 2, l = 1$ state is actually two states of slightly different energy, hydrogen atoms undergoing transitions from these states to the ground state emit two spectral lines slightly separated in wavelength. The term **fine structure** describes this and related spectral-line splittings. In the present example, the split spectral line is called a **doublet**. Figure 36.12 is an energy-level diagram showing the effect of spin-orbit splitting in hydrogen. Relativistic and other small corrections further alter the fine structure of hydrogen's spectrum.

The spin-orbit effect results from a magnetic field internal to the atom itself. But splitting of energy levels also occurs in an external magnetic field and is called the **Zeeman effect**. We showed an example of Zeeman splitting in Fig. 35.18.

Since it has zero orbital angular momentum, the ground state does not exhibit spin-orbit splitting. But interaction of the electron's magnetic moment with the nuclear magnetic dipole results in an even finer splitting known as **hyperfine structure**. The transition between the two hyperfine levels of the hydrogen ground state—corresponding physically to a change in the orientation of the electron spin vector—involves a photon of 21-cm wavelength. Radio astronomers use the 21-cm hydrogen radiation to map interstellar hydrogen in the cosmos.

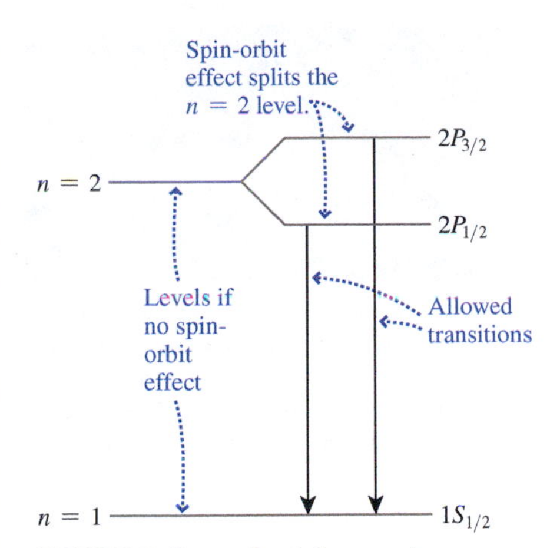

FIGURE 36.12 Energy-level diagram showing spin-orbit splitting of the $2P$ levels in hydrogen. Diagram is not to scale; the spacing between the $2P_{1/2}$ and $2P_{3/2}$ levels is actually about five millionths that of the $n = 1$ and $n = 2$ levels.

36.3 The Exclusion Principle

In trying to understand why atomic electrons distributed themselves as they did, Pauli in 1924 developed his **exclusion principle**, which, loosely, states that **two electrons cannot be in the same quantum state**. Since an electron's quantum state includes its spin orientation specified by m_s, the exclusion principle means that at most two electrons can occupy a state whose other quantum numbers n, l, and m_l are the same.

The Pauli exclusion principle has profound implications for multielectron systems, requiring that most electrons remain in high-energy states (Fig. 36.13). If the exclusion principle didn't hold, atomic electrons would collapse to the ground state and there would be no such thing as chemistry or life! The exclusion principle even manifests itself at the cosmic scale, as the Application shows.

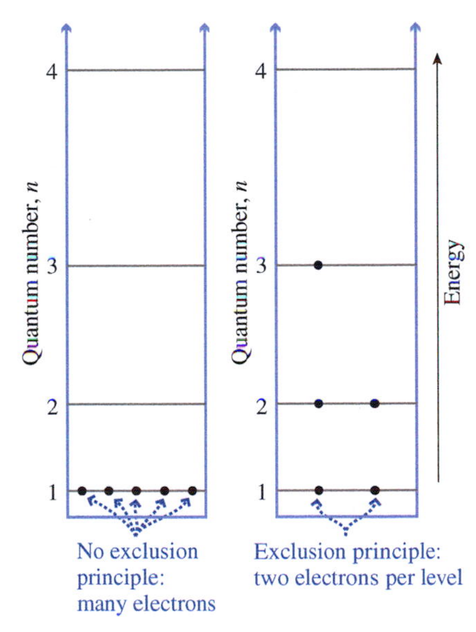

FIGURE 36.13 Particles in a square well, showing the exclusion principle's effect on electron distribution.

APPLICATION White Dwarfs and Neutron Stars

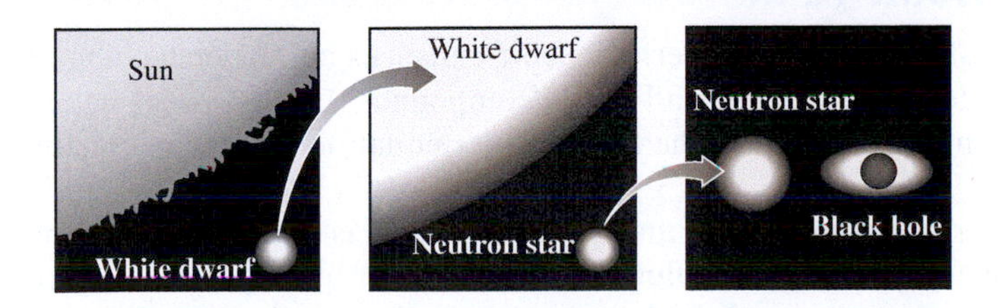

When a star exhausts its nuclear fuel, it collapses because there's no longer pressure to counter gravity. For a star with more than several times the Sun's mass, there's no force strong enough to halt the collapse, and the star becomes a black hole from which nothing can escape. But in less massive stars the collapse eventually halts because of a quantum-mechanical pressure associated with the exclusion principle.

When the Sun collapses some 5 billion years from now, its electrons will drop into the lowest available energy states. But as with the square well in Fig. 36.13, the exclusion principle requires that most of the Sun's 10^{57} electrons will end up in high-energy states. The associated **degenerate electron pressure**—independent of temperature, unlike the pressure of an ordinary gas—will stabilize the Sun as a **white dwarf**, about the size of Earth. For stars more massive than about 1.4 Sun masses, collapse proceeds until the protons and electrons merge to form neutrons. The neutrons, too, develop a degenerate pressure that stabilizes the resulting **neutron star**—an object with a mass exceeding the Sun's crammed into a 20-km sphere! The figure compares the sizes of these stellar endpoints.

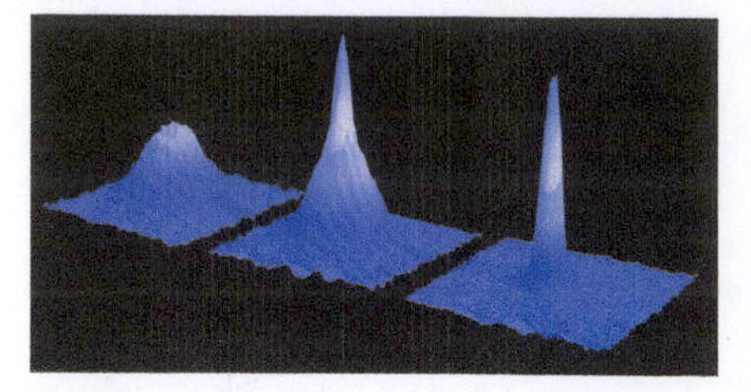

FIGURE 36.14 Velocity distribution of atoms in a Bose–Einstein condensate shows a large peak at the near-zero velocity of atoms all in their common ground state. The three peaks show the evolution from a normal gas to the condensate.

The Pauli exclusion principle quickly became fundamental to the developing quantum mechanics of the late 1920s. But physicists remained dissatisfied invoking this seemingly ad hoc rule with no theoretical basis. Late in the 1930s, following detailed analysis of relativistic quantum theories, Pauli finally showed that the exclusion principle, like spin, is ultimately grounded in the requirement of relativistic invariance. Pauli found that particles whose spin quantum number s is a half-integer (collectively called **fermions**) must necessarily obey the exclusion principle. On the other hand, particles with integer spin (called **bosons**) *do not* obey the exclusion principle. Photons, for example, are spin-1 particles and therefore an arbitrarily large number of them can occupy exactly the same quantum state. The laser, with its intense, coherent beam of light, is possible because the many photons that make up the beam are essentially all in the same state. In 1995 physicists at the University of Colorado first succeeded in producing an assemblage of bosonic matter all in the same quantum state (Fig. 36.14). This so-called **Bose–Einstein condensate** had been a goal of physicists since 1924, when the Indian physicist Satyendra Nath Bose first suggested the possibility. The Bose–Einstein condensate represents a truly new state of matter, in which thousands of atoms join quantum mechanically to behave as a single entity. Today physics labs around the world are experimenting with Bose–Einstein condensates, probing the fundamentals of quantum physics and developing applications including atom-beam analogs of the optical laser.

GOT IT? 36.3 If you put seven electrons in a quantum harmonic oscillator potential (recall Fig. 35.10), what will the total energy of the system be? (a) $\frac{7}{2}\hbar\omega$; (b) $\frac{13}{2}\hbar\omega$; (c) $7\hbar\omega$; (d) $\frac{25}{2}\hbar\omega$

36.4 Multielectron Atoms and the Periodic Table

Our modern understanding of the chemical elements developed in the late 18th century, when chemists first distinguished compounds, such as water, from elements, defined as substances that couldn't be decomposed by chemical means. From the formulas for various compounds, chemists could determine the relative atomic masses of the elements. The first attempts to organize the elements systematically used atomic mass, but a breakthrough occurred in 1869, when the Russian chemist Dmitri Mendeleev set up a table with the approximately 60 elements then known. He left blanks where necessary to maintain the periodic occurrence of similar chemical properties. Elements filling the blanks were soon discovered, validating Mendeleev's periodic table and suggesting an underlying order in the composition of atoms. Then, early in the 20th century, studies of X-ray spectra led to a table organized by **atomic number** Z, the number of protons in the nucleus. When this was done, a number of elements missing from earlier periodic tables were identified. The modern periodic table is shown in Fig. 36.15 and is printed with atomic weights inside the back cover.

Explaining the Periodic Table

The orderly arrangement of elements in the periodic table enhances our understanding of chemistry and our ability to formulate new and useful compounds. But why does nature exhibit this regularity? The answer lies in the Schrödinger equation and the exclusion principle.

Solution of the Schrödinger equation for multielectron atoms is complicated by the interactions among the electrons; analytic solutions like those for hydrogen aren't generally available. But qualitatively, we still find energy levels characterized by the principal quantum number n. Each such level is called a **shell**; for historical reasons, the shells $n = 1, 2, 3, \ldots$ are also labeled with the letters $K, L, M, \ldots$. As with hydrogen, an electron at the nth energy level can have any of the n values $l = 0, 1, 2, \ldots, n - 1$ for the orbital quantum number. The different angular momentum states within a shell are

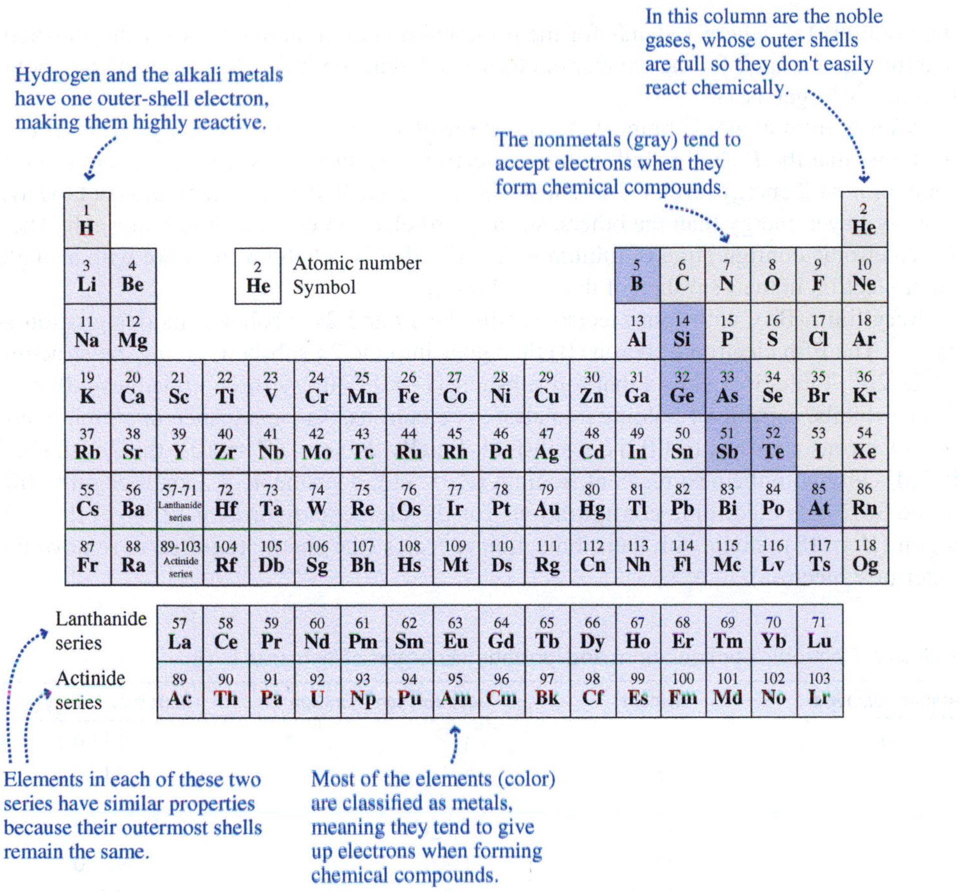

Hydrogen and the alkali metals have one outer-shell electron, making them highly reactive.

The nonmetals (gray) tend to accept electrons when they form chemical compounds.

In this column are the noble gases, whose outer shells are full so they don't easily react chemically.

1 H																	2 He
3 Li	4 Be			2 He	Atomic number Symbol							5 B	6 C	7 N	8 O	9 F	10 Ne
11 Na	12 Mg											13 Al	14 Si	15 P	16 S	17 Cl	18 Ar
19 K	20 Ca	21 Sc	22 Ti	23 V	24 Cr	25 Mn	26 Fe	27 Co	28 Ni	29 Cu	30 Zn	31 Ga	32 Ge	33 As	34 Se	35 Br	36 Kr
37 Rb	38 Sr	39 Y	40 Zr	41 Nb	42 Mo	43 Tc	44 Ru	45 Rh	46 Pd	47 Ag	48 Cd	49 In	50 Sn	51 Sb	52 Te	53 I	54 Xe
55 Cs	56 Ba	57-71 Lanthanide series	72 Hf	73 Ta	74 W	75 Re	76 Os	77 Ir	78 Pt	79 Au	80 Hg	81 Tl	82 Pb	83 Bi	84 Po	85 At	86 Rn
87 Fr	88 Ra	89-103 Actinide series	104 Rf	105 Db	106 Sg	107 Bh	108 Hs	109 Mt	110 Ds	111 Rg	112 Cn	113 Nh	114 Fl	115 Mc	116 Lv	117 Ts	118 Og
	Lanthanide series	57 La	58 Ce	59 Pr	60 Nd	61 Pm	62 Sm	63 Eu	64 Gd	65 Tb	66 Dy	67 Ho	68 Er	69 Tm	70 Yb	71 Lu	
	Actinide series	89 Ac	90 Th	91 Pa	92 U	93 Np	94 Pu	95 Am	96 Cm	97 Bk	98 Cf	99 Es	100 Fm	101 Md	102 No	103 Lr	

Elements in each of these two series have similar properties because their outermost shells remain the same.

Most of the elements (color) are classified as metals, meaning they tend to give up electrons when forming chemical compounds.

FIGURE 36.15 The periodic table. A larger version, with atomic weights, is inside the back cover; the names of the elements are in Appendix D.

termed **subshells**; subshells with the values $l = 0, 1, 2, 3, \ldots$ are labeled with the letters $s, p, d, f, \ldots$. Finally, for each subshell there are $2l + 1$ possible values of the magnetic orbital quantum number m_l, ranging from $-l$ to l. A state characterized by all three quantum numbers n, l, and m_l is called an **orbital**. Table 36.1 summarizes shell-structure notation; for completeness, the table also lists the spin quantum number m_s.

The structure of a multielectron atom is determined by the quantum states of its constituent electrons—their distribution among the shells, subshells, and orbitals. According to the exclusion principle, no two electrons can be in exactly the same quantum state; that is, they can't have the same values for all four quantum numbers n, l, m_l, and m_s. Since an atomic orbital is characterized by the three quantum numbers n, l, and m_l, the exclusion principle implies that at most two electrons can occupy a single orbital.

PhET: Build an Atom

We're now ready to understand the ground-state electronic structure of multielectron atoms. The simplest is helium (He), with two electrons. The K shell ($n = 1$) is the lowest possible energy level. As Table 36.1 shows, only the zero-angular-momentum s subshell is permitted within the K shell, and within that subshell there's only the single orbital corresponding to $m_l = 0$. But that orbital can accommodate two electrons. So in the ground state of helium, both electrons occupy the s subshell of the K shell. We describe this with

Table 36.1 Atomic Shell Structure

Quantum Number	Shell Notation	Allowed Values	Letter Labels	Number of States
n	Shell	$1, 2, 3, \ldots$	$K, L, M, \ldots$	Infinite
l	Subshell	$0, 1, 2, \ldots, n-1$	$s, p, d, f, \ldots$	n
m_l	Orbital	$-l, -l+1, \ldots, l-1, l$	—	$2l+1$
m_s	—	$-\frac{1}{2}, +\frac{1}{2}$		2

the notation $1s^2$, where 1 stands for the principal quantum number n, s for the subshell, and the superscript 2 for the number of electrons in that subshell. The corresponding notation for hydrogen is $1s^1$.

After helium comes lithium (Li), with three electrons. From our analysis of helium, we know that the K shell is full with two electrons. So the third electron goes into the L shell, or $n = 2$ energy level. Of the subshells in the L shell, the s subshell turns out to have slightly lower energy than the others, so the third electron occupies the s subshell. Then the electronic configuration of lithium is $1s^22s^1$—that is, a helium-like core with a single outer electron in the s subshell of the $n = 2$ level.

Beryllium (Be), with four electrons, fills the $1s$ and $2s$ subshells; its designation is $1s^22s^2$. The fifth electron of boron (B) then goes into the $2p$ subshell, giving the structure $1s^22s^22p^1$. Table 36.1 shows that a p subshell ($l = 1$) allows three m_l values—that is, three orbitals, capable of holding a total of six electrons. As we advance in atomic number, electrons continue to fill the p subshell. Finally, at neon ($Z = 10$), the $2p$ subshell is full. Only with the next element, sodium ($Z = 11$), does the $n = 3$ shell begin to fill. Table 36.2 lists electronic configurations for the elements hydrogen ($Z = 1$) through argon ($Z = 18$), along with their ionization energies (the energy required to remove the outermost electron).

Table 36.2 Electronic Configurations and Ionization Energies of Elements 1–18

Atomic Number, z	Element	Electronic Configuration	Ionization Energy (eV)
1	H	$1s^1$	13.60
2	He	$1s^2$	24.60
3	Li	$1s^22s^1$	5.390
4	Be	$1s^22s^2$	9.320
5	B	$1s^22s^22p^1$	8.296
6	C	$1s^22s^22p^2$	11.26
7	N	$1s^22s^22p^3$	14.55
8	O	$1s^22s^22p^4$	13.61
9	F	$1s^22s^22p^5$	17.42
10	Ne	$1s^22s^22p^6$	21.56
11	Na	$1s^22s^22p^63s^1$	5.138
12	Mg	$1s^22s^22p^63s^2$	7.644
13	Al	$1s^22s^22p^63s^23p^1$	5.984
14	Si	$1s^22s^22p^63s^23p^2$	8.149
15	P	$1s^22s^22p^63s^23p^3$	10.48
16	S	$1s^22s^22p^63s^23p^4$	10.36
17	Cl	$1s^22s^22p^63s^23p^5$	13.01
18	Ar	$1s^22s^22p^63s^23p^6$	15.76

Gaps mark ends of periodic-table rows.

An atom's chemical behavior is determined primarily by its outermost electrons, because these electrons interact most directly with nearby atoms and because they're most weakly bound to their nuclei. Table 36.2 shows that the outer-electron configurations for lithium through neon (Ne) are the same as the corresponding configurations for sodium (Na) through argon (Ar). The chemical properties of the corresponding atoms are therefore similar. Both lithium and sodium, for example, have a single electron in their outermost shell. As their relatively low ionization energies suggest, this electron is loosely bound, so it readily interacts with other atoms, giving these elements their extreme reactivity. Neon and argon, in contrast, both have completely filled outermost shells. All the outer-shell electrons have essentially the same energy; the corresponding ionization energy is high; and there's little tendency for these electrons to interact with other atoms.

As a result, argon and neon don't readily form chemical compounds, and at normal temperatures they're gases (collectively, the elements in the periodic-table column containing argon and neon are called the noble gases). Other element pairs also share similar properties. Fluorine and chlorine, for instance, each need only one more electron to achieve the energetically favorable noble-gas configuration. Consequently these elements readily accept electrons. Materials such as common salt, NaCl, owe their high melting points to the strong bond that results when reactive sodium gives up its outer electron to electron-accepting chlorine, and the resulting positive and negative ions bind strongly by the electrostatic force. We'll consider molecular bonding in the next chapter.

Beyond argon ($Z = 18$), shielding effects of the inner electrons result in the 4s states having lower energy than the 3d states. Potassium ($Z = 19$) thus has the electronic configuration $1s^22s^22p^63s^23p^64s^1$ rather than $1s^22s^22p^63s^23p^63d^1$. After potassium comes calcium, with two electrons in its single 4s orbital. But the 4p orbitals do have higher energy than the 3d, so elements beyond calcium begin filling the 3d orbitals. The next ten elements, scandium through zinc, have chemical properties that vary only slightly because their two outermost electrons remain 4s electrons; collectively, they're **transition elements**. (Chromium, $Z = 24$, and copper, $Z = 29$, are minor exceptions; in these an extra electron goes into the 3d orbitals, leaving only one 4s electron.) Finally, elements 31 (gallium) through 36 (krypton) repeat the pattern of aluminum through argon shown in Table 36.2, as their 4p orbitals fill with electrons. Krypton, with its outer p subshell full, is again a noble gas.

CONCEPTUAL EXAMPLE 36.1 The Periodic Table

Explain the general structure of the periodic table's first five rows.

EVALUATE Each row of the periodic table starts with an element whose outermost shell contains a single s electron; these elements include hydrogen and the highly reactive alkali metals. Each row ends with a noble gas, its outermost p subshell full. The first row involves filling the 1s orbital only; since this orbital holds at most two electrons, there are only two elements in the first row. The second row has eight elements, associated with the filling of the 2s and 2p orbitals, as shown in Table 36.2. The third row is like the second, but with the 3s and 3p orbitals filling. Because the 4s orbitals have lower energy than the 3d orbitals, the third row ends with a noble gas whose 3p orbitals are full, and the fourth row begins as the 4s orbital begins to fill. Then come the elements $Z = 21$ through $Z = 30$, in which the 3d orbitals are filling; these make for 10 additional elements in the fourth row. The fifth row is a repeat of the fourth, as first the 5s orbitals fill, then the higher-energy 4d orbitals, then the remaining 5p orbitals.

ASSESS Our explanation shows why elements in each column of the periodic table have similar chemical properties, while those properties generally change as we move across the table's rows.

MAKING THE CONNECTION Determine the electronic configuration of iron.

EVALUATE $Z = 26$ for iron, so we need to accommodate eight more electrons outside iron's argon-like core. Since iron is a transition element, its 4s orbitals fill before 3d. So two electrons go into 4s and the remaining six into 3d. Iron's electronic configuration is therefore $1s^22s^22p^63s^23p^63d^64s^2$.

The sixth and seventh rows don't quite fit our analysis in Conceptual Example 36.1. At element 57, lanthanum, the 4f orbitals begin filling while the outermost electron remains 6s. This continues through element 71, lutetium, giving elements 57–71 similar chemical properties. These elements constitute the **lanthanide series**, and they're printed separately below the main table. Row seven repeats this pattern, with the **actinide series**. Seventh-row elements beyond uranium (element 92) are radioactive with half-lives that are short compared to Earth's age. They're not found naturally but are produced in particle accelerators, fission reactors, and nuclear explosions. Elements beyond the actinides are very short-lived, although theory suggests that there may be an "island of stability" encompassing more massive nuclei than those yet produced; their lifetimes might range from minutes to possibly as much as millions of years. You'll learn more about nuclear lifetimes in Chapter 38.

Note the crucial role the exclusion principle plays in our discussion of the chemical elements. Without that principle, every atom in its ground state would have all its electrons in the 1s orbital. There would be no qualitative distinction among the elements, and the science of chemistry would not exist. Nor would there be any chemists or physicists, as life itself would be impossible without the rich diversity of chemical compounds formed from the different elements.

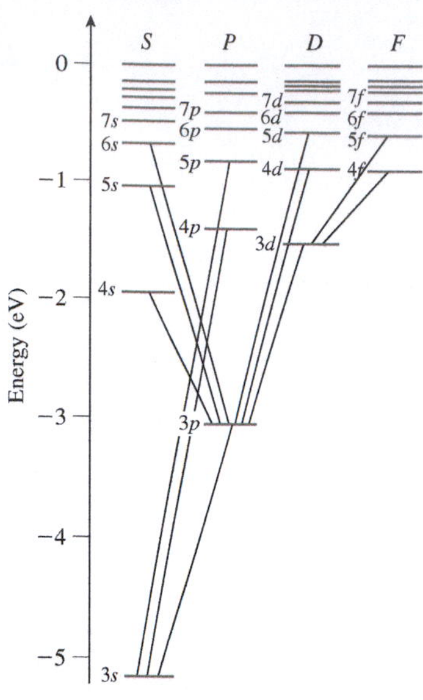

FIGURE 36.16 Energy-level diagram for sodium, neglecting spin-orbit splitting. Note the widely separated 4s and 4p levels, with 3d between them; this explains why the 4s orbital fills before 3d. Lines connecting levels show allowed transitions.

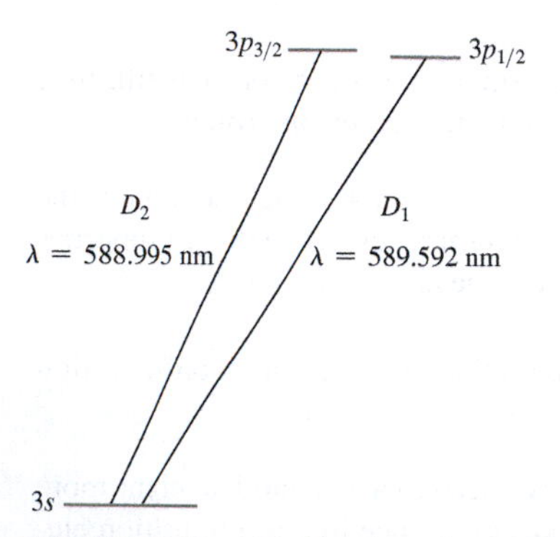

FIGURE 36.17 Magnified portion of sodium's energy-level diagram, showing spin-orbit splitting of the 3p level. Transitions from the two states result in the slightly different wavelengths of the sodium D doublet.

Video Tutor Demo | **Illuminating Sodium Vapor with Sodium and Mercury Lamps**

GOT IT? 36.4 What's the final term in the electron configuration for xenon (Xe)? Answer without working out the entire configuration. (a) $4p^6$; (b) $5p^6$; (c) $5p^5$; (d) $5p^7$; (e) $6s^1$

36.5 Transitions and Atomic Spectra

Emission and absorption of photons with specific energies provide the most direct manifestation of quantized atomic-energy levels, and give rise to the spectral lines that permit precise analysis of atomic systems from the laboratory to distant astrophysical objects. Even simple hydrogen exhibits myriad quantum states. In multielectron atoms, the possibilities for electronic excitation are even more numerous. The spectra of atoms reflect this rich array of available quantum states.

Selection Rules

Not all energy-level transitions are equally likely. So-called **selection rules** determine which are **allowed transitions**—those most likely to occur. One rule limits allowed transitions to those for which the orbital quantum number l changes by $\Delta l = \pm 1$; this and other selection rules are related to the conservation of angular momentum. Quantum mechanics also provides a way of calculating transition probabilities, and from them the mean lifetimes of excited states. For outer electrons, excited states that de-excite by allowed transitions have typical lifetimes on the order of 10^{-9} s.

Transitions that are not allowed by selection rules are called **forbidden transitions**; most are not strictly impossible but just extremely unlikely. States that can lose energy only by forbidden transitions are **metastable states**; their lifetimes are many orders of magnitude longer than the nanosecond timescale for allowed transitions. "Glow in the dark" phosphorescent materials emit light through the slow de-excitation of metastable states. Forbidden spectral lines are valuable probes of low-density astrophysical gases in which collisions are rare, and atoms can therefore remain in metastable states.

Optical Spectra

Spectral lines in or near the visible involve transitions among the incompletely filled outer atomic shells. The alkali metals, with a single outer s electron, therefore produce spectra qualitatively similar to that of hydrogen. However, the more complicated structure of a multielectron atom shifts some energy levels (Fig. 36.16). Many of the transitions in Fig. 36.16 are actually doublets or triplets resulting from spin-orbit splitting (Fig. 36.17). The energy-level structure is even more complicated for atoms with more than one outer-shell electron.

GOT IT? 36.5 Which of the transitions shown in Fig. 36.16 results in a photon of the shortest wavelength?

EXAMPLE 36.4 Atomic Spectra: The Sodium Doublet

Use Fig. 36.17 to determine the energy difference between the 3p states of sodium.

INTERPRET We're asked about the energy difference between two atomic states ($3p_{1/2}$ and $3p_{3/2}$), and we're given the wavelengths of photons emitted in transitions from those states to a common end state (3s). We know that those photons carry off energy equal to the difference between the energies of the starting and ending states.

DEVELOP The quantization condition $E = hf$ relates photon energy and frequency; since $f\lambda = c$, we also have $E = hc/\lambda$. Our plan is to use this expression to find the energies of the two transitions shown in Fig. 36.17, and then subtract to get the energy difference between the 3p levels.

EVALUATE We have

$$\Delta E_{3p} = \frac{hc}{588.995 \text{ nm}} - \frac{hc}{588.592 \text{ nm}} = 3.42\times10^{-22} \text{ J}.$$

ASSESS Our answer is about 2 meV, much lower than the eV-range energies associated with optical transitions themselves. That's expected, given the small separation between the 3p states evident in Fig. 36.17. In sodium, states below 3s are all full, so 3s is the lowest end state for optical transitions. ■

Spontaneous and Stimulated Transitions

What makes an electron jump between energy levels? In an upward transition, the electron must absorb the appropriate amount of energy. Generally, that energy is supplied by a photon whose energy is equal to the energy difference between the two levels; the process is called **stimulated absorption** (Fig. 36.18*a*). (Upward transitions can result from other processes as well, such as an energetic collision between two atoms or the interaction of a free electron with atomic electrons.)

For most downward transitions, however, there's no specific cause. An electron spontaneously jumps from a higher to a lower energy level and a photon is emitted; this is **spontaneous emission** (Fig. 36.18*b*). Although an individual spontaneous emission is a random event, quantum mechanics gives the probability per unit time for that event to occur; the inverse of that probability is the mean lifetime of the excited state.

In 1917 Einstein recognized a third possibility: Excited atoms can be stimulated to drop into lower energy states by the mere presence of a photon, again of energy appropriate to the transition. A second photon is emitted in the process, with the same energy and phase as the stimulating photon, and in the same direction. This process, **stimulated emission**, is the reverse of stimulated absorption (Fig. 36.18*c*).

Spontaneous emission, stimulated absorption, and stimulated emission play major roles in the transfer of radiation through gases. And stimulated emission is essential to the operation of lasers.

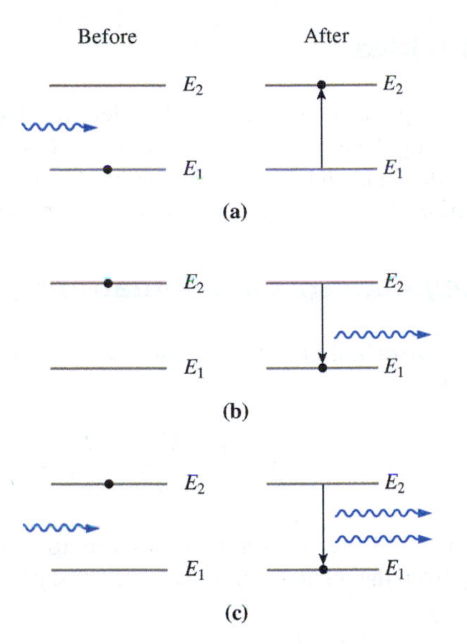

FIGURE 36.18 Interaction of photons with atomic electrons. Horizontal lines denote two atomic energy levels, and the wave is a photon with energy equal to the difference between the two levels. (a) Stimulated absorption; (b) spontaneous emission; (c) stimulated emission.

APPLICATION Lasers

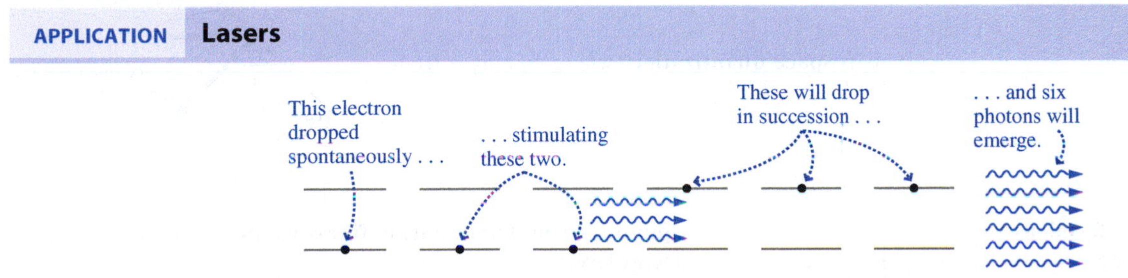

As Fig. 36.18c suggests, stimulated emission is a way to multiply photons with identical wavelength and phase. The **laser**, whose name derives from **l**ight **a**mplification by **s**timulated **e**mission of **r**adiation, exploits this effect to produce an intense beam of coherent light. The key to laser action is a **population inversion**, an unusual situation with many atoms in an excited state. The excited state is usually metastable, to prevent spontaneous emission from de-exciting the atoms. Atoms are first excited to a higher state from which they quickly drop by spontaneous emission to the metastable state, where they're "stuck" by the lack of allowed transitions downward. The excitation process is called **pumping**, and the excitation energy source is the **pump**. Laser pumps include flash lamps, sunlight, other lasers, electric currents, chemical reactions, and even nuclear explosions.

With a large number of excited atoms, it isn't long before one randomly de-excites even from the metastable state. It emits a photon that passes by other excited atoms, causing stimulated emission. That makes more photons and still more stimulated emission, as shown in the figure. The process snowballs, resulting in an intense beam of photons with the same wavelength and phase. In a laser, the radiating medium sits in a cavity with mirrors at the ends; as the photons reflect off the mirrors and traverse the medium, more and more stimulated emission results, building up the beam intensity. One mirror is only partially reflective to allow the laser beam to emerge. Some lasers produce a short burst of radiation before being pumped to prepare for another burst. Others are pumped continuously, resulting in a continuous beam.

The first laser, built in 1960, used a ruby rod as the lasing medium, surrounded by a coiled flash lamp for the pump. Since then myriad laser types have been developed. Almost anything can be used as the lasing medium, provided a population inversion is possible. Laser media include gases, solids, liquids, semiconductors, and ionized plasmas. Natural laser action occurs even in interstellar gas clouds. Some lasers, especially those using chemical dyes or temperature-sensitive semiconductors, are tunable over a range of wavelengths.

Laser light is monochromatic, since all photons have essentially the same energy. It's coherent because the photons all have the same phase. Coherence allows the beam to travel long distances with minimal spreading and enables very precise focusing. Finally, laser light can be made extremely intense, since stimulated emission extracts energy from many atoms simultaneously. Since photons are spin-1 particles that don't obey the exclusion principle, there's no limit to the number of photons in a laser beam. Small lasers like those used in laser pointers have power outputs in the submilliwatt range, while the most powerful continuous lasers exceed 1 MW. Pulsed lasers, in contrast, have produced outputs as great as 1 PW—1000 times the output of all the world's electric generating plants—but only with very brief pulses.

Today lasers are ubiquitous. They're used in commonplace applications like bar-code scanners and CD/DVD drives. Medical lasers correct vision (Application, "Laser Vision Correction," Chapter 31), whiten teeth, and perform bloodless surgery. Biologists use laser beams as "optical tweezers" to manipulate microscopic structures within cells. Lasers have replaced older technologies in surveying, leveling, and measuring instruments used in construction. Industrial lasers cut metal, shape gears, and harden surfaces. Semiconductor lasers drive the optical fibers that carry communications signals and Internet traffic. Military lasers lock on targets for precise weapons guidance. Ultrafast lasers probe chemical reactions that occur on femtosecond timescales (Application, "Femtosecond Photography of Chemical Reactions," Chapter 34). Lasers halt atoms' thermal motion, cooling materials to nanokelvin levels and enabling Bose–Einstein condensates. Laser beams reflected from the Moon measure its distance to within a few centimeters, testing Einstein's general relativity (Application, "Moon Distance," Chapter 30). Holograms capture interfering laser wavefronts, creating three-dimensional images. Future laser applications may include laser-driven spaceflight and the use of lasers to clear space debris.

CHAPTER 36 SUMMARY

Big Idea

The big idea here is that atomic electrons are quantum particles trapped in the three-dimensional potential well associated with the electric force. Solving the Schrödinger equation then leads to quantized energy levels. Considerations of electron spin and orbital angular momentum introduce subtle details into the atom's energy-level structure. The **exclusion principle** permits only one atomic electron per quantum state, and this fact underlies the shell structure of atoms and the periodic table of the elements.

Key Concepts and Equations

The **principal quantum number** n determines the energy levels in hydrogen:

$$E_n = -\frac{1}{n^2}\frac{\hbar^2}{2ma_0^2} = \frac{E_1}{n^2} = \frac{-13.6\text{ eV}}{n^2}$$

For $n = 1$ the electron is most likely to be found one Bohr radius a_0 from the nucleus; in higher-energy states it's likely to be farther away.

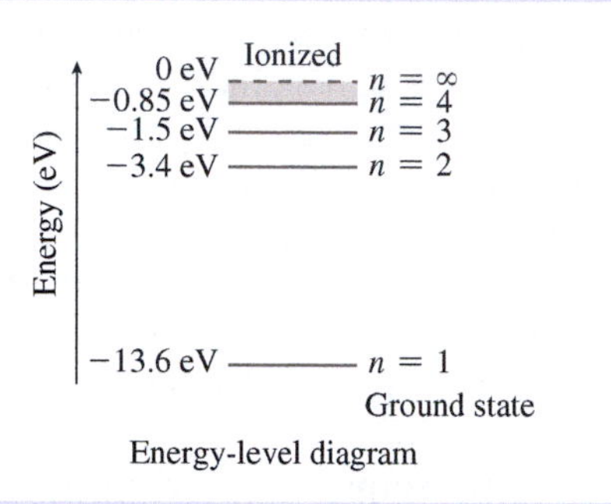

Energy-level diagram

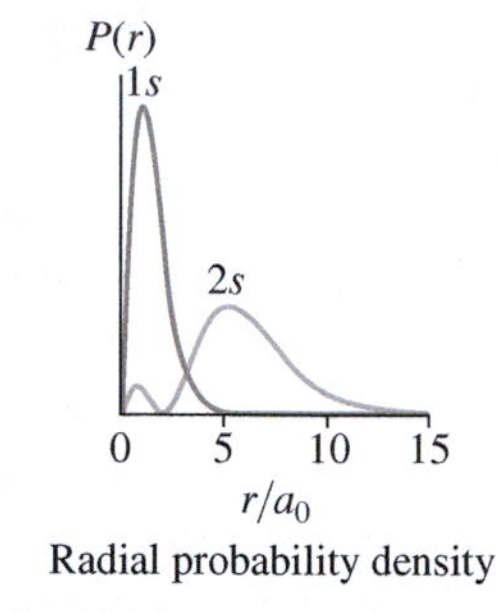

Radial probability density

The orbital quantum number l determines the angular momentum:

$$L = \sqrt{l(l+1)}\hbar$$

where l ranges from 0 to $n - 1$.

The orbital magnetic quantum number m_l determines the component of the angular momentum along any given axis:

$$L_z = m_l\hbar$$

This is **space quantization,** where m_l ranges from $-l$ to l.

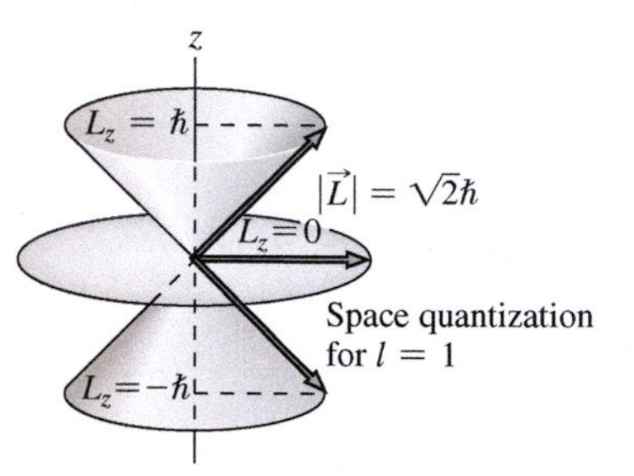

Electrons are **spin-$\frac{1}{2}$** particles or **fermions**; the component of their spin angular momentum on a given axis is $\pm\frac{1}{2}\hbar$.

Electron spin gives rise to the electron's intrinsic magnetic dipole moment, characterized by the **Bohr magneton**:

$$\mu_B = e\hbar/2m = 9.27\times10^{-24}\text{ A}\cdot\text{m}^2$$

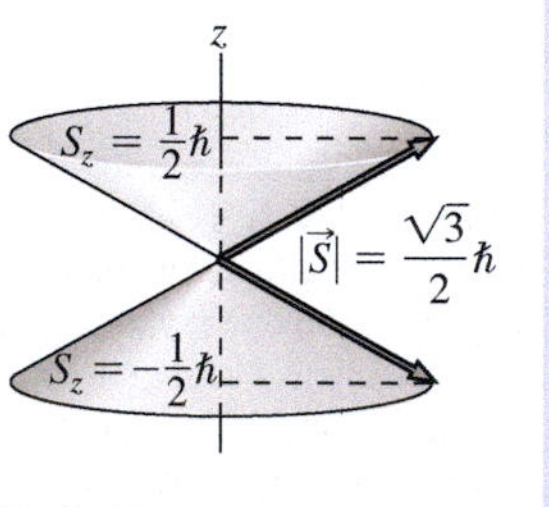

Spin-orbit coupling results in **fine-structure** splitting of atomic-energy levels.

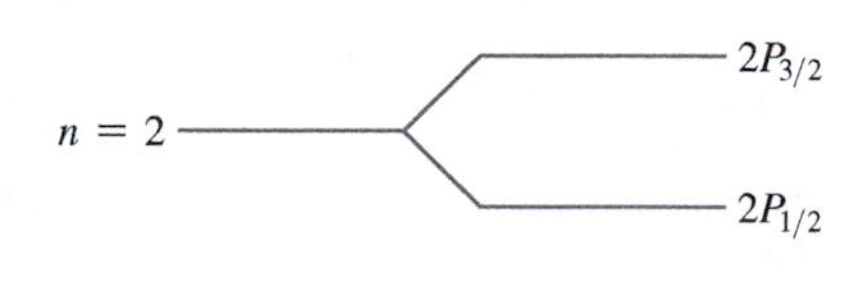

Spin angular momentum $\vec{S}$ and total angular momentum $\vec{J}$ obey quantization rules similar to those for orbital angular momentum.

Applications

Bosons are particles with integer spin. They don't obey the exclusion principle, allowing many particles to be in the same state, as happens in a **Bose–Einstein condensate** or a laser beam.

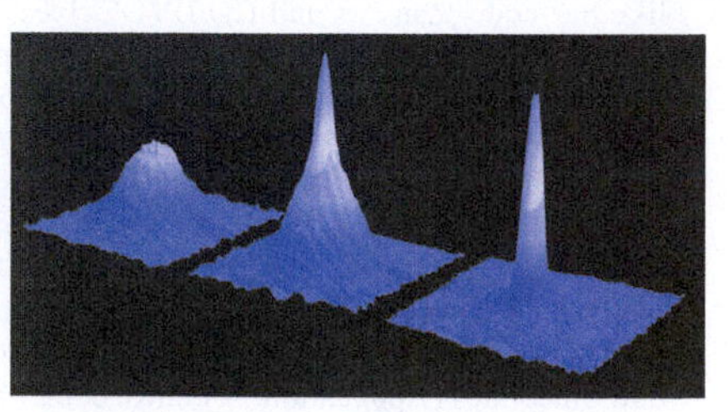

Forming a Bose–Einstein condensate

In stimulated absorption, an electron absorbs a photon and jumps to a higher energy level.

Electrons in excited states can drop to lower energy states by either **spontaneous** or **stimulated emission**.

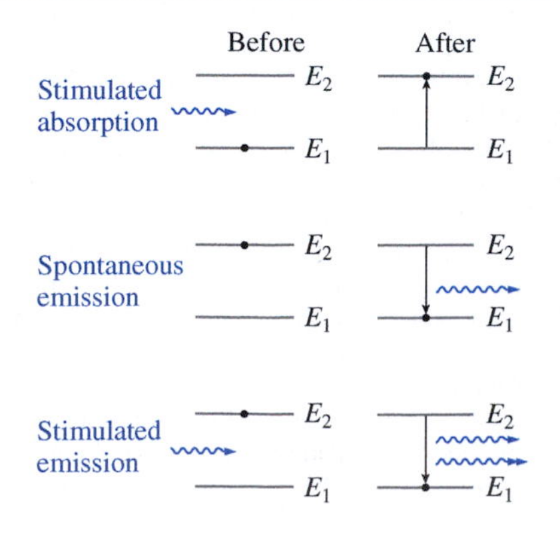

BIO *Biology and/or medicine-related problems* **DATA** *Data problems* **ENV** *Environmental problems* **CH** *Challenge problems*

For Thought and Discussion

1. The electron in a hydrogen atom is somewhat like a particle confined to a three-dimensional box. In the atom, what plays the role of the confining box?
2. A friend who hasn't studied physics asks you the size of a hydrogen atom. How do you answer?
3. How many quantum numbers are required to specify fully the state of a hydrogen atom?
4. Both the Bohr and Schrödinger theories predict the same ground-state energy for hydrogen. Do they agree about the angular momentum in the ground state? Explain.
5. Is it possible for a hydrogen atom to be in the 2*d* state? Explain.
6. Can the component of a quantized angular momentum measured on a given axis ever equal the magnitude of the angular momentum vector? Explain.
7. The electron is a spin-$\frac{1}{2}$ particle. Does this mean the electron's intrinsic angular momentum is $\frac{1}{2}\hbar$? Explain.
8. How does the Stern–Gerlach experiment provide convincing evidence for space quantization?
9. Why is there no spin-orbit splitting in hydrogen's ground state?
10. How does the exclusion principle explain the diversity of chemical elements?
11. Helium and lithium exhibit very different chemical behavior, yet they differ by only one unit of nuclear charge. Explain.
12. Why is stimulated emission essential for laser action?
13. What distinguishes a Bose–Einstein condensate from ordinary matter?

Exercises and Problems

Exercises

Section 36.1 The Hydrogen Atom

14. Using physical constants accurate to four significant figures (see inside front cover), verify the numerical values of the Bohr radius a_0 and the hydrogen ground-state energy E_1.
15. A group of hydrogen atoms is in the same excited state, and photons with at least 1.5-eV energy are required to ionize these atoms. What's the quantum number n for the initial excited state?
16. Find the maximum possible magnitude for the orbital angular momentum of an electron in the $n = 7$ state of hydrogen.
17. Which of the following is not a possible value for the magnitude of the orbital angular momentum in hydrogen: (a) $\sqrt{12}\hbar$; (b) $\sqrt{20}\hbar$; (c) $\sqrt{30}\hbar$; (d) $\sqrt{40}\hbar$; (e) $\sqrt{56}\hbar$?
18. The orbital angular momentum of the electron in a hydrogen atom has magnitude 2.585×10^{-34} J·s. Find its minimum possible energy.
19. What's the orbital quantum number for an electron whose orbital angular momentum has magnitude $L = \sqrt{30}\hbar$?
20. A hydrogen atom is in the 6*f* state. Find (a) its energy and (b) the magnitude of its orbital angular momentum.
21. Give a symbolic description for the state of the electron in a hydrogen atom with total energy -1.51 eV and orbital angular momentum $\sqrt{6}\hbar$.

Section 36.2 Electron Spin

22. Verify the value of the Bohr magneton μ_B in Equation 36.14.
23. Theories of quantum gravity predict a spin-2 particle called the *graviton*. What would be the magnitude of the graviton's spin angular momentum?
24. Some very short-lived particles known as delta resonances have spin $\frac{3}{2}$. Find (a) the magnitude of their spin angular momentum and (b) the number of possible spin states.
25. What are the possible j values for a hydrogen atom in the 3*D* state?

Section 36.3 The Exclusion Principle

26. An infinite square well contains nine electrons. Find the energy of the highest-energy electron in terms of the ground-state energy E_1.
27. A quantum harmonic oscillator with frequency ω contains 21 electrons. What's the energy of the highest-energy electron?

Section 36.4 Multielectron Atoms and the Periodic Table

28. Use shell notation to characterize rubidium's outermost electron.
29. Write the full electronic structure of scandium.
30. Write the full electronic structure of bromine.

Section 36.5 Transitions and Atomic Spectra

31. Show that the wavelength λ in nm of a photon with energy E in eV is $\lambda = 1240/E$.
32. The $4f \rightarrow 3p$ transition in sodium produces a spectral line at 567.0 nm. Find the energy difference between these two levels.
33. The $4p \rightarrow 3s$ transition in sodium produces a double spectral line at 330.237 nm and 330.298 nm. What's the energy splitting of the 4*p* level?

Problems

34. Adapt part (b) of Example 36.1 to find the probability that an electron in the hydrogen ground state will be found beyond two Bohr radii.
35. Determine the principal and orbital quantum numbers for a hydrogen atom whose electron has energy -0.850 eV and orbital angular momentum $L = \sqrt{12}\hbar$.
36. Find (a) the energy and (b) the magnitude of the orbital angular momentum for an electron in the 5*d* state of hydrogen.
37. Assuming the Moon's orbital angular momentum is quantized, estimate its orbital quantum number l.
38. The maximum possible angular momentum for a hydrogen atom in a certain state is $30\sqrt{11}\hbar$. Find (a) the principal quantum number and (b) the energy.
39. A hydrogen atom is in an $l = 2$ state. What are the possible angles its orbital angular momentum vector can make with a given axis?
40. A hydrogen atom has energy $E = -0.850$ eV. Find the maximum possible values for (a) its orbital angular momentum and (b) the component of that angular momentum on a given axis.
41. An electron in hydrogen is in the 5*f* state. What possible values, in units of $\hbar$, could a measurement of the orbital angular momentum component on a given axis yield?
42. Substitute Equation 36.3 into Equation 36.4 and carry out the differentiations to show that you get the first unnumbered equation following Equation 36.4.
43. Differentiate the radial probability density for the hydrogen ground state, and set the result to zero to show that the electron is most likely to be found at one Bohr radius.

44. Repeat Exercise 25 for the case where you know only that the principal quantum number is 3; that is, l might have any of its possible values.
45. A hydrogen atom is in the $4F_{5/2}$ state. Find (a) its energy in units of the ground-state energy, (b) its orbital angular momentum in units of $\hbar$, and (c) the magnitude of its total angular momentum in units of $\hbar$.
46. Suppose you put five electrons into an infinite square well of width L. Find an expression for the minimum energy of this system, consistent with the exclusion principle.
47. A harmonic oscillator potential of natural frequency ω contains eight electrons and is in its lowest-energy state. (a) What is its energy? (b) What would the lowest energy be if the electrons were replaced by spin-1 particles of the same mass?
48. You work for a nanotechnology company developing a new quantum device that operates essentially as a one-dimensional infinite square well of width 2.5 nm. You're asked to specify the maximum number of electrons in the device before the total electron energy exceeds 25 eV. Your answer?
49. Determine the electronic configuration of copper.
50. An electron in a highly excited state of hydrogen ($n_1 \gg 1$) drops into the state $n = n_2$. Find the lowest value of n_2 for which the emitted photon will be in the infrared.
51. A solid-state laser made from lead–tin selenide has a lasing transition at a wavelength of 30 μm. If its power output is 2.0 mW, how many lasing transitions occur each second?
52. For hydrogen, fine-structure splitting of the $2p$ state is only about 50 μeV. What percentage is this difference of the photon energy emitted in the $2p \rightarrow 1s$ transition? Your answer shows why it's hard to observe spin-orbit splitting in hydrogen.
53. Find the probability that the electron in the hydrogen ground state will be found in the radial-distance range $r = a_0 \pm 0.1a_0$.
54. **BIO** You've acquired a laser for your dental practice. It produces 400-mJ pulses at 2.94-μm wavelength. A patient wonders about the number of photons in each pulse, and where they lie in the EM spectrum. Your answer?
55. Singly ionized oxygen (so-called O-II) is a prevalent species in the tenuous gas between stars, and O-II emits a doublet spectral line at 372.60 nm and 372.88 nm. Astrophysicists analyze this line to learn, among other things, about the distribution of interstellar gas in distant galaxies. Find the energy splitting, in eV, that results in this doublet.
56. A harmonic oscillator potential with natural frequency ω contains a number of electrons and is in its state of lowest energy. If that energy is $6.5\hbar\omega$, (a) how many electrons are in the potential well and (b) what's the energy of the highest-energy electron?
57. A harmonic oscillator potential with natural frequency ω contains N electrons and is in its state of lowest energy. Find expressions for the energy of the highest-energy electron for (a) N even and (b) N odd.
58. **CH** A harmonic oscillator potential with natural frequency ω contains N electrons and is in its state of lowest energy. Find expressions for the total energy for (a) N even and (b) N odd.
59. An infinite square well containing a number of electrons is in its lowest possible energy state, which is 19 times the ground-state energy E_1 for the same well when it contains a single electron. (a) How many electrons are in the well? (b) What's the energy of the highest-energy electron?
60. **CH** Repeat Problem 58 for the case of an infinite square well containing N particles, rather than for a harmonic oscillator. Express your answer in terms of the ground-state energy E_1.
61. What's the most orbital angular momentum that could be added to an atomic electron initially in the $6d$ state without changing its principal quantum number? What would be the new state?
62. A hydrogen atom is in an F state. (a) Find the possible values for its total angular momentum. (b) For the state with the greatest angular momentum, find the number of possible values for the component of $\vec{J}$ on a given axis.
63. **CH** A hydrogen atom is in the $2s$ state. Find the probability that its electron will be found (a) beyond one Bohr radius and (b) beyond 10 Bohr radii.
64. Show that the maximum number of electrons in an atom's nth shell is $2n^2$.
65. Form the radial probability density $P_2(r)$ associated with the ψ_{2s} state of Equation 36.7, and find the electron's most probable radial position.
66. Substitute the wave function ψ_2 of Equation 36.7 into Equation 36.4 to verify that the equation is satisfied and that the energy is given by Equation 36.6 with $n = 2$.
67. (a) Verify Equation 36.8 by considering a single-electron atom with nuclear charge Ze instead of e. (b) Calculate the ionization energies for single-electron versions of helium, oxygen, lead, and uranium.
68. **BIO** Excimer lasers for vision correction generally use a combination of argon and fluorine to form a molecular complex that can exist only in an excited state. Stimulated de-excitation produces 6.42-eV photons, which form the laser's intense beam. What's the corresponding photon wavelength, and where in the spectrum does it lie?
69. A selection rule for the infinite square well allows only those transitions in which n changes by an odd number. Suppose an infinite square well of width 0.200 nm contains an electron in the $n = 4$ state. (a) Draw an energy-level diagram showing all allowed transitions that could occur as this electron drops toward the ground state, including transitions from lower levels that could be reached from $n = 4$. (b) Find all the possible photon energies emitted in these transitions.
70. An ensemble of one-electron square-well systems of width 1.17 nm all have their electrons in highly excited states. They undergo all possible transitions in dropping toward the ground state, obeying the selection rule that Δn must be odd. (a) What wavelengths of visible light are emitted? (b) Is there any infrared emission? If so, how many spectral lines lie in the infrared?
71. **CH** Use the radial probability density from Equation 36.5 and the normalized ground-state hydrogen wave function from Equation 36.3 and Example 36.1 to calculate the average radial distance r_{av} for an electron in the ground state. (*Note:* Because the probability-density curve isn't symmetric, the average radial distance isn't the same as the most probable distance shown in Fig. 36.3.)
72. **CH** Follow the procedure in Problem 71 to calculate the average radial distance for an electron in the $2s$ state of hydrogen.
73. The ratio of the magnetic moment, in units of the Bohr magneton μ_B, to the angular momentum, in units of $\hbar$, is called the g-factor. (a) Show that the classical orbital g-factor for an atomic electron in a circular Bohr orbit is $g_L = 1$. (b) Show that Equation 36.13 gives $g_S = 2$ for the g-factor associated with electron spin.
74. You work for a company that makes red helium–neon lasers widely used in physics experiments. Figure 36.19 shows an energy-level diagram for this laser. An electric current excites helium to a metastable level E_1 at 20.61 eV above the ground state. Collisions transfer energy to neon atoms, exciting them

to $E_2 = 20.66$ eV. The lasing transition drops the atoms to E_3, emitting a 632.8-nm photon in the process. You're asked to find the maximum possible efficiency for this laser—that is, the light energy emitted as a percentage of the energy supplied to excite the atoms. Your answer?

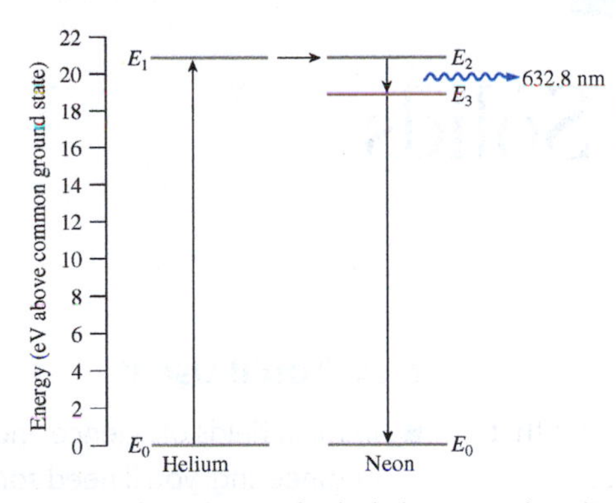

FIGURE 36.19 Energy-level diagram for the helium–neon laser (Problem 74).

75. Using the table below, make a plot of atomic volume versus atomic number, for the elements from $Z = 30$ to $Z = 59$ listed in the table. Comment on the structure of your graph in relation to the periodic table, the electronic structures of atoms, and their chemical properties. (Volumes are in units of 10^{-30} m^3.) DATA

Z	V	Z	V	Z	V
30	7.99	40	26.1	50	11.2
31	12.5	41	20.2	51	8.78
32	6.54	42	18.8	52	6.88
33	4.99	43	17.5	53	5.28
34	3.71	44	16.2	54	4.19
35	2.85	45	12.8	55	95.9
36	2.57	46	12.0	56	51.6
37	70.3	47	11.2	57	49.0
38	37.2	48	10.5	58	46.5
39	28.3	49	17.2	59	44.0

Passage Problems

With sufficient energy, it's possible to eject an electron from an inner atomic orbital. A higher-energy electron will then drop into the unoccupied state, emitting a photon with energy equal to the difference between the two levels. For inner-shell electrons, photon energies are in the keV range, putting them in the X-ray region of the spectrum. These **characteristic X rays** are labeled with the letter indicating the shell to which the electron drops, followed by a Greek letter indicating the higher level from which it drops; thus $K\alpha$ designates a transition from the L shell to the K shell.

Characteristic X rays provide scientists and physicians with an important diagnostic tool. Environmental scientists bombard pollution samples with high-energy electrons, knocking out inner-shell electrons and thus producing X-ray spectra that help identify contaminants (Fig. 36.20*a*). Geologists do the same with rocks. Medical radiologists reverse the process, exploiting the fact that X rays cause inner-shell transitions as well as complete ejection of inner-shell electrons. In particular, radiologists use the element barium in this way to produce high-contrast X-ray images of the intestinal tract (Fig. 36.20*b*).

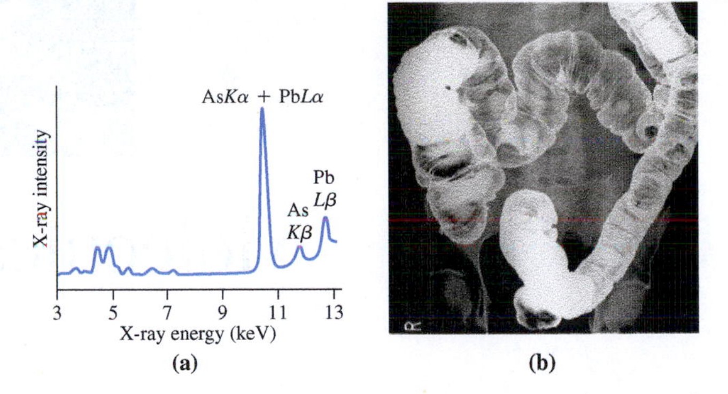

FIGURE 36.20 Passage Problems 76–79. (a) An X-ray spectrum from air pollutants trapped on a filter. The labeled peaks show the presence of lead (Pb) and arsenic (As), as evidenced by $K\alpha$, $K\beta$, $L\alpha$, and $L\beta$ characteristic X rays. (b) X-ray of an intestinal tract, made by coating the intestinal wall with X-ray-opaque barium.

76. Molybdenum's X-ray spectrum has its $K\alpha$ peak at 17.4 keV. The corresponding X-ray wavelength is closest to
    a. 1 pm.
    b. 100 pm.
    c. 1 nm.
    d. 100 nm.
77. In general, how should the energy of an element's $L\alpha$ X rays compare with the energy of its $K\alpha$ X rays?
    a. They have less energy.
    b. They have the same energy.
    c. They have greater energy.
    d. You can't tell without knowing the element.
78. Elements A and B have atomic numbers Z_A and $Z_B = 2Z_A$. How do you expect element B's $K\alpha$ X-ray energy to compare with that of element A?
    a. B's $K\alpha$ energy should be about one-fourth that of A.
    b. B's $K\alpha$ energy should be about half that of A.
    c. B's $K\alpha$ energy should be about twice that of A.
    d. B's $K\alpha$ energy should be about four times that of A.
79. Emission of characteristic X rays occurs in the context of multielectron atoms that generally have all but one of their electrons present. You should therefore expect the X-ray energies to be described
    a. quite accurately by Bohr's atomic theory.
    b. through hydrogen-like solutions to the Schrödinger equation.
    c. only approximately by Bohr's theory or hydrogenic solutions to the Schrödinger equation.

Answers to Chapter Questions

Answer to Chapter Opening Question

Only one electron is allowed in a given atomic quantum state, leading to the shell structure of atoms and to chemical properties based on the outermost atomic electrons.

Answers to GOT IT? Questions

36.1 (c)
36.2 (c)
36.3 (d)
36.4 (b)
36.5 $5p \rightarrow 3s$

Molecules and Solids

What You Know

- You've seen potential-energy curves, including curves for molecules, in Chapter 7.
- You understand the mechanics of rotation, including rotational inertia.
- You know how simple harmonic oscillations occur with linear restoring forces.
- You understand how quantized electron energies arise in atoms.
- You know the electrical properties of conductors and insulators.

What You're Learning

- You'll learn several common ways that atoms bond to form molecules.
- You'll learn to evaluate quantized energy levels associated with molecular rotation and vibration, and how these give rise to molecular spectra.
- You'll see how atoms bond to form crystalline solids.
- You'll learn to describe the electrical properties of insulators, conductors, and semiconductors using band theory.
- You'll develop a qualitative understanding of superconductivity.

How You'll Use It

- In most fields of science and engineering, you'll need some understanding of the properties of solids.
- Materials scientists are continually developing new materials, based on principles you'll learn in Chapter 37, and these materials will enrich your life no matter what your profession.
- Superconductors hold the possibility of stunning breakthroughs in electromagnetic technologies and, especially, energy efficiency.

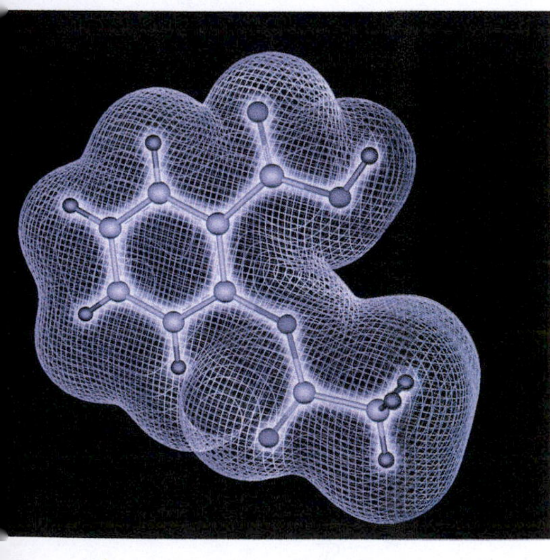

What equation determines the structure of a complex molecule, like the aspirin molecule shown here?

In principle, we could apply the Schrödinger equation to all the particles making up a molecule or even a solid, but in practice that's difficult for all but the simplest molecules. Ever-increasing computer power has brought more complex molecules within reach of structural calculations based in the Schrödinger equation, but in many cases it's appropriate to describe molecular and solid structure more qualitatively.

37.1 Molecular Bonding

The binding of atoms into molecules involves both electric forces and quantum-mechanical effects associated with the exclusion principle. Although individual atoms are electrically neutral, the distribution of charge within them gives rise to attractive or repulsive forces. When atoms are squeezed closely together, interactions involving the spins of their outermost electrons also result in attractive or repulsive interactions. For atoms with unfilled outer shells, it's energetically favorable for electrons to pair with opposite spins; this causes an attractive interaction. When the outer atomic shells are filled, the exclusion principle forces electrons from separate atoms into different states as the atoms are brought together. One or more electrons go into higher energy states, raising the overall energy and resulting in a repulsive interaction. Finally, if atoms get very close, the electrical repulsion of the nuclei becomes important. Ultimately, the balance of attractive and repulsive interactions determines the equilibrium configuration of a molecule. In energy terms, we can think of a stable molecule as a minimum-energy configuration of the electrons and nuclei making up two or more atoms (Fig. 37.1). Although such force and energy considerations ultimately govern all molecules, we distinguish several molecular binding mechanisms based on which interactions are most important.

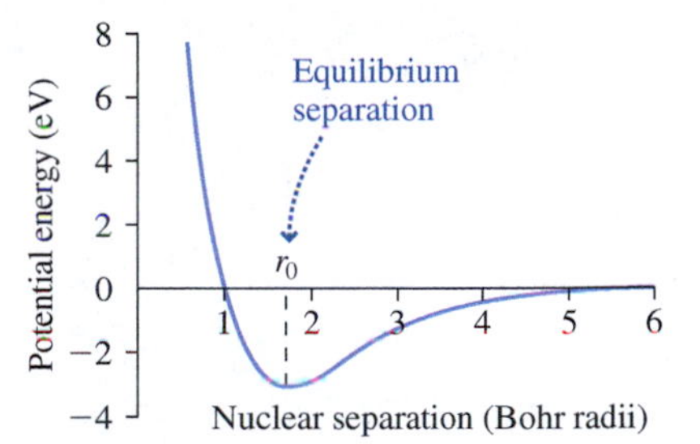

FIGURE 37.1 Potential energy of a pair of hydrogen atoms as a function of the distance between their nuclei.

Ionic Bonding

As we saw in Chapter 36, elements near the left side of the periodic table have few electrons in their outermost shells and correspondingly low ionization energies. In contrast, elements near the right side of the table have nearly filled shells and consequently strong affinities for electrons. When atoms from these different regions of the periodic table come together, it takes relatively little energy to transfer electrons between them. Sodium, for example, has an ionization energy of 5.1 eV, meaning it takes this much energy to make an Na^+ ion. Chlorine, at the opposite end of the periodic table, has such a strong electron affinity that the energy of a Cl^- ion is 3.8 eV below that of a neutral Cl atom. Thus an expenditure of only 1.3 eV $(5.1\text{ eV} - 3.8\text{ eV})$ is required to transfer the outermost electron from a sodium to a chlorine atom. The resulting ions are strongly attracted and come together until they reach equilibrium at an internuclear separation of about 0.24 nm. The total energy of the pair is then 4.2 eV below that of neutral chlorine and sodium atoms at large separation (Fig. 37.2). Since it would therefore take 4.2 eV to separate the atoms, this quantity is called the **dissociation energy**.

Because the minimum-energy sodium–chlorine structure consists of ions bound by the electrostatic force, the binding mechanism is termed **ionic bonding**. Ionic bonding generally occurs in crystalline solids. Because the building blocks of an ionically bound substance are electrically charged, each can bind to several of the opposite charge, resulting in a regular crystal pattern (Fig. 37.3). Because the electrostatic force is strong, ionic solids are tightly bound and therefore have high melting points (801°C for NaCl). And because all electrons are bound to individual nuclei, there are no free electrons and therefore ionic solids are electrical insulators.

MP

PhET: Quantum Bound States: Two Wells (Molecular Bonding)

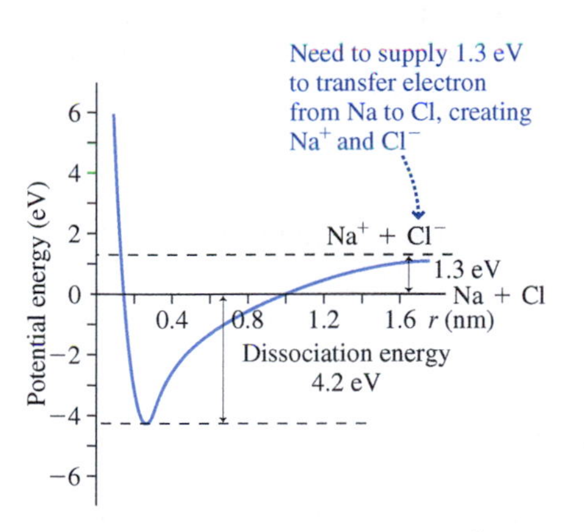

FIGURE 37.2 Potential-energy curve for Na^+ and Cl^- ions, with zero energy corresponding to infinite separation of neutral Na and Cl atoms.

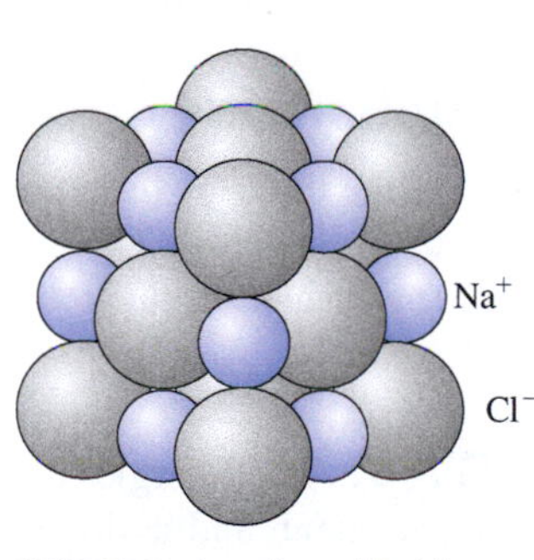

FIGURE 37.3 A sodium chloride crystal is a regular array of sodium and chlorine ions, bound by the electrostatic force.

Covalent Bonding

In an ionic bond, each electron is associated with only one ion. In a **covalent bond**, on the other hand, electrons are shared among atoms. Covalent bonds occur between atoms whose outermost shells aren't full, and whose outer electrons can therefore pair with opposite spins. The simplest example of a covalent bond is the hydrogen molecule, H_2. Since each hydrogen atom has a single 1s electron, each could accommodate in its 1s shell a second electron with opposite spin. When two hydrogen atoms join, quantum mechanics predicts a molecular ground state in which both electrons share a single orbital, with the highest probability of finding the electrons between the nuclei (Fig. 37.4). Dissociation energies for covalent bonds are, like those of ionic bonds, on the order of a few electronvolts.

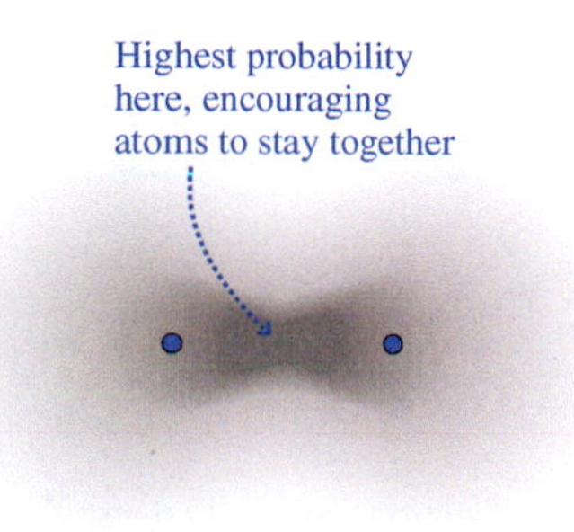

FIGURE 37.4 Probability density for finding electrons in the ground state of molecular hydrogen (H_2).

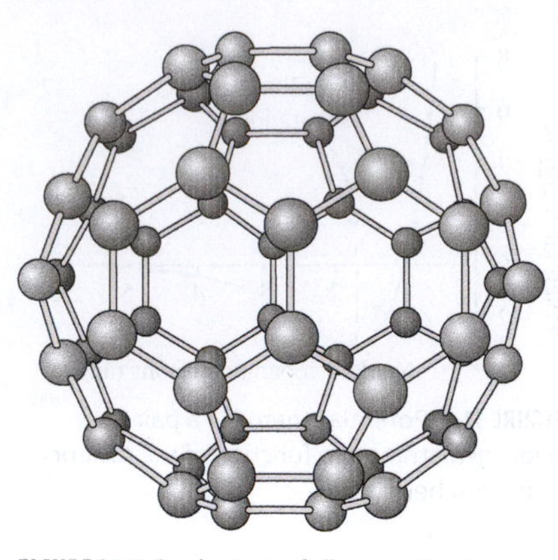

FIGURE 37.5 Buckminsterfullerene, C_{60}, is a symmetric arrangement of 60 carbon atoms. Discovered in the 1980s, C_{60} and related fullerenes hold promise in a wide range of technological applications.

With their outermost molecular orbitals full, covalently bonded molecules often have no room for another electron in their structures. For example, adding a third hydrogen to H_2 is impossible because the ground-state orbital already contains two electrons with opposite spins, so the exclusion principle requires that a third electron go into a higher energy state. The energy of that state is higher than that of an H_2 molecule and a distant H atom; for this reason H_3 isn't a stable molecule. Because their outermost molecular orbitals are full, covalent molecules interact only weakly, and as a result many common covalent materials—for example, H_2, CO, N_2, and H_2O—are either gases or liquids at ordinary temperatures. In other cases covalent bonds can form crystalline structures. A simple example is diamond, a pure-carbon solid formed when each carbon atom bonds covalently to its four nearest neighbors. A more dramatic covalent molecule is buckminsterfullerene, C_{60}, a soccer-ball configuration of 60 carbon atoms (Fig. 37.5).

Hydrogen Bonding

If water consists of covalently bonded molecules, why is it ever a solid? The answer lies in **hydrogen bonds** that form when the tiny, positively charged proton of a hydrogen nucleus nestles close to the negative parts of other molecules. In ice, hydrogen bonds link a proton in one H_2O molecule to the oxygen in another. Because covalent bonding within the water molecule leaves the oxygen only slightly negative and the hydrogen only slightly positive, hydrogen bonds are much weaker than ionic or covalent bonds; a typical hydrogen-bond energy is 0.1 eV. Hydrogen bonds are important in determining the overall configuration of complicated molecules. In DNA, for example, covalent bonds join atoms to form long chains; hydrogen bonds then link two chains into the double-helix structure.

Van der Waals Bonding

In Section 20.5 we mentioned the **van der Waals force** that arises from electrostatic interactions between induced dipole moments of otherwise nonpolar molecules. In gases, the van der Waals force causes deviations from the ideal-gas law that are most pronounced at high densities. As temperature drops, this weakly attractive force becomes effective in binding molecules into liquids and solids. Liquid and solid oxygen (O_2) and nitrogen (N_2), for example, are held together by van der Waals bonds.

Metallic Bonding

In a metal, the outermost atomic electrons aren't bound to individual nuclei, but move throughout the material. The metal forms a crystal lattice of positive ions, bound by this "electron gas." The free electrons give a metal its high electrical and thermal conductivities.

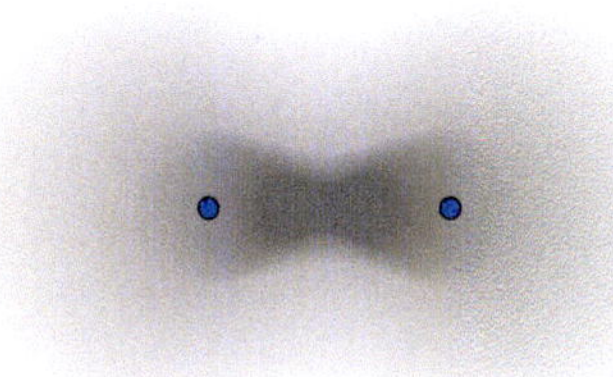

(a)

GOT IT? 37.1 Energies associated with molecular bonds are typically (a) several eV for ionic, covalent, and hydrogen bonds; (b) 10 eV for hydrogen bonds and a few eV for covalent and ionic bonds; or (c) a few eV for covalent and ionic bonds, and a fraction of an eV for hydrogen bonds.

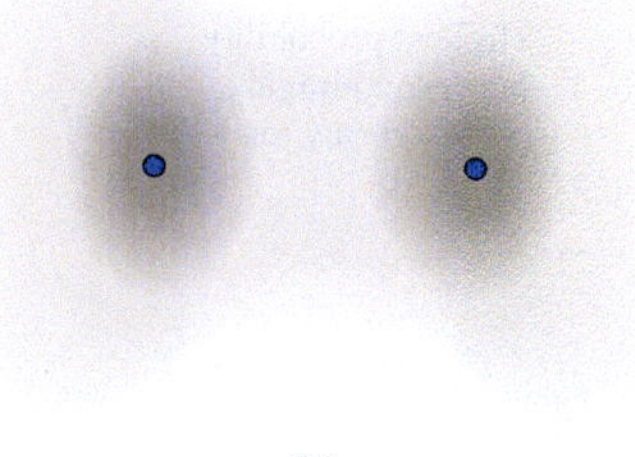

(b)

FIGURE 37.6 Electron probability densities for (a) the ground state and (b) the first excited state of hydrogen. Nuclei are farther apart in the excited state.

37.2 Molecular Energy Levels

In a molecule, electric forces bind electrons and nuclei into a single structure. Like any quantum-mechanical bound system, the energy levels of a molecule are quantized. As in atoms, differences among molecular energy levels are associated with different electronic configurations (Fig. 37.6). But molecules are more complex than atoms, and molecular energy can take additional forms.

In Chapter 18, you saw how a complete description of the specific heats of gases required that we consider the rotational and vibrational motions of individual molecules. We hinted at quantum mechanics, pointing out that each of these modes of molecular motion could absorb only certain discrete amounts of energy (see Fig. 18.17). Here, in a quantum-mechanical treatment of molecular energetics, we consider rotational and vibrational energy states as well as electronic configuration.

Rotational Energy Levels

If a molecule is rotating, it has angular momentum L whose magnitude, from Equation 11.4, is $L = I\omega$, where I is the rotational inertia and ω the angular speed. The quantization conditions that we found in Chapter 36 for atomic angular momenta also hold for the angular momentum of molecular rotation, so we have

$$L = \sqrt{l(l+1)}\hbar \qquad (37.1)$$

where the quantum number l takes on integer values 0, 1, 2, 3, ... But then the rotational energy, which from Equation 10.17 is $E_{rot} = \frac{1}{2}I\omega^2$, must also be quantized. Solving the equation $L = I\omega$ for ω allows us to write the energy as

$$E_{rot} = \tfrac{1}{2}I\left(\frac{L}{I}\right)^2 = \frac{L^2}{2I}$$

Applying the quantization condition 37.1 for L, we then have the quantized rotational energy levels:

$$E_{rot} = \frac{\hbar^2}{2I}l(l+1) \qquad \text{for } l = 0, 1, 2, 3, \ldots \qquad (37.2)$$

EXAMPLE 37.1 Molecular Rotation: Computing Molecular Size

A gas of HCl molecules shows spectral lines that result from transitions between pairs of adjacent rotational energy levels. The energy of the transition increases by 2.63 meV from one spectral line to the next. (a) Use this experimental result to determine the rotational inertia of the HCl molecule. (b) Approximating the more massive chlorine as being essentially fixed, find an expression for the rotational inertia in terms of the hydrogen mass and the separation of the two atomic nuclei. (c) Use the results of parts (a) and (b) to determine the internuclear separation in HCl.

INTERPRET We're asked to infer molecular properties from spectroscopic observations. We're given not the wavelength or energy of a given spectral line, but the energy difference associated with adjacent lines.

DEVELOP Spectral lines result from photons emitted as a molecule drops from one of the energy levels of Equation 37.2 to the next lower level. So the energy of such a photon is

$$\Delta E_{l\to l-1} = \frac{\hbar^2}{2I}[l(l+1) - (l-1)l] = \frac{\hbar^2 l}{I}$$

An adjacent spectral line would result from the transition $l-1 \to l-2$, giving $\Delta E_{l-1\to l-2} = \hbar^2(l-1)/I$. So our plan for (a) is to take the difference between these two transition energies, equate it to the observed energy difference of 2.63 meV between adjacent spectral lines, and solve for I. For (b) we can treat the molecule as having a fixed center (Cl) with a particle (H) in circular motion. Its rotational inertia is then given by Equation 10.12: $I = mR^2$. Equating the two expressions for I will let us solve for the internuclear separation R in (c).

EVALUATE (a) The energy difference $\Delta(\Delta E)$ between adjacent transition energies is

$$\Delta(\Delta E) = \frac{\hbar^2 l}{I} - \frac{\hbar^2(l-1)}{I} = \frac{\hbar^2}{I}$$

Setting this to the observed 2.63-meV spacing and solving for I gives $I = 2.65\times10^{-47}\,\text{kg}\cdot\text{m}^2$, where we had to convert meV to J to get the result in SI. (b) We've already shown that $I = mR^2$. (c) Equating our algebraic and numerical expressions for I and solving gives the internuclear distance:

$$R = \sqrt{\frac{I}{m}} = \sqrt{\frac{2.65\times10^{-47}\,\text{kg}\cdot\text{m}^2}{1.67\times10^{-27}\,\text{kg}}} = 0.126\ \text{nm}$$

where we approximated the hydrogen mass, m, with the proton mass listed inside the front cover.

ASSESS Our answer makes sense, since it's slightly larger than an isolated hydrogen atom. However, it's approximate because we took the chlorine as perfectly fixed and we also ignored the quantum-mechanical ground-state energy of molecular vibration, which stretches the molecule. Photon energies for low l values are on the order of that 2.63-meV spacing, corresponding to a wavelength of about 0.5 mm. This is in the microwave region of the spectrum, with much lower energy and longer wavelength than we've seen for transitions between atomic energy levels. ■

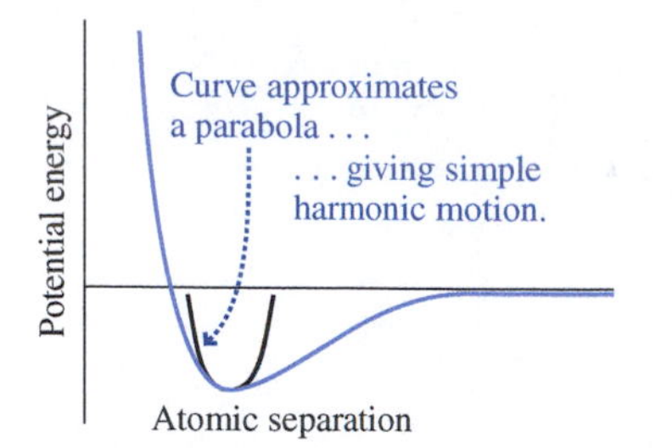

FIGURE 37.7 Near its minimum, the molecular potential-energy curve approximates a parabola.

Vibrational Energy Levels

The equilibrium configuration of a molecule corresponds to the minimum of the molecular potential-energy curve. In the vicinity of that minimum the curve is well approximated by a parabola (Fig. 37.7). In Chapter 13 we saw that parabolic potential-energy curves result in simple harmonic motion, and in Chapter 35 we used a parabolic potential-energy curve in the Schrödinger equation for the harmonic oscillator. There, we found that quantized vibrational energy levels are given by

$$E_{\text{vib}} = \left(n + \tfrac{1}{2}\right)\hbar\omega \tag{37.3}$$

where the quantum number n takes on integer values 0, 1, 2, 3, . . . , and where ω is the natural frequency for classical harmonic oscillations of the molecule. The selection rule for harmonic oscillators limits allowed transitions to those with $\Delta n = \pm 1$, so $\hbar\omega$ is the energy of photons emitted or absorbed in allowed transitions among vibrational energy levels. (Actually, the small-amplitude approximation is often justified for only the lower quantum states, so Equation 37.3 and the selection rule $\Delta n = \pm 1$ may apply to only these states.) For typical diatomic molecules, ω is on the order of $10^{14}\ \text{s}^{-1}$, in the infrared region of the spectrum. Consequently, study of molecular vibrations involves infrared spectroscopy.

As we found in Chapter 35, the minimum energy of a quantum harmonic oscillator is the ground-state energy $E_0 = \frac{1}{2}\hbar\omega$. Thus a molecule can never have zero vibrational energy, although Equation 37.2 shows that it *can* have zero rotational energy.

EXAMPLE 37.2 Molecular Energies: Rotational and Vibrational

An HCl molecule is in its vibrational ground state. Its classical vibration frequency is $f = 8.66\times10^{13}$ Hz. If its rotational and vibrational energies are nearly equal, what are its rotational quantum number and angular momentum?

INTERPRET We're being asked to compare energies associated with two different processes: vibration and rotation. We're told the vibrational energy state (the ground state, $n = 0$), and we need to find the rotational state with comparable energy.

DEVELOP With $n = 0$, Equation 37.3, $E_{\text{vib}} = \left(n + \frac{1}{2}\right)\hbar\omega$, gives the ground-state vibrational energy $E_{\text{vib}} = \frac{1}{2}\hbar\omega = \frac{1}{2}hf$, where we used $\hbar = h/2\pi$ and $\omega = 2\pi f$. Our plan is to equate this to the rotational energy of Equation 37.2, $E_{\text{rot}} = (\hbar^2/2I)l(l+1)$, and solve for the quantity $l(l+1)$. Then we'll use Equation 37.1 to get the angular momentum.

EVALUATE Equating the vibrational and rotational energies gives

$$\frac{\hbar^2}{2I}l(l+1) = \tfrac{1}{2}\hbar\omega = \hbar\pi f$$

where we used $\omega = 2\pi f$. So $l(l+1) = 2\pi f I/\hbar$, and Equation 37.1, $L = \sqrt{l(l+1)}\hbar$, gives

$$L = \sqrt{\frac{2\pi f I}{\hbar}}\hbar = \sqrt{2\pi f I\hbar} = \sqrt{fIh} = 1.23\times10^{-33}\ \text{J}\cdot\text{s}$$

where we used the rotational inertia $I = 2.65\times10^{-47}\ \text{kg}\cdot\text{m}^2$ that we found in Example 37.1.

ASSESS Our angular momentum is about 12 times $\hbar$; approximating $l(l+1)$ by l^2 in Equation 37.1 then shows that l is about 12. So we need a fairly high rotational quantum state to give the same energy as the vibrational ground state. That's consistent with transitions between adjacent rotational states involving microwave photons, while vibrational transitions involve infrared photons. ■

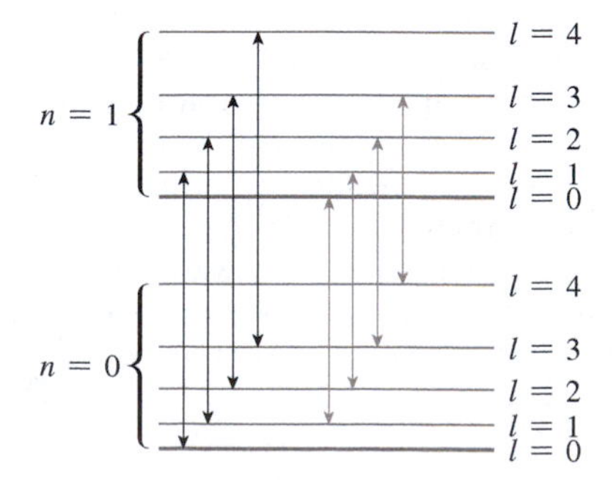

FIGURE 37.8 Energy-level diagram showing the ground state and first vibrational excited state of a diatomic molecule; also shown are four of the infinitely many rotational states for each n.

Molecular Spectra

A molecule with vibrational quantum number n and rotational quantum number l can undergo transitions obeying the selection rules $\Delta n = \pm 1$ and $\Delta l = \pm 1$. If molecules couldn't rotate, the molecular spectrum would consist of a single line at the classical vibration frequency, corresponding to transitions among adjacent vibrational states. But each vibrational level corresponds to an infinite number of rotational states. The resulting energy-level diagram is shown in Fig. 37.8. At typical temperatures, only the ground and first vibrational levels are significantly populated, but with energy distributed among many rotational levels. As a result, molecular spectra show a rich structure, with many lines corresponding to the different transitions of Fig. 37.8. Figure 37.9 is a spectrum of HCl, taken with a high-resolution infrared spectrometer that resolves the individual spectral lines. At lower resolution, the pattern shows up as a broad band, and we often speak of

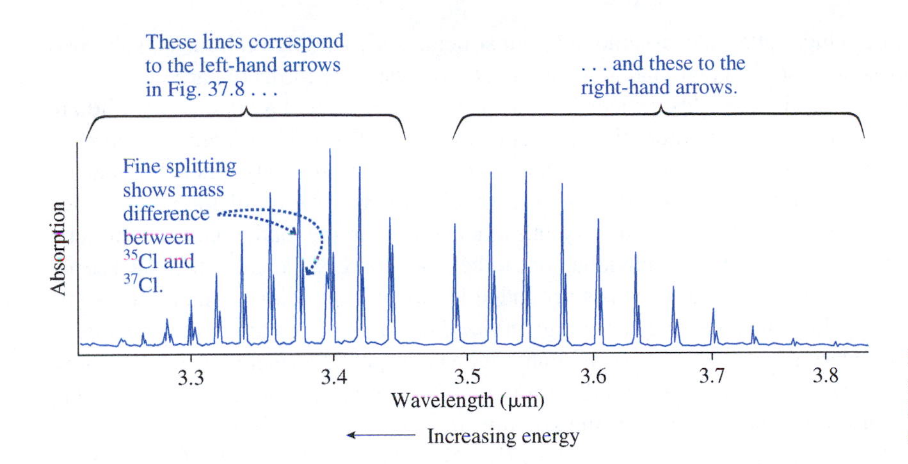

FIGURE 37.9 Absorption spectrum of HCl, showing lines that result from transitions between the $n = 1$ and $n = 0$ vibrational states.

infrared absorption bands in describing the effect of molecules on infrared radiation. For example, absorption bands of atmospheric water vapor and carbon dioxide limit the escape of infrared radiation from Earth, causing the greenhouse effect and global warming that we described in Chapter 16. Molecular energy levels are therefore at the heart of today's most global environmental concern. Figure 37.10 shows some of the molecular absorption bands that are most responsible for greenhouse warming.

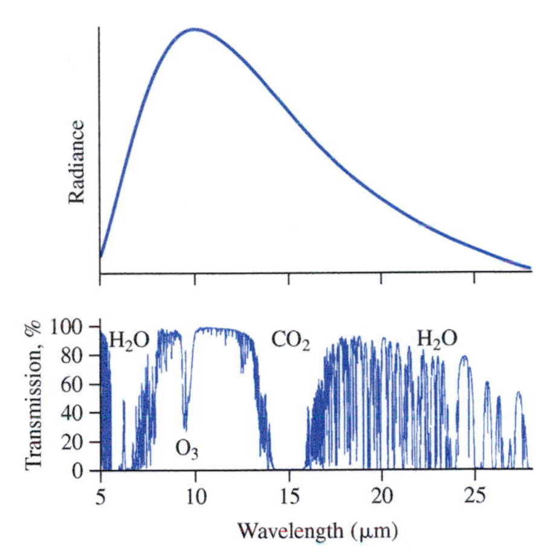

FIGURE 37.10 Upper graph shows the radiance of a blackbody at Earth's average surface temperature of 288 K, which is predominantly in the infrared with wavelengths from 5 μm to 25 μm. Lower graph shows the transmissivity of Earth's atmosphere in the same wavelength range, showing the effects of absorption by vibrational and rotational transitions in water vapor, carbon dioxide, and ozone. Where transmission is low, the atmosphere absorbs outgoing infrared. This absorption contributes to the greenhouse effect and global warming.

GOT IT? 37.2 If a scientist uses microwave technology to study molecular structure, what form of molecular energy is she most concerned with?

37.3 Solids

Bonding mechanisms can join relatively few atoms to form a molecule, or many to form a solid. In the lowest energy state, the atoms of a solid are arranged in a regular, repeating pattern; the solid is then **crystalline**. Sometimes solids form without their atoms having the opportunity to achieve a crystal structure; such solids are termed **amorphous**. Glass is a common amorphous solid. Amorphous materials are difficult to analyze due to their inherent randomness, so we concentrate here on crystalline solids.

Crystal Structure

The hallmark of a crystalline solid is the regular arrangement of atoms. Looking closely shows that a basic pattern repeats throughout the crystal (Fig. 37.11). This basic arrangement is the **unit cell**. Different crystalline materials have different unit cells (Fig. 37.11*a*, *c*). Sometimes

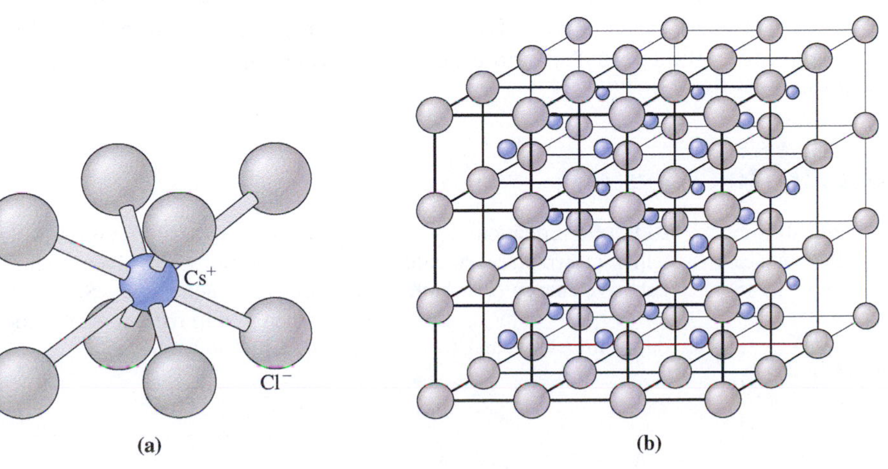

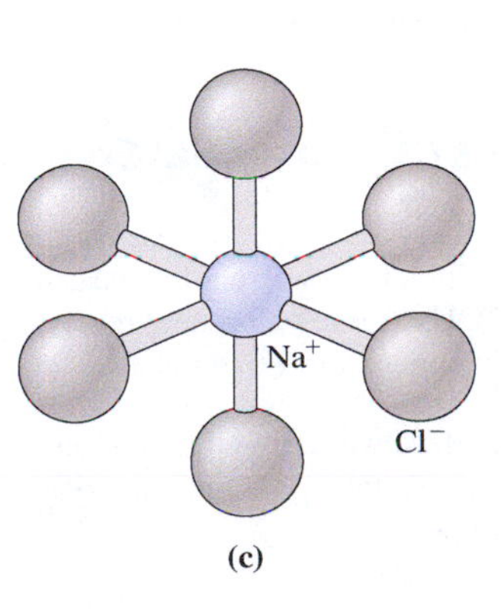

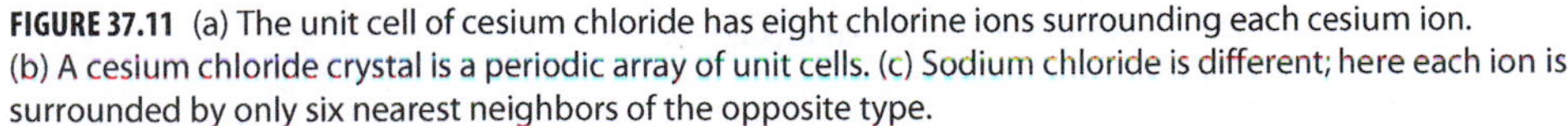

FIGURE 37.11 (a) The unit cell of cesium chloride has eight chlorine ions surrounding each cesium ion. (b) A cesium chloride crystal is a periodic array of unit cells. (c) Sodium chloride is different; here each ion is surrounded by only six nearest neighbors of the opposite type.

the same underlying matter may assume different structures, depending on how the solid was formed; this is the case with diamond and graphite, both crystalline forms of carbon.

As with individual molecules, properties like atomic separation in a crystalline solid are determined by the interplay of attractive and repulsive interactions. The situation is complicated, however, because an individual atom experiences forces from many other atoms in the crystal. With ionic bonding, those forces are electrical attraction and repulsion as described by Coulomb's law, and that makes ionic crystals amenable to simple mathematical treatment.

For ionic crystals, we can take individual ions to be point charges. Consider the NaCl structure of Fig. 37.11*c*. Each sodium ion is surrounded by six nearest chlorine ions, each some distance r away. The potential energy of a singly ionized positive sodium ion in the potential of each negative chlorine ion is $-ke^2/r$. So the contribution to the potential energy of the six nearest chlorines is $-6ke^2/r$, with the minus indicating an attractive interaction. But then there are 12 sodium ions a distance $\sqrt{2}r$ from the sodium; they give rise to a repulsive force and consequently a positive potential energy $+12ke^2/\sqrt{2}r$. At a distance of $\sqrt{3}r$ there are eight more chlorines, giving potential energy $-8ke^2/\sqrt{3}r$. The result is that the electrostatic potential energy of the sodium ion can be written as $U_1 = -\alpha(ke^2/r)$, where $\alpha = 6 - 12/\sqrt{2} + 8/\sqrt{3} - \cdots$; α is called the **Madelung constant**. Many terms in the series are required to compute α accurately, showing that the effect of distant ions is significant in determining the energy of an ion in the crystal. For the NaCl structure, α is approximately 1.748.

As ions are brought closer together, they experience the repulsive effect of the exclusion principle, as we discussed in Section 37.1. This repulsion is described approximately by a potential energy of the form $U_2 = A/r^n$, where A and n are constants. So the total potential energy of an ion in the crystalline solid is

$$U = U_1 + U_2 = -\alpha\frac{ke^2}{r} + \frac{A}{r^n}$$

At equilibrium the potential energy is a minimum (Fig. 37.12), corresponding to zero net force on the ion. Differentiating the potential energy with respect to r and setting dU/dr to zero to find the minimum, we have

$$0 = \frac{\alpha ke^2}{r_0^2} - \frac{nA}{r_0^{n+1}}$$

where r_0 designates the equilibrium separation. Solving for A gives $A = \alpha ke^2 r_0^{n-1}/n$, so the potential energy becomes

$$U = -\alpha\frac{ke^2}{r_0}\left[\frac{r_0}{r} - \frac{1}{n}\left(\frac{r_0}{r}\right)^n\right] \qquad (37.4)$$

The value of U at the equilibrium separation r_0 is designated U_0 and is called the **ionic cohesive energy**. The magnitude of U_0 represents the energy needed to remove an ion entirely from the crystal. The cohesive energy is sometimes given in kcal/mol, in which case its magnitude is the energy per mole needed to break an entire crystal into its constituent ions (see Exercise 23).

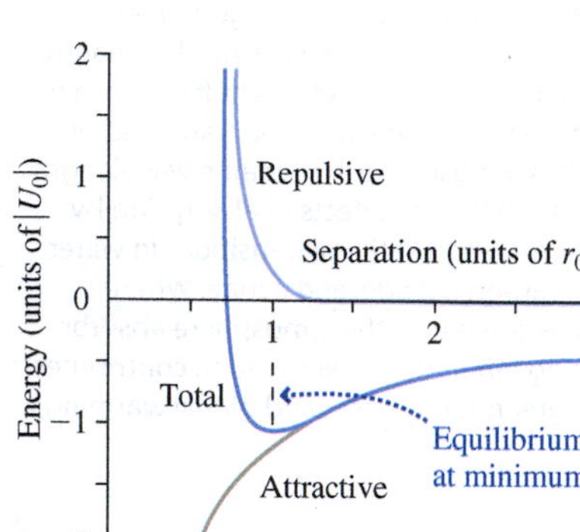
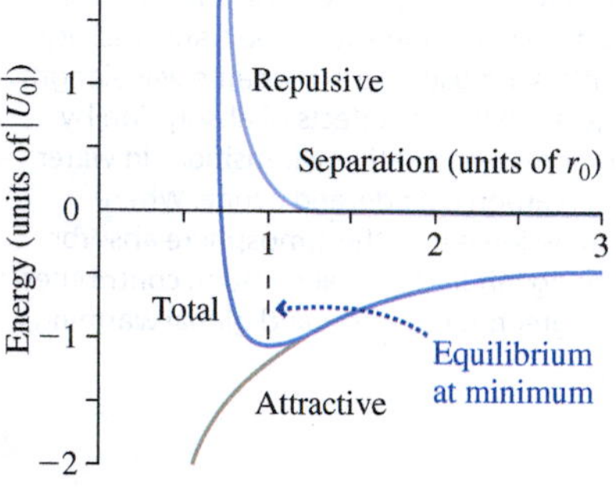

FIGURE 37.12 Potential-energy function for an ionic crystal, showing separate contributions of the attractive and repulsive terms.

EXAMPLE 37.3 Potential Energy in a Solid: The NaCl Crystal

The ionic cohesive energy for NaCl is −7.84 eV. The equilibrium separation, which follows from the measured density, is 0.282 nm (see Exercise 22). Use these values and the Madelung constant $\alpha = 1.748$ to find the exponent n in Equation 37.4 for NaCl.

INTERPRET Here we're given all but one of the quantities in the expression for a crystal's potential energy, and we're asked to solve for the one unknown, n.

DEVELOP Equation 37.4 for the potential energy U looks formidable, with n in two places, including an exponent. But we're given the ionic cohesive energy U_0, which is the value of U when $r = r_0$. So the two terms r_0/r become 1, and since $1^n = 1$, the n in the exponent drops out. With $r = r_0$ and $U = U_0$, Equation 37.4 simplifies to

$$U_0 = -\alpha\frac{ke^2}{r_0}\left(1 - \frac{1}{n}\right)$$

Our plan is to solve this equation for n.

EVALUATE Solving gives

$$n = \left(1 + \frac{U_0 r_0}{\alpha k e^2}\right)^{-1} = 8.22$$

with k the constant in Coulomb's law and e the elementary charge; other quantities are given in the problem statement.

ASSESS The large value of this exponent shows that the NaCl crystal is strongly resistant to compression. In Problem 46 you can calculate the associated repulsive force. ■

Band Theory

MP®

PhET: Band Structure

Quantum-mechanical analysis of a solid containing 10^{23} atoms or so might seem a hopeless task. But the regularity of a crystalline solid makes that problem, while not easy, at least amenable to mathematical treatment. The physical regularity of the solid is reflected mathematically in the properties of the wave function; specifically, the wave function for a crystalline solid in equilibrium is itself periodic. That's because equivalent points in different unit cells have exactly the same physical properties.

We won't solve the Schrödinger equation for a crystal, or even write the solutions. But you can see what some properties of those solutions must be. Consider two identical atoms, initially widely separated, as they're brought closer together. When the atoms are far apart, they're described by identical wave functions and associated energy-level diagrams; a given electron state, for example, has exactly the same energy in each atom. But as the atoms move closer, their wave functions begin to overlap to form a single wave function that characterizes the entire composite system. Because of the exclusion principle, two electrons that were in identical states in the two widely separated atoms can no longer be in the same state. This effect manifests itself as a separation of what were originally identical energy levels (Fig. 37.13*a*). As more and more atoms come together, initially identical energy levels split into ever more finely spaced levels (Fig. 37.13*b*). In a crystalline solid, there are so many atoms that each level splits into an essentially continuous **band** of allowed energies (Fig. 37.13*c*). **Band gaps** separate the bands arising from distinct single-atom states, as shown in Fig. 37.13*c*. An electron can have any energy between the top and bottom of a band, but energies in the band gaps are forbidden. The situation is like a single atom, where electrons are allowed only certain discrete energies, except now the discrete levels have broadened into bands.

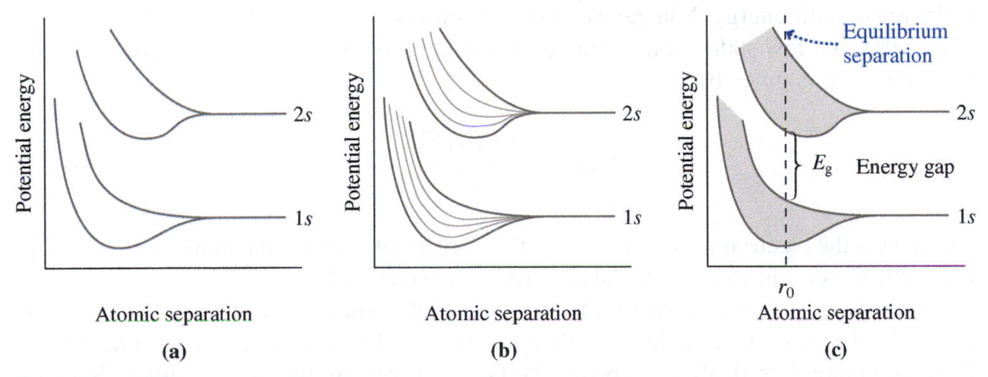

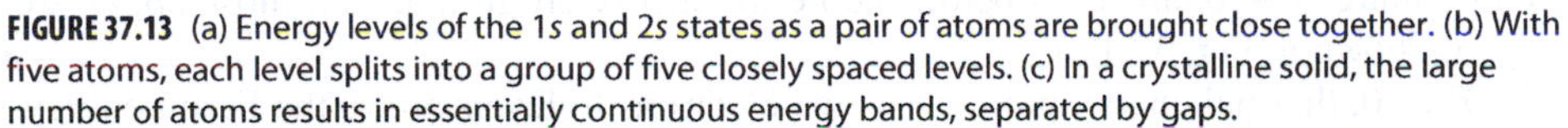

FIGURE 37.13 (a) Energy levels of the 1*s* and 2*s* states as a pair of atoms are brought close together. (b) With five atoms, each level splits into a group of five closely spaced levels. (c) In a crystalline solid, the large number of atoms results in essentially continuous energy bands, separated by gaps.

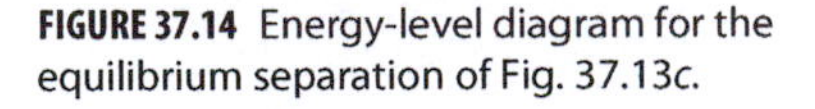

FIGURE 37.14 Energy-level diagram for the equilibrium separation of Fig. 37.13c.

We're usually interested in the properties of a solid at or near its equilibrium state, designated r_0 in Fig. 37.13*c*. There the solid is characterized by an energy-level diagram in which the energy levels are those of Fig. 37.13*c* at the value $r = r_0$ (Fig. 37.14).

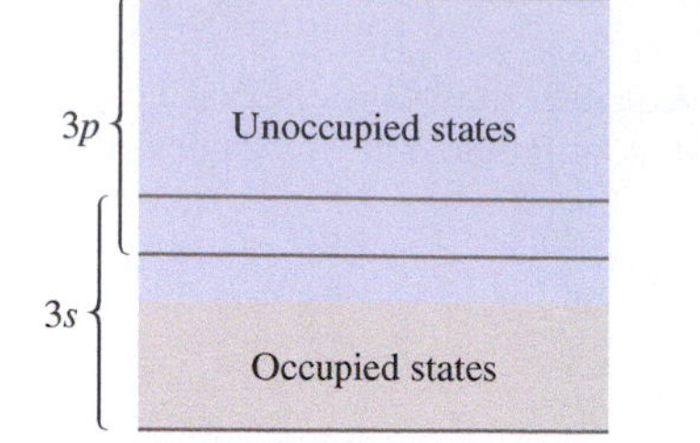

FIGURE 37.15 Band structure of metallic sodium, with gray indicating filled states and color unfilled states. Bands lower than 3*s* aren't shown; they correspond to inner electrons, whose levels aren't split significantly.

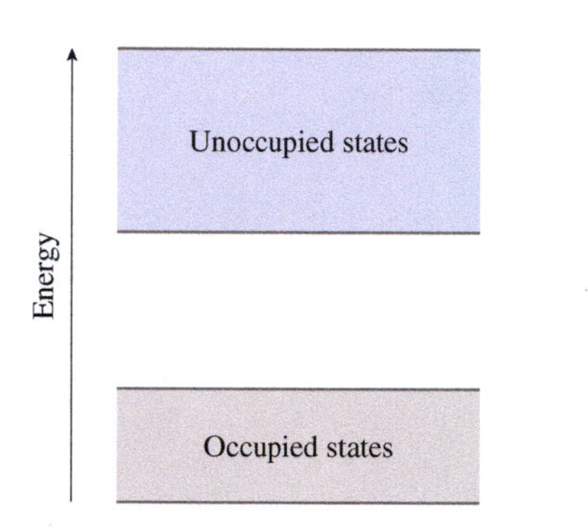

FIGURE 37.16 Band structure for an insulator.

Conductors, Insulators, and Semiconductors

Sometimes the splitting and shifting depicted in Fig. 37.13 result in overlapping bands. Figure 37.15 shows the band structure of sodium, in which the 3*s* and 3*p* bands overlap. Note that the high-energy band containing electrons—here the 3*s*/3*p* band—is not completely full, meaning that energy levels in the upper portion of the band aren't occupied by electrons.

We can determine which of the allowed energy levels of a solid will be occupied in the same way we established the electronic structure of atoms: by placing a given total number of electrons in the lowest possible levels consistent with the exclusion principle. In some materials, like sodium, that filling process results in the highest-energy occupied band being only partially full. But in others, like the material shown in Fig. 37.16, the highest-energy occupied band is completely full.

Figures 37.15 and 37.16 show the essential difference between conductors and insulators. A conductor is a material in which charges are free to move in response to an electric field. Classically, there's no problem with this: Apply an electric field, and if an electron is "free," it will accelerate and gain energy. But quantum mechanically, an electron can gain energy only by jumping into a higher allowed energy level. So there needs to be a higher unoccupied level available.

In sodium, the 3*s* atomic level contains a single electron, although it has room for two. Put N sodium atoms together to form a crystal, and the 3*s* band contains only N of the total $2N$ electrons it could hold. So the 3*s* band is only half full, as Fig. 37.15 shows, and therefore electrons near the top of the filled portion have available unoccupied states with only a little more energy. That makes it easy for an electric field to promote electrons to unoccupied levels. For that reason sodium is an electrical conductor.

In the material of Fig. 37.16, in contrast, one band is completely full and the next higher one empty. An electron in the filled band can't gain energy unless it's enough to jump the band gap. Electric fields of reasonable magnitude can't provide this energy, so the electrons are stuck in the filled bands. That makes the material an insulator.

Metallic Conductors

We found in Chapter 24 that classical physics can't account for the details of metallic conduction, in particular the temperature dependence of conductivity. Quantum mechanically, the conduction electrons in a metal are like electrons in the three-dimensional box of Section 35.4. They're free to move about inside the metal, but not to leave it. The number of states available to the electrons, per unit energy interval, turns out to increase with energy. You can see the beginnings of this trend in Fig. 35.17, which shows the first few states of the three-dimensional box. We won't do this count; the result, however, is given by

$$g(E) = \left(\frac{2^{7/2}\pi m^{3/2}}{h^3}\right)\sqrt{E} \tag{37.5}$$

where m is the electron mass and $g(E)$ is the **density of states**—the number of states per unit volume per unit energy interval centered on the energy E.

At absolute zero, electrons fill the lowest available states according to the exclusion principle. The energy of the highest filled level at absolute zero is the **Fermi energy**, E_F. At temperature $T = 0$, all states below the Fermi energy are full, and all those above are empty, as shown in Fig. 37.17*a*.

For $T > 0$, thermal energy promotes some electrons to levels above the Fermi energy, leaving some levels just below E_F vacant (Fig. 37.17*b*). Now, the Fermi energy in most metals is about 1–10 eV, much higher than the thermal energy at typical temperatures (0.025 eV at room temperature). So the electron distribution changes only slightly—and that means electrons near the Fermi energy carry essentially all the electric current, regardless of temperature. The mean electron speed is therefore quite different from the classical thermal speed (see Problem 50), and that makes the temperature dependence of electrical conductivity in metals very different from the classical prediction.

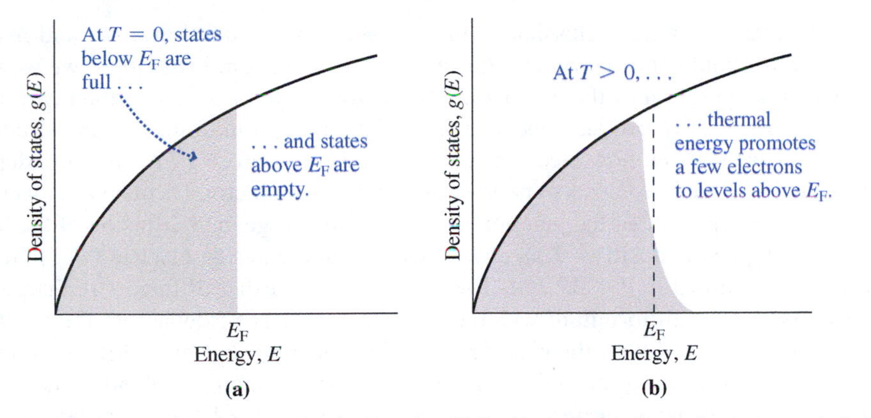

FIGURE 37.17 Density of states given by Equation 37.5, with shaded region indicating occupied energy levels. (a) $T = 0$; (b) $T > 0$.

GOT IT? 37.3 Both parts of Fig. 37.17 describe the same piece of metal. How do the shaded areas compare? (a) Area at left is greater; (b) area at right is greater; (c) areas are equal. Explain your answer.

Semiconductors

In Chapter 24 we gave a classical description of semiconductors—the materials at the heart of our modern electronic world. Here we see how band theory gives a quantum-mechanical explanation of semiconductors.

Our band diagram for an insulator (Fig. 37.16) is strictly correct only at absolute zero. Here the highest occupied band—the **valence band**—is full, and above it the **conduction band** is empty. At temperatures above absolute zero, though, random thermal energy may give an occasional electron enough energy to jump the gap into the conduction band, where it has plenty of nearby states available and can thus respond freely to an electric field. In good insulators, the band gap is many electronvolts and this effect is negligible. But in some materials, notably silicon and germanium, the band gap is on the order of 1 eV (see Table 37.1). At room temperature, thermal excitation promotes enough electrons into the conduction band that these materials conduct electricity, although their conductivity is much lower than in metallic conductors. Such a material is a **semiconductor**. Figure 37.18 compares the band structures for conductors, insulators, and semiconductors.

Table 37.1 Band-Gap Energies for Selected Semiconductors (at 300 K)

Semiconductor	Band-Gap Energy (eV)
Si	1.14
Ge	0.67
InAs	0.35
InP	1.35
GaP	2.26
GaAs	1.43
CdS	2.42
CdSe	1.74
ZnO	3.2
ZnS	3.6

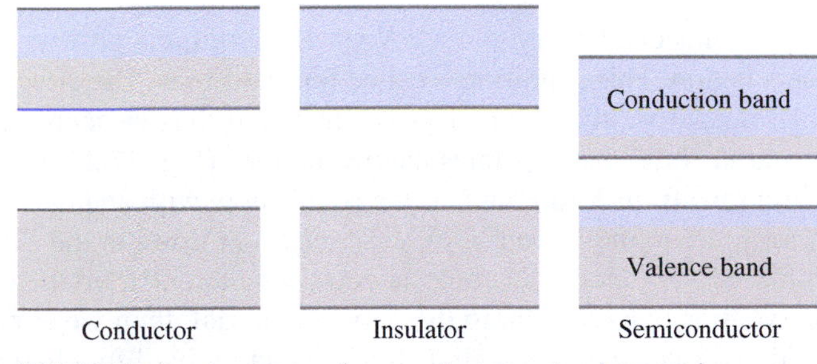

FIGURE 37.18 Band structures for a conductor, an insulator, and a semiconductor. Gray indicates occupied states.

In Chapter 24 we showed how doping—adding small quantities of impurities—could radically alter the electrical properties of semiconductors. In terms of band theory, a dopant such as phosphorus, with five valence electrons, adds **donor levels** just below the conduction band (Fig. 37.19*a*). Thermal energy readily promotes electrons from these levels into the conduction band, greatly increasing the conductivity. This makes an ***N*-type semiconductor** because its predominant charge carriers are electrons. A dopant like boron, in contrast, creates **acceptor levels** just above the valence band (Fig. 37.19*b*). Electrons promoted to these levels leave behind **holes** that act as positive charge carriers. The result is a ***P*-type semiconductor**.

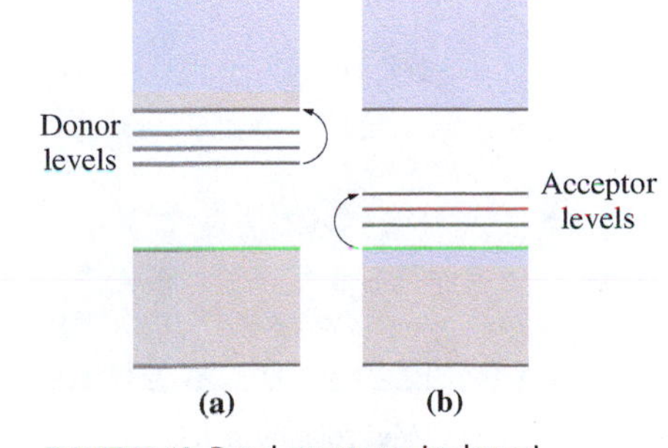

FIGURE 37.19 Band structures in doped semiconductors: (a) *N*-type; (b) *P*-type.

PhET: Semiconductors

In Chapter 24 we gave a classical explanation of how a junction of *P*- and *N*-type semiconductors conducts electric current in only one direction. From there we went on to describe the operation of the transistor—the semiconductor device at the heart of all modern electronics. We can also understand the *PN* junction in terms of band structure. In Fig. 24.11 we showed how electrons and holes diffuse across a *PN* junction, depleting the junction of charge carriers and making it a poor conductor. Diffusion of electrons also gives the *P*-type side of the junction a net negative charge, and diffusion of the holes makes the *N*-type side positive. This charge separation creates an electric field pointing from *N* to *P*, as shown in Fig. 37.20*a*. The field opposes further diffusion of charge and thus establishes an equilibrium in which there's no net charge flow across the junction. Because they've moved with the electric field, the electrons that have diffused into the *P*-type region have higher potential energy than those that remained behind in the *N*-type region. (Remember that electrons are negative, so their potential energy increases when they move in the *same* direction as an electric field.) As a result, the band-structure diagram for electrons in the *PN* junction takes the form shown in Fig. 37.20*b*.

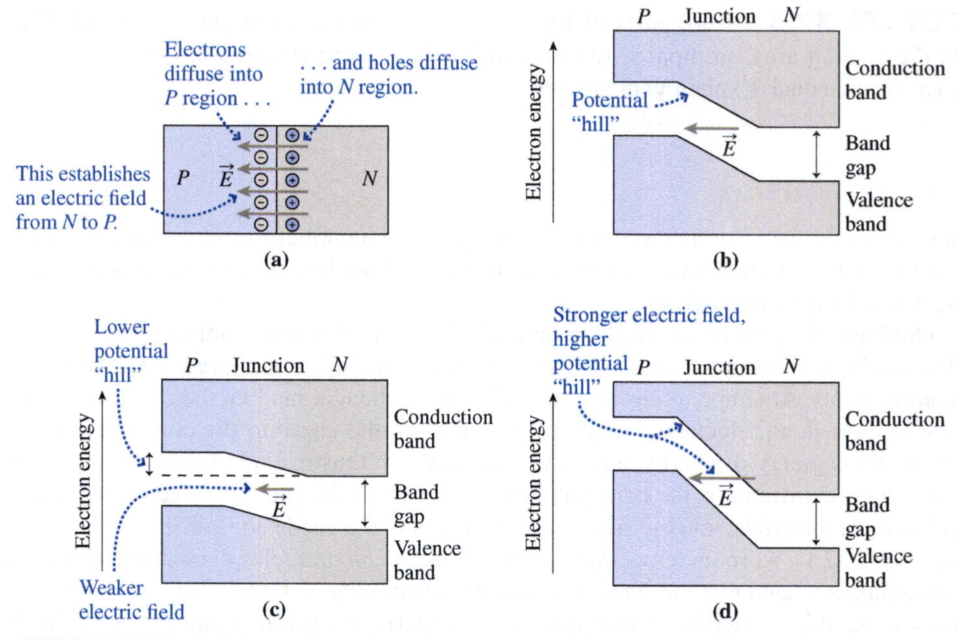

FIGURE 37.20 (a) Physical picture of an unbiased *PN* junction and (b) the corresponding band structure. (c) Band structure of a forward-biased junction and (d) a reverse-biased junction.

Now suppose we connect a battery to our *PN* junction, with the positive terminal to the *P*-type side of the junction. This condition is called **forward bias**. The effect is to make the *P*-type material less negative, the *N*-type less positive, and thus weaken the electric field and lower the potential "hill" that separates the two regions (Fig. 37.20*c*). It becomes easier for electrons to move from *N* to *P* and, as we could show with analogous diagrams for holes, it's easier for holes to move from *P* to *N*. So a current flows in the *P*-to-*N* direction, and the forward-biased *PN* junction becomes a good conductor. If, on the other hand, we connect the battery's positive terminal to the *N*-type material, then we strengthen the internal electric field and steepen the potential "hill," making it hard for charges to cross the junction (Fig. 37.20*d*). Now the *PN* junction is **reverse biased**, and it's a poor conductor.

As electrons and holes pour across a forward-biased junction, many recombine; that is, they drop from the conduction band into the valence band, releasing energy in the process. In light-emitting diodes (LEDs) and diode lasers, this energy appears as photons whose energy is close to that of the band gap. Because $E = hf$, the band gap determines the frequency and, equivalently, the wavelength and color of the emitted light. Development of semiconductor lasers with ever-larger band gaps enabled the evolution from CD to DVD to Blu-ray discs that we outlined in Chapter 32. Conversely, a material whose band gap corresponds to visible-light photons can absorb light energy, promoting electrons to the conduction band and driving current through an external circuit. Such **photovoltaic cells** are increasingly used to generate electric power, both on individual buildings and in large-scale solar power plants (Fig. 37.21).

FIGURE 37.21 Les Mées solar farm in France comprises 113,000 solar panels, giving a peak power output of 100 MW.

CONCEPTUAL EXAMPLE 37.1 **CD to Blu-ray: Engineering the Band Gap**

The amount of information stored on CDs, DVDs, and Blu-ray discs is limited in part by diffraction effects associated with the wavelength of the laser light used to "read" the disc (see Chapter 32's Application "Movies on Disc"). The lasers used in optical drives are semiconductor lasers, with wavelengths set by the semiconductors' band gaps. Compare the band gaps of lasers used for reading CDs and Blu-ray discs.

EVALUATE A CD holds 74 minutes of audio, yet a Blu-ray disc of the same physical size holds several hours of high-definition video. So Blu-ray data are stored at a smaller spatial scale and thus require a shorter wavelength to "read" the data. Since $E = hf = hc/\lambda$, that means higher photon energy and therefore a larger band gap.

ASSESS As Making the Connection shows, the band gap for Blu-ray is nearly twice that of a CD laser. In fact, the trade name "Blu-ray" comes from the blue wavelength used. Multiple-layer storage and better compression algorithms also contribute to Blu-ray's much greater capacity.

MAKING THE CONNECTION The lasers that "read" CDs, DVDs, and Blu-ray discs operate at 780 nm, 650 nm, and 405 nm, respectively. Find the corresponding band gaps.

EVALUATE Photon energy quantization $E = hf = hc/\lambda$ gives the photon energy and therefore the required band gap. Working in electron-volts gives 1.59 eV for CD, 1.91 eV for DVD, and 3.07 eV for Blu-ray.

37.4 Superconductivity

In Chapter 24 we introduced **superconductivity**—the complete loss of electrical resistance in some materials at low temperature. First discovered in mercury in 1911, superconductivity was for decades limited to a few elements and alloys below about 20 K. A breakthrough in 1986 brought a new class of metal-oxide superconductors with superconducting **transition temperatures** of about 100 K; today the highest transition temperatures exceed 200 K. The ultimate goal of a room-temperature superconductor, once thought beyond reach, may yet be achieved.

Superconductors find use in an ever-increasing range of applications, including high-strength electromagnets for MRI scanners, particle accelerators, materials separation, and research; compact, efficient motors for vehicle and marine propulsion; high-Q filters for cell-phone base stations; sensitive magnetic-field sensors for brain-wave imaging and physics research; underground power transmission in crowded cities; and so-called synchronous condensers for optimizing the power factor in AC power transmission (see Section 28.5). Other applications include superconducting electronic devices that promise orders-of-magnitude increases in computer speed, and magnetically levitated vehicles for ground transportation at speeds up to 500 km/h (see the Application on p. 714).

Superconductivity and Magnetism

The hallmark of a superconductor is zero electrical resistance. Another distinguishing characteristic is the **Meissner effect**, wherein a superconductor excludes magnetic flux from its interior (Fig. 37.22). Figure 37.22*c* shows why: Currents in the superconductor

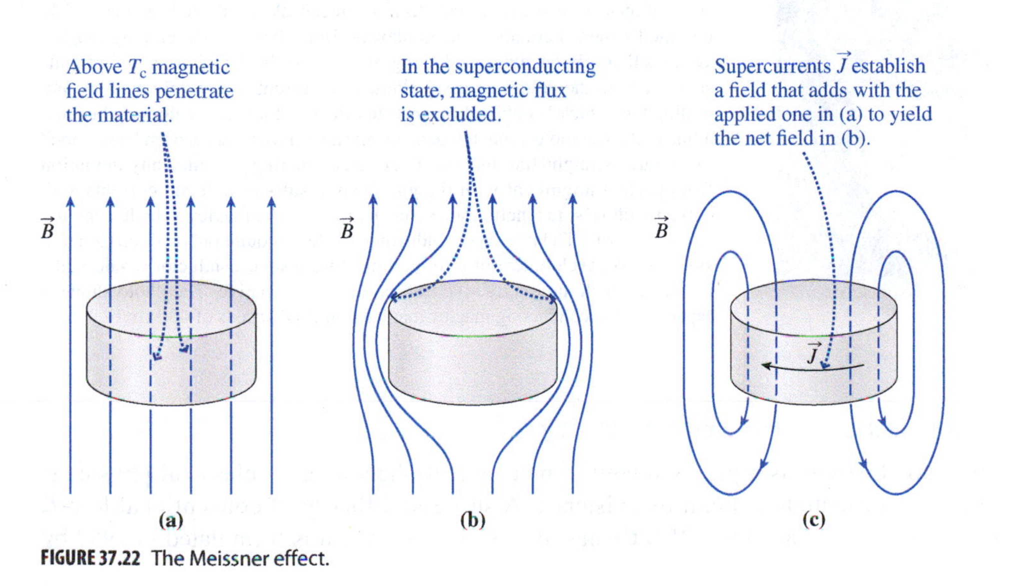

FIGURE 37.22 The Meissner effect.

create their own magnetic field that exactly cancels the field within the material. As we pointed out in Section 27.6, a superconductor's exclusion of magnetic flux makes it perfectly diamagnetic. The magnetic levitation shown in Fig. 27.34 is a manifestation of the Meissner effect, wherein a magnet is supported over a superconductor by mutual repulsion between the magnet and currents in the superconductor.

As the strength of an applied magnetic field increases, so do the currents and resulting magnetic field of the superconductor. But beyond a **critical field**, the external magnetic field alters the superconducting state, and the superconductor no longer excludes magnetic flux. In **type I superconductors**, superconductivity ceases abruptly at the critical field (Fig. 37.23*a*). **Type II superconductors**, in contrast, have upper and lower critical fields, between which superconductivity gradually diminishes (Fig. 37.23*b*). At the lower critical field the material begins to allow flux penetration, and a regular array of nonsuperconducting regions forms, centered on magnetic field lines. These grow with increasing field, until at the upper critical field the superconducting regions vanish altogether.

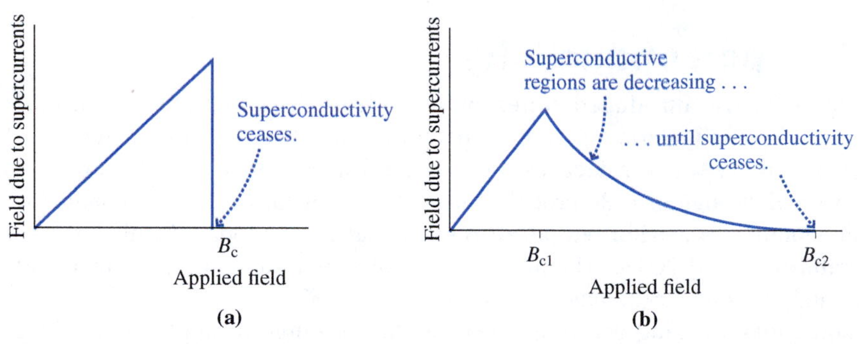

FIGURE 37.23 Responses of (a) type I and (b) type II superconductors to applied magnetic fields. B_c denotes the critical field.

Because electric currents generate magnetic fields, the critical field can limit the current-carrying capability of superconductors. Fortunately, type II superconductors have high enough upper critical fields to permit substantial currents. Type IIs tend to be alloys or complex mixtures, and include all the high-T superconductors. Critical fields of high-T superconductors are as high as 100 T; however, these materials are brittle ceramics and present engineering challenges to the fabrication of wires and other flexible conductors.

APPLICATION Maglev!

Passengers arriving at Shanghai's Pudong International Airport make the 30-km trip to a city metro station in only 7 minutes, "flying" on a magnetically levitated vehicle—a maglev—at speeds exceeding 400 km/h. The Shanghai system uses conventional electromagnets and electronic feedback circuits to keep the vehicle levitated a mere 1 cm above its guideway. But others are developing maglev systems that rely on superconducting magnets for both levitation and propulsion. Coils in the guideway carry alternating current, alternately pushing and pulling the vehicle's onboard magnet. In effect, vehicle and guideway become a linear electric motor, much like a conventional motor that's been "unwound" to produce straight-line motion. In superconducting systems, any deviation from perfect alignment with the guideway results in induced currents and, correspondingly, magnetic forces that act to keep the maglev vehicle centered in its guideway. Today's superconducting maglevs require onboard refrigeration systems, so development of a room-temperature superconductor would make maglev a much more attractive transportation alternative. The photo shows a Japanese superconducting maglev that has achieved speeds of 450 km/h.

Theories of Superconductivity

Superconductivity is a purely quantum-mechanical phenomenon; classical physics is totally inadequate to explain its existence. A successful theory of conventional low-T_c superconductors, called the **BCS theory** after its originators, was formulated in 1957 by

John Bardeen, Leon Cooper, and John Robert Schrieffer; the trio shared the 1972 Nobel Prize in Physics.

In BCS theory, superconductivity results from a quantum-mechanical pairing of electrons that leads to a lower-energy state in which electron pairs move through the crystal lattice with no energy loss to the ions, resulting in zero electrical resistance. The electron pairing involves one electron slightly deforming the ion lattice, with the second electron attracted by the slight positive charge of the deformed lattice (Fig. 37.24*a, b*). But the paired electrons aren't physically close; typically, a million other electrons, each paired with another distant electron, may lie between the two (Fig. 37.24*c*). The result of this long-range pairing is coherent motion of the conduction electrons that extends throughout the superconductor. Like well-choreographed dancers, the electrons all move together in a way that precludes energy loss to the ion lattice.

High-temperature superconductors aren't fully understood, although they almost certainly involve quantum-mechanical pairing of charge carriers. The mechanism of the pairing is less clear; one promising candidate involves magnetic interactions, although other mechanisms are under investigation. Superconductivity presents a continuing challenge to both theorists and experimentalists.

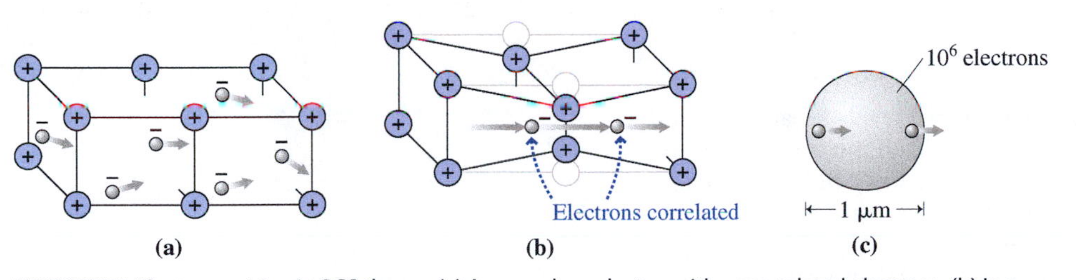

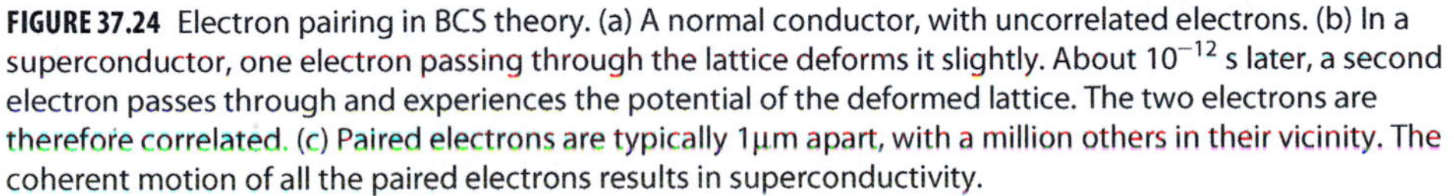

FIGURE 37.24 Electron pairing in BCS theory. (a) A normal conductor, with uncorrelated electrons. (b) In a superconductor, one electron passing through the lattice deforms it slightly. About 10^{-12} s later, a second electron passes through and experiences the potential of the deformed lattice. The two electrons are therefore correlated. (c) Paired electrons are typically 1μm apart, with a million others in their vicinity. The coherent motion of all the paired electrons results in superconductivity.

CHAPTER 37 SUMMARY

Big Idea

The big idea here is that quantum mechanics can explain the structure of molecules and solids as well as the atoms treated in Chapter 36. At this level we can't solve the Schrödinger equation for these many-particle systems, but we've argued—using energy and angular momentum quantization and the exclusion principle—that quantum effects are important in molecules and solids.

Key Concepts and Equations

Common types of molecular bonding include **ionic**, **covalent**, **hydrogen**, **van der Waals**, and **metallic** bonding. Whatever the bonding mechanism, a stable molecule is at the minimum of its potential-energy curve, shown below for H_2.

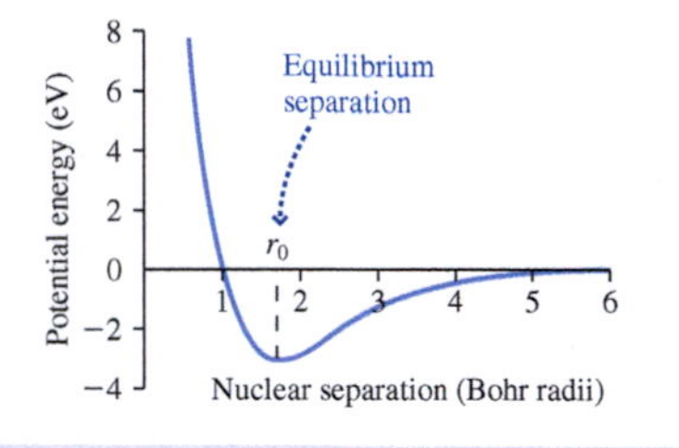

Molecules exhibit both rotational and vibrational energy, giving rise to a rich structure of quantized energy levels and the resulting spectra.

- Quantization of angular momentum leads to quantized rotational energy levels:

$$E_{\text{rot}} = \frac{\hbar^2}{2I} l(l + 1), l = 0, 1, 2, \ldots$$

- The vibrational energy levels are those of the harmonic oscillator:

$$E_{\text{vib}} = \left(n + \tfrac{1}{2}\right)\hbar\omega, n = 0, 1, 2, \ldots$$

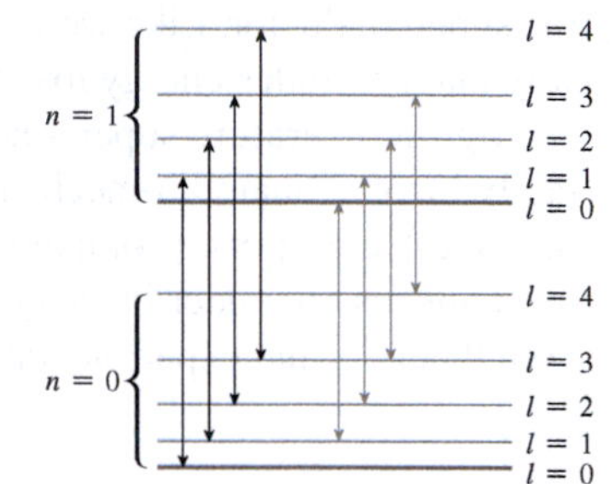

When atoms join to make solids, individual atomic energy levels separate to form **bands. Band theory** distinguishes conductors from insulators depending on whether the uppermost occupied band is partially or completely full, respectively. **Semiconductors** are like insulators, but with a much smaller band gap that permits thermal excitation of electrons into the conduction band.

In a metallic conductor, the energy of the highest occupied state at absolute zero is the **Fermi energy**.

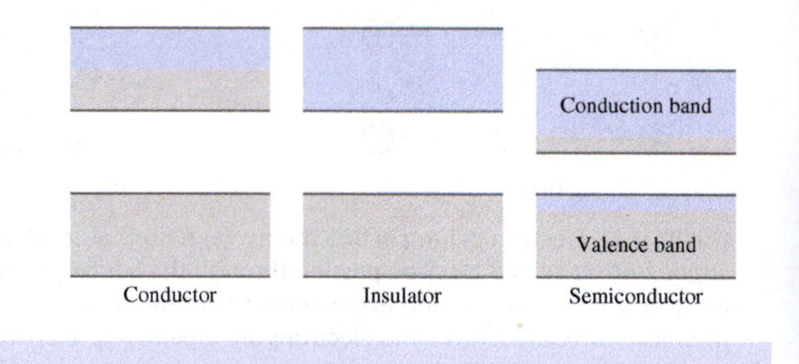

Applications

Superconductivity is a quantum-mechanical phenomenon that occurs at low temperatures and admits no classical explanation. Paired electrons move coherently through a superconductor without energy loss to the ion lattice, resulting in zero electrical resistance. Superconductors exclude magnetic fields—the **Meissner effect**—up to a **critical field** that destroys superconductivity, abruptly in **type I superconductors** and gradually in **type II superconductors**.

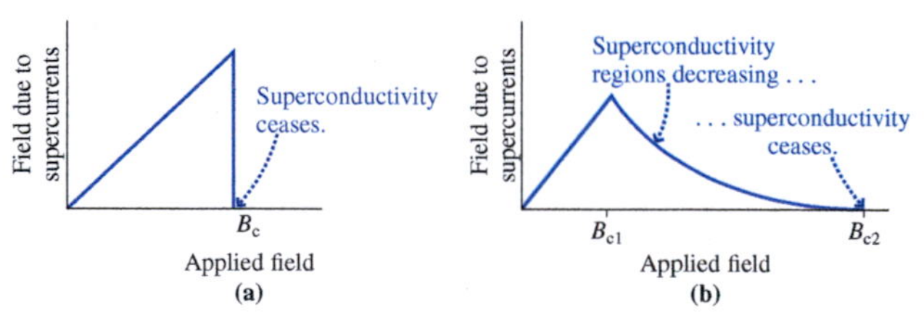

The band structure of doped semiconductors helps explain the one-way conduction of a *PN* junction, a phenomenon at the heart of modern electronics.

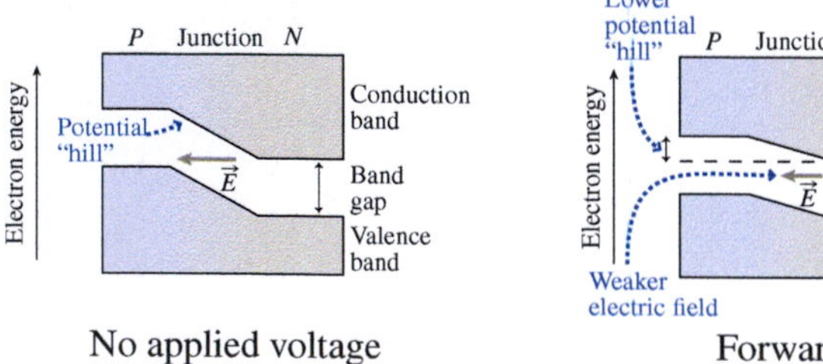

No applied voltage

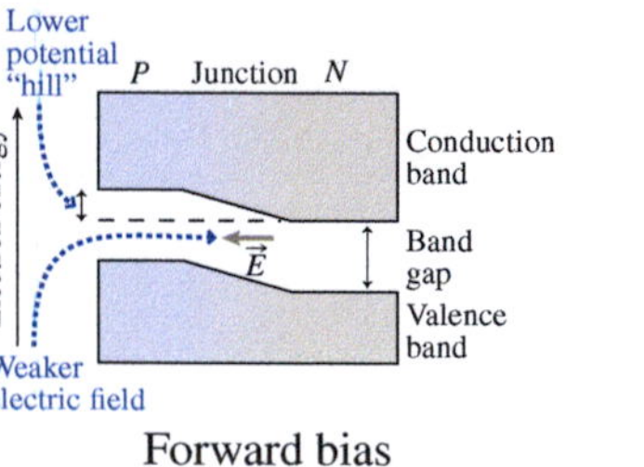

Forward bias

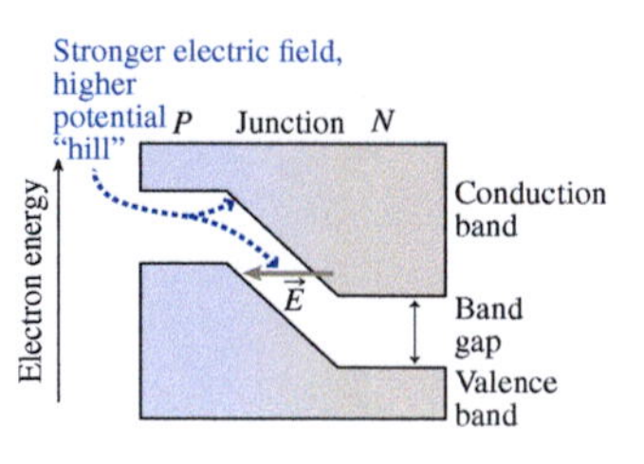

Reverse bias

MP For homework assigned on MasteringPhysics, go to www.masteringphysics.com

BIO *Biology and/or medicine-related problems* **DATA** *Data problems* **ENV** *Environmental problems* **CH** *Challenge problems*

For Thought and Discussion

1. If you push two atoms together to form a molecule, the exclusion principle results in a repulsive interaction between the atoms. How does this repulsion come about?
2. Why do ionically bonded materials have high melting points?
3. The electrostatic attraction between oppositely charged ions is what binds atoms in an ionic molecule. Is the electric force involved in covalent bonding? Explain.
4. Does it make sense to distinguish individual NaCl molecules in a salt crystal? What about individual H_2O molecules in an ice crystal? Explain.
5. Is it useful to think of the highest-energy electrons as "belonging" to individual atoms in an ionically bonded molecule? In a covalently bonded molecule?
6. What are the approximate relative magnitudes of the energies associated with electronic excitation of a molecule, with molecular vibration, and with molecular rotation?
7. Radio astronomers have discovered many complex organic molecules in interstellar space. Why were these discoveries made with radio telescopes and not optical telescopes?
8. In Fig. 18.17, why are rotational states excited at lower temperatures than vibrational states?
9. Would you expect solid hydrogen to conduct electricity? Why or why not?
10. The Fermi energy in metals is much higher than the thermal energy at typical temperatures. Why does this make the mean speed of conduction electrons nearly independent of temperature?
11. Why does the size of the band gap determine whether a material is an insulator or a semiconductor?
12. How would you expect the conductivity of an undoped semiconductor to depend on temperature? Why?
13. Name some technological innovations that might result from a room-temperature superconductor.
14. Suppose a room-temperature superconductor were discovered, but it had a very low critical field. In what way would this limit its practical applicability?
15. How do type I and type II superconductors differ?

Exercises and Problems

Exercises

Section 37.2 Molecular Energy Levels

16. Find the energies of the first four rotational states of the HCl molecule described in Example 37.1.
17. The rotational inertia of oxygen (O_2) is 1.95×10^{-46} kg·m^2. Find the wavelength of electromagnetic radiation needed to excite oxygen molecules to their first rotational excited state.
18. Find the wavelength of a photon emitted in the $l = 5$ to $l = 4$ transition of a molecule whose rotational inertia is 1.75×10^{-47} kg·m^2.
19. Photons of wavelength 1.68 cm excite transitions from the rotational ground state to the first rotational excited state in a gas. What's the rotational inertia of the gas molecules?
20. The classical vibration frequency for diatomic hydrogen (H_2) is 1.32×10^{14} Hz. Find the spacing between its vibrational energy levels.
21. The energy between adjacent vibrational levels in diatomic nitrogen is 0.293 eV. What's the classical vibration frequency of N_2?

Section 37.3 Solids

22. Use the 2.16-g/cm^3 density of NaCl to calculate the ionic spacing r_0 in the NaCl crystal. (*Hint:* Consult Appendix D.)
23. Express the 7.84-eV ionic cohesive energy of NaCl in kilocalories per mole of ions.
24. Lithium fluoride, LiF, has the same crystal structure as NaCl and therefore has essentially the same Madelung constant α. Its ionic cohesive energy is -10.5 eV and the value of n in Equation 37.4 is 6.25. Find the equilibrium ionic separation in LiF.
25. Find the wavelength of light emitted by a gallium phosphide (GaP) light-emitting diode. (*Hint:* See Table 37.1.)
26. What's the shortest wavelength of light that could be produced by electrons jumping the band gap in a material from Table 37.1? What is that material?
27. Which material in Table 37.1 would provide the longest wavelength of light in a light-emitting diode? What's that wavelength?
28. A common light-emitting diode is made with gallium arsenide phosphide (GaAsP) and emits red light at 650 nm. What's its band gap?

Problems

29. A molecule drops from the $l = 2$ to the $l = 1$ rotational level, emitting a 2.68-meV photon. If the molecule then drops to the rotational ground state, what energy photon will it emit?
30. A molecule absorbs a photon and jumps to the next higher rotational state. If the photon energy is three times what would be needed for a transition from the rotational ground state to the first rotational excited state, between what two levels is the transition?
31. Find an expression for the energy of a photon required for a transition from the $(l - 1)$th level to the lth level in a molecule with rotational inertia I.
32. A molecule with rotational inertia I undergoes a transition from the lth rotational level to the $(l - 1)$th level. Show that the wavelength of the emitted photon is $\lambda = 4\pi^2 Ic/hl$.
33. The rotational spectrum of diatomic oxygen shows spectral lines spaced 0.356 meV apart in energy. Find O_2's atomic separation. (*Hint:* See Example 37.1, and remember that the oxygen atoms have equal mass.)
34. Use the answer in the back of the book for Problem 59 to find the bond length in carbon monoxide (CO), given that excitation of the first rotational state requires photons of wavelength 2.59 mm.
35. For the HCl molecule of Example 37.2, determine (a) the energy of the vibrational ground state and (b) the energies of photons emitted in transitions among adjacent vibrational states, for the cases $\Delta l = +1$ and $\Delta l = -1$.
36. Diatomic deuterium has classical vibration frequency 9.35×10^{13} Hz and rotational inertia 9.17×10^{-48} kg·m^2. Find (a) the energy and (b) the wavelength of a photon emitted in a transition between the $n = 1, l = 1$ state and the $n = 0, l = 2$ state.

37. Carbon dioxide contributes to global warming because the triatomic CO_2 molecule exhibits many vibrational and rotational excited states, and transitions among them occur in the infrared region where Earth emits most of its radiation. Among the strongest IR-absorbing transitions is one that takes CO_2 from its ground state to the first excited state of a "bending" vibration and sets the molecule rotating in its first rotational excited state. The energy required for this transition is 82.96 meV. What IR wavelength does this transition absorb?
38. An oxygen molecule is in its vibrational and rotational ground states. It absorbs a photon of energy 0.19653 eV and jumps to the $n = 1, l = 1$ state. It then drops to $n = 0, l = 2$, emitting a 0.19546-eV photon. Find (a) the classical vibration frequency and (b) the rotational inertia of the molecule.
39. The internuclear spacing in diatomic hydrogen (H_2) is 74.14 pm. Find (a) the energy and (b) the wavelength of a photon emitted in a transition from the first rotational excited state to the ground state. (c) In what spectral region does this wavelength lie?
40. **BIO** Biological macromolecules are complex structures that exhibit many more vibrational modes than the diatomic molecules considered in this chapter. DNA has a low-frequency "breathing" mode whose associated photon wavelength is 330 μm. Find the corresponding (a) frequency and (b) photon energy in eV.
41. What wavelength of infrared radiation is needed to excite a transition between the $n = 0, l = 3$ state and the $n = 1, l = 2$ state in KCl, for which the rotational inertia is 2.43×10^{-45} kg·m^2 and the classical vibration frequency is 8.40 THz?
42. Find the wavelengths emitted in all allowed transitions between the first three rotational states in the $n = 1$ level to any states in the $n = 0$ level in H_2, whose rotational inertia and classical vibration frequency are 4.60×10^{-48} kg·m^2 and 3.69×10^{14} Hz, respectively.
43. Determine the constant n in Equation 37.4 for potassium chloride (KCl), which has the same crystal structure as NaCl and for which $r_0 = 0.315$ nm and $U_0 = -7.21$ eV.
44. A salt crystal contains 10^{21} sodium–chlorine pairs. How much energy would it take to compress the crystal to 90% of its normal size?
45. Lithium chloride, LiCl, has the same structure and therefore the same Madelung constant as NaCl. The equilibrium separation in LiCl is 0.257 nm, and $n = 7$ in Equation 37.4. Find the ionic cohesive energy of the LiCl crystal.
46. You're researching the possibility of storing radioactive waste in underground salt formations. In support of this idea, you'd like to demonstrate that salt is extremely resistant to compression. You differentiate Equation 37.4 to obtain an expression for the force on an ion in an ionic crystal, and then use your result to find the force on an ion in NaCl if the crystal were compressed to half its equilibrium spacing (see Example 37.3 for relevant parameters). You compare this with the electrostatic attraction at this compression. What do you find?
47. **CH** Integrating Equation 37.5 over all energies gives the total number of states per unit volume in a metal. Therefore, integrating from $E = 0$ to $E = E_F$—that is, over the occupied states only—gives the number of conduction electrons per unit volume. Carry out this integration to show that the electron number density is given by

$$n = \left(\frac{2^{9/2}\pi m^{3/2}}{3h^2}\right)E_F^{3/2}$$

48. The Fermi energy in aluminum is 11.6 eV. Use the result of Problem 47 to find the density of conduction electrons in aluminum.
49. Use the result of Problem 47 to determine the Fermi energy for calcium, which has 4.6×10^{28} conduction electrons per cubic meter.
50. Find (a) the speed associated with an electron whose kinetic energy is equal to copper's 7.00-eV Fermi and (b) the thermal speed of an electron at room temperature (293 K). (c) What does the difference between your calculated speeds tell you about whether a quantum or a classical model best describes copper's electrical conductivity?
51. The *Fermi temperature* is defined by equating the thermal energy kT to the Fermi energy, where k is Boltzmann's constant. Calculate the Fermi temperature for silver, for which $E_F = 5.48$ eV, and compare it with room temperature.
52. Photons with energy lower than a semiconductor's band gap aren't readily absorbed by the material, so a measurement of absorption versus wavelength gives the band gap. An absorption spectrum for silicon shows no absorption for wavelengths longer than 1090 nm. Use this information to calculate the band gap in silicon, and verify its value in Table 37.1.
53. Calculate the median wavelength λ_{median} for sunlight, treating the Sun as a 5800-K blackbody (see Equation 34.2b). Use your result to decide whether zinc selenide, with band gap 3.6 eV, would make a good photovoltaic cell.
54. Pure aluminum, which superconducts below 1.20 K, exhibits a critical field of 9.57 mT. Find the maximum current that can be carried in a 30-gauge (0.255-mm-diameter) aluminum superconducting wire without the field from that current exceeding the critical field. (*Hint:* Where is the field greatest? Consult Example 26.8.)
55. The critical field in a niobium–titanium superconductor is 15 T. What current in a 5000-turn solenoid 75 cm long will produce a field of this strength?
56. The transition from the ground state to the first rotational excited state in diatomic oxygen (O_2) requires 356 μeV. At what temperature would the thermal energy kT be sufficient to set diatomic oxygen into rotation? Would you ever find diatomic oxygen exhibiting the specific heat of a monatomic gas at normal pressure?
57. **BIO** *Green fluorescent protein* (GFP) is a substance that was first extracted from jellyfish; variants are used to "tag" biological molecules for study. The original "wild" GFP absorbs 395-nm light, undergoing an upward transition to an excited state. Movement of a proton within the protein then excites it to 2.44 eV above the ground state. Photons emitted in the subsequent downward transition to the ground state provide a visual indication of the GFP's location as seen in a microscope. What's the wavelength of these photons?
58. The density of rubidium iodide (RbI) is 3.55 g/cm^3, and its ionic cohesive energy is -145 kcal/mol. Determine (a) the equilibrium separation and (b) the exponent n in Equation 37.4 for RbI.
59. **CH** You're troubled that Example 37.1 neglects the mass of the hydrogen, and you wonder how much error this introduces. So you consider a diatomic molecule consisting of different atoms with masses m_1 and m_2, separated by a distance R, and derive an expression for the molecule's rotational inertia about its center of mass. You then calculate a more accurate value for the HCl bond length in Example 37.1. Your results?
60. What fraction of conduction electrons in a metal at absolute zero have energies less than half the Fermi energy?
61. **CH** The Madelung constant (Section 37.3) is notoriously difficult to calculate because it's the sum of an alternating series of nearly equal terms. But it can be calculated for a hypothetical one-dimensional crystal consisting of an evenly spaced line of

alternating positive and negative ions (Fig. 37.25). Show that the potential energy of an ion in this "crystal" can be written as

$$U = -\alpha \frac{ke^2}{r_0}$$

where the Madelung constant α has the value $2 \ln 2$.

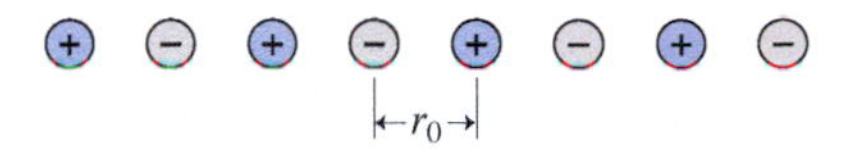

FIGURE 37.25 Problem 61

62. The lower-energy states in a covalently bound diatomic molecule can be found approximately from the so-called Morse potential $U(r) = U_0(e^{2(r-r_0)/a} - e^{-2(r-r_0)/a})$, where r is the atomic separation and U_0, r_0, and a are constants determined from experimental data. Calculate dU/dr and d^2U/dr^2 to show that U has a minimum, and find expressions for (a) U_{min} and (b) the separation r_{min} at the minimum energy.

63. **CH** (a) Count the number of electron states $N(E)$ with energy equal to or less than E in Equation 35.8 by finding the volume available to such states in the space with Cartesian coordinate axes n_x, n_y, n_z. (*Hint*: Consider each set of positive integers, at the corner of a unit cube, and that lies inside a radius $\sqrt{n_x^2 + n_y^2 + n_z^2}$, and remember that there are two spin values per state.) (b) Differentiate $N(E)$ with respect to E to obtain Equation 37.5.

64. **CH** Use Equation 37.5 to calculate the average energy of a conduction electron at $T = 0$ in terms of the Fermi energy.

65. **BIO** You're designing a new medical MRI imager. The design calls for a long solenoid wound with 75 turns per meter of niobium–titanium superconductor. The upper critical field for your particular Nb-Ti alloy is 12 T. To avoid a disastrous loss of superconductivity (see Example 27.9), you want to limit the actual field to half the upper critical field. What maximum current do you specify for your device?

66. **COMP** Squeezing a particular solid in all directions reduces the interatomic spacing to 76.6% of its equilibrium value; the result is that the solid's total potential energy becomes zero. Find the value of the exponent n in Equation 37.4 for this solid.

67. **DATA** The table below shows the wavelengths of photons emitted when identical molecules drop from the lth rotational level to the $(l-1)$th level. Find quantities that, when plotted, should yield a straight line. Make your plot, determine a best-fit line, and use the result to find the rotational inertia of the molecules.

Initial state, l	2	3	4	5	6
Wavelength, λ (mm)	0.24	0.17	0.12	0.095	0.078

Passage Problems

Photovoltaic (PV) cells convert sunlight energy directly into electricity, with no moving parts (recall Fig. 37.21). In a PV cell, photons incident on a semiconductor *PN* junction promote electrons to the conduction band, producing electron-hole pairs and driving current through an external circuit (Fig. 37.26). Commercially available PV cells are 15–20% efficient, meaning they convert this fraction of incident sunlight into electrical energy; the theoretical maximum efficiency is around 33% for silicon-based PV cells. An important limitation on PV efficiency is the relation between the solar spectrum and PV cells' semiconductor band-gap energy. For silicon, the band gap is 1.14 eV; photons with less energy can't promote electrons to the conduction zone and are thus unavailable for the PV energy conversion. Conversely, photons with more than the band-gap energy give up their excess energy as heat, also reducing PV efficiency.

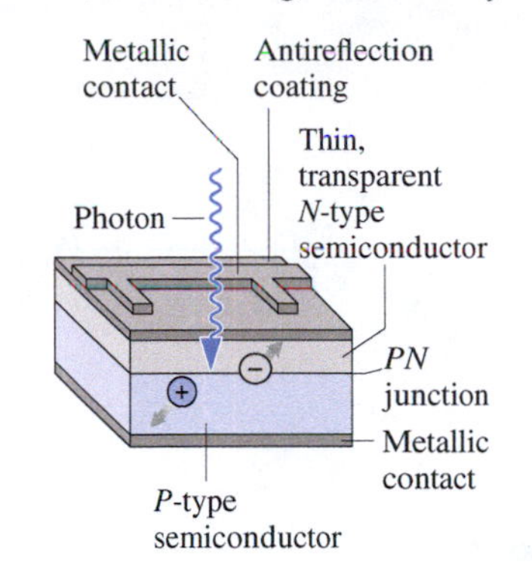

FIGURE 37.26 Operation of a photovoltaic cell, showing a solar photon producing an electron-hole pair at the *PN* junction (Passage Problems 68–71).

68. Problem 53 shows that the median wavelength in the solar spectrum is 710 nm, at the visible-IR boundary. What percentage of the incident solar energy can a silicon PV cell absorb? (*Hint*: See Exercise 36.31.)
    a. about 25%
    b. about 50%
    c. about 75%

69. How does the percentage of the number of incident solar photons that a PV cell absorbs compare with the energy percentage in the preceding problem?
    a. It's less than the energy percentage.
    b. It's the same as the energy percentage.
    c. It's more than the energy percentage.

70. Making PV cells with a semiconductor whose band gap is lower than silicon's will
    a. increase the fraction of solar energy absorbed while decreasing the amount of absorbed energy lost as heat.
    b. increase both the fraction of solar energy absorbed and the amount of absorbed energy lost as heat.
    c. decrease the fraction of solar energy absorbed while increasing the amount of absorbed energy lost as heat.
    d. decrease both the fraction of solar energy absorbed and the amount of absorbed energy lost as heat.

71. One way to improve PV efficiency is to make multilayer cells with several *PN* junctions using semiconductors with different band gaps. For a multilayer PV cell to be effective,
    a. the junction with the largest band gap should be closest to the top of the PV cell.
    b. the junction with the largest band gap should be closest to the bottom of the PV cell.
    c. the largest band gap should correspond to infrared wavelengths.
    d. the smallest band gap should correspond to ultraviolet wavelengths.

Answers to Chapter Questions

Answer to Chapter Opening Question

The Schrödinger equation.

Answers to GOT IT? Questions

37.1 (c)
37.2 rotational energy
37.3 (c) they're equal because both areas represent the total number of electrons

Nuclear Physics

What You Know

- You know that atoms are made of nuclei surrounded by electrons.
- You've seen the quantization of angular momentum.
- You understand the electric force between charges.
- You're familiar with energy in its various forms.
- You've seen relativistic mass–energy equivalence.

What You're Learning

- You'll learn to characterize nuclei by atomic number and mass number and to explain the difference between isotopes of the same element.
- You'll be able to distinguish between stable and unstable nuclei and interpret the chart of the nuclides.
- You'll be able to calculate the sizes and spin angular momenta of nuclei.
- You'll learn about three common types of nuclear reactions and to write equations describing each.
- You'll learn to quantify radioactivity and to describe its time dependence using half-life.
- You'll see the curve of binding energy and learn how it explains energy released in nuclear fission and nuclear fusion.
- You'll explore the technology of nuclear fission, including several types of nuclear reactors.
- You'll learn how nuclear fusion powers the Sun and other stars, and you'll explore current efforts to harness fusion as a terrestrial energy source.

How You'll Use It

- Depending on where you live, you may get a substantial portion of your electrical energy from nuclear fission.
- At some point in your life, you're likely to undergo medical diagnostic procedures based in nuclear physics.
- You're likely to eat foods that have been preserved by irradiation.
- Some decades in the future, humankind may learn to harness the almost inexhaustible energy resources of nuclear fusion.

In Chapters 36 and 37 we explored atomic structure and saw how atoms join to form molecules and solids. Now we turn inward, to the atomic nucleus. Since Ernest Rutherford and his colleagues discovered the nucleus in 1911, we've known that all the atom's positive charge and nearly all its mass are concentrated in a tiny nuclear region only about 10^{-5} of the atom's

Smoke pours from a damaged reactor at Japan's Fukushima Dai-ichi nuclear power plant. How dependent is humankind on nuclear power?

diameter. By 1920 Rutherford had proposed that nuclei beyond hydrogen contain neutral as well as positive particles, and today we know the nucleus is a composite of positive protons and neutral neutrons—collectively called **nucleons**. As we've seen, the uncertainty principle implies high minimum energies for particles confined in small regions, so we can infer that the nucleus is a huge energy repository. We'll conclude this chapter with a look at humankind's attempts to harness that energy.

PhET: Rutherford Scattering

38.1 Elements, Isotopes, and Nuclear Structure

You saw in Chapter 36 how the number of electrons determines an atom's shell structure and therefore its chemical behavior. It's the number of protons in the nucleus—the **atomic number**, Z—that, in turn, determines the number of electrons in a neutral atom. That means all nuclei with the same Z belong to the same element.

Isotopes and Nuclear Symbols

Nuclei of the same element can, however, have different numbers of neutrons. That's because neutrons don't affect the nuclear charge and therefore have negligible influence on chemical behavior. Nuclei of the same element with different numbers of neutrons are distinct **isotopes**. We call the total number of nucleons the **mass number**, A. Specification of the atomic number Z and mass number A then fully describes a nucleus. Figure 38.1 shows the conventional symbolism used in describing nuclei: the element symbol with a preceding subscript for Z and superscript for A. Actually, the atomic number and symbol are redundant. To be helium (He), for example, *means* to have two protons and therefore $Z = 2$; to be uranium *means* $Z = 92$. Sometimes, therefore, we write helium-4, He-4, or ^{4}He to mean the same thing as $^4_2$He.

Elements typically have several naturally occurring isotopes; a few are shown in Fig. 38.1. Most hydrogen has a single proton in its nucleus, but about one in 6500 hydrogen atoms is deuterium ($^2_1$H), whose nucleus contains a proton and a neutron. Most oxygen is $^{16}_8$O, but O-17 and O-18 also occur naturally; their ratios in polar ice cores provide valuable information about past climates. Most uranium is $^{238}_{92}$U, but 0.7% is the U-235 that's used in fission reactors and weapons—hence the great concern about the proliferation of uranium-enrichment facilities to increase the proportion of U-235. Incidentally, the atomic masses listed in the periodic table are averages that reflect the natural abundances of an element's several isotopes. Most elements also have short-lived radioactive isotopes that don't usually occur naturally but can be produced through nuclear reactions; more on these later.

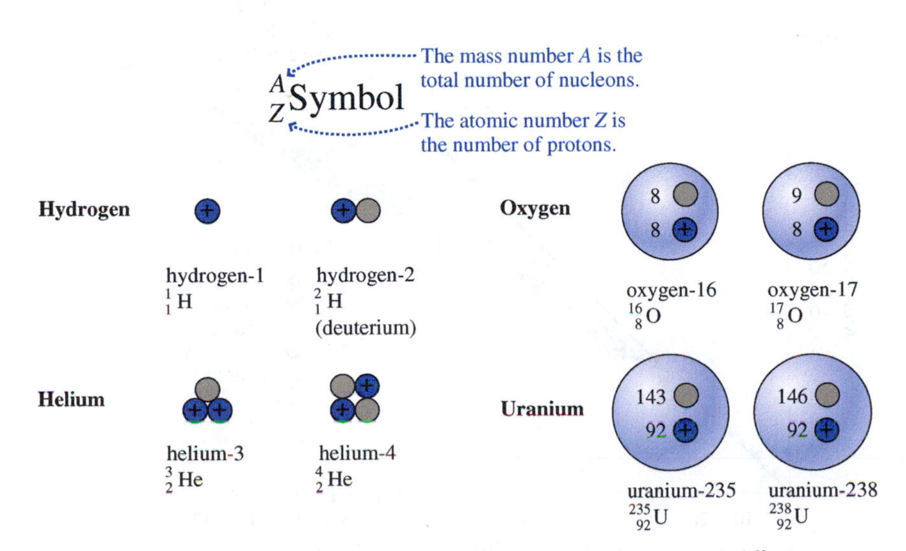

FIGURE 38.1 Isotopes of a given element have the same number of protons but different numbers of neutrons.

GOT IT? 38.1 Determine the number of protons and neutrons in these nuclei: (1) ${}^{12}_{6}\text{C}$; (2) ${}^{15}_{8}\text{O}$; (3) ${}^{57}_{26}\text{Fe}$; (4) ${}^{239}_{94}\text{Pu}$.

The Nuclear Force

Given the electrical repulsion of the protons, there must be another force acting attractively to bind the nuclear constituents. Throughout much of the 20th century, this **nuclear force** was considered fundamental, but we now recognize it as a manifestation of a more fundamental force between the quarks that make up neutrons and protons. We'll explore quarks and their interactions in Chapter 39.

The attractive nuclear force acts between all nucleons—neutrons and protons, protons and protons, neutrons and neutrons. It's very strong at distances of less than a few femtometers (10^{-15} m), but falls approximately exponentially with distance—more rapidly than the inverse-square falloff of the electric force. The attractive nuclear force therefore dominates between two neighboring protons, but electrical repulsion becomes dominant for more widely separated protons. The structure of the nucleus is determined, to a first approximation, by the interplay between the weaker but long-range electric force and the stronger but shorter-range nuclear force.

Stable Nuclei

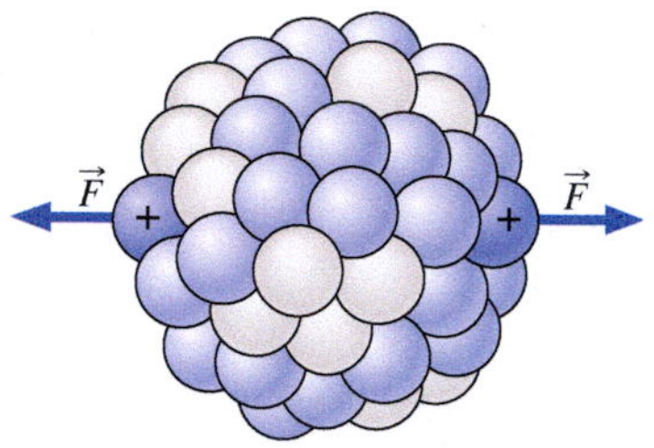

Not every combination of protons and neutrons will stick together indefinitely. Too many protons, and electrical repulsion wins out; sooner or later the nucleus decays by emitting a chunk of nuclear material (more details in Section 38.2). In larger nuclei most protons are far apart and therefore experience electrical repulsion more strongly than nuclear attraction (Fig. 38.2). To hold these nuclei together therefore requires more neutrons, which contribute attractive nuclear force but not electrical repulsion. So larger nuclei tend to have a higher ratio of neutrons to protons. Even this effect has its limits, though, and the result is that there are no stable nuclei for $Z > 83$.

FIGURE 38.2 Two widely separated protons in a large nucleus experience significant electrical repulsion and negligible nuclear attraction.

Too many neutrons also make a nucleus unstable. That's because the exclusion principle requires extra neutrons to go into higher energy states, making individual particles more likely to escape the nucleus. Furthermore, the neutron itself is an unstable particle; an isolated neutron decays spontaneously into a proton, an electron, and an elusive particle called a neutrino. This decay is suppressed in stable nuclei, but occurs if there are too many neutrons.

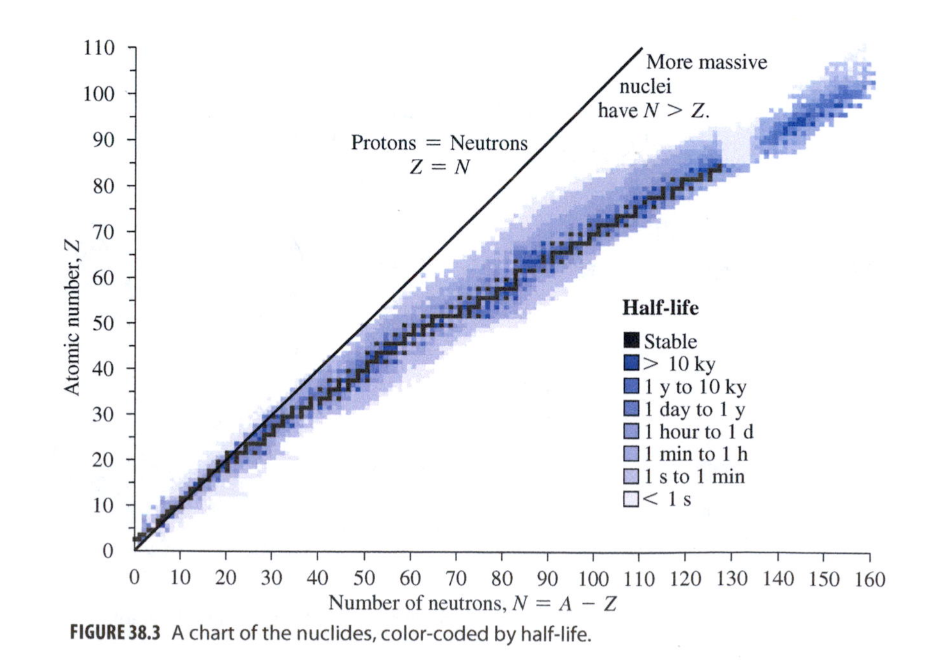

FIGURE 38.3 A chart of the nuclides, color-coded by half-life.

The delicate balance between neutrons and protons results in about 400 known stable nuclei, collectively called **nuclides**. Figure 38.3 is a **chart of the nuclides**, showing the stable nuclei, along with many unstable ones, on a chart of atomic number Z versus neutron number $N = A - Z$. The chart shows that lighter nuclei tend to have equal numbers of protons and neutrons, but that heavier nuclei invariably have more neutrons to compensate for the increasing electrical repulsion of their widely separated protons.

Nuclear Size

Unlike atomic electrons in their widely separated orbitals, nucleons pack tightly into the nucleus. Studies show that most nuclei are spherical, with the **nuclear radius**—defined as the radius at which the density has fallen to half its central value—given approximately by

$$R = R_0 A^{1/3} \tag{38.1}$$

where $R_0 = 1.2$ fm and A is the mass number. This cube-root dependence is what we should expect for a tightly packed sphere whose volume is proportional to the number A of its constituent particles, as suggested in Fig. 38.2. This tight packing also suggests that all nuclei have approximately the same density, on the order of $10^{17}\,\text{kg/m}^3$. A teaspoon of nuclear matter has a mass roughly equal to the mass of the Rock of Gibraltar! That absurdly high density reaffirms our picture of the complete atom as mostly empty space with its mass concentrated in a tiny nucleus.

Nuclear Spin

In Chapter 36 we noted the important role of electron spin in atomic structure. Protons and neutrons are, like electrons, spin-$\frac{1}{2}$ particles. The spins of individual nucleons, combined with any angular momentum associated with their motions within the nucleus, give the nucleus a quantized spin angular momentum I that obeys the same rules we've seen for other quantized angular momenta:

$$I = \sqrt{i(i+1)}\hbar \tag{38.2}$$

Here i, the nuclear spin quantum number, is a multiple of one-half. The component of I on a given axis is also quantized, just like other angular momenta, according to $I_z = m_i\hbar$, where m_i ranges from $-i$ to i in steps of 1.

The spin quantum number i is an even or odd multiple of one-half depending on whether the number of nucleons is even or odd. This means that nuclei with even values of A are bosons, particles with integer spin that don't obey the exclusion principle. Odd-A nuclei, in contrast, have half-integer spin and are fermions that do obey the exclusion principle. This distinction can lead to profound differences in physical behavior between isotopes of the same element. Helium-4, for example, becomes superfluid at low temperatures, meaning it flows without any viscosity. That's possible because helium-4 nuclei are bosons that can all occupy the same quantum state. Similar superfluidity doesn't occur in fermionic helium-3, although at extremely low temperatures He-3 nuclei themselves pair to form spin-1 particles that do make a superfluid.

The angular momentum of the nucleus results in a nuclear magnetic dipole moment, usually expressed in units of the **nuclear magneton**, $\mu_N = e\hbar/2m_p = 5.05\times10^{-27}$ J/T, where m_p is the proton mass. The proton itself has a magnetic moment whose component on a given axis takes either of the values $\pm 2.793\,\mu_N = \pm 1.41\times10^{-26}$ J/T—a value that's usually listed as "the magnetic moment of the proton," although it's actually the component. Interaction of the nuclear magnetic moment with magnetic fields alters very slightly the energy levels of the atom—although the effect is much smaller than with atomic electrons because the higher proton mass makes for a much smaller magnetic moment. In hydrogen, for example, the proton can have either of two spin orientations relative to the magnetic field due to the electron, and the result is **hyperfine splitting** of the ground state into two levels a mere 5.9 μeV apart (Fig. 38.4). Transitions between these levels result in a spectral line at a radio wavelength of 21 cm. Radio astronomers use this line to detect interstellar clouds of neutral hydrogen.

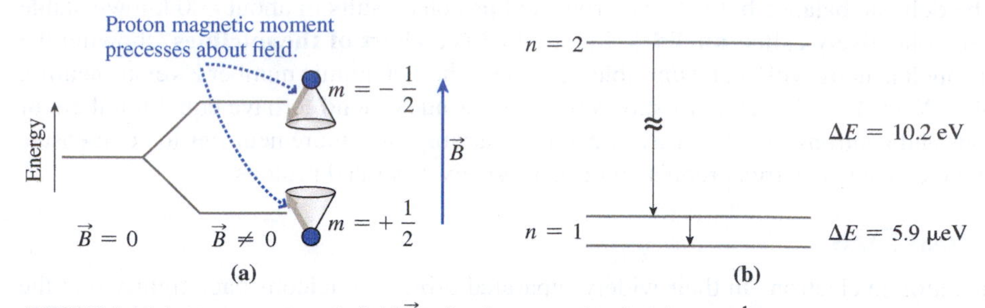

FIGURE 38.4 (a) A nonzero magnetic field $\vec{B}$ splits the energy level of the spin-$\frac{1}{2}$ proton into two levels. (b) The two possible orientations of the proton in the magnetic field of the electron split the hydrogen ground state into two levels 5.9 μeV apart.

MP

PhET: Simplified MRI

APPLICATION Nuclear Magnetic Resonance and MRI

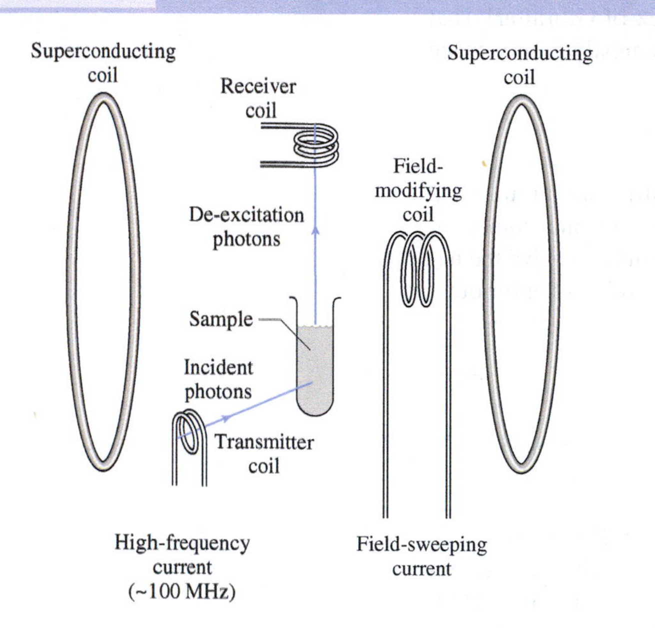

Putting nuclei in an external magnetic field creates two possible energy states, as suggested in Fig. 38.4*a*, depending on whether the nuclear magnetic moments are more nearly parallel or antiparallel to the field. Applying electromagnetic radiation with the appropriate photon energy will flip nuclei into the higher energy state. But because nuclei also experience magnetic fields from the electrons moving around them, the exact energy required is extremely sensitive to the details of the electron distribution—that is, to the surrounding molecular structure.

Nuclear magnetic resonance (NMR) uses this nuclear spin flipping to determine the structure of chemical compounds. In an NMR spectrometer, shown schematically in the figure, the sample under analysis is placed in a uniform magnetic field B, usually from superconducting coils. A smaller coil carries AC current at a frequency f corresponding to photon energy hf that would flip the spin of an isolated nucleus in the field B. The coil emits electromagnetic waves, and if the nuclei absorb the associated photons, then they flip into their higher states and drop back, emitting radiation of frequency f in the process. A receiver coil detects this radiation.

Because of the extra magnetic effect of the surrounding electrons, nuclei won't generally flip at the exact frequency and field B. So the field is varied until the superposition of the applied field and the electron-generated field is exactly right. This condition of magnetic resonance then produces the up/down spin flips that generate a signal in the receiver coil. Scanning the field through a range of values detects nuclei in different electron environments, and from this information scientists can deduce the molecular structure.

Nuclear magnetic resonance with protons (H nuclei) is the basis of **magnetic resonance imaging** (MRI), a widely used medical procedure. In MRI, a patient is placed inside a large solenoid whose field varies slightly with position. That makes the magnetic resonance frequency a function of position, and thus the resonance signal can be used to localize the protons undergoing magnetic resonance. A computer then uses the resonance information to construct an image. Most of the MRI signal comes from fat and water, making MRI especially good at imaging soft tissue that doesn't show well in X rays. The photo shows an MRI image of a human head and upper torso; soft-tissue structures including the brain are clearly visible.

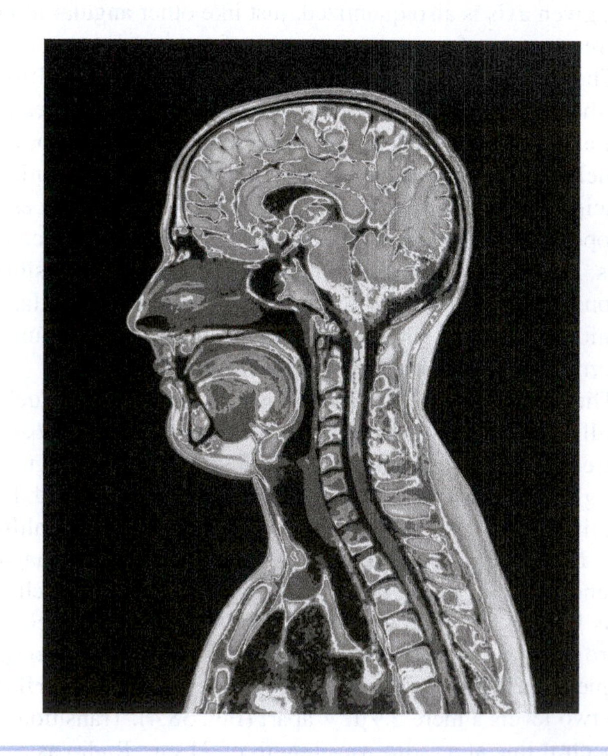

EXAMPLE 38.1 Nuclear Spins: Finding the MRI Frequency

The MRI solenoid of Example 26.10 produces a 1.50-T magnetic field. What frequency should be used to drive the transmitter coil in this MRI device?

INTERPRET MRI is an implementation of nuclear magnetic resonance using protons (see the Application), so we're being asked for the frequency corresponding to photons that will flip a proton in a 1.50-T magnetic field.

DEVELOP We need to calculate the necessary photon energy and then use $E = hf$ to find the corresponding frequency. We've just seen that the proton acts like a magnetic dipole whose component along the field is $\mu_p = \pm 1.41 \times 10^{-26}$ J/T. Equation 26.16 gives the energy of a magnetic dipole: $U = -\vec{\mu} \cdot \vec{B}$. Here we're given the component of the magnetic moment $\vec{\mu}$ along the field, so our two energies become $U = \pm \mu_p B$, where the signs correspond to the two possible spin orientations. A spin flip changes a proton's energy from $+\mu_p B$ to $-\mu_p B$, so our plan is to find the energy difference between these levels, equate it to the photon energy hf, and solve for f.

EVALUATE We have $E = \mu_p B - (-\mu_p B) = 2\mu_p B$, so

$$f = \frac{E}{h} = \frac{2\mu_p B}{h} = \frac{(2)(1.41 \times 10^{-26}\,\text{J/T})(1.50\,\text{T})}{6.63 \times 10^{-34}\,\text{J}\cdot\text{s}} = 63.8\,\text{MHz}$$

ASSESS This frequency is in the radio region of the electromagnetic spectrum, consistent with the diagram in the Application showing the use of coils and currents, and the approximate transmitter coil frequency of 100 MHz. ■

Models of Nuclear Structure

We've seen how the right ratio of neutrons to protons is essential for stable nuclei, and why that ratio increases for larger nuclei. Figure 38.3, the chart of the nuclides, summarizes this information. But take a closer look at that figure: There are more stable nuclei for even values of Z, and some—those with the so-called **magic numbers** 2, 8, 20, 28, 50, 82, and 126 protons or neutrons—have many more stable nuclei. Why?

Answering this question and explaining the decay mechanisms and lifetimes of unstable nuclei require a theory of nuclear structure. There still is no complete nuclear theory, analogous to the atomic theory of Chapter 36, that explains all aspects of all nuclei. Our still-imperfect knowledge of the nuclear force, and the tight packing of nucleons, render useless a simple two-particle model like the one we used for hydrogen. Instead, nuclear physicists resort to several models to explain different aspects of nuclear structure. Together, these models provide a good understanding of the nucleus and accurately predict nuclear properties, although not with the precision available in atomic physics.

The **liquid-drop model** provides a reasonable approximation for heavier nuclei, whose many nucleons behave somewhat like the molecules in a drop of liquid. A liquid-drop nucleus can rotate, vibrate, and change shape as long as its volume doesn't change, and the resulting quantized energy levels predict nuclear gamma-ray spectra that are in good agreement with observation. The liquid-drop model also helps explain nuclear fission, as you'll see in Section 38.4. But it can't account for the dramatic effects of small changes in nucleon number, particularly the role of the magic numbers.

The **nuclear shell model**, advanced in the late 1940s by physicists Maria Goeppert Mayer and J. Hans Jensen, gives the nucleus a shell structure similar to that of atoms. The shells occur because neutrons and protons obey the exclusion principle, and the magic numbers correspond to closed-shell configurations analogous to the electronic structure of inert gases. Closed-shell nucleons are tightly bound, making a magic nucleus particularly stable. Additional nucleons beyond a closed shell stay largely on the outskirts of the nucleus, where they're more readily excited to higher energy levels. Neutrons and protons behave independently in the shell model, and each has its own set of quantum numbers. Closed-shell structure therefore occurs with magic numbers of either protons or neutrons. Some nuclei, like $^{40}_{20}\text{Ca}$ ($Z = 20, N = 20$), are "doubly magic" and show exceptional stability.

The **collective model**, advanced by Niels Bohr's son Aage, combines aspects of the liquid-drop and shell models, emphasizing the collective quantum-mechanical behavior of the nucleons. One remarkable prediction of the collective model is that larger, nonmagical nuclei may be more stable if they take nonspherical shapes.

Active areas of nuclear-structure research involve the creation and exploration of exceptionally heavy or neutron-rich nuclei. The creation of elements 115 and 116 in the early 2000s and element 117 in 2010 suggests that physicists are approaching a region of longer-lived nuclei dubbed the "island of stability," which may be associated with a new magic neutron number of 184. And the recent creation of relatively stable silicon-42 implies that silicon's atomic number $Z = 14$ becomes magic in this neutron-bloated ($N = 28$) species. Until we have a complete nuclear theory, discoveries like these will continue to challenge physicists with nuclear surprises.

38.2 Radioactivity

MP

PhET: Radioactive Dating Game

In 1896 Henri Becquerel of Paris noticed that a photographic plate stored near uranium compounds became fogged, as though exposed to invisible rays. Becquerel had discovered **radioactivity**, wherein some substances spontaneously emit high-energy particles or photons. Marie and Pierre Curie promptly began a thorough exploration of the phenomenon, for which Marie Curie coined the name "radioactivity." The Curies shared the 1903 Nobel Prize in Physics with Becquerel, and Marie Curie won the 1911 Nobel Prize in Chemistry for her discovery of polonium and radium.

Decay Rate and Half-Life

Radioactivity results from the decay of unstable nuclei, a process that occurs at vastly differing rates in different isotopes. The number of decays per unit time is the **activity** of a radioactive sample; the SI unit of activity is the **becquerel** (Bq), equal to one decay per second. An older unit, the **curie** (Ci), is 3.7×10^{10} Bq and is approximately the activity of 1 gram of radium-226. For a given isotope, activity is proportional to the number N of nuclei present. N decreases as nuclei decay, so we can write

$$\frac{dN}{dt} = -\lambda N$$

where λ is the **decay constant**. As we've seen with discharging capacitors and decaying inductor currents, this differential equation is a prescription for exponential decay. We solve it the same way, multiplying both sides by dt/N and integrating:

$$\int_{N_0}^{N} \frac{dN}{N} = -\lambda \int_0^t dt$$

where N_0 is the initial number of nuclei at time $t = 0$. Evaluating the integrals gives $\ln(N/N_0) = -\lambda t$ or, exponentiating each side and using $e^{\ln x} = x$:

$$N = N_0 e^{-\lambda t} \tag{38.3a}$$

Equation 38.3a shows that the decay constant λ is a measure of the exponential decay rate. We can also interpret λ as the probability that a given atom will decay in a 1-s time interval. Another convenient measure of exponential decay is the **half-life**, $t_{1/2}$, defined as the time for half the nuclei in a given sample to decay. If we start with N_0 nuclei at time $t = 0$, then at a later time t the number of nuclei remaining will be

$$N = N_0 2^{-t/t_{1/2}} \quad \text{(radioactive decay)} \tag{38.3b}$$

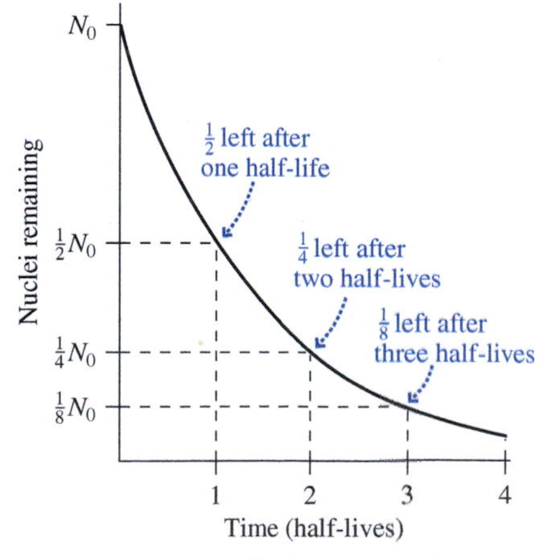

FIGURE 38.5 Exponential decay of a radioactive sample.

As you can show in Problem 54, $t_{1/2}$ and λ are related by $t_{1/2} = \ln 2/\lambda \simeq 0.693/\lambda$. Figure 38.5 is a graph of Equation 38.3b. Since activity and number of nuclei are proportional, both decline with the same half-life, as described in Equation 38.3b. Table 38.1 lists some significant radioisotopes and their half-lives.

Table 38.1 Selected Radioisotopes

Isotope	Half-life	Decay Mode	Comments
Carbon-14 ($^{14}_{6}C$)	5730 years	β^-	Used in radiocarbon dating
Iodine-131 ($^{131}_{53}I$)	8.04 days	β^-	Fission product abundant in fallout from nuclear weapons and reactor accidents; damages thyroid gland
Oxygen-15 ($^{15}_{8}O$)	2.03 minutes	β^+	Short-lived oxygen isotope used for PET scans
Potassium-40 ($^{40}_{19}K$)	1.25×10^9 years	β^-	Comprises 0.012% of natural potassium; dominant radiation source within the normal human body; used in radioisotope dating
Plutonium-239 ($^{239}_{94}Pu$)	24,110 years	α	Fissile isotope used in nuclear weapons
Radium-226 ($^{226}_{88}Ra$)	1600 years	α	Highly radioactive isotope discovered by Marie and Pierre Curie; results from decay of $^{238}_{92}U$
Radon-222 ($^{222}_{86}Rn$)	3.82 days	α	Radioactive gas formed naturally in decay of $^{226}_{88}Ra$; seeps into buildings, where it may cause serious radiation exposure
Strontium-90 ($^{90}_{38}Sr$)	29 years	β^-	Fission product that behaves chemically like calcium; readily absorbed into bones
Technetium-99m ($^{99m}_{43}Tc$)	6.006 hours	γ	Metastable excited state of Tc-99, widely used in medical diagnostics.
Tritium ($^{3}_{1}H$)	12.3 years	β^-	Hydrogen isotope used in biological studies and to enhance yields of nuclear weapons
Uranium-235 ($^{235}_{92}U$)	7.04×10^8 years	α	Fissile isotope comprising 0.72% of natural uranium; used as reactor fuel and in crude nuclear weapons
Uranium-238 ($^{238}_{92}U$)	4.46×10^9 years	α	Predominant uranium isotope; cannot sustain a chain reaction

EXAMPLE 38.2 Radioactive Decay: Fukushima Fallout

The 2011 tsunami-induced disaster at Japan's Fukushima Dai'ichi nuclear power plant spread radioactive fallout over the surrounding region and adjacent ocean. One isotope of particular concern was iodine-131, which is absorbed by the thyroid gland and can cause thyroid cancer. Shortly after the radiation releases began, the city of Iwaki, some 90 km from the Fukushima nuclear plant, recorded I-131 activity of 980 Bq/kg in milk. How long would such milk need to be held back from consumption in order to meet the 300-Bq/kg Japanese safety standard for milk?

INTERPRET This is a problem about radioactive decay. We're given the initial activity per kilogram of milk, and we need to find how long it takes for that to decay to the given safety standard. Using Table 38.1, we identify I-131's half-life as 8.04 days.

DEVELOP Equation 38.3b, $N = N_0 2^{-t/t_{1/2}}$, describes the decline in both the number of radioactive nuclei and their radioactivity. The equation shows that after n half-lives, activity drops to $1/2^n$ of its original level. So our plan is to find the number of half-lives n that will lower the milk's I-131 activity from 980 Bq/kg to 300 Bq/kg. Mathematically, we want $1/2^n = 300/980$. Our plan is to solve for n and then use the known half-life to get the actual time.

EVALUATE Inverting the expression $1/2^n = 300/980$ and taking logarithms of both sides gives

$$\ln(2^n) = \ln(980/300)$$

But $\ln(2^n) = n \ln 2$, so

$$n = \frac{\ln(980/300)}{\ln 2} = 1.71 \text{ half-lives}$$

With $t_{1/2} = 8.04$ days, this amounts to 13.7 days or just under 2 weeks.

ASSESS A quick check shows that our answer must be about right. The milk starts out contaminated with almost 1000 Bq/kg of I-131. After one half-life, the activity has dropped in half, to just under 500 Bq/kg. Another half-life, and it's about 250 Bq/kg—already under the safety standard. So the answer must be somewhere between one and two half-lives, and closer to two because the 250 Bq/kg reached after two half-lives isn't much less than the 300-Bq/kg safety standard. Note that the 2-week wait time depends not only on physics but also on policy—namely, the government's safety standard. Exercise 28 reworks this example using the lower international guideline of 100 Bq/kg, and Problem 55 considers a similar contamination situation following the Chernobyl nuclear accident. ■

✓TIP Half-Life and Powers of 2

After n half-lives, activity has dropped by a factor of $1/2^n$. When we estimate activity levels, it's useful to note that $2^{10} = 1024$, or very nearly 1000. Thus activity drops by a factor of about 1000 every 10 half-lives—and therefore by about 1 million in 20 half-lives.

GOT IT? 38.2 A PET-scan patient is injected with radioactive oxygen-15, whose half-life is 2 min. Approximately what fraction of the original ^{15}O remains undecayed an hour later? (a) 1/30; (b) 1/60; (c) one thousandth; (d) one millionth; (e) one billionth

APPLICATION Radiocarbon Dating and Fossil Carbon

Carbon-14 formed in the atmosphere is incorporated into a living organism through the food chain.

At death, ^{14}C uptake ceases.

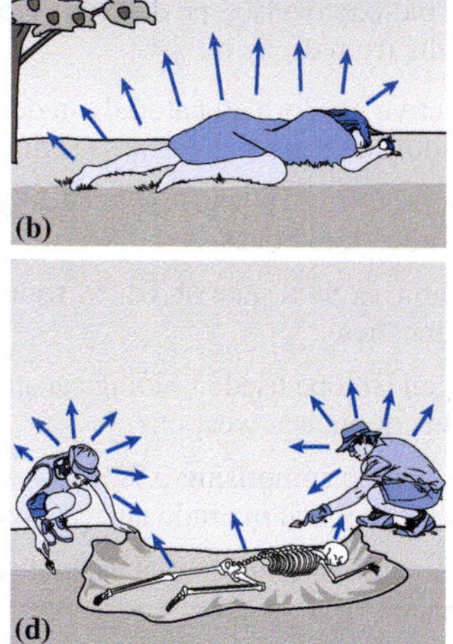

(c)

Much later, ^{14}C activity has decayed considerably.

Archaeologists excavate the long-dead remains. By measuring ^{14}C activity, they can infer the time since death. Note that the archaeologists, with their active ^{14}C intake, are more radioactive than their ancient ancestor.

Archaeologists, art historians, geologists, and others use radioactive decay to date ancient objects. For ages up to a few tens of thousands of years, the 5730-year isotope carbon-14 is especially useful. ^{14}C forms continuously in the atmosphere through reactions of cosmic rays with nitrogen. Living things take in ^{14}C and maintain a steady concentration through the balance between uptake and radioactive decay. At death, uptake ceases and the level of ^{14}C begins to drop. Measuring the ratio of ^{14}C to stable ^{12}C in a sample of once-living matter and comparing with the ratio found in living material then provides the age (see the figure and Example 38.3).

The cosmic-ray flux at Earth varies with solar activity, and so, therefore, does the atmospheric ^{14}C/^{12}C ratio. Scientists correct for this effect with data from growth rings in ancient trees, which provide an independent measure of age. Measuring the actual radioactivity takes a fairly large sample, so today the most sophisticated dating is done instead by counting individual C-14 atoms, separating them from ordinary C-12 using a mass spectrometer—a device we described in Example 26.2.

Radiocarbon dating is quite accurate to about 20,000 years and can be used back to about 50,000 years. For longer time spans, up to the billions of years characterizing the ages of rocks, ratios of longer-lived isotopes provide age information. Much knowledge of our own past, and our planet's and our solar system's, comes from radioisotope dating.

In the modern era, the atmospheric ^{14}C/^{12}C ratio provides evidence that the ongoing buildup of atmospheric CO_2 results from the combustion of fossil fuels (see the Application "The Greenhouse Effect and Global Warming," in Chapter 16). That ^{14}C/^{12}C ratio is dropping, showing that the added CO_2 is depleted in C-14. This is consistent with a carbon source that's been out of contact with the atmosphere for a long enough time that its C-14 has all decayed. Also decreasing is the ratio of the stable isotope ^{13}C to ^{12}C. Plants preferentially incorporate the lighter C-12, so taken together the decreasing ratios of C-14 and C-13 to C-12 point to a long-buried, plant-derived source of the new atmospheric carbon—namely, the fossil fuels.

EXAMPLE 38.3 Radioactive Decay: Archaeology

Archaeologists unearth charcoal from an ancient campfire and find its carbon-14 activity per unit mass to be 7.4% of the activity measured in living wood. Find the charcoal's age.

INTERPRET This is a problem about using the decay of carbon-14 to date a once-living material. We want the time it takes for ^{14}C activity to decline to 7.4% of its original level. From Table 38.1, we identify the half-life of ^{14}C as 5730 years.

DEVELOP Equation 38.3b, $N = N_0 2^{-t/t_{1/2}}$, shows that activity drops by $1/2^n$ in n half-lives, so our plan is to find the number of half-lives that makes the factor $1/2^n$ equal to 0.074. Then we can multiply by the half-life to get the actual time.

EVALUATE Solving as we did in Example 38.2 gives

$$n \ln 2 = \ln(1/0.074)$$

which gives $n = 3.76$ half-lives. With $t_{1/2} = 5730$ y, the age is then 21,500 years.

ASSESS Again a quick check suffices: One half-life drops activity to 50%; two half-lives to 25%, three to 12.5%, and four to just over 6%. So it must take a little less than four half-lives to get down to 7.4% of the original activity level. ■

Types of Radiation

Passing nuclear radiation through a magnetic field shows that there are three types: one positively charged, one negatively charged, and one neutral (Fig. 38.6). Early researchers named these alpha, beta, and gamma radiation, respectively. Today we know that alpha radiation consists of He-4 nuclei, beta radiation consists of high-energy electrons or positrons, and gamma rays are high-energy photons. They differ in penetrating power: A sheet of paper can stop alpha particles, several centimeters of matter stop most betas, and gamma rays can penetrate substantial thicknesses of concrete or lead. Different radioisotopes emit not only different types of radiation but also radiation of different energies.

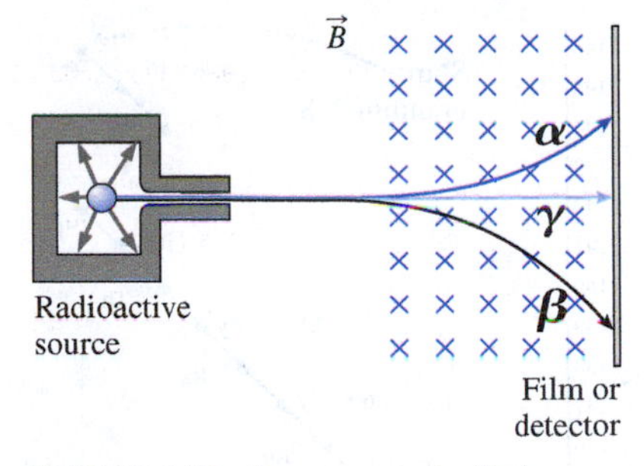

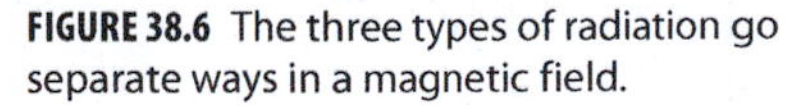
FIGURE 38.6 The three types of radiation go separate ways in a magnetic field.

Alpha Decay

Alpha emitters are nuclei with too much positive charge. They shed charge, and mass, by emitting a bundle of two protons and two neutrons—an alpha particle, ${}^4_2\text{He}$. Symbolically,

$$ {}^A_ZX \longrightarrow {}^{A-4}_{Z-2}Y + {}^4_2\text{He} \qquad \text{(alpha decay)} \tag{38.4} $$

Here X is the original or **parent** nucleus, and Y is the **daughter**. Note that the sums of the atomic numbers on both sides of this equation are equal, as are the mass numbers. Most of the energy released in the reaction appears as kinetic energy of the alpha particle. The alpha particle actually emerges with less energy than needed to overcome the nuclear potential barrier, and this provides one of the most direct confirmations of quantum tunneling—which is the only way the alpha particle can escape the nucleus.

PhET: Alpha Decay
PhET: Beta Decay

Beta Decay

Beta emitters have too many neutrons, one of which decays into an electron, a proton, and an elusive neutral particle called a neutrino (symbol ν). The electron exits at high energy to form beta radiation, leaving a nucleus with essentially the same mass but its atomic number increased because it has one more unit of positive charge:

$$ {}^A_ZX \longrightarrow {}_{Z+1}^{\;\;\;A}Y + e^- + \bar{\nu} \qquad \text{(beta decay)} \tag{38.5a} $$

In ordinary beta decay the neutrino is, in fact, an antineutrino—hence the bar over its symbol ν.

Beta decay is a manifestation of the weak nuclear force, and in the Sun it produces a steady stream of neutrinos that provide direct information on conditions in the solar core. That's because neutral, nearly massless neutrinos interact only rarely with matter; for example, they pass through the entire Earth with little probability of interaction. You'll see in Chapter 39 how neutrinos nonetheless are opening a new window on distant astrophysical events and the early universe.

A second type of beta decay converts a proton into a neutron, emitting both a positron (an anti-electron, e^+) and a neutrino:

$$ {}^A_ZX \longrightarrow {}_{Z-1}^{\;\;\;A}Y + e^+ + \nu \qquad \text{(beta decay, positron emission)} \tag{38.5b} $$

This reaction occurs in some short-lived isotopes of lighter elements like carbon and oxygen, and gamma rays from the subsequent annihilations of positrons are used in the medical imaging procedure known as positron emission tomography (PET; see the Application "PET Scans: Relativity in the Hospital," in Chapter 33.).

A third beta-decay process is **electron capture**, in which a nucleus captures an inner-shell atomic electron, converting a proton to a neutron and ejecting a neutrino:

$$ {}^A_ZX + e^- \longrightarrow {}_{Z-1}^{\;\;\;A}Y + \nu \qquad \text{(electron capture)} \tag{38.5c} $$

Gamma Decay

A nucleus in an excited state decays by emitting a photon, just like an atom. But the much higher energy associated with nuclear processes puts such photons in the gamma-ray region of the spectrum. Since the gamma-ray photon is neutral and massless, it doesn't change the type of nucleus; therefore, we write

$$ {}^A_ZX^* \longrightarrow {}^A_ZX + \gamma \qquad \text{(gamma decay)} \tag{38.6} $$

where X^* designates the excited state.

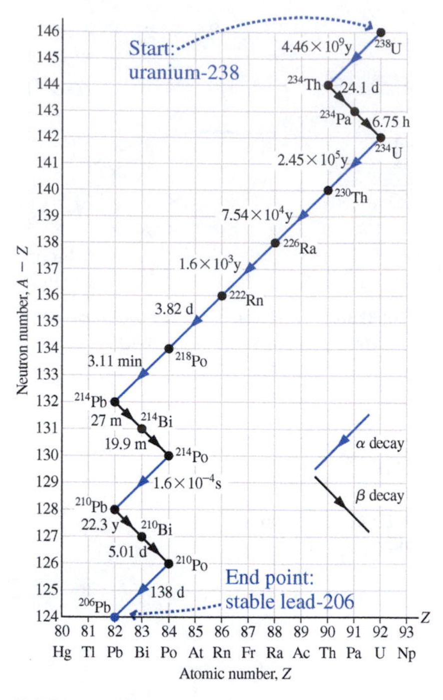

FIGURE 38.7 The decay of uranium-238 results in a series of shorter-lived nuclei. Times shown are half-lives.

Decay Series and Artificial Radioactivity

A few radioisotopes, like ^{40}K and ^{238}U, have half-lives comparable to Earth's age, so it's not surprising to find these in nature. But we also find shorter-lived species. Some, like cosmic-ray-produced ^{14}C, result from naturally occurring nuclear reactions. Many others arise in the decay of long-lived isotopes, while some we produce in particle accelerators, nuclear reactors, and nuclear explosions.

Figure 38.7 shows the **decay series** for uranium-238, whose 4.46-billion-year half-life ensures that there's still plenty of it around. The shorter-lived daughter products in this series are present wherever there's natural uranium. A balance between formation and decay establishes the abundance of each product in the decay series. One of the uranium daughters is radon-222, a radioactive gas that can be a serious health hazard in closed spaces.

In 1930 Marie Curie's daughter Irène and her husband Frédéric Joliot-Curie were the first to induce artificial radioactivity, by bombarding stable isotopes with alpha particles. Today we produce radioisotopes with particle beams or with neutrons from nuclear reactors, or by extracting them from the by-products of nuclear fission.

Uses of Radioactivity

Nuclear radiation has numerous beneficial uses in our technological society. Here we survey just a few:

- **Radioactive Tracers** "Tagging" molecules with radioactive atoms makes it easy to trace their flows through biological and physical systems. Biologists use radioactive tracers routinely to study the uptake and distribution of chemicals. Engineers use radioisotopes to study wear in mechanical parts. Physicians "tag" bone-seeking compounds with radioisotopes to image the skeletal system; the resulting "bone scans" reveal cancer and other diseases.
- **Cancer Treatment** Radiation destroys living cells, especially fast-dividing cancer cells. Early radiation treatment used gamma radiation; today, particle beams deliver radiation with less effect on surrounding tissue. Alternatively, radioisotope "seeds" are embedded directly into a tumor.
- **Food Preservation** High radiation doses destroy bacteria and enzymes that cause food spoilage, providing longer shelf life and a safer food supply. Though controversial, food irradiation is increasingly widespread, especially for spices, fruits, and some ground meats.
- **Insect Control** Radiation preferentially damages reproductive cells and can therefore cause sterility. Sterilizing large numbers of pest insects with radiation causes populations to collapse when the sterile insects mate with normal ones. The Mediterranean fruit fly, a serious pest of citrus crops, has been controlled in this way.
- **Fire Safety** Common smoke detectors contain americium-241, whose alpha radiation ionizes air, allowing it to carry electric current. Smoke particles interfere with the current, triggering the alarm. Exit signs containing radioactive tritium (^{3}H) glow without the need for electricity, providing another measure of fire safety in public buildings.
- **Activation Analysis** Bombarding materials with neutrons or other particles results in excited states or the production of unstable isotopes. Analyzing the resulting radiation helps identify unknown materials. Art historians use this technique to detect forgeries; environmental scientists identify the constituents of pollution; and airport luggage scanners search for the radiation "fingerprint" of chemical explosives.

Biological Effects of Radiation

Nuclear radiation has sufficiently high energy to ionize or otherwise disrupt biological molecules. Results include cell death, loss of biological functions, and mutations that lead to cancer or to genetic changes in future generations. Many early nuclear scientists, including Marie Curie and her daughter Irène, succumbed to leukemia and other cancers that undoubtedly resulted from radiation exposure.

The energy absorbed in a radiation dose is a rough measure of its biological danger. The SI unit of absorbed dose is the **gray** (Gy), defined as 1 J of energy per kg of absorbing material. A more appropriate measure is the **sievert** (Sv), which is weighted by the biological effectiveness of particular radiation types. Alpha particles, for example, cause more damage per unit energy than do gamma rays, so 1 Gy of alpha radiation is more harmful than 1 Gy of gamma radiation. But 1 Sv of alphas and 1 Sv of gammas cause essentially the same damage.

The biological effects of high radiation doses are well known; exposure to 4 Sv, for example, causes death in about half its human victims. But doses in the 0.1-Sv range and lower are more controversial. There are only a few cases of well-quantified exposures to populations large enough that small effects can be determined accurately. Even less certain are the effects of very low doses, such as the 1-mSv average dose to residents just outside the evacuation zone at Fukushima, or the even smaller 10-μSv dose to people living near the 1979 Three Mile Island nuclear accident. On the one hand, biological repair mechanisms may limit damage at low doses. On the other, even low doses may disproportionately affect the young and the unborn. A 2005 study by the U.S. National Academy of Sciences (NAS) suggests that the risk of cancer—the dominant health effect of low-level radiation—should scale linearly with dose. For a one-time dose of 1 mSv, the NAS study estimates a lifetime cancer risk of 1 in 10,000. This compares with a 42% lifetime chance of developing cancer from all causes.

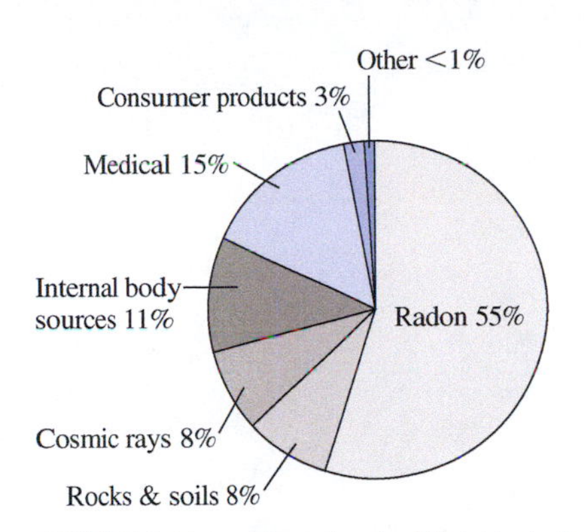

FIGURE 38.8 Natural (gray) and artificial (color) sources of radiation, as percentages of the U.S. average yearly dose of 3.6 mSv. "Other" includes nuclear power, radioactive waste, and weapons tests.

The average U.S. citizen receives about 3.6 mSv of radiation per year, most of it from natural sources (Fig. 38.8). The dominant source, at 55%, is the uranium decay product radon-222, which seeps into buildings from the decay of naturally occurring uranium in the ground and in building materials. Our own bodies account for some 11%, mostly from natural potassium-40. About 19% of our average exposure in the United States comes from artificial sources, mainly medical procedures.

Consumer products—mostly tobacco, drinking water, and building materials—account for about 3%. Less than 1% of our yearly radiation dose is from nuclear power and weapons. Radiation doses vary greatly with location and occupation; for example, residents of high-altitude Denver have greater exposure to cosmic rays, and airline flight crews' cosmic-radiation dose often exceeds the allowed dose for nuclear-plant workers. No matter what your exact dose, though, the risks to your health from radiation exposure pale compared with other risks you knowingly take.

38.3 Binding Energy and Nucleosynthesis

Disassembling a nucleus requires energy to overcome the strong nuclear force. The more tightly bound the nucleus, the higher this **binding energy**. The energies involved in nuclear interactions are high enough that Einstein's mass–energy equivalence is clearly evident, so accounting for energy conservation requires us to consider the rest energy of the particles. Then we can write

$$m_N c^2 + E_b = Zm_p c^2 + (A - Z)m_n c^2 \quad (38.7)$$

where the terms on the left are the rest energy of the nucleus, whose mass is m_N, and the binding energy E_b. The terms on the right are the rest energies of the Z individual protons and $A - Z$ neutrons that make up the nucleus. So Equation 38.7 shows that we can disassemble a nucleus into its constituent nucleons if we supply additional energy equal to the binding energy. Equivalently, E_b is the energy released if we assemble a nucleus from isolated nucleons.

Equation 38.7 shows that the nuclear mass m_N is *not* the sum of the constituent particles' masses; rather, it's less by the amount E_b/c^2. This is clear evidence for mass–energy equivalence. Again, as in Chapter 33, we emphasize that there's nothing uniquely nuclear about this so-called **mass defect**. The mass of a water molecule is also less than the sum of its constituent hydrogen and oxygen atoms—but with chemical binding the effect is so small as to be virtually immeasurable. It's the strength of the nuclear force that makes mass–energy equivalence more obvious in nuclear interactions.

It's convenient to measure nuclear and particle masses in **unified mass units**, u, currently defined as one-twelfth the mass of a neutral carbon-12 atom. The unified mass unit is very nearly 1.66054×10^{-27} kg, slightly less than the mass of the proton or neutron. High-energy physicists, ever cognizant of mass–energy equivalence, often express masses in MeV/c^2—a value numerically equal to the rest energy in MeV. Table 38.2 lists selected particle masses in kg, u, and MeV/c^2. In practice one often knows atomic rather than nuclear masses, but the difference is generally negligible because the extra mass of the electrons is so small.

Table 38.2 Selected Masses

	Mass (kg)	Mass (u)	Mass (MeV/c^2)
Electron	9.10939×10^{-31}	0.000548579	0.510999
Proton	1.67262×10^{-27}	1.007276	938.272
Neutron	1.67493×10^{-27}	1.008665	939.566
$^1_1$H atom	1.67353×10^{-27}	1.007825	938.783
α particle ($^4_2$He nucleus)	6.64466×10^{-27}	4.001506	3727.38
$^{12}_6$C atom	1.99265×10^{-26}	12	11,177.9
Unified mass unit (u)	1.66054×10^{-27}	1	931.494

EXAMPLE 38.4 Mass Defect in Helium: Powering the Sun

Use the appropriate masses from Table 38.2 to find the binding energy of $^4_2$He.

INTERPRET This is a question about binding energy—the energy difference between separate constituents of helium-4 and the helium-4 nucleus. We identify the constituent particles from the symbol $^4_2$He: $Z = 2$ protons and $N = A - Z = 2$ neutrons.

DEVELOP Equation 38.7 determines the binding energy in terms of the various masses:

$$E_b = Zm_pc^2 + (A - Z)m_nc^2 - m_Nc^2$$

EVALUATE Using our values for Z and $N = A - Z$, along with the proton, neutron, and alpha-particle (He-4 nucleus) masses from Table 38.2, gives

$$E_b = 2(938.272\text{ MeV}/c^2)c^2 + 2(939.566\text{ MeV}/c^2)c^2 - (3727.38\text{ MeV}/c^2)c^2$$
$$= 28.3\text{ MeV}$$

ASSESS Notice how easy it was to work with mass in units of MeV/c^2; the factor c^2 canceled and we didn't need to use the speed of light explicitly. The formation of helium through a sequence of nuclear reactions is what powers the Sun, and our 28.3-MeV result is very close to the actual 26.7 MeV released for each He-4 nucleus formed through the solar process. ■

The Curve of Binding Energy

Binding energy plays a crucial role in the formation of the elements and in nuclear energy. Figure 38.9 shows the **curve of binding energy**, a plot of binding energy *per nucleon* as a function of mass number A. The higher this quantity, the more tightly bound is the nucleus. The broad peak in the vicinity of $A = 60$ shows that nuclei with mass numbers around this value are most tightly bound. That means it's energetically favorable for two lighter nuclei to join through the process of **nuclear fusion**, making a middle-weight nucleus. But heavier nuclei can reach a lower energy state if they split or **fission** into two middle-weight nuclei. We'll discuss fission and fusion later in the chapter.

GOT IT? 38.3 Rank order these nuclei from the most to the least tightly bound: $^4_2$He; $^{238}_{92}$U; $^{57}_{26}$Fe; $^2_1$H; $^{132}_{54}$Xe.

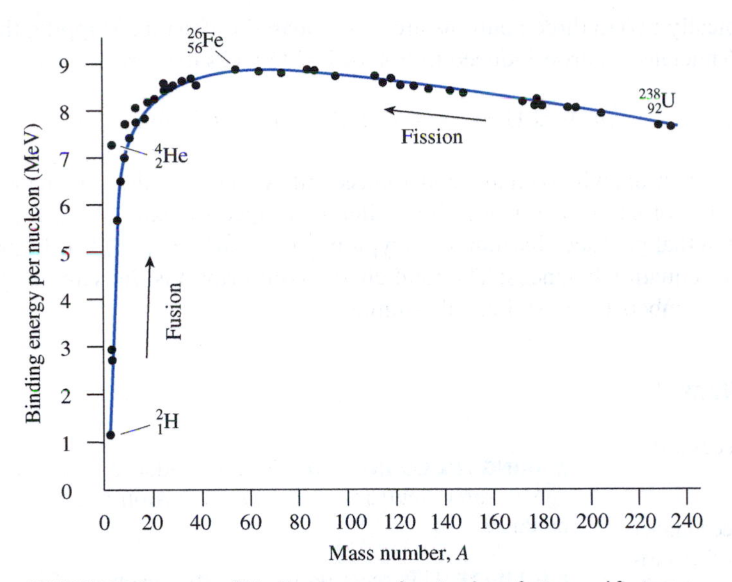

FIGURE 38.9 The curve of binding energy, showing how fusion and fission can result in the release of nuclear energy.

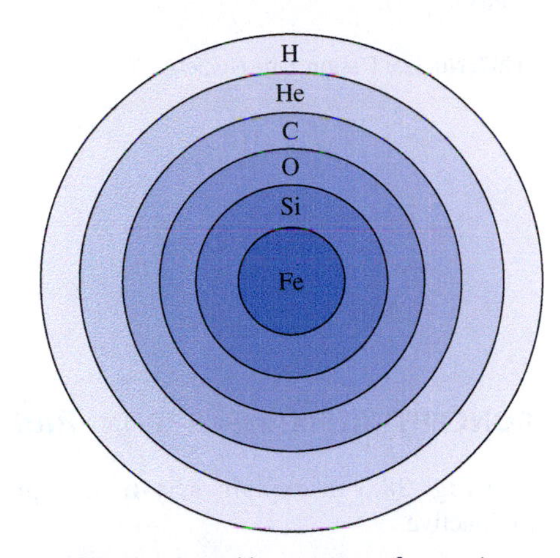

FIGURE 38.10 Onionlike structure of a massive star before it goes supernova. Successive stages of fusion reactions produce the elements shown, which accounts for their relative abundance.

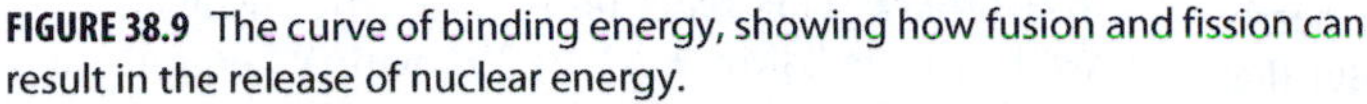

Nucleosynthesis and the Origin of the Elements

Since it's energetically favorable for light nuclei to fuse together, they'll do so if they have enough energy to overcome their electrical repulsion. This condition held in the high-temperature early universe, particularly from about 1 minute to 30 minutes after the start of the Big Bang. During that time, protons fused to form helium, leaving the universe with approximately its present composition of about 75% hydrogen and 25% helium, with traces of deuterium, lithium, beryllium, and boron. Hundreds of millions of years later the first stars formed, and in the interiors of more massive stars conditions were ripe for a two-step process that fused three helium nuclei to make carbon-12. From there fusion reactions led to the formation of isotopes up to those near the $A = 60$ peak in the curve of binding energy. In fact, the nuclei of essentially all the elements with $A < 60$—including most of the materials in our own bodies—were formed in the interiors of massive stars (Fig. 38.10). Some nuclei with $A > 60$ also formed inside massive stars; others formed in the violent supernova explosions that end such stars' lives. Those explosions spewed fusion-synthesized elements into the interstellar medium where, eons later, they're incorporated into new stars, planets, and even living things.

38.4 Nuclear Fission

Neutrons, first discovered in 1932, make excellent probes of the nucleus because they don't have to overcome electrical repulsion. In 1938 the German chemists Otto Hahn and Fritz Strassmann bombarded uranium with neutrons. They were puzzled to find among the reaction products radioactive versions of the much lighter elements barium and lanthanum. Physicist Lise Meitner and her nephew Otto Frisch interpreted these results to mean that uranium had split or, in their words, **fissioned** (Fig. 38.11). It was the eve of World War II, and the military implications were obvious and ominous: Nuclear fission represented an energy source orders of magnitude more potent than chemical reactions. The race for nuclear weapons was on. With the help of the international physics community, many of whom had fled fascism, the U.S. effort succeeded. A team led by the Italian Enrico Fermi built the first nuclear reactor under the stands of the University of Chicago stadium; it became operational in 1942. Three years later came the first nuclear weapons test, at Trinity Site in New Mexico, followed quickly by the nuclear destruction of Hiroshima and Nagasaki.

Although fission can occur spontaneously, it's much more likely when a neutron strikes a nucleus. Figure 38.11 shows U-235 absorbing a neutron to become U-236. This unstable nucleus undergoes dumbbell-shaped oscillations until electrical repulsion tears it apart. The resulting **fission products** are generally a pair of middle-weight nuclei with unequal

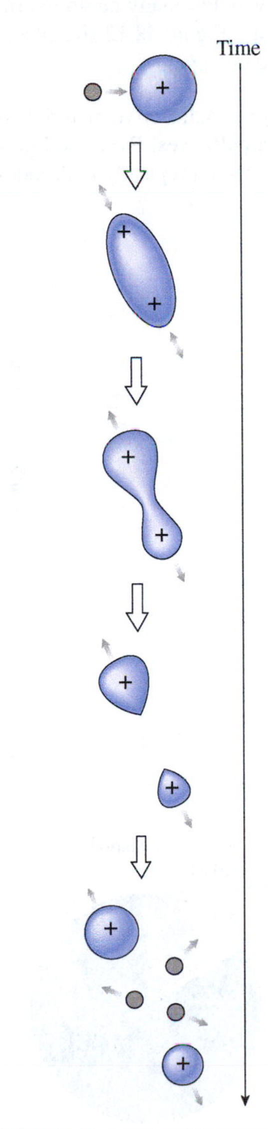

FIGURE 38.11 Neutron-induced fission of ^{235}U, showing three neutrons (gray) released in the process.

MP®

PhET: Nuclear Fission: One Nucleus

masses; typically two to three neutrons are also released in fission. Skipping the intermediate U-236 nucleus, neutron-induced fission of U-235 takes the form

$$ {}^{1}_{0}n + {}^{235}_{92}\text{U} \longrightarrow X + Y + b\,{}^{1}_{0}n \quad \text{(fission)} \tag{38.8}$$

Here ${}^{1}_{0}n$ is the neutron, with 0 charge and 1 mass unit; X and Y are the fission products; and b is the number of neutrons released immediately. A specific example of Equation 38.8 is ^{235}U fission that produces barium and krypton: ${}^{1}_{0}n + {}^{235}_{92}\text{U} \longrightarrow {}^{141}_{56}\text{Ba} + {}^{92}_{36}\text{Kr} + 3\,{}^{1}_{0}n$. Note how the equation balances: The total charge (subscripts) is the same on both sides, and the mass numbers (superscripts) also agree.

CONCEPTUAL EXAMPLE 38.1 Radioactive Waste!

Use Fig. 38.3 to explain why fission products are necessarily radioactive.

EVALUATE Figure 38.3 shows that more massive nuclei need higher ratios of neutrons to protons in order to overcome the protons' electrical repulsion. When uranium fissions, the resulting nuclei have nearly the same neutron-to-proton ratio as the original uranium. But that gives them way too many neutrons, making them highly radioactive via beta decay. Figure 38.12 shows a simplified chart of the nuclides to help make this point.

ASSESS Highly radioactive materials decay rapidly, giving them relatively short half-lives. Even the longer-lived fission products have half-lives measured typically in decades.

MAKING THE CONNECTION Neutron-induced fission of ^{235}U yields ${}^{102}_{42}$Mo, three neutrons, and another fission product. What's that product?

EVALUATE This reaction is a specific instance of Equation 38.8: ${}^{1}_{0}n + {}^{235}_{92}\text{U} \rightarrow {}^{102}_{42}\text{Mo} + {}^{Z}_{A}X + 3{}^{1}_{0}n$, with X being the unknown fission product. Balancing atomic and mass numbers gives $A = 131$ and $Z = 50$. The periodic table shows that $Z = 50$ is iodine, so X is I-131, a dangerous contaminant discussed in Example 38.2.

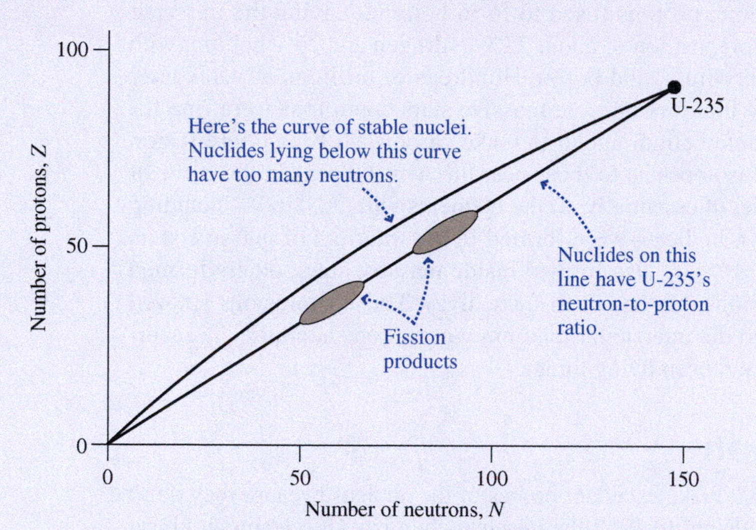

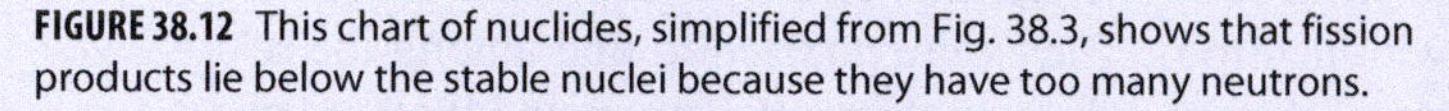
FIGURE 38.12 This chart of nuclides, simplified from Fig. 38.3, shows that fission products lie below the stable nuclei because they have too many neutrons.

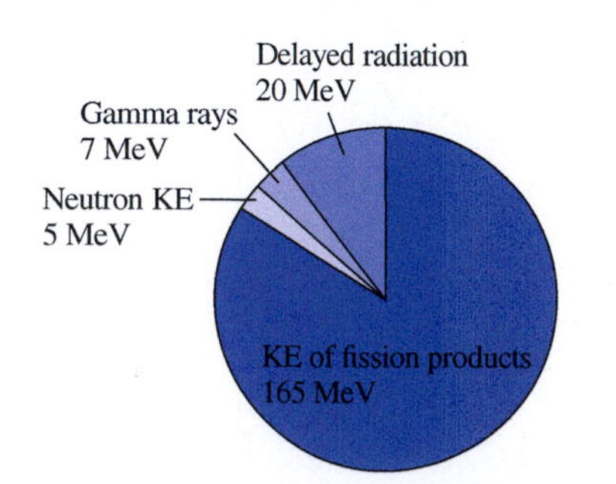

FIGURE 38.13 Fission energy is distributed among fission products, neutrons, and radiation.

Energy from Fission

Fission of a uranium nucleus releases about 200 MeV of energy, as shown in Fig. 38.13. Spontaneous fission is rare because of the energy barrier associated with forces on the outermost nucleons; rather, fission usually results when a nucleus absorbs a neutron, initiating the process shown in Fig. 38.11. Many heavy nuclei, including ^{238}U and ^{235}U, are **fissionable**, meaning they can undergo neutron-induced fission. **Fissile** nuclei will fission with neutrons of *any* energy, including thermal energy. Fissile nuclei are significant because they alone can sustain a nuclear chain reaction, and thus they're essential for both nuclear power and nuclear weapons. The three important fissile nuclei are uranium-233, uranium-235, and plutonium-239.

Uranium-235 presently constitutes only about 0.7% of natural uranium; nearly all the rest is ^{238}U. For most uses, uranium must be enriched in ^{235}U, to several percent for commercial power reactors and 80% or more for weapons. **Uranium enrichment** is difficult and expensive; since the isotopes ^{235}U and ^{238}U are chemically similar, enrichment techniques make use of their very slight mass difference. The technique of choice today involves spinning uranium hexafluoride gas in a sequence of high-speed centrifuges. Enrichment technology is highly sensitive because a nation possessing it can produce weapons-grade uranium.

Plutonium-239, with a 24,110-year half-life, does not occur in nature. It's produced by neutron bombardment of ^{238}U. The reaction forms ^{239}U, which undergoes two beta decays to produce first ^{239}Np and then the fissile ^{239}Pu:

$$
\begin{aligned}
{}^{1}_{0}n + {}^{238}_{92}\text{U} &\longrightarrow {}^{239}_{92}\text{U} \\
{}^{239}_{92}\text{U} &\longrightarrow {}^{239}_{93}\text{Np} + e^- + \bar{\nu} \\
{}^{239}_{93}\text{Np} &\longrightarrow {}^{239}_{94}\text{Pu} + e^- + \bar{\nu}
\end{aligned}
$$

Although ^{239}Pu is produced copiously in nuclear reactors (see Problem 80), **reprocessing** spent reactor fuel to extract plutonium is difficult and dangerous. Contamination with other plutonium isotopes further complicates the process. Like uranium enrichment, plutonium reprocessing is a sensitive technology, and the decision of several European countries and Japan to engage in commercial reprocessing for reactor fuel has made Pu-239 a commercial commodity.

EXAMPLE 38.5 Nuclear Fission: Uranium versus Coal

Assuming 200 MeV per fission, estimate the amount of pure ^{235}U that would provide the same energy as 1 metric ton (1000 kg) of coal.

INTERPRET We're asked to compare the energies of nuclear fission and the chemical burning of coal.

DEVELOP For coal, we can look up the energy released per unit mass in Appendix C's "Energy Content of Fuels" table. We can then find the number of fission events, at 200 MeV per fission, needed to release the same energy as 1000 kg of coal. Finally, we'll use the mass of a U-235 nucleus to find the corresponding mass of uranium.

EVALUATE Appendix C gives an energy content of 29 MJ/kg for coal, so burning 1000 kg of coal releases 29 GJ of energy. With 1.6×10^{-19} J/eV, each 200-MeV fission releases about 3.2×10^{-11} J. Then we need a total of

$$29\ \text{GJ}/3.2\times10^{-11}\ \text{J/fission} = 9.1\times10^{20}\ \text{fission events}$$

Each of the 9.1×10^{20} U-235 nuclei has a mass of approximately 235 u, so the total mass required is

$$(9.1\times10^{20}\ \text{nuclei})(235\ \text{u})(1.66\times10^{-27}\ \text{kg/u}) = 0.35\ \text{g}$$

ASSESS That's about one one-hundredth of an ounce of ^{235}U, packing as much energy as a ton of coal! Our result shows that U-235 contains about three million times as much energy as the same amount of coal. That's the reason nuclear power plants are fueled only about once a year, with a truckload or so of nuclear fuel, while coal-burning plants burn many 110-car trainloads of coal each week. It's also the reason for the immense destructive power of nuclear weapons. ■

The Chain Reaction

MP®

PhET: Nuclear Fission: Chain Reaction

Neutrons induce fission, and fission itself releases more neutrons. This makes possible a **chain reaction**, in which each fission event results in more fission. To sustain a chain reaction, each nucleus that fissions must, on average, cause at least one more fission event; otherwise, the reaction will fizzle to a halt. In a piece of material that's too small, most neutrons will escape without causing additional fission. For that reason there's a **critical mass** of nuclear fuel necessary to sustain a chain reaction. More than that amount is **supercritical** and results in an exponentially growing chain reaction (Fig. 38.14).

The size of the critical mass depends on the purity of the fissile material, its configuration, and surrounding materials. For plutonium it can be less than 5 kg, and as low as 15 kg for uranium. Those numbers are frighteningly small, and they show why we worry about city-destroying "suitcase bombs."

The **multiplication factor**, k, is the average number of neutrons from a fission event that cause additional fission. A critical mass has $k = 1$, and a supercritical mass has $k > 1$. The average time between successive fissions is the **generation time**. In a supercritical mass this can be as short as 10 ns, leading to the entire mass fissioning in about 1 μs.

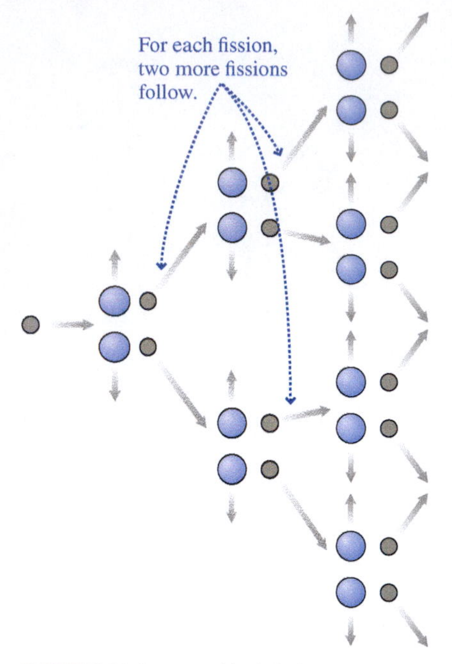

FIGURE 38.14 A supercritical chain reaction with multiplication factor $k = 2$.

MP

PhET: Nuclear Fission: Nuclear Reactor

Fission Weapons

A rapidly fissioning supercritical mass is a nuclear explosive. The major technological difficulty in producing a fission weapon is to assemble a supercritical mass so rapidly that the chain reaction consumes enough fissile material before it blows apart. With highly enriched uranium that's not an insurmountable challenge. The crude bomb that destroyed Hiroshima contained about 50 kg of enriched uranium, of which only about 1 kg actually fissioned. So confident were its developers that they never tested this design. Plutonium weapons present a greater challenge; neutrons from spontaneous fission make it more likely that the weapon will "pre-ignite" and blow itself apart.

Construction of a simple fission weapon is distressingly straightforward, but acquisition of weapons-grade fissile material is not. Again, that's why uranium enrichment and plutonium reprocessing technologies are so sensitive. We live in a dangerous and unstable world, and it's going to get more dangerous if fissile materials become widely available.

Nuclear Power

A **nuclear reactor** uses a controlled fission chain reaction with $k = 1$ to release energy at a steady rate. Since the average number of neutrons emitted in U-235 fission is about 2.5, reactors require that most neutrons don't cause fission. Commercial power reactors limit k in part by keeping the concentration of fissile U-235 low—typically a few percent—so that many neutrons are absorbed by U-238 instead of causing fission. **Control rods** made of neutron-absorbing material allow for active adjustment of k; these can be moved into and out of the nuclear fuel to provide precise control of the power level. A small fraction—about 0.65%—of fission-produced neutrons are emitted with delays from about 0.2 s to 1 min, and these **delayed neutrons** allow for relatively slow mechanical control of nuclear reactors. The next example explores this point.

EXAMPLE 38.6 Nuclear Fission: Delayed Neutrons and Reactor Control

A change in operating conditions makes a nuclear reactor slightly supercritical, with $k = 1.001$. Determine the time it would take the reactor power to double (a) if delayed neutrons establish a generation time $\tau = 0.1$ s, and (b) if prompt neutrons—those released immediately—sustain the reaction to give $\tau = 10^{-4}$ s.

INTERPRET We're asked to calculate the time until the reaction rate doubles, given the multiplication factor k and two different values for the generation time.

DEVELOP A multiplication factor $k = 1.001$ means the rate of fissioning increases by a factor of 1.001 with each generation time; after two generations it will have increased by k^2, and so forth. So our plan is to find the number n of generations that gives $k^n = 2$. Then we can multiply by the two different τ values to find the actual times: $t = n\tau$.

EVALUATE We set $k^n = 2$ and take the logarithm of both sides. With $\ln(k^n) = n \ln k$, we have $n \ln k = \ln 2$, or $n = \ln 2/\ln k = 693$ with $k = 1.001$. With $\tau = 0.1$ s that gives $t = n\tau = 69.3$ s or just over 1 min, but with $\tau = 10^{-4}$ s it's only 0.07 s.

ASSESS With delayed neutrons the doubling time is long enough for the reactor operators and their mechanical controls to take corrective action; with prompt neutrons there isn't time to prevent a serious nuclear accident. Delayed neutrons are crucial to reactor control! Loss of control to a reaction governed by prompt neutrons alone was a key factor in the 1986 Chernobyl nuclear accident. ■

High-energy fission neutrons aren't very effective at causing additional fission events, so in most reactor designs they must be slowed to roughly the mean thermal speed. A substance called the **moderator** effects this slowing through elastic collisions between neutrons and the moderator nuclei. In Chapter 9 you saw that the maximum energy transfer occurs when colliding particles have equal mass; therefore, the best moderators have low-mass nuclei. The choice of moderator is among the most significant distinguishing features of different reactor designs. Another important choice is the **coolant**, which carries off fission-generated heat.

Power reactors in the United States are **light-water reactors** (LWRs), using ordinary water with the protons of its hydrogen serving as the moderator nuclei. The same water acts as coolant and circulates through a pressure vessel containing uranium fuel rods and control rods. About one-third of the United States' roughly 100 power reactors are **boiling-water reactors** (BWRs), in which water boils in the reactor vessel to make steam that drives a turbine-generator (Fig. 38.15). The remainder are **pressurized-water reactors** (PWRs), in which liquid water under pressure transfers its energy to a secondary loop where water boils to make steam (Fig. 38.16). An advantage of this more complex system is that the steam loop doesn't become radioactive. Both types of light-water reactors have an intrinsic safety feature, in that a loss of coolant also means loss of moderator, and that brings the chain reaction to a halt. But light water has the disadvantage that ${}^{1}_{1}H$ readily absorbs neutrons, and therefore light-water reactor fuel must be enriched in ${}^{235}U$ in order to sustain the chain reaction. Refueling a LWR is also a big operation: The reactor must be shut down and the lid removed from the pressure vessel—a process that can take a month or longer.

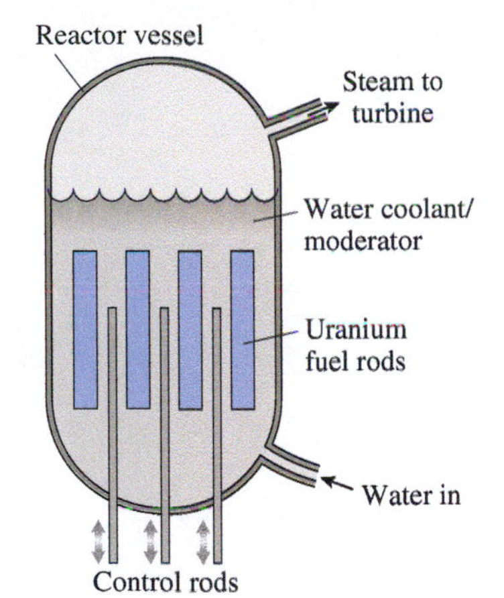

FIGURE 38.15 A boiling-water reactor, one of two types commonly used in the United States.

The Canadian CANDU design uses heavy water (${}^{2}_{1}H_2O$, or deuterium oxide) as moderator and coolant. Low neutron absorption means CANDU reactors can operate on natural uranium, eliminating the need for sensitive enrichment technology. And the CANDU design allows continuous refueling, although that increases another proliferation risk by making it easier to extract plutonium.

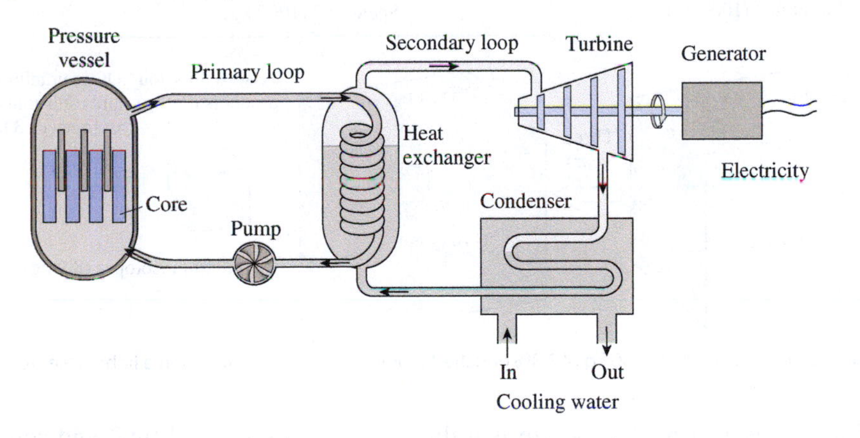

FIGURE 38.16 A complete power plant using a pressurized-water reactor, the most common type of power reactor in the United States.

An older Soviet-era design is the graphite-moderated, water-cooled RBMK reactor. Often built to provide both electric power and plutonium for weapons, this design suffered from the safety defect that loss of coolant not only didn't shut down the chain reaction but could actually accelerate it due to loss of neutron-absorbing hydrogen in the H_2O coolant. The disastrous 1986 Chernobyl accident involved an RBMK reactor. During a test of the emergency cooling system, operators inadvertently put the reactor in an unstable state where an increase in power boiled away more cooling water, resulting in a further increase. The power level soared by a factor of 4000 in 5 seconds, causing a steam explosion that blew the top off the reactor and ignited the flammable graphite moderator. Heavy smoke carried radioactive materials into the atmosphere, resulting in widespread contamination (see Problem 55). Today, thousands of square miles surrounding Chernobyl remain officially uninhabitable.

Other reactor designs include gas-cooled reactors that can operate at higher temperatures and therefore greater thermodynamic efficiencies, and **breeder reactors** designed specifically to "breed" plutonium from U-238 and therefore turn most of the nonfissile U-238 into fissile Pu-239. Breeders have no moderator, use liquid sodium coolant, and are critical with fast neutrons alone. Breeders are therefore less stable than so-called thermal reactors using slow neutrons, and widespread adoption of breeder technology entails international trafficking in fissile plutonium.

On the drawing board are a variety of advanced reactors, collectively termed Generation IV—although none has yet proven commercially viable. Some Gen-IV designs claim intrinsic safety, including the ability to survive a total loss of coolant. Others are designed to "burn" waste from current reactors, reducing humanity's long-term burden of nuclear waste. Even the decades-old light-water reactor has seen substantial improvements, including more robust safety systems. Had the 1970s-vintage reactors at Fukushima been, instead, the contemporary Generation III+ designs, the Fukushima power plant might well have survived the tsunami.

Nuclear Waste

We've seen that fission products are highly radioactive because they contain too many neutrons for stable middle-weight nuclei. Because of their high activity, fission products have relatively short half-lives, typically measured in decades. That makes fission-product waste dangerous for centuries to a few millennia. However, neutron absorption in fission reactors also produces plutonium and a host of other **transuranic** isotopes—those heavier than uranium—with much longer lifetimes. It's these substances that mean we'll have to safeguard nuclear waste for tens of thousands of years.

As fission proceeds, the concentration of fission products in the fuel increases. Before a reactor's ^{235}U is exhausted, fission products begin absorbing enough neutrons to interfere with the chain reaction. In U.S. LWRs, that requires about one-third of the fuel rods to be replaced annually. Older fuel is also rich in fissile plutonium, and at the end of a fuel rod's 3 years in the reactor, more than half the energy generation comes from fissioning plutonium rather than uranium. Figure 38.17 shows the evolution of nuclear fuel in an LWR.

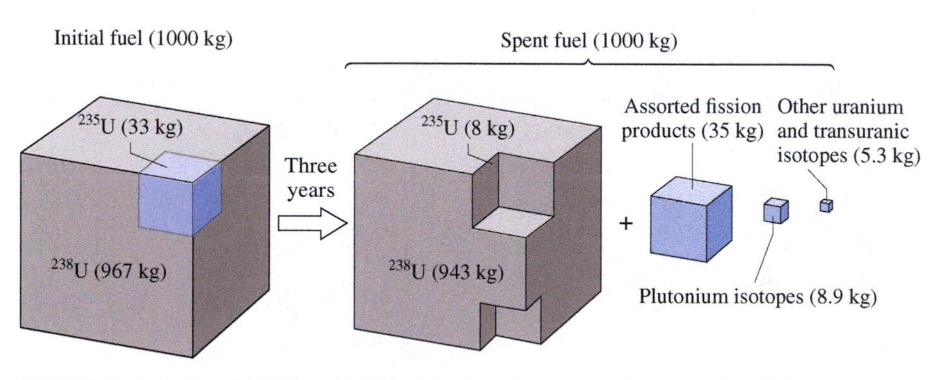

FIGURE 38.17 Evolution of 1000 kg of 3.3% enriched uranium over its 3-year stay in a light-water reactor.

The disposal of nuclear waste is a thorny issue, mixing political and scientific considerations. To date, the United States has no repository for commercial nuclear waste, which continues to accumulate at reactor sites. Lest you picture mountains of nuclear garbage, however, remember that factor-of-10^7 difference between nuclear and chemical energy sources. That translates into far less fuel needed for nuclear power plants, and far less waste produced. A 1-GW power reactor produces some 20 tons of high-level nuclear waste annually, while a comparable coal plant produces 1000 tons of carbon dioxide and 30 tons of solid waste *every hour*.

GOT IT? 38.4 Transportation and mining accidents involving coal are much more frequent than those involving uranium fuel. What's the fundamental reason for this?

Prospects for Nuclear Power

Today, nuclear power supplies some 11% of the world's electrical energy, a figure that's dropped from a high of 15% in 2006 due to economic factors and the Fukushima accident. Reliance on nuclear power varies widely; in nuclear-intensive France the figure is nearly 80%, and in the United States it's 20%. Dozens of new reactors are under construction, predominantly in Asia; most use advanced versions of light-water reactor designs. Recently, the United States has seen nuclear-plant license applications for the first time in 30 years. But worldwide, hundreds of older reactors are nearing the end their lifetimes, and without massive new construction, nuclear's share of the world's energy supply is unlikely to increase significantly.

Concern over climate change from fossil-fuel combustion has spurred a renewed interest in nuclear power, even among some environmentalists. New reactor designs promise greater reliability, economic viability, and, most important, safety based on intrinsic reactor design rather than complicated safety systems. Most physicists agree that the public's concern over nuclear power is exaggerated. We regularly accept much greater risks from other technologies—for example, some 13,000 premature deaths each year in the United States due to pollution from coal-fired power plants. Comparable estimates for nuclear plants range from about 10 to fewer than 1000 from even the most vigorously antinuclear groups.

Nevertheless, ongoing uncertainties about the risk of catastrophic nuclear accidents, long-term waste storage, terrorism, and weapons proliferation continue to haunt the nuclear power industry. Proliferation, especially, is a very real concern. Although nuclear power and weapons development are different enterprises, they share infrastructure and an educated technological elite that can be put to either purpose. If nuclear power is to advance, it will need to do so under strict international guarantees against diversion of materials and expertise to weapons production.

38.5 Nuclear Fusion

The curve of binding energy (Fig. 38.9) shows that fusion of light nuclei provides another approach to nuclear energy production. The curve is steepest at its left end, indicating that the most energy per nucleon comes from the fusion of hydrogen. Indeed, the fusion reactions powering the Sun and many other stars begin with the fusion of hydrogen to form deuterium. Also emitted in the process are a positron, a neutrino, and 0.42 MeV of energy:

$$ {}^{1}_{1}\text{H} + {}^{1}_{1}\text{H} \longrightarrow {}^{2}_{1}\text{H} + e^{+} + \nu \qquad (0.42\ \text{MeV}) \qquad (38.9\text{a}) $$

Deuterium then fuses with hydrogen to form helium-3 and a gamma ray:

$$ {}^{2}_{1}\text{H} + {}^{1}_{1}\text{H} \longrightarrow {}^{3}_{2}\text{He} + \gamma \qquad (5.49\ \text{MeV}) \qquad (38.9\text{b}) $$

Two helium-3 nuclei then react to form helium-4 and a pair of protons (${}^{1}_{1}\text{H}$), releasing 12.86 MeV:

$$ {}^{3}_{2}\text{He} + {}^{3}_{2}\text{He} \longrightarrow {}^{4}_{2}\text{He} + 2\,{}^{1}_{1}\text{H} \qquad (12.86\ \text{MeV}) \qquad (38.9\text{c}) $$

In addition, the positron from reaction 38.9a annihilates with an electron, forming two gamma rays with a total energy of $2mc^2$ or 1.022 MeV. Together, these reactions constitute the **proton–proton cycle**. In the full cycle, reactions 38.9a and b occur twice for each occurrence of reaction 33.9c. The net effect is to convert four protons and two electrons to a single He-4 nucleus, releasing 26.7 MeV (Fig. 38.18). In massive stars, ${}^{4}_{2}\text{He}$ then becomes a building block for still heavier elements, as we discussed earlier.

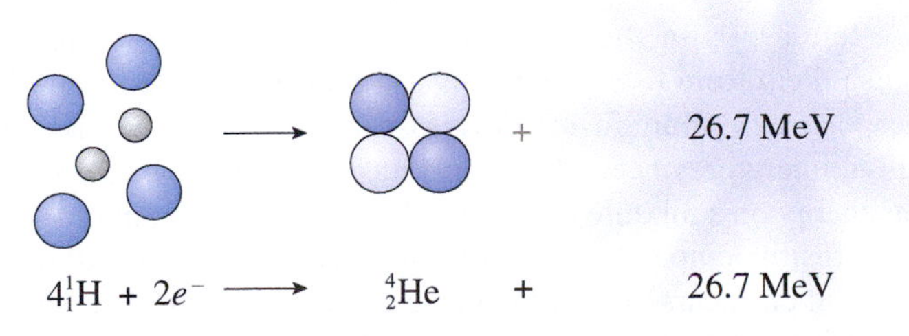

FIGURE 38.18 Net result of the proton–proton cycle of Equations 38.9.

Reaction 38.9a does not occur readily, and terrestrial fusion research has therefore focused on reactions involving the heavier hydrogen isotopes. Of immediate interest are deuterium–tritium (D-T) and deuterium–deuterium (D-D) reactions, listed below with the energy released in each:

$$ {}^{2}_{1}\text{H} + {}^{3}_{1}\text{H} \longrightarrow {}^{4}_{2}\text{He} + {}^{1}_{0}n \qquad (17.6\ \text{MeV; D-T reaction}) \qquad (38.10\text{a}) $$

$$ {}^{2}_{1}\text{H} + {}^{2}_{1}\text{H} \longrightarrow {}^{3}_{2}\text{He} + {}^{1}_{0}n \qquad (3.27\ \text{MeV; D-D reaction}) \qquad (38.10\text{b}) $$

$$ {}^{2}_{1}\text{H} + {}^{2}_{1}\text{H} \longrightarrow {}^{3}_{1}\text{H} + {}^{1}_{1}\text{H} \qquad (4.03\ \text{MeV; D-D reaction}) \qquad (38.10\text{c}) $$

The two outcomes of the D-D reaction have nearly equal probability.

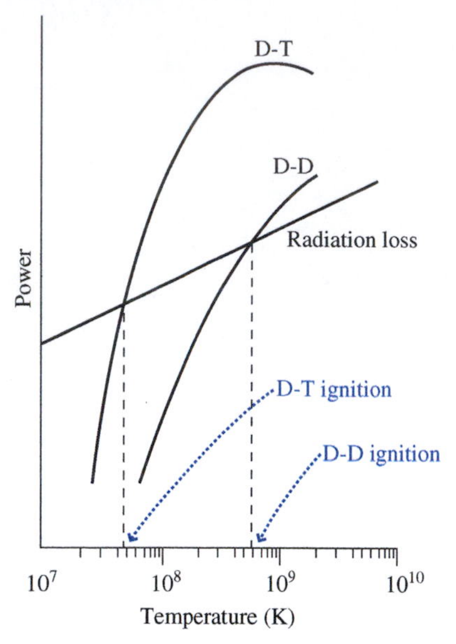

FIGURE 38.19 Power loss by radiation and power produced by D-D and D-T fusion reactions, as functions of temperature on a log-log plot.

The electrical repulsion between nuclei makes it difficult to get them close enough to fuse. Although quantum tunneling helps, it still takes very high nuclear speeds—corresponding to high temperatures—to initiate fusion. At fusion temperatures, atoms are stripped of their electrons and the fusing material constitutes a plasma. It's necessary somehow to contain this hot plasma. Stars achieve both ends with their immense gravity, which compresses stellar material to fusion temperatures and simultaneously provides confinement. In the Sun's core, for example, the temperature is some 15 MK, and fusing nuclei approach with energies on the order of 1 keV—although even under these conditions the fusion process isn't particularly efficient.

Terrestrial fusion requires still higher temperature, as high-energy particles undergo large accelerations that result in the plasma losing energy by radiation. The temperature at which fusion-generated power exceeds radiation loss is the **critical ignition temperature**. For the D-D reactions of Equations 38.10b and c, Fig. 38.19 shows that the ignition temperature is about 600 MK; for D-T it's a lower 50 MK. Net fusion-energy production requires not only high temperature but also confinement for long enough that the fusion energy produced exceeds the energy required to heat the plasma. The heat required depends on the number of nuclei or, on a volume basis, on the number density n. However, the rate of fusion-energy production depends on the *square* of the density. That's because doubling n doubles *both* the number of nuclei available to strike other nuclei and the number of nuclei available to be struck; the result is a quadrupling of the fusion rate. The total energy released therefore scales as $n^2\tau$, where τ is the **confinement time**. Meanwhile the radiation energy loss depends linearly on n, and as a result there's a minimum value of the product $n\tau$ necessary in an energy-producing fusion device. This condition is the **Lawson criterion**, given approximately by

$$\begin{aligned} n\tau &> 10^{22}\ \text{s/m}^3 \quad \text{(Lawson criterion, D-D fusion)} \\ n\tau &> 10^{20}\ \text{s/m}^3 \quad \text{(Lawson criterion, D-T fusion)} \end{aligned} \tag{38.11}$$

The factor-of-100 difference here shows that D-T fusion will be much easier to achieve.

Fusion technologies use two distinct approaches to the Lawson criterion. **Inertial confinement** strives for very high densities with short confinement times—so short that the particles' inertia alone is sufficient to prevent them from leaving the fusion site during the brief time needed. **Magnetic confinement** holds lower-density plasma in a "magnetic bottle" whose magnetic-field configuration minimizes the chance of escape during a relatively long confinement time. Neither approach has yet produced a sustained energy yield from fusion.

Inertial Confinement Fusion

Although peaceful fusion devices still elude us, inertial confinement has been used successfully since the 1950s in thermonuclear weapons—often called "hydrogen bombs" to distinguish them from fission explosives (incorrectly called "atomic bombs"). Thermonuclear weapons aren't pure fusion devices, though. They use a fission explosion to achieve the high temperatures needed to ignite fusion, and a clever arrangement for focusing the fission energy on a mixture of lithium deuteride and plutonium-239. The mixture is compressed to fusion temperatures, and fission neutrons convert lithium to helium and tritium. D-T fusion then occurs, providing the device with approximately half its explosive yield. The remainder comes from fission in an outer layer of natural uranium, whose nonfissile U-238 nevertheless fissions under the bombardment of high-energy neutrons. There's essentially no limit to the yield of a thermonuclear weapon, and devices as large as 58 megatons (Mt) TNT equivalent have been tested. That's 5000 times the energy of the fission bomb that destroyed Hiroshima. Today's missile-based thermonuclear weapons range from 40 kt to about 1 Mt, and there are still thousands of them in the world's arsenals.

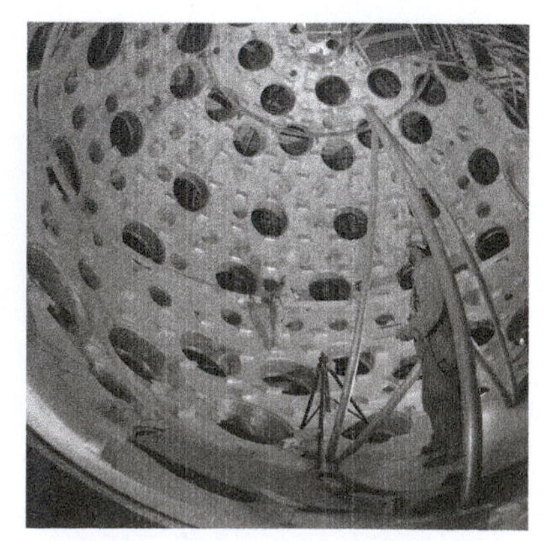

FIGURE 38.20 Target chamber of the National Ignition Facility is 11 m in diameter and weighs 130 tonnes. Holes are ports for the 192 laser beams that converge on the millimeter-size target.

Inertial confinement fusion (ICF) schemes for controlled fusion focus high-power laser beams on millimeter-size deuterium–tritium targets, producing miniature thermonuclear explosions. The most advanced ICF experiment is the National Ignition Facility (NIF) at California's Lawrence Livermore National Laboratory. NIF uses 192 laser beams to focus a 1.9-MJ pulse of some 20-ns duration on a gold chamber containing the D-T target (Fig. 38.20); the gold converts laser light energy into intense X rays that then compress the target to the temperature and density needed for fusion. During the brief pulse, the peak laser power is nearly 500 TW—about 30 times humanity's total energy-consumption rate. In 2013 NIF achieved a milestone of sorts, with the energy produced from fusion exceeding for the first time the energy absorbed by the target. But there's still a long way to go: Total laser beam energy was still 100 times the energy from fusion, and far more energy still was required to produce the laser beams. It takes several days to prepare NIF for a single laser "shot," while a fusion power plant would require some 15 shots each second to reach a commercially viable power output. Development of fusion energy is only one of NIF's three broad purposes; the others are to explore matter under extreme conditions and to simulate nuclear weapons explosions without carrying out actual nuclear tests.

Magnetic Confinement Fusion

In Chapter 26 we saw how charged particles in highly conducting plasma are essentially "frozen" to the magnetic field lines. Trapping of charged particles on magnetic field lines is the essence of magnetic confinement fusion schemes. The first job of magnetic confinement is to create a magnetic configuration that keeps plasma away from the relatively cool walls of the device. Plasma can escape to the walls in three general ways, as shown in Fig. 38.21.

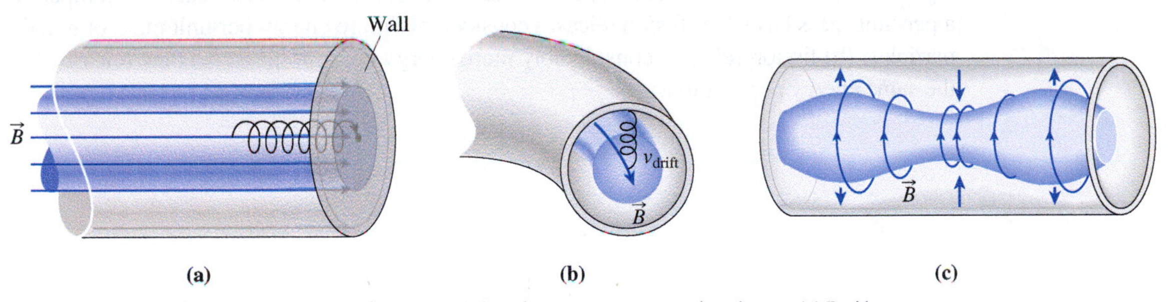

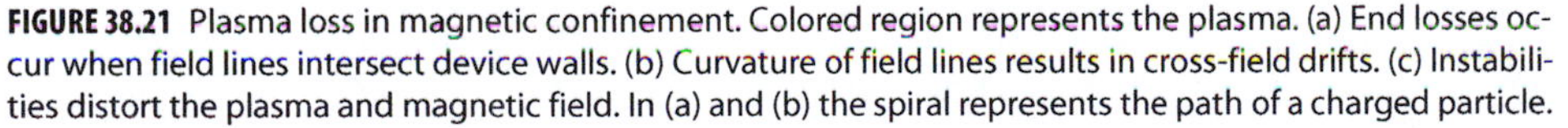

FIGURE 38.21 Plasma loss in magnetic confinement. Colored region represents the plasma. (a) End losses occur when field lines intersect device walls. (b) Curvature of field lines results in cross-field drifts. (c) Instabilities distort the plasma and magnetic field. In (a) and (b) the spiral represents the path of a charged particle.

The most promising magnetic fusion device is the **tokamak**, a Russian invention now used worldwide in fusion research. The tokamak has a toroidal configuration whose magnetic field lines never penetrate the walls, eliminating the end loss shown in Fig. 38.21*a*. Making the machine larger reduces field-line curvature, and with it the cross-field drift of Fig. 38.21*b*. Additional field components enhance confinement and suppress the instabilities of Fig. 38.21*c*. After smaller tokamaks paved the way, a vast international consortium is constructing the ITER fusion reactor in France (Fig. 38.22). ITER is scheduled to begin plasma experiments in the early 2020s and should begin fusion operations late in that decade. ITER is expected to be the first magnetic fusion system to produce net energy exceeding the energy used for plasma heating. ITER will operate at a plasma temperature higher than 100 MK and should generate 400 MW of fusion power from its 840 cubic meters of D-T plasma. ITER will use deuterium and lithium as fuel, with tritium (^3_1H) "bred" right in the reactor by neutron bombardment: $^6_3\text{Li} + ^1_0n \longrightarrow ^4_2\text{He} + ^3_1\text{H}$.

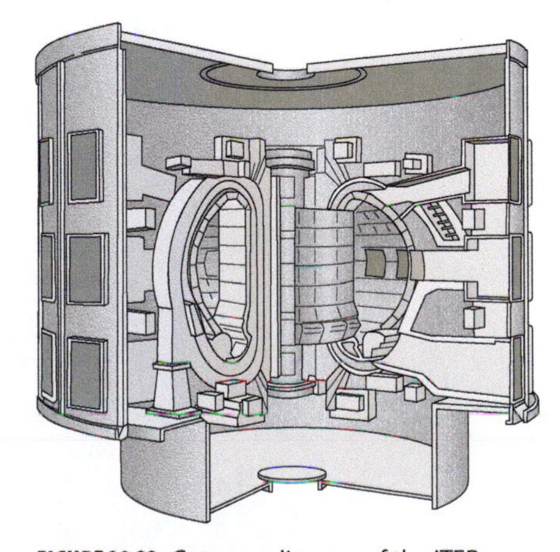

FIGURE 38.22 Cutaway diagram of the ITER fusion reactor. D-shaped structures are cross sections of the toroidal plasma chamber.

Prospects for Fusion Energy

When work on controlled fusion began in the 1950s, scientists confidently predicted that limitless fusion energy would be available in a few decades. More than half a century later, controlled fusion still appears decades away. But it's a goal worth pursuing: Problem 72 shows that with D-D fusion, a gallon of seawater is equivalent to some 300 gallons of gasoline, and Problem 73 shows that fusion energy resources could last far longer than the Sun will continue to shine!

Once controlled fusion proves scientifically feasible, there will be formidable engineering challenges in the design of a practical fusion power plant. Intense neutron fluxes from

D-T fusion degrade materials that form the reaction chamber. Neutron-capture reactions produce radioactive isotopes within the walls, although the associated radioactivity and that of the tritium fuel are much less than the radioactivity of fission waste.

The first practical fusion plants will likely use D-T fusion because its ignition temperature and Lawson criterion are lower than for D-D fusion; the resulting energy will probably run a conventional steam cycle. But the D-D reaction promises cleaner and more efficient power production. Because D-D reaction 38.10c produces charged protons ($^1_1$H) rather than neutral neutrons, there's the possibility of using a *magnetohydrodynamic* (MHD) *generator*, in which electromagnetic induction converts charged-particle kinetic energy directly to electricity. MHD generators would bypass the conventional steam cycle and greatly increase the thermodynamic efficiency of a fusion power plant.

There's one caveat to this rosy fusion future: Although fusion itself is relatively clean and produces no greenhouse gases, the availability of unlimited cheap energy would likely spur industrial growth at a level our planet might not tolerate. And ultimately, if fusion energy use grew exponentially, the waste heat from fusion power could itself have climate-changing consequences.

GOT IT? 38.5 A single D-T fusion reaction releases 17.6 MeV of energy, while fissioning of a single uranium nucleus releases some 200 MeV. How do these reactions compare on a per-unit-mass basis? (a) fusion releases considerably more energy per unit mass of reacting particles; (b) fission releases considerably more energy per unit mass; (c) they release about the same energy per unit mass

---

CHAPTER 38 SUMMARY

Big Idea

The big idea here is that the tiny but massive atomic nucleus is a repository of vast energy—on the order of 10^7 times the energy released in chemical reactions. The protons and neutrons that make up the nucleus can take many configurations, with the **atomic number**, Z, determining the element and the **mass number**, A, determining the particular **isotope**. We write ${}^A_Z X$ to describe the isotope with mass number A of the element whose atomic number is Z and whose symbol is X.

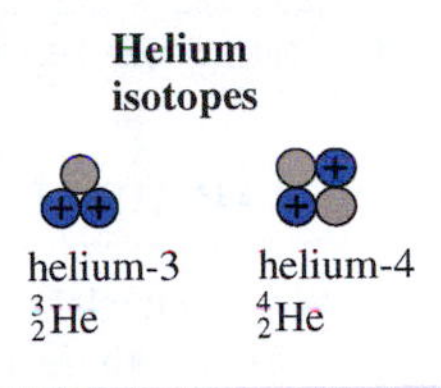

Stable isotopes require a delicate balance between protons and neutrons, with near equal numbers for lighter stable nuclei and more neutrons for heavier nuclei. **Unstable isotopes** are **radioactive** and decay by shedding particles. The **curve of binding energy** shows that energy can be released by either **fusion** of lighter nuclei or **fission** of heavier nuclei.

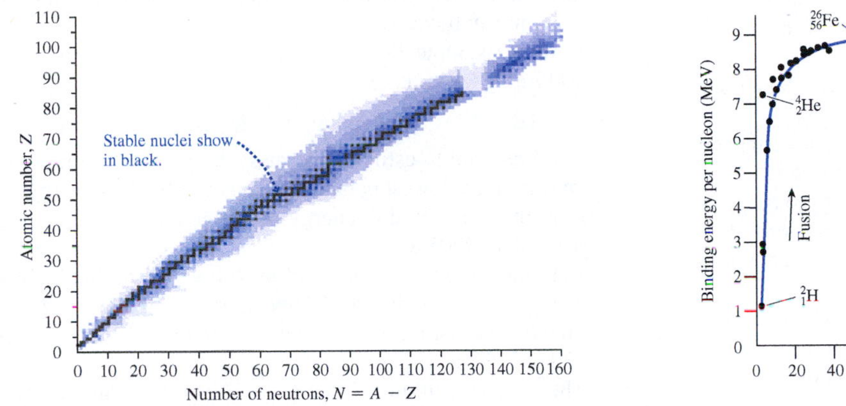

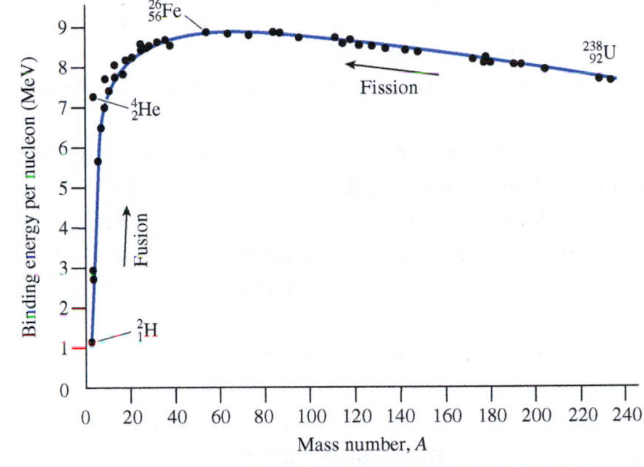

Key Concepts and Equations

Radioactive isotopes decay with a characteristic **half-life**, $t_{1/2}$: $N = N_0 2^{-t/t_{1/2}}$.

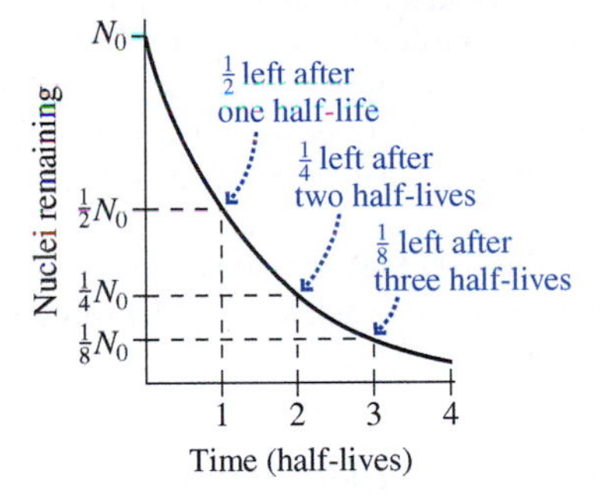

Alpha decay emits a helium-4 nucleus:

$$ {}^A_Z X \longrightarrow {}^{A-4}_{Z-2}Y + {}^4_2\text{He} $$

Beta decay emits an electron or a positron and an antineutrino or a neutrino:

$$ {}^A_Z X \longrightarrow {}_{Z+1}^{\ \ A}Y + e^- + \bar{\nu} $$
$$ {}^A_Z X \longrightarrow {}_{Z-1}^{\ \ A}Y + e^+ + \nu $$

Gamma decay emits a high-energy photon (gamma ray) as an excited nucleus drops to a lower energy state:

$$ X^* \longrightarrow X + \gamma $$

Applications

Radioactivity is measured in **becquerels**, with 1 Bq equal to one decay per second. **Sieverts** (Sv) measure the biological effects of radiation. Residents of the United States receive an average yearly radiation dose of about 3.6 mSv from both natural and artificial sources.

For fission, the most important isotopes are the **fissile** ${}^{235}_{92}$U and ${}^{239}_{94}$Pu, which can fission when struck by low-energy neutrons:

$$ {}^1_0 n + {}^{235}_{92}\text{U} \longrightarrow X + Y + b\,{}^1_0 n $$

The extra neutrons produced in fission can sustain a **chain reaction** provided there's a **critical mass** of fissile material. Exponentially growing chain reactions power fission weapons, while controlled fission occurs in **nuclear reactors** used for power generation.

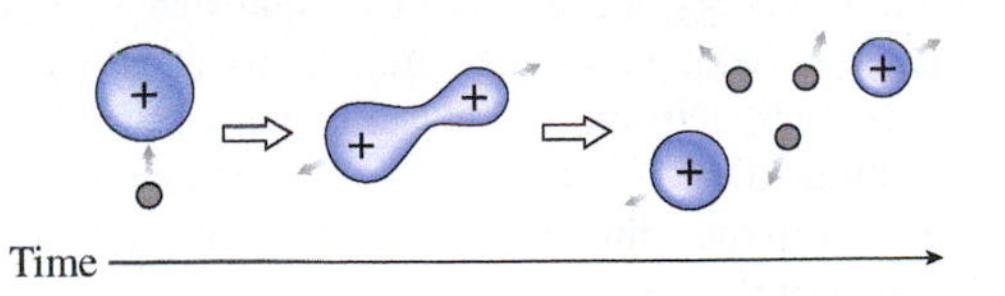

Fusion powers the Sun and stars but has proved elusive on Earth except in thermonuclear weapons. **Inertial confinement** or **magnetic confinement** fusion may one day provide us with nearly limitless energy.

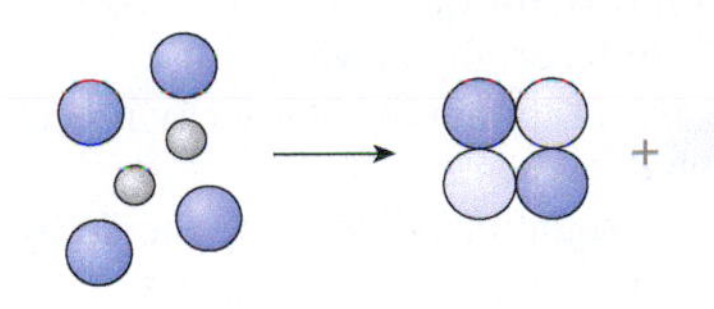

MP *For homework assigned on MasteringPhysics, go to www.masteringphysics.com*

BIO *Biology and/or medicine-related problems* **DATA** *Data problems* **ENV** *Environmental problems* **CH** *Challenge problems* **COMP** *Computer problems*

For Thought and Discussion

1. Why do nuclei contain neutrons?
2. Why are there no stable nuclei for sufficiently high atomic numbers?
3. Why might future archaeologists have problems dating samples from the second half of the 20th century?
4. Beta decay by positron emission is soon followed by a pair of 511-keV gamma rays. Why?
5. Why would it have been easier to make bombs fueled with uranium-235 a few billion years ago?
6. Why are iodine-131 and strontium-90 particularly dangerous radioisotopes?
7. Which model, liquid-drop or nuclear shell, does a better job explaining (a) nuclear fission and (b) gamma-ray spectra?
8. On an energy-release-per-unit-mass basis, by approximately what factor do nuclear reactions exceed chemical reactions?
9. Explain and distinguish the roles of the control rods and moderator in a nuclear reactor.
10. Why is a water-moderated reactor intrinsically safer in a loss-of-coolant accident than a graphite-moderated reactor?
11. Is ^{238}U fissionable? Is it fissile? Explain the distinction.
12. Why are fission fragments necessarily radioactive?
13. Nuclear waste comprises fission products and transuranics. Distinguish the two, including their implications for nuclear waste disposal.
14. What properties of fusion fuel require extreme values and thus present the greatest challenge to fusion energy technologies?
15. Explain the different approaches to the Lawson criterion taken by inertial-confinement and magnetic-confinement fusion schemes.
16. If you could extract all the deuterium from a gallon of seawater and use it as fusion fuel, how much gasoline would it be equivalent to in energy content?

Exercises and Problems

Exercises

Section 38.1 Elements, Isotopes, and Nuclear Structure

17. Three radon isotopes have 125, 134, and 136 neutrons. Write the symbol for each.
18. Write the symbol for the germanium isotope with 44 neutrons.
19. How do (a) the number of nucleons and (b) the nuclear charge compare in the two nuclei $^{35}_{17}$Cl and $^{35}_{19}$K?
20. Compare the radius of the proton (the $A = 1$ nucleus) with the Bohr radius of the hydrogen atom.
21. A uranium-235 nucleus splits into two roughly equal-size fragments. Find their common radius.

Section 38.2 Radioactivity

22. How many half-lives will it take for the activity of a radioactive sample to diminish to 10% of its original level?
23. Copper-64 can decay by any of the three beta-decay processes. Write the equation for each decay.
24. Referring to Fig. 38.7, write equations describing the decays of (a) radon-222 and (b) lead-214.
25. A milk sample shows iodine-131 activity of 450 pCi/L. What's its activity in Bq/L?
26. **BIO** Carbon-11–labeled acetate shows promise in PET scans for determining the extent of metastasized prostate cancer. (a) Given C-11's 20.4-min half-life, how long will it take an initial dose of 2.0 GBq to decay to 7.0 kBq (roughly the natural radioactivity of the human body)? (b) What nucleus remains after C-11 decays by positron emission?
27. Nuclear bomb tests of the 1950s deposited a layer of strontium-90 over Earth's surface. How long will it take from the time of the bomb tests for (a) 99% and (b) 99.9% of this radioactive contaminant to decay?
28. Rework Example 38.2, now using the international guideline of 100 Bq/kg for milk.

Section 38.3 Binding Energy and Nucleosynthesis

29. Use Fig. 38.9 to estimate the mass defect in deuterium, which appears at the lowest point on the curve of binding energy.
30. Find the total binding energy of oxygen-16, given its nuclear mass of 15.9905 u.
31. Determine the nuclear mass of nickel-60, given that its binding energy is very nearly 8.8 MeV/nucleon.
32. Find the nuclear mass of plutonium-239, given its atomic mass of 239.052157 u.
33. The mass of a lithium-7 nucleus is 7.01435 u. Find the binding energy per nucleon.

Section 38.4 Nuclear Fission

34. A ^{235}U nucleus undergoes neutron-induced fission, yielding ^{141}Cs, three neutrons, and another nucleus. What's that nucleus?
35. Neutron-induced fission of ^{235}U yields fission products iodine-139 and yttrium-95. How many neutrons are released?
36. Write a complete equation for neutron-induced fission of plutonium-239 that yields barium-143, two neutrons, and another nucleus.
37. Assuming 200 MeV per fission, determine the number of fission events occurring each second in a reactor whose thermal power output is 3.2 GW.

Section 38.5 Nuclear Fusion

38. Verify from Equations 38.9 that the proton–proton cycle yields net energy of 26.7 MeV.
39. In a magnetic-confinement fusion device with confinement time 0.5 s, what density is required to meet the Lawson criterion for D-T fusion?
40. The National Ignition Facility's 2013–2014 experiments achieving net fusion energy gain involved effective confinement times of about 150 ps. What corresponding density was required to meet the Lawson criterion for D-T fusion?
41. What confinement time is required for the D-T Lawson criterion in the ITER fusion reactor, given its plasma density of 10^{19} particles per cubic meter?

Problems

42. To what diameter would Earth have to collapse to be at nuclear density?
43. Find the energy needed to flip the spin state of a proton in Earth's magnetic field, whose magnitude is about 30 μT.
44. An NMR spectrometer is described as a "300-MHz instrument," meaning 3.00×10^8 Hz is the frequency supplied to its transmitter coil to flip the spin states of bare protons. What's the strength of its unperturbed magnetic field?

45. Iron-56, with nuclear mass 55.9206 u, is among the most tightly bound nuclei. Find the binding energy per nucleon, and check your answer against Fig. 38.9.
46. Find the atomic mass of iridium-193, whose binding energy is 7.94 MeV/nucleon.
47. As a geologist, you're assessing the feasibility of determining the ages of Earth's earliest rocks using radioactive dating. You estimate the number of half-lives that have passed for three different isotopes during Earth's 4.5-billion-year lifetime, and from that you determine the number of atoms remaining today from 10^6 atoms present at Earth's formation. The isotopes you consider are carbon-14, uranium-238, and potassium-40. What are your estimates, and which isotopes do you conclude are suitable for radioactive dating?
48. You measure the activity of a radioactive sample at 2.4 MBq. Thirty minutes later, the activity level is 1.9 MBq. Find the material's half-life.
49. You're a home inspector, and you find radon-222 activity of 23 pCi/L in the air inside a house, well above the EPA's "action" limit of 4 pCi/L. If radon infiltration were stopped but there were no significant ventilation, how long would it take for the radon activity to drop below the action limit?
50. **BIO** Nitrogen-13 is a 9.97-min-half-life isotope used to "tag" ammonia for PET scans, including quantification of myocardial infarction. Consider an intravenous injection incorporating 20.0 mCi of N-13. Plot a graph of N-13 activity versus time, with your vertical axis logarithmic and your horizontal axis linear. Why is the graph a straight line? What's the significance of its slope?
51. Thorium-232 is an α emitter with 14-billion-year half-life. Radium-228 is a β^- emitter with 5.75-year half-life. Actinium-228 is a β^- emitter with 6.13-hour half-life. (a) What's the third daughter in the thorium-232 decay series? (b) Make a chart similar to Fig. 38.7 showing the first three decays in the thorium series.
52. How much cobalt-60 ($t_{1/2} = 5.24$ years) must be used to make a laboratory source whose activity will exceed 1 GBq for 2 years?
53. Archaeologists unearth a bone and find its carbon-14 content is 34% of that in a living bone. How old is the archaeological find?
54. Show that the decay constant and half-life are related by $t_{1/2} = \ln 2/\lambda \simeq 0.693/\lambda$.
55. The table below lists reported levels of iodine-131 contamination in milk in four countries affected by the 1986 Chernobyl accident, along with each country's safety guideline. Given I-131's half-life of 8.04 days, how long did each country have to wait for I-131 levels to decline to a level deemed safe by its standards?

	Level (Bq/L)	
Country	**Reported**	**Safety Guideline**
Poland	2000	1000
Austria	1500	370
Germany	1184	500
Romania	2900	185

56. How many atoms are in a radioactive sample with activity 12 Bq and half-life 15 days?
57. Analysis of a Moon rock shows that 82% of its initial K-40 has decayed to Ar-40, a process with a half-life of 1.2×10^9 years. How old is the rock?
58. You're assessing the safety of an airport bomb-detection system in which neutron activation of the stable nitrogen isotope $^{15}_{7}N$ turns it into unstable $^{16}_{7}N$. The N-16 decays by beta emission with 7.13-s half-life. How long after activation will the N-16 activity have dropped by a factor of 1 million?
59. **BIO** *Brachytherapy* is a cancer treatment involving implantation of radioactive "seeds" at the tumor site. Iridium-192, often used for cancers of the head and neck, undergoes beta decay by electron capture with 74.2-day half-life. Inner-shell electrons drop to the orbital occupied by the captured electron, resulting in emission of gamma rays that kill surrounding tumor cells. What percentage of initial Ir-192 activity will remain one year after implant?
60. Today, uranium-235 comprises only 0.72% of natural uranium; essentially all the rest is U-238. Use the half-lives in Table 38.1 to determine the percentage of uranium-235 when Earth formed about 4.5 billion years ago.
61. You're a geologist assessing underground sites for nuclear waste storage. A ruling by the U.S. Environmental Protection Agency suggests that waste-storage facilities should be designed for a million years of radiation protection. What fraction of plutonium-239 initially in nuclear waste would remain after that time?
62. **BIO** Oxygen-15 ($t_{1/2} = 2.0$ min) is produced in a hospital's cyclotron. What should the initial activity concentration be if it takes 3.5 min to get the O-15 to a patient undergoing a PET scan requiring 0.50 mCi/L of activity?
63. How much ^{235}U would be needed to fuel the reactor of Exercise 37 for 1 year? (*Note:* Your answer is an overestimate because fission of ^{239}Pu also contributes to the power output.)
64. How much uranium-235 would be consumed in a fission bomb with a 25-kt explosive yield?
65. A neutron collides elastically head-on with a stationary deuteron in a reactor moderated by heavy water. How much of its kinetic energy is transferred to the deuteron? (*Hint:* Consult Chapter 9.)
66. A buildup of fission products "poisons" a reactor, dropping the multiplication factor to 0.992. How long will it take the reactor power to decrease by half, given a generation time of 0.10 s?
67. The total thermal power generated in a nuclear power reactor is 1.5 GW. How much U-235 does it consume in a year?
68. New Hampshire's Seabrook nuclear power plant produces electrical energy at the rate of 1.2 GW and consumes 1311 kg of U-235 each year. Assuming the plant operates continuously, find (a) its thermal power output and (b) its efficiency.
69. In the dangerous situation of prompt criticality in a fission reactor, the generation time drops to 100 μs as prompt neutrons sustain the chain reaction. If a reactor goes prompt critical with $k = 1.001$, how long does it take for a 100-fold increase in reactor power?
70. How much heavy water (deuterium oxide, ^2H_2O or D_2O) would be needed to fuel a 1000-MW D-D fusion power plant for 1 year?
71. The proton–proton cycle consumes four protons while producing 27 MeV of energy. (a) At what rate must the Sun consume protons to produce its power output of 4×10^{26} W? (b) The present phase of the Sun's life will end when it has consumed about 10% of its original protons. Estimate how long this phase will last, assuming the Sun's 2×10^{30}-kg mass was initially 71% hydrogen.
72. You're enthusiastic about fusion energy, and you want to convince others of the enormous fuel resource represented by the 0.015% of hydrogen nuclei that are actually deuterium. Using an average of 7.2 MeV per deuteron, you calculate the energy that would be released if all the deuterium in a gallon of seawater underwent fusion, and you compare your result with the energy in a gallon of gasoline (see Appendix C). What do you find for the gasoline equivalent of a gallon of seawater?
73. In a further effort to convince others of the benefits of fusion energy, you use the data from Problem 72 to estimate how long the deuterium in the world's oceans (average depth 3 km) could supply humanity's energy needs at the current consumption rate of about 16 TW. You then compare this with the Sun's remaining lifetime, about 5 billion years. What do you find?

74. The atomic masses of uranium-238 and thorium-234 are 238.050784 u and 234.043593 u, respectively. Find the energy released in the alpha decay of U-238.
75. Bismuth-209 and chromium-54 combine to form a heavy nucleus plus a neutron. Identify the heavy nucleus.
76. It's possible, but difficult, to realize alchemists' dreams of synthesizing gold. One approach bombards mercury-198 with neutrons to produce, for each neutron captured, a gold-197 nucleus and another particle. Write the equation for this reaction.
77. Nickel-65 beta decays by electron emission with decay constant $\lambda = 0.275\ \text{h}^{-1}$. (a) Identify the daughter nucleus. (b) In a sample of initially pure Ni-65, find the time when there are twice as many daughter nuclei as parents.
78. **BIO** The dominant naturally occurring radioisotopes in the typical human body include 16 mg of ^{40}K and 16 ng of ^{14}C. Using half-lives from Table 38.1, estimate the body's natural radioactivity.
79. A laser-fusion fuel pellet has mass 1.0 mg and consists of equal parts (by mass) of deuterium and tritium. (a) If half the deuterons and an equal number of tritons participate in D-T fusion, how much energy is released? (b) At what rate must pellets be fused in a power plant with 3000-MW thermal power output? (c) What mass of fuel would be needed to run the plant for 1 year? Compare your answer with the 3.6×10^6 tons of coal needed to fuel a comparable coal-burning power plant.
80. Of the neutrons emitted in each fission event in a light-water reactor, an average of 0.6 neutron is absorbed by ^{238}U, leading to the formation of ^{239}Pu. (a) Assuming 200 MeV per fission, how much ^{239}Pu forms each year in a 30%-efficient nuclear plant whose electric power output is 1.0 GW? (b) With careful design, a fission explosive can be made from 5 kg of ^{239}Pu. How many potential bombs are produced each year in the power plant of part (a)?
81. **BIO** A family member is about to have a brain scan using technetium-99m, an excited isotope with 6.01-hour half-life. The hospital makes Tc-99m from the decay of molybdenum-99 ($t_{1/2} = 2.7$ days), then delivers it to the nuclear medicine department. You're told that the Tc-99m will arrive 90 minutes after production, and that there must be 10 mg of it. The technician says she will produce 12 mg of Tc-99m. Is that sufficient?
82. **DATA CH** A mix of two isotopes, one of them from Table 38.1, is observed over a period of 15 days, and the total radioactivity is tabulated below. Determine a quantity that, when plotted against time, should yield one or more straight lines. Make your plot and use it to determine the half-lives of the isotopes. Identify the isotope from Table 38.1.

Time (days)	0	0.25	0.5	0.75	1.0
Activity (kBq)	200	103	54	29	17

Time (days)	1.3	5.0	10	15
Activity (kBq)	4.6	3.2	2.1	1.4

83. **CH** The probability that a radioactive nucleus will have lifetime t is the probability that it will survive from time 0 to time t multiplied by the probability that it will decay in the interval from t to $t + dt$. Use this to show that the average lifetime of a nucleus is equal to the inverse of the decay constant in Equation 38.3a.
84. **CH** Nucleus A decays into B with decay constant λ_A and B decays into a stable product C with decay constant λ_B. A pure sample starts with N_0 nuclei A at $t = 0$. Find an expression for the total activity of the sample at time t.
85. **CH** (a) Example 38.6 explains that the number of fission events in a chain reaction increases by a factor k with each generation. Show that the total number of fission events in n generations is $N = (k^{n+1} - 1)/(k - 1)$. (b) In a typical nuclear explosive, k is about 1.5 and the generation time is about 10 ns. Use the result from (a) to calculate the time for all the nuclei in a 10-kg mass ^{235}U to fission. *Hint*: Sum a series in part (a), and neglect 1 compared with N in part (b).

Passage Problems

In 1972, a worker at a nuclear fuel plant in France found that uranium from a mine in Oklo, in the African Republic of Gabon, had less U-235 than the normal 0.7%—a quantity known from meteorites and Moon rocks to be constant throughout the solar system. Further analysis showed the presence of isotopes that would result from the decay of fission products. Scientists drew the remarkable conclusion that a natural nuclear fission reaction had occurred some 2 billion years ago, lasting for about 100,000 years. Water, mixing with rich uranium ore, provided the moderator that enabled the chain reaction. More significantly, U-235's 700-My half-life means that 2 billion years ago there was a higher abundance of U-235 in natural uranium.

86. At the time of the Oklo fission reaction, the actual *amount* of U-235 present was
    a. about the same as today.
    b. about twice as much as today.
    c. about four times as much as today.
    d. about eight times as much as today.
87. Given U-238's 4.5-billion-year half-life, the percentage of U-235 in natural uranium 2 billion years ago was
    a. about 1%.
    b. about 4%.
    c. about 10%.
    d. nearly 100%.
88. The power output from fission at Oklo was 10 kW to 100 kW. If at some point that power had been sufficient to boil away the water at the reaction site, the chain reaction would have
    a. ceased.
    b. continued, but more slowly.
    c. been unaffected.
    d. sped up.
89. At the Oklo site today, you would expect to find measurable amounts of
    a. strontium-90.
    b. cesium-137.
    c. plutonium-239.
    d. none of the above.

Answers to Chapter Questions

Answer to Chapter Opening Question

Nuclear power supplies about 9% of the world's total energy, and more than 11% of electrical energy.

Answers to GOT IT? Questions

38.1 (a) $Z = 6, N = 6$; (b) $Z = 8, N = 7$; (c) $Z = 26, N = 31$; (d) $Z = 94, N = 145$
38.2 (e)
38.3 $^{57}_{26}$Fe, $^{132}_{54}$Xe, $^{238}_{92}$U, $^{4}_{2}$He, $^{2}_{1}$H
38.4 Energy is vastly more concentrated in nuclear fuels, which means far less nuclear fuel is mined and transported.
38.5 (a)

39

From Quarks to the Cosmos

What You Know

- You're familiar with the concepts of energy and force.
- You've studied gravity and electromagnetism, which you know are among the fundamental forces.
- You're familiar with the most common of the "elementary" particles, including the electron, the proton, and the neutron.
- You've studied several conservation laws.

What You're Learning

- You'll learn how particles mediate interactions in the quantum description of forces.
- You'll learn to evaluate conserved quantities in particle interactions.
- You'll be able to describe the standard model of particles and forces, including the role of the newly discovered Higgs boson.
- You'll learn that baryons (including the proton and neutron) aren't fundamental particles but are composed of quarks.
- You'll get a glimpse of physicists' current efforts to unify the fundamental forces.
- You'll learn the latest evidence for the Big Bang origin of the universe and the ongoing cosmic expansion, emphasizing especially the importance of the cosmic microwave background radiation.
- You'll be able to outline the overall history of the universe.

How You'll Use It

- This chapter covers the frontiers of physics. After you've completed your physics course, keep watching for news of new discoveries in particle physics, fundamental forces, and cosmology!

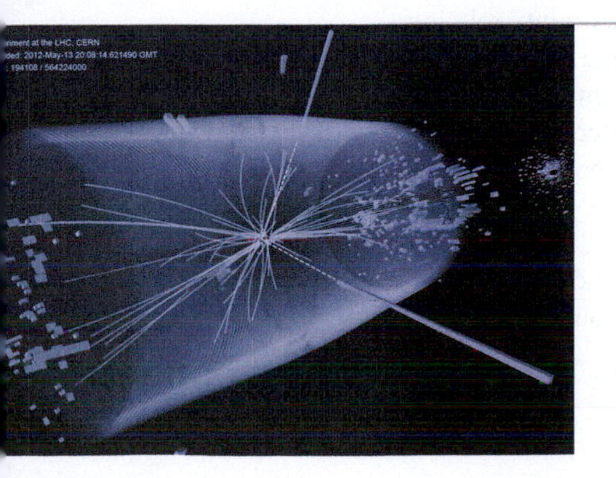

The past five chapters have extended the realm of physics to the scales of atoms and molecules and then down to the atomic nucleus. Here we go further still, probing the structure of nucleons themselves and trying to make sense of the host of subatomic particles nature reveals. We'll be asking questions about the ultimate nature of matter at the smallest scales, but in the process we'll find a remarkable connection with questions of the largest scale—questions about the origin and ultimate fate of the entire universe.

A Higgs boson decays to two photons in this image from the Large Hadron Collider. Discovery of the Higgs in 2012 completed physicists' standard model of particles and fields. What role does the Higgs play in particle physics?

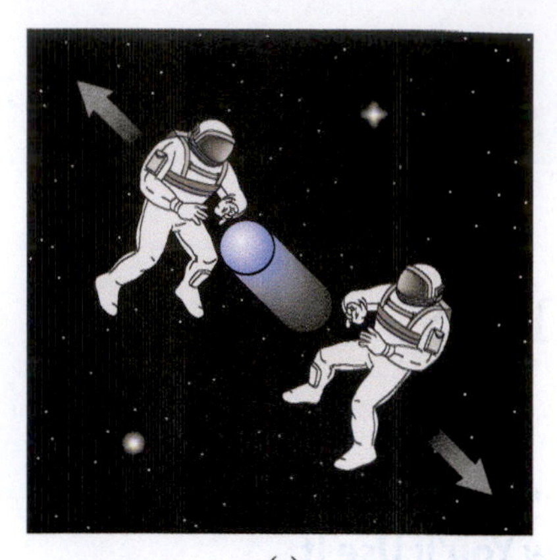
(a)

(b)

FIGURE 39.1 Analogs for particle-mediated forces: (a) repulsive and (b) attractive. The ball represents a photon.

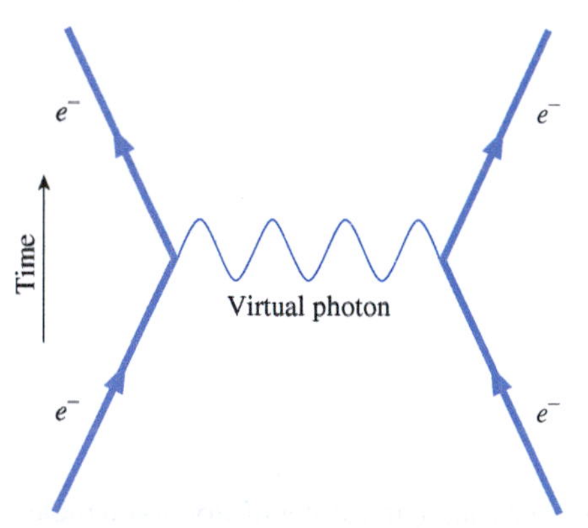

FIGURE 39.2 A Feynman diagram, showing the interaction of two electrons through the exchange of a virtual photon. The diagram provides a quantum description of the electrons' Coulomb repulsion.

39.1 Particles and Forces

By 1932 four "elementary" particles of matter were known: the electron, the proton, the neutron, and the neutrino. In addition, there were the positron, antiparticle to the electron, and the photon of electromagnetic radiation. There were also the seemingly fundamental forces—gravity, the electromagnetic force, the nuclear force, and the weak force that manifests itself in beta decay.

In Chapter 34 you saw how the interaction of electromagnetic waves with matter ultimately involves individual photons—the quanta of the electromagnetic field. In the quantum-mechanical view of electromagnetism, the force between two charged particles also involves photons, now exchanged between the interacting particles. Imagine two astronauts tossing a ball back and forth (Fig. 39.1*a*). Catching or throwing the ball, one astronaut gains momentum in a direction away from the other, so the exchange results in a net average repulsive force. If the two astronauts struggle for possession of the ball, then the ball mediates what appears as an attractive interaction (Fig. 39.1*b*). Figure 39.1 gives classical analogs for the attractive and repulsive electrical interactions involving photon exchange.

You know that photons are emitted when a particle jumps into a lower energy state, with the photon carrying off energy equal to the energy difference between the two states. This process conserves energy. But now we're saying that a single, free electron emits photons that it exchanges with another particle to produce what we call the electromagnetic force. How can that process conserve energy? The energy–time uncertainty relation (Equation 34.16) says that an energy measured in a time Δt is necessarily uncertain by an amount $\Delta E \geq \hbar/\Delta t$. The photon exchanged by two particles lasts only a short time, and therefore its energy is uncertain. So we can't really say that energy conservation is violated. A photon created in this way and lasting for only the short time it takes to exchange with another particle is called a **virtual photon**. We never "see" the virtual photon, since it's emitted by one particle and absorbed by the other.

The quantum theory of the electromagnetic interaction is called **quantum electrodynamics** (QED). Although begun by Paul Dirac, it was brought to consistent form in 1948 by Richard Feynman, Sin-Itiro Tomonaga, and Julian Schwinger. The fundamental event in QED is the interaction of a photon with an electrically charged particle. Two such events joined by a common virtual photon give the quantum electrodynamical description of the electromagnetic force (Fig. 39.2). The predictions of quantum electrodynamics have been confirmed experimentally to a remarkably high precision, and today QED is our best-verified theory of physical reality.

Mesons

In 1935 the Japanese physicist Hideki Yukawa proposed that the nuclear force should, like the electromagnetic force, be mediated by exchange of a particle. Yukawa called his hypothetical particle a **meson**. Because the range of the nuclear force is limited, Yukawa argued, the meson should have nonzero mass. The reason for this connection between mass and range again lies in the energy–time uncertainty relation.

The electromagnetic force falls off as $1/r^2$ and thus has an infinite range. Two particles can be very far apart and still interact electromagnetically. Since photons travel at the finite speed of light, the time Δt for a photon interaction can be arbitrarily long. The energy–time uncertainty relation $\Delta E\,\Delta t \geq \hbar$ thus shows that the energy uncertainty ΔE can be arbitrarily small. Thus, the possible energies for virtual photons must extend downward toward zero—and that can happen only if photons have zero rest energy.

The nuclear force, in contrast, has a finite range of about 1.5 fm. At close to the speed of light, the longest a particle mediating this force would need to exist is a time Δt given by

$$\Delta t = \frac{\Delta x}{c} = \frac{1.5\times10^{-15}\,\text{m}}{3.0\times10^{8}\,\text{m/s}} = 5.0\times10^{-24}\,\text{s}$$

Then energy–time uncertainty gives

$$\Delta E \geq \frac{\hbar}{\Delta t} = \frac{1.05\times10^{-34}\,\text{J}\cdot\text{s}}{5.0\times10^{-24}\,\text{s}} = 2.1\times10^{-11}\,\text{J} = 130\,\text{MeV}$$

Yukawa therefore proposed a new particle with mass 130 MeV/c^2, about 250 times that of the electron. Yukawa's prediction was eventually confirmed—but not before physicists found yet another particle.

39.2 Particles and More Particles

In the 1930s the most available source of high-energy particles was cosmic radiation—high-energy protons and other particles of extraterrestrial origin. In 1937 the American physicist Carl Anderson (who had earlier discovered the positron) and his colleagues identified in cosmic rays a particle with a mass 207 times that of the electron. Now called the **muon**, this particle had the same charge and spin as the electron and seemed to behave much like a heavier version of the electron. Two muons were found: the negatively charged μ^- and its antiparticle μ^+. Although the muon mass was close to Yukawa's prediction, the muon interacted only weakly with nuclei and, therefore, could not be the mediator of the nuclear force.

The real Yukawa particle was discovered 10 years later in 1947, again in cosmic rays, and turned out to have a mass about 270 times that of the electron. This time there were three related particles, now called **pions**: positive π^+, negative π^-, and neutral π^0.

The new particles are all unstable, undergoing decays that ultimately result in well-known stable particles. The negative pion, for example, decays with a mean lifetime of 26 ns to a negative muon and an antineutrino:

$$\pi^- \longrightarrow \mu^- + \overline{\nu}$$

The muon then decays with a 2.2-μs lifetime to an electron and a neutrino–antineutrino pair:

$$\mu^- \longrightarrow e^- + \nu + \overline{\nu}$$

APPLICATION Particle Detection

Despite their small size, we can, remarkably, follow the trajectories of individual subatomic particles. Early particle detectors included the **cloud chamber** and the **bubble chamber**, in which particles ionize vapor or liquid, causing visible condensation or bubble formation along the particle tracks. More recent is the **multiwire proportional chamber**, in which criss-crossed wire grids record current pulses from electrons liberated as particles pass through a gas-filled chamber (see the photo, which shows a multiwire chamber at the Stanford Linear Accelerator Center). Analyzing the pulse distribution then reveals particle trajectories. Applying a magnetic field in a particle detector curves the trajectories, enabling scientists to determine the particles' charge-to-mass ratios. **Scintillation detectors** give off light flashes as particles pass through them, and the flash intensity provides a measure of particle energy. **Calorimeters**, consisting of layers of scintillators and energy-absorbing material, analyze the showers of secondary particles produced by a single high-energy particle to determine the original particle's energy. Modern detectors are huge agglomerations of several basic detector types, arranged to extract the maximum information from particle interactions. Computer analysis of detector output allows the identification of events so rare they may occur only once in a million interactions.

Classifying Particles

The availability of increasingly powerful particle accelerators led to an upsurge in particle discoveries. By 1980 there were more than 100 "elementary" particles. Early attempts to classify particles distinguished them by mass, but a more enlightening approach is based on the fundamental forces. We'll now use this approach to outline three particle classes.

Leptons are particles that don't experience the strong force. They include the familiar electron, the muon, a more massive particle called the tau, and three types of neutrinos, one associated with each of the charged leptons. The neutrinos were long thought to be massless, but recent experiments show that neutrinos have small nonzero mass and that they "oscillate" among the three types. Each of the leptons has an antiparticle as well. There are thus a total of six lepton–antilepton pairs, and experimental evidence strongly suggests that no others can exist. The leptons all have spin $\frac{1}{2}$ and are therefore **fermions**, which obey the Pauli exclusion principle. Leptons are believed to be true elementary point particles with zero size and no internal structure.

Hadrons are particles that do experience the strong force. They fall into two subclasses: mesons and baryons. **Mesons** have integer spin and are therefore **bosons** that don't obey the exclusion principle. Mesons include Yukawa's pions and a host of others; all are unstable. **Baryons** have half-integer spins and are therefore fermions. They include the familiar proton and neutron and similar but more massive particles. The baryons are grouped into pairs, triplets, and higher-multiple groupings of closely related particles. The neutron and proton, for example, form a pair that differ in charge and very slightly in mass. Each baryon has an antiparticle, as do most mesons, but some neutral mesons are their own antiparticles.

The third class of particles comprises the **field particles** or **gauge bosons**, quanta of the different force fields and "carriers" of those forces. These include the familiar photon for the electromagnetic force; three particles called the W^+, W^-, and Z for the weak force; the **gluon** for the strong force and a hypothetical **graviton** that would carry the gravitational force in an as yet incomplete theory of quantum gravity. All the field particles are bosons, carrying spin 1 or, for the graviton, spin 2. You might think Yukawa's meson should be in this category in its role as carrier of the nuclear force. That it doesn't appear here is a hint that the nuclear force isn't really fundamental; as we'll soon see, it's the gluon that plays the more fundamental role.

Table 39.1 lists some of the particles known even before full confirmation of today's elementary particle theories.

Particle Properties and Conservation Laws

Many new particles can be characterized by known properties such as mass, spin, and electric charge. Of these, spin and charge are associated with important conservation laws—conservation of angular momentum and conservation of electric charge. Allowed interactions among particles *must* conserve these quantities. The annihilation of an electron–positron pair, for example, is allowed because the initial particles have no net charge and neither do the resulting photons. Similarly, beta decay of the neutron produces an electron, a proton, and a neutral antineutrino and thus conserves charge:

$$n \longrightarrow p + e^- + \bar{\nu}_e$$

Here the subscript on the antineutrino indicates that it's an electron antineutrino, as opposed to the muon or tau variety.

Other particle properties appear to be conserved as well. Associated with each baryon or antibaryon is its **baryon number** (B), assigned the value $+1$ for a baryon and -1 for an antibaryon. All experimental evidence to date points to conservation of baryon number: In all observed particle reactions, the sums of the baryon numbers before and after the reaction are equal. An example is, again, beta decay of the neutron: The process starts with a neutron of baryon number $B = 1$ and ends with a proton $(B = 1)$, an electron, and an antineutrino. The last two are leptons, so their baryon numbers are zero, and thus baryon number is conserved. Some theories, which we describe shortly, suggest that baryon number is only approximately conserved. If that's so, then the proton itself is an unstable particle with a mean lifetime in excess of 10^{35} years.

Lepton number seems also to be conserved. Again, beta decay provides an example: The neutron and proton, being baryons, have lepton number zero, while the resulting electron and *anti*neutrino have lepton numbers $+1$ and -1, respectively.

In the late 1950s particles called K, Λ, Σ, and Ξ were discovered. Strange characteristics of these particles' decays could be explained by introducing a new fundamental property, called **strangeness** (s), with the new particles having $s = \pm 1$ or ± 2. Strangeness is conserved in strong and electromagnetic interactions, but in weak interactions its value can change. We'll soon see that several other new properties are needed to characterize matter fully.

Table 39.1 Selected Particles*

Category/Particle	Symbol (Particle/ Antiparticle)	Spin	Mass (MeV/c^2)	Baryon Number, B	Lepton Number, L	Strangeness, s	Lifetime (s)
Field particles							
Photon	γ, γ	1	0	0	0	0	Stable
Z^0	Z^0, Z^0	1	91,188	0	0	0	$\sim 10^{-25}$
Leptons							
Electron	e^-, e^+	$\frac{1}{2}$	0.511	0	+1	0	Stable
Muon	μ^-, μ^+	$\frac{1}{2}$	105.7	0	+1	0	2.2×10^{-6}
Tau	τ^-, τ^+	$\frac{1}{2}$	1777	0	+1	0	2.9×10^{-13}
Electron neutrino	$\nu_e, \bar{\nu}_e$	$\frac{1}{2}$	$<3\times10^{-6}$	0	+1	0	Stable
Muon neutrino	$\nu_\mu, \bar{\nu}_\mu$	$\frac{1}{2}$	<0.19	0	+1	0	Stable
Tau neutrino	$\nu_\tau, \bar{\nu}_\tau$	$\frac{1}{2}$	<18.2	0	+1	0	Stable
Hadrons							
Mesons							
Pion	π^+, π^-	0	139.6	0	0	0	2.6×10^{-8}
Pion	π^0, π^0	0	135.0	0	0	0	8.4×10^{-17}
Eta	η^0, η^0	0	547.8	0	0	0	$\sim 5\times10^{-19}$
Rho	ρ^0, ρ^0	1	775.8	0	0	0	$\sim 4\times10^{-24}$
Kaon	K^+, K^-	0	493.7	0	0	1	1.2×10^{-8}
Kaon	K^0, K^0	0	497.6	0	0	1	0.895×10^{-10} 5.18×10^{-8}†
Baryons							
Proton	$p, \bar{p}$	$\frac{1}{2}$	938.3	1	0	0	Stable
Neutron	$n, \bar{n}$	$\frac{1}{2}$	939.6	1	0	0	885.7
Lambda	$\Lambda^0, \bar{\Lambda}^0$	$\frac{1}{2}$	1115.7	1	0	-1	2.6×10^{-10}
Sigma	$\Sigma^+, \bar{\Sigma}^-$	$\frac{1}{2}$	1189.4	1	0	-1	0.80×10^{-10}
Sigma	$\Sigma^0, \bar{\Sigma}^0$	$\frac{1}{2}$	1192.6	1	0	-1	7.4×10^{-20}
Sigma	$\Sigma^-, \bar{\Sigma}^+$	$\frac{1}{2}$	1197.4	1	0	-1	1.5×10^{-10}
Omega	$\Omega^-, \bar{\Omega}^+$	$\frac{3}{2}$	1672.45	1	0	-3	0.82×10^{-10}

*Does not include any hadrons with c, b, or t quarks.

†The neutral kaon exists as a quantum-mechanical superposition of states with two different lifetimes.

CONCEPTUAL EXAMPLE 39.1 Conservation Laws: Evaluating a Particle Interaction

A pion collides with a proton to produce a neutral kaon and a lambda particle:

$$\pi^- + p \longrightarrow K^0 + \Lambda^0$$

(a) Which of the following are conserved: electric charge, baryon number, lepton number, strangeness? (b) Could another possible result of a pion–proton collision be an electron and a proton?

EVALUATE (a) Table 39.1 shows that all the particles are hadrons, so the lepton number is zero on both sides of the equation. On the left we have a positive and a negative particle and on the right two neutrals, so the electric charge is zero on both sides. The pion is a meson and the proton a baryon, so the baryon number on the left is 1; similarly, the kaon is a meson and the Λ^0 a baryon, so the baryon number is conserved. Finally, neither pion nor proton is strange, so the total strangeness on the left is zero. Table 39.1 lists the K^0 with $s = +1$ and the Λ^0 with $s = -1$, so strangeness, too, is conserved. (b) Having an electron and a proton as the final state would conserve charge (0), strangeness (0), and baryon number (1). But it wouldn't conserve lepton number, which was originally 0 but would become 1.

ASSESS This example shows how conservation laws restrict the possible outcomes of particle interactions.

MAKING THE CONNECTION What minimum kinetic energy is required for the pion and proton together in order for this reaction to occur?

EVALUATE Table 39.1 gives the rest masses of the pion as 139.6 MeV/c^2 and the proton as 938.3 MeV/c^2; therefore, their rest energies mc^2 are 139.6 MeV and 938.3 MeV, for a total of 1078 MeV. But the tabulated masses of the K^0 and Λ^0 show that the sum of their rest energies is 1613 MeV, so we need an additional $1613 - 1078 = 535$ MeV of energy to drive the reaction. This is the minimum value for the initial kinetic energy—the sum of the pion and proton kinetic energies.

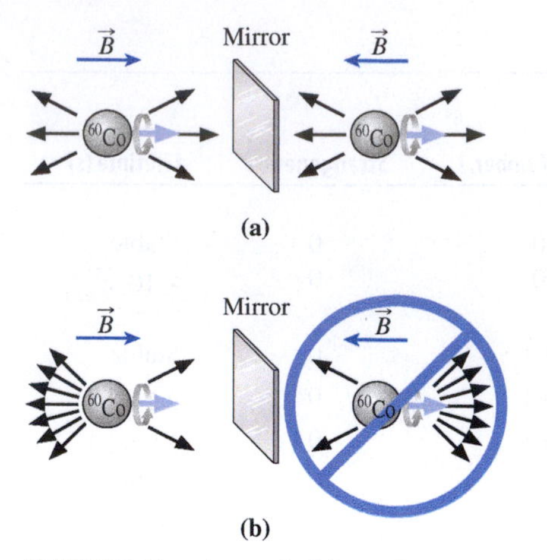

FIGURE 39.3 Experimental evidence for nonconservation of parity. At left of the mirror, a ^{60}Co nucleus has its spin aligned with a magnetic field. (a) Reflected in the mirror, the spin vector still points to the right, even though the magnetic field is reversed. If the mirror image were equally likely, beta emission (arrows) would occur with equal probability along and opposite the spin direction. (b) Experiment shows that beta emission occurs preferentially opposite the spin direction, so the mirror-image situation at the right in (b) does not occur.

Symmetries

Watch a physical process in a mirror, and you expect the image to be a possible physical process; that is, the laws of physics should exhibit **symmetry** with respect to mirror reflection. At the subatomic level, the statement that a process and its mirror image are equally likely is called **conservation of parity**. Mathematically, a system has parity $+1$ if its wave function is unchanged on reflection through the origin—that is, on a coordinate change $x \rightarrow -x, y \rightarrow -y, z \rightarrow -z$. If the wave function changes sign, then the parity is -1. Parity is conserved if its value is unchanged in a particle interaction.

In 1957 theoretical physicists Tsung-dao Lee and Chen Ning Yang pointed out that parity conservation had not been tested for the weak force. They made the revolutionary suggestion that parity need not be conserved—tantamount to suggesting that nature can distinguish right- from left-handed systems that are otherwise identical. A group led by Chien-Shiung Wu took up the challenge. Wu studied the beta decay of cobalt-60 in a magnetic field that established a left–right symmetry. Her experiments showed a preferential beta emission opposite the field direction—a clear violation of parity conservation (Fig. 39.3).

Although parity might not be conserved, theorists held that a combination of parity reversal (P) and charge conjugation (C)—changing particles into antiparticles—would result in indistinguishable physical behavior. But in 1964 a violation of this CP conservation was found in a rare decay of the neutral kaon to a pion–antipion pair. The Russian physicist Andrei Sakharov suggested that this asymmetric decay might account for the preponderance of matter over antimatter in today's universe.

It still appears that CPT conservation holds; that is, a combination of mirror reflection, charge conjugation, and reversal of the time coordinate makes a new physical process indistinguishable from the original. There may be a deep philosophical connection here with the direction of time, but the full implications of CPT symmetry and the failure of its individual components aren't fully understood.

39.3 Quarks and the Standard Model

The proliferation of particles distressed physicists used to finding an underlying simplicity in nature. Were all those particles really "elementary," or was there a more fundamental, simpler level? In 1961 physicists Murray Gell-Mann and Yuval Ne'eman independently noticed patterns in the then-known particles. They called their patterns the **Eightfold Way**, after Buddhist principles for right living. An empty spot in one pattern led Gell-Mann to predict a new particle with strangeness -3. Experimentalists soon found the particle, now known as the Ω^-, in bubble-chamber photographs from earlier experiments.

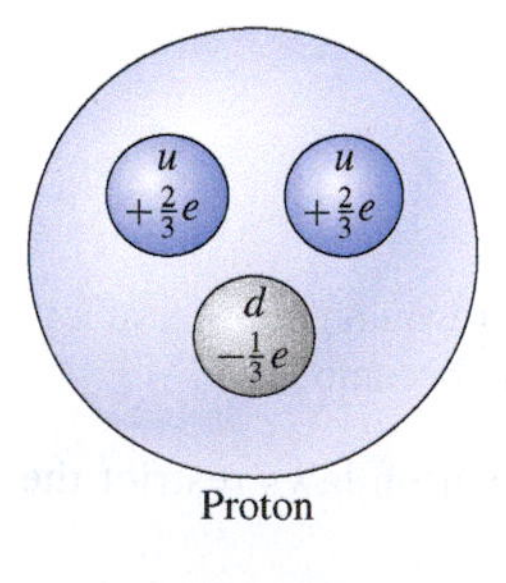

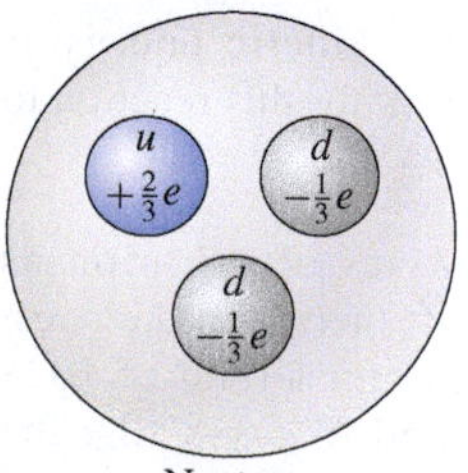

FIGURE 39.4 Protons and neutrons consist of the quark combinations *uud* and *udd*, respectively.

Quarks

Success of the Eightfold Way convinced physicists that many "elementary" particles weren't really elementary. In 1964 Gell-Mann and his colleague George Zweig independently proposed a set of three particles called **quarks** that combined to form the then-known hadrons. These became known as the **up quark**, the **down quark**, and the **strange quark**. For each there was a corresponding antiquark.

One surprising thing about quarks is that they carry fractional electric charges. The two least massive quarks, the up and the down, carry $+\frac{2}{3}e$ and $-\frac{1}{3}e$, respectively; their antiparticles have the opposite charges. The quarks combine in pairs or triplets to make the two classes of hadrons. Baryons, like the proton and the neutron, consist of three quarks (Fig. 39.4). Mesons contain quark–antiquark pairs (Fig. 39.5). The quarks all have spin $\frac{1}{2}$, which explains why the three-quark baryons all have odd half-integer spin, and why the two-quark mesons have integer spin.

The Pauli exclusion principle precludes three particles having the same quantum numbers, so there must be an additional property distinguishing quarks. Called **color**, this property is a kind of "charge"—not to be confused with electric charge—that can take on any of three values called, whimsically, red, green, and blue. The force binding quarks of different colors is the **strong force**, and the quark theory is known as **quantum chromodynamics (QCD)**. In QCD, **gluons** are the field particles that play the role of photons in quantum electrodynamics, binding particles subject to the strong force. Particles formed from quarks—the mesons and the baryons—are always **colorless**. This is true of mesons because they contain a quark of one color and another of its anticolor. It's true of baryons because they contain three quarks of different colors, which combine to give the baryon zero net color charge. The nuclear force, once thought to be fundamental, is actually a residual manifestation of the strong force, acting between the quarks in colorless particles—in much the same way that the van der Waals force between neutral gas molecules is a "residue" of the stronger electric force among the particles that make up the molecules.

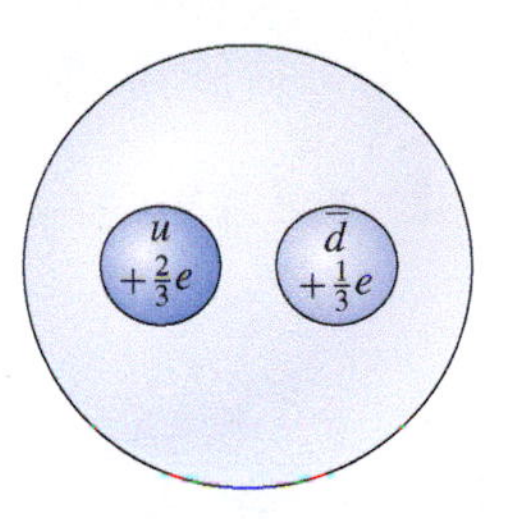

FIGURE 39.5 Mesons consist of a quark and an antiquark. The meson here is the π^+, made from an up quark and a down antiquark.

Photons mediate the electromagnetic force between charged particles but are themselves uncharged. In contrast, gluons, like the quarks they bind, carry color charge. There are eight different gluons; six carry combinations like red–antiblue ($R\overline{B}$), green–antired ($G\overline{R}$), and so on; the other two are colorless. Exchange of a colored gluon, unlike photon exchange in quantum electrodynamics, thus changes the colors of the particles involved.

Another surprising aspect of quarks is that the strong force doesn't decrease with separation. For that reason it appears impossible to isolate a single quark (Fig. 39.6). As a result we never see individual free particles with fractional electric charge.

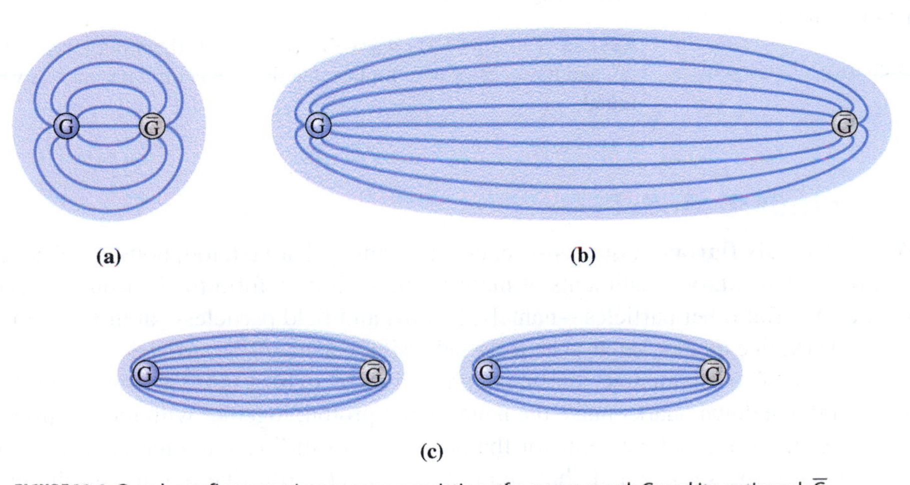

FIGURE 39.6 Quark confinement in a meson consisting of a green quark G and its antiquark $\overline{G}$. (a) Field lines represent the "color field" that joins the two. (b) The field remains confined as the quarks are moved apart, so the field strength stays essentially constant. (c) Pulling the quarks far apart builds up enough energy to create another quark–antiquark pair, rather than isolating individual parts.

The up, down, and strange quarks soon proved insufficient to account for all the observed particles. Theorist Sheldon Glashow argued for a fourth quark, called the **charmed quark**. Ten years later, following intensive searches, experimental teams at Brookhaven National Laboratory and the Stanford Linear Accelerator Center announced the discovery of a particle that implies the existence of the charmed quark. The charmed and strange quarks form a related pair, similar to the up/down quark pair.

There's still one more quark pair. A 1977 experiment confirmed the existence of the **bottom quark**, and in 1995 Fermilab announced the discovery of the **top quark**. The more exotic quarks are more massive, and therefore, through mass–energy equivalence, it takes more energy to produce particles containing them. This need for higher energy is what drives the push for ever more powerful and expensive particle accelerators. (See Application, p. 756.) Table 39.2 lists some properties of the six quarks.

Table 39.2 Matter Particles of the Standard Model

Quark Name	Symbol	Approximate Mass (MeV/c^2)*	Charge	Corresponding Leptons (Symbol, Mass in MeV/c^2)
Down	d	5.0	$-\frac{1}{3}e$	Electron $(e, 0.511)$, electron neutrino (ν_e)
Up	u	2.4	$+\frac{2}{3}e$	
Strange	s	100	$-\frac{1}{3}e$	Muon $(\mu, 106)$, muon neutrino (ν_μ)
Charmed	c	1300	$+\frac{2}{3}e$	
Bottom	b	4200	$-\frac{1}{3}e$	Tau $(\tau, 1777)$, tau neutrino (ν_τ)
Top	t	1.75×10^5	$+\frac{2}{3}e$	

*Quark masses cannot be measured directly and are not precisely determinable; rather, they are calculated based on both experimental results and a particular theoretical framework.

EXAMPLE 39.1 Quarks: Particle Composition and Properties

Given that the strange quark has strangeness $s = -1$, find the charge and strangeness of the Λ^0 particle, which has quark composition *uds*.

INTERPRET We're asked to find how the charge and strangeness of three individual quarks combine in a composite particle, the Λ^0.

DEVELOP Our plan is to sum the values of the quark's charge and strangeness to get the parameters of the Λ^0.

EVALUATE Table 39.2 gives the charges of the u, d, and s quarks as $+\frac{2}{3}e$, $-\frac{1}{3}e$, and $-\frac{1}{3}e$, respectively. These sum to zero net charge, so the Λ^0 is electrically neutral. The up and down quarks aren't strange, so they have $s = 0$. With $s = -1$ for the strange quark, the Λ^0 must have strangeness -1.

ASSESS Table 39.1 shows we're right about the strangeness of the Λ^0, and its superscript 0 implies that this is a neutral particle, as we've found. ■

The Standard Model

We now have six **flavors** of quarks—up, down, strange, charmed, top, bottom—that seem to be truly elementary constituents of matter. Quarks join to form the hadrons—baryons and mesons. But other particles—namely, leptons and field particles—aren't made from quarks. They, like quarks, seem to be truly indivisible, elementary particles.

In this "zoo" of elementary particles, physicists recognize three distinct "families." The up and the down quarks make the neutron and proton; together with the electron and its related neutrino, they account for the properties of ordinary matter. A second family consists of the strange and charmed quarks, the electron-like muon, and the muon neutrino. The quarks of this family are more massive than the up and down quarks, and the muon is more massive than the electron. More massive still are the particles of the third family, consisting of the top and bottom quarks, the electron-like tau particle, and the tau neutrino. Table 39.2 lists these three families, from which all known matter is constructed.

You may be expecting that future editions of this book will tell of additional quarks, and thus of additional families of matter. Whether such additional families exist was an open question until physicists at the Large Electron Positron Collider in Geneva, Switzerland, examined more than half a million particle-decay events and concluded that the number of different types of neutrinos that can exist is 2.99 ± 0.06—at least for low-mass neutrinos like those we've so far detected. Since there's presumably a neutrino type for each family, this result seems to preclude the existence of additional families.

The theory that currently describes elementary particles and their interactions is called the **standard model**. In addition to the matter particles listed in Table 39.2, the standard model includes the gauge bosons responsible for particle interactions. Table 39.3 shows these bosons, which include the photons that mediate the electromagnetic interaction, the gluons of the strong force, the W and Z particles that mediate the weak force, and the Higgs boson—a particle first proposed in 1964 but not discovered until 2012. Interaction of elementary matter particles—quarks and leptons—with the Higgs and its associated field is believed to be responsible for the masses of these particles. The Higgs mechanism,

Table 39.3 Field Particles and Forces

Particle	Mass (GeV/c^2)	Electric and Color Charges*	Force Mediated	Range	Approximate Strength at 1 fm (Relative to Strong Force)
Graviton	0	0, 0	Gravity	Infinite	10^{-38}
$W^{\pm}$	80.2	± 1, 0	Weak	$<2.4\times10^{-18}$ m	10^{-13}
Z^0	91.2	0, 0			
Photon, γ	0	0, 0	Electromagnetic	Infinite	10^{-2}
Gluon, g (8 varieties)	0	0, 6 color–anticolor combinations, 2 colorless	Strong	Infinite†	1
Higgs boson, H^0	126	0, 0	Interaction with the Higgs field gives quarks and leptons their mass.		

*Color is a quark property analogous to, but more complicated than, electric charge.
†The nuclear force is the residual strong force between colorless particles and has a range of about 1 fm (10^{-15} m).

however, makes only a small contribution to the masses of composite particles like baryons, whose mass is largely associated with virtual quarks and gluons.

The standard model is successful in explaining the phenomena of particle physics, but it leaves many fundamental questions unanswered. Why, for example, do the quarks and leptons have the particular masses they do? Why are there only three families of elementary particles? Why are leptons and quarks distinct? Are these particles really elementary, or are there even smaller structures at hitherto unexplored scales? Continuing theoretical work and experiments at ever-higher energies may someday answer these questions.

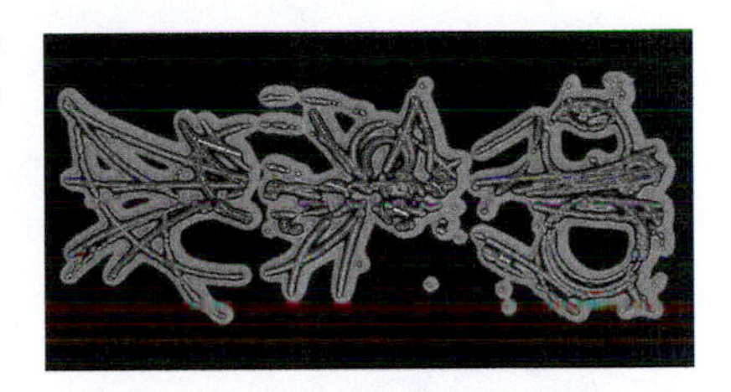

FIGURE 39.7 Particle tracks from the decay of a Z particle help confirm electroweak unification.

39.4 Unification

We introduced the fundamental forces of nature in Chapter 4: gravity, the electroweak force, and the strong force. It was not always that simple, though. In Chapters 20–29 we studied the electric and magnetic forces, first separately but then with the realization that they fall under the single umbrella of electromagnetism. The unification of electricity and magnetism was a major step forward in our understanding of physical reality. Physicists continue to strive for further unification, with the ultimate hope that someday all the forces will be understood as a manifestation of a single common interaction.

Electroweak Unification

In the 1960s and early 1970s, a century after Maxwell formalized the unification of electromagnetism, physicists Steven Weinberg, Abdus Salam, and Sheldon Glashow proposed that the electromagnetic force and the weak force are really aspects of the same thing. Their theory predicted the existence of the particles W^+, Z^0, and W^-, the "carriers" of the unified electroweak force. In 1983 a huge international consortium headed by Carlo Rubbia discovered the W and Z particles (Fig. 39.7), using advances in accelerator technology developed by Simon van der Meer. That discovery confirmed the electroweak unification, and Rubbia and van der Meer joined a long list of physicists who had won the Nobel Prize for contributions to our understanding of the structure of matter.

Further Unification

Electroweak unification led to the present situation in physics, with the strong force, the electroweak force, and gravity comprising the **fundamental forces** that describe all interactions of matter. A further step, the **grand unification theories** (GUTs), attempts to merge the electroweak and strong forces. Some versions of GUT predict that the proton should decay on the very long timescale of some 10^{36} years. We can't wait that long, but we can put 10^{34} protons together in the form of tens of thousands of tons of water and watch for proton decay (Fig. 39.8). So far, such experiments have not found the predicted decays. Nevertheless, many physicists believe that some form of grand unification will soon be achieved.

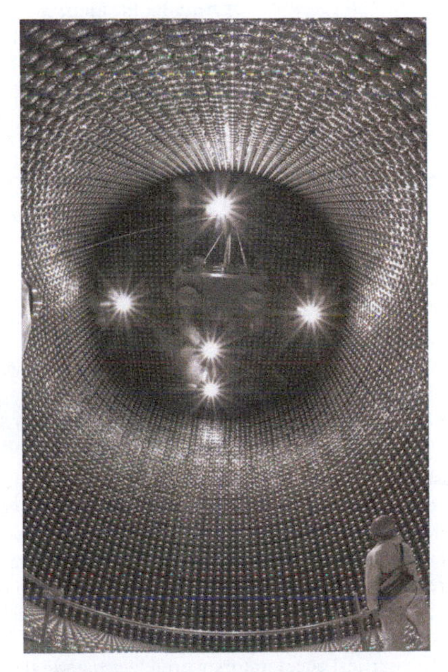

FIGURE 39.8 Japan's Super Kamiokande experiment consists of 50,000 tons of pure water in an underground chamber, surrounded by some 10,000 photomultiplier tubes to detect flashes from rare nuclear reactions including neutrino interactions and hypothetical proton decays.

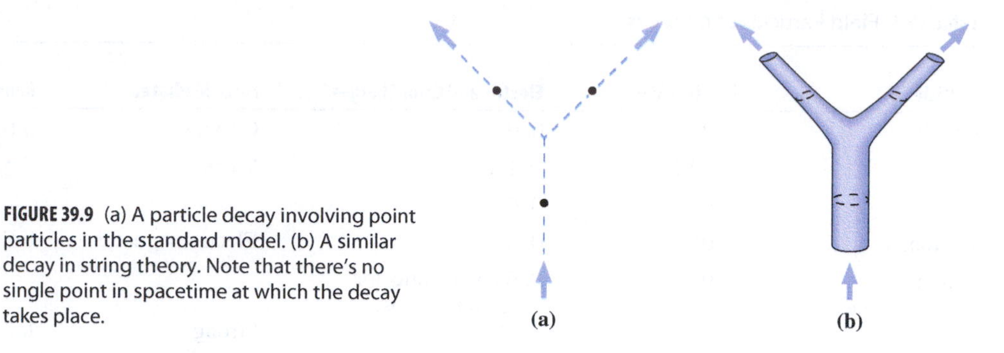

FIGURE 39.9 (a) A particle decay involving point particles in the standard model. (b) A similar decay in string theory. Note that there's no single point in spacetime at which the decay takes place.

Even grand unification would still leave two forces, one of them gravity. Attempts to reconcile our current theory of gravity—Einstein's general theory of relativity—with quantum mechanics have so far made little progress. Yet such a reconciliation is a necessary prerequisite for a final unification of all known forces. A possible candidate is **string theory**, which pictures elementary particles as vibration modes on stringlike structures that may be as short as 10^{-35} m (Fig. 39.9). String theory is set not in the four-dimensional spacetime to which we're accustomed, but in a spacetime with 10 or more dimensions. The extra dimensions are "compactified" in a way that makes them undetectable in normal interactions. To some physicists, string theories hold the promise of a "theory of everything," explaining all our observations about the behavior of the universe. To others they're another in a long line of unsuccessful attempts at a comprehensive explanation of physical phenomena. Only further research will tell.

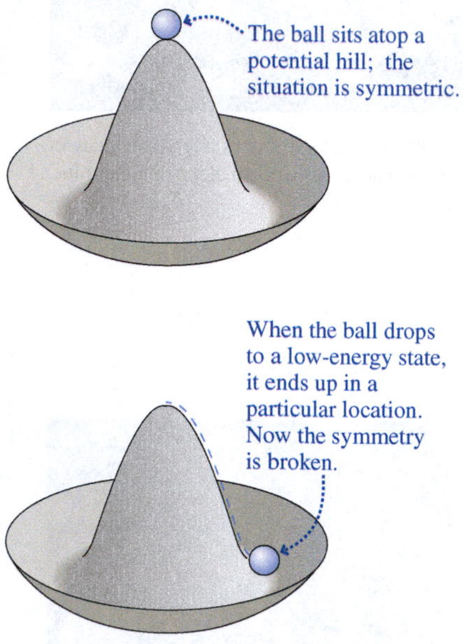

The ball sits atop a potential hill; the situation is symmetric.

When the ball drops to a low-energy state, it ends up in a particular location. Now the symmetry is broken.

FIGURE 39.10 A mechanical analogy for symmetry breaking, showing a ball subject to a hat-shaped potential-energy curve.

Symmetry Breaking

Unification theories predict that phenomena that appear distinct under normal conditions will be seen as one at sufficiently high energies. The observed unification represents a kind of symmetry that's "broken" as the energy level drops. Figure 39.10 shows a mechanical analogy for such **symmetry breaking**. With high energy, a ball sits atop a potential "hill," and the situation is symmetric. But when the ball drops to a low-energy state, it ends up at a particular angular position, and the symmetry is broken. Analogously, at energies above 100 GeV, what we call the electromagnetic and weak forces are one and the same. But at lower energies the symmetry is broken, and we see two distinct forces. Particle accelerators now being planned will exceed the energy of electroweak symmetry breaking, allowing us to explore that interaction in its fundamental simplicity. But the energy at which symmetry breaking occurs increases to some 10^{15} GeV as we move from electroweak to grand unification, making it unlikely that we'll achieve that energy in the foreseeable future. And the energy at which gravity, too, would join a single unified force is an even more remote 10^{19} GeV.

APPLICATION Particle Accelerators

Most particles are more massive than the proton, and some, like the weak-force mediators $W^{\pm}$ and Z, are extremely massive. Since the more massive particles are all unstable, discovering them involves first creating them—and that requires energy of at least mc^2, with m the particle mass. That energy requirement, along with the hint of new phenomena such as force unification, drives particle physicists' seemingly insatiable desire for particle accelerators of ever-higher energies.

The earliest accelerators were electrostatic devices that established large potential differences between conducting electrodes, and they used the associated electric field to accelerate charged particles. But such accelerators are limited to maximum energies of about 20 MeV because of the difficulties of handling high voltages. We saw in Chapter 26 how this problem is cleverly circumvented in the cyclotron, an accelerator that uses a magnetic field to keep particles in circular orbits so they can gain energy on each orbit from a modest electric field. But cyclotrons work for only nonrelativistic particles, for which the cyclotron frequency is independent of particle energy. Today's high-energy experiments call for ultrarelativistic particles, whose speeds differ only minutely from the speed of light. As a result, today's accelerators are primarily variations on the **synchrotron**, a device in which the magnetic field increases with the particle energy to maintain particles in a circular orbit of fixed radius. An alternative to the synchrotron is the **linear accelerator**, the largest of which is the 3-km-long Stanford Linear Accelerator (which was shown in Fig. 33.9).

A head-on collision between two cars is much more damaging than a collision of a moving car with a stationary one, since in the former case all the energy goes into damaging the cars while in the latter a great deal of energy goes into accelerating the initially stationary "target" car. For the same reason head-on collisions of high-energy particles make much more energy available for the creation of new particles. As a result, most of the highest-energy accelerators today are colliders, with particle beams accelerated in opposite directions and brought to collide inside elaborate detectors. The largest accelerator today

Aerial view showing the location of the 4.3-km-diameter Large Hadron Collider on the Swiss–French border. The accelerator itself is in a tunnel 50 m–175 m below ground. The device accelerates protons to 14 TeV. The smaller ring accelerates the protons to 450 GeV before they're injected into the main accelerator.

is the Large Hadron Collider (LHC) at CERN, the European Laboratory for Particle Physics, at Geneva, Switzerland. With a 27-km circumference, LHC collides proton beams at energies up to 28 TeV. Both LHC and the Relativistic Heavy Ion Collider at Brookhaven National Laboratory in the United States create conditions that existed when the universe was only 10^{-12} s old; more on this in the next section.

39.5 The Evolving Universe

We come at the end to the broadest possible questions about physical reality: How did the universe begin? What is its overall structure? What will its future bring? Remarkably, these cosmological questions are closely linked with the questions of particle physics. Even more remarkably, we now have precise, quantitative answers to these and other questions that once seemed highly speculative.

Expansion of the Universe

Early in the 20th century, astronomers argued about the nature of certain fuzzy patches visible in telescope photographs. Many thought they were gas clouds scattered among the visible stars, but others made a more radical proposal: that some of these "nebulae" were gravitationally bound systems containing billions of stars and that they were almost inconceivably distant.

In the 1920s, the opening of the 2.5-m telescope at California's Mt. Wilson Observatory finally resolved the issue. There, astronomer Edwin Hubble proved that some nebulae were indeed distant galaxies like our own Milky Way, each containing billions of stars. Today cosmologists think of galaxies as individual "point particles" whose distribution traces the overall structure of the universe.

Hubble continued to study the galaxies throughout the 1920s, and by analyzing their spectra he made a remarkable discovery: Spectral lines from distant galaxies were shifted toward the red, with the amount of shift dependent on the distance to the galaxies. The most reasonable explanation was a redshift caused by the Doppler effect (see Section 14.8). Then the implication of Hubble's work is that the distant galaxies are receding from us at speeds proportional to their distances. This result is known as **Hubble's law**:

$$v = H_0 d \quad (39.1)$$

where v is the recession speed, d the distance, and H_0 the **Hubble constant**, whose value is now known to be very nearly 20.8 kilometers per second per million light years of distance. Astronomers now use the Hubble relation to find the distances to remote galaxies, measuring their redshifts and using Equation 39.1 to infer their distances. Although the Hubble relation is written in terms of velocity, a more sophisticated view describes the Hubble expansion not as galactic motion but as a stretching of space itself—a process that also stretches light waves, giving the observed wavelength increase.

It may sound like Hubble's law puts us right in the center of things, in grotesque violation of modern science's view that Earth and its inhabitants don't occupy a favored

FIGURE 39.11 A portion of the Hubble Deep Field, a Hubble Space Telescope image showing distant galaxies whose redshifts provide information about cosmic expansion.

position in the universe. But actually the inhabitants of any other galaxy would observe the same thing: All the distant galaxies would be receding from them at speeds proportional to their distances. As long as the universe is infinite in extent, none can claim to be at the center. And if it's not infinite, then Einstein's general theory of relativity gives it a closed-curve shape that still has no center.

Extrapolating Hubble's law backward in time suggests there was a time when all the galaxies were together. This implies that the universe had a definite beginning, in the form of a colossal explosion that flung matter into an expansion that continues today. Based on additional evidence that we'll describe shortly, scientists are quite certain that the universe began with such a **Big Bang**. Some of that evidence comes from the Hubble Space Telescope, named to honor Hubble's pioneering studies (Fig. 39.11).

EXAMPLE 39.2 Hubble's Law: Calculating the Age of the Universe

Using $H_0 = 20.8\,\mathrm{km/s/Mly}$ and assuming the expansion rate has been constant, find out how long the universe has been expanding.

INTERPRET We're given the Hubble constant H_0 and asked to extrapolate back to the time when all the galaxies were together.

DEVELOP If a galaxy has been moving with constant speed for time t, then its distance from us today is $d = vt$. But Hubble's law gives its speed in terms of distance: $v = H_0 d$. So our plan is to substitute $H_0 d$ for v and solve for the time t.

EVALUATE We have $t = d/v = d/H_0 d = 1/H_0$. To evaluate this expression we need to convert from the mixed units used for the Hubble constant:

$$t = \frac{1}{H_0} = \frac{1}{(20.8\ \mathrm{km/s/Mly})/[3.00\times10^5(\mathrm{km/s})/(\mathrm{ly/y})]} = 14.4\ \mathrm{Gy}$$

ASSESS Our calculation shows that the universe is about 14 billion years old—on the assumption that the expansion rate hasn't changed. We'll soon see that this assumption isn't quite correct, but our result remains a good estimate for the age of the universe (the actual number is close to 13.8 Gy). Note that we used $c = 3.00\times10^5$ km/s in converting from km/s to ly/y so our answer would come out in years. ■

The Cosmic Background Radiation

In 1965 Arno Penzias and Robert Wilson at Bell Laboratories found a faint "noise" of microwave radiation in a satellite communications antenna they were testing. The noise seemed to come from all directions in the sky. Theorists at Princeton identified Penzias and Wilson's "noise" as radiation dating to a much earlier era in the universe. This **cosmic microwave background radiation** is the strongest evidence yet for the Big Bang.

The Big Bang theory suggests that the universe started very hot and then cooled as it expanded, doing work against its own gravitation. At first it was so hot that any atoms that formed would be dashed apart by collisions at the high thermal energy prevailing. Thus, in its early times the universe was populated by individual charged particles. These interacted readily with electromagnetic radiation, making the universe opaque. But by 380,000 years the temperature had dropped to some 3000 K, and at that point atoms of hydrogen and helium could form. Since neutral atoms interact much more weakly with electromagnetic radiation, the universe became transparent, and photons emitted as the atoms formed could travel throughout the universe with little chance of being subsequently absorbed. Those photons became the cosmic background radiation, permeating the entire universe.

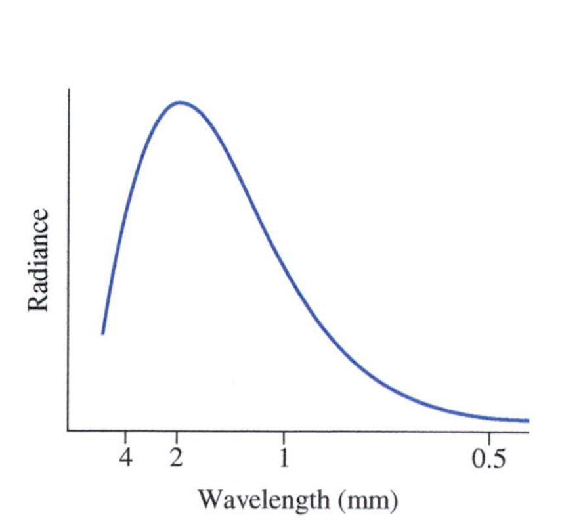

FIGURE 39.12 Spectrum of the cosmic microwave background matches perfectly that of a blackbody at 2.726 K.

Measurements of the cosmic microwave background show a near-perfect fit to a 2.7-K blackbody spectrum (Fig. 39.12). Applying the "stretching of space" interpretation of the Hubble expansion shows that the universe has expanded about 1000-fold since the background radiation formed, dropping the temperature from 3000 K to about 3 K, and stretching the radiation's wavelength by the same factor. That's why radiation that initially had μm wavelengths now peaks in the microwave region with mm wavelengths. Thus the cosmic microwave background is a direct reflection of the conditions when it formed 380,000 years after the universe began.

The cosmic microwave background (CMB) is remarkably uniform, but not perfectly so. Tiny spatial variations in the effective temperature of the radiation (about one part in 10^5;

see Fig. 39.13) provide a wealth of information about the early universe and confirm not only the universe's Big Bang origin but also many precise details of its composition and evolution. A look at Fig. 39.13 shows that fluctuations in the CMB temperature occur on a variety of angular scales, with a scale on the order of about 1° being most prominent. These variations resulted, ultimately, from quantum fluctuations in the primordial density. Those fluctuations led, in turn, to oscillations driven by a combination of gravity and pressure, resulting in sound waves that propagated through the early universe. The sound waves were "frozen" at the time of formation of the cosmic background, when radiation and matter effectively decoupled. As a result, a spectrum of the size scales of the CMB fluctuations carries a wealth of information about the early universe.

FIGURE 39.13 All-sky map of the cosmic microwave background shows minute spatial variations in the radiation intensity and therefore in the associated temperature. The image spans the entire sky—360°—and thus the most prominent fluctuations are on angular scales of around 1°. This image was made with the European Space Agency's *Planck* spacecraft.

Figure 39.14 is such a spectrum. On the horizontal axis is the angular scale of CMB fluctuations, and on the vertical axis is the "power" at each scale—roughly, the strength of the fluctuations at that scale. A smooth curve represents the theoretical prediction of our current cosmological theory, with a number of cosmological parameters adjusted for a best fit to the data. That the data points lie precisely on the curve is a strong indication that the theory is correct, and the fitting provides precise values for cosmological parameters, such as the Hubble constant introduced in Equation 39.1.

Figure 39.13, from which the data in Fig. 39.14 are derived, reveals only angular sizes of the CMB fluctuations. But the physics of sound waves gives the corresponding physical sizes. Comparing physical and angular sizes then reveals the geometry of the universe—which, as you saw in Section 33.9's brief introduction to general relativity, is associated with gravity. In particular, the location of the prominent peak in Fig. 39.14 at just under 1° scale shows that the universe is flat overall—that is, Euclidean geometry governs the overall universe, despite deviations in the curved spacetime near massive objects. The second peak in Fig. 39.14 is related to the density of baryons in the early universe, and the third peak is associated with all forms of nonrelativistic matter—all particles with velocities substantially less than c. All in all, data like those shown in Fig. 39.14 have taken cosmology from an inexact science to a precise description of the overall universe and its history, with many of its characteristic parameters now measured to several significant figures.

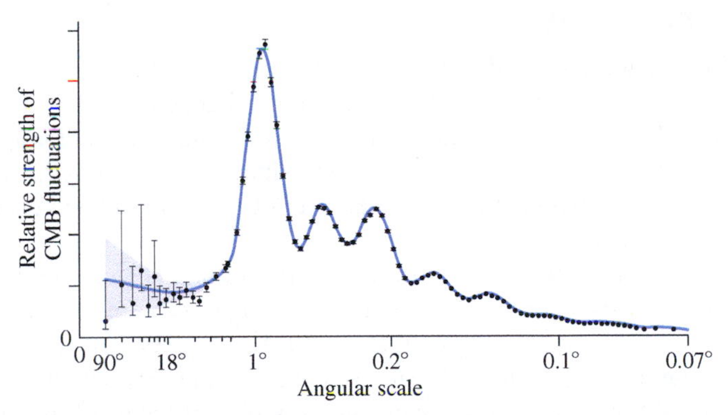

FIGURE 39.14 Spectrum of the angular scales of fluctuations in the cosmic microwave background. Points are data from the *Planck* spacecraft, and the smooth curve is a theoretical fit. Note the error bars, which are tiny except at very large angular scales. These data yield values of cosmological parameters to several significant figures.

The Earliest Times

The cosmic microwave background shows us the universe as it was some 380,000 years after the Big Bang. Nuclear physics takes us back even further, to the time from about 1 second to 30 minutes when the lightest nuclei were forming. The first composite nuclei were the simplest: deuterium, consisting of a proton and a neutron. The rate of deuterium formation is critically sensitive to the expansion rate of the universe. Measurements of deuterium abundance, based on spectral lines from interstellar deuterium, therefore provide direct evidence for conditions early in the Big Bang.

Evidence for still earlier times comes from particle physics, which predicts the interactions and particle populations under the hot, high-energy conditions that existed a fraction of a second after the Big Bang. In 2005, experiments using the Relativistic Heavy Ion Collider (RHIC) at Brookhaven National Laboratory produced a so-called quark–gluon plasma similar to the state of the universe just microseconds after the Big Bang. Thus RHIC and other high-energy particle accelerators can act as "time machines," allowing us to study conditions that prevailed in the very early universe. Figure 39.15 summarizes our understanding of the universe's evolution, showing that the phenomena of particle physics and unification are inextricably tied with cosmic expansion.

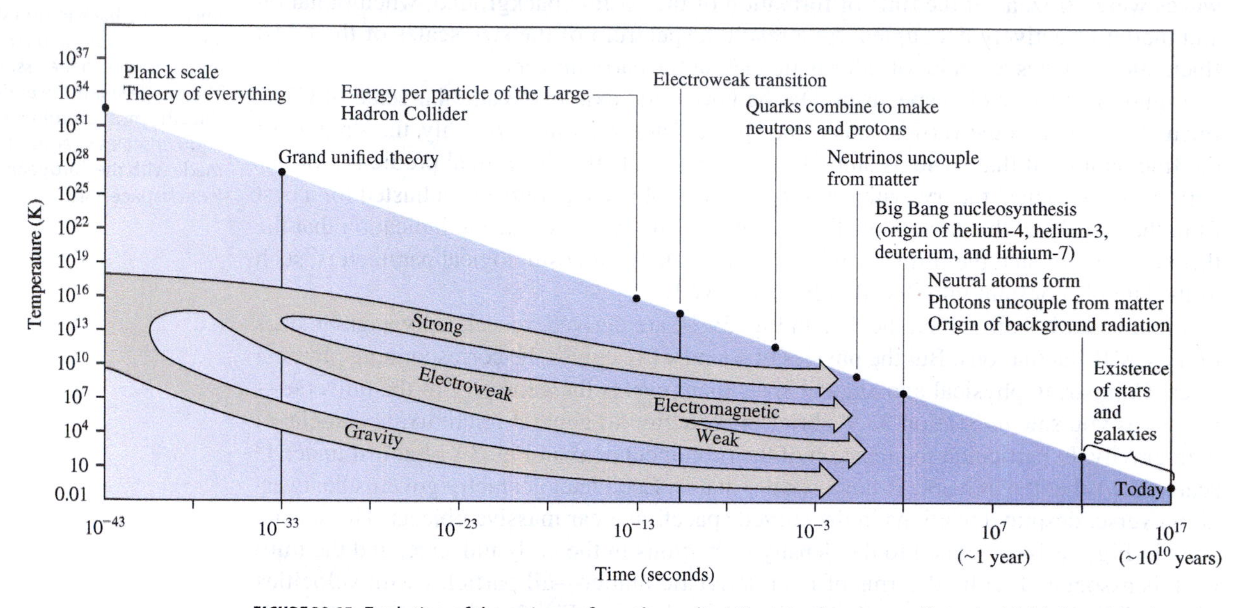

FIGURE 39.15 Evolution of the universe from the earliest times to the present. Note the highly logarithmic scales.

The Inflationary Universe

The original Big Bang theory had difficulty explaining several features of the observed universe. Why, for example, does the universe seem homogeneous and in thermodynamic equilibrium on the largest scales, when the most distant regions are so far apart that light could not have traveled between them in the time since the beginning? And why is the overall geometry of the universe flat, as evidenced in Fig. 39.14, when general relativity allows for curved space?

The solution to these conundrums is **inflation**, an idea first advanced by Alan Guth of MIT. Guth's theory holds that the universe underwent a period of exponential expansion beginning at about 10^{-35} s and lasting until about 10^{-32} s (Fig. 39.16). The expansion was the result of a delay in the symmetry breaking that made the fundamental forces appear distinct. Because of the tremendous expansion, now-distant locations would once have been close enough to reach the thermodynamic equilibrium that we now observe. Furthermore, the inflationary expansion would "flatten out" any overall curvature, giving us the flat universe we see today.

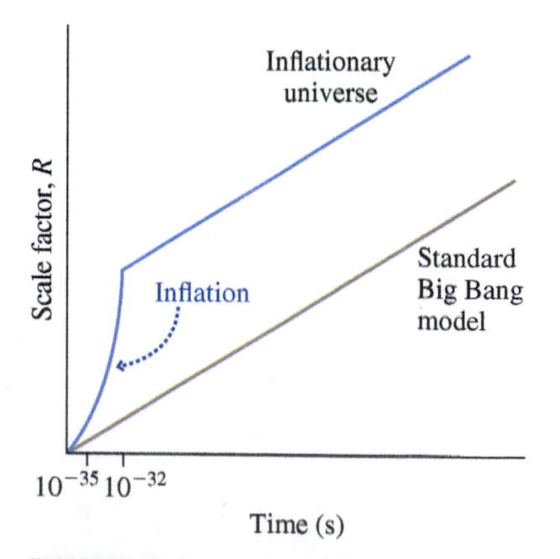

FIGURE 39.16 Expansion of the universe in Big Bang theories, both with and without inflation. The scale factor R measures the amount of expansion.

It might seem impossible to find observational evidence for an event that happened only 10^{-35} s into the history of the universe, but, remarkably, inflation would have left a "fingerprint" in the form of gravitational waves—"ripples" in spacetime whose existence (see Section 33.9) is predicted by general relativity. Those waves would have imposed a characteristic pattern on the polarization of the cosmic microwave background. Such polarization was tentatively detected in 2014, using the microwave telescope BICEP2 (for Background Imaging of Cosmic Extragalactic Polarization) at the South Pole. If confirmed, these results will provide direct evidence for inflation in the early universe.

Dark Matter, Dark Energy, and the Future of the Universe

Will the universe expand forever, or will the expansion eventually stop and reverse? That's like asking whether a spacecraft will escape Earth forever or ultimately return, and the answer is the same: If the system's kinetic energy exceeds the magnitude of its (negative) potential energy, then expansion will proceed forever. Thus a single parameter—designated Ω (capital Greek omega)—determines the fate of the universe:

$$\Omega = \frac{|\text{potential energy of the universe}|}{\text{kinetic energy of the universe}}$$

where $\Omega > 1$ implies eventual collapse, and $\Omega < 1$ means continued expansion. An alternative way of expressing the same dichotomy is in terms of the average density. A simple Newtonian calculation based on kinetic and gravitational potential energies of particles in an expanding universe gives Ω in terms of the average density ρ and the Hubble constant: $\Omega = 8\pi G\rho/3H_0^2$—a result that turns out to be identical to the correct general relativistic calculation. Setting $\Omega = 1$ gives the **critical density** that divides eternal expansion from eventual collapse:

$$\rho_c = \frac{3H_0^2}{8\pi G} \quad \text{(critical density)} \tag{39.2}$$

Analysis of the cosmic microwave background fluctuations strongly suggests that the actual universe is at the critical density, with $\Omega = 1$. But when we total the visible matter in the universe, we come up with far less than the critical density. Elementary particle theories corroborate this observational result, suggesting that there can be only enough ordinary matter—protons, neutrons, nuclei, and atoms—to make up about 4% of the critical density. Furthermore, the motions of stars in galaxies and of galaxies in clusters suggest the presence of a great deal more mass than is visible. All this implies the existence of **dark matter** whose composition is unknown and which can't be the sort of matter—made mostly from quarks—with which we're familiar.

Another approach to the cosmic density is to study the most distant galaxies, whose light has taken so long to reach us that we're seeing them as they were in the early universe. One might expect that cosmic expansion was faster in earlier times, and slowed as the galaxies did work against their mutual gravitational attraction. Comparing the recession speeds of ancient, distant galaxies with the speeds of nearer ones should then give the rate of cosmic deceleration—which, in turn, should depend on cosmic density. But observations in 1998 gave a surprising and unexpected result: The cosmic expansion is actually accelerating!

Cosmic acceleration implies a kind of "antigravity" operating on the largest scales. Ironically, Einstein had proposed just such a phenomenon in his original formulation of general relativity. Einstein needed this **cosmological constant** in his theory in order to keep the universe static—which, in 1916, astronomers thought it was. When Hubble then showed that the universe is expanding, Einstein dropped the cosmological constant and called it "the greatest blunder of my life." Now it appears that Einstein had the right idea in the first place.

The source of the cosmic acceleration is **dark energy**, which may be just another name for Einstein's cosmological constant or may be a different phenomenon with the same "antigravity" effect. At this point we don't know quite what it is. But we do know how much of it there is: Dark energy is fully 69% of the "stuff" that makes up our universe. Another 26% is dark matter. That means only 5% of the universe is in the form of familiar matter made from baryons and thus ultimately from quarks (Fig. 39.17). These numbers come from a confluence of recent observations of the cosmic microwave background, distant supernovae, and surveys of distant galaxies; together, they tightly constrain the relative amounts of matter and dark energy. Here, at the end of a long physics course, it's sobering to realize that much of our universe consists of "stuff" about which we know so very little!

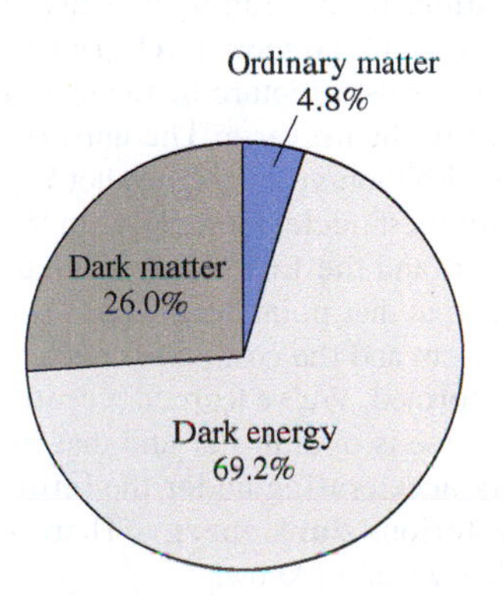

FIGURE 39.17 Composition of the universe, from the *Planck* spacecraft supplemented with additional data. Only a tiny fraction (colored wedge) is "stuff" we understand!

Understanding the Universe

With this chapter's brief survey, we've reached the limits of present understanding of the universe. On the way we've seen that particle physics and cosmology are inextricably linked. To understand our universe, we need to understand all its aspects from the largest to the smallest—and that means understanding all the forces, from gravity to the weak force; all the physical laws, from Newton's and Maxwell's to the laws of quantum mechanics; and the nature of the elementary particles. In this text we've touched on all these topics, and we hope we've given you a foundation for further understanding and appreciation of the richness and diversity of the physical universe.

CHAPTER 39 SUMMARY

Big Idea

The big idea here is that the structure of ordinary matter and its interactions can be explained in terms of a handful of particles: **quarks**, which make up the familiar proton and neutron and a host of other **hadrons**; **leptons**, including the familiar electrons and elusive neutrinos; and **gauge bosons**, which mediate the fundamental forces and include the photon for the electromagnetic force. The **standard model** describes these elementary particles and their interactions; it sheds light not only on the structure of matter we see today but on the early universe as well. But there's a humbling caveat: Only about 5% of the universe consists of familiar matter. The rest is **dark energy** and **dark matter**, about which we know very little.

Key Concepts and Equations

Quarks, which come in six flavors and three colors, join in threes to form **baryons**, including protons and neutrons, and in twos to form **mesons**. Quarks carry fractional electric charges, but because the **strong force** between quarks doesn't decrease with distance, it appears impossible to have an isolated quark.

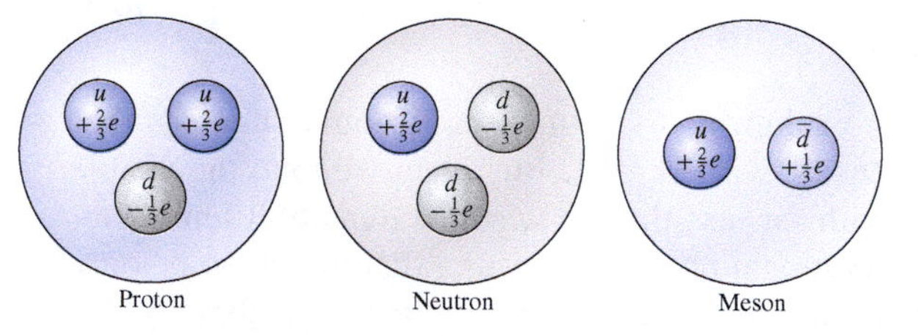

Leptons are the other class of elementary matter particles, and include the electron and its more massive cousins, the muon and tau, and the three types of neutrinos.

The three **fundamental forces**—strong, electroweak, and gravity—are believed to be manifestations of the same fundamental interactions that appear unified at high enough energies.

At typical energies in today's universe, though, the electroweak force separates into the electromagnetic and weak forces. The forces differ greatly in strength:

Force	Relative strength at 1 fm (approximate)
Gravity	10^{-38}
Weak	10^{-13}
Electromagnetic	10^{-2}
Strong	1

Exchange of **gauge bosons** explains the forces between particles.

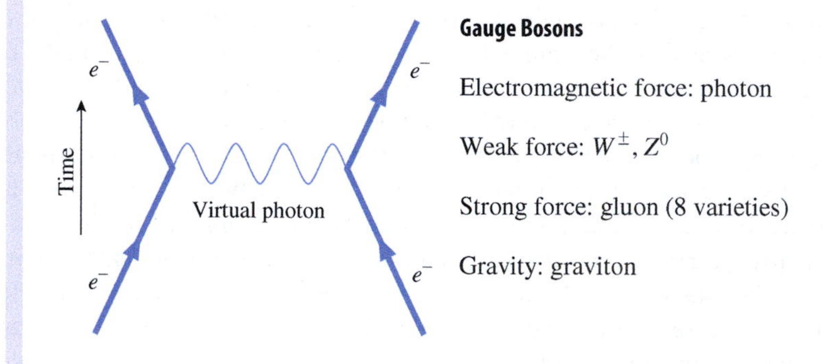

Applications

Our knowledge of particles and their interactions, combined with general relativity and with observations of the **Hubble expansion** and the **cosmic microwave background radiation**, gives us a picture of the origin and structure of the universe. The universe began some 14 billion years ago in a hot **Big Bang**. The simplest nuclei formed within the first half-hour, and the first atoms at about 380,000 years; at that point the universe became transparent and the cosmic microwave background formed. We've learned recently that the universe is overall flat and that its expansion is accelerating under the influence of mysterious **dark energy**. There's still much that we don't know!

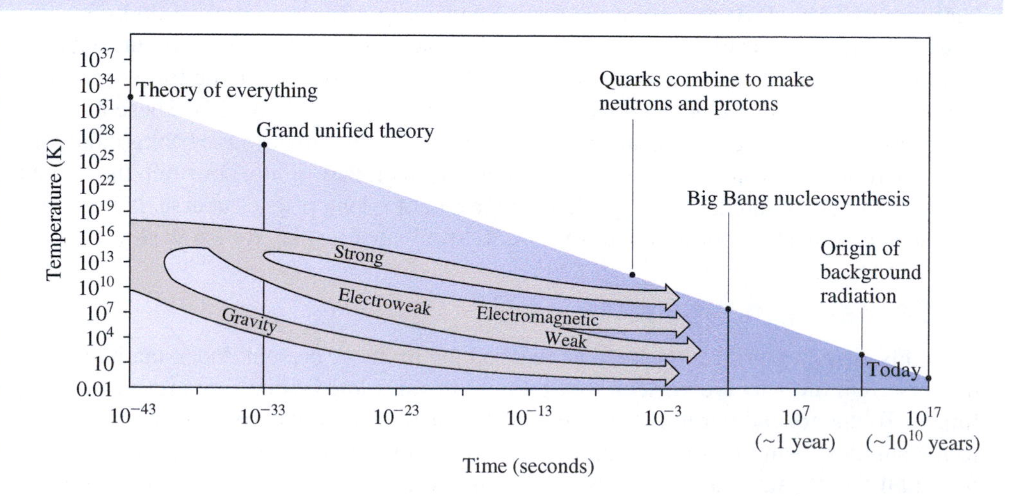

BIO *Biology and/or medicine-related problems* **DATA** *Data problems* **ENV** *Environmental problems* **CH** *Challenge problems* **COMP** *Computer problems*

For Thought and Discussion

1. Why did Yukawa conclude that the particle mediating the strong force should have nonzero mass?
2. How can we follow the tracks of individual particles?
3. How are baryons fundamentally different from leptons?
4. What coordinates are changed under the inversion processes P and T?
5. Why are we unlikely to observe an isolated quark?
6. Describe the relation between the strong force and the nuclear force.
7. What's the role of gluons?
8. Classify (a) mesons and (b) baryons as fermions or bosons, and relate your classification to the particles' quark compositions.
9. Name the fundamental force involved in (a) binding of a proton and a neutron to make a deuterium nucleus; (b) decay of a neutron to a proton, an electron, and a neutrino; (c) binding of an electron and a proton to make a hydrogen atom.
10. What forces are unified in the electroweak theory?
11. What forces would be unified by GUTs?
12. Why do we need higher-energy particle accelerators to explore fully the standard model?
13. How can Hubble's law hold without the universe having a center?
14. Is it possible for a charged particle to be its own antiparticle?
15. Describe the origin of the cosmic microwave background.
16. Explain how particle accelerators can help us understand the early universe.
17. **BIO** What medical diagnostic procedure makes use of the fact that every particle has an antiparticle? What particle/antiparticle pair is involved?
18. The radiation that we observe as the cosmic microwave background started out largely as infrared. Why is it now the *microwave* background?
19. If the hypothetical graviton is ever discovered, what should its mass be? Use Yukawa's argument (Section 39.1) and the fact that the gravitational force falls as $1/r^2$.
20. What data provide the most robust confirmation of cosmological theory as well as the most precise values for cosmological parameters?

Exercises and Problems

Exercises

Section 39.1 Particles and Forces

21. How long could a virtual photon of 633-nm red laser light exist without violating conservation of energy?
22. Some scientists have speculated on a possible "fifth force," with a range of about 100 m. Following Yukawa's reasoning, what would be the mass of the field particle mediating such a force?

Section 39.2 Particles and More Particles

23. Write the equation for the decay of a positive pion to a muon and a neutrino, being sure to label the type of neutrino. (*Hint:* The positive muon is an antiparticle.)
24. Use Table 39.1 to find the total strangeness before and after the decay $\Lambda^0 \rightarrow \pi^- + p$, and use your answer to determine which force is involved in this reaction.
25. The η^0 particle is a neutral nonstrange meson that can decay to a positive pion, a negative pion, and a neutral pion. Write the reaction for this decay, and verify that it conserves charge, baryon number, and strangeness.
26. Are either or both of these decay schemes possible for the tau particle: (a) $\tau^- \rightarrow e^- + \bar{\nu}_e + \nu_\tau$; (b) $\tau^- \rightarrow \pi^- + \pi^0 + \nu_\tau$?
27. Is the interaction $p + p \rightarrow p + \pi^+$ allowed? If not, what conservation law does it violate?

Section 39.3 Quarks and the Standard Model

28. Determine the quark composition of the π^-.
29. The Eightfold Way led Gell-Mann to predict a baryon with strangeness −3. Determine this particle's quark composition.
30. The Σ^+ and Σ^- have quark compositions *uus* and *dds*, respectively. Are the Σ^+ and Σ^- each other's antiparticles? If not, give the quark compositions of their antiparticles.

Section 39.4 Unification

31. Estimate the volume of the 50,000 tons of water used in the Super Kamiokande experiment shown in Fig. 39.8.
32. Estimate the temperature in a gas of particles such that the thermal energy kT is high enough to make electromagnetism and the weak force appear as a single phenomenon.
33. Repeat Exercise 32 for the 10^{15}-GeV energy of grand unification.

Section 39.5 The Evolving Universe

34. Express the Hubble constant in SI units.
35. Find the distance to a galaxy whose redshift reveals that it's receding from us at 2.5×10^4 km/s.
36. Find the recession speed of a galaxy 360 Mly from Earth.
37. Before the high-precision cosmological data from the *Planck* spacecraft became available in 2013, the best estimate for the Hubble constant was 22.7 km/s/Mly. Repeat Example 39.2 using this value for H_0.

Problems

38. The mass of the photon is assumed to be zero, but experiments put only an upper limit of 5×10^{-63} kg on the photon mass. What would the range of the electromagnetic force be if the photon mass were actually at this upper limit?
39. Which of the following reactions is not possible, and why not? (a) $\Lambda^0 \rightarrow \pi^+ + \pi^-$; (b) $K^0 \rightarrow \pi^+ + \pi^-$
40. Both the neutral kaon and the neutral ρ meson can decay to a pion–antipion pair. Which of these decays is mediated by the weak force? How can you tell?
41. Some grand unification theories suggest that the decay $p \rightarrow \pi^0 + e^+$ may be possible, in which case all matter may eventually become radiation. Are (a) baryon number and (b) electric charge conserved in this hypothetical proton decay?
42. Consider systems described by wave functions that are proportional to the terms (a) xy^2z, (b) x^2yz, and (c) xyz, where x, y, and z are the spatial coordinates. Which pairs of these systems could be transformed into each other under a parity-conserving interaction?
43. The J/ψ particle is an uncharmed meson that nevertheless includes charmed quarks. Determine its quark composition.
44. List all the possible quark triplets formed from any combination of up, down, and charmed quarks, along with the charge of each.
45. The Tevatron at Fermilab accelerates protons to energy of 1 TeV. (a) How much is this in joules? (b) How far would a 1-g mass have to fall in Earth's gravitational field to gain this much energy?
46. (a) Find the relativistic factor γ for a 14-TeV proton in the Large Hadron Collider. (b) Find the proton's speed, expressed as a

decimal fraction of c, and accurate to 10 significant figures (you might need the binomial theorem here).

47. In 2015 the proton energy in the Large Hadron Collider was doubled, from 7 TeV to 14 TeV. In working this problem, keep just two significant figures. (a) How did this energy change affect the protons' speed? (b) What is the new speed? (c) How long does it take a 14-TeV proton to circle the LHC's 27-km circumference?
48. Estimate the critical density of the universe.
49. Estimate the diameter to which the Sun would have to be expanded for its average density to be the critical density found in Problem 48.
50. A baryon called the neutral lambda particle has mass 1116 MeV/c^2. Find the minimum speed necessary for the particles in a proton–antiproton collider to produce lambda–antilambda pairs.
51. A so-called muonic atom is a hydrogen atom with the electron replaced by a muon; the muon's mass is 207 times the electron's. Find (a) the size and (b) the ground-state energy of a muonic atom.
52. (a) By what factor must the magnetic field in a proton synchrotron be increased as the proton energy increases by a factor of 10? Assume the protons are highly relativistic, so $\gamma \gg 1$. (b) By what factor must the diameter of the accelerator be increased to raise the energy by a factor of 10 without changing the magnetic field?
53. A galaxy's hydrogen-β spectral line, normally at 486.1 nm, appears at 495.4 nm. (a) Use the Doppler shift of Chapter 14 to find the galaxy's recession speed, and (b) infer the distance to the galaxy. Is it appropriate to use Chapter 14's nonrelativistic Doppler formulas in this case?
54. At the time the cosmic microwave background radiation originated, the temperature of the universe was about 3000 K. What were (a) the median wavelength of the newly formed radiation (Equation 34.2b) and (b) the corresponding photon energy?
55. Many particles are far too short-lived for their lifetimes to be measured directly. Instead, tables of particle properties often list "width," measured in energy units and indicating the width of the distribution of measured rest energies. For example, the Z^0 has mass 91.18 GeV and width 2.5 GeV. Use the energy–time uncertainty relation to estimate its corresponding lifetime.
56. A mix of particles starts with equal numbers of the three types of sigma particles listed in Table 39.1. Find the relative portion of each after (a) 5×10^{-20} s and (b) 5×10^{-10} s. Give your answer in a reference frame in which the particles are at rest.
57. You pick up an old astronomy book and read that the Hubble constant is 17 km/s/Mly. You know that today's more accurate value is 20.8 km/s/Mly. Use the simplified reasoning of Example 39.2 to compare the ages for the universe implied by these two values of H_0.
58. A friend believes that the universe is less than 10,000 years old. Based on Hubble's law, how would you argue that the universe is older? What would the Hubble constant be for a 10,000-year-old universe?
59. Your roommate is writing a science-fiction novel set very far in the future, 60 Gy after the Big Bang. One of the characters is a cosmologist, and your roommate wants to know what the cosmologist will measure for the Hubble constant. What's your answer, assuming a steady expansion rate?
60. DATA When physicists "discover" a new particle, it isn't by finding the particle itself in their detectors. Rather, they look for events that might indicate the particle's decay. For a given type of event, they plot the frequency of events (number per unit energy interval) versus energy. The particle shows up as a bump or deviation from an otherwise smooth background curve. You can explore this indirect approach to particle detection by plotting the data in the table below. These data are from the Large Hadron Collider and show events producing a pair of photons versus the energy of the pair. You should see a modest bump whose energy you can roughly determine. That bump is evidence for the Higgs boson!

Energy (GeV)	Events per 1.5 GeV
116	1031
118	965
119	866
121	811
122	829
124	818
125	820
127	743
128	668
130	612
131	567
133	549

Passage Problems

Pions are the lightest mesons, with mass some 270 times that of the electron. Charged pions decay typically into a muon and a neutrino or antineutrino. This makes pion beams useful for producing beams of neutrinos, which physicists use to study those elusive particles. In a medical application during the late 20th century, accelerator centers installed "biomedical beam lines" to test pions for cancer therapy. In these experiments, pions attached themselves to atomic nuclei within cancer cells. The nuclei would literally explode, delivering a "pion star" of cancer-killing nuclear debris. Unfortunately, results were not as encouraging as hoped, and enthusiasm for this technique has waned.

61. The negative pion usually decays into a negative muon and one other particle. The other particle could be
    a. a proton.
    b. an antineutrino.
    c. a neutrino.
    d. an up quark.
62. In the cancer-treatment experiments described in the passage, for which pions is it energetically easiest to be captured by a nucleus?
    a. π^+
    b. π^0
    c. π^-
    d. Energetically, capture is equally likely for all three pions.
63. The lifetime of charged pions is 26 ns. The length of an accelerator's biomedical beam line, from the point where pions are created to the patient, could be at most about
    a. 800 m long.
    b. 80 m long.
    c. 8 m long.
    d. 80 cm long.
64. The quark composition of the negative pion is
    a. uud.
    b. $d\bar{u}$.
    c. ud.
    d. $c\bar{c}$.

Answer to Chapter Question

Answer to Chapter Opening Question

Other particles acquire their masses through interactions with the Higgs and its associated field.

PART SIX SUMMARY Modern Physics

Modern physics, developed since the year 1900, contrasts with the **classical physics** that came before. Modern physics is essential in understanding physical reality at the atomic scale, at very low temperatures, at very high relative velocities, in regions of very strong gravity, and in the evolution and large-scale structure of the universe.

The two big ideas in modern physics are **relativity** and **quantum mechanics**. Relativity is based on a simple principle but drastically alters our commonsense notions of space and time, matter and energy, and the nature of gravity. Quantum mechanics replaces Newtonian determinism with a statistical description in which matter and energy exhibit both wave-like and particle-like behaviors.

Einstein's **special theory of relativity** is based in the statement that the laws of physics are the same for all observers in uniform motion. Therefore, Maxwell's prediction that there should be electromagnetic waves propagating at the speed of light c is valid in all uniformly moving reference frames. So measures of space and time cannot be absolute, but depend on one's reference frame.

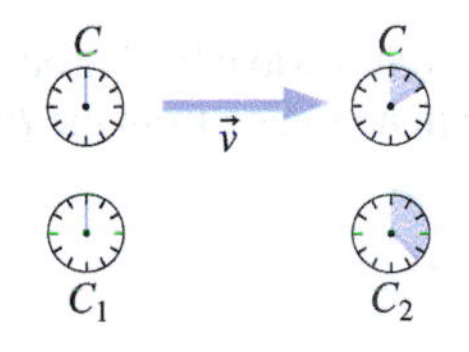

The time between two events is shortest in a reference frame where events occur at the same place; here that's the reference frame of clock C.

$\vec{v}$

An object's length is longest in a reference frame in which it's at rest.

Energy E, momentum p, and mass m in relativity are related by $E^2 = p^2c^2 + (mc^2)^2$.

For an object at rest with respect to an observer, this gives $E = mc^2$, showing the relativistic equivalence of mass and energy.

General relativity is Einstein's theory of gravity, which explains gravity as the geometric curvature of spacetime. General relativity is central to modern astrophysics and cosmology, describing phenomena from black holes to the overall structure of the universe.

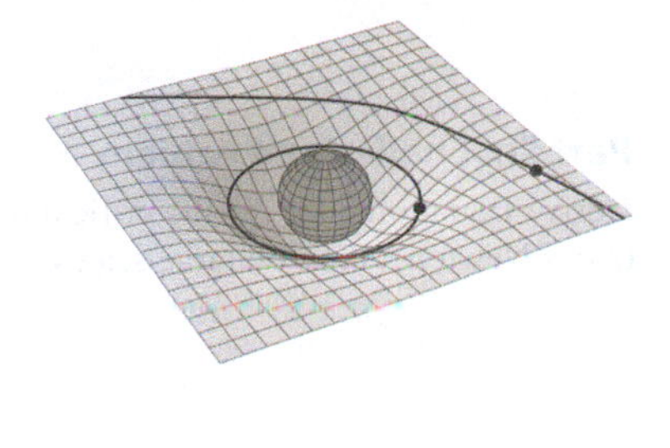

Quantum physics arose from attempts to explain several phenomena observed around the turn of the 20th century. These include blackbody radiation from hot objects, the photoelectric effect, and the existence and spectra of atoms.

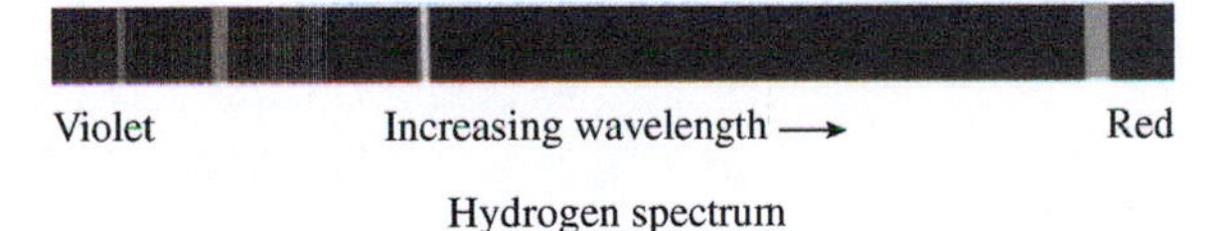

Hydrogen spectrum

Wave–particle duality is at the heart of quantum physics. The energy in electromagnetic radiation of frequency f is concentrated in particle-like "bundles" called photons. Thus electromagnetic energy is quantized, with each photon carrying energy $E = hf$, where $h = 6.63\times10^{-34}$ J·s is Planck's constant. Conversely, matter exhibits wave-like behavior. The de Broglie wavelength associated with a particle of momentum p is $\lambda = h/p$.

Quantization of atomic angular momentum leads to the **Bohr model** for the atom, with quantized energy levels that help explain atomic spectra.

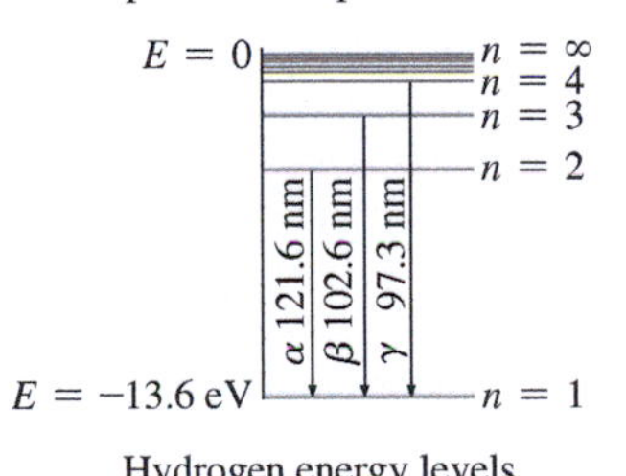

Hydrogen energy levels

Wave–particle duality is closely related to the **Heisenberg uncertainty principle**, which states that it's impossible to measure a particle's position and momentum simultaneously with perfect accuracy. Rather, the uncertainties Δx and Δp in position and momentum must obey the inequality

$$\Delta x\, \Delta p \geq \hbar \text{ (uncertainty principle), where } \hbar = h/2\pi$$

Quantum mechanics describes phenomena at the atomic scale. The **Schrödinger equation** gives the **wave function**, ψ, for a particle of mass m with potential energy U and total energy E:

$$-\frac{\hbar^2}{2m}\frac{d^2\psi(x)}{dx^2} + U(x)\psi = E\psi(x)$$

ψ^2 is the **probability density** for finding the particle. Application of the Schrödinger equation to bound systems results in quantized energy levels.

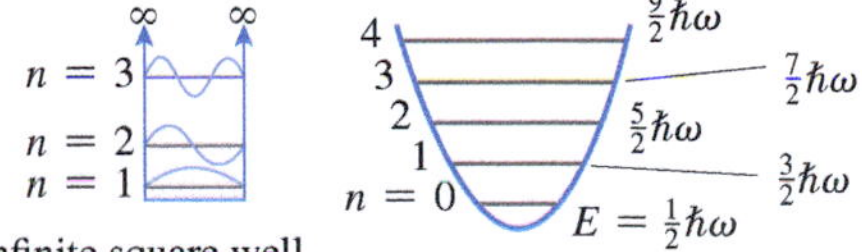

Infinite square well

Harmonic oscillator

Application of the Schrödinger equation to atoms and molecules explains atomic and molecular structure, the organization of the chemical elements in the periodic table, and the behavior of crystalline solids.

(continued)

Nuclear physics plunges into the heart of the atom and shows that larger nuclei require more neutrons than protons in order for the strong nuclear force to overcome the repulsive electric interaction between protons. Nuclear physics describes such phenomena as radioactivity and the production of energy by nuclear fission and fusion.

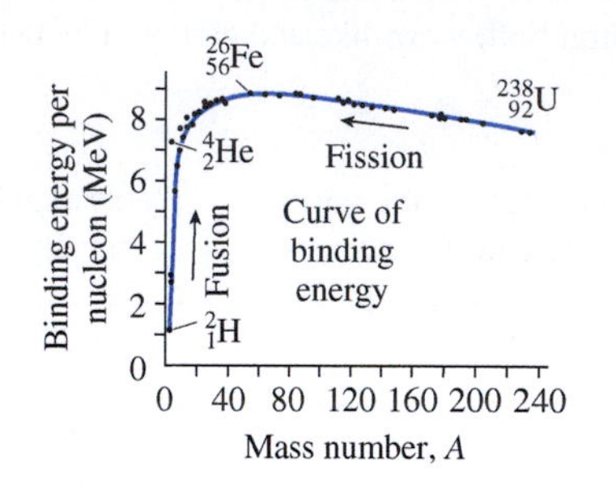

Applying the principles of physics from the subatomic scale of quantum and particle physics to the largest scales described by general relativity gives us our modern-day picture of the origin and evolution of the universe.

The cosmic microwave background radiation, shown here in an image from the *Planck* spacecraft.

Part Six Challenge Problem

Derive Equation 39.2 for the critical density of the universe on the assumption that the universe is spherically symmetric, is homogeneous, and obeys Newton's laws on large scales.

Mathematics

APPENDIX A

A-1 Algebra and Trigonometry

Quadratic Formula

If $ax^2 + bx + c = 0$, then $x = \dfrac{-b \pm \sqrt{b^2 - 4ac}}{2a}$.

Circumference, Area, Volume

Where $\pi \simeq 3.14159\ldots$:

circumference of circle	$2\pi r$
area of circle	πr^2
surface area of sphere	$4\pi r^2$
volume of sphere	$\frac{4}{3}\pi r^3$
area of triangle	$\frac{1}{2}bh$
volume of cylinder	$\pi r^2 l$

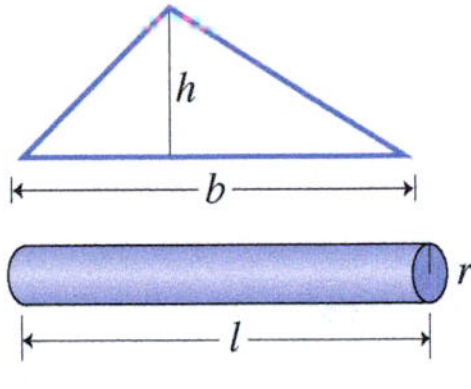

Trigonometry

definition of angle (in radians): $\theta = \dfrac{s}{r}$

2π radians in complete circle

1 radian $\simeq 57.3°$

Trigonometric Functions

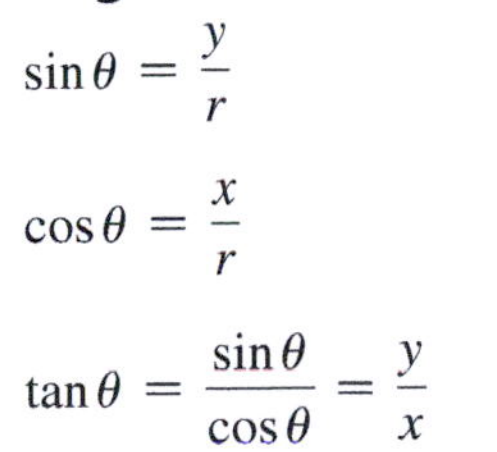

$\sin\theta = \dfrac{y}{r}$

$\cos\theta = \dfrac{x}{r}$

$\tan\theta = \dfrac{\sin\theta}{\cos\theta} = \dfrac{y}{x}$

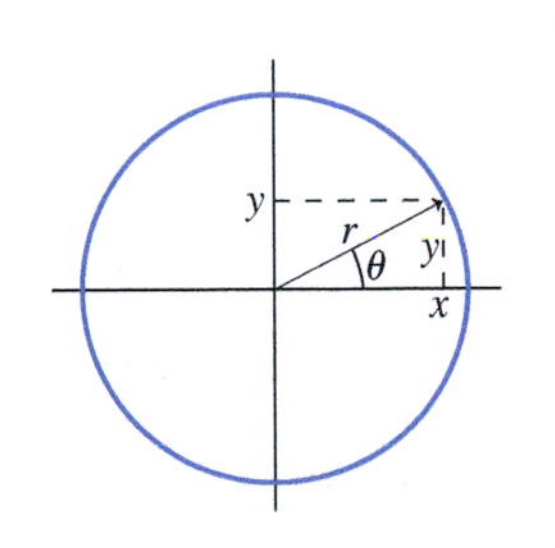

Values at Selected Angles

$\theta \rightarrow$	0	$\frac{\pi}{6}$ (30°)	$\frac{\pi}{4}$ (45°)	$\frac{\pi}{3}$ (60°)	$\frac{\pi}{2}$ (90°)
$\sin\theta$	0	$\frac{1}{2}$	$\frac{\sqrt{2}}{2}$	$\frac{\sqrt{3}}{2}$	1
$\cos\theta$	1	$\frac{\sqrt{3}}{2}$	$\frac{\sqrt{2}}{2}$	$\frac{1}{2}$	0
$\tan\theta$	0	$\frac{\sqrt{3}}{3}$	1	$\sqrt{3}$	∞

Graphs of Trigonometric Functions

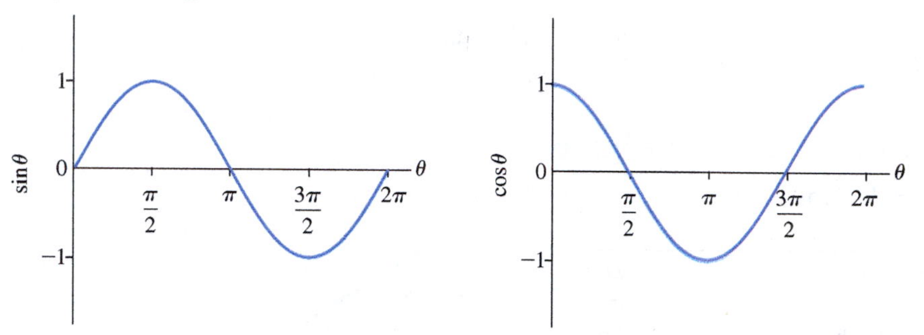

Trigonometric Identities

$$\sin(-\theta) = -\sin\theta$$

$$\cos(-\theta) = \cos\theta$$

$$\sin\left(\theta \pm \frac{\pi}{2}\right) = \pm\cos\theta$$

$$\cos\left(\theta \pm \frac{\pi}{2}\right) = \mp\sin\theta$$

$$\sin^2\theta + \cos^2\theta = 1$$

$$\sin 2\theta = 2\sin\theta\cos\theta$$

$$\cos 2\theta = \cos^2\theta - \sin^2\theta = 1 - 2\sin^2\theta = 2\cos^2\theta - 1$$

$$\sin(\alpha \pm \beta) = \sin\alpha\cos\beta \pm \cos\alpha\sin\beta$$

$$\cos(\alpha \pm \beta) = \cos\alpha\cos\beta \mp \sin\alpha\sin\beta$$

$$\sin\alpha \pm \sin\beta = 2\sin\left[\tfrac{1}{2}(\alpha \pm \beta)\right]\cos\left[\tfrac{1}{2}(\alpha \mp \beta)\right]$$

$$\cos\alpha + \cos\beta = 2\cos\left[\tfrac{1}{2}(\alpha + \beta)\right]\cos\left[\tfrac{1}{2}(\alpha - \beta)\right]$$

$$\cos\alpha - \cos\beta = -2\sin\left[\tfrac{1}{2}(\alpha + \beta)\right]\sin\left[\tfrac{1}{2}(\alpha - \beta)\right]$$

Laws of Cosines and Sines

Where A, B, C are the sides of an arbitrary triangle and α, β, γ the angles opposite those sides:

Law of cosines

$$C^2 = A^2 + B^2 - 2AB\cos\gamma$$

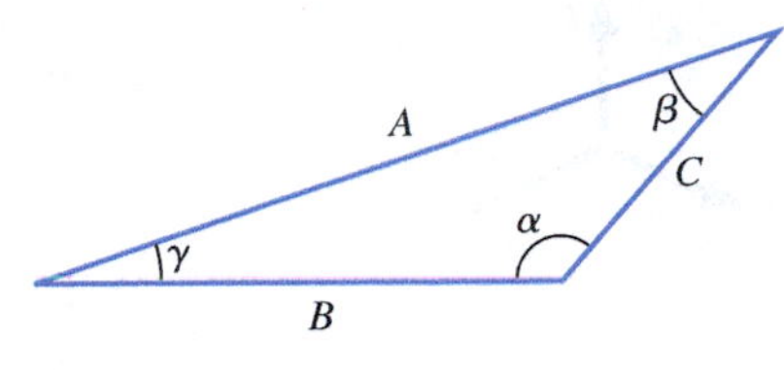

Law of sines

$$\frac{\sin\alpha}{A} = \frac{\sin\beta}{B} = \frac{\sin\gamma}{C}$$

Exponentials and Logarithms

$$e^{\ln x} = x, \quad \ln e^x = x \quad e = 2.71828\ldots$$

$$a^x = e^{x\ln a} \qquad \ln(xy) = \ln x + \ln y$$

$$a^x a^y = a^{x+y} \qquad \ln\left(\frac{x}{y}\right) = \ln x - \ln y$$

$$(a^x)^y = a^{xy} \qquad \ln\left(\frac{1}{x}\right) = -\ln x$$

$$\log x \equiv \log_{10} x = \ln(10)\ln x \simeq 2.3\ln x$$

Approximations

For $|x| \ll 1$, the following expressions provide good approximations to common functions:

$$e^x \simeq 1 + x$$

$$\sin x \simeq x$$

$$\cos x \simeq 1 - \tfrac{1}{2}x^2$$

$$\ln(1 + x) \simeq x$$

$$(1 + x)^p \simeq 1 + px \quad \text{(binomial approximation)}$$

Expressions that don't have the forms shown may often be put in the appropriate form. For example:

$$\frac{1}{\sqrt{a^2 + y^2}} = \frac{1}{a\sqrt{1 + \dfrac{y^2}{a^2}}} = \frac{1}{a}\left(1 + \frac{y^2}{a^2}\right)^{-1/2} \simeq \frac{1}{a}\left(1 - \frac{y^2}{2a}\right) \quad \text{for} \quad y^2/a^2 \ll 1, \text{ or } y^2 \ll a^2$$

Vector Algebra

Vector Products

$$\vec{A}\cdot\vec{B} = AB\cos\theta$$

$|\vec{A}\times\vec{B}| = AB\sin\theta$, with direction of $\vec{A}\times\vec{B}$ given by the right-hand rule:

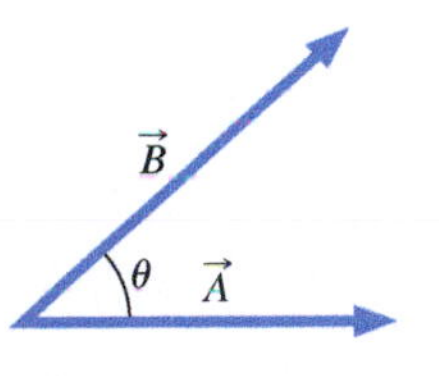

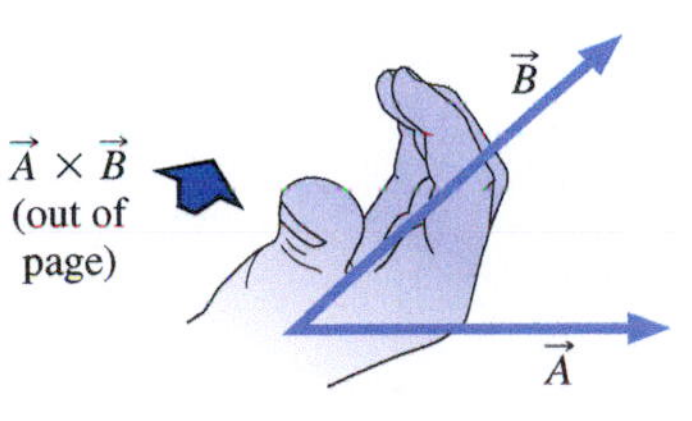

Unit Vector Notation

An arbitrary vector $\vec{A}$ may be written in terms of its components A_x, A_y, A_z and the unit vectors $\hat{\imath}, \hat{\jmath}, \hat{k}$ that have magnitude 1 and lie along the x-, y-, z-axes:

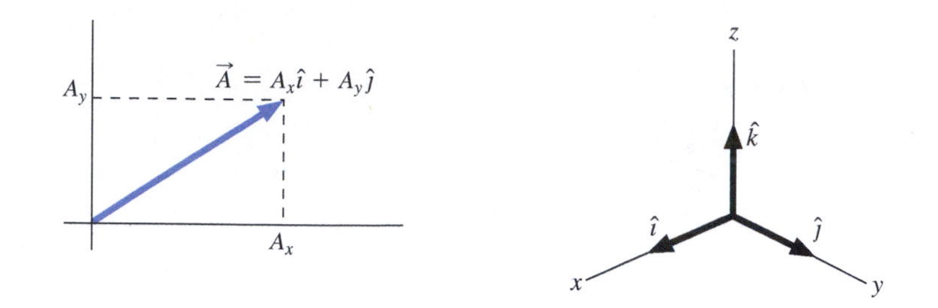

In unit vector notation, vector products become

$$\vec{A} \cdot \vec{B} = A_xB_x + A_yB_y + A_zB_z$$

$$\vec{A} \times \vec{B} = (A_yB_z - A_zB_y)\hat{\imath} + (A_zB_x - A_xB_z)\hat{\jmath} + (A_xB_y - A_yB_x)\hat{k}$$

Vector Identities

$$\vec{A} \cdot \vec{B} = \vec{B} \cdot \vec{A}$$

$$\vec{A} \times \vec{B} = -\vec{B} \times \vec{A}$$

$$\vec{A} \cdot (\vec{B} \times \vec{C}) = \vec{B} \cdot (\vec{C} \times \vec{A}) = \vec{C} \cdot (\vec{A} \times \vec{B})$$

$$\vec{A} \times (\vec{B} \times \vec{C}) = (\vec{A} \cdot \vec{C})\vec{B} - (\vec{A} \cdot \vec{B})\vec{C}$$

A-2 Calculus

Derivatives

Definition of the Derivative

If y is a function of x, then the **derivative of y with respect to x** is the ratio of the change Δy in y to the corresponding change Δx in x, in the limit of arbitrarily small Δx:

$$\frac{dy}{dx} = \lim_{\Delta x \to 0} \frac{\Delta y}{\Delta x}$$

Algebraically, the derivative is the rate of change of y with respect to x; geometrically, it is the slope of the y versus x graph—that is, of the tangent line to the graph at a given point:

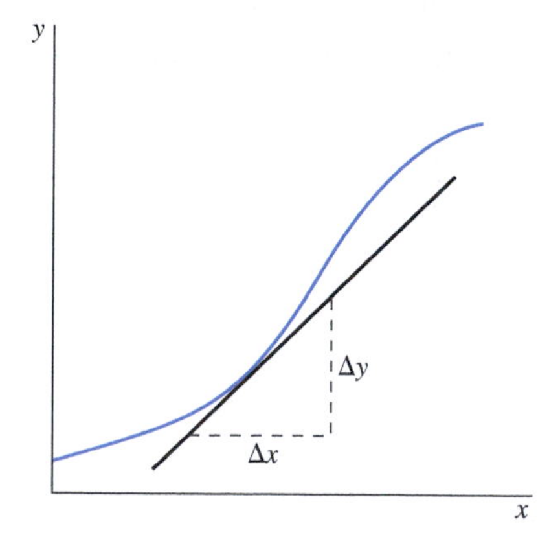

Derivatives of Common Functions

$$\frac{da}{dx} = 0 \quad (a \text{ is a constant})$$

$$\frac{dx^n}{dx} = nx^{n-1} \quad (n \text{ need not be an integer})$$

$$\frac{d}{dx}\sin x = \cos x$$

$$\frac{d}{dx}\cos x = -\sin x$$

$$\frac{d}{dx}\tan x = \frac{1}{\cos^2 x}$$

$$\frac{de^x}{dx} = e^x$$

$$\frac{d}{dx}\ln x = \frac{1}{x}$$

Derivatives of Sums, Products, and Functions of Functions

1. Derivative of a constant times a function

$$\frac{d}{dx}[af(x)] = a\frac{df}{dx} \qquad (a \text{ is a constant})$$

2. Derivative of a sum

$$\frac{d}{dx}[f(x) + g(x)] = \frac{df}{dx} + \frac{dg}{dx}$$

3. Derivative of a product

$$\frac{d}{dx}[f(x)g(x)] = g\frac{df}{dx} + f\frac{dg}{dx}$$

Examples

$$\frac{d}{dx}(x^2\cos x) = \cos x\frac{dx^2}{dx} + x^2\frac{d}{dx}\cos x = 2x\cos x - x^2\sin x$$

$$\frac{d}{dx}(x\ln x) = \ln x\frac{dx}{dx} + x\frac{d}{dx}\ln x = (\ln x)(1) + x\left(\frac{1}{x}\right) = \ln x + 1$$

4. Derivative of a quotient

$$\frac{d}{dx}\left[\frac{f(x)}{g(x)}\right] = \frac{1}{g^2}\left(g\frac{df}{dx} - f\frac{dg}{dx}\right)$$

Example

$$\frac{d}{dx}\left(\frac{\sin x}{x^2}\right) = \frac{1}{x^4}\left(x^2\frac{d}{dx}\sin x - \sin x\frac{dx^2}{dx}\right) = \frac{\cos x}{x^2} - \frac{2\sin x}{x^3}$$

5. Chain rule for derivatives

If f is a function of u and u is a function of x, then

$$\frac{df}{dx} = \frac{df}{du}\frac{du}{dx}$$

Examples

a. Evaluate $\frac{d}{dx}\sin(x^2)$. Here $u = x^2$ and $f(u) = \sin u$, so

$$\frac{d}{dx}\sin(x^2) = \frac{d}{du}\sin u\frac{du}{dx} = (\cos u)\frac{dx^2}{dx} = 2x\cos(x^2)$$

b. $\frac{d}{dt}\sin\omega t = \frac{d}{d\omega t}\sin\omega t\frac{d}{dt}\omega t = \omega\cos\omega t \qquad (\omega \text{ is a constant})$

c. Evaluate $\dfrac{d}{dx}\sin^2 5x$. Here $u = \sin 5x$ and $f(u) = u^2$, so

$$\frac{d}{dx}\sin^2 5x = \frac{d}{du}u^2\frac{du}{dx} = 2u\frac{du}{dx} = 2\sin 5x\frac{d}{dx}\sin 5x$$
$$= (2)(\sin 5x)(5)(\cos 5x) = 10\sin 5x\cos 5x = 5\sin 2x$$

Second Derivative

The second derivative of y with respect to x is defined as the derivative of the derivative:

$$\frac{d^2y}{dx^2} = \frac{d}{dx}\left(\frac{dy}{dx}\right)$$

Example

If $y = ax^3$, then $dy/dx = 3ax^2$, so

$$\frac{d^2y}{dx^2} = \frac{d}{dx}3ax^2 = 6ax$$

Partial Derivatives

When a function depends on more than one variable, then the partial derivatives of that function are the derivatives with respect to each variable, taken with all other variables held constant. If f is a function of x and y, then the partial derivatives are written

$$\frac{\partial f}{\partial x} \quad \text{and} \quad \frac{\partial f}{\partial y}$$

Example

If $f(x, y) = x^3 \sin y$, then

$$\frac{\partial f}{\partial x} = 3x^2 \sin y \quad \text{and} \quad \frac{\partial f}{\partial y} = x^3 \cos y$$

Integrals

Indefinite Integrals

Integration is the inverse of differentiation. The **indefinite integral**, $\int f(x)\,dx$, is defined as a function whose derivative is $f(x)$:

$$\frac{d}{dx}\left[\int f(x)\,dx\right] = f(x)$$

If $A(x)$ is an indefinite integral of $f(x)$, then because the derivative of a constant is zero, the function $A(x) + C$ is also an indefinite integral of $f(x)$, where C is any constant. Inverting the derivatives of common functions listed in the preceding section gives the integrals that follow (a more extensive table appears at the end of this appendix).

$$\int a\,dx = ax + C \qquad \int \cos x\,dx = \sin x + C$$

$$\int x^n\,dx = \frac{x^{n+1}}{n+1} + C, \quad n \neq -1 \qquad \int e^x\,dx = e^x + C$$

$$\int \sin x\,dx = -\cos x + C \qquad \int x^{-1}\,dx = \ln x + C$$

Definite Integrals

In physics we're most often interested in the **definite integral**, defined as the sum of a large number of very small quantities, in the limit as the number of quantities grows arbitrarily large and the size of each arbitrarily small:

$$\int_{x_1}^{x_2} f(x)\,dx \equiv \lim_{\substack{\Delta x \to 0 \\ N \to \infty}} \sum_{i=1}^{N} f(x_i)\,\Delta x$$

where the terms in the sum are evaluated at values x_i between the limits of integration x_1 and x_2; in the limit $\Delta x \to 0$, the sum is over all values of x in the interval.

The key to evaluating the definite integral is provided by the **fundamental theorem of calculus**. The theorem states that, if $A(x)$ is an *indefinite* integral of $f(x)$, then the *definite integral* is given by

$$\int_{x_1}^{x_2} f(x)\,dx = A(x_2) - A(x_1) \equiv A(x)\Big|_{x_1}^{x_2}$$

Geometrically, the definite integral is the area under the graph of $f(x)$ between the limits x_1 and x_2:

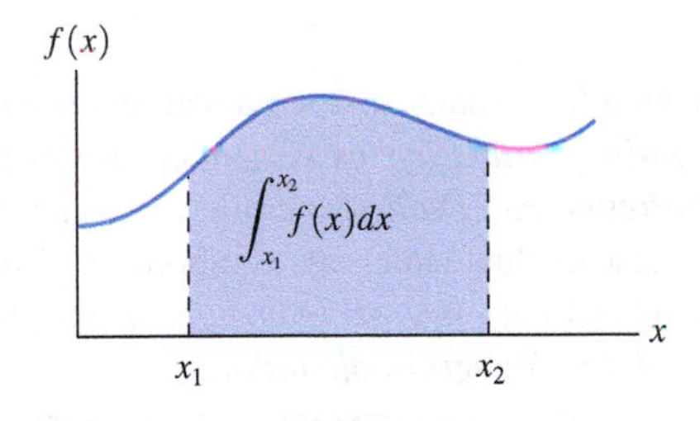

Evaluating Integrals

The first step in evaluating an integral is to express all varying quantities within the integral in terms of a single variable; Tactics 9.1 in Chapter 9 outlines a general strategy for setting up an integral. Once you've set up an integral, you can evaluate it yourself or look it up in tables. Two common techniques can help you evaluate integrals or convert them to forms listed in tables:

1. **Change of variables**
   An unfamiliar integral can often be put into familiar form by defining a new variable. For example, it is not obvious how to integrate the expression

$$\int \frac{x\,dx}{\sqrt{a^2 + x^2}}$$

where a is a constant. But let $z = a^2 + x^2$. Then

$$\frac{dz}{dx} = \frac{da^2}{dx} + \frac{dx^2}{dx} = 0 + 2x = 2x$$

so $dz = 2x\,dx$. Then the quantity $x\,dx$ in our unfamiliar integral is just $\frac{1}{2}dz$, while the quantity $\sqrt{a^2 + x^2}$ is just $z^{1/2}$. So the integral becomes

$$\int \frac{1}{2} z^{-1/2}\,dz = \frac{\frac{1}{2} z^{1/2}}{\frac{1}{2}} = \sqrt{z}$$

where we have used the standard form for the integral of a power of the independent variable. Substituting back $z = a^2 + x^2$ gives

$$\int \frac{x\,dx}{\sqrt{a^2 + x^2}} = \sqrt{a^2 + x^2}$$

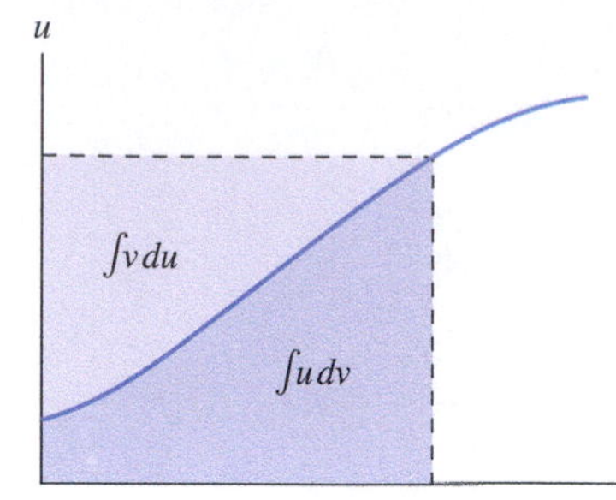

2. Integration by parts
The quantity $\int u\,dv$ is the area under the curve of u as a function of v between specified limits. In the figure, that area can also be expressed as the area of the rectangle shown minus the area under the curve of v as a function of u. Mathematically, this relation among areas may be expressed as a relation among integrals:

$$\int u\,dv = uv - \int v\,du \qquad \text{(integration by parts)}$$

This expression may often be used to transform complicated integrals into simpler ones.

Example

Evaluate $\int x\cos x\,dx$. Here let $u = x$, so $du = dx$. Then $dv = \cos x\,dx$, so we have $v = \int dv = \int \cos x\,dx = \sin x$. Integrating by parts then gives

$$\int x\cos x\,dx = (x)(\sin x) - \int \sin x\,dx = x\sin x + \cos x$$

where the + sign arises because $\int \sin x\,dx = -\cos x$.

Table of Integrals

More extensive tables are available in many mathematical and scientific handbooks; see, for example, ***Handbook of Chemistry and Physics*** (Chemical Rubber Co.) or Dwight, ***Tables of Integrals and Other Mathematical Data*** (Macmillan). Some math software, including *Mathematica* and *Maple*, can also evaluate integrals symbolically. Wolfram Research provides *Mathematica*-based integration both at *integrals.wolfram.com* and through WolframAlpha at *www.wolframalpha.com/calculators/integral-calculator*.

In the expressions below, a and b are constants. An arbitrary constant of integration may be added to the right-hand side.

$$\int e^{ax}\,dx = \frac{e^{ax}}{a}$$

$$\int \sin ax\,dx = -\frac{\cos ax}{a}$$

$$\int \cos ax\,dx = \frac{\sin ax}{a}$$

$$\int \tan ax\,dx = -\frac{1}{a}\ln(\cos ax)$$

$$\int \sin^2 ax\,dx = \frac{x}{2} - \frac{\sin 2ax}{4a}$$

$$\int \cos^2 ax\,dx = \frac{x}{2} + \frac{\sin 2ax}{4a}$$

$$\int x\sin ax\,dx = \frac{1}{a^2}\sin ax - \frac{1}{a}x\cos ax$$

$$\int x\cos ax\,dx = \frac{1}{a^2}\cos ax + \frac{1}{a}x\sin ax$$

$$\int \frac{dx}{\sqrt{a^2 - x^2}} = \sin^{-1}\left(\frac{x}{a}\right)$$

$$\int \frac{dx}{\sqrt{x^2 \pm a^2}} = \ln\left(x + \sqrt{x^2 \pm a^2}\right)$$

$$\int \frac{dx}{x^2 + a^2} = \frac{1}{a}\tan^{-1}\left(\frac{x}{a}\right)$$

$$\int \frac{x\,dx}{\sqrt{a^2 - x^2}} = -\sqrt{a^2 - x^2}$$

$$\int \frac{x\,dx}{\sqrt{x^2 \pm a^2}} = \sqrt{x^2 \pm a^2}$$

$$\int \frac{dx}{(x^2 \pm a^2)^{3/2}} = \frac{\pm x}{a^2\sqrt{x^2 \pm a^2}}$$

$$\int x e^{ax}\,dx = \frac{e^{ax}}{a^2}(ax - 1)$$

$$\int x^2 e^{ax}\,dx = \frac{x^2 e^{ax}}{a} - \frac{2}{a}\left[\frac{e^{ax}}{a^2}(ax - 1)\right]$$

$$\int \frac{dx}{a + bx} = \frac{1}{b}\ln(a + bx)$$

$$\int \frac{dx}{(a + bx)^2} = -\frac{1}{b(a + bx)}$$

$$\int \ln ax\,dx = x\ln ax - x$$

The International System of Units (SI)

APPENDIX B

The International System of Units (SI) system is undergoing a major overhaul, which should be completed by 2017. The new SI will give explicit-constant definitions to six of the seven base units, defining those units by setting exact values for appropriate physical constants. Here we list SI unit definitions informally to reflect this ongoing transition.

length (meter): The meter is defined so that the speed of light in vacuum is exactly 299,792,458 m/s. In effect since 1983, this definition is reworded but not otherwise changed.

mass (kilogram): In the new SI, the definition of the kilogram changes from one based on a physical standard, the international prototype kilogram, to an explicit-constant definition based on a defined value for Planck's constant h.

time (second): The second is defined as the duration of 9,192,631,770 periods of the radiation corresponding to the transition between the two hyperfine levels of the ground state of the cesium-133 atom. In effect since 1967, this definition is reworded but not otherwise changed.

electric current (ampere): Since 1948 the ampere has been defined in terms of the force between two current-carrying wires, but that changes to a definition based on an exact value for the elementary charge.

temperature (kelvin): Since 1967 the kelvin has been defined such that the triple point of water is at 273.16 K. That changes to a definition based on an exact value for Boltzmann's constant k.

amount of substance (mole): The 1971 definition of the mole in terms of carbon-12 atoms changes to one based on an exact value for Avogadro's number.

luminous intensity (candela): The 1979 definition defines the candela as the luminous intensity of a 540-THz source emitting (1/683) watt per steradian. This is reworded but not otherwise changed.

SI Base and Supplementary Units

	SI Unit	
Quantity	**Name**	**Symbol**
Base Unit		
Length	meter	m
Mass	kilogram	kg
Time	second	s
Electric current	ampere	A
Thermodynamic temperature	kelvin	K
Amount of substance	mole	mol
Luminous intensity	candela	cd
Supplementary Units		
Plane angle	radian	rad
Solid angle	steradian	sr

SI Prefixes

Factor	Prefix	Symbol
10^{24}	yotta	Y
10^{21}	zetta	Z
10^{18}	exa	E
10^{15}	peta	P
10^{12}	tera	T
10^{9}	giga	G
10^{6}	mega	M
10^{3}	kilo	k
10^{2}	hecto	h
10^{1}	deka	da
10^{0}	—	—
10^{-1}	deci	d
10^{-2}	centi	c
10^{-3}	milli	m
10^{-6}	micro	μ
10^{-9}	nano	n
10^{-12}	pico	p
10^{-15}	femto	f
10^{-18}	atto	a
10^{-21}	zepto	z
10^{-24}	yocto	y

Some SI Derived Units with Special Names

			SI Unit	
Quantity	**Name**	**Symbol**	**Expression in Terms of Other Units**	**Expression in Terms of SI Base Units**
Frequency	hertz	Hz		s^{-1}
Force	newton	N		$m \cdot kg \cdot s^{-2}$
Pressure, stress	pascal	Pa	N/m^2	$m^{-1} \cdot kg \cdot s^{-2}$
Energy, work, heat	joule	J	$N \cdot m$	$m^2 \cdot kg \cdot s^{-2}$
Power	watt	W	J/s	$m^2 \cdot kg \cdot s^{-3}$
Electric charge	coulomb	C		$s \cdot A$
Electric potential, potential difference, electromotive force	volt	V	J/C	$m^2 \cdot kg \cdot s^{-3} \cdot A^{-1}$
Capacitance	farad	F	C/V	$m^{-2} \cdot kg^{-1} \cdot s^4 \cdot A^2$
Electric resistance	ohm	Ω	V/A	$m^2 \cdot kg \cdot s^{-3} \cdot A^{-2}$
Magnetic flux	weber	Wb	$T \cdot m^2$, $V \cdot s$	$m^2 \cdot kg \cdot s^{-2} \cdot A^{-1}$
Magnetic field	tesla	T	Wb/m^2	$kg \cdot s^{-2} \cdot A^{-1}$
Inductance	henry	H	Wb/A	$m^2 \cdot kg \cdot s^{-2} \cdot A^{-2}$
Radioactivity	becquerel	Bq	1 decay/s	s^{-1}
Absorbed radiation dose	gray	Gy	J/kg, 100 rad	$m^2 \cdot s^{-2}$
Radiation dose equivalent	sievert	Sv	J/kg, 100 rem	$m^2 \cdot s^{-2}$

Conversion Factors

APPENDIX C

The listings below give the SI equivalents of non-SI units. To convert from the units shown to SI, multiply by the factor given; to convert the other way, divide. For conversions within the SI system, see the table of SI prefixes in Appendix B, Chapter 1, or the inside front cover. Conversions that are not exact by definition are given to, at most, four significant figures.

Length

1 inch (in) = 0.0254 m

1 foot (ft) = 0.3048 m

1 yard (yd) = 0.9144 m

1 mile (mi) = 1609 m

1 nautical mile = 1852 m

1 angstrom (Å) = 10^{-10} m

1 light year (ly) = 9.46×10^{15} m

1 astronomical unit (AU) = 1.496×10^{11} m

1 parsec = 3.09×10^{16} m

1 fermi = 10^{-15} m = 1 fm

Mass

1 slug = 14.59 kg

1 metric ton (tonne; t) = 1000 kg

1 unified mass unit (u) = 1.661×10^{-27} kg

Force units in the English system are sometimes used (incorrectly) for mass. The units given below are actually equal to the number of kilograms multiplied by *g*, the acceleration of gravity.

1 pound (lb) = weight of 0.454 kg

1 ton = 2000 lb = weight of 908 kg

1 ounce (oz) = weight of 0.02835 kg

Time

1 minute (min) = 60 s

1 hour (h) = 60 min = 3600 s

1 day (d) = 24 h = 86,400 s

1 year (y) = 365.2422 d* = 3.156×10^{7} s

Area

1 hectare (ha) = 10^4 m^2

1 square inch (in^2) = 6.452×10^{-4} m^2

1 square foot (ft^2) = 9.290×10^{-2} m^2

1 acre = 4047 m^2

1 barn = 10^{-28} m^2

1 shed = 10^{-30} m^2

Volume

1 liter (L) = 1000 cm^3 = 10^{-3} m^3

1 cubic foot (ft^3) = 2.832×10^{-2} m^3

1 cubic inch (in^3) = 1.639×10^{-5} m^3

1 fluid ounce = 1/128 gal = 2.957×10^{-5} m^3

1 barrel (bbl) = 42 gal = 0.1590 m^3

1 gallon (U.S.; gal) = 3.785×10^{-3} m^3

1 gallon (British) = 4.546×10^{-3} m^3

Angle, Phase

1 degree (°) = $\pi/180$ rad = 1.745×10^{-2} rad

1 revolution (rev) = 360° = 2π rad

1 cycle = 360° = 2π rad

*The length of the year changes very slowly with changes in Earth's orbital period.

Speed, Velocity

1 km/h = (1/3.6) m/s = 0.2778 m/s
1 mi/h (mph) = 0.4470 m/s
1 ft/s = 0.3048 m/s
1 ly/y = 3.00×10^8 m/s

Angular Speed, Angular Velocity, Frequency, and Angular Frequency

1 rev/s = 2π rad/s = 6.283 rad/s (s^{-1})
1 Hz = 1 cycle/s = 2π s^{-1}
1 rev/min (rpm) = 0.1047 rad/s (s^{-1})

Force

1 dyne = 10^{-5} N
1 pound (lb) = 4.448 N

Pressure

1 dyne/cm^2 = 10^{-1} Pa
1 atmosphere (atm) = 1.013×10^5 Pa
1 torr = 1 mm Hg at 0°C = 133.3 Pa
1 bar = 10^5 Pa = 0.987 atm
1 lb/in^2 (psi) = 6.895×10^3 Pa
1 in H_2O (60°F) = 248.8 Pa
1 in Hg (60°F) = 3.377×10^3 Pa

Energy, Work, Heat

1 erg = 10^{-7} J
1 calorie* (cal) = 4.184 J
1 electronvolt (eV) = 1.602×10^{-19} J
1 foot-pound (ft·lb) = 1.356 J
1 Btu* = 1.054×10^3 J
1 kWh = 3.6×10^6 J
1 megaton (explosive yield; Mt) = 4.18×10^{15} J

Power

1 erg/s = 10^{-7} W
1 horsepower (hp) = 746 W
1 Btu/h (Btuh) = 0.293 W
1 ft·lb/s = 1.356 W

Magnetic Field

1 gauss (G) = 10^{-4} T
1 gamma (γ) = 10^{-9} T

Radiation

1 curie (ci) = 3.7×10^{10} Bq
1 rad = 10^{-2} Gy
1 rem = 10^{-2} Sv

Energy Content of Fuels

Energy Source	Energy Content
Coal	29 MJ/kg = 7300 kWh/ton = 25×10^6 Btu/ton
Oil	43 MJ/kg = 39 kWh/gal = 1.3×10^5 Btu/gal
Gasoline	44 MJ/kg = 36 kWh/gal = 1.2×10^5 Btu/gal
Natural gas	55 MJ/kg = 30 kWh/100 ft^3 = 1000 Btu/ft^3
Uranium (fission)	
Normal abundance	5.8×10^{11} J/kg = 1.6×10^5 kWh/kg
Pure U-235	8.2×10^{13} J/kg = 2.3×10^7 kWh/kg
Hydrogen (fusion)	
Normal abundance	7×10^{11} J/kg = 3.0×10^4 kWh/kg
Pure deuterium	3.3×10^{14} J/kg = 9.2×10^7 kWh/kg
Water	1.2×10^{10} J/kg = 1.3×10^4 kWh/gal = 340 gal gasoline/gal water
100% conversion, matter to energy	9.0×10^{16} J/kg = 931 MeV/u = 2.5×10^{10} kWh/kg

*Values based on the thermochemical calorie; other definitions vary slightly.

The Elements

APPENDIX D

The atomic weights of stable elements reflect the abundances of different isotopes; values given here apply to elements as they exist naturally on Earth. For stable elements, parentheses express uncertainties in the last decimal place given. For elements with no stable isotopes (indicated in **boldface**), at most three isotopes are given; for elements 99 and beyond, only the longest-lived isotope is given. (Exceptions are the unstable elements thorium, protactinium, and uranium, for which atomic weights reflect natural abundances of long-lived isotopes.) See also the periodic table inside the back cover.

Atomic Number	Names	Symbol	Atomic Weight
1	Hydrogen	H	1.00794 (7)
2	Helium	He	4.002602 (2)
3	Lithium	Li	6.941 (2)
4	Beryllium	Be	9.012182 (3)
5	Boron	B	10.811 (5)
6	Carbon	C	12.011 (1)
7	Nitrogen	N	14.00674 (7)
8	Oxygen	O	15.9994 (3)
9	Fluorine	F	18.9984032 (9)
10	Neon	Ne	20.1797 (6)
11	Sodium (Natrium)	Na	22.989768 (6)
12	Magnesium	Mg	24.3050 (6)
13	Aluminum	Al	26.981539 (5)
14	Silicon	Si	28.0855 (3)
15	Phosphorus	P	30.973762 (4)
16	Sulfur	S	32.066 (6)
17	Chlorine	Cl	35.4527 (9)
18	Argon	Ar	39.948 (1)
19	Potassium (Kalium)	K	39.0983 (1)
20	Calcium	Ca	40.078 (4)
21	Scandium	Sc	44.955910 (9)
22	Titanium	Ti	47.88 (3)
23	Vanadium	V	50.9415 (1)
24	Chromium	Cr	51.9961 (6)
25	Manganese	Mn	54.93805 (1)
26	Iron	Fe	55.847 (3)
27	Cobalt	Co	58.93320 (1)
28	Nickel	Ni	58.69 (1)
29	Copper	Cu	63.546 (3)
30	Zinc	Zn	65.39 (2)
31	Gallium	Ga	69.723 (1)
32	Germanium	Ge	72.61 (2)
33	Arsenic	As	74.92159 (2)

(continued)

Atomic Number	Names	Symbol	Atomic Weight
34	Selenium	Se	78.96 (3)
35	Bromine	Br	79.904 (1)
36	Krypton	Kr	83.80 (1)
37	Rubidium	Rb	85.4678 (3)
38	Strontium	Sr	87.62 (1)
39	Yttrium	Y	88.90585 (2)
40	Zirconium	Zr	91.224 (2)
41	Niobium	Nb	92.90638 (2)
42	Molybdenum	Mo	95.94 (1)
43	**Technetium**	**Tc**	**97, 98, 99**
44	Ruthenium	Ru	101.07 (2)
45	Rhodium	Rh	102.90550 (3)
46	Palladium	Pd	106.42 (1)
47	Silver	Ag	107.8682 (2)
48	Cadmium	Cd	112.411 (8)
49	Indium	In	114.82 (1)
50	Tin	Sn	118.710 (7)
51	Antimony (Stibium)	Sb	121.75 (3)
52	Tellurium	Te	127.60 (3)
53	Iodine	I	126.90447 (3)
54	Xenon	Xe	131.29 (2)
55	Cesium	Cs	132.90543 (5)
56	Barium	Ba	137.327 (7)
57	Lanthanum	La	138.9055 (2)
58	Cerium	Ce	140.115 (4)
59	Praseodymium	Pr	140.90765 (3)
60	Neodymium	Nd	144.24 (3)
61	**Promethium**	**Pm**	**145, 147**
62	Samarium	Sm	150.36 (3)
63	Europium	Eu	151.965 (9)
64	Gadolinium	Gd	157.25 (3)
65	Terbium	Tb	158.92534 (3)
66	Dysprosium	Dy	162.50 (3)
67	Holmium	Ho	164.93032 (3)
68	Erbium	Er	167.26 (3)
69	Thulium	Tm	168.93421 (3)
70	Ytterbium	Yb	173.04 (3)
71	Lutetium	Lu	174.967 (1)
72	Hafnium	Hf	178.49 (2)
73	Tantalum	Ta	180.9479 (1)
74	Tungsten (Wolfram)	W	183.85 (3)
75	Rhenium	Re	186.207 (1)
76	Osmium	Os	190.2 (1)
77	Iridium	Ir	192.22 (3)
78	Platinum	Pt	195.08 (3)
79	Gold	Au	196.96654 (3)
80	Mercury	Hg	200.59 (3)
81	Thallium	Tl	204.3833 (2)
82	Lead	Pb	207.2 (1)
83	Bismuth	Bi	208.98037 (3)

Atomic Number	Names	Symbol	Atomic Weight
84	Polonium	Po	209, 210
85	Astatine	At	210, 211
86	Radon	Rn	211, 220, 222
87	Francium	Fr	223
88	Radium	Ra	223, 224, 226
89	Actinium	Ac	227
90	Thorium	Th	232.0381 (1)
91	Protactinium	Pa	231.03588 (2)
92	Uranium	U	238.0289 (1)
93	Neptunium	Np	237, 239
94	Plutonium	Pu	239, 242, 244
95	Americium	Am	241, 243
96	Curium	Cm	245, 247, 248
97	Berkelium	Bk	247, 249
98	Californium	Cf	249, 250, 251
99	Einsteinium	Es	252
100	Fermium	Fm	257
101	Mendelevium	Md	258
102	Nobelium	No	259
103	Lawrencium	Lr	262
104	Rutherfordium	Rf	263
105	Dubnium	Db	268
106	Seaborgium	Sg	266
107	Bohrium	Bh	272
108	Hassium	Hs	277
109	Meitnerium	Mt	276
110	Darmstadtium	Ds	281
111	Roentgenium	Rg	280
112	Copernicium	Cn	285
113	Nihonium	Nh	284
114	Flerovium	Fl	289
115	Moscovium	Mc	288
116	Livermorium	Lv	292
117	Tennessine	Ts	294
118	Oganesson	Og	294

APPENDIX E

Astrophysical Data

Sun, Planets, Principal Satellites

Body	Mass (10^{24} kg)	Mean Radius (10^6 m except as noted)	Surface Gravity (m/s^2)	Escape Speed (km/s)	Sidereal Rotation Period* (days)	Mean Distance from Central Body† (10^6 km)	Orbital Period	Mean Orbital Speed (km/s)
Sun	1.99×10^6	696	274	618	36 at poles 27 at equator	2.6×10^{11}	200 My	250
Planets								
Mercury	0.330	2.44	3.70	4.25	58.6	57.9	88.0 d	47.4
Venus	4.87	6.05	8.87	10.4	−243	108	225 d	35.0
Earth	5.97	6.37	9.81	11.2	0.997	149.6	365.2 d	29.8
Moon	0.0735	1.74	1.62	2.38	27.3	0.3844	27.3 d	1.02
Mars	0.642	3.39	3.71	5.03	1.03	228	1.88 y	24.1
Phobos	1.07×10^{-8}	9–13 km	0.0057	0.0114	0.319	9.4×10^{-3}	0.319 d	2.14
Deimos	1.48×10^{-9}	5–8 km	0.003	0.00556	1.26	23×10^{-3}	1.26 d	1.35
Jupiter	1.90×10^3	69.9	24.8	60.2	0.414	778	11.9 y	13.1
Io	0.0893	1.82	1.80	2.38	1.77	0.422	1.77 d	17.3
Europa	0.480	1.56	1.32	2.03	3.55	0.671	3.55 d	13.7
Ganymede	0.148	2.63	1.43	2.74	7.15	1.07	7.15 d	10.9
Callisto	0.108	2.41	1.24	2.44	16.7	1.88	16.7 d	8.20
and 13 smaller satellites								
Saturn	568	58.2	10.4	36.1	0.444	1.43×10^3	29.5 y	9.69
Tethys	0.0007	0.53	0.2	0.4	1.89	0.294	1.89 d	11.3
Dione	0.00015	0.56	0.3	0.6	2.74	0.377	2.74 d	10.0
Rhea	0.0025	0.77	0.3	0.5	4.52	0.527	4.52 d	8.5
Titan	0.135	2.58	1.35	2.64	15.9	1.22	15.9 d	5.6
and 12 smaller satellites								
Uranus	86.8	25.4	8.87	21.4	−0.720	2.87×10^3	84.0 y	6.80
Ariel	0.0013	0.58	0.3	0.4	2.52	0.19	2.52 d	5.5
Umbriel	0.0013	0.59	0.3	0.4	4.14	0.27	4.14 d	4.7
Titania	0.0018	0.81	0.2	0.5	8.70	0.44	8.70 d	3.7
Oberon	0.0017	0.78	0.2	0.5	13.5	0.58	13.5 d	3.1
and 11 smaller satellites								
Neptune	102	24.6	11.2	23.5	0.673	4.50×10^3	165 y	5.43
Triton	0.134	1.9	2.5	3.1	5.88	0.354	5.88 d	4.4
and 7 smaller satellites								
Dwarf Planets								
Ceres	0.000945	0.476	0.27	0.51	0.38	414	4.60 y	17.9
Pluto	0.0131	1.20	0.58	1.2	−6.39	5.91×10^3	248 y	4.67
Charon	0.00162	0.604	0.278	0.580	−6.39	0.00196	6.39 d	0.23
and 4 smaller satellites								
Eris	0.0167	1.16	0.827	1.38	1.1	1.02×10^4	560 y	3.43
and 1 small satellite, Dysnomia								

*Negative rotation period indicates retrograde motion, in opposite sense from orbital motion. Periods are sidereal, meaning the time for the body to return to the same orientation relative to the distant stars rather than the Sun.

†Central body is galactic center for Sun, Sun for planets, and planet for satellites.

Answers to Odd-Numbered Problems

Chapter 1

13. 10^5
15. 108.783 ps
17. 10^8
19. 0.62 rad = 35°
21. 30 g
23. 10^6
25. 8.6 m^2/L
27. 3.6 km/h
29. 57.3°
31. 24 Zm
33. 7.4×10^6 m/s^2
35. 4×10^6
37. 41 m
39. (a) 5.18 (b) 5.20
41. 3×10^6
43. About 0.08%
45. (a) ~3×10^3 m^3 (b) ~100 days
47. 10^5
49. ~ 250 μm
51. (a) 40 nm (b) 5×10^5 calculations per second
53. $\Delta = 100(\pm 0.05/N)\%$
55. in the U.S.
57. about 2000
59. about 1–2 m^2
61. (a) 1.0 m (b) 0.001 m^2 (c) 0.0 m (d) 1.0
63. $10.10
65. slope = 4.09 g/cm^3
67. b
69. c

Chapter 2

13. (a) 375 yd/min (b) 5.72 min
15. 21 h
17. (a) 3.0×10^4 m/s (b) 19 mi/s
21. (a) $v = b - 2ct$ (b) 8.4 s
23. 0.35 m/s^2
25. falling: 9.82 m/s^2, stopping: 84.0 m/s^2
27. 17 m/s^2
29. $v = dx/dt = d/dt(x_0 + v_0t + at^2/2) = v_0 + at$
31. (a) 46 m/s^2 (b) 61 s
33. 27 ft/s^2
35. 15 s
37. 95 m
39. (a) 123 m (b) 39 m/s, 40 m (c) 9.8 m/s, 100 m; (d) −20 m/s, 100 m
41. 11 m/s
43. 48 mi/h
45. 2.2 s
47. (a) 9.82 m/s (b) 9.34 m/s (c) 9.18 m/s (d) 9.18 m/s
49. (a) 39.95 m/s (b) 0.13%
51. 4.3 m/s^2
53. 2.75 s
55. 55%
57. (a) 0.014 s (b) 51 cm
59. 0.89 km
61. (a) 25 m/s (b) 180 m
63. 0.0051 m/s^2
65. 11 m/s
67. 270 m
69. $-\frac{1}{2}\sqrt{hg}$
71. (a) 7.88 m/s, 7.67 m/s (b) 0.162 s
73. 3.9 s, 6.2 m/s
75. 4.8 m/s (17 km/h)
77. (a) $\bar{v} = (v_1 + v_2)/2$ (b) $\bar{v} = (2v_1v_2)/(v_1 + v_2)$ (c) in the first case
79. 70.7 %
81. −0.3 m/s
83. $\frac{h}{4}\left(\frac{2h}{g\Delta t^2}\right)\left(\frac{g\Delta t^2}{2h} - 1\right)^2$
85. 15 s^{-1}
89. (a) $-(a_0/\omega)\sin\omega t$ (b) $(a_0/\omega^2)\cos\omega t$ (c) $v_{max} = a_0/\omega, x_{max} = a_0/\omega^2$
91. (a) $v_0 > \sqrt{gh_0/2}$ (b) $h_0 - gh_0^2/2v_0$
93. c
95. c

Chapter 3

13. 270 m, 150°
15. 700 km, 110°
17. $105\hat{\imath} + 58\hat{\jmath}$ km
19. 1.414, $\theta = 45°$
21. (−14 m/s, −12 m/s)
23. $3ct^2\hat{\imath}$
25. (a) $\vec{v} = -2.2\times10^{-6}\hat{\jmath}$ m/s (b) $\vec{a} = -3.2\times10^{-10}\,\hat{\imath}$ m/s^2
27. $\vec{v}_2 = 1.3\hat{\imath} + 2.3\hat{\jmath}$ m/s
29. (a) 26° upstream (b) 53.9 s
31. 42.8° west of south
33. 49 m, 6.4° to your original direction
35. (a) 1.3 s (b) 15 m
37. 34 nm
39. 1090 m
41. 2.28×10^{-7}m/s^2
43. 2.8 mm/s^2
45. (a) $A\sqrt{5}$ (b) $A\sqrt{10}$
47. $\vec{C} = -15\hat{\imath} + 9\hat{\jmath} - 18\hat{k}$
49. (a) $4c/3d$ (b) $c/3d$
51. 96 m
53. (a) 0.249 m/s (b) 7.00×10^{-4} m/s^2 (c) 7.21×10^{-4} m/s^2, about 3% difference
55. $A = B$
57. 0.50 m/s^2
59. 5.7 m/s
61. (a) $x_1 = x_2$ implies $y_1 = h\left(1 - \frac{gh}{v_0^2}\right) = y_2$ (b) $v_0 \geq \sqrt{gh}$
63. 8.3 m/s, 61°
67. semi-circle of radius 2.5 cm; 6.54 m/s, 17.1 m/s^2
69. Yes
71. 66°
73. (a) $v\sqrt{2/\sqrt{3}} \approx 1.07v$ (b) $\sqrt[4]{3}t \approx 1.3t$
75. 2.3 km
77. $2h$
79. 19 m
81. $dx/d\theta_0 = 2v_0^2/g\cos(2\theta_0) = 0 \Rightarrow \theta_0 = 45°$.
83. $\frac{1}{2}\cos^{-1}(1/(1 + v_0^2/gh))$
87. $2va_t/r$
89. c
91. c

Chapter 4

15. (a) 2.0 m/s^2 (b) 0.082 m/s^2
17. −13 kN
19. 2.0×10^6 m/s^2
21. 22 cm
25. 210 kg
27. 9000 kg
29. 490 N
31. 380 N
33. Required thrust 11 MN; force on astronaut 1.3 kN
35. 55 kN
37. 130 N
39. 19 cm
41. 2.94 m/s^2, downward
43. 4.9 m/s^2
45. 0.53 s
47. 6.0 N
49. 1.62×10^{-7}N/m
51. (a) 5.3 kN (b) 1.1 kN (c) 0.49 kN (d) 0.59 kN

53. 680 m
55. 0.96 m
57. 950 N
59. (a) $-0.40mg$ (b) $2.40mg$ (c) $1.40mg$
61. F-35A: yes, 0.81 m/s^2; A-380: no
65. 1.96 m/s^2
67. (a) $(m_f - m_s)g/m_s$ (b) $\dfrac{m_f a_s h_0}{(m_f - m_s)(g + a_s)}$
69. (a) 60.0 m/s (b) 0.672 m
71. 11.8 m/s^2
73. 0.92 kg, 1.4 kg
75. $\omega F_0/M$
77. a
79. b

Chapter 5

13. $5.40\hat{\imath} + 11.0\hat{\jmath}$ N
15. 22.2°
17. 880 N
19. 34°
21. $m_R/m_L = 2.5$
23. (a) 3.9 m/s^2 (b) 530 N
27. Train was speeding at 71.4 km/h
29. 490 km/h
31. 0.18
33. 0.12
35. 0.43 m
37. about 2.62 times
39. $T = m_2 g$, $\tau = 2\pi\sqrt{(m_1 R)/(m_2 g)}$
41. 310 N downward (b) $-m_{SB}v^2/R$ (c) nothing
43. 8.5 km
45. 0.15
47. $a = 0.19$ m/s^2; $t = 23$ s
49. Yes
51. $0.23 \leq \mu_s \leq 0.30$
53. 4.2 m/s^2
55. 0.62
57. (a) 9.8 cm (b) no
59. 100 km/h
63. 17 min^{-1}
65. Brake, don't swerve
67. 28 cm
69. $T' = u_k/\sqrt{1 + \mu_k^2}$
73. Yes
75. 7.6 km
77. a
79. b

Chapter 6

13. 900 J
15. 150 kJ
17. 190 MN
19. $\vec{A}\cdot(\vec{B} + \vec{C}) = AB\cos(\theta_{AB}) + AC\cos(\theta_{AC}) = \vec{A}\cdot\vec{B} + \vec{A}\cdot\vec{C}$
21. 1.9 m
23. (a) 1 J (b) 3 J
25. 30 cm
27. 7.5 GJ
29. ± 120 km/h
31. 110 m/s
33. 97 W
35. (a) 60 kW (b) 1 kW (c) 41.7 W
37. 9.4×10^6 J
39. 0 W
41. 22 s
43. (a) 400 J (b) 31 kg
45. (a) 76,000 (b) 14 kW
47. 25°
49. (a) 0 (b) 90°
51. 622 J
53. $k_B = 8k_A$
55. $W = F_0\left(x - \dfrac{x^2}{2L_0} + \dfrac{L_0^2}{L_0 + x} - L_0\right)$
57. $v_2 = \pm 2v_1$
59. (a) 1.3×10^{-17}W (b) 1.4×10^{-14}J
61. 9.6 kW
63. $F_0 x_0/3$
65. 70.5°
67. 370×10^6 gal/day
69. 26 m/s
71. 0.60
73. (a) 0.45 kW (b) 7.99 kJ
75. 42 kJ
77. 6.0 years
81. $W_{x_1\to x_2} = 2b(\sqrt{x_2} - \sqrt{x_1})$, $W(x_1 = 0) = 2b\sqrt{x_2}$
83. (a) $\frac{1}{2}kL_0^2 + \frac{1}{3}bL_0^3 + \frac{1}{4}cL_0^4 + \frac{1}{5}dL_0^5$ (b) 12 kJ
85. 135 J
87. 30 people
89. Stopping force is 35 times weight of leg
91. c
93. c

Chapter 7

11. $W_a = W_b = -mgL$
13. (a) 1.3 MJ (b) -59 kJ
15. 840 m
17. 55 cm
19. ± 22 m/s, ± 35 m/s
21. 92 m
23. 2.3 kN/m
25. 0.75
27. ± 2.0 m
29. (a) 4.4×10^{13}J (b) 11 h
31. (a) 1.07 J (b) 1.12 J
33. 778 J, 4.90%
35. 2.5 J
37. $U(x) = -\frac{1}{3}ax^3 - bx$
39. $r = \dfrac{kx^2}{2mg\sin\theta}$
43. (a) -11 cm (b) ± 4 m/s
45. $h \geq 5R/2$
49. (a) $U(x) = -\dfrac{a}{2}x^2 + \dfrac{b}{4}x^4$ (b) 0.7 m and 2 m
51. 20 m/s, 30 m/s
53. 1.4 m
55. 62.5 cm
57. 2.9 m
59. 14 m
61. $v = 2x^{3/4}\sqrt{\dfrac{a}{3m}}$
63. 5.8 s
65. $\dfrac{mgh}{2d}\sqrt{2g(h - d)}$
67. 185 N/m
69. d
71. b

Chapter 8

11. $R_P = R_E/\sqrt{2}$
13. 57.5%
15. 8.6 kg
17. 542 m
19. 3070 m/s
21. 1.77 d
23. 0.28×10^6 m
25. 3.17 GJ
27. 4.29 km/s
29. -2.64×10^{33} J
31. (a) 2.44 km/s (b) 2.10×10^8 m/s
33. 10 m/s^2
35. $g(h)/g(0) = 0.414$
37. 2.73×10^{-3} m/s^2, $a_c/g = 2.78\times10^{-4}$
39. 60.5 min
41. 2.6×10^{41} kg
43. $T^2 = \dfrac{4\pi^2L^3}{3GM}$
45. 2.79 AU
47. $E > 0$, hyperbolic path
51. The comet is going faster than the escape velocity from the Sun, so it will not return to Earth's vicinity.
55. (a) 2.06×10^6 m (b) 0.805×10^6 m
57. (a) 4.59 km/s (b) 14.2 km/s
59. 4.17 km/s
61. 4.60×10^{10} m
63. 1.42×10^3 km
65. 1.58×10^{16} kg
67. 3.8 m/century
69. No danger, since the puck needs at least 6100 km/h to go into orbit.
71. 1.5×10^6 km
73. d
75. d

Chapter 9

15. $2m$
17. $(0, 0.289L)$
19. $\vec{v}_2 = -67\hat{\imath}$ cm/s
21. 0.268 Mm/s
23. 1.21 J
25. The impulse imparted by gravity is 0.08% of the collision impulse.
27. 41.8 s
31. The second truck's load was about 1100 kg—well below the legal limit.
33. 46 m/s

35. $v_{1f} = -11$ Mm/s, $v_{2f} = +6.9$ Mm/s, velocities are exchanged
37. (0, 0.115 (a)
39. $\vec{r}_{cm} = (2t^2 + 4t + \frac{8}{3})\hat{\imath} + (\frac{5}{3}t + \frac{4}{3})\hat{\jmath}$; $\vec{v}_{cm} = (4t + 4)\hat{\imath} + (\frac{5}{3})\hat{\jmath}$; $\vec{a}_{cm} = 4\hat{\imath}$
41. $m_b = 4m_m$
43. $(0, 0, h/4)$
45. (a) 0.99 m (b) 3.9 m/s
47. (a) $\vec{a}_c = \frac{v_0}{M}\left(\frac{dm}{dt}\right)\hat{\imath}$ (b) v_0
51. (a) (0, 0, 13 m) (b) (0, 0, 11 m)
53. $\vec{v}_3 = 4.4\hat{\imath} + 3.0\hat{\jmath}$ m/s
55. 9.4 m/s
57. $\frac{2}{5}v$; $\frac{7}{5}v$
59. (a) 37.7° (b) −65.8 cm/s
61. 5.8 s
63. 0.92 m/s
65. If $v_{Buick} = 55$ km/h, then $v_{Toyota} = 90$ km/h; if $v_{Toyata} = 55$ km/h, then $v_{Buick} = 65$ km/h
67. 120°
69. 5.83
73. 18.6%
75. $J = 2F_0/a$
77. (a) 12.0 m/s (b) 15.4 m/s
79. $v_1 = v/6$, $v_2 = 5v/6$
81. 8.3 kg
83. The peak force of 327 kN occurs at 165 ms.
87. The center of mass lies along line through the middle of the slice, at a distance of $(4R/3\theta)\sin(\frac{1}{2}\theta)$ from the tip.
89. 3.75 min
91. (a) $\frac{M}{1+a}$; (b) $\frac{1+a}{2+a}L$ (c) M and $\frac{1}{2}L$
93. 3 collisions, final speeds $0.26v_0$ and $0.31v_0$
95. b
97. a

Chapter 10

13. (a) $7.27\times10^{-5}\,\text{s}^{-1}$ (b) $1.75\times10^{-3}\,\text{s}^{-1}$ (c) $1.45\times10^{-4}\,\text{s}^{-1}$ (d) $31.4\,\text{s}^{-1}$
15. (a) 75 rad/s (b) 2.4×10^{-4} rad/s (c) 6×10^3 rad/s (d) 2×10^{-7} rad/s
17. (a) 0.068 rpm/s (b) $7.1\times10^{-3}\,\text{s}^{-2}$
19. (a) 0.16 rev (b) 0.07 rad/s
21. 1.2 m
23. 7.9×10^{-2} N·m
25. (a) $2mL^2$ (b) mL^2
27. (a) 4.4×10^{-4} kg·m² (b) 3.7×10^{-3} N·m
29. (a) mL^2 (b) $\frac{1}{2}mL^2$
31. (a) 3.2×10^{38} kg·m² (b) 1.8×10^{34} N·m
33. 20 min
35. 12,000 y
37. (a) 1.6×10^8 J (b) 16 MW
39. 1/3
41. (a) 6.9 rad/s (b) 3.7 s
43. (a) 1.1 rad/s (b) 1.1 m/s
45. (a) $170\,\text{s}^{-2}$ (b) 2.9 m/s² (c) 150 revolutions
47. 570 rev
49. (a) $2ML^2/3$ (b) $2ML^2/3$ (c) $4ML^2/3$
51. $Ma^2/3$
53. 33 pN
55. (a) 7.2 h (b) 1900 rev
57. 0.36
59. ±2.1 rad/s
61. $v = \sqrt{\frac{6}{5}gd\sin\theta}$
63. 17%
65. $0.494\,MR^2$
67. 33 m
69. (a) $M = \frac{2\pi\rho_0 wR^2}{3}$ (b) $I = 3MR^2/5$
71. yes for spin-up time (53 s), but no for efficiency (94%)
73. $3MR^2/10$
75. $\tau = \frac{1}{2}MGL\sin\theta$
77. The specs are incorrect. The storage capacity is 3 MJ below what's claimed.
79. 5.2×10^{-5} kg·m²
81. a
83. b

Chapter 11

15. $\vec{\omega} = 63\,\text{s}^{-1}$ west
17. (a) $0.524\,\text{s}^{-2}$ (b) −37°
19. (a) $-12\hat{k}$ N·m (b) $36\hat{k}$ N·m (c) $12\hat{\imath} - 36\hat{\jmath}$ N·m
21. 3.1 N·m, out of the page
23. 414 kg·m²·s⁻¹
25. 2.3 J·s along axis
27. 17.4 rpm
29. 2.5 days.
31. $-9.0\hat{k}$ N·m
33. 1600 N·m
35. 37 J·s
37. 2.66×10^5 J·s, out of plane of figure
39. 3.1×10^{-16} J·s
41. 0.21 kg·m²
43. 63%
45. 5.5 m/s
47. 3.1 rpm
51. (a) 0.25 rad/s (b) 6.4 kJ
53. (a) $\omega d(\frac{1}{2} - I/2md^2)$ (b) ωd (c) $\omega d(2 + I/md^2)$
55. 2.8%, orbital angular momentum of Jupiter
57. (a) 140 rpm (b) 21%
59. (a) $2\omega_0/7$ (b) $t = \frac{2R\omega_0}{\mu_k g}$
63. 9.2×10^{26}N·m
65. d
67. d

Chapter 12

15. (a) $\tau = mgL/2$ (b) $\tau = 0$ (c) $\tau = -mgL/2$
17. 16 m relative to the wall
19. (a) 0.61 m from left end (b) 1.42 m from left end
21. 480 N
23. −0.797 m, unstable; 1.46 m, stable
25. (a) 40 N·m (b) 1.3 kN
27. 500 N
29. 79 kg
31. 1.4 W
35. 87 kg
37. $\tan^{-1}(L/w)$
39. (a) $\frac{mg}{2}[L\sin\theta - w(1 - \cos\theta)]$ (b) $\tan^{-1}(L/w)$ (c) concave down, unstable
41. 74 kg
43. $0.366\,mgs$
47. $F_{app} = Mg\tan(\theta/2)$
49. $\mu_s < \tan\alpha = 1/2$
53. $\mu \geq \frac{\tan\theta}{2 + \tan^2\theta}$
55. 840 N
57. 170 N
59. (a) $F = G\frac{M_E m}{R_E^2}(1.229)$, 21.3° (b) $\tau = G\frac{M_E m}{R_E}(-0.0356)$
61. The tie beam will not hold under the 10 kN of tension.
63. stable equilibrium at ~6 nm and ~14 nm, unstable equilibrium at ~11 nm
65. a
67. b

Chapter 13

17. 2.27×10^{-3} s
19. (a) $x(t) = (12.5\text{ cm})\cos[(42.0\text{ s}^{-1})t]$ (b) $x(t) = (2.15\text{ cm})\sin[(4.63\text{ s}^{-1})t]$
21. 22 ms
23. 0.59 Hz; 1.7 s
25. (a) $10.6\,\text{s}^{-1}$ (b) 16.5 cm (c) 38.6 N/m
27. (a) 2.2 rad/s (b) 2.8 s (c) 0.63 m
29. 1.21 s
31. 1.6 s
33. 7 oscillations in x direction for 4 oscillations in the y direction
35. ±1.7 rad, ±15 rad/s
37. 0.25 s
39. 65 km/h
41. 0.70 s
43. (a) $t = \pi\sqrt{m/k}$ (b) $A = v_0\sqrt{m/k}$
45. 71 min
47. (a) 67 μN/m (b) 3.4×10^{-10} kg
51. 821 kg
53. (a) $|\vec{r}| = A$ (b) $\vec{v} = (\omega A\cos\omega t)\hat{\imath} - (\omega A\sin\omega t)\hat{\jmath}$ (c) $|\vec{v}| = \omega A$ (d) ω
55. 0.147%
57. (a) 7.9 N/m (b) 0.80 kg
59. $\omega = \sqrt{(k_1 + k_2)/m}$
63. $\omega = \sqrt{2k/3M}$
65. 34
67. (a) 6.5 cm (b) 0.51 s
69. $f = \frac{1}{2\pi}\sqrt{2a/m}$
71. (a) $E_1 = 4E_2$ (b) $a_{max,1} = 4a_{max,2}$
73. 27°
75. $T = 2\pi\sqrt{7l/(10ga)}$

77. 0.54 Hz; 22 cm; −0.11 rad
79. c) $T = 2\pi\sqrt{\frac{r_0^3}{GM}}$ (d) 16 days
83. 2.1 m/s^2
85. (a) up (b) 0.46 turn
87. c
89. c

Chapter 14

15. (a) 0.19 s (b) 6.5 cm
17. (a) 300 m (b) 1.58 m (c) 3.0 cm (d) 8 μm (e) 500 nm (f) 3.0 Å
19. (a) 0.19 mm (b) 0.43 mm
21. (a) 1.3 cm (b) 9.1 cm (c) 0.20 s (d) 45 cm/s (e) $-x$ direction
23. (a) 12.8 rad/s (b) 0.336 cm^{-1} (c) $(2.34 \text{ cm})\cos[(0.336 \text{ cm}^{-1})x + (12.8 \text{ s}^{-1})t]$
25. 250 m/s
27. 7.6 N
29. 9.9 W
31. 343 m/s
33. 420 m/s
35. 940 Hz
37. 5.4 m
39. (a) 280 Hz (b) 70 Hz (c) 210 Hz
41. 14 cm
43. 93 Hz
45. Galaxy receding
47. 30 m/s
49. $\bar{E} = \frac{4\pi^2 F A^2}{\lambda}$
51. 1.0×10^2 W
55. $v = \sqrt{\frac{kL(L - L_0)}{m}}$
57. 10 m
59. $L_0 = 5L_1/7$
61. 440 mph
65. 6.3 m
71. 7.3 km
73. 41 m/s
75. radar worked properly
77. Not sufficient: The minimum measurable speed is 5.4 km/h.
79. 3.9 kg
81. b
83. c

Chapter 15

15. 1.2 kg
17. (a) 180 kg/m^3 (b) 7.3 m^3
19. 249 Pa
21. 322 kPa
23. 1.7×10^3 kg/m^3
25. 92 m
27. 2.4%
29. 46 kg
31. 0.75%
33. 2.8 m/s
35. (a) 1.8×10^4 m^3/s (b) 1.5 m/s
37. 1.8 cm/s
39. 830 cm^2
41. (a) 620 Pa (b) 1.2 kPa
43. 3.6 mm
45. 8100 kg
47. The accused apparently drank 51 oz.
49. 27 m
51. (a) 49 kg (b) 2500 kg
53. 14 kPa
55. 14 m
57. (a) 1.5 m/s (b) 0.47 L/s
59. 70%
61. (a) 98% less (b) 17 cm
63. (a) 603 Pa (b) 11.0 km
65. 15 kg
67. Yes, the wind farm could produce 1 GW of power.
69. $t = \frac{A_0}{A_1}\sqrt{\frac{2h}{g}}$
71. (b) 5.8 km
73. $\frac{M}{4\pi R^3(1 - 2e^{-1})}$
75. 2.1×10^{12} N·m
77. Yes
79. $\rho_{H_2O} L \tan\frac{\theta}{2}(h_0^2 - h_1^2)$
81. c
83. e

Chapter 16

15. 2.5°F to 5.6°F
17. 20°C
19. −40°C = −40°F
21. 102°F
23. 32 kJ
25. 100 W
27. (a) 170 J/K (b) 480 J/(kg·K)
29. 0.293 W
31. 55 kW
33. 4 W
35. R_{air} = 0.98 m^2·K/W, $R_{concrete}$ = 0.03 m^2·K/W, $R_{fiberglass}$ = 0.60 m^2·K/W, R_{glass} = 0.03 m^2·K/W, $R_{Styrofoam}$ = 0.88 m^2·K/W, R_{pine} = 0.23 m^2·K/W
37. 2.2 kW
39. 2×10^{-5} m^2
41. (a) 138 kPa (b) 33.4 kPa (c) 233 kPa
43. 263 K = −10°C
45. 364 g
47. (a) 23.2 kJ (b) 337 kJ (c) 65.2 kJ
49. 138 s
51. 0.56 kg
53. 1.8 kg
55. 9.2 K
57. 0.20 kg
59. 2.0×10^2 W
61. The house will remain at a comfortable 19°C
63. (a) 1.1×10^3 K (b) 700 K
65. 24°C
67. 1200 K
69. (a) $319/month (b) $37.58/month
71. 44 K
73. 418.76 kJ, 0.09% higher
75. Mars: 207K vs. ~210 K measured; Venus: 301 K vs. ~740 K measured
77. The solar increase accounts for only 4% of recent warming.
79. The hutch temperature will be −2.5°C, so the water will freeze.
81. 10 h
83. c
85. a

Chapter 17

17. 1.8 m^3
19. 1.8×10^6 Pa
21. (a) 27 L (b) 330 K
23. 3.16 km/s
25. 22 kJ
27. 3.9 kg
29. 6.0 MJ
31. 0.987 L
33. 263°C
35. 1×10^{15} m^{-3}, which is over 10 billion times less dense than Earth's atmosphere
37. (a) 235 mol (b) 5.65 m^3
39. (a) 1.27 atm (b) 0.980 mol (c) 0.786 atm
41. 27.6 min
43. 14 min
45. 43.9 min
47. 10°C
49. 46.1°C
51. 177 g
53. 4.9°C
55. 19 kW
57. 56 min
59. 251 K
61. 307 K
63. $d = \frac{L_0}{2}\sqrt{2\alpha\Delta T + \alpha^2\Delta T^2}$
65. (a) 61 h (b) 52 h
67. 3.97°C
69. 34.1 km
73. (a) $y^2 = \frac{1}{4}(L_0^2 - d^2) + \frac{1}{2}L_0^2\alpha\Delta T$ (b) $\alpha = 2.35 \times 10^{-5}$/C°, d = 80.00 cm (c) aluminium
77. (a) 244 K (b) 247 K
79. c
81. c

Chapter 18

15. 29.3 kJ
17. 250 J
19. −14 kW
21. $2p_1V_1$
23. (a) 4/3 (b) 220 J
25. 0.177
27. 2.1 MJ
29. 57.7%
31. (a) 200 K (b) 120 K
33. 380 W
35. (a) 1.49 mm (b) 10.7 μJ
37. 1.35
39. (a) 300 kPa (b) 240 J
41. 440°C

43. (a) 810 K (b) 25.8 atm
45. 354
47. (a) 1.27 (b) internal energy to raise gas temperature
49. (a) 255 K, 1.75 kJ (b) 279 K (c) 272 K, 500 J
51. (a) 40 kPa (b) 83 kPa (c) 80 kJ
53. 930 J
57. The temperature rises 75°C, missing the criteria.
59. 57 kJ
61. 330 K
63. (a) 202 J (b) 500 J transferred out of the gas
65. 20 mol
67. 140 atm
73. 2.0 mJ
75. $4p_1V_1/3$
77. Yes
79. 18%
81. a
83. c

Chapter 19

13. (a) 26.8% (b) 7.05% (c) 77.0%
15. 0.948 K
17. 9.10
19. No
21. 8.8 kJ/K
23. 21.9 kg
25. (a) 1/64 (b) 5/16
27. 52.1% (winter), 47.7% (summer)
29. (a) 1.75 GW (b) 43.0% (c) 232°C
31. 2×10^{11} kg/s
33. (a) 8.53 (b) 1.10×10^3 kg
35. $68
37. 2.83
39. (a) 5.7 (b) 3.5 kW (c) pump: 54¢/h; oil furnace; $2.40/h
41. (a) 17.4% (b) 83.3%
43. 140 MJ/K
47. (a) 86.0 J/K (b) 120 J/K (c) 0
49. 0
51. 160 J/K
53. 12.1 kJ/K
55. $1-r^{1-\gamma}$
59. 61%
61. $C_0(1 - T_0/T_1)$
63. $W = CT_h(\ln x - 1 + 1/x)$
65. $\Delta S_{tot} = mc\ln\left(1 + \dfrac{(T_h - T_c)^2}{4T_hT_c}\right)$
67. 36.2 J/K
69. 62%
71. c
73. c

Chapter 20

13. 3 C, or about 0.05 C/kg
15. (a) *uud* (b) *udd*
17. 1.1×10^9
19. 5.1 m
21. (a) $\hat{\jmath}$ (b) $-\hat{\imath}$ (c) $0.316\hat{\imath} + 0.949\hat{\jmath}$
23. 3.8×10^9 N/C
25. (a) 2.2×10^6 N/C (b) 77 N
27. $-1.6\hat{\imath}$ pN
29. (a) 26 MN/C, to the left; (b) 5.2 MN/C, to the right; (c) 58 MN/C, to the right
31. 1.1 kN/C
33. $E = kQ/(\sqrt{8}a^2)$
35. 5.1×10^4 N/C
37. 980 N/C
39. (a) 22.3 μC (b) no
41. $16\hat{\imath} - 9.1\hat{\jmath}$ N
43. $4q$
45. (a) 20 μC (b) 1.6 N
47. (a) $2.3\hat{\imath}$ MN/C (b) $0.83\hat{\imath} + 0.83\hat{\jmath}$ MN/C (c) $\vec{E} = 0.30\hat{\imath} - 0.89\hat{\jmath}$ MN/C
49. $-4e$
51. (a) $8.0\hat{\jmath}$ GN/C (b) $190\hat{\jmath}$ MN/C (c) $220\hat{\jmath}$ kN/C
53. 0
55. $q_1 = \pm 40\ \mu C, q_2 = \mp 6.9\ \mu C$
59. −14 μC/m
61. The device doesn't work because its two halves depend on charge-to-mass ratio in the same way.
63. 1.3×10^{-30} C·m
65. (a) $2kQqa/x^2$ (b) $2kQqa/x^3$ (c) upward
67. 0.4*e*, 0.03*e*
69. (a) $\vec{E}(x) = 2kqa^2\dfrac{(3x^2 - a^2)}{x^2(x^2 - a)^2}\hat{\imath}$ (b) $\vec{E}(x) \approx \dfrac{6kqa^2}{x^4}\hat{\imath}$
71. (a) 2.5 μC/m (b) 300 kN/C (c) 1.8 N/C
73. (b) $dq = 2\pi\sigma r dr$ (c) $dE_x = \dfrac{2\pi k\sigma xr}{(x^2 + r^2)^{3/2}}dr$
77. $y = a/\sqrt{2}$
79. $E = -\dfrac{k\lambda_0\hat{\imath}}{L}\left[\dfrac{1}{2} + 2\ln(2)\right]$
81. mdv^2/qL^2
83. a
85. a

Chapter 21

17. 3 μC
19. $Q_C = 2Q = -Q_B$
21. 650 kN/C
23. ±1.5 kN·m²/C
25. 69 N·m²/C
27. (a) $-q/\epsilon_0$ (b) $-2q/\epsilon_0$ (c) 0 (d) 0
29. 49 kN·m²/C
31. (a) 1.2 MN/C (b) 2.0 MN/C (c) 50×10^4 N/C
33. Line symmetry
35. 49×10^3 N/C
37. (a) 5.1×10^6 N/C (b) 34 N/C
39. (a) 2.0×10^6 N/C (b) 7.2×10^3 N/C
41. (a) 0 (b) 4.0×10^{-3} C/m²
43. 1.8 MN/C
45. $\pm E_0a^2/2$
47. 7.0 MN/C; 17 MN/C
49. (a) 2.8 cm (b) 3.5 nC
51. ±154 nC
53. (a) $3.6\hat{r}$ MN/C (b) $3.8\hat{r}$ MN/C (c) $7.8\hat{r}$ MN/C
55. (a) $20\hat{r}$ kNC (b) $1.7\hat{r}$ kN/C
57. 6.3 μC/m³
59. (a) $\rho x/\epsilon_0$ (b) $\rho d/2\epsilon_0$; away from the center plane of slab if $\rho > 0$, toward center plane if $\rho < 0$
61. 18 N/C
63. (b) $-Q$
67. (a) $Q = \pi\rho_0a^3$ (b) $E(r) = \rho_0r^2/(4\epsilon_0a)$
69. $a = 5\rho_0/(3R^2)$
71. $R\dfrac{\rho_0}{\epsilon_0}(e - 2)$
73. $\dfrac{\rho_0r^2}{3\epsilon_0R}$
75. $E_{in} = \dfrac{\rho_0x^2}{2\epsilon_0d}, E_{out} = \dfrac{\rho_0d}{8\epsilon_0}$
77. c
79. d

Chapter 22

15. 600 μJ
17. 3.0 kV
19. 910 V
21. Proton, ionized He atom: 1.6×10^{-17} J, proton: 3.2×10^{-17} J
23. $-E_0y$
25. 53 nC
27. (a) 440 kV, 9.2×10^6 m/s
31. (a) 4 V (b) $E_x = 1$ V/m, $E_y = -12$ V/m, $E_z = 3$ V/m
33. 3 kV
35. 5.6 kV/m
37. 4.5 V
41. 6.1 μC
43. $\sqrt{2keQ/(mR)}$
45. kQ/R
47. $-ax^2/2$
49. −52 nC/m
51. $-a/2, a/4$
53. (a) 2.6 kV (b) 1.8 kV (c) 0
55. $V = 2kQ/R$
57. $2\pi k\sigma(\sqrt{x^2 + b^2} - \sqrt{x^2 + a^2})$
61. $(V/R)\hat{r}$
63. (a) 43 kV (b) 1.7 MN/C (c) 540 V (d) 0
65. $-E_0R/3$
67. (a) 7.2 kV (b) 14 kV
69. 14 cm, 1.7 nC
71. 0.12 J
73. $\omega = 232$ nC/m², $q = 3.75$ nC, $r = 7.18$ cm
77. (a) $\pi k\sigma_0a[\sqrt{1 + (x/a)^2} - (x/a)^2\ln(a/x + \sqrt{1 + (a/x)^2})]$
79. $-\dfrac{k\lambda_0}{L^2}\left[Lx + x^2\ \ln\left(\dfrac{2x - L}{2x + L}\right)\right]$
81. 8.0 mm
83. d
85. b

Chapter 23

13. 4.4 kJ
15. 0
17. −48.5 eV
19. (a) 1.4 J (b) 4.2 J
21. 22 nF
23. 740 pF
25. 1.5 J
27. 3.0 μF, 0.67 μF
29. (a) 1.20 μF (b) $Q_1 = 14.4$ μC, $Q_2 = 4.80$ μC, $Q_3 = 9.60$ μC (c) $V_1 = 7.2$ V, $V_2 = V_3 = 4.8$ V
31. 8.2×10^5 V/m
33. No
35. $Q_y = 4Q_0/(\sqrt{2} + 1) \approx 1.66Q_0$
37. 2.8 μC
41. (a) 4.4 kV (b) 120 kW
43. 129 F
45. 0.86 μF
49. (a) 4.1 nF (b) 1.3 kV
51. 2.7 nm
53. 24 μJ
55. $U = kQ^2/(2R)$
57. 6.0×10^{-4} J
59. 13 min
61. $C = \dfrac{4\pi\epsilon_0 ab}{b - a}$
63. $\dfrac{1}{6}$
65. (b) $\dfrac{C_0V_0^2}{2}\left(\dfrac{\kappa x + L - x}{L}\right)$ (c) $\dfrac{C_0V_0^2(\kappa - 1)}{2L}$
67. $\dfrac{\pi\rho^2R^4}{8\epsilon_0}$
69. (b) 4.3 μF
71. a
73. c

Chapter 24

13. 9.4×10^{18}
15. 1.9×10^{11}
17. 3.2×10^6 A/m^2
19. 6.8 cm
21. (a) $5.95\times10^7(\Omega\cdot\text{m})^{-1}$ (b) $4.55\ (\Omega\cdot\text{m})^{-1}$
23. 360 V
25. 32 mΩ
27. 4R
29. (a) 6.0 V (b) 8.0 Ω
31. 230 V
33. 300 Ω
35. (a) 0.12 mA (b) no
37. (a) 420 A/mm^2 (b) 0.24 A/mm^2
39. greater in Cu by factors of (a) 7.6; (b) 4
41. 9.7 μC
43. (a) 5.8 MA/m^2 (b) 97 mV/m
45. Ge
47. 50 ft
49. $R_1 = 388$ μΩ, $R_2 = 0.971$ μΩ, and $R_3 = 0.243$ μΩ
51. (a) 81 miles (b) 7.3 h at 3.3 kW, 3.6 h at 6.6 kW, 33 min at 44 kW (c) 203A
53. ~ 2 TW
55. 2.8 min
57. $d_1 = \sqrt{2}d_2$.
59. 0.63 A
61. Aluminum at $3.30/m is more economical than copper at $14/m.
63. 2.5 A
65. 250°C
69. $2\pi J_0a^2/3$
71. 19°
73. a
75. c

Chapter 25

17. 1.4 h
19. 43 kΩ
21. 10 V
23. 50 Ω
25. $I_1 = 2$ A, $I_2 = 0.2$ A, $I_3 = 2$ A
27. 0 A
29. −0.66%
35. $\mathcal{E}R_2/(R_1 + R_2)$
37. 1.5 mA
39. 30 A
41. 14 W
43. 120 mA, so yes, possibility fatal
45. (a) $\mathcal{E}R_1/(R + 2R_1)$
47. 2.4 W
49. 7R/5
51. (a) 48 V (b) 57 V (c) 60 V
53. (a) 0.992 A (b) 0.83%
55. 360 μF; 1200 V
57. 3.4 μJ
59. a. $V_C = 0, I_1 = 25$ mA, $I_2 = 0$
    b. $V_C = 60$ V, $I_1 = I_2 = 10$ mA
    c. $V_C = 60$ V, $I_1 = 0, I_2 = 10$ mA
    d. $V_C = 0, I_1 = I_2 = 0$
61. (a) 5.015 V (b) 66.53 Ω
63. 1.07 A, left to right
65. 2.15 μF
67. 80 μs
69. 8 Ω; 89 W
71. (a) R_1 (b) R_1 (c) R_1
75. (a) 9 V (b) 1.5 ms (c) 0.3 μF
77. 220 mV
79. $\tau = \dfrac{R_1R_2C}{R_1 + R_2}$
81. Yes
83. b
85. c

Chapter 26

15. (a) 16 G (b) 23 G
17. (a) 2.0×10^{-14} N (b) 1.0×10^{-14} N (c) 0
19. 400 km/s
21. 360 ns
23. (a) 87.6 mT (b) 1.25 keV
25. 0.373 N
27. 12,500 lb, so clamping down the bar is a good idea.
29. (a) 9.85 cm (b) 14.8 μT
31. 1.2 mT
33. 5 mN/m
35. 4.05×10^{-2} A·m^2 (b) 7.78×10^{-2} N·m
37. 7.0 A
39. (a) 0.569 mT (b) 3.90 mT (c) 2.85 mT
41. 17 T
43. 2.3×10^{27} A·m^2
45. 3.8 GA
49. (a) 71 μm (b) 440 μm
51. 0.53 A
53. 8.5×10^{22} cm^{-3}
55. (a) 4.6 A·m^2 (b) 0.43 N·m
57. 0.021 N, 45° above horizontal
59. $(1 + \pi)\dfrac{\mu_0I}{2\pi a}$, out of page
61. $\dfrac{\mu_0I}{4a}$, into page
63. 16 μN, toward long wire
65. (a) 0 (b) $B = \mu_0I/(2\pi r)$
67. (a) 2300 (b) 3.3 kW
71. (a) 8.0 μT (b) 4.0 μT (c) 0
73. $\dfrac{\mu_0J_sx}{d}$
75. (a) $B \approx \dfrac{\mu_0I}{2w}$ (b) $B \approx \dfrac{\mu_0I}{2\pi r}$
77. (a) $\pi R^2J_0/3$ (b) $B = \dfrac{\mu_0J_0R^2}{6r}$ (c) $B = \dfrac{\mu_0J_0r}{2}\left(1 - \dfrac{2r}{3R}\right)$
79. Since $\tau \propto 1/N$, more torque from a 1-turn loop.
81. $\dfrac{\mu_0I^2}{2\pi w}\ln\left(\dfrac{a + w}{a}\right)$
83. $\mu_0nIl/\sqrt{l^2 + 4a^2}$
85. No; the force between each meter of the two conductors is 150 N.
87. The hall potential is 10,000 times smaller than bioelectric potentials.
89. d
91. d

Chapter 27

15. 1.2×10^{-4} Wb
17. 160 T/s
19. 6.5 mH
21. 42 kV
23. 330 mH
25. 3.1 kJ
27. 66 mJ
31. 4.4 T
33. $-rbl/2$
35. (a) −0.30 A (b) −0.20 A
37. 15 mT
39. (a) 3 s (b) clockwise
41. (a) 2.0 mA (b) 4.4 mA
43. −42 mA, clockwise
45. 130
47. (a) Upper bar (b) 0

49. (a) 25 mA (b) 1.3 mN (c) 2.5 mW (d) 2.5 mW
51. 58 T/ms
55. 0.76 s
57. 20 s
59. (a) 5 Ω (b) 500 J
61. (a) 1.0 A (b) 0.43 A (c) −1.7 A
63. 190 mΩ
65. 3.4×10^{21} J/m^3
67. $\dfrac{\mu_0 I^2}{16\pi}$
69. 3×10^8 m/s (speed of light)
71. (a) $-br/(2\rho)$ (b) $\dfrac{\pi b^2 h a^4}{8\rho}$
73. 3.69 H
75. $v(t) = \dfrac{FR}{B^2 l^2}\left[1 - \exp\left(-\dfrac{B^2 l^2}{Rm}t\right)\right]$
77. (a) $\dfrac{\mu_0 I^2}{4\pi}\ln(b/a)$
81. c
83. a

Chapter 28

15. (a) 294 V (b) 2.51×10^3 s^{-1}
17. (a) $V(0) \approx V_p/\sqrt{2}$, 45°
    (b) $V(0) = 0, \phi_b = 0$
    (c) $V(0) = V_p, \phi_c = 90°$
    (d) $V(0) = 0, \phi_d = \pm\pi$
    (e) $V(0) = -V_p, \phi_e = -90°$
19. $I_{R,\text{rms}} = 13$ mA, $I_{C,\text{rms}} = 24$ mA, $I_{L,\text{rms}} = 22$ mA
21. (a) 250 V (b) 15 V
23. 16 kHz
25. 78.1 H
27. (a) 32 mH (b) 1.0 V
29. 3.5 kΩ
31. 5.0 mA
33. 390 mA
35. 1
37. (a) 150 mA (b) 330 mA
41. 4.3 kHz
43. (a) 52 nF (b) 350 Hz
45. 0.199 μH
47. (a) $1/\sqrt{2}$ (b) 1/2 (c) $-1/\sqrt{2}$ (d) 1/2
49. 50
51. 6.2 Ω
53. (a) Above resonance; (b) ∼50°
55. (a) 0.369 (b) 6.43 W
57. (a) 5.5% (b) 9.1%
59. 3.7 mF
61. 2.7 V
63. 1620 Hz
67. $R = 400\,\Omega, L = 68$ mH, $C = 94$ nF
71. 910 Hz, 36 V
73. a
75. d

Chapter 29

13. 1.3 nA
15. $-\hat{k}$
19. 11.2 km
21. 2.57 s
23. 5.00×10^6 m
25. x-direction
27. 12%
29. 1×10^{10} W/m^2
31. The radio has a minimum intensity of 0.27 nW/m^2, so it will work at the cabin.
33. 20 kW
35. 3.1 cm
37. 0.94 PHz -1.0 PHz
39. 1.07 pT
41. 91%
43. 19%
45. 0.00004%
47. Quasar power is greater by factor of 4×10^{10}
49. (a) 4.6 kW (b) 53 mV/m
51. (a) $1/r$ (b) $1/r^2$
53. (a) 8.9×10^6 W/m^2 (b) 58×10^3 V/m
55. 6.2×10^3 y
57. 2.52 kPa
59. (a) 1.12 MV/m (b) 4.14 mm (c) 91.0 mJ (d) 3.03×10^{-10} kg·m/s (e) 86.0 W
61. 6
65. 2.75 m
67. 2.2 km
69. (a) 51 MV/m (b) 0.17 T (c) 96 TW
71. b
73. d

Chapter 30

11. 15°
13. 0.5°
15. Ice
17. 77.7°
19. 14.2°
21. 1.9
23. 79.1°
25. 1.66
27. 6.41°
29. (a) 18° (b) 390 nm
31. Ethyl alcohol
33. 1.83
35. 5.1 m
37. 139 nm
39. 35°
41. Diagonal face, 23°
43. 1.07
47. 63.8°
49. red: 72.3°, violet: total internal reflection
51. 2.7 m
53. 1.9 m
57. c) 50.9°
61. $\dfrac{d}{c}\left(\dfrac{2}{3}n_1 + \dfrac{1}{3}n_2\right)$
63. c
65. b

Chapter 31

15. 35°
17. (a) −1/4 (b) real, inverted
19. (a) $3f$ (b) $3f/2$ (c) real
21. −2
23. 21 cm
25. 27 cm
27. 40 cm
29. 0.86 mm
31. 2.2 diopters
33. −1.3 diopters
35. −200
37. (a) −24 cm (b) 29 mm (c) virtual, upright, enlarged
39. 18 cm
41. 16 cm
43. 7.59 cm
45. 12 cm
47. (a) −7.7 cm, inverted, real (b) +7.7 cm, upright, virtual
49. 29 cm or 41 cm
51. 11 cm
53. $s' = 1.1$ m, inverted real image
55. −67.9 cm
57. 2.0
59. 2
61. Choose plastic, because it meets requirements and is cheaper.
63. (a) Real, inverted image (b) −2.82
65. 3.3 diopters
67. 0.3°
69. 72 cm
79. (a) $dn = -\dfrac{2c}{\lambda^3}d\lambda$ (b) 0.858 mm
81. c
83. d

Chapter 32

11. 1.7 cm
13. 420 nm
15. 4
17. (a) 4.8°, 9.7° (b) 2.9°, 6.8°
19. (a) 2 (b) 1
21. 103 nm
23. 594 nm, 424 nm
25. The top 1.5-cm of the film
27. 29.3°
29. 1.62%
31. 37 cm
33. 3×10^{-4} rad
35. 96°
37. (a) 38 (b) 3
39. 44 μm
43. 2
45. Not feasible because a 2-km-diameter telescope is needed
47. 3.3 Å
49. 5
51. 236
53. 128.8 μm
55. $1 + 2.93 \times 10^{-4}$
57. 34 m
59. 2.0 μm
61. 6.9 km
63. Rep is correct, but microscope won't resolve rhinovirus.
65. $n_{\text{gas}} = 1 + \dfrac{m\lambda}{2L}$

67. $\Delta y = D\lambda/2d$
69. 54
71. c
73. a

Chapter 33

13. (a) 4.50 h (b) 4.56 h (c) 4.62 h
15. 33 ly
17. 40 m
19. $0.14c$
21. (a) 2.0 (b) 2.5
23. $0.14c$
25. (a) 2.1 MeV (b) 1.6 MeV
29. (a) $0.86c$ (b) 9.7 min
31. $c/\sqrt{2}$
33. Twin A = 83.2 years old, twin B = 39.7 years old
35. $0.96c$
39. Civilization B, 3.8×10^5 y
41. yes, from A toward B at $0.45c$
43. earlier by 5.2 min
47. $0.94c$
49. (a) 10 ly, 13 y (b) 0 ly, 7.5 y
53. (a) 4.2 ly (b) −2.4 ly
55. (a) $0.758c$ (b) 1.09 GeV/c
57. 25 h
59. (a) 0.26 eV (b) 1.3 keV (c) 3.1 MeV
65. $0.866c$
67. $0.95c$
71. $\gamma ma\left[1+\left(\frac{\gamma u}{c}\right)^2\right]$
73. 0.31 c; 27 kV
75. (a) 2.976×10^8 m/s (b) 9.46×10^{-31} kg, 4% higher than known electron mass (Answers are very sensitive to the precise values used for constants and conversion factors.)
77. a
79. a

Chapter 34

15. 16
17. $\lambda_{peak} = 10.1\ \mu m$, $\lambda_{median} = 14.3\ \mu m$
19. (a) 500.0 nm (b) 708.6 nm
21. 2.8×10^{-19} J to 5.0×10^{-19} J
23. 1.44
25. 122 nm, 103 nm, 97.2 nm
27. 91.2 nm
29. (a) 3.7×10^{-63} m (b) 73 nm
31. The electron moves 1836 times faster than the proton.
33. 6×10^7 m/s
35. 130 nm
37. 23 keV
39. UV is smaller by factor 5.4×10^{-2}
41. (a) 5.19×10^3 K (b) 0.748
43. (a) $1.7\times10^{28}\ s^{-1}$ (b) $3.2\times10^{15}\ s^{-1}$ (c) $1.3\times10^{18}\ s^{-1}$
45. (a) 1.12×10^{15} Hz (b) 2.79 eV
47. (a) 2.9 eV and 1.9 eV (b) Plants absorb blue and red, reflect green.
49. 440 nm
51. (a) 154 pm (b) 222 eV
53. No
55. (a) 313 m/s (b) 96 km/s
57. 0.22 meV
59. (a) 26.4 cm (b) 4.70 μeV
61. 229
63. 3.40 eV
65. (a) 0.0265 nm (b) 40.8 eV
67. 1.62 km/s
69. 2.5 km/s
71. 1 ps
75. $E_0 = \frac{1}{2}m_ec^2[(\gamma-1)+\sqrt{(\gamma-1)(\gamma+3)}]$
83. (a) 6.65×10^{-34} J·s (b) 2.3 eV (c) potassium
85. b
87. c

Chapter 35

13. (a) 0 (b) $\pm a\sqrt{\ln 2/2}$
15. 5
17. 3.8 meV
19. (a) 1.6 eV (b) 6.5 eV
21. Electron
23. 0.2 MeV
25. 33 eV
27. 8.0 eV
29. $E \rightarrow E/4$
31. 930 pm
35. (a) 2.2 eV (b) 570 nm
37. 21 μm
39. (a) 6 (b) $\lambda_{4\rightarrow1} = 153$ nm, $\lambda_{4\rightarrow2} = 191$ nm, $\lambda_{4\rightarrow3} = 328$ nm, $\lambda_{3\rightarrow1} = 287$ nm, $\lambda_{3\rightarrow2} = 459$ nm, $\lambda_{2\rightarrow1} = 765$ nm (c) UV, visible, and IR
41. (a) $\psi_{n-\text{odd}}(x) = \sqrt{\frac{2}{L}}\cos\left(\frac{n\pi x}{L}\right)$, $\psi_{n-\text{even}}(x) = \sqrt{\frac{2}{L}}\sin\left(\frac{n\pi x}{L}\right)$ (b) $E_n = n^2h^2/(8mL^2)$
43. 0.759 nm
45. 2.5×10^{-17} eV; quantization is insignificant
47. (a) 0.30 (b) 0.15
53. 4
55. (c) $A_0 = (\alpha^2/\pi)^{1/4}$
57. 2.23 nm
61. b
63. a

Chapter 36

15. 3
17. d
19. 5
21. 3d
23. 2.58×10^{-34} J·s
25. 3/2, 5/2
27. $11.5\hbar\omega$
29. $1s^22s^22p^63s^23p^64s^23d^1$
33. 0.69 meV
35. $n = 4, l = 3$
37. 2.67×10^{68}
39. 90°, 65.9°, 114°, 35.3°, 145°
41. 0, ±1, ±2, ±3
45. (a) $E_1/16$ (b) $\sqrt{12}\hbar$ (c) $\frac{1}{2}\sqrt{35}\hbar$
47. (a) $16\hbar\omega$ (b) $4\hbar\omega$
49. $1s^22s^22p^63s^23p^64s^13d^{10}$
51. 3.0×10^{17}
53. 0.1
55. 2.5 meV
57. even N: $\hbar\omega(N-1)/2$; odd N: $\hbar\omega N/2$
59. (a) 5 (b) $9E_1$
61. $3\hbar$, $6h$
63. (a) 0.966 (b) 0.0595
65. $P(r)dr = 4\pi r^2\psi_{2s}^2 dr$, $3+\sqrt{5}$
67. (b) 54.4 eV, 870 eV, 91.4 keV, 115 keV
69. (b) 141 eV, 65.8 eV, 47.0 eV, 28.2 eV
71. $3a_0/2$
77. a
79. c

Chapter 37

17. 3.48 mm
19. 9.41×10^{-46} kg·m²
21. 7.08×10^{13} Hz
23. 181 kcal/mol
25. 549 nm
27. 3.54 μm
29. 1.34 meV
31. $l\hbar^2/I$
33. 0.121 nm
35. (a) 0.179 eV (b) 0.358 eV
37. 14.95 μm
39. (a) 15.09 meV (b) 82.22 μm (c) far infrared
41. 35.8 μm
43. 10.2
45. −8.40 eV
49. 4.68 eV
51. 6.36×10^4 K, ~200 times room temperature
53. 709 nm, no
55. 1.8 kA
57. 508 nm
59. $I = m_1m_2R^2/(m_1+m_2)$; 0.128 nm
63. (a) $(2^{9/2}\pi m^{3/2}L^3/3h^3)\,E^{3/2}$
65. 64 kA
67. 2.75×10^{-47} kg·m²
69. a
71. a

Chapter 38

17. $^{211}_{86}$Ra, $^{220}_{86}$Ra, and $^{222}_{86}$Ra.
19. (a) $A = 35$ for both (b) $Z_K = Z_{Cl} + 2$
21. 5.9 fm
23. $^{64}_{29}\text{Cu} \rightarrow {}^{64}_{30}\text{Zn} + e^- + \bar{\nu}$
$^{64}_{29}\text{Cu} \rightarrow {}^{64}_{28}\text{Ni} + e^+ + \nu$
$^{64}_{29}\text{Cu} + e^- \rightarrow {}^{64}_{28}\text{Ni} + \nu$
25. 17 Bq/L

27. (a) 190 y (b) 290 y
29. 4×10^{-30} kg
31. 59.930 u
33. 5.612 MeV
35. 2
37. $1.0 \times 10^{20}\ s^{-1}$
39. $2 \times 10^{20}\ m^{-3}$
41. 10^3 s
43. 5.3×10^{-12} eV
45. 8.80 MeV
47. 0 atoms; 5×10^5 atoms; 8×10^4 atoms; U-238 and K-40 are suitable
49. 9.6 d
51. (a) $^{228}_{90}Th$
53. 8.9×10^3 y
55. Poland: 8.04 d; Austria: 16.2 d, Germany: 10.0 d
57. 3.0×10^9 y
59. 3.31%
61. 3×10^{-13}
63. 1.2×10^3 kg
65. 88.9%
67. 580 kg
69. 0.461 s
71. (a) $4 \times 10^{38}\ s^{-1}$ (b) 7×10^9 y
73. 8×10^{17} s, which is about 20 billion years longer than the Sun will shine
75. Bohrium-262 ($^{262}_{107}Bh$)
77. (a) $^{65}_{29}Cu$ (b) 4 h
79. (a) 210 MJ (b) $14\ s^{-1}$ (c) 450 kg
81. Yes
85. (b) 1.4 μs
87. b
89. d

Chapter 39

21. 0.336 fs
23. $\pi^+ \rightarrow \mu^+ + \nu_\mu$
25. $\eta \rightarrow \pi^+ + \pi^- + \pi^0$
27. No, violates conservation of baryon number and angular momentum
29. *sss*
31. 4.54×10^7 L
33. 10^{28} K
35. 1.2 Gly
37. 1.32×10^{10} yr
39. Reaction (a) is not possible because it violates conservation of baryon number and angular momentum.
41. (a) No (b) yes
43. $c\bar{c}$
45. (a) 0.16 μJ (b) 0.02 mm
47. (a) essentially no change (b) essentially c (c) 90 μs
49. 313 ly
51. (a) 256 fm (b) −2.81 keV
53. (a) 5.740×10^3 km/s (b) 276 Mly
55. 2.6×10^{-25} s
57. older value gives 22% larger age, 17.6 Gy vs. 14.4 Gy
59. 5.0 km/s/Mly
61. b
63. c

Credits

Front Matter

Page i: (volume 1): 68/Ocean/Corbis. Page i: (volume 2): Argus/Shutterstock. Page vi: Robert Keren.

Chapter 1

Page 1: David Parker/Science Source. Page 3: NASA. Page 5: The Photos/Fotolia. Page 6: NASA. Page 6: Andrew Syred/ Science Source. Page 9: cesc_assawin/Shutterstock. Page 11: NASA Images. Page 11: cesc_assawin/Shutterstock. Page 14: F1online digitale Bildagentur GmbH/Alamy. Page 14: Paul Fleet/Alamy.

Chapter 2

Page 15: Bob Thomas/Getty Images. Page 21: DARPA. Page 24: Richard Megna/Fundamental Photographs. Page 26: National Institute of Standards and Technology.

Chapter 3

Page 32: Martha Holmes/Nature Picture Library. Page 38: Richard Megna/Fundamental Photographs. Page 39: Paul Whitfield/Rough Guides/Dorling Kindersley ltd. Page 41: Jamie Squire/Getty Images.

Chapter 4

Page 51: NASA/JPL-Caltech. Page 56: Dudarev Mikhail/ Shutterstock. Page 57: Marvin Dembinsky Photo Associates/ Alamy. Page 61: NASA. Page 64: David Scharf/Science Source. Page 69: Source: From WP 3e Fig05-40.

Chapter 5

Page 71: Kevin Maskell/Alamy. Page 76: NIC BOTHMA/ EPA /Landov Media. Page 86: Source: From EUP 2e ISM CH5 Exercise 17 Solution. Page 89: AP Photo/Lionel Cironneau.

Chapter 6

Page 90: PAOLO COCCO/AFP/Getty Images. Page 92: Source: K. Trenberth et al., Bulletin of American Meteorological Society. March 2009. Page 314. Page 107: Jeffrey Liao/Shutterstock.

Chapter 7

Page 109: Kennan Ward/Encyclopedia/Corbis. Page 113: United States Geological Survey. Page 118: Pearson Education.

Chapter 8

Page 129: NASA/JPL-Caltech. Page 130: Lowell Observatory. Page 134: David Hardy/Science Source. Page 137: Yekaterina Pustynnikova/Chelyabinsk.ru/AP Images.

Chapter 9

Page 144: Peter Muller/Cultura RM/Alamy. Page 148: Wally McNamee/Corbis Sports/Corbis. Page 153: NASA/Bill Ingalls. Page 166: Fundamental Photographs.

Chapter 10

Page 168: IM_photo/Shutterstock. Page 179: Mark Flynn/ University of Texas-Austin.

Chapter 11

Page 189: Goddard Space Flight Center/NASA. Page 196: Courtesy of Moog, Inc. Page 196: Courtesy of Chipworks. Page 198: Stefano Rellandini/Reuters.

Chapter 12

Page 204: Michael Jenner/Robert Harding Picture Library Ltd/ Alamy.

Chapter 13

Page 221: Tischenko Irina/Shutterstock. Page 222: Tommy/ Fotolia. Page 222: Sheila Terry/Science Source. Page 227: Peter Tsai Photography/Alamy. Page 227: Shutterstock. Page 236: AP Images. Page 236: Marc Wuchner/Corbis. Page 242: NASA Johnson Space Center (NASA-JSC).

Chapter 14

Page 243: AGE Fotostock. Page 251: Gene Chutka/E+/ Getty Images. Page 253: Science Source. Page 255: Richard Megna/Fundamental Photographs. Page 257: Michael Freeman Photography. Page 260: SVSimagery/Shutterstock. Page 260: Drazen Vukelic/E+/Getty Images. Page 264: David Rydevik.

Chapter 15

Page 265: Ralph A. Clevenger/Crush/Corbis. Page 269: Bojan Pavlukovic/Fotolia. Page 271: culture-images/MPI/Frank Herzog/Alamy. Page 273: Pascal Rossignol/Reuters. Page 275: Pearson Science/Pearson Education. Page 277: Maciej Noskowski/Getty Images. Page 277: Alekss/Fotolia. Page 281: Richard Wolfson. Page 283: Science Source.

Chapter 16

Page 284: Mark Antman/The Image Works. Page 284: Arogant/Shutterstock. Page 285: Ted Kinsman/Science Source. Page 292: Branko Miokovic/E+/Getty Images. Page 293: Scott Camazine/Alamy. Page 297: Source: This is Figure 15.12 from the W.W. Norton book Energy, Environment, and Climate 2e by Rich Wolfson. It was adapted from Meehl et al, "Climate System Response to External Forcings and Climate Change Projections in CCSM4", J. Climate, 25, 3661-3683, doi: 10.1175/JCLI-D-11-00240.1. Page 302: Branko Miokovic/E+/Getty Images.

Chapter 17

Page 303: MIMOTITO/Getty Images. Page 310: AP Photo. Page 311: Brian J. Skerry/National Geographic/Getty Images. Page 315: samoshkin/Shutterstock.

Chapter 18

Page 317: Don Farrall/Photodisc/Getty Images. Page 322: Steven Coling/Shutterstock.com. Page 324: Walter Bibikow/Agency Jon Arnold Images/AGE Fotostock.

Chapter 19

Page 334: James Hardy/PhotoAlto/AGE Fotostock. Page 337: Encyclopaedia Britannica/Universal Images Group Limited/Alamy. Page 341: Peter Bowater/Photo Researchers, Inc.

Chapter 20

Page 354: NASA. Page 355: Anna Omelchenko/Shutterstock. Page 356: Fundamental Photographs, NYC. Page 360: Steven Puetzer/Getty Images. Page 369: Helen Sessions/Alamy. Page 369: Courtesy of Bohdan Senyuk. Page 369: CB2/ZOB/WENN.com/Newscom.

Chapter 21

Page 375: Peter Menzel/Science Source. Page 393: Fox Photos/Stringer/Hulton Archive/. Page 398: Eric Schrader/Pearson Education.

Chapter 22

Page 399: Mark Graham/AP Images. Page 412: Kevin Cruff/Getty Images.

Chapter 23

Page 418: Rob Kim/Stringer/Getty Images Entertainment/Getty Images. Page 422: Pearson Education. Page 425: Cindy Charles/PhotoEdit. Page 431: Lawrence Livermore National Laboratory.

Chapter 24

Page 432: Daniel Sambraus/Science Source. Page 443: AP Images. Page 444: Pearson Education.

Chapter 25

Page 449: villorejo/Shutterstock. Page 452: Pearson Education.

Chapter 26

Page 469: NASA. Page 470: Cordelia Molloy/Science Source. Page 473: Photo by Ivan Massar. Page 485: Jerry Lodriguss/Science Source. Page 490: 1990 Richard Megna/Fundamental Photographs.

Chapter 27

Page 497: NASA Images. Page 499: Source: Albert Einstein. Page 516: NASA Images. Page 519: Copyright: © 1988 Richard Megna-Fundamental Photographs. Page 521: Source: A. Einstein, ON THE ELECTRODYNAMICS OF MOVING BODIES, June 30, 1905.

Chapter 28

Page 525: Li Ding/Alamy. Page 533: Pearson Education. Page 537: B.S.P.I./Documentary/Corbis. Page 537: Larry Lawhead/Getty Images. Page 537: ROBERT ROBINSON/Getty Images. Page 537: jlsohio/Getty Images. Page 537: EricVega/E+/Getty Images. Page 537: Arogant/Shutterstock. Page 537: TebNad/Shutterstock.

Chapter 29

Page 543: Arthur S. Aubry/Photodisc/Getty Images. Page 543: Brent Bossom/Getty Images. Page 543: Neustockimages/Getty Images. Page 553: Fundamental Photographs, NYC. Page 555: Babak Tafreshi/Science Source. Page 555: NASA E/PO, Sonoma State University, Aurore Simonnet. Page 563: NASA.

Chapter 30

Page 565: Craig Tuttle/Documentary/Corbis. Page 566: Charles Hood/Alamy. Page 567: D. Scott/NASA. Page 568: Paul Carstairs/Alamy. Page 571: Universal Images Group Limited/Alamy. Page 571: Richard Megna /Fundamental Photographs. Page 572: Spencer Grant/Science Source. Page 574: Richard Megna/Fundamental Photographs. Page 574: Library of Congress Prints and Photographs Division Washington, D.C. 20540 USA http://hdl.loc.gov/loc.pnp/pp.print. Page 575: Universal Images Group Limited/Alamy.

Chapter 31

Page 579: Hunckstock inc./Alamy. Page 580: Fundamental Photographs, NYC. Page 581: Space Telescope Science Institute/NASA. Page 582: Pearson Education. Page 582: Scott Leigh/Getty Images. Page 584: Science Source. Page 587: Pearson Education. Page 592: Rolf Vennenbernd/AP Images. Page 594: TMT Observatory Corporation. Page 598: ajt /Shutterstock.

Chapter 32

Page 599: PRNewsFoto/GeoEye, Inc/AP Images. Page 601: Andrew Syred/Science Source. Page 602: Courtesy of Dr. Christopher Jones. Page 604: Courtesy of Dr. Christopher Jones. Page 608: Jay M. Pasachoff. Page 609: Fundamental Photographs, NYC. Page 609: Pearson Education. Page 610: Ames Research Center/Ligo Project/NASA. Page 613: M. Cagnet et al. Atlas of Otical Phenomena, Springer-Verlag, 1962. Page 613: M. Cagnet et al. Atlas of Otical Phenomena, Springer-Verlag, 1962. Page 613: Courtesy of Dr. Christopher Jones. Page 614: Courtesy of Dr. Christopher Jones. Page 615: Courtesy of the Blu-ray Disc Association. Page 615: Courtesy of the Blu-ray Disc Association. Page 615: Courtesy of the Blu-ray Disc Association. Page 616: Courtesy of Dr. Christopher Jones. Page 616: M. Cagnet et al. Atlas of Otical Phenomena, Springer-Verlag, 1962. Page 616: Courtesy of Dr. Christopher Jones. Page 616: Courtesy of Dr. Christopher Jones. Page 616: Courtesy of Dr. Christopher Jones. Page 616: Courtesy of Dr. Christopher Jones. Page 620: M. Cagnet et al. Atlas of Otical Phenomena, Springer-Verlag, 1962. Page 620: Courtesy of Dr. Christopher Jones. Page 620: Courtesy of Dr. Christopher Jones. Page 620: Courtesy of Dr. Christopher Jones. Page 620: Courtesy of Dr. Christopher Jones.

Chapter 33

Page 621: Delft University of Technology/Science Source. Page 622: ASSOCIATED PRESS/AP Photos. Page 623: Source: Albert Einstein; Leopold Infeld, The evolution of physics : the growth of ideas from the early concepts to relativity and quanta, Cambridge : Cambridge University Press, 1938. Page 625: The Hebrew University of Jerusalem. Page 631: SLAC National Accelerator Laboratory/AP Images. Page 638: WDCN/Univ. College London/Science Source. Page 642: NASA.

Chapter 34

Page 647: Dr. Wolfgang Ketterle. Page 658: Andrew Syred/Science Source. Page 659: Millie LeBlanc/Courtesy of The Educational Development Center. Page 661: Kim Steele/Stockbyte/Getty Images. Page 661: Source: Werner Heisenberg, Physics and Philosophy: The Revolution in Modern Science (New York: Harper & Brothers, 1962).

Chapter 35

Page 667: IBM Corporation. Page 669: RIchard Wolfson. Page 671: Source: Albert Einstein, Max Born, The Born-Einstein Letters 1916-55, Palgrave Macmillan, 1971. Page 677: IBM Corporation. Page 683: Courtesy of Dr. Stephan Diez.

Chapter 36

Page 684: Fundamental Photographs. Page 692: Wolfgang Ketterle. Page 698: Wolfgang Ketterle. Page 701: Medical Body Scans/Science Source.

Chapter 37

Page 702: Alfred Pasieka/Science Source. Page 712: Hemis/Alamy. Page 714: Koichi Kamoshida/Getty Images.

Chapter 38

Page 720: TEPCO/AFLO/Newscom. Page 724: David Job/The Image Bank/Getty Images. Page 740: LLNL/Science Source.

Chapter 39

Page 747: CMS Experiment at the LHC, CERN. Page 749: Brookhaven National Laboratory. Page 755: European Organization for Nuclear Research. Page 755: Kyodo News/Newscom. Page 757: CERN/European Organization for Nuclear Research. Page 758: NASA. Page 759: European Space Agency (ESA). Page 759: Source: http://arxiv.org/pdf/1303.5062v1.pdf. Page 765: Richard Megna/Fundamental Photographs. Page 766: European Space Agency (ESA).

Index

D

N

O

P

Q

R

S

X

Y

Z